FORMULAS/EQUATIONS

Distance Formula
If $P_1 = (x_1, y_1)$ and $P_2 = (x_2, y_2)$, the distance from P_1 to P_2 is
$$d(P_1, P_2) = \sqrt{(x_2 - x_1)^2 + (y_2 - y_1)^2}$$

Slope of a Line
The slope m of the line containing the points $P_1 = (x_1, y_1)$ and $P_2 = (x_2, y_2)$ is
$$m = \frac{y_2 - y_1}{x_2 - x_1} \qquad \text{if } x_1 \neq x_2$$
$$m \text{ is undefined} \qquad \text{if } x_1 = x_2$$

Point-Slope Equation of a Line
The equation of a line with slope m containing the point (x_1, y_1) is
$$y - y_1 = m(x - x_1)$$

Slope-Intercept Equation of a Line
The equation of a line with slope m and y-intercept $(0, b)$ is
$$y = mx + b$$

Quadratic Formula
The solutions of the equation $ax^2 + bx + c = 0$, $a \neq 0$, are
$$x = \frac{-b \pm \sqrt{b^2 - 4ac}}{2a}$$

If $b^2 - 4ac > 0$, there are two real unequal solutions.
If $b^2 - 4ac = 0$, there is a repeated real solution.
If $b^2 - 4ac < 0$, there are no real solutions.

FUNCTIONS

Constant Function
$f(x) = b$

Linear Function
$f(x) = mx + b$, $\quad m$ is slope, $(0, b)$ is y-intercept

Quadratic Function
$f(x) = ax^2 + bx + c$

Polynomial Function
$f(x) = a_n x^n + a_{n-1} x^{n-1} + \cdots + a_1 x + a_0$

Rational Function
$R(x) = \dfrac{p(x)}{q(x)} = \dfrac{a_n x^n + a_{n-1} x^{n-1} + \cdots + a_1 x + a_0}{b_m x^m + b_{m-1} x^{m-1} + \cdots + b_1 x + b_0}$

Exponential Function
$f(x) = a^x$, $\quad a > 0, a \neq 1$

Logarithmic Function
$f(x) = \log_a x$, $\quad a > 0, a \neq 1$

Exponential Function (base e)
$f(x) = e^x$

Natural Logarithm Function
$f(x) = \ln x$

Mathematics
An Applied Approach

Mathematics

An
Applied
Approach

Sixth Edition

Abe Mizrahi
Indiana University Northwest

Michael Sullivan
Chicago State University

John Wiley & Sons, Inc.
New York / Chichester / Brisbane / Toronto / Singapore

ACQUISITIONS EDITOR: Ruth Baruth
DEVELOPMENTAL EDITOR: Madalyn Stone
MARKETING MANAGER: Debra Riegert
PRODUCTION SERVICE: Phyllis Niklas
TEXT AND COVER DESIGNER: Karin Gerdes Kincheloe
MANUFACTURING MANAGER: Mark Cirillo
ILLUSTRATION COORDINATOR: Rosa Bryant
COVER ILLUSTRATION: Roy Wiemann

This book was set in Times Roman by Progressive Information Technologies and printed and bound by Courier/Westford. The cover was printed by Lehigh Press.

Library of Congress Cataloging in Publication Data:
Mizrahi, Abe.
     Mathematics, an applied approach / Abe Mizrahi, Michael Sullivan.
   —6th ed.
        p.   cm.
     Rev. ed. of: Mathematics for business, life sciences, and social sciences. 5th ed. c1993.
     Includes bibliographical references and index.
     ISBN 0-471-10701-8 (cloth : alk. paper)
     1. Business mathematics.   2. Business mathematics—Problems, exercises, etc.   3. Social sciences—Mathematics.   I. Sullivan, Michael, 1942–   .   II. Mizrahi, Abe. Mathematics for business, life sciences, and social sciences.   III. Title.
HF5691.M59   1996   95-42833
650′.01′513—dc20   CIP

Printed in the United States of America

10 9 8 7 6 5 4

To Our Families

About the Authors

Abe Mizrahi received his doctorate in mathematics from the Illinois Institute of Technology in 1965. He is currently Professor of Mathematics at Indiana University Northwest. Dr. Mizrahi is a member of the Mathematical Association of America. Articles he has written explore topics in math education and the applications of mathematics to economics. Professor Mizrahi has served on many CUPM committees and has been a panel member on the CUPM Committee on Applied Mathematics in the Undergraduate Curriculum. Professor Mizrahi is a recipient of many NSF grants and has served as a consultant to a number of businesses and federal agencies.

Michael Sullivan received his doctorate in mathematics from the Illinois Institute of Technology in 1967. Since 1965, he has been Professor of Mathematics and Computer Science at Chicago State University. Dr. Sullivan is a member of the American Mathematical Society, the Mathematical Association of America, and the Text and Academic Authors Association. He has served on CUPM curriculum committees and is a member of the Illinois Section MAA High School Lecture Committee. Professor Sullivan has written a variety of scholarly articles and has served as a curriculum consultant to high schools, colleges, professional organizations, and government agencies.

Drs. Sullivan and Mizrahi are also the coauthors of *Finite Mathematics: An Applied Approach,* Seventh Edition, John Wiley & Sons, Inc. 1996.

Preface

The first edition of *Mathematics* was published in 1976. At that time our purpose was to present an accessible approach to the mathematics required in business and the social sciences, while giving emphasis to real-world applications from these fields. We achieved this goal of making mathematics accessible to students through clear exposition, appropriate pacing, timely suggestions for interaction, and motivating examples, exercises, and applications. In subsequent editions this has remained our purpose, and in this edition we have enlarged our range of applications to include those using a graphing calculator.

DEVELOPMENT

This edition of *Mathematics: An Applied Approach* has undergone extensive development to ensure its pedagogical soundness and mathematical accuracy. Surveying users of previous editions, monitoring curriculum changes, reviewing proposed organizational changes, reviewing first-draft and final-draft manuscript, reading all phases of manuscript for accuracy, monitoring the selection of new examples and exercises, and working the answers and solutions to all exercises are some of the activities comprising the development process.

Reviewing

Current and former users, and users of other texts, reacted to critical components such as quality of examples and exercises, level, pacing, sequencing and organization, pedagogical approach, and mathematical accuracy. Based on the results of this survey and the current trends in curriculum, a reorganized table of contents was proposed. This was reviewed by professors to gain their reactions to scope, sequencing, pacing, emphasis, and new topics. Reactions were analyzed and incorporated. The first draft of the manuscript was reviewed by a group of finite mathematics professors and a group of calculus professors. They provided additional reactions to pedagogical approach, accuracy, choice of examples and exercises, and organization. In turn, their suggestions were analyzed and incorporated into the next phase of the manuscript.

Accuracy Checks

Maintaining accuracy through manuscript development is an abiding concern of both authors and publisher. For this reason, each phase of the manuscript was read by yet another group of professors to ensure mathematical accuracy in features such as use of notation and computational detail of examples. Professors also monitored exercises to ensure that they gradually increased in difficulty, coordinated well with instruction, and were of interest to students. Equally important, the authors and publishers endeavored to impart to users of this edition the confidence that everything that is humanly possible has been undertaken to ensure the accuracy of answers and solutions to problems. This is evidenced by the special attention given through the working, and reworking, of every problem of this text by an experienced professor of mathematics.

Prepublication Reactions

Just prior to publication, the manuscript was reviewed by professors from a cross section of colleges and universities. These professors are familiar with previous editions of this text and profess a knowledge of competing texts. Their highly positive reaction to exposition, approach, examples, exercises, applications, reorganization, and new topics attests to the pedagogical superiority of this Sixth Edition of *Mathematics: An Applied Approach*.

FEATURES FOR THE STUDENT

A variety of features work together to help students use this book.

Style and Pedagogy

The writing style is clear and concise. Students can easily follow the straightforward explanations. At the same time, the text is mathematically accurate and pedagogically sound. Topics are introduced in simple algebraic settings so that students can more easily concentrate on the underlying concepts.

Examples

An abundance of examples is used to thoroughly exemplify concepts. Some exhibit a basic idea, whereas others gradually raise the complexity of the idea. Where appropriate, examples are used to illustrate applications of the material. In all cases, sufficient detail is provided so that the mathematics is easy to follow.

Interaction

To encourage students to interact with the text, the instruction ''Now Work Problem XX'' appears in the margin in selected places in each section. These references direct students to problems in the exercise set at the end of the section that relate to the concepts and examples just introduced. By working a related problem in the exercises, the student is able to discover whether the idea just presented is understood, or whether it needs further study before he or she continues to the next concept.

Exercises

At the end of each section, problems are given that progress from easy (to instill confidence) to ones that are more difficult (to challenge better students). Motivation is sustained through application-type exercises that deal with business and the life and social sciences.

Technology

At the end of most sections, Technology Exercises have been included that demonstrate the usefulness of a graphing utility for solving certain types of exercises. These are used as an instructional tool in order to facilitate the crucial combination of graphical, numerical, and algebraic viewpoints. These exercises, which require a programmable and/or graphing calculator or graphing software, have been placed in appropriate exercise sets.

Appendix B contains examples of solving linear programming problems using the LINDO software package.

Chapter Review

Each chapter concludes with a summary of important terms and formulas, followed by their page reference. A variety of problems affords another check as to whether the material of the chapter has been mastered.

Professional Exams

When appropriate, mathematical questions from CPA, CMA, and Actuary exams are given. This feature adds to the realism of the text by providing students with a preview of those questions that might appear on professional exams.

Functional Appendix

Prerequisite material from algebra and geometry, including exercise sets, is presented in Appendix A. This review focuses precisely on topics from earlier courses that are used and needed in this course. Timely references to Appendix A appear at appropriate places in the text.

Purposeful Design

Important terms are given in boldface type when first introduced. Those requiring more emphasis are set off with their definitions in boldface type. Formulas and properties are enclosed in boxes and highlighted with color. Procedures are also highlighted in boxes. Numerous illustrations have been used, each utilizing a second color for clarity and emphasis.

Answers

A section of Answers to Odd-Numbered Problems appears at the end of the book.

CHANGES AND IMPROVEMENTS TO THIS EDITION

Much of the narrative has been rewritten to improve clarity and to heighten understanding. New examples have been added where appropriate. The exercise sets have been improved so that the early problems help build student confidence, while later problems challenge the better student. Many new applied problems have been added for student motivation, and technology exercises have been added at the ends of appropriate sections. All the exercises have been reworked to ensure accuracy in the answer section and in the solutions manuals.

x   Preface

Organizational and Content Changes

Chapter 1 is now streamlined to improve its usefulness as a review chapter. The sections on functions and graphs of equations have been relocated to a new review chapter, Chapter 11, Precalculus—A Review, which immediately precedes the calculus part of the book. Applications of linear equations now appear in a separate section (Section 1.3) to improve the flexibility of coverage. Discussions on perpendicular lines, a summary about pairs of lines, and applications involving prediction have been added. More applied exercises have been added throughout.

In Chapter 2, applications involving the Leontief model, cryptography, and accounting now appear in a separate section (Section 2.7) at the end of the chapter to improve the flexibility of coverage. New applied exercises have been added to Sections 2.1 and 2.2, and new applied examples have been used to motivate systems of equations in Section 2.1. A new applied example has been added to Section 2.5 to motivate matrix multiplication, and new technology exercises have been added.

Chapters 3 and 4 now cover linear programming, which was covered in one chapter (Chapter 3) in the previous edition. Chapter 3 covers the geometric approach, while Chapter 4 covers the simplex method. The section using LINDO to solve linear programming problems now appears as Appendix B.

In Chapter 3, a new example (and matching exercise) has been added to provide motivation for graphing systems of inequalities (Section 3.1). Many new applied exercises have been added to Sections 3.1 and 3.2.

In Chapter 4, the steps for pivoting are explained in terms of row operations, and proper warning is now given. The current values of the objective function and corresponding variables are provided after each pivot. Row operations used to pivot are cited. A new example was added in Section 4.2 to improve the clarity of the discussion, and many new applied exercises were added in Sections 4.2 and 4.4. Section 4.3, on the standard minimum problem and the duality principle, is entirely new to this edition. This section is optional for those who cover the more general Phase I/Phase II method, which now appears in Section 4.4.

Chapter 5, Finance, was repositioned to improve the flow of material. (It was Chapter 8 in the previous edition.) More emphasis is given to using a calculator throughout the chapter, and technology exercises have been added to Sections 5.2, 5.3, and 5.4. Section 5.2 was rewritten to improve clarity, and several new examples were added, including one on the time required to double an investment. In Section 5.3, the connection between an annuity and the sum of a geometric series is made, and new examples on computing an annuity and a sinking fund have been added. In Section 5.4, a discussion on the connection between present value and the sum of a geometric series has been added.

In Chapter 6, the introduction to the discussion on combinations now begins with two examples that provide the basis for the derivation of the formula. A new section, Section 6.5, More Counting Problems, contains several challenging examples as well as many new exercises.

Chapters 7 and 8 now cover probability, which was covered in one chapter (Chapter 5) in the previous edition. This is a less imposing presentation for the students.

Chapter 7, Probability, contains the first five sections from Chapter 5 of the previous edition. These sections have been completely rewritten and reorganized to improve the clarity of the exposition. The discussion on equally likely outcomes appears much earlier, and new examples have been added. To improve the discussion on conditional probability, a new example has been added to help students use the formulas correctly; the use of Venn diagrams is more extensive than before; and tree diagrams are used early and extensively to solve Product Rule problems. The exercises were expanded to include problems involving Venn diagrams and tree diagrams.

Chapter 8, Additional Probability Topics, contains the last three sections from Chapter 5 of the previous edition. In the discussion on Bayes' Formula, tree diagrams are now used to motivate the formula, as well as to provide an alternative technique of solution. This allows for simplification of the explanation, while also using previous experience. In the discussion on expected value, a new example has been added to demonstrate using expectation in decision-making situations. A new section on random variables has been added. Section 8.4, Applications to Operations Research, was a subsection in the previous edition and is now a separate, stand-alone section, which allows for more flexibility of coverage.

Chapter 9, Statistics, was formerly Chapter 7 in the previous edition. New sections on pie charts and bar graphs have been added, as well as new exercises and examples.

Chapter 10, Markov Chains; Games, was Chapter 6 in the previous edition. Technology exercises have been added.

Chapter 11 introduces quadratic functions and the exponential and logarithm functions. Many new examples and exercises are introduced, especially applied problems.

In Chapter 12, Section 12.1 has been completely rewritten to introduce left and right limits earlier. Section 12.4 includes a new discussion of cases for which $f'(x)$ does not exist. Section 12.5 is a new section for this edition.

Chapter 13 has been completely reorganized, with implicit differentiation and higher-order derivatives discussed at the end of the chapter. The derivative of x^n, n a real number, is now introduced earier. Relative and percentage rates of change are introduced in Section 13.2. The section on the chain rule has been completely rewritten; composite functions are introduced first, then the chain rule, and then the power rule. The section on implicit differentiation has been rewritten. The derivatives of the exponential and logarithm functions are introduced in this chapter, with new examples and exercises. Many more new examples and exercises appear throughout the chapter.

In Chapter 14, many examples have been rewritten to make them easier to understand. The discussion on concavity has been rewritten. The section on related rates has been completely rewritten, with new examples and exercises added.

Chapters 15 and 16 have been refined and new exercises added.

Chapter 17 includes a new section on the double integral and a detailed discussion of production functions.

Appendix A contains three new sections for review: A.1, Basic Algebra; A.2, Exponents and Logarithms; and A.3, Geometric Sequences.

TEACHING OPTIONS

Based on an extensive survey of what topics are actually taught, this edition was reorganized to meet the varied needs of different curricular requirements. In this edition, you may now teach any of the following clusters of material in any order you wish:

Systems of Equations, Matrices	Finance	Calculus
Linear Programming	Probabilty/Statistics	

Within the calculus cluster, after the derivative formulas have been established, you may choose to do applications of the derivative or integral calculus as you wish.

The flowchart that follows shows how the various chapters can be selected and sequenced to meet your specific requirements.

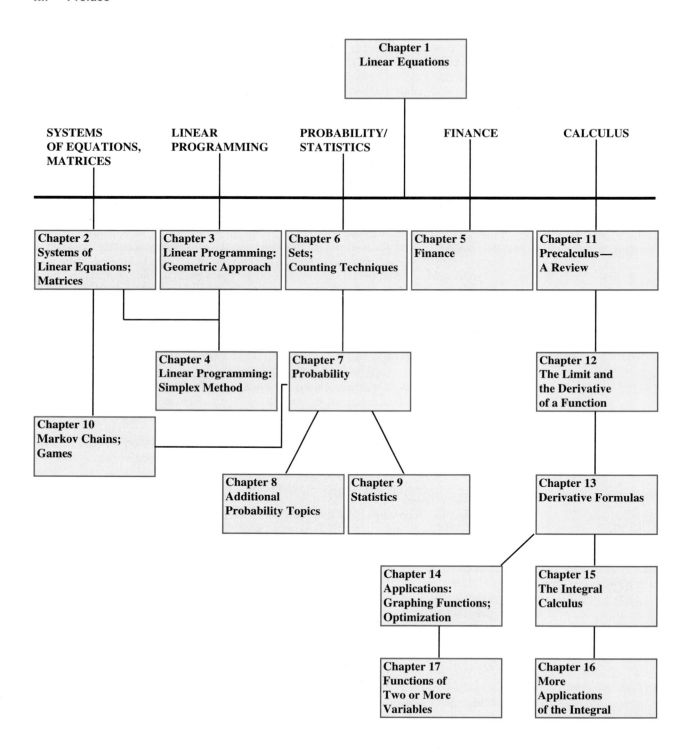

Supplements

Discovering Finite Mathematics and Calculus: A Laboratory Approach with examples on the TI-82 and TI-85 to Accompany *Mathematics: An Applied Approach*, Sixth Edition.
Prepared by Donna Pirich and Patricia Bigliani
0-471-14264-6

Computerized Testbank Macintosh $3\frac{1}{2}$ to Accompany *Mathematics: An Applied Approach*, Sixth Edition.
0-471-13680-8

Computerized Testbank IBM $3\frac{1}{2}$ to Accompany *Mathematics: An Applied Approach*, Sixth Edition.
0-471-13682-4

Testbank to Accompany *Mathematics: An Applied Approach*, Sixth Edition (print version).
Prepared by Deborah Betthauser Britt
0-471-13679-4

Instructor's Solutions Manual to Accompany *Mathematics: An Applied Approach*, Sixth Edition.
Prepared by Stephen L. Davis
0-471-13678-6

Student Solutions Manual to Accompany *Mathematics: An Applied Approach*, Sixth Edition.
Prepared by Stephen L. Davis
0-471-13677-8

Acknowledgments

W e thank the following individuals who made valuable contributions to this edition.

Reviewers

Pasquale J. Arpaia, *St. John Fisher College*
Donna Christy, *Rhode Island College*
Charles E. Cleaver, *The Citadel*
Lucinda Gallagher, *Florida State University*
Melinda Gann, *Mississippi College*
Richard Hobbs, *Mission College*
V. J. Klassen, *California State University—Fullerton*
Melvin Lax, *California State University—Long Beach*
David Minda, *University of Cincinnati*
Gilbert F. Orr, *University of Southern Colorado*
Barbara Rath, *Moorehead State University*
Steven Rodi, *Austin Community College*
Marlene Sims, *Kennesaw State College*
Samuel W. Spero, *Cuyahoga Community College*

Accuracy Readers and Accuracy Checkers

Irl C. Bivens, *Davidson College*
Stephen L. Davis, *Davidson College*
Susan L. Friedman, *Bernard M. Baruch College, CUNY*
Sharon O'Donnell, *Chicago State University*
Stella Pudar-Hozo, *Indiana University Northwest*

Other Acknowledgments

We also appreciate the patience and skill of Candy Myers, who typed this revision. Recognition and thanks are especially due the following individuals for their invaluable assistance in the preparation of this edition: Ruth Baruth for her patience and support; Madalyn Stone for her commitment to excellence and her encouragement; Sharon O'Donnell, Georgia Kamvosoulis, and Stella Pudar-Hozo for their careful check of the answer section; Stephen Davis for his careful work on the solutions manuals; Carl White, Charlotte Hyland, and the Wiley sales staff for their confidence in this project. Special thanks to Professor Iztok Hozo, Mathematics Department, Indiana University Northwest, for his work in preparing the technology problems.

Abe Mizrahi
Michael Sullivan

Contents

Part One

Linear Algebra

Chapter 1

Linear Equations

Much of the material presented here will already be familiar to you. We begin with a review of rectangular coordinates, followed by a discussion of lines, including parallel and intersecting lines. We close the chapter with some applications of lines.

1.1 RECTANGULAR COORDINATES; LINES

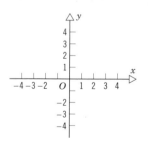

Figure 1

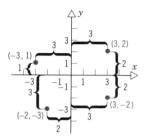

Figure 2

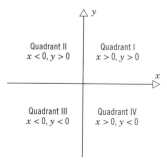

Figure 3

Now Work Problem 1

Rectangular Coordinates

We begin with two lines located in the same plane: one horizontal and the other vertical. We call the horizontal line the **x-axis,** the vertical line the **y-axis,** and the point of intersection the **origin O.** We assign coordinates to every point on these lines as shown in Figure 1, using a convenient scale. (The scales are usually, but not necessarily, the same on both axes.) Also, the origin O has a value of 0 on both the x-axis and the y-axis. We follow the usual convention that points on the x-axis to the right of O are associated with positive real numbers, and those to the left of O are associated with negative real numbers. Those on the y-axis above O are associated with positive real numbers, and those below O are associated with negative real numbers. In Figure 1, the x-axis and y-axis are labeled as x and y, respectively, and we have used an arrow at the end of each axis to denote the positive direction.

The coordinate system described here is called a **rectangular,** or **Cartesian, coordinate system.** The plane containing the x-axis and y-axis is sometimes called the **xy-plane,** and the x-axis and y-axis are referred to as the **coordinate axes.**

Any point P in the xy-plane can then be located by using an **ordered pair** (x, y) of real numbers, also called the **coordinates** of P. The number x denotes the signed distance of P from the y-axis (*signed* in the sense that if P is to the right of the y-axis, then $x > 0$, and if P is to the left of the y-axis, then $x < 0$); y denotes the signed distance of P from the x-axis. If (x, y) are the coordinates of a point P, then x is called the **x-coordinate,** or **abscissa,** of P and y is the **y-coordinate,** or **ordinate,** of P. We identify the point P by its coordinates (x, y) by writing $P = (x, y)$. Usually, we will simply say "the point (x, y)" rather than "the point whose coordinates are (x, y)."

To locate the point $(-3, 1)$, we go 3 units along the x-axis to the left of O and then go straight up 1 unit. We **plot** this point by placing a dot at this location. See Figure 2, in which the points with coordinates $(-3, 1)$, $(-2, -3)$, $(3, -2)$, and $(3, 2)$ are plotted.

The origin has coordinates $(0, 0)$. Any point on the x-axis has coordinates of the form $(x, 0)$, and any point on the y-axis has coordinates of the form $(0, y)$.

The coordinate axes divide the xy-plane into four sections, called **quadrants,** as shown in Figure 3. In quadrant I both the x-coordinate and the y-coordinate are positive; in quadrant II x is negative and y is positive; in quadrant III both x and y are negative; and in quadrant IV x is positive and y is negative. Points on the coordinate axes belong to no quadrant.

Graphs of Linear Equations in Two Variables

A **linear equation in two variables** is of the form

$$Ax + By = C \tag{1}$$

where A and B are not both zero. Examples of linear equations are

$$3x - 5y - 6 = 0 \quad \text{This equation can be written as}$$
$$3x - 5y = 6 \qquad A = 3, B = -5, C = 6$$

$$-3x = 2y - 1 \quad \text{This equation can be written as}$$
$$-3x - 2y = -1 \qquad A = -3, B = -2, C = -1$$

$$y = \tfrac{3}{4}x - 5$$

Here we can write
$-\tfrac{3}{4}x + y = -5$ or $3x - 4y = 20$
$A = 3, B = -4, C = 20$

$$y = -5$$

Here we can write
$0 \cdot x + y = -5$ $A = 0, B = 1, C = -5$

$$x = 4$$

Here we can write
$x + 0 \cdot y = 4$ $A = 1, B = 0, C = 4$

The **graph** of an equation is the set of all points (x, y) whose coordinates satisfy the equation. For example, $(0, 4)$ is a point on the graph of the equation $3x + 4y = 16$, because when x is replaced by zero and y is replaced by 4 in the equation, we get

$$3 \cdot 0 + 4 \cdot 4 = 16$$

which is a true statement.

It can be shown that if A, B, C are real numbers, with A and B not both zero, then the graph of the equation

$$Ax + By = C$$

is a *line*. This is the reason we call it a *linear* equation.

Conversely, any line is the graph of an equation of this form.

Since any line can be written as an equation in the form $Ax + By = C$, we call this form the **general equation** of a line.

Given a linear equation, we can obtain its graph by plotting two points that satisfy its equation and connecting them with a line. The easiest two points to plot are the *intercepts*.

INTERCEPTS

The points at which the graph of a line crosses the axes are called *intercepts*. The *x-intercept* is the point at which the graph crosses the x-axis; the *y-intercept* is the point at which the graph crosses the y-axis.

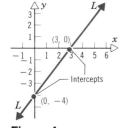

Figure 4

See Figure 4. The line has the intercepts $(0, -4)$ and $(3, 0)$. To find the intercepts, follow these steps:

Step 1 Set $y = 0$ and solve for x. This determines the x-intercept of the line.

Step 2 Set $x = 0$ and solve for y. This determines the y-intercept of the line.

EXAMPLE 1

Find the intercepts of the equation $2x + 3y = 6$. Graph the equation.

SOLUTION

Step 1 To find the x-intercept, we need to find the number x for which $y = 0$. Thus we set $y = 0$ to get

$$2x + 3(0) = 6$$
$$2x = 6$$
$$x = 3$$

The x-intercept is $(3, 0)$.

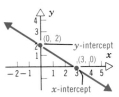

Figure 5

Step 2 To find the y-intercept, we set $x = 0$ and solve for y:

$$2(0) + 3y = 6$$
$$3y = 6$$
$$y = 2$$

The y-intercept is $(0, 2)$.

Since the equation is a linear equation, its graph is a line. We use the two intercepts $(3, 0)$ and $(0, 2)$ to graph it. See Figure 5.

········

EXAMPLE 2

Graph the equation: $y = 2x + 5$

SOLUTION

This equation can be written as

$$-2x + y = 5$$

This is a linear equation, so its graph is a line. The intercepts are $(0, 5)$ and $\left(-\frac{5}{2}, 0\right)$, which you should verify. For reassurance we'll find a third point. Arbitrarily, we let $x = 10$. Then $y = 2(10) + 5 = 25$, so $(10, 25)$ is a point on the graph. See Figure 6.

Figure 6

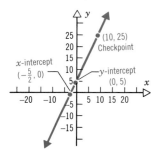

x	y
0	5
$-\frac{5}{2}$	0
10	25

········

EXAMPLE 3

Graph the equation: $-x + y = 0$

SOLUTION

This is a linear equation, so its graph is a line. The only intercept is $(0, 0)$. To locate another point on the graph, let $x = 3$. Then $y = 3$ and $(3, 3)$ is a point on the graph. See Figure 7.

Figure 7

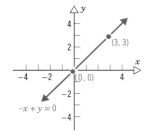

Now Work Problem 5

········

EXAMPLE 4

Graph the equation: $x = 3$

SOLUTION

We are looking for all points (x, y) in the plane for which $x = 3$. Thus, no matter what y-coordinate is used, the corresponding x-coordinate always equals 3. Consequently, the graph of the equation $x = 3$ is a vertical line with x-intercept $(3, 0)$ as shown in Figure 8.

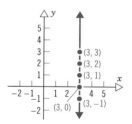

Figure 8 $x = 3$

As suggested by Example 4, we have the following result:

Equation of a Vertical Line

A vertical line is given by an equation of the form

$$x = a$$

where $(a, 0)$ is the x-intercept.

Slope of a Line

An important characteristic of a line, called its *slope,* is best defined by using rectangular coordinates.

SLOPE OF A LINE

Let $P = (x_1, y_1)$ and $Q = (x_2, y_2)$ be two distinct points with $x_1 \neq x_2$. The *slope m* of the nonvertical line L containing P and Q is defined by the formula

$$m = \frac{y_2 - y_1}{x_2 - x_1} \qquad x_1 \neq x_2 \qquad (2)$$

If $x_1 = x_2$, L is a vertical line and the slope m of L is *undefined* (since this results in division by 0).

Figure 9(a) provides an illustration of the slope of a nonvertical line; Figure 9(b) illustrates a vertical line.

Figure 9

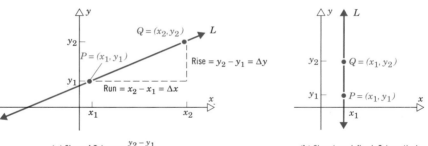

(a) Slope of L is $m = \frac{y_2 - y_1}{x_2 - x_1}$

(b) Slope is undefined; L is vertical

As Figure 9(a) illustrates, the slope m of a nonvertical line may be viewed as

$$m = \frac{y_2 - y_1}{x_2 - x_1} = \frac{\text{Rise}}{\text{Run}} = \frac{\text{Change in } y}{\text{Change in } x}$$

The change in y may be denoted by Δy, read "delta y," and the change in x by Δx. Using this notation, the slope m is

$$m = \frac{\Delta y}{\Delta x}$$

That is, the slope m of a nonvertical line L is the ratio of the change in the y-coordinates from P to Q to the change in the x-coordinates from P to Q.

Two comments about computing the slope of a nonvertical line may prove helpful:

1. Any two distinct points on the line can be used to compute the slope of the line. (See Figure 10 for justification.)
2. The slope of a line may be computed from $P = (x_1, y_1)$ to $Q = (x_2, y_2)$ or from Q to P because

$$\frac{y_2 - y_1}{x_2 - x_1} = \frac{y_1 - y_2}{x_1 - x_2}$$

Figure 10   Triangles ABC and PQR are similar (equal angles). Hence, ratios of corresponding sides are proportional. Thus:

Slope using P and $Q = \dfrac{y_2 - y_1}{x_2 - x_1}$ = Slope using A and B = $\dfrac{d(B, C)}{d(A, C)}$

where $d(B, C)$ denotes the distance from B to C and $d(A, C)$ denotes the distance from A to C.

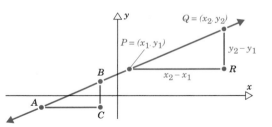

To get a better idea of the meaning of the slope m of a line L, consider the following example.

EXAMPLE 5

Compute the slopes of the lines L_1, L_2, L_3, and L_4 containing the following pairs of points. Graph all four lines on the same set of coordinate axes.

$$
\begin{aligned}
L_1: \quad & P = (2, 3) \qquad Q_1 = (-1, -2) \\
L_2: \quad & P = (2, 3) \qquad Q_2 = (3, -1) \\
L_3: \quad & P = (2, 3) \qquad Q_3 = (5, 3) \\
L_4: \quad & P = (2, 3) \qquad Q_4 = (2, 5)
\end{aligned}
$$

SOLUTION

Let m_1, m_2, m_3, and m_4 denote the slopes of the lines L_1, L_2, L_3, and L_4, respectively. Then

$$m_1 = \frac{-2-3}{-1-2} = \frac{-5}{-3} = \frac{5}{3} \qquad \text{A rise of 5 divided by a run of 3}$$

$$m_2 = \frac{-1-3}{3-2} = \frac{-4}{1} = -4 \qquad \text{A rise (drop) of } -4 \text{ divided by a run of 1}$$

$$m_3 = \frac{3-3}{5-2} = \frac{0}{3} = 0 \qquad \text{A rise of 0 divided by a run of 3}$$

m_4 is undefined

The graphs of these lines are given in Figure 11 below.

· ·

As Figure 11 illustrates, when the slope m of a line is positive, the line slants upward from left to right (L_1); when the slope m is negative, the line slants downward from left to right (L_2); when the slope m is 0, the line is horizontal (L_3); and when the slope m is undefined, the line is vertical (L_4).

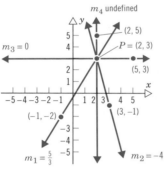

Figure 11

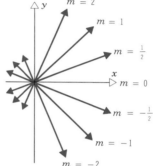

Figure 12

Now Work Problem 13

Figure 12 illustrates the slopes of several lines. Note the pattern. Do you see that the slope of a line is an indicator of its steepness?

EXAMPLE 6

Draw a graph of the line that passes through the point (3, 2) and has a slope of

(a) $\frac{3}{4}$ (b) $-\frac{4}{5}$

SOLUTION

(a) Slope = rise/run. The fact that the slope is $\frac{3}{4}$ means that for every horizontal movement (run) of 4 units to the right, there will be a vertical movement (rise) of 3 units. If we start at the given point (3, 2) and move 4 units to the right and 3 units up, we reach the point (7, 5). By drawing the line through this point and the point (3, 2), we have the graph (see Figure 13, p. 10).

(b) The fact that the slope is $-\frac{4}{5} = \frac{-4}{5}$ means that for every horizontal movement of 5 units to the right, there will be a corresponding vertical movement of -4 units (a downward movement). If we start at the given point (3, 2) and move 5 units to the right and then 4 units down, we arrive at the point $(8, -2)$. By drawing the line through these points, we have the graph (see Figure 14).

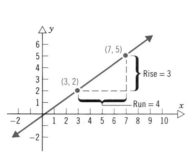

Figure 13   Slope $= \frac{3}{4}$ **Figure 14**   Slope $= -\frac{4}{5}$

Alternatively, we can set $-\frac{4}{5} = \frac{4}{-5}$ so that for every horizontal movement of -5 units (a movement to the left), there will be a corresponding vertical movement of 4 units (upward). This approach brings us to the point $(-2, 6)$, which is also on the graph shown in Figure 14.

Now Work Problem 23

· ·

Other Forms of the Equation of a Line

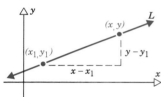

Figure 15

Let L be a nonvertical line with slope m and containing the point (x_1, y_1). See Figure 15. Any two distinct points on L can be used to compute slope. Thus, for any other point (x, y) on L, we have

$$m = \frac{y - y_1}{x - x_1} \qquad \text{or} \qquad y - y_1 = m(x - x_1)$$

Point–Slope Form of an Equation of a Line

An equation of a nonvertical line of slope m that passes through the point (x_1, y_1) is

$$y - y_1 = m(x - x_1) \tag{3}$$

EXAMPLE 7

An equation of the line with slope 4 and passing through the point $(1, 2)$ can be found by using the point–slope form with $m = 4$, $x_1 = 1$, and $y_1 = 2$:

$$y - y_1 = m(x - x_1)$$
$$y - 2 = 4(x - 1)$$
$$y = 4x - 2$$

See Figure 16.

Figure 16 $y = 4x - 2$

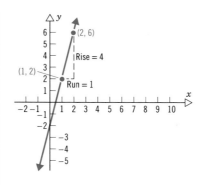

EXAMPLE 8

Find an equation of the horizontal line passing through the point $(3, 2)$.

SOLUTION

The slope of a horizontal line is 0. To get an equation, we use the point–slope form with $m = 0$, $x_1 = 3$, and $y_1 = 2$:

$$y - y_1 = m(x - x_1)$$
$$y - 2 = 0 \cdot (x - 3)$$
$$y - 2 = 0$$
$$y = 2$$

See Figure 17 for the graph.

Figure 17 $y = 2$

As suggested by Example 8, we have the following result:

Equation of a Horizontal Line

A horizontal line is given by an equation of the form

$$y = b$$

where $(0, b)$ is the y-intercept.

EXAMPLE 9

Find the equation of the line L passing through the points $(2, 3)$ and $(-4, 5)$. Graph the line L.

SOLUTION

Since two points are given, we first compute the slope of the line:

$$m = \frac{5 - 3}{-4 - 2} = \frac{2}{-6} = \frac{-1}{3}$$

We use the point $(2, 3)$ and the fact that the slope $m = -\frac{1}{3}$ to get the point–slope form of the equation of the line:

$$y - 3 = -\tfrac{1}{3}(x - 2)$$

See Figure 18 for the graph.

Figure 18

$y - 3 = -\frac{1}{3}(x - 2)$

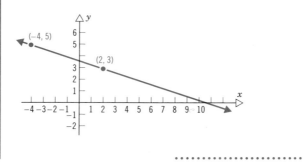

In the solution in Example 9 we could have used the point $(-4, 5)$ instead of the point $(2, 3)$. The equation that results, although it looks different, is equivalent to the equation we obtained in the example. (Try it for yourself.)

The general form of the equation of the line in Example 9 can be obtained by multiplying both sides of the point–slope equation by 3 and collecting terms:

$$y - 3 = -\tfrac{1}{3}(x - 2)$$
$$3(y - 3) = 3(-\tfrac{1}{3})(x - 2) \quad \text{Multiply by 3.}$$
$$3y - 9 = -1(x - 2)$$
$$3y - 9 = -x + 2$$
$$x + 3y = 11$$

This is the general form of the equation.

Now Work Problem 35   Another useful equation of a line is obtained when the slope m and y-intercept $(0, b)$ are known. In this case we know both the slope m of the line and a point $(0, b)$ on the line; thus, we may use the point–slope form, Equation (3), to obtain the following equation:

$$y - b = m(x - 0) \qquad \text{or} \qquad y = mx + b$$

Slope–Intercept Form of an Equation of a Line

An equation of a line L with slope m and y-intercept $(0, b)$ is

$$y = mx + b \tag{4}$$

When an equation of a line is written in slope–intercept form, it is easy to find the slope m and y-intercept $(0, b)$ of the line. For example, suppose the equation of the line is

$$y = -2x + 3$$

Compare it to $y = mx + b$:

$$y = -2x + 3$$
$$\uparrow \quad \uparrow$$
$$y = \ \ mx + b$$

The slope of this line is -2 and its y-intercept is $(0, 3)$.
Let's look at another example.

EXAMPLE 10

Find the slope m and y-intercept $(0, b)$ of the line $2x + 4y - 8 = 0$. Graph the line.

SOLUTION

To obtain the slope and y-intercept, we transform the equation into its slope–intercept form. Thus we need to solve for y:

$$2x + 4y - 8 = 0$$
$$4y = -2x + 8$$
$$y = -\tfrac{1}{2}x + 2$$

The coefficient of x, $-\tfrac{1}{2}$, is the slope, and the y-intercept is $(0, 2)$.
We can graph the line in either of two ways:

1. Use the fact that the y-intercept is $(0, 2)$ and the slope is $-\tfrac{1}{2}$. Then, starting at the point $(0, 2)$, go to the right 2 units and then down 1 unit to the point $(2, 1)$.
2. Locate the intercepts. The y-intercept is $(0, 2)$. To obtain the x-intercept, we let $y = 0$ and solve for x. When $y = 0$, we have

$$2x + 4 \cdot 0 - 8 = 0$$
$$2x - 8 = 0$$
$$x = 4$$

Thus, the intercepts are $(4, 0)$ and $(0, 2)$.
See Figure 19.

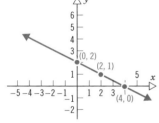

Figure 19
$2x + 4y - 8 = 0$

·······················

[*Note:* The second method, locating the intercepts, will not work when the line passes through the origin. In this case some other point on the line must be found. Refer back to Example 3.]

Now Work Problem 51

EXAMPLE 11

Daily Cost of Production   A certain factory has daily fixed overhead expenses of $2000, while each item produced costs $100. Find an equation that relates the daily cost C to the number x of items produced each day.

SOLUTION

The fixed overhead expense of $2000 represents the fixed cost, the cost incurred no matter how many items are produced. Since each item produced costs $100, the variable cost of producing x items is $100x$. Thus the total daily cost C of production is

$$C = 100x + 2000$$

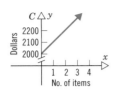

Figure 20

The graph of this equation is given by the line in Figure 20. Notice that the fixed cost $2000 is represented by the y-intercept, while the $100 cost of producing each item is the slope. Also notice that a different scale is used on each axis.

....

Summary

The graph of a linear equation, $Ax + By = C$, where A and B are not both zero, is a line. In this form it is referred to as the general equation of a line.

1. Given the general equation of a line, information can be found about the line:
   (a) Place the equation in slope–intercept form $y = mx + b$ to find the slope m and y-intercept $(0, b)$.
   (b) Let $x = 0$ and solve for y to find the y-intercept.
   (c) Let $y = 0$ and solve for x to find the x-intercept.

2. Given information about a straight line, an equation of the line can be found. The form of the equation depends on the given information.

Given	Use	Equation
Point (x_1, y_1), slope m	Point–slope form	$y - y_1 = m(x - x_1)$
Two points $(x_1, y_1), (x_2, y_2)$	If $x_1 = x_2$: The line is vertical If $x_1 \neq x_2$: Find the slope m, $m = \dfrac{y_2 - y_1}{x_2 - x_1}$ Then use the point–slope form	$x = x_1$ $y - y_1 = m(x - x_1)$
Slope m, y-intercept $(0, b)$	Slope–intercept form	$y = mx + b$

EXERCISE 1.1   Answers to odd-numbered problems begin on page AN-1.

1. Give the coordinates of each point in the following figure.

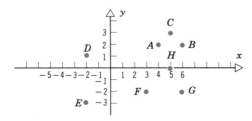

2. Plot each point in the xy-plane. Tell in which quadrant or on what coordinate axis each point lies.

   (a) $A = (-3, 2)$ (b) $B = (6, 0)$
   (c) $C = (-2, -2)$ (d) $D = (6, 5)$
   (e) $E = (0, -3)$ (f) $F = (6, -3)$

3. Plot the points $(2, 0)$, $(2, -3)$, $(2, 4)$, $(2, 1)$, and $(2, -1)$. Describe the collection of all points of the form $(2, y)$, where y is a real number.

4. Plot the points $(0, 3)$, $(1, 3)$, $(-2, 3)$, $(5, 3)$, and $(-4, 3)$. Describe the collection of all points of the form $(x, 3)$, where x is a real number.

In Problems 5–8 copy the tables given and fill in the missing values using the given equations. Use these points to graph each equation.

5. $y = x - 3$

x	0		2	-2	4	-4
y		0				

6. $y = -3x + 3$

x	0		2	-2	4	-4
y		0				

7. $2x - y = 6$

x	0		2	-2	4	-4
y		0				

8. $x + 3y = 9$

x	0		2	-2	4	-4
y		0				

In Problems 9–12 find the slope of the line.

9.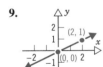

10.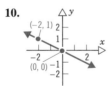

11.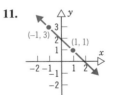

12.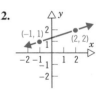

In Problems 13–22 plot each pair of points and determine the slope of the line containing them. Graph the line.

13. $(2, 3)$; $(1, 0)$

14. $(1, 2)$; $(3, 4)$

15. $(-2, 3)$; $(2, 1)$

16. $(-1, 1)$; $(2, 3)$

17. $(-3, -1)$; $(2, -1)$

18. $(4, 2)$; $(-5, 2)$

19. $(-1, 2)$; $(-1, -2)$

20. $(2, 0)$; $(2, 2)$

21. $(\sqrt{2}, 3)$; $(1, \sqrt{3})$

22. $(-2\sqrt{2}, 0)$; $(4, \sqrt{5})$

In Problems 23–30 graph the line passing through the point P and having slope m.

23. $P = (1, 2)$; $m = 2$

24. $P = (2, 1)$; $m = 3$

25. $P = (2, 4)$; $m = -\frac{3}{4}$

26. $P = (1, 3)$; $m = -\frac{2}{5}$

27. $P = (-1, 3)$; $m = 0$

28. $P = (2, -4)$; $m = 0$

29. $P = (0, 3)$;   slope undefined

30. $P = (-2, 0)$;   slope undefined

In Problems 31–34 find a general equation of each line. Write the equation in the form Ax + By = C.

31.

32.

33.

34.

In Problems 35–46 find a general equation for the line with the given properties. Write the equation in the form Ax + By = C.

35. Slope $= 2$; passing through $(-2, 3)$

36. Slope $= 3$; passing through $(4, -3)$

37. Slope $= -\frac{2}{3}$; passing through $(1, -1)$

38. Slope $= \frac{1}{2}$; passing through $(3, 1)$

39. Passing through $(1, 3)$ and $(-1, 2)$

40. Passing through $(-3, 4)$ and $(2, 5)$

41. Slope $= -3$; y-intercept $= (0, 3)$

42. Slope $= -2$; y-intercept $= (0, -2)$

43. x-intercept $= (2, 0)$; y-intercept $= (0, -1)$

44. x-intercept $= (-4, 0)$; y-intercept $= (0, 4)$

45. Slope undefined; passing through $(1, 4)$

46. Slope undefined; passing through $(2, 1)$

In Problems 47–62 find the slope and y-intercept of each line. Graph the line.

47. $y = 2x + 3$

48. $y = -3x + 4$

49. $\frac{1}{2}y = x - 1$

50. $\frac{1}{3}x + y = 2$

51. $2x - 3y = 6$

52. $3x + 2y = 6$

53. $x + y = 1$

54. $x - y = 2$

55. $x = -4$

56. $y = -1$

57. $y = 5$

58. $x = 2$

59. $y - x = 0$

60. $x + y = 0$

61. $2y - 3x = 0$

62. $3x + 2y = 0$

63. Find a general equation of the x-axis.

64. Find a general equation of the y-axis.

65. Temperature Conversion The relationship between Celsius (°C) and Fahrenheit (°F) degrees for measuring temperature is linear. Find an equation relating °C and °F if 0°C corresponds to 32°F and 100°C corresponds to 212°F. Use the equation to find the Celsius measure of 70°F.

66. Temperature Conversion The Kelvin (K) scale for measuring temperature is obtained by adding 273 to the Celsius temperature.

(a) Write an equation relating K and °C.
(b) Write an equation relating K and °F (see Problem 65).

67. Profit from Selling Newspapers Each Sunday a newspaper agency sells x copies of a certain newspaper for $1.00 per copy. The cost to the agency of each newspaper is $0.50. The agency pays a fixed cost for storage, delivery, and so on, of $100 per Sunday. Write an equation that relates the profit P, in dollars, to the number x of copies sold. Graph this equation.

68. Repeat Problem 67 if the cost to the agency is $0.45 per copy and the fixed cost is $125 per Sunday.

69. Electricity Rates Commonwealth Edison Company supplies electricity in the summer months to residential customers for a monthly customer charge of $9.06 plus 10.189 cents per kilowatt-hour for all kilowatt-hours supplied in the month.* Write an equation that relates the monthly charge C, in dollars, to the number x of kilowatt-hours used in a month. Graph this equation. What is the monthly charge for using 300 kilowatt-hours? For using 900 kilowatt-hours?

* *Source:* Commonwealth Edison Rates for Residential Service, 1991.

70. Cost of Operating a Car The average cost of operating a standard-size car is $0.45 per mile. Write an equation that relates the average cost C of operating a standard-size car and the number x of miles it is driven.

71. Cost of Renting a Truck The cost of renting a truck is $280 per week plus a charge of $0.20 per mile driven. Write an equation that relates the cost C for a weekly rental in which the truck is driven x miles.

72. Wages of a Car Salesperson Dan receives $300 per week for selling used cars. As part of his weekly salary he also receives 10% of the sales he generates. Write an equation that relates Dan's weekly salary S when he generates sales of x.

73. Weight–Height Relation Suppose the weights (w) of male college students are linearly related to their heights (h). If a student 62 inches tall weighs 120 pounds and a student 72 inches tall weighs 170 pounds, write an equation to express weight in terms of height.

74. Disease Propagation Research indicates that in a controlled environment, the number of diseased mice will increase linearly each day after one of the mice in the cage is infected with a particular type of disease-causing germ. There were 8 diseased mice 4 days after the first exposure and 14 diseased mice after 6 days. Write an equation that will give the number of diseased mice after any given number of days. If there were 40 mice in the cage, how long will it take until they are all infected?

75. Water Preservation Since the beginning of the month a local reservoir has been losing water at a constant rate. On the 10th of the month, the reservoir held 300 million gallons of water and on the 18th it held only 262 million gallons.

(a) Write an equation that will give the amount of water in the reservoir at any time.

(b) How much water was in the reservoir on the 14th of the month?

76. Product Promotion A cereal company finds that the number of people who will buy one of its products the first month it is introduced is linearly related to the amount of money it spends on advertising. If it spends $400,000 on advertising, 100,000 boxes of cereal will be sold, and if it spends $600,000, 140,000 boxes will be sold.

(a) Write an equation describing the relation between the amount spent on advertising and the number of boxes sold.

(b) How much advertising is needed to sell 200,000 boxes of cereal?

Technology Exercises
In Problems 1–8 graph each linear equation. Be sure to use a viewing rectangle that shows the intercepts. Then locate each intercept correct to two decimal places.

1. $1.2x + 0.8y = 20$

2. $-1.3x + 2.7y = 81.1$

3. $215x - 0.1y = 53$

4. $0.5x - 313y = 82$

5. $\frac{4}{17}x + \frac{6}{23}y = \frac{2}{3}$

6. $\frac{9}{14}x - \frac{43}{8}y = \frac{22}{7}$

7. $\pi x - \sqrt{3}\, y = \sqrt{6}$

8. $x + \pi y = \sqrt{15}$

9. Seeing the Concept On the same screen graph all the following lines:

(a) $y = 0$ (slope 0)

(b) $y = \frac{1}{2}x$ (slope $\frac{1}{2}$)

(c) $y = x$ (slope 1)

(d) $y = 2x$ (slope 2)

10. Seeing the Concept On the same screen graph all the following lines:

(a) $y = 0$ (slope 0)

(b) $y = -\frac{1}{2}x$ (slope $-\frac{1}{2}$)

(c) $y = -x$ (slope -1)

(d) $y = -2x$ (slope -2)

11. Exploration On the same screen graph all the following lines:

(a) $y = 2x - 3$

(b) $y = 2x - 1$

(c) $y = 2x$

(d) $y = 2x + 2$

What do you conclude about lines having the same slope?

1.2 PARALLEL AND INTERSECTING LINES

Let L and M be two lines. Exactly one of the following three relationships must hold for these two lines:

1. All the points on L are the same as the points on M.
2. L and M have no points in common.
3. L and M have exactly one point in common.

If the first relationship holds, the lines L and M are identical. Such lines are called **coincident lines.**

Coincident Lines

Coincident lines that are vertical have undefined slope and the same x-intercept. Coincident lines that are nonvertical have the same slope and the same y-intercept.

EXAMPLE 1

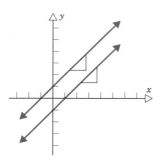

Figure 21

Show that the lines given by the equations below are coincident.

$$L: \quad 2x - y = 5 \qquad M: \quad -4x + 2y = -10$$

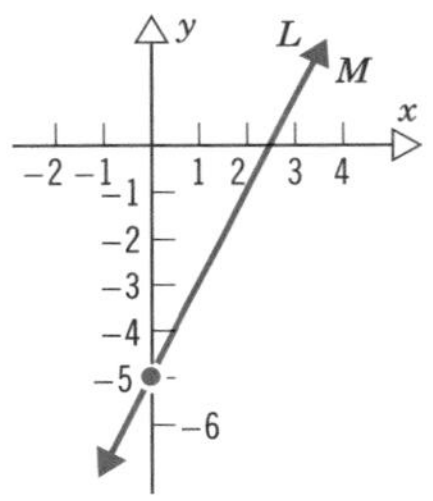

SOLUTION

We put each equation into slope–intercept form:

$$
\begin{array}{ll}
L: \quad 2x - y = 5 & M: \quad -4x + 2y = -10 \\
\qquad -y = -2x + 5 & \qquad 2y = 4x - 10 \\
\qquad y = 2x - 5 & \qquad y = 2x - 5
\end{array}
$$

The lines L and M have the same slope 2 and the same y-intercept $(0, -5)$. Hence, they are coincident. See Figure 21.

$\cdots\cdots\cdots\cdots\cdots\cdots\cdots$

If two lines L and M (in the same plane) have no points in common, they are called **parallel lines.**

Look at Figure 22. For the two nonvertical parallel lines, equal runs result in equal rises. As a result, nonvertical parallel lines have equal slopes. Since they also have no points in common, they will also have different y-intercepts.

Figure 22

Parallel Lines

Parallel lines that are vertical have undefined slope and different x-intercepts. Parallel lines that are nonvertical have the same slope but different y-intercepts.

EXAMPLE 2

Show that the lines given by the equations below are parallel.

$$L: \quad 2x + 3y = 6 \qquad M: \quad 4x + 6y = 0$$

SOLUTION

To see if these lines have equal slopes, we put each equation into slope–intercept form:

$$L:\quad 2x + 3y = 6 \qquad\qquad M:\quad 4x + 6y = 0$$
$$3y = -2x + 6 \qquad\qquad\qquad 6y = -4x$$
$$y = \frac{-2}{3}x + 2 \qquad\qquad\qquad y = \frac{-2}{3}x$$
$$\text{Slope} = \frac{-2}{3} \qquad\qquad\qquad \text{Slope} = \frac{-2}{3}$$
$$y\text{-intercept} = (0, 2) \qquad\qquad y\text{-intercept} = (0, 0)$$

Since each has slope $-\frac{2}{3}$ and different y-intercepts, the lines are parallel. See Figure 23.

· ·

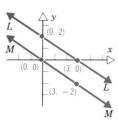

Figure 23

If two lines L and M have exactly one point in common, they are said to **intersect,** and the common point is called the **point of intersection.** Necessarily, the slopes of intersecting lines are unequal.

Intersecting Lines

Intersecting lines have different slopes.

EXAMPLE 3

Show that the lines given by the equations below intersect.

$$L:\quad 2x - y = 5 \qquad M:\quad x + y = 4$$

SOLUTION

We put each equation into slope–intercept form.

$$L:\quad 2x - y = 5 \qquad\qquad M:\quad x + y = 4$$
$$-y = -2x + 5 \qquad\qquad\qquad y = -x + 4$$
$$y = 2x - 5$$

The slope of L is 2 and the slope of M is -1. Hence, the lines intersect.

Now Work Problem 1

· ·

EXAMPLE 4

Find the point of intersection of the two lines

$$L:\quad 2x - y = 5 \qquad M:\quad x + y = 4$$

SOLUTION

The slope–intercept form of each line is

$$L:\quad y = 2x - 5 \qquad M:\quad y = -x + 4$$

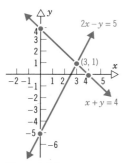

Figure 24

Now Work Problem 11

If (x_0, y_0) denotes the point of intersection, then (x_0, y_0) is a point on both L and M. As a result, we must have

$$y_0 = 2x_0 - 5 \quad \text{and} \quad y_0 = -x_0 + 4$$

By setting these equal and solving, we obtain

$$2x_0 - 5 = -x_0 + 4$$
$$3x_0 = 9$$
$$x_0 = 3$$

Using $x_0 = 3$ in $y_0 = 2x_0 - 5$ (or in $y_0 = -x_0 + 4$), we find

$$y_0 = 2x_0 - 5 = 2(3) - 5 = 1$$

Thus the point of intersection of L and M is (3, 1). See Figure 24.

Check: We verify that the point (3, 1) is on both L and M.

$$L: \quad 2x - y = 2(3) - 1 = 6 - 1 = 5 \qquad M: \quad x + y = 3 + 1 = 4$$

Perpendicular Lines

When two lines intersect and form a right angle, they are said to be **perpendicular.** For example, a vertical line and a horizontal line are perpendicular.

> **Perpendicular Lines**
>
> Two distinct nonvertical lines L and M with slopes m_1 and m_2, respectively, are **perpendicular** if and only if $m_1 m_2 = -1$. That is, the product of their slopes is -1.

EXAMPLE 5

Show that the lines given by the equations below are perpendicular.

$$L: x - 2y = 6 \qquad M: 2x + y = 1$$

SOLUTION

To see if these lines are perpendicular, find the slope of each:

$$x - 2y = 6 \qquad\qquad 2x + y = 1$$
$$-2y = -x + 6 \qquad\qquad y = -2x + 1$$
$$y = \tfrac{1}{2}x - 3 \qquad \text{Slope} = m_2 = -2$$
$$\text{Slope} = m_1 = \tfrac{1}{2}$$

Since $m_1 m_2 = \tfrac{1}{2} \cdot (-2) = -1$, the lines are perpendicular.

EXAMPLE 6

Given the line $x - 4y = 8$, find an equation for the line that passes through $(2, 1)$ and is

(a) Parallel to the given line   (b) Perpendicular to the given line

SOLUTION

First find the slope of the line $x - 4y = 8$ by putting it in slope–intercept form $y = mx + b$:

$$x - 4y = 8$$
$$-4y = -x + 8$$
$$y = \tfrac{1}{4}x - 2$$

The slope of the line is $\tfrac{1}{4}$.

(a) We seek a line parallel to the given line that contains the point $(2, 1)$. The slope of this line must be $\tfrac{1}{4}$. (Do you see why?) Using the point–slope form of the equation of a line, we have

$$y - y_1 = m(x - x_1) \qquad m = \tfrac{1}{4},\ x_1 = 2,\ y_1 = 1$$
$$y - 1 = \tfrac{1}{4}(x - 2)$$
$$y = \tfrac{1}{4}x + \tfrac{1}{2}$$

(b) We seek a line perpendicular to the given line, whose slope is $\tfrac{1}{4}$. The slope m of this line obeys

$$m \cdot \tfrac{1}{4} = -1$$
$$m = -4$$

Since the line we seek contains the point $(2, 1)$, we have

$$y - y_1 = m(x - x_1)$$
$$y - 1 = -4(x - 2)$$
$$y = -4x + 9$$

Figure 25 illustrates these solutions.

Figure 25

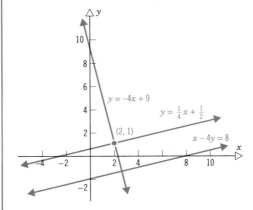

Now Work Problems 31 and 37

Summary

Pair of Lines	Conclusion
Both vertical	Coincident, if they have the same x-intercept
	Parallel, if they have different x-intercepts
One vertical, one nonvertical	Intersect
	Perpendicular, if the nonvertical line is horizontal
Neither vertical	Write the equation of each line in slope–intercept form:
	$y = m_1 x + b_1$, $y = m_2 x + b_2$
	Coincident, if $m_1 = m_2$, $b_1 = b_2$
	Parallel, if $m_1 = m_2$, $b_1 \neq b_2$
	Intersecting, if $m_1 \neq m_2$
	Perpendicular, if $m_1 m_2 = -1$

EXERCISE 1.2   Answers to odd-numbered problems begin on page AN-4.

In Problems 1–10 determine whether the given pairs of lines are parallel, coincident, or intersecting.

1. L: $x + y = 10$
    M: $3x + 3y = 6$

2. L: $x - y = 5$
    M: $-2x + 2y = 8$

3. L: $x + y = 5$
    M: $3x - y = 7$

4. L: $2x + y = 7$
    M: $x - y = -4$

5. L: $-x + y = 2$
    M: $2x - 2y = -4$

6. L: $x + y = -4$
    M: $3x + 3y = -12$

7. L: $2x - 3y = -8$
    M: $6x - 9y = -2$

8. L: $4x - 2y = -7$
    M: $-2x + y = -2$

9. L: $3x - 4y = 1$
    M: $x - 2y = -4$

10. L: $4x + 3y = 2$
    M: $2x - y = -1$

In Problems 11–20 the given pairs of lines intersect. Find the point of intersection. Graph each pair of lines.

11. L: $x + y = 5$
    M: $3x - y = 7$

12. L: $2x + y = 7$
    M: $x - y = -4$

13. L: $x - y = 2$
    M: $2x + y = 7$

14. L: $2x - y = -1$
    M: $x + y = 4$

15. L: $4x + 2y = 4$
    M: $4x - 2y = 4$

16. L: $4x - 2y = 8$
    M: $6x + 3y = 0$

17. L: $3x - 4y = 2$
    M: $x + 2y = 4$

18. L: $4x + 3y = 2$
    M: $2x - y = -1$

19. L: $3x - 2y = -5$
    M: $3x + y = -2$

20. L: $4x + y = 6$
    M: $4x - 2y = 0$

In Problems 21–26 show that the lines are perpendicular.

21. $x - 3y = 2$
    $6x + 2y = 5$

22. $2x + 3y = 4$
    $9x - 6y = 1$

23. $x + 2y = 7$
    $2x - y = 15$

24. $4x + 12y = 3$
    $15x - 5y = -1$

25. $3x + 12y = 2$
    $4x - y = -2$

26. $20x - 2y = -7$
    $x + 10y = 8$

In Problems 27–30, find an equation for the line L. Express the answer using the general form or the slope–intercept form, whichever you prefer.

27.

$y = 2x \mid L$
L is parallel to $y = 2x$

28.

$y = -x$
L is parallel to $y = -x$

29.
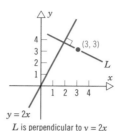
$y = 2x$
L is perpendicular to $y = 2x$

30.
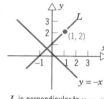
$y = -x$
L is perpendicular to $y = -x$

In Problems 31–40 find an equation for the line with the given properties. Express the answer using the general form or the slope–intercept form, whichever you prefer.

31. Parallel to the line $y = 3x$;   passing through $(-1, 2)$

32. Parallel to the line $y = -2x$;   passing through $(-1, 2)$

33. Parallel to the line $2x - y + 2 = 0$; passing through $(0, 0)$

34. Parallel to the line $x - 2y + 5 = 0$; passing through $(0, 0)$

35. Parallel to the line $x = 5$;   passing through $(4, 2)$

36. Parallel to the line $y = 5$;   passing through $(4, 2)$

37. Perpendicular to the line $y = 2x - 5$; passing through $(-1, -2)$

38. Perpendicular to the line $6x - 2y + 5 = 0$; passing through $(-1, -2)$

39. Perpendicular to the line $y = 2x - 5$; passing through $(\frac{-1}{3}, \frac{4}{5})$

40. Perpendicular to the line $y = 3x - 15$; passing through $(\frac{-2}{3}, \frac{3}{5})$

41. Find t so that $tx - 4y + 3 = 0$ is perpendicular to the line $2x + 2y - 5 = 0$.

42. Find t so that $(1, 2)$ is a point on the line $tx - 3y + 4 = 0$.

43. Find the equation of the line passing through $(-2, -5)$ and perpendicular to the line through $(-2, 9)$ and $(3, -10)$.

44. Find the equation of the line passing through $(-2, -5)$ and perpendicular to the line through $(-4, 5)$ and $(2, -1)$.

45. Find the equation of the horizontal line passing through $(-1, -3)$.

46. Find the equation of the vertical line passing through $(-2, 5)$.

1.3 APPLICATIONS

Prediction

Linear equations are sometimes used as predictors of future results. Let's look at an example.

EXAMPLE 1

Predicting the Cost of a Home   In 1990 the cost of an average home in the Midwest was \$75,000.* In 1994 the cost was \$87,000. Assuming that the relationship between time and cost is linear, develop a formula for predicting the cost of an average home in 1999.

* *Source:* National Association of Realtors.

SOLUTION

We agree to let x represent the year and y represent the cost. We seek a relationship between x and y. Two points on the graph of the equation relating x and y are

$$(1990, 75{,}000) \qquad \text{and} \qquad (1994, 87{,}000)$$

The assumption is that the equation relating x and y is linear. The slope of this line is

$$\frac{87{,}000 - 75{,}000}{1994 - 1990} = 3000$$

Using this fact and the point $(1990, 75{,}000)$, the point–slope form of the equation of the line is

$$y - 75{,}000 = 3000(x - 1990)$$

For $x = 1999$ we find the cost of an average home to be

$$\begin{aligned}
y &= 75{,}000 + 3000(x - 1990) \\
&= 75{,}000 + 3000(1999 - 1990) \\
&= 75{,}000 + 3000(9) \\
&= 102{,}000
\end{aligned}$$

Figure 26 illustrates the situation.

Figure 26

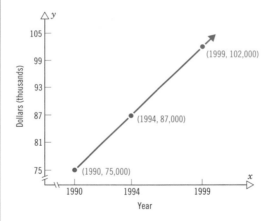

This prediction of future cost is based on the assumption that annual increases remain constant. If this assumption is not accurate, our predictions will be incorrect.

Now Work Problem 1

· ·

Break-Even Point

In many businesses the cost C of production and the number x of items produced can be expressed as a linear equation. Similarly, sometimes the revenue R obtained from sales and the number x of items produced can be expressed as a linear equation. When the cost C of production exceeds the revenue R from the sales, the business is operating at a loss; when the revenue R exceeds the cost C, there is a profit; and when the revenue R and the cost C are equal, there is no profit or loss. The point at which $R = C$, that is, the point of intersection of the two lines, is usually referred to as the **break-even point** in business.

EXAMPLE 2

Finding the Break-Even Point   Sweet Delight Candies, Inc., has daily fixed costs from salaries and building operations of $300. Each pound of candy produced costs $1 and is sold for $2.

(a) Find the cost C of production for x pounds of candy.
(b) Find the revenue R from selling x pounds of candy.
(c) What is the break-even point? That is, how many pounds of candy must be sold daily to guarantee no loss and no profit?

SOLUTION

(a) The cost C of production is the fixed cost plus the variable cost of producing x pounds of candy at $1 per pound. Thus

$$C = \$1 \cdot x + \$300 = x + 300$$

(b) The revenue R realized from the sale of x pounds of candy at $2 per pound is

$$R = \$2 \cdot x = 2x$$

(c) The break-even point is the point where $R = C$. Setting $R = C$, we find

$$2x = x + 300$$
$$x = 300$$

That is, 300 pounds of candy must be sold to break even.

....

In Figure 27 we see a graphical interpretation of the break-even point for Example 2. Note that for $x > 300$, the revenue R always exceeds the cost C so that a profit results. Similarly, for $x < 300$, the cost exceeds the revenue, resulting in a loss.

Figure 27

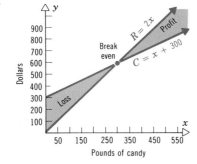

EXAMPLE 3

Analyzing Break-Even Points   After negotiations with employees of Sweet Delight Candies and an increase in the price of sugar, the daily cost C of production for x pounds of candy is

$$C = \$1.05x + \$330$$

(a) If each pound of candy is sold for $2.00, how many pounds must be sold daily to make a profit?

(b) If the selling price is increased to $2.25 per pound, what is the break-even point?

(c) If it is known that at least 325 pounds of candy can be sold daily, what price should be charged per pound to guarantee no loss?

SOLUTION

(a) If each pound is sold for $2.00, the revenue R from sales is

$$R = \$2x$$

where x represents the number of pounds sold. When we set $R = C$, we find that the break-even point is the solution of

$$2x = 1.05x + 330$$
$$0.95x = 330$$
$$x = \frac{330}{0.95} = 347.37$$

Thus, if 347 pounds of candy are sold, a loss is incurred; if 348 pounds or more are sold, a profit results.

(b) If the selling price is increased to $2.25 per pound, the revenue R from sales is

$$R = \$2.25x$$

The break-even point is the solution of

$$2.25x = 1.05x + 330$$
$$1.2x = 330$$
$$x = \frac{330}{1.2} = 275$$

With the new selling price, the break-even point is 275 pounds.

(c) If we know that at least 325 pounds of candy will be sold daily, the price per pound p needed to guarantee no loss (that is, to guarantee at worst a break-even point) is the solution of

$$325p = (1.05)(325) + 330$$
$$325p = 671.25$$
$$p = \$2.07$$

We should charge at least $2.07 per pound to guarantee no loss, provided at least 325 pounds will be sold.

....

EXAMPLE 4

Analyzing Break-Even Points   A producer sells items for $0.30 each.

(a) If the cost for production is

$$C = \$0.15x + \$105$$

where x is the number of items sold, find the break-even point.
(b) If the cost can be changed to

$$C = \$0.12x + \$110$$

would it be advantageous?

SOLUTION

The revenue R received is

$$R = \$0.3x$$

(a) If the cost for production is $C = \$0.15x + \$105$, then the break-even point is the solution of the equation

$$0.3x = 0.15x + 105$$
$$0.15x = 105$$
$$x = 700$$

Thus the break-even point is 700 items.
(b) If the revenue received remains at $R = \$0.3x$, but the cost for production changes to $C = \$0.12x + \$110$, then the break-even point is the solution of the equation

$$0.3x = 0.12x + 110$$
$$0.18x = 110$$
$$x = 611.11$$

The break-even point for the cost in (a) was 700 items. Since the cost in (b) will require fewer items to be sold in order to break even, management should probably change over to the new cost. See Figure 28.

Figure 28

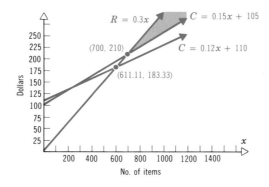

Now Work Problem 13

Mixture Problems

EXAMPLE 5

Mixing Peanuts   A store that specializes in selling nuts sells cashews for $5 per pound and peanuts for $2 per pound. At the end of the month the manager finds that the peanuts are not selling well. In order to sell 30 pounds of peanuts more quickly, the manager decides to mix the 30 pounds of peanuts with some cashews and sell the mixture of peanuts and cashews for $3 a pound. How many pounds of cashews should be mixed with the peanuts so that the revenue remains the same?

SOLUTION

There are two unknowns: the number of pounds of cashews (call this x) and the number of pounds of the mixture (call this y). Since we know that the number of pounds of cashews plus 30 pounds of peanuts equals the number of pounds of the mixture, we can write

$$y = x + 30$$

Also, in order to keep revenue the same, we must have

$$\begin{pmatrix} \text{Price} \\ \text{per} \\ \text{pound} \end{pmatrix} \cdot \begin{pmatrix} \text{Pounds} \\ \text{of} \\ \text{cashews} \end{pmatrix} + \begin{pmatrix} \text{Price} \\ \text{per} \\ \text{pound} \end{pmatrix} \cdot \begin{pmatrix} \text{Pounds} \\ \text{of} \\ \text{peanuts} \end{pmatrix} = \begin{pmatrix} \text{Price} \\ \text{per} \\ \text{pound} \end{pmatrix} \cdot \begin{pmatrix} \text{Pounds} \\ \text{of} \\ \text{mixture} \end{pmatrix}$$

That is,

$$5x + 2(30) = 3y$$
$$\tfrac{5}{3}x + 20 = y$$

Thus we have the two equations

$$y = x + 30 \qquad \text{and} \qquad y = \tfrac{5}{3}x + 20$$

Since the number of pounds of the mixture is the same in each case, we have

$$\tfrac{5}{3}x + 20 = x + 30$$
$$\tfrac{2}{3}x = 10$$
$$x = 15$$

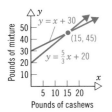

Figure 29

The manager should mix 15 pounds of cashews with 30 pounds of peanuts. See Figure 29. Notice that the point of intersection $(15, 45)$ represents the pounds of cashews (15) in the mixture (45 pounds).

Now Work Problem 15

....

Economics

The **supply equation** in economics is used to specify the amount of a particular commodity that sellers are willing to offer in the market at various prices. The **demand**

equation specifies the amount of a particular commodity that buyers are willing to purchase at various prices.

An increase in price p usually causes an increase in the supply S and a decrease in demand D. On the other hand, a decrease in price brings about a decrease in supply and an increase in demand. The **market price** is defined as the price at which supply and demand are equal (the point of intersection).

Figure 30 illustrates a typical supply/demand situation.

Figure 30

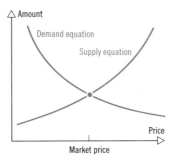

EXAMPLE 6

Supply and Demand   The supply and demand for flour have been estimated as being given by the equations

$$S = 0.8p + 0.5 \qquad D = -0.4p + 1.5$$

where p is measured in dollars and S and D are measured in pound units of flour. Find the market price and graph the supply and demand equations.

SOLUTION

The market price p is the solution of the equation

$$S = D$$
$$0.8p + 0.5 = -0.4p + 1.5$$
$$1.2p = 1$$
$$p = 0.83$$

Thus, at a price of $0.83 per pound, supply and demand for flour are equal.

The graphs are shown in Figure 31.

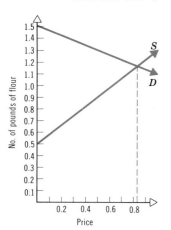

Figure 31

Now Work Problem 27

·······················

EXERCISE 1.3   Answers to odd-numbered problems begin on page AN-4.

Problems 1–6 involve prediction.

1. Suppose the sales of a company are given by

$$S = \$5000x + \$80,000$$

where x is measured in years and $x = 0$ corresponds to the year 1996.

(a) Find S when $x = 0$.
(b) Find S when $x = 3$.
(c) Find the predicted sales in 2001, assuming this trend continues.
(d) Find the predicted sales in 2004, assuming this trend continues.

2. Rework Problem 1 if the sales of the company are given by

$$S = \$3000x + \$60,000$$

3. **Predicting the Cost of a Compact Car**   In 1992, the cost of a compact car averaged $7000. In 1995, the cost of a compact car averaged $8500. Assuming that the relationship between time and cost is linear, develop a formula for predicting the average cost of a compact car in the future. What do you predict the average cost of a compact car will be in 1997?

4. **Oil Depletion**   Oil is pumped from an oil field at a constant rate each year, so that its oil reserves have been decreasing linearly with time. Geologists estimate that the field reserves were 400,000 barrels in 1980 and 320,000 barrels in 1990.

   (a) Write an equation describing the amount of oil left in the field at any time.
   (b) If the trend continues, when will the oil well dry out?

5. **SAT Scores**   The average SAT scores of incoming freshmen at a midwestern college have been declining at a constant rate in recent years. In 1990 the average SAT score was 592, while in 1994 the average SAT score was only 564.

   (a) Write an equation that will give the average SAT score at any time.
   (b) If the trends continue, what will the average SAT score of incoming freshmen be in 1998? Do you think this is why the scoring was changed in 1995?

6. **College Degrees**   The percent of people over 55 who have college degrees is summarized in the following table:

Year (t)	1970	1975	1980	1985	1990
Percent with college degree (y)	30	34	38	42	46

   (a) Find an equation of the line through the points (1970, 30) (1990, 46).
   (b) If the trend continues, estimate the percentage of people over 55 who will have a college degree by the year 2000.

Problems 7–14 involve break-even points. In Problems 7–10 find the break-even point for the cost C of production and the revenue R. Graph each result.

7. $C = \$10x + \$600$ $R = \$30x$

8. $C = \$5x + \$200$ $R = \$8x$

9. $C = \$0.2x + \$50$ $R = \$0.3x$

10. $C = \$1800x + \$3000$ $R = \$2500x$

11. **Break-Even Point**   A manufacturer produces items at a daily cost of $0.75 per item and sells them for $1 per item. The daily operational overhead is $300. What is the break-even point? Graph your result.

12. **Break-Even Point**   If the manufacturer of Problem 11 is able to reduce the cost per item to $0.65, but with a result-

ant increase to $350 in operational overhead, is it advantageous to do so? Graph your result.

13. **Profit from Selling Newspapers**   Each Sunday, a newspaper agency sells x copies of a certain newspaper for $1.00 per copy. The cost to the agency for each newspaper is $0.50. The agency pays a fixed cost for storage, delivery, and so on, of $100 per Sunday. How many newspapers need to be sold for the agency to break even?

14. **Profit from Selling Newspapers**   Repeat Problem 13 if the cost to the agency is $0.45 per copy and the fixed cost is $125 per Sunday.

Problems 15–22 involve mixtures.

15. **Mixture Problem**   Sweet Delight Candies sells boxes of candy consisting of creams and caramels. Each box sells for $8.00 and holds 50 pieces of candy (all pieces are the same size). If the caramels cost $0.10 to produce and the creams cost $0.20 to produce, how many caramels and creams should be in each box for no profit or loss? Would you increase or decrease the number of caramels in order to obtain a profit?

16. **Mixture Problem**   The manager of Nutt's Nuts regularly sells cashews for $4.50 per pound, pecans for $4.80 per pound, and peanuts for $1.80 per pound. How many pounds of cashews and pecans should be mixed with 40 pounds of peanuts to obtain a mixture of 100 pounds that will sell for $4.25 a pound so that the revenue is unchanged?

17. **Investment Problem**   Mr. Nicholson has just retired and needs $10,000 per year in supplementary income. He has $150,000 to invest and can invest in AA bonds at 10% annual interest or in Savings and Loan Certificates at 5% interest per year. How much money should be invested in each so that he realizes exactly $10,000 in extra income per year?

18. **Investment Problem**   Mr. Nicholson finds after 2 years that because of inflation he now needs $12,000 per year in supplementary income. How should he transfer his funds to achieve this amount? (Use the data from Problem 17.)

19. **Theater Attendance Problem**   The Star Theater wants to know whether the majority of its patrons are adults or children. During a week in July 2600 tickets were sold and the receipts totaled $11,874. The adult admission is $6 and the children's admission is $3. How many adult patrons were there?

20. **Mixture Problem**   A coffee manufacturer wants to market a new blend of coffee that will cost $2.90 per pound by mixing $2.75 per pound coffee and $3.00 per pound coffee. What amounts of the $2.75 per pound coffee and $3.00 per pound coffee should be blended to obtain the desired mixture? [*Hint:* Assume the total weight of the desired blend is 100 pounds.]

21. **Mixture Problem**   One solution is 15% acid and another is 5% acid. How many cubic centimeters of each should be mixed to obtain 100 cubic centimeters of a solution that is 8% acid?

22. **Investment Problem**   A bank loaned $10,000, some at an annual rate of 8% and some at an annual rate of 12%. If the income from these loans was $1000, how much was loaned at 8%? How much at 12%?

Problems 23–30 involve economics. In Problems 23–26 find the market price for each pair of supply and demand equations.

23. $S = p + 1$ $D = 3 - p$

24. $S = 2p + 3$ $D = 6 - p$

25. $S = 20p + 500$ $D = 1000 - 30p$

26. $S = 40p + 300$ $D = 1000 - 30p$

27. **Market Price of Sugar**   The supply and demand equations for sugar have been estimated to be given by the equations

$$S = 0.7p + 0.4 \qquad D = -0.5p + 1.6$$

Find the market price. What quantity of supply is demanded at this market price? Graph both the supply and demand equations. Interpret the point of intersection of the two lines.

28. **Supply and Demand Problem**   The market price for a certain product is $5.00 per unit and occurs when 14,000 units are produced. At a price of $1, no units are manu-

factured and, at a price of $19.00, no units will be purchased. Find the supply and demand equations, assuming they are linear.

29. **Supply and Demand Problem**   For a certain commodity the supply equation is given by

$$S = 2p + 5$$

At a price of $1, 19 units of the commodity are demanded. If the demand equation is linear and the market price is $3, find the demand equation.

30. **Supply and Demand Problem**   For a certain commodity the demand equation is given by

$$D = -3p + 20$$

At a price of $1, four units of the commodity are supplied. If the supply equation is linear and the market price is $4, find the supply equation.

CHAPTER REVIEW

IMPORTANT TERMS AND CONCEPTS

rectangular coordinates 4	point–slope form 10	perpendicular lines 20
linear equation 4	slope–intercept form 12	break-even point 25
finding intercepts 5	parallel lines 18	supply and demand equations 28
slope of a line 7	intersecting lines 19	

IMPORTANT FORMULAS

Linear Equation, General Form

$$Ax + By = C \quad A, B \text{ not both zero}$$

Slope of a Line

$$m = \frac{y_2 - y_1}{x_2 - x_1} \quad \text{if } x_1 \neq x_2$$

Point–Slope Form of the Equation of a Line

$$y - y_1 = m(x - x_1)$$

Slope–Intercept Form of the Equation of a Line

$$y = mx + b$$

Pair of Lines	*Conclusion*
Both vertical	Coincident, if they have the same x-intercept Parallel, if they have different x-intercepts
One vertical, one nonvertical	Intersect Perpendicular, if the non-vertical line is horizontal
Neither vertical	Write the equation of each line in slope–intercept form: $y = m_1 x + b_1$, $y = m_2 x + b_2$ Coincident, if $m_1 = m_2$, $b_1 = b_2$ Parallel, if $m_1 = m_2$, $b_1 \neq b_2$ Intersecting, if $m_1 \neq m_2$ Perpendicular, if $m_1 m_2 = -1$

TRUE–FALSE ITEMS   Answers are on page AN-5.

T_____ F_____ **1.** In the slope–intercept equation of a line, $y = mx + b$, m is the slope and $(0, b)$ is the x-intercept.

T_____ F_____ **2.** The graph of the equation $Ax + By = C$, where A, B, C are real numbers and A, B are not both zero, is a straight line.

T_____ F_____ **3.** The y-intercept of the line $2x - 3y + 6 = 0$ is $(0, 2)$.

T_____ F_____ **4.** The slope of the line $2x - 4y + 7 = 0$ is $-\frac{1}{2}$.

T_____ F_____ **5.** Parallel lines always have the same intercepts.

T_____ F_____ **6.** Intersecting lines have different slopes.

T_____ F_____ **7.** Perpendicular lines have slopes that are reciprocals of each other.

T_____ F_____ **8.** A linear relation between two variables can always be graphed as a line.

T_____ F_____ **9.** All lines with equal slopes are distinct.

T_____ F_____ **10.** All vertical lines have positive slope.

FILL IN THE BLANKS   Answers are on page AN-5.

1. If (x, y) are rectangular coordinates of a point, the number x is called the _____ and y is called the _____ .

2. The slope of a vertical line is _____ ; the slope of a horizontal line is _____ .

3. If a line slants downward as it moves from left to right, its slope will be a _____ number.

4. If two lines have the same slope but different y-intercepts, they are _____ .

5. If two lines have the same slope and the same y-intercept, they are said to be _____ .

7. Distinct   lines   that   have   different   slopes   are _____ lines.

6. Lines that intersect at right angles are said to be _____ to each other.

REVIEW EXERCISES   Answers to odd-numbered problems begin on page AN-5.

In Problems 1–4 graph each equation.

1. $y = -2x + 3$ **2.** $y = 6x - 2$ **3.** $2y = 3x + 6$ **4.** $3y = 2x + 6$

In Problems 5–8 find a general equation for the line containing each pair of points.

5. $P = (1, 2)$ $Q = (-3, 4)$ **6.** $P = (-1, 3)$ $Q = (1, 1)$

7. $P = (0, 0)$ $Q = (-2, 3)$ **8.** $P = (-2, 3)$ $Q = (0, 0)$

In Problems 9–18 find an equation of the line having the given characteristics. Express the answer using the general form or the slope–intercept form, whichever you prefer.

9. Slope $= -2$;   passing through $(2, -1)$

10. Slope $= 0$;   passing through $(-3, 4)$

11. Slope undefined;   passing through $(-3, 4)$

12. x-intercept $= (2, 0)$;   passing through $(4, -5)$

13. y-intercept $= (0, -2)$;   passing through $(5, -3)$

14. Passing through $(3, -4)$ and $(2, 1)$

15. Parallel to the line $2x - 3y + 4 = 0$;   passing through $(-5, 3)$

16. Parallel to the line $x + y - 2 = 0$;   passing through $(1, -3)$

17. Perpendicular to the line $2x - 3y + 4 = 0$;   passing through $(-5, 3)$

18. Perpendicular to the line $x + y - 2 = 0$;   passing through $(1, -3)$

In Problems 19–22 find the slope and y-intercept of each line. Graph each line.

19. $-9x - 2y + 18 = 0$ **20.** $-4x - 5y + 20 = 0$ **21.** $4x + 2y - 9 = 0$ **22.** $3x + 2y - 8 = 0$

In Problems 23–28 determine whether the two lines are parallel, coincident, or intersecting.

23. $3x - 4y + 12 = 0$
$6x - 8y + 9 = 0$

24. $2x + 3y + 5 = 0$
$4x + 6y + 10 = 0$

25. $x - y + 2 = 0$
$3x - 4y + 12 = 0$

26. $2x + 3y - 5 = 0$
$x + y - 2 = 0$

27. $4x + 6y + 12 = 0$
$2x + 3y + 6 = 0$

28. $3x - y = 0$
$6x - 2y + 5 = 0$

In Problems 29–34 the given pair of lines intersect. Find the point of intersection. Graph the lines.

29. $L:$ $x - y = 4$
$M:$ $x + 2y = 7$

30. $L:$ $x + y = 4$
$M:$ $x - 2y = 1$

31. $L:$ $x - y = -2$
$M:$ $x + 2y = 7$

32. L: $2x + 4y = 4$
    M: $2x - 4y = 8$

33. L: $2x - 4y = -8$
    M: $3x + 6y = 0$

34. L: $3x + 4y = 2$
    M: $x - 2y = 1$

35. Investment Problem Mr. and Mrs. Byrd have just retired and find that they need $10,000 per year to live on. Fortunately, they have a nest egg of $90,000, which they can invest in somewhat risky B-rated bonds at 12% interest per year or in a well-known bank at 5% per year. How much money should they invest in each so that they realize exactly $10,000 in interest income each year?

36. Mixture Problem One solution is 20% acid and another is 12% acid. How many cubic centimeters of each solution should be mixed to obtain 100 cubic centimeters of a solution that is 15% acid?

37. Car Sales The annual sales of Motors, Inc., over a period of 5 years are listed in the table.

Year	Units Sold (in thousands)
1988	3400
1989	3200
1990	3100
1991	2800
1992	2200

(a) Graph this information using the x-axis for years and the y-axis for units sold. (For convenience, use different scales on the axes.)

(b) Draw a line L that passes through two of the points and comes close to passing through the remaining points.
(c) Find the equation of this line L.
(d) Using this equation of the line, what is your estimate for units sold in 1993?

38. Attendance at a Dance A church group is planning a dance in the school auditorium to raise money for its school. The band they will hire charges $500; the advertising costs are estimated at $100; and food will be supplied at the rate of $5 per person. The church group would like to clear at least $900 after expenses.

(a) Determine how many people need to attend the dance for the group to break even if tickets are sold at $10 each.
(b) Determine how many people need to attend in order to achieve the desired profit if tickets are sold for $10 each.
(c) Answer the above two questions if the tickets are sold for $12 each.

Mathematical Questions from Professional Exams

1. CPA Exam The Oliver Company plans to market a new product. Based on its market studies, Oliver estimates that it can sell 5500 units in 1992. The selling price will be $2 per unit. Variable costs are estimated to be 40% of the selling price. Fixed costs are estimated to be $6000. What is the break-even point?

(a) 3750 units   (b) 5000 units
(c) 5500 units   (d) 7500 units

2. CPA Exam The Breiden Company sells rodaks for $6 per unit. Variable costs are $2 per unit. Fixed costs are $37,500. How many rodaks must be sold to realize a profit before income taxes of 15% of sales?

(a) 9375 units   (b) 9740 units
(c) 11,029 units   (d) 12,097 units

3. CPA Exam Given the following notations, what is the break-even sales level in units?

$$SP = \text{Selling price per unit}$$
$$FC = \text{Total fixed cost}$$
$$VC = \text{Variable cost per unit}$$

(a) $\dfrac{SP}{FC \div VC}$ (b) $\dfrac{FC}{VC \div SP}$

(c) $\dfrac{VC}{SP - FC}$ (d) $\dfrac{FC}{SP - VC}$

4. CPA Exam At a break-even point of 400 units sold, the variable costs were $400 and the fixed costs were $200. What will the 401st unit sold contribute to profit before income taxes?

(a) $0   (b) $0.50   (c) $1.00   (d) $1.50

5. CPA Exam A graph is set up with "depreciation expense" on the vertical axis and "time" on the horizontal axis. Assuming linear relationships, how would the graphs for straight-line and sum-of-the-year's-digits depreciation, respectively, be drawn?

(a) Vertically and sloping down to the right
(b) Vertically and sloping up to the right
(c) Horizontally and sloping down to the right
(d) Horizontally and sloping up to the right

The following statement applies to Questions 6–8:

In analyzing the relationship of total factory overhead with changes in direct labor hours, the following relationship was found to exist: $Y = \$1000 + \$2X.$

6. CMA Exam The relationship as shown above is

(a) Parabolic
(b) Curvilinear
(c) Linear
(d) Probabilistic
(e) None of the above

7. CMA Exam Y in the above equation is an estimate of

(a) Total variable costs
(b) Total factory overhead
(c) Total fixed costs
(d) Total direct labor hours
(e) None of the above

8. CMA Exam The $2 in the equation is an estimate of

(a) Total fixed costs
(b) Variable costs per direct labor hour
(c) Total variable costs
(d) Fixed costs per direct labor hour
(e) None of the above

Chapter 2

Systems of Linear Equations; Matrices

In this chapter we take up the problem of solving systems of linear equations containing two or more variables. As the section titles suggest, there are various ways to solve such problems. The *method of substitution* for solving equations in several unknowns goes back to ancient times. The *method of elimination,* though it had existed for centuries, was put into systematic order by Karl Friedrich Gauss (1777–1855) and by Camille Jordan (1838–1922). This method led to the *matrix method* that is now used for solving large systems by computer.

The theory of *matrices* was developed in 1857 by Arthur Cayley (1821–1895), though only later were matrices used as we use them in this chapter. Matrices have become a very flexible instrument, invaluable in almost all areas of mathematics.

2.1* SYSTEMS OF LINEAR EQUATIONS: SUBSTITUTION; ELIMINATION

We begin with an example.

EXAMPLE 1

A movie theater sells adult tickets for $8 each and child tickets for $5 each. One Saturday the theater took in $3255 in revenue. If x represents the number of adult tickets sold and y the number of child tickets sold, write an equation that relates these variables.

SOLUTION

Each adult ticket brings in $8, so x adult tickets will bring in $8x$ dollars. Similarly, y child tickets bring in $5y$ dollars. If the total brought in is $3255, then we must have

$$8x + 5y = 3255$$

....

The equation found in Example 1 is an example of a **linear equation in two variables.** Some other examples of linear equations are

$$2x + 3y = 2 \qquad 5x - 2y + 3z = 10 \qquad 8x_1 + 8x_2 - 2x_3 + 5x_4 = 0$$
$$\text{2 variables} \qquad\qquad \text{3 variables} \qquad\qquad \text{4 variables}$$

Notice that each term contains at most one variable and that each variable has exponent 1.

In general, an equation in n variables is said to be **linear** if it can be written in the form

$$a_1x_1 + a_2x_2 + \cdots + a_nx_n = b$$

where $x_1, x_2, \ldots, x_n$ are n distinct variables, $a_1, a_2, \ldots, a_n, b$ are constants, and at least one of the a_i's is not 0.

In Example 1 suppose we also know that 525 tickets were sold on that Saturday. Then we have another equation relating the variables x and y, namely,

$$x + y = 525$$

The two linear equations

$$x + y = 525$$
$$8x + 5y = 3255$$

form a *system* of linear equations. In general, a **system of linear equations** is a collection of two or more linear equations, each containing one or more variables. Example 2 gives some samples of systems of linear equations.

---

* Based on material from *Precalculus,* 4th ed., by Michael Sullivan. Used here with the permission of the author and Prentice-Hall, Inc.

EXAMPLE 2

(a) $\begin{cases} 2x + y = 5 & (1) \\ -4x + 6y = -2 & (2) \end{cases}$ Two equations containing two variables, x and y

(b) $\begin{cases} x + y + z = 6 & (1) \\ 3x - 2y + 4z = 9 & (2) \\ x - y - z = 0 & (3) \end{cases}$ Three equations containing three variables, x, y, and z

(c) $\begin{cases} x + y + z = 5 & (1) \\ x - y = 2 & (2) \end{cases}$ Two equations containing three variables, x, y, and z

(d) $\begin{cases} x + y + z = 6 & (1) \\ 2x + 2z = 4 & (2) \\ y + z = 2 & (3) \\ x = 4 & (4) \end{cases}$ Four equations containing three variables, x, y, and z

(e) $\begin{cases} x_1 - 2x_2 + x_3 - x_4 = 5 & (1) \\ 3x_1 + x_2 - x_3 - 5x_4 = 2 & (2) \end{cases}$ Two equations containing four variables x_1, x_2, x_3, and x_4

• •

We use a brace, as shown above, to remind us that we are dealing with a system of equations. We also will find it convenient to number each equation in the system.

A **solution** of a system of equations consists of values for the variables that result in each equation of the system being a true statement. To **solve** a system of equations means to find all solutions of the system.

For example, $x = 2$, $y = 1$ is a solution of the system in Example 2(a) because

$$2(2) + 1 = 5 \qquad \text{and} \qquad -4(2) + 6(1) = -2$$

A solution of the system in Example 2(b) is $x = 3$, $y = 2$, $z = 1$ because

$$\begin{cases} 3 + 2 + 1 = 6 & (1) \\ 3(3) - 2(2) + 4(1) = 9 & (2) \\ 3 - 2 - 1 = 0 & (3) \end{cases}$$

Note that $x = 3$, $y = 3$, $z = 0$ is not a solution of the system in Example 2(b):

$$\begin{cases} 3 + 3 + 0 = 6 & (1) \\ 3(3) - 2(3) + 4(0) = 3 \neq 9 & (2) \\ 3 - 3 - 0 = 0 & (3) \end{cases}$$

Although these values satisfy Equations (1) and (3), they do not satisfy Equation (2). Any solution of the system must satisfy *each* equation of the system.

When a system of equations has at least one solution, it is said to be **consistent;** otherwise, it is called **inconsistent.**

Now Work Problem 1

Two Linear Equations Containing Two Variables

We can view the problem of solving a system of two linear equations containing two variables as a geometry problem. The graph of each equation in such a system is a straight line. Thus, a system of two equations containing two variables represents a pair of lines. The lines either (1) are parallel or (2) are intersecting or (3) are coincident (that is, identical).

1. If the lines are parallel, then the system of equations has no solution, because the lines never intersect. The system is **inconsistent.**
2. If the lines intersect, then the system of equations has one solution, given by the point of intersection. The system is **consistent** and the equations are **independent.**
3. If the lines are coincident, then the system of equations has infinitely many solutions, represented by the totality of points on the line. The system is **consistent** and the equations are **dependent.**

Thus a system of equations is either

    (I)   Inconsistent; has no solution

or

    (II)   Consistent; with

          (a)   One solution (equations are independent)

          or

          (b)   Infinitely many solutions (equations are dependent)

Figure 1 illustrates these conclusions.

Figure 1

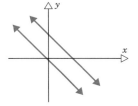 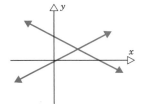 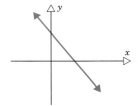

(a) Parallel lines; system has no solution and is inconsistent

(b) Intersecting lines; system has one solution and is consistent; the equations are independent

(c) Coincident lines; system has infinitely many solutions and is consistent; the equations are dependent

EXAMPLE 3

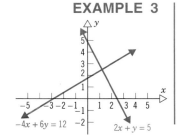

Figure 2

Graph the system: $\begin{cases} 2x + y = 5 & (1) \\ -4x + 6y = 12 & (2) \end{cases}$

SOLUTION

Equation (1) is a line with x-intercept $(\frac{5}{2}, 0)$ and y-intercept $(0, 5)$. Equation (2) is a line with x-intercept $(-3, 0)$ and y-intercept $(0, 2)$.

    Figure 2 shows their graphs.

· ·

From the graph in Figure 2 we see that the lines intersect, so the system is consistent and the equations are independent. We can also use the graph as a means of approximating the solution. For this system the solution would appear to be close to the point $(1, 3)$. The actual solution, which you should verify, is $(\frac{9}{8}, \frac{11}{4})$.

To obtain the exact solution, we use algebraic methods. The first algebraic method we take up is the *method of substitution.*

Method of Substitution

We illustrate the method of substitution by solving the system in Example 3.

EXAMPLE 4

Solve: $\begin{cases} 2x + y = 5 & (1) \\ -4x + 6y = 12 & (2) \end{cases}$

SOLUTION

We solve the first equation for y, obtaining

$$y = -2x + 5$$

We substitute this result for y in the second equation. This results in an equation containing one variable, which we can solve.

$$-4x + 6y = 12$$
$$-4x + 6(-2x + 5) = 12$$
$$-4x - 12x + 30 = 12$$
$$-16x = -18$$
$$x = \tfrac{-18}{-16} = \tfrac{9}{8}$$

Once we know $x = \frac{9}{8}$, we can easily find the value of y by **back-substitution,** that is, by substituting $\frac{9}{8}$ for x in one of the original equations or an equivalent form of one of them. We will substitute $x = \frac{9}{8}$ into the equation $y = -2x + 5$. The result is

$$y = -2x + 5 = -2(\tfrac{9}{8}) + 5 = -\tfrac{9}{4} + \tfrac{20}{4} = \tfrac{11}{4}$$

The solution of the system is $x = \frac{9}{8}$, $y = \frac{11}{4}$.

Check:
$$2x + y = 2(\tfrac{9}{8}) + \tfrac{11}{4} = \tfrac{9}{4} + \tfrac{11}{4} = \tfrac{20}{4} = 5$$
$$-4x + 6y = -4(\tfrac{9}{8}) + 6(\tfrac{11}{4}) = -\tfrac{18}{4} + \tfrac{66}{4} = \tfrac{48}{4} = 12$$

· ·

The method used to solve the system in Example 4 is called **substitution.** The steps used are outlined in the box at the top of the next page.

> **Steps for Solving by Substitution**
>
> **Step 1** Pick one of the equations and solve for one of the variables in terms of the remaining variables.
>
> **Step 2** Substitute the result in the remaining equations.
>
> **Step 3** If one equation in one variable results, solve this equation. Otherwise, repeat Steps 1 and 2 until a single equation with one variable remains.
>
> **Step 4** Find the values of the remaining variables by back-substitution.
>
> **Step 5** Check the solution found.

EXAMPLE 5

Solve: $\begin{cases} 3x - 2y = 5 & (1) \\ 5x - y = 6 & (2) \end{cases}$

SOLUTION

Step 1 After looking at the two equations, we conclude that it is easiest to solve for the variable y in Equation (2):

$$5x - y = 6$$
$$y = 5x - 6$$

Step 2 We substitute this result into Equation (1) and simplify:

$$3x - 2y = 5$$
$$3x - 2(5x - 6) = 5$$

Step 3
$$-7x + 12 = 5$$
$$-7x = -7$$
$$x = 1$$

Step 4 Knowing $x = 1$, we can find y from the equation

$$y = 5x - 6 = 5(1) - 6 = -1$$

Step 5 *Check:* $\begin{cases} 3(1) - 2(-1) = 3 + 2 = 5 \\ 5(1) - (-1) = 5 + 1 = 6 \end{cases}$

The solution of the system is $x = 1$, $y = -1$.

Now Work Problem 7 Using the Method of Substitution

· ·

Method of Elimination

A second method for solving a system of linear equations is the *method of elimination.* This method is usually preferred over substitution if substitution leads to fractions or if the system contains more than two variables. Elimination also provides the necessary motivation for solving systems using matrices (the subject of the next section).

The idea behind the method of elimination is to keep replacing the original equations in the system with equivalent equations until a system of equations with an obvious solution is reached. When we proceed in this way, we obtain **equivalent systems of equations,** namely, systems of equations that have the same solutions. The rules for obtaining equivalent equations are given below:

Rules for Obtaining an Equivalent System of Equations

1. Interchange any two equations of the system.
2. Multiply (or divide) each side of an equation by the same nonzero constant.
3. Replace any equation in the system by the sum (or difference) of that equation and any other equation in the system.

An example will give you the idea. As you work through the example, pay particular attention to the pattern being followed.

EXAMPLE 6

Solve: $\begin{cases} 2x + 3y = 1 & (1) \\ -x + y = -3 & (2) \end{cases}$

SOLUTION

We multiply each side of Equation (2) by 2 so that the coefficients of x in the two equations are negatives of one another. The result is the equivalent system

$$\begin{cases} 2x + 3y = 1 & (1) \\ -2x + 2y = -6 & (2) \end{cases}$$

Now we add the two equations and solve the resulting equation for y.

$$\begin{cases} 2x + 3y = 1 \\ -2x + 2y = -6 \end{cases}$$
$$5y = -5$$
$$y = -1$$

We use this value of y in Equation (1) and simplify, to get

$$2x + 3(-1) = 1$$
$$2x = 4$$
$$x = 2$$

Thus the solution of the original system is $x = 2$, $y = -1$. We leave it to you to check the solution.

The procedure used in Example 6 is called the **method of elimination.** Notice the pattern of the solution. First, we eliminated the variable x from the two equations. Then we back-substituted; that is, we substituted the value found for y back into the first equation to find x.

Steps for Using the Method of Elimination

Step 1 Select two equations from the system and eliminate a variable from them.

Step 2 If there are additional equations in the system, pair off equations and eliminate the same variable from them.

Step 3 Continue Steps 1 and 2 on successive systems until one equation containing one variable remains.

Step 4 Solve for this variable and back-substitute in previous equations until all the variables have been found.

Let's do another example.

EXAMPLE 7

Use the method of elimination to solve the system of equations:

$$\begin{cases} \frac{1}{3}x + 5y = -4 & (1) \\ 2x + 3y = 3 & (2) \end{cases}$$

SOLUTION

We begin by multiplying each side of Equation (1) by 3 in order to remove the fraction $\frac{1}{3}$:

$$\begin{cases} \frac{1}{3}x + 5y = -4 & (1) \\ 2x + 3y = 3 & (2) \end{cases}$$

$$\begin{cases} x + 15y = -12 & (1) \\ 2x + 3y = 3 & (2) \end{cases}$$

Thinking ahead, we decide to multiply each side of Equation (1) above by -2, because then the sum of the two equations will result in an equation with the variable x eliminated. [Note that we also could multiply each side of Equation (1) by 2 and then replace Equation (1) by the difference of the two equations. However, because subtracting is more likely to result in a calculation error, we follow the safer practice of adding equations.]

$$\begin{cases} -2x - 30y = 24 & \text{Multiply Equation (1) by } -2. \\ 2x + 3y = 3 \end{cases}$$

$$-27y = 27 \quad \text{Add the two equations.}$$
$$y = -1 \quad \text{Solve for } y.$$

Now substitute $y = -1$ in Equation (2). Then,

$$2x + 3(-1) = 3$$
$$2x = 6$$
$$x = 3$$

The solution of the original system is $x = 3$, $y = -1$, which you should check.

Now Work Problem 7 Using
the Method of Elimination

....

We have already seen several examples in which the system of equations has exactly one solution. The next example illustrates what happens when the method of substitution is used to solve a system of equations that has no solution.

EXAMPLE 8

Solve: $\begin{cases} 2x + y = 5 & (1) \\ 4x + 2y = 8 & (2) \end{cases}$

SOLUTION

We choose to solve Equation (1) for y:

$$2x + y = 5$$
$$y = 5 - 2x$$

Substituting in Equation (2), we get

$$4x + 2y = 8$$
$$4x + 2(5 - 2x) = 8$$
$$4x + 10 - 4x = 8$$
$$0 \cdot x = -2$$

This equation has no solution. Thus we conclude that the system itself has no solution and is therefore inconsistent.

....

Figure 3 illustrates the pair of lines whose equations form the system in Example 8. Notice that the graphs of the two equations are lines, each with slope -2; one line has y-intercept $(0, 5)$, the other has y-intercept $(0, 4)$. Thus, the lines are parallel and have no point of intersection. This geometric statement is equivalent to the algebraic statement that the system has no solution.

Figure 3

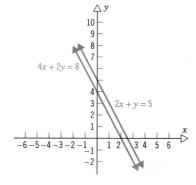

The next example is an illustration of a system with infinitely many solutions.

EXAMPLE 9

Solve: $\begin{cases} 2x + y = 4 & (1) \\ -6x - 3y = -12 & (2) \end{cases}$

SOLUTION

We choose to use the method of elimination:

$\begin{cases} 2x + y = 4 & (1) \\ -6x - 3y = -12 & (2) \end{cases}$

$\begin{cases} 6x + 3y = 12 & (1) \\ -6x - 3y = -12 & (2) \end{cases}$ Multiply each side of Equation (1) by 3.

$\begin{cases} 6x + 3y = 12 & (1) \\ 0 = 0 & (2) \end{cases}$ Replace Equation (2) by the sum of Equations (1) and (2).

The original system is thus equivalent to a system containing one equation. This means that any values of x and y for which $6x + 3y = 12$ (or, equivalently, $2x + y = 4$) are solutions. For example, $x = 2$, $y = 0$; $x = 0$, $y = 4$; $x = -2$, $y = 8$; $x = 4$, $y = -4$; and so on, are solutions. There are, in fact, infinitely many values of x and y for which $2x + y = 4$, so the original system has infinitely many solutions. We will write the solutions of the original system either as

$$y = 4 - 2x$$

where x can be any real number, or as

$$x = 2 - \tfrac{1}{2}y$$

where y can be any real number.

· ·

Figure 4 illustrates the situation presented in Example 9. Notice that the graphs of the two equations are lines, each with slope -2 and each with y-intercept $(0, 4)$. Thus the lines are coincident. Notice also that Equation (2) in the original system is just -3 times Equation (1). This indicates that the two equations are dependent.

Figure 4

$\begin{cases} 2x + y = 4 \\ -6x - 3y = -12 \end{cases}$

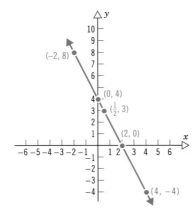

For the system in Example 9 we can find some of the infinite number of solutions by assigning values to x and then finding $y = 4 - 2x$. Thus:

If $x = 4$, then $y = -4$.   This is the point $(4, -4)$ on the graph.

If $x = 0$, then $y = 4$.   This is the point $(0, 4)$ on the graph.

If $x = \frac{1}{2}$, then $y = 3$.   This is the point $(\frac{1}{2}, 3)$ on the graph.

Now Work Problems 15 and 19

EXAMPLE 10

The movie theater of Example 1 charges $8 for each adult admission and $5 for each child. One Saturday 525 tickets were sold, bringing in a total of $3255. How many of each type of ticket were sold?

SOLUTION

If x represents the number of adult tickets and y the number of child tickets, then, as mentioned at the beginning of this chapter, the given information leads to the system of equations

$$\begin{cases} x + y = 525 \\ 8x + 5y = 3255 \end{cases}$$

We use elimination and multiply the first equation by -5 and then add the two equations:

$$\begin{cases} -5x - 5y = -2625 \\ 8x + 5y = 3255 \end{cases}$$
$$3x = 630$$
$$x = 210$$

Since $x + y = 525$, then $y = 525 - x = 525 - 210 = 315$. Thus 210 adult tickets and 315 child tickets were sold that Saturday.

Now Work Problem 55

....

Three Linear Equations Containing Three Variables

Just as a system of two linear equations containing two variables, a system of three linear equations containing three variables also has either no solution, or one solution, or infinitely many solutions.

Let's see how elimination works on a system of three equations containing three variables.

EXAMPLE 11

Use the method of elimination to solve the system of equations:

$$\begin{cases} x + y - z = -1 & (1) \\ 4x - 3y + 2z = 16 & (2) \\ 2x - 2y - 3z = 5 & (3) \end{cases}$$

SOLUTION

For a system of three equations we attempt to eliminate one variable at a time, using pairs of equations. Our plan of attack on this system will be to first eliminate the variable x from Equations (1) and (3) and then from (1) and (2). Next we will eliminate the variable y from the resulting equations, leaving one equation containing only the variable z. Back-substitution can then be used to obtain the values of x and y.

We begin by multiplying each side of Equation (1) by -2, in anticipation of eliminating the variable x by adding Equations (1) and (3):

$$\begin{cases} -2x - 2y + 2z = 2 & (1) \\ 4x - 3y + 2z = 16 & (2) \\ 2x - 2y - 3z = 5 & (3) \end{cases}$$ Multiply each side of Equation (1) by -2.

$$\begin{cases} -2x - 2y + 2z = 2 & (1) \\ 4x - 3y + 2z = 16 & (2) \\ -4y - z = 7 & (3) \end{cases}$$ Replace Equation (3) by the sum of Equations (1) and (3).

We now eliminate the same variable x from Equation (2):

$$\begin{cases} -4x - 4y + 4z = 4 & (1) \\ 4x - 3y + 2z = 16 & (2) \\ -4y - z = 7 & (3) \end{cases}$$ Multiply each side of Equation (1) by 2.

$$\begin{cases} -4x - 4y + 4z = 4 & (1) \\ -7y + 6z = 20 & (2) \\ -4y - z = 7 & (3) \end{cases}$$ Replace Equation (2) by the sum of Equations (1) and (2).

We now eliminate y from Equation (3):

$$\begin{cases} -4x - 4y + 4z = 4 & (1) \\ -28y + 24z = 80 & (2) \\ 28y + 7z = -49 & (3) \end{cases}$$ Multiply each side by 4. Multiply each side by -7.

$$\begin{cases} -4x - 4y + 4z = 4 & (1) \\ -28y + 24z = 80 & (2) \\ 31z = 31 & (3) \end{cases}$$ Replace Equation (3) by the sum of Equations (2) and (3).

$$\begin{cases} -4x - 4y + 4z = 4 & (1) \\ -28y + 24z = 80 & (2) \\ z = 1 & (3) \end{cases}$$ Multiply each side by $\frac{1}{31}$.

$$\begin{cases} -4x - 4y + 4 = 4 & (1) \\ -28y + 24 = 80 & (2) \\ z = 1 & (3) \end{cases}$$ Back-substitute; replace z by 1 in Equations (1) and (2).

$$\begin{cases} -4x - 4y \quad\quad = \quad 0 \quad (1) \\ \quad\quad y \quad\quad = -2 \quad (2) \\ \quad\quad\quad z = \quad 1 \quad (3) \end{cases}$$ Solve Equation (2) for y.

$$\begin{cases} -4x + 8 \quad\quad = \quad 0 \quad (1) \\ \quad\quad y \quad\quad = -2 \quad (2) \\ \quad\quad\quad z = \quad 1 \quad (3) \end{cases}$$ Back-substitute; replace y by -2.

$$\begin{cases} x = \quad 2 \quad (1) \\ y = -2 \quad (2) \\ z = \quad 1 \quad (3) \end{cases}$$

The solution of the original system is $x = 2$, $y = -2$, $z = 1$. (You should check this.)

····················

Look back over the solution given in Example 11. Note the pattern of making Equation (3) contain only the variable z, followed by making Equation (2) contain only the variable y and Equation (1) contain only the variable x. Although which variables to isolate is your choice, the method remains the same for all systems.

Now Work Problem 37

EXERCISE 2.1   Answers to odd-numbered problems begin on page AN-6.

In Problems 1–6 verify that the values of the variables listed are solutions of the system of equations.

1. $\begin{cases} 2x - y = 5 \\ 5x + 2y = 8 \end{cases}$
   $x = 2, y = -1$

2. $\begin{cases} 3x + 2y = \quad 2 \\ x - 7y = -30 \end{cases}$
   $x = -2, y = 4$

3. $\begin{cases} 3x - 4y = 4 \\ x - 3y = \frac{1}{2} \end{cases}$
   $x = 2, y = \frac{1}{2}$

4. $\begin{cases} 2x + y = \quad 0 \\ 5x - 4y = -\frac{13}{2} \end{cases}$
   $x = -\frac{1}{2}, y = 1$

5. $\begin{cases} 3x + 3y + 2z = \quad 4 \\ x - y - z = \quad 0 \\ 2y - 3z = -8 \end{cases}$
   $x = 1, y = -1, z = 2$

6. $\begin{cases} 4x - z = 7 \\ 8x + 5y - z = 0 \\ -x - y + 5z = 6 \end{cases}$
   $x = 2, y = -3, z = 1$

In Problems 7–42 solve each system of equations. If the system has no solution, say it is inconsistent. Use either <u>substitution or elimination</u>.

7. $\begin{cases} x + y = 8 \\ x - y = 4 \end{cases}$

8. $\begin{cases} x + 2y = 5 \\ x + y = 3 \end{cases}$

9. $\begin{cases} 5x - y = 13 \\ 2x + 3y = 12 \end{cases}$

10. $\begin{cases} x + 3y = \quad 5 \\ 2x - 3y = -8 \end{cases}$

11. $\begin{cases} 3x = 24 \\ x + 2y = 0 \end{cases}$

12. $\begin{cases} 4x + 5y = -3 \\ -2y = -4 \end{cases}$

13. $\begin{cases} 3x - 6y = 24 \\ 5x + 4y = 12 \end{cases}$

14. $\begin{cases} 2x + 4y = \quad 16 \\ 3x - 5y = -9 \end{cases}$

15. $\begin{cases} 2x + y = 1 \\ 4x + 2y = 6 \end{cases}$

16. $\begin{cases} x - y = 5 \\ -3x + 3y = 2 \end{cases}$

17. $\begin{cases} 2x - 4y = -2 \\ 3x + 2y = 3 \end{cases}$

18. $\begin{cases} 3x + 3y = 3 \\ 4x + 2y = \frac{8}{3} \end{cases}$

19. $\begin{cases} x + 2y = 4 \\ 2x + 4y = 8 \end{cases}$

20. $\begin{cases} 3x - y = 7 \\ 9x - 3y = 21 \end{cases}$

21. $\begin{cases} 2x - 3y = -1 \\ 10x + 10y = 5 \end{cases}$

22. $\begin{cases} 3x - 2y = 0 \\ 5x + 10y = 4 \end{cases}$

23. $\begin{cases} 2x + 3y = 6 \\ x - y = \frac{1}{2} \end{cases}$

24. $\begin{cases} \frac{1}{2}x + y = -2 \\ x - 2y = 8 \end{cases}$

25. $\begin{cases} 2x + 3y = 5 \\ 4x + 6y = 10 \end{cases}$

26. $\begin{cases} 2x + 3y = 5 \\ 4x + 6y = 6 \end{cases}$

27. $\begin{cases} 3x - 5y = 3 \\ 15x + 5y = 21 \end{cases}$

28. $\begin{cases} 2x - y = -1 \\ x + \frac{1}{2}y = \frac{3}{2} \end{cases}$

29. $\begin{cases} x - y = 6 \\ 2x - 3z = 16 \\ 2y + z = 4 \end{cases}$

30. $\begin{cases} 2x + y = -4 \\ -2y + 4z = 0 \\ 3x - 2z = -11 \end{cases}$

31. $\begin{cases} x - 2y + 3z = 7 \\ 2x + y + z = 4 \\ -3x + 2y - 2z = -10 \end{cases}$

32. $\begin{cases} 2x + y - 3z = 0 \\ -2x + 2y + z = -7 \\ 3x - 4y - 3z = 7 \end{cases}$

33. $\begin{cases} x - y - z = 1 \\ 2x + 3y + z = 2 \\ 3x + 2y = 0 \end{cases}$

34. $\begin{cases} 2x - 3y - z = 0 \\ -x + 2y + z = 5 \\ 3x - 4y - z = 1 \end{cases}$

35. $\begin{cases} x - y - z = 1 \\ -x + 2y - 3z = -4 \\ 3x - 2y - 7z = 0 \end{cases}$

36. $\begin{cases} 2x - 3y - z = 0 \\ 3x + 2y + 2z = 2 \\ x + 5y + 3z = 2 \end{cases}$

37. $\begin{cases} 2x - 2y + 3z = 6 \\ 4x - 3y + 2z = 0 \\ -2x + 3y - 7z = 1 \end{cases}$

38. $\begin{cases} 3x - 2y + 2z = 6 \\ 7x - 3y + 2z = -1 \\ 2x - 3y + 4z = 0 \end{cases}$

39. $\begin{cases} x + y - z = 6 \\ 3x - 2y + z = -5 \\ x + 3y - 2z = 14 \end{cases}$

40. $\begin{cases} x - y + z = -4 \\ 2x - 3y + 4z = -15 \\ 5x + y - 2z = 12 \end{cases}$

41. $\begin{cases} x + 2y - z = -3 \\ 2x - 4y + z = -7 \\ -2x + 2y - 3z = 4 \end{cases}$

42. $\begin{cases} x + 4y - 3z = -8 \\ 3x - y + 3z = 12 \\ x + y + 6z = 1 \end{cases}$

43. The sum of two numbers is 81. The difference of twice one and three times the other is 62. Find the two numbers.

44. The difference of two numbers is 40. Six times the smaller one less the larger one is 5. Find the two numbers.

45. Dimensions of a Floor The perimeter of a rectangular floor is 90 feet. Find the dimensions of the floor if the length is twice the width.

46. Dimensions of a Field The length of fence required to enclose a rectangular field is 3000 meters. What are the dimensions of the field if it is known that the difference between its length and width is 50 meters?

47. Cost of Fast Food Four large cheeseburgers and two chocolate shakes cost a total of $7.90. Two shakes cost 15 cents more than one cheeseburger. What is the cost of a cheeseburger? A shake?

48. Movie Theater Tickets A movie theater charges $7 for adults and $4 for children under 12. On a day when 325 people paid for admission, the total receipts were $1900. How many who paid were adults? How many were under 12?

49. Investment Mr. Nicholson has just retired and needs $6000 per year in income to live on. He has $50,000 to invest and can invest in AA bonds at 15% annual interest or in Savings and Loan Certificates at 7% interest per year. How much money should be invested in each so that he realizes exactly $6000 in income per year?

50. Investment Mr. Nicholson finds after 2 years that because of inflation he now needs $7000 per year to live on. How should he transfer his funds to achieve this amount? (Use the data from Problem 49.)

51. Joan has $1.65 in her piggy bank. She knows she placed only nickels and quarters in the bank and she knows that, in all, she put 13 coins in the bank. Can she find out how many nickels she has without breaking her bank?

52. Mixture A coffee manufacturer wants to market a new blend of coffee that will cost $5 per pound by mixing $3.75 per pound coffee and $8 per pound coffee. What amounts of the $3.75 per pound coffee and $8 per pound coffee should be blended to obtain the desired mixture? [*Hint:* Assume the total weight of the desired blend is 100 pounds.]

53. Mixture One solution is 30% acid and another is 10% acid. How many cubic centimeters of each should be mixed to obtain 100 cc of a solution that is 18% acid?

54. Investment   A bank loaned $10,000, some at an annual rate of 8% and some at an annual rate of 18%. If the income from these loans was $1000, how much was loaned at 8%? How much at 18%?

55. Receipts   The Star Theater wants to know whether the majority of its patrons are adults or children. During a week in July 5200 tickets were sold and the receipts totaled $33,840. The adult admission is $8 and the children's admission is $4. How many adult patrons were there?

56. Mixture   A candy manufacturer sells a candy mix for $1.10 a pound. It is composed of two candies, one worth $0.90 a pound, the other worth $1.50 a pound. How many pounds of each candy is in 60 pounds of mixture?

57. Finance   Two investments are made totaling $50,000. In 1 year the first investment yields a profit of 10%, whereas the second yields a profit of 12%. Total profit for this year is $5250. Find the amount initially put into each investment.

58. Finance   Laura invested part of her money at 8% and the rest at 10%. The income from both investments totaled $3200. If she interchanged her investments, her in-

come would have totaled $2800. How much did she have in each investment?

59. Agriculture   The Smith farm has 1000 acres of land to be used to raise corn and soybeans. The cost of cultivating the corn and the soybeans is $62 and $44 per acre, respectively. Mr. Smith has a budget of $45,800 to use for cultivating these crops. If Mr. Smith wishes to use all the land and all the money budgeted, how many acres of each crop should he plant?

60. Diet Preparation   A farmer prepares feed for livestock by combining two grains. Each unit of the first grain contains 2 units of protein and 5 units of iron, while each unit of the second grain contains 4 units of protein and 1 unit of iron. Determine the number of units of each kind of grain the farmer needs to feed each animal daily if each animal must have 10 units of protein and 16 units of iron each day.

61. Purchasing   David owns a restaurant and wants to order 200 dinner sets. One design costs $20 per set and the other costs $15 per set. He has $3200 with which to buy the sets. How many of each type of set should he buy if he is to use all the money and acquire 200 sets?

Technology Exercises

In Problems 1–4 graph each system of linear equations. Be sure to use a viewing rectangle that shows the point of intersection. Then locate the point of intersection correct to two decimal places.

1. $\begin{cases} 1.2x + 0.8y = 20 \\ -1.3x + 2.7y = 81.1 \end{cases}$

2. $\begin{cases} 215x - 0.1y = 53 \\ 0.5x - 313y = 82 \end{cases}$

3. $\begin{cases} \frac{4}{17}x + \frac{6}{23}y = \frac{2}{3} \\ \frac{9}{14}x - \frac{43}{8}y = \frac{22}{7} \end{cases}$

4. $\begin{cases} \pi x - \sqrt{3}\,y = \sqrt{6} \\ x + \pi y = \sqrt{15} \end{cases}$

2.2 SYSTEMS OF LINEAR EQUATIONS: MATRIX METHOD

In the previous section we discussed two methods for solving systems of linear equations. In this section and the next we present a third method, using matrices, for solving a system of linear equations.

A **matrix** can be defined as a rectangular array of numbers. Some examples of matrices are

(a) $\begin{bmatrix} 8 & 0 \\ 1 & 3 \\ -2 & 4 \end{bmatrix}$ (b) $\begin{bmatrix} 4 & 1 & -3 \\ 2 & 1 & 2 \end{bmatrix}$ (c) $\begin{bmatrix} 4 & 3 \end{bmatrix}$

The square brackets are commonly used to enclose the numbers in a matrix.

The numbers in a matrix are called the **entries** of the matrix. Matrix (a) above has three rows and two columns; matrix (b) has two rows and three columns; matrix (c) has one row and two columns. For example, in matrix (a), the entry 4 is in row 3, column 2; in matrix (b), the entry -3 is in row 1, column 3.

Matrix Representation of a System of Linear Equations

Consider the following two systems of two linear equations containing two variables

$$\begin{cases} x + 4y = 14 \\ 3x - 2y = 0 \end{cases} \quad \text{and} \quad \begin{cases} u + 4v = 14 \\ 3u - 2v = 0 \end{cases}$$

We observe that, except for the symbols used to represent the variables, these two systems are identical. That is, it is the coefficients of the variables and the numbers that appear to the right of the equal sign that distinguish one system from another. As a result, we can dispense altogether with the letters used to symbolize the variables, provided we have some means of keeping track of them. A matrix serves us well in this regard.

When a matrix is used to represent a system of linear equations, it is called the **augmented matrix** of the system. For example, the system

$$\begin{cases} x + 4y = 14 \\ 3x - 2y = 0 \end{cases}$$

can be represented by the augmented matrix

	Column 1 (x)	Column 2 (y)	Column 3 (right-hand side)
Row 1	1	4	14
Row 2	3	-2	0

Here it is understood that column 1 contains the coefficients of the variable x, column 2 contains the coefficients of the variable y, and column 3 contains the numbers to the right of the equal sign. Each row of the matrix represents an equation of the system. Although not required, it has become customary to place a vertical bar in the matrix as a reminder of the equal sign.

In this book we shall follow the practice of using x and y to denote the variables for systems containing two variables. We will use x, y, and z for systems containing three variables; we will use subscripted variables (x_1, x_2, x_3, x_4, etc.) for systems containing four or more variables.

EXAMPLE 1

System of Linear Equations   *Augmented Matrix*

(a) $\begin{cases} 5x - 2y = 5 \\ 2x - y = -4 \end{cases}$ $\begin{bmatrix} 5 & -2 & | & 5 \\ 2 & -1 & | & -4 \end{bmatrix}$

(b) $\begin{cases} 3x + 4y + 3z = 10 \\ x + y - z = 1 \\ x + 6y = 4 \end{cases}$ $\begin{bmatrix} 3 & 4 & 3 & | & 10 \\ 1 & 1 & -1 & | & 1 \\ 1 & 6 & 0 & | & 4 \end{bmatrix}$

(c) $\begin{cases} 3x_1 - x_2 + x_3 + x_4 = 5 \\ 2x_1 + 6x_3 = 2 \end{cases}$ $\begin{bmatrix} 3 & -1 & 1 & 1 & | & 5 \\ 2 & 0 & 6 & 0 & | & 2 \end{bmatrix}$

Now Work Problem 1

Given an augmented matrix, we can write the corresponding system of equations.

EXAMPLE 2

Write a system of linear equations for each augmented matrix.

(a) $\begin{bmatrix} 4 & -1 & | & 6 \\ 3 & 1 & | & 2 \end{bmatrix}$ (b) $\begin{bmatrix} 3 & 2 & 1 & 1 & | & 5 \\ -1 & 0 & 1 & 4 & | & 0 \end{bmatrix}$

SOLUTION

(a) If we use x and y as variables, the system of equations is

$$\begin{cases} 4x - y = 6 \\ 3x + y = 2 \end{cases}$$

(b) If we use x_1, x_2, x_3, and x_4 as variables, the system of equations is

$$\begin{cases} 3x_1 + 2x_2 + x_3 + x_4 = 5 \\ -x_1 + x_3 + 4x_4 = 0 \end{cases}$$

....

The use of matrices to solve a system of linear equations requires the use of *row operations* on the augmented matrix of the system.

Row Operations

We begin by going through the solution of a system of two linear equations in two variables. Although this system is more easily solved using the methods of substitution or elimination (Section 2.1), the *pattern* of the solution shown below will lead to the formulation of a third method for solving systems of linear equations.

EXAMPLE 3

Solve: $\begin{cases} x + 4y = 14 \\ 3x - 2y = 0 \end{cases}$ (1)

SOLUTION

We multiply the first equation by -3, obtaining $-3x - 12y = -42$, and then add it to the second equation.

$$\begin{cases} -3x - 12y = -42 \\ \underline{3x - 2y = 0} \\ -14y = -42 \end{cases}$$

The original system of Equations (1) may now be written as the equivalent system

$$\begin{cases} x + 4y = 14 \\ -14y = -42 \end{cases}$$

Dividing the second equation by -14 or, equivalently, multiplying by $-\frac{1}{14}$, we get

$$\begin{cases} x + 4y = 14 \\ y = 3 \end{cases}$$

To find x, we multiply the second equation by -4 and add it to the first. The result is

$$\begin{cases} x = 2 \\ y = 3 \end{cases} \tag{2}$$

This system of equations has the obvious solution $x = 2$, $y = 3$ and is equivalent to the original system (1), so that the solution of the original system (1) is also $x = 2$, $y = 3$.

······················

We obtained the final system (2) from the original system (1) by a series of operations. We chose the above operations because the *pattern* of the solution has certain advantages:

1. It is algorithmic in character. This means it consists of repetitive steps that can be programmed on a computer.
2. It works on any system of linear equations.

Let's repeat the steps we took to solve Example 3, except now we will manipulate the augmented matrix instead of the equations. For convenience, the equations are listed next to the augmented matrix.

$$\begin{bmatrix} 1 & 4 & | & 14 \\ 3 & -2 & | & 0 \end{bmatrix} \qquad \begin{cases} x + 4y = 14 \\ 3x - 2y = 0 \end{cases}$$

Multiplying the first *equation* by -3 and adding it to the second *equation* corresponds to multiplying each entry in the first *row* of the matrix by -3 and adding the result to each corresponding entry in the second *row*. The result is the matrix

$$\begin{bmatrix} 1 & 4 & | & 14 \\ 0 & -14 & | & -42 \end{bmatrix} \qquad \begin{cases} x + 4y = 14 \\ -14y = -42 \end{cases}$$

Next, divide the second row of this last matrix by -14 or, equivalently, multiply it by $-\frac{1}{14}$, to obtain

$$\begin{bmatrix} 1 & 4 & | & 14 \\ 0 & 1 & | & 3 \end{bmatrix} \qquad \begin{cases} x + 4y = 14 \\ y = 3 \end{cases}$$

Finally, multiply the second row of this matrix by -4 and add it to the first row to obtain

$$\begin{bmatrix} 1 & 0 & | & 2 \\ 0 & 1 & | & 3 \end{bmatrix} \qquad \begin{cases} x = 2 \\ y = 3 \end{cases}$$

When manipulations such as the above are performed on a matrix, they are called *row operations*. We now list the three basic types of row operations:

> **Row Operations**
>
> 1. The interchange of any two rows of a matrix.
> 2. The replacement of any row of a matrix by a nonzero constant multiple of itself.
> 3. The replacement of any row of a matrix by the sum of itself and a constant multiple of some other row.

An example of each type of row operation is given below.

EXAMPLE 4

$$A = \begin{bmatrix} 3 & 4 & -3 \\ 7 & -\frac{1}{2} & 0 \end{bmatrix}$$

1. The matrix obtained by interchanging the first and second rows of A is

$$\begin{bmatrix} 7 & -\frac{1}{2} & 0 \\ 3 & 4 & -3 \end{bmatrix}$$

2. The matrix obtained by multiplying row 2 of A by 5 is

$$\begin{bmatrix} 3 & 4 & -3 \\ 35 & -\frac{5}{2} & 0 \end{bmatrix}$$

We denote this operation by writing $R_2 = 5r_2$, where r_2 denotes the "old" row 2 and R_2 denotes the "new" row 2.

3. The matrix obtained from A by replacing row 2 of A by 3 times row 1 plus row 2 is

$$\begin{bmatrix} 3 & 4 & -3 \\ 3 \cdot 3 + 7 & 3 \cdot 4 + (-\frac{1}{2}) & 3 \cdot (-3) + 0 \end{bmatrix} = \begin{bmatrix} 3 & 4 & -3 \\ 16 & \frac{23}{2} & -9 \end{bmatrix}$$

We denote this operation by writing $R_2 = 3r_1 + r_2$.

Now Work Problem 11

· ·

Compare the three types of row operations to the Rules for Obtaining an Equivalent System of Equations (see page 43). The three types of row operations, when applied to the augmented matrix of a system of equations, result in a new augmented matrix that represents a system of equations equivalent to the original. With this in mind, let's see how row operations are used to solve a system of equations.

EXAMPLE 5

Solve: $\begin{cases} x - y = 2 \\ 2x - 3y = 2 \end{cases}$

SOLUTION

First, we write the augmented matrix:

$$\begin{bmatrix} 1 & -1 & | & 2 \\ 2 & -3 & | & 2 \end{bmatrix} \qquad \begin{cases} x - y = 2 \\ 2x - 3y = 2 \end{cases}$$

Perform the row operation

$$R_2 = -2r_1 + r_2$$

(This has the effect of leaving row 1 unchanged and getting a 0 in row 2, column 1.)

$$\begin{bmatrix} 1 & -1 & \bigm| & 2 \\ 0 & -1 & \bigm| & -2 \end{bmatrix} \qquad \begin{cases} x - y = & 2 \\ -y = & -2 \end{cases}$$

Perform the row operation

$$R_2 = -r_2$$

(This has the effect of getting a 1 in row 2, column 2.)

$$\begin{bmatrix} 1 & -1 & \bigm| & 2 \\ 0 & 1 & \bigm| & 2 \end{bmatrix} \qquad \begin{cases} x - y = 2 \\ y = 2 \end{cases}$$

Perform the row operation

$$R_1 = r_2 + r_1$$

(This has the effect of leaving row 2 unchanged and getting a 0 in row 1, column 2.)

$$\begin{bmatrix} 1 & 0 & \bigm| & 4 \\ 0 & 1 & \bigm| & 2 \end{bmatrix} \qquad \begin{cases} x = 4 \\ y = 2 \end{cases}$$

The solution of the system is $x = 4$, $y = 2$, which you should check.

· ·

The pattern of steps used in the previous examples can be followed to solve any system of linear equations. Let's list those steps.

Steps for Solving a System of Linear Equations Using Matrices

Step 1 Write the augmented matrix corresponding to the system. Remember that the constants must be to the right of the equal sign.

Step 2 Perform row operations to obtain the entry 1 in row 1, column 1.

Step 3 Perform row operations that leave the entry 1 obtained in Step 2 unchanged while getting 0s in the rest of column 1.

Step 4 Perform row operations to get a 1 in row 2, column 2 (if possible) without changing column 1. If this is not possible, move to the first nonzero entry in row 2 and get a 1 there without changing column 1.

Step 5 Perform row operations to get 0s in the rest of this column without changing the entries of the preceding columns.

Step 6 Repeat Step 4, to make the first nonzero entry in row 3 a 1. Continue in this way up to and including the last row of the matrix.

Now Work Problem 21

Here is an example of three linear equations containing three variables.

EXAMPLE 6

Solve: $\begin{cases} x + y + z = 6 \\ 3x + 2y - z = 4 \\ 3x + y + 2z = 11 \end{cases}$

SOLUTION

Step 1 The augmented matrix corresponding to the system is

$$\left[\begin{array}{ccc|c} 1 & 1 & 1 & 6 \\ 3 & 2 & -1 & 4 \\ 3 & 1 & 2 & 11 \end{array}\right]$$

Step 2 Since a 1 appears in row 1, column 1, we can skip to Step 3.

Step 3 Perform the row operations*

$$R_2 = -3r_1 + r_2$$
$$R_3 = -3r_1 + r_3$$

Notice that these operations leave row 1 unchanged, while getting 0s in the rest of column 1:

$$\left[\begin{array}{ccc|c} 1 & 1 & 1 & 6 \\ 0 & -1 & -4 & -14 \\ 0 & -2 & -1 & -7 \end{array}\right]$$

Step 4 To get a 1 in row 2, column 2, we use

$$R_2 = (-1)r_2$$

Notice that column 1 remains unchanged:

$$\left[\begin{array}{ccc|c} 1 & 1 & 1 & 6 \\ 0 & 1 & 4 & 14 \\ 0 & -2 & -1 & -7 \end{array}\right]$$

Step 5 To get 0s in the rest of column 2, we use

$$R_1 = -r_2 + r_1$$
$$R_3 = 2r_2 + r_3$$

Notice that row 2 is left unchanged:

$$\left[\begin{array}{ccc|c} 1 & 0 & -3 & -8 \\ 0 & 1 & 4 & 14 \\ 0 & 0 & 7 & 21 \end{array}\right]$$

* You should convince yourself that doing both of these simultaneously is the same as doing the first followed by the second.

Step 6 Continuing, we seek a 1 in row 3, column 3, so we use

$$R_3 = \tfrac{1}{7}r_3$$

Notice that this is the only choice available to us, since any other choice would change the 0s and 1s already obtained:

$$\begin{bmatrix} 1 & 0 & -3 & | & -8 \\ 0 & 1 & 4 & | & 14 \\ 0 & 0 & 1 & | & 3 \end{bmatrix}$$

To get 0s in the rest of column 3, we use

$$R_1 = 3r_3 + r_1$$
$$R_2 = -4r_3 + r_2$$

The result is

$$\begin{bmatrix} 1 & 0 & 0 & | & 1 \\ 0 & 1 & 0 & | & 2 \\ 0 & 0 & 1 & | & 3 \end{bmatrix}$$

The solution of the system is $x = 1$, $y = 2$, $z = 3$, which you should check.

Now Work Problem 37

· ·

EXAMPLE 7

FoodPerfect Corporation manufactures three models of the Perfect Foodprocessor. Each Model X processor requires 30 minutes of electrical assembly, 40 minutes of mechanical assembly, and 30 minutes of testing; each Model Y requires 20 minutes of electrical assembly, 50 minutes of mechanical assembly, and 30 minutes of testing; and each Model Z requires 30 minutes of electrical assembly, 30 minutes of mechanical assembly, and 20 minutes of testing. If 2500 minutes of electrical assembly, 3500 minutes of mechanical assembly, and 2400 minutes of testing are used in one day, how many of each model will be produced?

SOLUTION

The table below summarizes the given information:

	Model			Time Used
	X	Y	Z	
Electrical Assembly	30	20	30	2500
Mechanical Assembly	40	50	30	3500
Testing	30	30	20	2400

We assign variables to represent the unknowns:

$$x = \text{Number of Model X produced}$$
$$y = \text{Number of Model Y produced}$$
$$z = \text{Number of Model Z produced}$$

Based on the table, we obtain the following system of equations:

$$\begin{cases} 30x + 20y + 30z = 2500 \\ 40x + 50y + 30z = 3500 \\ 30x + 30y + 20z = 2400 \end{cases} \quad \text{or} \quad \begin{cases} 3x + 2y + 3z = 250 & (1) \\ 4x + 5y + 3z = 350 & (2) \\ 3x + 3y + 2z = 240 & (3) \end{cases}$$

The augmented matrix of this system is

$$\left[\begin{array}{ccc|c} 3 & 2 & 3 & 250 \\ 4 & 5 & 3 & 350 \\ 3 & 3 & 2 & 240 \end{array}\right]$$

We could obtain a 1 in row 1, column 1 by using the row operation $R_1 = \frac{1}{3}r_1$, but the introduction of fractions is best avoided. Instead, we use

$$R_2 = (-1)r_1 + r_2$$

to get a 1 in row 2, column 1. The result is

$$\left[\begin{array}{ccc|c} 3 & 2 & 3 & 250 \\ 1 & 3 & 0 & 100 \\ 3 & 3 & 2 & 240 \end{array}\right]$$

Next, interchange row 1 and row 2:

$$\left[\begin{array}{ccc|c} 1 & 3 & 0 & 100 \\ 3 & 2 & 3 & 250 \\ 3 & 3 & 2 & 240 \end{array}\right]$$

Use $R_2 = (-3)r_1 + r_2$, $R_3 = (-3)r_1 + r_3$ to obtain

$$\left[\begin{array}{ccc|c} 1 & 3 & 0 & 100 \\ 0 & -7 & 3 & -50 \\ 0 & -6 & 2 & -60 \end{array}\right]$$

We use $R_2 = (-1)r_2$ followed by $R_2 = r_3 + r_2$:

$$\left[\begin{array}{ccc|c} 1 & 3 & 0 & 100 \\ 0 & 7 & -3 & 50 \\ 0 & -6 & 2 & -60 \end{array}\right] \qquad \left[\begin{array}{ccc|c} 1 & 3 & 0 & 100 \\ 0 & 1 & -1 & -10 \\ 0 & -6 & 2 & -60 \end{array}\right]$$

Next, use $R_1 = (-3)r_2 + r_1$, $R_3 = 6r_2 + r_3$ to obtain

$$\left[\begin{array}{ccc|c} 1 & 0 & 3 & 130 \\ 0 & 1 & -1 & -10 \\ 0 & 0 & -4 & -120 \end{array}\right]$$

Next, use $R_3 = -\frac{1}{4}r_3$. The result is

$$\left[\begin{array}{ccc|c} 1 & 0 & 3 & 130 \\ 0 & 1 & -1 & -10 \\ 0 & 0 & 1 & 30 \end{array}\right]$$

Finally, let $R_1 = (-3)r_3 + r_1$, $R_2 = r_3 + r_2$:

$$\begin{bmatrix} 1 & 0 & 0 & | & 40 \\ 0 & 1 & 0 & | & 20 \\ 0 & 0 & 1 & | & 30 \end{bmatrix}$$

The solution of the system is $x = 40$, $y = 20$, $z = 30$. Thus in one day 40 Model X, 20 Model Y, and 30 Model Z processors were produced.

· ·

The matrix method also reveals systems that have no solution or an infinite number of solutions.

EXAMPLE 8

Solve: $\begin{cases} 3x - 6y = 4 \\ 6x - 12y = 5 \end{cases}$

SOLUTION

The augmented matrix representing this system is

$$\begin{bmatrix} 3 & -6 & | & 4 \\ 6 & -12 & | & 5 \end{bmatrix}$$

To get a 1 in row 1, column 1, we use $R_1 = \frac{1}{3}r_1$. (Note that no other choice is possible, since 6 is a multiple of 3.)

$$\begin{bmatrix} 1 & -2 & | & \frac{4}{3} \\ 6 & -12 & | & 5 \end{bmatrix}$$

To get 0 in row 2, column 1, we use $R_2 = -6r_1 + r_2$:

$$\begin{bmatrix} 1 & -2 & | & \frac{4}{3} \\ 0 & 0 & | & -3 \end{bmatrix}$$

Let's stop here to look at the actual system of equations:

$$\begin{cases} x - 2y = \frac{4}{3} \\ 0x + 0y = -3 \end{cases}$$

The second equation can never be true—no matter what the choice of x and y. Hence, there are no numbers x and y that can obey both equations. That is, the system has no solution.

· ·

EXAMPLE 9

Solve: $\begin{cases} 2x - 3y = 5 \\ 4x - 6y = 10 \end{cases}$

SOLUTION

The augmented matrix representing this system is

$$\left[\begin{array}{cc|c} 2 & -3 & 5 \\ 4 & -6 & 10 \end{array}\right]$$

To get a 1 in row 1, column 1, we use $R_1 = \frac{1}{2}r_1$:

$$\left[\begin{array}{cc|c} 1 & -\frac{3}{2} & \frac{5}{2} \\ 4 & -6 & 10 \end{array}\right]$$

To get a 0 in column 1, row 2, we use $R_2 = -4r_1 + r_2$:

$$\left[\begin{array}{cc|c} 1 & -\frac{3}{2} & \frac{5}{2} \\ 0 & 0 & 0 \end{array}\right]$$

The system of equations looks like

$$\begin{cases} x - \frac{3}{2}y = \frac{5}{2} \\ 0x + 0y = 0 \end{cases}$$

The second equation is true for any choice of x and y. Hence, all numbers x and y that obey the first equation are solutions of the system. Since any point on the line $x - \frac{3}{2}y = \frac{5}{2}$ is a solution, there are an infinite number of solutions.

We can list some of these solutions by assigning values to y and then calculating x from the equation $x - \frac{3}{2}y = \frac{5}{2}$.

If $y = 0$, then $x = \frac{5}{2}$. Thus $x = \frac{5}{2}$, $y = 0$ is a solution.

If $y = 1$, then $x = 4$. Thus $x = 4$, $y = 1$ is a solution.

If $y = 5$, then $x = 10$. Thus $x = 10$, $y = 5$ is a solution.

If $y = -3$, then $x = -2$. Thus $x = -2$, $y = -3$ is a solution.

And so on.

Now Work Problem 47

· ·

We close with a capsule summary of what to expect from any system of two equations containing two variables.

Summary

After completing the steps outlined earlier, one of the following matrices will result for a system of two linear equations containing two variables.

$$\left[\begin{array}{cc|c} 1 & 0 & c \\ 0 & 1 & d \end{array}\right] \qquad \textbf{Unique solution: } x = c, y = d$$

$$\left[\begin{array}{cc|c} a & b & c \\ 0 & 0 & 0 \end{array}\right] \qquad \textbf{Infinite number of solutions: } ax + by = c$$

$$\left[\begin{array}{cc|c} a & b & c \\ 0 & 0 & \text{nonzero number} \end{array}\right] \qquad \textbf{No solution}$$

EXERCISE 2.2   Answers to odd-numbered problems begin on page AN-7.

In Problems 1–10 write the augmented matrix of each system.

1. $\begin{cases} 2x - 3y = 5 \\ x - y = 3 \end{cases}$

2. $\begin{cases} 4x + y = 5 \\ 2x + y = 5 \end{cases}$

3. $\begin{cases} 2x + y + 6 = 0 \\ x + y = -1 \end{cases}$

4. $\begin{cases} x - y = -3 \\ 4x - y + 2 = 0 \end{cases}$

5. $\begin{cases} 2x - y - z = 0 \\ x - y - z = 1 \\ 3x - y = 2 \end{cases}$

6. $\begin{cases} x + y + z = 3 \\ 2x + z = 0 \\ 3x - y - z = 1 \end{cases}$

7. $\begin{cases} 2x - 3y + z - 7 = 0 \\ x + y - z = 1 \\ 2x + 2y - 3z + 4 = 0 \end{cases}$

8. $\begin{cases} 5x - 3y + 6z + 1 = 0 \\ -x - y + z = 1 \\ 2x + 3y + 5 = 0 \end{cases}$

9. $\begin{cases} 4x_1 - x_2 + 2x_3 - x_4 = 4 \\ x_1 + x_2 + 6 = 0 \\ 2x_2 - x_3 + x_4 = 5 \end{cases}$

10. $\begin{cases} 3x - 5 = 0 \\ x - y + z = 6 \\ 2y + z + 4 = 0 \end{cases}$

In Problems 11–20 perform, in order, (a), followed by (b), followed by (c), on the given augmented matrix.

11. $\left[\begin{array}{ccc|c} 1 & -3 & -5 & -2 \\ 2 & -5 & -4 & 5 \\ -3 & 5 & 4 & 6 \end{array}\right]$ (a) $R_2 = -2r_1 + r_2$ (b) $R_3 = 3r_1 + r_3$ (c) $R_3 = 4r_2 + r_3$

12. $\left[\begin{array}{ccc|c} 1 & -3 & -3 & -3 \\ 2 & -5 & 2 & -4 \\ -3 & 2 & 4 & 6 \end{array}\right]$ (a) $R_2 = -2r_1 + r_2$ (b) $R_3 = 3r_1 + r_3$ (c) $R_3 = 7r_2 + r_3$

13. $\left[\begin{array}{ccc|c} 1 & -3 & 4 & 3 \\ 2 & -5 & 6 & 6 \\ -3 & 3 & 4 & 6 \end{array}\right]$ (a) $R_2 = -2r_1 + r_2$ (b) $R_3 = 3r_1 + r_3$ (c) $R_3 = 6r_2 + r_3$

14. $\left[\begin{array}{ccc|c} 1 & -3 & 3 & -5 \\ 2 & -5 & -3 & -5 \\ -3 & -2 & 4 & 6 \end{array}\right]$ (a) $R_2 = -2r_1 + r_2$ (b) $R_3 = 3r_1 + r_3$ (c) $R_3 = 11r_2 + r_3$

15. $\left[\begin{array}{ccc|c} 1 & -3 & 2 & -6 \\ 2 & -5 & 3 & -4 \\ -3 & -6 & 4 & 6 \end{array}\right]$ (a) $R_2 = -2r_1 + r_2$ (b) $R_3 = 3r_1 + r_3$ (c) $R_3 = 15r_2 + r_3$

16. $\left[\begin{array}{ccc|c} 1 & -3 & -4 & -6 \\ 2 & -5 & 6 & -6 \\ -3 & 1 & 4 & 6 \end{array}\right]$ (a) $R_2 = -2r_1 + r_2$ (b) $R_3 = 3r_1 + r_3$ (c) $R_3 = 8r_2 + r_3$

17. $\left[\begin{array}{ccc|c} 1 & -3 & 1 & -2 \\ 2 & -5 & 6 & -2 \\ -3 & 1 & 4 & 6 \end{array}\right]$ (a) $R_2 = -2r_1 + r_2$ (b) $R_3 = 3r_1 + r_3$ (c) $R_3 = 8r_2 + r_3$

18. $\left[\begin{array}{ccc|c} 1 & -3 & -1 & 2 \\ 2 & -5 & 2 & 6 \\ -3 & -6 & 4 & 6 \end{array}\right]$ (a) $R_2 = -2r_1 + r_2$ (b) $R_3 = 3r_1 + r_3$ (c) $R_3 = 15r_2 + r_3$

19. $\left[\begin{array}{ccc|c} 1 & -3 & -2 & 3 \\ 2 & -5 & 2 & -1 \\ -3 & -2 & 4 & 6 \end{array}\right]$ (a) $R_2 = -2r_1 + r_2$ (b) $R_3 = 3r_1 + r_3$ (c) $R_3 = 11r_2 + r_3$

20. $\left[\begin{array}{ccc|c} 1 & -3 & 5 & -3 \\ 2 & -5 & 1 & -4 \\ -3 & 3 & 4 & 6 \end{array}\right]$ (a) $R_2 = -2r_1 + r_2$ (b) $R_3 = 3r_1 + r_3$ (c) $R_3 = 6r_2 + r_3$

In Problems 21–46 solve each system of equations using the method of row operations.

21. $\begin{cases} x + y = 6 \\ 2x - y = 0 \end{cases}$

22. $\begin{cases} x - y = 2 \\ 2x + y = 1 \end{cases}$

23. $\begin{cases} 2x + y = 5 \\ x - y = 1 \end{cases}$

24. $\begin{cases} 3x + 2y = 7 \\ x + y = 3 \end{cases}$

25. $\begin{cases} 2x + 3y = 7 \\ 3x - y = 5 \end{cases}$

26. $\begin{cases} 2x - 3y = 5 \\ 3x + y = 2 \end{cases}$

27. $\begin{cases} 5x - 7y = 31 \\ 3x + 2y = 0 \end{cases}$

28. $\begin{cases} 2x + 8y = 17 \\ 3x - y = 1 \end{cases}$

29. $\begin{cases} 2x - 3y = 0 \\ 4x + 9y = 5 \end{cases}$

30. $\begin{cases} 3x - 4y = 3 \\ 6x + 2y = 1 \end{cases}$

31. $\begin{cases} 4x - 3y = 4 \\ 2x + 6y = 7 \end{cases}$

32. $\begin{cases} 3x - 5y = 3 \\ 6x + 10y = 10 \end{cases}$

33. $\begin{cases} \frac{1}{2}x + \frac{1}{3}y = 2 \\ x + y = 5 \end{cases}$

34. $\begin{cases} x - \frac{1}{4}y = 0 \\ \frac{1}{2}x + \frac{1}{2}y = \frac{5}{2} \end{cases}$

35. $\begin{cases} x + y = 1 \\ 3x - 2y = \frac{4}{3} \end{cases}$

36. $\begin{cases} 4x - y = \frac{11}{4} \\ 3x + y = \frac{5}{2} \end{cases}$

37. $\begin{cases} 2x + y + z = 6 \\ x - y - z = -3 \\ 3x + y + 2z = 7 \end{cases}$

38. $\begin{cases} x + y + z = 5 \\ 2x - y + z = 2 \\ x + 2y - z = 3 \end{cases}$

39. $\begin{cases} x + y - z = -2 \\ 3x + y + z = 0 \\ 2x - y + 2z = 1 \end{cases}$

40. $\begin{cases} 2x - y - z = -5 \\ x + y + z = 2 \\ x + 2y + 2z = 5 \end{cases}$

41. $\begin{cases} 2x + y - z = 2 \\ x + 3y + 2z = 1 \\ x + y + z = 2 \end{cases}$

42. $\begin{cases} 2x + 2y + z = 6 \\ x - y - z = -2 \\ x - 2y - 2z = -5 \end{cases}$

43. $\begin{cases} x + y - z = 0 \\ 2x + 4y - 4z = -1 \\ 2x + y + z = 2 \end{cases}$

44. $\begin{cases} x + y - z = 0 \\ 4x + 2y - 4z = 0 \\ x + 2y + z = 0 \end{cases}$

45. $\begin{cases} 3x + y - z = \frac{2}{3} \\ 2x - y + z = 1 \\ 4x + 2y = \frac{8}{3} \end{cases}$

46. $\begin{cases} x + y = 1 \\ 2x - y + z = 1 \\ x + 2y + z = \frac{8}{3} \end{cases}$

In Problems 47–54 discuss each system of equations. Determine whether the system has a unique solution, no solution, or infinitely many solutions. Use matrix techniques.

47. $\begin{cases} x - y = 5 \\ 2x - 2y = 6 \end{cases}$

48. $\begin{cases} 4x + y = 5 \\ 8x + 2y = 10 \end{cases}$

49. $\begin{cases} 2x - 3y = 6 \\ 4x - 6y = 12 \end{cases}$

50. $\begin{cases} 2x - 3y = 6 \\ 4x - 6y = 8 \end{cases}$

51. $\begin{cases} 5x - 6y = 1 \\ -10x + 12y = 0 \end{cases}$

52. $\begin{cases} 3x + 4y = 7 \\ x - y = 2 \end{cases}$

53. $\begin{cases} 2x + 3y = 5 \\ 4x + 4y = 8 \end{cases}$

54. $\begin{cases} 2x - y = 0 \\ 4x - 2y = 0 \end{cases}$

55. Mixture Problem A store sells cashews for $5 per pound and peanuts for $1.50 per pound. The manager decides to mix 30 pounds of peanuts with some cashews and sell the mixture for $3 per pound. How many pounds of cashews should be mixed with the peanuts so that the mixture will produce the same revenue as would selling the nuts separately?

56. Mixture Problem A store sells almonds for $6 per pound, cashews for $5 per pound, and peanuts for $2 per pound. One week the manager decides to prepare 100 16-ounce packages of nuts by mixing 40 pounds of peanuts with some almonds and cashews. Each package will be sold for $4. How many pounds of almonds and cashews should be mixed with the peanuts so that the mixture will produce the same revenue as selling the nuts separately?

57. Laboratory Work Stations A chemistry laboratory can be used by 38 students at one time. The laboratory has 16 work stations, some set up for 2 students each and the others set up for 3 students each. How many are there of each kind of work station?

58. Cost of Fast Food One group of people purchased 10 hot dogs and 5 soft drinks at a cost of $12.50. A second group bought 7 hot dogs and 4 soft drinks at a cost of $9. What is the cost of a single hot dog? A single soft drink?

59. Refunding a Purchase The grocery store we use does not mark prices on its goods. My wife went to this store, bought three 1-pound packages of bacon and two cartons of eggs, and paid a total of $7.45. Not knowing she went to the store, I also went to the same store, purchased two 1-pound packages of bacon and three cartons of eggs, and paid a total of $6.45. Now we want to return two 1-pound packages of bacon and two cartons of eggs. How much will be refunded?

60. Coin Collections A coin collection consists of 37 coins —nickels, dimes, and quarters. If the collection has a face value of $3.25 and there are 5 more dimes than there are nickels, how many of each coin are in the collection?

61. Theater Seating A Broadway theater has 500 seats, divided into orchestra, main, and balcony seating. Orchestra seats sell for $75, main seats for $50, and balcony seats for $35. If all the seats are sold, the gross revenue to the theater is $23,000. If all the main and balcony seats are sold, but only half the orchestra seats are sold, the gross revenue is $21,500. How many are there of each kind of seat?

62. Investment Problem   An amount of $5000 is put into three investments at rates of 6%, 7%, and 8% per annum, respectively. The total annual income is $358. The income from the first two investments is $70 more than the income from the third investment. Find the amount of each investment.

63. Investment Problem   An amount of $6500 is placed in three investments at rates of 6%, 8%, and 9% per annum, respectively. The total annual income is $480. If the income from the third investment is $60 more than the income from the second investment, find the amount of each investment.

64. Mixture   Sally's Girl Scout troop is selling cookies for the Christmas season. There are three different kinds of cookies in three different containers: *bags* that hold 1 dozen chocolate chip and 1 dozen oatmeal; *gift boxes* that hold 2 dozen chocolate chip, 1 dozen mints, and 1 dozen oatmeal; and *cookie tins* that hold 3 dozen mints and 2 dozen chocolate chip. Sally's mother is having a Christmas party and wants 6 dozen oatmeal cookies, 10 dozen mints, and 14 dozen chocolate chip cookies. How can Sally fill her mother's order?

65. Production   A citrus company completes the preparation of its products by cleaning, filling, and labeling bottles. Each case of orange juice requires 10 minutes in the cleaning machine, 4 minutes in the filling machine, and 2 minutes in the labeling machine. For each case of tomato juice, the times are 12 minutes of cleaning, 4 minutes for filling, and 1 minute to label. Pineapple juice requires 9 minutes of cleaning, 6 minutes of filling, and 1 minute of labeling per case. If the company runs the cleaning machine for 398 minutes, the filling machine for 164 minutes, and the labeling machine for 58 minutes, how many cases of each type of juice are prepared?

66. Production   The manufacture of an automobile requires painting, drying, and polishing. The Rome Motor Company produces three types of cars: the Centurion, the Tribune, and the Senator. Each Centurion requires 8 hours for painting, 2 hours for drying, and 1 hour for polishing. A Tribune needs 10 hours for painting, 3 hours for drying, and 2 hours for polishing. It takes 16 hours of painting, 5 hours of drying, and 3 hours of polishing to prepare a Senator. If the company uses 240 hours for painting, 69 hours for drying, and 41 hours for polishing in a given month, how many of each type of car are produced?

67. Inventory Control   An art teacher finds that colored paper can be bought in three different packages. The first package has 20 sheets of white paper, 15 sheets of blue paper, and 1 sheet of red paper. The second package has 3 sheets of blue paper and 1 sheet of red paper. The last package has 40 sheets of white paper and 30 sheets of blue paper. Suppose he needs 200 sheets of white paper, 180 sheets of blue paper, and 12 sheets of red paper. How many of each type of package should he order?

68. Inventory Control   An interior decorator has ordered 12 cans of sunset paint, 35 cans of brown, and 18 cans of fuchsia. The paint store has special pair packs, containing 1 can each of sunset and fuchsia; darkening packs containing 2 cans of sunset, 5 cans of brown, and 2 cans of fuchsia; and economy packs, containing 3 cans of sunset, 15 cans of brown, and 6 cans of fuchsia. How many of each type of pack should the paint store send to the interior decorator?

69. Diet Preparation   A hospital dietician is planning a meal consisting of three foods whose ingredients are summarized as follows:

	Units of		
	Food I	Food II	Food III
Units of Protein	10	5	15
Units of Carbohydrates	3	6	3
Units of Iron	4	4	6

Determine the number of units of each food needed to create a meal containing 100 units of protein, 50 units of carbohydrates, and 50 units of iron.

70. Production   A luggage manufacturer produces three types of luggage: economy, standard, and deluxe. The company produces 1000 pieces of luggage at a cost of $20, $25, and $30 for the economy, standard, and deluxe luggage, respectively. The manufacturer has a budget of $20,700. Each economy luggage requires 6 hours of labor, each standard luggage requires 10 hours of labor, and each deluxe model requires 20 hours of labor. The manufacturer has a maximum of 6800 hours of labor available. If the manufacturer sells all the luggage, consumes the entire budget, and uses all the available labor, how many of each type of luggage should be produced?

71. Packaging   A recreation center wants to purchase albums to be used in the center. There is no requirement as to the artists. The only requirement is that they purchase 40 rock albums, 32 western albums, and 14 blues albums. There are three different shipping packages offered by the record company. They are an *assorted* carton, containing 2 rock albums, 4 western albums, and 1 blues album; a *mixed* carton containing 4 rock and 2 western albums; and a *single* carton containing 2 blues albums. What combination of these packages is needed to fill the center's order?

72. Mixture   Suppose that a store has three sizes of cans of nuts. The *large* size contains 2 pounds of peanuts and 1 pound of cashews. The *mammoth* size contains 1 pound of walnuts, 6 pounds of peanuts, and 2 pounds of cashews. The *giant* size contains 1 pound of walnuts, 4 pounds of peanuts, and 2 pounds of cashews. Suppose that the store receives an order for 5 pounds of walnuts, 26 pounds of peanuts, and 12 pounds of cashews. How can it fill this order with the given sizes of cans?

73. Mixture   Suppose that the store in Problem 72 receives a new order for 6 pounds of walnuts, 34 pounds of peanuts, and 15 pounds of cashews. How can this order be filled with the given cans?

Technology Exercises
Some graphing calculators are able to perform elementary row operations on matrices stored in the calculator memory. This can be of great assistance, especially in large systems of equations. For example, the TI-81 can swap rows, add two rows together, and multiply a row by a scalar value. Check your user's manual to see how to perform these functions with your graphing calculator.

In Problems 1–4 use the elementary row operations available on your graphing calculator to perform the given operations on the matrix

$$A = \begin{bmatrix} 1 & 0 & 4 & | & 0 \\ -2 & 1 & -4 & | & 1 \\ 3 & -1 & 0 & | & 1 \end{bmatrix}$$

1. $R_1 = -2r_1$

2. $R_1 = -2r_1 + r_3$

3. $R_2 = 3r_1 + r_2$

4. $R_3 = r_1 - 2r_3$

In Problems 5–10 use the elementary row operations available on your graphing calculator to solve each system.

5. $\begin{cases} 2x - 2y + z = 2 \\ x - \frac{1}{2}y + 2z = 1 \\ 2x + \frac{1}{3}y - z = 0 \end{cases}$

6. $\begin{cases} x + y = -1 \\ x - z = 0 \\ y - z = 1 \end{cases}$

7. $\begin{cases} x + y + z = 4 \\ x - y - z = 0 \\ y - z = -4 \end{cases}$

8. $\begin{cases} 2x + y + z = 6 \\ x - y - z = -3 \\ 3x + y + 2z = 7 \end{cases}$

9. $\begin{cases} x_1 + x_2 + x_3 + x_4 = 20 \\ x_2 + x_3 + x_4 = 0 \\ x_3 + x_4 = 13 \\ x_2 - 2x_4 = -5 \end{cases}$

10. $\begin{cases} x_1 - 2x_2 + 3x_3 - 4x_4 = 40 \\ 4x_2 + 6x_4 = -10 \\ x_3 - x_4 = 12 \\ x_2 + 2x_4 = -10 \end{cases}$

2.3 SYSTEMS OF m LINEAR EQUATIONS CONTAINING n VARIABLES

We learned in the previous two sections that a system of linear equations will have either one solution, no solution, or infinitely many solutions. As it turns out, no matter how many equations are in a system of linear equations and no matter how many variables a system has, only these three possibilities can arise.

Thus, for example, the system of three linear equations containing four variables

$$x_1 + 3x_2 + 5x_3 + x_4 = 2$$
$$2x_1 + 3x_2 + 4x_3 + 2x_4 = 1$$
$$x_1 + 2x_2 + 3x_3 + x_4 = 1$$

will have either no solution, one solution, or infinitely many solutions.

SYSTEM OF m EQUATIONS CONTAINING n VARIABLES

A *system of m linear equations containing n variables $x_1, x_2, \ldots, x_n$ is of the form*

$$\begin{cases} a_{11}x_1 + a_{12}x_2 + \cdots + a_{1n}x_n = b_1 \\ a_{21}x_1 + a_{22}x_2 + \cdots + a_{2n}x_n = b_2 \\ a_{31}x_1 + a_{32}x_2 + \cdots + a_{3n}x_n = b_3 \\ \vdots \qquad \vdots \qquad\quad \vdots \qquad\quad \vdots \\ a_{i1}x_1 + a_{i2}x_2 + \cdots + a_{in}x_n = b_i \\ \vdots \qquad \vdots \qquad\quad \vdots \qquad\quad \vdots \\ a_{m1}x_1 + a_{m2}x_2 + \cdots + a_{mn}x_n = b_m \end{cases}$$

where a_{ij} and b_i are real numbers, $i = 1, 2, \ldots, m, j = 1, 2, \ldots, n.$

A **solution** of a system of m equations containing n variables $x_1, x_2, \ldots, x_n$ is any ordered set $(x_1, x_2, \ldots, x_n)$ of real numbers for which *each* of the m equations of the system is satisfied.

Reduced Row-Echelon Form

A system of m linear equations containing n variables will have either no solution, one solution, or infinitely many solutions. We can determine which of these possibilities occurs and, if solutions exist, find them by performing row operations on the augmented matrix of the system until we arrive at a matrix that has the following configuration:

Conditions for the Reduced Row-Echelon Form of a Matrix

1. Any rows that consist entirely of 0s are located at the bottom of the matrix.

2. The first nonzero element in each row is 1 and it has 0s above it and below it.

3. The leftmost 1 in any row is to the right of the leftmost 1 in the row above.

The matrix having this configuration is unique and is called the **reduced row-echelon form** of the augmented matrix.

EXAMPLE 1 | The matrix

$$\left[\begin{array}{ccc|c} 1 & 0 & 0 & 3 \\ 0 & 1 & 0 & 8 \\ 0 & 0 & 1 & -4 \end{array}\right]$$

is the reduced row-echelon form of a system of three linear equations containing three variables. If the variables are x, y, z, this matrix represents the system of equations

$$\begin{cases} x = 3 \\ y = 8 \\ z = -4 \end{cases}$$

Thus the system has the one solution: $x = 3, y = 8, z = -4$.

· ·

EXAMPLE 2

The matrix

$$\left[\begin{array}{ccc|c} 1 & 0 & 3 & 0 \\ 0 & 1 & 2 & 0 \\ 0 & 0 & 0 & 1 \\ 0 & 0 & 0 & 0 \end{array}\right]$$

is the reduced row-echelon form of a system of four equations containing three variables. If the variables are x, y, z, the equation represented by the third row is

$$0 \cdot x + 0 \cdot y + 0 \cdot z = 1 \qquad \text{or} \qquad 0 = 1$$

Since $0 = 1$ is a contradiction, we conclude the system has no solution.

· ·

EXAMPLE 3

The matrix

$$\left[\begin{array}{cccc|c} 1 & 0 & 0 & 2 & 5 \\ 0 & 1 & 0 & 1 & 2 \\ 0 & 0 & 1 & 3 & 4 \end{array}\right]$$

is the reduced row-echelon form of a system of three equations containing four variables. If x_1, x_2, x_3, x_4 are the variables, the system of equations is

$$\begin{cases} x_1 + 2x_4 = 5 \\ x_2 + x_4 = 2 \\ x_3 + 3x_4 = 4 \end{cases}$$

Thus the system has infinitely many solutions—the variable x_4 can take on any value, from which x_1, x_2, x_3, can be calculated. Some of the possibilities are

If $x_4 = 0$, then $x_1 = 5, x_2 = 2, x_3 = 4$.

If $x_4 = 1$, then $x_1 = 3, x_2 = 1, x_3 = 1$.

If $x_4 = 2$, then $x_1 = 1, x_2 = 0, x_3 = -2$.

And so on.

· ·

EXAMPLE 4

The following matrices are not in reduced row-echelon form:

$$\left[\begin{array}{cc|c} 1 & 0 & 0 \\ 0 & 0 & 0 \\ 0 & 1 & 0 \end{array}\right]$$

The second row contains all 0s and the third does not—this violates the rule that states that any rows containing all 0s are at the bottom.

$$\left[\begin{array}{ccc|c} 1 & 0 & 2 & 4 \\ 0 & 0 & 2 & 3 \\ 0 & 0 & 0 & 1 \end{array}\right]$$

The first nonzero element in row 2 is not a 1.

$$\left[\begin{array}{cc|c} 1 & 0 & 0 \\ 0 & 0 & 1 \\ 0 & 1 & 0 \end{array}\right]$$

The leftmost 1 in the third row is not to the right of the leftmost 1 in the row above it.

Now Work Problems 1
and 11

• •

Let's review the procedure for getting the reduced row-echelon form of a matrix.

EXAMPLE 5

Find the reduced row-echelon form of

$$A = \left[\begin{array}{cc|c} 1 & -1 & 2 \\ 2 & -3 & 2 \\ 3 & -5 & 2 \end{array}\right]$$

SOLUTION

The entry in row 1, column 1 is 1. Thus we proceed to obtain a matrix in which all the remaining entries in column 1 are 0s. We can obtain such a matrix by performing the row operations

$$R_2 = -2r_1 + r_2$$
$$R_3 = -3r_1 + r_3$$

The new matrix is

$$\left[\begin{array}{cc|c} 1 & -1 & 2 \\ 0 & -1 & -2 \\ 0 & -2 & -4 \end{array}\right]$$

We want the entry in row 2, column 2 (namely, -1), to be 1. By multiplying row 2 by -1, we obtain

$$\left[\begin{array}{cc|c} 1 & -1 & 2 \\ 0 & 1 & 2 \\ 0 & -2 & -4 \end{array}\right]$$

Now we want the entry in row 1, column 2 and in row 3, column 2 to be 0. This can be accomplished by applying the row operations

$$R_1 = r_2 + r_1$$
$$R_3 = 2r_2 + r_3$$

The new matrix is

$$\left[\begin{array}{cc|c} 1 & 0 & 4 \\ 0 & 1 & 2 \\ 0 & 0 & 0 \end{array}\right]$$

This is the reduced row-echelon form of A.

· ·

EXAMPLE 6

Solve: $\begin{cases} x - y = 2 \\ 2x - 3y = 2 \\ 3x - 5y = 2 \end{cases}$

SOLUTION

The augmented matrix of this system is

$$\left[\begin{array}{cc|c} 1 & -1 & 2 \\ 2 & -3 & 2 \\ 3 & -5 & 2 \end{array}\right]$$

Following the solution to Example 5, we place this matrix in reduced row-echelon form.

$$\left[\begin{array}{cc|c} 1 & 0 & 4 \\ 0 & 1 & 2 \\ 0 & 0 & 0 \end{array}\right]$$

We conclude that the system has the solution $x = 4$, $y = 2$.

· ·

EXAMPLE 7

Solve: $\begin{cases} x - y + 2z = 2 \\ 2x - 3y + 2z = 1 \\ 3x - 5y + 2z = -3 \\ -4x + 12y + 8z = 10 \end{cases}$

SOLUTION

We need to find the reduced row-echelon form of the augmented matrix of this system, namely,

$$\left[\begin{array}{ccc|c} 1 & -1 & 2 & 2 \\ 2 & -3 & 2 & 1 \\ 3 & -5 & 2 & -3 \\ -4 & 12 & 8 & 10 \end{array}\right]$$

The entry 1 is already present in row 1, column 1. To obtain 0s elsewhere in column 1, we use the row operations

$$R_2 = -2r_1 + r_2 \qquad R_3 = -3r_1 + r_3 \qquad R_4 = 4r_1 + r_4$$

The new matrix is

$$\left[\begin{array}{rrr|r} 1 & -1 & 2 & 2 \\ 0 & -1 & -2 & -3 \\ 0 & -2 & -4 & -9 \\ 0 & 8 & 16 & 18 \end{array}\right]$$

To obtain the entry 1 in row 2, column 2, we use $R_2 = -r_2$, obtaining

$$\left[\begin{array}{rrr|r} 1 & -1 & 2 & 2 \\ 0 & 1 & 2 & 3 \\ 0 & -2 & -4 & -9 \\ 0 & 8 & 16 & 18 \end{array}\right]$$

To obtain 0s elsewhere in column 2, we use

$$R_1 = r_2 + r_1 \qquad R_3 = 2r_2 + r_3 \qquad R_4 = -8r_2 + r_4$$

The new matrix is

$$\left[\begin{array}{rrr|r} 1 & 0 & 4 & 5 \\ 0 & 1 & 2 & 3 \\ 0 & 0 & 0 & -3 \\ 0 & 0 & 0 & -6 \end{array}\right]$$

We can stop here even though the matrix is not yet row-reduced. The third row yields the equation

$$0 \cdot x + 0 \cdot y + 0 \cdot z = -3$$

Thus we conclude the system has no solution.

Now Work Problem 23

$\cdots\cdots\cdots\cdots\cdots\cdots\cdots\cdots$

Infinite Number of Solutions: Parameters

We have seen several examples of systems of linear equations that have an infinite number of solutions. Let's look at a few more examples.

EXAMPLE 8   Solve: $\begin{cases} x + y + z = 7 \\ x - y - 3z = 1 \end{cases}$

SOLUTION

The augmented matrix of the system is

$$\left[\begin{array}{rrr|r} 1 & 1 & 1 & 7 \\ 1 & -1 & -3 & 1 \end{array}\right]$$

The reduced row-echelon form (as you should verify) is

$$\left[\begin{array}{rrr|r} 1 & 0 & -1 & 4 \\ 0 & 1 & 2 & 3 \end{array}\right]$$

The system of equations represented by this matrix is

$$\begin{cases} x - z = 4 \\ y + 2z = 3 \end{cases} \tag{1}$$

We can assign any value to z and use it to compute values of x and y. Thus the system has infinitely many solutions.

. . . .

If we rearrange the equations in (1) in the form

$$\begin{cases} x = 4 + z \\ y = 3 - 2z \end{cases} \tag{2}$$

it is easier to see the role the variable z plays. In fact, because the remaining variables are expressed in terms of z, we call z a **parameter.**

We check the solution to Example 8 as follows:

$$x + y + z = (4 + z) + (3 - 2z) + z = 7 + z - 2z + z = 7$$
$$x - y - 3z = (4 + z) - (3 - 2z) - 3z = 1 + z + 2z - 3z = 1$$

The solution is verified.

The next example illustrates a system having an infinite number of solutions with two parameters.

EXAMPLE 9

Solve: $\begin{cases} x_1 + x_2 + 2x_3 + 2x_4 = 2 \\ x_1 + x_3 + x_4 = 0 \\ x_2 + x_3 + x_4 = 2 \end{cases}$

SOLUTION

The augmented matrix of the system is

$$\begin{bmatrix} 1 & 1 & 2 & 2 & | & 2 \\ 1 & 0 & 1 & 1 & | & 0 \\ 0 & 1 & 1 & 1 & | & 2 \end{bmatrix}$$

The reduced row-echelon form (as you should verify) is

$$\begin{bmatrix} 1 & 0 & 1 & 1 & | & 0 \\ 0 & 1 & 1 & 1 & | & 2 \\ 0 & 0 & 0 & 0 & | & 0 \end{bmatrix}$$

The equations represented by this system are

$$\begin{cases} x_1 + x_3 + x_4 = 0 \\ x_2 + x_3 + x_4 = 2 \end{cases}$$

We can rewrite this system in the form

$$\begin{cases} x_1 = -x_3 - x_4 \\ x_2 = 2 - x_3 - x_4 \end{cases} \tag{3}$$

The system has infinitely many solutions, obtained by assigning the two parameters x_3 and x_4 arbitrary values. Some choices are shown in the table.

x_3	x_4	x_1	x_2	Solution (x_1, x_2, x_3, x_4)
0	0	0	2	$(0, 2, 0, 0)$
1	0	-1	1	$(-1, 1, 1, 0)$
0	2	-2	0	$(-2, 0, 0, 2)$

· ·

The variables used as parameters are not unique. We could have chosen x_1 and x_4 as parameters by rewriting Equations (3) in the following manner.

From the first equation

$$x_3 = -x_1 - x_4$$

We can replace the parameter x_3 in the second equation by the above to produce

$$x_2 = 2 - x_3 - x_4 = 2 + x_1 + x_4 - x_4 = 2 + x_1$$

We then get the system

$$\begin{cases} x_2 = 2 + x_1 \\ x_3 = -x_1 - x_4 \end{cases}$$

showing the role of x_1 and x_4 as parameters.

Now Work Problem 27

Summary

To solve a system of m linear equations containing n variables:

Steps for Solving a System of m Linear Equations Containing n Variables

Step 1 Write its augmented matrix.

Step 2 Compute the reduced row-echelon form of the augmented matrix.

Step 3 If this matrix has a row of zeros, except for a nonzero entry in the last column, the system has no solution.

Step 4 Otherwise determine from this matrix whether the system has a unique solution or infinitely many solutions.

Application

EXAMPLE 10

Mixing Chemicals   In a chemistry laboratory one solution contains 10% hydrochloric acid (HCl), a second solution contains 20% HCl, and a third contains 40% HCl. How many liters of each should be mixed to obtain 100 liters of 25% HCl?

SOLUTION

Let x, y, and z represent the number of liters of 10%, 20%, and 40% solutions of HCl, respectively. Since we want 100 liters in all and the amount of HCl obtained from each solution must sum to 25% of 100, or 25 liters, we must have

$$x + y + z = 100$$
$$0.1x + 0.2y + 0.4z = 25$$

Thus our problem is to solve a system of two equations containing three variables. By matrix techniques we obtain the solution

$$x = 2z - 50$$
$$y = -3z + 150$$

(4)

where z can represent any real number. Now the practical considerations of this problem lead us to the conditions that $x \geq 0$, $y \geq 0$, $z \geq 0$. From (4) we see that we must have $z \geq 25$ and $z \leq 50$, since otherwise $x < 0$ or $y < 0$. Some possible solutions are listed in the table. The final determination by the chemistry laboratory will more than likely be based on the amount and availability of one acid solution versus the others.

No. of Liters 10% Solution	No. of Liters 20% Solution	No. of Liters 40% Solution	No. of Liters 10% Solution	No. of Liters 20% Solution	No. of Liters 40% Solution
0	75	25	26	36	38
10	60	30	30	30	40
12	57	31	36	21	43
16	51	33	38	18	44
20	45	35	46	6	48
25	37.5	37.5	50	0	50

Now Work Problem 45

• •

Matrices in Practice

Systems of linear equations arise in business, economics, sociology, chemistry — in fact, in any field that has a quantitative side to it. The systems encountered in practice are often quite large, with 100 equations containing 100 variables not unusual, making hand calculations with such systems out of the question. Thus computer routines are used to implement work such as row-reducing the augmented matrix. A very popular collection of such routines is the LINPACK package. In more advanced treatments it is shown that solving n equations containing n variables requires roughly n^3 multiplications and additions. Thus a 100 by 100 system will require $(100)^3 = 10^6$ arithmetic operations. But if we have access to a machine that can perform 100,000 operations per second, the task seems less formidable since our 100 by 100 system would be solved in 10 seconds. The availability of high-speed computing has greatly enhanced the applicability of matrices since they can now be used in large-scale problems. See the article by Kolata in *Science*.*

There is an aspect of computer-performed matrix calculations that can at times be potentially troublesome in applications. Computers by their nature can perform arithmetic only on decimals that have finitely many nonzero terms. Decimals that are infinite in length are rounded off. For example, $\frac{2}{3}$ has the nonterminating decimal expansion 0.6666. . . . A machine would round this off and store it as, say, 0.6666667 (the actual number of significant places would vary with the machine). This can have consequences for matrix calculations, as we show in the following example.

EXAMPLE 11

Suppose we wish to find the row-reduced form of the matrix

$$A = \begin{bmatrix} 1 & \frac{1}{3} \\ 2 & \frac{2}{3} \end{bmatrix}$$

A direct hand calculation shows that the row-reduced form is

$$\begin{bmatrix} 1 & \frac{1}{3} \\ 0 & 0 \end{bmatrix}$$

Now assume we did this on a computer that rounded off and stored, say, two significant digits.

Our matrix A would then be represented in the machine as

$$A = \begin{bmatrix} 1 & 0.33 \\ 2 & 0.67 \end{bmatrix}$$

If a program were now called upon to row-reduce A, the following steps would result:

$$R_2 = -2r_1 + r_2 \qquad \begin{bmatrix} 1 & 0.33 \\ 0 & 0.01 \end{bmatrix}$$

$$R_2 = 100r_2 \qquad \begin{bmatrix} 1 & 0.33 \\ 0 & 1 \end{bmatrix}$$

$$R_1 = -0.33r_2 + r_1 \qquad \begin{bmatrix} 1 & 0 \\ 0 & 1 \end{bmatrix}$$

Note that the end result of the computer calculation differs drastically from our own computed row-reduced form. The problem clearly lies in the rounded representation of the numbers $\frac{1}{3}$ and $\frac{2}{3}$.

. . . .

* Gina Kolata, "Solving Linear Systems Faster." *Science* (June 14, 1985).

Though the above example is a bit simplistic, errors in matrix calculations introduced by round-off or truncation can and do occur and can have disastrous consequences. Were the matrix in the example above the coefficient matrix of a system, then the computer program would conclude that the system has a unique solution, while in reality the system has either infinitely many solutions or no solution. The subject of error propagation when doing matrix arithmetic on a computer is of great importance to people who use matrices in practice, since they want to be assured that their results are meaningful. It is also a subject where much current work is being done by computer scientists and mathematicians.

EXERCISE 2.3   Answers to odd-numbered problems begin on page AN-8.

In Problems 1–10 tell whether the given matrix is in reduced row-echelon form.

1. $\begin{bmatrix} 1 & 2 & 3 \\ 0 & 0 & 0 \\ 0 & 0 & 1 \end{bmatrix}$ NO

2. $\begin{bmatrix} 1 & 2 & 3 \\ 0 & 0 & 0 \\ 0 & 0 & 0 \end{bmatrix}$ yes

3. $\begin{bmatrix} 1 & 1 & 0 \\ 0 & 1 & 0 \end{bmatrix}$ yes

4. $\begin{bmatrix} 1 & 0 & 3 \\ 0 & 1 & 0 \end{bmatrix}$ yes

$\begin{bmatrix} 1 & 0 \\ 0 & 1 \end{bmatrix}$

5. $\begin{bmatrix} 0 & 1 \\ 1 & 0 \end{bmatrix}$ yes

6. $\begin{bmatrix} 0 & 1 & 0 \\ 0 & 0 & 1 \\ 0 & 0 & 0 \end{bmatrix}$ yes

7. $\begin{bmatrix} 0 & 0 & 1 \\ 0 & 0 & 0 \end{bmatrix}$ yes

8. $\begin{bmatrix} 0 & 0 \\ 0 & 0 \end{bmatrix}$ no

9. $\begin{bmatrix} 1 & 0 & 0 & 0 & 0 \\ 0 & 0 & 1 & 2 & 0 \\ 0 & 0 & 0 & 0 & 1 \\ 0 & 0 & 0 & 0 & 0 \end{bmatrix}$ yes

10. $\begin{bmatrix} 1 & 1 & 0 \\ 0 & 0 & 2 \end{bmatrix}$ no

In Problems 11–20 the reduced row-echelon form of the augmented matrix of a system of linear equations is given. Tell whether the system has one solution, no solution, or infinitely many solutions.

11. $\left[\begin{array}{cc|c} 1 & 1 & 1 \\ 0 & 0 & 0 \end{array}\right]$

12. $\left[\begin{array}{cc|c} 1 & 0 & 0 \\ 0 & 0 & 1 \end{array}\right]$

13. $\left[\begin{array}{ccc|c} 1 & 0 & 0 & 0 \\ 0 & 1 & 0 & 0 \\ 0 & 0 & 1 & 6 \end{array}\right]$

14. $\left[\begin{array}{ccc|c} 1 & 0 & -2 & 6 \\ 0 & 1 & 3 & 1 \end{array}\right]$

15. $\left[\begin{array}{ccc|c} 1 & 2 & 0 & 1 \\ 0 & 0 & 1 & 2 \\ 0 & 0 & 0 & 0 \end{array}\right]$

16. $\left[\begin{array}{ccc|c} 1 & 2 & 0 & 0 \\ 0 & 0 & 1 & 0 \\ 0 & 0 & 0 & 1 \end{array}\right]$

17. $\left[\begin{array}{cccc|c} 1 & 0 & 1 & -1 & 0 \\ 0 & 1 & 2 & 1 & 1 \\ 0 & 0 & 0 & 0 & 0 \end{array}\right]$

18. $\left[\begin{array}{ccc|c} 1 & 0 & 0 & -1 \\ 0 & 1 & 0 & 3 \\ 0 & 0 & 1 & 4 \\ 0 & 0 & 0 & 0 \end{array}\right]$

19. $\left[\begin{array}{ccc|c} 1 & 0 & -1 & 1 \\ 0 & 1 & 2 & 1 \end{array}\right]$

20. $\left[\begin{array}{cc|c} 1 & 0 & 1 \\ 0 & 1 & 1 \\ 0 & 0 & 0 \end{array}\right]$

In Problems 21–44 solve each system of equations by finding the reduced row-echelon form of the augmented matrix. If there is no solution, say the system is inconsistent.

21. $\begin{cases} x + y = 3 \\ 2x - y = 3 \end{cases}$

22. $\begin{cases} x - y = 5 \\ 2x + 3y = 15 \end{cases}$

23. $\begin{cases} 3x - 3y = 12 \\ 3x + 2y = -3 \\ 2x + y = 4 \end{cases}$

24. $\begin{cases} 6x + y = 8 \\ x - 3y = -5 \\ 2x + y = 2 \end{cases}$

25. $\begin{cases} 2x - 4y = 8 \\ x - 2y = 4 \\ -x + 2y = -4 \end{cases}$

26. $\begin{cases} 3x + y = 8 \\ 6x + 2y = 16 \\ -9x - 3y = -24 \end{cases}$

27. $\begin{cases} 2x + y + 3z = -1 \\ -x + y + 3z = 8 \\ 2x - 2y - 6z = -16 \end{cases}$

28. $\begin{cases} x + 2y + 3z = 5 \\ -2x + 6y + 4z = 0 \\ 2x + 4y + 6z = 10 \end{cases}$

29. $\begin{cases} x - y = 1 \\ y - z = 6 \\ x + z = -1 \end{cases}$

30. $\begin{cases} 2x - y + 3z = 0 \\ x + 2y - z = 5 \\ 2y + z = 1 \end{cases}$

31. $\begin{cases} x_1 + x_2 = 7 \\ x_2 - x_3 + x_4 = 5 \\ x_1 - x_2 + x_3 + x_4 = 6 \\ x_2 - x_4 = 10 \end{cases}$

32. $\begin{cases} x_1 + x_2 + x_3 + x_4 = 0 \\ 2x_1 - x_2 - x_3 + x_4 = 0 \\ x_1 - x_2 - x_3 + x_4 = 0 \\ x_1 + x_2 - x_3 - x_4 = 0 \end{cases}$

33. $\begin{cases} x_1 + 2x_2 + 3x_3 - x_4 = 0 \\ 3x_1 - x_4 = 4 \\ x_2 - x_3 - x_4 = 2 \end{cases}$

34. $\begin{cases} 2x - 3y + 4z = 7 \\ x - 2y + 3z = 2 \end{cases}$

35. $\begin{cases} x - y + z = 5 \\ 2x - 2y + 2z = 8 \end{cases}$

36. $\begin{cases} x + y + z = 3 \\ x - y + z = 7 \\ x - y - z = 1 \end{cases}$

37. $\begin{cases} 3x - y + 2z = 3 \\ 3x + 3y + z = 3 \\ 3x - 5y + 3z = 12 \end{cases}$

38. $\begin{cases} x + y - z = 12 \\ 3x - y = 1 \\ 2x - 3y + 4z = 3 \end{cases}$

39. $\begin{cases} x_1 + x_2 + x_3 + x_4 = 4 \\ 2x_1 - x_2 + x_3 = 0 \\ 3x_1 + 2x_2 + x_3 - x_4 = 6 \\ x_1 - 2x_2 - 2x_3 + 2x_4 = -1 \end{cases}$

40. $\begin{cases} x_1 + x_2 + x_3 + x_4 = 4 \\ -x_1 + 2x_2 + x_3 = 0 \\ 2x_1 + 3x_2 + x_3 - x_4 = 6 \\ -2x_1 + x_2 - 2x_3 + 2x_4 = -1 \end{cases}$

41. $\begin{cases} 2x - y - z = 0 \\ x - y - z = 1 \\ 3x - y - z = 2 \end{cases}$

42. $\begin{cases} x + y + z = 3 \\ 2x + y + z = 0 \\ 3x + y + z = 1 \end{cases}$

43. $\begin{cases} 2x - y + z = 6 \\ 3x - y + z = 6 \\ 4x - 2y + 2z = 12 \end{cases}$

44. $\begin{cases} x - y + z = 2 \\ 2x - 3y + z = 0 \\ 3x - 3y + 3z = 6 \end{cases}$

45. Mixing Chemicals A chemistry laboratory has available three kinds of hydrochloric acid (HCl): 10%, 30%, and 50% solutions. How many liters of each should be mixed to obtain 100 liters of 25% HCl? Provide a table showing at least six of the possible solutions.

46. Repeat Problem 45 if the mixture is to be 100 liters of 40% HCl.

47. Cost of Fast Food   One group of customers bought 8 deluxe hamburgers, 6 orders of large fries, and 6 large colas for $26.10. A second group ordered 10 deluxe hamburgers, 6 large fries, and 8 large colas and paid $31.60. Is there sufficient information to determine the price of each food item? If not, construct a table showing the various possibilities. Assume the hamburgers cost between $1.75 and $2.25, the fries between $0.75 and $1.00, and the colas between $0.60 and $0.90.

48. Use the information given in Problem 47 and add a third group that purchased 3 deluxe hamburgers, 2 large fries, and 4 colas for $10.95. Is there now sufficient information to determine the price of each food item?

49. Investment Goals   A retired couple has $25,000 available to invest. They require a return on their investment of $2000 per year. As their financial consultant, you recommend they invest some money in Treasury bills that yield 7%, some money in corporate bonds that yield 9%, and some in junk bonds that yield 11%. Prepare a table showing the various ways this couple can achieve their goal.

50. (a) Rework Problem 49 if the couple has only $20,000 to invest.
   (b) Rework Problem 49 if the couple has $30,000 to invest.

(c) What general conclusions can you make regarding the amount to invest and the choices available to the couple?

51. (a) Rework Problem 49 if the couple requires a return of only $1500 per year.
   (b) Rework Problem 49 if the couple requires a return of $2500 per year.
   (c) What general conclusions can you make regarding the required return per year and the choices available to the couple?

52. Inventory Control   Three species of bacteria will be kept in one test tube and will feed on three resources. Each member of the first species consumes three units of the first resource and one unit of the third. Each bacterium of the second type consumes one unit of the first resource and two units each of the second and third. Each bacterium of the third type consumes two units of the first resource and four each of the second and third. If the test tube is supplied daily with 12,000 units of the first resource, 12,000 units of the second resource, and 14,000 units of the third, how many of each species can coexist in equilibrium in the test tube so that all of the supplied resources are consumed?

Technology Exercises

In Problems 1–6 use the elementary row operations on your graphing calculator to find the reduced row-echelon form of the augmented matrix, and solve the system. If there is no solution, say the system is inconsistent.

1.
$$\begin{cases} 2x_1 + x_2 + x_3 + x_4 = 6 \\ x_1 - x_2 - 2x_4 = 2 \\ -x_1 + x_4 = 0 \\ x_2 - x_3 + 2x_4 = -1 \end{cases}$$

2.
$$\begin{cases} x_1 + x_2 + x_3 + x_4 = 20 \\ -x_1 + 2x_2 + x_4 = 0 \\ 2x_1 + x_4 = 0 \\ x_2 + 2x_4 = -5 \end{cases}$$

3.
$$\begin{cases} x_1 + x_2 - x_3 - x_4 = 6 \\ x_1 + x_2 = -4 \\ -2x_1 - x_2 - x_3 + x_4 = 0 \\ 2x_3 + x_4 = -1 \end{cases}$$

4.
$$\begin{cases} x_1 + x_3 = 6 \\ x_2 + x_4 = 4 \\ -x_3 + x_4 = 0 \\ x_2 - x_3 = -2 \end{cases}$$

5.
$$\begin{cases} x_1 - 2x_2 - 5x_3 = -25 \\ x_2 + x_3 = 15 \\ 2x_2 + 3x_3 + 4x_4 = 15 \\ 3x_2 - 2x_4 = -10 \end{cases}$$

6.
$$\begin{cases} x_1 + x_2 - 5x_3 + x_5 = -25 \\ x_2 + x_3 - x_5 = 15 \\ 2x_2 + 3x_3 + 4x_4 = 15 \\ 3x_2 - 2x_4 - x_5 = -10 \\ x_1 + x_4 - 4x_5 = 18 \end{cases}$$

2.4 MATRIX ALGEBRA

Matrices can be added, subtracted, and multiplied. They also possess many of the algebraic properties of real numbers. Matrix algebra is the study of these properties. Its importance lies in the fact that many situations in both pure and applied mathematics involve rectangular arrays of numbers. In fact, in many branches of business and the biological and social sciences, it is necessary to express and use a set of numbers in a rectangular array. Let's look at an example.

EXAMPLE 1

Motors, Inc., produces three models of cars: a sedan, a hardtop, and a station wagon. If the company wishes to compare the units of raw material and the units of labor involved in 1 month's production of each of these models, the rectangular array displayed below may be used to present the data:

	Sedan Model	Hardtop Model	Station Wagon Model
Units of Material	23	16	10
Units of Labor	7	9	11

The same information may be written concisely as the matrix

$$\begin{bmatrix} 23 & 16 & 10 \\ 7 & 9 & 11 \end{bmatrix}$$

This matrix has 2 rows (the units) and 3 columns (the models). The first row represents units of material and the second row represents units of labor. The first, second, and third columns represent the sedan, hardtop, and station wagon models, respectively.

· ·

A formal definition of a matrix as a rectangular array of numbers is given next.

MATRIX

A *matrix* A is a rectangular array of numbers a_{ij} of the form

$$\begin{array}{c} j\text{th column} \\ \downarrow \end{array}$$

$$A = \begin{bmatrix} a_{11} & a_{12} & \cdots & a_{1j} & \cdots & a_{1n} \\ a_{21} & a_{22} & \cdots & a_{2j} & \cdots & a_{2n} \\ \cdot & \cdot & & \cdot & & \cdot \\ \cdot & \cdot & & \cdot & & \cdot \\ \cdot & \cdot & & \cdot & & \cdot \\ a_{i1} & a_{i2} & \cdots & a_{ij} & \cdots & a_{in} \\ \cdot & \cdot & & \cdot & & \cdot \\ \cdot & \cdot & & \cdot & & \cdot \\ \cdot & \cdot & & \cdot & & \cdot \\ a_{m1} & a_{m2} & \cdots & a_{mj} & \cdots & a_{mn} \end{bmatrix} \leftarrow i\text{th row}$$

This matrix contains $m \cdot n$ numbers.

Each number a_{ij} of the matrix A has two indices: the **row index,** i, and the **column index,** j. The symbols $a_{i1}, a_{i2}, \ldots, a_{in}$ represent the numbers in the ith row, and the symbols $a_{1j}, a_{2j}, \ldots, a_{mj}$ represent the numbers in the jth column. The numbers a_{ij} of a matrix are referred to as the **entries** (or **components** or **elements**) of the matrix. The matrix A, above, which has m rows and n columns, can be abbreviated by

$$A = [a_{ij}] \qquad i = 1, 2, \ldots, m; \qquad j = 1, 2, \ldots, n$$

DIMENSION

The *dimension of a matrix A* is determined by the number of rows and the number of columns of the matrix. If a matrix A has m rows and n columns, we denote the dimension of A by $m \times n$, read as "*m by n*."

For a 2×3 matrix, remember that the first number 2 denotes the number of rows and the second number 3 is the number of columns. A matrix with 3 rows and 2 columns is of dimension 3×2.

SQUARE MATRIX

If a matrix A has the same number of rows as it has columns, it is called a *square matrix*.

In a square matrix $A = [a_{ij}]$ the entries for which $i = j$, namely $a_{11}, a_{22}, a_{33}, a_{44}$, and so on, are the **diagonal entries** of A.

EXAMPLE 2

In the recent U.S. census the following figures were obtained with regard to the city of Glenwood. Each year 7% of city residents move to the suburbs and 1% of the people in the suburbs move to the city. This situation can be represented by the matrix

$$P = \begin{matrix} \text{City} \\ \text{Suburbs} \end{matrix} \begin{matrix} \text{City} & \text{Suburbs} \\ \begin{bmatrix} 0.93 & 0.07 \\ 0.01 & 0.99 \end{bmatrix} \end{matrix}$$

Here, the entry in row 1, column 2—0.07—indicates that 7% of city residents move to the suburbs. The matrix P is a square matrix and its dimension is 2×2. The diagonal entries are 0.93 and 0.99.

. . . .

ROW AND COLUMN MATRIX

A *row matrix* is a matrix with 1 row of entries. A *column matrix* is a matrix with 1 column of entries. Row matrices and column matrices are also referred to as *row vectors* and *column vectors,* respectively.

EXAMPLE 3

The matrices

$$A = [23 \quad 16 \quad 10] \qquad B = [7 \quad 9] \qquad C = \begin{bmatrix} 23 \\ -1 \\ 7 \end{bmatrix}$$

$$D = \begin{bmatrix} 16 \\ 9 \end{bmatrix} \qquad E = [9]$$

have the following dimensions: A, 1×3; B, 1×2; C, 3×1; D, 2×1; E, 1×1. Here A and B are row vectors and C and D are column vectors. E is both a row vector and a column vector.

· ·

The matrix $E = [9]$ in Example 3 is a 1×1 matrix and, as such, can be treated simply as a number.

Now Work Problem 3

Equality of Matrices

As with most mathematical quantities, we now want to define various relationships between two matrices. We might ask, "When, if at all, are two matrices equal?"

Let's try to arrive at a sound definition for equality of matrices by requiring equal matrices to have certain desirable properties. First, it would seem necessary that two equal matrices have the same dimension — that is, that they both be $m \times n$ matrices. Next, it would seem necessary that their entries be identical numbers. With these two restrictions, we define equality of matrices.

EQUALITY OF MATRICES

Two matrices A and B are *equal* if they are of the same dimension and if corresponding entries are equal. In this case we write $A = B$, read as "matrix A is equal to matrix B."

EXAMPLE 4

In order for the two matrices

$$\begin{bmatrix} p & q \\ 1 & 0 \end{bmatrix} \quad \text{and} \quad \begin{bmatrix} 2 & 4 \\ n & 0 \end{bmatrix}$$

to be equal, we must have $p = 2$, $q = 4$, and $n = 1$.

· ·

EXAMPLE 5

Let A and B be two matrices given by

$$A = \begin{bmatrix} x + y & 6 \\ 2x - 3 & 2 - y \end{bmatrix} \qquad B = \begin{bmatrix} 5 & 5x + 2 \\ y & x - y \end{bmatrix}$$

Find x and y so that A and B are equal (if possible).

SOLUTION

Both A and B are 2×2 matrices. Thus $A = B$ if

(a) $x + y = 5$ (b) $6 = 5x + 2$
(c) $2x - 3 = y$ (d) $2 - y = x - y$

Here we have four equations containing the two variables x and y. From Equation (d) we see that $x = 2$. Using this value in Equation (a), we obtain $y = 3$. But $x = 2$, $y = 3$ do not satisfy either (b) or (c). Hence, there are *no* values for x and y satisfying all four equations. This means A and B can never be equal.

Now Work Problem 9

$\cdots\cdots\cdots\cdots\cdots\cdots$

Addition of Matrices

Can two matrices be added? And, if so, what is the rule or law for addition of matrices?

Let's return to Example 1. In that example we recorded 1 month's production of Motors, Inc., by the matrix

$$A = \begin{bmatrix} 23 & 16 & 10 \\ 7 & 9 & 11 \end{bmatrix}$$

Suppose the next month's production is

$$B = \begin{bmatrix} 18 & 12 & 9 \\ 14 & 6 & 8 \end{bmatrix}$$

in which the pattern of recording units and models remains the same.

The total production for the 2 months can be displayed by the matrix

$$C = \begin{bmatrix} 41 & 28 & 19 \\ 21 & 15 & 19 \end{bmatrix}$$

since the number of units of material for sedan models is $41 = 23 + 18$, the number of units of material for hardtop models is $28 = 16 + 12$, and so on.

This leads us to define the sum $A + B$ of two matrices A and B as the matrix consisting of the sum of corresponding entries from A and B.

ADDITION OF MATRICES

Let $A = [a_{ij}]$ and $B = [b_{ij}]$ be two $m \times n$ matrices. The *sum* $A + B$ is defined as the $m \times n$ matrix $[a_{ij} + b_{ij}]$.

EXAMPLE 6

(a) $\begin{bmatrix} 23 & 16 & 10 \\ 7 & 9 & 11 \end{bmatrix} + \begin{bmatrix} 18 & 12 & 9 \\ 14 & 6 & 8 \end{bmatrix} = \begin{bmatrix} 23 + 18 & 16 + 12 & 10 + 9 \\ 7 + 14 & 9 + 6 & 11 + 8 \end{bmatrix}$

$\qquad\qquad\qquad = \begin{bmatrix} 41 & 28 & 19 \\ 21 & 15 & 19 \end{bmatrix}$

(b) $\begin{bmatrix} 0.6 & 0.4 \\ 0.1 & 0.9 \end{bmatrix} + \begin{bmatrix} 2.3 & 0.6 \\ 1.8 & 5.2 \end{bmatrix} = \begin{bmatrix} 0.6 + 2.3 & 0.4 + 0.6 \\ 0.1 + 1.8 & 0.9 + 5.2 \end{bmatrix}$

$\qquad\qquad\qquad = \begin{bmatrix} 2.9 & 1.0 \\ 1.9 & 6.1 \end{bmatrix}$

$\cdots\cdots\cdots\cdots\cdots\cdots$

Notice that it is possible to add two matrices only if their dimensions are the same. Also, the dimension of the sum of two matrices is the same as that of the two original matrices.

The following pairs of matrices cannot be added since they are of different dimensions:

$$A = \begin{bmatrix} 1 & 2 \\ 7 & 2 \end{bmatrix} \qquad \text{and} \qquad B = \begin{bmatrix} 1 \\ -3 \end{bmatrix}$$

$$A = [2 \quad 3] \qquad \text{and} \qquad B = [1 \quad 1 \quad 1]$$

$$A = \begin{bmatrix} -1 & 7 & 0 \\ 2 & \frac{1}{2} & 0 \end{bmatrix} \qquad \text{and} \qquad B = \begin{bmatrix} -1 & 2 \\ 3 & 0 \\ 1 & 5 \end{bmatrix}$$

Now Work Problem 17

It turns out that the usual rules for the addition of real numbers (such as the commutative property and associative property) are also valid for matrix addition.

EXAMPLE 7

Let

$$A = \begin{bmatrix} 1 & 5 \\ 7 & -3 \end{bmatrix} \qquad \text{and} \qquad B = \begin{bmatrix} 3 & -2 \\ 4 & 1 \end{bmatrix}$$

Then

$$A + B = \begin{bmatrix} 1 & 5 \\ 7 & -3 \end{bmatrix} + \begin{bmatrix} 3 & -2 \\ 4 & 1 \end{bmatrix} = \begin{bmatrix} 1 + 3 & 5 + (-2) \\ 7 + 4 & -3 + 1 \end{bmatrix}$$

$$= \begin{bmatrix} 4 & 3 \\ 11 & -2 \end{bmatrix}$$

$$B + A = \begin{bmatrix} 3 & -2 \\ 4 & 1 \end{bmatrix} + \begin{bmatrix} 1 & 5 \\ 7 & -3 \end{bmatrix} = \begin{bmatrix} 4 & 3 \\ 11 & -2 \end{bmatrix}$$

· ·

This leads us to formulate the following property for addition.

Commutative Property for Addition

If A and B are two matrices of the same dimension, then

$$A + B = B + A$$

The associative property for addition of matrices is also true. Thus:

Associative Property for Addition

If A, B, and C are three matrices of the same dimension, then

$$A + (B + C) = (A + B) + C$$

Now Work Problems 35 and 36

The fact that addition of matrices is associative means that the notation $A + B + C$ is *not* ambiguous, since $(A + B) + C = A + (B + C)$.

ZERO MATRIX

A matrix in which all entries are 0 is called a *zero matrix*. We use the symbol 0 to represent a zero matrix of any dimension.

For real numbers, 0 has the property that $x + 0 = x$ for any x. An important property of a zero matrix is that $A + \mathbf{0} = A$, provided the dimension of $\mathbf{0}$ is the same as that of A.

EXAMPLE 8

Let
$$A = \begin{bmatrix} 3 & 4 & -\frac{1}{2} \\ \sqrt{2} & 0 & 3 \end{bmatrix}$$

Then
$$A + \mathbf{0} = \begin{bmatrix} 3 & 4 & -\frac{1}{2} \\ \sqrt{2} & 0 & 3 \end{bmatrix} + \begin{bmatrix} 0 & 0 & 0 \\ 0 & 0 & 0 \end{bmatrix}$$
$$= \begin{bmatrix} 3+0 & 4+0 & -\frac{1}{2}+0 \\ \sqrt{2}+0 & 0+0 & 3+0 \end{bmatrix} = \begin{bmatrix} 3 & 4 & -\frac{1}{2} \\ \sqrt{2} & 0 & 3 \end{bmatrix} = A$$

If A is any matrix, the **additive inverse** of A, denoted by $-A$, is the matrix obtained by replacing each entry in A by its negative.

EXAMPLE 9

If
$$A = \begin{bmatrix} -3 & 0 \\ 5 & -2 \\ 1 & 3 \end{bmatrix} \qquad \text{then} \qquad -A = \begin{bmatrix} 3 & 0 \\ -5 & 2 \\ -1 & -3 \end{bmatrix}$$

Additive Inverse Property

For any matrix A, we have the property that
$$A + (-A) = \mathbf{0}$$

Now Work Problem 37

Subtraction of Matrices

Now that we have defined the sum of two matrices and the additive inverse of a matrix, it is natural to ask about the *difference* of two matrices. As you will see, subtracting matrices and subtracting real numbers are much the same kind of process.

DIFFERENCE OF TWO MATRICES

Let $A = [a_{ij}]$ and $B = [b_{ij}]$ be two $m \times n$ matrices. The *difference* $A - B$ is defined as the $m \times n$ matrix $[a_{ij} - b_{ij}]$.

EXAMPLE 10

Let
$$A = \begin{bmatrix} 2 & 3 & 4 \\ 1 & 0 & 2 \end{bmatrix} \quad \text{and} \quad B = \begin{bmatrix} -2 & 1 & -1 \\ 3 & 0 & 3 \end{bmatrix}$$

Then
$$A - B = \begin{bmatrix} 2 & 3 & 4 \\ 1 & 0 & 2 \end{bmatrix} - \begin{bmatrix} -2 & 1 & -1 \\ 3 & 0 & 3 \end{bmatrix}$$
$$= \begin{bmatrix} 2 - (-2) & 3 - 1 & 4 - (-1) \\ 1 - 3 & 0 - 0 & 2 - 3 \end{bmatrix} = \begin{bmatrix} 4 & 2 & 5 \\ -2 & 0 & -1 \end{bmatrix}$$

Notice that the difference $A - B$ is nothing more than the matrix formed by subtracting the entries in B from the corresponding entries in A.

Using the matrices A and B from Example 10, we find that
$$B - A = \begin{bmatrix} -2 & 1 & -1 \\ 3 & 0 & 3 \end{bmatrix} - \begin{bmatrix} 2 & 3 & 4 \\ 1 & 0 & 2 \end{bmatrix}$$
$$= \begin{bmatrix} -2 - 2 & 1 - 3 & -1 - 4 \\ 3 - 1 & 0 - 0 & 3 - 2 \end{bmatrix} = \begin{bmatrix} -4 & -2 & -5 \\ 2 & 0 & 1 \end{bmatrix}$$

Observe that $A - B \neq B - A$, illustrating that matrix subtraction, like subtraction of real numbers, is not commutative.

Multiplying a Matrix by a Number

Let's return to the production of Motors Inc., during the month specified in Example 1. The matrix A describing this production is
$$A = \begin{bmatrix} 23 & 16 & 10 \\ 7 & 9 & 11 \end{bmatrix}$$

Let's assume that for 3 consecutive months, the monthly production remained the same. Then the total production for the 3 months is simply the sum of the matrix A taken 3 times. If we represent the total production by the matrix T, then
$$T = \begin{bmatrix} 23 & 16 & 10 \\ 7 & 9 & 11 \end{bmatrix} + \begin{bmatrix} 23 & 16 & 10 \\ 7 & 9 & 11 \end{bmatrix} + \begin{bmatrix} 23 & 16 & 10 \\ 7 & 9 & 11 \end{bmatrix}$$
$$= \begin{bmatrix} 23 + 23 + 23 & 16 + 16 + 16 & 10 + 10 + 10 \\ 7 + 7 + 7 & 9 + 9 + 9 & 11 + 11 + 11 \end{bmatrix}$$
$$= \begin{bmatrix} 3 \cdot 23 & 3 \cdot 16 & 3 \cdot 10 \\ 3 \cdot 7 & 3 \cdot 9 & 3 \cdot 11 \end{bmatrix} = \begin{bmatrix} 69 & 48 & 30 \\ 21 & 27 & 33 \end{bmatrix}$$

In other words, when we add the matrix A 3 times, we multiply each entry of A by 3. This leads to the following definition.

SCALAR MULTIPLICATION

Let $A = [a_{ij}]$ be an $m \times n$ matrix and let c be a real number, called a *scalar*. The product of the matrix A by the scalar c, called *scalar multiplication*, is the $m \times n$ matrix $cA = [ca_{ij}]$.

When multiplying a matrix by a real number, each entry of the matrix is multiplied by the number. Notice that the dimension of A and the dimension of the product cA are the same.

EXAMPLE 11

For
$$A = \begin{bmatrix} 2 \\ 5 \\ -7 \end{bmatrix} \quad \text{and} \quad B = \begin{bmatrix} 20 & 0 \\ 18 & 8 \end{bmatrix}$$

compute   (a) $3A$ (b) $\frac{1}{2}B$

SOLUTION

(a) $3A = 3 \begin{bmatrix} 2 \\ 5 \\ -7 \end{bmatrix} = \begin{bmatrix} 3 \cdot 2 \\ 3 \cdot 5 \\ 3 \cdot (-7) \end{bmatrix} = \begin{bmatrix} 6 \\ 15 \\ -21 \end{bmatrix}$

(b) $\frac{1}{2}B = \frac{1}{2} \begin{bmatrix} 20 & 0 \\ 18 & 8 \end{bmatrix} = \begin{bmatrix} \frac{1}{2} \cdot 20 & \frac{1}{2} \cdot 0 \\ \frac{1}{2} \cdot 18 & \frac{1}{2} \cdot 8 \end{bmatrix} = \begin{bmatrix} 10 & 0 \\ 9 & 4 \end{bmatrix}$

....

EXAMPLE 12

For
$$A = \begin{bmatrix} 3 & 1 \\ 4 & 0 \\ 2 & -3 \end{bmatrix} \quad \text{and} \quad B = \begin{bmatrix} 2 & -3 \\ -1 & 1 \\ 1 & 0 \end{bmatrix}$$

compute   (a) $A - B$ (b) $A + (-1) \cdot B$

SOLUTION

(a) $A - B = \begin{bmatrix} 1 & 4 \\ 5 & -1 \\ 1 & -3 \end{bmatrix}$

(b) $A + (-1) \cdot B = \begin{bmatrix} 3 & 1 \\ 4 & 0 \\ 2 & -3 \end{bmatrix} + \begin{bmatrix} -2 & 3 \\ 1 & -1 \\ -1 & 0 \end{bmatrix} = \begin{bmatrix} 1 & 4 \\ 5 & -1 \\ 1 & -3 \end{bmatrix}$

....

The above example illustrates the result that

$$A - B = A + (-1) \cdot B = A + (-B)$$

Now Work Problem 27

We continue our study of scalar multiplication by listing some of its properties in the box at the top of the next page.

Properties of Scalar Multiplication

Let k and h be two real numbers and let A and B be two matrices of dimension $m \times n$. Then

$$\textbf{(I)} \qquad k(hA) = (kh)A$$

$$\textbf{(II)} \qquad (k + h)A = kA + hA$$

$$\textbf{(III)} \qquad k(A + B) = kA + kB$$

Properties I–III are illustrated in the following example.

EXAMPLE 13

For $\qquad A = \begin{bmatrix} 2 & -3 & -1 \\ 5 & 6 & 4 \end{bmatrix} \qquad$ and $\qquad B = \begin{bmatrix} -3 & 0 & 4 \\ 2 & -1 & 5 \end{bmatrix}$

show that

(a) $5[2A] = 10A$ (b) $(4 + 3)A = 4A + 3A$ (c) $3[A + B] = 3A + 3B$

SOLUTION

(a) $5[2A] = 5 \begin{bmatrix} 4 & -6 & -2 \\ 10 & 12 & 8 \end{bmatrix} = \begin{bmatrix} 20 & -30 & -10 \\ 50 & 60 & 40 \end{bmatrix}$

$\qquad 10A = \begin{bmatrix} 20 & -30 & -10 \\ 50 & 60 & 40 \end{bmatrix}$

(b) $(4 + 3)A = 7A = \begin{bmatrix} 14 & -21 & -7 \\ 35 & 42 & 28 \end{bmatrix}$

$\qquad 4A + 3A = \begin{bmatrix} 8 & -12 & -4 \\ 20 & 24 & 16 \end{bmatrix} + \begin{bmatrix} 6 & -9 & -3 \\ 15 & 18 & 12 \end{bmatrix}$

$\qquad\qquad = \begin{bmatrix} 14 & -21 & -7 \\ 35 & 42 & 28 \end{bmatrix}$

(c) $3[A + B] = 3 \begin{bmatrix} -1 & -3 & 3 \\ 7 & 5 & 9 \end{bmatrix} = \begin{bmatrix} -3 & -9 & 9 \\ 21 & 15 & 27 \end{bmatrix}$

$\qquad 3A + 3B = \begin{bmatrix} 6 & -9 & -3 \\ 15 & 18 & 12 \end{bmatrix} + \begin{bmatrix} -9 & 0 & 12 \\ 6 & -3 & 15 \end{bmatrix}$

$\qquad\qquad = \begin{bmatrix} -3 & -9 & 9 \\ 21 & 15 & 27 \end{bmatrix}$

EXERCISE 2.4   Answers to odd-numbered problems begin on page AN-9.

In Problems 1–8 write the dimension of each matrix.

1. $\begin{bmatrix} 3 & 2 \\ -1 & 3 \end{bmatrix}$
2. $\begin{bmatrix} -1 & 0 \\ 0 & 5 \end{bmatrix}$
3. $\begin{bmatrix} 2 & 1 & -3 \\ 1 & 0 & -1 \end{bmatrix}$
4. $\begin{bmatrix} 1 & 2 \\ 2 & 1 \\ 0 & -3 \end{bmatrix}$

5. $\begin{bmatrix} 4 \\ 1 \end{bmatrix}$
    6. $[2 \quad 1 \quad -3]$
    7. $[2]$
    8. $[0]$

In Problems 9–16 determine whether the given statements are true or false. If false, tell why.

9. $\begin{bmatrix} 0 \\ 1 \end{bmatrix} = [0 \quad 1]$

10. $\begin{bmatrix} 3 & 2 \\ -1 & 0 \end{bmatrix} = \begin{bmatrix} 3 & 2 \\ -1 & 4 \end{bmatrix}$

11. $\begin{bmatrix} 5 & 0 \\ 0 & 1 \end{bmatrix}$ is square

12. $\begin{bmatrix} 3 & 2 & 1 \\ 4 & -1 & 0 \end{bmatrix}$ is 3 × 2

13. $\begin{bmatrix} x & 2 \\ 4 & 0 \end{bmatrix} = \begin{bmatrix} 3 & 2 \\ 4 & 0 \end{bmatrix}$ if x = 3

14. $\begin{bmatrix} x & y \\ 0 & 0 \end{bmatrix} = [x \quad y]$

15. $\begin{bmatrix} 5 & 0 \\ 1 & 1 \end{bmatrix} = \begin{bmatrix} 2+3 & 0 \\ 1 & 1 \end{bmatrix}$

16. $\begin{bmatrix} 1 & 0 \\ 0 & 1 \end{bmatrix} = \begin{bmatrix} 3-2 & 3-3 \\ 3-3 & 3-2 \end{bmatrix}$

In Problems 17–24 perform the indicated operations. Express your answer as a single matrix.

17. $\begin{bmatrix} 3 & -1 \\ 4 & 2 \end{bmatrix} + \begin{bmatrix} -2 & 2 \\ 1 & 3 \end{bmatrix}$

18. $\begin{bmatrix} 2 & 4 \\ 3 & -1 \end{bmatrix} - \begin{bmatrix} 8 & 1 \\ 0 & 1 \end{bmatrix}$

19. $3 \begin{bmatrix} 2 & 6 & 0 \\ 4 & -2 & 1 \end{bmatrix}$

20. $-3 \begin{bmatrix} 2 & 1 \\ -2 & 1 \\ 0 & 3 \end{bmatrix}$

21. $2 \begin{bmatrix} 1 & -1 & 8 \\ 2 & 4 & 1 \end{bmatrix} - 3 \begin{bmatrix} 0 & -2 & 8 \\ 1 & 4 & 1 \end{bmatrix}$

22. $6 \begin{bmatrix} 2 & 1 \\ 3 & 1 \\ -1 & 0 \end{bmatrix} + 4 \begin{bmatrix} 6 & -4 \\ -2 & -3 \\ 0 & 1 \end{bmatrix}$

23. $3 \begin{bmatrix} a & 8 \\ b & 1 \\ c & -2 \end{bmatrix} + 5 \begin{bmatrix} 2a & 6 \\ -b & -2 \\ -c & 0 \end{bmatrix}$

24. $2 \begin{bmatrix} 2x & y & z \\ 2 & -4 & 8 \end{bmatrix} - 3 \begin{bmatrix} -3x & 4y & 2z \\ 6 & -1 & 4 \end{bmatrix}$

In Problems 25–40 use the matrices below. For Problems 25-34 perform the indicated operation(s); for Problems 35–40 verify the indicated property.

$$A = \begin{bmatrix} 2 & -3 & 4 \\ 0 & 2 & 1 \end{bmatrix} \quad B = \begin{bmatrix} 1 & -2 & 0 \\ 5 & 1 & 2 \end{bmatrix} \quad C = \begin{bmatrix} -3 & 0 & 5 \\ 2 & 1 & 3 \end{bmatrix}$$

25. $A + B$

26. $B + C$

27. $2A - 3C$

28. $3C - 4B$

29. $(A + B) - 2C$

30. $4C + (A - B)$

31. $3A + 4(B + C)$

32. $(A + B) + 3C$

33. $2(A - B) - C$

34. $2A - 5(B + C)$

35. Verify the commutative property for addition by finding $A + B$ and $B + A$.

36. Verify the associative property for addition by finding $(A + B) + C$ and $A + (B + C)$.

37. Verify the additive inverse property by showing that $A + (-A) = 0$.

38. Verify Property I of scalar multiplication by finding $2(3A)$ and $6A$.

39. Verify Property II of scalar multiplication by finding $2B + 3B$ and $5B$.

40. Verify Property III of scalar multiplication by finding $2(A + C)$ and $2A + 2C$.

41. Find x and z so that

$$\begin{bmatrix} x \\ z \end{bmatrix} = \begin{bmatrix} 4 \\ 3 \end{bmatrix}$$

42. Find x, y, and z so that

$$\begin{bmatrix} x+y & 2 \\ 4 & 0 \end{bmatrix} = \begin{bmatrix} 6 & x-y \\ 4 & z \end{bmatrix}$$

43. Find x and y so that

$$\begin{bmatrix} x-2y & 0 \\ -2 & 6 \end{bmatrix} = \begin{bmatrix} 3 & 0 \\ -2 & x+y \end{bmatrix}$$

44. Find x, y, and z so that

$$\begin{bmatrix} x-2 & 3 & 2z \\ 6y & x & 2y \end{bmatrix} = \begin{bmatrix} y & z & 6 \\ 18z & y+2 & 6z \end{bmatrix}$$

45. Find x, y, and z so that

$$[2 \quad 3 \quad -4] + [x \quad y \quad z] = [6 \quad -8 \quad 2]$$

46. Find x and y so that

$$\begin{bmatrix} 3 & -2 & 2 \\ 1 & 0 & -1 \end{bmatrix} + \begin{bmatrix} x-y & 2 & -2 \\ 4 & x & 6 \end{bmatrix}$$
$$= \begin{bmatrix} 6 & 0 & 0 \\ 5 & 2x+y & 5 \end{bmatrix}$$

47. Nail Production XYZ Company produces steel and aluminum nails. One week 25 gross of $\frac{1}{2}$-inch steel nails and 45 gross of 1-inch steel nails were produced. Suppose 13 gross of $\frac{1}{2}$-inch aluminum nails, 20 gross of 1-inch aluminum nails, 35 gross of 2-inch steel nails, and 23 gross of 2-inch aluminum nails were also made. Write a 2×3 matrix depicting this. Could you also write a 3×2 matrix for this situation?

48. Katy, Mike, and Danny go to the candy store. Katy buys 5 sticks of gum, 2 ice cream cones, and 10 jelly beans. Mike buys 2 sticks of gum, 15 jelly beans, and 2 candy bars. Danny buys 1 stick of gum, 1 ice cream cone, and 4 candy bars. Write a matrix depicting this situation.

49. Use a matrix to display the information given at the top of the next column, which was obtained in a survey of voters. Label the rows and columns.

351   Democrats earning under $25,000
271   Republicans earning under $25,000
 73   Independents earning under $25,000
203   Democrats earning $25,000 or more
215   Republicans earning $25,000 or more
 55   Independents earning $25,000 or more

50. Listing Stocks One day on the New York Stock Exchange, 800 issues went up and 600 went down. Of the 800 up issues, 200 went up more than $1 per share. Of the 600 down issues, 50 went down more than $1 per share. Express this information in a 2×2 matrix. Label the rows up and down and the columns more than $1 and less than $1.

51. Surveys In a survey of 1000 college students, the following information was obtained: 400 were Liberal Arts and Sciences (LAS) majors, of which 50% were female; 300 were Engineering (ENG) majors, of which 75% were male; and the remaining were Education (EDUC) majors, of which 60% were female. Express this information using a 2×3 matrix. Label the rows male and female and the columns LAS, ENG, and EDUC.

52. Sales of Cars The sales figures for two car dealers during June showed that dealer A sold 100 compacts, 50 intermediates, and 40 full-size cars, while dealer B sold 120 compacts, 40 intermediates, and 35 full-size cars. During July, dealer A sold 80 compacts, 30 intermediates, and 10 full-size cars, while dealer B sold 70 compacts, 40 intermediates, and 20 full-size cars. Total sales over the 3-month period of June–August revealed that dealer A sold 300 compacts, 120 intermediates, and 65 full-size cars. In the same 3-month period, dealer B sold 250 compacts, 100 intermediates, and 80 full-size cars.

(a) Write 2×3 matrices summarizing sales data for June, July, and the 3-month period for each dealer.
(b) Use matrix addition to find the sales over the 2-month period for June and July for each dealer.
(c) Use matrix subtraction to find the sales in August for each dealer.

Technology Exercises
Most graphing calculators are able to store and manipulate matrices. For example, the TI-81 can store up to three 6×6 or smaller matrices. Check your user's manual to see how to enter a matrix into memory and how to copy one matrix into another.

In Problems 1–6 use your graphing calculator to perform the indicated operations on the matrices below.

$$A = \begin{bmatrix} -1 & -1 & 3 & 0 \\ 2 & 6 & 2 & 2 \\ -4 & 2 & 3 & 2 \\ 7 & 0 & 5 & -1 \end{bmatrix} \qquad B = \begin{bmatrix} -1 & 2 & 4 & 5 \\ 2 & 0 & 5 & 3 \\ 0.5 & 6 & -7 & 11 \\ 5 & -1 & 2 & 7 \end{bmatrix} \qquad C = \begin{bmatrix} 13 & -8 & 7 & 0 \\ 0 & 5 & 0 & -2 \\ 5 & 0 & 7 & 0 \\ 7 & 7 & 7 & 7 \end{bmatrix}$$

1. $A + B$

2. $3C - 2B$

3. $C - 3(A + B)$

4. $2(A - B) + \frac{1}{2}C$

5. $3(B + C) - A$

6. $\frac{1}{3}(A + 2C) - B$

2.5 MULTIPLICATION OF MATRICES

While addition of matrices and scalar multiplication are fairly straightforward, defining the *product* $A \cdot B$ of the two matrices A and B requires a bit more detail.

We explain first what we mean by the product of a row vector with a column vector.

Let's look at a simple example. In a given month suppose 23 units of material and 7 units of labor were required to manufacture 4-door sedans. We can represent this by the column vector

$$\begin{bmatrix} 23 \\ 7 \end{bmatrix}$$

Also, suppose the cost per unit of material is $450 and the cost per unit of labor is $600. We represent these costs by the row vector

$$[450 \quad 600]$$

The total cost of producing the sedans is calculated as follows:

$$\begin{aligned} \text{Total cost} &= (\text{Cost per unit of material}) \times (\text{Units of material}) \\ &\quad + (\text{Cost per unit of labor}) \times (\text{Units of labor}) \\ &= 450 \times 23 + 600 \times 7 = 14{,}550 \end{aligned}$$

In terms of the matrix representations,

$$\text{Total cost} = [450 \quad 600] \begin{bmatrix} 23 \\ 7 \end{bmatrix} = 450 \times 23 + 600 \times 7 = 14{,}550$$

This leads us to formulate the following definition for multiplying a row vector times a column vector.

VECTOR PRODUCT

If $R = [r_1\ r_2\ \ldots\ r_n]$ is a row vector and $C = \begin{bmatrix} c_1 \\ c_2 \\ \vdots \\ c_n \end{bmatrix}$ is a column vector, then by

the *product of R and C* we mean the number

$$r_1 c_1 + r_2 c_2 + r_3 c_3 + \ldots + r_n c_n$$

EXAMPLE 1

If
$$R = [1 \quad 5 \quad 3] \quad \text{and} \quad C = \begin{bmatrix} 2 \\ -1 \\ 4 \end{bmatrix}$$

then the product of R and C is

$$R \cdot C = 1 \cdot 2 + 5 \cdot (-1) + 3 \cdot 4 = 9$$

· ·

Notice that for the product of a row vector R and a column vector C to make sense, if R is a $1 \times n$ row vector, then C must have dimension $n \times 1$.

EXAMPLE 2

Let
$$R = [1 \quad 0 \quad 1] \quad \text{and} \quad C = \begin{bmatrix} 0 \\ -11 \\ 0 \end{bmatrix}$$

Then the product of R and C is

$$R \cdot C = 1 \cdot 0 + 0 \cdot (-11) + 1 \cdot 0 = 0$$

Now Work Problem 1

· ·

Given two matrices A and B, the rows of A can be thought of as row vectors, while the columns of B can be thought of as column vectors. This observation will be used in the following main definition.

MATRIX MULTIPLICATION

Let $A = [a_{ij}]$ be a matrix of dimension $m \times r$ and let $B = [b_{ij}]$ be a matrix of dimension $r \times n$. The *product $A \cdot B$ is the matrix of dimension $m \times n$, whose ijth entry is the sum of the products of corresponding elements of the ith row of A and the jth column of B. That is, the ijth entry of $A \cdot B$ is

$$a_{i1}b_{1j} + a_{i2}b_{2j} + a_{i3}b_{3j} + \cdots + a_{ir}b_{rj}$$

The rule for multiplication of matrices is best illustrated by an example.

EXAMPLE 3

Find the product $A \cdot B$ if

$$A = \begin{bmatrix} 1 & 3 & -2 \\ 4 & -1 & 5 \end{bmatrix} \quad \text{and} \quad B = \begin{bmatrix} 2 & -3 & 4 & 1 \\ -1 & 2 & 2 & 0 \\ 4 & 5 & 1 & 1 \end{bmatrix}$$

SOLUTION

Since A is 2×3 and B is 3×4, the product $A \cdot B$ will be 2×4. To obtain, for example, the entry in row 2, column 3 of $A \cdot B$, we form the product of the second row of A with the third column of B. That is,

$$\begin{bmatrix} 1 & 3 & -2 \\ 4 & -1 & 5 \end{bmatrix} \quad \begin{bmatrix} 2 & -3 & 4 & 1 \\ -1 & 2 & 2 & 0 \\ 4 & 5 & 1 & 1 \end{bmatrix}$$

We compute

$$4 \cdot 4 + (-1) \cdot 2 + 5 \cdot 1 = 19$$

Thus far, we have

$$A \cdot B = \begin{bmatrix} - & - & - & - \\ - & - & 19 & - \end{bmatrix}$$

To obtain the entry in row 1, column 2, we compute

$$[1 \quad 3 \quad -2] \begin{bmatrix} -3 \\ 2 \\ 5 \end{bmatrix} = 1 \cdot (-3) + 3 \cdot 2 + (-2) \cdot 5 = -3 + 6 - 10 = -7$$

The other entries of $A \cdot B$ are obtained in a similar fashion. The final result is shown here. You should verify it.

$$A \cdot B = \begin{bmatrix} -9 & -7 & 8 & -1 \\ 29 & 11 & 19 & 9 \end{bmatrix}$$

· ·

EXAMPLE 4

One month's production at Motors, Inc., may be given in matrix form as

$$A = \begin{array}{c} \\ \text{Sedan} \quad \text{Hardtop} \quad \begin{array}{c} \text{Station} \\ \text{Wagon} \end{array} \\ \begin{bmatrix} 23 & 16 & 10 \\ 7 & 9 & 11 \end{bmatrix} \begin{array}{l} \text{Units of material} \\ \text{Units of labor} \end{array} \end{array}$$

Suppose that in this month's production, the cost for each unit of material is \$450 and the cost for each unit of labor is \$600. What is the total cost to manufacture the sedans, the hardtops, and the station wagons?

SOLUTION

For sedans, the cost is 23 units of material at \$450 each, plus 7 units of labor at \$600 each, for a total cost of

$$23 \cdot \$450 + 7 \cdot \$600 = 10{,}350 + 4200 = \$14{,}550$$

Similarly, for hardtops, the total cost is

$$16 \cdot \$450 + 9 \cdot \$600 = 7200 + 5400 = \$12{,}600$$

Finally, for station wagons, the total cost is

$$10 \cdot \$450 + 11 \cdot \$600 = 4500 + 6600 = \$11{,}100$$

If we represent the cost of units of material and units of labor by the 1×2 matrix

$$U = \begin{matrix} \text{Unit cost of} & \text{Unit cost of} \\ \text{material} & \text{labor} \\ [450 & 600] \end{matrix}$$

then the total cost is the product $U \cdot A$.

$$U \cdot A = [450 \quad 600] \begin{bmatrix} 23 & 16 & 10 \\ 7 & 9 & 11 \end{bmatrix}$$

$$= [450 \cdot 23 + 600 \cdot 7 \quad 450 \cdot 16 + 600 \cdot 9 \quad 450 \cdot 10 + 600 \cdot 11]$$

$$= [14{,}550 \quad 12{,}600 \quad 11{,}100]$$

....

Properties of Matrix Multiplication

Let's look at some consequences of the definition of matrix multiplication. If A is a matrix of dimension $m \times r$ (which has r columns) and B is a matrix of dimension $p \times n$ (which has p rows) and if $r \neq p$, the product $A \cdot B$ is not defined. That is,

> Multiplication of matrices is possible only if the number of columns of the **matrix on the left** equals the number of rows of the **matrix on the right.**

If A is of dimension $m \times r$ and B is of dimension $r \times n$, then the product $A \cdot B$ is of dimension $m \times n$. See Figure 5.

Figure 5

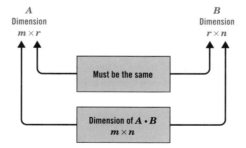

In Example 3, A is of dimension 2×3, B is of dimension 3×4, and we found the product $A \cdot B$ to be of dimension 2×4. Observe that the product $B \cdot A$ is not defined.

Now Work Problem 11

EXAMPLE 5

For

$$A = \begin{bmatrix} 2 & 0 \\ 1 & 5 \end{bmatrix} \quad \text{and} \quad B = \begin{bmatrix} 3 & 2 \\ 1 & 4 \end{bmatrix}$$

compute $A \cdot B$ and $B \cdot A$.

SOLUTION

We observe that the products $A \cdot B$ and $B \cdot A$ are both defined, and so

$$A \cdot B = \begin{bmatrix} 6 & 4 \\ 8 & 22 \end{bmatrix} \qquad B \cdot A = \begin{bmatrix} 8 & 10 \\ 6 & 20 \end{bmatrix}$$

○ ●

Note from the above example that even if both $A \cdot B$ and $B \cdot A$ are defined, **they may not be equal.** We conclude that

---

Matrix multiplication is not commutative. That is, in general,

$A \cdot B$ **is not always equal to** $B \cdot A$

In listing some of the properties of matrix multiplication in this book, **we agree to** follow the usual convention and write $A \cdot B$ as AB.

Associative Property of Multiplication

Let A be a matrix of dimension $m \times r$, let B be a matrix of dimension $r \times p$, and let C be a matrix of dimension $p \times n$. Then matrix multiplication is **associative.** That is,

$$A(BC) = (AB)C$$

The resulting matrix ABC is of dimension $m \times n$.

Notice the limitations that are placed on the dimensions of the matrices in **order for** multiplication to be associative.

Distributive Property

Let A be a matrix of dimension $m \times r$. Let B and C be matrices of dimension $r \times n$. Then the **distributive property** states that

$$A(B + C) = AB + AC$$

Now Work Problems 35 and 36

The resulting matrix $AB + AC$ is of dimension $m \times n$.

Identity Matrix

A special type of square matrix is the **identity matrix,** which is denoted by I_n. It is the matrix whose diagonal entries are 1s and all other entries are 0s. Thus,

$$I_n = \begin{bmatrix} 1 & 0 & \cdots & 0 & 0 \\ 0 & 1 & \cdots & 0 & 0 \\ \cdot & \cdot & \cdot & \cdot & \cdot \\ \cdot & \cdot & \cdot & \cdot & \cdot \\ \cdot & \cdot & \cdot & \cdot & \cdot \\ 0 & 0 & \cdots & 1 & 0 \\ 0 & 0 & \cdots & 0 & 1 \end{bmatrix}$$

where the subscript n implies that I_n is of dimension $n \times n$.

EXAMPLE 6

For
$$A = \begin{bmatrix} 3 & 2 \\ -4 & 5 \end{bmatrix}$$

compute   (a) AI_2 (b) I_2A

SOLUTION

(a) $AI_2 = \begin{bmatrix} 3 & 2 \\ -4 & 5 \end{bmatrix}\begin{bmatrix} 1 & 0 \\ 0 & 1 \end{bmatrix} = \begin{bmatrix} 3 & 2 \\ -4 & 5 \end{bmatrix} = A$

(b) $I_2A = \begin{bmatrix} 1 & 0 \\ 0 & 1 \end{bmatrix}\begin{bmatrix} 3 & 2 \\ -4 & 5 \end{bmatrix} = \begin{bmatrix} 3 & 2 \\ -4 & 5 \end{bmatrix} = A$

In general, if A is a square matrix of dimension $n \times n$, then $AI_n = I_nA = A$. Thus for square matrices the identity matrix plays the role that the number 1 plays in the set of real numbers.

When the matrix A is not square, care must be taken when forming the products AI and IA. For example, if

$$A = \begin{bmatrix} 1 & 2 \\ 3 & 2 \\ 1 & 1 \end{bmatrix}$$

then A is of dimension 3×2 and

$$AI_2 = A \begin{bmatrix} 1 & 0 \\ 0 & 1 \end{bmatrix} = \begin{bmatrix} 1 & 2 \\ 3 & 2 \\ 1 & 1 \end{bmatrix}\begin{bmatrix} 1 & 0 \\ 0 & 1 \end{bmatrix} = \begin{bmatrix} 1 & 2 \\ 3 & 2 \\ 1 & 1 \end{bmatrix} = A$$

Although the product I_2A is not defined, we can calculate the product I_3A as follows:

$$I_3A = \begin{bmatrix} 1 & 0 & 0 \\ 0 & 1 & 0 \\ 0 & 0 & 1 \end{bmatrix}\begin{bmatrix} 1 & 2 \\ 3 & 2 \\ 1 & 1 \end{bmatrix} = \begin{bmatrix} 1 & 2 \\ 3 & 2 \\ 1 & 1 \end{bmatrix} = A$$

This example can be generalized as follows:

Identity Property

If A is a matrix of dimension $m \times n$ and if I_n denotes the identity matrix of dimension $n \times n$, and I_m denotes the identity matrix of dimension $m \times m$, then

$$I_m A = A \qquad \text{and} \qquad A I_n = A$$

The Method of Least Squares*

The method of least squares refers to a technique that is often used in data analysis to find the "best" linear equation that fits a given collection of experimental data. (We shall see a little later just what we mean by the word *best*.) It is a technique employed by statisticians, economists, business forecasters, and those who try to interpret and analyze data. Before we begin our discussion, we will need to define the *transpose* of a matrix.

TRANSPOSE

Let A be a matrix of dimension $m \times n$. The *transpose* of A, written A^T, is the $n \times m$ matrix obtained from A by interchanging the rows and columns of A.

Thus, the first row of A^T is the first column of A; the second row of A^T is the second column of A; and so on.

EXAMPLE 7

If we let

$$A = \begin{bmatrix} 1 & 2 & 3 \\ 0 & -1 & 2 \end{bmatrix} \qquad B = \begin{bmatrix} 1 & 1 \\ 0 & 1 \\ 2 & 3 \end{bmatrix} \qquad C = [1 \quad 0 \quad -1]$$

then

$$A^T = \begin{bmatrix} 1 & 0 \\ 2 & -1 \\ 3 & 2 \end{bmatrix} \qquad B^T = \begin{bmatrix} 1 & 0 & 2 \\ 1 & 1 & 3 \end{bmatrix} \qquad C^T = \begin{bmatrix} 1 \\ 0 \\ -1 \end{bmatrix}$$

· ·

Note how dimensions are reversed when computing A^T: A has the dimension 2×3, while A^T has the dimension 3×2, and so on.

Now Work Problem 49

---

* This subsection may be omitted without loss of continuity.

We begin our discussion of the method of least squares by an example.

Suppose a product has been sold over time at various prices and that we have some data that show the demand for the product (in thousands of units) in terms of its price (in dollars). If we use x to represent price and y to represent demand, the data might look like the information in the following table.

Price, x	4	5	9	12
Demand, y	9	8	6	3

Suppose also that we have reason to assume that y and x are *linearly related*. That is, we assume we can write

$$y = mx + b$$

for some, as yet unknown, m and b. In other words, we are assuming that when demand is graphed against price, the resulting graph will be a straight line. Our belief that y and x are linearly related might be based on past experience or economic theory. However, if we plot the data from the above table (see Figure 6), it seems pretty clear that no single straight line passes through the plotted points.

This may be due to several reasons:

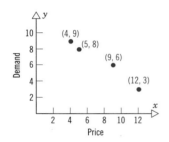

Figure 6

1. The reporting of the data may not have been accurate, or
2. Our assumption that $y = mx + b$ may not be totally warranted.

That is, we are looking for an ideal answer to a question about something in the real world where what happens may only roughly (approximately) fit an ideal design. Refer again to Figure 6. Though no straight line will pass through all the points, we might still ask: "Is there a straight line that 'best fits' the points?" That is, we seek a line of the sort in Figure 7 that provides a good straight-line approximation to the data. Finding such a line would give us at least an approximate feel for how the demand y is related to the price x.

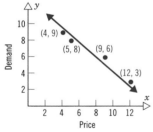

Figure 7

OUR PROBLEM

Given a set of noncollinear points, find the straight line that "best fits" these points.

In the process we will need to clarify our use of the phrase "best fits." First, though, suppose there are four data points, labeled from left to right as (x_1, y_1), (x_2, y_2), and so on. See Figure 8. Now suppose the equation of the straight line L in Figure 8 is

$$y = mx + b$$

where we do not know what m and b are. If (x_1, y_1) is actually on the line, it would be true that

$$y_1 = mx_1 + b$$

Figure 8

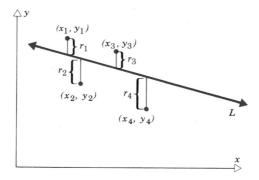

But, since we can't be sure (x_1, y_1) lies on L, the above equation may not be true. That is, there may be a nonzero difference between y_1 and $mx_1 + b$. We designate this difference by r_1. That is,

$$r_1 = y_1 - (mx_1 + b)$$

or

$$y_1 = mx_1 + b + r_1$$

What we have just said about (x_1, y_1) we can repeat for the remaining data points (x_i, y_i), $i = 2, 3,$ and 4. Therefore, since we can't expect y_i to be equal to $mx_i + b$, we will measure the difference by

$$r_i = y_i - (mx_i + b)$$

or, equivalently, by

$$y_i = mx_i + b + r_i \qquad i = 1, 2, 3, 4 \tag{1}$$

Geometrically, the r_i's measure the vertical distance between the sought-after line L and the data points.

Recall that our problem is to find a line L that "best fits" the data points (x_i, y_i). Intuitively, we would seek a line L for which the r_i's are all simultaneously small. Since some r_i's may be positive and some may be negative, it will not do to just add them up. To eliminate the signs, we square each r_i. The method of least squares consists of finding a line L for which the sum of the squares of the r_i's is as small as possible. Since finding the line $y = mx + b$ is the same as finding m and b, we restate the least squares problem for this example as follows:

LEAST SQUARES PROBLEM

Given the data points (x_i, y_i), $i = 1, 2, 3, 4$, find m and b satisfying

$$y_i = mx_i + b + r_i \qquad i = 1, 2, 3, 4 \tag{1}$$

so that $r_1^2 + r_2^2 + r_3^2 + r_4^2$ is minimized.

A solution to the problem can be compactly expressed using matrix language. However, since a derivation that this does, in fact, give the solution would require either calculus or more advanced linear algebra, we omit the proof and simply present the solution:

Solution to Least Squares Problem

The line of best fit

$$y = mx + b \tag{2}$$

to a set of four points (x_1, y_1), (x_2, y_2), (x_3, y_3), (x_4, y_4) is obtained by solving the system of two equations in two unknowns

$$A^TAX = A^TY \tag{3}$$

for m and b, where

$$A = \begin{bmatrix} x_1 & 1 \\ x_2 & 1 \\ x_3 & 1 \\ x_4 & 1 \end{bmatrix} \qquad X = \begin{bmatrix} m \\ b \end{bmatrix} \qquad Y = \begin{bmatrix} y_1 \\ y_2 \\ y_3 \\ y_4 \end{bmatrix} \tag{4}$$

EXAMPLE 8   Use least squares to find the line of best fit for the points

$$(4, 9), (5, 8), (9, 6), (12, 3)$$

SOLUTION

We use Equations (4) to set up the matrices A and Y:

$$A = \begin{bmatrix} 4 & 1 \\ 5 & 1 \\ 9 & 1 \\ 12 & 1 \end{bmatrix} \qquad Y = \begin{bmatrix} 9 \\ 8 \\ 6 \\ 3 \end{bmatrix}$$

Equation (3), $A^TAX = A^TY$, becomes

$$\begin{bmatrix} 4 & 5 & 9 & 12 \\ 1 & 1 & 1 & 1 \end{bmatrix} \begin{bmatrix} 4 & 1 \\ 5 & 1 \\ 9 & 1 \\ 12 & 1 \end{bmatrix} \begin{bmatrix} m \\ b \end{bmatrix} = \begin{bmatrix} 4 & 5 & 9 & 12 \\ 1 & 1 & 1 & 1 \end{bmatrix} \begin{bmatrix} 9 \\ 8 \\ 6 \\ 3 \end{bmatrix}$$

This reduces to a system of two equations in two unknowns:

$$\begin{cases} 266m + 30b = 166 \\ 30m + 4b = 26 \end{cases}$$

The solution, which you can verify, is given by

$$m = -\frac{29}{41} \qquad \text{and} \qquad b = \frac{484}{41}$$

Hence, the least squares solution to the problem is given by the straight line

$$y = -\frac{29}{41}x + \frac{484}{41}$$

· ·

Note, for example, that corresponding to the value $x = 5$, the above line of "best fit" has $y = -\frac{29}{41} \cdot 5 + \frac{484}{41} = 8.27$, while the experimentally observed demand had value 8. Were we to use our computed line to approximate the connection between price and demand, we would predict that a selling price of $x = 6$ would yield a demand of $y = -\frac{29}{41} \cdot 6 + \frac{484}{41} = 7.56$.

THE GENERAL LEAST SQUARES PROBLEM

We presented the method of least squares using a particular example with four data points. It is easy to extend this analysis.

The general least squares problem consists of finding the best straight-line fit to a given set of data points:

$$(x_1, y_1), (x_2, y_2), \ldots , (x_n, y_n) \tag{5}$$

If we let

$$A = \begin{bmatrix} x_1 & 1 \\ x_2 & 1 \\ \cdot & \cdot \\ \cdot & \cdot \\ \cdot & \cdot \\ x_n & 1 \end{bmatrix} \qquad X = \begin{bmatrix} m \\ b \end{bmatrix} \qquad Y = \begin{bmatrix} y_1 \\ y_2 \\ \cdot \\ \cdot \\ \cdot \\ y_n \end{bmatrix}$$

then the line of best fit to the data points in (5) is

$$y = mx + b$$

where $X = \begin{bmatrix} m \\ b \end{bmatrix}$ is found by solving the system of two equations in two unknowns,

$$A^T A X = A^T Y \tag{6}$$

It can be shown that the system given by Equation (6) always has a unique solution provided the data points do not all lie on the same vertical line.

The least squares techniques presented here along with more elaborate variations are frequently used today by statisticians and researchers.

EXERCISE 2.5   Answers to odd-numbered problems begin on page AN-10.

In Problems 1–10 find the product.

1. $[1 \quad 3] \begin{bmatrix} 2 \\ 4 \end{bmatrix}$

2. $[1 \quad -2 \quad 3] \begin{bmatrix} 0 \\ 1 \\ 2 \end{bmatrix}$

3. $[1 \quad 3] \begin{bmatrix} 2 & 0 \\ 4 & -2 \end{bmatrix}$

4. $[1 \quad -2 \quad 3] \begin{bmatrix} 0 & 1 \\ 1 & 2 \\ 2 & 3 \end{bmatrix}$

5. $\begin{bmatrix} 2 & 0 \\ 4 & -2 \end{bmatrix} \begin{bmatrix} 2 \\ 3 \end{bmatrix}$

6. $\begin{bmatrix} 1 & -2 & 3 \\ 4 & 0 & 6 \end{bmatrix} \begin{bmatrix} 0 \\ 1 \\ 2 \end{bmatrix}$

7. $\begin{bmatrix} 2 & 0 \\ 4 & -2 \end{bmatrix} \begin{bmatrix} 2 & 1 \\ 3 & -2 \end{bmatrix}$

8. $\begin{bmatrix} 1 & -2 & 3 \\ 4 & 0 & 6 \end{bmatrix} \begin{bmatrix} 0 & -2 \\ 1 & 0 \\ 2 & 4 \end{bmatrix}$

9. $\begin{bmatrix} 2 & 0 \\ 4 & -2 \\ 6 & -1 \end{bmatrix} \begin{bmatrix} 2 & 1 \\ 3 & -2 \end{bmatrix}$

10. $\begin{bmatrix} 1 & -2 & 3 \\ 4 & 0 & 6 \end{bmatrix} \begin{bmatrix} -1 & 2 & 1 \\ 1 & 3 & 0 \\ 0 & 4 & 1 \end{bmatrix}$

In Problems 11–20 let A be of dimension 3 × 4, B be of dimension 3 × 3, C be of dimension 2 × 3, and D be of dimension 3 × 2. Determine which of the following expressions are defined and, for those that are, give the dimension.

11. BA

12. CD

13. AB

14. DC

15. $(BA)C$

16. $A(CD)$

17. $BA + A$

18. $CD + BA$

19. $DC + B$

20. $CB - A$

In Problems 21–36 use the matrices below. For Problems 21–34 perform the indicated operation(s); for Problems 35 and 36 verify the indicated property.

$$A = \begin{bmatrix} 1 & 2 \\ 0 & 4 \end{bmatrix} \quad B = \begin{bmatrix} 1 & 2 & 3 \\ -1 & 4 & -2 \end{bmatrix} \quad C = \begin{bmatrix} 3 & 1 \\ 4 & -1 \\ 0 & 2 \end{bmatrix} \quad D = \begin{bmatrix} 1 & 0 & 4 \\ 0 & 1 & 2 \\ 0 & -1 & 1 \end{bmatrix} \quad E = \begin{bmatrix} 3 & -1 \\ 4 & 2 \end{bmatrix}$$

21. AB

22. DC

23. BC

24. AA

25. $(D + I_3)C$

26. $DC + C$

27. EI_2

28. I_3D

29. $(2E)B$

30. $E(2B)$

31. $-5E + A$

32. $3A + 2E$

33. $3CB + 4D$

34. $2EA - 3BC$

35. Verify the associative property of matrix multiplication by finding $D(CB)$ and $(DC)B$.

36. Verify the distributive property by finding $(D + B)C$ and $DC + BC$.

37. For

$$A = \begin{bmatrix} 1 & -1 \\ 2 & 0 \end{bmatrix} \quad \text{and} \quad B = \begin{bmatrix} 3 & 2 \\ -1 & 4 \end{bmatrix}$$

find AB and BA. Notice that $AB \neq BA$.

38. Show that, for all values a, b, c, and d, the matrices

$$A = \begin{bmatrix} a & b \\ -b & a \end{bmatrix} \quad \text{and} \quad B = \begin{bmatrix} c & d \\ -d & c \end{bmatrix}$$

are commutative; that is, $AB = BA$.

39. If possible, find a matrix A such that

$$A \begin{bmatrix} 0 & 1 \\ 2 & -1 \end{bmatrix} = \begin{bmatrix} 2 & 1 \\ -1 & 0 \end{bmatrix}$$

Hint: Let $A = \begin{bmatrix} a & b \\ c & d \end{bmatrix}$.

40. For what numbers x will the following be true?

$$[x \quad 4 \quad 1] \begin{bmatrix} 2 & 1 & 0 \\ 1 & 0 & 2 \\ 0 & 2 & 4 \end{bmatrix} \begin{bmatrix} x \\ -7 \\ \frac{5}{4} \end{bmatrix} = 0$$

41. Let

$$A = \begin{bmatrix} 1 & 2 & 5 \\ 2 & 4 & 10 \\ -1 & -2 & -5 \end{bmatrix}$$

Show that $A \cdot A = A^2 = \mathbf{0}$. Thus the rule in the real number system that if $a^2 = 0$, then $a = 0$ does not hold for matrices.

42. What must be true about a, b, c, and d, if we demand that $AB = BA$ for the following matrices?

$$A = \begin{bmatrix} a & b \\ c & d \end{bmatrix} \qquad B = \begin{bmatrix} 1 & 1 \\ -1 & 1 \end{bmatrix}$$

Assume that

$$\begin{bmatrix} a & b \\ c & d \end{bmatrix} \ne \begin{bmatrix} 1 & 0 \\ 0 & 1 \end{bmatrix}$$

43. Let

$$A = \begin{bmatrix} a & b \\ b & a \end{bmatrix}$$

Find a and b such that $A^2 + A = \mathbf{0}$, where $A^2 = AA$.

44. For the matrix

$$A = \begin{bmatrix} a & 1-a \\ 1+a & -a \end{bmatrix}$$

show that $A^2 = AA = I_2$.

45. Find the vector $[x_1 \quad x_2]$ for which

$$[x_1 \quad x_2] \begin{bmatrix} \frac{1}{2} & \frac{1}{2} \\ \frac{1}{4} & \frac{3}{4} \end{bmatrix} = [x_1 \quad x_2]$$

under the condition that $x_1 + x_2 = 1$. Here, the vector $[x_1 \quad x_2]$ is called a **fixed vector** of the matrix

$$\begin{bmatrix} \frac{1}{2} & \frac{1}{2} \\ \frac{1}{4} & \frac{3}{4} \end{bmatrix}$$

46. Department Store Purchases   Lee went to a department store and purchased 6 pairs of pants, 8 shirts, and 2 jackets. Chan purchased 2 pairs of pants, 5 shirts, and 3 jackets. If pants are $25 each, shirts are $18 each, and jackets are $39 each, use matrix multiplication to find the amounts spent by Lee and Chan.

47. Factory Production   Suppose a factory is asked to produce three types of products, which we will call P_1, P_2, P_3. Suppose the following purchase order was received: $P_1 = 7$, $P_2 = 12$, $P_3 = 5$. Represent this order by a row vector and call it P:

$$P = [7 \quad 12 \quad 5]$$

To produce each of the products, raw material of four kinds is needed. Call the raw material M_1, M_2, M_3, and M_4. The matrix below gives the amount of material needed for each product:

$$Q = \begin{array}{c} \\ P_1 \\ P_2 \\ P_3 \end{array} \begin{array}{cccc} M_1 & M_2 & M_3 & M_4 \\ \begin{bmatrix} 2 & 3 & 1 & 12 \\ 7 & 9 & 5 & 20 \\ 8 & 12 & 6 & 15 \end{bmatrix} \end{array}$$

Suppose the cost for each of the materials M_1, M_2, M_3, and M_4 is $10, $12, $15, and $20, respectively. The cost vector is

$$C = \begin{bmatrix} 10 \\ 12 \\ 15 \\ 20 \end{bmatrix}$$

Compute each of the following and interpret each one:

(a) PQ (b) QC (c) PQC

48. For a square matrix A it is possible to find $A \cdot A = A^2$. It is also clear that we can compute

$$A^n = \underbrace{A \cdot A \cdot \ \ldots \ \cdot A}_{n \text{ factors}}$$

Find A^2, A^3, and A^4 for each of the following square matrices:

(a) $A = \begin{bmatrix} 1 & 0 \\ 3 & 2 \end{bmatrix}$ (b) $A = \begin{bmatrix} 3 & 1 \\ -2 & -1 \end{bmatrix}$

(c) $A = \begin{bmatrix} 1 & 0 \\ 0 & 1 \end{bmatrix}$ (d) $A = \begin{bmatrix} \frac{1}{2} & \frac{1}{2} \\ \frac{1}{4} & \frac{3}{4} \end{bmatrix}$

Can you guess what A^n looks like for Part (c)? For Part (d)?

In Problems 49–54 compute A^T.

49. $A = \begin{bmatrix} 4 & 1 & 2 \\ 3 & 1 & 0 \end{bmatrix}$ **50.** $A = \begin{bmatrix} 5 & 2 & -1 \\ 1 & 3 & 6 \\ 1 & -1 & 2 \end{bmatrix}$ **51.** $A = \begin{bmatrix} 1 & 11 \\ 0 & 12 \\ 1 & 4 \end{bmatrix}$

52. $A = [-1 \quad 6 \quad 4]$ **53.** $A = \begin{bmatrix} 8 \\ 6 \\ 3 \end{bmatrix}$ **54.** $A = \begin{bmatrix} 5 & 3 \\ 3 & 7 \end{bmatrix}$

Problems 55–58 involve the method of least squares.

55. Supply Equation   The following table shows the supply (in thousands of units) of a product at various prices (in dollars):

Price, x	3	5	6	7
Supply, y	10	13	15	16

(a) Find the least squares line of best fit to the above data.
(b) Use the equation of this line to estimate the supply of the product at a price of $8.

56. Study Time vs. Performance   Data giving the number of hours a person had studied compared to his or her performance on an exam are given in the table:

Hours Studied x	Exam Score y
0	50
2	74
4	85
6	90
8	92

(a) Find a least squares line of best fit to the above data.
(b) What prediction would this line make for a student who studied 9 hours?

57. Advertising vs. Sales   A business would like to determine the relationship between the amount of money spent on advertising and its total weekly sales. Over a period of 5 weeks it gathers the following data:

Amount Spent on Advertising (in thousands) x	Weekly Sales Volume (in thousands) y
10	50
17	61
11	55
18	60
21	70

Find a least squares line of best fit to the above data.

58. Drug Concentration vs. Time   The following data show the connection between the number of hours a drug has been in a person's body and its concentration in the body.

Number of Hours	Drug Concentration (parts per million)
2	2.1
4	1.6
6	1.4
8	1.0

(a) Fit a least squares line to the above data.
(b) Use the equation of the line to estimate the drug concentration after 5 hours.

59. A matrix is **symmetric** if $A^T = A$. Which of the following matrices are symmetric?

(a) $\begin{bmatrix} 1 & 1 & 2 \\ 1 & 0 & 1 \\ 3 & 2 & 3 \end{bmatrix}$ (b) $\begin{bmatrix} 0 & 1 & 3 \\ 1 & 4 & 7 \\ 3 & 7 & 5 \end{bmatrix}$

(c) $\begin{bmatrix} 1 & 2 & 3 & 0 \\ 2 & 4 & 5 & 0 \\ 3 & 5 & 1 & 0 \end{bmatrix}$

Need a symmetric matrix be square?

60. Show that the matrix $A^T A$ is always symmetric.

Technology Exercises
Most graphing calculators are able to store and manipulate matrices. For example, the TI-81 can store up to three 6×6 or smaller matrices. Check your user's manual to see how to enter a matrix into memory and how to copy one matrix into another.

In Problems 1–8 use your graphing calculator to perform the indicated operations on the matrices below.

$$A = \begin{bmatrix} -1 & -1 & 3 & 0 \\ 2 & 6 & 2 & 2 \\ -4 & 2 & 3 & 2 \\ 7 & 0 & 5 & -1 \end{bmatrix} \quad B = \begin{bmatrix} -1 & 2 & 4 & 5 \\ 2 & 0 & 5 & 3 \\ 0.5 & 6 & -7 & 11 \\ 5 & -1 & 2 & 7 \end{bmatrix} \quad C = \begin{bmatrix} 13 & -8 & 7 & 0 \\ 0 & 5 & 0 & -2 \\ 5 & 0 & 7 & 0 \\ 7 & 7 & 7 & 7 \end{bmatrix}$$

1. AB

2. BA

3. $(AB)C$

4. $A(BC)$

5. $B(A + C)$

6. $(A + C)B$

7. $A(2B - 3C)$

8. $(A + B)(A - B)$

9. For a square matrix A it is possible to find $A \cdot A = A^2$. It is also clear that we can compute

$$A^n = \underbrace{A \cdot A \cdot \ldots \cdot A}_{n \text{ factors}}$$

Use your graphing calculator to compute A^2, A^{10}, and A^{15} for each of the following matrices:

$$(\text{a}) \ A = \begin{bmatrix} -0.5 & -1 & 0.3 & 0 & 0.3 \\ 2 & 1.6 & 1 & -1 & 0.4 \\ -4 & 2 & 0.7 & 2 & 0.2 \\ 1 & 0 & 0 & -1 & 0 \\ 0 & 0 & -0.9 & 0 & 0 \end{bmatrix}$$

$$(\text{b}) \ A = \begin{bmatrix} -1 & 0.02 & 0.24 & 0 \\ 2 & 0 & 0 & 1.3 \\ 0.5 & 6 & -0.7 & 1.1 \\ 2.5 & -1 & 0.02 & 0.7 \end{bmatrix}$$

$$(\text{c}) \ A = \begin{bmatrix} 0 & 1 & 0 \\ 1 & 0 & 1 \\ 1 & 1 & 1 \end{bmatrix}$$

Problem 10 involves the method of least squares.

10. Study Time vs. Performance Data giving the number of hours a person had studied compared to his or her performance on an exam are given in the table:

Hours Studied	Exam Score
x	y
0	50
2	74
4	85
6	90
8	92

(a) Use your graphing calculator (or a computer) to define the matrix

$$A = \begin{bmatrix} x_1 & 1 \\ x_2 & 1 \\ x_3 & 1 \\ x_4 & 1 \\ x_5 & 1 \end{bmatrix}$$

Most graphing calculators are able to find the transpose of a matrix with a single operation. Find A^T using your graphing calculator, and calculate matrix $B = A^TA$.

(b) Define the matrix

$$Y = \begin{bmatrix} y_1 \\ y_2 \\ y_3 \\ y_4 \\ y_5 \end{bmatrix}$$

Use your graphing calculator to solve $BX = A^TY$.

(c) Find a least squares line of best fit to the above data.

2.6 INVERSE OF A MATRIX

The *inverse* of a matrix, if it exists, plays the role in matrix algebra that the reciprocal of a number plays in the set of real numbers. The product of a real number and its reciprocal equals 1, the multiplicative identity. The product of a matrix and its inverse equals the identity matrix.

INVERSE

Let A be a matrix of dimension $n \times n$. A matrix B of dimension $n \times n$ is called the *inverse* of A if $AB = BA = I_n$. We denote the inverse of a matrix A, if it exists, by A^{-1}.

EXAMPLE 1

Show that $\begin{bmatrix} \frac{1}{2} & -\frac{1}{2} \\ 0 & 1 \end{bmatrix}$ is the inverse of $\begin{bmatrix} 2 & 1 \\ 0 & 1 \end{bmatrix}$.

SOLUTION

Since

$$\begin{bmatrix} 2 & 1 \\ 0 & 1 \end{bmatrix} \begin{bmatrix} \frac{1}{2} & -\frac{1}{2} \\ 0 & 1 \end{bmatrix} = \begin{bmatrix} 1 & 0 \\ 0 & 1 \end{bmatrix}$$

and

$$\begin{bmatrix} \frac{1}{2} & -\frac{1}{2} \\ 0 & 1 \end{bmatrix} \begin{bmatrix} 2 & 1 \\ 0 & 1 \end{bmatrix} = \begin{bmatrix} 1 & 0 \\ 0 & 1 \end{bmatrix}$$

the required condition is met.

Now Work Problem 1

· ·

The next example provides a technique for finding the inverse of a matrix. Although this technique is not the one we shall ultimately use, it is illustrative.

EXAMPLE 2

Find the inverse of the matrix $A = \begin{bmatrix} 2 & 1 \\ 0 & 1 \end{bmatrix}$.

SOLUTION

We begin by assuming that this matrix has an inverse of the form

$$A^{-1} = \begin{bmatrix} a & b \\ c & d \end{bmatrix}$$

Then the product of A and A^{-1} must be the identity matrix:

$$\begin{bmatrix} 2 & 1 \\ 0 & 1 \end{bmatrix} \begin{bmatrix} a & b \\ c & d \end{bmatrix} = \begin{bmatrix} 1 & 0 \\ 0 & 1 \end{bmatrix}$$

Multiplying the matrices on the left side, we get

$$\begin{bmatrix} 2a + c & 2b + d \\ c & d \end{bmatrix} = \begin{bmatrix} 1 & 0 \\ 0 & 1 \end{bmatrix}$$

The condition for equality requires that

$$2a + c = 1 \qquad 2b + d = 0 \qquad c = 0 \qquad d = 1$$

Thus

$$a = \tfrac{1}{2} \qquad b = -\tfrac{1}{2} \qquad c = 0 \qquad d = 1$$

Hence, the inverse of

$$A = \begin{bmatrix} 2 & 1 \\ 0 & 1 \end{bmatrix} \qquad \text{is} \qquad A^{-1} = \begin{bmatrix} \tfrac{1}{2} & -\tfrac{1}{2} \\ 0 & 1 \end{bmatrix}$$

• •

Sometimes a square matrix does not have an inverse.

EXAMPLE 3

Show that the matrix below does not have an inverse.

$$A = \begin{bmatrix} 0 & 1 \\ 0 & 0 \end{bmatrix}$$

SOLUTION

We proceed as in Example 2 by assuming that A does have an inverse. It will be of the form

$$A^{-1} = \begin{bmatrix} a & b \\ c & d \end{bmatrix}$$

The product of A and A^{-1} must be the identity matrix. Thus

$$\begin{bmatrix} 0 & 1 \\ 0 & 0 \end{bmatrix} \begin{bmatrix} a & b \\ c & d \end{bmatrix} = \begin{bmatrix} 1 & 0 \\ 0 & 1 \end{bmatrix}$$

Performing the multiplication on the left side, we have

$$\begin{bmatrix} c & d \\ 0 & 0 \end{bmatrix} = \begin{bmatrix} 1 & 0 \\ 0 & 1 \end{bmatrix}$$

But these two matrices can never be equal (since $0 \neq 1$). We conclude that our assumption that A has an inverse is false. That is, A does not have an inverse.

Now Work Problem 21

• •

So far, we have shown that a square matrix may or may not have an inverse. What about nonsquare matrices? Can they have inverses? The answer is ''No.'' By definition, whenever a matrix has an inverse, it will commute with its inverse under multiplication. So if the nonsquare matrix A had the alleged inverse B, then AB would have to be equal to BA. But the fact that A is not square causes AB and BA to have different dimensions and prevents them from being equal. So such a B could not exist.

> A nonsquare matrix has no inverse.

The procedure used in Example 2 to find the inverse, if it exists, of a square matrix becomes quite involved as the dimension of the matrix gets larger. A more efficient method is provided next.

Reduced Row-Echelon Technique

We will introduce this technique by looking at an example.

EXAMPLE 4

Find the inverse of the matrix

$$A = \begin{bmatrix} 4 & 2 \\ 3 & 1 \end{bmatrix}$$

SOLUTION

Assuming A has an inverse, we will denote it by

$$X = \begin{bmatrix} x_1 & x_2 \\ x_3 & x_4 \end{bmatrix}$$

Then the product of A and X is the identity matrix of dimension 2×2. That is,

$$AX = I_2$$

$$\begin{bmatrix} 4 & 2 \\ 3 & 1 \end{bmatrix} \begin{bmatrix} x_1 & x_2 \\ x_3 & x_4 \end{bmatrix} = \begin{bmatrix} 1 & 0 \\ 0 & 1 \end{bmatrix}$$

Performing the multiplication on the left yields

$$\begin{bmatrix} 4x_1 + 2x_3 & 4x_2 + 2x_4 \\ 3x_1 + x_3 & 3x_2 + x_4 \end{bmatrix} = \begin{bmatrix} 1 & 0 \\ 0 & 1 \end{bmatrix}$$

This matrix equation can be written as the following system of four equations containing four variables:

$$\begin{cases} 4x_1 + 2x_3 = 1 & \quad 4x_2 + 2x_4 = 0 \\ 3x_1 + x_3 = 0 & \quad 3x_2 + x_4 = 1 \end{cases}$$

We find the solution to be

$$x_1 = -\tfrac{1}{2} \qquad x_2 = 1 \qquad x_3 = \tfrac{3}{2} \qquad x_4 = -2$$

Thus the inverse of A is

$$A^{-1} = \begin{bmatrix} -\tfrac{1}{2} & 1 \\ \tfrac{3}{2} & -2 \end{bmatrix}$$

• •

Let's look at this example more closely. The system of four equations containing four variables can be written as two systems:

(a) $\begin{cases} 4x_1 + 2x_3 = 1 \\ 3x_1 + x_3 = 0 \end{cases}$ (b) $\begin{cases} 4x_2 + 2x_4 = 0 \\ 3x_2 + x_4 = 1 \end{cases}$

Their augmented matrices are

(a) $\begin{bmatrix} 4 & 2 & | & 1 \\ 3 & 1 & | & 0 \end{bmatrix}$ (b) $\begin{bmatrix} 4 & 2 & | & 0 \\ 3 & 1 & | & 1 \end{bmatrix}$

Since the matrix A appears in both (a) and (b), any row operation we perform on (a) and (b) can be performed more easily on the single augmented matrix that combines the two right-hand columns. We denote this matrix by $A|I_2$ and write

$$[A|I_2] = \begin{bmatrix} 4 & 2 & | & 1 & 0 \\ 3 & 1 & | & 0 & 1 \end{bmatrix}$$

If we perform row operations on $[A|I_2]$, just as if we were computing the reduced row-echelon form of A, we get

$$\begin{bmatrix} 1 & 0 & | & -\frac{1}{2} & 1 \\ 0 & 1 & | & \frac{3}{2} & -2 \end{bmatrix}$$

The 2×2 matrix on the right-hand side of the vertical bar is A^{-1}.

This example illustrates the general procedure:

To find the inverse, if it exists, of a square matrix of dimension $n \times n$, follow these steps:

Steps for Finding the Inverse of a Matrix

Step 1 Write the augmented matrix $[A|I_n]$.

Step 2 Using row operations, write $[A|I_n]$ in reduced row-echelon form.

Step 3 If the resulting matrix is of the form $[I_n|B]$, that is, if the identity matrix appears on the left side of the bar, then B is the inverse of A. Otherwise, A has no inverse.

Let's work another example.

EXAMPLE 5

Find the inverse of

$$A = \begin{bmatrix} 1 & 1 & 2 \\ 2 & 1 & 0 \\ 1 & 2 & 2 \end{bmatrix}$$

SOLUTION

Since A is of dimension 3×3, we use the identity matrix I_3. The matrix $[A|I_3]$ is

$$\begin{bmatrix} 1 & 1 & 2 & | & 1 & 0 & 0 \\ 2 & 1 & 0 & | & 0 & 1 & 0 \\ 1 & 2 & 2 & | & 0 & 0 & 1 \end{bmatrix}$$

We proceed to transform this matrix, using row operations:

Use $R_2 = -2r_1 + r_2$
$R_3 = -r_1 + r_3$ to obtain
$$\left[\begin{array}{ccc|ccc} 1 & 1 & 2 & 1 & 0 & 0 \\ 0 & -1 & -4 & -2 & 1 & 0 \\ 0 & 1 & 0 & -1 & 0 & 1 \end{array}\right]$$

Use $R_2 = (-1)r_2$ to obtain
$$\left[\begin{array}{ccc|ccc} 1 & 1 & 2 & 1 & 0 & 0 \\ 0 & 1 & 4 & 2 & -1 & 0 \\ 0 & 1 & 0 & -1 & 0 & 1 \end{array}\right]$$

Use $R_1 = -r_2 + r_1$
$R_3 = -r_2 + r_3$ to obtain
$$\left[\begin{array}{ccc|ccc} 1 & 0 & -2 & -1 & 1 & 0 \\ 0 & 1 & 4 & 2 & -1 & 0 \\ 0 & 0 & -4 & -3 & 1 & 1 \end{array}\right]$$

Use $R_3 = (-\frac{1}{4})r_3$ to obtain
$$\left[\begin{array}{ccc|ccc} 1 & 0 & -2 & -1 & 1 & 0 \\ 0 & 1 & 4 & 2 & -1 & 0 \\ 0 & 0 & 1 & \frac{3}{4} & -\frac{1}{4} & -\frac{1}{4} \end{array}\right]$$

Use $R_1 = 2r_3 + r_1$
$R_2 = -4r_3 + r_2$ to obtain
$$\left[\begin{array}{ccc|ccc} 1 & 0 & 0 & \frac{1}{2} & \frac{1}{2} & -\frac{1}{2} \\ 0 & 1 & 0 & -1 & 0 & 1 \\ 0 & 0 & 1 & \frac{3}{4} & -\frac{1}{4} & -\frac{1}{4} \end{array}\right]$$

Since the identity matrix I_3 appears on the left side, the matrix appearing on the right is the inverse. That is,

$$A^{-1} = \left[\begin{array}{ccc} \frac{1}{2} & \frac{1}{2} & -\frac{1}{2} \\ -1 & 0 & 1 \\ \frac{3}{4} & -\frac{1}{4} & -\frac{1}{4} \end{array}\right]$$

· ·

You should verify that, in fact, $AA^{-1} = I_3$.

EXAMPLE 6

Show that the matrix given below has no inverse.
$$\left[\begin{array}{cc} 3 & 2 \\ 6 & 4 \end{array}\right]$$

SOLUTION

We set up the matrix
$$\left[\begin{array}{cc|cc} 3 & 2 & 1 & 0 \\ 6 & 4 & 0 & 1 \end{array}\right]$$

Use $R_1 = \frac{1}{3}r_1$ to obtain
$$\left[\begin{array}{cc|cc} 1 & \frac{2}{3} & \frac{1}{3} & 0 \\ 6 & 4 & 0 & 1 \end{array}\right]$$

Use $R_2 = -6r_1 + r_2$ to obtain
$$\left[\begin{array}{cc|cc} 1 & \frac{2}{3} & \frac{1}{3} & 0 \\ 0 & 0 & -2 & 1 \end{array}\right]$$

The 0s in row 2 tell us we cannot get the identity matrix. This, in turn, tells us the original matrix has no inverse.

Now Work Problem 33

· ·

Solving a System of n Linear Equations Containing n Variables Using Inverses

The inverse of a matrix can also be used to solve a system of n linear equations containing n variables. Let's look at an example.

EXAMPLE 7

Solve the system of equations:
$$\begin{cases} x + y + 2z = 1 \\ 2x + y = 2 \\ x + 2y + 2z = 3 \end{cases}$$

SOLUTION

If we let

$$A = \begin{bmatrix} 1 & 1 & 2 \\ 2 & 1 & 0 \\ 1 & 2 & 2 \end{bmatrix} \qquad X = \begin{bmatrix} x \\ y \\ z \end{bmatrix} \qquad B = \begin{bmatrix} 1 \\ 2 \\ 3 \end{bmatrix}$$

the above system can be written as

$$AX = B$$

From Example 5 we know A has an inverse, A^{-1}. If we multiply both sides of the equation by A^{-1}, we obtain

$$A^{-1}(AX) = A^{-1}B$$
$$(A^{-1}A)X = A^{-1}B$$
$$I_3X = A^{-1}B$$
$$X = A^{-1}B$$

$$X = \begin{bmatrix} \frac{1}{2} & \frac{1}{2} & -\frac{1}{2} \\ -1 & 0 & 1 \\ \frac{3}{4} & -\frac{1}{4} & -\frac{1}{4} \end{bmatrix} \begin{bmatrix} 1 \\ 2 \\ 3 \end{bmatrix} = \begin{bmatrix} 0 \\ 2 \\ -\frac{1}{2} \end{bmatrix}$$

Thus the solution is $x = 0$, $y = 2$, $z = -\frac{1}{2}$.

. . . .

Based on the discussion given above, a system of equations

$$AX = B$$

for which A is a square matrix and A^{-1} exists, always has a unique solution, given by

$$X = A^{-1}B$$

If, for a system of equations $AX = B$, the matrix A has no inverse, then the system may be consistent or inconsistent. In such cases, the methods discussed in Section 2.3 must be used to investigate the system.

The method used in Example 7 for solving a system of equations is particularly useful for applications in which the constants appearing to the right of the equal sign change while the coefficients of the variables on the left side do not. See Problems 39–50 for an illustration. See also the discussion of Leontief models in Section 2.7.

EXERCISE 2.6   Answers to odd-numbered problems begin on page AN-12.

In Problems 1–6 show that the given matrices are inverses of each other.

1. $\begin{bmatrix} 1 & 2 \\ 2 & 3 \end{bmatrix} \begin{bmatrix} -3 & 2 \\ 2 & -1 \end{bmatrix}$

2. $\begin{bmatrix} 1 & 5 \\ 2 & 0 \end{bmatrix} \begin{bmatrix} 0 & \frac{1}{2} \\ \frac{1}{5} & -\frac{1}{10} \end{bmatrix}$

3. $\begin{bmatrix} -1 & -2 \\ 3 & 4 \end{bmatrix} \begin{bmatrix} 2 & 1 \\ -\frac{3}{2} & -\frac{1}{2} \end{bmatrix}$

4. $\begin{bmatrix} 1 & 3 \\ 2 & -1 \end{bmatrix} \begin{bmatrix} \frac{1}{7} & \frac{3}{7} \\ \frac{2}{7} & -\frac{1}{7} \end{bmatrix}$

5. $\begin{bmatrix} 1 & 2 & 3 \\ 2 & 3 & 4 \\ 1 & 2 & 1 \end{bmatrix} \begin{bmatrix} -\frac{5}{2} & 2 & -\frac{1}{2} \\ 1 & -1 & 1 \\ \frac{1}{2} & 0 & -\frac{1}{2} \end{bmatrix}$

6. $\begin{bmatrix} 1 & 3 & 3 \\ 1 & 4 & 3 \\ 1 & 3 & 4 \end{bmatrix} \begin{bmatrix} 7 & -3 & -3 \\ -1 & 1 & 0 \\ -1 & 0 & 1 \end{bmatrix}$

In Problems 7–20 find the inverse of each matrix using the reduced row-echelon technique.

7. $\begin{bmatrix} 2 & 5 \\ 1 & 3 \end{bmatrix}$

8. $\begin{bmatrix} 4 & 1 \\ 3 & 1 \end{bmatrix}$

9. $\begin{bmatrix} 1 & -1 \\ 3 & -4 \end{bmatrix}$

10. $\begin{bmatrix} 5 & 3 \\ 3 & 2 \end{bmatrix}$

11. $\begin{bmatrix} 2 & 1 \\ 4 & 3 \end{bmatrix}$

12. $\begin{bmatrix} 2 & 3 \\ 2 & -1 \end{bmatrix}$

13. $\begin{bmatrix} 0 & 0 & 1 \\ 0 & 1 & 0 \\ 1 & 0 & 0 \end{bmatrix}$

14. $\begin{bmatrix} -1 & 1 & 0 \\ 1 & 0 & 2 \\ 3 & 1 & 0 \end{bmatrix}$

15. $\begin{bmatrix} 1 & 1 & -1 \\ 3 & -1 & 0 \\ 2 & -3 & 4 \end{bmatrix}$

16. $\begin{bmatrix} 1 & 1 & 1 \\ 2 & 1 & 1 \\ 1 & 1 & 2 \end{bmatrix}$

17. $\begin{bmatrix} 1 & 1 & -1 \\ 2 & 1 & 1 \\ 1 & 0 & 1 \end{bmatrix}$

18. $\begin{bmatrix} 2 & 3 & -1 \\ 1 & 1 & 1 \\ 0 & 2 & -1 \end{bmatrix}$

19. $\begin{bmatrix} 1 & 1 & 0 & 0 \\ 0 & 1 & -1 & 1 \\ 1 & -1 & 1 & 1 \\ 0 & 1 & 0 & -1 \end{bmatrix}$

20. $\begin{bmatrix} 1 & 2 & -3 & -2 \\ 0 & 1 & 4 & -2 \\ 3 & -1 & 4 & 0 \\ 2 & 1 & 0 & 3 \end{bmatrix}$

In Problems 21–26 show that each matrix has no inverse.

21. $\begin{bmatrix} 4 & 6 \\ 2 & 3 \end{bmatrix}$

22. $\begin{bmatrix} -1 & 2 \\ 3 & -6 \end{bmatrix}$

23. $\begin{bmatrix} -8 & 4 \\ -4 & 2 \end{bmatrix}$

24. $\begin{bmatrix} 2 & 10 \\ 1 & 5 \end{bmatrix}$

25. $\begin{bmatrix} 1 & 1 & 1 \\ 3 & -4 & 2 \\ 0 & 0 & 0 \end{bmatrix}$

26. $\begin{bmatrix} -1 & 2 & 3 \\ 5 & 2 & 0 \\ 2 & -4 & -6 \end{bmatrix}$

In Problems 27–34 find the inverse, if it exists, of each matrix.

27. $\begin{bmatrix} 1 & 1 \\ 1 & 2 \end{bmatrix}$

28. $\begin{bmatrix} 2 & 1 \\ 1 & 1 \end{bmatrix}$

29. $\begin{bmatrix} 3 & -2 \\ 0 & 2 \end{bmatrix}$

30. $\begin{bmatrix} 4 & -1 \\ -1 & 0 \end{bmatrix}$

31. $\begin{bmatrix} 3 & 2 \\ 6 & 4 \end{bmatrix}$

32. $\begin{bmatrix} 4 & 2 \\ 2 & 1 \end{bmatrix}$

33. $\begin{bmatrix} 1 & -2 & -1 \\ -2 & 5 & 4 \\ 3 & -8 & -5 \end{bmatrix}$

34. $\begin{bmatrix} 1 & 1 & -1 \\ -2 & -1 & 4 \\ 3 & 2 & -8 \end{bmatrix}$

35. Find the inverse of both

$$A = \begin{bmatrix} 1 & 2 \\ 2 & -1 \end{bmatrix} \quad \text{and} \quad B = \begin{bmatrix} 1 & 4 \\ 3 & 1 \end{bmatrix}$$

to determine $A^{-1} - B^{-1}$.

36. Find the inverse of both

$$A = \begin{bmatrix} 1 & -4 \\ 2 & -1 \end{bmatrix} \quad \text{and} \quad B = \begin{bmatrix} 1 & -2 \\ 3 & -2 \end{bmatrix}$$

to determine $A^{-1} - B^{-1}$.

37. Write the matrix product $A^{-1}B$ used to find the solution to the system $AX = B$.

$$\begin{cases} x + 3y + 2z = 2 \\ 2x + 7y + 3z = 1 \\ x \quad\quad + 6z = 3 \end{cases}$$

38. Write the matrix product $A^{-1}B$ used to find the solution to the system $AX = B$.

$$\begin{cases} x + 2y + 2z = 3 \\ 2x + 5y + 7z = 2 \\ 2x + y - 4z = 4 \end{cases}$$

In Problems 39–50 solve each system of equations by the method of Example 7.

39. $\begin{cases} 3x + 7y = 10 \\ 2x + 5y = 7 \end{cases}$

40. $\begin{cases} 3x + 7y = -4 \\ 2x + 5y = -3 \end{cases}$

41. $\begin{cases} 3x + 7y = 13 \\ 2x + 5y = 9 \end{cases}$

42. $\begin{cases} 3x + 7y = 20 \\ 2x + 5y = 14 \end{cases}$

43. $\begin{cases} 3x + 7y = 12 \\ 2x + 5y = -4 \end{cases}$

44. $\begin{cases} 3x + 7y = -2 \\ 2x + 5y = 10 \end{cases}$

45. $\begin{cases} x + y - z = 3 \\ 3x - y = -4 \\ 2x - 3y + 4z = 6 \end{cases}$

46. $\begin{cases} x + y - z = 6 \\ 3x - y = 8 \\ 2x - 3y + 4z = -2 \end{cases}$

47. $\begin{cases} x + y - z = 12 \\ 3x - y = -4 \\ 2x - 3y + 4z = 16 \end{cases}$

48. $\begin{cases} x + y - z = -8 \\ 3x - y = 12 \\ 2x - 3y + 4z = -2 \end{cases}$

49. $\begin{cases} x + y - z = 9 \\ 3x - y = -8 \\ 2x - 3y + 4z = -6 \end{cases}$

50. $\begin{cases} x + y - z = 21 \\ 3x - y = 12 \\ 2x - 3y + 4z = 14 \end{cases}$

51. Show that the inverse of

$$A = \begin{bmatrix} a & b \\ c & d \end{bmatrix}$$

is given by the formula

$$A^{-1} = \begin{bmatrix} \dfrac{d}{\Delta} & \dfrac{-b}{\Delta} \\ \dfrac{-c}{\Delta} & \dfrac{a}{\Delta} \end{bmatrix}$$

where $\Delta = ad - bc \neq 0$. The number Δ is called the **determinant** of A.

In Problems 52–54 use the result of Problem 51 to find the inverse of each matrix.

52. $\begin{bmatrix} 1 & 2 \\ 2 & 3 \end{bmatrix}$

53. $\begin{bmatrix} 1 & 5 \\ 2 & 0 \end{bmatrix}$

54. $\begin{bmatrix} -1 & -2 \\ 3 & 4 \end{bmatrix}$

Technology Exercises
A graphing calculator can be used to find the inverse of a matrix. If no inverse
exists, an error message is given.

In Problems 1–4 use a graphing calculator to find the inverse, if it exists, of each matrix.

1. $\begin{bmatrix} 25 & 61 & -12 \\ 18 & -2 & 4 \\ 8 & 35 & 21 \end{bmatrix}$

2. $\begin{bmatrix} 18 & -3 & 4 \\ 6 & -20 & 14 \\ 10 & 25 & -15 \end{bmatrix}$

3. $\begin{bmatrix} 44 & 21 & 18 & 6 \\ -2 & 10 & 15 & 5 \\ 21 & 12 & -12 & 4 \\ -8 & -16 & 4 & 9 \end{bmatrix}$

4. $\begin{bmatrix} 16 & 22 & -3 & 5 \\ 21 & -17 & 4 & 8 \\ 2 & 8 & 27 & 20 \\ 5 & 15 & -3 & -10 \end{bmatrix}$

Graphing calculators also can be used to multiply and add matrices. Use the idea
behind Example 7 and a graphing calculator to solve the following systems of equations.

5. $\begin{cases} 25x + 61y - 12z = 10 \\ 18x - 12y + 7z = -9 \\ 3x + 4y - z = 12 \end{cases}$

6. $\begin{cases} 25x + 61y - 12z = 5 \\ 18x - 12y + 7z = -3 \\ 3x + 4y - z = 12 \end{cases}$

7. $\begin{cases} 25x + 61y - 12z = 21 \\ 18x - 12y + 7z = 7 \\ 3x + 4y - z = -2 \end{cases}$

8. $\begin{cases} 25x + 61y - 12z = 25 \\ 18x - 12y + 7z = 10 \\ 3x + 4y - z = -4 \end{cases}$

9. Use your graphing calculator to find the inverse of the matrix

$$A = \begin{bmatrix} 3 & 0 & 2 & -1 & 3 \\ -2 & 1 & 2 & 3 & 0 \\ 2 & 2 & 1 & 1 & -1 \\ 1 & 2 & 0 & 2 & -3 \\ 4 & 0 & -1 & 1 & -1 \end{bmatrix}$$

Check your result by multiplying your answer by matrix
A. What should the result be?

10. Try to find the inverse of the matrix

$$A = \begin{bmatrix} 0 & 0 & 2 & -1 & 3 \\ -2 & 0 & 2 & 3 & 0 \\ 2 & 2 & 0 & 0 & -1 \\ 1 & 2 & 0 & 2 & -3 \\ 4 & 4 & 0 & 0 & -2 \end{bmatrix}$$

How does your calculator (or a computer) respond?

2.7 APPLICATIONS: LEONTIEF MODEL; CRYPTOGRAPHY; ACCOUNTING*

Application 1: Leontief Models

The Leontief models in economics are named after Wassily Leontief, who received the Nobel prize in economics in 1973. (See the October 29, 1973, issue of *Newsweek*, p. 94.) These models can be characterized as a description of an economy in which input equals output or, in other words, consumption equals production. That is, the models assume that whatever is produced is always consumed.

Leontief models are of two types: closed, in which the entire production is consumed by those participating in the production; and open, in which some of the production is

* This section is optional, and the applications given are independent of one another.

consumed by those who produce it and the rest of the production is consumed by external bodies.

In the *closed model* we seek the relative income of each participant in the system. In the *open model* we seek the amount of production needed to achieve a forecast demand, when the amount of production needed to achieve current demand is known.

THE CLOSED MODEL

We begin with an example to illustrate the idea.

EXAMPLE 1

Three homeowners, Juan, Luis, and Carlos, each with certain skills, agreed to pool their talents to make repairs on their houses. As it turned out, Juan spent 20% of his time on his own house, 40% of his time on Luis' house, and 40% on Carlos' house. Luis spent 10% of his time on Juan's house, 50% of his time on his own house, and 40% on Carlos' house. Of Carlos' time, 60% was spent on Juan's house, 10% on Luis', and 30% on his own. Now that the projects are finished, they need to figure out how much money each should get for his work, including the work performed on his own house, so that the amount paid by each person equals the amount received by each one. They agreed in advance that the payment to each one should be approximately $3000.00.

SOLUTION

We place the information given in the problem in a 3×3 matrix, as follows:

	Work done by		
	Juan	Luis	Carlos
Proportion of work done on Juan's house	0.2	0.1	0.6
Proportion of work done on Luis' house	0.4	0.5	0.1
Proportion of work done on Carlos' house	0.4	0.4	0.3

Next, we define the variables:

$$x = \text{Juan's wages}$$
$$y = \text{Luis' wages}$$
$$z = \text{Carlos' wages}$$

We require that the total amount paid out by each one equals the total amount received by each one. Let's analyze this requirement, by looking just at the work done on Juan's house. Juan's wages are x. Juan's expenditures for work done on his house are $0.2x + 0.1y + 0.6z$. Juan's wages and expenditures are required to be equal, so

$$x = 0.2x + 0.1y + 0.6z$$

Similarly,

$$y = 0.4x + 0.5y + 0.1z$$
$$z = 0.4x + 0.4y + 0.3z$$

These three equations can be written compactly as

$$\begin{bmatrix} x \\ y \\ z \end{bmatrix} = \begin{bmatrix} 0.2 & 0.1 & 0.6 \\ 0.4 & 0.5 & 0.1 \\ 0.4 & 0.4 & 0.3 \end{bmatrix} \begin{bmatrix} x \\ y \\ z \end{bmatrix}$$

Some computation reduces the system to

$$\begin{cases} 0.8x - 0.1y - 0.6z = 0 \\ -0.4x + 0.5y - 0.1z = 0 \\ -0.4x - 0.4y + 0.7z = 0 \end{cases}$$

Solving for x, y, z, we find that

$$x = \tfrac{31}{36}z \qquad y = \tfrac{32}{36}z$$

where z is the parameter. To get solutions that fall close to \$3000, we set $z = 3600$.*
The wages to be paid out are therefore

$$x = \$3100 \qquad y = \$3200 \qquad z = \$3600$$

· ·

The matrix in Example 1, namely,

$$\begin{bmatrix} 0.2 & 0.1 & 0.6 \\ 0.4 & 0.5 & 0.1 \\ 0.4 & 0.4 & 0.3 \end{bmatrix}$$

is called an **input–output matrix.**

In the general closed model we have an economy consisting of n components. Each component produces an *output* of some goods or services, which, in turn, is completely used up by the n components. The proportionate use of each component's output by the economy makes up the input–output matrix of the economy. The problem is to find suitable pricing levels for each component so that total income equals total expenditure.

Closed Leontief Model

In general, an input–output matrix for a closed Leontief model is of the form

$$A = [a_{ij}] \qquad i, j = 1, 2, \ldots, n$$

where the a_{ij} represent the fractional amount of goods or services used by i and produced by j. For a closed model the sum of each column equals 1 (this is the condition that all production is consumed internally) and $0 \le a_{ij} \le 1$ for all entries (this is the restriction that each entry is a fraction).

If A is the input–output matrix of a closed system with n components and X is a column vector representing the price of each output of the system, then

$$X = AX$$

represents the requirement that total income equal expenditure.

---

* Other choices for z are, of course, possible. The choice of which value to use is up to the homeowners. No matter what choice is made, the amount paid by each one equals the amount received by each one.

For example, the first entry of the matrix equality $X = AX$ requires that

$$x_1 = a_{11} x_1 + a_{12} x_2 + \cdots + a_{1n} x_n$$

The right side represents the price paid by component 1 for the goods it uses, while x_1 represents the income of component 1; we are requiring they be equal.

We can rewrite the equation $X = AX$ as

$$X - AX = \mathbf{0}$$
$$I_n X - AX = \mathbf{0}$$
$$(I_n - A)X = \mathbf{0}$$

This matrix equation, which represents a system of equations in which the right-hand side is always $\mathbf{0}$, is called a **homogeneous system of equations.** It can be shown that if the entries in the input–output matrix A are positive and if the sum of each column of A equals 1, then this system has a one-parameter solution; that is, we can solve for $n - 1$ of the variables in terms of the remaining one, which serves as the parameter. This parameter serves as a "scale factor."

Now Work Problem 1

THE OPEN MODEL

For the open model, in addition to internal consumption of goods produced, there is an outside demand for the goods produced. This outside demand may take the form of exportation of goods or may be the goods needed to support consumer demand. Again, however, we make the assumption that whatever is produced is also consumed.

For example, suppose an economy consists of three industries R, S, and T, and suppose each one produces a single product. We assume that a portion of R's production is used by each of the three industries, while the remainder is used up by consumers. The same is true of the production of S and T. To organize our thoughts, we construct a table that describes the interaction of the use of R, S, and T's production over some fixed period of time. See Table 1.

Table 1

	R	S	T	Consumer	Total
R	50	20	40	70	180
S	20	30	20	90	160
T	30	20	20	50	120

All entries in the table are in appropriate units, say, in dollars. The first row (row R) represents the production in dollars of industry R (input). Out of the total of $180 worth of goods produced, R, S, and T use $50, $20, and $40, respectively, for the production of their goods, while consumers purchase the remaining $70 for their consumption (output). Observe that input equals output since everything produced by R is used up by R, S, T, and consumers.

The second and third rows are interpreted in the same way.

An important observation is that the goal of R's production is to produce \$70 worth of goods, since this is the demand of consumers. In order to meet this demand, R must produce a total of \$180, since the difference, \$110, is required internally by R, S, and T.

Suppose, however, that consumer demand is expected to change. To effect this change, how much should each industry now produce? For example, in Table 1, current demand for R, S, and T can be represented by a **demand vector:**

$$D_0 = \begin{bmatrix} 70 \\ 90 \\ 50 \end{bmatrix}$$

But suppose marketing forecasts predict that in 3 years the demand vector will be

$$D_3 = \begin{bmatrix} 60 \\ 110 \\ 60 \end{bmatrix}$$

Here, the demand for item R has decreased; the demand for item S has significantly increased, and the demand for item T is higher. Given the current total output of R, S, and T at 180, 160, and 120, respectively, what must it be in 3 years to meet this projected demand?

In using input–output analysis to obtain a solution to such a forecasting problem, we take into account the fact that the output of any one of these industries is affected by changes in the other two, since the total demand for, say, R in 3 years depends not only on consumer demand for R, but also on consumer demand for S and T. That is, the industries are interrelated.

The solution of this type of forecasting problem is derived from the *open Leontief model* in input–output analysis.

To obtain the solution, we need to determine how much of each of the three products R, S, and T is required to produce 1 unit of R. For example, to obtain 180 units of R requires the use of 50 units of R, 20 units of S, and 30 units of T (the entries in column 1). Forming the ratios, we find that to produce 1 unit of R requires $\frac{50}{180} = 0.278$ of R, $\frac{20}{180} = 0.111$ of S, and $\frac{30}{180} = 0.167$ of T. If we want, say, x units of R, we will require $0.278x$ units of R, $0.111x$ units of S, and $0.167x$ units of T.

Continuing in this way, we can construct the matrix

$$A = \begin{array}{c} \\ R \\ S \\ T \end{array} \begin{array}{c} \quad R \qquad\quad S \qquad\quad T \\ \begin{bmatrix} 0.278 & 0.125 & 0.333 \\ 0.111 & 0.188 & 0.167 \\ 0.167 & 0.125 & 0.167 \end{bmatrix} \end{array}$$

Observe that column 1 represents the amounts of R, S, T required for 1 unit of R; column 2 represents the amounts of R, S, and T required for 1 unit of S; and column 3 represents the amounts of R, S, and T required for 1 unit of T. For example, the entry in row 3, column 2 (0.125), represents the amount of T needed to produce 1 unit of S.

As a result of placing the entries this way, if

$$X = \begin{bmatrix} x \\ y \\ z \end{bmatrix}$$

represents the total output required to obtain a given demand, the product AX represents the amount of R, S, and T required for internal consumption. The condition that production = consumption requires that

$$\text{Internal consumption} + \text{Consumer demand} = \text{Total output}$$

In terms of the matrix A, the total output X, and the demand vector D, this requirement is equivalent to the equation

$$AX + D = X$$

In this equation we seek to find X for a prescribed demand D. The matrix A is calculated as above for some initial production process.*

EXAMPLE 2

For the data given in Table 1, find the total output X required to achieve a future demand of

$$D_3 = \begin{bmatrix} 60 \\ 110 \\ 60 \end{bmatrix}$$

SOLUTION

We need to solve for X in

$$AX + D_3 = X$$

Simplifying, we have

$$[I - A]X = D_3$$

Solving for X, we have

$$X = [I - A]^{-1} \cdot D_3$$

$$= \begin{bmatrix} 0.722 & -0.125 & -0.333 \\ -0.111 & 0.812 & -0.167 \\ -0.167 & -0.125 & 0.833 \end{bmatrix}^{-1} \begin{bmatrix} 60 \\ 110 \\ 60 \end{bmatrix}$$

$$= \begin{bmatrix} 1.6048 & 0.3568 & 0.7131 \\ 0.2946 & 1.3363 & 0.3857 \\ 0.3660 & 0.2721 & 1.4013 \end{bmatrix} \begin{bmatrix} 60 \\ 110 \\ 60 \end{bmatrix}$$

$$= \begin{bmatrix} 178.322 \\ 187.811 \\ 135.969 \end{bmatrix}$$

* The entries in A can be checked by using the requirement that $AX + D = X$, for $D =$ Initial demand and $X =$ Total output. For our example, it must happen that

$$\underset{AX}{\underbrace{\begin{bmatrix} 0.278 & 0.125 & 0.333 \\ 0.111 & 0.188 & 0.167 \\ 0.167 & 0.125 & 0.167 \end{bmatrix} \begin{bmatrix} 180 \\ 160 \\ 120 \end{bmatrix}}} + \underset{D}{\underbrace{\begin{bmatrix} 70 \\ 90 \\ 50 \end{bmatrix}}} = \underset{X}{\underbrace{\begin{bmatrix} 180 \\ 160 \\ 120 \end{bmatrix}}}$$

$$\text{Internal consumption} \quad + \quad \text{Consumer demand} = \text{Total output}$$

Thus the total output of R, S, and T required for the forecast demand D_3 is

$$x = 178.322 \qquad y = 187.811 \qquad z = 135.969$$

....

The general open model can be described as follows: Suppose there are n industries in the economy. Each industry produces some goods or services, which are partially consumed by the n industries, while the rest are used to meet a prescribed current demand. Given the output required of each industry to meet current demand, what should the output of each industry be to meet some different future demand?

Open Leontief Model

The matrix $A = [a_{ij}]$, $i, j = 1, \ldots, n$, of the open model is defined to consist of entries a_{ij}, where a_{ij} is the amount of output of industry j required for one unit of output of industry i. If X is a column vector representing the production of each industry in the system and D is a column vector representing future demand for goods produced in the system, then

$$X = AX + D$$

From the equation above we find

$$[I_n - A]X = D$$

It can be shown that the matrix $I_n - A$ has an inverse, provided each entry in A is positive and the sum of each column in A is less than 1. Under these conditions we may solve for X to get

$$X = [I_n - A]^{-1} \cdot D$$

This form of the solution is particularly useful since it allows us to find X for a variety of demands D by doing one calculation: $[I_n - A]^{-1}$.

We conclude by noting that the use of an input–output matrix to solve forecasting problems assumes that each industry produces a single commodity and that no technological advances take place in the period of time under investigation (in other words, the proportions found in the matrix A are fixed).

Application 2: Cryptography

Our second application is to *cryptography,* the art of writing or deciphering secret codes. We begin by giving examples of elementary codes.

EXAMPLE 3

A message can be encoded by associating each letter of the alphabet with some other letter of the alphabet according to a prescribed pattern. For example, we might have

A B C D E F G H I J K L M N O P Q R S T U V W X Y Z
↓ ↓
C D E F G H I J K L M N O P Q R S T U V W X Y Z A B

With the above code the word *BOMB* would become DQOD.

· ·

EXAMPLE 4

Another code may associate numbers with the letters of the alphabet. For example, we might have

A B C D E F G H I J K L M N O P Q R S T U V W X Y Z
↓ ↓
26 25 24 23 22 21 20 19 18 17 16 15 14 13 12 11 10 9 8 7 6 5 4 3 2 1

In this code the word *PEACE* looks like 11 22 26 24 22.

· ·

Both the above codes have one important feature in common. The association of letters with the coding symbols is made using a one-to-one correspondence so that no possible ambiguities can arise.

Suppose we want to encode the following message:

$$\text{BEWARE   THE   IDES   OF   MARCH}$$

If we decide to divide the message into pairs of letters, the message becomes

$$\text{BE   WA   RE   TH   EI   DE   SO   FM   AR   CH}$$

(If there is a letter left over, we arbitrarily assign Z to the last position.) Using the correspondence of letters to numbers given in Example 4, and writing each pair of letters as a column vector, we obtain

$$\begin{bmatrix} B \\ E \end{bmatrix} = \begin{bmatrix} 25 \\ 22 \end{bmatrix} \quad \begin{bmatrix} W \\ A \end{bmatrix} = \begin{bmatrix} 4 \\ 26 \end{bmatrix} \quad \begin{bmatrix} R \\ E \end{bmatrix} = \begin{bmatrix} 9 \\ 22 \end{bmatrix} \quad \begin{bmatrix} T \\ H \end{bmatrix} = \begin{bmatrix} 7 \\ 19 \end{bmatrix} \quad \text{etc.}$$

Next, we arbitrarily choose a 2×2 matrix A, which we know has an inverse A^{-1} (the reason for this is seen later). Suppose we choose

$$A = \begin{bmatrix} 2 & 3 \\ 1 & 2 \end{bmatrix}$$

Its inverse is

$$A^{-1} = \begin{bmatrix} 2 & -3 \\ -1 & 2 \end{bmatrix}$$

Now, we transform the column vectors representing the message by multiplying each of them on the left by the matrix A:

$$A \begin{bmatrix} B \\ E \end{bmatrix} = A \begin{bmatrix} 25 \\ 22 \end{bmatrix} = \begin{bmatrix} 116 \\ 69 \end{bmatrix}$$

$$A \begin{bmatrix} W \\ A \end{bmatrix} = A \begin{bmatrix} 4 \\ 26 \end{bmatrix} = \begin{bmatrix} 86 \\ 56 \end{bmatrix}$$

$$A \begin{bmatrix} R \\ E \end{bmatrix} = A \begin{bmatrix} 9 \\ 22 \end{bmatrix} = \begin{bmatrix} 84 \\ 53 \end{bmatrix} \quad \text{etc.}$$

The coded message is

$$116 \quad 69 \quad 86 \quad 56 \quad 84 \quad 53 \quad \text{etc.}$$

To decode or unscramble the above message, pair the numbers in 2×1 column vectors. Multiply each of these column vectors by A^{-1} on the left:

$$A^{-1} \begin{bmatrix} 116 \\ 69 \end{bmatrix} = \begin{bmatrix} 25 \\ 22 \end{bmatrix}$$

$$A^{-1} \begin{bmatrix} 86 \\ 56 \end{bmatrix} = \begin{bmatrix} 4 \\ 26 \end{bmatrix}$$

By reassigning letters to these numbers, we obtain the original message.

EXAMPLE 5

The message to be encoded is

$$\text{THE} \quad \text{END} \quad \text{IS} \quad \text{NEAR}$$

We agree to associate numbers to letters as follows:

A B C D E F G H I J K L M N O P Q R S T U V W X Y Z
↓ ↓
1 2 3 4 5 6 7 8 9 10 11 12 13 14 15 16 17 18 19 20 21 22 23 24 25 26

The encoded message is to be formed of triplets of numbers.

SOLUTION

This time we must divide the message into triplets of letters, obtaining

$$\text{THE} \quad \text{END} \quad \text{ISN} \quad \text{EAR}$$

in order for the encoded message to have triplets of numbers. (If the message required additional letters to complete the triplet, we would have used Z or YZ.)

Now we choose a 3×3 matrix such as

$$A = \begin{bmatrix} 1 & 0 & 0 \\ 3 & 1 & 5 \\ -2 & 0 & 1 \end{bmatrix}$$

Its inverse is

$$A^{-1} = \begin{bmatrix} 1 & 0 & 0 \\ -13 & 1 & -5 \\ 2 & 0 & 1 \end{bmatrix}$$

The encoded message is obtained by multiplying the matrix A times each column vector of the original message:

$$A \begin{bmatrix} T \\ H \\ E \end{bmatrix} = \begin{bmatrix} 1 & 0 & 0 \\ 3 & 1 & 5 \\ -2 & 0 & 1 \end{bmatrix} \begin{bmatrix} 20 \\ 8 \\ 5 \end{bmatrix} = \begin{bmatrix} 20 \\ 93 \\ -35 \end{bmatrix}$$

$$A \begin{bmatrix} E \\ N \\ D \end{bmatrix} = \begin{bmatrix} 1 & 0 & 0 \\ 3 & 1 & 5 \\ -2 & 0 & 1 \end{bmatrix} \begin{bmatrix} 5 \\ 14 \\ 4 \end{bmatrix} = \begin{bmatrix} 5 \\ 49 \\ -6 \end{bmatrix}$$

$$A \begin{bmatrix} I \\ S \\ N \end{bmatrix} = \begin{bmatrix} 1 & 0 & 0 \\ 3 & 1 & 5 \\ -2 & 0 & 1 \end{bmatrix} \begin{bmatrix} 9 \\ 19 \\ 14 \end{bmatrix} = \begin{bmatrix} 9 \\ 116 \\ -4 \end{bmatrix}$$

$$A \begin{bmatrix} E \\ A \\ R \end{bmatrix} = \begin{bmatrix} 1 & 0 & 0 \\ 3 & 1 & 5 \\ -2 & 0 & 1 \end{bmatrix} \begin{bmatrix} 5 \\ 1 \\ 18 \end{bmatrix} = \begin{bmatrix} 5 \\ 106 \\ 8 \end{bmatrix}$$

The coded message is

$$20 \quad 93 \quad -35 \quad 5 \quad 49 \quad -6 \quad 9 \quad 116 \quad -4 \quad 5 \quad 106 \quad 8$$

. . . .

To decode the message in Example 5, form 3×1 column vectors of the numbers in the coded message and multiply on the left by A^{-1}.

The above are elementary examples of encoding and decoding. Modern-day cryptography uses sophisticated computer-implemented codes that depend on higher-level mathematics. For an interesting survey of cryptographic techniques see the article ''The Mathematics of Public-Key Cryptography'' in the August 1979 issue of *Scientific American*.

Now Work Problem 11

Application 3: Accounting

Consider a firm that has two types of departments, production and service. The production departments produce goods that can be sold in the market, and the service departments provide services to the production departments. A major objective of the cost accounting process is the determination of the full cost of manufactured products on a per unit basis. This requires an allocation of indirect costs, first, from the service department (where they are incurred) to the producing department in which the goods are manufactured and, second, to the specific goods themselves. For example, an accounting department usually provides accounting services for service departments, as well as for the production departments. Thus the indirect costs of service rendered by a service department must be determined in order to correctly assess the production departments. The total costs of a service department consist of its direct costs (salaries, wages, and materials) and its indirect costs (charges for the services it receives from other service departments). The nature of the problem and its solution are illustrated by the following example.

EXAMPLE 6

Consider a firm with two production departments, P_1 and P_2, and three service departments, S_1, S_2, and S_3. These five departments are listed in the leftmost column of Table 2. The total monthly costs of these departments are unknown and are

Table 2

Department	Total Costs	Direct Costs, Dollars	Indirect Costs for Services from Departments		
			S_1	S_2	S_3
S_1	x_1	600	$0.25x_1$	$0.15x_2$	$0.15x_3$
S_2	x_2	1100	$0.35x_1$	$0.20x_2$	$0.25x_3$
S_3	x_3	600	$0.10x_1$	$0.10x_2$	$0.35x_3$
P_1	x_4	2100	$0.15x_1$	$0.25x_2$	$0.15x_3$
P_2	x_5	1500	$0.15x_1$	$0.30x_2$	$0.10x_3$
Totals		5900	x_1	x_2	x_3

denoted by x_1, x_2, x_3, x_4, x_5. The direct monthly costs of the five departments are shown in the third column of the table. The fourth, fifth, and sixth columns show the allocation of charges for the services of S_1, S_2, and S_3 to the various departments. Since the total cost for each department is its direct costs plus its indirect costs, the first three rows of the table yield the total costs for the three service departments:

$$x_1 = \ \ \ 600 + 0.25x_1 + 0.15x_2 + 0.15x_3$$
$$x_2 = 1100 + 0.35x_1 + 0.20x_2 + 0.25x_3$$
$$x_3 = \ \ \ 600 + 0.10x_1 + 0.10x_2 + 0.35x_3$$

Let X, C, and D denote the following matrices:

$$X = \begin{bmatrix} x_1 \\ x_2 \\ x_3 \end{bmatrix} \quad C = \begin{bmatrix} 0.25 & 0.15 & 0.15 \\ 0.35 & 0.20 & 0.25 \\ 0.10 & 0.10 & 0.35 \end{bmatrix} \quad D = \begin{bmatrix} 600 \\ 1100 \\ 600 \end{bmatrix}$$

Then the system of equations above can be written in matrix notation as

$$X = D + CX$$

which is equivalent to

$$[I_3 - C]X = D$$

The total costs of the three service departments can be obtained by solving this matrix equation for X:

$$X = [I_3 - C]^{-1}D$$

Now

$$[I_3 - C] = \begin{bmatrix} 0.75 & -0.15 & -0.15 \\ -0.35 & 0.80 & -0.25 \\ -0.10 & -0.10 & 0.65 \end{bmatrix}$$

from which it can be verified that

$$[I_3 - C]^{-1} = \begin{bmatrix} 1.57 & 0.36 & 0.50 \\ 0.79 & 1.49 & 0.76 \\ 0.36 & 0.28 & 1.73 \end{bmatrix}$$

It is significant that the inverse of $[I_3 - C]$ exists, and that all of its entries are nonnegative. Because of this and the fact that the matrix D contains only nonnegative entries, the matrix X will also have only nonnegative entries. This means there is a meaningful solution to the accounting problem:

$$X = \begin{bmatrix} 1638.00 \\ 2569.00 \\ 1562.00 \end{bmatrix}$$

Thus, $x_1 = \$1638.00$, $x_2 = \$2569.00$, and $x_3 = \$1562.00$. All direct and indirect costs can now be determined by substituting these values in Table 2, as shown in Table 3.

From Table 3 we learn that department P_1 pays $\$1122.25$ for the services it receives from S_1, S_2, S_3, and P_2 pays $\$1172.60$ for the services it receives from these departments. The procedure we have followed charges the direct costs of the service departments to the production departments, and each production department is charged according to the services it utilizes. Furthermore, the total cost for P_1 and P_2 is $\$5894.85$, and this figure approximates the sum of the direct costs of the three service departments and the two production departments. The results are consistent with conventional accounting procedure. Discrepancies that occur are due to rounding off.

Table 3

Department	Total Costs, Dollars	Direct Costs, Dollars	Indirect Costs for Services from Departments, Dollars		
			S_1	S_2	S_3
S_1	1629.15	600	409.50	385.35	234.30
S_2	2577.60	1100	573.30	513.80	390.50
S_3	1567.40	600	163.80	256.90	546.70
P_1	3222.25	2100	245.70	642.25	234.30
P_2	2672.60	1500	245.70	770.70	156.20

Finally, a comment should be made about the allocation of charges for services as shown in Table 2. How is it determined that 25% of the total cost x_1 of S_1 should be charged to S_1, 35% to S_2, 10% to S_3, 15% to P_1, and 15% to P_2? The services of each department can be measured in some suitable unit, and each department can be charged according to the number of these units of service it receives. If 20% of the accounting items concern a given department, that department is charged 20% of the total cost of the accounting department. When services are not readily measurable, the allocation basis is subjectively determined.

Now Work Problem 15

························

EXERCISE 2.7   Answers to odd-numbered problems begin on page AN-13.

Problems 1–9 involve the Leontief model. In Problems 1–4 find the relative wages of each person for the given closed input–output matrix. In each case take the wages of C to be the parameter and use z = C's wages = $30,000.

1.
$$\begin{array}{c} \\ A \\ B \\ C \end{array} \begin{array}{ccc} A & B & C \\ \left[\begin{array}{ccc} \frac{1}{2} & \frac{1}{3} & \frac{1}{4} \\ \frac{1}{4} & \frac{1}{3} & \frac{1}{4} \\ \frac{1}{4} & \frac{1}{3} & \frac{1}{2} \end{array}\right] \end{array}$$

2.
$$\begin{array}{c} \\ A \\ B \\ C \end{array} \begin{array}{ccc} A & B & C \\ \left[\begin{array}{ccc} \frac{1}{4} & \frac{2}{3} & \frac{1}{2} \\ \frac{1}{2} & \frac{1}{6} & \frac{1}{4} \\ \frac{1}{4} & \frac{1}{6} & \frac{1}{4} \end{array}\right] \end{array}$$

3.
$$\begin{array}{c} \\ A \\ B \\ C \end{array} \begin{array}{ccc} A & B & C \\ \left[\begin{array}{ccc} 0.2 & 0.3 & 0.1 \\ 0.6 & 0.4 & 0.2 \\ 0.2 & 0.3 & 0.7 \end{array}\right] \end{array}$$

4.
$$\begin{array}{c} \\ A \\ B \\ C \end{array} \begin{array}{ccc} A & B & C \\ \left[\begin{array}{ccc} 0.4 & 0.3 & 0.2 \\ 0.2 & 0.3 & 0.3 \\ 0.4 & 0.4 & 0.5 \end{array}\right] \end{array}$$

5. For the three industries R, S, and T in the open Leontief model of Table 1 on page 115, compute the total output vector X if the forecast demand vector is

$$D_2 = \begin{bmatrix} 80 \\ 90 \\ 60 \end{bmatrix}$$

6. Rework Problem 5 if the forecast demand vector is

$$D_4 = \begin{bmatrix} 100 \\ 80 \\ 60 \end{bmatrix}$$

7. **Closed Leontief Model**   A society consists of four individuals: a farmer, a builder, a tailor, and a rancher (who produces meat products). Of the food produced by the farmer, $\frac{3}{10}$ is used by the farmer, $\frac{2}{10}$ by the builder, $\frac{2}{10}$ by the tailor, and $\frac{3}{10}$ by the rancher. The builder's production is utilized 30% by the farmer, 30% by the builder, 10% by the tailor, and 30% by the rancher. The tailor's production is used in the ratios $\frac{3}{10}$, $\frac{3}{10}$, $\frac{1}{10}$, and $\frac{3}{10}$ by the farmer, builder, tailor, and rancher, respectively. Finally, meat products are used 20% by each of the farmer, builder, and tailor,

and 40% by the rancher. What is the relative income of each if the rancher's income is scaled at $25,000?

8. If in Problem 7 the meat production utilization changes so that it is used equally by all four individuals, while everyone else's production utilization remains the same, what are the relative incomes?

9. **Open Leontief Model**   Suppose the interrelationships between the production of two industries R and S in a given year are given in the table:

	R	S	Current Consumer Demand	Total Output
R	30	40	60	130
S	20	10	40	70

If the forecast demand in 2 years is

$$D_2 = \begin{bmatrix} 80 \\ 40 \end{bmatrix}$$

what should the total output X be?

Problems 10–14 involve cryptography, using the correspondence

A B C D E F G H I J K L M N O P Q R S T U V W X Y Z
↓ ↓
1 2 3 4 5 6 7 8 9 10 11 12 13 14 15 16 17 18 19 20 21 22 23 24 25 26

10. Use the matrices

$$(I) \quad A = \begin{bmatrix} 2 & 3 \\ 1 & 2 \end{bmatrix} \qquad (II) \quad A = \begin{bmatrix} 1 & 0 & 0 \\ 3 & 1 & 5 \\ -2 & 0 & 1 \end{bmatrix}$$

to encode the following messages:

(a) MEET ME AT THE CASBAH
(b) TOMORROW NEVER COMES
(c) THE MISSION IS IMPOSSIBLE   2 2

11. Use the matrix

$$A = \begin{bmatrix} 2 & 3 \\ 1 & 2 \end{bmatrix}$$

to decode the following messages:

(a) 51   30   27   16   75   47   19   10   48   26
(b) 70   45   103   62   58   38   102   61   88   57

12. Use the matrix

$$A = \begin{bmatrix} 1 & 0 & 0 \\ 3 & 1 & 5 \\ -2 & 0 & 1 \end{bmatrix}$$

to decode the message

25   195   − 29   6   135   9   14   183   − 2

13. The matrix A used to encode a message has as its inverse the matrix

$$A^{-1} = \begin{bmatrix} 2 & -3 \\ -1 & 2 \end{bmatrix}$$

Decode the message

11 7 84 51 51 28 66 43 44 29 107 65 64 41

14. Use the matrix

$$A = \begin{bmatrix} 1 & 0 & 0 \\ 3 & 1 & 5 \\ -2 & 0 & 1 \end{bmatrix}$$

to encode the message: SELL THE COMPANY

Problems 15 and 16 concern accounting.

15. Consider the accounting problem described by the data in the table:

Department	Total Costs	Direct Costs, Dollars	Indirect Costs	
			S_1	S_2
S_1	x_1	2,000	$\frac{1}{9}x_1$	$\frac{3}{9}x_2$
S_2	x_2	1,000	$\frac{3}{9}x_1$	$\frac{1}{9}x_2$
P_1	x_3	2,500	$\frac{1}{9}x_1$	$\frac{2}{9}x_2$
P_2	x_4	1,500	$\frac{3}{9}x_1$	$\frac{1}{9}x_2$
P_3	x_5	3,000	$\frac{1}{9}x_1$	$\frac{2}{9}x_2$
Totals		10,000	x_1	x_2

Determine whether this accounting problem has a solution. If it does, find the total costs. Prepare a table similar to Table 3. Show that the total of the service charges allocated to P_1, P_2, and P_3 is equal to the sum of the direct costs of the service departments S_1 and S_2.

16. Follow the directions of Problem 15 for the accounting problem described by the following data:

Department	Total Costs	Direct Costs, Dollars	Indirect Costs for Services from Departments		
			S_1	S_2	S_3
S_1	x_1	500	$0.20x_1$	$0.10x_2$	$0.10x_3$
S_2	x_2	1000	$0.40x_1$	$0.15x_2$	$0.30x_3$
S_3	x_3	500	$0.10x_1$	$0.05x_2$	$0.30x_3$
P_1	x_4	2000	$0.20x_1$	$0.35x_2$	$0.20x_3$
P_2	x_5	1500	$0.10x_1$	$0.35x_2$	$0.10x_3$
Totals		5500	x_1	x_2	x_3

Technology Exercises

Problems 1–5 involve the open Leontief model. In Problems 1–4 use a graphing calculator (or a computer) to solve each open Leontief model. Find the production vector X for given input–output matrix A and demand vector D, using the formula
$$X = [I_n - A]^{-1} \cdot D.$$

1. $A = \begin{bmatrix} 0 & 0.1 & 0.2 & 0.1 \\ 0.2 & 0 & 0.4 & 0.3 \\ 0.2 & 0.7 & 0 & 0 \\ 0.1 & 0.2 & 0 & 0.2 \end{bmatrix}$; $D = \begin{bmatrix} 3 \\ 0 \\ 8 \\ 1 \end{bmatrix}$

2. $A = \begin{bmatrix} 0 & 0 & 0.2 & 0.1 & 0.3 \\ 0.2 & 0 & 0.2 & 0.3 & 0 \\ 0.2 & 0.2 & 0 & 0 & 0.1 \\ 0.1 & 0.2 & 0 & 0.2 & 0.3 \\ 0.4 & 0 & 0.1 & 0 & 0.2 \end{bmatrix}$; $D = \begin{bmatrix} 3 \\ 0 \\ 7 \\ 5 \\ 1 \end{bmatrix}$

3. $A = \begin{bmatrix} 0 & 0.2 & 0.1 & 0.3 \\ 0 & 0.2 & 0.3 & 0 \\ 0.2 & 0 & 0 & 0.1 \\ 0 & 0 & 0.2 & 0.3 \end{bmatrix}$; $D = \begin{bmatrix} 5 \\ 7 \\ 2 \\ 0 \end{bmatrix}$

4. $A = \begin{bmatrix} 0 & 0 & 0.02 & 0 & 0.3 \\ 0.2 & 0 & 0 & 0.3 & 0 \\ 0.2 & 0 & 0 & 0 & 0.01 \\ 0.1 & 0.7 & 0 & 0.01 & 0.03 \\ 0.04 & 0.4 & 0 & 0 & 0.2 \end{bmatrix}$; $D = \begin{bmatrix} 7 \\ 0 \\ 6 \\ 8 \\ 2 \end{bmatrix}$

5. Open Leontief Model Suppose the interrelationships between the production of five industries (manufacturing,

electric power, petroleum, transportation, and textiles) in a given year are as listed in the table:

	Manu.	E.P.	Petr.	Tran.	Text.
Manu.	0.2	0.12	0.15	0.18	0.1
E.P.	0.17	0.11	0	0.19	0.28
Petr.	0.11	0.11	0.12	0.46	0.12
Tran.	0.1	0.14	0.18	0.17	0.19
Text.	0.16	0.18	0.02	0.1	0.3

If the demand vector is given by
$$D = \begin{bmatrix} 100 \\ 100 \\ 200 \\ 100 \\ 100 \end{bmatrix}$$

what should the production vector X be? [*Hint:* Use your graphing calculator to find the inverse $[I_n - A]^{-1}$.]

CHAPTER REVIEW

IMPORTANT TERMS AND CONCEPTS

TRUE–FALSE ITEMS   Answers are on page AN-13.

T_____ F_____ **1.** Matrices of the same dimension can always be added.

T_____ F_____ **2.** Matrices of the same dimension can always be multiplied.

T_____ F_____ **3.** A square matrix will always have an inverse.

T_____ F_____ **4.** The reduced row-echelon form of a matrix A is unique.

T_____ F_____ **5.** If A and B are each of dimension 4×4, then $AB = BA$.

T_____ F_____ **6.** Matrix addition is always defined.

T_____ F_____ **7.** Matrix multiplication is commutative.

FILL IN THE BLANKS   Answers are on page AN-13.

1. If matrix A is of dimension 3×4 and matrix B is of dimension 4×2, then AB is of dimension _____ .

2. A system of three linear equations containing three variables has either _____ solution, or no solution, or _____ _____ solutions.

3. If A is a matrix of dimension 3×4, the 3 tells the number of _____ and the 4 tells the number of _____ .

4. If $AB = I$, the identity matrix, then B is called the _____ of A.

5. If B is a 2×3 matrix and BA^2 is defined, then A is a _____ matrix.

6. If A is a 4×5 matrix and AB^3 is defined, then B is a _____ matrix.

REVIEW EXERCISES   Answers to odd-numbered problems begin on page AN-13.

In Problems 1–14 perform the indicated operations using

$$A = \begin{bmatrix} -2 & 0 & 7 \\ 1 & 8 & 3 \\ 2 & 4 & 21 \end{bmatrix} \quad B = \begin{bmatrix} 1 & 3 & 9 \\ 2 & 7 & 5 \\ 3 & 6 & 8 \end{bmatrix} \quad C = \begin{bmatrix} 0 & 1 & 2 \\ 0 & 5 & 1 \\ 8 & 7 & 9 \end{bmatrix}$$

1. $A + B$

2. $B + A$

3. $3(A + B)$

4. $3A + 3B$

5. $3A - 3B$

6. $B - C$

7. $2(5A)$

8. $\frac{3}{2}A$

9. $2A + \frac{1}{2}B - 3C$

10. $A - 2B + 3C$

11. AB

12. BA

13. $(B - A)C$

14. $BC - AC$

In Problems 15–22 find the inverse, if it exists, of each matrix.

15. $\begin{bmatrix} 3 & 0 \\ -2 & 1 \end{bmatrix}$

16. $\begin{bmatrix} 4 & 1 \\ 3 & 1 \end{bmatrix}$

17. $\begin{bmatrix} 1 & 2 & 3 \\ 2 & 4 & 5 \\ 3 & 5 & 6 \end{bmatrix}$

18. $\begin{bmatrix} -1 & 2 & 0 \\ 3 & 2 & -1 \\ 4 & 0 & 3 \end{bmatrix}$

19. $\begin{bmatrix} 4 & 3 & -1 \\ 0 & 2 & 2 \\ 3 & -1 & 0 \end{bmatrix}$

20. $\begin{bmatrix} -6 & 6 & 2 \\ 13 & 3 & 1 \\ 8 & -8 & 8 \end{bmatrix}$

21. $\begin{bmatrix} 1 & 2 & -3 \\ 4 & 6 & 2 \\ -3 & -6 & 9 \end{bmatrix}$

22. $\begin{bmatrix} 9 & 6 & -3 \\ 2 & -6 & 4 \\ -3 & 2 & 1 \end{bmatrix}$

In Problems 23–40 find the solution, if it exists, of each system of linear equations. If the system has infinitely many solutions, list at least three solutions.

23. $\begin{cases} -5x + 2y = -2 \\ -3x + 3y = \ \ \ 4 \end{cases}$

24. $\begin{cases} -3x + 2y = 3 \\ -3x - 4y = 4 \end{cases}$

25. $\begin{cases} x + 2y + \ \ 5z = \ \ 6 \\ 3x + 7y + 12z = 23 \\ x + 4y \qquad\quad = 25 \end{cases}$

26. $\begin{cases} x + 2y - \ \ z = -4 \\ 3x + 7y - 6z = -21 \\ x + 4y - 6z = -17 \end{cases}$

27. $\begin{cases} x + 2y + \ \ 7z = \ \ \ \ 2 \\ 3x + 7y + 18z = \ \ -1 \\ x + 4y + \ \ 2z = -13 \end{cases}$

28. $\begin{cases} x + 2y - \ \ 7z = -1 \\ 3x + 7y - 24z = \ \ 6 \\ x + 4y - 12z = 26 \end{cases}$

29. $\begin{cases} 2x - y + z = 1 \\ x + y - z = 2 \\ 3x - y + z = 0 \end{cases}$

30. $\begin{cases} 2x + 3y - z = 5 \\ x - y + z = 1 \\ 3x - 3y + 3z = 3 \end{cases}$

31. $\begin{cases} y - 2z = \ \ 6 \\ 3x + 2y - z = \ \ 2 \\ 4x + 3z = -1 \end{cases}$

32. $\begin{cases} 2x - y + 3z = \ \ \ 5 \\ x + 2z = \ \ \ 0 \\ 3x + 2y + z = -3 \end{cases}$

33. $\begin{cases} x - 3y = 5 \\ 3y + z = 0 \\ 2x - y + 2z = 2 \end{cases}$

34. $\begin{cases} x - z = 2 \\ 2x - y = 4 \\ x + y + z = 6 \end{cases}$

35. $\begin{cases} 3x + y - 2z = 3 \\ x - 2y + z = 4 \end{cases}$

36. $\begin{cases} 2x - y - 3z = 0 \\ x - 2y + z = 4 \end{cases}$

37. $\begin{cases} x + 2y - z = 5 \\ 2x - y + 2z = 0 \end{cases}$

38. $\begin{cases} x - y + 2z = \ \ 6 \\ 2x + 2y - z = -1 \end{cases}$

39. $\begin{cases} 2x - y = 6 \\ x - 2y = 0 \\ 3x - y = 6 \end{cases}$

40. $\begin{cases} x - 2y = \ \ 0 \\ 2x + y = \ \ 5 \\ x - 3y = -3 \end{cases}$

41. What must be true about x, y, z, w, if we require $AB = BA$ for the matrices

$$A = \begin{bmatrix} x & y \\ z & w \end{bmatrix} \quad \text{and} \quad B = \begin{bmatrix} 1 & 1 \\ -1 & 1 \end{bmatrix}$$

42. Let $t = [t_1 \ \ t_2]$, with $t_1 + t_2 = 1$, and let $A = \begin{bmatrix} \frac{1}{4} & \frac{3}{4} \\ \frac{2}{3} & \frac{1}{3} \end{bmatrix}$. Find t such that $tA = t$.

43. Mixture Sweet Delight Candies, Inc., sells boxes of candy consisting of creams and caramels. Each box sells for $4 and holds 50 pieces of candy (all pieces are the same size). If the caramels cost $0.05 to produce and the creams cost $0.10 to produce, how many caramels and creams should be in each box for no profit and no loss? Would you increase or decrease the number of caramels in order to obtain a profit?

44. Cookie Orders A cookie company makes three kinds of cookies, oatmeal raisin, chocolate chip, and shortbread, packaged in small, medium, and large boxes. The small box contains 1 dozen oatmeal raisin and 1 dozen chocolate chip; the medium box has 2 dozen oatmeal raisin, 1 dozen chocolate chip, and 1 dozen shortbread; the large box contains 2 dozen oatmeal raisin, 2 dozen chocolate chip, and 3 dozen shortbread. If you require exactly 15 dozen oatmeal raisin, 10 dozen chocolate chip, and 11 dozen shortbread cookies, how many of each size box should you buy?

45. Mixture Problem A store sells almonds for $6 per pound, cashews for $5 per pound, and peanuts for $2 per pound. One week the manager decides to prepare 100 16-ounce packages of nuts by mixing the peanuts, almonds, and cashews. Each package will be sold for $4. The mixture is to produce the same revenue as selling the nuts separately. Prepare a table that shows some of the possible ways the manager can prepare the mixture.

46. Financial Planning Three retired couples each require an additional annual income of $1800 per year. As their financial consultant, you recommend they invest some money in Treasury bills that yield 6%, some money in corporate bonds that yield 8%, and some money in junk bonds that yield 10%. Prepare a table for each couple showing the various ways their goal can be achieved
(a) If the first couple has $20,000 to invest
(b) If the second couple has $25,000 to invest
(c) If the third couple has $30,000 to invest

47. Financial Planning A retired couple has $40,000 to invest. As their financial consultant, you recommend they invest some money in Treasury bills that yield 6%, some money in corporate bonds that yield 8%, and some money in junk bonds that yield 10%. Prepare a table showing the various ways this couple can achieve the following goals:
(a) They want $2500 per year in income.
(b) They want $3000 per year in income.
(c) They want $3500 per year in income.

Mathematical Questions from Professional Exams
Use the following information to answer Problems 1–4:

Akron, Inc. owns 80% of the capital stock of Benson Company and 70% of the capital stock of Cashin, Inc. Benson Company owns 15% of the capital stock of Cashin, Inc. Cashin, Inc., in turn, owns 25% of the capital stock of Akron, Inc. These ownership interrelationships are illustrated in the following diagram:

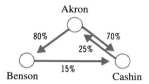

Net income before adjusting for interests in intercompany net income for each corporation follows:

Akron, Inc.	$190,000
Benson Co.	$170,000
Cashin, Inc.	$230,000

Ignore all income tax considerations.

A_e = *Akron's consolidated net income;*
that is, its net income plus its share of the consolidated net income of Benson and Cashin

B_e = *Benson's consolidated net income;*
that is, its net income plus its share of the consolidated net income of Cashin

C_e = *Cashin's consolidated net income;*
that is, its net income plus its share of the consolidated net income of Akron

1. CPA Exam   The equation, in a set of simultaneous equations, which computes A_e is

(a) $A_e = .75(190,000 + .8B_e + .7C_e)$
(b) $A_e = 190,000 + .8B_e + .7C_e$
(c) $A_e = .75(190,000) + .8(170,000) + .7(230,000)$
(d) $A_e = .75(190,000) + .8B_e + .7C_e$

2. CPA Exam   The equation, in a set of simultaneous equations, which computes B_e is

(a) $B_e = 170,000 + .15C_e - .75A_e$
(b) $B_e = 170,000 + .15C_e$
(c) $B_e = .2(170,000) + .15(230,000)$
(d) $B_e = .2(170,000) + .15C_e$

3. CPA Exam   Cashin's minority interest in consolidated net income is

(a) $.15(230,000)$ (b) $230,000 + .25A_e$
(c) $.15(230,000) + .25A_e$ (d) $.15C_e$

4. CPA Exam   Benson's minority interest in consolidated net income is

(a) $34,316   (b) $25,500   (c) $45,755   (d) $30,675

Chapter 3

Linear Programming: Geometric Approach

Whenever the analysis of a problem leads to minimizing or maximizing a linear expression in which the variables must obey a collection of linear inequalities, a solution may be obtained using linear programming techniques.

Historically, linear programming problems evolved out of the need to solve problems involving resource allocation by the U.S. Army during World War II. Among those who worked on such problems was George Dantzig, who later gave a general formulation of the linear programming problem and offered a method for solving it, called the *simplex method*. This method is discussed in Chapter 4.

In this chapter we study ways to solve linear programming problems that involve only two variables. As a result, we can use a geometric approach utilizing linear inequalities to solve the problem.

3.1 LINEAR INEQUALITIES

We have already discussed linear equations (linear equalities) in two variables x and y (Section 1.1). These are equations of the form

$$Ax + By = C \qquad (1)$$

where A, B, C are real numbers and A and B are not both zero. If in Equation (1) we replace the equal sign by an inequality symbol, namely, one of the symbols $<, >, \leq, \geq$, we obtain a **linear inequality in two variables** x and y. For example, the expressions

$$3x + 2y \geq 4 \qquad 2x - 3y < 0 \qquad 3x + 5y > -8$$

are each linear inequalities in two variables. The first of these is called a **nonstrict inequality** since the inequality symbol $\geq$ is nonstrict; the remaining two linear inequalities are **strict.***

The Graph of a Linear Inequality

The **graph of a linear inequality** in two variables x and y is the set of all points (x, y) for which the inequality holds.
Let's look at an example.

EXAMPLE 1

Graph the inequality: $2x + 3y \geq 6$

SOLUTION

The inequality $2x + 3y \geq 6$ is equivalent to $2x + 3y > 6$ or $2x + 3y = 6$. So we begin by graphing the line $2x + 3y = 6$, noting that any point on the line must satisfy the inequality $2x + 3y \geq 6$. See Figure 1(a).

Figure 1

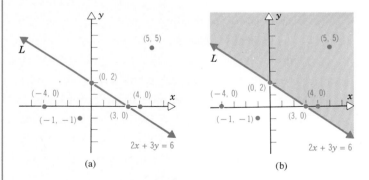

(a)   (b)

Now let's test a few points, such as $(-1, -1)$, $(5, 5)$, $(4, 0)$, $(-4, 0)$, to see if they satisfy the inequality. We do this by substituting the coordinates of each point into the left member of the inequality and determining whether the result is ≥ 6 or < 6.

---

* A review of inequalities may be found in Appendix A, Section A.1.

$$2x \quad + 3y \qquad\qquad\qquad Conclusion$$

	$2x + 3y$	Conclusion
$(-1, -1)$:	$2(-1) + 3(-1) = -2 - 3 = -5 < 6$	Not part of graph
$(5, 5)$:	$2(5) \quad + 3(5) \quad = 25 > 6$	Part of graph
$(4, 0)$:	$2(4) \quad + 3(0) \quad = 8 > 6$	Part of graph
$(-4, 0)$:	$2(-4) + 3(0) \quad = -8 < 6$	Not part of graph

Notice that the two points $(4, 0)$ and $(5, 5)$ that are part of the graph both lie on one side of L, while the points $(-4, 0)$ and $(-1, -1)$ (not part of the graph) lie on the other side of L. This is not an accident. The graph of the inequality consists of all points on the same side of L as $(4, 0)$ and $(5, 5)$. The shaded region of Figure 1(b) illustrates the graph of the inequality.

⋯⋯⋯⋯⋯⋯⋯⋯⋯⋯⋯⋯

The inequality in Example 1 was *nonstrict,* so the *corresponding line was part of the graph of the inequality.* If the inequality is *strict,* the *corresponding line is not part of the graph of the inequality.* We will indicate the latter by using dashes to graph the line. Let's outline the procedure for graphing a linear inequality:

Steps for Graphing a Linear Inequality

Step 1 Graph the corresponding linear equation, a line L. If the inequality is nonstrict, graph L using a solid line; if the inequality is strict, graph L using dashes.

Step 2 Select a test point P not on the line L.

Step 3 Substitute the coordinates of the test point P into the given inequality. If the coordinates of this point P satisfy the linear inequality, then all points on the same side of L as the point P satisfy the inequality. If the coordinates of the point P do not satisfy the linear inequality, then all points on the opposite side of L from P satisfy the inequality.

EXAMPLE 2

Graph the linear inequality: $2x - y < -4$

SOLUTION

The corresponding linear equation is the line

$$L: \quad 2x - y = -4$$

Since the inequality is strict, points on L are not part of the graph of the linear inequality. When we graph L, we use a dashed line to indicate this fact. See Figure 2(a) on the next page.

We select a point not on the line L to be tested, for example, $(0, 0)$:

$$2(0) - 0 = 0 > -4$$

Figure 2

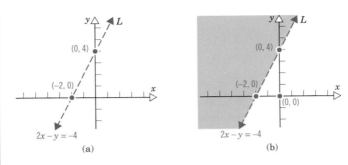

(a)   (b)

Since $(0, 0)$ does not satisfy the inequality, all points on the opposite side of L from $(0, 0)$ are on the graph. The graph of $2x - y < -4$ is the shaded region of Figure 2(b).

· ·

EXAMPLE 3   Graph: (a) $x \leq 3$ (b) $2x \leq y$

SOLUTION

(a) The corresponding linear equation is $x = 3$, a vertical line. If we choose $(0, 0)$ as the test point, we find that it satisfies the inequality. Thus all points to the left of, and on, the vertical line are on the graph. See Figure 3(a).

(b) The corresponding linear equation is $2x = y$. Since it passes through $(0, 0)$, we choose the point $(0, 2)$ as the test point. The inequality is satisfied by $(0, 2)$, so it is on the graph. See Figure 3(b).

Figure 3

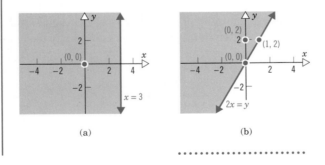

(a)   (b)

· ·

The set of points belonging to the graph of a linear inequality [for example, the shaded region in Figure 3(b)] is sometimes called a **half-plane.**

Now Work Problem 5

Systems of Linear Inequalities

A **system of linear inequalities** is a collection of two or more linear inequalities. To **graph** a system of two inequalities we locate all points whose coordinates satisfy each of the linear inequalities of the system.

EXAMPLE 4

Which of the points $P_1 = (7, 5)$, $P_2 = (9, 12)$, $P_3 = (3, 1)$ are part of the graph of the following system?

$$\begin{cases} 10x - y \geq 0 \\ -x + 2y \geq 0 \\ x + y \leq 15 \end{cases}$$

SOLUTION

For $P_1 = (7, 5)$ to be part of the graph, it must satisfy each of the linear inequalities. Since $10(7) - 5 = 65 \geq 0$, the first inequality is satisfied. Since $-7 + 2(5) = 3 \geq 0$, so is the second. And since $7 + 5 = 12 \leq 15$, so is the third. Thus P_1 is part of the graph.

For $P_2 = (9, 12)$ we have

$$10(9) - 12 = 78 \geq 0 \qquad -9 + 2(12) = 15 \geq 0 \qquad 9 + 12 = 21 \text{ is not } \leq 15$$

Thus P_2 is not part of the graph.

For $P_3 = (3, 1)$ we have

$$10(3) - 1 = 29 \geq 0 \qquad -3 + 2(1) = -1 \text{ is not } \geq 0$$

Thus P_3 is not part of the graph.

Now Work Problem 11

· ·

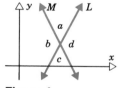

Figure 4

Several possible graphs can result from a system of two linear inequalities in two variables. For example, suppose L and M are the lines corresponding to two linear inequalities, and suppose L and M intersect. See Figure 4. Then the two lines L and M divide the plane into four regions, a, b, c, and d. One of these regions is the solution of the system.

EXAMPLE 5

Graph the system: $\begin{cases} 2x - y \leq -4 \\ x + y \geq -1 \end{cases}$

SOLUTION

First we graph each inequality separately. See Figures 5(a) and 5(b).

Figure 5

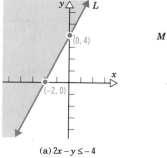

(a) $2x - y \leq -4$

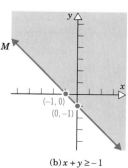

(b) $x + y \geq -1$

The solution of the system consists of all points common to these two half-planes. The dark blue shaded region in Figure 6 represents the solution of the system.

Figure 6

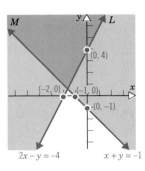

$2x - y = -4$ $x + y = -1$

Now Work Problem 17

$\cdots\cdots\cdots\cdots\cdots\cdots\cdots\cdots\cdots$

If the lines L and M are parallel, the system of linear inequalities may or may not have a solution. Examples of such situations are given below.

EXAMPLE 6

Graph the system: $\begin{cases} 2x - y \leq -4 \\ 2x - y \leq -2 \end{cases}$

SOLUTION

First we graph each inequality separately. See Figures 7(a) and 7(b).

Figure 7

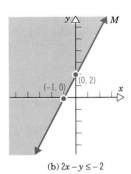

(a) $2x - y \leq -4$ (b) $2x - y \leq -2$

The dark blue shaded region in Figure 8 represents the solution of the system.

$\cdots\cdots\cdots\cdots\cdots\cdots\cdots\cdots\cdots$

Notice that the solution of this system is the same as that of the single linear inequality $2x - y \leq -4$.

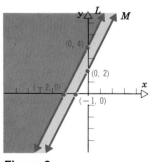

Figure 8

EXAMPLE 7

The solution of the system

$$\begin{cases} 2x - y \geq -4 \\ 2x - y \leq -2 \end{cases}$$

is the dark blue shaded region in Figure 9.

Figure 9

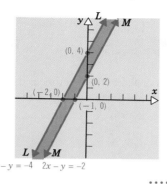

$$2x - y = -4 \quad 2x - y = -2$$

EXAMPLE 8

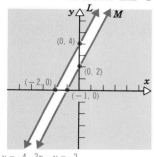

$$2x - y = -4 \quad 2x - y = -2$$

Figure 10

The system

$$\begin{cases} 2x - y \le -4 \\ 2x - y \ge -2 \end{cases}$$

has no solution, as Figure 10 indicates, because the two half-planes have no points in common.

Until now, we have considered systems of only two linear inequalities. The next example is of a system of four linear inequalities. As we shall see, the technique for graphing such systems is the same as that used for graphing systems of two linear inequalities in two variables.

EXAMPLE 9

Graph the system:
$$\begin{cases} x + y \ge 2 \\ 2x + y \ge 3 \\ x \ge 0 \\ y \ge 0 \end{cases}$$

SOLUTION

Again we first graph the four lines:

$$\begin{aligned} L_1: & \quad x + y = 2 \\ L_2: & \quad 2x + y = 3 \\ L_3: & \quad x = 0 \\ L_4: & \quad y = 0 \end{aligned}$$

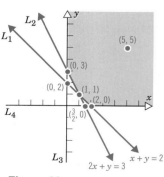

$$2x + y = 3 \qquad x + y = 2$$

Figure 11

The lines L_1 and L_2 intersect at the point (1, 1). (Do you see why?) The inequalities $x \ge 0$ and $y \ge 0$ indicate that the graph of the system lies in quadrant I. Thus the graph of the system consists of that part of the graph of the inequalities $x + y \ge 2$ and $2x + y \ge 3$ that lies in the first quadrant. Since (5, 5) is a point that satisfies each of these inequalities, we obtain Figure 11.

EXAMPLE 10

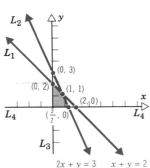

Figure 12

Graph the system: $\begin{cases} x + y \leq 2 \\ 2x + y \leq 3 \\ x \geq 0 \\ y \geq 0 \end{cases}$

SOLUTION

Since the lines associated with these linear inequalities are the same as those of the previous example, we proceed directly to the graph. See Figure 12.

· ·

Some Terminology

Compare the graphs of the systems of linear inequalities given in Figures 11 and 12. The graph in Figure 11 is said to be **unbounded** in the sense that it extends infinitely far in some direction. The graph in Figure 12 is **bounded** in the sense that it can be enclosed by some circle of sufficiently large radius. See Figure 13.

Figure 13

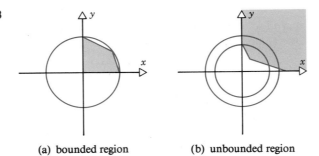

(a) bounded region   (b) unbounded region

The boundary of each of the graphs in Figures 11 and 12 consists of line segments. In fact, the graph of any system of linear inequalities will have line segments as boundaries. The point of intersection of two line segments that form the boundary is called a **corner point** of the graph. For example, the graph of the system given in Example 9 has the corner points $(0, 3)$, $(1, 1)$, and $(2, 0)$. See Figure 11. The graph of the system given in Example 10 has the corner points $(0, 2)$, $(0, 0)$, $(\frac{3}{2}, 0)$, $(1, 1)$. See Figure 12.

We shall soon see that the corner points of the graph of a system of linear inequalities play a major role in the procedure for solving linear programming problems.

Now Work Problem 25

Application

EXAMPLE 11

Nutt's Nuts has 75 pounds of cashews and 120 pounds of peanuts. These are to be mixed in 1-pound packages as follows: a low-grade mixture that contains 4 ounces of cashews and 12 ounces of peanuts and a high-grade mixture that contains 8 ounces of cashews and 8 ounces of peanuts.

(a) Use x to denote the number of packages of the low-grade mixture and use y to denote the number of packages of the high-grade mixture and write a system of linear inequalities that describes the possible number of each kind of package.
(b) Graph the system and list its corner points.

SOLUTION

(a) We begin by naming the variables:

$$x = \text{Number of packages of low-grade mixture}$$
$$y = \text{Number of packages of high-grade mixture}$$

First, we note that the only meaningful values for x and y are nonnegative values. Thus we must restrict x and y so that

$$x \geq 0 \qquad \text{and} \qquad y \geq 0$$

Next, we note that there is a limit to the number of pounds of cashews and peanuts available. That is, the total number of pounds of cashews cannot exceed 75 pounds (1200 ounces), and the number of pounds of peanuts cannot exceed 120 pounds (1920 ounces). This means that

$$\begin{pmatrix} \text{Ounces of} \\ \text{cashews} \\ \text{required} \\ \text{for low-grade} \\ \text{mixture} \end{pmatrix} \begin{pmatrix} \text{Number of} \\ \text{packages of} \\ \text{low-grade} \\ \text{mixture} \end{pmatrix} + \begin{pmatrix} \text{Ounces of} \\ \text{cashews} \\ \text{required} \\ \text{for high-} \\ \text{grade} \\ \text{mixture} \end{pmatrix} \begin{pmatrix} \text{Number of} \\ \text{packages} \\ \text{of high-} \\ \text{grade} \\ \text{mixture} \end{pmatrix} \begin{matrix} \text{cannot} \\ \text{exceed} \end{matrix} 1200$$

$$\begin{pmatrix} \text{Ounces of} \\ \text{peanuts} \\ \text{required} \\ \text{for low-grade} \\ \text{mixture} \end{pmatrix} \begin{pmatrix} \text{Number of} \\ \text{packages of} \\ \text{low-grade} \\ \text{mixture} \end{pmatrix} + \begin{pmatrix} \text{Ounces of} \\ \text{peanuts} \\ \text{for high-} \\ \text{grade} \\ \text{mixture} \end{pmatrix} \begin{pmatrix} \text{Number of} \\ \text{packages of} \\ \text{high-grade} \\ \text{mixture} \end{pmatrix} \begin{matrix} \text{cannot} \\ \text{exceed} \end{matrix} 1920$$

In terms of the data given and the variables introduced, we can write these statements compactly as

$$4x + 8y \leq 1200$$
$$12x + 8y \leq 1920$$

The system of linear inequalities that gives the possible values x and y can take on is

$$\begin{cases} 4x + 8y \leq 1200 \\ 12x + 8y \leq 1920 \\ \quad\quad\; x \geq \quad 0 \\ \quad\quad\; y \geq \quad 0 \end{cases}$$

(b) The system of linear inequalities given above can be simplified to the equivalent form

$$\begin{cases} x + 2y \leq 300 \\ 3x + 2y \leq 480 \\ \quad\; x \geq \quad 0 \\ \quad\; y \geq \quad 0 \end{cases}$$

The graph of the system is given in Figure 14. The corner points of the graph are the points of intersection of the lines ① and ②, ① and ③, ② and ④, and ③ and ④ as shown in Figure 14. The last three are easy to identify by inspection: the first one requires that we solve the system of equations

$$\begin{cases} x + 2y = 300 \\ 3x + 2y = 480 \end{cases}$$

Subtracting the first equation from the second gives $2x = 180$ or $x = 90$. Using this in the first equation, we find

$$2y = 300 - x \quad \text{or} \quad 2y = 210 \quad \text{or} \quad y = 105$$

Therefore, (90, 105) is a corner point. The four corner points are

$$(0, 0), (0, 150), (160, 0), \text{ and } (90, 105)$$

Figure 14

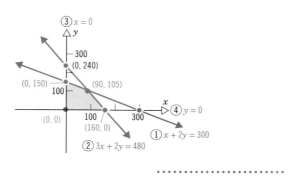

EXERCISE 3.1   Answers to odd-numbered problems begin on page AN-14.

In Problems 1–10 graph each inequality.

1. $x \geq 0$

2. $y \geq 0$

3. $x \geq 0, y \geq 0$

4. $x \leq 0, y \leq 0$

5. $2x - 3y \leq -6$

6. $3x + 2y \geq 6$

7. $5x + y \leq -10$

8. $x - 2y > -4$

9. $x \geq 5$

10. $y \leq -2$

11. Which of the points $P_1 = (3, 8), P_2 = (12, 9), P_3 = (5, 1)$ are part of the graph of the following system?

$$\begin{cases} -10x + 3y \leq 0 \\ -3x + 2y \geq 0 \\ 2x + y \leq 15 \end{cases}$$

13. Which of the points $P_1 = (5, -8), P_2 = (10, 10), P_3 = (5, 1)$ are part of the graph of the following system?

$$\begin{cases} 10x + 3y \geq 0 \\ 3x + 2y \geq 0 \\ x + y \leq 15 \end{cases}$$

12. Which of the points $P_1 = (9, -5), P_2 = (12, -4), P_3 = (4, 1)$ are part of the graph of the following system?

$$\begin{cases} 10x + y \leq 0 \\ -x + 2y \geq 0 \\ 4x + y \leq 15 \end{cases}$$

14. Which of the points $P_1 = (2, 6), P_2 = (12, 4), P_3 = (4, 1)$ are part of the graph of the following system?

$$\begin{cases} -10x + y \leq 0 \\ 2x - 5y \leq 0 \\ x + 3y \leq 15 \end{cases}$$

15. Which of the points $P_1 = (5, 3)$, $P_2 = (6, 12)$, $P_3 = (6, 1)$ are part of the graph of the following system?

$$\begin{cases} 2y - 10x \geq 0 \\ 2y - \quad x \geq 0 \\ \quad y + 6x \geq 15 \end{cases}$$

16. Which of the points $P_1 = (1, -4)$, $P_2 = (10, 6)$, $P_3 = (6, -2)$ are part of the graph of the following system?

$$\begin{cases} 3y - 10x \leq 0 \\ 2y + 5x \geq 0 \\ 4y + \quad x \leq 15 \end{cases}$$

In Problems 17–24 determine which region a, b, c, or d represents the graph of the given system of linear inequalities. The regions a, b, c, and d are nonoverlapping regions bounded by the indicated lines.

17. $\begin{cases} 5x - 4y \leq 8 \\ 2x + 5y \leq 23 \end{cases}$

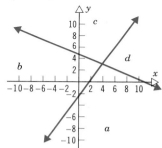

18. $\begin{cases} 4x - 5y \leq 0 \\ 4x + 2y \leq 28 \end{cases}$

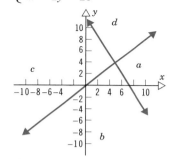

19. $\begin{cases} 2x - 3y \leq -3 \\ 4x + 6y \leq 30 \end{cases}$

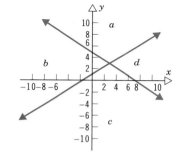

20. $\begin{cases} 6x - 5y \leq 5 \\ 2x + 4y \leq 30 \end{cases}$

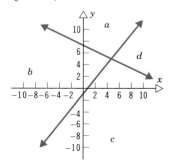

21. $\begin{cases} 5x - 3y \leq 3 \\ 2x + 6y \leq 30 \end{cases}$

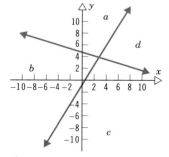

22. $\begin{cases} 6x - 5y \leq 10 \\ 6x + 4y \leq 46 \end{cases}$

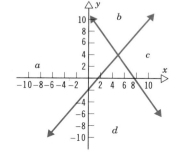

23. $\begin{cases} 5x - 4y \leq 0 \\ 2x + 4y \leq 28 \end{cases}$

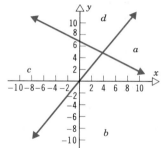

24. $\begin{cases} 2x - 5y \leq -5 \\ 3x + 5y \leq 30 \end{cases}$

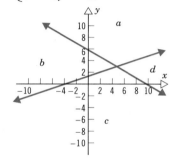

In Problems 25–34 graph each system of linear inequalities. Tell whether the graph is bounded or unbounded and list each corner point of the graph.

25. $\begin{cases} x + y \le 2 \\ x \ge 0 \\ y \ge 0 \end{cases}$

26. $\begin{cases} 2x + 3y \le 6 \\ x \ge 0 \\ y \ge 0 \end{cases}$

27. $\begin{cases} x + y \ge 2 \\ 2x + 3y \le 6 \\ x \ge 0 \\ y \ge 0 \end{cases}$

28. $\begin{cases} x + y \ge 2 \\ 2x + 3y \le 12 \\ 3x + 2y \le 12 \\ x \ge 0 \\ y \ge 0 \end{cases}$

29. $\begin{cases} 2 \le x + y \\ x + y \le 8 \\ 2x + y \le 10 \\ x \ge 0 \\ y \ge 0 \end{cases}$

30. $\begin{cases} 2 \le x + y \\ x + y \le 8 \\ 1 \le x + 2y \\ x \ge 0 \\ y \ge 0 \end{cases}$

31. $\begin{cases} x + y \ge 2 \\ 2x + 3y \le 12 \\ 3x + y \le 12 \\ x \ge 0 \\ y \ge 0 \end{cases}$

32. $\begin{cases} 2 \le x + y \\ x + y \le 10 \\ 2x + y \le 3 \\ x \ge 0 \\ y \ge 0 \end{cases}$

33. $\begin{cases} 1 \le x + 2y \\ x + 2y \le 10 \\ x \ge 0 \\ y \ge 0 \end{cases}$

34. $\begin{cases} 1 \le x + 2y \\ x + 2y \le 10 \\ 2 \le x + y \\ x + y \le 8 \\ x \ge 0 \\ y \ge 0 \end{cases}$

35. Rework Example 11 if 60 pounds of cashews and 90 pounds of peanuts are available.

36. Rework Example 11 if the high-grade mixture contains 10 ounces of cashews and 6 ounces of peanuts.

37. **Manufacturing**   Mike's Famous Toy Trucks company manufactures two kinds of toy trucks—a dumpster and a tanker. In the manufacturing process, each dumpster requires 3 hours of grinding and 4 hours of finishing, while each tanker requires 2 hours of grinding and 3 hours of finishing. The company has two grinders and three finishers, each of whom works at most 40 hours per week.

   (a) Using x to denote the number of dumpsters and y to denote the number of tankers, write a system of linear inequalities that describes the possible numbers of each truck that can be manufactured.
   (b) Graph the system and list its corner points.

38. **Manufacturing**   Repeat Problem 37 if one grinder and two finishers, each of whom works at most 40 hours per week, are available.

39. **Financial Planning**   A retired couple has up to $25,000 to invest. As their financial adviser, you recommend they place at least $15,000 in Treasury bills yielding 6% and at most $10,000 in corporate bonds yielding 9%.

   (a) Using x to denote the amount of money invested in Treasury bills and y to denote the amount invested in corporate bonds, write a system of inequalities that describes this situation.
   (b) Graph the system and list its corner points.
   (c) Interpret the meaning of each corner point in relation to the investments it represents.

40. **Financial Planning**   Use the information supplied in Problem 39, along with the fact that the couple will invest at least $15,000, to answer Parts (a), (b), and (c).

41. **Nutrition**   A farmer prepares feed for livestock by combining two types of grain. Each unit of the first grain contains 1 unit of protein and 5 units of iron while each unit of the second grain contains 2 units of protein and 1 unit of iron. Each animal must receive at least 5 units of protein and 16 units of iron each day.

   (a) Write a system of linear inequalities that describes the possible amounts of each grain the farmer needs to prepare.
   (b) Graph the system and list the corner points.

42. **Investment Strategy**   Laura wishes to invest up to a total of $40,000 in class AA bonds and stocks. Furthermore, she believes that the amount invested in class AA

bonds should be at most one-third of the amount invested in stocks.

(a) Write a system of linear inequalities that describes the possible amount of investments in each security.

(b) Graph the system and list the corner points.

43. Nutrition   To maintain an adequate daily diet, nutritionists recommend the following: at least 85 g of carbohydrate, 70 g of fat, and 50 g of protein. An ounce of food A contains 5 g of carbohydrate, 3 g of fat, and 2 g of protein, while an ounce of food B contains 4 g of carbohydrate, 3 g of fat, and 3 g of protein.

(a) Write a system of linear inequalities that describes the possible quantities of each food.

(b) Graph the system and list the corner points.

44. Transportation   A microwave company has two plants, one on the East Coast and one in the Midwest. It takes 25 hours (packing, transportation, and so on) to transport an order of microwaves from the Eastern plant to its central warehouse and 20 hours from the Midwest plant to its central warehouse. It costs $80 to transport an order from the Eastern plant to the central warehouse and $40 from the Midwestern to its central warehouse. There are 1000 work-hours available for packing, transportation, and so on, and $3000 for transportation cost.

(a) Write a system of linear inequalities that describes the transportation system.

(b) Graph the system and list the corner points.

Technology Exercises

Some graphing calculators can graph systems of inequalities and shade the region representing the solution to the system on the screen. Consult your user's manual, and use your graphing calculator to solve the systems of inequalities in Problems 1–6. If your calculator does not graph inequalities, graph the boundary line of each inequality. Use TRACE to find the corner points.

1. $\begin{cases} 1 < x + y \\ x \le 2 \\ y \le 2 \end{cases}$ **2.** $\begin{cases} x + 2y \le 4 \\ x \ge 0 \\ y \ge 0 \end{cases}$ **3.** $\begin{cases} 2x + 3y \le 6 \\ x \ge 1 \\ y \ge 0 \end{cases}$

4. $\begin{cases} 3x + 4y \le 12 \\ x \ge 0 \\ y \ge 0 \end{cases}$ **5.** $\begin{cases} y - x + 1 \ge 0 \\ y + x \le 5 \\ x \ge 0 \end{cases}$ **6.** $\begin{cases} 2x - 3y \ge 6 \\ x \ge 0 \\ y \le 2 \\ y \ge 0 \end{cases}$

3.2 A GEOMETRIC APPROACH TO LINEAR PROGRAMMING PROBLEMS

We begin by restating a portion of Example 11 given in the previous section: Nutt's Nuts has 75 pounds of cashews and 120 pounds of peanuts. These are to be mixed in 1-pound packages as follows: a low-grade mixture that contains 4 ounces of cashews and 12 ounces of peanuts and a high-grade mixture that contains 8 ounces of cashews and 8 ounces of peanuts.

Suppose that in addition to the information given above, we also know what the profit will be on each type of mixture. For example, suppose the profit is $0.25 on each package of the low-grade mixture and is $0.45 on each package of the high-grade mixture. The question of importance to the manager is "How many packages of each type of mixture should be prepared to maximize the profit?"

If P symbolizes the profit, x the number of packages of low-grade mixture, and y the number of high-grade packages, then the question can be restated as "What are the values of x and y so that the expression

$$P = \$0.25x + \$0.45y$$

is a maximum?"

This problem is typical of a **linear programming problem.** It requires that a certain linear expression, the profit, be maximized. This linear expression is called the **objective function.** Furthermore, the problem requires that the maximum profit be achieved under certain restrictions or **constraints,** each of which are linear inequalities involving the variables. The linear programming problem may be restated as

Maximize

$$P = \$0.25x + \$0.45y \quad \text{Objective function}$$

subject to the conditions that

$$x + 2y \leq 300 \quad \text{Cashew constraint}$$
$$3x + 2y \leq 480 \quad \text{Peanut constraint}$$
$$x \geq 0 \quad \text{Nonnegativity constraint}$$
$$y \geq 0 \quad \text{Nonnegativity constraint}$$

In general, every linear programming problem has two components:

1. A linear objective function to be maximized or minimized.
2. A collection of linear inequalities that must be satisfied simultaneously.

LINEAR PROGRAMMING PROBLEM

A *linear programming problem* in two variables, x and y, consists of *maximizing* or *minimizing* an *objective function*

$$z = Ax + By$$

where A and B are given real numbers, subject to certain conditions or *constraints* expressible as a system of linear inequalities in x and y.

Let's look at this definition more closely. To maximize (or minimize) the quantity $z = Ax + By$ means to locate the points (x, y) that result in the largest (or smallest) value of z. But not all points (x, y) are eligible. Only the points that obey *all* the constraints are potential solutions. Hence, we refer to such points as **feasible points.**

In a linear programming problem we want to find the feasible point that maximizes (or minimizes) the objective function.

LINEAR PROGRAMMING PROBLEM SOLUTION

By a *solution* to a linear programming problem we mean a feasible point (x, y), together with the value of the objective function at that point, which maximizes (or minimizes) the objective function.

If none of the feasible points maximizes (or minimizes) the objective function, or if there are no feasible points, then the linear programming problem has no solution.

EXAMPLE 1

Minimize the quantity

$$z = x + 2y$$

subject to the constraints

$$x + y \geq 1$$
$$x \geq 0$$
$$y \geq 0$$

SOLUTION

The objective function to be minimized is $z = x + 2y$. The constraints are the linear inequalities

$$x + y \geq 1$$
$$x \geq 0$$
$$y \geq 0$$

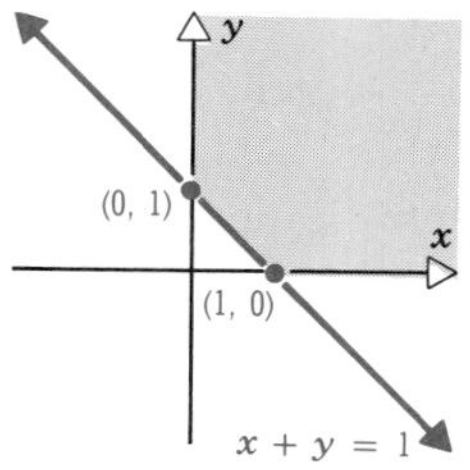

Figure 15

The shaded portion of Figure 15 illustrates the set of feasible points.

To see if there is a smallest z, we graph $z = x + 2y$ for some choice of z, say, $z = 3$. See Figure 16. By moving the line $x + 2y = 3$ parallel to itself, we can observe what happens for different values of z. Since we want a minimum value for z, we try to move $z = x + 2y$ down as far as possible while keeping some part of the line within the set of feasible points. The "best" solution is obtained when the line just touches a corner point of the set of feasible points. If you refer to Figure 16, you will see that the best solution is $x = 1$, $y = 0$, which yields $z = 1$. There is no other feasible point for which z is smaller.

Figure 16

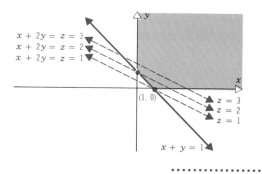

In Example 1 we can see that the feasible point that minimizes z occurs at a corner point. This is not an unusual situation. If there is a feasible point minimizing (or maximizing) the objective function, it is usually located at a corner point of the set of feasible points.

However, it is possible for a feasible point that is not a corner point to minimize (or maximize) the objective function. For example, if the slope of the objective function is the same as the slope of one of the boundaries of the set of feasible points and if the two adjacent corner points are solutions, then so are all the points on the line segment joining them. The following example illustrates this situation.

EXAMPLE 2   Minimize the quantity

$$z = x + 2y$$

subject to the constraints

$$x + y \geq 1$$
$$2x + 4y \geq 3$$
$$x \geq 0$$
$$y \geq 0$$

SOLUTION

Again we first graph the constraints. The shaded portion of Figure 17 illustrates the set of feasible points.

Figure 17

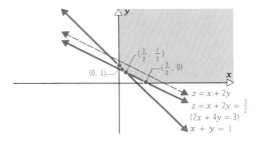

If we graph the objective equation $z = x + 2y$ for some choice of z and move it down, we find that a minimum is reached when $z = \frac{3}{2}$. In fact, any point on the line $2x + 4y = 3$ between the adjacent corner points $(\frac{1}{2}, \frac{1}{2})$ and $(\frac{3}{2}, 0)$ and including these corner points will minimize the objective function. Of course, the reason any feasible point on $2x + 4y = 3$ minimizes the objective equation $z = x + 2y$ is that these two lines each have slope $-\frac{1}{2}$. Thus this linear programming problem has infinitely many solutions.

· ·

Now Work Problem 1

The next example illustrates a linear programming problem that has no solution.

EXAMPLE 3   Maximize the quantity

$$z = x + 2y$$

subject to the constraints

$$x + y \geq 1$$
$$x \geq 0$$
$$y \geq 0$$

SOLUTION

First, we graph the constraints. The shaded portion of Figure 18 illustrates the set of feasible points.

The graphs of the objective function $z = x + 2y$ for $z = 2$, $z = 8$, and $z = 12$ are also shown in Figure 18. Observe that we continue to get larger values for z by moving the graph of the objective function upward. But there is no feasible point that will make z *largest*. No matter how large a value is assigned to z, there is a feasible point that will give a larger value. Since there is no feasible point that makes z largest, we conclude that this linear programming problem has no solution.

Figure 18

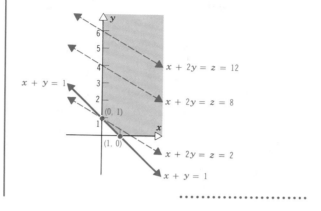

We now state the conditions that will tell when the solution of a linear programming problem does exist.

Existence of a Solution

Consider a linear programming problem with the set R of feasible points and objective function $z = ax + by$.

1. If R is bounded, then z has both a maximum and a minimum value on R.
2. If R is unbounded and $a \geq 0$, $b \geq 0$, and the constraints include $x \geq 0$ and $y \geq 0$, then z has a minimum value on R but not a maximum value (see Example 3).
3. If R is the empty set, then the linear programming problem has no solution and z has neither a maximum nor a minimum value.

For any linear programming problem that has a solution, the following general result is true:

Fundamental Theorem of Linear Programming

If a linear programming problem has a solution, it is located at a corner point of the set of feasible points; if a linear programming problem has multiple solutions, at least one of them is located at a corner point of the set of feasible points. In either case the corresponding value of the objective function is unique.

Since the objective function attains its maximum or minimum value at the corner points of the set of feasible points, we can outline a procedure for solving a linear programming problem provided that it has a solution.

Steps for Solving a Linear Programming Problem

If a linear programming problem has a solution, follow these steps to find it:

Step 1 Write an expression for the quantity that is to be maximized or minimized (the objective function).

Step 2 Determine all the constraints and graph the set of feasible points.

Step 3 List the corner points of the set of feasible points.

Step 4 Determine the value of the objective function at each corner point.

Step 5 Select the optimal solution, that is, the maximum or minimum value of the objective function.

Let's look at some examples.

EXAMPLE 4

Maximize and minimize the objective function

$$z = x + 5y$$

subject to the constraints

$$x + 4y \leq 12 \quad ①$$
$$x \leq 8 \quad ②$$
$$x + y \geq 2 \quad ③$$
$$x \geq 0 \quad ④$$
$$y \geq 0 \quad ⑤$$

SOLUTION

The objective function and the constraints (numbered for convenience) are given (this will not be the case when we do word problems), so we can proceed to graph the constraints. The shaded portion of Figure 19 illustrates the set of feasible points. Since this set is bounded, we know a solution exists.

Figure 19

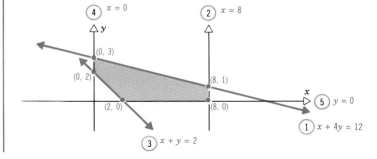

Now we locate the corner points of the set of feasible points at the points of intersection of lines ① and ④, ① and ②, ② and ⑤, ③ and ⑤, and ③ and ④. Using methods discussed earlier, we find that the corner points are

$$(0, 3) \qquad (8, 1) \qquad (8, 0) \qquad (2, 0) \qquad (0, 2)$$

To find the maximum and minimum value of $z = x + 5y$, we set up a table:

Corner Point (x, y)	Value of Objective Function $z = x + 5y$
(0, 3)	$z = 0 + 5(3) = 15$
(8, 1)	$z = 8 + 5(1) = 13$
(8, 0)	$z = 8 + 5(0) = 8$
(2, 0)	$z = 2 + 5(0) = 2$
(0, 2)	$z = 0 + 5(2) = 10$

The maximum value of z is 15, and it occurs at the point (0, 3). The minimum value of z is 2, and it occurs at the point (2, 0).

Now Work Problems 13 and 25

. . . .

Applications

Now let's solve the problem of the cashews and peanuts.

EXAMPLE 5

Maximizing Profit   Maximize

$$P = 0.25x + 0.45y$$

subject to the constraints

$$x + 2y \le 300 \quad ①$$
$$3x + 2y \le 480 \quad ②$$
$$x \ge 0 \quad ③$$
$$y \ge 0 \quad ④$$

SOLUTION

Before applying the method of this chapter to solve this problem, let's discuss a solution that might be suggested by intuition. Namely, since the profit is higher for the high-grade mixture, you might think that Nutt's Nuts should prepare as many packages of the high-grade mixture as possible. If this were done, then there would be a total of 150 packages (8 ounces divides into 75 pounds of cashews exactly 150 times) and the total profit would be

$$150(0.45) = \$67.50$$

As we shall see, this is not the best solution to the problem.

To obtain the maximum profit, we use linear programming. The graph of the set of feasible points is given in Figure 20.

Figure 20

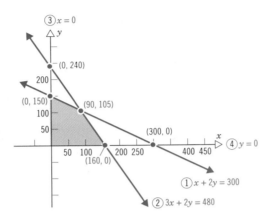

Since this set is bounded, we proceed to locate its corner points. The corner points of the set of feasible points are the points of intersection of lines ③ and ④, ① and ③, ② and ④, and ① and ②:

$$(0, 0) \qquad (0, 150) \qquad (160, 0) \qquad (90, 105)$$

(Notice that the points of intersection of lines ① and ④ and lines ② and ③ are not feasible points.) It remains only to evaluate the objective equation at each corner point:

Corner Point (x, y)	Value of Objective Function $P = (\$0.25)x + (\$0.45)y$
$(0, 0)$	$P = (0.25)(0) + (0.45)(0) = 0$
$(0, 150)$	$P = (0.25)(0) + (0.45)(150) = \$67.50$
$(160, 0)$	$P = (0.25)(160) + (0.45)(0) = \$40.00$
$(90, 105)$	$P = (0.25)(90) + (0.45)(105) = \$69.75$

Thus a maximum profit is obtained if 90 packages of low-grade mixture and 105 packages of high-grade mixture are made. The maximum profit obtainable under the conditions described is $69.75.

Now Work Problem 41

・・・・・・・・・・・・・・・・・・・・・・・

EXAMPLE 6

Maximizing Profit   Mike's Famous Toy Trucks manufactures two kinds of toy trucks—a standard model and a deluxe model. In the manufacturing process each standard model requires 2 hours of grinding and 2 hours of finishing, and each deluxe model needs 2 hours of grinding and 4 hours of finishing. The company has two grinders and three finishers, each of whom works at most 40 hours per week. Each standard model toy truck brings a profit of $3 and each deluxe model a profit of $4. Assuming that every truck made will be sold, how many of each should be made to maximize profits?

SOLUTION

First, we name the variables:

$$x = \text{Number of standard models made}$$
$$y = \text{Number of deluxe models made}$$

The quantity to be maximized is the profit, which we denote by P:

$$P = \$3x + \$4y$$

This is the objective function. To manufacture one standard model requires 2 grinding hours and to make one deluxe model requires 2 grinding hours. Thus, the number of grinding hours of x standard and y deluxe models is

$$2x + 2y$$

But the total amount of grinding time available is 80 hours per week. This means we have the constraint

$$2x + 2y \leq 80 \quad \text{Grinding time constraint}$$

Similarly, for the finishing time we have the constraint

$$2x + 4y \leq 120 \quad \text{Finishing time constraint}$$

By simplifying each of these constraints and adding the nonnegativity constraints $x \geq 0$ and $y \geq 0$, we may list all the constraints for this problem:

$$x + y \leq 40 \quad \text{①}$$
$$x + 2y \leq 60 \quad \text{②}$$
$$x \geq 0 \quad \text{③}$$
$$y \geq 0 \quad \text{④}$$

Figure 21 illustrates the set of feasible points, which is bounded.

Figure 21

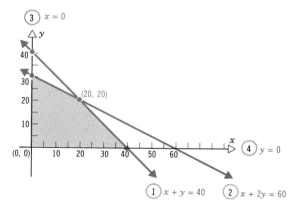

The corner points of the set of feasible points are

$$(0, 0) \qquad (0, 30) \qquad (40, 0) \qquad (20, 20)$$

The table lists the corresponding values of the objective equation:

Corner Point (x, y)	Value of Objective Function $P = \$3x + \$4y$
(0, 0)	$P = 0$
(0, 30)	$P = \$120$
(40, 0)	$P = \$120$
(20, 20)	$P = 3(20) + 4(20) = \$140$

Thus a maximum profit is obtained if 20 standard trucks and 20 deluxe trucks are manufactured. The maximum profit is $140.

· ·

EXAMPLE 7

Financial Planning   A retired couple has up to $30,000 to invest in fixed-income securities. Their broker recommends investing in two bonds: one a AAA bond yielding 8%; the other a B^+ bond paying 12%. After some consideration, the couple decides to invest at most $12,000 in the B^+-rated bond and at least $6000 in the AAA bond. They also want the amount invested in the AAA bond to exceed or equal the amount invested in the B^+ bond. What should the broker recommend if the couple (quite naturally) wants to maximize the return on their investment?

SOLUTION

First, we name the variables:

$$x = \text{Amount invested in the AAA bond}$$
$$y = \text{Amount invested in the } B^+ \text{ bond}$$

The quantity to be maximized, the couple's return on investment, which we denote by P, is

$$P = 0.08x + 0.12y$$

This is the objective function. The conditions specified by the problem are

Up to $30,000 available to invest	$x + y \leq 30,000$
Invest at most $12,000 in the B^+ bond	$y \leq 12,000$
Invest at least $6000 in the AAA bond	$x \geq 6000$
Amount in the AAA bond must exceed or equal amount in the B^+ bond	$x \geq y$

In addition, we must have the conditions $x \geq 0$ and $y \geq 0$. The total list of constraints is

$$x + y \leq 30,000 \quad ①$$
$$y \leq 12,000 \quad ②$$
$$x \geq 6000 \quad ③$$
$$x \geq y \quad ④$$
$$x \geq 0 \quad ⑤$$
$$y \geq 0 \quad ⑥$$

Figure 22 illustrates the set of feasible points, which is bounded. The corner points of the set of feasible points are

(6000, 0)   (6000, 6000)   (12,000, 12,000)   (18,000, 12,000)   (30,000, 0)

Figure 22

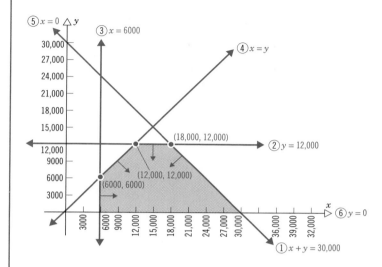

The corresponding return on investment at each corner point is

$$P = 0.08(6000) + 0.12(0) = \$480$$
$$P = 0.08(6000) + 0.12(6000) = 480 + 720 = \$1200$$
$$P = 0.08(12,000) + 0.12(12,000) = 960 + 1440 = \$2400$$
$$P = 0.08(18,000) + 0.12(12,000) = 1440 + 1440 = \$2880$$
$$P = 0.08(30,000) + 0.12(0) = \$2400$$

Thus the maximum return on investment is $2880, obtained by placing $18,000 in the AAA bond and $12,000 in the B⁺ bond.

Now Work Problem 47

· ·

EXAMPLE 8*

Land Reclamation   This example concerns reclaimed land and its allocation into two major uses—agricultural and urban (or nonagricultural). The reclamation of land for urban purposes costs $400 per acre and for agricultural uses, $300. The reclamation agency wishes to minimize the total cost C of reclaiming the land:

$$C = \$400x + \$300y$$

where x = the number of acres of urban land and y = the number of acres of agricultural land. Although this equation can be minimized by setting both x and y at

---

* This example is adapted from Maurice Yeates, *An Introduction to Quantitative Analysis in Economic Geography.* New York: McGraw-Hill, 1968.

zero, that is, reclaiming nothing, the problem derives from a number of constraints due to three different groups.

The first is an urban group, which insists that at least 4000 acres of land be reclaimed for urban purposes. The second group is concerned with agriculture and says that at least 5000 acres of land must be reclaimed for agricultural uses. Finally, the third group is concerned only with reclamation and is quite uninterested in the use to which the land will be put. The third group, however, says that at least 10,000 acres of land must be reclaimed. The problem and the constraints can, therefore, be written in full as follows:

Minimize

$$C = \$400x + \$300y$$

subject to the constraints

$$x \geq 4000 \quad \text{①}$$
$$y \geq 5000 \quad \text{②}$$
$$x + y \geq 10,000 \quad \text{③}$$
$$x \geq 0 \quad \text{④}$$
$$y \geq 0 \quad \text{⑤}$$

Figure 23 illustrates the set of feasible points, which is not bounded. However, this minimum problem has a solution since the coefficients of x and y in the objective function are positive and the constraints include $x \geq 0$ and $y \geq 0$. See Condition 2 for existence of a solution (p. 147).

Figure 23

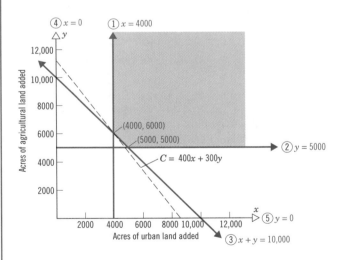

The combination of urban and agricultural land at the corner point (4000, 6000) reveals that if 4000 acres are devoted to urban purposes and 6000 acres to agricultural purposes, the cost is a minimum and is

$$C = (\$400)(4000) + (\$300)(6000) = \$3,400,000$$

. . . .

Application: Pollution Control

The following model is taken from a paper by Robert E. Kohn.* In this paper a linear programming model is proposed to help determine what air pollution controls should be adopted in an airshed. The basic premise is that air quality goals should be achieved at the least possible cost. Advantages of the model are its simplicity, its emphasis on economic efficiency, and its appropriateness for the kind of data that are already available.

To illustrate the model, consider a hypothetical airshed with a single industry, cement manufacturing. Annual production is 2,500,000 barrels of cement. Although the kilns are equipped with mechanical collectors for air pollution control, they are still emitting 2 pounds of dust for every barrel of cement produced. The industry can be required to replace the mechanical collectors with four-field electrostatic precipitators, which would reduce emissions to 0.5 pound of dust per barrel of cement or with five-field electrostatic precipitators, which would reduce emissions to 0.2 pound per barrel. If the capital and operating costs of the four-field precipitator are $0.14 per barrel of cement produced and of the five-field precipitator are $0.18 per barrel, what control methods should be required of this industry? Assume that, for this hypothetical airshed, it has been determined that particulate emissions (which now total 5,000,000 pounds per year) should be reduced by 4,200,000 pounds per year.

If C represents the cost of control, x the number of barrels of annual cement production subject to the four-field electrostatic precipitator (cost is $0.14 a barrel of cement produced and pollutant reduction is $2 - 0.5 = 1.5$ pounds of particulates per barrel of cement produced), and y the number of barrels of annual cement production subject to the five-field electrostatic precipitator (cost is $0.18 a barrel and pollutant reduction is $2 - 0.2 = 1.8$ pounds per barrel of cement produced), then the problem can be stated as follows:

Minimize

$$C = \$0.14x + \$0.18y$$

subject to

$$x + y \leq 2{,}500{,}000 \quad \text{①}$$
$$1.5x + 1.8y \geq 4{,}200{,}000 \quad \text{②}$$
$$x \geq 0 \quad \text{③}$$
$$y \geq 0 \quad \text{④}$$

The first equation states that our objective is to minimize air pollution control costs; the first constraint states that barrels of cement production subject to the two control methods cannot exceed the annual production; the second constraint states that the particulate reduction from the two methods must be greater than or equal to the particulate reduction target; and the last two inequalities mean that we cannot have negative quantities of cement. Figure 24 (p. 156) illustrates a graphic solution to the problem.

*R. E. Kohn, ''A Mathematical Programming Model for Air Pollution Control,'' *School Science and Mathematics* (June 1969), pp. 487–499.

Figure 24

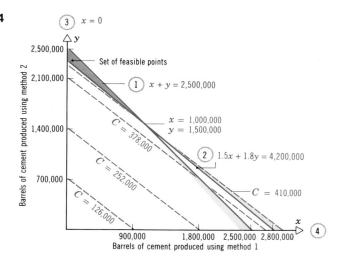

The least costly solution would be to install the four-field precipitator on kilns producing 1,000,000 ($x = 1,000,000$) and the five-field precipitator on kilns producing 1,500,000 ($y = 1,500,000$) barrels of cement at a cost of $C = \$410,000$.

EXERCISE 3.2   Answers to odd-numbered problems begin on page AN-16.

In Problems 1–6 the given figure illustrates the graph of the set of feasible points of a linear programming problem. Find the maximum and minimum values of each objective function.

1. $z = 2x + 3y$

2. $z = 3x + 27y$

3. $z = x + 8y$

4. $z = 3x + y$

5. $z = x + 6y$

6. $z = x + 5y$

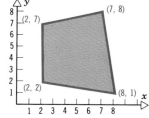

In Problems 7–12 list the corner points for each collection of constraints of a linear programming problem.

7.
$x \leq 13$ ①
$4x + 3y \geq 12$ ②
$x \geq 0$ ③
$y \geq 0$ ④

8.
$x \leq 8$ ①
$2x + 3y \geq 6$ ②
$x \geq 0$ ③
$y \geq 0$ ④

9.
$y \leq 10$ ①
$x + y \leq 15$ ②
$x \geq 0$ ③
$y \geq 0$ ④

10.
$y \leq 8$ ①
$2x + y \geq 10$ ②
$x \geq 0$ ③
$y \geq 0$ ④

11.
$x \leq 10$ ①
$y \leq 8$ ②
$4x + 3y \geq 12$ ③
$x \geq 0$ ④
$y \geq 0$ ⑤

12.
$x \leq 9$ ①
$y \leq 12$ ②
$2x + 3y \leq 24$ ③
$x \geq 0$ ④
$y \geq 0$ ⑤

In Problems 13–20 maximize (if possible) the quantity $z = 5x + 7y$ subject to the given constraints.

13. $x + y \leq 2$
$\quad\quad y \geq 1$
$\quad\quad x \geq 0$
$\quad\quad y \geq 0$

14. $2x + 3y \leq 6$
$\quad\quad\quad x \leq 2$
$\quad\quad\quad x \geq 0$
$\quad\quad\quad y \geq 0$

15. $\quad x + y \geq 2$
$\quad 2x + 3y \leq 6$
$\quad\quad\quad x \geq 0$
$\quad\quad\quad y \geq 0$

16. $\quad x + y \geq 2$
$\quad 2x + 3y \leq 12$
$\quad 3x + 2y \leq 12$
$\quad\quad\quad x \geq 0$
$\quad\quad\quad y \geq 0$

17. $\quad 2 \leq x + y$
$\quad\quad x + y \leq 8$
$\quad 2x + y \leq 10$
$\quad\quad\quad x \geq 0$
$\quad\quad\quad y \geq 0$

18. $\quad 2 \leq x + y$
$\quad\quad x + y \leq 8$
$\quad\quad 1 \leq x + 2y$
$\quad\quad x + 2y \leq 10$
$\quad\quad\quad x \geq 0$
$\quad\quad\quad y \geq 0$

19. $x + y \leq 10$
$\quad\quad x \geq 6$
$\quad\quad x \geq 0$
$\quad\quad y \geq 0$

20. $x + y \leq 8$
$\quad\quad y \geq 2$
$\quad\quad x \geq 0$
$\quad\quad y \geq 0$

In Problems 21–26 minimize (if possible) the quantity $z = 2x + 3y$ subject to the given constraints.

21. $x + y \geq 2$
$\quad\quad y \leq x$
$\quad\quad x \geq 0$
$\quad\quad y \geq 0$

22. $2x + y \geq 3$
$\quad\quad\quad y \geq x$
$\quad\quad\quad x \geq 0$
$\quad\quad\quad y \geq 0$

23. $\quad x + y \geq 2$
$\quad\quad x + 3y \leq 12$
$\quad\quad 3x + y \leq 12$
$\quad\quad\quad\quad x \geq 0$
$\quad\quad\quad\quad y \geq 0$

24. $\quad 2 \leq x + y$
$\quad\quad x + y \leq 10$
$\quad 2x + 3y \leq 6$
$\quad\quad\quad x \geq 0$
$\quad\quad\quad y \geq 0$

25. $\quad 2y \leq x$
$\quad x + 2y \leq 10$
$\quad\quad\quad x \geq 0$
$\quad\quad\quad y \geq 0$

26. $\quad 1 \leq x + 2y$
$\quad x + 2y \leq 10$
$\quad\quad\quad y \geq 2x$
$\quad\quad x + y \leq 8$
$\quad\quad\quad x \geq 0$
$\quad\quad\quad y \geq 0$

In Problems 27–32 find the maximum and minimum values (if possible) of the given objective function subject to the constraints

$$x + y \leq 10$$
$$2x + y \geq 10$$
$$x + 2y \geq 10$$
$$x \geq 0$$
$$y \geq 0$$

27. $z = x + y$

28. $z = 2x + 3y$

29. $z = 5x + 2y$

30. $z = x + 2y$

31. $z = 3x + 4y$

32. $z = 3x + 6y$

33. Find the maximum and minimum values of
$z = 18x + 33y$ subject to the constraints $3y + 3x \geq 9$,
$-x + 4y \leq 12$, and $4x - y \leq 12$.

34. Find the maximum and minimum values of
$z = 20x + 26y$ subject to the constraints $3y + 4x \geq 12$,
$-2x + 4y \leq 16$, and $6x - y \leq 18$.

35. Find the maximum and minimum values of $z = 7x + 36y$
subject to the constraints $3y + 2x \geq 6$, $-3x + 4y \leq 8$,
and $5x - y \leq 15$.

36. Find the maximum and minimum values of $z = 6x + 30y$
subject to the constraints $2y + 2x \geq 4$, $-x + 5y \leq 10$,
and $3x - 3y \leq 6$.

37. Maximize $z = -20x + 30y$ subject to the constraints
$0 \leq x \leq 15$, $0 \leq y \leq 10$, $3y + 5x \geq 15$, and
$3y - 3x \leq 21$.

38. Maximize $z = -10x + 10y$ subject to the constraints
$0 \leq x \leq 15$, $0 \leq y \leq 10$, $y + 6x \geq 6$, and $y - 3x \leq 7$.

39. Maximize $z = -12x + 24y$ subject to the constraints
$0 \leq x \leq 15$, $0 \leq y \leq 10$, $3y + 3x \geq 9$, and
$2y - 3x \leq 14$.

40. Maximize $z = -20x + 10y$ subject to the constraints
$0 \leq x \leq 15$, $0 \leq y \leq 10$, $3y + 4x \geq 12$, and
$y - 3x \leq 7$.

41. In Example 5, if the profit on the low-grade mixture is
$0.30 per package and the profit on the high-grade mix-
ture is $0.40 per package, how many packages of each
mixture should be made for a maximum profit?

42. Using the information supplied in Example 6, suppose the
profit on each standard model is $4 and the profit on each
deluxe model is also $4. How many of each should be
manufactured in order to maximize profit?

43. **Optimal Land Use**   A farmer has 70 acres of land available on which to grow some soybeans and some corn. The
cost of cultivation per acre, the workdays needed per acre,
and the profit per acre are indicated in the table:

	Soybeans	Corn	Total Available
Cultivation Cost per Acre	$60	$30	$1800
Days of Work per Acre	3 days	4 days	120 days
Profit per Acre	$300	$150	

As indicated in the last column, the acreage to be culti-
vated is limited by the amount of money available for
cultivation costs and by the number of working days that
can be put into this part of the business. Find the number
of acres of each crop that should be planted in order to
maximize the profit.

44. **Investment Strategy**   An investment broker wants to
invest up to $20,000. She can purchase a type A bond
yielding a 10% return on the amount invested and she can
purchase a type B bond yielding a 15% return on the
amount invested. She also wants to invest at least as much
in the type A bond as in the type B bond. She will also
invest at least $5000 in the type A bond and no more than
$8000 in the type B bond. How much should she invest in
each type of bond to maximize her return?

45. **Manufacturing**   A factory manufactures two products,
each requiring the use of three machines. The first machine can be used at most 70 hours; the second machine at
most 40 hours; and the third machine at most 90 hours.
The first product requires 2 hours on machine 1, 1 hour on
machine 2, and 1 hour on machine 3; the second product
requires 1 hour each on machines 1 and 2, and 3 hours on
machine 3. If the profit is $40 per unit for the first product
and $60 per unit for the second product, how many units
of each product should be manufactured to maximize
profit?

46. **Diet Problem**   A diet is to contain at least 400 units of
vitamins, 500 units of minerals, and 1400 calories. Two
foods are available: F_1, which costs $0.05 per unit, and
F_2, which costs $0.03 per unit. A unit of food F_1 contains
2 units of vitamins, 1 unit of minerals, and 4 calories; a
unit of food F_2 contains 1 unit of vitamins, 2 units of
minerals, and 4 calories. Find the minimum cost for a diet
that consists of a mixture of these two foods and also
meets the minimal nutrition requirements.

47. **Investment Strategy**   A financial consultant wishes to
invest up to a total of $30,000 in two types of securities,
one that yields 10% per year and another that yields 8%
per year. Furthermore, she believes that the amount invested in the first security should be at most one-third of
the amount invested in the second security. What investment program should the consultant pursue in order to
maximize income?

48. **Scheduling**   Blink appliances has a sale on microwaves
and stoves. Each microwave requires 2 hours to unpack
and set up, and each stove requires 1 hour. The storeroom
space is limited to 50 items. The budget of the store

allows only 80 hours of employee time for unpacking and setup. Microwaves sell for $300 each, and stoves sell for $200 each. How many of each should the store order to maximize revenue?

49. **Cost Control** An appliance repair shop has 5 vacuum cleaners, 12 TV sets, and 18 VCRs to be repaired. The store employs two part-time repairmen. One repairman can repair one vacuum cleaner, three TV sets, and three VCRs in 1 week, while the second repairman can repair one vacuum cleaner, two TV sets and six VCRs in 1 week. The first employee is paid $250 a week and the second employee is paid $220 a week. To minimize the cost, how many weeks should each of the two repairmen be employed?

50. **Transportation** An appliance company has a warehouse and two terminals. To minimize shipping costs, the manager must decide how many appliances should be shipped to each terminal. There is a total supply of 1200 units in the warehouse and a demand for 400 units in terminal A and 500 units in terminal B. It costs $12 to ship each unit to terminal A and $16 to ship to terminal B. How many units should be shipped to each terminal in order to minimize cost?

51. **Pollution Control** A chemical plant produces two items A and B. For each item A produced, 2 cubic feet of carbon monoxide and 6 cubic feet of sulfur dioxide are emitted into the atmosphere, whereas to produce item B, 4 cubic feet of carbon monoxide and 3 cubic feet of sulfur dioxide are emitted into the atmosphere. Government pollution standards permit the manufacturer to emit a maximum of 3000 cubic feet of carbon monoxide and 5400 cubic feet of sulfur dioxide per week. The manufacturer can sell all of the items that it produces and make a profit of $1.50 per unit for item A and $1.00 per unit for item B. Determine the number of units of each item to be produced each week to maximize profit without exceeding government standards.

52. **Production Scheduling** A company produces two types of steel. Type 1 requires 2 hours of melting, 4 hours of cutting, and 10 hours of rolling per ton. Type 2 requires 5 hours of melting, 1 hour of cutting, and 5 hours of rolling per ton. Forty hours are available for melting, 20 for cutting, and 60 for rolling. Each ton of Type 1 produces $240 profit, and each ton of Type 2 yields $80 profit. Find the maximum profit and the production schedule that will produce this profit.

53. **Diet Problem** Danny's Chicken Farm is a producer of frying chickens. In order to produce the best fryers possi-

ble, the regular chicken feed is supplemented by four vitamins. The minimum amount of each vitamin required per 100 ounces of feed is: vitamin 1, 50 units; vitamin 2, 100 units; vitamin 3, 60 units; vitamin 4, 180 units. Two supplements are available: supplement I costs $0.03 per ounce and contains 5 units of vitamin 1 per ounce, 25 units of vitamin 2 per ounce, 10 units of vitamin 3 per ounce, and 35 units of vitamin 4 per ounce. Supplement II costs $0.04 per ounce and contains 25 units of vitamin 1 per ounce, 10 units of vitamin 2 per ounce, 10 units of vitamin 3 per ounce, and 20 units of vitamin 4 per ounce. How much of each supplement should Danny buy to add to each 100 ounces of feed in order to minimize his cost, but still have the desired vitamin amounts present?

54. **Maximizing Income** J. B. Rug Manufacturers has available 1200 square yards of wool and 1000 square yards of nylon for the manufacture of two grades of carpeting: high-grade, which sells for $500 per roll, and low-grade, which sells for $300 per roll. Twenty square yards of wool and 40 square yards of nylon are used in a roll of high-grade carpet, and 40 square yards of nylon are used in a roll of low-grade carpet. Forty work-hours are required to manufacture each roll of the high-grade carpet, and 20 work-hours are required for each roll of the low-grade carpet, at an average cost of $6 per work-hour. A maximum of 800 work-hours are available. The cost of wool is $5 per square yard and the cost of nylon is $2 per square yard. How many rolls of each type of carpet should be manufactured to maximize income? [*Hint:* Income = revenue from sale − (production cost for material + labor)]

55. The rug manufacturer in Problem 54 finds that maximum income occurs when no high-grade carpet is produced. If the price of the low-grade carpet is kept at $300 per roll, in what price range should the high-grade carpet be sold so that income is maximized by selling some rolls of each type of carpet? Assume all other data remain the same.

56. Maximize
$$P = 2x + y + 3z$$
subject to
$$x + 2y + z \leq 25 \quad \text{①}$$
$$3x + 2y + 3z \leq 30 \quad \text{②}$$
$$x \geq 0 \quad \text{③}$$
$$y \geq 0 \quad \text{④}$$
$$z \geq 0 \quad \text{⑤}$$

[*Hint:* Solve the constraints three at a time, find the feasible points, and test each of them in the objective function. Assume a solution exists.]

Technology Exercises
Some graphing calculators can graph systems of inequalities and shade the region representing the solution to the system on the screen. Consult your user's manual, and use your graphing calculator to solve the systems of inequalities in Problems 1–6. If your calculator does not graph inequalities, graph the boundary line of each inequality.

For each system in Problems 1–6, find the minimum and the maximum of the objective function

$$z = 3.5x + 1.25y$$

within the region representing the solution to the system of inequalities. Use TRACE to find the corner points.

1. $\begin{cases} y \le x + 1 \\ y + x \le 9 \\ x + y \ge 3 \\ y + 3 \ge x \end{cases}$

2. $\begin{cases} x + y \le 7 \\ y \le 2x + 2 \\ y + \frac{1}{2}x \ge 2 \\ y + 3 \ge x \end{cases}$

3. $\begin{cases} y \le \frac{1}{2}x + 5 \\ y + 2x \le 15 \\ 5x + y \ge 8 \\ y + 1 \ge x \end{cases}$

4. $\begin{cases} y \le 1.3x + 4 \\ y + 1.17x \le 12.33 \\ y \ge 3 \\ x \ge 0 \end{cases}$

5. $\begin{cases} y \le \frac{1}{2}x + 5 \\ y + 2x \le 15 \\ 5x + y \ge 8 \\ y + 5 \ge x \\ y \ge 0 \end{cases}$

6. $\begin{cases} y \le 6 \\ y + 2x \le 6 \\ x + y \ge 0 \\ x \le 4 \end{cases}$

CHAPTER REVIEW

IMPORTANT TERMS AND CONCEPTS

graph of a linear inequality 132
steps for graphing a linear inequality 133
systems of linear inequalities 134
linear programming problem 144
solution to a linear programming
   problem 144

existence of solutions to a
   linear programming
   problem 147
fundamental theorem of
   linear programming 147

steps for solving a
   linear programming
   problem 148

TRUE–FALSE ITEMS   Answers are on page AN-17.

T_____ F_____ **1.** The graph of a system of linear inequalities may be bounded or unbounded.

T_____ F_____ **2.** The graph of the set of constraints of a linear programming problem, under certain conditions, could have a circle for a boundary.

T_____ F_____ **3.** The objective function of a linear programming problem is always a linear equation involving the variables.

T_____ F_____ **4.** In a linear programming problem, there may be more than one point that maximizes or minimizes the objective function.

T_____ F_____ **5.** Some linear programming problems will have no solution.

T_____ F_____ **6.** If a linear programming problem has a solution, it is located at the center of the set of feasible points.

FILL IN THE BLANKS   Answers are on page AN-17.

1. The graph of a linear inequality in two variables is called a _____.

2. In a linear programming problem the quantity to be maximized or minimized is referred to as the _____ function.

3. The points that obey the collection of constraints of a linear programming problem are called _____ points.

4. A linear programming problem will always have a solution if the set of feasible points is _____.

5. If a linear programming problem has a solution, it is located at a _____ of the set of feasible points.

REVIEW EXERCISES   Answers to odd-numbered problems begin on page AN-17.

In Problems 1–4 graph each linear inequality.

1. $x - 3y < 0$ **2.** $4x + y \geq 8$ **3.** $5x + y \geq 10$ **4.** $2x + 3y > 6$

5. Which of the points $P_1 = (4, -3)$, $P_2 = (2, -6)$, $P_3 = (8, -3)$ are part of the graph of the following system?

$$\begin{cases} 7y + 10x \leq 0 \\ 2y + 9x \geq 0 \\ y + 3x \leq 15 \end{cases}$$

6. Which of the points $P_1 = (8, 6)$, $P_2 = (2, 5)$, $P_3 = (4, 1)$ are part of the graph of the following system?

$$\begin{cases} y - 10x \leq 0 \\ 2y - 3x \geq 0 \\ y + x \leq 15 \end{cases}$$

In Problems 7–8 determine which region, a, b, c, or d, represents the graph of the given system of linear inequalities.

7. $\begin{cases} 6x - 4y \leq 12 \\ 3x + 2y \leq 18 \end{cases}$

8. $\begin{cases} 6x - 5y \leq 5 \\ 6x + 6y \leq 60 \end{cases}$

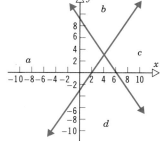

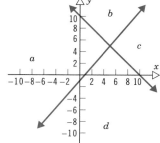

In Problems 9–14 graph each system of linear inequalities. Locate the corner points and tell whether the graph is bounded or unbounded.

9. $\begin{cases} 3x + 2y \leq 12 \\ x + y \geq 1 \\ x \geq 0 \quad y \geq 0 \end{cases}$

10. $\begin{cases} x + y \leq 8 \\ 2x + y \geq 2 \\ x \geq 0 \quad y \geq 0 \end{cases}$

11. $\begin{cases} x + 2y \geq 4 \\ 3x + y \geq 6 \\ x \geq 0 \quad y \geq 0 \end{cases}$

12. $\begin{cases} 2x + y \geq 4 \\ 3x + 2y \geq 6 \\ x \geq 0 \qquad y \geq 0 \end{cases}$ **13.** $\begin{cases} 3x + 2y \geq 6 \\ 3x + 2y \leq 12 \\ x + 2y \leq 8 \\ x \geq 0 \qquad y \geq 0 \end{cases}$ **14.** $\begin{cases} x + 2y \geq 2 \\ x + 2y \leq 10 \\ 2x + y \leq 10 \\ x \geq 0 \qquad y \geq 0 \end{cases}$

In Problems 15–22 use the constraints below to solve each linear programming problem.

$$x + 2y \leq 40$$
$$2x + y \leq 40$$
$$x + y \geq 10$$
$$x \geq 0 \qquad y \geq 0$$

15. Maximize $z = x + y$ **16.** Maximize $z = 2x + 3y$ **17.** Minimize $z = 5x + 2y$

18. Minimize $z = 3x + 2y$ **19.** Maximize $z = 2x + y$ **20.** Maximize $z = x + 2y$

21. Minimize $z = 2x + 5y$ **22.** Minimize $z = x + y$

In Problems 23–26 maximize and minimize (if possible) the quantity $z = 15x + 20y$ subject to the given constraints.

23. $\begin{aligned} x &\leq 5 \\ y &\leq 8 \\ 3x + 4y &\geq 12 \\ x \geq 0 \quad y &\geq 0 \end{aligned}$ **24.** $\begin{aligned} x &\leq 6 \\ y &\leq 6 \\ 3x + 2y &\geq 6 \\ x \geq 0 \quad y &\geq 0 \end{aligned}$ **25.** $\begin{aligned} 2x + 3y &\leq 22 \\ x &\leq 5 \\ y &\leq 6 \\ x \geq 0 \quad y &\geq 0 \end{aligned}$ **26.** $\begin{aligned} x + 2y &\leq 20 \\ x + 10y &\geq 36 \\ 5x + 2y &\geq 36 \\ x \geq 0 \quad y &\geq 0 \end{aligned}$

In Problems 27–30 solve each linear programming problem.

27. Maximize

$$z = 2x + 1.2y$$

subject to the constraints

$$0 \leq x \leq 9$$
$$0 \leq y \leq 8$$
$$-3x + 3y \leq 15$$
$$2x + 4y \leq 42$$
$$2x - 7y \leq 4$$

28. Maximize

$$z = 4x + 1.11y$$

subject to the constraints

$$0 \leq x \leq 7$$
$$0 \leq y \leq 8$$
$$-3x + 3y \leq 15$$
$$4x + 3y \leq 40$$
$$2x - 5y \leq 4$$

29. Maximize

$$z = x + 1.48y$$

subject to the constraints

$$0 \leq x \leq 8$$
$$0 \leq y \leq 8$$
$$-3x + 3y \leq 15$$
$$x + 4y \leq 36$$
$$2x - 6y \leq 4$$

30. Maximize

$$z = x + 1.36y$$

subject to the constraints

$$0 \leq x \leq 8$$
$$0 \leq y \leq 7$$
$$-2x + 3y \leq 15$$
$$x + 4y \leq 32$$
$$2x - 6y \leq 4$$

31. **Diet Problem**   Katy needs at least 60 units of carbohydrates, 45 units of protein, and 30 units of fat each month. From each pound of food A, she receives 5 units of carbohydrates, 3 of protein, and 4 of fat. Food B contains 2 units of carbohydrates, 2 units of protein, and 1 unit of fat per pound. If food A costs $1.30 per pound and food B costs $0.80 per pound, how many pounds of each food should Katy buy each month to keep costs at a minimum?

32. **Minimizing Cost**   The ACE Meat Market makes up a combination package of ground beef and ground pork for meat loaf. The ground beef is 75% lean (75% beef, 25% fat) and costs the market 70¢ per pound. The ground pork is 60% lean (60% pork, 40% fat) and costs the market 50¢ per pound. If the meat loaf is to be at least 70% lean, how much ground beef and ground pork should be mixed to keep cost at a minimum?

33. **Maximizing Profit**   A ski manufacturer makes two types of skis: downhill and cross-country. Using the information given in the table below, how many of each type of ski should be made for a maximum profit to be achieved? What is the maximum profit?

	Downhill	Cross-Country	Maximum Time Available
Manufacturing Time per Ski	2 hours	1 hour	40 hours
Finishing Time per Ski	1 hour	1 hour	32 hours
Profit per Ski	$70	$50	

34. Rework Problem 33 if the manufacturing unit has a maximum of 48 hours available.

35. **Maximizing Profit**   A company makes two explosives: type I and type II. Due to storage problems, a maximum of 100 pounds of type I and 150 pounds of type II can be mixed and packaged each week. One pound of type I takes 60 hours to mix and 70 hours to package; 1 pound of type II takes 40 hours to mix and 40 hours to package. The mixing department has at most 7200 work-hours available each week, and packaging has at most 7800 work-hours available. If the profit for 1 pound of type I is $60 and for 1 pound of type II is $40, what is the maximum profit possible each week?

36. **Production Scheduling**   A company sells two types of shoes. The first uses 2 units of leather and 2 units of synthetic material and yields a profit of $8 per pair. The second type requires 5 units of leather and 1 unit of synthetic material and gives a profit of $10 per pair. If there are 40 units of leather and 16 units of synthetic material available, how many pairs of each type of shoe should be sold to maximize profit? What is the maximum profit?

37. **Mixture**   A company makes two kinds of animal food, A and B, which contain two food supplements. It takes 2 pounds of the first supplement and one pound of the second to make a dozen cans of food A, and 4 pounds of the first supplement and 5 pounds of the second to make a dozen cans of food B. On a certain day 80 pounds of the first supplement and 70 pounds of the second is available. Maximize company profits if the profit on a dozen cans of food A is $3.00 and the profit on a dozen cans of food B is $10.00.

Mathematical Questions from Professional Exams
Use the following information to answer Problems 1–3:

CPA Exam   *The Random Company manufactures two products, Zeta and Beta. Each product must pass through two processing operations. All materials are introduced at the start of process 1. There are no work-in-process inventories. Random may produce either one product exclusively or various combinations of both products subject to the following constraints:*

	Process No. 1	Process No. 2	Contribution Margin per Unit
Hours required to produce one unit of			
Zeta	1 hour	1 hour	$4.00
Beta	2 hours	3 hours	5.25
Total capacity in hours per day	1000 hours	1275 hours	

A shortage of technical labor has limited Beta production to 400 units per day. There are no constraints on the production of Zeta other than the hour constraints in the above schedule. Assume that all relationships between capacity and production are linear, and that all of the above data and relationships are deterministic rather than probabilistic.

1. Given the objective to maximize total contribution margin, what is the production constraint for process 1?

   (a) Zeta + Beta ≤ 1000
   (b) Zeta + 2 Beta ≤ 1000
   (c) Zeta + Beta ≥ 1000
   (d) Zeta + 2 Beta ≥ 1000

2. Given the objective to maximize total contribution margin, what is the labor constraint for production of Beta?

   (a) Beta ≤ 400   (b) Beta ≥ 400
   (c) Beta ≤ 425   (d) Beta ≥ 425

3. What is the objective function of the data presented?

   (a) Zeta + 2 Beta = $9.25
   (b) ($4.00)Zeta + 3($5.25)Beta = Total contribution margin
   (c) ($4.00)Zeta + ($5.25)Beta = Total contribution margin
   (d) 2($4.00)Zeta + 3($5.25)Beta = Total contribution margin

4. **CPA Exam**   Williamson Manufacturing intends to produce two products, X and Y. Product X requires 6 hours of time on machine 1 and 12 hours of time on machine 2. Product Y requires 4 hours of time on machine 1 and no time on machine 2. Both machines are available for 24 hours. Assuming that the objective function of the total contribution margin is $2X + $1Y$, what product mix will produce the maximum profit?

   (a) No units of product X and 6 units of product Y.
   (b) 1 unit of product X and 4 units of product Y.
   (c) 2 units of product X and 3 units of product Y.
   (d) 4 units of product X and no units of product Y.

5. **CPA Exam**   Quepea Company manufactures two products, Q and P, in a small building with limited capacity. The selling price, cost data, and production time are given below:

	Product Q	Product P
Selling price per unit	$20	$17
Variable costs of producing and selling a unit	$12	$13
Hours to produce a unit	3	1

   Based on this information, the profit maximization objective function for a linear programming solution may be stated as

   (a) Maximize $20Q + $17P.
   (b) Maximize $12Q + $13P.
   (c) Maximize $3Q + $1P.
   (d) Maximize $8Q + $4P.

6. **CPA Exam**   Patsy, Inc., manufactures two products, X and Y. Each product must be processed in each of three departments: machining, assembling, and finishing. The hours needed to produce one unit of product per department and the maximum possible hours per department follow:

Department	Production Hours per Unit X	Y	Maximum Capacity in Hours
Machining	2	1	420
Assembling	2	2	500
Finishing	2	3	600

Other restrictions follow:

$$X \geq 50 \qquad Y \geq 50$$

The objective function is to maximize profits where profit = $4X + $2Y. Given the objective and constraints, what is the most profitable number of units of X and Y, respectively, to manufacture?

(a) 150 and 100 (b) 165 and 90
(c) 170 and 80   (d) 200 and 50

7. CPA Exam Milford Company manufactures two models, medium and large. The contribution margin expected is $12 for the medium model and $20 for the large model. The medium model is processed 2 hours in the machining department and 4 hours in the polishing department. The large model is processed 3 hours in the machining department and 6 hours in the polishing department. How would the formula for determining the maximization of total contribution margin be expressed?

(a) $5X + 10Y$ (b) $6X + 9Y$
(c) $12X + 20Y$ (d) $12X(2 + 4) + 20Y(3 + 6)$

8. CPA Exam Hale Company manufactures products A and B, each of which requires two processes, polishing and grinding. The contribution margin is $3 for product A and $4 for product B. The illustration shows the maximum number of units of each product that may be processed in the two departments.

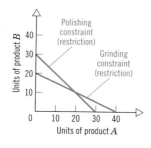

Considering the constraints (restrictions) on processing, which combination of products A and B maximizes the total contribution margin?

(a) 0 units of A and 20 units of B.
(b) 20 units of A and 10 units of B.
(c) 30 units of A and 0 units of B.
(d) 40 units of A and 0 units of B.

9. CPA Exam Johnson, Inc., manufactures product X and product Y, which are processed as follows:

	Machine A	Machine B
Product X	6 hours	4 hours
Product Y	9 hours	5 hours

The contribution margin is $12 for product X and $7 for product Y. The available time daily for processing the two products is 120 hours for machine A and 80 hours for machine B. How would the restriction (constraint) for machine B be expressed?

(a) $4X + 5Y$ (b) $4X + 5Y \leq 80$
(c) $6X + 9Y \leq 120$ (d) $12X + 7Y$

10. CMA Exam A small company makes only two products, with the following two production constraints representing two machines and their maximum availability:

$$2X + 3Y \leq 18$$
$$2X + \ Y \leq 10$$

where X = Units of the first product
       Y = Units of the second product

If the profit equation is Z = $4X + $2Y, the maximum possible profit is

(a) $20   (b) $21   (c) $18   (d) $24
(e) Some profit other than those given above

CMA Exam *Questions 11–13 are based on the Jarten Company, which manufactures and sells two products. Demand for the two products has grown to such a level that Jarten can no longer meet the demand with its facilities. The company can work a total of 600,000 direct labor-hours annually using three shifts. A total of 200,000 hours of machine time is available annually. The company plans to use linear programming to determine a production schedule that will maximize its net return.*

The company spends $2,000,000 in advertising and promotion and incurs $1,000,000 for general and administrative costs. The unit sale price for model A is

$27.50; model B sells for $75.00 each. The unit manufacturing requirements and unit cost data are as shown below. Overhead is assigned on a machine-hour (MH) basis.

	Model A		*Model B*	
Raw material		$ 3		$ 7
Direct labor	1 DLH @ $8	8	1.5 DLH @ $8	12
Variable overhead	0.5 MH @ $12	6	2.0 MH @ $12	24
Fixed overhead	0.5 MH @ $4	2	2.0 MH @ $4	8
		$19		$51

11. The objective function that would maximize Jarten's net income is

   (a) $10.50A + 32.00B$ (b) $8.50A + 24.00B$
   (c) $27.50A + 75.00B$ (d) $19.00A + 51.00B$
   (e) $17.00A + 43.00B$

12. The constraint function for the direct labor is

   (a) $1A + 1.5B \le 200,000$
   (b) $8A + 12B \le 600,000$
   (c) $8A + 12B \le 200,000$
   (d) $1A + 1.5B \le 4,800,000$
   (e) $1A + 1.5B \le 600,000$

13. The constraint function for the machine capacity is

   (a) $6A + 24B \le 200,000$
   (b) $(1/0.5)A + (1.5/2.0)B \le 800,000$

   (c) $0.5A + 2B \le 200,000$
   (d) $(0.5 + 0.5)A + (2 + 2)B \le 200,000$
   (e) $(0.5 \times 1) + (1.5 \times 2.00) \le (200,000 \times 600,000)$

14. **CPA Exam**   Boaz Company manufactures two models, medium (X) and large (Y). The contribution margin expected is $24 for the medium model and $40 for the large model. The medium model is processed 2 hours in the machining department and 4 hours in the polishing department. The large model is processed 3 hours in the machining department and 6 hours in the polishing department. If total contribution margin is to be maximized, using linear programming, how would the objective function be expressed?

   (a) $24X(2 + 4) + 40Y(3 + 6)$ (b) $24X + 40Y$
   (c) $6X + 9Y$ (d) $5X + 10Y$

Chapter 4

Linear Programming: Simplex Method

In Chapter 3 we described a geometrical method (using graphs) for solving linear programming problems. Unfortunately, this method is useful only when there are no more than two variables and the number of constraints is small.

If we have a large number of either variables or constraints, it is still true that the optimal solution will be found at a corner point of the set of feasible points. In fact, we could find these corner points by writing all the equations corresponding to the inequalities of the problem and then proceeding to solve all possible combinations of these equations. We would, of course, have to discard any solutions that are not feasible (because they do not satisfy one or more of the constraints). Then we could evaluate the objective function at the remaining corner points. After all this, we might discover that the problem has no optimal solution after all.

Just how difficult is this procedure? Well, if there were just 4 variables and 7 constraints, we would have to solve all possible combinations of 4 equations chosen from a set of 7 equations—that would be 35 solutions in all. Each of these solutions would then have to be tested for feasibility. So even for this relatively small number of variables and constraints, the work would be quite tedious. In the real world of applications, it is fairly common to encounter problems with *hundreds,* even *thousands,* of variables and constraints. Of course, such problems must be solved by computer. Even so, choosing a more efficient problem-solving strategy than the geometrical method might reduce the computer's running time from hours to seconds, or, for very large problems, from years to hours.

A more systematic approach would involve choosing a solution at one corner point of the feasible set, then moving from there to another corner point at which the objective function has a better value, and continuing in this way until the best possible value is found. One very efficient and popular way of doing this is the subject of the present chapter: the *simplex method.*

A discussion of LINDO, a software package that mimics the simplex method, may be found in Appendix B.

4.1 THE SIMPLEX TABLEAU; PIVOTING

Standard Form of a Maximum Problem

A linear programming problem in which the objective function is to be maximized is referred to as a **maximum linear programming problem.** Such problems are said to be in **standard form** provided two conditions are met:

Standard Form of a Maximum Linear Programming Problem

Condition 1. All the variables are nonnegative.

Condition 2. All other constraints are written as a linear expression that is less than or equal to a positive constant.

EXAMPLE 1   Determine which of the following maximum linear programming problems are in standard form.*

(a) Maximize

$$z = 5x_1 + 4x_2$$

---

* Due to the nature of solving linear programming problems using the simplex method, we shall find it convenient to use subscripted variables for the remainder of this chapter.

subject to the constraints

$$3x_1 + 4x_2 \leq 120$$
$$4x_1 + 3x_2 \leq 20$$
$$x_1 \geq 0, x_2 \geq 0$$

(b) Maximize

$$z = 8x_1 + 2x_2 + 3x_3$$

subject to the constraints

$$4x_1 + 8x_2 \qquad \leq 120$$
$$3x_2 + 4x_3 \leq 120$$
$$x_1 \geq 0, x_2 \geq 0$$

(c) Maximize

$$z = 6x_1 - 8x_2 + x_3$$

subject to the constraints

$$3x_1 + x_2 \qquad \leq \quad 10$$
$$4x_1 - x_2 \qquad \leq \quad 5$$
$$x_1 - x_2 - x_3 \geq -3$$
$$x_1 \geq 0, x_2 \geq 0, x_3 \geq 0$$

(d) Maximize

$$z = 8x_1 + x_2$$

subject to the constraints

$$3x_1 + 4x_2 \geq 2$$
$$x_1 + x_2 \leq 6$$
$$x_1 \geq 0, x_2 \geq 0$$

SOLUTION

(a) This is a maximum problem containing two variables x_1 and x_2. Since both variables are nonnegative and since the other constraints

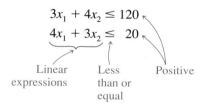

Linear   Less   Positive
expressions   than or
equal

are each written as linear expressions less than or equal to a positive constant, we conclude the maximum problem is in standard form.

(b) This is a maximum problem containing three variables x_1, x_2, and x_3. Since the variable x_3 is not given as nonnegative, the maximum problem is not in standard form.

(c) This is a maximum problem containing three variables x_1, x_2, and x_3. Each variable is nonnegative. The set of constraints

$$
\begin{aligned}
3x_1 + x_2 &\le 10 \\
4x_1 - x_2 &\le 5 \\
x_1 - x_2 - x_3 &\ge -3
\end{aligned}
$$

contains $x_1 - x_2 - x_3 \ge -3$, which is not a linear expression that is less than or equal to a positive constant. Thus the maximum problem is not in standard form. Notice, however, that by multiplying this constraint by -1, we get

$$-x_1 + x_2 + x_3 \le 3$$

which is in the desired form. Thus, although the maximum problem as stated is not in standard form, it can easily be modified to conform to the requirements of the standard form.

(d) The maximum problem contains two variables x_1 and x_2, each of which is nonnegative. Of the other constraints, the first one, $3x_1 + 4x_2 \ge 2$ does not conform. Thus the maximum problem is not in standard form. Notice that we cannot modify this problem to place it in standard form. Even though multiplying by -1 will change the $\ge$ to $\le$, in so doing the 2 will change to -2.

Now Work Problem 1

. . . .

Slack Variables and the Simplex Tableau

In order to apply the simplex method to a maximum problem, we need to first

1. Introduce *slack variables*.
2. Construct the *initial simplex tableau*.

We will show how these steps are done by working with a specific maximum problem in standard form. (This problem is Example 6 of Section 3.2, page 150.)

Maximize

$$P = 3x_1 + 4x_2$$

subject to the constraints

$$
\begin{aligned}
2x_1 + 4x_2 &\le 120 \\
2x_1 + 2x_2 &\le 80 \\
x_1 \ge 0, \, x_2 &\ge 0
\end{aligned}
$$

This problem is in standard form for a maximum linear programming problem. When we say that $2x_1 + 4x_2 \le 120$, we mean that there is a number greater than or equal to 0, which we designate by s_1, such that

$$2x_1 + 4x_2 + s_1 = 120$$

This number s_1 is a variable. It must be nonnegative since it is the difference between 120 and a number that is less than or equal to 120. We call it a **slack variable** since it "takes up the slack" between the left and right sides of the inequality.

Similarly, when we say that $2x_1 + 2x_2 \leq 80$, we are saying that there is a slack variable s_2 such that

$$2x_1 + 2x_2 + s_2 = 80$$

Furthermore, the objective function $P = 3x_1 + 4x_2$ can be rewritten as

$$P - 3x_1 - 4x_2 = 0$$

In effect, we have now replaced our original system of constraints and the objective function by a system of three equations containing five variables, P, x_1, x_2, s_1, and s_2:

$$
\begin{aligned}
P - 3x_1 - 4x_2 \qquad\qquad &= 0 \qquad \text{Objective function}\\
2x_1 + 4x_2 + s_1 \qquad &= 120 \;\Big\}\\
2x_1 + 2x_2 \qquad + s_2 &= 80 \;\Big\} \quad \text{Constraints}
\end{aligned}
$$

$$x_1 \geq 0 \qquad x_2 \geq 0 \qquad s_1 \geq 0 \qquad s_2 \geq 0$$

To solve the maximum problem is to find the particular solution (P, x_1, x_2, s_1, s_2) that gives the largest possible value for P. The augmented matrix for this system is given below:

$$
\begin{array}{ccccc}
P & x_1 & x_2 & s_1 & s_2\\
\end{array}
$$
$$
\left[\begin{array}{ccccc|c}
1 & -3 & -4 & 0 & 0 & 0\\
0 & 2 & 4 & 1 & 0 & 120\\
0 & 2 & 2 & 0 & 1 & 80
\end{array}\right]
$$

If we write the augmented matrix in the form given next, we have the **initial simplex tableau** for the maximum problem:

BV	P	x_1	x_2	s_1	s_2	RHS
P	1	-3	-4	0	0	0
s_1	0	2	4	1	0	120
s_2	0	2	2	0	1	80

(1)

The first row of the initial simplex tableau represents the objective function and is called the **objective row.** The remaining rows represent the constraints. We separate the objective row from these rows with a horizontal rule. Notice that we have written the symbol for each variable above the column in which its coefficients appear. The notation **BV** stands for **basic variables.** These are the variables that have a coefficient of 1 and 0 elsewhere in their column. The notation **RHS** stands for **right-hand side,** that is, the numbers to the right of the equal sign in each equation.

So far, we have seen this much of the simplex method:

Standard Form of a Maximum Problem

For a maximum problem in standard form:

1. The constraints are changed from inequalities to equations by the introduction of extra variables—one for each constraint and all nonnegative—called **slack variables.**

2. These equations, together with one that describes the objective function, are placed in the **initial simplex tableau.**

EXAMPLE 2

The following maximum problems are in standard form. For each one introduce slack variables and set up the initial simplex tableau.

(a) Maximize

$$P = 3x_1 + 2x_2 + x_3$$

subject to the constraints

$$3x_1 + x_2 + x_3 \leq 30$$
$$5x_1 + 2x_2 + x_3 \leq 24$$
$$x_1 + x_2 + 4x_3 \leq 20$$
$$x_1 \geq 0 \qquad x_2 \geq 0 \qquad x_3 \geq 0$$

(b) Maximize

$$P = x_1 + 4x_2 + 3x_3 + x_4$$

subject to the constraints

$$2x_1 + x_2 \qquad\qquad \leq 10$$
$$3x_1 + x_2 + x_3 + 2x_4 \leq 18$$
$$x_1 + x_2 + x_3 + x_4 \leq 14$$
$$x_1 \geq 0 \qquad x_2 \geq 0 \qquad x_3 \geq 0 \qquad x_4 \geq 0$$

SOLUTION

(a) We write the objective function in the form

$$P - 3x_1 - 2x_2 - x_3 = 0$$

For each constraint we introduce a nonnegative slack variable to obtain the following equations:

$$3x_1 + x_2 + x_3 + s_1 \qquad\qquad = 30$$
$$5x_1 + 2x_2 + x_3 \qquad + s_2 \qquad = 24$$
$$x_1 + x_2 + 4x_3 \qquad\qquad + s_3 = 20$$
$$x_1 \geq 0 \qquad x_2 \geq 0 \qquad x_3 \geq 0$$
$$s_1 \geq 0 \qquad s_2 \geq 0 \qquad s_3 \geq 0$$

These equations, together with the objective function P, give the initial simplex tableau:

BV	P	x_1	x_2	x_3	s_1	s_2	s_3	RHS
P	1	-3	-2	-1	0	0	0	0
s_1	0	3	1	1	1	0	0	30
s_2	0	5	2	1	0	1	0	24
s_3	0	1	1	4	0	0	1	20

(b) We write the objective function in the form

$$P - x_1 - 4x_2 - 3x_3 - x_4 = 0$$

For each constraint we introduce a nonnegative slack variable to obtain the equations

$$2x_1 + x_2 \qquad\qquad + s_1 \qquad\qquad = 10$$
$$3x_1 + x_2 + x_3 + 2x_4 \qquad + s_2 \qquad = 18$$
$$x_1 + x_2 + x_3 + x_4 \qquad\qquad + s_3 = 14$$
$$x_1 \geq 0 \qquad x_2 \geq 0 \qquad x_3 \geq 0 \qquad x_4 \geq 0$$
$$s_1 \geq 0 \qquad s_2 \geq 0 \qquad s_3 \geq 0$$

These equations, together with the objective function P, give the initial simplex tableau:

BV	P	x_1	x_2	x_3	x_4	s_1	s_2	s_3	RHS
P	1	-1	-4	-3	-1	0	0	0	0
s_1	0	2	1	0	0	1	0	0	10
s_2	0	3	1	1	2	0	1	0	18
s_3	0	1	1	1	1	0	0	1	14

Notice that in each initial simplex tableaux an identity matrix appears under the columns headed by P and the slack variables. Notice too that the right-hand column (RHS) will always contain nonnegative constants.

Now Work Problem 17

The Pivot Operation

Before going any further in our discussion of the simplex method, we need to discuss the matrix operation known as *pivoting*. The first thing one does in a pivot operation is to choose a *pivot element*. However, for now the pivot element will be specified in advance. The method of selecting pivot elements in the simplex tableau will be shown in the next section.

PIVOTING

To *pivot* **a matrix about a given element, called the *pivot element*, is to apply row operations so that the pivot element is replaced by a 1 and all other entries in the same column, called the *pivot column,* become 0s.**

Steps for Pivoting

The correct sequence of steps is

Step 1 In the *pivot row* (where the pivot element appears), divide each entry by the *pivot element* (we assume it is not 0).

Step 2 Obtain 0s elsewhere in the *pivot column* by performing row operations using the pivot row.

The steps for pivoting utilize two variations of the three row operations for matrices, namely:

Step 1 Replace the pivot row by a positive multiple of that same row.

Step 2 Replace a row by the sum of that row and a multiple of the pivot row.

Warning! Notice that Step 2 requires row operations that must involve the pivot row.

We continue with the initial simplex tableau given in Display (1), page 171, to illustrate the pivot operation.

EXAMPLE 3

Perform a pivot operation on the initial simplex tableau given in (1) and repeated below in (2), where the pivot element is circled, and the pivot row and pivot column are marked by arrows:

$$
\begin{array}{c|ccccc|c}
\text{BV} & P & x_1 & x_2 & s_1 & s_2 & \text{RHS} \\
\hline
P & 1 & -3 & -4 & 0 & 0 & 0 \\
\rightarrow\; s_1 & 0 & 2 & ④ & 1 & 0 & 120 \\
s_2 & 0 & 2 & 2 & 0 & 1 & 80
\end{array}
\tag{2}
$$

SOLUTION

In this tableau the pivot column is the column corresponding to the variable x_2 and the pivot row is the row corresponding to the variable s_1. Step 1 of the pivoting procedure tells us to divide the pivot row by 4. This is the row operation

$$R_2 = \tfrac{1}{4} r_2$$

$$
\begin{array}{c|ccccc|c}
 & P & x_1 & x_2 & s_1 & s_2 & \text{RHS} \\
\hline
 & 1 & -3 & -4 & 0 & 0 & 0 \\
 & 0 & \tfrac{1}{2} & ① & \tfrac{1}{4} & 0 & 30 \\
 & 0 & 2 & 2 & 0 & 1 & 80
\end{array}
$$

For Step 2 the pivot row is row 2. To obtain 0s elsewhere in the pivot column, we multiply row 2 by 4 and add it to row 1; in addition, we multiply row 2 by -2 and add it to row 3. The row operations specified are

$$R_1 = 4r_2 + r_1 \qquad R_3 = -2r_2 + r_3$$

The new tableau looks like this:

$$
\begin{array}{c|ccccc|c}
\text{BV} & P & x_1 & x_2 & s_1 & s_2 & \text{RHS} \\
\hline
P & 1 & -1 & 0 & 1 & 0 & 120 \\
x_2 & 0 & \tfrac{1}{2} & ① & \tfrac{1}{4} & 0 & 30 \\
s_2 & 0 & 1 & 0 & -\tfrac{1}{2} & 1 & 20
\end{array}
\tag{3}
$$

This completes the pivot operation since the pivot column (column 3) now appears as

$$
\begin{array}{c}
0 \\
1 \\
0
\end{array}
$$

. . . .

This process should look familiar. It is exactly the one used in Chapter 2 to obtain the reduced row-echelon form of a matrix.

Just what has the pivot operation done? To see, we look again at the initial simplex tableau (2). Observe that the entries in columns P, s_1, and s_2 form an identity matrix (I_3, to be exact). This makes it easy to solve for P, s_1, and s_2 using the other variables as parameters:

$$
\begin{aligned}
P - 3x_1 - 4x_2 \quad\quad &= \quad 0 \quad &\text{or} \quad P = \quad 3x_1 + 4x_2 \\
2x_1 + 4x_2 + s_1 \quad\quad &= 120 \quad &\text{or} \quad s_1 = -2x_1 - 4x_2 + 120 \\
2x_1 + 2x_2 \quad + s_2 &= \ 80 \quad &\text{or} \quad s_2 = -2x_1 - 2x_2 + \ 80
\end{aligned}
$$

The variables P, s_1, and s_2 are the original basic variables (BV) listed in the tableau.

After pivoting, we obtain the tableau given in (3). Notice that in this form, it is easy to solve for P, x_2, and s_2 in terms of x_1 and s_1.

$$
\begin{aligned}
P &= \quad\ x_1 - \quad s_1 + 120 \\
x_2 &= -\tfrac{1}{2}x_1 - \tfrac{1}{4}s_1 + \ 30 \quad\quad (4)\\
s_2 &= \quad -x_1 + \tfrac{1}{2}s_1 + \ 20
\end{aligned}
$$

The variables P, x_2, and s_2 are the new basic variables of the tableau. The variables x_1 and s_1 are the **nonbasic variables.** Thus the result of pivoting is that x_2 becomes a basic variable, while s_1 becomes a nonbasic variable.

Notice in Equations (4) that if we let the value of the nonbasic variables x_1 and s_1 equal 0, then the basic variables P, x_2, and s_2 equal the entries across from them in the right-hand side (RHS) of the tableau (3). Thus for the tableau in Display (3) the current value of the objective function is $P = 120$, obtained for $x_1 = 0$, $s_1 = 0$. The values of x_2 and s_2 are $x_2 = 30$, $s_2 = 20$. Because $P = x_1 - s_1 + 120$ and $x_1 \geq 0$, the value of P can be increased beyond 120 when $x_1 > 0$ and $s_1 = 0$. So we have not maximized P yet.

We summarize this discussion below.

Analyzing a Tableau

To obtain the current values of the objective function and the basic variables in a tableau, follow these steps:

Step 1 From the tableau write the equations corresponding to each row.

Step 2 Solve the first equation for P and the remaining equations for the basic variables.

Step 3 Set each nonbasic variable equal to zero to obtain the current values of P and the basic variables.

Now Work Problem 25

EXAMPLE 4 | Perform another pivot operation on the tableau given in Display (3). Use the circled pivot element in the tableau at the top of the next page.

$\downarrow$

BV	P	x_1	x_2	s_1	s_2	RHS
P	1	-1	0	1	0	120
x_2	0	$\frac{1}{2}$	1	$\frac{1}{4}$	0	30
$\rightarrow s_2$	0	①	0	$-\frac{1}{2}$	1	20

SOLUTION

Since the pivot element happens to be a 1 in this case, we skip Step 1. For Step 2 we perform the row operations

$$R_1 = r_3 + r_1 \qquad R_2 = -\tfrac{1}{2}r_3 + r_2$$

The result is

BV	P	x_1	x_2	s_1	s_2	RHS
P	1	0	0	$\frac{1}{2}$	1	140
x_2	0	0	1	$\frac{1}{2}$	$-\frac{1}{2}$	20
x_1	0	1	0	$-\frac{1}{2}$	1	20

(5)

In the tableau given in (5), the new basic variables are P, x_2, and x_1. The variables s_1 and s_2 are the nonbasic variables. The result of pivoting caused x_1 to become a basic variable and s_2 to become a nonbasic variable. Finally, the equations represented by (5) can be written as

$$P = -\tfrac{1}{2}s_1 - s_2 + 140$$
$$x_2 = -\tfrac{1}{2}s_1 + \tfrac{1}{2}s_2 + 20$$
$$x_1 = \tfrac{1}{2}s_1 - s_2 + 20$$

If we let the nonbasic variables s_1 and s_2 equal 0, then $P = 140$, $x_2 = 20$, and $x_1 = 20$. Thus the current values of the basic variables are $P = 140$, $x_2 = 20$, and $x_1 = 20$. The pivot process has improved the value of P. Because $P = -\tfrac{1}{2}s_1 - s_2 + 140$ and $s_1 \geq 0$ and $s_2 \geq 0$, the value of P cannot increase beyond 140 (any values of s_1 and s_2, other than 0, reduce the value of P). So we have maximized P.

We'll stop here to do some problems before continuing our discussion of the simplex method.

EXERCISE 4.1   Answers to odd-numbered problems begin on page AN-18.

In Problems 1–10 determine which maximum linear programming problems are in standard form. Do not attempt to solve them!

1. Maximize

$$P = 2x_1 + x_2$$

subject to the constraints

$$x_1 + x_2 \leq 5$$
$$2x_1 + 3x_2 \leq 2$$
$$x_1 \geq 0 \qquad x_2 \geq 0$$

2. Maximize

$$P = 3x_1 + 4x_2$$

subject to the constraints

$$3x_1 + x_2 \leq 6$$
$$x_1 + 4x_2 \leq 74$$
$$x_1 \geq 0 \qquad x_2 \geq 0$$

3. Maximize
$$P = 3x_1 + x_2 + x_3$$
subject to the constraints
$$x_1 + x_2 + x_3 \le 6$$
$$2x_1 + 3x_2 + 4x_3 \le 10$$
$$x_1 \ge 0$$

4. Maximize
$$P = 2x_1 + x_2 + 4x_3$$
subject to the constraints
$$2x_1 + x_2 + x_3 \le 10$$
$$x_2 \ge 0$$

5. Maximize
$$P = 3x_1 + x_2 + x_3$$
subject to the constraints
$$x_1 + x_2 + x_3 \le 8$$
$$2x_1 + x_2 + 4x_3 \ge 6$$
$$x_1 \ge 0 \qquad x_2 \ge 0$$

6. Maximize
$$P = 2x_1 + x_2 + 4x_3$$
subject to the constraints
$$2x_1 + x_2 + x_3 \le -1$$
$$x_1 \ge 0 \qquad x_2 \ge 0$$

7. Maximize
$$P = 2x_1 + x_2$$
subject to the constraints
$$x_1 + x_2 \ge -6$$
$$2x_1 + x_2 \le 4$$
$$x_1 \ge 0 \qquad x_2 \ge 0$$

8. Maximize
$$P = 3x_1 + x_2$$
subject to the constraints
$$x_1 + 3x_2 \le 4$$
$$2x_1 - x_2 \ge 1$$
$$x_1 \ge 0 \qquad x_2 \ge 0$$

9. Maximize
$$P = 2x_1 + x_2 + 3x_3$$
subject to the constraints
$$x_1 + x_2 - x_3 \le 10$$
$$x_2 + x_3 \le 4$$
$$x_1 \ge 0 \qquad x_2 \ge 0 \qquad x_3 \ge 0$$

10. Maximize
$$P = 2x_1 + 2x_2 + 3x_3$$
subject to the constraints
$$x_1 - x_2 + x_3 \le 6$$
$$x_1 \le 4$$
$$x_1 \ge 0 \qquad x_2 \ge 0 \qquad x_3 \ge 0$$

In Problems 11–16 each maximum problem is not in standard form. Determine if the problem can be modified so as to be in standard form. If it can, write the modified version.

11. Maximize
$$P = x_1 + x_2$$
subject to the constraints
$$3x_1 - 4x_2 \le -6$$
$$x_1 + x_2 \le 4$$
$$x_1 \ge 0 \qquad x_2 \ge 0$$

12. Maximize
$$P = 2x_1 + 3x_2$$
subject to the constraints
$$-4x_1 + 2x_2 \ge -8$$
$$x_1 - x_2 \le 6$$
$$x_1 \ge 0 \qquad x_2 \ge 0$$

13. Maximize
$$P = x_1 + x_2 + x_3$$
subject to the constraints
$$x_1 + x_2 + x_3 \le 6$$
$$4x_1 + 3x_2 \ge 12$$
$$x_1 \ge 0 \qquad x_2 \ge 0 \qquad x_3 \ge 0$$

14. Maximize
$$P = 2x_1 + x_2 + 3x_3$$
subject to the constraints
$$x_1 + x_2 + x_3 \ge -8$$
$$x_1 - x_2 \le -6$$
$$x_1 \ge 0 \qquad x_2 \ge 0 \qquad x_3 \ge 0$$

15. Maximize
$$P = 2x_1 + x_2 + 3x_3$$
subject to the constraints
$$-x_1 + x_2 + x_3 \geq -6$$
$$2x_1 - 3x_2 \geq -12$$
$$x_3 \leq 2$$
$$x_1 \geq 0 \qquad x_2 \geq 0 \qquad x_3 \geq 0$$

16. Maximize
$$P = x_1 + x_2 + x_3$$
subject to the constraints
$$2x_1 - x_2 + 3x_3 \leq 8$$
$$x_1 - x_2 \geq 6$$
$$x_3 \leq 4$$
$$x_1 \geq 0 \qquad x_2 \geq 0 \qquad x_3 \geq 0$$

In Problems 17–24 each maximum problem is in standard form. For each one introduce slack variables and set up the initial simplex tableau.

17. Maximize
$$P = 2x_1 + x_2 + 3x_3$$
subject to the constraints
$$5x_1 + 2x_2 + x_3 \leq 20$$
$$6x_1 + x_2 + 4x_3 \leq 24$$
$$x_1 + x_2 + 4x_3 \leq 16$$
$$x_1 \geq 0 \qquad x_2 \geq 0 \qquad x_3 \geq 0$$

18. Maximize
$$P = 3x_1 + 2x_2 + x_3$$
subject to the constraints
$$3x_1 + 2x_2 - x_3 \leq 10$$
$$x_1 - x_2 + 3x_3 \leq 12$$
$$2x_1 + x_2 + x_3 \leq 6$$
$$x_1 \geq 0 \qquad x_2 \geq 0 \qquad x_3 \geq 0$$

19. Maximize
$$P = 3x_1 + 5x_2$$
subject to the constraints
$$2.2x_1 - 1.8x_2 \leq 5$$
$$0.8x_1 + 1.2x_2 \leq 2.5$$
$$x_1 + x_2 \leq 0.1$$
$$x_1 \geq 0 \qquad x_2 \geq 0$$

20. Maximize
$$P = 2x_1 + 3x_2$$
subject to the constraints
$$1.2x_1 - 2.1x_2 \leq 0.5$$
$$0.3x_1 + 0.4x_2 \leq 1.5$$
$$x_1 + x_2 \leq 0.7$$
$$x_1 \geq 0 \qquad x_2 \geq 0$$

21. Maximize
$$P = 2x_1 + 3x_2 + x_3$$
subject to the constraints
$$x_1 + x_2 + x_3 \leq 50$$
$$3x_1 + 2x_2 + x_3 \leq 10$$
$$x_1 \geq 0 \qquad x_2 \geq 0 \qquad x_3 \geq 0$$

22. Maximize
$$P = x_1 + 4x_2 + 2x_3$$
subject to the constraints
$$3x_1 + x_2 + x_3 \leq 10$$
$$x_1 + x_2 + 3x_3 \leq 5$$
$$x_1 \geq 0 \qquad x_2 \geq 0 \qquad x_3 \geq 0$$

23. Maximize
$$P = 3x_1 + 4x_2 + 2x_3$$
subject to the constraints
$$3x_1 + x_2 + 4x_3 \leq 5$$
$$x_1 + x_2 \leq 5$$
$$2x_1 - x_2 + x_3 \leq 6$$
$$x_1 \geq 0 \qquad x_2 \geq 0 \qquad x_3 \geq 0$$

24. Maximize
$$P = 2x_1 + x_2 + 3x_3$$
subject to the constraints
$$2x_1 + x_2 + x_3 \leq 2$$
$$x_1 - x_2 \leq 4$$
$$2x_1 + x_2 - x_3 \leq 5$$
$$x_1 \geq 0 \qquad x_2 \geq 0 \qquad x_3 \geq 0$$

In Problems 25–29 perform a pivot operation on each tableau. The pivot element is circled. Using the new tableau obtained, write the corresponding system of equations. Indicate the current values of the objective function and the basic variables.

25.

BV	P	x_1	x_2	s_1	s_2	RHS
P	1	-1	-2	0	0	0
s_1	0	1	(2)	1	0	300
s_2	0	3	2	0	1	480

26.

BV	P	x_1	x_2	s_1	s_2	RHS
P	1	-2	-1	0	0	0
s_1	0	1	4	1	0	100
s_2	0	(2)	5	0	1	50

27.

BV	P	x_1	x_2	x_3	s_1	s_2	s_3	RHS
P	1	-1	-2	-3	0	0	0	0
s_1	0	1	2	4	1	0	0	24
s_2	0	2	-1	1	0	1	0	32
s_3	0	3	2	(4)	0	0	1	18

28.

BV	P	x_1	x_2	x_3	s_1	s_2	s_3	RHS
P	1	-1	-2	-3	0	0	0	0
s_1	0	1	2	1	1	0	0	6
s_2	0	2	3	1	0	1	0	12
s_3	0	1	-2	(3)	0	0	1	0

29.

BV	P	x_1	x_2	x_3	x_4	s_1	s_2	s_3	s_4	RHS
P	1	-1	-2	-3	-4	0	0	0	0	0
s_1	0	-3	0	1	0	1	0	0	0	20
s_2	0	2	0	0	(1)	0	1	0	0	24
s_3	0	0	-3	1	0	0	0	1	0	28
s_4	0	0	-3	0	1	0	0	0	1	24

Technology Exercises

In Problems 1–4 use the elementary row operations on your graphing calculator to perform a pivot operation on each tableau. The pivot element is circled.

1.

BV	P	x_1	x_2	s_1	s_2	RHS
P	1	-2	-3	0	0	0
s_1	0	3	2	1	0	200
s_2	0	1	(3)	0	1	150

2.

BV	P	x_1	x_2	s_1	s_2	RHS
P	1	-5	-2	0	0	0
s_1	0	1	3	1	0	300
s_2	0	(3)	4	0	1	480

3.

BV	P	x_1	x_2	s_1	s_2	RHS
P	1	-2	-3	0	0	0
s_1	0	1	(4)	1	0	100
s_2	0	2	5	0	1	135

4.

BV	P	x_1	x_2	s_1	s_2	RHS
P	1	-2.75	-1.45	0	0	0
s_1	0	(3.25)	2.12	1	0	125
s_2	0	1.50	3.35	0	1	75

4.2 THE SIMPLEX METHOD: SOLVING MAXIMUM PROBLEMS IN STANDARD FORM

We are now ready to state the details of the simplex method for solving a maximum linear programming problem. This method requires that the problem be in standard form and that the problem be placed in an initial simplex tableau with slack variables.

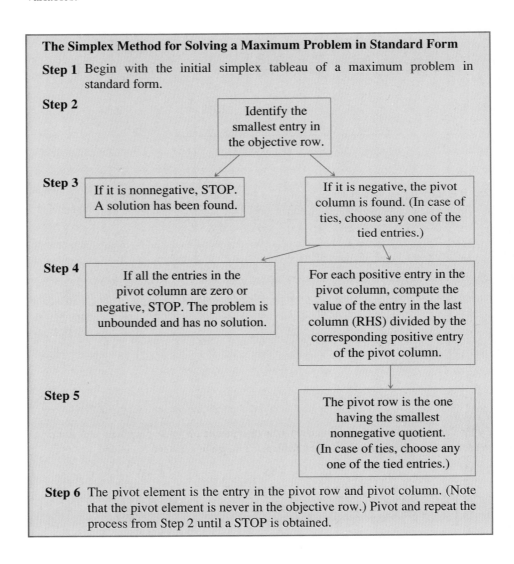

The Simplex Method for Solving a Maximum Problem in Standard Form

Step 1 Begin with the initial simplex tableau of a maximum problem in standard form.

Step 2   Identify the smallest entry in the objective row.

Step 3   If it is nonnegative, STOP. A solution has been found.

If it is negative, the pivot column is found. (In case of ties, choose any one of the tied entries.)

Step 4   If all the entries in the pivot column are zero or negative, STOP. The problem is unbounded and has no solution.

For each positive entry in the pivot column, compute the value of the entry in the last column (RHS) divided by the corresponding positive entry of the pivot column.

Step 5   The pivot row is the one having the smallest nonnegative quotient. (In case of ties, choose any one of the tied entries.)

Step 6 The pivot element is the entry in the pivot row and pivot column. (Note that the pivot element is never in the objective row.) Pivot and repeat the process from Step 2 until a STOP is obtained.

Let's go through the process.

EXAMPLE 1

Maximize
$$P = 3x_1 + 4x_2$$
subject to the constraints
$$2x_1 + 4x_2 \leq 120$$
$$2x_1 + 2x_2 \leq 80$$
$$x_1 \geq 0 \qquad x_2 \geq 0$$

SOLUTION

Step 1 This is a maximum problem in standard form. To obtain the initial simplex tableau, we proceed as follows: The objective function is written in the form
$$P - 3x_1 - 4x_2 = 0$$
After introducing slack variables s_1 and s_2, the constraints take the form
$$2x_1 + 4x_2 + s_1 \qquad = 120$$
$$2x_1 + 2x_2 \qquad + s_2 = \quad 80$$
$$x_1 \geq 0 \qquad x_2 \geq 0$$
$$s_1 \geq 0 \qquad s_2 \geq 0$$
The initial simplex tableau is

BV	P	x_1	x_2	s_1	s_2	RHS
P	1	−3	−4	0	0	0
s_1	0	2	4	1	0	120
s_2	0	2	2	0	1	80

Notice that the objective row contains the negatives of the coefficients in the objective function.

Step 2 The smallest entry in the objective row is −4.

Step 3 Since −4 is negative, the pivot column is the column corresponding to x_2.

Step 4 For each positive entry in the pivot column, form the quotient of the corresponding RHS entry by the positive entry.

RHS	120	80
Positive entry	4	2
Quotient	$\frac{120}{4} = 30$	$\frac{80}{2} = 40$

Step 5 The smallest nonnegative value is 30, so the pivot row is the row corresponding to the basic variable s_1.

The tableau below summarizes this process.

BV	P	x_1	x_2	s_1	s_2	RHS		
P	1	−3	−4	0	0	0	$P = 3x_1 + 4x_2$	$P = 0$
s_1	0	2	④	1	0	120	$120 \div 4 = 30$	$s_1 = 120$
s_2	0	2	2	0	1	80	$80 \div 2 = 40$	$s_2 = 80$

Current values

(1)

Step 6 The pivot element is 4. Next, we pivot by using the row operations:

1. $R_2 = \frac{1}{4}r_2$
2. $R_1 = r_1 + 4r_2, \quad R_3 = r_3 + (-2)r_2$

After pivoting, we have this tableau:

BV	P	x_1	x_2	s_1	s_2	RHS	Current values
P	1	-1	0	1	0	120	$P = 120$
x_2	0	$\frac{1}{2}$	1	$\frac{1}{4}$	0	30	$x_2 = 30$
s_2	0	1	0	$-\frac{1}{2}$	1	20	$s_2 = 20$

(2)

Notice that x_2 is now a basic variable, replacing s_1. The objective row, $P = x_1 - s_1 + 120$, has a current value of 120. Because $P = x_1 - s_1 + 120$ and $x_1 \geq 0$, the value of P can be increased beyond 120, when $x_1 > 0$ and $s_1 = 0$. So we have not maximized P yet.

We continue with the simplex method at Step 2.

Step 2 The smallest entry in the objective row is -1.

Step 3 Since -1 is negative, the pivot column is the column corresponding to x_1.

Step 4 For each positive entry in the pivot column, form the quotient of the corresponding RHS entry by the positive entry.

RHS		30	20
Positive entry		$\frac{1}{2}$	1
Quotient		$\frac{30}{\frac{1}{2}} = 60$	$\frac{20}{1} = 20$

Step 5 The smallest nonnegative value is 20, so the pivot row is the row corresponding to the basic variable s_2.

The tableau below summarizes this process.

BV	P	x_1↓	x_2	s_1	s_2	RHS		Current values
P	1	-1	0	1	0	120	$P = x_1 - s_1 + 120$	$P = 120$
x_2	0	$\frac{1}{2}$	1	$\frac{1}{4}$	0	30	$30 \div \frac{1}{2} = 60$	$x_2 = 30$
$\to s_2$	0	①	0	$-\frac{1}{2}$	1	20	$20 \div 1 = 20$	$s_2 = 20$

Step 6 The pivot element is 1. Next, we pivot by using the row operations:

$$R_1 = r_1 + r_3 \qquad R_2 = r_2 + (-\tfrac{1}{2})r_3$$

The result is this tableau:

BV	P	x_1	x_2	s_1	s_2	RHS	Current values
P	1	0	0	$\frac{1}{2}$	1	140	$P = 140$
x_2	0	0	1	$\frac{1}{2}$	$-\frac{1}{2}$	20	$x_2 = 20$
x_1	0	1	0	$-\frac{1}{2}$	1	20	$x_1 = 20$

(3)

We continue with the simplex method at Step 2.

Step 2 The smallest entry in the objective row is 0.

Step 3 Since it is nonnegative, we STOP. To see why we STOP, we write the equations from the objective row, namely,

$$P = -\tfrac{1}{2}s_1 - s_2 + 140$$

Since $s_1 \geq 0$ and $s_2 \geq 0$, any positive value of s_1 or s_2 would make the value of P less than 140. By choosing $s_1 = 0$ and $s_2 = 0$, we obtain the largest possible value for P, namely, 140. If we write the equations from the second and third rows, substituting 0 for s_1 and s_2, we have

$$x_2 = -\tfrac{1}{2}s_1 + \tfrac{1}{2}s_2 + 20 = 20$$
$$x_1 = \tfrac{1}{2}s_1 - s_2 + 20 = 20$$

In other words, we have found the optimal solution,

$$P = 140$$

the maximum, which occurs at

$$x_1 = 20 \qquad x_2 = 20 \qquad s_1 = 0 \qquad s_2 = 0$$

The tableau in Display (3) is called the **final tableau** because with this tableau an optimal solution is found.

To summarize, each tableau obtained by applying the simplex method provides information about the current status of the solution:

1. The RHS entry in the objective row gives the current value of the objective function.
2. The remaining entries in the RHS column give the current values of the corresponding basic variables.

EXAMPLE 2 Determine whether each tableau

1. is a final tableau. If it is, give the solution.
2. requires additional pivoting. If so, identify the pivot element.
3. indicates no solution.

(a)

BV	P	x_1	x_2	s_1	s_2	RHS
P	1	0	-2	0	1	20
s_1	0	0	0	1	1	40
x_1	0	1	-1	0	1	20

(b)

BV	P	x_1	x_2	s_1	s_2	RHS
P	1	0	0	1	2	220
x_1	0	1	0	2	-1	40
x_2	0	0	1	-1	1	20

SOLUTION

(a) The smallest entry in the objective row is -2, which is negative. The pivot column is the column corresponding to the variable x_2. Since all the entries in the pivot column are zero or negative, the problem is unbounded and has no solution.

(b) The objective row contains no negative entries, so this is a final tableau. The solution is

$$P = 220 \qquad \text{when} \qquad x_1 = 40 \quad x_2 = 20 \quad s_1 = 0 \quad s_2 = 0$$

Now Work Problem 1

· ·

EXAMPLE 3

Maximize

$$P = 6x_1 + 8x_2 + x_3$$

subject to the constraints

$$3x_1 + 5x_2 + 3x_3 \le 20$$
$$x_1 + 3x_2 + 2x_3 \le 9$$
$$6x_1 + 2x_2 + 5x_3 \le 30$$
$$x_1 \ge 0 \qquad x_2 \ge 0 \qquad x_3 \ge 0$$

SOLUTION

Note that the problem is in standard form. By introducing slack variables s_1, s_2, and s_3, the constraints take the form

$$3x_1 + 5x_2 + 3x_3 + s_1 \qquad\qquad = 20$$
$$x_1 + 3x_2 + 2x_3 \qquad + s_2 \qquad = 9$$
$$6x_1 + 2x_2 + 5x_3 \qquad\qquad + s_3 = 30$$
$$x_1 \ge 0 \qquad x_2 \ge 0 \qquad x_3 \ge 0$$
$$s_1 \ge 0 \qquad s_2 \ge 0 \qquad s_3 \ge 0$$

Since

$$P - 6x_1 - 8x_2 - x_3 = 0$$

the initial simplex tableau is

BV	P	x_1	x_2	x_3	s_1	s_2	s_3	RHS		Current values
P	1	-6	-8	-1	0	0	0	0		$P = 0$
s_1	0	3	5	3	1	0	0	20	$20 \div 5 = 4$	$s_1 = 20$
s_2	0	1	③	2	0	1	0	9	$9 \div 3 = 3$	$s_2 = 9$
s_3	0	6	2	5	0	0	1	30	$30 \div 2 = 15$	$s_3 = 30$

The pivot column is found by locating the column containing the smallest entry in the objective row (-8 in the column corresponding to the variable x_2). The pivot row is obtained by dividing each entry in the RHS column by the corresponding entry in the pivot column and selecting the smallest nonnegative quotient. Thus the pivot row is the row corresponding to the basic variable s_2. The pivot element is 3, which is circled. After pivoting, the new tableau is

BV	P	x_1	x_2	x_3	s_1	s_2	s_3	RHS	
P	1	$-\frac{10}{3}$	0	$\frac{13}{3}$	0	$\frac{8}{3}$	0	24	
$\rightarrow$ s_1	0	$\left(\frac{4}{3}\right)$	0	$-\frac{1}{3}$	1	$-\frac{5}{3}$	0	5	
x_2	0	$\frac{1}{3}$	1	$\frac{2}{3}$	0	$\frac{1}{3}$	0	3	
s_3	0	$\frac{16}{3}$	0	$\frac{11}{3}$	0	$-\frac{2}{3}$	1	24	

Current values

$P = 24$

$5 \div \frac{4}{3} = 3.75$ $s_1 = 5$

$3 \div \frac{1}{3} = 9$ $x_2 = 3$

$24 \div \frac{16}{3} = 4.5$ $s_3 = 24$

The value of P has improved to 24, but the negative entry, $-\frac{10}{3}$, in the objective row indicates further improvement is possible. By the same procedure as before, we determine the next pivot element to be $\frac{4}{3}$. After pivoting, we get

BV	P	x_1	x_2	x_3	s_1	s_2	s_3	RHS
P	1	0	0	$\frac{7}{2}$	$\frac{5}{2}$	$-\frac{3}{2}$	0	$\frac{73}{2}$
x_1	0	1	0	$-\frac{1}{4}$	$\frac{3}{4}$	$-\frac{5}{4}$	0	$\frac{15}{4}$
x_2	0	0	1	$\frac{3}{4}$	$-\frac{1}{4}$	$\frac{3}{4}$	0	$\frac{7}{4}$
$\rightarrow$ s_3	0	0	0	5	-4	$\left(6\right)$	1	4

Current values

$P = \frac{73}{2}$

$x_1 = \frac{15}{4}$

$\frac{7}{4} \div \frac{3}{4} = \frac{7}{3}$ $x_2 = \frac{7}{4}$

$4 \div 6 = \frac{2}{3}$ $s_3 = 4$

The value of P has improved to $\frac{73}{2}$, but, since we still observe a negative entry in the objective row, we pivot again. (Remember, in finding the pivot row, we ignore the objective row and any rows in which the pivot column contains a negative number or zero—in this case, $-\frac{5}{4}$.) The new tableau is

BV	P	x_1	x_2	x_3	s_1	s_2	s_3	RHS
P	1	0	0	$\frac{19}{4}$	$\frac{3}{2}$	0	$\frac{1}{4}$	$\frac{75}{2}$
x_1	0	1	0	$\frac{19}{24}$	$-\frac{1}{12}$	0	$\frac{5}{24}$	$\frac{55}{12}$
x_2	0	0	1	$\frac{1}{8}$	$\frac{1}{4}$	0	$-\frac{1}{8}$	$\frac{5}{4}$
s_2	0	0	0	$\frac{5}{6}$	$-\frac{2}{3}$	1	$\frac{1}{6}$	$\frac{2}{3}$

Current values

$P = \frac{75}{2}$

$x_1 = \frac{55}{12}$

$x_2 = \frac{5}{4}$

$s_2 = \frac{2}{3}$

This is a final tableau since all the entries in the objective row are nonnegative. The objective (top) row yields the equation

$$P = -\tfrac{19}{4} x_3 - \tfrac{3}{2} s_1 - \tfrac{1}{4} s_3 + \tfrac{75}{2}$$

Thus P is a maximum when $x_3 = 0$, $s_1 = 0$, and $s_3 = 0$, giving $P = \frac{75}{2}$. From rows 2, 3, and 4 of the final tableau we have the equations

$$x_1 = -\tfrac{19}{24} x_3 + \tfrac{1}{12} s_1 - \tfrac{5}{24} s_3 + \tfrac{55}{12}$$
$$x_2 = -\tfrac{1}{8} x_3 - \tfrac{1}{4} s_1 + \tfrac{1}{8} s_3 + \tfrac{5}{4}$$
$$s_2 = -\tfrac{5}{6} x_3 + \tfrac{2}{3} s_1 - \tfrac{1}{6} s_3 + \tfrac{2}{3}$$

Using $x_3 = 0$, $s_1 = 0$, and $s_3 = 0$ in these equations, we find

$$x_1 = \tfrac{55}{12} \qquad x_2 = \tfrac{5}{4} \qquad s_2 = \tfrac{2}{3} \tag{4}$$

Thus the optional solution is $P = \frac{75}{2}$, obtained when $x_1 = \frac{55}{12}$, $x_2 = \frac{5}{4}$, and $x_3 = 0$.

. . . .

Note that the information in (4) may be easily obtained by looking at the current values column of the final tableau.

Now Work Problem 7

EXAMPLE 4   Maximize
$$P = 4x_1 + 4x_2 + 3x_3$$

subject to the constraints

$$2x_1 + 3x_2 + x_3 \le 6$$
$$x_1 + 2x_2 + 3x_3 \le 6$$
$$x_1 + x_2 + x_3 \le 5$$
$$x_1 \ge 0 \qquad x_2 \ge 0 \qquad x_3 \ge 0$$

SOLUTION

The problem is in standard form. We introduce slack variables s_1, s_2, s_3, and write the constraints as

$$2x_1 + 3x_2 + x_3 + s_1 \qquad\qquad = 6$$
$$x_1 + 2x_2 + 3x_3 \qquad + s_2 \qquad = 6$$
$$x_1 + x_2 + x_3 \qquad\qquad + s_3 = 5$$
$$x_1 \ge 0 \qquad x_2 \ge 0 \qquad x_3 \ge 0$$
$$s_1 \ge 0 \qquad s_2 \ge 0 \qquad s_3 \ge 0$$

Since $P - 4x_1 - 4x_2 - 3x_3 = 0$, the initial simplex tableau is

BV	P	x_1	x_2	x_3	s_1	s_2	s_3	RHS	
P	1	-4	-4	-3	0	0	0	0	
s_1	0	②	3	1	1	0	0	6	$6 \div 2 = 3$
s_2	0	1	2	3	0	1	0	6	$6 \div 1 = 6$
s_3	0	1	1	1	0	0	1	5	$5 \div 1 = 5$

Since -4 is the smallest negative entry in the objective row, we have a tie for pivot column between the columns corresponding to the variables x_1 and x_2. We choose (arbitrarily) as pivot column the column corresponding to the variable x_1. The pivot row is the row corresponding to the variable s_1. (Do you see why?) The pivot element is 2, which is circled. After pivoting, we obtain the following tableau:

BV	P	x_1	x_2	x_3	s_1	s_2	s_3	RHS		Current values
P	1	0	2	-1	2	0	0	12		$P = 12$
x_1	0	1	$\frac{3}{2}$	$\frac{1}{2}$	$\frac{1}{2}$	0	0	3	$3 \div \frac{1}{2} = 6$	$x_1 = 3$
s_2	0	0	$\frac{1}{2}$	⑤⁄₂	$-\frac{1}{2}$	1	0	3	$3 \div \frac{5}{2} = \frac{6}{5}$	$s_2 = 3$
s_3	0	0	$-\frac{1}{2}$	$\frac{1}{2}$	$-\frac{1}{2}$	0	1	2	$2 \div \frac{1}{2} = 4$	$s_3 = 2$

The pivot element is $\frac{5}{2}$, which is circled. After pivoting, we obtain

BV	P	x_1	x_2	x_3	s_1	s_2	s_3	RHS	Current values
P	1	0	$\frac{11}{5}$	0	$\frac{9}{5}$	$\frac{2}{5}$	0	$\frac{66}{5}$	$P = \frac{66}{5}$
x_1	0	1	$\frac{7}{5}$	0	$\frac{3}{5}$	$-\frac{1}{5}$	0	$\frac{12}{5}$	$x_1 = \frac{12}{5}$
x_3	0	0	$\frac{1}{5}$	1	$-\frac{1}{5}$	$\frac{2}{5}$	0	$\frac{6}{5}$	$x_3 = \frac{6}{5}$
s_3	0	0	$-\frac{3}{5}$	0	$-\frac{2}{5}$	$-\frac{1}{5}$	1	$\frac{7}{5}$	$s_3 = \frac{7}{5}$

This is a final tableau. The solution is $P = \frac{66}{5}$, obtained when $x_1 = \frac{12}{5}$, $x_2 = 0$, $x_3 = \frac{6}{5}$.

·····················

The reader is encouraged to solve Example 4 again, this time choosing the pivot column corresponding to x_2 for the first pivot.

EXAMPLE 5

Maximizing Profit   Mike's Famous Toy Trucks specializes in making four kinds of toy trucks: a delivery truck, a dump truck, a garbage truck, and a gasoline truck. Three machines—a metal casting machine, a paint spray machine, and a packaging machine—are used in the production of these trucks. The time, in hours, each machine works to make each type of truck and the profit for each truck are given in Table 1. The maximum time available per week for each machine is given: metal casting, 4000 hours; paint spray, 1800 hours; and packaging, 1000 hours. How many of each type truck should be produced to maximize profit? Assume that every truck made is sold.

Table 1

	Delivery Truck	Dump Truck	Garbage Truck	Gasoline Truck	Maximum Time
Metal Casting	2 hours	2.5 hours	2 hours	2 hours	4000 hours
Paint Spray	1 hour	1.5 hours	1 hour	2 hours	1800 hours
Packaging	0.5 hour	0.5 hour	1 hour	1 hour	1000 hours
Profit	$0.50	$1.00	$1.50	$2.00	

SOLUTION

Let x_1, x_2, x_3, and x_4 denote the number of delivery trucks, dump trucks, garbage trucks, and gasoline trucks, respectively, to be made. If P denotes the profit to be maximized, we have this problem:

Maximize

$$P = 0.5x_1 + x_2 + 1.5x_3 + 2x_4$$

subject to the constraints

$$2x_1 + 2.5x_2 + 2x_3 + 2x_4 \leq 4000$$
$$x_1 + 1.5x_2 + x_3 + 2x_4 \leq 1800$$
$$0.5x_1 + 0.5x_2 + x_3 + x_4 \leq 1000$$
$$x_1 \geq 0 \qquad x_2 \geq 0 \qquad x_3 \geq 0 \qquad x_4 \geq 0$$

Since this problem is in standard form, we introduce slack variables s_1, s_2, and s_3, write the initial simplex tableau, and solve:

BV	P	x_1	x_2	x_3	x_4	s_1	s_2	s_3	RHS	Current values
P	1	-0.5	-1	-1.5	-2	0	0	0	0	$P = 0$
s_1	0	2	2.5	2	2	1	0	0	4000	$s_1 = 4000$
s_2	0	1	1.5	1	②	0	1	0	1800	$s_2 = 1800$
s_3	0	0.5	0.5	1	1	0	0	1	1000	$s_3 = 1000$

BV	P	x_1	x_2	x_3	x_4	s_1	s_2	s_3	RHS	Current values
P	1	0.5	0.5	-0.5	0	0	1	0	1800	$P = 1800$
s_1	0	1	1	1	0	1	-1	0	2200	$s_1 = 2200$
x_4	0	0.5	0.75	0.5	1	0	0.5	0	900	$x_4 = 900$
s_3	0	0	-0.25	⑤.5	0	0	-0.5	1	100	$s_3 = 100$

BV	P	x_1	x_2	x_3	x_4	s_1	s_2	s_3	RHS	Current values
P	1	0.5	0.25	0	0	0	0.5	1	1900	$P = 1900$
s_1	0	1	1.5	0	0	1	0	-2	2000	$s_1 = 2000$
x_4	0	0.5	1	0	1	0	1	-1	800	$x_4 = 800$
x_3	0	0	-0.5	1	0	0	-1	2	200	$x_3 = 200$

This is a final tableau. The maximum profit is $P = \$1900$, and it is attained for

$$x_1 = 0 \qquad x_2 = 0 \qquad x_3 = 200 \qquad x_4 = 800$$

....

The practical considerations of the situation described in Example 5 are that delivery trucks and dump trucks are too costly to produce or too little profit is being gained from their sale. Since the slack variable s_1 has a value of 2000 for maximum P and since s_1 represents the number of hours the metal casting machine is printing no truck (that is, the time the machine is idle), it may be possible to release this machine for other duties. Also note that both the paint spray and packaging machines are operating at full capacity. This means that to increase productivity, more paint spray and packaging capacity is required.

Now Work Problem 23

Analyzing the Simplex Method

To justify some of the steps of the simplex method, we return to Example 1 and analyze more carefully what we did. Look back to Display (1).

The reason we choose the most negative entry in the objective row is that it is the negative of the *largest* coefficient in the objective function:

$$P = 3x_1 + 4x_2$$

If we were to set $x_1 = x_2 = 0$, we would obtain $P = 0$ as a first approximation for the profit P. Of course, this is not a very good approximation; it can easily be improved by increasing either x_1 or x_2. But the profit per unit of x_2 is \$4, while the profit per unit of x_1 is only \$3. Thus it is more effective to increase x_2 than x_1. But what is the largest amount by which x_2 can be increased?

We can answer this question by referring back to Display (2), the tableau that resulted after the first pivot. The corresponding equations are

$$P = \quad x_1 - \quad s_1 + 120$$
$$x_2 = -\tfrac{1}{2}x_1 - \tfrac{1}{4}s_1 + \quad 30$$
$$s_2 = \quad - \quad x_1 + \tfrac{1}{2}s_1 + \quad 20$$

This suggests that x_2 can be as large as 30, if we take both x_1 and s_1 to be 0, in which case P will be 120.

So we chose the pivot column for Display (1) because we wanted to increase x_2. But why did we choose the pivot row? Let's see what would have happened if instead we had chosen the row corresponding to the variable s_2 as the pivot row. The result would have been

BV	P	x_1	x_2	s_1	s_2	RHS
P	1	1	0	0	2	160
s_1	0	-2	0	1	-2	-40
x_2	0	1	1	0	$\tfrac{1}{2}$	40

However, this is not acceptable because of the negative number in the last column. (In effect, this matrix tells us that we could get P to be as large as 160, by setting $x_1 = 0$ and $s_2 = 0$. Then $x_2 = 40$ and $s_1 = -40$; but this is not a feasible point because s_1 is supposed to be greater than or equal to 0.)

Each iteration of the simplex method consists in choosing an "entering" variable from the nonbasic ones and a "leaving" variable from the basic ones, using a selective criterion so that the value of the objective function is not decreased (sometimes the value may remain unchanged). The iterative procedure stops when no variable can enter from the nonbasic variables.

So the reasoning behind the simplex method for standard maximum problems is not too complicated, and the process of "moving to a better solution" is made quite easy simply by following the rules. Briefly, the pivoting strategy works like this:

> The choice of the pivot column forces us to pivot the variable that apparently improves the value of the objective function most effectively.
>
> The choice of the pivot row prevents us from making this variable *too large* to be feasible.

Geometry of the Simplex Method

The maximum value (provided it exists) of the objective function will occur at one of the corner points of the feasible region. The simplex method is designed to move from corner point to corner point of the feasible region, at each stage improving the value of the objective function until a solution is found. More precisely, the geometry behind the simplex method is outlined at the top of the next page.

1. A given tableau corresponds to a corner point of the feasible region.
2. The operation of pivoting moves us to an adjacent corner point, where the objective function has a value at least as large as it did at the previous corner point.
3. The process continues until the final tableau is reached—which produces a corner point that maximizes the objective function.

Though drawings of this can be rendered in only two or three dimensions (that is, when the objective function has two or three variables in it), the same interpretation can be shown to hold regardless of the number of variables involved. Let's look at an example.

EXAMPLE 6

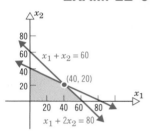

Figure 1

Maximize

$$P = 3x_1 + 5x_2$$

subject to the constraints

$$x_1 + x_2 \leq 60$$
$$x_1 + 2x_2 \leq 80$$
$$x_1 \geq 0 \qquad x_2 \geq 0$$

SOLUTION

The feasible region is shown in Figure 1.

Below, on the left, we apply the simplex method, indicating on the right the corner point corresponding to each tableau and the value of the objective function there. You should supply the details.

Tableau							Corner Point (x_1, x_2)	Value of $P = 3x_1 + 5x_2$ at the Corner Point
BV	P	x_1	x_2	s_1	s_2	RHS		
P	1	-3	-5	0	0	0	(0, 0)	0
s_1	0	1	1	1	0	60		
s_2	0	1	2	0	1	80		
			(Pivot)					
P	1	$-\frac{1}{2}$	0	0	$\frac{5}{2}$	200	(0, 40)	200
s_1	0	$\frac{1}{2}$	0	1	$-\frac{1}{2}$	20		
x_2	0	$\frac{1}{2}$	1	0	$\frac{1}{2}$	40		
			(Pivot)					
P	1	0	0	1	2	220	(40, 20)	220
x_1	0	1	0	2	-1	40		
x_2	0	0	1	-1	1	20		
			(Final tableau)					(Maximum value)

· ·

The Unbounded Case

So far in our discussion, it has always been possible to continue to choose pivot elements until the problem has been solved. But it may turn out that all the entries in a column of a tableau are 0 or negative at some stage. If this happens, it means that the problem is *unbounded* and a maximum solution does not exist.

For example, consider the tableau

BV	P	x_1	x_2	s_1	s_2	RHS
P	1	-1	-1	0	0	0
s_1	0	-1	1	1	0	2
s_2	0	1	-1	0	1	2

When the smallest negative entries in the objective (top) row are equal, you may choose either column as the pivot column. Suppose we arbitrarily choose column 2 to pivot, and the tableau becomes

BV	P	x_1	x_2	s_1	s_2	RHS
P	1	0	-2	0	1	2
s_1	0	0	0	1	1	4
x_1	0	1	-1	0	1	2

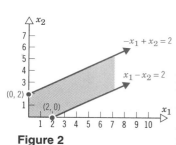

Figure 2

Now the only negative entry in the objective (top) row is in column 3, and it is impossible to choose a pivot element in that column. This implies that the objective function is unbounded. Indeed, it is easy to see that if the only constraints are $-x_1 + x_2 \leq 2$, $x_1 - x_2 \leq 2$, $x_1 \geq 0$, and $x_2 \geq 0$, then $P = x_1 + x_2$ has no maximum. The feasible region is shown in Figure 2.

Summary of the Simplex Method

The general procedure for solving a maximum linear programming problem in standard form using the simplex method can be outlined as follows:

1. The maximum problem is stated in standard form as

Maximize

$$P = c_1 x_1 + c_2 x_2 + \cdots + c_n x_n$$

subject to the constraints

$$a_{11} x_1 + a_{12} x_2 + \cdots + a_{1n} x_n \leq b_1$$
$$a_{21} x_1 + a_{22} x_2 + \cdots + a_{2n} x_n \leq b_2$$
$$\vdots$$
$$a_{m1} x_1 + a_{m2} x_2 + \cdots + a_{mn} x_n \leq b_m$$
$$x_1 \geq 0, x_2 \geq 0, \ldots, x_n \geq 0$$

in which $b_1 > 0, b_2 > 0, \ldots, b_m > 0$.

2. Introduce slack variables $s_1, s_2, \ldots, s_m$ so that the constraints take the form of equalities:

$$a_{11}x_1 + a_{12}x_2 + \cdots + a_{1n}x_n + s_1 = b_1$$
$$a_{21}x_1 + a_{22}x_2 + \cdots + a_{2n}x_n + s_2 = b_2$$
$$\vdots \qquad\qquad\qquad\qquad \vdots$$
$$a_{m1}x_1 + a_{m2}x_2 + \cdots + a_{mn}x_n + s_m = b_m$$
$$x_1 \geq 0, x_2 \geq 0, \ldots, x_n \geq 0$$
$$s_1 \geq 0, s_2 \geq 0, \ldots, s_m \geq 0$$

3. Write the objective function in the form

$$P - c_1x_1 - c_2x_2 - \cdots - c_nx_n = 0$$

4. Set up the initial simplex tableau

BV	P	x_1	x_2	$\cdots$	x_n	s_1	s_2	s_m	RHS
P	1	$-c_1$	$-c_2$	$\cdots -c_n$		0	$0 \cdots 0$		0
s_1	0	a_{11}	a_{12}	$\cdots$	a_{1n}	1	$0 \cdots 0$		b_1
s_2	0	a_{21}	a_{22}	$\cdots$	a_{2n}	0	$1 \cdots 0$		b_2
$\vdots$	$\vdots$	$\vdots$	$\vdots$		$\vdots$	$\vdots$	$\vdots$		$\vdots$
s_m	0	a_{m1}	a_{m2}	$\cdots$	a_{mn}	0	$0 \cdots 1$		b_m

5. Pivot until
   (a) All the entries in the objective row are nonnegative. This is a final tableau from which a solution can be read.

   Or until

   (b) The selection of the pivot column is a column whose entries are negative or zero. In this case the problem is unbounded and there is no solution.

The flowchart in Figure 3 illustrates the steps to be used in solving standard maximum linear programming problems.

Other Linear Programming Techniques

Since its introduction in the 1940s, linear programming has gained ever wider acceptance and application. Today airlines, brokerage houses, and oil companies routinely construct large-scale linear programming problems involving thousands, even tens of thousands, of variables and constraints. To date, the method of choice for solving these problems has been the simplex algorithm developed some 50 years ago by George Dantzig.

Computer implementation of the simplex algorithm has been relatively successful, yet practitioners in the field have always asked if perhaps some algorithm other than the simplex method might not be more appropriate since there were already scattered instances of problems that required too much computer time to solve with the simplex method. Also, the demands of industry could easily envision problems involving hundreds of thousands of variables for which a quicker algorithm would be more cost efficient and even necessary.

Figure 3

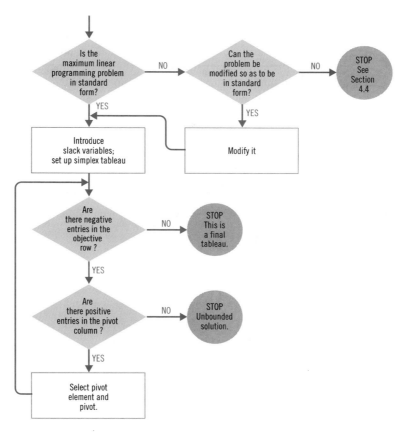

In 1979 a young Soviet mathematician named Leonid Khatchian proposed an algorithm, called the *ellipsoid method,* which proceeded to find the optimal solution by successively surrounding the feasible region by ellipsoids. Though this algorithm did better on some problems, its practical significance was diminished by the fact that it performed no better on the average than the simplex method.

In 1984 Narendra Karmarkar, a researcher at AT&T Bell Laboratories, announced a linear programming technique that speeds up the handling of certain problems. Unlike the simplex method, which moves along the edges of the feasible region from corner point to corner point, both the Khatchian and the Karmarkar algorithms deal with points in the interior of the feasible region in their search for the solution.

Both Karmarkar's method and the ellipsoid method have been shown to be **polynomial time algorithms.** This implies that if a linear programming problem of size n is solved by either Karmarkar's method or the ellipsoid method, there exist positive numbers a and b such that the problem can be solved in a time of at most an^b.

In contrast to the Karmarkar and the ellipsoid methods, the simplex method is an **exponential time algorithm** for solving linear programming problems. If a linear programming problem of size n is solved by the simplex method, then there exists a positive number c such that the simplex method will find the optimal solution in a time of at most $c2^n$. For large enough n (for positive a, b, and c), $c2^n > an^b$. This means that, in theory, a polynomial time algorithm is superior to an exponential time algorithm. Extensive testing of the ellipsoid method has shown it to require much more computer time than the simplex method. However, experimentation with Karmarkar's method has shown that for large linear programming problems arising in actual applications, Karmarkar's method may be up to 50 times faster than the simplex method.

EXERCISE 4.2   Answers to odd-numbered problems begin on page AN-20.

In Problems 1–6 determine which of the following statements is true about each tableau:
(a) It is the final tableau.
(b) It requires additional pivoting.
(c) It indicates no solution to the problem.
If the answer is (a), write down the solution; if the answer is (b), indicate the pivot element.

1.

BV	P	x_1	x_2	s_1	s_2	RHS
P	1	-1	0	0	1	120
s_1	0	1	0	1	$-\frac{1}{2}$	20
x_2	0	$\frac{1}{2}$	1	0	$\frac{1}{4}$	30

2.

BV	P	x_1	x_2	s_1	s_2	RHS
P	1	0	0	1	$\frac{1}{2}$	140
x_1	0	1	0	1	$-\frac{1}{2}$	20
x_2	0	0	1	$-\frac{1}{2}$	$\frac{1}{2}$	20

3.

BV	P	x_1	x_2	s_1	s_2	RHS
P	1	0	$\frac{12}{7}$	0	$\frac{32}{7}$	$\frac{256}{7}$
s_1	0	0	$\frac{1}{14}$	1	$-\frac{1}{7}$	$\frac{186}{21}$
x_1	0	1	$\frac{12}{7}$	0	$\frac{4}{7}$	$\frac{32}{7}$

4.

BV	P	x_1	x_2	s_1	s_2	RHS
P	1	-8	-12	0	0	0
s_1	0	$\frac{1}{4}$	$\frac{1}{2}$	1	0	10
s_2	0	$\frac{7}{4}$	3	0	1	8

5.

BV	P	x_1	x_2	s_1	s_2	RHS
P	1	5	-10	12	4	20
x_1	0	1	-2	0	4	24
s_1	0	0	-2	1	4	36

6.

BV	P	x_1	x_2	s_1	s_2	RHS
P	1	-2	-5	0	0	0
s_1	0	1	3	1	0	30
s_2	0	2	1	0	1	12

In Problems 7–22 use the simplex method to solve each maximum linear programming problem.

7. Maximize
$$P = 5x_1 + 7x_2$$
subject to
$$2x_1 + 3x_2 \leq 12$$
$$3x_1 + x_2 \leq 12$$
$$x_1 \geq 0 \qquad x_2 \geq 0$$

8. Maximize
$$P = x_1 + 5x_2$$
subject to
$$2x_1 + x_2 \leq 10$$
$$x_1 + 2x_2 \leq 10$$
$$x_1 \geq 0 \qquad x_2 \geq 0$$

9. Maximize
$$P = 5x_1 + 7x_2$$
subject to
$$x_1 + 2x_2 \leq 2$$
$$2x_1 + x_2 \leq 2$$
$$x_1 \geq 0 \qquad x_2 \geq 0$$

10. Maximize

$$P = 5x_1 + 4x_2$$

subject to

$$x_1 + x_2 \le 2$$
$$2x_1 + 3x_2 \le 6$$
$$x_1 \ge 0 \qquad x_2 \ge 0$$

11. Maximize

$$P = 3x_1 + x_2$$

subject to

$$x_1 + x_2 \le 2$$
$$2x_1 + 3x_2 \le 12$$
$$3x_1 + x_2 \le 12$$
$$x_1 \ge 0 \qquad x_2 \ge 0$$

12. Maximize

$$P = 3x_1 + 5x_2$$

subject to

$$2x_1 + x_2 \le 4$$
$$x_1 + 2x_2 \le 6$$
$$x_1 \ge 0 \qquad x_2 \ge 0$$

13. Maximize

$$P = 2x_1 + x_2 + x_3$$

subject to

$$-2x_1 + x_2 - 2x_3 \le 4$$
$$x_1 - 2x_2 + x_3 \le 2$$
$$x_1 \ge 0 \qquad x_2 \ge 0 \qquad x_3 \ge 0$$

14. Maximize

$$P = 4x_1 + 2x_2 + 5x_3$$

subject to

$$x_1 + 3x_2 + 2x_3 \le 30$$
$$2x_1 + x_2 + 3x_3 \le 12$$
$$x_1 \ge 0 \qquad x_2 \ge 0 \qquad x_3 \ge 0$$

15. Maximize

$$P = 2x_1 + x_2 + 3x_3$$

subject to

$$x_1 + 2x_2 + x_3 \le 25$$
$$3x_1 + 2x_2 + 3x_3 \le 30$$
$$x_1 \ge 0 \qquad x_2 \ge 0 \qquad x_3 \ge 0$$

16. Maximize

$$P = 6x_1 + 3x_2 + 2x_3$$

subject to

$$2x_1 + 2x_2 + 3x_3 \le 30$$
$$2x_1 + 2x_2 + x_3 \le 12$$
$$x_1 \ge 0 \qquad x_2 \ge 0 \qquad x_3 \ge 0$$

17. Maximize

$$P = 2x_1 + 4x_2 + x_3 + x_4$$

subject to

$$2x_1 + x_2 + 2x_3 + 3x_4 \le 12$$
$$2x_2 + x_3 + 2x_4 \le 20$$
$$2x_1 + x_2 + 4x_3 \le 16$$
$$x_1 \ge 0 \qquad x_2 \ge 0 \qquad x_3 \ge 0 \qquad x_4 \ge 0$$

18. Maximize

$$P = 2x_1 + 4x_2 + x_3$$

subject to

$$-x_1 + 2x_2 + 3x_3 \le 6$$
$$-x_1 + 4x_2 + 5x_3 \le 5$$
$$-x_1 + 5x_2 + 7x_3 \le 7$$
$$x_1 \ge 0 \qquad x_2 \ge 0 \qquad x_3 \ge 0$$

19. Maximize

$$P = 2x_1 + x_2 + x_3$$

subject to

$$x_1 + 2x_2 + 4x_3 \le 20$$
$$2x_1 + 4x_2 + 4x_3 \le 60$$
$$3x_1 + 4x_2 + x_3 \le 90$$
$$x_1 \ge 0 \qquad x_2 \ge 0 \qquad x_3 \ge 0$$

20. Maximize

$$P = x_1 + 2x_2 + 4x_3$$

subject to

$$8x_1 + 5x_2 - 4x_3 \le 30$$
$$-2x_1 + 6x_2 + x_3 \le 5$$
$$-2x_1 + 2x_2 + x_3 \le 15$$
$$x_1 \ge 0 \qquad x_2 \ge 0 \qquad x_3 \ge 0$$

21. Maximize

$$P = x_1 + 2x_2 + 4x_3 - x_4$$

subject to

$$5x_1 + 4x_3 + 6x_4 \le 20$$
$$4x_1 + 2x_2 + 2x_3 + 8x_4 \le 40$$
$$x_1 \ge 0 \qquad x_2 \ge 0 \qquad x_3 \ge 0 \qquad x_4 \ge 0$$

22. Maximize

$$P = x_1 + 2x_2 - x_3 + 3x_4$$

subject to

$$2x_1 + 4x_2 + 5x_3 + 6x_4 \le 24$$
$$4x_1 + 4x_2 + 2x_3 + 2x_4 \le 4$$

$$x_1 \ge 0 \qquad x_2 \ge 0 \qquad x_3 \ge 0 \qquad x_4 \ge 0$$

23. **Process Utilization**   A jean manufacturer makes three types of jeans, each of which goes through three manufacturing phases—cutting, sewing, and finishing. The number of minutes each type of product requires in each of the three phases is given below:

Jean	Cutting	Sewing	Finishing
I	8	12	4
II	12	18	8
III	18	24	12

There are 5200 minutes of cutting time, 6000 minutes of sewing time, and 2200 minutes of finishing time each day. The company can sell all the jeans it makes and make a profit of $3 on each Jean I, $4.50 on each Jean II, and $6 on each Jean III. Determine the number of jeans in each category it should make each day to maximize its profits.

24. **Process Utilization**   A company manufactures three types of toys A, B, and C. Each requires rubber, plastic, and aluminum as listed below:

Toy	Rubber	Plastic	Aluminum
A	2	2	4
B	1	2	2
C	1	2	4

The company has available 600 units of rubber, 800 units of plastic, and 1400 units of aluminum. The company makes a profit of $4, $3, and $2 on toys A, B, and C, respectively. Assuming all toys manufactured can be sold, determine a production order so that profit is maximum.

25. **Scheduling**   Products A, B, and C are sold door-to-door. Product A costs $3 per unit, takes 10 minutes to sell (on the average), and costs $0.50 to deliver to the customer. Product B costs $5, takes 15 minutes to sell, and is left with the customer at the time of sale. Product C costs $4, takes 12 minutes to sell, and costs $1.00 to deliver. During any week a salesperson is allowed to draw up to $500 worth of A, B, and C (at cost) and is allowed delivery expenses not to exceed $75. If a salesperson's selling time is not expected to exceed 30 hours (1800 minutes) in a week, and if the salesperson's profit (net after all ex-

penses) is $1 each on a unit of A or B and $2 on a unit of C, what combination of sales of A, B, and C will lead to maximum profit and what is this maximum profit?

26. **Resource Allocations**   Suppose that a large hospital classifies its surgical operations into three categories according to their length and charges a fee of $600, $900, and $1200, respectively, for each of the categories. The average time of the operations in the three categories is 30 minutes, 1 hour, and 2 hours, respectively; and the hospital has four operating rooms, each of which can be used for 10 hours per day. If the total number of operations cannot exceed 60, how many of each type should the hospital schedule to maximize its revenues?

27. **Mixture**   The Lee refinery blends high and low octane gasoline into three intermediate grades: regular, premium, and super premium. The regular grade consists of 60% high octane and 40% low octane, the premium consists of 70% high octane and 30% low octane, and the super premium consists of 80% high octane and 20% low octane. The company has available 140,000 gallons of high octane and 120,000 gallons of low octane, but can mix only 225,000 gallons. Regular gas sells for $1.20 per gallon, premium sells for $1.30 per gallon, and super premium sells for $1.40 per gallon. How many gallons of each grade should the company mix in order to maximize revenues?

28. **Mixture**   Repeat Problem 27 under the assumption that the combined total number of gallons produced by the refinery cannot exceed 200,000 gallons.

29. **Investment**   A financial consultant has at most $90,000 to invest in stocks, corporate bonds, and municipal bonds. The average yields for stocks, corporate bonds, and municipal bonds is 10%, 8%, and 6%, respectively. Determine how much she should invest in each security to maximize the return on her investments, if she has decided that her investment in stocks should not exceed half her funds, and that twice her investment in corporate bonds should not exceed her investment in municipal bonds.

30. **Investment**   Repeat Problem 29 under the assumption that no more than $25,000 can be invested in stocks.

31. **Crop Planning**   A farmer has at most 200 acres of farm-land suitable for cultivating crops A, B, and C. The costs for cultivating crops A, B, and C are $40, $50, and $30 per acre, respectively. The farmer has a maximum of $18,000 available for land cultivation. Crops A, B, and C require 20, 30, and 15 hours per acre of labor, respectively, and there is a maximum of 4200 hours of labor available. If the farmer expects to make a profit of $70,

$90, and $50 per acre on crops A, B, and C, respectively, how many acres of each crop should he plant in order to maximize his profit?

32. Crop Planning   Repeat Problem 31 if the farmer modified his allocations as follows:

	Cost of Cultivating			Maximum Available
	Crop A	**Crop B**	**Crop C**	
Cost	$30	$40	$20	$12,000
Hours	10	20	18	3,600
Profit	$50	$60	$40	

33. Mixture Problem   Nutt's Nut Company has 500 pounds of peanuts, 100 pounds of pecans, and 50 pounds of cashews on hand. They package three types of 5-pound cans of nuts: can I contains 3 pounds peanuts, 1 pound pecans, and 1 pound cashews; can II contains 4 pounds peanuts, $\frac{1}{2}$ pound pecans, and $\frac{1}{2}$ pound cashews; and can III contains 5 pounds peanuts. The selling price is $28 for can I, $24 for can II, and $20 for can III. How many cans of each kind should be made to maximize revenue?

34. Maximizing Profit   One of the methods used by the Alexander Company to separate copper, lead, and zinc from ores is the flotation separation process. This process consists of three steps: oiling, mixing, and separation. These steps must be applied for 2, 2, and 1 hour, respectively, to produce 1 unit of copper; 2, 3, and 1 hour, respectively, to produce 1 unit of lead; and 1, 1, and 3 hours, respectively, to produce 1 unit of zinc. The oiling and separation phases of the process can be in operation for a maximum of 10 hours a day, while the mixing phase can be in operation for a maximum of 11 hours a day. The Alexander Company makes a profit of $45 per unit of copper, $30 per unit of lead, and $35 per unit of zinc. The demand for these metals is unlimited. How many units of each metal should be produced daily by use of the flotation process to achieve the highest profit?

35. Maximizing Profit   A wood cabinet manufacturer produces cabinets for television consoles, stereo systems, and radios, each of which must be assembled, decorated, and crated. Each television console requires 3 hours to assemble, 5 hours to decorate, and 0.1 hour to crate and returns a profit of $10. Each stereo system requires 10 hours to assemble, 8 hours to decorate, and 0.6 hour to crate and returns a profit of $25. Each radio requires 1 hour to assemble, 1 hour to decorate, and 0.1 hour to crate and returns a profit of $3. The manufacturer has 30,000, 40,000, and 120 hours available weekly for assembling, decorating, and crating, respectively. How many units of each product should be manufactured to maximize profit?

36. Maximizing Profit   The finishing process in the manufacture of cocktail tables and end tables requires sanding, staining, and varnishing. The time in minutes required for each finishing process is given below:

	Sanding	Staining	Varnishing
End table	8	10	4
Cocktail table	4	4	8

The equipment required for each process is used on one table at a time and is available for 6 hours each day. If the profit on each cocktail table is $20 and on each end table is $15, how many of each should be manufactured each day in order to maximize profit?

37. Maximizing Profit   A large TV manufacturer has warehouse facilities for storing its 25-inch color TVs in Chicago, New York, and Denver. Each month the city of Atlanta is shipped at most four hundred 25-inch TVs. The cost of transporting each TV to Atlanta from Chicago, New York, and Denver averages $20, $20, and $40, respectively, while the cost of labor required for packing averages $6, $8, and $4, respectively. Suppose $10,000 is allocated each month for transportation costs and $3000 is allocated for labor costs. If the profit on each TV made in Chicago is $50, in New York is $80, and Denver is $40, how should monthly shipping arrangements be scheduled to maximize profit?

Technology Exercises

A graphing calculator can be programmed to perform pivoting using elementary row operations. In Problems 1–4 use your graphing calculator to solve each maximum linear programming problem by the simplex method.

1. Maximize

$$P = 5x_1 + 3x_2$$

subject to

$$2x_1 + 3x_2 \le 125$$
$$3x_1 + x_2 \le 75$$
$$x_1 \ge 0$$
$$x_2 \ge 0$$

2. Maximize

$$P = x_1 + 5x_2$$

subject to

$$x_1 + 3x_2 \leq 50$$
$$3x_1 + x_2 \leq 75$$
$$x_1 \geq 0$$
$$x_2 \geq 0$$

3. Maximize

$$P = 5x_1 + 7x_2$$

subject to

$$x_1 + 2x_2 \leq 2$$
$$2x_1 + x_2 \leq 2$$
$$x_1 \geq 0$$
$$x_2 \geq 0$$

4. Maximize

$$P = 3.25x_1 + 4.25x_2$$

subject to

$$1.25x_1 + 1.75x_2 \leq 2.50$$
$$3.75x_1 + 1.50x_2 \leq 7.50$$
$$x_1 \geq 0$$
$$x_2 \geq 0$$

4.3 SOLVING MINIMUM PROBLEMS IN STANDARD FORM; THE DUALITY PRINCIPLE*

Standard Form of a Minimum Problem

A linear programming problem in which the objective function is to be minimized is referred to as a **minimum linear programming problem.** Such problems are said to be in **standard form** provided the following three conditions are met:

Standard Form of a Minimum Linear Programming Problem

Condition 1 All the variables are nonnegative.

Condition 2 All other constraints are written as linear expressions that are greater than or equal to a constant.

Condition 3 The objective function must be expressed as a linear expression with nonnegative coefficients.

EXAMPLE 1

Determine which of the following minimum problems are in standard form.

SOLUTION

(a) Minimize

$$C = 2x_1 + 3x_2$$

subject to the constraints

$$x_1 + 3x_2 \geq 24$$
$$2x_1 + x_2 \geq 18$$
$$x_1 \geq 0 \qquad x_2 \geq 0$$

(a) Since all three conditions are met, this minimum problem is in standard form.

---

* The solution of general minimum problems is discussed in Section 4.4. If you plan to cover Section 4.4, this section may be omitted without loss of continuity.

SOLUTION

(b) Minimize

$$C = 3x_1 - x_2 + 4x_3$$

subject to the constraints

$$3x_1 + x_2 + x_3 \geq 12$$
$$x_1 + x_2 + x_3 \geq 8$$
$$x_1 \geq 0 \qquad x_2 \geq 0 \qquad x_3 \geq 0$$

(b) Conditions 1 and 2 are met, but Condition 3 is not, since the coefficient of x_2 in the objective function is negative. Thus, this minimum problem is not in standard form.

(c) Minimize

$$C = 2x_1 + x_2 + x_3$$

subject to the constraints

$$x_1 - 3x_2 + x_3 \leq 12$$
$$x_1 + x_2 + x_3 \geq 1$$
$$x_1 \geq 0 \qquad x_2 \geq 0 \qquad x_3 \geq 0$$

(c) Conditions 1 and 3 are met, but Condition 2 is not, since the first constraint

$$x_1 - 3x_2 + x_3 \leq 12$$

is not written with a $\geq$ sign. Thus, the minimum problem as stated is not in standard form. Notice, however, that by multiplying by -1, we can write this constraint as

$$-x_1 + 3x_2 - x_3 \geq -12$$

Written in this way, the minimum problem is in standard form.

(d) Minimize

$$C = 2x_1 + x_2 + 3x_3$$

subject to the constraints

$$-x_1 + 2x_2 + x_3 \geq -2$$
$$x_1 + x_2 + x_3 \geq 6$$
$$x_1 \geq 0 \qquad x_2 \geq 0 \qquad x_3 \geq 0$$

(d) Conditions 1, 2, and 3 are each met, so this minimum problem is in standard form.

Now Work Problem 1

· · · · · · · · · · · · · · · · · · · ·

The Duality Principle

One technique for solving a minimum problem in standard form was developed by John von Neumann and others. The solution (if it exists) is found by solving a related maximum problem, called the **dual problem.** The next example illustrates how to obtain the dual problem.

EXAMPLE 2

Obtain the dual problem of the following minimum problem:

Minimize

$$C = 300x_1 + 480x_2$$

subject to the constraints

$$x_1 + 3x_2 \geq 0.25$$
$$2x_1 + 2x_2 \geq 0.45$$
$$x_1 \geq 0 \qquad x_2 \geq 0$$

SOLUTION

First notice that the minimum problem is in standard form. We begin by writing down a matrix that represents the constraints and the objective function:

$$
\begin{array}{c}
 \\
\text{Constraint} \\
\text{Constraint} \\
\text{Objective function}
\end{array}
\begin{array}{cc}
x_1 & x_2 \\
\end{array}
\left[
\begin{array}{cc|c}
1 & 3 & 0.25 \\
2 & 2 & 0.45 \\
300 & 480 & 0
\end{array}
\right]
$$

Now form the matrix that has as columns the rows of the above matrix:

$$
\left[
\begin{array}{cc|c}
1 & 2 & 300 \\
3 & 2 & 480 \\
0.25 & 0.45 & 0
\end{array}
\right]
$$

This matrix is called the **transpose** of the first matrix. From this matrix, create the following maximum problem:

Maximize

$$P = 0.25y_1 + 0.45y_2$$

subject to the conditions

$$y_1 + 2y_2 \leq 300$$
$$3y_1 + 2y_2 \leq 480$$
$$y_1 \geq 0 \qquad y_2 \geq 0$$

This maximum problem is the dual of the given minimum problem.

· ·

Notice that the dual of a minimum problem in standard form is a maximum problem in standard form that can be solved by using techniques discussed in the previous section. The significance of this is expressed in the following principle:

VON NEUMANN DUALITY PRINCIPLE

The optimal solution, if it exists, of a minimum linear programming problem in standard form has the same value as the optimal solution of the dual problem, a maximum problem in standard form.

So, one way to solve a minimum problem in standard form is to form the dual problem and solve it. Another way to solve minimum problems (even those not in standard form) is given in Section 4.4.

The steps to use for obtaining the dual problem are listed at the top of the next page.

> **Steps for Obtaining the Dual Problem**
>
> **Step 1** Write the minimum problem in standard form.
>
> **Step 2** Construct a matrix that represents the constraints and the objective function.
>
> **Step 3** Interchange the rows and columns to form the matrix of the dual problem.
>
> **Step 4** Translate this matrix into a maximum problem in standard form.

EXAMPLE 3

Find the dual of the following minimum problem:

Minimize

$$C = 2x_1 + 3x_2$$

subject to

$$2x_1 + x_2 \geq 6$$
$$x_1 + 2x_2 \geq 4$$
$$x_1 + x_2 \geq 5$$
$$x_1 \geq 0 \qquad x_2 \geq 0$$

SOLUTION

Observe that the minimum problem is in standard form. The matrix that represents the constraints and the objective function is

$$\begin{bmatrix} 2 & 1 & | & 6 \\ 1 & 2 & | & 4 \\ 1 & 1 & | & 5 \\ 2 & 3 & | & 0 \end{bmatrix}$$

Interchanging rows and columns, we obtain the matrix

$$\begin{bmatrix} 2 & 1 & 1 & | & 2 \\ 1 & 2 & 1 & | & 3 \\ 6 & 4 & 5 & | & 0 \end{bmatrix}$$

This matrix represents the following maximum problem:

Maximize

$$P = 6y_1 + 4y_2 + 5y_3$$

subject to

$$2y_1 + y_2 + y_3 \leq 2$$
$$y_1 + 2y_2 + y_3 \leq 3$$
$$y_1 \geq 0 \qquad y_2 \geq 0 \qquad y_3 \geq 0$$

This maximum problem is in standard form and is the dual problem of the minimum problem.

. . . .

Some observations about Example 3:

1. The variables (x_1, x_2) of the minimum problem are different from the variables of its dual problem (y_1, y_2, y_3).
2. The minimum problem has three constraints and two variables, while the dual problem has two constraints and three variables. (In general, if a minimum problem has m constraints and n variables, its dual problem will have n constraints and m variables.)
3. The inequalities defining the constraints are $\geq$ for the minimum problem and $\leq$ for the maximum problem.
4. Since the coefficients in the minimal objective function are positive, the dual problem has nonnegative numbers to the right of the $\leq$ signs.
5. We follow the custom of denoting an objective function by C (for *Cost*) if it is to be minimized and P (for *Profit*) if it is to be maximized.

Now Work Problem 7

EXAMPLE 4

Solve the maximum problem of Example 3 by the simplex method and thereby obtain the solution for the minimum problem.

SOLUTION

We introduce slack variables s_1 and s_2 to get

$$2y_1 + y_2 + y_3 + s_1 \qquad = 2$$
$$y_1 + 2y_2 + y_3 \qquad + s_2 = 3$$

The initial simplex tableau is

BV	P	y_1	y_2	y_3	s_1	s_2	RHS
P	1	-6	-4	-5	0	0	0
s_1	0	②	1	1	1	0	2
s_2	0	1	2	1	0	1	3

The pivot element, 2, is circled. After pivoting, we obtain this tableau:

BV	P	y_1	y_2	y_3	s_1	s_2	RHS
P	1	0	-1	-2	3	0	6
y_1	0	1	$\frac{1}{2}$	$(\frac{1}{2})$	$\frac{1}{2}$	0	1
s_2	0	0	$\frac{3}{2}$	$\frac{1}{2}$	$-\frac{1}{2}$	1	2

The pivot element, $\frac{1}{2}$, is circled. After pivoting, we obtain this tableau:

BV	P	y_1	y_2	y_3	s_1	s_2	RHS
P	1	4	1	0	5	0	10
y_3	0	2	1	1	1	0	2
s_2	0	-1	1	0	-1	1	1

This is a final tableau so an optimal solution has been obtained. We read from it that the solution to the maximum problem is

$$P = 10 \qquad y_1 = 0 \qquad y_2 = 0 \qquad y_3 = 2$$

The duality principle states that the minimum value of the objective function in the original problem is the same as the maximum value in the dual; that is,

$$C = 10$$

But which values of x_1 and x_2 will yield this minimum value? There are some details of the duality principle and its application that we will omit here. These concern the relationships between the variables of the original problem and the slack variables used in the solution of the dual problem. As a consequence of these relationships, the entire minimal solution can be read from the right end of the objective row of the final tableau:

$$x_1 = 5 \qquad x_2 = 0 \qquad C = 10$$

Notice in the solution to Example 4 that the value of x_1 is found in the objective row in the column corresponding to s_1, and x_2 is similarly found in the objective row in the column corresponding to s_2.

We summarize how to solve a minimum linear programming problem below.

Solving a Minimum Problem in Standard Form

Step 1 Write the dual (maximum) problem.

Step 2 Solve this maximum problem by the simplex method.

Step 3 Read the optimal solution for the maximum problem from the objective row of the final simplex tableau.

Step 4 The minimum value of the objective function (C) will appear in the upper right corner of the final tableau; it is equal to the maximum value of the dual objective function (P).

EXAMPLE 5

Minimize

$$C = 6x_1 + 8x_2 + x_3$$

subject to

$$3x_1 + 5x_2 + 3x_3 \geq 20$$
$$x_1 + 3x_2 + 2x_3 \geq 9$$
$$6x_1 + 2x_2 + 5x_3 \geq 30$$
$$x_1 \geq 0 \qquad x_2 \geq 0 \qquad x_3 \geq 0$$

SOLUTION

This minimum problem is in standard form. The matrix representing this problem is

$$\begin{bmatrix} 3 & 5 & 3 & | & 20 \\ 1 & 3 & 2 & | & 9 \\ 6 & 2 & 5 & | & 30 \\ 6 & 8 & 1 & | & 0 \end{bmatrix}$$

We interchange rows and columns to get

$$\begin{bmatrix} 3 & 1 & 6 & | & 6 \\ 5 & 3 & 2 & | & 8 \\ 3 & 2 & 5 & | & 1 \\ 20 & 9 & 30 & | & 0 \end{bmatrix}$$

The dual problem is:

Maximize

$$P = 20y_1 + 9y_2 + 30y_3$$

subject to

$$3y_1 + y_2 + 6y_3 \le 6$$
$$5y_1 + 3y_2 + 2y_3 \le 8$$
$$3y_1 + 2y_2 + 5y_3 \le 1$$
$$y_1 \ge 0 \qquad y_2 \ge 0 \qquad y_3 \ge 0$$

We introduce slack variables s_1, s_2, and s_3. The initial tableau for this problem is

BV	P	y_1	y_2	y_3	s_1	s_2	s_3	RHS
P	1	-20	-9	-30	0	0	0	0
s_1	0	3	1	6	1	0	0	6
s_2	0	5	3	2	0	1	0	8
s_3	0	3	2	5	0	0	1	1

The final tableau (as you should verify) is

BV	P	y_1	y_2	y_3	s_1	s_2	s_3	RHS
P	1	0	$\frac{13}{3}$	$\frac{10}{3}$	0	0	$\frac{20}{3}$	$\frac{20}{3}$
s_1	0	0	-1	1	1	0	-1	5
s_2	0	0	$-\frac{1}{3}$	$-\frac{19}{3}$	0	1	$-\frac{5}{3}$	$\frac{19}{3}$
y_1	0	1	$\frac{2}{3}$	$\frac{5}{3}$	0	0	$\frac{1}{3}$	$\frac{1}{3}$

The solution to the maximum problem is

$$P = \tfrac{20}{3} \qquad y_1 = \tfrac{1}{3} \qquad y_2 = 0 \qquad y_3 = 0$$

For the minimum problem, the values of x_1, x_2, and x_3 are read as the entries in the objective row in the columns under s_1, s_2, and s_3, respectively. Hence, the solution to the minimum problem is

$$x_1 = 0 \qquad x_2 = 0 \qquad x_3 = \tfrac{20}{3}$$

and the minimum value is $C = \tfrac{20}{3}$.

....

EXERCISE 4.3   Answers to odd-numbered problems begin on page AN-20.

In Problems 1–6 determine which of the given minimum problems are in standard form.

1. Minimize
$$C = 2x_1 + 3x_2$$
subject to the constraints
$$4x_1 - x_2 \geq 2$$
$$x_1 + x_2 \geq 1$$
$$x_1 \geq 0 \qquad x_2 \geq 0$$

2. Minimize
$$C = 3x_1 + 5x_2$$
subject to the constraints
$$3x_1 - x_2 \geq 4$$
$$x_1 - 2x_2 \geq 3$$
$$x_1 \geq 0 \qquad x_2 \geq 0$$

3. Minimize
$$C = 2x_1 - x_2$$
subject to the constraints
$$2x_1 - x_2 \geq 1$$
$$-2x_1 \geq -3$$
$$x_1 \geq 0 \qquad x_2 \geq 0$$

4. Minimize
$$C = 2x_1 + 3x_2$$
subject to the constraints
$$x_1 - x_2 \leq 3$$
$$2x_1 + 3x_2 \geq 4$$
$$x_1 \geq 0 \qquad x_2 \geq 0$$

5. Minimize
$$C = 3x_1 + 7x_2 + x_3$$
subject to the constraints
$$x_1 + x_3 \leq 6$$
$$2x_1 + x_2 \geq 4$$
$$x_1 \geq 0 \qquad x_2 \geq 0 \qquad x_3 \geq 0$$

6. Minimize
$$C = x_1 - x_2 + x_3$$
subject to the constraints
$$x_1 + x_2 \geq 6$$
$$2x_1 - x_3 \geq 4$$
$$x_1 \geq 0 \qquad x_2 \geq 0 \qquad x_3 \geq 0$$

In Problems 7–10 write the dual problem for each minimum linear programming problem.

7. Minimize
$$C = 2x_1 + 3x_2$$
subject to
$$x_1 + x_2 \geq 2$$
$$2x_1 + 3x_2 \geq 6$$
$$x_1 \geq 0 \qquad x_2 \geq 0$$

8. Minimize
$$C = 3x_1 + 4x_2$$
subject to
$$2x_1 + x_2 \geq 2$$
$$2x_1 + x_2 \geq 6$$
$$x_1 \geq 0 \qquad x_2 \geq 0$$

9. Minimize
$$C = 3x_1 + x_2 + x_3$$
subject to
$$x_1 + x_2 + x_3 \geq 5$$
$$2x_1 + x_2 \geq 4$$
$$x_1 \geq 0 \qquad x_2 \geq 0 \qquad x_3 \geq 0$$

10. Minimize
$$C = 2x_1 + x_2 + x_3$$
subject to
$$2x_1 + x_2 + x_3 \geq 4$$
$$x_1 + 2x_2 + x_3 \geq 6$$
$$x_1 \geq 0 \qquad x_2 \geq 0 \qquad x_3 \geq 0$$

In Problems 11–18 solve each minimum linear programming problem.

11. Minimize

$$C = 6x_1 + 3x_2$$

subject to

$$x_1 + x_2 \geq 2$$
$$2x_1 + 6x_2 \geq 6$$
$$x_1 \geq 0 \qquad x_2 \geq 0$$

12. Minimize

$$C = 3x_1 + 4x_2$$

subject to

$$x_1 + x_2 \geq 3$$
$$2x_1 + x_2 \geq 4$$
$$x_1 \geq 0 \qquad x_2 \geq 0$$

13. Minimize

$$C = 6x_1 + 3x_2$$

subject to

$$x_1 + x_2 \geq 4$$
$$3x_1 + 4x_2 \geq 12$$
$$x_1 \geq 0 \qquad x_2 \geq 0$$

14. Minimize

$$C = 2x_1 + 3x_2 + 4x_3$$

subject to

$$x_1 - 2x_2 - 3x_3 \geq -2$$
$$x_1 + x_2 + x_3 \geq 2$$
$$2x_1 \qquad + x_3 \geq 3$$
$$x_1 \geq 0 \qquad x_2 \geq 0 \qquad x_3 \geq 0$$

15. Minimize

$$C = x_1 + 2x_2 + x_3$$

subject to

$$x_1 - 3x_2 + 4x_3 \geq 12$$
$$3x_1 + x_2 + 2x_3 \geq 10$$
$$x_1 - x_2 - x_3 \geq -8$$
$$x_1 \geq 0 \qquad x_2 \geq 0 \qquad x_3 \geq 0$$

16. Minimize

$$C = x_1 + 2x_2 + 4x_3$$

subject to

$$x_1 - x_2 + 3x_3 \geq 4$$
$$2x_1 + 2x_2 - 3x_3 \geq 6$$
$$-x_1 + 2x_2 + 3x_3 \geq 2$$
$$x_1 \geq 0 \qquad x_2 \geq 0 \qquad x_3 \geq 0$$

17. Minimize

$$C = x_1 + 4x_2 + 2x_3 + 4x_4$$

subject to

$$x_1 + x_3 \geq 1$$
$$x_2 + x_4 \geq 1$$
$$-x_1 - x_2 - x_3 - x_4 \geq -3$$
$$x_1 \geq 0 \qquad x_2 \geq 0 \qquad x_3 \geq 0 \qquad x_4 \geq 0$$

18. Minimize

$$C = x_1 + 2x_2 + 3x_3 + 4x_4$$

subject to

$$x_1 + x_3 \geq 1$$
$$x_2 + x_4 \geq 1$$
$$-x_1 - x_2 - x_3 - x_4 \geq -3$$
$$x_1 \geq 0 \qquad x_2 \geq 0 \qquad x_3 \geq 0 \qquad x_4 \geq 0$$

19. Diet Preparation Mr. Jones needs to supplement his diet with at least 50 mg calcium and 8 mg iron daily. The minerals are available in two types of vitamin pills, P and Q. Pill P contains 5 mg calcium and 2 mg iron, while Pill Q contains 10 mg calcium and 1 mg iron. If each P pill costs 3 cents and each Q pill costs 4 cents, how could Mr. Jones minimize the cost of adding the minerals to his diet? What would the daily minimum cost be?

20. Production Schedule A company owns two mines. Mine A produces 1 ton of high-grade ore, 3 tons of medium-grade ore, and 5 tons of low-grade ore each day. Mine B produces 2 tons of each grade ore per day. The company needs at least 80 tons of high-grade ore, at least 160 tons of medium-grade ore, and at least 200 tons of low-grade ore. How many days should each mine be operated to minimize costs if it costs $2000 per day to operate each mine?

21. Production Schedule Argus Company makes three products: A, B, and C. Each unit of A costs $4, each unit of B costs $2, and each unit of C costs $1 to produce. Argus must produce at least 20 As, 30 Bs, and 40 Cs, and cannot produce fewer than 200 total units of As, Bs, and Cs combined. Minimize Argus's costs.

22. Diet Planning A health clinic dietician is planning a meal consisting of three foods whose ingredients are summarized as follows:

	One Unit of		
	Food I	**Food II**	**Food III**
Units of protein	10	15	20
Units of carbohydrates	1	2	1
Units of iron	4	8	1
Calories	80	120	100

The dietician wishes to determine the number of units of each food to use to create a balanced meal containing at least 40 units of protein, 6 units of carbohydrates, and 12 units of iron, with as few calories as possible.

23. Menu Planning Fresh Starts Catering offers the following lunch menu:

Menu		
Lunch #1	Soup, salad, sandwich	$6.20
Lunch #2	Salad, pasta	$7.40
Lunch #3	Salad, sandwich, pasta	$9.10

Mrs. Mintz and her friends would like to order 4 bowls of soup, 9 salads, 6 sandwiches, and 5 orders of pasta and keep the cost as low as possible. Compose their order.

24. Inventory Control A department store stocks three brands of toys: A, B, and C. Each unit of brand A occupies 1 square foot of shelf space, each unit of brand B occupies 2 square feet, and each unit of brand C occupies 3 square feet. The store has 120 square feet available for storage. Surveys show that the store should have on hand at least 12 units of brand A and at least 30 units of A and B combined. Each unit of brand A costs the store $8, each unit of brand B $6, and each unit of brand C $10. Minimize the cost to the store.

Technology Exercises
A graphing calculator can be programmed to perform pivoting using elementary row operations. In Problems 1–4 use your graphing calculator to solve each minimum linear programming problem by the simplex method.

1. Minimize
$$C = 5x_1 + 3x_2$$
subject to
$$2x_1 + 3x_2 \geq 125$$
$$3x_1 + x_2 \geq 75$$
$$x_1 \geq 0$$
$$x_2 \geq 0$$

2. Minimize
$$C = x_1 + 5x_2$$
subject to
$$x_1 + 3x_2 \geq 53$$
$$3x_1 + x_2 \geq 74$$
$$x_1 \geq 0$$
$$x_2 \geq 0$$

3. Minimize
$$C = 5x_1 + 7x_2$$
subject to
$$x_1 + 2x_2 \geq 2$$
$$2x_1 + x_2 \geq 2$$
$$x_1 \geq 0$$
$$x_2 \geq 0$$

4. Minimize
$$C = 3.25x_1 + 4.25x_2$$
subject to
$$1.25x_1 + 1.75x_2 \geq 2.50$$
$$3.75x_1 + 1.50x_2 \geq 7.50$$
$$x_1 \geq 0$$
$$x_2 \geq 0$$

4.4 THE SIMPLEX METHOD WITH MIXED CONSTRAINTS: PHASE I/PHASE II

Thus far, we have developed the simplex method only for solving linear programming problems in standard form. In this section we develop the simplex method for linear programming problems that cannot be written in standard form.

The Simplex Method with Mixed Constraints

Recall that for a maximum problem in standard form each constraint must be of the form

$$a_1x_1 + a_2x_2 + \cdots + a_nx_n \le b_1 \qquad b_1 > 0$$

That is, each is a linear expression *less than or equal to a positive constant.* When the constraints are of any other form (greater than or equal to, or equal to) we have what are called **mixed constraints.** The following example illustrates the simplex method for solving these problems with mixed constraints.

EXAMPLE 1

Maximize

$$P = 20x_1 + 15x_2$$

subject to the constraints

$$\begin{aligned}
x_1 + x_2 &\ge 7 \\
9x_1 + 5x_2 &\le 45 \\
2x_1 + x_2 &\ge 8 \\
x_1 \ge 0 \qquad x_2 &\ge 0
\end{aligned}$$

SOLUTION

We first observe this is a maximum problem that is not in standard form. Second, it cannot be modified so as to be in standard form.

> **Step 1** Write each constraint except the nonnegative constraints as an inequality with the variables on the left side of a $\le$ sign.

To do this, we merely multiply the first and third inequality by -1. The result is that the constraints become

$$\begin{aligned}
-x_1 - x_2 &\le -7 \\
9x_1 + 5x_2 &\le 45 \\
-2x_1 - x_2 &\le -8 \\
x_1 \ge 0 \qquad x_2 &\ge 0
\end{aligned}$$

> **Step 2** Introduce nonnegative slack variables on the left side of each inequality to form an equality.

To do this, we will use the slack variables s_1, s_2, s_3 to obtain

$$\begin{aligned}
-x_1 - x_2 + s_1 \qquad\qquad &= -7 \\
9x_1 + 5x_2 \qquad + s_2 \qquad &= 45 \\
-2x_1 - x_2 \qquad\qquad + s_3 &= -8 \\
s_1 \ge 0 \qquad s_2 \ge 0 \qquad s_3 &\ge 0
\end{aligned}$$

Step 3 Set up the initial simplex tableau.

BV	P	x_1	x_2	s_1	s_2	s_3	RHS
P	1	-20	-15	0	0	0	0
s_1	0	-1	-1	1	0	0	-7
s_2	0	9	5	0	1	0	45
s_3	0	-2	-1	0	0	1	-8

This initial tableau represents the solution $x_1 = 0$, $x_2 = 0$, $s_1 = -7$, $s_2 = 45$, $s_3 = -8$. This is not a feasible point. The reason for this lies in the existence of the two negative entries in the right-hand column. That is, this tableau represents a solution that causes two of the variables to be negative, in violation of the non-negativity requirement. Whenever this occurs, the simplex algorithm consists of two phases.

Phase I or Phase II

Step 4 Determine whether Phase I or Phase II applies. Phase I is used whenever negative entries appear in the right-hand column (RHS) of the constraint equations; Phase II is used whenever all the entries in the right-hand column (RHS) of the constraint equations are nonnegative. In determining whether Phase I or Phase II applies, *the objective row is ignored.*

Select the Pivot Element

Step 5 *Phase I:* The pivot row is the row with the most negative value in the right column (RHS) of the constraint equations. (The objective row is ignored and ties are possible.) If all the other entries in the pivot row are nonnegative, there are no feasible solutions, and the problem has no solution.

For each column with a negative entry in the pivot row, form the quotient of the entry in the last column (RHS) divided by the corresponding negative entry in the pivot row. The smallest of these quotients is the pivot column.

Phase II: Follow the pivoting strategy given on page 180.

For our example we use Phase I. The pivot row corresponds to the variable s_3 because of the -8 in the RHS column. To find the pivot column, we form the quotients $(-8)/(-2) = 4$ and $(-8)/(-1) = 8$, the smallest being 4. Thus the pivot column corresponds to the variable x_1. As a result of using this pivot element, x_1 will become a basic variable and s_3 will become a nonbasic variable.

Step 6 Go back to Step 4 unless a final tableau has been reached.

For our ex ample, after pivoting we get the tableau

BV	P	x_1	x_2	s_1	s_2	s_3	RHS
P	1	0	-5	0	0	-10	80
s_1	0	0	$\left(-\frac{1}{2}\right)$	1	0	$\left(-\frac{1}{2}\right)$	-3
s_2	0	0	$\frac{1}{2}$	0	1	$\frac{9}{2}$	9
x_1	0	1	$\frac{1}{2}$	0	0	$-\frac{1}{2}$	4

$$\frac{-3}{-\frac{1}{2}} = 6$$

Step 4 Phase I applies because of the negative entry, -3, in the RHS column.

Step 5 The pivot row corresponds to the variable s_1; the pivot column corresponds to either the variable x_2 or s_3. We'll select x_2. After pivoting we get the tableau

BV	P	x_1	x_2	s_1	s_2	s_3	RHS
P	1	0	0	-10	0	-5	110
x_2	0	0	1	-2	0	1	6
s_2	0	0	0	1	1	4	6
x_1	0	1	0	(1)	0	-1	1

$$\frac{6}{1} = 6$$
$$\frac{1}{1} = 1$$

Step 4 Phase II applies because there are no negative entries in the RHS column.

Step 5 The pivot column corresponds to the variable s_1; the pivot row corresponds to the variable x_1. After pivoting we get the tableau

BV	P	x_1	x_2	s_1	s_2	s_3	RHS
P	1	10	0	0	0	-15	120
x_2	0	2	1	0	0	-1	8
s_2	0	-1	0	0	1	(5)	5
s_1	0	1	0	1	0	-1	1

$$\frac{5}{5} = 1$$

Step 4 Phase II applies because there are no negative entries in the RHS column.

Step 5 The pivot column corresponds to the variable s_3; the pivot row corresponds to the variable s_2. After pivoting we get the tableau

BV	P	x_1	x_2	s_1	s_2	s_3	RHS
P	1	7	0	0	3	0	135
x_2	0	$\frac{9}{5}$	1	0	$\frac{1}{5}$	0	9
s_3	0	$-\frac{1}{5}$	0	0	$\frac{1}{5}$	1	1
s_1	0	$\frac{4}{5}$	0	1	$\frac{1}{5}$	0	2

This is a final tableau. The maximum value of P is 135, and it is achieved when $x_1 = 0$, $x_2 = 9$, $s_1 = 2$, $s_2 = 0$, $s_3 = 1$.

Now Work Problem 1

The Minimum Problem

In general, a minimum problem can be changed to a maximum problem by realizing that in order to minimize z we can instead maximize $-z$. The following example illustrates this method.

EXAMPLE 2

Minimize
$$z = 5x_1 + 6x_2$$
subject to the constraints
$$x_1 + x_2 \leq 10$$
$$x_1 + 2x_2 \geq 12$$
$$2x_1 + x_2 \geq 12$$
$$x_1 \geq 3$$
$$x_1 \geq 0 \qquad x_2 \geq 0$$

SOLUTION

We change our problem from minimizing $z = 5x_1 + 6x_2$ to maximizing $P = -z = -5x_1 - 6x_2$.

Step 1 Write each constraint with $\leq$.
$$x_1 + x_2 \leq 10$$
$$-x_1 - 2x_2 \leq -12$$
$$-2x_1 - x_2 \leq -12$$
$$-x_1 \leq -3$$

Step 2 Introduce nonnegative slack variables to form equalities:
$$x_1 + x_2 + s_1 \qquad\qquad = 10$$
$$-x_1 - 2x_2 \qquad + s_2 \qquad\qquad = -12$$
$$-2x_1 - x_2 \qquad\qquad + s_3 \qquad = -12$$
$$-x_1 \qquad\qquad\qquad + s_4 = -3$$
$$s_1 \geq 0 \qquad s_2 \geq 0 \qquad s_3 \geq 0 \qquad s_4 \geq 0$$

Step 3 Set up the initial simplex tableau

BV	P	x_1	x_2	s_1	s_2	s_3	s_4	RHS
P	1	5	6	0	0	0	0	0
s_1	0	1	1	1	0	0	0	10
s_2	0	-1	$\boxed{-2}$	0	1	0	0	-12
s_3	0	-2	-1	0	0	1	0	-12
s_4	0	-1	0	0	0	0	1	-3

$\frac{-12}{-1} = 12; \frac{-12}{-2} = 6$

$\frac{-12}{-2} = 6; \frac{-12}{-1} = 12$

Step 4 Phase I applies because of the negative entries in the RHS column.

Step 5 The pivot row corresponds to the variable s_2 or s_3. Using s_2, the pivot column corresponds to the variable x_2; using s_3, the pivot column corresponds to the variable x_1. We'll use row s_2, column x_2 as the pivot element. After pivoting we get the tableau

BV	P	x_1	x_2	s_1	s_2	s_3	s_4	RHS
P	1	2	0	0	3	0	0	-36
s_1	0	$\frac{1}{2}$	0	1	$\frac{1}{2}$	0	0	4
x_2	0	$\frac{1}{2}$	1	0	$-\frac{1}{2}$	0	0	6
s_3	0	$\left(-\frac{3}{2}\right)$	0	0	$-\frac{1}{2}$	1	0	-6
s_4	0	-1	0	0	0	0	1	-3

$$\frac{-6}{-\frac{3}{2}} = 4; \quad \frac{-6}{-\frac{1}{2}} = 12$$

Step 4 Phase I applies because of the negative entries in the RHS column.

Step 5 The pivot row corresponds to the variable s_3; the pivot column corresponds to the variable x_1. After pivoting we get the tableau

BV	P	x_1	x_2	s_1	s_2	s_3	s_4	RHS
P	1	0	0	0	$\frac{7}{3}$	$\frac{4}{3}$	0	-44
s_1	0	0	0	1	$\frac{1}{3}$	$\frac{1}{3}$	0	2
x_2	0	0	1	0	$-\frac{2}{3}$	$\frac{1}{3}$	0	4
x_1	0	1	0	0	$\frac{1}{3}$	$-\frac{2}{3}$	0	4
s_4	0	0	0	0	$\frac{1}{3}$	$-\frac{2}{3}$	1	1

This is a final tableau. The maximum value of P is -44, so the minimum value of z is 44. This occurs when $x_1 = 4$, $x_2 = 4$, $s_1 = 2$, $s_2 = 0$, $s_3 = 0$, $s_4 = 1$.

· ·

Thus, to solve a minimum linear programming problem, change it to a maximum linear programming problem as follows:

Steps for Solving a Minimum Problem

Step 1 If z is to be minimized, let $P = -z$.

Step 2 Solve the linear programming problem: Maximize P subject to the same constraints as the minimum problem.

Step 3 Use the principle that

$$\text{Minimum of } z = -\text{Maximum of } P$$

Now Work Problem 5

Equality Constraints

So far, all our constraints used $\leq$ or $\geq$. What can be done if one of the constraints is an equality? One method is to replace the $=$ constraint with the two constraints $\leq$ and $\geq$. The next example illustrates this method.

EXAMPLE 3   Minimize

$$z = 7x_1 + 5x_2 + 6x_3$$

subject to the constraints

$$\begin{aligned} x_1 + x_2 + x_3 &= 10 \\ x_1 + 2x_2 + 3x_3 &\leq 19 \\ 2x_1 + 3x_2 &\geq 21 \\ x_1 \geq 0 \qquad x_2 \geq 0 \qquad x_3 &\geq 0 \end{aligned}$$

SOLUTION

We wish to maximize $P = -z = -7x_1 - 5x_2 - 6x_3$ subject to the constraints

$$\begin{aligned} x_1 + x_2 + x_3 &\leq 10 \\ x_1 + x_2 + x_3 &\geq 10 \\ x_1 + 2x_2 + 3x_3 &\leq 19 \\ 2x_1 + 3x_2 &\geq 21 \\ x_1 \geq 0 \qquad x_2 \geq 0 \qquad x_3 &\geq 0 \end{aligned}$$

Step 1 Rewrite the constraints with $\leq$:

$$\begin{aligned} x_1 + x_2 + x_3 &\leq 10 \\ -x_1 - x_2 - x_3 &\leq -10 \\ x_1 + 2x_2 + 3x_3 &\leq 19 \\ -2x_1 - 3x_2 &\leq -21 \end{aligned}$$

Step 2 Introduce nonnegative slack variables:

$$\begin{aligned} x_1 + x_2 + x_3 + s_1 &= 10 \\ -x_1 - x_2 - x_3 + s_2 &= -10 \\ x_1 + 2x_2 + 3x_3 + s_3 &= 19 \\ -2x_1 - 3x_2 + s_4 &= -21 \\ s_1 \geq 0 \qquad s_2 \geq 0 \qquad s_3 \geq 0 \qquad s_4 &\geq 0 \end{aligned}$$

Step 3 Set up the initial simplex tableau:

BV	P	x_1	x_2	x_3	s_1	s_2	s_3	s_4	RHS
P	1	7	5	6	0	0	0	0	0
s_1	0	1	1	1	1	0	0	0	10
s_2	0	-1	-1	-1	0	1	0	0	-10
s_3	0	1	2	3	0	0	1	0	19
s_4	0	-2	$\boxed{-3}$	0	0	0	0	1	-21

$\frac{-21}{-2} = 10.5; \frac{-21}{-3} = 7$

Step 4 Phase I applies because of the negative entries in the RHS column.

Step 5 The pivot row corresponds to the variable s_4; the pivot column corresponds to the variable x_2. By pivoting we obtain the tableau

BV	P	x_1	x_2	x_3	s_1	s_2	s_3	s_4	RHS
P	1	$\frac{11}{3}$	0	6	0	0	0	$\frac{5}{3}$	-35
s_1	0	$\frac{1}{3}$	0	1	1	0	0	$\frac{1}{3}$	3
s_2	0	$-\frac{1}{3}$	0	$\boxed{-1}$	0	1	0	$-\frac{1}{3}$	-3
s_3	0	$-\frac{1}{3}$	0	3	0	0	1	$\frac{2}{3}$	5
x_2	0	$\frac{2}{3}$	1	0	0	0	0	$-\frac{1}{3}$	7

$$\frac{-3}{-\frac{1}{3}} = 9; \quad \frac{-3}{-1} = 3$$

Step 4 Phase I applies because of the negative entry in the RHS column.

Step 5 The pivot row corresponds to the variable s_2; the pivot column corresponds to the variable x_3. By pivoting we obtain the tableau

BV	P	x_1	x_2	x_3	s_1	s_2	s_3	s_4	RHS
P	1	$\frac{5}{3}$	0	0	0	6	0	$-\frac{1}{3}$	-53
s_1	0	0	0	0	1	1	0	0	0
x_3	0	$\frac{1}{3}$	0	1	0	-1	0	$\frac{1}{3}$	3
s_3	0	$\boxed{-\frac{4}{3}}$	0	0	0	3	1	$-\frac{1}{3}$	-4
x_2	0	$\frac{2}{3}$	1	0	0	0	0	$-\frac{1}{3}$	7

$$\frac{-4}{-\frac{4}{3}} = 3; \quad \frac{-4}{-\frac{1}{3}} = 12$$

Step 4 Phase I applies because of the negative entry in the RHS column.

Step 5 The pivot row corresponds to the variable s_3; the pivot column corresponds to the variable x_1. By pivoting we obtain the tableau

BV	P	x_1	x_2	x_3	s_1	s_2	s_3	s_4	RHS
P	1	0	0	0	0	$\frac{39}{4}$	$\frac{5}{4}$	$-\frac{3}{4}$	-58
s_1	0	0	0	0	1	1	0	0	0
x_3	0	0	0	1	0	$-\frac{1}{4}$	$\frac{1}{4}$	$\boxed{\frac{1}{4}}$	2
x_1	0	1	0	0	0	$-\frac{9}{4}$	$-\frac{3}{4}$	$\frac{1}{4}$	3
x_2	0	0	1	0	0	$\frac{3}{2}$	$\frac{1}{2}$	$-\frac{1}{2}$	5

$$\frac{2}{\frac{1}{4}} = 8$$
$$\frac{3}{\frac{1}{4}} = 12$$

Step 4 Phase II applies because there are no negative entries in the RHS column.

Step 5 The pivot column corresponds to the variable s_4; the pivot row corresponds to the variable x_3. By pivoting we get the tableau

BV	P	x_1	x_2	x_3	s_1	s_2	s_3	s_4	RHS
P	1	0	0	3	0	9	2	0	-52
s_1	0	0	0	0	1	1	0	0	0
s_4	0	0	0	4	0	-1	1	1	8
x_1	0	1	0	-1	0	-2	-1	0	1
x_2	0	0	1	2	0	1	1	0	9

This is a final tableau. The maximum value of P is -52, so the minimum value of z is 52. This occurs when $x_1 = 1$, $x_2 = 9$, $x_3 = 0$, $s_1 = 0$, $s_2 = 0$, $s_3 = 0$, $s_4 = 8$.

Now Work Problem 7

. . . .

EXAMPLE 4

Minimizing Cost   The Red Tomato Company operates two plants for canning its tomatoes and has two warehouses for storing the finished products until they are purchased by retailers. The schedule shown in the table represents the per case shipping costs from plant to warehouse.

		Warehouse	
		A	B
Plant	I	$0.25	$0.18
	II	$0.25	$0.14

Each week plant I can produce at most 450 cases and plant II can produce no more than 350 cases of tomatoes. Also, each week warehouse A requires at least 300 cases and warehouse B requires at least 500 cases. If we represent the number of cases shipped from plant I to warehouse A by x_1, from plant I to warehouse B by x_2, and so on, the above data can be represented by the following table:

		Warehouse		Maximum Available
		A	B	
Plant	I	x_1	x_2	450
	II	x_3	x_4	350
Minimum Demand		300	500	

The company wants to arrange its shipments from the plants to the warehouses so that the requirements of the warehouses are met and shipping costs are kept at a minimum. How should the company proceed?

SOLUTION

The linear programming problem is stated as follows:

Minimize the cost equation

$$C = 0.25x_1 + 0.18x_2 + 0.25x_3 + 0.14x_4$$

subject to

$$
\begin{aligned}
x_1 + x_2 &\leq 450 \\
x_3 + x_4 &\leq 350 \\
x_1 + x_3 &\geq 300 \\
x_2 + x_4 &\geq 500 \\
x_1 \geq 0 \quad x_2 \geq 0 \quad x_3 &\geq 0 \quad x_4 \geq 0
\end{aligned}
$$

Thus we shall maximize

$$P = -C = -0.25x_1 - 0.18x_2 - 0.25x_3 - 0.14x_4$$

subject to the same constraints.

Step 1 Write each constraint with $\le$.

$$
\begin{aligned}
x_1 + x_2 &&&& \le && 450 \\
&& x_3 + x_4 && \le && 350 \\
-x_1 && - x_3 && &\le& - 300 \\
- x_2 && - x_4 && &\le& - 500
\end{aligned}
$$

Step 2 Introduce nonnegative slack variables to form equalities:

$$
\begin{aligned}
x_1 + x_2 && && + s_1 &&&& &=& 450 \\
&& x_3 + x_4 && && + s_2 && &=& 350 \\
-x_1 && - x_3 && && && + s_3 &=& - 300 \\
- x_2 && - x_4 && && && + s_4 &=& - 500 \\
\end{aligned}
$$

$$s_1 \ge 0 \qquad s_2 \ge 0 \qquad s_3 \ge 0 \qquad s_4 \ge 0$$

Step 3 Set up the initial simplex tableau:

BV	P	x_1	x_2	x_3	x_4	s_1	s_2	s_3	s_4	RHS	
P	1	0.25	0.18	0.25	0.14	0	0	0	0	0	
s_1	0	1	1	0	0	1	0	0	0	450	
s_2	0	0	0	1	1	0	1	0	0	350	
s_3	0	-1	0	-1	0	0	0	1	0	-300	
s_4	0	0	(-1)	0	-1	0	0	0	1	-500	$\frac{-500}{-1} = 500$

Step 4 Phase I applies because of the negative entries in the RHS column.

Step 5 The pivot row corresponds to the variable s_4; the pivot column corresponds to the variable x_2 or x_4. We'll use x_2. By pivoting we obtain

BV	P	x_1	x_2	x_3	x_4	s_1	s_2	s_3	s_4	RHS	
P	1	0.25	0	0.25	-0.04	0	0	0	0.18	-90	
s_1	0	1	0	0	-1	1	0	0	1	-50	
s_2	0	0	0	1	1	0	1	0	0	350	
s_3	0	-1	0	(-1)	0	0	0	1	0	-300	$\frac{-300}{-1} = 300$
x_2	0	0	1	0	1	0	0	0	-1	500	

Step 4 Phase I applies because of the negative entries in the RHS column.

Step 5 The pivot row corresponds to the variable s_3; we'll use the column corresponding to x_3 as the pivot column. By pivoting we obtain

BV	P	x_1	x_2	x_3	x_4	s_1	s_2	s_3	s_4	RHS	
P	1	0	0	0	-0.04	0	0	0.25	0.18	-165	
s_1	0	1	0	0	(-1)	1	0	0	1	-50	$\frac{-50}{-1} = 50$
s_2	0	-1	0	0	1	0	1	1	0	50	
x_3	0	1	0	1	0	0	0	-1	0	300	
x_2	0	0	1	0	1	0	0	0	-1	500	

Step 4 Phase I applies because of the negative entry in the RHS column.

Step 5 The pivot row corresponds to the variable s_1; the pivot column corresponds to the variable x_4. By pivoting we obtain

BV	P	x_1	x_2	x_3	x_4	s_1	s_2	s_3	s_4	RHS	
P	1	-0.04	0	0	0	-0.04	0	0.25	0.14	-163	
x_4	0	-1	0	0	1	-1	0	0	-1	50	
s_2	0	0	0	0	0	①	1	1	1	0	$\frac{0}{1}=0$
x_3	0	1	0	1	0	0	0	-1	0	300	$\frac{300}{1}=300$
x_2	0	1	1	0	0	1	0	0	0	450	$\frac{450}{1}=450$

Step 4 Phase II applies because there are no negative entries in the RHS column.

Step 5 The pivot column corresponds to the variables x_1 or s_1; we choose s_1. The pivot row corresponds to the variable s_2. By pivoting we obtain

BV	P	x_1	x_2	x_3	x_4	s_1	s_2	s_3	s_4	RHS	
P	1	-0.04	0	0	0	0	0.04	0.29	0.18	-163	
x_4	0	-1	0	0	1	0	1	1	0	50	
s_1	0	0	0	0	0	1	1	1	1	0	
x_3	0	①	0	1	0	0	0	-1	0	300	$\frac{300}{1}=300$
x_2	0	1	1	0	0	0	-1	-1	-1	450	$\frac{450}{1}=450$

Step 4 Phase II applies because there are no negative entries in the RHS column.

Step 5 The pivot column corresponds to the variable x_1; the pivot row corresponds to the variable x_3. By pivoting we obtain

BV	P	x_1	x_2	x_3	x_4	s_1	s_2	s_3	s_4	RHS
P	1	0	0	0.04	0	0	0.04	0.25	0.18	-151
x_4	0	0	0	1	1	0	1	0	0	350
s_1	0	0	0	0	0	1	1	1	1	0
x_1	0	1	0	1	0	0	0	-1	0	300
x_2	0	0	1	-1	0	0	-1	0	-1	150

This is a final tableau. The maximum value of P is -151, so the minimum cost C is \$151. This occurs when $x_1 = 300, x_2 = 150, x_3 = 0, x_4 = 350$. This means plant I should deliver 300 cases to warehouse A and 150 cases to warehouse B; and plant II should deliver 350 cases to warehouse B to keep costs at the minimum (\$151).

· ·

EXERCISE 4.4   Answers to odd-numbered problems begin on page AN-21.

In Problems 1–8 use the Phase I/Phase II method to solve each linear programming problem.

1. Maximize

$$P = 3x_1 + 4x_2$$

subject to the constraints

$$x_1 + x_2 \leq 12$$
$$5x_1 + 2x_2 \geq 36$$
$$7x_1 + 4x_2 \geq 14$$
$$x_1 \geq 0 \qquad x_2 \geq 0$$

2. Maximize

$$P = 5x_1 + 2x_2$$

subject to the constraints

$$x_1 + x_2 \geq 11$$
$$2x_1 + 3x_2 \geq 24$$
$$x_1 + 3x_2 \leq 18$$
$$x_1 \geq 0 \qquad x_2 \geq 0$$

3. Maximize

$$P = 3x_1 + 2x_2 - x_3$$

subject to the constraints

$$x_1 + 3x_2 + x_3 \leq 9$$
$$2x_1 + 3x_2 - x_3 \geq 2$$
$$3x_1 - 2x_2 + x_3 \geq 5$$
$$x_1 \geq 0 \qquad x_2 \geq 0 \qquad x_3 \geq 0$$

4. Maximize

$$P = 3x_1 + 2x_2 - x_3$$

subject to the constraints

$$2x_1 - x_2 - x_3 \leq 2$$
$$x_1 + 2x_2 + x_3 \geq 2$$
$$x_1 - 3x_2 - 2x_3 \leq -5$$
$$x_1 \geq 0 \qquad x_2 \geq 0 \qquad x_3 \geq 0$$

5. Minimize

$$z = 6x_1 + 8x_2 + x_3$$

subject to the constraints

$$3x_1 + 5x_2 + 3x_3 \geq 20$$
$$x_1 + 3x_2 + 2x_3 \geq 9$$
$$6x_1 + 2x_2 + 5x_3 \geq 30$$
$$x_1 + x_2 + x_3 \leq 10$$
$$x_1 \geq 0 \qquad x_2 \geq 0 \qquad x_3 \geq 0$$

6. Minimize

$$z = 2x_1 + x_2 + x_3$$

subject to the constraints

$$3x_1 - x_2 - 4x_3 \leq -12$$
$$x_1 + 3x_2 + 2x_3 \geq 10$$
$$x_1 - x_2 + x_3 \leq 8$$
$$x_1 \geq 0 \qquad x_2 \geq 0 \qquad x_3 \geq 0$$

7. Maximize

$$P = 3x_1 + 2x_2$$

subject to the constraints

$$2x_1 + x_2 \leq 4$$
$$x_1 + x_2 = 3$$
$$x_1 \geq 0 \qquad x_2 \geq 0$$

8. Maximize

$$P = 45x_1 + 27x_2 + 18x_3 + 36x_4$$

subject to the constraints

$$5x_1 + x_2 + x_3 + 8x_4 = 30$$
$$2x_1 + 4x_2 + 3x_3 + 2x_4 = 30$$
$$x_1 \geq 0 \qquad x_2 \geq 0 \qquad x_3 \geq 0 \qquad x_4 \geq 0$$

9. Shipping Private Motors, Inc., has two plants, M1 and M2, which manufacture engines; the company also has two assembly plants, A1 and A2, which assemble the cars. M1 can produce at most 600 engines per week. M2 can produce at most 400 engines per week. A1 needs at least 500 engines per week and A2 needs at least 300 engines per week. Following is a table of charges to ship engines to assembly plants.

	A1	**A2**
M1	$400	$100
M2	$200	$300

How many engines should be shipped each week from each engine plant to each assembly plant? [*Hint:* Consider four variables: $x_1 =$ number of units shipped from M1 to A1, $x_2 =$ number of units shipped from M1 to A2, $x_3 =$ number of units shipped from M2 to A1, and $x_4 =$ number of units shipped from M2 to A2.]

10. Minimizing Materials Quality Oak Tables, Inc., has an individual who does all its finishing work, and it wishes to use him in this capacity at least 6 hours each day. The assembly area can be used at most 8 hours each day. The company has three models of oak tables, T1, T2, T3. T1 requires 1 hour for assembly, 2 hours for finishing, and 9 board feet of oak. T2 requires 1 hour for assembly, 1 hour for finishing, and 9 board feet of oak. T3 requires 2 hours for assembly, 1 hour for finishing, and 3 board feet of oak. If we wish to minimize the board feet of oak used, how many of each model should be made?

11. Mixture Problem Minimize the cost of preparing the following mixture, which is made up of three foods, I, II,

III. Food I costs $2 per unit, food II costs $1 per unit, and food III costs $3 per unit. Each unit of food I contains 2 ounces of protein and 4 ounces of carbohydrate; each unit of food II has 3 ounces of protein and 2 ounces of carbohydrate; and each unit of food III has 4 ounces of protein and 2 ounces of carbohydrate. The mixture must contain at least 20 ounces of protein and 15 ounces of carbohydrate.

12. **Advertising**   A local appliance store has decided on an advertising campaign utilizing newspaper and radio. Each dollar spent on newspaper advertising is expected to reach 50 people in the "Under $25,000" and 40 in the "Over $25,000" bracket. Each dollar spent on radio advertising is expected to reach 70 people in the "Under $25,000" and 20 people in the "Over $25,000" bracket. If the store wants to reach at least 100,000 people in the "Under $25,000" and at least 120,000 in the "Over $25,000" bracket, how should it proceed so that the cost of advertising is minimized?

13. **Shipping Schedule**   A television manufacturer must fill orders from two retailers. The first retailer, R_1, has ordered 55 television sets, while the second retailer, R_2, has ordered 75 sets. The manufacturer has the television sets stored in two warehouses, W_1 and W_2. There are 100 sets in W_1 and 120 sets in W_2. The shipping costs per television set are: $8 from W_1 to R_1; $12 from W_1 to R_2; $13

from W_2 to R_1; $7 from W_2 to R_2. Find the number of television sets to be shipped from each warehouse to each retailer if the total shipping cost is to be a minimum. What is this minimum cost?

14. **Shipping Schedule**   A motorcycle manufacturer must fill orders from two dealers. The first dealer, D_1, has ordered 20 motorcycles, while the second dealer, D_2, has ordered 30 motorcycles. The manufacturer has the motorcycles stored in two warehouses, W_1 and W_2. There are 40 motorcycles in W_1 and 15 in W_2. The shipping costs per motorcycle are as follows: $15 from W_1 to D_1; $13 from W_1 to D_2; $14 from W_2 to D_1; $16 from W_2 to D_2. Under these conditions, find the number of motorcycles to be shipped from each warehouse to each dealer if the total shipping cost is to be held to a minimum. What is this minimum cost?

15. **Production Control**   RCA manufacturing received an order for a machine. The machine is to weigh 150 pounds. The two raw materials used to produce the machine are A, with a cost of $4 per unit, and B, with a cost of $8 per unit. At least 14 units of B and no more than 20 units of A must be used. Each unit of A weighs 5 pounds; each unit of B weighs 10 pounds. How much of each type of raw material should be used for each unit of final product if we wish to minimize cost?

Technology Exercises

A graphing calculator can be programmed to perform pivoting using elementary row operations. In Problems 1–4 use your graphing calculator to solve each linear programming problem by the simplex method.

1. Maximize
$$P = 1.50x_1 + 3x_2$$
subject to
$$0.25x_1 + 2.75x_2 \leq 2.45$$
$$2.33x_1 + 1.85x_2 \leq 2.66$$
$$x_1 \geq 0$$
$$x_2 \geq 0$$

2. Maximize
$$P = 2.45x_1 + 7.85x_2$$
subject to
$$10.76x_1 + 62.75x_2 \leq 254.55$$
$$52.65x_1 + 10.85x_2 \leq 405.88$$
$$x_1 \geq 0$$
$$x_2 \geq 0$$

3. Minimize
$$C = 5.45x_1 + 2.59x_2$$
subject to
$$12.45x_1 + 45.75x_2 \geq 53.33$$
$$30.88x_1 + 8.12x_2 \geq 74.45$$
$$x_1 \geq 0$$
$$x_2 \geq 0$$

4. Minimize
$$C = 17.88x_1 + 62.59x_2$$
subject to
$$44.45x_1 + 11.75x_2 \geq 153.33$$
$$26.88x_1 + 83.12x_2 \geq 174.45$$
$$x_1 \geq 0$$
$$x_2 \geq 0$$

Graphing calculators that are able to operate with larger matrices and can be programmed to perform pivoting using elementary row operations are especially useful in solving the following type of problem. Check your user's manual to see if your graphing calculator can deal with large matrices. In Problems 5 and 6 use your graphing calculator or a computer to solve each linear programming problem by the simplex method.

5. Maximize
$$P = 1.50x_1 + 2.15x_2 + 1.75x_3 + 5.48x_4$$
subject to
$$2.86x_1 + 2.45x_2 \leq 24.5$$
$$1.75x_3 + 5.48x_4 \leq 26.6$$
$$3.46x_1 + 1.99x_3 \leq 68.9$$
$$3.95x_2 + 1.68x_4 \leq 24.5$$
$$x_1 \geq 0$$
$$x_2 \geq 0$$
$$x_3 \geq 0$$
$$x_4 \geq 0$$

6. Minimize
$$C = 5.66x_1 + 1.89x_2 + 5.88x_3 + 5.48x_4$$
subject to
$$3.89x_1 + 2.69x_2 + 7.78x_3 \geq 9.13$$
$$3.77x_2 + 1.65x_3 + 5.48x_4 \geq 8.44$$
$$0.85x_1 + 6.05x_3 + 3.77x_4 \geq 6.33$$
$$3.95x_2 + 0.33x_3 + 1.68x_4 \geq 0.26$$
$$x_1 \geq 0$$
$$x_2 \geq 0$$
$$x_3 \geq 0$$
$$x_4 \geq 0$$

CHAPTER REVIEW

IMPORTANT TERMS AND CONCEPTS

standard form of a maximum
  problem 168
initial simplex tableau 171
pivoting 173
the simplex method 180

flowchart to solve a
  maximum linear
  programming problem 193
standard form of a minimum
  problem 198
duality principle 200

steps for obtaining the dual problem
  201
simplex method with mixed
  constraints 208
steps for solving a minimum
  problem 212

TRUE–FALSE ITEMS   Answers are on page AN-21.

T_____ F_____ **1.** In a maximum problem written in standard form, each of the constraints, with the exception of the nonnegativity constraints, is written with a $\leq$ symbol.

T_____ F_____ **2.** In a maximum problem written in standard form, the slack variables are sometimes negative.

T_____ F_____ **3.** Once the pivot element is identified in a tableau, the pivot operation causes the pivot element to become a 1 and causes the remaining entries in the pivot column to become 0s.

T_____ F_____ **4.** The pivot element is sometimes in the objective row.

T_____ F_____ **5.** One way to solve a minimum problem is to first solve its dual, which is a maximum problem.

T_____ F_____ **6.** Another way to solve a minimum problem is to solve the maximum problem whose objective function is the negative of the minimum problem's objective function.

FILL IN THE BLANKS   Answers are on page AN-21.

1. The constraints of a maximum problem in standard form are changed from an inequality to an equation by introducing _____ _____.

2. For a maximum problem in standard form, the pivot _____ is located by selecting the most negative entry in the objective row.

3. For a minimum problem to be in standard form all the constraints must be written with _____ signs.

4. The _____ _____ _____ principle states that the optimal solution of a minimum linear programming problem, if it exists, has the same value as the optimal solution of the maximum problem, which is its dual.

5. To solve a maximum linear programming problem that cannot be placed in standard form, the _____ method is used.

REVIEW EXERCISES   Answers to odd-numbered problems begin on page AN-21.

In Problems 1–9 use the simplex method.

1. Maximize
$$P = 100x_1 + 200x_2 + 50x_3$$
subject to the constraints
$$5x_1 + 5x_2 + 10x_3 \leq 1000$$
$$10x_1 + 8x_2 + 5x_3 \leq 2000$$
$$10x_1 + 5x_2 \qquad \leq 500$$
$$x_1 \geq 0 \qquad x_2 \geq 0 \qquad x_3 \geq 0$$

2. Maximize
$$P = x_1 + 2x_2 + x_3$$
subject to the constraints
$$3x_1 + x_2 + x_3 \leq 3$$
$$x_1 - 10x_2 - 4x_3 \leq 20$$
$$x_1 \geq 0 \qquad x_2 \geq 0 \qquad x_3 \geq 0$$

3. Maximize
$$P = 40x_1 + 60x_2 + 50x_3$$
subject to the constraints
$$2x_1 + 2x_2 + x_3 \leq 8$$
$$x_1 - 4x_2 + 3x_3 \leq 12$$
$$x_1 \geq 0 \qquad x_2 \geq 0 \qquad x_3 \geq 0$$

4. Maximize
$$P = 2x_1 + 8x_2 + 10x_3 + x_4$$
subject to the constraints
$$x_1 + 2x_2 + x_3 + x_4 \leq 50$$
$$3x_1 + x_2 + 2x_3 + x_4 \leq 100$$
$$x_1 \geq 0 \qquad x_2 \geq 0 \qquad x_3 \geq 0 \qquad x_4 \geq 0$$

5. Minimize
$$z = 2x_1 + x_2$$
subject to the constraints
$$2x_1 + 2x_2 \leq 8$$
$$x_1 - x_2 \leq 2$$
$$x_1 \geq 0 \qquad x_2 \geq 0$$

6. Minimize
$$z = 4x_1 + 2x_2$$
subject to the constraints
$$x_1 + 2x_2 \leq 4$$
$$x_1 + 4x_2 \leq 6$$
$$x_1 \geq 0 \qquad x_2 \geq 0$$

7. Minimize
$$z = 5x_1 + 4x_2 + 3x_3$$
subject to the constraints
$$x_1 + x_2 + x_3 \geq 100$$
$$2x_1 + x_2 \qquad \geq 50$$
$$x_1 \geq 0 \qquad x_2 \geq 0 \qquad x_3 \geq 0$$

8. Minimize
$$z = 2x_1 + x_2 + 3x_3 + x_4$$
subject to the constraints
$$x_1 + x_2 + x_3 + x_4 \geq 50$$
$$3x_1 + x_2 + 2x_3 + x_4 \geq 100$$
$$x_1 \geq 0 \qquad x_2 \geq 0 \qquad x_3 \geq 0 \qquad x_4 \geq 0$$

9. Maximize

$$P = 300x_1 + 200x_2 + 450x_3$$

subject to the constraints

$$x_1 \geq 0 \qquad x_2 \geq 0 \qquad x_3 \geq 0$$
$$4x_1 + 3x_2 + 5x_3 \leq 140$$
$$x_1 + x_2 + x_3 = 30$$

10. **Mixture**   A brewery manufactures three types of beer —lite, regular, and dark. Each vat of lite beer requires 6 bags of barley, 1 bag of sugar, and 1 bag of hops. Each vat of regular beer requires 4 bags of barley, 3 bags of sugar, and 1 bag of hops. Each vat of dark beer requires 2 bags of barley, 2 bags of sugar, and 4 bags of hops. Each day the brewery has 800 bags of barley, 600 bags of sugar, and 300 bags of hops available. The brewery realizes a profit of $10 per vat of lite beer, $20 per vat of regular beer, and $30 per vat of dark beer. How many vats of lite, regular, and dark beer should be brewed in order to maximize profits? What is the maximum profit?

11. **Management**   The manager of a supermarket meat department finds that there are 160 pounds of round steak, 600 pounds of chuck steak, and 300 pounds of pork in stock on Saturday morning. From experience, the manager knows that half of these quantities can be sold as straight cuts. The remaining meat will have to be ground into hamburger patties and picnic patties for which there is a large weekend demand. Each pound of hamburger patties contains 20% ground round and 60% ground chuck. Each pound of picnic patties contains 30% ground pork and 50% ground chuck. The remainder of each product consists of an inexpensive nonmeat filler that the store has in unlimited quantities. How many pounds of each product should be made if the objective is to maximize the amount of meat used to make the patties?

12. **Scheduling**   An automobile manufacturer must fill orders from two dealers. The first dealer, D_1, has ordered 40 cars, while the second dealer, D_2, has ordered 25 cars. The manufacturer has the cars stored in two locations, W_1 and W_2. There are 30 cars in W_1 and 50 cars in W_2. The shipping costs per car are as follows: $180 from W_1 to D_1; $150 from W_1 to D_2; $160 from W_2 to D_1; $170 from W_2 to D_2. Under these conditions, how many cars should be shipped from each storage location to each dealer so as to minimize the total shipping costs? What is this minimum cost?

13. **Optimal Land Use**   A farmer has 1000 acres of land on which corn, wheat, or soybeans can be grown. Each acre of corn costs $100 for preparation, requires 7 days of labor, and yields a profit of $30. An acre of wheat costs $120 to prepare, requires 10 days of labor, and yields $40 profit. An acre of soybeans costs $70 to prepare, requires 8 days of labor, and yields $40 profit. If the farmer has $10,000 for preparation and can count on enough workers to supply 8000 days of labor, how many acres should be devoted to each crop to maximize profits?

Mathematical Questions from Professional Exams
Use the following information to answer Problems 1–4.

CPA Exam   *The Ball Company manufactures three types of lamps, which are labeled A, B, and C. Each lamp is processed in two departments—I and II. Total available work-hours per day for departments I and II are 400 and 600, respectively. No additional labor is available. Time requirements and profit per unit for each lamp type are as follows:*

	A	B	C
Work-hours required in department I	2	3	1
Work-hours required in department II	4	2	3
Profit per unit (sales price less all variable costs)	$5	$4	$3

The company has assigned you, as the accounting member of its profit planning committee, to determine the number of types of A, B, and C lamps that it should produce in order to maximize its total profit from the sale of lamps. The following questions relate to a linear programming model that your group has developed.

1. The coefficients of the objective function would be

   (a) 4, 2, 3   (b) 2, 3, 1   (c) 5, 4, 3   (d) 400, 600

2. The constraints in the model would be

   (a) 2, 3, 1   (b) 5, 4, 3   (c) 4, 2, 3   (d) 400, 600

3. The constraint imposed by the available work-hours in department I could be expressed as

   (a) $4X_1 + 2X_2 + 3X_3 \leq 400$
   (b) $4X_1 + 2X_2 + 3X_3 \geq 400$
   (c) $2X_1 + 3X_2 + 1X_3 \leq 400$
   (d) $2X_1 + 3X_2 + 1X_3 \geq 400$

4. The most types of lamps that would be included in the optimal solution would be

   (a) 2   (b) 1   (c) 3   (d) 0

5. **CPA Exam**   In a system of equations for a linear programming model, what can be done to equalize an inequality such as $3X + 2Y \leq 15$?

   (a) Nothing.
   (b) Add a slack variable.
   (c) Add a tableau.
   (d) Multiply each element by -1.

Use the following information to answer Problems 6 and 7.

CPA Exam   *The Golden Hawk Manufacturing Company wants to maximize the profits on products A, B, and C. The contribution margin for each product follows:*

Product	Contribution Margin
A	$2
B	$5
C	$4

The production requirements and departmental capacities, by departments, are as follows:

Department	Production Requirements by Product (Hours)		
	A	B	C
Assembling	2	3	2
Painting	1	2	2
Finishing	2	3	1

Department	Departmental Capacity (Total Hours)
Assembling	30,000
Painting	38,000
Finishing	28,000

6. What is the profit maximization formula for the Golden Hawk Company?

   (a) $\$2A + \$5B + \$4C = X$ (where X = Profit)
   (b) $5A + 8B + 5C \leq 96,000$
   (c) $\$2A + \$5B + \$4C \leq X$ (where X = Profit)
   (d) $\$2A + \$5B + \$4C = 96,000$

7. What is the constraint for the painting department of the Golden Hawk Company?

   (a) $1A + 2B + 2C \geq 38,000$
   (b) $\$2A + \$5B + \$4C \geq 38,000$
   (c) $1A + 2B + 2C \leq 38,000$
   (d) $2A + 3B + 2C \leq 30,000$

8. **CPA Exam**   Watch Corporation manufactures products A, B, and C. The daily production requirements are shown below.

Product	Profit per Unit	Hours Required per Unit per Department		
		Machining	Plating	Polishing
A	$10	1	1	1
B	$20	3	1	2
C	$30	2	3	2
Total Hours per Day per Department		16	12	6

What is Watch's objective function in determining the daily production of each unit?

(a) $A + B + C \leq \$60$

(b) $\$3A + \$6B + \$7C = \$60$

(c) $A + B + C \leq$ Profit

(d) $\$10A + \$20B + \$30C =$ Profit

CPA Exam   *Problems 9–11 are based on a company that uses a linear programming model to schedule the production of three products. The per-unit selling prices, variable costs, and labor time required to produce these products are presented below. Total labor time available is 200 hours.*

Product	Selling Price	Variable Cost	Labor (Hours)
A	$4.00	$1.00	2
B	$2.00	$ .50	2
C	$3.50	$1.50	3

9. The objective function to maximize the company's gross profit (Z) is

(a) $4A + 2B + 3.5C = Z$

(b) $2A + 2B + 3C = Z$

(c) $5A + 2.5B + 5C = Z$

(d) $3A + 1.5B + 2C = Z$

(e) $A + B + C = Z$

10. The constraint of labor time available is represented by

(a) $2A + 2B + 3C \leq 200$

(b) $2A + 2B + 3C \geq 200$

(c) $A + B + C \geq 200$

(d) $4A + 2B + 3.5C = 200$

(e) $A/2 + B/2 + C/3 = 200$

11. A linear programming model produces an optimal solution by

(a) Ignoring resource constraints.

(b) Minimizing production costs.

(c) Minimizing both variable production costs and labor costs.

(d) Maximizing the objective function subject to resource constraints.

(e) Finding the point at which various resource constraints intersect.

Chapter 5

Finance

In this chapter we discuss several types of problems in finance, such as compound interest, annuities, sinking funds, and mortgage payments.

5.1 INTEREST

Interest is money paid for the use of money. The total amount of money borrowed, whether by an individual from a bank in the form of a loan or by a bank from an individual in the form of a savings account, is called the **principal.**

The **rate of interest** is the amount charged for the use of the principal for a given length of time, usually on a yearly, or *per annum,* basis. Rates of interest are usually expressed as percents: 10% per annum, 14% per annum, $7\frac{1}{2}$% per annum, and so on.

The word **percent** means ''per hundred.'' The familiar symbol % thus means to divide by one hundred. For example,

$$1\% = \frac{1}{100} = 0.01 \qquad 12\% = \frac{12}{100} = 0.12 \qquad 0.3\% = \frac{0.3}{100} = \frac{3}{1000} = 0.003$$

By reversing these ideas, we can write decimals as percents.

$$0.35 = \frac{35}{100} = 35\% \qquad 1.25 = \frac{125}{100} = 125\% \qquad 0.005 = \frac{5}{1000} = \frac{0.5}{100}$$
$$= 0.5\% = \tfrac{1}{2}\%$$

Now Work Problems 1 and 9

EXAMPLE 1

(a) Find 12% of 80.   (b) What percent of 40 is 18?
(c) 15% of what number is 8?

SOLUTION

(a) The English word ''of'' translates to ''multiply'' when percents are involved. Thus

$$12\% \text{ of } 80 = 12\% \text{ times } 80 = (0.12) \cdot (80) = 9.6$$

(b) Let x represent the unknown percent. Then

$$x\% \text{ of } 40 = 18$$
$$\frac{x}{100} \cdot 40 = 18$$
$$40x = 1800$$
$$x = \frac{1800}{40} = 45$$

Thus 45% of 40 is 18.

(c) Let x represent the number. Then

$$15\% \text{ of } x = 8$$
$$0.15x = 8$$
$$x = \frac{8}{0.15} = 53.33$$

Now Work Problems 17, 23, and 27

Thus 15% of 53.33 is 8.

. . . .

EXAMPLE 2

A resident of Illinois has base income, after adjustment for deductions, of $18,000. The state income tax on this base income is 3%. What tax is due?

SOLUTION

We must find 3% of $18,000. We convert 3% to its decimal equivalent and then multiply by $18,000.

$$3\% \text{ of } \$18{,}000 = (0.03)(\$18{,}000) = \$540$$

The state income tax is $540.00.

· ·

Simple Interest

The easiest type of interest to deal with is called *simple interest.*

SIMPLE INTEREST

Simple interest **is interest computed on the principal for the entire period it is borrowed.**

Simple Interest Formula

If a principal P is borrowed at a simple interest rate of r per annum (where r is a decimal) for a period of t years, the interest charge I is

$$I = Prt$$

The **amount** A owed at the end of t years is the sum of the principal borrowed and the interest charge:

$$A = P + I = P + Prt = P(1 + rt) \tag{1}$$

EXAMPLE 3

A loan of $250 is made for 9 months at a simple interest rate of 10% per annum. What is the interest charge? What amount is due after 9 months?

SOLUTION

The actual period the money is borrowed for is 9 months, which is $\frac{3}{4}$ of a year. The interest charge is the product of the amount borrowed, $250, the annual rate of interest, 0.10, and the length of time in years, $\frac{3}{4}$. Thus

$$I = Prt$$

$$\text{Interest charge} = (\$250)(0.10)(\tfrac{3}{4}) = \$18.75$$

The amount A due after 9 months is

$$A = P + I = 250 + 18.75 = \$268.75$$

Now Work Problem 31

· ·

EXAMPLE 4

A person borrows $1000 for a period of 6 months. What simple interest rate is being charged if the amount A that must be repaid after 6 months is $1045?

SOLUTION

The principal P is $1000, the period is $\frac{1}{2}$ year (6 months), and the amount A owed after 6 months is $1045. We need to solve for r in this equation:

$$A = P + Prt$$
$$1045 = 1000 + 1000r(\tfrac{1}{2})$$
$$45 = 500r$$
$$r = \frac{45}{500} = 0.09$$

The per annum rate of interest is 9%.

Now Work Problem 37

· ·

EXAMPLE 5

A bank borrows $1,000,000 for 1 month at a simple interest rate of 9% per annum. How much must the bank pay back at the end of 1 month?

SOLUTION

The principal P is $1,000,000, the period t is $\frac{1}{12}$ year, and the rate r is 0.09.

$$A = P(1 + rt)$$
$$A = 1,000,000\,[1 + 0.09(\tfrac{1}{12})]$$
$$= 1,000,000(1.0075)$$
$$= \$1,007,500$$

At the end of 1 month, the bank must pay back $1,007,500.

· ·

Discounted Loans

If a lender deducts the interest from the amount of the loan at the time the loan is made, the loan is said to be **discounted.** The interest deducted from the amount of the loan is the **discount.** The amount the borrower receives is called the **proceeds,** and the amount to be repaid is called the **maturity value.**

Discounted Loans

If r is the per annum rate of interest, t the time in years, and A the amount repaid, the maturity value P, the proceeds, is given by

$$P = A - Art = A(1 - rt) \tag{2}$$

where Art is the discount, the interest deducted from the amount of the loan.

EXAMPLE 6

A borrower signs a note for a discounted loan and agrees to pay $1000 in 9 months at a 10% rate of interest. How much does this borrower receive?

SOLUTION

The maturity value, the amount to be repaid, is $A = 1000$. The rate of interest is $r = 10\% = 0.10$. The time is $t = 9$ months $= \frac{9}{12}$ year. The discount is

$$Art = \$1000(0.10)(\tfrac{9}{12}) = \$75$$

The discount is deducted from the maturity value of $1000, so that the proceeds, the amount the borrower receives, is

$$P = A - Art = 1000 - 75 = \$925$$

Now Work Problem 43

· ·

EXAMPLE 7

What simple rate of interest is the borrower in Example 6 paying on the $925 that was borrowed for 9 months?

SOLUTION

The principal P is $925, t is $\frac{9}{12} = \frac{3}{4}$ of a year, and the amount A is $1000. If r is the simple rate of interest, then, from (1),

$$A = P + Prt$$
$$1000 = 925 + 925r(\tfrac{3}{4})$$
$$75 = 693.75r$$
$$r = \frac{75}{693.75} = 0.108108$$

The simple rate of interest is 10.81%.

· ·

EXAMPLE 8

You wish to borrow $10,000 for 3 months. If the person you are borrowing from offers a discounted loan at 8%, how much must you repay at the end of 3 months?

SOLUTION

The principal P you borrow (the proceeds) is $10,000, r is 0.08, and the time t is $\frac{3}{12} = \frac{1}{4}$ year. From Equation (2), the amount A you repay obeys

$$P = A(1 - rt)$$
$$10,000 = A[1 - 0.08(\tfrac{1}{4})]$$
$$10,000 = 0.98A$$
$$A = \frac{10,000}{0.98} = \$10,204.08$$

· ·

Treasury Bills

Treasury bills (T-Bills) are short-term securities issued by the Federal Reserve. The bills do not specify a rate of interest. They are sold at public auction with financial institutions making competitive bids. For example, a financial institution may bid $982,400 for a 3-month $1 million treasury bill. At the end of 3 months the financial institution receives $1 million, which covers the interest earned and the cost of the T-bill. This is basically a simple discount transaction.

EXAMPLE 9

Bidding on Treasury Bills How much should a bank bid to earn 7.65% simple interest on a 6-month $1 million treasury bill?

SOLUTION

The maturity value, the amount to be repaid to the bank by the government, is $A = \$1,000,000$. The rate of interest is $r = 7.65\% = 0.0765$. The time is $t = 6$ months $= \frac{1}{2}$ year. The proceeds P to the government are

$$P = A(1 - rt) = \$1,000,000[1 - 0.0765(\tfrac{1}{2})]$$
$$= 1,000,000(0.96175)$$
$$= 961,750$$

The bank should bid $961,750.

· ·

EXERCISE 5.1 Answers to odd-numbered problems begin on page AN-21.

In Problems 1–8 write each decimal as a percent.

1. 0.45	**2.** 0.85	**3.** 1.12	**4.** 1.25
5. 0.06	**6.** 0.07	**7.** 0.0025	**8.** 0.0015

In Problems 9–16 write each percent as a decimal.

9. 42%	**10.** 7.25%	**11.** 0.2%	**12.** 300%
13. 0.001%	**14.** 4.3%	**15.** 73.4%	**16.** 92%

In Problems 17–30 calculate the indicated quantity.

17. 15% of 1000	**18.** 20% of 500	**19.** 18% of 100
20. 10% of 50	**21.** 210% of 50	**22.** 135% of 1000

23. What percent of 80 is 4?   **24.** What percent of 60 is 5?

25. What percent of 5 is 8?   **26.** What percent of 25 is 45?

27. 8% of what number is 20?

28. 12% of what number is 25?

29. 15% of what number is 50?

30. 18% of what number is 40?

In Problems 31–36 find the interest due on each loan.

31. $1000 is borrowed for 3 months at 4% simple interest.

32. $100 is borrowed for 6 months at 8% simple interest.

33. $500 is borrowed for 9 months at 12% simple interest.

34. $800 is borrowed for 8 months at 5% simple interest.

35. $1000 is borrowed for 18 months at 10% simple interest.

36. $100 is borrowed for 24 months at 12% simple interest.

In Problems 37–42 find the simple interest rate for each loan.

37. $1000 is borrowed; the amount owed after 6 months is $1050.

38. $500 is borrowed; the amount owed after 8 months is $600.

39. $300 is borrowed; the amount owed after 12 months is $400.

40. $600 is borrowed; the amount owed after 9 months is $660.

41. $900 is borrowed; the amount owed after 10 months is $1000.

42. $800 is borrowed; the amount owed after 3 months is $900.

In Problems 43–46 find the proceeds for each discounted loan.

43. $1200 repaid in 6 months at 10%.

44. $500 repaid in 8 months at 9%.

45. $2000 repaid in 24 months at 8%.

46. $1500 repaid in 18 months at 10%.

In Problems 47–50 find the amount you must repay for each discounted loan.

47. You borrow $1200 for 6 months at 10%.

48. You borrow $500 for 8 months at 9%.

49. You borrow $2000 for 24 months at 8%.

50. You borrow $1500 for 18 months at 10%.

In Problems 51 and 52 determine at which rate the borrower pays the least interest.

51. A discounted loan for 6 months at 9% or a loan using a simple interest rate of 10%.

52. A discounted loan for 9 months at 8% or a loan using a simple interest rate of $8\frac{1}{2}$%.

53. Madalyn wants to buy a $500 stereo set in 9 months. How much should she invest at 8% simple interest to have the money then?

54. Mike borrows $10,000 for a period of 3 years at a simple interest rate of 10%. Determine the interest due on the loan.

55. Tami borrowed $600 at 8% simple interest. The amount of interest paid was $156. What was the length of the loan?

56. The owner of a restaurant would like to borrow $12,000 from a bank to buy some equipment. The bank will give the owner a discounted loan at an 11% rate of interest for 9 months. What maturity value should be used so that the owner will receive $12,000?

57. Ruth would like to borrow $2000 for one year from a bank. She is given a choice of a simple interest loan at 12.3% or a discounted loan at 12.1%. What should she do?

58. A bank wants to earn 8.5% simple interest on a 3-month $1 million treasury bill. How much should they bid?

59. A bank paid $979,000 for a 3-month $1 million treasury bill. What simple interest was earned?

60. How much should a bank bid on a 6-month $3 million treasury bill to earn 7.715% simple interest?

5.2 COMPOUND INTEREST

Compound Interest Formula

In working with problems involving interest we use the term **payment period** as follows:

Annually	Once per year
Semiannually	Twice per year
Quarterly	4 times per year
Monthly	12 times per year
Daily	365 times per year*

If the interest due at the end of a payment period is added to the principal, so that the interest computed for the next payment period is based on this new amount of the old principal plus interest, then the interest is said to have been **compounded.** That is, **compound interest** is interest paid on the initial principal and previously earned interest.

EXAMPLE 1

A bank pays 6% per annum compounded quarterly. If $200 is placed in a savings account and the quarterly interest is left in the account, how much money is in the account after 1 year?

SOLUTION

At the first quarter (3 months) the interest earned is

$$I = Prt = (\$200)(\tfrac{1}{4})(0.06) = \$3.00$$

The new principal is $P + I = \$203$. The interest on this principal at the second quarter is

$$I = (\$203)(\tfrac{1}{4})(0.06) = \$3.05$$

The interest at the third quarter on the principal of $203 + $3.05 = $206.05 is

$$I = (\$206.05)(\tfrac{1}{4})(0.06) = \$3.09$$

The interest for the fourth quarter is

$$I = (\$209.14)(\tfrac{1}{4})(0.06) = \$3.14$$

Thus after 1 year the total in the savings account is $212.28.
    These results are shown in Figure 1.

....

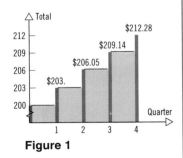

Figure 1

* Some banks use 360 times per year.

Let's develop a formula for computing the amount when interest is compounded. Suppose P is the principal and r is the per annum rate of interest compounded each payment period. Then the rate of interest per payment period is

$$i = \frac{\text{Per annum rate of interest}}{\text{Number of payment periods}}$$

For example, if the annual rate of interest is 10% and the compounding is monthly, then there are 12 payment periods per year and

$$i = \frac{0.10}{12} = 0.00833$$

If 18% is the annual rate compounded daily (365 payment periods), then

$$i = \frac{0.18}{365} = 0.000493$$

At the end of the first payment period the amount A_1 is

$$A_1 = P + Pi = P(1 + i)$$

At the end of the second payment period, and subsequent ones, the amounts are

$$A_2 = A_1 + A_1 i = A_1(1 + i) = P(1 + i)(1 + i) = P(1 + i)^2$$
$$A_3 = A_2 + A_2 i = A_2(1 + i) = P(1 + i)^2(1 + i) = P(1 + i)^3$$
$$.$$
$$.$$
$$.$$
$$A_n = A_{n-1} + A_{n-1} i = A_{n-1}(1 + i) = P(1 + i)^{n-1}(1 + i) = P(1 + i)^n$$

Compound Interest Formula

The amount A_n accrued on a principal P after n payment periods at i interest per payment period is

$$A_n = P(1 + i)^n$$

In working with the compound interest formula, we use a calculator with a $\boxed{y^x}$ key. To use this key, enter the value of y, press $\boxed{y^x}$, enter the value of x, and press $\boxed{=}$.

EXAMPLE 2

If $1000 is invested at an annual rate of interest of 10%, what is the amount after 5 years if the compounding takes place

(a) Annually?   (b) Monthly?   (c) Daily?

How much interest is earned in each case?

SOLUTION

The principal is $P = \$1000$.

(a) We use a calculator. For annual compounding, $i = 0.10$ and $n = 5$. The amount A is

$$A = P(1 + i)^n = (\$1000)(1 + 0.10)^5 = (\$1000)(1.61051) = \$1610.51$$

The interest earned is

$$A - P = \$1610.51 - \$1000.00 = \$610.51$$

(b) For monthly compounding, there are $5 \cdot 12 = 60$ payment periods over 5 years. The interest rate per payment period is $i = 0.10/12$. Using a calculator, the amount A is

$$A = P(1 + i)^n = (\$1000)\left(1 + \frac{0.10}{12}\right)^{60} = (\$1000)(1.64531) = \$1645.31$$

The interest earned is

$$A - P = \$1645.31 - \$1000.00 = \$645.31$$

(c) For daily compounding, there are $5 \cdot 365 = 1825$ payment periods over 5 years. The interest rate per payment period is $i = 0.10/365$. Using a calculator, the amount A is

$$A = P(1 + i)^n = (\$1000)\left(1 + \frac{0.10}{365}\right)^{1825} = (\$1000)(1.64861) = \$1648.61$$

The interest earned is

$$A - P = \$1648.61 - \$1000.00 = \$648.61$$

. . . .

The results of Example 2 are summarized in Table 1.

Table 1

Per Annum Rate	Compounding Method	Interest Rate per Payment Period	Initial Principal	Amount after 5 Years	Interest Earned
10%	Annual	0.10	$1000	$1610.51	$610.51
10%	Monthly	0.00833	$1000	$1645.31	$645.31
10%	Daily	0.000274	$1000	$1648.61	$648.61

Notice the substantial increase in interest earned between annual compounding and monthly compounding ($34.80), compared to the slight increase between monthly compounding and daily compounding ($3.30). It can be shown that as the number of compoundings per year gets larger and larger, the interest earned does not increase indefinitely, but instead approaches a limit. This type of compounding is referred to as *continuous compounding.*

Now Work Problem 1

Let's look at the effect of various compounding periods on a principal of $1000 after 1 year using a rate of interest of 8% per annum.

Annual compounding:	$A = \$1000(1 + 0.08)^1$	$= \$1080.00$
Semiannual compounding:	$A = \$1000(1 + 0.04)^2$	$= \$1081.60$
Monthly compounding:	$A = \$1000 \left(1 + \dfrac{0.08}{12}\right)^{12}$	$= \$1083.00$
Daily compounding:	$A = \$1000 \left(1 + \dfrac{0.08}{365}\right)^{365}$	$= \$1083.28$

With semiannual compounding the amount $1081.60 could have been obtained with a simple interest of 8.16%. We describe this by saying that the **effective rate of interest** of 8% compounded semiannually is 8.16%. Thus the effective rate of interest of 8% compounded monthly is 8.3%, and the effective rate of interest of 8% compounded daily is 8.328%.

EXAMPLE 3

What is the effective rate of interest of 10% compounded monthly?

SOLUTION

To find the effective rate of interest, we need to compute the amount after 1 year, using a principal of, say, $1000. For monthly compounding at 10%, there are 12 payment periods, and the interest rate per period is $i = 0.10/12$. Using a calculator, the amount A is

$$A = P(1 + i)^n = \$1000 \left(1 + \frac{0.10}{12}\right)^{12} = \$1104.71$$

The effective rate of interest of 10% compounded monthly is 10.471%.

· ·

EXAMPLE 4

Three local banks offer the following 1-year Certificates of Deposit (CDs):

(a) Simple interest of 5.2% per annum
(b) 5% per annum compounded monthly
(c) $4\frac{3}{4}$% per annum compounded daily

Which CD results in the most interest?

SOLUTION

We use the effective rate of interest to compare the three CDs. Thus we compute the amount $1000 will grow to in each case.

(a) At simple interest of 5.2%, $1000 will grow to

$$A = P + Prt = \$1000 + \$1000(0.052)(1) = \$1052.00$$

The effective rate of interest is 5.2%.

(b) There are 12 payment periods and the rate of interest per payment period is $i = 0.05/12$. The amount A of $1000 is

$$A = P(1 + i)^n = \$1000 \left(1 + \frac{0.05}{12}\right)^{12} = \$1051.16$$

The effective rate of interest is 5.116%.

(c) There are 365 payment periods and the rate of interest per payment period is $i = 0.0475/365$. The amount A of $1000 is

$$A = P(1 + i)^n = \$1000 \left(1 + \frac{0.0475}{365}\right)^{365} = \$1048.64$$

The effective rate of interest is 4.864%.

The CD offering 5.2% simple interest is the best choice.

Now Work Problem 31

∙∙∙∙∙∙∙∙∙∙∙∙∙∙∙∙∙∙∙∙∙∙∙

Present Value

The compound interest formula states that principal P earning an interest rate per payment period i will, after n payment periods, be worth the amount A, where

$$A = P(1 + i)^n$$

If we solve for P, we obtain

$$P = \frac{A}{(1 + i)^n} = A(1 + i)^{-n}$$

In this formula P is called the **present value** of the amount A due at the end of n interest periods at i interest per payment period. In other words, P is the amount that must be invested for n interest periods at i interest per payment period in order to accumulate the amount A.

Values for $(1 + i)^{-n}$ can be found using a calculator with a $\boxed{y^x}$ key.

The compound interest formula and the present value formula can be used to solve many different kinds of problems. The examples below illustrate some of these applications.

EXAMPLE 5

How much money should be invested at 8% per annum so that after 2 years the amount will be $10,000 when the interest is compounded

(a) Annually?   (b) Monthly?   (c) Daily?

SOLUTION

In this problem we want to find the principal P when we know that the amount A after 2 years is going to be $10,000. That is, we want to find the present value of $10,000.

(a) Since compounding is once per year for 2 years, $n = 2$. We find

$$P = A(1 + i)^{-n} = 10,000(1 + 0.08)^{-2} = 10,000(0.8573388) = \$8573.39$$

(b) Since compounding is 12 times per year for 2 years, $n = 24$. We find

$$P = A(1 + i)^{-n} = 10,000 \left(1 + \frac{0.08}{12} \right)^{-24} = 10,000(0.852596) = \$8525.96$$

(c) Since compounding is 365 times per year for 2 years, $n = 730$. We find

$$P = A(1 + i)^{-n} = 10,000 \left(1 + \frac{0.08}{365} \right)^{-730} = 10,000(0.8521587)$$

$$= \$8521.59$$

Now Work Problem 7

· ·

EXAMPLE 6

What annual rate of interest compounded annually should you seek if you want to double your investment in 5 years?

SOLUTION

If P is the principal and we want P to double, the amount A will be $2P$. We use the compound interest formula with $n = 5$ to find i:

$$2P = P(1 + i)^5$$
$$2 = (1 + i)^5$$
$$1 + i = \sqrt[5]{2}$$
$$i = \sqrt[5]{2} - 1 = 1.148698 - 1 = 0.148698$$

$$\uparrow$$
$$\sqrt[5]{2} = 2^{1/5} = 2^{0.2}$$
Use the $\boxed{y^x}$ key on your calculator

The annual rate of interest needed to double the principal in 5 years is 14.87%.

· ·

EXAMPLE 7*

(a) How long will it take for an investment to double in value if it earns 5% compounded monthly?
(b) How long will it take to triple at this rate?

SOLUTION

(a) If P is the initial investment and we want P to double, the amount A will be $2P$. We use the compound interest formula with $i = 0.05/12$. Then

$$A = P(1 + i)^n$$

$$2P = P \left(1 + \frac{0.05}{12} \right)^n$$

$$2 = (1.0041667)^n$$

$$n = \log_{1.0041667} 2 = \frac{\log_{10} 2}{\log_{10} 1.0041667} = 166.7 \text{ months}$$

$$\uparrow \qquad\qquad\qquad\qquad \uparrow$$
Change-of-   Payment period
base formula   measured in months

* Requires a knowledge of logarithms, especially the change-of-base formula. See Appendix A for a review.

It will take about 13 years 11 months to double the investment.

(b) To triple, we have

$$3P = P \left(1 + \frac{0.05}{12} \right)^n$$
$$3 = (1.0041667)^n$$
$$n = \log_{1.0041667} 3 = \frac{\log_{10} 3}{\log_{10} 1.0041667} = 264.2 \text{ months} = 22 \text{ years}$$

·······················

EXERCISE 5.2   Answers to odd-numbered problems begin on page AN-21.

In Problems 1–6 find the amount owed.

1. $1000 is borrowed at 10% compounded monthly for 36 months.

2. $100 is borrowed at 6% compounded monthly for 36 months.

3. $500 is borrowed at 9% compounded annually for 1 year.

4. $200 is borrowed at 10% compounded annually for 10 years.

5. $800 is borrowed at 12% compounded daily for 200 days.

6. $400 is borrowed at 7% compounded daily for 180 days.

In Problems 7–10 find the principal needed now to get each amount.

7. To get $100 in 6 months at 10% compounded monthly

8. To get $500 in 1 year at 12% compounded annually

9. To get $500 in 1 year at 9% compounded daily

10. To get $800 in 2 years at 5% compounded monthly

11. If $1000 is invested at 9% compounded

(a) Annually   (b) Semiannually
(c) Quarterly   (d) Monthly

what is the amount after 3 years? How much interest is earned?

12. If $2000 is invested at 12% compounded

(a) Annually   (b) Semiannually
(c) Quarterly   (d) Monthly

what is the amount after 5 years? How much interest is earned?

13. If $1000 is invested at 12% compounded quarterly, what is the amount after

(a) 2 years?   (b) 3 years?   (c) 4 years?

14. If $2000 is invested at 8% compounded quarterly, what is the amount after

(a) 2 years?   (b) 3 years?   (c) 4 years?

15. If a bank pays 6% compounded semiannually, how much should be deposited now to have $5000

(a) 4 years later?   (b) 8 years later?

16. If a bank pays 8% compounded quarterly, how much should be deposited now to have $10,000

(a) 5 years later?   (b) 10 years later?

17. Find the effective rate of interest for money invested at

(a) 8% compounded semiannually
(b) 12% compounded monthly

18. Find the effective rate of interest for money invested at

(a) 6% compounded monthly
(b) 14% compounded semiannually

19. What annual rate of interest compounded annually is required to double an investment in 3 years?

20. What annual rate of interest compounded annually is required to double an investment in 10 years?

21. Approximately how long will it take to triple an investment at 10% compounded annually?

22. Approximately how long will it take to triple an investment at 9% compounded annually?

23. Mr. Nielsen wants to borrow $1000 for 2 years. He is given the choice of (a) a simple interest loan of 12% or (b) a loan at 10% compounded monthly. Which loan results in less interest due?

24. Rework Problem 23 if the simple interest loan is 15% and the other loan is at 14% compounded daily.

25. What principal is needed now to get $1000 1 year from today and $1000 2 years from today at 9% compounded annually?

26. Repeat Problem 25 using 9% compounded daily.

27. Find the effective rate of interest for $5\frac{1}{4}$% compounded quarterly.

28. Repeat Problem 27 using 6% compounded quarterly.

29. What interest rate compounded quarterly will give an effective interest rate of 7%?

30. Repeat Problem 29 using 10%.

In Problems 31–34 which of the two rates would yield the larger amount in 1 year? [Hint: Start with a principal of $10,000 in each instance.].

31. 6% compounded quarterly or $6\frac{1}{4}$% compounded annually

32. 9% compounded quarterly or $9\frac{1}{4}$% compounded annually

33. 9% compounded monthly or 8.8% compounded daily

34. 8% compounded semiannually or 7.9% compounded daily

35. If the price of homes rises an average of 5% per year for the next 4 years, what will be the selling price of a home that is selling for $90,000 today 4 years from today? Express your answer rounded to the nearest hundred dollars.

36. A department store charges 1.25% per month on the unpaid balance for customers with charge accounts (interest is compounded monthly). A customer charges $200 and does not pay her bill for 6 months. What is the bill at that time?

37. A major credit card company has a finance charge of 1.5% per month on the outstanding indebtedness. Caryl charged $600 and did not pay her bill for 6 months. What is the bill at that time?

38. Laura wishes to have $8000 available to buy a car in 3 years. How much should she invest in a savings account now so that she will have enough if the bank pays 8% interest compounded quarterly?

39. Tami and Todd will need $40,000 for a down payment on a house in 4 years. How much should they invest in a savings account now so that they will be able to do this? The bank pays 8% compounded quarterly.

40. A newborn child receives a $3000 gift toward a college education. How much will the $3000 be worth in 17 years if it is invested at 10% compounded quarterly?

41. A child's grandparents have opened a $6000 savings account for the child on the day of her birth. The account pays 8% compounded semiannually. The child will be allowed to withdraw the money when she becomes 25 years old. What will the account be worth at that time?

42. What will a $90,000 house cost 5 years from now if the inflation rate over that period averages 5% compounded annually? Express your answer rounded to the nearest hundred dollars.

43. A town increased in population 2% per year for 8 years. If the population was 17,000 at the beginning, what is the size of the population at the end of 8 years?

44. Omega Company can invest its earnings at (a) 7% per year compounded monthly, at (b) 7.15% per year compounded semiannually, or at (c) 7.20% per year compounded annually. Which rate should Omega choose?

45. Jack is considering buying 1000 shares of a stock that sells at $15 per share. The stock pays no dividends. From the history of the stock, Jack is certain that he will be able to sell it 4 years from now at $20 per share. Jack's goal is not to make any investment unless it returns at least 7% compounded quarterly. Should Jack buy the stock?

46. Repeat Problem 45 if Jack requires a return of at least 14% compounded quarterly.

47. An Individual Retirement Account (IRA) has $2000 in it, and the owner decides not to add any more money to the account other than the interest earned at 9% compounded quarterly. How much will be in the account 25 years from the day the account was opened?

48. If Jack sold a stock for $35,281.50 (net) that cost him $22,485.75 three years ago, what annual compound rate of return did Jack make on his investment?

*For Problems 49–51 zero coupon bonds are used. A **zero coupon bond** is a bond
that is sold now at a discount and will pay its face value at some time in the future
when it matures; no interest payments are made.*

49. Tami's grandparents are considering buying a $40,000
face value zero coupon bond at birth so that she will have
enough money for her college education 17 years later. If
money is worth 8% compounded annually, what should
they pay for the bond?

50. How much should a $10,000 face value zero coupon
bond, maturing in 10 years, be sold for now if its rate of
return is to be 8% compounded annually?

51. If you pay $12,485.52 for a $25,000 face value zero coupon bond that matures in 8 years, what is your annual
compound rate of return?

52. A bank advertises that it pays interest on saving accounts
at the rate of 6.25% compounded daily. Find the effective
rate if the bank uses (a) 360 days, or (b) 365 days in
determining the daily rate.

Problems 53 and 54 require logarithms.

53. How many years will it take for an initial investment of
$10,000 to grow to $25,000? Assume a rate of interest of
6% compounded daily.

54. How many years will it take for an initial investment of
$25,000 to grow to $80,000? Assume a rate of interest of
7% compounded daily.

Technology Exercises
*A business or financial calculator is preprogrammed to use the compound interest
formula. If you have one, consult your manual and use your calculator to
check your answers to Problems 9, 11(a), 13, 19, 20, 53, and 54 above.
A graphing calculator can be programmed to do what a financial calculator is
preprogrammed to do.*

1. Write a program that will calculate the amount after n years
if a principal P is invested at r% per annum compounded
quarterly. Use it to verify your answers to Problems 11(c)
and 13 above.

2. Write a program that will calculate the principal needed
now to get the amount A in n years at r% per annum compounded daily. Use it to verify your answer to Problem 9
above.

3. Write a program that will calculate the rate of interest
required to double an investment in n years. Use it to verify
your answers to Problems 19 and 20 above.

4. Write a program that will calculate the number of months
required for an initial investment of x dollars to grow to y
dollars at r% per annum compounded daily. Use it to verify
your answer to Problems 53 and 54 above.

5.3 ANNUITIES; SINKING FUNDS

Annuity

In the previous sections we saw how to compute the future value of an investment when
a fixed amount of money is deposited in an account that pays interest compounded
periodically. Often, however, people and financial institutions do not deposit money
and then sit back and watch it grow. Rather, money is invested in small amounts at
periodic intervals. Examples of such investments are annual life insurance premiums,
monthly deposits in a bank, installment loan payments, and dollar averaging in the
stock market.

An **annuity** is a sequence of equal periodic deposits. When the deposits are made at the same time the interest is credited, the annuity is termed **ordinary.** We shall concern ourselves only with ordinary annuities in this book.

The payment period can be annual, semiannual, quarterly, monthly, or any other fixed length of time.

AMOUNT OF AN ANNUITY

The *amount of an annuity* is the sum of all deposits made plus all interest accumulated.

EXAMPLE 1

Find the amount of an annuity after 5 deposits if each deposit is equal to $100 and is made on an annual basis at an interest rate of 10% per annum compounded annually.

SOLUTION

After 5 deposits the first $100 deposit will have accumulated interest compounded annually at 10% for 4 years. Its value A_1 after 4 years is

$$A_1 = \$100(1 + 0.10)^4 = \$100(1.4641) = \$146.41$$

The second deposit of $100, made 1 year after the first deposit, will accumulate interest compounded at 10% for 3 years. Its value A_2 after the fifth deposit is

$$A_2 = \$100(1 + 0.10)^3 = \$100(1.331) = \$133.10$$

Similarly, the third, fourth, and fifth deposits will have the values

$$A_3 = \$100(1 + 0.10)^2 = \$100(1.21) = \$121.00$$
$$A_4 = \$100(1 + 0.10)^1 = \$100(1.10) = \$110.00$$
$$A_5 = \$100$$

The amount of the annuity after 5 deposits is

$$A_1 + A_2 + A_3 + A_4 + A_5 = \$146.41 + \$133.10 + \$121.00$$
$$+ \$110.00 + \$100.00$$
$$= \$610.51$$

· ·

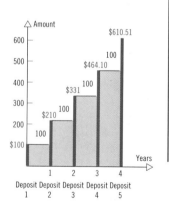

Figure 2

Figure 2 illustrates the annuity formula for the situation described in Example 1.

To develop a formula for the amount of an annuity, suppose $\$P$ is the deposit for an annuity at an interest rate of i percent per payment period (in decimal form) and with a term of n payment periods. Since deposits are made at the end of each period, the first deposit will accumulate the value A_1 compounded over $n - 1$ periods at i percent per payment period. The second deposit will accumulate the value A_2 compounded over $n - 2$ periods at i percent per payment period, and so on. Thus

$$A_1 = \$P(1 + i)^{n-1}, \quad A_2 = \$P(1 + i)^{n-2}, \ldots, \quad A_n = \$P(1 + i)^0 = \$P$$

The total amount A of the annuity after n payment periods is

$$A = A_1 + \cdots + A_n = P(1 + i)^{n-1} + P(1 + i)^{n-2} + \cdots + P(1 + i) + P$$
$$= P[1 + (1 + i) + \cdots + (1 + i)^{n-1}]$$

The expression in brackets is the sum of a geometric sequence* with n terms and common ratio $1 + i$. As a result,

$$1 + (1 + i) + \cdots + (1 + i)^{n-1} = \frac{1 - (1 + i)^n}{1 - (1 + i)} = \frac{1 - (1 + i)^n}{-i} = \frac{(1 + i)^n - 1}{i}$$

We have established the following result.

Amount of an Annuity

If P represents the deposit in dollars made at each payment period for an annuity at i percent interest per payment period, the amount A of the annuity after n payment periods is

$$A = P \cdot \frac{(1 + i)^n - 1}{i}$$

Table I, Amount of an Annuity, in the back of the book, gives values for

$$\frac{(1 + i)^n - 1}{i}$$

for per annum rates of 8%, 10%, and 12% compounded annually and monthly. For most of our work, a calculator will be used.

EXAMPLE 2

Find the amount of an annuity if a deposit of $100 per year is made for 5 years at 10% compounded annually. How much interest is earned?

SOLUTION

The deposit is $P = \$100$. The number of payment periods is $n = 5$ years, and the interest per payment period is $i = 0.10$. The amount A after 5 years is

$$A = 100\left[\frac{(1 + 0.10)^5 - 1}{0.10} \right] = \$100(6.1051) = \$610.51$$

The interest accrued is the amount after 5 years less the 5 annual payments of $100 each:

$$\text{Interest accrued} = A - 500 = 610.51 - 500 = \$110.50$$

Now Work Problem 1

....

---

* The sum of the first n terms of the geometric sequence with common ratio r is

$$1 + r + r^2 + \cdots + r^{n-1} = \frac{1 - r^n}{1 - r}$$

See Appendix A for a more detailed discussion.

EXAMPLE 3

Mary decides to put aside $100 every month in an insurance fund that pays 8% compounded monthly. After making 8 deposits, how much money does Mary have?

SOLUTION

This is an annuity with $P = \$100$, $n = 8$, and $i = 0.08/12$. The amount A after 8 deposits is

$$A = 100\left[\frac{\left(1 + \frac{0.08}{12}\right)^8 - 1}{\frac{0.08}{12}}\right] = \$100(8.1892) = \$818.92$$

....

EXAMPLE 4

Saving for College   To save for his son's college education, Mr. Graff decides to put $50 aside every month in a credit union account paying 10% interest compounded monthly. If he begins this savings program when his son is 3 years old, how much will he have saved by the time his son is 18 years old?

SOLUTION

When his son is 18 years old, Mr. Graff will have made his 180th payment (15 years $\times$ 12 payments per year). This is an annuity with $P = \$50$, $n = 180$, and $i = 0.10/12$. The amount A saved is

$$A = 50\left[\frac{\left(1 + \frac{0.10}{12}\right)^{180} - 1}{\frac{0.10}{12}}\right] = \$50(414.4703) = \$20,723.52$$

....

EXAMPLE 5

Funding an IRA   Joe, at age 35, decides to invest in an IRA. He will put aside $2000 per year for the next 30 years. How much will he have at age 65 if his rate of return is assumed to be 10% per annum?

SOLUTION

This is an annuity with $P = \$2000$, $n = 30$, and $i = 0.10$. The amount A in Joe's IRA after 30 years is

$$A = 2000\left[\frac{(1 + 0.10)^{30} - 1}{0.10}\right] = \$2000(164.49402) = \$328,988.05$$

....

EXAMPLE 6

Funding an IRA   If, in Example 5, Joe had begun his IRA at age 25, instead of 35, what would his IRA be worth at age 65?

SOLUTION

$$A = 2000\left[\frac{(1 + 0.10)^{40} - 1}{0.10}\right] = \$2000(442.59256) = \$885,185.11$$

....

EXAMPLE 7＊

How long does it take to save $500,000 if you place $500 per month in an account paying 6% compounded monthly?

SOLUTION

This is an annuity in which $A = 500,000$, $P = 500$, and $i = 0.06/12$. Thus

$$A = P \cdot \frac{(1 + i)^n - 1}{i}$$

$$500,000 = 500 \cdot \frac{\left(1 + \dfrac{0.06}{12}\right)^n - 1}{\dfrac{0.06}{12}}$$

$$5 = \left(1 + \frac{0.06}{12}\right)^n - 1$$

$$6 = (1.005)^n$$

$$n = \log_{1.005} 6 = \frac{\log_{10} 6}{\log_{10} 1.005} = 359 \text{ months}$$

                ↑   ↑

    Change-of-base Payment period
       formula   in months

· ·

Sinking Fund

A person with a debt may decide to accumulate sufficient funds to pay off the debt by agreeing to set aside enough money each month (or quarter, or year) so that when the debt becomes payable, the money set aside each month plus the interest earned will equal the debt. The fund created by such a plan is called a **sinking fund.**

We shall limit our discussion of sinking funds to those in which equal payments are made at equal time intervals at the end of each period.

EXAMPLE 8

Funding a Bond Obligation   The Board of Education received permission to issue $4,000,000 in bonds to build a new high school. The board is required to make payments every 6 months into a sinking fund paying 8% compounded semiannually. At the end of 12 years the bond obligation will be retired. What should each payment be?

SOLUTION

This is an example of a sinking fund. The payment P required twice a year to accumulate $4,000,000 in 12 years (24 payments at a rate of interest of $i = 0.08/2$ per payment period) obeys the formula

＊ This example requires a knowledge of logarithms, especially the change-of-base formula. A review of logarithms appears in Appendix A.

$$A = P \cdot \frac{(1 + i)^n - 1}{i}$$

$$4{,}000{,}000 = P \cdot \frac{\left(1 + \dfrac{0.08}{2}\right)^{24} - 1}{\dfrac{0.08}{2}}$$

$$4{,}000{,}000 = P(39.082604)$$

$$P = \$102{,}347.33$$

· ·

EXAMPLE 9

Funding a Loan   A woman borrows $3000, which will be paid back to the lender in one payment at the end of 5 years. She agrees to pay interest monthly at an annual rate of 12%. At the same time she sets up a sinking fund in order to repay the loan at the end of 5 years. She decides to make monthly deposits into her sinking fund, which earns 8% interest compounded monthly.

(a) What is the monthly sinking fund deposit?
(b) Construct a table that shows how the sinking fund grows over time.
(c) How much does she need each month to be able to pay the interest on the loan and make the sinking fund deposit?

SOLUTION

(a) The sinking fund deposit is the value of P in the formula

$$A = P \frac{(1 + i)^n - 1}{i}$$

where A equals the amount to be accumulated, namely, $A = \$3000$, $n = 60$, and $i = 0.08/12$. The sinking fund deposit is therefore

$$P = 3000 \left[\frac{\left(1 + \dfrac{0.08}{12}\right)^{60} - 1}{\dfrac{0.08}{12}} \right]^{-1} = 3000(0.0136097) = \$40.83$$

(b) Table 2 (p. 246) shows the growth of the sinking fund over time. The entries for payment number 12 are obtained by using the amount of an annuity formula for a monthly payment of $40.83 made for 12 months at 8% compounded monthly:

$$\text{Total} = \$40.83 \left[\frac{\left(1 + \dfrac{0.08}{12}\right)^{12} - 1}{\dfrac{0.08}{12}} \right] = \$40.83(12.449926) = \$508.33$$

The deposit for payment number 60, the final payment, is only $40.77 because a deposit of $40.83 results in a total of $3000.06.

Table 2

Payment Number	Sinking Fund Deposit, $	Cumulative Deposits	Accumulated Interest, $	Total, $
1	40.83	40.83	0	40.83
12	40.83	489.96	18.37	508.33
24	40.83	979.92	78.93	1058.85
36	40.83	1469.88	185.19	1655.07
48	40.83	1959.84	340.93	2300.77
60	40.77	2449.74	550.26	3000.00

(c) The monthly interest payment due on the loan of $3000 at 12% interest is found using the simple interest formula.

$$I = 3000(0.12)(\tfrac{1}{12}) = \$30$$

Thus the woman needs to be able to pay $40.83 + $30 = $70.83 each month.

Now Work Problem 7

· ·

EXAMPLE 10

Depletion Investment   A gold mine is expected to yield an annual net return of $200,000 for the next 10 years, after which it will be worthless. An investor wants an annual return on his investment of 18%. If he can establish a sinking fund earning 10% annually, how much should he be willing to pay for the mine?

SOLUTION

Let p denote the purchase price. Then $0.18p$ represents an 18% annual return on investment. The annual sinking fund contribution needed to obtain the amount p in 10 years is

$$p\left[\frac{(1 + i)^n - 1}{i}\right]^{-1}$$

where $n = 10$ and $i = 0.10$. The investor should be willing to pay an amount p so that

$$\left(\begin{array}{c}\text{Annual return}\\\text{on investment}\end{array}\right) + \left(\begin{array}{c}\text{Annual sinking}\\\text{fund requirement}\end{array}\right) = \text{Annual return}$$

$$0.18p + p\left[\frac{(1 + 0.10)^{10} - 1}{0.10}\right]^{-1} = \$200,000$$

$$0.18p + (0.0627453)p = \$200,000$$

$$0.2427453p = \$200,000$$

$$p = \$823,909$$

A purchase price of $823,909 will achieve the investor's goals.

· ·

EXERCISE 5.3   Answers to odd-numbered problems begin on page AN-22.

In Problems 1–6 find the amount of each annuity.

1. The deposit is $100 annually for 10 years at 10% compounded annually.

2. The deposit is $200 monthly for 1 year at 5% compounded monthly.

3. The deposit is $400 monthly for 1 year at 12% compounded monthly.

4. The deposit is $1000 annually for 5 years at 10% compounded annually.

5. The deposit is $200 monthly for 3 years at 6% compounded monthly.

6. The deposit is $2000 semiannually for 20 years at 5% compounded semiannually.

In Problems 7–12 find the payment required for each sinking fund.

7. The amount required is $10,000 after 5 years at 5% compounded monthly.

8. The amount required is $5000 after 180 days at 4% compounded daily.

9. The amount required is $20,000 after $2\frac{1}{2}$ years at 6% compounded quarterly.

10. The amount required is $50,000 after 10 years at 7% compounded semiannually.

11. The amount required is $25,000 after 6 months at $5\frac{1}{2}$% compounded monthly.

12. The amount required is $100,000 after 25 years at $4\frac{1}{4}$% compounded annually.

13. **Market Value of a Mutual Fund**   Caryl invests $2500 a year in a mutual fund for 15 years. If the market value of the fund increases on the average 7% per year, what will be the value of the fund at the end of the 15th year?

14. **Saving for a Car**   Laura wants to invest an amount at the end of every 3 months so that she will have $12,000 in 3 years to buy a new car. The account pays 8% compounded quarterly. How much should she deposit each quarter?

15. Todd and Tami pay $300 at the end of each 3 months for 6 years into an ordinary annuity paying 8% compounded quarterly. What is the value of the annuity at the end of 6 years?

16. **Saving for a House**   In 4 years Colleen and Bill would like to have $30,000 for a down payment on a house. How much should they deposit each month into an account paying 9% compounded monthly?

17. **Funding a Pension**   Dan wishes to have $350,000 in a pension fund 20 years from now. How much should he

deposit each month in an account paying 9% compounded monthly?

18. **Funding a Keogh Plan**   Pat has a Keogh retirement plan (this type of plan is tax-deferred until money is withdrawn). If deposits of $7500 are made each year into an account paying 8% compounded annually, how much will be in the account after 25 years?

19. A company establishes a sinking fund to provide for the payment of a $100,000 debt maturing in 4 years. Contributions to the fund are to be made at the end of every year. Find the amount of each annual deposit if interest is 8% per annum.

20. A state has $5,000,000 worth of bonds that are due in 20 years. A sinking fund is established to pay off the debt. If the state can earn 10% annually on its money, what is the annual sinking fund deposit needed?

21. **Depletion Investment**   An investor wants to know the amount she should pay for an oil well expected to yield an annual return of $30,000 for the next 30 years, after which the well will be dry. Find the amount she should pay to yield a 14% annual return if a sinking fund earns 10% annually.

22. If you deposit $10,000 every year in an account paying 8% compounded annually, how long will it take to accumulate $1,000,000?

23. A city has issued bonds to finance a new library. The bonds have a total face value of $1,000,000 and are payable in 10 years. A sinking fund has been opened to retire the bonds. If the interest rate on the fund is 8% compounded quarterly, what will the quarterly payments be?

24. (a) Laura invested $2000 per year in an IRA each year for 10 years earning 8% compounded annually. At

the end of 10 years she ceased the IRA payments, but continued to invest her accumulated amount at 8% compounded annually, for the next 30 years. What was the value of her IRA investment at the end of 10 years? What was the value of her investment at the end of the next 30 years?

(b) Tami started her IRA investment in the 11th year and invested $2000 per year for the next 30 years at 8% compounded annually. What was the value of her investment at the end of 30 years?

(c) Who had more money at the end of the period?

Problems 25 and 26 require logarithms.

25. How many years will it take to save $1,000,000 if you place $600 per month in an account that earns 7% compounded monthly?

26. How many years will it take to save $1,000,000 if you place $1000 per month in an account that earns 6% compounded monthly?

Technology Exercises
A business or financial calculator is preprogrammed to use the annuity formula. If you have one, consult your manual and use your calculator to check your answers to Problems 1, 4, 7, 11, 25, and 26 above. A graphing calculator can be programmed to do what a financial calculator is preprogrammed to do.

1. Write a program that will calculate the amount of an annuity after n years if a principal P is invested at $r\%$ per annum compounded annually. Use it to verify your answers to Problems 1 and 4 above.

2. Write a program that will calculate the monthly payment needed to get the amount A after n years at $r\%$ per annum

compounded monthly. Use it to verify your answers to Problems 7 and 11 above.

3. Write a program that will calculate the number of months required for a monthly investment of x dollars to grow to y dollars at $r\%$ per annum compounded monthly. Use it to verify your answers to Problems 25 and 26 above.

5.4 PRESENT VALUE OF AN ANNUITY; AMORTIZATION

In Section 5.2 we defined present value as it relates to the compound interest formula as the amount of money needed now to obtain an amount A in the future. A similar idea is used for periodic withdrawals.

Suppose you want to withdraw $10,000 per year each year for the next 5 years for a retirement account that earns 10% compounded annually. How much money is required initially in this account for this to happen? In fact, the amount is the present value of each of the $10,000 withdrawals. This leads to the following definition.

PRESENT VALUE OF AN ANNUITY
The *present value* of an annuity is the sum of the present values of the payments.

In other words, the present value of an annuity is the amount of money needed now so that if it is invested at i percent, n equal dollar amounts can be withdrawn without any money left over.

EXAMPLE 1 | Compute the amount of money required to pay out $10,000 per year for 5 years at 10% compounded annually.

SOLUTION

For the first $10,000 withdrawal the present value V_1 (the dollars needed now to do this) is

$$V_1 = \$10,000(1 + 0.10)^{-1} = \$10,000(0.9090909) = \$9090.91$$

For the second $10,000 withdrawal the present value V_2 is

$$V_2 = \$10,000(1 + 0.10)^{-2} = \$10,000(0.826446) = \$8264.46$$

Similarly,

$$V_3 = \$10,000(1 + 0.10)^{-3} = \$10,000(0.7513148) = \$7513.15$$
$$V_4 = \$10,000(1 + 0.10)^{-4} = \$10,000(0.6830134) = \$6830.13$$
$$V_5 = \$10,000(1 + 0.10)^{-5} = \$10,000(0.6209213) = \$6209.21$$

The present value V for 5 withdrawals of $10,000 each is

$$
\begin{aligned}
V &= V_1 + V_2 + V_3 + V_4 + V_5 \\
&= \$9090.91 + \$8264.46 + \$7513.15 + \$6830.13 + \$6209.21 \\
&= \$37,907.86
\end{aligned}
$$

Thus a person would need $37,907.86 now invested at 10% per annum in order to withdraw $10,000 per year for the next 5 years.

Table 3(a) summarizes these results. Table 3(b) lists the amount at the beginning of each year.

Table 3(a)

Withdrawal	Present Value
1st	$10,000(1.10)^{-1} = \$\ 9,090.91$
2nd	$10,000(1.10)^{-2} = \$\ 8,264.46$
3rd	$10,000(1.10)^{-3} = \$\ 7,513.15$
4th	$10,000(1.10)^{-4} = \$\ 6,830.13$
5th	$10,000(1.10)^{-5} = \$\ 6,209.21$
Total	$37,907.86

Table 3(b)

Year	Amount at the Beginning of the Year	Add Interest	Subtract Withdrawal
1	37,907.86	3,790.79	10,000.00
2	31,698.65	3,169.87	10,000.00
3	24,868.52	2,486.85	10,000.00
4	17,355.37	1,735.54	10,000.00
5	9,090.91	909.09	10,000.00
6	0		

To derive a formula for present value, suppose P is the withdrawal per payment period for an annuity at an interest rate of i percent per payment period (in decimal form) and with a term of n payment periods. Then the present value V_1 of the first withdrawal is

$$V_1 = P(1 + i)^{-1}$$

The present value V_2 for the second withdrawal is

$$V_2 = P(1 + i)^{-2}$$

The present value V_n for the nth withdrawal is

$$V_n = P(1 + i)^{-n}$$

The total present value V of the annuity is

$$V = V_1 + \cdots + V_n = P(1 + i)^{-1} + \cdots + P(1 + i)^{-n}$$
$$= P(1 + i)^{-n}[1 + (1 + i) + \cdots + (1 + i)^{n-1}]$$

The expression in brackets is the sum of the first n terms of a geometric sequence, whose ratio is $1 + i$. As a result,

$$V = P(1 + i)^{-n} \frac{1 - (1 + i)^n}{1 - (1 + i)} = P \cdot \frac{(1 + i)^{-n} - 1}{-i} = P \cdot \frac{1 - (1 + i)^{-n}}{i}$$

Present Value of an Annuity

If P represents the withdrawal per payment period, the present value V of an annuity at a rate i per payment period for n payment periods is

$$V = P \cdot \frac{1 - (1 + i)^{-n}}{i}$$

Table II, Present Value of an Annuity, in the back of the book, gives values for

$$\frac{1 - (1 + i)^{-n}}{i}$$

for per annum rates of 8%, 10%, and 12% compounded annually and monthly. For most of our work, a calculator will be used.

Now Work Problem 1

EXAMPLE 2

True Cost of a Car   A man agrees to pay $300 per month for 48 months to pay off a car loan. If interest of 12% per annum is charged monthly, how much did the car originally cost? How much interest was paid?

SOLUTION

This is the same as asking for the present value V of an annuity of $300 per month at 12% for 48 months. The original cost of the car is

$$V = 300 \left[\frac{1 - \left(1 + \dfrac{0.12}{12}\right)^{-48}}{\dfrac{0.12}{12}} \right] = \$300(37.9739595) = \$11,392.19$$

The total payment is ($300)(48) = $14,400. Thus the interest paid is

$$14,400 - \$11,392.19 = \$3007.81$$

·······················

Amortization

We can look at Example 2 differently. What it also says is that if the man pays $300 per month for 48 months with an interest of 12% compounded monthly, then the car is his. In other words, he *amortized* the debt in 48 equal monthly payments. (The Latin word *mort* means "death." Paying off a loan is regarded as "killing" it.) A loan with a fixed rate of interest is said to be **amortized** if both principal and interest are paid by a sequence of equal payments made over equal periods of time.

When a loan of V dollars is amortized at a rate of interest i per payment period over n payment periods, the question is, "What is the payment P?" In other words, in amortization problems, we want to find the amount of payment P that, after n payment periods at the rate of interest i per payment period, gives us a present value equal to the amount of the loan. Thus, we need to find P in the formula

$$V = P\left[\frac{1 - (1 + i)^{-n}}{i}\right]$$

Thus

$$P = V\left[\frac{1 - (1 + i)^{-n}}{i}\right]^{-1}$$

Amortization

The monthly payment P required to pay off a loan of V dollars borrowed for n payment periods at a rate of interest i per payment period is

$$P = V\left[\frac{1 - (1 + i)^{-n}}{i}\right]^{-1}$$

Table II, Present Value of an Annuity, in the back of the book, gives values for

$$\left[\frac{1 - (1 + i)^{-n}}{i}\right]^{-1}$$

for per annum rates of 8%, 10%, and 12% compounded annually and monthly. For most of our work, a calculator will be used.

EXAMPLE 3

What monthly payment is necessary to pay off a loan of $800 at 10% per annum

(a) In 2 years?   (b) In 3 years?
(c) What total amount is paid out for each loan?

SOLUTION

(a) For the 2-year loan $V = \$800$, $n = 24$, and $i = 0.10/12$. The monthly payment P is

$$P = 800 \left[\frac{1 - \left(1 + \dfrac{0.10}{12}\right)^{-24}}{\dfrac{0.10}{12}} \right]^{-1} = \$800(0.04614493) = \$36.92$$

(b) For the 3-year loan $V = \$800$, $n = 36$, and $i = 0.10/12$. The monthly payment P is

$$P = 800 \left[\frac{1 - \left(1 + \dfrac{0.10}{12}\right)^{-36}}{\dfrac{0.10}{12}} \right]^{-1} = \$800(0.03226719) = \$25.81$$

(c) For the 2-year loan, the total amount paid out is $(36.92)(24) = \$886.08$; for the 3-year loan, the total amount paid out is $(\$25.81)(36) = \$929.16$.

Now Work Problem 9

· ·

EXAMPLE 4

Mortgage Payments   Mr. and Mrs. Corey have just purchased a $70,000 house and have made a down payment of $15,000. They can amortize the balance ($55,000) at 9% for 25 years.

(a) What are the monthly payments?
(b) What is their total interest payment?
(c) After 20 years, what equity do they have in their house (that is, what is the sum of the down payment and the amount paid on the loan)?

SOLUTION

(a) The monthly payment P needed to pay off the loan of $55,000 at 9% for 25 years (300 months) is

$$P = \$55,000 \left[\frac{1 - \left(1 + \dfrac{0.09}{12}\right)^{-300}}{\dfrac{0.09}{12}} \right]^{-1} = \$55,000(0.008392) = \$461.56$$

(b) The total paid out for the loan is

$$(\$461.56)(300) = \$138,468.00$$

Thus the interest on this loan amounts to

$$\$138,468 - \$55,000 = \$83,468.00$$

(c) After 20 years (240 months) there remains 5 years (or 60 months) of payments. The present value of the loan is the present value of a monthly payment of $461.56 for 60 months at 9%, namely,

$$($461.56)V = $461.56 \left[\frac{1 - \left(1 + \frac{0.09}{12} \right)^{-60}}{\frac{0.09}{12}} \right]$$

$$= (461.56)(48.17337) = $22,234.90$$

The amount paid on the loan is

$$\left(\begin{array}{c} \text{Original loan} \\ \text{amount} \end{array} \right) - \left(\begin{array}{c} \text{Present} \\ \text{value} \end{array} \right) = $55,000 - $22,234.90 = $32,765.10$$

Thus the equity after 20 years is

$$\left(\begin{array}{c} \text{Down} \\ \text{payment} \end{array} \right) + \left(\begin{array}{c} \text{Amount paid} \\ \text{on loan} \end{array} \right) = $15,000 + $32,765.10 = $47,765.10$$

....

Table 4 gives a partial schedule of payments for the loan in Example 4. It is interesting to observe how slowly the amount paid on the loan increases early in the payment schedule, with very little of the payment used to reduce principal, and how quickly the amount paid on the loan increases during the last 5 years.

Table 4

Payment Number	Monthly Payment	Principal	Interest	Amount Paid on Loan
1	$461.56	$49.06	$412.50	$49.06
60	$461.56	$75.94	$385.62	$3,699.94
120	$461.56	$119.27	$342.29	$9,493.23
180	$461.56	$186.85	$274.71	$18,563.67
240	$461.56	$292.56	$169.00	$32,765.10
300	$461.56	$458.16	$3.40	$55,000.00

Now Work Problem 11

EXAMPLE 5

Inheritance   When Mr. Nicholson died, he left an inheritance of $15,000 for his family to be paid to them over a 10-year period in equal amounts at the end of each year. If the $15,000 is invested at 10% per annum, what is the annual payout to the family?

SOLUTION

This example asks what annual payment is needed at 10% for 10 years to disperse $15,000. That is, we can think of the $15,000 as a loan amortized at 10% for 10 years. The payment needed to pay off the loan is the yearly amount Mr. Nicholson's family will receive. Thus the yearly payout P is

$$P = \$15,000 \left[\frac{1 - (1 + 0.10)^{-10}}{0.10} \right]^{-1}$$

$$= \$15,000(0.16274539) = \$2441.18$$

. . . .

EXAMPLE 6

Joan is 20 years away from retiring and starts saving $100 a month in an account paying 6% compounded monthly. When she retires, she wishes to withdraw a fixed amount each month for 25 years. What will this fixed amount be?

SOLUTION

After 20 years the amount accumulated in her account is the amount of an annuity with a monthly payment of $100 and an interest rate of $i = 0.06/12$. Thus

$$A = 100 \left[\frac{\left(1 + \frac{0.06}{12} \right)^{240} - 1}{\frac{0.06}{12}} \right]$$

$$= 100 \cdot (462.0408951) = \$46,204.09$$

The amount P she can withdraw for 300 payments (25 years) at 6% compounded monthly is

$$P = 46,204.09 \left[\frac{1 - \left(1 + \frac{0.06}{12} \right)^{-300}}{\frac{0.06}{12}} \right]^{-1} = \$297.69$$

. . . .

EXERCISE 5.4   Answers to odd-numbered problems begin on page AN-22.

In Problems 1–6 find the present value of each annuity.

1. The withdrawal is to be $500 per month for 36 months at 10% compounded monthly.

2. The withdrawal is to be $1000 per year for 3 years at 8% compounded annually.

3. The withdrawal is to be $100 per month for 9 months at 12% compounded monthly.

4. The withdrawal is to be $400 per month for 18 months at 5% compounded monthly.

5. The withdrawal is to be $10,000 per year for 20 years at 10% compounded annually.

6. The withdrawal is to be $2000 per month for 3 years at 4% compounded monthly.

7. A husband and wife contribute $4000 per year to an IRA paying 10% compounded annually for 20 years. What is the value of their IRA? How much can they withdraw each year for 25 years at 10% compounded annually?

8. Rework Problem 7 if the interest rate is 8%.

9. What monthly payment is needed to pay off a loan of $10,000 amortized at 12% compounded monthly for 2 years?

10. What monthly payment is needed to pay off a loan of $500 amortized at 12% compounded monthly for 2 years?

11. In Example 4 if Mr. and Mrs. Corey amortize the $55,000 loan at 10% for 20 years, what is their monthly payment?

12. In Example 4 if Mr. and Mrs. Corey amortize their $55,000 loan at 12% for 15 years, what is their monthly payment?

13. In Example 5 if Mr. Nicholson left $15,000 to be paid over 20 years in equal yearly payments and if this amount were invested at 12%, what would the annual payout be?

14. Joan has a sum of $30,000 that she invests at 10% compounded monthly. What equal monthly payments can she receive over a 10-year period? Over a 20-year period?

15. Mr. Doody, at age 65, can expect to live for 20 years. If he can invest at 10% per annum compounded monthly, how much does he need now to guarantee himself $250 every month for the next 20 years?

16. Sharon, at age 65, can expect to live for 25 years. If she can invest at 10% per annum compounded monthly, how much does she need now to guarantee herself $300 every month for the next 25 years?

17. **House Mortgage** A couple wishes to purchase a house for $120,000 with a down payment of $25,000. They can amortize the balance either at 8% for 20 years or at 9% for 25 years. Which monthly payment is greater? For which loan is the total interest paid greater? After 10 years, which loan provides the greater equity?

18. **House Mortgage** A couple has decided to purchase a $100,000 house using a down payment of $20,000. They can amortize the balance at 8% for 25 years. (a) What is their monthly payment? (b) What is the total interest paid? (c) What is their equity after 5 years? (d) What is the equity after 20 years?

19. John is 45 years old and wants to retire at 65. He wishes to make monthly deposits in an account paying 9% compounded monthly so when he retires he can withdraw $300 a month for 30 years. How much should John deposit each month?

20. The grand prize in the Illinois lottery is $6,000,000 paid out in 20 equal yearly payments of $300,000 each. How much should the state deposit in an account paying 8% compounded annually to achieve this goal?

21. Dan works during the summer to help with expenses at school the following year. He is able to save $100 each week for 12 weeks, and he invests it at 6% compounded weekly.

   (a) How much has he saved after 12 weeks?
   (b) When school starts, Dan will begin to withdraw equal amounts from this account each week. What is the most Dan can withdraw each week for 34 weeks?

22. Repeat Problem 21 if the rate of interest is 8%. What if it is only 4%?

23. Mike and Yola have just purchased a town house for $76,000. They obtain financing with the following terms: a 20% down payment and the balance to be amortized over 30 years at 9%.

   (a) What is their down payment?
   (b) What is the loan amount?
   (c) How much is their monthly payment on the loan?
   (d) How much total interest do they pay over the life of the loan?
   (e) If they pay an additional $100 each month toward the loan, when will the loan be paid?
   (f) With the $100 additional monthly payment, how much total interest is paid over the life of the loan?

24. Mr. Smith obtained a 25-year mortgage on a house. The monthly payments are $749.19 (principal and interest) and are based on a 7% interest rate. How much did Mr. Smith borrow? How much interest will be paid?

25. A car costs $12,000. You put 20% down and amortize the rest with equal monthly payments over a 3-year period at 15% to be compounded monthly. What will the monthly payment be?

26. Jay pays $320 per month for 36 months for furniture, making no down payment. If the interest charged is 0.5% per month on the unpaid balance, what was the original cost of the furniture? How much interest did he pay?

27. A restaurant owner buys equipment costing $20,000. If the owner pays 10% down and amortizes the rest with equal monthly payments over 4 years at 12% compounded monthly, what will be the monthly payments? How much interest is paid?

28. A house that was bought 8 years ago for $50,000 is now worth $100,000. Originally the house was financed by paying 20% down with the rest financed through a 25-year mortgage at 10.5% interest. The owner (after making 96 equal monthly payments) is in need of cash, and would like to refinance the house. The finance company is willing to loan 80% of the new value of the house amortized over 25 years with the same interest rate. How much cash will the owner receive after paying the balance of the original loan?

29. A home buyer is purchasing a $140,000 house. The down payment will be 20% of the price of the house, and the remainder will be financed by a 30-year mortgage at a rate of 9.8% interest compounded monthly. What will the

monthly payment be? Compare the monthly payments and the total amounts of interest paid if a 15-year mortgage is chosen instead of a 30-year mortgage.

30. Mr. and Mrs. Hoch are interested in building a house that

will cost $180,000. They intend to use the $60,000 equity in their present house as a down payment on the new one and will finance the rest with a 25-year mortgage at an interest rate of 10.2% compounded monthly. How large will their monthly payment be on the new house?

Problems 31 and 32 require logarithms.

31. How long will it take to exhaust an IRA of $100,000 if you withdraw $2000 every month? Assume a rate of interest of 5% compounded monthly.

32. How long will it take to exhaust an IRA of $200,000 if you withdraw $3000 every month? Assume a rate of interest of 4% compounded monthly.

Technology Exercises
A business or financial calculator is preprogrammed to use the present value of an annuity formula. If you have one, consult your manual and use your calculator to check your answers to Problems 2, 5, 9, 11, 31, and 32 above. A graphing calculator can be programmed to do what a financial calculator is preprogrammed to do.

1. Write a program that will calculate the present value of an annuity after n years if P dollars is withdrawn each year at $r\%$ per annum compounded annually. Use it to verify your answers to Problems 2 and 5 above.

2. Write a program that will calculate the monthly payment needed to pay off a loan of V dollars borrowed for n months

at $r\%$ per annum compounded monthly. Use it to verify your answers to Problems 9 and 11 above.

3. Write a program that will calculate the number of months it will take to exhaust A dollars if you withdraw x dollars every month. Assume a rate of interest of $r\%$ per annum compounded monthly. Use it to verify your answers to Problems 31 and 32 above.

5.5 APPLICATIONS: LEASING; CAPITAL EXPENDITURE; BONDS

Leasing

EXAMPLE 1

Lease or Purchase   A corporation may obtain a particular machine either by leasing it for 4 years (the useful life) at an annual rent of $1000 or by purchasing the machine for $3000.

(a) Which alternative is preferable if the corporation can invest money at 10% per annum?
(b) What if it can invest at 14% per annum?

SOLUTION

(a) Suppose the corporation may invest money at 10% per annum. The present value of an annuity of $1000 for 4 years at 10% equals $3169.87, which exceeds the purchase price. Therefore, purchase is preferable.

(b) Suppose the corporation may invest at 14% per annum. The present value of an annuity of $1000 for 4 years at 14% equals $2913.71, which is less than the purchase price. Hence, leasing is preferable.

Now Work Problem 1

....

Capital Expenditure

EXAMPLE 2

Selecting Equipment   A corporation is faced with a choice between two machines, both of which are designed to improve operations by saving on labor costs. Machine A costs $8000 and will generate an annual labor savings of $2000. Machine B costs $6000 and will save $1800 annually. However, machine A has a useful life of 7 years while machine B has a useful life of only 5 years. Assuming that the time value of money (the investment opportunity rate) of the corporation is 10% per annum, which machine is preferable? (Assume annual compounding and that the savings is realized at the end of each year.)

SOLUTION

Machine A costs $8000 and has a life of 7 years. Since an annuity of $1 for 7 years at 10% interest has a present value of $4.87, the cost of machine A may be considered the present value of an annuity:

$$\frac{\$8000}{4.87} = \$1642.71$$

The $1642.71 may be termed the *equivalent annual cost* of machine A. Similarly, the equivalent annual cost of machine B may be calculated by reference to the present value of an annuity of 5 years, namely,

$$\frac{\$6000}{3.79} = \$1583.11$$

Thus the net annual savings of each machine is as follows:

	A	B
Labor savings	$2000.00	$1800.00
Equivalent annual cost	1642.71	1583.11
Net savings	$ 357.29	$ 216.89

Therefore, machine A is preferable.

Now Work Problem 3

....

Bonds

We begin our third application with some definitions of terms concerning corporate bonds. A calculator will be needed for the example that follows.

FACE AMOUNT (FACE VALUE OR PAR VALUE)

The denomination of a bond. This is normally $1000. Generally, the *face amount* is the amount paid to the bondholder at maturity. It is also the amount usually paid by the bondholder when the bond is originally issued.

NOMINAL INTEREST (COUPON RATE)

The contractual periodic interest payments on a bond.

Nominal interest is normally quoted as an annual percentage of the face amount. Nominal interest payments are conventionally made semiannually, so semiannual periods are used for compound interest calculations. For example, if a bond has a face amount of $1000 and a coupon rate of 8%, then every 6 months the owner of the bond would receive $(1000)(0.08)(\frac{1}{2}) = \$40$.

But, because of market conditions, such as the current prime rate of interest, or the discount rate set by the Federal Reserve Board, or changes in the credit rating of the company issuing the bond, the price of a bond will fluctuate. When the bond price is higher than the face amount, it is trading at a **premium;** when it is lower, it is trading at a **discount.** Thus, for example, a bond with a face amount of $1000 and a coupon rate of 8% may trade in the marketplace at a price of $1100, which means the **true yield** is less than 8%.

To obtain the **true interest rate** of a bond, we view the bond as a combination of an annuity of semiannual interest payments plus a single future amount payable at maturity. The price of a bond is therefore the sum of the present value of the annuity of semiannual interest payments plus the present value of the single future payment at maturity. This present value is calculated by discounting at the true interest rate and assuming semiannual discounting periods.

EXAMPLE 3

Pricing Bonds   A bond has a face amount of $1000 and matures in 10 years. The nominal interest rate is 8.5%. What is the price of the bond to yield a true interest rate of 8%?

SOLUTION

Step 1 Calculate the amount of each semiannual interest payment:

$$(\$1000)(\tfrac{1}{2})(0.085) = \$42.50$$

Step 2 Calculate the present value of the annuity of semiannual payments:

Amount of each payment from Step 1	$ 42.50
Number of payments (2 × 10 years): 20	
True interest rate per period	
(half of stated true rate): 4%	
Factor from formula for V (use a calculator)	13.5903
Present value of interest payments	$ 577.59

Step 3 Calculate the present value of the amount payable at maturity:

Amount payable at maturity	$ 1000
Number of semiannual compounding periods before maturity: 20	
True interest rate per period: 4%	
Factor from formula for $(1 + i)^{-n}$ (use a calculator)	0.45639
Present value of maturity value	$ 456.39

Step 4 Determine the price of the bond:

Present value of interest payments	$ 577.59
Present value of maturity payment	456.39
Price of bond	$1033.98

· ·

EXERCISE 5.5   Answers to odd-numbered problems begin on page AN-22.

1. **Leasing Problem**   A corporation may obtain a machine either by leasing it for 5 years (the useful life) at an annual rent of $2000 or by purchasing the machine for $8100. If the corporation can borrow money at 10% per annum, which alternative is preferable?

2. If the corporation in Problem 1 can borrow money at 14% per annum, which alternative is preferable?

3. **Capital Expenditure Analysis** Machine A costs $10,000 and has a useful life of 8 years, and machine B costs $8000 and has a useful life of 6 years. Suppose machine A generates an annual labor savings of $2000 while machine B generates an annual labor savings of $1800. Assuming the time value of money (investment opportunity rate) is 10% per annum, which machine is preferable?

4. In Problem 3 if the time value of money is 14% per annum, which machine is preferable?

5. **Corporate Bonds**   A bond has a face amount of $1000 and matures in 15 years. The nominal interest rate is 9%. What is the price of the bond that will yield an effective interest rate of 8%?

6. For the bond in Problem 5 what is the price of the bond to yield an effective interest rate of 10%?

CHAPTER REVIEW

IMPORTANT TERMS AND CONCEPTS

simple interest formula 227
discounted loans 228
compound interest formula 233

present value 236
amount of an annuity 242
sinking fund 244

present value of an annuity 250
amortization 251

IMPORTANT FORMULAS

Simple Interest Formula

$$I = Prt$$

Discounted Loans

$$P = A - Art$$

Compound Interest Formula
$$A_n = P(1 + i)^n$$

Amount of an Annuity
$$A = P\frac{(1 + i)^n - 1}{i}$$

Present Value of an Annuity; Amortization
$$V = P \cdot \frac{1 - (1 + i)^{-n}}{i}$$

TRUE–FALSE ITEMS   Answers are on page AN-22.

T_____ F_____ **1.** Simple interest is interest computed on the principal for the entire period it is borrowed.

T_____ F_____ **2.** The amount A accrued on a principal P after five payment periods at i interest per payment period is
$$A = P(1 + i)^5$$

T_____ F_____ **3.** The effective rate of interest for 10% compounded daily is about 9.58%.

T_____ F_____ **4.** The present value of an annuity is the sum of all the present values of the payment less any interest.

FILL IN THE BLANKS   Answers are on page AN-22.

1. For a discounted loan the amount the borrower receives is called the _____ .

2. In the formula $P = A(1 + i)^{-n}$, P is called the _____ _____ of the amount A due at the end of n interest periods at i interest per payment period.

3. The amount of an _____ is the sum of all deposits made plus all interest accumulated.

4. A loan with a fixed rate of interest is _____ if both principal and interest are paid by a sequence of equal payments over equal periods of time.

REVIEW EXERCISES   Answers to odd-numbered problems begin on page AN-22.

1. Find the interest (I) and amount (A) if $400 is borrowed for 9 months at 12% simple interest.

2. Dan borrows $500 at 9% per annum simple interest for 1 year and 2 months. What is the interest charged, and what is the amount of the loan?

3. Find the amount of an investment of $100 after 2 years and 3 months at 10% compounded monthly.

4. Mike places $200 in a savings account that pays 4% per annum compounded monthly. How much is in his account after 9 months?

5. A car dealer offers Mike the choice of two loans:

(a) $3000 for 3 years at 12% per annum simple interest
(b) $3000 for 3 years at 10% per annum compounded monthly

Which loan costs Mike the least?

6. A mutual fund pays 9% per annum compounded monthly. How much should I invest now so that 2 years from now I will have $100 in the account?

7. Katy wants to buy a bicycle that costs $75 and will purchase it in 6 months. How much should she put in her savings account for this if she can get 10% per annum compounded monthly?

8. Saving for a Car   Mike decides he needs $500 1 year from now to buy a used car. If he can invest at 8% compounded monthly, how much should he save each month to buy the car?

9. Saving for a House   Mr. and Mrs. Corey are newlyweds and want to purchase a home, but they need a down payment of $10,000. If they want to buy their home in 2 years, how much should they save each month in their savings account that pays 3% per annum compounded monthly?

10. **True Cost of a Car**   Mike has just purchased a used car and will make equal payments of $50 per month for 18 months at 12% per annum charged monthly. How much did the car actually cost? Assume no down payment.

11. **House Mortgage**   Mr. and Mrs. Ostedt have just purchased an $80,000 home and made a 25% down payment. The balance can be amortized at 10% for 25 years.

    (a) What are the monthly payments?
    (b) How much interest will be paid?
    (c) What is their equity after 5 years?

12. An inheritance of $25,000 is to be paid in equal amounts over a 5-year period at the end of each year. If the $25,000 can be invested at 10% per annum, what is the annual payment?

13. **House Mortgage**   A mortgage of $125,000 is to be amortized at 9% per annum for 25 years. What are the monthly payments? What is the equity after 10 years?

14. A state has $8,000,000 worth of construction bonds that are due in 25 years. What annual sinking fund deposit is needed if the state can earn 10% per annum on its money?

15. **Depletion Problem**   How much should Mr. Graff pay for a gold mine expected to yield an annual return of $20,000 and to have a life expectancy of 20 years, if he wants to have a 15% annual return on his investment and he can set up a sinking fund that earns 10% a year?

16. Mr. Doody, at age 70, is expected to live for 15 years. If he can invest at 12% per annum compounded monthly, how much does he need now to guarantee himself $300 every month for the next 15 years?

17. **Depletion Problem**   An oil well is expected to yield an annual net return of $25,000 for the next 15 years, after which it will run dry. An investor wants a return on his investment of 20%. He can establish a sinking fund earning 10% annually. How much should he pay for the oil well?

18. Hal deposited $100 a month in an account paying 9% per annum compounded monthly for 25 years. What is the largest amount he may withdraw monthly for 35 years?

19. Mr. Jones wants to save for his son's education. If he deposits $500 every 6 months at 6% compounded semiannually, how much will he have on hand at the end of 8 years?

20. How much money should be invested at 8% compounded quarterly in order to have $20,000 in 6 years?

21. Bill borrows $1000 at 5% compounded annually. He is able to establish a sinking fund to pay off the debt and interest in 7 years. The sinking fund will earn 8% compounded quarterly. What should be the size of the quarterly payments into the fund?

22. A man has $50,000 invested at 12% compounded quarterly at the time he retires. If he wishes to withdraw money every 3 months for the next 7 years, what will be the size of his withdrawals?

23. How large should monthly payments be to amortize a loan of $3000 borrowed at 12% compounded monthly for 2 years?

24. A school board issues bonds in the amount of $20,000,000 to be retired in 25 years. How much must be paid into a sinking fund at 6% compounded annually to pay off the total amount due?

25. What effective rate of interest corresponds to a nominal rate of 9% compounded monthly?

26. If an amount was borrowed 5 years ago at 6% compounded quarterly, and $6000 is owed now, what was the original amount borrowed?

27. John is the beneficiary of a trust fund set up for him by his grandparents. If the trust fund amounts to $20,000 earning 8% compounded semiannually and he is to receive the money in equal semiannual installments for the next 15 years, how much will he receive each 6 months?

28. For 20 years a person has had on deposit $4000. At 6% annual interest compounded monthly, how much will have been saved at the end of the 20 years?

29. An employee gets paid at the end of each month and $60 is withheld from her paycheck for a retirement fund. The fund pays 1% per month (equivalent to 12% annually compounded monthly). What amount will be in the fund at the end of 30 months?

30. $6000 is borrowed at 10% compounded semiannually. The amount is to be paid back in 5 years. If a sinking fund is established to repay the loan and interest in 5 years, and the fund earns 8% compounded quarterly, how much will have to be paid into the fund every 3 months?

31. A student borrowed $4000 from a credit union toward purchasing a car. The interest rate on such a loan is 14% compounded quarterly, with payments due every quarter. The student wants to pay off the loan in 4 years. Find the quarterly payment.

Mathematical Questions from Professional Exams

1. **CPA Exam**   Which of the following should be used to calculate the amount of the equal periodic payments that could be equivalent to an outlay of $3000 at the time of the last payment?

   (a) Amount of 1
   (b) Amount of an annuity of 1
   (c) Present value of an annuity of 1
   (d) Present value of 1

2. **CPA Exam**   A businessman wants to withdraw $3000 (including principal) from an investment fund at the end of each year for 5 years. How should he compute his required initial investment at the beginning of the first year if the fund earns 6% compounded annually?

   (a) $3000 times the amount of an annuity of $1 at 6% at the end of each year for 5 years
   (b) $3000 divided by the amount of an annuity of $1 at 6% at the end of each year for 5 years
   (c) $3000 times the present value of an annuity of $1 at 6% at the end of each year for 5 years
   (d) $3000 divided by the present value of an annuity of $1 at 6% at the end of each year for 5 years

3. **CPA Exam**   A businesswoman wants to invest a certain sum of money at the end of each year for 5 years. The investment will earn 6% compounded annually. At the end of 5 years, she will need a total of $30,000 accumulated. How should she compute the required annual investment?

   (a) $30,000 times the amount of an annuity of $1 at 6% at the end of each year for 5 years
   (b) $30,000 divided by the amount of an annuity of $1 at 6% at the end of each year for 5 years
   (c) $30,000 times the present value of an annuity of $1 at 6% at the end of each year for 5 years
   (d) $30,000 divided by the present value of an annuity of $1 at 6% at the end of each year for 5 years

4. **CPA Exam**   Shaid Corporation issued $2,000,000 of 6%, 10-year convertible bonds on June 1, 1993, at 98 plus accrued interest. The bonds were dated April 1, 1993, with interest payable April 1 and October 1. Bond discount is amortized semiannually on a straight-line basis.

   On April 1, 1994, $500,000 of these bonds were converted into 500 shares of $20 par value common stock. Accrued interest was paid in cash at the time of conversion.

   What was the effective interest rate on the bonds when they were issued?

   (a) 6%   (b) Above 6%   (c) Below 6%
   (d) Cannot be determined from the information given.

CPA Exam   *Items 5–7 apply to the appropriate use of present value tables. Given below are the present value factors for $1.00 discounted at 8% for 1 to 5 periods. Each of the following items is based on 8% interest compounded annually from day of deposit to day of withdrawal.*

Periods	Present Value of $1 Discounted at 8% per Period
1	0.926
2	0.857
3	0.794
4	0.735
5	0.681

5. What amount should be deposited in a bank today to grow to $1000 3 years from today?

   (a) $\dfrac{\$1000}{0.794}$ (b) $\$1000 \times 0.926 \times 3$

   (c) $(\$1000 \times 0.926) + (\$1000 \times 0.857) + (\$1000 \times 0.794)$
   (d) $\$1000 \times 0.794$

6. What amount should an individual have in her bank account today before withdrawal if she needs $2000 each year for 4 years with the first withdrawal to be made today and each subsequent withdrawal at 1-year intervals? (She is to have exactly a zero balance in her bank account after the fourth withdrawal.)

(a) $2000 + ($2000 × 0.926) + ($2000 × 0.857) + ($2000 × 0.794)

(b) $\dfrac{\$2000}{0.735} \times 4$

(c) ($2000 × 0.926) + ($2000 × 0.857) + ($2000 × 0.794) + ($2000 × 0.735)

(d) $\dfrac{\$2000}{0.926} \times 4$

7. If an individual put $3000 in a savings account today, what amount of cash would be available 2 years from today?

(a) $3000 × 0.857

(b) $3000 × 0.857 × 2

(c) $\dfrac{\$3000}{0.857}$

(d) $\dfrac{\$3000}{0.926} \times 2$

Part Two

Probability

Chapter 6

Sets; Counting Techniques

This chapter discusses techniques and formulas for counting the number of objects in a set, a part of the branch of mathematics called *combinatorics.* These formulas are used in computer science to analyze algorithms and recursive functions and to study stacks and queues. They are also used to determine *probabilities,* the likelihood that a certain outcome of a random experiment will occur, which is the subject of Chapter 7. The final section explores the binomial theorem and some of its uses.

6.1 SETS

Set Properties and Set Notation

When we want to treat a collection of distinct objects as a whole, we use the idea of a **set**. For example, the set of **digits** consists of the collection of numbers 0, 1, 2, 3, 4, 5, 6, 7, 8, and 9. If we use the symbol D to denote the set of digits, then we can write

$$D = \{0, 1, 2, 3, 4, 5, 6, 7, 8, 9\}$$

In this notation the braces { } are used to enclose the objects, or **elements,** in the set. This method of denoting a set is called the **roster method.** A second way to denote a set is to use **set-builder notation,** where the set D of digits is written as

$$D = \quad \{ \qquad x \qquad | \qquad x \text{ is a digit}\}$$
$$\uparrow \uparrow \quad \uparrow \qquad \uparrow \qquad \quad \uparrow \qquad \quad \uparrow$$

Read as ''D is the set of all x such that x is a digit.''

EXAMPLE 1

(a) $E = \{x | x \text{ is an even digit}\} = \{0, 2, 4, 6, 8\}$
(b) $O = \{x | x \text{ is an odd digit}\} = \{1, 3, 5, 7, 9\}$

· ·

EXAMPLE 2

(a) Let E denote the set that consists of all possible outcomes resulting from tossing a coin three times. If we let H denote ''heads'' and T denote ''tails,'' then the set E can be written as

$$E = \{TTT, TTH, THT, THH, HHH, HTH, HHT, HTT \}$$

where, for instance, THT means the first toss resulted in tails, the second toss in heads, and the third toss in tails.
(b) Let F denote the set consisting of all possible arrangements of the digits without repetition. Some typical elements of F are

1478906532   4875326019   3214569870

The number of elements in F is very large, so listing all of them is impractical. Later in this chapter we will study a technique to compute the number of elements in F.

· ·

A set that has no elements, called the **empty set** or **null set,** is denoted by the symbol $\emptyset$. Also, the elements of a set are not repeated. That is, we do not write $\{3, 2, 2\}$ but rather write $\{3, 2\}$. Finally, the order in which the elements of a set are listed does not make any difference. Thus the three sets

$$\{3, 2, 4\} \qquad \{2, 3, 4\} \qquad \{4, 3, 2\}$$

are different listings of the same set. The *elements* of a set distinguish the set—not the *order* in which the elements are written.

EQUALITY OF SETS

Let A and B be two sets. We say that A *is equal to B,* written as

$$A = B$$

if and only if A and B have the same elements. If two sets A and B are not equal, we write

$$A \neq B$$

SUBSET

Let A and B be two sets. We say that A *is a subset of B* or that A *is contained in B,* written as

$$A \subseteq B$$

if and only if every element of A is also an element of B. If a set A is not a subset of a set B, we write

$$A \nsubseteq B$$

When we say that A is a subset of B, we can also say "there are no elements in set A that are not also elements in set B." Of course, $A \subseteq B$ if and only if whenever x is in A, then x is in B for all x. This latter way of interpreting the meaning of $A \subseteq B$ is useful for obtaining various laws that sets obey.

From the definition it follows that for any set A, $A \subseteq A$; that is, every set is a subset of itself.

PROPER SUBSET

Let A and B be two sets. We say that A *is a proper subset of B* or that A *is properly contained in B,* written as

$$A \subset B$$

if and only if every element in the set A is also in the set B, but there is at least one element in set B that is *not* in set A.

Notice that "A is a proper subset of B" means that there are *no* elements of A that are not also elements of B, but there is at least one element of B that is not in A.

If a set A is *not* a proper subset of a set B, we write

$$A \not\subset B$$

From the definition it follows that $A \not\subset A$; that is, a set is never a proper subset of itself.

The following example illustrates some uses of the three relationships, $=$, $\subseteq$, and $\subset$, just defined.

EXAMPLE 3   Consider three sets A, B, and C given by

$$A = \{1, 2, 3\} \qquad B = \{1, 2, 3, 4, 5\} \qquad C = \{3, 2, 1\}$$

Some of the relationships between pairs of these sets are

(a) $A = C$ (b) $A \subseteq B$ (c) $A \subseteq C$ (d) $A \subset B$ (e) $C \subseteq A$

....

In comparing the two definitions of *subset* and *proper subset,* you should notice that if a set A is a subset of a set B, then either A is a proper subset of B or else A equals B. That is,

$$A \subseteq B \qquad \text{if and only if either} \qquad A \subset B \text{ or } A = B$$

Also, if A is a proper subset of B, we can infer that A is a subset of B, but A does not equal B. That is,

$$A \subset B \qquad \text{if and only if} \qquad A \subseteq B \text{ and } A \neq B$$

The distinction that is made between *subset* and *proper subset* is rather subtle but quite important.

We can think of the relationship $\subset$ as a refinement of $\subseteq$. On the other hand, the relationship $\subseteq$ is an extension of $\subset$, in the sense that $\subseteq$ may include equality whereas with $\subset$, equality cannot be included.

Let A denote any set. Since the empty set $\varnothing$ has no elements, there is no element of the set $\varnothing$ that is not also in A. That is,

$$\varnothing \subseteq A$$

Also, if A is any nonempty set, that is, any set having at least one element, then

$$\varnothing \subset A$$

Now Work Problems 1 and 3

In applications the elements that may be considered are usually limited to some specific all-encompassing set. For example, in discussing students eligible to graduate from Midwestern University, the discussion would be limited to students enrolled at the university.

UNIVERSAL SET

The *universal set* U is defined as the set consisting of all elements under consideration.

Thus if A is any set and if U is the universal set, then every element in A must be in U (since U consists of all elements under consideration). Hence, we may write

$$A \subseteq U$$

for *any* set A.

It is convenient to represent a set as the interior of a circle. Two or more sets may be depicted as circles enclosed in a rectangle, which represents the universal set. The circles may or may not overlap, depending on the situation. Such diagrams of sets are called **Venn diagrams.** See Figure 1.

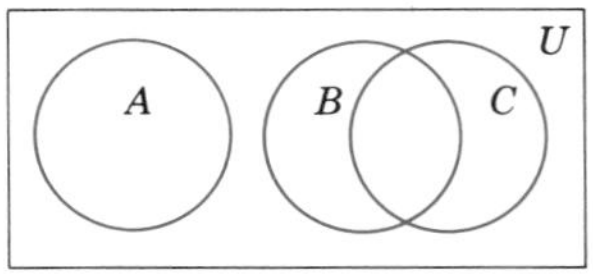

Figure 1

Operations on Sets

Next, we introduce operations that may be performed on sets.

UNION OF TWO SETS

Let A and B be any two sets. The *union* of A with B, written as

$$A \cup B$$

and read as "A union B" or as "A or B," is defined to be the set consisting of those elements either in A or in B or in both A and B. That is,

$$A \cup B = \{x \mid x \text{ is in } A \text{ or } x \text{ is in } B\}$$

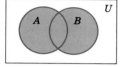

Figure 2

Warning! In English, the word *or* has two meanings: the **inclusive or** "A or B" means A or B or both. The **exclusive or** "A or B" means A or B but *not* both. Thus A or B in mathematics means use the *inclusive or* from English.

In the Venn diagram in Figure 2 the shaded area corresponds to $A \cup B$.

INTERSECTION OF TWO SETS

Let A and B be any two sets. The *intersection* of A with B, written as

$$A \cap B$$

and read as "A intersect B" or as "A and B," is defined as the set consisting of those elements that are in both A and B. That is,

$$A \cap B = \{x \mid x \text{ is in } A \text{ and } x \text{ is in } B\}$$

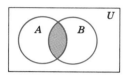

Figure 3

In other words, to find the intersection of two sets A and B means to find the elements *common* to A *and* B. In the Venn diagram in Figure 3 the shaded region is $A \cap B$.

For any set A it follows that

$$A \cup \varnothing = A \qquad A \cap \varnothing = \varnothing$$

EXAMPLE 4

Use the sets

$$A = \{1, 3, 5\} \qquad B = \{3, 4, 5, 6\} \qquad C = \{6, 7\}$$

to find   (a) $A \cup B$ (b) $A \cap B$ (c) $A \cap C$

SOLUTION

(a) $A \cup B = \{1, 3, 5\} \cup \{3, 4, 5, 6\} = \{1, 3, 4, 5, 6\}$
(b) $A \cap B = \{1, 3, 5\} \cap \{3, 4, 5, 6\} = \{3, 5\}$
(c) $A \cap C = \{1, 3, 5\} \cap \{6, 7\} = \varnothing$

EXAMPLE 5

Let T be the set of all taxpayers and let S be the set of all people over 65 years of age. Describe $T \cap S$.

SOLUTION

Now Work Problems 11 and 13

$T \cap S$ is the set of all taxpayers who are also over 65 years of age.

DISJOINT SETS

If two sets A and B have no elements in common, that is, if

$$A \cap B = \varnothing$$

then A and B are called *disjoint sets*.

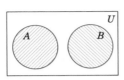

Figure 4

Two disjoint sets A and B are illustrated in the Venn diagram in Figure 4. Notice that the areas corresponding to A and B do not overlap anywhere because $A \cap B$ is empty.

EXAMPLE 6

Suppose that a die* is tossed. Let A be the set of outcomes in which an even number turns up; let B be the set of outcomes in which an odd number shows. Find $A \cap B$.

SOLUTION

$$A = \{2, 4, 6\} \qquad B = \{1, 3, 5\}$$

We note that A and B have no elements in common since an even number and an odd number cannot occur simultaneously on a single toss of a die. Therefore, $A \cap B = \varnothing$ and the sets A and B are disjoint sets.

. . . .

Suppose we consider all the employees of some company as our universal set U. Let A be the subset of employees who smoke. Then all the nonsmokers will make up the subset of U that is called the *complement* of the set of smokers.

COMPLEMENT OF A SET

Let A be any set. The *complement* of A, written as

$$\overline{A} \qquad \text{(or } A'\text{)}$$

is defined as the set consisting of elements in the universe U that are not in A. Thus

$$\overline{A} = \{x \mid x \text{ is not in } A\}$$

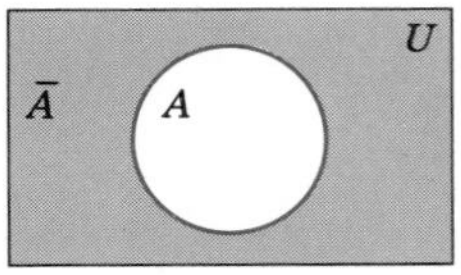

The shaded region in Figure 5 illustrates the complement, $\overline{A}$.
For any set A it follows that

$$A \cup \overline{A} = U \qquad A \cap \overline{A} = \varnothing \qquad \overline{\overline{A}} = A$$

Figure 5

EXAMPLE 7

Use the sets

$$U = \{a, b, c, d, e, f\} \qquad A = \{a, b, c\} \qquad B = \{a, c, f\}$$

to list the elements of the following sets:

(a) $\overline{A}$ (b) $\overline{B}$ (c) $\overline{A \cup B}$
(d) $\overline{A} \cap \overline{B}$ (e) $\overline{A \cap B}$ (f) $\overline{A} \cup \overline{B}$

* A **die** (*plural* **dice**) is a cube with 1, 2, 3, 4, 5, or 6 dots showing on the six faces.

SOLUTION

(a) $\overline{A}$ consists of all the elements in U that are not in A, namely, $\overline{A} = \{d, e, f\}$.

(b) Similarly, $\overline{B} = \{b, d, e\}$.

(c) To determine $\overline{A \cup B}$, we first determine the elements in $A \cup B$:

$$A \cup B = \{a, b, c, f\}$$

The complement of the set $A \cup B$ is then

$$\overline{A \cup B} = \{d, e\}$$

(d) From Parts (a) and (b) we find that

$$\overline{A} \cap \overline{B} = \{d, e\}$$

(e) As in Part (c) we first determine the elements in $A \cap B$:

$$A \cap B = \{a, c\}$$

Then

$$\overline{A \cap B} = \{b, d, e, f\}$$

(f) From Parts (a) and (b) we find that

$$\overline{A} \cup \overline{B} = \{b, d, e, f\}$$

Now Work Problem 23(d)
and (e)

The answers to Parts (c) and (d) in Example 7 are the same, and so are the results from Parts (e) and (f). This is no coincidence. There are two important properties involving intersections and unions of complements of sets. They are known as *De Morgan's properties:*

De Morgan's Properties

Let A and B be any two sets. Then

(a) $\overline{A \cup B} = \overline{A} \cap \overline{B}$ **(b)** $\overline{A \cap B} = \overline{A} \cup \overline{B}$

De Morgan's properties state that all we need to do to form the complement of a union (or intersection) of sets is to form the complements of the individual sets and then change the union symbol to an intersection (or the intersection to a union). We shall employ Venn diagrams to verify De Morgan's properties.

(a) First we draw two diagrams, as shown in Figure 6.

Figure 6

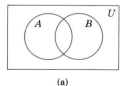

(a)

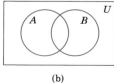

(b)

We will use the diagram on the left for $\overline{A} \cap \overline{B}$ and the one on the right for $\overline{A \cup B}$. Figure 7 illustrates the completed Venn diagrams of these sets.

Figure 7

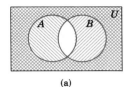
(a)

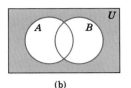
(b)

Thus in Figure 7(a) $\overline{A} \cap \overline{B}$ is represented by the cross-hatched region, and in Figure 7(b) $\overline{A \cup B}$ is represented by the shaded region. Since these regions correspond, this illustrates that the two sets $\overline{A} \cap \overline{B}$ and $\overline{A \cup B}$ are equal.

(b) This verification is left to you. See Problem 28, part (c).

EXAMPLE 8

Use a Venn diagram to illustrate the set $(A \cup B) \cap C$.

SOLUTION

First we construct Figure 8(a). Then we hatch $A \cup B$ and C as in Figure 8(b). The cross-hatched region of Figure 8(b) is the set $(A \cup B) \cap C$.

Figure 8

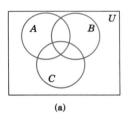
(a)

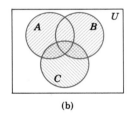
(b)

....

EXAMPLE 9

Use a Venn diagram to illustrate the following equality:

$$A \cup B = (A \cap \overline{B}) \cup (A \cap B) \cup (\overline{A} \cap B)$$

SOLUTION

First we construct Figure 9. Now hatch the regions $A \cap \overline{B}$, $A \cap B$, and $\overline{A} \cap B$, as shown in Figure 10. The union of the three regions in Figure 10 is the set $A \cup B$. See Figure 11.

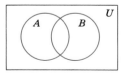

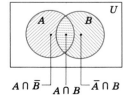
$A \cap \overline{B}$ $A \cap B$ $\overline{A} \cap B$

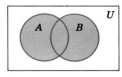

Now Work Problem 27(a) and (g)

Figure 9   **Figure 10**   **Figure 11**

....

The Number of Elements in a Set

When we count objects, what we are actually doing is taking each object to be counted and matching each of these objects exactly once to the counting numbers 1, 2, 3, and so on, until *no* objects remain. Even before numbers had names and symbols assigned to them, this method of counting was used. Prehistoric peoples used rocks to determine how many cattle did not return from pasture. As each cow left, a rock was placed aside. As each cow returned, a rock was removed from the pile. If rocks remained after all the cows returned, it was then known that some cows were missing.

If A is any set, we will denote by $c(A)$ the number of elements in A. Thus, for example, for the set L of letters in the alphabet,

$$L = \{a, b, c, d, e, f, \ldots, x, y, z\}$$

we write $c(L) = 26$ and say "the number of elements in L is 26."

Also, for the set

$$N = \{1, 2, 3, 4, 5\}$$

we write $c(N) = 5$.

The empty set $\varnothing$ has no elements, and we write

$$c(\varnothing) = 0$$

If the number of elements in a set is zero or a positive integer, we say that the set is **finite.** Otherwise, the set is said to be **infinite.** The area of mathematics that deals with the study of finite sets is called **finite mathematics.**

EXAMPLE 10

A survey of a group of people indicated there were 25 with brown eyes and 15 with black hair. If 10 people had both brown eyes and black hair and 23 people had neither, how many people were interviewed?

SOLUTION

Let A denote the set of people with brown eyes and B the set of people with black hair. Then the data given tell us

$$c(A) = 25 \qquad c(B) = 15 \qquad c(A \cap B) = 10$$

The number of people with either brown eyes or black hair cannot be $c(A) + c(B)$, since those with both would be counted twice. The correct procedure then would be to subtract those with both. That is,

$$c(A \cup B) = c(A) + c(B) - c(A \cap B) = 25 + 15 - 10 = 30$$

The sum of people found either in A or in B and those found neither in A nor in B is the total interviewed. Thus the number of people interviewed is

$$30 + 23 = 53$$

Figure 12 illustrates how a Venn diagram can be used to represent this situation.

· ·

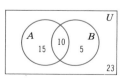

Figure 12

In Example 10 we discovered the following important relationship:

> **Counting Formula**
>
> Let A and B be two finite sets. Then
>
> $$c(A \cup B) = c(A) + c(B) - c(A \cap B) \tag{1}$$

EXAMPLE 11

Let $A = \{a, b, c, d, e\}$, $B = \{a, e, g, u, w, z\}$. Find $c(A)$, $c(B)$, $c(A \cap B)$, and $c(A \cup B)$.

SOLUTION

$c(A) = 5$ and $c(B) = 6$. To find $c(A \cap B)$, we note that $A \cap B = \{a, e\}$ so that $c(A \cap B) = 2$. Since $A \cup B = \{a, b, c, d, e, g, u, w, z\}$, we have $c(A \cup B) = 9$. This checks with the formula

$$c(A \cup B) = c(A) + c(B) - c(A \cap B) = 5 + 6 - 2 = 9$$

Now Work Problem 41

· ·

Applications

EXAMPLE 12

Consumer Survey   In a survey of 75 consumers, 12 indicated they were going to buy a new car, 18 said they were going to buy a new refrigerator, and 24 said they were going to buy a new stove. Of these, 6 were going to buy both a car and a refrigerator, 4 were going to buy a car and a stove, and 10 were going to buy a stove and refrigerator. One person indicated he was going to buy all three items.

(a) How many were going to buy none of these items?
(b) How many were going to buy only a car?
(c) How many were going to buy only a stove?
(d) How many were going to buy only a refrigerator?
(e) How many were going to buy a stove and refrigerator, but not a car?

SOLUTION

We denote the sets of people buying cars, refrigerators, and stoves by C, R, and S, respectively. Then we know from the data given that

$$c(C) = 12 \qquad c(R) = 18 \qquad c(S) = 24$$
$$c(C \cap R) = 6 \qquad c(C \cap S) = 4 \qquad c(S \cap R) = 10$$
$$c(C \cap R \cap S) = 1$$

We use the information given above in the reverse order and put it into a Venn diagram. Thus, beginning with the fact that $c(C \cap R \cap S) = 1$, we place a 1 in that set, as shown in Figure 13(a). Now $c(C \cap R) = 6$, $c(C \cap S) = 4$, and $c(S \cap R) = 10$. Thus we place $6 - 1 = 5$ in the proper region (giving a total of 6 in the set $C \cap R$). Similarly, we place 3 and 9 in the proper regions for the sets $C \cap S$ and $S \cap R$. See Figure 13(b). Now $c(C) = 12$ and 9 of these 12 are already accounted for. Also, $c(R) = 18$ with 15 accounted for and $c(S) = 24$ with 13 accounted for.

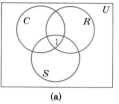
(a)

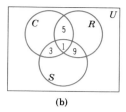
(b)

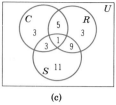
(c)

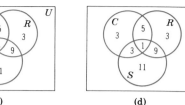
(d)

Figure 13

See Figure 13(c). Finally, the number in $\overline{C \cup R \cup S}$ is the total of 75 less those accounted for in C, R, and S, namely, $3 + 5 + 1 + 3 + 3 + 9 + 11 = 35$. Thus

$$c(\overline{C \cup R \cup S}) = 75 - 35 = 40$$

See Figure 13(d). From this figure, we can see that (a) 40 were going to buy none of the items; (b) 3 were going to buy only a car; (c) 11 were going to buy only a stove; (d) 3 were going to buy only a refrigerator, and (e) 9 were going to buy a stove and refrigerator, but not a car.

Now Work Problem 51

· ·

EXAMPLE 13

Analyzing Data   In a survey of 10,281 people restricted to those who were either black or male or over 18 years of age, the following data were obtained:

Black:	3490	Black males:	1745	Black male over 18:	239
Male:	5822	Over 18 and male:	859		
Over 18:	4722	Over 18 and black:	1341		

The data are inconsistent. Why?

SOLUTION

We denote the set of people who were black by B, male by M, and over 18 by H. Then we know that

$$c(B) = 3490 \qquad c(M) = 5822 \qquad c(H) = 4722$$
$$c(B \cap M) = 1745 \qquad c(H \cap M) = 859 \qquad c(H \cap B) = 1341$$
$$c(H \cap M \cap B) = 239$$

Since $H \cap M \cap B \neq \varnothing$, we use the Venn diagram shown in Figure 14. This means that

$$239 + 1102 + 620 + 1506 + 3457 + 643 + 2761 = 10,328$$

people were interviewed. However, it is given that only 10,281 were interviewed. This means the data are inconsistent.

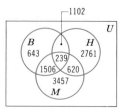

Figure 14

· ·

EXERCISE 6.1   Answers to odd-numbered problems begin on page AN-23.

In Problems 1–10 tell whether the given statement is true or false.

1. $\{1, 3, 2\} = \{2, 1, 3\}$ **2.** $\{2, 3\} \subseteq \{1, 2, 3\}$ **3.** $\{6, 7, 9\} \subseteq \{1, 6, 9\}$

4. $\{4, 8, 2\} = \{2, 4, 6, 8\}$ **5.** $\{6, 7, 9\} \subset \{1, 6, 9\}$ **6.** $\{4, 8, 2\} \subset \{2, 4, 6, 8\}$

7. $\{1, 2\} \cap \{2, 3, 4\} = \{2\}$ **8.** $\{2, 3\} \cap \{2, 3, 4\} = \{2, 3, 4\}$ **9.** $\{4, 5\} \cap \{1, 2, 3, 4\} \subseteq \{4\}$

10. $\{1, 4\} \cup \{2, 3\} \subseteq \{1, 2, 3, 4\}$

In Problems 11–18 write each expression as a single set.

11. $\{1, 2, 3\} \cap \{2, 3, 4, 5\}$ **12.** $\{0, 1, 3\} \cup \{2, 3, 4\}$ **13.** $\{1, 2, 3\} \cup \{2, 3, 4, 5\}$

14. $\{0, 1, 2\} \cap \{2, 3, 4\}$ **15.** $\{2, 4, 6, 8\} \cap \{1, 3, 5, 7\}$ **16.** $\{2, 4, 6\} \cup \{1, 3, 5\}$

17. $\{a, b, e\} \cup \{d, e, f, q\}$ **18.** $\{a, e, m\} \cup \{p, o, m\}$

In Problems 19–22 find the number of elements in each set.

19. $\{2, 4, 6, 8\}$ **20.** $\{0, 1, 2\}$ **21.** $\{0, 1, 2, 3, 4, 5, 6, 7, 8, 9\}$ **22.** $\{10\}$

23. If U = Universal set = $\{0, 1, 2, 3, 4, 5, 6, 7, 8, 9\}$ and if $A = \{0, 1, 5, 7\}$, $B = \{2, 3, 5, 8\}$, $C = \{5, 6, 9\}$, find

(a) $A \cup B$ (b) $B \cap C$
(c) $A \cap B$ (d) $\overline{A \cap B}$
(e) $\overline{A} \cap \overline{B}$ (f) $A \cup (B \cap A)$
(g) $(C \cap A) \cap (\overline{A})$ (h) $(A \cap B) \cup (B \cap C)$

24. If U = Universal set = $\{1, 2, 3, 4, 5\}$ and if $A = \{3, 5\}$, $B = \{1, 2, 3\}$, $C = \{2, 3, 4\}$, find

(a) $\overline{A} \cap \overline{C}$ (b) $(A \cup B) \cap C$
(c) $A \cup (B \cap C)$ (d) $(A \cup B) \cap (A \cup C)$
(e) $\overline{A} \cap C$ (f) $\overline{A \cup B}$
(g) $\overline{A} \cap \overline{B}$ (h) $(A \cap B) \cup C$

25. Let U = {All letters of the alphabet}, $A = \{b, c, d\}$, and $B = \{c, e, f, g\}$. List the elements of the sets:

(a) $A \cup B$ (b) $A \cap B$
(c) $\overline{A} \cap \overline{B}$ (d) $\overline{A \cup B}$

26. Let $U = \{a, b, c, d, e, f\}$, $A = \{b, c\}$, and $B = \{c, d, e\}$. List the elements of the sets:

(a) $A \cup B$ (b) $A \cap B$
(c) $\overline{A}$ (d) $\overline{B}$
(e) $\overline{A} \cap B$ (f) $\overline{A \cup B}$

27. Use Venn diagrams to illustrate the following sets:

(a) $\overline{A} \cap B$ (b) $(\overline{A} \cap \overline{B}) \cup C$
(c) $A \cap (A \cup B)$
(d) $A \cup (A \cap B)$
(e) $(A \cup B) \cap (A \cup C)$
(f) $A \cup (B \cap C)$
(g) $A = (A \cap B) \cup (A \cap \overline{B})$
(h) $B = (A \cap B) \cup (\overline{A} \cap B)$

28. Use Venn diagrams to illustrate the following properties:

(a) $A \cap (B \cup C) = (A \cap B) \cup (A \cap C)$
     (Distributive property)
(b) $A \cap (A \cup B) = A$ (Absorption property)
(c) $\overline{A \cap B} = \overline{A} \cup \overline{B}$ (De Morgan's property)
(d) $(A \cup B) \cup C = A \cup (B \cup C)$
     (Associative property)

In Problems 29–32 use

$$A = \{x \mid x \text{ is a customer of IBM}\}$$

$$B = \{x \mid x \text{ is a secretary employed by IBM}\}$$

$$C = \{x \mid x \text{ is a computer operator at IBM}\}$$

$$D = \{x \mid x \text{ is a stockholder of IBM}\}$$

$$E = \{x \mid x \text{ is a member of the Board of Directors of IBM}\}$$

to describe each set in words.

29. $A \cap E$ **30.** $B \cap D$ **31.** $A \cup D$ **32.** $C \cap E$

In Problems 33–36 use

$$U = \{\text{All college students}\}$$
$$M = \{\text{All male students}\}$$
$$S = \{\text{All students who smoke}\}$$

to describe each set in words.

33. $M \cap S$ **34.** $\overline{M}$ **35.** $\overline{M} \cap \overline{S}$ **36.** $M \cup S$

In Problems 37–40 use the sets A = {1, 2, 3, 4, 5, 6} and B = {2, 4, 6, 8} to find the number of elements in each set.

37. $A \cap B$ **38.** $A \cup B$ **39.** $(A \cap B) \cup A$ **40.** $(B \cap A) \cup B$

41. Find $c(A \cup B)$, given that $c(A) = 4$, $c(B) = 3$, and $c(A \cap B) = 2$.

42. Find $c(A \cup B)$, given that $c(A) = 14$, $c(B) = 11$, and $c(A \cap B) = 6$.

43. Find $c(A \cap B)$, given that $c(A) = 5$, $c(B) = 4$, and $c(A \cup B) = 7$.

44. Find $c(A \cap B)$, given that $c(A) = 8$, $c(B) = 9$, and $c(A \cup B) = 16$.

45. Find $c(A)$, given that $c(B) = 8$, $c(A \cap B) = 4$, and $c(A \cup B) = 14$.

46. Find $c(B)$, given that $c(A) = 10$, $c(A \cap B) = 5$, and $c(A \cup B) = 29$.

47. Motors, Inc., manufactured 325 cars with automatic transmissions, 216 with power steering, and 89 with both these options. How many cars were manufactured if every car has at least one option?

48. Suppose that out of 1500 first-year students at a certain college, 350 are taking history, 300 are taking mathematics, and 270 are taking both history and mathematics. How many first-year students are taking history or mathematics?

In Problems 49–58 use the data in the figure to answer each question.

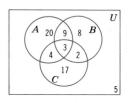

49. How many elements are in set A?   **50.** How many elements are in set B?

51. How many elements are in A or B?   **52.** How many elements are in B or C?

53. How many elements are in A but not B?   **54.** How many elements are in B but not C?

55. How many elements are in A or B or C?   **56.** How many elements are in neither A nor B nor C?

57. How many elements are in A and B and C?   **58.** How many elements are in U?

59. Voting Patterns   Suppose the influence of religion and age on voting preference is given by the following table:

	Age		
	Below 35	35–54	Over 54
Protestant Voting Republican	82	152	111
Protestant Voting Democratic	42	33	15
Catholic Voting Republican	27	33	7
Catholic Voting Democratic	44	47	33

Find

(a) The number of voters who are Catholic or Republican or both.
(b) The number of voters who are Catholic or over 54 or both.
(c) The number voting Democratic below 35 or over 54.

60. The Venn diagram illustrates the number of seniors (S), female students (F), and students on the dean's list (D) at a small western college. Describe each number in terms of the sets S, F, or D.

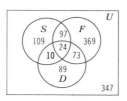

61. At a small midwestern college:

31	female seniors were on the dean's list
62	women were on the dean's list who were not seniors
45	male seniors were on the dean's list
87	female seniors were not on the dean's list
96	male seniors were not on the dean's list
275	women were not seniors and were not on the dean's list
89	men were on the dean's list who were not seniors
227	men were not seniors and were not on the dean's list

(a) How many were seniors?
(b) How many were women?

(c) How many were on the dean's list?
(d) How many were seniors on the dean's list?
(e) How many were female seniors?
(f) How many were women on the dean's list?
(g) How many were students at the college?

62. Survey Analysis   In a survey of 75 college students, it was found that of the three weekly news magazines *Time, Newsweek*, and *U.S. News and World Report*:

23	read *Time*
18	read *Newsweek*
14	read *U.S. News and World Report*
10	read *Time* and *Newsweek*
9	read *Time* and *U.S. News and World Report*
8	read *Newsweek* and *U.S. News and World Report*
5	read all three

(a) How many read none of these three magazines?
(b) How many read *Time* alone?
(c) How many read *Newsweek* alone?
(d) How many read *U.S. News and World Report* alone?
(e) How many read neither *Time* nor *Newsweek*?
(f) How many read *Time* or *Newsweek* or both?

63. Car Sales   Of the cars sold during the month of July, 90 had air conditioning, 100 had automatic transmissions, and 75 had power steering. Five cars had all three of these extras. Twenty cars had none of these extras. Twenty cars had only air conditioning; 60 cars had only automatic transmissions; and 30 cars had only power steering. Ten cars had both automatic transmission and power steering.

(a) How many cars had both power steering and air conditioning?
(b) How many had both automatic transmission and air conditioning?
(c) How many had neither power steering nor automatic transmission?
(d) How many cars were sold in July?
(e) How many had automatic transmission or air conditioning or both?

64. Incorrect Information   A staff member at a large engineering school was presenting data to show that the students there received a liberal education as well as a scientific one. "Look at our record," she said. "Out of our senior class of 500 students, 281 are taking English, 196 are taking English and history, 87 are taking history and a foreign language, 143 are taking a foreign language and English, and 36 are taking all of these." She was fired. Why?

65. Blood Classification   Blood is classified as being either Rh-positive or Rh-negative and according to type. If blood contains an A antigen, it is type A; if it has a B antigen, it is type B; if it has both A and B antigens, it is type AB; and if it has neither antigen, it is type O. Use a Venn diagram to illustrate these possibilities. How many different possibilities are there?

66. Survey Analysis   A survey of 52 families from a suburb of Chicago indicated that there was a total of 241 children below the age of 18. Of these, 109 were male; 132 were below the age of 11; 143 had played Little League; 69 males were below the age of 11; 45 females under 11 had played Little League; and 30 males under 11 had played Little League.

(a) How many children over 11 and under 18 had played Little League?

(b) How many females under 11 did not play Little League?

67. Survey Analysis   Of 100 personal computer users surveyed: 27 use IBM; 35 use Apple; 35 use AT&T; 10 use both IBM and Apple; 10 use both IBM and AT&T; 10 use both Apple and AT&T; 3 use all three; and 30 use another computer brand. How many people exclusively use one of the three brands mentioned, that is, only IBM or only Apple or only AT&T?

68. List all the subsets of $\{a, b, c\}$. How many are there?

69. List all the subsets of $\{a, b, c, d\}$. How many are there?

6.2 THE MULTIPLICATION PRINCIPLE

In this section we introduce a general principle of counting, called the *Multiplication Principle*. We begin with two examples.

EXAMPLE 1

In traveling from New York to Los Angeles, Mr. Doody wishes to stop over in Chicago. If he has five different routes to choose from in driving from New York to Chicago and has three routes to choose from in driving from Chicago to Los Angeles, in how many ways can Mr. Doody travel from New York to Los Angeles?

SOLUTION

The task of traveling from New York to Los Angeles is composed of two consecutive operations:

Choose a route from New York to Chicago Task 1	Choose a route from Chicago to Los Angeles Task 2

In Figure 15 we see after the five routes from New York to Chicago there are three routes from Chicago to Los Angeles. These different routes can be enumerated as

$1A, 1B, 1C \qquad 2A, 2B, 2C \qquad 3A, 3B, 3C \qquad 4A, 4B, 4C \qquad 5A, 5B, 5C$

Thus, in all, there are $5 \cdot 3 = 15$ different routes.

Figure 15

Notice that the total number of ways the trip can be taken is simply the product of the number of ways of doing Task 1 with the number of ways of doing Task 2.

....

The different routes in Example 1 can also be depicted in a **tree diagram.** See Figure 16.

Figure 16

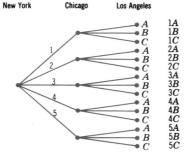

Now Work Problem 1

EXAMPLE 2

In a city election there are four candidates for mayor, three candidates for vice-mayor, six candidates for treasurer, and two for secretary. In how many ways can these four offices be filled?

SOLUTION

The task of filling an office can be divided into four consecutive operations:

| Select a mayor | Select a vice-mayor | Select a treasurer | Select a secretary |

Corresponding to each of the four possible mayors, there are three vice-mayors. These two offices can be filled in $4 \cdot 3 = 12$ different ways. Also, corresponding to each of these 12 possibilities, we have six different choices for treasurer—giving $12 \cdot 6 = 72$ different possibilities. Finally, to each of these 72 possibilities there correspond two choices for secretary. Thus, in all, these offices can be filled in $4 \cdot 3 \cdot 6 \cdot 2 = 144$ different ways. A partial illustration is given by the tree diagram in Figure 17.

Figure 17

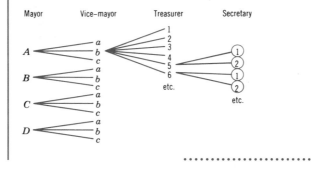

The examples just solved demonstrate a general type of counting problem, which can be solved by the Multiplication Principle.

> **The Multiplication Principle**
>
> If we can perform a first task in p different ways, a second task in q different ways, a third task in r different ways, . . . , then the total act of performing the first task, followed by performing the second task, and so on, can be done in $p \cdot q \cdot r \cdot \ \cdot \ \cdot$ different ways.

EXAMPLE 3

A particular type of combination lock has 10 numbers on it. How many sequences of four numbers can be formed to open the lock?

SOLUTION

Each of the four numbers can be chosen in 10 ways. Therefore, there are

$$10 \cdot 10 \cdot 10 \cdot 10 = 10,000$$

different sequences.

. . . .

EXAMPLE 4

(a) How many possible batting orders can the manager of a baseball team construct from nine available players?
(b) If the manager adheres to the rule that the pitcher always bats last and the star homerun hitter is always fourth (cleanup), then how many batting orders are possible from nine given players?

SOLUTION

(a) In the first slot any of the nine players can be chosen. In the second slot any one of the remaining eight can be chosen, and so on, so that there are

$$9 \cdot 8 \cdot 7 \cdot 6 \cdot 5 \cdot 4 \cdot 3 \cdot 2 \cdot 1 = 362,880$$

possible batting orders.
(b) Since two of the players have designated spots in the batting order, there are seven remaining positions to be filled. The first position can be filled in any one of seven ways, the second in any of six ways, the third in any of five ways, the fourth in four ways, the fifth in three ways, the sixth in two ways, and the seventh in one way, so that here there are

$$7 \cdot 6 \cdot 5 \cdot 4 \cdot 3 \cdot 2 \cdot 1 = 5040$$

different batting orders of nine players after two have been designated.

Now Work Problem 9

. . . .

EXAMPLE 5

License plates in the state of Maryland consist of three letters of the alphabet followed by three digits.

(a) The Maryland system will allow how many possible license plates?
(b) Of these, how many will have all their digits distinct?

6 5 4 3

☐ ☐ ☐ ☐

SOLUTION

(a) There are six positions on the plate to be filled, the first three by letters and the last three by digits. The positions 1, 2, and 3 can be filled in any one of 26 ways, while the remaining positions can each be filled in any of 10 ways. The total number of plates, by the Multiplication Principle, is then

$$26 \cdot 26 \cdot 26 \cdot 10 \cdot 10 \cdot 10 = 17,576,000$$

(b) Here, the tasks involved in filling the digit positions are slightly different. The first digit can be any one of 10, but the second digit can be only any one of 9 (we cannot duplicate the first digit); there are only 8 choices for the third digit (we cannot duplicate either the first or the second). Thus there are

$$26 \cdot 26 \cdot 26 \cdot 10 \cdot 9 \cdot 8 = 12,654,720$$

plates with no repeated digit.

....

EXERCISE 6.2   Answers to odd-numbered problems begin on page AN-24.

1. There are 2 roads between towns A and B. There are 4 roads between towns B and C. How many different routes may one travel between towns A and C?

2. A woman has 4 blouses and 5 skirts. How many different outfits can she wear?

3. XYZ Company wants to build a complex consisting of a factory, office building, and warehouse. If the building contractor has 3 different kinds of factories, 2 different office buildings, and 4 different warehouses, how many models must be built to show all possibilities to XYZ Company?

4. Cars, Inc., has 3 different car models and 6 color schemes. If you are one of the dealers, how many cars must you display to show each possibility?

5. A man has 3 pairs of shoes, 8 pairs of socks, 4 pairs of slacks, and 9 sweaters. How many outfits can he wear?

6. There are 14 teachers in a math department. A student is asked to indicate her favorite and her least favorite. In how many ways is this possible?

7. A house has 3 outside doors and 12 windows. In how many ways can a person enter the house through a window and exit through a door?

8. A woman has 4 pairs of gloves. In how many ways can she select a right-hand glove and a left-hand glove that do not match?

9. A corporation has a board of directors consisting of 12 members. The board must select from its members a chairman, vice-chairman, and a secretary. In how many ways can this be done?

10. How many license plates consisting of 2 letters followed by 2 digits are possible?

11. A restaurant offers 6 different salads, 5 different main courses, 10 different desserts, and 4 different drinks. How many different lunches—each consisting of a salad, a main course, a dessert, and a drink—are possible?

12. Five different mathematics books are to be arranged on a student's desk. How many arrangements are possible?

13. How many ways can 6 people be seated in a row of 6 seats? 8 people be seated in a row of 8 seats?

14. How many different arrangements can be formed using the 5 letters of the word CLIPS?

15. How many 4-letter code words are possible from the first 6 letters of the alphabet with no letters repeated? How many when letters are allowed to repeat?

16. Find the number of 7-digit telephone numbers

   (a) With no repeated digits (lead 0 is allowed)
   (b With no repeated digits (lead 0 not allowed)
   (c) With repeated digits allowed including a lead 0

17. How many ways are there to rank 7 candidates who apply for a job?

18. (a) How many different ways are there to arrange the 7 letters in the word PROBLEM?
   (b) If we insist that the letter P comes first, how many ways are there?
   (c) If we insist that the letter P comes first and the letter M last, how many ways are there?

19. Testing On a math test there are 10 multiple-choice questions with 4 possible answers and 15 true–false questions. In how many possible ways can the 25 questions be answered?

20. Using the digits 1, 2, 3, and 4, how many different 4-digit numbers can be formed?

21. License Plate Possibilities How many different license plate numbers can be made using 2 letters followed by 4 digits selected from the digits 0 through 9, if

   (a) Letters and digits may be repeated?
   (b) Letters may be repeated, but digits are not repeated?
   (c) Neither letters nor digits may be repeated?

22. Security A system has 7 switches, each of which may be either open or closed. The state of the system is described by indicating for each switch whether it is open or closed. How many different states of the system are there?

23. Product Choice An automobile manufacturer produces 3 different models. Models A and B can come in any of 3 body styles; model C can come in only 2 body styles. Each car also comes in either black or green. How many distinguishable car types are there?

24. Using the digits 0, 1, 2, . . . , 9, how many 7-digit numbers can be formed if the first digit cannot be 0 or 9 and if the last digit is greater than or equal to 2 and less than or equal to 3? Repeated digits are allowed.

25. Home Choices A contractor constructs homes with 5 different choices of exterior finish, 3 different roof arrangements, and 4 different window designs. How many different types of homes can be built?

26. License Plate Possibilities A license plate consists of 1 letter, excluding O and I, followed by a 4-digit number that cannot have a 0 in the lead position. How many different plates are possible?

27. Using only the digits 0 and 1, how many different numbers consisting of 8 digits can be formed?

28. Stock Portfolios As a financial planner, you are asked to select one stock each from the following group: 8 DOW stocks, 15 NASDAQ stocks, and 4 global stocks. How many different portfolios are possible?

29. Combination Locks A combination lock has 50 numbers on it. To open it, you turn counterclockwise to a number, then rotate clockwise to a second number, and then counterclockwise to the third number. How many different lock combinations are there?

30. Opinion Polls An opinion poll is to be conducted among college students. Eight multiple-choice questions, each with 3 possible answers, will be asked. In how many different ways can a student complete the poll, if exactly one response is given to each question?

31. Path Selection in a Maze The maze below is constructed so that a rat must pass through a series of one-way doors. How many different paths are there from start to finish?

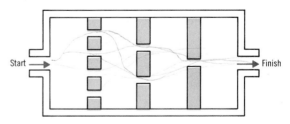

32. How many 3-letter code words are possible using the first 10 letters of the alphabet if

   (a) No letter can be repeated?
   (b) Letters can be repeated?
   (c) Adjacent letters cannot be the same?

6.3 PERMUTATIONS

In the next two sections we use the Multiplication Principle to discuss two general types of counting problems, called *permutations* and *combinations*. These concepts arise often in applications, especially in probability.

Factorial

Before discussing permutations, we introduce a useful shorthand notation—the *factorial symbol.*

FACTORIAL

The symbol $n!$, read as *"n factorial,"* is defined as

$$0! = 1$$
$$1! = 1$$
$$2! = 2 \cdot 1 \qquad = 2$$
$$3! = 3 \cdot 2 \cdot 1 \qquad = 6$$
$$4! = 4 \cdot 3 \cdot 2 \cdot 1 = 24$$

and, in general, for $n \geq 1$ an integer,

$$n! = n \cdot (n - 1) \cdot (n - 2) \cdot \ \cdot \ \cdot \ \cdot 3 \cdot 2 \cdot 1$$

Thus to compute $n!$, we find the product of all consecutive integers from n down to 1, inclusive. Remember that by definition, $0! = 1$.

A formula we shall find useful is

$$(n + 1)! = (n + 1) \cdot n!$$

EXAMPLE 1

(a) $4! = (4)(3)(2)(1) = 24$

(b) $\dfrac{5!}{4!} = \dfrac{5 \cdot 4!}{4!} = 5$

(c) $\dfrac{52!}{5!47!} = \dfrac{52 \cdot 51 \cdot 50 \cdot 49 \cdot 48 \cdot 47!}{5 \cdot 4 \cdot 3 \cdot 2 \cdot 1 \cdot 47!} = 2,598,960$

(d) $\dfrac{7!}{(7 - 5)!5!} = \dfrac{7!}{2!5!} = \dfrac{7 \cdot 6 \cdot 5!}{2!5!} = \dfrac{7 \cdot 6}{2} = 21$

(e) $\dfrac{50 \cdot 49 \cdot 48 \cdot 47 \cdot 46}{50!} = \dfrac{50 \cdot 49 \cdot 48 \cdot 47 \cdot 46}{50 \cdot 49 \cdot 48 \cdot 47 \cdot 46 \cdot 45!} = \dfrac{1}{45!}$

Now Work Problem 1

. . . .

Factorials grow very quickly. Compare the following:

$$5! = 120$$
$$10! = 3,628,800$$
$$15! = 1,307,674,368,000$$

In fact, if your calculator has a factorial key, you will find that $69! = 1.71 \cdot 10^{98}$, while $70!$ produces an error message—indicating you have exceeded the range of the calculator. Because of this, it is important to "cancel out" factorials whenever possible so that "out of range" errors may be avoided. Thus to calculate $100!/95!$, we write

$$\frac{100!}{95!} = \frac{100 \cdot 99 \cdot 98 \cdot 97 \cdot 96 \cdot 95!}{95!} = 9.03 \cdot 10^9$$

Permutations

We start with an example.

EXAMPLE 2

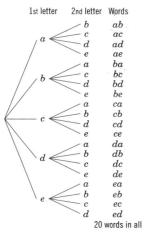

1st letter   2nd letter   Words

20 words in all

Figure 18

Suppose we are setting up a code of two-letter "words" and have five different letters, *a*, *b*, *c*, *d*, *e*, from which to choose. If the code must not repeat any letter and if such words as *ab* and *ba* are considered different, how many different words can be formed?

SOLUTION

We solve the problem by using the Multiplication Principle. In selecting a first letter, we have five choices. Since whatever letter is chosen cannot be repeated, we have four choices available for the second letter. In all, then, there are $5 \cdot 4 = 20$ words of two letters that can be formed. See Figure 18 for a tree diagram of this solution.

· ·

A way of rephrasing the question posed in Example 2 would be to ask: How many ordered arrangements using two distinct letters can be formed from the five letters *a* through *e*?

In general, we could ask: Given *n* distinct objects, how many ordered arrangements can be formed using *r* of the objects where $r \leq n$?

PERMUTATION

Ordered arrangements of distinct objects are called *permutations*. An *r* permutation of a set of *n* distinct objects is an ordered arrangement using *r* of the *n* objects. *P*(*n*, *r*) is defined to be the *number* of *r* permutations of a set of *n* distinct objects.

$P(n, r)$ is also referred to as the number of **permutations of *n* different objects taken *r* at a time.**

Thus $P(n, r)$ is the number of ordered arrangements that can be formed using *r* objects chosen from a set of *n* distinct objects.

In computing $P(n, r)$, we want to find the number of possible different arrangements of *r* objects that are chosen from *n* distinct objects in which no item is repeated and order is important. Thus the symbol $P(5, 2)$, means to rearrange two items chosen from five items. Since there are five choices for the first item and four for the second (no repetitions allowed), we find that

$$P(5, 2) = \underbrace{5 \cdot 4}_{\text{2 factors}} = 20$$

Other examples are

$$P(7, 3) = \underbrace{7 \cdot 6 \cdot 5}_{\text{3 factors}} = 210 \qquad P(5, 5) = \underbrace{5 \cdot 4 \cdot 3 \cdot 2 \cdot 1}_{\text{5 factors}} = 5! = 120$$

Now Work Problem 13

To find a formula for $P(n, r)$, we note that the first entry can be filled by any one of the *n* possibilities, the second by any one of the remaining $(n - 1)$ possibilities, the third by any one of the now remaining $(n - 2)$ possibilities, and so on. Since there are r positions to be filled, the number of possibilities is

$$P(n, r) = \underbrace{n(n - 1)(n - 2) \cdot \,\cdots}_{r \text{ factors}}$$

To obtain the last factor in the expression for $P(n, r)$, we observe the following pattern:

First factor is n

Second factor is $n - 1$

Third factor is $n - 2$

.

.

.

rth factor is $n - (r - 1) = n - r + 1$

Thus

$$P(n, r) = n(n - 1) \cdot \,\cdots\, \cdot (n - r + 1)$$

Multiplying the right side by 1 in the form of $(n - r)!/(n - r)!$, we obtain

$$P(n, r) = n(n - 1)(n - 2) \cdot \,\cdots\, \cdot (n - r + 1) \frac{(n - r)!}{(n - r)!}$$

Thus since $n(n - 1)(n - 2) \cdot \,\cdots\, \cdot (n - r + 1)(n - r)! = n!$, we obtain the following alternate formula for $P(n, r)$:

$$P(n, r) = \frac{n!}{(n - r)!}$$

Permutation Formula

The number of different arrangements using r objects chosen from n objects in which

1. The n objects are all different

2. No object is repeated in an arrangement

3. Order is important

is given by the formula*

$$P(n, r) = n(n - 1) \cdot \,\cdots\, \cdot (n - r + 1) = \frac{n!}{(n - r)!}$$

---

* Some calculators have a key for computing $P(n, r)$. If yours has such a key, consult your manual to find out how to use it.

Here is a list of short problems with their solutions given in $P(n, r)$ notation.

Problem	Solution
Find the number of ways of choosing five people from a group of 10 and arranging them in a line.	$P(10, 5)$
Find the number of six-letter ''words'' that can be formed with no letter repeated.	$P(26, 6)$
Find the number of seven-digit telephone numbers, with no repeated digit (allow 0 for a first digit).	$P(10, 7)$
Find the number of ways of arranging eight people in a line.	$P(8, 8)$

Note that in all of the above examples, *order is important.*

EXAMPLE 3

You own eight different mathematics books. How many ways can a shelf arrangement of five of the mathematics books be formed?

SOLUTION

We are seeking the number of arrangements using five of the eight books. The answer is given by

$$P(8, 5) = \frac{8!}{(8 - 5)!} = \frac{8!}{3!} = \frac{8 \cdot 7 \cdot 6 \cdot 5 \cdot 4 \cdot 3!}{3!} = 8 \cdot 7 \cdot 6 \cdot 5 \cdot 4 = 6720$$

Now Work Problem 25

EXAMPLE 4

A student has six questions on an examination and is allowed to answer the questions in any order. In how many different orders could the student answer these questions?

SOLUTION

The student wants the number of ordered arrangements of the six questions using all six of them. The number is given by

$$P(6, 6) = \frac{6!}{(6 - 6)!} = \frac{6!}{0!} = \frac{6!}{1} = 720$$

In general, the number of permutations (arrangements) of n different objects using all n of them is given by

$$P(n, n) = n!$$

So, for example, in a class of n students, there are $n!$ ways of positioning all the students in a line.

Now Work Problem 33

EXAMPLE 5

You own eight mathematics books and six computer science books and wish to fill seven positions on a shelf. If the first four positions are to be occupied by math books and the last three by computer science books, in how many ways can this be done?

SOLUTION

We think of the problem as consisting of two tasks. Task 1 is to fill the first four positions with four of the eight mathematics books. This can be done in $P(8, 4)$ ways. Task 2 is to fill the remaining three positions with three of six computer books. This can be done in $P(6, 3)$ ways. By the Multiplication Principle the seven positions can be filled in

$$P(8, 4) \cdot P(6, 3) = \frac{8!}{4!} \cdot \frac{6!}{3!} = 8 \cdot 7 \cdot 6 \cdot 5 \cdot 6 \cdot 5 \cdot 4 = 201{,}600 \text{ ways}$$

··············

EXERCISE 6.3   Answers to odd-numbered problems begin on page AN-24.

In Problems 1–20 evaluate each expression.

1. $\dfrac{5!}{2!}$

2. $\dfrac{8!}{2!}$

3. $\dfrac{6!}{3!}$

4. $\dfrac{9!}{3!}$

5. $\dfrac{10!}{8!}$

6. $\dfrac{11!}{9!}$

7. $\dfrac{9!}{8!}$

8. $\dfrac{10!}{9!}$

9. $\dfrac{8!}{2!6!}$

10. $\dfrac{9!}{3!6!}$

11. $P(7, 2)$

12. $P(5, 1)$

13. $P(8, 1)$

14. $P(6, 6)$

15. $P(5, 0)$

16. $P(6, 4)$

17. $\dfrac{8!}{(8-3)!3!}$

18. $\dfrac{9!}{(9-5)!5!}$

19. $\dfrac{6!}{(6-6)!6!}$

20. $\dfrac{7!}{(0-0)!7!}$

21. (a) How many different ways are there to arrange the 6 letters in the word SUNDAY?
    (b) If we insist that the letter S come first, how many ways are there?
    (c) If we insist that the letter S come first and the letter Y be last, how many ways are there?

22. How many ways are there to rank 8 candidates who apply for a job?

23. From a pool of 10 job applicants, a list ranking the top 4 must be made. How many such lists are possible?

24. How many different 5-letter ''words'' (sequences of letters) can be formed from the standard alphabet if repeated letters are not allowed? If repeated letters are allowed?

25. A station wagon has 9 seats. In how many different ways can 5 people be seated in it?

26. There are 5 different French books and 5 different Spanish books. How many ways are there to arrange them on a shelf if

(a) Books of the same language must be grouped together, French on the left, Spanish on the right?
(b) French and Spanish books must alternate in the grouping, beginning with a French book?

27. In how many ways can 8 different books be distributed to 12 children if no child gets more than one book?

28. A computer must assign each of 4 outputs to one of 8 different printers. In how many ways can it do this provided no printer gets more than one output?

29. **Lottery Tickets**   From 1500 lottery tickets that are sold, 3 tickets are to be selected for first, second, and third prizes. How many possible outcomes are there?

30. **Personnel Assignment**   A salesperson is needed in each of 7 different sales territories. If 10 equally qualified persons apply for the jobs, in how many ways can the jobs be filled?

31. **Slate Assignment**   A club has 15 members. In how many ways can a slate of 4 officers consisting of a president, vice-president, secretary, and treasurer be chosen?

32. **Psychology Testing**   In an ESP experiment a person is asked to select and arrange 3 cards from a set of 6 cards labeled *A, B, C, D, E,* and *F*. Without seeing the card, a second person is asked to give the arrangement. Determine the number of possible responses by the second person if he simply guesses.

33. How many ways are there to arrange 5 people in a line?

34. How many ways are there to seat 4 people in a 6-passenger automobile?

6.4 COMBINATIONS

Permutations focus on the order in which objects are arranged. However, in many cases, order is not important. For example, in a draw poker hand, the order in which you receive the cards is not relevant—all that matters is what cards are received. That is, with poker hands, we are concerned only with what *combination* of cards we have—not the particular order of the cards.

Combinations

The following example illustrates the distinction between selections in which order is important and those for which order is not important.

EXAMPLE 1

From the four letters *a, b, c, d,* choose two without repeating any letter

(a) If order is important   (b) If order is not important

SOLUTION

(a) If order is important, there are $P(4, 2) = 4 \cdot 3 = 12$ possible selections, namely,

$$ab \quad ac \quad ad \quad ba \quad bc \quad bd \quad ca \quad cb \quad cd \quad da \quad db \quad dc$$

See Figure 19.

(b) If order is not important, only 6 of the 12 selections found in Part (a) are listed, namely,

$$ab \quad ac \quad ad \quad bc \quad bd \quad cd$$

Notice that the number of ordered selections, 12, is $2! = 2$ times the number of unordered selections, 6. The reason is that each unordered selection consists of two letters that allow for 2! rearrangements. For example, the selection *ab* in the unordered list gives rise to *ab* and *ba* in the ordered list.

· ·

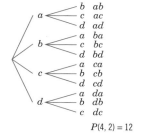

$P(4, 2) = 12$

Figure 19

EXAMPLE 2

From the four letters *a, b, c, d,* choose three without repeating any letter

(a) If order is important   (b) If order is not important

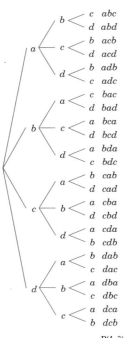

$P(4, 3) = 24$

Figure 20

SOLUTION

(a) If order is important, there are $P(4, 3) = 4 \cdot 3 \cdot 2 = 24$ possible selections, namely,

abc	abd	acb	acd	adb	adc	bac	bad	bca	bcd	bda	bdc
cab	cad	cba	cbd	cda	cdb	dab	dac	dba	dbc	dca	dcb

See Figure 20.

(b) If order is not important, only 4 of the 24 selections found in Part (a) are listed, namely,

$$abc \qquad abd \qquad acd \qquad bcd$$

Notice that the number of ordered selections, 24, is $3! = 6$ times the number of unordered selections, 4. The reason is that each unordered selection consists of three letters that allow for 3! rearrangements. For example, the selection *abc* in the unordered list gives rise to *abc, acb, bac, bca, cab, cba* in the ordered list.

• •

Unordered selections are called *combinations.*

COMBINATIONS

The *combination $C(n, r)$* is defined to be the *number* of ways of choosing r distinct objects from a set of n distinct objects, without regard to the order of the selection.

$C(n, r)$ is also referred to as the number of **combinations of n objects taken r at a time.**

To obtain a formula for $C(n, r)$, we first observe that each unordered selection of r objects will give rise to $r!$ ordered selections. Thus the number of ordered selections, $P(n, r)$, is $r!$ times the number of unordered selections, $C(n, r)$. That is,

$$P(n, r) = r!C(n, r)$$

Using the previously developed formula for $P(n, r)$, we find that

$$C(n, r) = \frac{P(n, r)}{r!} = \frac{n!}{r!(n - r)!}$$

Combination Formula

The number of different selections of r objects chosen from n objects in which

1. The n objects are all different

2. No object is repeated

3. Order is not important

is given by the formula*

$$C(n, r) = \frac{n!}{r!(n - r)!}$$

---

* Some calculators have a key for computing $C(n, r)$. If yours has such a key, consult your manual to find out how to use it.

EXAMPLE 3

Compute the following combinations:

(a) $C(50, 2)$ (b) $C(7, 5)$ (c) $C(7, 7)$ (d) $C(7, 0)$

SOLUTION

(a) $C(50, 2) = \dfrac{50!}{2!(50 - 2)!} = \dfrac{50!}{2!48!} = \dfrac{50 \cdot 49 \cdot 48!}{2 \cdot 48!} = 1225$

(b) $C(7, 5) = \dfrac{7!}{5!(7 - 5)!} = \dfrac{7!}{5!2!} = \dfrac{7 \cdot 6 \cdot 5!}{5! \cdot 2} = 21$

(c) $C(7, 7) = \dfrac{7!}{7!(7 - 7)!} = \dfrac{7!}{7!0!} = 1$

(d) $C(7, 0) = \dfrac{7!}{0!(7 - 0)!} = \dfrac{7!}{0!7!} = 1$

Now Work Problem 1

· ·

Here are some problems whose solutions are given in $C(n, r)$ notation.

Problem	Solution
Find the number of ways of selecting four people from a group of six	$C(6, 4)$
Find the number of five-card unordered poker hands	$C(52, 5)$
Find the number of committees of six that can be formed from the U.S. Senate (100 members)	$C(100, 6)$
Find the number of ways of selecting five courses from a catalog containing 200	$C(200, 5)$

EXAMPLE 4

From a deck of 52 cards a hand of 5 cards is dealt. How many different hands are possible?

SOLUTION

Such a hand is an unordered selection of 5 cards from a deck of 52. So the number of different hands is

$$C(52, 5) = \frac{52!}{5!47!} = \frac{52 \cdot 51 \cdot 50 \cdot 49 \cdot 48 \cdot 47!}{5 \cdot 4 \cdot 3 \cdot 2 \cdot 1 \cdot 47!} = 2{,}598{,}960$$

· ·

EXAMPLE 5

A bit is a 0 or a 1. A six-bit string is a sequence of length six consisting of 0s and 1s. How many six-bit strings contain

(a) Exactly one 1?   (b) Exactly two 1s?

SOLUTION

(a) To form a six-bit string having one 1, we need but specify where the single 1 is located (the other positions are 0s). The location for the 1 can be chosen in

$$C(6, 1) = 6 \text{ ways}$$

(b) Here, we must choose two of the six positions to contain 1s. Hence, there are

$$C(6, 2) = 15 \text{ such strings}$$

....

EXAMPLE 6

A sociologist needs a sample of 12 welfare recipients located in a large metropolitan area. He divides the city into four areas—northwest, northeast, southwest, southeast. Each section contains 25 welfare recipients. The sociologist may select the 12 recipients in any way he wants. How many different groups of 12 recipients are there?

SOLUTION

Since order of selection is not important and since the selection is of 12 things from a possible $4 \cdot 25 = 100$ things, there are $C(100, 12)$ different groups. That is, there are

$$C(100, 12) = \frac{100!}{12!88!} \approx 1.05 \cdot 10^{15}$$

$$\uparrow$$
$$\text{Use a}$$
$$\text{calculator}$$

different groups.

....

EXAMPLE 7

From five faculty members and four students a committee of four is to be chosen that includes two students and two faculty members. In how many ways can this be done?

SOLUTION

The faculty members can be chosen in $C(5, 2)$ ways. The students can be chosen in $C(4, 2)$ ways. By the Multiplication Principle there are then

$$C(5, 2) \cdot C(4, 2) = \frac{5!}{2!3!} \cdot \frac{4!}{2!2!} = 10 \cdot 6 = 60 \text{ different ways}$$

Now Work Problem 17

....

EXAMPLE 8

From six women and four men a committee of three is to be formed. The committee must include at least two women. In how many ways can this be done?

SOLUTION

A committee of three that includes at least two women will contain either exactly two women and one man, or exactly three women and zero men. Since no committee can contain exactly two women and simultaneously exactly three women, once we have counted the number of ways a committee of exactly two women and the number of ways a committee of exactly three women can be formed, their sum will give the number of ways exactly two women or exactly three women are on the committee. (Refer to Equation (1) on page 276, noting that the sets are disjoint.)

A committee of exactly two women and one man can be formed from six women, four men in $C(6, 2) \cdot C(4, 1)$ ways, while a committee of exactly three women, zero men can be formed in $C(6, 3) \cdot C(4, 0)$ ways. Thus a committee of three consisting of at least two women can be formed in

$$C(6, 2) \cdot C(4, 1) + C(6, 3) \cdot C(4, 0) = \frac{6!}{4!2!} \cdot \frac{4!}{3!1!} + \frac{6!}{3!3!} \cdot \frac{4!}{4!0!}$$

$$= 15 \cdot 4 + 20 \cdot 1 = 60 + 20 = 80 \text{ ways}$$

. . . .

Pascal's Triangle

Sometimes the notation $\binom{n}{r}$, read as "from n choose r," is used in place of $C(n, r)$. $\binom{n}{r}$ is called a **binomial coefficient** because of its connection with the binomial theorem (discussed in Section 6.6). A triangular display of $\binom{n}{r}$ for $n = 0$ to $n = 6$ is given in Figure 21. This triangular display is called **Pascal's triangle.**

For example, $\binom{5}{2} = 10$ is found in the row marked $n = 5$ and on the diagonal marked $r = 2$.

In the Pascal triangle successive entries can be obtained by adding the two nearest entries in the row above it. The shaded triangles in Figure 21 illustrate this. For example, $10 + 5 = 15$, etc.

Figure 21

The Pascal triangle, as the figure indicates, is symmetric. Thus when n is even, the largest entry occurs in the middle, and corresponding entries on either side are equal. When n is odd, there are two equal middle entries with corresponding equal entries on either side.

The reasons behind these properties of Pascal's triangle as well as other properties of binomial coefficients are discussed in Section 6.6.

EXERCISE 6.4   Answers to odd-numbered problems begin on page AN-24.

In Problems 1–8 find the value of each expression.

1. $C(6, 4)$ **2.** $C(5, 4)$ **3.** $C(7, 2)$ **4.** $C(8, 7)$ **5.** $\binom{5}{1}$

6. $\binom{8}{1}$ **7.** $\binom{8}{6}$ **8.** $\binom{8}{4}$

9. In how many ways can a committee of 3 senators be selected from a group of 8 senators?

10. In how many ways can a committee of 4 representatives be selected from a group of 9 representatives?

11. A math department is allowed to tenure 4 of 17 eligible teachers. In how many ways can the selection for tenure be made?

12. How many different hands are possible in a bridge game? (A bridge hand consists of 13 cards dealt from a deck of 52 cards.)

13. There are 25 students in the Math Club. In how many ways can a subcommittee of 3 members be formed?

14. How many different relay teams of 4 persons can be chosen from a group of 10 runners?

15. **Basketball Teams**   A basketball team has 6 players who play guard (2 of 5 starting positions). How many different teams are possible, assuming the remaining 3 positions are filled and it is not possible to distinguish a left guard from a right guard?

16. **Basketball Teams**   On a basketball team of 12 players, 2 play only center, 3 play only guard, and the rest play forward (5 players on a team: 2 forwards, 2 guards, and 1 center). How many different teams are possible, assuming it is not possible to distinguish left and right guards and left and right forwards?

17. The Student Affairs Committee has 3 faculty members, 2 administration members, and 5 students on it. In how many ways can a subcommittee of 1 faculty, 1 administrator, and 2 students be formed?

18. **Stock Trading**   Of 1352 stocks traded in 1 day on the New York Stock Exchange, 641 advanced, 234 declined, and the remainder were unchanged. In how many ways can this happen?

19. How many different ways can an offensive football team be formed from a squad that consists of 20 linemen, 3 quarterbacks, 8 halfbacks, and 4 fullbacks? This football team must have 1 quarterback, 2 halfbacks, 1 fullback, and 7 linemen.

20. **Baseball Teams**   How many different ways can a baseball team be made up from a squad of 15 players, if 3 players are used only as pitchers and the remaining players can be placed at any position except pitcher (9 players on a team)?

21. A little girl has 1 penny, 1 nickel, 1 dime, 1 quarter, and 1 half dollar in her purse. If she pulls out 3 coins, how many different sums are possible?

22. How many 8-bit strings contain exactly two 1s? Exactly three 1s?

23. **Lottery Tickets**   A state of Maryland million dollar lottery ticket consists of 6 distinct numbers chosen from the range 00 through 99. The order in which the numbers appear on the ticket is irrelevant. How many distinct lottery tickets can be issued?

24. How many 5-card poker hands contain all spades? (A deck of 52 cards contains 13 spades.)

25. How many committees of 5 can be formed from members of the U.S. Senate?

26. A test has 3 parts. In part 1 a student must do 3 of 5 questions, in part 2 a student must choose 2 of 4 questions, and in part 3 a student must pick 3 of 6 questions. In how many different ways can a student complete the test?

In Problems 27–36 use the Multiplication Principle, permutations, or combinations, as appropriate.

27. **Test Panel Selection**   A sample of 8 persons is selected for a test from a group containing 40 smokers and 15 nonsmokers. In how many ways can the 8 persons be selected?

28. **Resource Allocation**   A trucking company has 8 trucks and 6 drivers available when requests for 4 trucks are received. How many different ways are there of selecting the trucks and the drivers to meet these requests?

29. **Group Selection**   From a group of 5 people we are required to select a different person to participate in each of 3 different tests. In how many ways can the selections be made?

30. **Congressional Committees**   In the U.S. Congress a conference committee is to be composed of 5 senators and 4 representatives. In how many ways can this be done? (There are 435 representatives and 100 senators.)

31. **Quality Control**   A box contains 24 light bulbs. The quality control engineer will pick a sample of 4 light bulbs for inspection. How many different samples are there?

32. **Mating**   A horse stable has 12 mares and 4 stallions. In how many different ways can they be mated?

33. **Rating**   A sportswriter makes a preseason guess of the top 15 university basketball teams (in order) from among

50 major university teams. How many different possibilities are there?

34. Packaging   A manufacturer produces 8 different items. He packages assortments of equal parts of 3 different items. How many different assortments can be packaged?

35. The digits 0 through 9 are written on 10 cards. Four different cards are drawn, and a 4-digit number is formed. How many different 4-digit numbers can be formed in this way?

36. Investment Selection   An investor is going to invest $21,000 in 3 stocks from a list of 12 prepared by his broker. How many different investments are possible if

(a) $7000 is to be invested in each stock?
(b) $10,000 is to be invested in one stock, $6000 in another, and $5000 in the third?
(c) $8000 is to be invested in each of 2 stocks and $5000 in a third stock?

6.5 MORE COUNTING PROBLEMS

In this section we consider some counting problems that will be useful in our discussion of probability.

Examples

The first example deals with a coin-tossing experiment in which a coin is tossed a fixed number of times. There are exactly two possible outcomes on each trial or toss (heads, H, or tails, T). For instance, in tossing a coin three times, one possible outcome is HTH—heads on the first toss, tails on the second toss, and heads on the third toss.

EXAMPLE 1

Suppose an experiment consists of tossing a coin four times.

(a) How many different outcomes are possible?
(b) How many different outcomes have exactly 3 tails?
(c) How many outcomes have at most 2 tails?
(d) How many outcomes have at least 1 tail?

SOLUTION

(a) Each outcome of the experiment consists of a sequence of four letters H or T, where the first letter records the result of the first toss, the second letter the result of the second toss, and so forth. Thus the process can be visualized as the tree diagram in Figure 22.

Figure 22

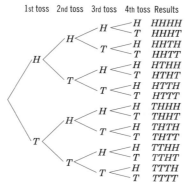

We can also visualize the process as filling an empty box at each toss with either an H or a T.

1st toss | T | | | |

2nd toss | T | H | | |

3rd toss | T | H | H | |

4th toss | T | H | H | H |

Since each box can be filled in two ways, by the Multiplication Principle the sequence of four boxes can be filled in

$$2 \cdot 2 \cdot 2 \cdot 2 = 2^4 = 16 \text{ ways}$$

$$\underbrace{\qquad\qquad}_{4 \text{ factors}}$$

Thus there are $2^4 = 16$ different possible outcomes.

(b) Any sequence that contains exactly $3 T$s must contain 1 H. A particular outcome is determined as soon as we decide where to place the Ts in the four boxes. The three boxes to receive the Ts can be selected from the four boxes in $C(4, 3)$ different ways. So the number of outcomes with exactly 3 tails is

$$C(4, 3) = \frac{4!}{3!1!} = 4$$

(c) The outcomes with at most 2 tails correspond to the sequences with 0, 1, or 2 Ts:

0 T Only one outcome is possible, namely, $HHHH$.
1 T These outcomes, which include $THHH$ and $HTHH$, are determined by selecting one box out of four in which to place the single T. This can be done in $C(4, 1) = 4$ ways.
2 Ts   These outcomes, which include $THHT$ and $HTTH$, are determined by selecting two boxes out of four in which to place the two Ts. This can be done in $C(4, 2) = 6$ ways.

Thus the number of outcomes with at most 2 tails is just the sum of all these results, which is $1 + 4 + 6 = 11$.

(d) The outcomes with at least 1 tail are the results with 1, 2, 3, or 4 tails. The total number of such outcomes is

$$C(4, 1) + C(4, 2) + C(4, 3) + C(4, 4)$$

But there is a simpler way of obtaining the answer. If we start with the total number of outcomes obtained in Part (a) and subtract the number of outcomes of at most 0 tails, namely $C(4, 0) = 1$, we get the number of outcomes with at least 1 tail:

$$16 - 1 = 15$$

．．．．．．．．．．．．．．．．．．．．．．

Part (d) of Example 1 illustrates a counting technique that is often useful: Count the outcomes that are not favorable to you and subtract from the total number of outcomes. This "backdoor" approach can be easier at times than a direct attack. Here is another example of its use.

Now Work Problem 1

EXAMPLE 2

Find the number of seven-digit telephone numbers that have at least one repeated digit. Leading 0s are allowed.

SOLUTION

A direct application of the Multiplication Principle shows that there are 10^7 total possible seven-digit telephone numbers. The number of telephone numbers that have *no* repeated digits is

$$P(10, 7) = \frac{10!}{3!} = 604{,}800$$

Hence, the number that have at least one repeated digit is

$$10^7 - \frac{10!}{3!} = 9{,}395{,}200$$

· ·

EXAMPLE 3

An urn contains 8 white balls and 4 red balls. Four balls are selected. In how many ways can the 4 balls be drawn from the total of 12 balls

(a) If 3 balls are white and 1 is red?   (b) If all 4 balls are white?
(c) If all 4 balls are red?

SOLUTION

(a) The desired answer involves two operations: first, the selection of 3 white balls from 8; and second, the selection of 1 red ball from 4:

Select 3 white balls Operation 1	Select 1 red ball Operation 2

The first operation can be performed in $C(8, 3)$ ways; the second operation can be performed in $C(4, 1)$ ways. By the Multiplication Principle, the answer is

$$C(8, 3) \cdot C(4, 1) = \frac{8!}{3!5!} \cdot \frac{4!}{1!3!} = 224$$

(b) Since all 4 balls must be selected from the 8 that are white, the answer is

$$C(8, 4) = \frac{8!}{4!4!} = \frac{8 \cdot 7 \cdot 6 \cdot 5}{4 \cdot 3 \cdot 2 \cdot 1} = 70 \text{ ways}$$

(c) Since the 4 red balls must be selected from 4 red balls, the answer is 1 way.

Now Work Problem 3

· ·

Permutations with Repetition

Our previous discussion of permutations required that the objects we were rearranging be distinct. We now examine what happens when repetitions are allowed. The following example shows that allowing repetition of some objects introduces modifications.

EXAMPLE 4   How many three-letter words (real or imaginary) can be formed from the letters in the word

(a) MAD?   (b) DAD?

SOLUTION

(a) The three distinct letters in MAD can be rearranged in

$$P(3, 3) = 3! = 6 \text{ ways}$$

(b) Straightforward listing shows that there are only three ways of rearranging the letters in the word DAD:

$$\text{DAD, DDA, and ADD}$$

· ·

The word DAD in Example 4 contains 2 Ds, and it is this duplication that results in fewer rearrangements for DAD than for MAD. In the next example we describe a way of dealing with the problem of duplication.

EXAMPLE 5   How many distinct ''words'' can be formed using all the letters of the six-letter word

$$\text{M A M M A L ?}$$

SOLUTION

Any such word will have 6 letters formed from 3 Ms, 2 As, and 1 L. To form a word, think of six blank positions that will have to be filled by the above letters.

$$\underline{\quad} \; \underline{\quad} \; \underline{\quad} \; \underline{\quad} \; \underline{\quad} \; \underline{\quad}$$
$$1 \quad 2 \quad 3 \quad 4 \quad 5 \quad 6$$

We separate the construction of a word into three tasks.

   Task 1 Choose 3 of the positions for the Ms.
   Task 2 Choose 2 of the remaining positions for the As.
   Task 3 Choose the remaining position for the L.

Doing this sequence of tasks will result in a word and, conversely, every rearrangement of MAMMAL can be interpreted as resulting from this sequence of tasks.

Task 1 can be done in $C(6, 3)$ ways. There are now three positions left for the 2 As, so Task 2 can be done in $C(3, 2)$ ways. Five blanks have been filled, so that the L must go in the remaining blank. That is, Task 3 can be done in $C(1, 1)$ way. The Multiplication Principle says that the number of rearrangements is

$$C(6, 3) \cdot C(3, 2) \cdot C(1, 1) = \frac{6!}{3!3!} \cdot \frac{3!}{2!1!} \cdot \frac{1!}{1!}$$

$$= \frac{6!}{3!2!1!}$$

· ·

The form of the answer in Example 5 is suggestive. Had the letters in MAMMAL been distinct, there would have been 6! rearrangements possible. The presence of 3Ms, 2As, and 1 L yielded the denominator above. The very same reasoning used in Example 5 can be used to derive the following general result.

Permutation with Repetition

The number of distinct permutations of n things, of which n_1 are of one kind, n_2 of a second kind, . . . , n_k of a kth kind, is

$$\frac{n!}{n_1! \cdot n_2! \cdot \cdots \cdot n_k!}$$

where $n_1 + n_2 + \cdots + n_k = n$.

EXAMPLE 6

How many different vertical arrangements are possible for 10 flags, if 2 are white, 3 are red, and 5 are blue?

SOLUTION

Here we want the different arrangements of 10 objects, which are not all different. Following the result above, we have

$$\frac{10!}{2!3!5!} = \frac{10 \cdot 9 \cdot 8 \cdot 7 \cdot 6 \cdot 5!}{2 \cdot 3 \cdot 2 \cdot 5!} = 2520 \text{ different arrangements}$$

· ·

EXAMPLE 7

How many different 11-letter words (real or imaginary) can be formed from the word below?

MISSISSIPPI

SOLUTION

Here we want the number of distinct 11-letter words with 4 Is, 4 Ss, 2 Ps, and 1 M, so that the total number of 11-letter words is

$$\frac{11!}{1!2!4!4!} = \frac{39{,}916{,}800}{1152} = 34{,}650$$

· ·

Now Work Problem 9

The ideas above can be adapted to problems involving assignments of objects to locations.

EXAMPLE 8

A sorority house has three bedrooms and 10 students. One bedroom has three beds, the second has two beds, and the third has five beds. In how many different ways can the students be assigned rooms?

SOLUTION

Designate the bedrooms as *A*, *B*, and *C*. We think of the 10 students as standing in a row and we hand each student a letter corresponding to her assigned bedroom. An assignment of rooms can then be visualized as a sequence of length 10 (the number of students) containing 3 *A*s, 2 *B*s, and 5 *C*s (the capacity of the rooms). For example, the sequence

$$A \quad B \quad B \quad C \quad . \quad . \quad .$$

would have the first student in room *A*, students 2 and 3 in room *B*, and so on. There are

$$\frac{10!}{3!2!5!} = 2520$$

such sequences and, hence, room assignments.

· ·

EXERCISE 6.5   Answers to odd-numbered problems begin on page AN-24.

1. An experiment consists of tossing a coin 10 times.

(a) How many different outcomes are possible?
(b) How many different outcomes have exactly 4 heads?
(c) How many different outcomes have at most 2 heads?
(d) How many different outcomes have at least 3 heads?

2. An experiment consists of tossing a coin six times.

(a) How many different outcomes are possible?
(b) How many different outcomes have exactly 3 heads?
(c) How many different outcomes have at least 2 heads?
(d) How many different outcomes have 4 heads or 5 heads?

3. An urn contains 7 white balls and 3 red balls. Three balls are selected. In how many ways can the 3 balls be drawn from the total of 10 balls

(a) If 2 balls are white and 1 is red?
(b) If all 3 balls are white?
(c) If all 3 balls are red?

4. An urn contains 15 red balls and 10 white balls. Five balls are selected. In how many ways can the 5 balls be drawn from the total of 25 balls

(a) If all balls are red?
(b) If 3 balls are red and 2 are white?
(c) If at least 4 are red balls?

5. In the World Series the American League team (*A*) and the National League team (*N*) play until one team wins four games. If the sequence of winners is designated by letters (for example, *NAAAA* means the National League team won the first game and the American League team won the next four), how many different sequences are possible?

6. How many different ways can 3 red, 4 yellow, and 5 blue bulbs be arranged in a string of Christmas tree lights with 12 sockets?

7. In how many ways can 3 apple trees, 4 peach trees, and 2 plum trees be arranged along a fence line if one does not distinguish between trees of the same kind?

8. How many different 9-letter words (real or imaginary) can be formed from the letters in the word ECONOMICS?

9. How many different 11-letter words (real or imaginary) can be formed from the letters in the word MATHEMATICS?

10. The U.S. Senate has 100 members. Suppose it is desired to place each senator on exactly 1 of 7 possible committees. The first committee has 22 members, the second has 13, the third has 10, the fourth has 5, the fifth has 16, and the sixth and seventh have 17 apiece. In how many ways can these committees be formed?

11. In how many ways can 12 children be placed on 3 distinct teams of 3, 5, and 4 members?

12. A group of 9 people is going to be formed into committees of 4, 3, and 2 people. How many committees can be formed if

   (a) A person can serve on any number of committees?
   (b) No person can serve on more than one committee?

13. A group consists of 5 men and 8 women. A committee of 4 is to be formed from this group, and policy dictates that at least 1 woman be on this committee.

   (a) How many committees can be formed that contain exactly 1 man?
   (b) How many committees can be formed that contain exactly 2 women?
   (c) How many committees can be formed that contain at least 1 man?

14. How many distinct seven-dig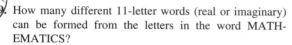 formed

   (a) If the digits 1, 1, 2, 2, 5, 5, 5 are used
   (b) If it is required that the first digit be a 5?

15. In how many ways can 30 diplomats be assigned to countries, with each country receiving an equal number of diplomats?

16. How many rearrangements of the letters in the word SUCCESS have the U before the E?

17. An experiment consists of tossing a coin 8 times. How many outcomes have more heads than tails?

18. Eight couples (husband and wife) are present at a meeting where a committee of 3 is to be chosen. How many ways can this be done so that the committee

   (a) Contains a couple?   (b) Contains no couple?

19. In how many ways can a committee of 4 be selected from 6 men and 8 women if the committee must contain at least 2 women?

20. A man wants to invite 1 or more of his 4 friends to dinner. In how many ways can he do this?

21. How many 8-bit strings contain an even number of 1s? An odd number of 1s?

22. An office manager must locate 12 secretaries into three offices that hold, respectively, 6, 4, and 2 secretaries. In how many ways can the three groups be chosen to occupy the three offices?

6.6 THE BINOMIAL THEOREM

The *binomial theorem* deals with the problem of expanding an expression of the form $(x + y)^n$, where n is a positive integer.

Expressions such as $(x + y)^2$ and $(x + y)^3$ are not too difficult to expand. For example,

$$(x + y)^2 = x^2 + 2xy + y^2$$
$$(x + y)^3 = (x + y)^2(x + y) = (x^2 + 2xy + y^2)(x + y) = x^3 + 3x^2y + 3xy^2 + y^3$$

However, expanding expressions such as $(x + y)^6$ or $(x + y)^8$ by the normal process of multiplication would be tedious and time consuming. It is here that the binomial theorem is especially useful.

Recall that

$$C(n, r) = \binom{n}{r} = \frac{n!}{r!(n - r)!}$$

now 2 Pascal

$\binom{2}{1} = 2$, and $\binom{2}{2} = 1$, we can write the expression

$$(x + y)^2 = x^2 + 2xy + y^2$$

rm

$$(x + y)^2 = \binom{2}{0} x^2 + \binom{2}{1} xy + \binom{2}{2} y^2$$

Since $\binom{3}{0} = 1$, $\binom{3}{1} = 3$, $\binom{3}{2} = 3$, and $\binom{3}{3} = 1$, the expansion of $(x + y)^3$ can be written as

$$(x + y)^3 = x^3 + 3x^2y + 3xy^2 + y^3 = \binom{3}{0} x^3 + \binom{3}{1} x^2y + \binom{3}{2} xy^2 + \binom{3}{3} y^3$$

Similarly, the expansion of $(x + y)^4$ can be written as

$$(x + y)^4 = x^4 + 4x^3y + 6x^2y^2 + 4xy^3 + y^4$$
$$= \binom{4}{0} x^4 + \binom{4}{1} x^3y + \binom{4}{2} x^2y^2 + \binom{4}{3} xy^3 + \binom{4}{4} y^4$$

The binomial theorem generalizes this pattern.

Binomial Theorem

If n is a positive integer,

$$(x + y)^n = \binom{n}{0} x^n + \binom{n}{1} x^{n-1}y + \binom{n}{2} x^{n-2}y^2$$
$$+ \cdots + \binom{n}{k} x^{n-k}y^k + \cdots + \binom{n}{n} y^n \qquad (1)$$

Observe that the powers of x begin at n and decrease by 1, while the powers of y begin with 0 and increase by 1. Also, the coefficient of the term involving y^k is always $\binom{n}{k}$.

Let's get some practice using the binomial theorem.

EXAMPLE 1

Expand $(x + y)^6$ using the binomial theorem.

SOLUTION

$$(x + y)^6 = \binom{6}{0} x^6 + \binom{6}{1} x^5y + \binom{6}{2} x^4y^2 + \binom{6}{3} x^3y^3$$
$$+ \binom{6}{4} x^2y^4 + \binom{6}{5} xy^5 + \binom{6}{6} y^6$$
$$= x^6 + 6x^5y + 15x^4y^2 + 20x^3y^3 + 15x^2y^4 + 6xy^5 + y^6$$

. . . .

Now Work Problem 1
Note that the coefficients in the expansion of $(x + y)^6$ are the entries in the Pascal triangle for $n = 6$. (See Figure 21, page 295.)

EXAMPLE 2

Find the coefficient of x^3y^4 in the expansion of $(x + y)^7$.

SOLUTION

The expansion of $(x + y)^7$ is

$$(x + y)^7 = \binom{7}{0} x^7 + \binom{7}{1} x^6y + \binom{7}{2} x^5y^2 + \binom{7}{3} x^4y^3 + \binom{7}{4} x^3y^4$$

$$+ \binom{7}{5} x^2y^5 + \binom{7}{6} xy^6 + \binom{7}{7} y^7$$

Thus the coefficient of x^3y^4 is

$$\binom{7}{4} = \frac{7 \cdot 6 \cdot 5}{3 \cdot 2 \cdot 1} = 35$$

Now Work Problem 7

EXAMPLE 3

Expand $(x + 2y)^4$ using the binomial theorem.

SOLUTION

Here, we let "$2y$" play the role of "y" in the binomial theorem. We then get

$$(x + 2y)^4 = \binom{4}{0} x^4 + \binom{4}{1} x^3(2y) + \binom{4}{2} x^2(2y)^2$$

$$+ \binom{4}{3} x(2y)^3 + \binom{4}{4} (2y)^4$$

$$= x^4 + 8x^3y + 24x^2y^2 + 32xy^3 + 16y^4$$

Now Work Problem 3

To explain why the binomial theorem is true, we take a close look at what happens when we compute $(x + y)^3$. Think of $(x + y)^3$ as the product of three factors, namely,

$$(x + y)^3 = (x + y)(x + y)(x + y)$$

Were we to multiply these three factors together without any attempt at simplification or collecting of terms, we would get

$$\underbrace{(x + y)}_{\substack{\text{Factor}\\1}} \underbrace{(x + y)}_{\substack{\text{Factor}\\2}} \underbrace{(x + y)}_{\substack{\text{Factor}\\3}} = xxx + xyx + yxx + yyx + xxy + xyy + yxy + yyy$$

Notice that the terms on the right yield all possible products that can be formed by picking either an x or y from each of the three factors on the left. Thus, for example,

xyx results from choosing an x from factor 1,
a y from factor 2, and an x from factor 3

Now, the number of terms on the right that will simplify to, say, xy^2, will be those terms that resulted from choosing ys from two of the factors and an x from the remaining factor. How many such terms are there? There are as many as there are ways of choosing two of the three factors to contribute ys—that is, there are $C(3, 2) = \binom{3}{2}$ such terms. This is why the coefficient of xy^2 in the expansion of $(x + y)^3$ is $\binom{3}{2}$.

In general, if we think of $(x + y)^n$ as the product of n factors,

$$(x + y)^n = \underbrace{(x + y) \cdot (x + y) \cdot \cdots \cdot (x + y)}_{n \text{ factors}}$$

then, upon multiplying out and simplifying, there will be as many terms of the form $x^{n-k} y^k$ as there are ways of choosing k of the n factors to contribute ys (and the remaining $n - k$ to contribute xs). There are $C(n, k)$ ways of making this choice. So the coefficient of $x^{n-k} y^k$ is thus $\binom{n}{k}$, and this is the assertion of the binomial theorem.

Binomial Identities

Binomial coefficients have some interesting properties. We discuss some of these in the examples below.

EXAMPLE 4

Show that

$$\binom{n}{k} = \binom{n}{n - k}$$

SOLUTION

By definition,

$$\binom{n}{k} = \frac{n!}{k!(n - k)!}$$

while

$$\binom{n}{n - k} = \frac{n!}{(n - k)![n - (n - k)]!}$$

Since $n - (n - k) = k$, a comparison of the expressions shows that they are equal.

· ·

Another way of explaining the equality would be to imagine that we wanted to pick a team of k players from n people. Then choosing those k who will play is the same as choosing those $n - k$ who will not. So the number of ways of choosing the players equals the number of ways of choosing the nonplayers. The players can be chosen in $C(n, k) = \binom{n}{k}$ ways, while those to be left out can be chosen in $\binom{n}{n-k}$ ways, and the equality follows.

Thus

$$\binom{5}{3} = \binom{5}{2} \qquad \binom{10}{2} = \binom{10}{8}$$

and so on. This identity accounts for the symmetry in the rows of Pascal's triangle.

EXAMPLE 5

Show that

$$\binom{n}{k} = \binom{n-1}{k} + \binom{n-1}{k-1}$$

SOLUTION

We could expand both sides of the above identity using the definition of binomial coefficients and, after some algebra, demonstrate the equality. But we choose the route of posing a problem that we solve two different ways. Equating the two solutions will prove the identity.

A committee of k is to be chosen from n people. The total number of ways this can be done is $C(n, k) = \binom{n}{k}$.

We now count the total number of committees a different way. Assume that one of the n people is Jennifer. We compute

(1) those committees not containing Jennifer

and

(2) those committees containing Jennifer

The number of committees of type 1 is $\binom{n-1}{k}$ since the k people must be chosen from the $n-1$ people who are not Jennifer. The number of committees of type 2 will correspond to the number of ways we can choose the $k-1$ people other than Jennifer to be on the committee. So the number of committees of type 2 is given by $\binom{n-1}{k-1}$. Since the number of committees of type 1 plus the number of committees of type 2 equals the total number of committees, our identity follows.

....

For example, $\binom{8}{5} = \binom{7}{5} + \binom{7}{4}$. It is precisely this identity that explains the reason why an entry in Pascal's triangle can be obtained by adding the nearest two entries in the row above it. Due to its recursive character, the identity in Example 5 is sometimes used in computer programs that evaluate binomial coefficients.

Now Work Problem 21

EXAMPLE 6

Explain why

$$\binom{6}{3} = \binom{2}{2} + \binom{3}{2} + \binom{4}{2} + \binom{5}{2}$$

SOLUTION

We make repeated use of the identity in Example 5. So

$$\binom{6}{3} = \binom{5}{3} + \binom{5}{2}$$

$$= \left[\binom{4}{3} + \binom{4}{2} \right] + \binom{5}{2} \qquad \text{Apply the identity to } \binom{5}{3}.$$

$$= \left[\binom{3}{3} + \binom{3}{2} \right] + \binom{4}{2} + \binom{5}{2} \qquad \text{Apply the identity to } \binom{4}{3}.$$

$$= \binom{2}{2} + \binom{3}{2} + \binom{4}{2} + \binom{5}{2} \qquad \text{Since } \binom{2}{2} = \binom{3}{3} = 1$$

....

EXAMPLE 7

Show that

$$\binom{n}{0} + \binom{n}{1} + \binom{n}{2} + \cdots + \binom{n}{n} = 2^n$$

SOLUTION

We make use of the binomial theorem. Since the binomial theorem is valid for all x and y, we may set $x = y = 1$ in Equation (1). This gives

$$2^n = (1 + 1)^n = \binom{n}{0} + \binom{n}{1} + \binom{n}{2} + \cdots + \binom{n}{n}$$

· ·

This shows, for example, that the sum of the elements in the row marked $n = 6$ of Pascal's triangle is $2^6 = 64$. The result in Example 7 can be used to find the number of subsets of a set with n elements. $\binom{n}{0}$ gives the number of subsets with 0 elements; $\binom{n}{1}$ the number of subsets with 1 element; $\binom{n}{2}$ the number of subsets with 2 elements; and so on. The sum $\binom{n}{0} + \binom{n}{1} + \cdots + \binom{n}{n}$ is thus the total number of subsets of a set with n elements. The result in Example 7 can then be rephrased as follows:

A set with n elements has 2^n subsets.

Now Work Problem 11

Thus a set with 5 elements has $2^5 = 32$ subsets.

EXAMPLE 8

Show that

$$\binom{n}{0} - \binom{n}{1} + \binom{n}{2} - \cdots + (-1)^n \binom{n}{n} = 0$$

(The last term will be preceded by a plus or minus sign depending on whether n is even or odd.)

SOLUTION

We again make use of the binomial theorem. This time we let $x = 1$ and $y = -1$ in Equation (1). This produces

$$(1 - 1)^n = \binom{n}{0} + \binom{n}{1}(-1) + \binom{n}{2}(-1)^2 + \binom{n}{3}(-1)^3$$

$$+ \cdots + \binom{n}{n}(-1)^n$$

$$0 = \binom{n}{0} - \binom{n}{1} + \binom{n}{2} - \cdots + (-1)^n \binom{n}{n}$$

· ·

We mention an interpretation of this identity by examining the instance where $n = 5$. The identity gives

$$\binom{5}{0} - \binom{5}{1} + \binom{5}{2} - \binom{5}{3} + \binom{5}{4} - \binom{5}{5} = 0$$

Rearranging some terms yields

$$\binom{5}{0} + \binom{5}{2} + \binom{5}{4} = \binom{5}{1} + \binom{5}{3} + \binom{5}{5}$$

This says that a set with 5 elements has as many subsets containing an even number of elements as it has subsets containing an odd number of elements.

EXERCISE 6.6 Answers to odd-numbered problems begin on page AN-24.

In Problems 1–6 use the binomial theorem to expand each expression.

 1. $(x + y)^5$ **2.** $(x + y)^4$ **3.** $(x + 3y)^3$ **4.** $(2x + y)^3$ **5.** $(2x - y)^4$ **6.** $(x - y)^4$

7. What is the coefficient of x^2y^3 in the expansion of $(x + y)^5$?

8. What is the coefficient of x^2y^6 in the expansion of $(x + y)^8$?

9. What is the coefficient of x^8 in the expansion of $(x + 3)^{10}$?

10. What is the coefficient of x^3 in the expansion of $(x + 2)^5$?

11. How many different subsets can be chosen from a set with 5 elements?

12. How many different subsets can be chosen from a set with 50 elements?

13. How many nonempty subsets does a set with 10 elements have?

14. How many subsets with an even number of elements does a set with 10 elements have?

15. How many subsets with an odd number of elements does a set with 10 elements have?

16. Explain why

$$\binom{8}{5} = \binom{4}{4} + \binom{5}{4} + \binom{6}{4} + \binom{7}{4}$$

17. Explain why

$$\binom{10}{7} = \binom{6}{6} + \binom{7}{6} + \binom{8}{6} + \binom{9}{6}$$

18. Show that

$$\binom{7}{1} + \binom{7}{3} + \binom{7}{5} + \binom{7}{7} = 2^6$$

19. Replace $\binom{11}{6} + \binom{11}{5}$ by a single binomial coefficient.

20. Replace $\binom{8}{8} + \binom{9}{8} + \binom{10}{8}$ by a single binomial coefficient.

21. Show that

$$k \binom{n}{k} = n \binom{n-1}{k-1}$$

CHAPTER REVIEW

IMPORTANT TERMS AND CONCEPTS

set 268
subset 269
proper subset 269
universal set 270
Venn diagram 270
union of sets 271
intersection of sets 271

disjoint sets 272
complement of a set 272
De Morgan's properties 273
counting formula 276
Multiplication Principle 283
factorial 286

permutation 287
combination 292
binomial coefficient 295
Pascal's triangle 295
permutation with repetition 301
binomial theorem 304

IMPORTANT FORMULAS

Counting Formula

$$c(A \cup B) = c(A) + c(B) - c(A \cap B)$$

Permutation Formula

$$P(n, r) = n(n - 1) \cdot \cdots \cdot (n - r + 1) = \frac{n!}{(n - r)!}$$

Combination Formula

$$C(n, r) = \binom{n}{r} = \frac{n!}{r!(n - r)!} = \frac{P(n, r)}{r!}$$

Binomial Theorem

$$(x + y)^n = \binom{n}{0} x^n + \binom{n}{1} x^{n-1}y + \binom{n}{2} x^{n-2}y^2$$
$$+ \cdots + \binom{n}{k} x^{n-k}y^k + \cdots + \binom{n}{n} y^n$$

TRUE–FALSE ITEMS   Answers are on page AN-25.

T_____ F_____ **1.** If $A \cup B = A \cap B$, then $A = B$.

T_____ F_____ **2.** If A and B are disjoint sets, then $c(A \cup B) = c(A) + c(B)$.

T_____ F_____ **3.** The number of permutations of 4 different objects taken 4 at a time is 12.

T_____ F_____ **4.** $C(5, 3) = 20$

T_____ F_____ **5.** $5! = 120$

T_____ F_____ **6.** $\dfrac{7!}{6!} = \dfrac{7}{6}$

T_____ F_____ **7.** In the binomial expansion of $(x + 1)^7$, the coefficient of x^4 is 4.

FILL IN THE BLANKS   Answers are on page AN-25.

1. Two sets that have no elements in common are called _____ .

2. The number of different arrangements of r objects from n objects in which (a) the n objects are different, (b) no object is repeated more than once in an arrangement, and (c) order is important is called a _____ .

3. If in 2 above, condition (c) is replaced by "order is not important," we have a _____ .

4. A triangular display of combinations is called the _____ triangle.

5. The combinations $\binom{n}{r}$ are sometimes called _____ _____ .

6. To expand an expression such as $(x + y)^n$, n a positive integer, we can use the _____ _____ .

7. The coefficient of x^3 in the expansion of $(x + 2)^5$ is _____ .

REVIEW EXERCISES   Answers to odd-numbered problems begin on page AN-25.

In Problems 1–16 replace the asterisk by any of the symbol(s) $\subset$, $\subseteq$, $=$ that result in a true statement. If none result in a true statement, write "None of these." More than one answer may be possible.

1. $\{0\} * \varnothing$

2. $\{0\} * \{1, 0, 3\}$

3. $\{5, 6\} \cap \{2, 6\} * \{8\}$

4. $\{2, 3\} \cup \{3, 4\} * \{3\}$

5. $\{8, 9\} * \{9, 10, 11\}$

6. $\{1\} * \{1, 3, 5\} \cup \{3, 4\}$

7. $\{5\} * \{0, 5\}$

8. $\varnothing * \{1, 2, 3\}$

9. $\varnothing * \{1, 2\} \cap \{3, 4, 5\}$

10. $\{2, 3\} * \{3, 4\}$

11. $\{1, 2\} * \{1\} \cup \{3\}$

12. $\{5\} * \{1\} \cup \{2, 3\}$

13. $\{4, 5\} \cap \{5, 6\} * \{4, 5\}$

14. $\{6, 8\} * \{8, 9, 10\}$

15. $\{6, 7, 8\} \cap \{8\} * \{6\}$

16. $\{4\} * \{6, 8\} \cap \{4, 8\}$

17. For the sets

$$A = \{1, 3, 5, 6, 8\} \quad B = \{2, 3, 6, 7\} \quad C = \{6, 8, 9\}$$

find

(a) $(A \cap B) \cup C$ (b) $(A \cap B) \cap C$
(c) $(A \cup B) \cap B$

18. For the sets U = universal set = $\{1, 2, 3, 4, 5, 6, 7\}$ and

$$A = \{1, 3, 5, 6\} \quad B = \{2, 3, 6, 7\} \quad C = \{4, 6, 7\}$$

find:

(a) $\overline{A \cap B}$ (b) $(B \cap C) \cap A$ (c) $\overline{B} \cup \overline{A}$

19. If A and B are sets and if $c(A) = 24$, $c(A \cup B) = 33$, $c(B) = 12$, find $c(A \cap B)$.

20. Car Options During June, Colleen's Motors sold 75 cars with air conditioning, 95 with power steering, and 100 with automatic transmissions. Twenty cars had all three options, 10 cars had none of these options, and 10 cars were sold that had only air conditioning. In addition, 50 cars had both automatic transmissions and power steering, and 60 cars had both automatic transmissions and air conditioning.

(a) How many cars were sold in June?
(b) How many cars had only power steering?

21. Student Survey In a survey of 125 college students, it was found that of three newspapers, the *Wall Street Journal, New York Times,* and *Chicago Tribune:*

60	read the *Chicago Tribune*
40	read the *New York Times*
15	read the *Wall Street Journal*
25	read the *Chicago Tribune* and *New York Times*
8	read the *New York Times* and *Wall Street Journal*
3	read the *Chicago Tribune* and *Wall Street Journal*
1	read all three

(a) How many read none of these papers?
(b) How many read only the *Chicago Tribune?*
(c) How many read neither the *Chicago Tribune* nor the *New York Times?*

22. If U = universal set = $\{1, 2, 3, 4, 5\}$ and $B = \{1, 4, 5\}$, find all sets A for which $A \cap B = \{1\}$.

23. Compute $P(6, 3)$.

24. Compute $C(6, 2)$.

25. In how many different ways can a committee of 3 people be formed from a group of 5 people?

26. In how many different ways can 4 people line up?

27. In how many different ways can 3 books be placed on a shelf?

28. In how many different ways can 3 people be seated in 4 chairs?

29. How many house styles are possible if a contractor offers 3 choices of roof designs, 4 choices of window designs, and 6 choices of brick?

30. How many different answers are possible in a true–false test consisting of 10 questions?

31. You are to set up a code of 2-digit words using the digits 1, 2, 3, 4 without using any digit more than once. What is the maximum number of words in such a language? If all words of the form *ab* and *ba* are the same, how many words are possible?

32. You are to set up a code of 3-digit words using the digits 1, 2, 3, 4, 5, 6 without using any digit more than once in the same word. What is the maximum number of words in such a language? If the words 124, 142, etc., designate the same word, how many different words are possible?

33. A small town consists of a north side and a south side. The north side has 16 houses and the south side has 10 houses. A pollster is asked to visit 4 houses on the north side and 3 on the south side. In how many ways can this be done?

34. Program Selection A ceremony is to include 7 speeches and 6 musical selections.

(a) How many programs are possible?
(b) How many programs are possible if speeches and musical selections are to be alternated?

35. There are 7 boys and 6 girls willing to serve on a committee. How many 7-member committees are possible if a committee is to contain:

(a) 3 boys and 4 girls?
(b) At least one member of each sex?

36. Juan's Ice Cream Parlor offers 31 different flavors to choose from and specializes in double dip cones.

(a) How many different cones are there to choose from if you may select the same flavor for each dip?

(b) How many different cones are there to choose from if you cannot repeat any flavor? Assume that a cone with vanilla on top of chocolate is different from a cone with chocolate on top of vanilla.

(c) How many different cones are there if you consider any cone having chocolate on top and vanilla on the bottom the same as having vanilla on top and chocolate on the bottom?

37. A person has 4 history, 5 English, and 6 mathematics books. How many ways can the books be arranged on a shelf if books on the same subject must be together?

38. Five people are to line up for a group photograph. If two of them refuse to stand next to each other, in how many ways can the photograph be taken?

39. In how many ways can a committee of 8 boys and 5 girls be formed if there are 10 boys and 11 girls eligible to serve on the committee?

40. A football squad has 7 linemen, 11 linebackers, and 9 safeties. How many different teams composed of 5 linemen, 3 linebackers, and 3 safeties can be formed?

41. In how many ways can we choose three words, one each from five 3-letter words, six 4-letter words, and eight 5-letter words?

42. A newborn child can be given 1, 2, or 3 names. In how many ways can a child be named if we can choose from 100 names?

43. In how many ways can 5 girls and 3 boys be divided into 2 teams of 4 if each team is to include at least 1 boy?

44. A meeting is to be addressed by 5 speakers, A, B, C, D, E. In how many ways can the speakers be ordered if B must not precede A?

45. What is the answer to Problem 44 if B is to speak immediately after A?

46. License Plate Numbers   An automobile license number contains 1 or 2 letters followed by a 4-digit number. Compute the maximum number of different licenses.

47. There are 5 rotten plums in a crate of 25 plums. How many samples of 4 of the 25 plums contain

(a) Only good plums?
(b) Three good plums and 1 rotten plum?
(c) One or more rotten plums?

48. An admissions test given by a university contains 10 true–false questions. Eight or more of the questions must be answered correctly in order to be admitted.

(a) How many different ways can the answer sheet be filled out?

(b) How many different ways can the answer sheet be filled out so that 8 or more questions are answered correctly?

49. The figure below indicates the locations of two houses, A and B, in a city, where the lines represent streets. A person at A wishes to reach B and can travel in only two directions, to the right and up. How many different paths are there from A to B?

50. A car driver picks up a passenger at point A (see the figure below) whose destination is point B. After completing the trip, the driver is to proceed to the garage at point C. If the cab must travel to the right or up, how many different routes are there from A to C?

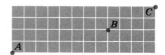

51. Expand $(x + 2)^4$.

52. Expand $(x - 1)^5$.

53. What is the coefficient of x^3 in the expansion of $(x + 2)^7$?

54. What is the coefficient of x^4 in the expansion of $(2x + 1)^6$?

Chapter 7

Probability

Probability theory is a part of mathematics that is useful for discovering and investigating the *regular* features of *random events.* Although it is not really possible to give a precise and simple definition of what is meant by the words *random* and *regular,* we hope that the explanation and the examples given below will help you understand these concepts.

Certain phenomena in the real world may be considered *chance phenomena.* These phenomena do not always produce the same observed outcome, and the outcome of any given observation of the phenomena may not be predictable. But they have a long-range behavior known as *statistical regularity.*

In some cases we are familiar enough with the phenomenon under investigation to feel justified in making *exact* predictions with respect to the result of each individual observation. For example, if you want to know the time when the sun will set, you can find the exact time in the weather section of a newspaper.

However, in many cases our knowledge is not precise enough to allow exact predictions in particular situations. Some examples of such cases, called *random events,* follow.

Tossing a fair coin gives a result that is either a head or a tail. For any one throw, we cannot predict the result, although it is obvious that it is determined by definite causes (such as the initial velocity of the coin, the initial angle of throw, and the surface on which the coin rests). Even though some of these causes can be controlled, we cannot predetermine the result of any particular toss. Thus the result of tossing a coin is a *random event.*

Although we cannot predict the result of any particular toss of the coin, if we perform a long sequence of tosses, we will notice that the number of times heads occurs is approximately equal to the number of times tails appears. That is, it seems *reasonable* to say that in any toss of this fair coin, a head or a tail is *equally likely* to occur. As a result, we might *assign a probability* of $\frac{1}{2}$ for obtaining a head (or tail) on a particular toss.

In a series of throws with an ordinary die, each throw results in one of the numbers 1, 2, 3, 4, 5, or 6. Thus the result of throwing a die is a *random event.*

The appearance of any particular face of the die is an outcome of the event. However, if we perform a long series of tosses, any face is as *equally likely* to occur as any other, provided the die is fair. Here we might *assign a probability* of $\frac{1}{6}$ for obtaining a particular face.

The sex of a newborn baby is either male or female. This, too, is an example of a *random event.*

Table 1*

Year of Birth	Number of Births		Total Number of Births	Ratio of Births	
	Boys b	Girls g	$b + g$	$\dfrac{b}{b + g}$	$\dfrac{g}{b + g}$
1989	2,069,490	1,971,468	4,040,958	.512	.488
1988	2,002,424	1,907,086	3,909,510	.512	.488
1987	1,951,153	1,858,241	3,809,394	.512	.488
1986	1,924,868	1,831,679	3,756,547	.512	.488
1985	1,927,983	1,832,578	3,760,561	.513	.487
1984	1,879,490	1,789,651	3,669,141	.512	.488
1983	1,865,553	1,773,380	3,638,933	.513	.487
1982	1,885,676	1,794,861	3,680,537	.512	.488
1981	1,860,272	1,768,966	3,629,238	.513	.487
1980	1,852,616	1,759,642	3,612,258	.513	.487

**Source:* U.S. Department of Health and Human Services, *Monthly Vital Statistics Report,* December 1991.

Our intuition tells us that a boy baby and a girl baby are *equally likely* to occur. If we follow this reasoning, we might *assign a probability* of $\frac{1}{2}$ to having a girl baby. However, if we consult the data about births in the United States found in Table 1, we see that it might be more accurate to assign a probability of .488 to having a girl baby.

These examples demonstrate that in studying a sequence of random experiments it is not possible to forecast individual results. These are subject to irregular, random fluctuations that cannot be exactly predicted. However, if the number of observations is large—that is, if we deal with a *mass phenomenon*—some regularity appears.

7.1 SAMPLE SPACES AND ASSIGNMENT OF PROBABILITIES

In studying probability we are concerned with experiments, real or conceptual, and their outcomes. In this study we try to formulate in a precise manner a mathematical theory that closely resembles the experiment in question. The first stage of development of such a mathematical theory is the building of what is termed a *probability model.* This model is then used to analyze and predict outcomes of the experiment. The purpose of this section is to learn how a probability model can be constructed.

Sample Spaces

We begin by writing the associated *sample space* of an experiment; that is, we write all outcomes that can occur as a result of the experiment.

EXAMPLE 1

If the experiment consists of flipping a coin, we would agree that the only possible outcomes are heads (H) and tails (T). Therefore, a sample space for the experiment is the set $\{H, T\}$.

. . . .

EXAMPLE 2

Consider an experiment in which, for the sake of simplicity, one die is green and the other is red. When the dice are rolled, the set of outcomes consists of all the different ways that the dice may come to rest, that is, the set of all *possibilities* that can occur as a result of the experiment.

An application of the Multiplication Principle reveals that the number of outcomes of this experiment is 36 since there are 6 possible ways for the green die to come up and 6 ways for the red die to come up.

We can use a tree diagram to obtain a list of the 36 outcomes. See Figure 1 (p. 316). Figure 2 gives a graphical representation of the 36 outcomes.

Figure 1

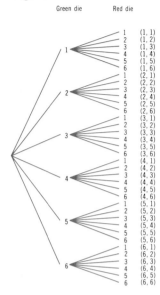

Figure 2

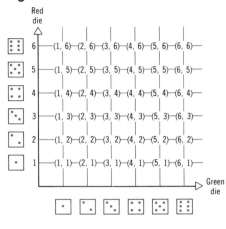

SAMPLE SPACE; OUTCOME

A *sample space S,* associated with a real or conceptual experiment, is the set of all possibilities that can occur as a result of the experiment. Each element of a sample space S is called an *outcome.*

We list below some experiments and their sample spaces.

Experiment	*Sample Space*
(a) An experiment consists of tossing two coins, a penny and a nickel, and observing whether each coin falls heads (H) or tails (T).	$\{HH, HT, TH, TT\}$
(b) An experiment consists of tossing two coins, a penny and a nickel, and observing the number of heads that appear.	$\{0, 1, 2\}$
(c) An experiment consists of tossing two coins, a penny and a nickel, and observing whether the coins match (M) or do not match (D).	$\{M, D\}$
(d) An experiment consists of selecting three manufactured parts from the production process and observing whether they are acceptable (A) or defective (D).	$\{AAA, AAD, ADA, ADD, DAA, DAD, DDA, DDD\}$
(e) An experiment consists of polling customers at a supermarket whether they *like* (L) a certain product or *dislike* (D) it and recording their response.	$\{L, D\}$
(f) A spinner is marked from 1 to 8. An experiment consists of spinning the dial once.	$\{1, 2, 3, 4, 5, 6, 7, 8\}$

Experiment	*Sample Space*
(g) An experiment consists of surveying whether a customer entering a fast food restaurant orders a hamburger (H), french fries (F), both (B), or neither (N).	$\{H, F, B, N\}$

Now Work Problem 1

The sample space of an experiment plays the same role in probability as the universal set does in set theory for all questions concerning the experiment.

In this chapter we confine our attention to those cases for which the sample space is finite, that is, to those situations in which it is possible to have only a finite number of outcomes.

Notice that in our definition we say *a* sample space, rather than *the* sample space, since an experiment can be described in many different ways. In general, it is a safe guide to include as much detail as possible in the description of the outcomes of the experiment in order to answer all pertinent questions concerning the result of the experiment.

EXAMPLE 3

Consider the set of all different types of families with three children. Describe a sample space for the experiment of selecting one family from the set of all possible three-child families.

SOLUTION

One way of describing a sample space is by denoting the number of girls in the family. The only possibilities are members of the set

$$\{0, 1, 2, 3\}$$

That is, a three-child family can have 0, 1, 2, or 3 girls.

This sample space has four outcomes. A disadvantage of describing the experiment using this sample space is that a question such as ''Was the second child a girl?'' cannot be answered. Thus this method of classifying the outcomes may not provide enough wanted information.

Another way of describing a sample space is by first defining B and G as ''boy'' and ''girl,'' respectively. Then the sample space would be given as

$$\{BBB, BBG, BGB, BGG, GBB, GBG, GGB, GGG\}$$

where BBB means first child is a boy, second child is a boy, third child is a boy, and so on. This experiment can be depicted by the tree diagram in Figure 3. The experiment has $2 \cdot 2 \cdot 2 = 8$ possible outcomes, as the Multiplication Principle indicates.

Figure 3

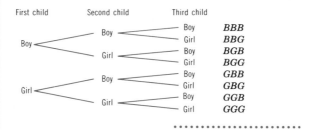

· ·

An advantage of the classification given by the second sample space in Example 3 is that each outcome of the experiment corresponds to exactly one element in this sample space.

Assignment of Probabilities

We are now ready to give a definition of the probability of an outcome of a sample space.

PROBABILITY OF AN OUTCOME

Let S denote a sample space. To each outcome e of S, we assign a real number, $P(e)$, called the *probability of the outcome e,* which has two properties:

(I) $P(e) \geq 0$ for each outcome e in S.
(II) The sum of the probabilities of all the outcomes in S equals 1.

Thus if the sample space S is given by

$$S = \{e_1, e_2, \ldots, e_n\}$$

then

(I) $P(e_1) \geq 0, P(e_2) \geq 0, \ldots, P(e_n) \geq 0$

(II) $P(e_1) + P(e_2) + \cdots + P(e_n) = 1$ (1)

EXAMPLE 4

Let a die be thrown. A sample space S is then

$$S = \{1, 2, 3, 4, 5, 6\}$$

There are six outcomes in S: 1, 2, 3, 4, 5, 6.

One acceptable assignment of probabilities is

$$P(1) = \tfrac{1}{6} \quad P(2) = \tfrac{1}{6} \quad P(3) = \tfrac{1}{6} \quad P(4) = \tfrac{1}{6} \quad P(5) = \tfrac{1}{6} \quad P(6) = \tfrac{1}{6}$$

This choice is consistent with the definition since the probability assigned each outcome is nonnegative and their sum is 1. This assignment is made when the die is **fair.**

Another assignment that is consistent with the definition is

$$P(1) = 0 \quad P(2) = 0 \quad P(3) = \tfrac{1}{3} \quad P(4) = \tfrac{2}{3} \quad P(5) = 0 \quad P(6) = 0$$

This assignment, although unnatural, is made when the die is "loaded" in such a way that only a 3 or a 4 can occur and the 4 is twice as likely as the 3 to occur.

Many other assignments can also be made that are consistent with the definition.

Now Work Problem 23

....

EXAMPLE 5

A coin is weighted so that heads (H) is 5 times more likely to occur than tails (T). What probability should we assign to heads? To tails?

SOLUTION

Let x denote the probability that tails occurs. Then

$$P(T) = x \quad \text{and} \quad P(H) = 5x$$

Since the sum of the probabilities of all the outcomes must equal 1, we have

$$P(H) + P(T) = 5x + x = 1$$
$$6x = 1$$
$$x = \tfrac{1}{6}$$

Thus we assign the probabilities

$$P(H) = \tfrac{5}{6} \qquad P(T) = \tfrac{1}{6}$$

Now Work Problem 27

The sample space and the assignment of probabilities to each outcome of an experiment constitutes a *probability model* for the experiment.

Constructing a Probability Model

To construct a probability model requires two steps:

Step 1 Find a sample space. List all the possible outcomes of the experiment, or, if this is not easy to do, determine the number of outcomes of the experiment.

Step 2 Assign to each outcome e a probability $P(e)$ so that
(a) $P(e) \geq 0$
(b) The sum of all the probabilities assigned to the outcomes equals 1.

EXAMPLE 6

A fair coin is tossed. If it comes up heads (H), then a fair die is rolled. If it comes up tails (T), then the coin is tossed once more. Construct a probability model for this experiment.

SOLUTION

We begin by constructing a tree diagram (Figure 4) that reflects the experiment.

Figure 4

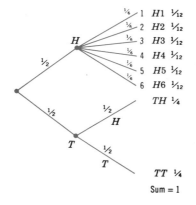

Based on the tree diagram, we can list the possible outcomes of the experiment. Thus a sample space for the experiment is

$$S = \{H1,\ H2,\ H3,\ H4,\ H5,\ H6,\ TH,\ TT\}$$

where $H1$ indicates heads for the coin and then 1 for the die, and so on.

Next, since the coin is fair, the probability is $\frac{1}{2}$ for a head H and $\frac{1}{2}$ for a tail T. Also, since the die is fair, the probability for any face to occur is $\frac{1}{6}$. Refer again to Figure 4. Based on all this, we assign the following probabilities to each outcome in S:

$$P(H1) = P(H2) = P(H3) = P(H4) = P(H5) = P(H6) = \tfrac{1}{12}$$
$$P(TH) = P(TT) = \tfrac{1}{4}$$

The above discussion constitutes a probability model, or **stochastic model,** for the experiment.

. . . .

Notice that the probability assignments of Example 6 are obtained by multiplying along the branches of the tree. We'll have more to say about this later.

Events and Simple Events

EVENT; SIMPLE EVENT

An *event* is any subset of the sample space. If an event has exactly one element, that is, consists of only one outcome, it is called a *simple event*.

Every event can be written as the union of simple events. For example, consider the sample space of Example 3. The event E that the family consists of exactly two boys is

$$E = \{BBG,\ BGB,\ GBB\}$$

The event E is the union of the three simple events $\{BBG\}$, $\{BGB\}$, $\{GBB\}$. That is,

$$E = \{BBG\} \cup \{BGB\} \cup \{GBB\}$$

Since a sample space S is also an event, we can express a sample space as the union of simple events. Thus if the sample space S consists of n outcomes,

$$S = \{e_1, e_2,\ \ldots\ ,\ e_n\}$$

then

$$S = \{e_1\} \cup \{e_2\} \cup \cdots \cup \{e_n\}$$

Suppose probabilities have been assigned to each outcome of S. We now raise the question, "What is the probability of an event?" Let S be a sample space and let E be any event of S. It is clear that either $E = \varnothing$ or E is a simple event or E is the union of two or more simple events.

PROBABILITY OF AN EVENT

If $E = \varnothing$, the event E is *impossible*. We define the *probability of $\varnothing$* as

$$P(\varnothing) = 0$$

If $E = \{e\}$ is a simple event, then $P(E) = P(e)$; that is, $P(E)$ equals the probability assigned to the outcome e.

If E is the union of r simple events $\{e_1\}, \{e_2\}, \ldots, \{e_r\}$, we define the *probability of* E to be the sum of the probabilities assigned to each simple event in E. That is,

$$P(E) = P(e_1) + P(e_2) + \cdots + P(e_r) \tag{2}$$

In particular, if the sample space S is given by

$$S = \{e_1, e_2, \ldots, e_n\}$$

we must have

$$P(S) = P(e_1) + \cdots + P(e_n) = 1$$

Thus the probability of S, the sample space, is 1.

EXAMPLE 7

In an experiment with two dice, the following are events:

(a) The sum of the faces is 3.   (b) The sum of the faces is 7.
(c) The sum of the faces is 7 or 3.   (d) The sum of the faces is 7 and 3.

Find the probability of these events, assuming the dice are fair.

SOLUTION

The tree diagram shown in Figure 5 illustrates the experiment. Since the dice are fair, we assign a probability of $\frac{1}{6}$ along each branch. The probability assigned to each outcome in S is therefore $\frac{1}{36}$.

Figure 5

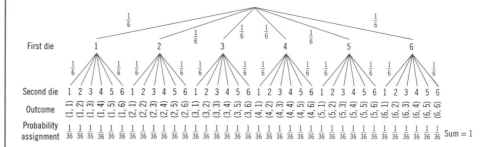

(a) The sum of the faces is 3 if and only if the outcome is an element of the event $A = \{(1, 2), (2, 1)\}$. Thus

$$P(A) = \tfrac{1}{36} + \tfrac{1}{36} = \tfrac{2}{36} = \tfrac{1}{18}$$

(b) The sum of the faces is 7 if and only if the outcome is an element of the event $B = \{(1, 6), (2, 5), (3, 4), (4, 3), (5, 2), (6, 1)\}$. Thus

$$P(B) = \tfrac{1}{36} + \tfrac{1}{36} + \tfrac{1}{36} + \tfrac{1}{36} + \tfrac{1}{36} + \tfrac{1}{36} = \tfrac{6}{36} = \tfrac{1}{6}$$

(c) The sum of the faces is 7 or 3 if and only if the outcome is an element of $A \cup B$, as defined in Parts (a) and (b):

$$A \cup B = \{(2, 1), (1, 2), (1, 6), (2, 5), (3, 4), (4, 3), (5, 2), (6, 1)\}$$

Thus,

$$P(A \cup B) = \tfrac{1}{36} + \tfrac{1}{36} + \tfrac{1}{36} + \tfrac{1}{36} + \tfrac{1}{36} + \tfrac{1}{36} + \tfrac{1}{36} + \tfrac{1}{36} = \tfrac{8}{36} = \tfrac{2}{9}$$

(d) The sum of the faces is 3 and 7 if and only if the outcome is an element of $A \cap B$. Since $A \cap B = \varnothing$, the event is impossible. That is, $P(A \cap B) = 0$.

....

EXAMPLE 8

For the model described in Example 6 let E be the event

$$E = \{H2, H4, TH\}$$

The probability of event E is

$$P(E) = P(H2) + P(H4) + P(TH)$$
$$= \tfrac{1}{12} + \tfrac{1}{12} + \tfrac{1}{4} = \tfrac{5}{12}$$

Now Work Problem 37

....

EXAMPLE 9

Let two coins be tossed. A sample space S is

$$S = \{HH, TH, HT, TT\}$$

Let E be the event that both coins show heads or both show tails. Compute the probability for this event E using the following two assignments of probabilities:

(a) $P(HH) = P(TH) = P(HT) = P(TT) = \tfrac{1}{4}$
(b) $P(HH) = \tfrac{1}{9}$, $P(TH) = \tfrac{2}{9}$, $P(HT) = \tfrac{2}{9}$, $P(TT) = \tfrac{4}{9}$

SOLUTION

The event E is $\{HH, TT\}$.

(a) $P(E) = P(HH) + P(TT) = \tfrac{1}{4} + \tfrac{1}{4} = \tfrac{1}{2}$
(b) $P(E) = P(HH) + P(TT) = \tfrac{1}{9} + \tfrac{4}{9} = \tfrac{5}{9}$

....

The fact that we obtained different probabilities for the same event in Example 9 is not unexpected since it results from our original assignment of probabilities to the outcomes of the experiment. Any assignment that conforms to the restrictions given in (1) is mathematically correct. The question of which assignment to make is not a mathematical question, but is one that depends on the real world situation to which the theory is applied. In this example the coins were fair in case (a) and were loaded in case (b).

EXERCISE 7.1   Answers to odd-numbered problems begin on page AN-25.

1. A nickel and a dime are tossed. List the elements of the sample space

   (a) If we are interested in whether the dime falls heads (*H*) or tails (*T*).
   (b) If we are interested only in the number of heads that appear on a single toss of the two coins.

   (c) If we are interested in whether the coins match (*M*) or do not match (*D*).

2. A card is selected from a regular deck of cards.* List the elements of the sample space

   (a) If we are interested in the color of the card
   (b) If we are interested in the suit of the card

In Problems 3–8 describe a sample space associated with each experiment. List the outcomes of each sample space. In each experiment we are interested in whether the coin falls heads (H) or tails (T).

3. Tossing 2 coins

4. Tossing 1 coin twice

5. Tossing a coin 3 times

6. Tossing 3 coins

7. Tossing a coin 2 times and then a die

8. Tossing a coin and then a die

In Problems 9–16 use the pictured spinners to list the outcomes of a sample space associated with each experiment.

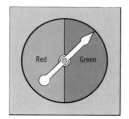 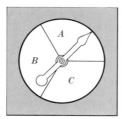 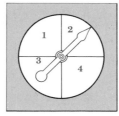

Spinner 1   **Spinner 2**   **Spinner 3**

9. First spinner 1 is spun and then spinner 2 is spun.

10. First spinner 3 is spun and then spinner 1 is spun.

11. Spinner 1 is spun twice.

12. Spinner 3 is spun twice.

13. Spinner 2 is spun twice and then spinner 3 is spun.

14. Spinner 3 is spun once and then spinner 2 is spun twice.

15. Spinners 1, 2, and 3 are each spun once in this order.

16. Spinners 3, 2, and 1 are each spun once in this order.

In Problems 17–22 find the number of outcomes of a sample space associated with each random experiment.

17. Tossing a coin 4 times

18. Tossing a coin 5 times

19. Tossing 3 dice

20. Tossing 2 dice and then a coin

21. Selecting 2 cards (without replacement) from a regular deck of 52 cards (Assume order is not important.)

22. Selecting 3 cards (without replacement) from a regular deck of 52 cards (Assume order is not important.)

* A regular deck of cards has 52 cards. There are four suits of 13 cards each. The suits are called *clubs* (black), *diamonds* (red), *hearts* (red), and *spades* (black). In each suit the 13 cards are labeled A (ace), 2, 3, 4, 5, 6, 7, 8, 9, 10, J (jack), Q (queen), and K (king).

In Problems 23–26 consider the experiment of tossing a coin twice. The table lists six possible assignments of probabilities for this experiment:

	Outcome	HH	HT	TH	TT
			Sample Space		
Assignments	1	$\frac{1}{4}$	$\frac{1}{4}$	$\frac{1}{4}$	$\frac{1}{4}$
	2	0	0	0	1
	3	$\frac{3}{16}$	$\frac{5}{16}$	$\frac{5}{16}$	$\frac{3}{16}$
	4	$\frac{1}{2}$	$\frac{1}{2}$	$-\frac{1}{2}$	$\frac{1}{2}$
	5	$\frac{1}{8}$	$\frac{1}{4}$	$\frac{1}{4}$	$\frac{1}{8}$
	6	$\frac{1}{9}$	$\frac{2}{9}$	$\frac{2}{9}$	$\frac{4}{9}$

Using this table, answer the following four questions.

23. Which of the assignments of probabilities are consistent with the definition of the probability of an outcome?

24. Which of the assignments of probabilities should be used if the coin is known to be fair?

25. Which of the assignments of probabilities should be used if the coin is known to always come up tails?

26. Which of the assignments of probabilities should be used if tails is twice as likely as heads to occur?

27. A coin is weighted so that heads is three times as likely as tails to occur. What probability should we assign to heads? to tails?

28. A coin is weighted so that tails is twice as likely as heads to occur. What probability should we assign to heads? to tails?

29. A die is weighted so that an odd-numbered face is twice as likely as an even-numbered face. What probability should we assign to each face?

30. A die is weighted so that a 6 cannot appear. If the other faces each have the same probability, what probability should we assign to each face?

In Problems 31–36 the random experiment consists of tossing 2 fair dice. Construct a probability model for this experiment and find the probability of each given event.

31. $A = \{(1, 2), (2, 1)\}$

32. $B = \{(1, 5), (2, 4), (3, 3), (4, 2), (5, 1)\}$

33. $C = \{(1, 4), (2, 4), (3, 4), (4, 4)\}$

34. $D = \{(1, 2), (2, 1), (2, 4), (4, 2), (3, 6), (6, 3)\}$

35. $E = \{(1, 1), (2, 2), (3, 3), (4, 4), (5, 5), (6, 6)\}$

36. $F = \{(6, 6)\}$

In Problems 37–42 the random experiment consists of tossing a fair die and then a fair coin. Construct a probability model for this experiment. Then find the probability of each given event.

37. A: The coin comes up heads

38. B: The die comes up 1

39. C: The die does *not* come up 1

40. D: The die comes up 5 or 6

41. E: The die comes up 3, 4, or 5

42. F: The coin comes up heads and the die comes up a number less than 4

In Problems 43–46 assign valid probabilities to the outcomes of each random experiment.

43. Tossing a fair coin twice

44. Tossing a fair coin 3 times

45. Tossing a fair die and then a fair coin

46. Tossing a fair die and then 2 fair coins

In Problems 47–52 the random experiment consists of tossing a fair coin 4 times.

47. List the outcomes of the sample space and assign probabilities to each one.

48. Write the elements of the event, "The first 2 tosses are heads."

49. Write the elements of the event, "The last 3 tosses are tails."

50. Write the elements of the event, "Exactly 3 tosses come up tails."

51. Write the elements of the event, "The number of heads exceeds 1 but is fewer than 4."

52. Write the elements of the event, "The first 2 tosses are heads and the second 2 are tails."

53. In a T-maze a mouse may turn to the right (R) and receive a mild shock, or to the left (L) and get a piece of cheese. Its behavior in making such "choices" is studied by psychologists. Suppose a mouse runs a T-maze 3 times. List the set of all possible outcomes and assign valid probabilities to each outcome under the assumption that the first two times the maze is run the mouse chooses equally between left and right, but on the third run, the mouse is twice as likely to choose cheese. Find the probability of each of the events listed.

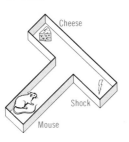

(a) E: Run to the right 2 consecutive times
(b) F: Never run to the right
(c) G: Run to the left on the first trial
(d) H: Run to the right on the second trial

54. Let $S = \{e_1, e_2, e_3, e_4, e_5, e_6, e_7\}$ be a given sample space. Let the probabilities assigned to each outcome be given as follows:

$$P(e_1) = \ P(e_2) = P(e_6)$$
$$P(e_3) = 2P(e_4) = \tfrac{1}{2}P(e_1)$$
$$P(e_5) = \tfrac{1}{2}P(e_7) = \tfrac{1}{4}P(e_1)$$

(a) Find $P(e_1)$, $P(e_2)$, $P(e_3)$, $P(e_4)$, $P(e_5)$, $P(e_6)$, $P(e_7)$.
(b) If $A = \{e_1, e_2\}$, $B = \{e_2, e_3, e_4\}$, $C = \{e_5, e_6, e_7\}$, and $D = \{e_1, e_5, e_6\}$, find $P(A)$, $P(B)$, $P(C)$, $P(D)$, $P(A \cup B)$, $P(A \cap D)$, $P(D \cap B)$, and $P(A \cap \overline{B})$.

55. Three cars, C_1, C_2, C_3, are in a race. If the probability of C_1 winning is p, that is, $P(C_1) = p$, and $P(C_1) = \tfrac{1}{2}P(C_2)$ and $P(C_3) = \tfrac{1}{2}P(C_2)$, find $P(C_1)$, $P(C_2)$, and $P(C_3)$.

56. Consider an experiment with a loaded die such that the probability of any of the faces appearing in a toss is equal to that face times the probability that a 1 will occur. That is, $P(6) = 6 \cdot P(1)$, and so on.

(a) Describe the sample space.
(b) Find $P(1)$, $P(2)$, $P(3)$, $P(4)$, $P(5)$, and $P(6)$.
(c) Let the events A, B, and C be described as

$\quad A$: Even-numbered face
$\quad B$: Odd-numbered face
$\quad C$: Prime number on face (2, 3, 5 are prime)

Find $P(A)$, $P(B)$, $P(C)$, $P(A \cup B)$, and $P(A \cup \overline{C})$.

7.2 PROPERTIES OF THE PROBABILITY OF AN EVENT

In this section we state and prove results involving the probability of an event after the probability model has been constructed. The main tool we employ is set theory.

Mutually Exclusive Events

MUTUALLY EXCLUSIVE EVENTS

Two or more events of a sample space S are said to be *mutually exclusive* **if and only if they have no outcomes in common.**

Thus if we treat mutually exclusive events as sets, they are disjoint.

The following result gives us a way of computing probabilities for mutually exclusive events.

Probability of E or F if E and F Are Mutually Exclusive

Let E and F be two events of a sample space S. If E and F are mutually exclusive, that is, if $E \cap F = \varnothing$, then the probability of E or F is the sum of their probabilities, namely,

$$P(E \cup F) = P(E) + P(F) \qquad \textbf{(1)}$$

Since E and F can be written as a union of simple events in which no simple event of E appears in F and no simple event of F appears in E, the result follows.

EXAMPLE 1

In the experiment of tossing two fair dice, what is the probability of obtaining either a sum of 7 or a sum of 11?

SOLUTION

Let E and F be the events

$$E: \quad \text{Sum is 7} \qquad F: \quad \text{Sum is 11}$$

Since the dice are fair, and a sum of 7 can occur in six different ways, we have

$$P(E) = P(\text{Sum is 7}) = \tfrac{1}{36} + \tfrac{1}{36} + \tfrac{1}{36} + \tfrac{1}{36} + \tfrac{1}{36} + \tfrac{1}{36} = \tfrac{6}{36}$$

Similarly,

$$P(F) = \tfrac{2}{36}$$

The two events E and F are mutually exclusive. Therefore, by (1), the probability that the sum is 7 or 11 is

$$P(E \cup F) = P(E) + P(F) = \tfrac{6}{36} + \tfrac{2}{36} = \tfrac{8}{36} = \tfrac{2}{9}$$

Now Work Problem 17

....

Additive Rule

The following result, called the **Additive Rule,** provides a technique for finding the probability of the union of two events whether they are mutually exclusive or not.

> **Additive Rule**
>
> For any two events E and F of a sample space S,
>
> $$P(E \cup F) = P(E) + P(F) - P(E \cap F) \qquad (2)$$

This result concerning probability is closely related to the counting formula discussed in the previous chapter (page 276). A proof is outlined in Problem 57.

EXAMPLE 2

If $P(E) = .30$, $P(F) = .20$, and $P(E \cup F) = .40$, find $P(E \cap F)$.

SOLUTION

By the Additive Rule we know that

$$P(E \cup F) = P(E) + P(F) - P(E \cap F)$$

Using the data given, we have

$$.40 = .30 + .20 - P(E \cap F)$$

Thus

$$P(E \cap F) = .30 + .20 - .40 = .10$$

....

EXAMPLE 3

Consider the two events

E: A shopper spends at least \$40 for food
F: A shopper spends at least \$15 for meat

Because of recent studies, we might assign

$$P(E) = .56 \qquad P(F) = .63$$

Suppose the probability that a shopper spends at least \$40 for food and \$15 for meat is .33. What is the probability that a shopper spends at least \$40 for food or at least \$15 for meat?

SOLUTION

Since we are looking for the probability of $E \cup F$, we use the Additive Rule and find that

$$P(E \cup F) = P(E) + P(F) - P(E \cap F)$$
$$= .56 + .63 - .33 = .86$$

....

Now Work Problem 5

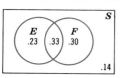

Figure 6

Often a Venn diagram is helpful in solving probability problems. A Venn diagram depicting the information of Example 3 is given in Figure 6. To obtain the diagram, we begin with the fact that $P(E \cap F) = .33$. Since $P(E) = .56$ and $P(F) = .63$, we fill in E with $.56 - .33 = .23$ and we fill in F with $.63 - .33 = .30$. Since $P(S) = 1$, we complete Figure 6 by entering $1 - (.23 + .33 + .30) = .14$. Now it is easy to see that the probability of E, but not F, is .23. The probability of neither E nor F is .14.

EXAMPLE 4

In an experiment with two fair dice, consider the events

E: The sum of the faces is 8
F: Doubles are thrown

What is the probability of obtaining E or F?

SOLUTION

$$E = \{(2, 6), (3, 5), (4, 4), (5, 3), (6, 2)\}$$
$$F = \{(1, 1), (2, 2), (3, 3), (4, 4), (5, 5), (6, 6)\}$$
$$E \cap F = \{(4, 4)\}$$

Also,

$$P(E) = \tfrac{5}{36} \qquad P(F) = \tfrac{6}{36} \qquad P(E \cap F) = \tfrac{1}{36}$$

Thus the probability of E or F is

$$P(E \cup F) = \tfrac{5}{36} + \tfrac{6}{36} - \tfrac{1}{36} = \tfrac{10}{36} = \tfrac{5}{18}$$

• •

We could also have solved Example 4 by actually finding $E \cup F$. Then

$$E \cup F = \{(2, 6), (3, 5), (4, 4), (5, 3), (6, 2), (1, 1), (2, 2), (3, 3), (5, 5), (6, 6)\}$$

Since $c(E \cup F) = 10$, we have

$$P(E \cup F) = \underbrace{\frac{1}{36} + \frac{1}{36} + \cdots + \frac{1}{36}}_{10 \text{ times}} = \frac{10}{36} = \frac{5}{18}$$

as before.

The probability of any outcome of a sample space S is nonnegative. Furthermore, since any event E of S is the union of outcomes in S, and since $P(S) = 1$, it follows that

$$0 \le P(E) \le 1$$

Properties of the Probability of an Event

To summarize, the probability of an event E of a sample space S has the following three properties:

(I) $0 \le P(E) \le 1$ for every event E of S

(II) $P(\varnothing) = 0$ and $P(S) = 1$

(III) $P(E \cup F) = P(E) + P(F) - P(E \cap F)$ for any two events E and F of S

Complement of an Event

Let E be an event of a sample space S. The complement of E is the event "Not E" in S. The next result gives a relationship between their probabilities.

Probability of the Complement of an Event

Let E be an event of a sample space S. Then

$$P(\overline{E}) = 1 - P(E) \qquad\qquad (3)$$

where $\overline{E}$ is the complement of E.

Proof   We know that

$$S = E \cup \overline{E} \qquad E \cap \overline{E} = \varnothing$$

Since E and $\overline{E}$ are mutually exclusive,

$$P(S) = P(E) + P(\overline{E})$$

Now we apply Property II, setting $P(S) = 1$. It follows that

$$P(\overline{E}) = 1 - P(E) \qquad\qquad\blacksquare$$

This result gives us a tool for finding the probability that an event does not occur if we know the probability that it does occur. Thus the probability $P(\overline{E})$ that E does not occur is obtained by subtracting from 1 the probability $P(E)$ that E does occur. We will see shortly that it is sometimes easier to find $P(E)$ by finding $P(\overline{E})$ and using (3), than it is to proceed directly.

EXAMPLE 5

A study of people over 40 with an MBA degree shows that it is reasonable to assign a probability of .756 that such a person will have annual earnings in excess of $80,000. The probability that such a person will earn $80,000 or less is then

$$1 - .756 = .244$$

Now Work Problem 1

· ·

EXAMPLE 6

In an experiment using two fair dice find

(a) The probability that the sum of the faces is less than or equal to 7
(b) The probability that the sum of the faces is greater than 7

SOLUTION

(a) The number of outcomes in the event E, "the sum of the faces is less than or equal to 7," is 21. The number of outcomes of the sample space S is 36. Thus, because the dice are fair,

$$P(E) = \tfrac{21}{36} = \tfrac{7}{12}$$

(b) We need to find $P(\overline{E})$. From (3),

$$P(\overline{E}) = 1 - P(E) = 1 - \tfrac{7}{12} = \tfrac{5}{12}$$

That is, the probability that the sum of the faces is greater than 7 is $\tfrac{5}{12}$.

· ·

Equally Likely Outcomes

So far, we have seen that in many cases it is reasonable to assign the same probability to each outcome of a sample space. Such outcomes are termed *equally likely outcomes*.

Equally likely outcomes often occur when items are selected randomly. For example, in randomly selecting 1 person from a group of 10, the probability of selecting a particular individual is $\tfrac{1}{10}$. If a card is chosen randomly from a deck of 52 cards, the probability of drawing a particular card is $\tfrac{1}{52}$. In general, if a sample space S has n equally likely outcomes, the probability assigned to each outcome is $\tfrac{1}{n}$.

Let a sample space S be given by

$$S = \{e_1, e_2, \ldots, e_n\}$$

Suppose each of the outcomes $e_1, \ldots, e_n$ is equally likely to occur. If E contains m of these n outcomes, then

$$P(E) = \frac{m}{n} \qquad\qquad (4)$$

A proof of Equation (4) is outlined in Problem 58.

For example, if a person is chosen randomly from a group of 100 people, 60 female and 40 male, the probability a male is chosen is

$$\underbrace{\tfrac{1}{100} + \tfrac{1}{100} + \cdots + \tfrac{1}{100}}_{40 \text{ times}} = \tfrac{40}{100}$$

This leads to the following formulation:

Probability of an Event E in a Sample Space with Equally Likely Outcomes

If an experiment has n equally likely outcomes, among which the event E occurs m times, then the probability of event E, written as $P(E)$, is m/n. That is,

$$P(E) = \frac{\text{Number of possible ways the event } E \text{ can take place}}{\text{Number of outcomes in } S} = \frac{c(E)}{c(S)} \qquad (5)$$

Thus, to compute the probability of an event E in which the outcomes are equally likely, count the number $c(E)$ of outcomes in E, and divide by the total number $c(S)$ of outcomes in the sample space.

EXAMPLE 7

A jar contains 10 marbles; 5 are solid color, 4 are speckled, and 1 is clear.

(a) If one marble is picked at random, what is the probability it is speckled?
(b) If one marble is picked at random, what is the probability it is clear or a solid color?

SOLUTION

The experiment is an example of one in which the outcomes are equally likely, that is, no one marble is more likely to be picked than another. If S is the sample space, then there are 10 possible outcomes in S, so $c(S) = 10$.

(a) Define the event E: A speckled marble is picked. There are 4 ways E can occur. Thus

$$P(E) = \frac{c(E)}{c(S)} = \frac{4}{10} = .4$$

(b) Define the events F: A clear marble is picked and G: A solid color marble is picked. Then there is 1 way for F to occur and 5 ways for G to occur. Thus

$$P(F) = \frac{c(F)}{c(S)} = \frac{1}{10} \qquad P(G) = \frac{c(G)}{c(S)} = \frac{5}{10}$$

We seek $P(F \cup G)$. Since F and G are mutually exclusive,

$$P(F \cup G) = P(F) + P(G) = \tfrac{1}{10} + \tfrac{5}{10} = \tfrac{6}{10} = .6$$

Now Work Problem 23

· ·

Odds

In many instances the probability of an event may be expressed as *odds*—either *odds for* an event or *odds against* an event.

If E is an event:

The **odds for E** are $\qquad \dfrac{P(E)}{P(\overline{E})} \qquad$ or $\qquad P(E)$ to $P(\overline{E})$

The **odds against E** are $\qquad \dfrac{P(\overline{E})}{P(E)} \qquad$ or $\qquad P(\overline{E})$ to $P(E)$

EXAMPLE 8

Suppose the probability of the event

$$E: \quad \text{It will rain}$$

is .3. The odds for rain are

$$\frac{.3}{.7} \qquad \text{or} \qquad 3 \text{ to } 7$$

The odds against rain are

$$\frac{.7}{.3} \qquad \text{or} \qquad 7 \text{ to } 3$$

· ·

To obtain the probability of the event E when either the odds for E or the odds against E are known, we use the following formulas:

If the odds for E are a to b, then

$$P(E) = \frac{a}{a+b} \qquad (6)$$

If the odds against E are a to b, then

$$P(E) = \frac{b}{a+b}$$

The proof of Equation (6) is outlined in Problem 59.

EXAMPLE 9

(a) The odds for a Republican victory in the next Presidential election are 7 to 5. What is the probability that a Republican victory occurs?
(b) The odds against the Chicago Cubs winning the league pennant are 200 to 1. What is the probability that the Cubs win the pennant?

SOLUTION

(a) The event E is "A Republican victory occurs." The odds for E are 7 to 5. Thus,

$$P(E) = \frac{7}{7+5} = \frac{7}{12} \approx .583$$

(b) The event F is "The Cubs win the pennant." The odds against F are 200 to 1. Thus

$$P(F) = \frac{1}{200+1} = \frac{1}{201} \approx .00498$$

Now Work Problem 41

EXERCISE 7.2   Answers to odd-numbered problems begin on page AN-26.

In Problems 1–6 find the probability of the indicated event if $P(A) = .20$ and $P(B) = .30$.

1. $P(\overline{A})$

2. $P(\overline{B})$

3. $P(A \cup B)$ if A, B are mutually exclusive

4. $P(A \cap B)$ if A, B are mutually exclusive

5. $P(A \cup B)$ if $P(A \cap B) = .15$

6. $P(A \cap B)$ if $P(A \cup B) = .40$

In Problems 7–16 a card is drawn at random from a regular deck of 52 cards. Calculate the probability of each event.

7. The ace of hearts is drawn.

8. An ace is drawn.

9. A spade is drawn.

10. A red card is drawn.

11. A picture card (J, Q, K) is drawn.

12. A number card (A, 2, 3, 4, 5, 6, 7, 8, 9, 10) is drawn.

13. A card with a number less than 6 is drawn (count A as 1).

14. A card with a value of 10 or higher is drawn.

15. A card that is not an ace is drawn.

16. A card that is either a queen or king of any suit is drawn.

17. In tossing 2 fair dice, are the events "Sum is 2" and "Sum is 12" mutually exclusive? What is the probability of obtaining either a 2 or a 12?

18. In tossing 2 fair dice, are the events "Sum is 6" and "Sum is 8" mutually exclusive? What is the probability of obtaining either a 6 or an 8?

In Problems 19–26 a ball is picked at random from a box containing 3 white, 5 red, 8 blue, and 7 green balls. Find the probability of each event.

19. White ball is picked.

20. Blue ball is picked.

21. Green ball is picked.

22. Red ball is picked.

23. White or red ball is picked.

24. Green or blue ball is picked.

25. Neither red nor green ball is picked.

26. Red or white or blue ball is picked.

27. In a throw of 2 fair dice, what is the probability that the number on one die is double the number on the other?

28. In a throw of 2 fair dice, what is the probability that one die gives a 5 and the other die a number less than 5?

29. The Chicago Bears football team has a probability of winning of .65 and of tying of .05. What is their probability of losing?

30. The Chicago Black Hawks hockey team has a probability of winning of .6 and a probability of losing of .25. What is the probability of a tie?

31. Jenny is taking courses in both mathematics and English. She estimates her probability of passing mathematics at .4 and English at .6, and she estimates her probability of passing at least one of them at .8. What is her probability of passing both courses?

32. After midterm exams, the student in Problem 31 reassesses her probability of passing mathematics to .7. She feels her probability of passing at least one of these courses is still .8, but she has a probability of only .1 of passing both courses. If her probability of passing English is less than .4, she will drop English. Should she drop English? Why?

33. Let A and B be events of a sample space S and let $P(A) = .5$, $P(B) = .3$, and $P(A \cap B) = .1$. Find the probabilities of each of the following events:

(a) A or B
(b) A but not B
(c) B but not A
(d) Neither A nor B

34. If A and B represent two mutually exclusive events such that $P(A) = .35$ and $P(B) = .50$, find each of the follow-

ing probabilities:

(a) $P(A \cup B)$
(b) $P(\overline{A \cup B})$
(c) $P(\overline{B})$
(d) $P(\overline{A})$
(e) $P(A \cap B)$

35. At the Milex tune-up and brake repair shop, the manager has found that a car will require a tune-up with a probability of .6, a brake job with a probability of .1, and both with a probability of .02.

(a) What is the probability that a car requires either a tune-up or a brake job?

(b) What is the probability that a car requires a tune-up but not a brake job?

(c) What is the probability that a car requires neither type of repair?

36. A factory needs two raw materials, say, E and F. The probability of not having an adequate supply of material E is .06, whereas the probability of not having an adequate supply of material F is .04. A study shows that the probability of a shortage of both E and F is .02. What is the probability of the factory being short of either material E or F?

37. In a survey of the number of TV sets in a house, the following probability table was constructed:

Number of TV sets	0	1	2	3	4 or more
Probability	.05	.24	.33	.21	.17

Find the probability of a house having

(a) 1 or 2 TV sets
(b) 1 or more TV sets
(c) 3 or fewer TV sets
(d) 3 or more TV sets
(e) Fewer than 2 TV sets
(f) Not even 1 TV set
(g) 1, 2, or 3 TV sets
(h) 2 or more TV sets

38. Through observation it has been determined that the probability for a given number of people waiting in line at a particular checkout register of a supermarket is as shown in the table:

Number waiting in line	0	1	2	3	4 or more
Probability	.10	.15	.20	.24	.31

Find the probability of

(a) At most 2 people in line
(b) At least 2 people in line
(c) At least 1 person in line

39. From a sales force of 150 people, 1 person will be chosen to attend a special sales meeting. If 52 are single, 72 are college graduates, and, of the 52 who are single, $\frac{3}{4}$ are college graduates, what is the probability that a salesperson selected at random will be neither single nor a college graduate?

40. In an election two amendments were proposed. The results indicate that of 850 people eligible to cast a ballot, 480 voted in favor of Amendment I, 390 voted for Amendment II, 120 voted for both, and 100 approved of neither. If an eligible voter is selected at random (that is, any one is as likely to be chosen as another), compute the following probabilities:

(a) The voter is in favor of I, but not II.
(b) The voter is in favor of II, but not I.

In Problems 41–46 determine the probability of E for the given odds.

41. 3 to 1 for E **42.** 4 to 1 against E **43.** 7 to 5 against E

44. 2 to 9 for E **45.** 1 to 1 for E (even)   **46.** 50 to 1 for E

In Problems 47–50 determine the odds for and against each event for the given probability.

47. $P(E) = .7$ **48.** $P(H) = \frac{1}{3}$ **49.** $P(F) = \frac{4}{5}$ **50.** $P(G) = .01$

51. If two fair dice are thrown, what are the odds of obtaining a 7? An 11? A 7 or 11?

52. If the odds for event A are 1 to 5 and the odds for event B are 1 to 3, what are the odds for the event A or B, assuming the event A and B is impossible?

53. The probability of a person getting a job interview is .54; what are the odds against getting the interview?

54. If the probability of war is .6, what are the odds against war?

55. In a track contest the odds that A will win are 1 to 2, and the odds that B will win are 2 to 3. Find the probability and the odds that A or B wins the race, assuming a tie is impossible.

56. It has been estimated that in 70% of the fatal accidents involving two cars, at least one of the drivers is drunk. If you hear of a two-car fatal accident, what odds should you give a friend for the event that at least one of the drivers was drunk?

57. Prove the Additive Rule, Equation (2) on page 327.
[*Hint:* From Example 9 in Section 6.1 (page 274) we have

$$E \cup F = (E \cap \bar{F}) \cup (E \cap F) \cup (\bar{E} \cap F)$$

Since $E \cap \bar{F}, E \cap F$, and $\bar{E} \cap F$ are pairwise disjoint, we can write

$$P(E \cup F) = P(E \cap \bar{F}) + P(E \cap F) + P(\bar{E} \cap F) \quad \text{(a)}$$

We may write the sets E and F in the form

$$E = (E \cap F) \cup (E \cap \bar{F})$$
$$F = (E \cap F) \cup (\bar{E} \cap F)$$

Since $E \cap F$ and $E \cap \bar{F}$ are disjoint and $E \cap F$ and $\bar{E} \cap F$ are disjoint, we have

$$P(E) = P(E \cap F) + P(E \cap \bar{F})$$
$$P(F) = P(E \cap F) + P(\bar{E} \cap F) \quad \text{(b)}$$

Now combine (a) and (b).]

58. Prove Equation (4) on page 330.
[*Hint:*   If $S = \{e_1, e_2, \ldots, e_n\}$, then

$$S = \{e_1\} \cup \{e_2\} \cup \cdots \cup \{e_n\}$$

and

$$P(S) = P(e_1) + P(e_2) + \cdots + P(e_n) = 1$$

If $P(e_1) = P(e_2) = \cdots = P(e_n)$, show that the common value is $1/n$. Next, if
$$E = \{e_1, e_2, \ldots, e_m\}$$
then
$$E = \{e_1\} \cup \{e_2\} \cup \cdots \cup \{e_m\}$$
Use the definition of probability for an event to calculate $P(E) = m/n$.]

59. Prove Equation (6) on page 332.
[*Hint:* If the odds for E are a to b, then, by the definition of odds,
$$\frac{P(E)}{P(\overline{E})} = \frac{a}{b}$$

But $P(\overline{E}) = 1 - P(E)$. So
$$\frac{P(E)}{1 - P(E)} = \frac{a}{b}$$
Now solve for $P(E)$.]

60. Generalize the Additive Rule by showing the probability of the occurrence of at least one of the three events A, B, C is given by
$$P(A \cup B \cup C) = P(A) + P(B) + P(C)$$
$$- P(A \cap B) - P(A \cap C)$$
$$- P(B \cap C) + P(A \cap B \cap C)$$

Technology Exercises

Most graphing calculators have a random number function (usually RAND or RND) generating numbers between 0 and 1. Check your user's manual to see how to use this function.

Sometimes experiments are simulated using a random number function instead of actually performing the experiment. In Problems 1–6 use a random number function to simulate each experiment.

1. **Tossing a Fair Coin**   Consider an experiment of tossing a fair coin. Simulate the experiment using a random number function on your calculator, considering a toss to be tails (T) if the result is less than 0.5, and considering a toss to be heads (H) if the result is more than or equal to 0.5. [*Note:* Most calculators repeat the action of the last entry if you simply press the ENTER, or EXE, key again.] Repeat the experiment 10 times. Using these 10 outcomes of the experiment you can estimate the probabilities $P(H)$ and $P(T)$ by the ratios
$$P(H) \approx \frac{\text{Number of times } H \text{ occurred}}{10}$$
$$P(T) \approx \frac{\text{Number of times } T \text{ occurred}}{10}$$
What are the actual probabilities? How close are the results of the experiment to the actual values?

2. **Urn and Balls**   Consider an experiment of choosing a ball from an urn containing 18 red and 12 white balls. Simulate the experiment using a random number function on your calculator, considering a selection to be a red ball (R) if the result is less than 0.6, and considering a toss to be a white ball (W) if the result is more than or equal to 0.6. [*Note:* Most calculators repeat the action of the last entry if you simply press the ENTER, or EXE, key again.] Repeat the experiment 10 times. Using these 10 outcomes of the

experiment you can estimate the probabilities $P(R)$ and $P(W)$ by the ratios
$$P(R) \approx \frac{\text{Number of times } R \text{ occurred}}{10}$$
$$P(W) \approx \frac{\text{Number of times } W \text{ occurred}}{10}$$
What are the actual probabilities? How close are the results of the experiment to the actual values?

3. **Tossing a Loaded Coin**   Consider an experiment of tossing a loaded coin. Simulate the experiment using a random number function on your calculator, considering a toss to be tails (T) if the result is less than 0.25, and considering a toss to be heads (H) if the result is more than or equal to 0.25. [*Note:* Most calculators repeat the action of the last entry if you simply press the ENTER, or EXE, key again.] Repeat the experiment 20 times. Using these 20 outcomes of the experiment you can estimate the probabilities $P(H)$ and $P(T)$ by the ratios
$$P(H) \approx \frac{\text{Number of times } H \text{ occurred}}{20}$$
$$P(T) \approx \frac{\text{Number of times } T \text{ occurred}}{20}$$
What are the actual probabilities? How close are the results of the experiment to the actual values?

4. Urn and Balls Consider an experiment of choosing a ball from an urn containing 15 red and 35 white balls. Simulate the experiment using a random number function on your calculator, considering a selection to be a red ball (*R*) if the result is less than 0.3, and considering a toss to be a white ball (*W*) if the result is more than or equal to 0.3. [*Note:* Most calculators repeat the action of the last entry if you simply press the ENTER, or EXE, key again.] Repeat the experiment 10 times. Using these 10 outcomes of the experiment you can estimate the probabilities $P(R)$ and $P(W)$ by the ratios

$$P(R) \approx \frac{\text{Number of times } R \text{ occurred}}{10}$$

$$P(W) \approx \frac{\text{Number of times } W \text{ occurred}}{10}$$

What are the actual probabilities? How close are the results of the experiment to the actual values?

5. Jar and Marbles Consider an experiment of choosing a marble from a jar containing 5 red, 2 yellow, and 8 white marbles. Simulate the experiment using a random number function on your calculator, considering a selection to be a red marble (*R*) if the result is less than or equal to 0.33, a yellow marble (*Y*) if the result is between 0.33 and 0.47, and a white marble (*W*) if the result is more than or equal to 0.47. [*Note:* Most calculators repeat the action of the last entry if you simply press the ENTER, or EXE, key again.] Repeat the experiment 10 times. Using these 10 outcomes of the experiment you can estimate the probabilities $P(R)$, $P(Y)$, and $P(W)$ by the ratios

$$P(R) \approx \frac{\text{Number of times } R \text{ occurred}}{10}$$

$$P(Y) \approx \frac{\text{Number of times } Y \text{ occurred}}{10}$$

$$P(W) \approx \frac{\text{Number of times } W \text{ occurred}}{10}$$

What are the actual probabilities? How close are the results of the experiment to the actual values?

6. Poker Game The probabilities in a game of poker that Adam, Beatrice, or Cathy win are .22, .60, .18, respectively. Simulate the game using a random number function on your calculator, considering that Adam won (*A*) if the result is less than or equal to 0.22, that Beatrice won (*B*) if the result is between 0.22 and 0.82, and that Cathy won (*C*) if the result is more than or equal to 0.82. [*Note:* Most calculators repeat the action of the last entry if you simply press the ENTER, or EXE, key again.] Repeat the experiment 20 times; that is, simulate 20 games. Using these 20 outcomes of the experiment you can estimate the probabilities $P(A)$, $P(B)$, and $P(C)$ by the ratios

$$P(A) \approx \frac{\text{Number of times } A \text{ won}}{20}$$

$$P(B) \approx \frac{\text{Number of times } B \text{ won}}{20}$$

$$P(C) \approx \frac{\text{Number of times } C \text{ won}}{20}$$

What are the actual values? How close are the results of the experiment to the actual values?

7.3 PROBABILITY PROBLEMS USING COUNTING TECHNIQUES

We have seen that many problems in probability require the ability to count the elements of a set. In this section we look at several examples of probability problems and the related counting problems.

EXAMPLE 1 What is the probability that a four-digit telephone extension has one or more repeated digits? Assume no one digit is more likely than another to be used.

SOLUTION

There are $10^4 = 10,000$ distinct four-digit telephone extensions; this, therefore, is the number of outcomes in the sample space.

We wish to find the probability that a four-digit telephone extension has one or more repeated digits. Since it is difficult to count the outcomes of this event, we first compute the probability of the complement, namely,

$$E: \quad \text{No repeated digits in four-digit extension}$$

Notice that the event $\overline{E}$ is one or more repeated digits. By the Multiplication Principle, the number of four-digit extensions that have *no* repeated digits is

$$10 \cdot 9 \cdot 8 \cdot 7 = 5040$$

Hence,

$$P(E) = \frac{5040}{10{,}000} = .504$$

Therefore, the probability of one or more repeated digits is

$$P(\overline{E}) = 1 - .504 = .496$$

Now Work Problem 1

· ·

EXAMPLE 2

A box contains 12 light bulbs, of which 5 are defective. All bulbs look alike and have equal probability of being chosen. Pick 3 light bulbs and place them in a box.

(a) What is the probability that all 3 are defective?
(b) What is the probability that exactly 2 are defective?
(c) What is the probability that at least 2 are defective?

SOLUTION

(a) The number of elements in the sample space S is equal to the number of combinations of 12 light bulbs taken 3 at a time, namely,

$$C(12, 3) = \frac{12!}{3!9!} = 220$$

Define E as the event, "3 bulbs are defective." Then E can occur in $C(5, 3)$ ways, that is, the number of ways in which 3 defective bulbs can be chosen from 5 defectives ones. The probability $P(E)$ is

$$P(E) = \frac{C(5, 3)}{C(12, 3)} = \frac{\dfrac{5!}{3!2!}}{220} = \frac{10}{220} = .04545$$

(b) Define F as the event, "2 bulbs are defective." To obtain 2 defective bulbs when 3 are chosen requires that we select 2 defective bulbs from the 5 available defective ones and 1 good bulb from the 7 good ones. By the Multiplication Principle this can be done in the following number of ways:

$$C(5, 2) \cdot C(7, 1) \qquad = \frac{5!}{2!3!} \cdot \frac{7!}{1!6!} = 10 \cdot 7 = 70$$

Number of ways to select 2 defectives from 5 defectives

Number of ways to select 1 good bulb from 7 good ones

The probability of selecting exactly 2 defective bulbs is therefore

$$P(F) = P(\text{Exactly 2 defectives}) = \frac{C(5, 2) \cdot C(7, 1)}{C(12, 3)} = \frac{70}{220} = .31818$$

(c) Define G as the event, "At least 2 are defective." The event G is equivalent to asking for the probability of selecting either exactly 2 or exactly 3 defective bulbs. Since these events are mutually exclusive, the sum of their probabilities will give the probability of G. Therefore,

$$P(G) = P(\text{Exactly 2 defectives}) + P(\text{Exactly 3 defectives})$$
$$= P(F) + P(E) = .31818 + .04545 = .36363$$

Now Work Problem 9

. . . .

EXAMPLE 3

A fair coin is tossed 10 times.

(a) What is the probability of obtaining exactly 5 heads and 5 tails?
(b) What is the probability of obtaining between 4 and 6 heads, inclusive?

SOLUTION

The number of elements in the sample space S is found by using the Multiplication Principle. Each toss results in a head (H) or a tail (T). Since the coin is tossed 10 times, we have

$$c(S) = \underbrace{2 \cdot 2 \cdot \cdots \cdot 2}_{10 \text{ twos}} = 2^{10}$$

The outcomes are equally likely since the coin is fair.

(a) Any sequence that contains 5 heads and 5 tails is determined once the position of the 5 heads (or 5 tails) is known. The number of ways we can position 5 heads in a sequence of 10 slots is $C(10, 5)$. The probability of the event E: Exactly 5 heads, 5 tails is

$$P(E) = \frac{c(E)}{c(S)} = \frac{C(10, 5)}{2^{10}} = .2461$$

(b) Let F be the event: Between 4 and 6 heads, inclusive. To obtain between 4 and 6 heads is equivalent to the event: Exactly 4 heads or exactly 5 heads or exactly 6 heads. Since these are mutually exclusive (for example, it is impossible to obtain exactly 4 heads and exactly 5 heads when tossing a coin 10 times), we have

$$P(F) = P(4 \text{ heads or 5 heads or 6 heads})$$
$$= P(4 \text{ heads}) + P(5 \text{ heads}) + P(6 \text{ heads})$$

The probabilities on the right are obtained as in Part (a). Thus,

$$P(F) = \frac{C(10, 4)}{2^{10}} + \frac{C(10, 5)}{2^{10}} + \frac{C(10, 6)}{2^{10}} = .2051 + .2461 + .2051$$

$$= .6563$$

Now Work Problem 5

. . . .

EXAMPLE 4

Birthday Problem   An interesting problem, called the **birthday problem,** is to find the probability that in a group of r people there are at least two people who have the same birthday (the same month and day of the year).

SOLUTION

We first determine the number of outcomes in the sample space. There are 365 possibilities for each person's birthday (we exclude February 29 for simplicity). Since there are r people in the group, there are 365^r possibilities for the birthdays. [For one person in the group, there are 365 days on which his or her birthday can fall; for two people, there are $(365)(365) = 365^2$ pairs of days; and, in general, using the Multiplication Principle, for r people there are 365^r possibilities.]

Next, we assume a person is no more likely to be born on one day than another, so that we assign the probability $1/365^r$ to each outcome.

We wish to find the probability that at least two people have the same birthday. It is difficult to count the elements in this set; it is much easier to count the elements of the event

$$E:\quad \text{No two people have the same birthday}$$

Notice that the event $\overline{E}$ is that at least two people have the same birthday. To find the probability of E, we proceed as follows: Choose one person at random. There are 365 possibilities for his or her birthday. Choose a second person. There are 364 possibilities for this birthday, if no two people are to have the same birthday. Choose a third person. There are 363 possibilities left for this birthday. Finally, we arrive at the rth person. There are $365 - (r - 1)$ possibilities left for this birthday. By the Multiplication Principle, the total number of possibilities is $365 \cdot 364 \cdot 363 \cdot \cdots \cdot (365 - r + 1)$.

Hence, the probability of event E is

$$P(E) = \frac{365 \cdot 364 \cdot 363 \cdot \cdots \cdot (365 - r + 1)}{365^r}$$

The probability of two or more people having the same birthday is then $P(\overline{E}) = 1 - P(E)$.

$\bullet$

For example, in a group of eight people, the probability that two or more will have the same birthday is

$$P(\overline{E}) = 1 - \frac{365 \cdot 364 \cdot 363 \cdot 362 \cdot 361 \cdot 360 \cdot 359 \cdot 358}{365^8}$$
$$= 1 - .93$$
$$= .07$$

The table on p. 340 gives the probabilities for two or more people having the same birthday. Notice that the probability is better than $\frac{1}{2}$ for any group of 23 or more people.

	Number of People														
	5	**10**	**15**	**20**	**21**	**22**	**23**	**24**	**25**	**30**	**40**	**50**	**60**	**70**	**80**	**90**
Probability That 2 or More Have Same Birthday	.027	.117	.253	.411	.444	.476	.507	.538	.569	.706	.891	.970	.994	.99916	.99991	.99999

Now Work Problem 13

EXAMPLE 5

Find the probability of obtaining (a) a straight and (b) a flush in a poker hand.* (A poker hand is a set of 5 cards chosen at random from a regular deck of 52 cards.)

SOLUTION

The sample space contains $C(52, 5)$ simple events, each equally likely to occur.

(a) A straight consists of 5 consecutive cards, not all of the same suit. Now, for the straight 4, 5, 6, 7, 8, the 4 can be drawn in 4 different ways, as can the 5, the 6, the 7, and the 8, for a total of 4^5 ways. There is a total of 10 different kinds of straights (A, 2, 3, 4, 5; 2, 3, 4, 5, 6; . . . ; 9, 10, J, Q, K; 10, J, Q, K, A). Thus, all together, there are $10 \cdot 4^5$ ways to obtain 5 consecutive cards. However, among these are the 36 straight flushes (A, 2, 3, 4, 5; 2, 3, 4, 5, 6; 3, 4, 5, 6, 7; . . . ; 9, 10, J, Q, K all of the same suit) and the 4 royal flushes (10, J, Q, K, A, all of the same suit). Thus there are

$$10 \cdot 4^5 - 36 - 4 = 10{,}240 - 40 = 10{,}200 \text{ straights}$$

The probability of drawing a straight is therefore

$$\frac{10{,}200}{C(52, 5)} = .0039$$

(b) A flush consists of 5 nonconsecutive cards in a single suit. The number of ways of obtaining 5 cards, all of the same suit, is $C(13, 5) = 1287$, and there are 4 different suits, for a total of $4(1287) = 5148$ ways. However, straight flushes (36) and the 4 royal flushes are not included. Thus there are

$$5148 - 36 - 4 = 5108 \text{ flushes}$$

The probability of drawing a flush is therefore

$$\frac{5108}{C(52, 5)} = .0020$$

Now Work Problem 25

. . . .

---

* Poker hands of 5 cards are characterized by: *royal flush*—A, K, Q, J, 10 of the same suit (all clubs, all diamonds, all hearts, or all spades); *straight flushes*—A, 2, 3, 4, 5; 2, 3, 4, 5, 6; 3, 4, 5, 6, 7; 4, 5, 6, 7, 8; . . . ; 9, 10, J, Q, K (all of the same suit); *flushes*—5 nonconsecutive cards, all of same suit (example: 2, 5, 9, K, A of hearts); *straights*—5 consecutive cards, not all of same suit (example: 4, 5, 6, 7, 8, not all of same suit).

EXAMPLE 6

If the letters in the word MISTER are randomly scrambled, what is the probability that the resulting rearrangement has the I preceding the E?

SOLUTION

The 6 letters in the word MISTER can be rearranged in $P(6, 6) = 6!$ ways. We need to count the number of arrangements in which the letter I precedes the letter E. We can view the problem as consisting of two tasks. Task 1 is to choose two of the six positions for the I and E—the I would then of necessity have to occupy the leftmost position of the two chosen.

Task 1 can be done in $C(6, 2)$ ways.

Task 2 would be to arrange the letters MSTR in the remaining four positions. This can be done in $4!$ ways. Hence, by the Multiplication Principle, the number of arrangements with the I preceding the E is

$$C(6, 2) \cdot 4!$$

So the desired probability is

$$\frac{C(6, 2) \cdot 4!}{6!} = \frac{\frac{6!}{2!4!} \cdot 4!}{6!} = \frac{1}{2!} = .5$$

The answer should be intuitively plausible—we would expect one half of the arrangements to have I before E and the other half to have E before I.

EXERCISE 7.3 Answers to odd-numbered problems begin on page AN-26.

1. What is the probability that a seven-digit phone number has one or more repeated digits?

2. What is the probability that a seven-digit phone number contains the number 7?

3. Five letters, with repetition allowed, are selected from the alphabet. What is the probability that none is repeated?

4. Four letters, with repetition allowed, are selected from the alphabet. What is the probability that none of them is a vowel (a, e, i, o, u)?

5. A fair coin is tossed 5 times.

(a) Find the probability that exactly 3 heads appear.
(b) Find the probability that no heads appear.

6. A fair coin is tossed 6 times.

(a) Find the probability that exactly 1 tail appears.
(b) Find the probability that no more than 1 tail appears.

7. A pair of fair dice are tossed 3 times.

(a) Find the probability that the sum of seven appears 3 times.
(b) Find the probability that a sum of 7 or 11 appears at least twice.

8. A pair of fair dice are tossed 5 times.

(a) Find the probability that the sum is never 2.
(b) Find the probability that the sum is never 7.

9. Through a mix-up on the production line, 6 defective refrigerators were shipped out with 44 good ones. If 5 are selected at random, what is the probability that all 5 are defective? What is the probability that at least 2 of them are defective?

10. In a shipment of 50 transformers 10 are known to be defective. If 30 transformers are picked at random, what is the probability that all 30 are nondefective? Assume that all transformers look alike and have an equal probability of being chosen.

11. A rare coins dealer has 37 distinctly valued silver dollars in a bag, one of which is valued at more than $10,000. The winner of a certain contest is given the opportunity to reach into the bag, while blindfolded, and pull out 4 of the coins. What is the probability that one of the 4 coins is the one valued at more than $10,000?

12. A person is dealt 3 cards from a regular deck of 52 cards. What is the probability that they are all hearts, all diamonds, or all spades?

13. What is the probability that, in a group of 3 people, at least 2 were born in the same month (disregard day and year)?

14. What is the probability that, in a group of 6 people, at least 2 were born in the same month (disregard day and year)?

15. A box contains 100 slips of paper numbered from 1 to 100. If 3 slips are drawn in succession with replacement, what is the probability that at least 2 of them have the same number?

16. If, in Problem 15, 10 slips are drawn with replacement, what is the probability that at least 2 of them have the same number?

17. Use the idea behind Example 4 to find the approximate probability that 2 or more U.S. senators have the same birthday. (There are 100 Senators.)

18. Follow the directions of Problem 17 for the House of Representatives. (There are more than 365 representatives.)

19. If the five letters in the word VOWEL are rearranged, what is the probability the L will precede the E?

20. If the four letters in the word MATH are rearranged, what is the probability the word begins with the letters TH?

21. If the five letters in the word VOWEL are rearranged, what is the probability the word will begin with L?

22. If the seven letters in the word DEFAULT are rearranged, what is the probability the word will end in E and begin with T?

23. A spinner has 26 equally spaced wedges, each labeled consecutively, 1, 2, 3, . . . , 26. What is the probability, from one spin, of landing on an even integer or on any of the last 19 integers?

24. Five cards are dealt at random from a regular deck of 52 playing cards. Find the probability that

    (a) All are hearts.   (b) Exactly 4 are spades.
    (c) Exactly 2 are clubs.

25. In a game of bridge find the probability that a hand of 13 cards consists of 5 spades, 4 hearts, 3 diamonds, and 1 club.

26. Find the probability of obtaining each of the following poker hands:

    (a) Royal flush (10, J, Q, K, A all of the same suit)
    (b) Straight flush (5 cards in sequence in a single suit, but not a royal flush)
    (c) Four of a kind (4 cards of the same face value)
    (d) Full house (one pair and one triple of the same face values)

27. **Elevator Problem***   An elevator starts with 5 passengers and stops at 8 floors. Find the probability that no 2 passengers leave at the same floor. Assume that all arrangements of discharging the passengers have the same probability.

7.4 CONDITIONAL PROBABILITY

Recall that whenever we compute the probability of an event, we do it relative to the entire sample space in question. Thus when we ask for the probability $P(E)$ of the event E, this probability $P(E)$ represents an appraisal of the likelihood that a chance experiment will produce an outcome in the set E relative to a sample space S.

* William Feller, *An Introduction to Probability Theory and Its Applications,* 3rd ed. New York: Wiley, 1968.

However, sometimes we would like to compute the probability of an event E of a sample space relative to another event F of the same sample space. That is, if we have *prior* information that the outcome must be in a set F, this information should be used to reappraise the likelihood that the outcome will also be in E. This reappraised probability is denoted by $P(E|F)$, and is read as the *conditional probability of E given F*. It represents the answer to the question, "How probable is E, given that F has occurred?"

Let's work some examples to illustrate conditional probability before giving the definition.

EXAMPLE 1

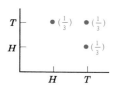

Figure 7

Consider the experiment of flipping two fair coins. As we have previously seen, the sample space S is

$$S = \{HH, HT, TH, TT\}$$

Figure 7 illustrates the sample space and, for convenience, the probability of each outcome.

Suppose the experiment is performed by another person and we have no knowledge of the result, but we are informed that at least one tail was tossed. This information means the outcome HH could not have occurred. But the remaining outcomes HT, TH, TT are still possible. How does this alter the probabilities of the remaining outcomes?

For instance, we might be interested in calculating the probability of the event $\{TT\}$. The three outcomes TH, HT, TT were each assigned the probability $\frac{1}{4}$ *before* we knew the information that at least one tail occurred, so it is not reasonable to assign them this same probability now. Since only three equally likely outcomes are now possible, we assign to each of them the probability $\frac{1}{3}$. See Figure 8.

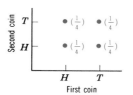

Figure 8

· ·

EXAMPLE 2

Suppose a population of 1000 people includes 70 accountants and 520 females. There are 40 females who are accountants.

Suppose a person is chosen at random, and we are told the person is female. The probability the person is an accountant, given that the person is female, is $\frac{40}{520}$. The ratio $\frac{40}{520} = \frac{1}{13}$ represents the *conditional probability* of the event E (accountant) assuming the event F (the person chosen is female).

· ·

We shall use the symbol $P(E|F)$, read "the probability of E given F," to denote conditional probability. Thus, for Example 2, if we let E be the event "A person chosen at random is an accountant" and we let F be the event "A person chosen at random is female," we would write

$$P(E|F) = \tfrac{40}{520} = \tfrac{1}{13}$$

Figure 9 illustrates that in computing $P(E|F)$ in Example 2, we form the ratio of the numbers of those entries in E and in F with the numbers that are in F. Since $c(E \cap F) = 40$ and $c(F) = 520$, then

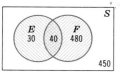

Figure 9

$$P(E|F) = \frac{c(E \cap F)}{c(F)} = \frac{40}{520} = \frac{1}{13}$$

Notice that

$$P(E|F) = \frac{\dfrac{c(E \cap F)}{c(S)}}{\dfrac{c(F)}{c(S)}} = \frac{P(E \cap F)}{P(F)}$$

With this result in mind we choose to define conditional probability as follows:

CONDITIONAL PROBABILITY

Let E and F be events of a sample space S and suppose $P(F) > 0$. The *conditional probability of event E, assuming the event F*, denoted by $P(E|F)$, is defined as

$$P(E|F) = \frac{P(E \cap F)}{P(F)} \tag{1}$$

For sample spaces involving equally likely outcomes, this formula can be proved. Suppose E and F are two events for a particular experiment. Assume that the sample space S for this experiment has n equally likely outcomes. Suppose event F has m outcomes, while $E \cap F$ has k outcomes ($k \le m$). Since the outcomes are equally likely, we have

$$P(F) = \frac{\text{Number of outcomes in } F}{n} = \frac{m}{n}$$

and

$$P(E \cap F) = \frac{\text{Number of outcomes in } E \cap F}{n} = \frac{k}{n}$$

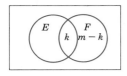

Figure 10

We wish to compute $P(E|F)$, the probability that E occurs given that F has occurred. Since we assume F has occurred, look only at the m outcomes in F. Of these m outcomes, there are k outcomes where E also occurs since $E \cap F$ has k outcomes. See Figure 10. Thus,

$$P(E|F) = \frac{k}{m}$$

Divide numerator and denominator by n to get

$$P(E|F) = \frac{\dfrac{k}{n}}{\dfrac{m}{n}} = \frac{P(E \cap F)}{P(F)}$$

EXAMPLE 3

Consider a three-child family for which the sample space S is

$$S = \{BBB, BBG, BGB, BGG, GBB, GBG, GGB, GGG\}$$

We assume that each outcome is equally likely, so that each is assigned a probability of $\frac{1}{8}$. Let E be the event, ''The family has exactly two boys'' and let F be the event ''The first child is a boy.'' What is the probability that the family has two boys, given that the first child is a boy?

SOLUTION

We want to find $P(E|F)$. The events E and F are

$$E = \{BBG, BGB, GBB\} \qquad F = \{BBB, BBG, BGB, BGG\}$$

Since $E \cap F = \{BBG, BGB\}$, we have

$$P(E \cap F) = \tfrac{1}{4} \qquad P(F) = \tfrac{1}{2}$$

$$P(E|F) = \frac{P(E \cap F)}{P(F)} = \frac{\tfrac{1}{4}}{\tfrac{1}{2}} = \frac{1}{2}$$

·······················

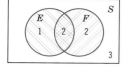

Figure 11

Now Work Problem 3

Since the sample space S of Example 3 consists of equally likely outcomes, we can also compute $P(E|F)$ using the Venn diagram in Figure 11. Then

$$P(E|F) = \frac{c(E \cap F)}{c(F)} = \frac{2}{4} = \frac{1}{2}$$

EXAMPLE 4

A regular deck of playing cards consists of 52 cards: 13 clubs (black), 13 diamonds (red), 13 hearts (red), and 13 spades (black). The 13 cards are labeled A(ace), 2, 3, 4, 5, 6, 7, 8, 9, 10, J(jack), Q(queen), K(king). Consider the experiment of drawing a single card. We are interested in the event E consisting of the outcome that a black ace is drawn. Since we may assume that there are 52 equally likely possible outcomes and there are 2 black aces in the deck, we have

$$P(E) = \tfrac{2}{52}$$

However, suppose a card is drawn and we are informed that it is a spade. How should this information be used to *reappraise* the likelihood of the event E?

SOLUTION

Let F denote the event "A spade is drawn." The probability of a black ace being drawn, given that a spade has been drawn, is the conditional probability $P(E|F)$. The probability of the event $E \cap F$, "A black ace and a spade is drawn," is $P(E \cap F) = \tfrac{1}{52}$, since $E \cap F$ can occur in only one way. Since there are 13 spades, $P(F) = \tfrac{13}{52}$. Thus,

$$P(E|F) = \frac{P(E \cap F)}{P(F)} = \frac{\tfrac{1}{52}}{\tfrac{13}{52}} = \frac{1}{13}$$

The result is that $P(E|F) = \tfrac{1}{13}$.

·······················

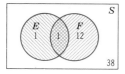

Figure 12

Let's analyze the situation in Example 4 more carefully. See Figure 12. The event E is "A black ace is drawn." We have computed the probability of event E, knowing event F has occurred. This means we are computing a probability relative to a *new sample space F*. That is, F is treated as the universal set and we should consider only that part of E that is included in F, namely, $E \cap F$. Thus if E is any subset of the sample space S, then $P(E|F)$ provides the reappraisal of the likelihood that an outcome of the experiment will be in the set E if we have prior information that it must be in the set F.

EXAMPLE 5

In a certain experiment the events E and F have the characteristics

$$P(E) = .7 \qquad P(F) = .8 \qquad P(E \cap F) = .6$$

Find

(a) $P(E \cup F)$ (b) $P(E|F)$ (c) $P(F|E)$ (d) $P(\overline{E}|\overline{F})$

SOLUTION

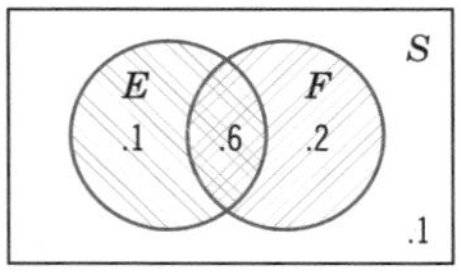

We can visualize the situation by using a Venn diagram. See Figure 13.

Figure 13

(a) To find $P(E \cup F)$, we use the Additive Rule.

$$P(E \cup F) = P(E) + P(F) - P(E \cap F) = .7 + .8 - .6 = .9$$

(b) To find the conditional probability $P(E|F)$, we use (1).

$$P(E|F) = \frac{P(E \cap F)}{P(F)} = \frac{.6}{.8} = .75$$

(c) To find the conditional probability $P(F|E)$, we use a variation of (1).

$$P(F|E) = \frac{P(F \cap E)}{P(E)} = \frac{P(E \cap F)}{P(E)} = \frac{.6}{.7} = .857$$

(d) To find the conditional probability $P(\overline{E}|\overline{F})$, we use a variation of (1).

$$P(\overline{E}|\overline{F}) = \frac{P(\overline{E} \cap \overline{F})}{P(\overline{F})}$$

Since $\overline{E} \cap \overline{F} = \overline{E \cup F}$ (De Morgan's property), we have

$$P(\overline{E}|\overline{F}) = \frac{P(\overline{E \cup F})}{P(\overline{F})} = \frac{1 - P(E \cup F)}{1 - P(F)} = \frac{1 - .9}{1 - .8} = \frac{.1}{.2} = .50$$

Now Work Problems 23
and 25

· ·

Product Rule

If in Formula (1) we multiply both sides of the equation by $P(F)$, we obtain the following useful relationship, which is referred to as the **Product Rule:**

Product Rule

$$P(E \cap F) = P(F) \cdot P(E|F) \qquad\qquad (2)$$

The next example illustrates how the Product Rule is used to compute the probability of an event that is itself a sequence of two events.

EXAMPLE 6

Two cards are drawn at random (without replacement) from a regular deck of 52 cards. What is the probability that the first card is a diamond and the second is red?

SOLUTION

We seek the probability of an event that is a sequence of two events, namely,

A: The first card is a diamond
B: The second card is red

Since there are 52 cards in the deck, of which 13 are diamonds, it follows that

$$P(A) = \tfrac{13}{52} = \tfrac{1}{4}$$

If A occurred, it means that there are only 51 cards left in the deck, of which 25 are red, so

$$P(B|A) = \tfrac{25}{51}$$

By the Product Rule,

$$P(B \cap A) = P(A) \cdot P(B|A) = \tfrac{1}{4} \cdot \tfrac{25}{51} = \tfrac{25}{204}$$

· ·

A tree diagram is helpful for problems like that of Example 6. See Figure 14. The branch leading to A: Diamond has probability $\tfrac{13}{52} = \tfrac{1}{4}$; the branch leading to Heart also has probability $\tfrac{13}{52} = \tfrac{1}{4}$; and the branch leading to Club or Spade has probability $\tfrac{26}{52} = \tfrac{1}{2}$. These must add up to 1 since no other possibilities (branches) are possible. The branch from A to Red is the conditional probability of drawing a red card on the second draw after a diamond on the first draw, P (Red|Diamond), which is $\tfrac{25}{51}$. The branch from A to Black is the conditional probability P (Black|Diamond), which is $\tfrac{26}{51}$ (26 black cards and 51 total cards remain after a diamond on the first draw). The remaining entries are obtained similarly. Notice that the probability the first card is a diamond and the second a red corresponds to tracing the top branch of the tree. The product rule then tells us to multiply the branch probabilities.

Figure 14

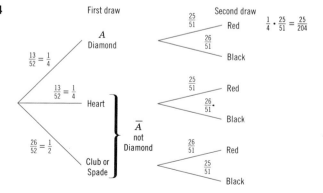

Based on Figure 14, we can draw the following conclusions:

Probability the first card is a heart $\frac{1}{4} \cdot \frac{26}{51}$
and the second is black
Probability the first card is black $\frac{1}{2} \cdot \frac{26}{51}$
and the second is red
Probability the first card is black $\frac{1}{2} \cdot \frac{25}{51}$
and the second is black

EXAMPLE 7

From a box containing four white, three yellow, and one green ball, two balls are drawn one at a time without replacing the first before the second is drawn. Use a tree diagram to find the probability that one white and one yellow ball are drawn.

SOLUTION

To fix our ideas, we define the events

W: White ball drawn Y: Yellow ball drawn G: Green ball drawn

The tree diagram for this experiment is given in Figure 15.

Figure 15

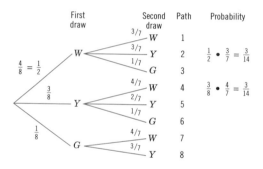

The event of drawing one white ball and one yellow ball can occur in two ways: drawing a white ball first and then a yellow ball (path 2 of the tree diagram in Figure 15), or drawing a yellow ball first and then a white ball (path 4).

Consider path 2. Since four of the eight balls are white, on the first draw we have

$$P(W) = P(W \text{ on 1st}) = \tfrac{4}{8} = \tfrac{1}{2}$$

Since one white ball has been removed, leaving seven balls in the box, of which three are yellow, on the second draw we have

$$P(Y|W) = P(Y \text{ on 2nd}|W \text{ on 1st}) = \tfrac{3}{7}$$

Thus for path 2 we have

$$P(W) \cdot P(Y|W) = P(W \text{ on 1st}) \cdot P(Y \text{ on 2nd}|W \text{ on 1st}) = \tfrac{1}{2} \cdot \tfrac{3}{7} = \tfrac{3}{14}$$

Similarly, for path 4 we have

$$P(Y) \cdot P(W|Y) = P(Y \text{ on 1st}) \cdot P(W \text{ on 2nd}|Y \text{ on 1st}) = \tfrac{3}{8} \cdot \tfrac{4}{7} = \tfrac{3}{14}$$

Since the two events are mutually exclusive, the probability of drawing one white ball and one yellow ball is the sum of these two probabilities:

$$\tfrac{3}{14} + \tfrac{3}{14} = \tfrac{6}{14} = \tfrac{3}{7}$$

Now Work Problem 33

• • • • • • • • • • • • • • • • • • • •

EXAMPLE 8

Motors, Inc., has two plants to manufacture cars. Plant I manufacturers 80% of the cars and plant II manufactures 20%. At plant I, 85 out of every 100 cars are rated standard quality or better. At plant II, only 65 out of every 100 cars are rated standard quality or better. We would like to answer the following questions:

(a) What is the probability that a customer obtains a standard quality car if he buys a car from Motors, Inc.?

(b) What is the probability that the car came from plant I if it is known that the car is of standard quality?

SOLUTION

We begin with a tree diagram. See Figure 16.

Figure 16

```
        .85
.80           Standard or better   (.80)(.85) = .68
    Plant I
        .15   Not

        .65
.20           Standard or better   (.20)(.65) = .13
    Plant II
        .35   Not
```

We define the following events:

$$\text{I: Car came from plant I} \qquad \text{II: Car came from plant II}$$
$$A: \text{Car is of standard quality}$$

(a) There are two ways a standard car can be obtained: either it is standard and came from plant I, or else it is standard and came from plant II. By the Product Rule,

$$P(A \cap \text{I}) = P(\text{I}) \cdot P(A|\text{I}) = (.8)(.85) = .68$$
$$P(A \cap \text{II}) = P(\text{II}) \cdot P(A|\text{II}) = (.2)(.65) = .13$$

Since $A \cap \text{I}$ and $A \cap \text{II}$ are mutually exclusive, we have

$$P(A) = P(A \cap \text{I}) + P(A \cap \text{II}) = .68 + .13 = .81$$

(b) To compute $P(\text{I}|A)$, we use the definition of conditional probability.

$$P(\text{I}|A) = \frac{P(\text{I} \cap A)}{P(A)} = \frac{.68}{.81} = .8395$$

• •

EXERCISE 7.4   Answers to odd-numbered problems begin on page AN-27.

In Problems 1–8 use the Venn diagram below to find the specified probabilities.

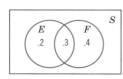

1. $P(E)$ **2.** $P(F)$ **3.** $P(E|F)$ **4.** $P(F|E)$

5. $P(E \cap F)$ **6.** $(E \cup F)$ **7.** $P(\overline{E})$ **8.** $P(\overline{F})$

9. If E and F are events with $P(E) = .4$, $P(F) = .2$, and $P(E \cap F) = .1$, find the probability of E given F. Find $P(F|E)$.

10. If E and F are events with $P(E) = .6$, $P(F) = .5$, and $P(E \cap F) = .3$, find the probability of E given F. Find $P(F|E)$.

11. If E and F are events with $P(E \cap F) = .1$ and $P(E|F) = .2$, find $P(F)$.

12. If E and F are events with $P(E \cap F) = .2$ and $P(E|F) = .5$, find $P(F)$.

13. If E and F are events with $P(F) = \frac{5}{13}$ and $P(E|F) = \frac{4}{5}$, find $P(E \cap F)$.

14. If E and F are events with $P(F) = .38$ and $P(E|F) = .46$, find $P(E \cap F)$.

15. If E and F are events with $P(E \cap F) = \frac{1}{3}$, $P(E|F) = \frac{1}{2}$, and $P(F|E) = \frac{2}{3}$, find

(a) $P(E)$ (b) $P(F)$

16. If E and F are events with $P(E \cap F) = .1$, $P(E|F) = .25$, and $P(F|E) = .125$, find

(a) $P(E)$ (b) $P(F)$

In Problems 17–22 find the indicated probabilities by referring to the following tree diagram:

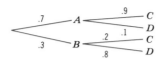

17. $P(C)$ **18.** $P(D)$ **19.** $P(C|A)$ **20.** $P(D|A)$

21. $P(C|B)$ **22.** $P(D|B)$

In Problems 23–28 E and F are events in a sample space S for which

$$P(E) = .5 \qquad P(F) = .4 \qquad P(E \cup F) = .8$$

Find the indicated probabilities.

23. $P(E \cap F)$ **24.** $P(E|F)$ **25.** $P(F|E)$

26. $P(\overline{E}|F)$ **27.** $P(E|\overline{F})$ **28.** $P(\overline{E}|\overline{F})$

29. For a 3-child family, find the probability of exactly 2 girls, given that the first child is a girl.

30. For a 3-child family, find the probability of exactly 1 girl, given that the first child is a boy.

31. A fair coin is tossed 4 successive times. Find the probability of obtaining 4 heads. Does the probability change if we are told that the second throw resulted in a head?

32. A pair of fair dice is thrown and we are told that at least one of them shows a 2. If we know this, what is the probability that the total is 7?

33. Two cards are drawn at random (without replacement) from a regular deck of 52 cards. What is the probability that the first card is a heart and the second is red?

34. From a box containing 3 white, 2 green, and 1 yellow ball, 2 balls are drawn at a time without replacing the first before the second is drawn. Find the probability that 1 white and 1 yellow ball are drawn.

35. If 2 cards are drawn from a regular deck of 52 cards without replacement, what is the probability that the second card is a queen?

36. A box contains 2 red, 4 green, 1 black, and 8 yellow marbles. If 2 marbles are selected without replacement, what is the probability that 1 is red and 1 is green?

37. A card is drawn at random from a regular deck of 52 cards. What is the probability that

(a) The card is a red ace?
(b) The card is a red ace if it is known an ace was picked?
(c) The card is a red ace if it is known a red card was picked?

38. A card is drawn at random from a regular deck of 52 cards. What is the probability that

(a) The card is a black jack?
(b) The card is a black jack if it is known a jack was picked?

(c) The card is a black jack if it is known a black card was picked?

39. In a small town it is known that 25% of the families have no children, 25% have 1 child, 18% have 2 children, 16% have 3 children, 8% have 4 children, and 8% have 5 or more children. Find the probability that a family has more than 2 children if it is known that it has at least 1 child.

40. A sequence of 2 cards is drawn from a regular deck of 52 cards (without replacement). What is the probability that the first card is red and the second is black?

In Problems 41–52 use the table to obtain probabilities for events in a sample space S.

	E	F	G	Totals
H	.10	.06	.08	.24
I	.30	.14	.32	.76
Totals	.40	.20	.40	1.00

For Problems 41–48, read each probability directly from the table:

41. $P(E)$ **42.** $P(G)$ **43.** $P(H)$ **44.** $P(I)$

45. $P(E \cap H)$ **46.** $P(E \cap I)$ **47.** $P(G \cap H)$ **48.** $P(G \cap I)$

For Problems 49–52 use Equation (1) and the appropriate values from the above table to compute the conditional probability.

49. $P(E|H)$ **50.** $P(E|I)$ **51.** $P(G|H)$ **52.** $P(G|I)$

53. A recent poll of residents in a certain community revealed the following information about voting preferences:

	Democrat	Republican	Independent
Male	50	40	30
Female	60	30	25

Events $M, F, D, R,$ and I are defined as follows:

M: Resident is male
F: Resident is female
D: Resident is a Democrat
R: Resident is a Republican
I: Resident is an Independent

Find

(a) $P(F|I)$ (b) $P(R|F)$ (c) $P(M|D)$
(d) $P(D|M)$ (e) $P(M|R \cup I)$ (f) $P(I|M)$

54. Let E be the event, "A person is an executive" and let F be the event "A person earns over $65,000 per year." State in words what is expressed by each of the following probabilities:

(a) $P(E|F)$ (b) $P(F|E)$
(c) $P(\bar{E}|\bar{F})$ (d) $P(\bar{E}|F)$

55. The following table summarizes the graduating class of a midwestern university:

	Arts and Sciences A	Education E	Business B	Total
Male, M	342	424	682	1448
Female, F	324	102	144	570
Total	666	526	826	2018

A student is selected at random from the graduating class. Find the probability that the student

(a) Is male
(b) Is receiving an arts and sciences degree
(c) Is a female receiving a business degree
(d) Is a female, given that the student is receiving an education degree
(e) Is receiving an arts and sciences degree, given that the student is a male

(f) Is a female, given that the student is receiving an arts and science degree or an education degree
(g) Is not receiving a business degree and is male
(h) Is female, given that an education degree is not received

56. The following data are the result of a survey conducted by a marketing company among students for their beer preferences.

	Do Not Drink Beer N	Light Beer L	Regular Beer R	Total
Female, F	224	420	622	1266
Male, M	196	512	484	1192
Total	420	932	1106	2458

A student is selected at random from this class. Find the probability that

(a) The student does not drink beer.
(b) The student is a female.
(c) The student is a female who prefers regular beer.
(d) The student prefers regular beer, given that the student is male.
(e) The student is male, given that the student prefers regular beer.
(f) The student is female, given that the student prefers regular beer or does not drink beer.

57. In a sample survey it is found that 35% of the men and 70% of the women weigh less than 160 pounds. Assume that 50% of the sample are men. If a person is selected at random and this person weighs less than 160 pounds, what is the probability that this person is a woman?

58. If the probability that a married man will vote in a given election is .50, the probability that a married woman will vote in the election is .60, and the probability that a woman will vote in the election, given that her husband votes, is .90, find

(a) The probability that a husband and wife will both vote in the election.

(b) The probability that a married man will vote in the election, given that at least one member of the married couple will vote.

59. Of the first-year students in a certain college, it is known that 40% attended private secondary schools and 60% attended public schools. The registrar reports that 30% of all students who attended private schools maintain an A average in their first year at college and that 24% of all first-year students had an A average. At the end of the year, one student is chosen at random from the class. If the student has an A average, what is the conditional probability that the student attended a private school? [*Hint:* Use a tree diagram.]

60. In a rural area in the north, registered Republicans outnumber registered Democrats by 3 to 1. In a recent election all Democrats voted for the Democratic candidate and enough Republicans also voted for the Democratic candidate so that the Democrat won by a ratio of 5 to 4. If a voter is selected at random, what is the probability he or she is Republican? What is the probability a voter is Republican, if it is known that he or she voted for the Democratic candidate?

61. "Temp Help" uses a preemployment test to screen applicants for the job of programmer. The test is passed by 70% of the applicants. Among those who pass the test, 85% complete training successfully. In an experiment a random sample of applicants who do not pass the test is also employed. Training is successfully completed by only 40% of this group. If no preemployment test is used, what percentage of applicants would you expect to complete the training successfully?

62. If E and F are two events with $P(E) > 0$ and $P(F) > 0$, show that
$$P(F) \cdot P(E|F) = P(E) \cdot P(F|E)$$

63. Show that $P(E|E) = 1$ when $P(E) \neq 0$.

64. Show that $P(E|F) + P(\overline{E}|F) = 1$.

65. If S is the sample space, show that $P(E|S) = P(E)$.

66. If $P(E) > 0$ and $P(E|F) = P(E)$, show that $P(F|E) = P(F)$.

7.5 INDEPENDENT EVENTS

One of the most important concepts in probability is that of independence. In this section we define what is meant by two events being *independent*. First, however, we try to develop an intuitive idea of the meaning of independent events.

EXAMPLE 1

Consider a group of 36 students. Suppose that E and F are two properties that each student either has or does not have. For example, the events E and F might be

E: Student has blue eyes F: Student is female

With regard to these two properties, suppose it is found that the 36 students are distributed as follows:

	Blue Eyes E	Not Blue Eyes $\overline{E}$	Totals
Female, F	12	12	24
Male, $\overline{F}$	6	6	12
Totals	18	18	36

What is the probability $P(E|F)$?

SOLUTION

If we choose a student at random, the following probabilities can be obtained from the table:

$$P(E) = \tfrac{18}{36} = \tfrac{1}{2} \qquad P(F) = \tfrac{24}{36} = \tfrac{2}{3}$$
$$P(E \cap F) = \tfrac{12}{36} = \tfrac{1}{3}$$

Then we find that

$$P(E|F) = \frac{P(E \cap F)}{P(F)} = \frac{\tfrac{1}{3}}{\tfrac{2}{3}} = \frac{1}{2} = P(E)$$

· ·

In Example 1 the probability of E given F equals the probability of E. This situation can be described by saying that the information that the event F has occurred does not affect the probability of the event E. If this is the case, we say that E *is independent of F*.

E IS INDEPENDENT OF F

Let E and F be two events of a sample space S with $P(F) > 0$. *The event E is independent of the event F if and only if*

$$P(E|F) = P(E)$$

Theorem

Let E, F be events for which $P(E) > 0$ and $P(F) > 0$. If E is independent of F, then F is independent of E.

The proof of this result is outlined in Problem 41.

INDEPENDENT EVENTS

If two events E and F have positive probabilities and if the event E is independent of F, then F is also independent of E. In this case E and F are called *independent events.*

We have the following result concerning independent events:

> **Test for Independence**
>
> Two events E and F of a sample space S are independent events if and only if
> $$P(E \cap F) = P(E) \cdot P(F) \qquad (1)$$
> That is, the probability of E *and* F is equal to the product of the probability of E and the probability of F.

The proof of this result is outlined in Problem 42.

The Test for Independence can be used to determine whether two events are independent.

EXAMPLE 2

Suppose $P(E) = \frac{1}{4}$, $P(F) = \frac{2}{3}$, and $P(E \cap F) = \frac{1}{6}$. Show that E and F are independent.

SOLUTION

$$P(E) \cdot P(F) = \frac{1}{4} \cdot \frac{2}{3} = \frac{1}{6}$$
$$= P(E \cap F)$$

By the Test for Independence (1), E and F are independent.

. . . .

If E and F are two independent events, the Test for Independence can also be used to find $P(E \cap F)$, the probability of E and F.

Warning! Two events E and F must be independent in order to use (1) to find $P(E \cap F)$.

EXAMPLE 3

Suppose E and F are independent events with $P(E) = .4$ and $P(F) = .3$. Find $P(E \cap F)$.

SOLUTION

Since E and F are independent,

$$P(E \cap F) = P(E) \cdot P(F) \quad \text{By (1)}$$
$$= (.4) \cdot (.3)$$
$$= .12$$

Now Work Problem 1

. . . .

EXAMPLE 4

Suppose a red die and a green die are thrown. Let event E be "Throw a 5 with the red die," and let event F be "Throw a 6 with the green die." Show that E and F are independent events.

SOLUTION

In this experiment the events E and F are

$$E = \{(5, 1), (5, 2), (5, 3), (5, 4), (5, 5), (5, 6)\}$$
$$F = \{(1, 6), (2, 6), (3, 6), (4, 6), (5, 6), (6, 6)\}$$

Then

$$P(E) = \tfrac{6}{36} = \tfrac{1}{6} \qquad P(F) = \tfrac{6}{36} = \tfrac{1}{6}$$

Also, the event E and F is

$$E \cap F = \{(5, 6)\}$$

so that

$$P(E \cap F) = \tfrac{1}{36}$$

Since $P(E) \cdot P(F) = \tfrac{1}{6} \cdot \tfrac{1}{6} = \tfrac{1}{36} = P(E \cap F)$, E and F are independent events.

Now Work Problem 7

· ·

EXAMPLE 5

In a T-maze a mouse may run to the right, R, or to the left, L. Suppose its behavior in making such "choices" is random so that R and L are equally likely outcomes. Suppose a mouse is put through the T-maze three times. Define events E, G, and H as

E: Run to the right 2 consecutive times
G: Run to the left on first trial
H: Run to the right on second trial

(a) Show that E and G are not independent.
(b) Show that G and H are independent.

SOLUTION

(a) The events E and G are

$$E = \{RRL, LRR, RRR\}$$
$$G = \{LLL, LLR, LRL, LRR\}$$

The sample space S has eight elements, so that

$$P(E) = \tfrac{3}{8} \qquad P(G) = \tfrac{1}{2}$$

Also, the event E and G is

$$E \cap G = \{LRR\}$$

so that

$$P(E \cap G) = \tfrac{1}{8}$$

Since $P(E \cap G) \neq P(E) \cdot P(G)$, the events E and G are not independent.

(b) The event H is
$$H = \{\text{RRL, RRR, LRL, LRR}\}$$
so that
$$P(H) = \tfrac{1}{2}$$
The event G and H and its probability are
$$G \cap H = \{\text{LRL, LRR}\} \qquad P(G \cap H) = \tfrac{1}{4}$$
Since $P(G \cap H) = P(G) \cdot P(H)$, the events G and H are independent.

· ·

Example 5 illustrates that the question of whether two events are independent can be answered simply by determining whether Formula (1) is satisfied. Although we may often suspect two events E and F as being independent, our intuition must be checked by computing $P(E)$, $P(F)$, and $P(E \cap F)$ and determining whether $P(E \cap F) = P(E) \cdot P(F)$.

Now Work Problem 13   For some probability models, an assumption of independence is made. In such instances when two events are independent, Formula (1) may be used to compute the probability that both events occur. The following example illustrates such a situation.

EXAMPLE 6   In a group of seeds, $\tfrac{1}{4}$ of which should produce white plants, the best germination that can be obtained is 75%. If one seed is planted, what is the probability that it will grow into a white plant?

SOLUTION

Let G and W be the events

G: The plant will grow
W: The seed will produce a white plant

Assume that W and G are independent events.
Then the probability that the plant grows and is white, namely, $P(W \cap G)$ is
$$P(W \cap G) = P(W) \cdot P(G) = \tfrac{1}{4} \cdot \tfrac{3}{4} = \tfrac{3}{16}$$
A white plant will grow 3 out of 16 times.

· ·

There is a danger that mutually exclusive events and independent events may be confused. A source of this confusion is the common expression, ''Events are independent if they have nothing to do with each other.'' This expression provides a description of independence when applied to everyday events; but when it is applied to sets, it suggests nonoverlapping. Nonoverlapping sets are mutually exclusive but in general are not independent. See Problem 36.

Independence for More than Two Events

The concept of independence can be applied to more than two events:

INDEPENDENCE

A set $E_1, E_2, \ldots , E_n$ of events is called *independent* if the occurrence of one or more of them does not change the probability of any of the others. It can be shown that, for such events,

$$P(E_1 \cap E_2 \cap \cdots \cap E_n) = P(E_1) \cdot P(E_2) \cdots \cdots P(E_n) \tag{2}$$

EXAMPLE 7

A new skin cream can cure skin infection 90% of the time. If five randomly selected people with skin infections use this cream, assuming independence, what is the probability that

(a) All five are cured?   (b) All five still have the infection?

SOLUTION

(a) Let

E_1: First person does not have infection
E_2: Second person does not have infection
E_3: Third person does not have infection
E_4: Fourth person does not have infection
E_5: Fifth person does not have infection

Then, since events E_1, E_2, E_3, E_4, E_5 are given to be independent,

$$P(\text{All 5 are cured}) = P(E_1 \cap E_2 \cap E_3 \cap E_4 \cap E_5)$$
$$= P(E_1) \cdot P(E_2) \cdot P(E_3) \cdot P(E_4) \cdot P(E_5)$$
$$= (.9)^5$$
$$= .59$$

(b) Let

$\overline{E}_i$: ith person has an infection, $i = 1, 2, 3, 4, 5$

Then

$$P(\overline{E}_i) = 1 - .9 = .1$$
$$P(\text{All 5 are infected}) = P(\overline{E}_1 \cap \overline{E}_2 \cap \overline{E}_3 \cap \overline{E}_4 \cap \overline{E}_5)$$
$$= P(\overline{E}_1) \cdot P(\overline{E}_2) \cdot P(\overline{E}_3) \cdot P(\overline{E}_4) \cdot P(\overline{E}_5)$$
$$= (.1)^5$$
$$= .00001$$

EXERCISE 7.5   Answers to odd-numbered problems begin on page AN-27.

1. If E and F are independent events and if $P(E) = .3$ and $P(F) = .5$, find $P(E \cap F)$.

2. If E and F are independent events and if $P(E) = .6$ and $P(E \cap F) = .3$, find $P(F)$.

3. If E and F are independent events, find $P(F)$ if $P(E) = .2$ and $P(E \cup F) = .3$.

4. If E and F are independent events, find $P(E)$ if $P(F) = .3$ and $P(E \cup F) = .6$.

5. Suppose E and F are two events such that $P(E) = \frac{4}{21}$, $P(F) = \frac{7}{12}$, and $P(E \cap F) = \frac{2}{9}$. Are E and F independent?

6. If E and F are two events such that $P(E) = .25$, $P(F) = .36$, and $P(E \cap F) = .09$, then are E and F independent?

7. If E and F are two independent events with $P(E) = .3$, and $P(F) = .5$, find

   (a) $P(E|F)$ (b) $P(F|E)$
   (c) $P(E \cap F)$ (d) $P(E \cup F)$

8. If E and F are independent events with $P(E) = .32$, $P(F) = .41$, find

   (a) $P(E|F)$ (b) $P(F|E)$
   (c) $P(E \cap F)$ (d) $P(E \cup F)$

9. If E, F, and G are three independent events with $P(E) = \frac{2}{3}$, $P(F) = \frac{3}{7}$, and $P(G) = \frac{2}{21}$, then find $P(E \cap F \cap G)$.

10. If E_1, E_2, E_3, and E_4 are four independent events with $P(E_1) = .62$, $P(E_2) = .35$, $P(E_3) = .58$, and $P(E_4) = .41$, find $P(E_1 \cap E_2 \cap E_3 \cap E_4)$.

11. If $P(E) = .3$, $P(F) = .2$, and $P(E \cup F) = .4$, what is $P(E|F)$? Are E and F independent?

12. If $P(E) = .4$, $P(F) = .6$, and $P(E \cup F) = .7$, what is $P(E|F)$? Are E and F independent?

13. A fair die is rolled. Let E be the event "1, 2, or 3 is rolled" and let F be the event "3, 4, or 5 is rolled." Are E and F independent?

14. A loaded die is rolled. The probabilities for this die are $P(1) = P(2) = P(4) = P(5) = \frac{1}{8}$ and $P(3) = P(6) = \frac{1}{4}$. Are the events defined in Problem 13 independent in this case?

15. For a 3-child family let E be the event "The family has at most 1 boy" and let F be the event "The family has children of each sex." Are E and F independent events?

16. For a 2-child family are the events E and F as defined in Problem 15 independent?

17. A first card is drawn at random from a regular deck of 52 cards and is then put back into the deck. A second card is drawn. What is the probability that

   (a) The first card is a club?
   (b) The second card is a heart, given that the first is a club?
   (c) The first card is a club and the second is a heart?

18. For the situation described in Problem 17 what is the probability that

   (a) The first card is an ace?
   (b) The second card is a king, given that the first card is an ace?
   (c) The first card is an ace and the second is a king?

19. In the T-maze of Example 5 are the two events E and H independent?

20. A fair coin is tossed twice. Define the events E and F to be

   E: A head turns up on the first throw of a fair coin

   F: A tail turns up on the second throw of a fair coin

Show that E and F are independent events.

21. A die is loaded so that

$$P(1) = P(2) = P(3) = \frac{1}{4} \qquad P(4) = P(5) = P(6) = \frac{1}{12}$$

If $A = \{1, 2\}$, $B = \{2, 3\}$, $C = \{1, 3\}$, show that any pair of these events is independent.

22. Determine whether the events E, F, and G defined below for the experiment of tossing 2 fair dice are independent.

   E: The first die shows a 6

   F: The second die shows a 3

   G: The sum of the 2 dice is 7

23. Cardiovascular Disease   Records show that a child of parents with heart disease has a probability of $\frac{3}{4}$ of inheriting the disease. Assuming independence, what is the probability that, for a couple with heart disease that have two children:

   (a) Both children have heart disease.
   (b) Neither child has heart disease.
   (c) Exactly one child has heart disease.

24. Hitting a Target   A marksman hits a target with probability $\frac{4}{5}$. Assuming independence for successive firings, find the probabilities of getting

   (a) One miss followed by two hits
   (b) Two misses and one hit (in any order)

25. A coin is loaded so that tails is three times as likely as heads. If the coin is tossed three times, find the probability of getting

   (a) All tails
   (b) Two heads and one tail (in any order)

26. Sex of Newborns   The probability of a newborn baby being a girl is .49. Assuming that the sex of one baby is independent of the sex of all other babies—that is, the events are independent—what is the probability that all four babies born in a certain hospital on one day are girls?

27. Recovery Rate The recovery rate from a flu is .8. If 4 people have this flu, what is the probability (assume independence) that

(a) All will recover?   (b) Exactly 2 will recover?
(s) At least 2 will recover?

28. Germination In a group of seeds, $\frac{1}{3}$ of which should produce violets, the best germination that can be obtained is 60%. If one seed is planted, what is the probability that it will grow into violets?

29. A box has 9 marbles in it, 5 red and 4 white. Suppose we draw a marble from the box, replace it, and then draw another. Find the probability that

(a) Both marbles are red.
(b) Just one of the two marbles is red.

30. Survey In a survey of 100 people, categorized as drinkers or nondrinkers, with or without a liver ailment, the following data were obtained:

	Liver Ailment F	No Liver Ailment $\overline{F}$
Drinkers, E	52	18
Nondrinkers, $\overline{E}$	8	22

(a) Are the events E and F independent?
(b) Are the events $\overline{E}$ and $\overline{F}$ independent?
(c) Are the events E and $\overline{F}$ independent?

31. Insurance By examining the past driving records over a period of 1 year of 840 randomly selected drivers, the following data were obtained.

	Under 25 U	Over 25 $\overline{U}$	Totals
Accident, A	40	5	45
No Accident, $\overline{A}$	285	510	795
Totals	325	515	840

(a) What is the probability of a driver having an accident, given that the person is under 25?
(b) What is the probability of a driver having an accident, given that the person is over 25?

(c) Are events U and A independent?
(d) Are events U and $\overline{A}$ independent?
(e) Are events $\overline{U}$ and A independent?
(f) Are events $\overline{U}$ and $\overline{A}$ independent?

32. Voting Patterns The following data show the number of voters in a sample of 1000 from a large city, categorized by religion and their voting preference.

	Democrat D	Republican R	Independent	Totals
Catholic, C	160	150	90	400
Protestant, P	220	220	60	500
Jewish, J	20	30	50	100
Totals	400	400	200	1000

(a) Find the probability a person is a Democrat.
(b) Find the probability a person is a Catholic.
(c) Find the probability a person is Catholic, knowing the person is a Democrat.
(d) Are the events R and D independent?
(e) Are the events P and R independent?

33. Election A candidate for office believes that $\frac{4}{5}$ of registered voters in her district will vote for her in the next election. If two registered voters are independently selected at random, what is the probability that

(a) Both of them will vote for her in the next election?
(b) Neither will vote for her in the next election?
(c) Exactly one of them will vote for her in the next election?

34. A woman has 10 keys but only 1 fits her door. She tries them successively (without replacement). Find the probability that a key fits in exactly 5 tries.*

35. Chevalier de Mere's Problem Which of the following random events do you think is more likely to occur?

(a) To obtain a 1 on at least one die in a simultaneous throw of four fair dice.
(b) To obtain at least one pair of 1s in a series of 24 throws of a pair of fair dice.
[*Hint:* Part (a): $P(\text{No 1s are obtained}) = 5^4/6^4 = \frac{625}{1296} = .4823$. Part (b): The probability of not obtaining a double 1 on any given toss is $\frac{35}{36}$. Thus $P(\text{No double 1s are obtained}) = (\frac{35}{36})^{24} = .509$.]

---

* William Feller, *An Introduction to Probability Theory and Its Applications*, 3rd ed. New York: Wiley, 1968.

36. Give an example of two events that are

(a) Independent, but not mutually exclusive
(b) Not independent, but mutually exclusive
(c) Not independent and not mutually exclusive

37. Show that whenever two events are both independent and mutually exclusive, then at least one of them is impossible.

38. Let E be any event. If F is an impossible event, show that E and F are independent.

39. Show that if E and F are independent events, so are $\overline{E}$ and $\overline{F}$. [*Hint:* Use De Morgan's properties.]

40. Show that if E and F are independent events and if $P(E) \neq 0$, $P(F) \neq 0$, then E and F are not mutually exclusive.

41. Suppose $P(E) > 0$, $P(F) > 0$ and E is independent of F. Show that F is independent of E.
[*Hint:* First we note that

$$P(F) \cdot P(E|F) = P(F) \cdot \frac{P(E \cap F)}{P(F)} = P(E \cap F)$$

and

$$P(E) \cdot P(F|E) = P(E) \cdot \frac{P(F \cap E)}{P(E)} = P(E \cap F)$$

Then $P(F) \cdot P(E|F) = P(E) \cdot P(F|E)$. Now use the fact that E is independent of F—that is, $P(E|F) = P(E)$ —to show that F is independent of E—that is, $P(F|E) = P(F)$.]

42. Prove Equation (1) on page 354.
[*Hint:* If E and F are independent events, then

$$P(E|F) = \frac{P(E \cap F)}{P(F)} \quad \text{and} \quad P(E|F) = P(E)$$

Use these relationships to show that $P(E \cap F) = P(E) \cdot P(F)$. Conversely, suppose $P(E \cap F) = P(E) \cdot P(F)$. Use the fact that

$$P(E|F) = \frac{P(E \cap F)}{P(F)}$$

to show that $P(E|F) = P(E)$.]

CHAPTER REVIEW

IMPORTANT TERMS AND CONCEPTS

sample space 316
outcome 316
probability of an outcome 318
constructing a probability model 319
event; simple event 320
probability of an event 320
mutually exclusive events 326
probability of E or F, if E, F are
    mutually exclusive 326

Additive Rule 327
properties of the probability
    of an event 328
probability of the complement
    of an event 329
probability of an event
    with equally likely
    outcomes 330

odds 332
conditional probability 344
Product Rule 346
independent events 354
Test for Independence 354
independence of more than
    two events 357

IMPORTANT FORMULAS

Mutually Exclusive Events

$$P(E \cup F) = P(E) + P(F)$$

Additive Rule

$$P(E \cup F) = P(E) + P(F) - P(E \cap F)$$

Complementary Events

$$P(\overline{E}) = 1 - P(E)$$

Equally Likely Outcomes

$$P(E) = \frac{c(E)}{c(S)}$$

Odds for E Are a to b

$$P(E) = \frac{a}{a + b}$$

Conditional Probability

$$P(E|F) = \frac{P(E \cap F)}{P(F)}$$

Product Rule

$$P(E \cap F) = P(F) \cdot P(E|F)$$

E Is Independent of F

$$P(E|F) = P(E)$$

E, F Are Independent

if and only if $P(E \cap F) = P(E) \cdot P(F)$

TRUE–FALSE ITEMS   Answers are on page AN-27.

T____ F____ **1.** If the odds for an event E are 2 to 1, then $P(E) = \frac{2}{3}$.

T____ F____ **2.** The conditional probability of E given F is
$$P(E|F) = \frac{P(E \cap F)}{P(E)}$$

T____ F____ **3.** If two events in a sample space have no outcomes in common, they are said to be independent.

T____ F____ **4.** $P(E|F) = P(F|E)$ for any events E and F.

T____ F____ **5.** $P(E) + P(\overline{E}) = 1$

T____ F____ **6.** If $P(E) = .4$ and $P(F) = .3$, then $P(E \cup F)$ must be .7.

T____ F____ **7.** If $P(E \cap F) = 0$, then E and F are said to be mutually exclusive.

T____ F____ **8.** If $P(E \cup F) = .7$, $P(E) = .4$, and $P(F) = .3$, then $P(E \cap F) = 0$.

FILL IN THE BLANKS   Answers are on page AN-27.

1. If $S = \{a, b, c, d\}$ is a sample space, the outcomes are equally likely, and $E = \{a, b\}$, then $P(\overline{E}) = $ _____.

2. If a coin is tossed five times, the number of outcomes in the sample space of this experiment is _____.

3. If an event is certain to occur, then its probability is _____. If an event is impossible, its probability is _____.

4. If $P(\overline{E}) = .2$, then $P(E) = $ _____.

5. If $P(E) = .6$, the odds _____ E are 3 to 2.

6. When each outcome in a sample space is assigned the same probability, the outcomes are termed _____ _____.

7. If two events in a sample space have no outcomes in common, they are said to be _____ _____.

REVIEW EXERCISES   Answers to odd-numbered problems begin on page AN-27.

1. A survey of families with 2 children is made, and the sexes of the children are recorded. Describe the sample space and draw a tree diagram of this random experiment.

2. A fair coin is tossed three times.
   (a) Construct a probability model corresponding to this experiment.
   (b) Find the probabilities of the following events:

(i)   The first toss is tails.
(ii)   The first toss is heads.
(iii)   Either the first toss is tails or the third toss is heads.
(iv)   At least one of the tosses is heads.
(v)   There are at least 2 tails.
(vi)   No tosses are heads.

3. A jar contains 3 white marbles, 2 yellow marbles, 4 red marbles, and 5 blue marbles. Two marbles are picked at random. What is the probability that

(a) Both are blue?   (b) Exactly 1 is blue?
(c) At least 1 is blue?

4. Let A and B be events with $P(A) = .3$, $P(B) = .5$, and $P(A \cap B) = .2$. Find the probability that

(a) A or B happens.   (b) A does not happen.
(c) Neither A nor B happens.
(d) Either A does not happen or B does not happen.

5. A loaded die is rolled 400 times, and the following outcomes are recorded:

Face	1	2	3	4	5	6
No. Times Showing	32	45	84	74	92	73

Estimate the probability of rolling a

(a) 3   (b) 5   (c) 6

6. A survey of a group of criminals shows that 65% came from low-income families, 40% from broken homes, and 30% came from low-income and broken families. Define

    E: Criminal came from low income

    F: Criminal came from broken home

If a criminal is selected at random, find

(a) The probability that the criminal is not from a low-income family.
(b) The probability that the criminal comes from a broken home or a low-income family.
(c) Are E and F mutually exclusive?

7. If E and F are events with $P(E \cup F) = \frac{5}{8}$, $P(E \cap F) = \frac{1}{3}$, and $P(E) = \frac{1}{2}$, find

(a) $P(\overline{E})$ (b) $P(F)$ (c) $P(\overline{F})$

8. If E and F represent mutually exclusive events, $P(E) = .30$, and $P(F) = .45$, find each of the probabilities:

(a) $P(\overline{E})$ (b) $P(\overline{F})$
(c) $P(E \cap F)$ (d) $P(E \cup F)$
(e) $P(\overline{E} \cap F)$ (f) $P(\overline{E \cup F})$
(g) $P(\overline{E} \cup \overline{F})$ (h) $P(\overline{E} \cap \overline{F})$

9. Consider the experiment of spinning the spinner shown in the figure at the top of the next column 3 times. (Assume the spinner cannot fall on a line.)

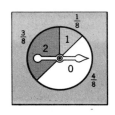

(a) Are all outcomes equally likely?
(b) If not, which of the outcomes has the highest probability?
(c) Let F be the event, "Each digit will occur exactly once." Find $P(F)$.

10. What are the odds in favor of a 5 when a fair die is thrown?

11. A bettor is willing to give 7 to 6 odds that the Bears will win the NFL title. What is the probability of the Bears winning?

12. A biased coin is such that the probability of heads (H) is $\frac{1}{4}$ and the probability of tails (T) is $\frac{3}{4}$. Show that in flipping this coin twice the events E and F defined below are independent.

    E: A head turns up in the first throw

    F: A tail turns up in the second throw

13. The records of Midwestern University show that in one semester, 38% of the students failed mathematics, 27% of the students failed physics, and 9% of the students failed mathematics and physics. A student is selected at random.

(a) If a student failed physics, what is the probability that he or she failed mathematics?
(b) If a student failed mathematics, what is the probability that he or she failed physics?
(c) What is the probability that he or she failed mathematics or physics?

14. A pair of fair dice is thrown 3 times. What is the probability that on the first toss the sum of the 2 dice is even, on the second toss the sum is less than 6, and on the third toss the sum is 7?

15. In a certain population of people, 25% are blue-eyed and 75% are brown-eyed. Also, 10% of the blue-eyed people are left-handed and 5% of the brown-eyed people are left-handed.

(a) What is the probability that a person chosen at random is blue-eyed and left-handed?
(b) What is the probability that a person chosen at random is left-handed?

(c) What is the probability that a person is blue-eyed, given that the person is left-handed?

16. Score Distribution Two forms of a standardized math exam were given to 100 students. The following are the results.

	Form A	Form B	Total
Over 80%	8	12	20
Under 80%	32	48	80
Totals	40	60	100

(a) What is the probability that a student who scored over 80% took form A?
(b) What is the probability that a student who took form A scored over 80%?
(c) Show that the events ''scored over 80%'' and ''took form A'' are independent.
(d) Are the events ''scored over 80%'' and ''took form B'' independent?

17. ACT Scores The following data compare ACT scores of students with their performance in the classroom (based on a 4.0).

Average	Below 21	22–27	Above 28	Total
3.6–4	8	56	104	168
3.0–3.5	47	70	30	147
Below 3	47	34	4	85
Totals	102	160	138	400

Mathematical Questions from Professional Exams

1. Actuary Exam—Part I If P and Q are events having positive probability in the sample space S such that $P \cap Q = \varnothing$, then all of the following pairs are independent EXCEPT

(a) $\varnothing$ and P (b) P and Q
(c) P and S (d) P and $P \cap Q$
(e) $\varnothing$ and the complement of P

2. Actuary Exam—Part I A box contains 12 varieties of candy and exactly 2 pieces of each variety. If 12 pieces of candy are selected at random, what is the probability that a given variety is represented?

A graduating student is selected at random. Find the probability that

(a) The student scored above 28.
(b) The student point average is 3.6–4.
(c) The student scored above 28 with an average 3.6–4.
(d) The student ACT score was in the 22–27 range.
(e) The student had a 3.0–3.5 average.
(f) Show that ''ACT above 28'' and ''average below 3'' are not independent.

18. Three envelopes are addressed for 3 secret letters written in invisible ink. A secretary randomly places each of the letters in an envelope and mails them. What is the probability that at least 1 person receives the correct letter?

19. Jones lives at O (see the figure). He owns 5 gas stations located 4 blocks away (dots). Each afternoon he checks on one of his gas stations. He starts at O. At each intersection he flips a fair coin. If it shows heads, he will head north (N); otherwise, he will head toward the east (E). What is the probability that he will end up at gas station G before coming to one of the other stations?

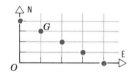

(a) $\dfrac{2^{12}}{(12!)^2}$ (b) $\dfrac{2^{12}}{24!}$ (c) $\dfrac{2^{12}}{\binom{24}{12}}$ (d) $\dfrac{11}{46}$ (e) $\dfrac{35}{46}$

3. Actuary Exam—Part II What is the probability that a 3-card hand drawn at random and without replacement from a regular deck consists entirely of black cards?

(a) $\frac{1}{17}$ (b) $\frac{2}{17}$ (c) $\frac{1}{8}$ (d) $\frac{3}{17}$ (e) $\frac{4}{17}$

4. Actuary Exam—Part II Events S and T are independent with $\Pr(S) < \Pr(T)$, $\Pr(S \cap T) = \frac{6}{25}$, and $\Pr(S|T) + \Pr(T|S) = 1$. What is $\Pr(S)$?

(a) $\frac{1}{25}$ (b) $\frac{1}{5}$ (c) $\frac{5}{25}$ (d) $\frac{2}{5}$ (e) $\frac{3}{5}$

5. Actuary Exam—Part II What is the least number of independent times that an unbiased die must be thrown to make the probability that all throws do not give the same result greater than .999?

(a) 3   (b) 4   (c) 5   (d) 6   (e) 7

6. Actuary Exam—Part II In a group of 20,000 men and 10,000 women, 6% of the men and 3% of the women have a certain affliction. What is the probability that an afflicted member of the group is a man?

(a) $\frac{3}{5}$ (b) $\frac{2}{3}$ (c) $\frac{3}{4}$ (d) $\frac{4}{5}$ (e) $\frac{8}{9}$

7. Actuary Exam—Part II An unbiased die is thrown 2 independent times. Given that the first throw resulted in an even number, what is the probability that the sum obtained is 8?

(a) $\frac{5}{36}$ (b) $\frac{1}{6}$ (c) $\frac{4}{21}$ (d) $\frac{7}{36}$ (e) $\frac{1}{3}$

8. Actuary Exam—Part II If the events S and T have equal probability and are independent with $\Pr(S \cap T) = p > 0$, then $\Pr(S) =$

(a) $\sqrt{p}$ (b) p^2 (c) $\frac{p}{2}$ (d) p (e) $2p$

9. Actuary Exam—Part II The probability that both S and T occur, the probability that S occurs and T does not, and the probability that T occurs and S does not are all equal to p. What is the probability that either S or T occurs?

(a) p (b) $2p$ (c) $3p$ (d) $3p^2$ (e) p^3

10. Actuary Exam—Part II What is the probability that a bridge hand contains 1 card of each denomination (i.e., 1 ace, 1 king, 1 queen, . . . , 1 three, 1 two)?

(a) $\dfrac{13!}{13^{13}}$ (b) $\dfrac{4^{13}}{\binom{52}{13}}$ (c) $\dfrac{\binom{52}{4}}{\binom{52}{13}}$

(d) $\left(\dfrac{1}{13}\right)^{13}$ (e) $\dfrac{13^4}{\binom{52}{13}}$

Chapter 8

Additional Probability Topics

This chapter deals with various applications of probability. Bayes' formula involves sample spaces that can be divided into mutually exclusive events. The binomial probability model investigates experiments that are repetitive. Expected value uses probability to assess the value of a situation. The section on operations research applies expected value. The chapter closes with a brief description of random variables.

8.1 BAYES' FORMULA

In this section we consider experiments with sample spaces that will be divided or partitioned into two (or more) mutually exclusive events. This study involves a further application of conditional probabilities and leads us to the famous Bayes' formula, named after Thomas Bayes, who first published it in 1763.

We begin by considering the following example.

EXAMPLE 1

Given two urns, suppose urn I contains four black and seven white balls. Urn II contains three black, one white, and four yellow balls. We select an urn at random and then draw a ball. What is the probability that we obtain a black ball?

SOLUTION A

Let U_I and U_{II} stand for the events "Urn I is chosen" and "Urn II is chosen," respectively. Similarly, let B, W, Y stand for the event that "A black," "A white," or "A yellow" ball is chosen, respectively.

A solution to Example 1 can be depicted using the tree diagram shown in Figure 1.

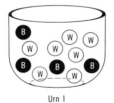 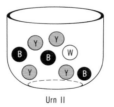 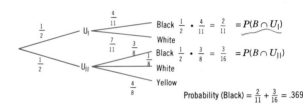

Urn I   Urn II

Figure 1

$$\text{Probability (Black)} = \tfrac{2}{11} + \tfrac{3}{16} = .369$$

Then

$$P(B) = \tfrac{2}{11} + \tfrac{3}{16} = \tfrac{65}{176} = .369$$

SOLUTION B

Let's see how we can solve Example 1 without using a tree diagram. First we observe that

$$P(U_I) = P(U_{II}) = \tfrac{1}{2}$$
$$P(B|U_I) = \tfrac{4}{11} \qquad P(B|U_{II}) = \tfrac{3}{8}$$

The event B can be written as

$$B = (B \cap U_I) \cup (B \cap U_{II})$$

Since $B \cap U_I$ and $B \cap U_{II}$ are disjoint, we add their probabilities. Then

$$P(B) = P(B \cap U_I) + P(B \cap U_{II}) \qquad (1)$$

Using the Product Rule, we have

$$P(B \cap U_I) = P(U_I) \cdot P(B|U_I) \qquad P(B \cap U_{II}) = P(U_{II}) \cdot P(B|U_{II}) \qquad (2)$$

Combining (1) and (2), we have

$$P(B) = P(U_I) \cdot P(B|U_I) + P(U_{II}) \cdot P(B|U_{II})$$
$$= \tfrac{1}{2} \cdot \tfrac{4}{11} + \tfrac{1}{2} \cdot \tfrac{3}{8} = \tfrac{2}{11} + \tfrac{3}{16} = .369$$

· ·

Partitions

The preceding discussion leads to the following generalization. Suppose A_1 and A_2 are two nonempty, mutually exclusive events of a sample space S and the union of A_1 and A_2 is S; that is,

$$A_1 \neq \varnothing \qquad A_2 \neq \varnothing \qquad A_1 \cap A_2 = \varnothing \qquad S = A_1 \cup A_2$$

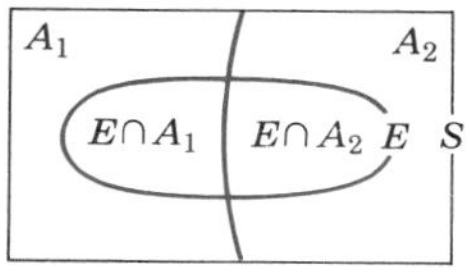

In this case we say that A_1 and A_2 form a **partition** of S. See Figure 2. Now if we let E be any event in S, we may write the set E in the form

$$E = (E \cap A_1) \cup (E \cap A_2)$$

The sets $E \cap A_1$ and $E \cap A_2$ are disjoint since

$$(E \cap A_1) \cap (E \cap A_2) = (E \cap E) \cap (A_1 \cap A_2) = E \cap \varnothing = \varnothing$$

Using the Product Rule, the probability of E is therefore

$$P(E) = P(E \cap A_1) + P(E \cap A_2)$$
$$= P(A_1) \cdot P(E|A_1) + P(A_2) \cdot P(E|A_2) \qquad (3)$$

Formula (3) is used to find the probability of an event E of a sample space when the sample space is partitioned into two sets A_1 and A_2. See Figure 3 for a tree diagram depicting Formula (3).

Figure 2

A_1 | A_2 | $E \cap A_1$ | $E \cap A_2$ | E | S

Figure 3

$P(A_1)$ — A_1 — $P(E|A_1)$ — E / $\bar{E}$
$P(A_2)$ — A_2 — $P(E|A_2)$ — E / $\bar{E}$
$P(E) = P(A_1) \cdot P(E|A_1) + P(A_2) \cdot P(E|A_2)$

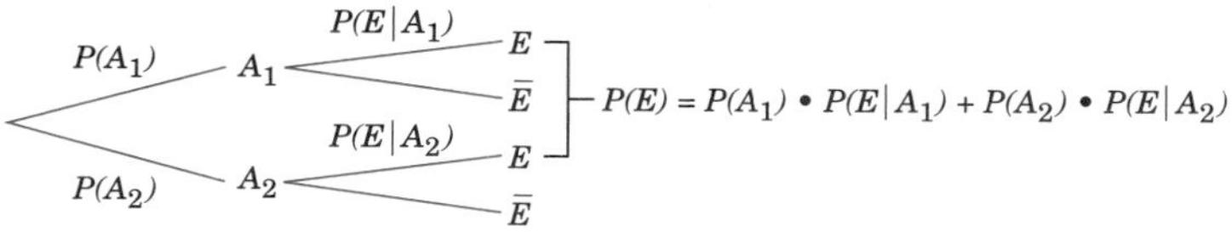

EXAMPLE 2

Admissions Tests for Medical School   Of the applicants to a medical school, 80% are eligible to enter and 20% are not. To aid in the selection process, an admissions test is administered that is designed so that an eligible candidate will pass 90% of the time, while an ineligible candidate will pass only 30% of the time. What is the probability that an applicant for admission will pass the admissions test?

SOLUTION A

Figure 4 provides a tree diagram solution.

SOLUTION B

The sample space S consists of the applicants for admission, and S can be partitioned into the following two events:

$$A_1\text{:} \quad \text{Eligible applicant} \qquad A_2\text{:} \quad \text{Ineligible applicant}$$

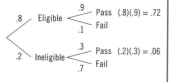

Probability of passing = .72 + .06 = .78

Figure 4

Figure 5

These two events are disjoint, and their union is S. See Figure 5.

The event E is

$$E: \quad \text{Applicant passes admissions test}$$

Now

$$P(A_1) = .8 \qquad P(A_2) = .2$$
$$P(E|A_1) = .9 \qquad P(E|A_2) = .3$$

Using Formula (3), we have

$$P(E) = P(A_1) \cdot P(E|A_1) + P(A_2) \cdot P(E|A_2) = (.8)(.9) + (.2)(.3) = .78$$

Thus the probability that an applicant will pass the admissions test is .78.

Now Work Problem 15

· ·

If we partition a sample space S into three sets A_1, A_2, and A_3 so that

$$S = A_1 \cup A_2 \cup A_3$$
$$A_1 \cap A_2 = \varnothing \qquad A_2 \cap A_3 = \varnothing \qquad A_1 \cap A_3 = \varnothing$$
$$A_1 \neq \varnothing \qquad A_2 \neq \varnothing \qquad A_3 \neq \varnothing$$

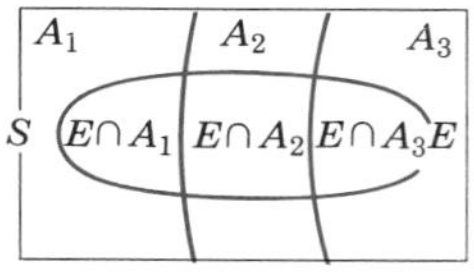

we may write any set E in S in the form

$$E = (E \cap A_1) \cup (E \cap A_2) \cup (E \cap A_3)$$

Figure 6

See Figure 6.

Since $E \cap A_1$, $E \cap A_2$, and $E \cap A_3$ are disjoint, the probability of event E is

$$\begin{aligned}P(E) &= P(E \cap A_1) + P(E \cap A_2) + P(E \cap A_3) \\ &= P(A_1) \cdot P(E|A_1) + P(A_2) \cdot P(E|A_2) + P(A_3) \cdot P(E|A_3)\end{aligned} \tag{4}$$

Formula (4) is used to find the probability of an event E of a sample space when the sample space is partitioned into three sets A_1, A_2, and A_3. See Figure 7 for a tree diagram depicting Formula (4).

Figure 7

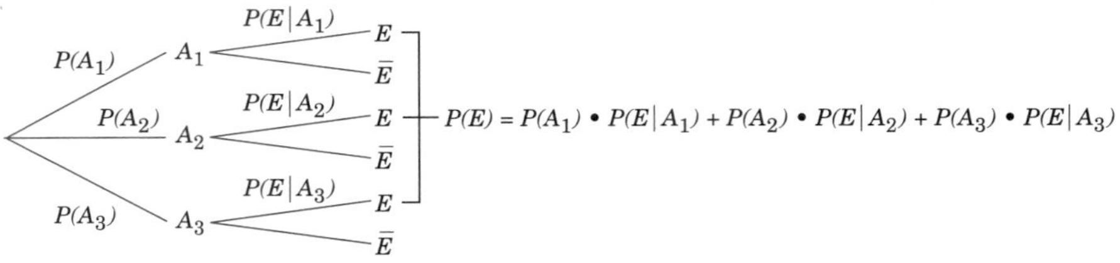

$$P(E) = P(A_1) \bullet P(E|A_1) + P(A_2) \bullet P(E|A_2) + P(A_3) \bullet P(E|A_3)$$

EXAMPLE 3

Three machines, I, II, and III, manufacture .4, .5, and .1 of the total production in a plant, respectively. The percentage of defective items produced by I, II, and III is 2%, 4%, and 1%, respectively. For an item chosen at random, what is the probability that it is defective?

SOLUTION A

Figure 8 gives a tree diagram solution.

Figure 8

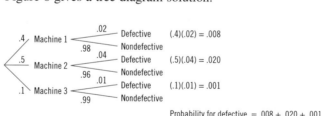

Probability for defective $= .008 + .020 + .001 = .029$

SOLUTION B

The sample space S is partitioned into three events $A_1, A_2,$ and A_3 defined as follows:

A_1: Item produced by machine I
A_2: Item produced by machine II
A_3: Item produced by machine III

Clearly, the events $A_1, A_2,$ and A_3 are mutually exclusive, and their union is S. Define the event E in S to be

E: Item is defective

Now

$$P(A_1) = .4 \qquad P(A_2) = .5 \qquad P(A_3) = .1$$
$$P(E|A_1) = .02 \qquad P(E|A_2) = .04 \qquad P(E|A_3) = .01$$

Thus, using Formula (4), we see that

$$P(E) = (.4)(.02) + (.5)(.04) + (.1)(.01)$$
$$= .008 + .020 + .001 = .029$$

Now Work Problem 17

• •

To generalize Formulas (3) and (4) to a sample space S partitioned into n subsets, we first introduce the following definition:

PARTITION

A sample space S is *partitioned* into n subsets $A_1, A_2, \ldots, A_n$, provided:

(a) The intersection of any two of the subsets is empty.
(b) Each subset is nonempty.
(c) $A_1 \cup A_2 \cup \cdots \cup A_n = S$

Let S be a sample space and let $A_1, A_2, A_3, \ldots, A_n$ be n events that form a partition of the set S. If E is any event in S, then

$$E = (E \cap A_1) \cup (E \cap A_2) \cup \cdots \cup (E \cap A_n)$$

Since $E \cap A_1, E \cap A_2, \ldots, E \cap A_n$ are mutually exclusive events, we have

$$P(E) = P(E \cap A_1) + P(E \cap A_2) + \cdots + P(E \cap A_n) \qquad (5)$$

In (5) replace $P(E \cap A_1), P(E \cap A_2), \ldots, P(E \cap A_n)$ using the Product Rule. Then we obtain the formula

$$P(E) = P(A_1) \cdot P(E|A_1) + P(A_2) \cdot P(E|A_2) + \cdots + P(A_n) \cdot P(E|A_n) \qquad (6)$$

EXAMPLE 4

Admissions Tests for Medical School   In Example 2 suppose an applicant passes the admissions test. What is the probability that he or she was among those eligible; that is, what is the probability $P(A_1|E)$?

SOLUTION

By the definition of conditional probability,

$$P(A_1|E) = \frac{P(A_1 \cap E)}{P(E)} = \frac{P(A_1) \cdot P(E|A_1)}{P(E)}$$

But $P(E)$ is given by (6) when $n = 2$, or by (3). Thus

$$P(A_1|E) = \frac{P(A_1) \cdot P(E|A_1)}{P(A_1) \cdot P(E|A_1) + P(A_2) \cdot P(E|A_2)} \qquad (7)$$

Using the information supplied in Example 2, we find

$$P(A_1|E) = \frac{(.8)(.9)}{.78} = \frac{.72}{.78} = .923$$

The admissions test is a reasonably effective device. Less than 8% of the students passing the test are ineligible.

Now Work Problem 19

· ·

Bayes' Formula

Equation (7) is a special case of **Bayes' formula** when the sample space is partitioned into two sets A_1 and A_2. The general formula is given below.

Bayes' Formula

Let S be a sample space partitioned into n events, $A_1, \ldots, A_n$. Let E be any event of S for which $P(E) > 0$. The probability of the event A_j ($j = 1, 2, \ldots, n$), given the event E, is

$$P(A_j|E) = \frac{P(A_j) \cdot P(E|A_j)}{P(E)} \qquad (8)$$

$$= \frac{P(A_j) \cdot P(E|A_j)}{P(A_1) \cdot P(E|A_1) + P(A_2) \cdot P(E|A_2) + \cdots + P(A_n) \cdot P(E|A_n)}$$

The proof is left as an exercise (see Problem 42).

The following example illustrates a use for Bayes' formula when the sample space S is partitioned into three events.

EXAMPLE 5

Source of Defective Cars   Motors, Inc., has three plants. Plant I produces 35% of the car output, plant II produces 20%, and plant III produces the remaining 45%. One percent of the output of plant I is defective, as is 1.8% of the output of plant II, and 2% of the output of plant III. The annual total output of Motors, Inc., is 1,000,000 cars. A car is chosen at random from the annual output and it is found to be defective. What is the probability that it came from plant I? Plant II? Plant III?

SOLUTION

To answer these questions, let's define the following events:

$$E: \quad \text{Car is defective}$$
$$A_1: \quad \text{Car produced by plant I}$$
$$A_2: \quad \text{Car produced by plant II}$$
$$A_3: \quad \text{Car produced by plant III}$$

Also, $P(A_1|E)$ indicates the probability that a car is produced by plant I, given that it was defective; $P(A_2|E)$ and $P(A_3|E)$ are similarly defined. To find these probabilities, we first need to find $P(E)$.

From the data given in the problem we can determine the following:

$$\begin{aligned} P(A_1) &= .35 & P(E|A_1) &= .010 \\ P(A_2) &= .20 & P(E|A_2) &= .018 \\ P(A_3) &= .45 & P(E|A_3) &= .020 \end{aligned} \tag{9}$$

Now,

$$A_1 \cap E \text{ is the event ``Produced by plant I and is defective''}$$
$$A_2 \cap E \text{ is the event ``Produced by plant II and is defective''}$$
$$A_3 \cap E \text{ is the event ``Produced by plant III and is defective''}$$

From the Product Rule we find

$$P(A_1 \cap E) = P(A_1) \cdot P(E|A_1) = (.35)(.010) = .0035$$
$$P(A_2 \cap E) = P(A_2) \cdot P(E|A_2) = (.20)(.018) = .0036$$
$$P(A_3 \cap E) = P(A_3) \cdot P(E|A_3) = (.45)(.020) = .0090$$

Since $E = (A_1 \cap E) \cup (A_2 \cap E) \cup (A_3 \cap E)$, we have

$$\begin{aligned} P(E) &= P(A_1 \cap E) + P(A_2 \cap E) + P(A_3 \cap E) \\ &= .0035 + .0036 + .0090 \\ &= .0161 \end{aligned}$$

Thus the probability that a defective car is chosen is .0161. See Figure 9.

Figure 9

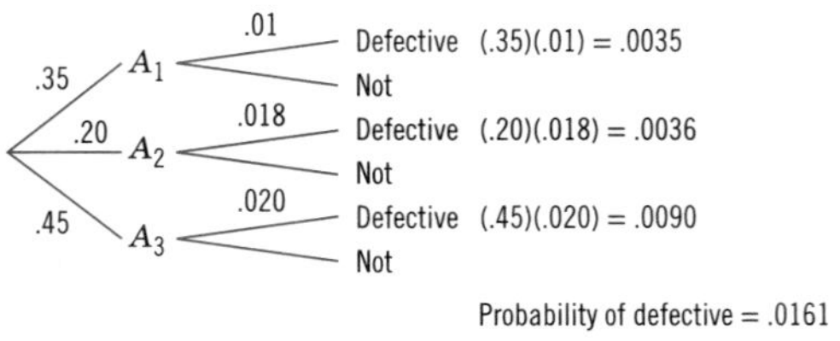

Probability of defective = .0161

Given that the car chosen is defective, the probability that it came from plant I is $P(A_1|E)$, from plant II is $P(A_2|E)$, and from plant III is $P(A_3|E)$. To compute these probabilities, we use Bayes' formula:

$$P(A_1|E) = \frac{P(A_1) \cdot P(E|A_1)}{P(A_1) \cdot P(E|A_1) + P(A_2) \cdot P(E|A_2) + P(A_3) \cdot P(E|A_3)}$$

$$= \frac{P(A_1) \cdot P(E|A_1)}{P(E)} = \frac{(.35)(.01)}{.0161} = .217$$

$$P(A_2|E) = \frac{P(A_2) \cdot P(E|A_2)}{P(E)} = \frac{.0036}{.0161} = .224 \qquad (10)$$

$$P(A_3|E) = \frac{P(A_3) \cdot P(E|A_3)}{P(E)} = \frac{.0090}{.0161} = .559$$

Now Work Problem 24

•••••••••••••••••••••

A Priori, A Posteriori Probabilities

In Bayes' formula the probabilities $P(A_j)$ are referred to as *a priori* probabilities, while the $P(A_j|E)$ are called *a posteriori* probabilities. We use Example 5 to explain the reason for this terminology. Knowing nothing else about a car, the probability that it was produced by plant I is given by $P(A_1)$. Thus $P(A_1)$ can be regarded as a "before the fact," or a priori, probability. With the additional information that the car is defective, we reassess the likelihood of whether it came from plant I and compute $P(A_1|E)$. Thus $P(A_1|E)$ can be viewed as an "after the fact," or a posteriori, probability.

Note that $P(A_1) = .35$, while $P(A_1|E) = .217$. So the knowledge that a car is defective decreases the chances that it came from plant I.

EXAMPLE 6

Testing for Cancer   The residents of a community are examined for cancer. The examination results are classified as positive ($+$) if a malignancy is suspected, and as negative ($-$) if there are no indications of a malignancy. If a person has cancer, the probability of a positive result from the examination is .98. If a person does not have cancer, the probability of a positive result is .15. If 5% of the community has cancer, what is the probability of a person not having cancer if the examination is positive?

SOLUTION

Let us define the following events:

$$A_1: \quad \text{Person has cancer}$$
$$A_2: \quad \text{Person does not have cancer}$$
$$E: \quad \text{Examination is positive}$$

We want to know the probability of a person not having cancer if it is known that the examination is positive; that is, we wish to find $P(A_2|E)$. Now

$$P(A_1) = .05 \qquad P(A_2) = .95$$
$$P(E|A_1) = .98 \qquad P(E|A_2) = .15$$

Using Bayes' formula, we get

$$P(A_2|E) = \frac{P(A_2) \cdot P(E|A_2)}{P(A_1) \cdot P(E|A_1) + P(A_2) \cdot P(E|A_2)}$$

$$= \frac{(.95)(.15)}{(.05)(.98) + (.95)(.15)} = .744$$

Thus even if the examination is positive, the person examined is more likely not to have cancer than to have cancer. The reason the test is designed this way is that it is better for a healthy person to be examined more thoroughly than for someone with cancer to go undetected. Simply stated, the test is useful because of the high probability (.98) that a person with cancer will not go undetected.

Now Work Problem 29

. . . .

EXAMPLE 7

Car Repair Diagnosis The manager of a car repair shop knows from past experience that when a call is received from a person whose car will not start, the probabilities for various troubles (assuming no two can occur simultaneously) are as follows:

Event	Trouble	Probability
A_1	Flooded	.3
A_2	Battery cable loose	.2
A_3	Points bad	.1
A_4	Out of gas	.3
A_5	Something else	.1

The manager also knows that if the person will hold the gas pedal down and try to start the car, the probability that it will start (E) is

$$P(E|A_1) = .9 \qquad P(E|A_2) = 0 \qquad P(E|A_3) = .2$$
$$P(E|A_4) = 0 \qquad P(E|A_5) = .2$$

(a) If a person has called and is instructed to "hold the pedal down . . . ," what is the probability that the car will start?

(b) If the car does start after holding the pedal down, what is the probability that the car was flooded?

SOLUTION

(a) We need to compute $P(E)$. Using Formula (6) for $n = 5$ (the sample space is partitioned into five disjoint sets), we have

$$P(E) = P(A_1) \cdot P(E|A_1) + P(A_2) \cdot P(E|A_2) + P(A_3) \cdot P(E|A_3)$$
$$+ P(A_4) \cdot P(E|A_4) + P(A_5) \cdot P(E|A_5)$$
$$= (.3)(.9) + (.2)(0) + (.1)(.2) + (.3)(0) + (.1)(.2)$$
$$= .27 + .02 + .02 = .31$$

See Figure 10.

Figure 10

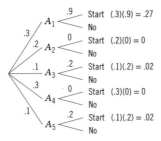

Probability of starting = .31

(b) We use Bayes' formula to compute the a posteriori probability $P(A_1|E)$:

$$P(A_1|E) = \frac{P(A_1) \cdot P(E|A_1)}{P(E)} = \frac{(.3)(.9)}{.31} = .87$$

Thus the probability that the car was flooded, after it is known that holding down the pedal started the car, is .87.

....

EXERCISE 8.1   Answers to odd-numbered problems begin on page AN-28.

In Problems 1–14 find the indicated probabilities by referring to the tree diagram below and using Bayes' formula.

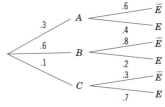

$(.3 \cdot .4)$

1. $P(E|A)$ $(.3 \cdot .1)$ 2. $P(\bar{E}|A)$ 3. $P(E|B)$ 4. $P(\bar{E}|B)$ 5. $P(E|C)$

6. $P(\bar{E}|C)$ 7. $P(E)$ 8. $P(\bar{E})$ 9. $P(A|E)$ 10. $P(B|\bar{E})$

11. $P(C|E)$ $(.7 \cdot .1)$ 12. $P(A|\bar{E})$ 13. $P(B|E)$ 14. $P(C|\bar{E})$ $(.3 \cdot .1)$

15. Events A_1 and A_2 form a partition of a sample space S with $P(A_1) = .3$ and $P(A_2) = .7$. If E is an event in S with $P(E|A_1) = .01$ and $P(E|A_2) = .02$, compute $P(E)$.

16. Events A_1 and A_2 form a partition of a sample space S with $P(A_1) = .4$ and $P(A_2) = .6$. If E is an event in S with $P(E|A_1) = .03$ and $P(E|A_2) = .01$, compute $P(E)$.

17. Events A_1, A_2, and A_3 form a partition of a sample space S with $P(A_1) = .5$, $P(A_2) = .3$, and $P(A_3) = .2$. If E is an event in S with $P(E|A_1) = .01$, $P(E|A_2) = .03$, and $P(E|A_3) = .02$, compute $P(E)$.

18. Events A_1, A_2, and A_3 form a partition of a sample space S with $P(A_1) = .3$, $P(A_2) = .3$, and $P(A_3) = .4$. If E is an

event in S with $P(E|A_1) = .01$, $P(E|A_2) = .02$, and $P(E|A_3) = .02$, compute $P(E)$.

19. Use the information in Problem 15 to find $P(A_1|E)$ and $P(A_2|E)$.

20. Use the information in Problem 16 to find $P(A_1|E)$ and $P(A_2|E)$.

21. Use the information in Problem 17 to find $P(A_1|E)$, $P(A_2|E)$, and $P(A_3|E)$.

22. Use the information in Problem 18 to find $P(A_1|E)$, $P(A_2|E)$, and $P(A_3|E)$.

23. In Example 3 (page 368) suppose it is known that a defective item was produced. Find the probability that it came from machine I. From machine II. From machine III.

24. In Example 5 (page 371) suppose $P(A_1) = P(A_2) = P(A_3) = \frac{1}{3}$. Find $P(A_2|E)$ and $P(A_3|E)$.

25. In Example 7 (page 373), compute the a posteriori probabilities $P(A_2|E)$, $P(A_3|E)$, $P(A_4|E)$, and $P(A_5|E)$.

26. In Example 6 (page 372) compute $P(A_1|E)$.

27. Three jars contain colored balls as follows:

Jar	Red, R	White, W	Blue, B
I	5	6	5
II	3	4	9
III	7	5	4

One jar is chosen at random and a ball is withdrawn. The ball is red. What is the probability that it came from jar I? From jar II? From jar III? [*Hint:* Define the events E: Ball selected is red; U_I: jar I selected; U_{II}: jar II selected; and U_{III}: jar III selected. Determine $P(U_I|E)$, $P(U_{II}|E)$, and $P(U_{III}|E)$ by using Bayes' formula.]

28. **TB Screening** Suppose that if a person with tuberculosis is given a TB screening, the probability that his or her condition will be detected is .90. If a person without tuberculosis is given a TB screening, the probability that he or she will be diagnosed incorrectly as having tuberculosis is .3. Suppose, further, that 11% of the adult residents of a certain city have tuberculosis. If one of these adults is diagnosed as having tuberculosis based on the screening, what is the probability that he or she actually has tuberculosis? Interpret your result.

29. **Car Production** Cars are being produced by two factories, but factory I produces twice as many cars as factory II in a given time. Factory I is known to produce 2% defectives and factory II produces 1% defectives. A car is examined and found to be defective. What are the a priori and a posteriori probabilities that the car was produced by factory I?

30. An absent-minded nurse is to give Mr. Brown a pill each day. The probability that the nurse forgets to administer the pill is $\frac{2}{3}$. If he receives the pill, the probability that Brown will die is $\frac{1}{3}$. If he does not get his pill, the probability that he will die is $\frac{3}{4}$. Mr. Brown died. What is the probability that the nurse forgot to give Brown the pill?

31. **Color-Blind** In a human population 51% are male and 49% are female. 5% of the males and 0.3% of the females are color-blind. If a person randomly chosen from the population is found to be color-blind, what is the probability that the person is a male?

32. **Medical Diagnosis** In a certain small town 16% of the population developed lung cancer. If 45% of the population are smokers, and 85% of those developing lung cancer are smokers, what is the probability that a smoker in this population will develop lung cancer?

33. **Voting Pattern** In Cook County 55% of the registered voters are Democrats, 30% are Republicans, and 15% are independents. During a recent election, 35% of the Democrats voted, 65% of the Republicans voted, and 75% of the independents voted. What is the probability that someone who voted is a Democrat? Republican? Independent?

34. **Quality Control** A computer manufacturer has three assembly plants. Records show that 2% of the sets shipped from plant A turn out to be defective, as compared to 3% of those that come from plant B and 4% of those that come from plant C. In all, 30% of the manufacturer's total production comes from plant A, 50% from plant B, and 20% from plant C. If a customer finds that his computer is defective, what is the probability it came from plant B?

35. **Oil Drilling** An oil well is to be drilled in a certain location. The soil there is either rock (probability .53), clay (probability .21), or sand. If it is rock, a geological test gives a positive result with 35% accuracy; if it is clay, this test gives a positive result with 48% accuracy; and if it is sand, the test gives a positive result with 75% accuracy. Given that the test is positive, what is the probability that the soil is rock? What is the probability that the soil is clay? What is the probability that the soil is sand?

36. **Oil Drilling** A geologist is using seismographs to test for oil. It is found that if oil is present, the test gives a positive result 95% of the time, and if oil is not present, the test gives a positive result 2% of the time. Oil is actu-

ally present in 1% of the cases tested. If the test shows positive, what is the probability that oil is present?

37. **Political Polls**   In conducting a political poll, a pollster divides the United States into four sections: Northeast (N), containing 40% of the population; South (S), containing 10% of the population; Midwest (M), containing 25% of the population; and West (W), containing 25% of the population. From the poll it is found that in the next election 40% of the people in the Northeast say they will vote for Republicans, in the South 56% will vote Republican, in the Midwest 48% will vote Republican, and in the West 52% will vote Republican. What is the probability that a person chosen at random will vote Republican? Assuming a person votes Republican, what is the probability that he or she is from the Northeast?

38. **Marketing**   To introduce a new beer, a company conducted a survey. It divided the United States into four regions, eastern, northern, southern, and western. The company estimates that 35% of the potential customers for the beer are in the eastern region, 30% are in the northern region, 20% are in the southern region, and 15% are in the western region. The survey indicates that 50% of the potential customers in the eastern region, 40% of the potential customers in the northern region, 36% of the potential customers in the southern region, and 42% of those in the western region will buy the beer. If a potential customer chosen at random indicates that he or she will buy the beer, what is the probability that the customer is from the southern region?

39. **College Majors**   Data collected by the Office of Admissions of a large midwestern university indicate the following choices made by the members of the freshman class regarding their majors:

Major	Percentage of Freshmen Choosing Major	Female (in percent)	Male (in percent)
Engineering	26	40	60
Business	30	35	65
Education	9	80	20
Social science	12	52	48
Natural science	12	56	44
Humanities	9	65	35
Other	2	51	49

What is the probability that a female student selected at random from the freshman class is majoring in engineering?

40. **Testing for HIV**   An article in the *New York Times* some time ago reported that college students are beginning to routinely ask to be tested for the AIDS virus. The standard test for the HIV virus is the Elias test, which tests for the presence of HIV antibodies. It is estimated that this test has a 99.8% sensitivity and a 99.8% specificity. A 99.8% sensitivity means that, in a large-scale screening test, for every 1000 people tested who have the virus we can expect 998 people to test positive and 2 to have a false negative test. A 99.8% specificity means that, in a large-scale screening test, for every 1000 people tested who do not have the virus we can expect 998 people to have a negative test and 2 to have a false positive test.

(a) The *New York Times* article remarks that it is estimated that about 2 in every 1000 college students have the HIV virus. Assume that a large group of randomly chosen college students, say 100,000, are tested by the Elias test. If a student tests positive, what is the chance that this student has the HIV virus?

(b) What would this probability be for a population at high risk, where 5% of the population has the HIV virus?

(c) Suppose Jack tested positive on an Elias test. Another Elias test* is performed and the results are positive again. Assuming that the tests are independent, what is the probability that Jack has the HIV virus?

41. **Medical Test**   A scientist designed a medical test for a certain disease. Among 100 patients who have the disease, the test will show the presence of the disease in 97 cases out of 100, and will fail to show the presence of the disease in the remaining 3 cases out of a 100. Among those who do not have the disease, the test will erroneously show the presence of the disease in 4 cases out of 100, and will show that there is no disease in the remaining 96 cases out of 100.

(a) What is the probability that a patient who tested positive on this test actually has the disease, if it is estimated that 20% of the population has the disease?

(b) What is the probability that a patient who tested positive on this test actually has the disease, if it is estimated that 4% of the population has the disease?

(c) What is the probability that a patient who took the test

---

* Actually, in practice, if a person tests positive on an Elias test, then two more Elias tests are carried out. If either is positive, then one more confirmatory test, called the Western blot test, is carried out. If this is positive, the person is assumed to have the HIV virus.

twice and tested positive both times actually has the disease, if it is estimated that 4% of the population has the disease?

42. Prove Bayes' formula (8).

43. Show that $P(E|F) = 1$ if F is a subset of E and $P(F) \neq 0$.

8.2 THE BINOMIAL PROBABILITY MODEL

Bernoulli Trials

In this section we study situations that can be analyzed by using a simple probability model called the *binomial probability model*. The model was first studied by J. Bernoulli about 1700 and, for this reason, the model is sometimes referred to as a *Bernoulli trial.*

The binomial probability model is a sequence of trials, each of which consists of repetition of a single experiment. We assume the outcome of one experiment does not affect the outcome of any other one; that is, we assume the trials to be independent. Furthermore, we assume that there are only two possible outcomes for each trial and label them A, for *success,* and F, for *failure*. The probability of success, $p = P(A)$, remains the same from trial to trial. In addition, since there are only two outcomes in each trial, the probability of failure must be $1 - p$, and we write

$$q = 1 - p = P(F)$$

Thus $p + q = 1$.

Any random experiment for which the binomial probability model is appropriate is called a *Bernoulli trial.*

BERNOULLI TRIAL

Random experiments are called *Bernoulli trials* if

(a) The same experiment is repeated several times.
(b) There are only two possible outcomes, success and failure, on each trial.
(c) The repeated trials are independent.
(d) The probability of each outcome remains the same for each trial.

Many real world situations have the characteristics of the binomial probability model. For example, in repeatedly running a subject through a T-maze, we may label a turn to the left by A and a turn to the right by F. The assumption of independence of each trial is equivalent to presuming the subject has no memory.

In opinion polls one person's response is independent of any other person's response, and we may designate the answer "Yes" by an A and any other answer ("No" or "Don't know") by an F.

In testing TVs, we have a sequence of independent trials (each test of a particular TV is a trial), and we label a nondefective TV with an A and a defective one with an F.

In determining whether 9 out of 12 persons will recover from a tropical disease, we assume that each of the 12 persons has the same chance of recovery from the disease and that their recoveries are independent (they are not treated by the same doctor or in the same hospital). We may designate "recovery" by A and "nonrecovery" by F.

Next we consider an experiment that will lead us to formulate a general expression for the probability of obtaining k successes in a sequence of n Bernoulli trials ($k \leq n$).

The experiment is to toss a coin three times. We define success as heads (H) and failure as tails (T). The coin may be fair or loaded, so we let p be the probability of heads and q the probability of tails.

The tree diagram in Figure 11 lists all the possible outcomes and their respective probabilities.

Figure 11

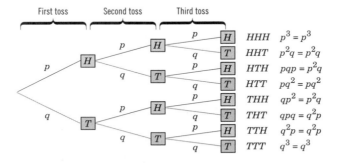

The outcomes are

$$HHH, \quad HHT, \quad HTH, \quad HTT, \quad THH, \quad THT, \quad TTH, \quad TTT$$

Here, for example, HHT means that the first two tosses are heads and the third is tails.

The outcome HHH, in which all three tosses are heads, has a probability $ppp = p^3$ since each Bernoulli trial has a probability p of resulting in heads.

If we wish to calculate the probability that exactly two heads appear, we consider only the outcomes

$$HHT, \quad HTH, \quad THH$$

The outcome HHT has probability $ppq = p^2q$ since the two heads each have probability p and the tail has probability q. In the same way the other two outcomes, HTH and THH, in which there are two heads and one tail, also have probabilities p^2q. Therefore, the probability that exactly two heads appear is equal to the sum of the probabilities of the three outcomes and, hence, is given by $3p^2q$.

In a similar way we compute the following probabilities:

$$P \text{ (Exactly 1 head and 2 tails)} = 3pq^2$$
$$P \text{ (All 3 heads)} = p^3$$
$$P \text{ (All 3 tails)} = q^3$$

Binomial Probabilities

When considering tosses in excess of three, it would be extremely tedious, to say the least, to solve problems of this type using a tree diagram. This is why it is desirable to find a general formula.

Suppose the probability of a success in a Bernoulli trial is p, and suppose we wish to find the probability of exactly k successes in n repeated trials. One possible outcome is

$$\underbrace{AAA \cdots A}_{k \text{ successes}} \cdot \underbrace{FF \cdots F}_{n-k \text{ failures}} \qquad (1)$$

where k successes come first, followed by $n - k$ failures. The probability of this outcome is

$$\underbrace{ppp \cdots p}_{k \text{ factors}} \cdot \underbrace{qq \cdots q}_{n-k \text{ factors}} = p^k q^{n-k}$$

The k successes could also be obtained by rearranging the letters in (1) above. The number of such sequences must be equal to the number of ways of choosing k of the n trials to contain successes—namely, $C(n, k) = \binom{n}{k}$. If we multiply this number by the probability of obtaining any one such sequence, we arrive at the following result:

Formula for $b(n, k; p)$

In a Bernoulli trial the probability of exactly k successes in n trials is given by

$$b(n, k; p) = \binom{n}{k} p^k \cdot q^{n-k} = \frac{n!}{k!(n-k)!} p^k \cdot q^{n-k} \qquad (2)$$

where $q = 1 - p$.

The symbol $b(n, k; p)$, which represents the probability of exactly k successes in n trials, is called a **binomial probability.**

Now Work Problem 1

EXAMPLE 1

A common example of a Bernoulli trial is a coin-flipping experiment:

1. There are exactly two possible mutually exclusive outcomes on each trial or toss (heads or tails).
2. The probability of a particular outcome (say, heads) remains constant from trial to trial (toss to toss).
3. The outcome on any trial (toss) is independent of the outcome on any other trial (toss).

We would like to compute the probability of obtaining exactly one tail in six tosses of a fair coin.

SOLUTION

Let T denote the outcome "Tail shows" and let H denote the outcome "Head shows." Using Formula (2), in which $k = 1$, $n = 6$, and $p = \frac{1}{2} = P(T)$, we obtain

$$P(\text{Exactly 1 success}) = b\left(6, 1; \frac{1}{2}\right) = \binom{6}{1}\left(\frac{1}{2}\right)^1 \left(\frac{1}{2}\right)^{6-1} = \frac{6}{64} = .0938$$

Now Work Problem 15

EXAMPLE 2

Baseball   A baseball pitcher gives up a hit on the average of once every fifth pitch. If nine pitches are thrown, what is the probability that

(a) Exactly three of them result in hits?   (b) None of them results in a hit?
(c) No more than seven result in hits?

SOLUTION

In this example $P(\text{Success}) = P(\text{Allowing a hit}) = \frac{1}{5} = .2$. The number of trials n is the number of pitches (9), and k is the number of pitches that result in hits.

(a) $n = 9$, $k = 3$, and $q = 1 - P(\text{Success}) = 1 - .2 = .8$.

$$P(\text{Exactly 3}) = b(9, 3; .2) = \binom{9}{3}(.2)^3(.8)^6 = .1762$$

(b) If none of the nine pitches resulted in a hit, $k = 0$.

$$P(\text{Exactly 0}) = b(9, 0; .2) = \binom{9}{0}(.2)^0(.8)^9 = .1342$$

For Part (c), as in many other applications, it is necessary to compute the probability not of exactly k successes, but of *at least* or *at most* k successes. To obtain such probabilities, we have to compute all the individual probabilities and add them.

(c) "No more than seven pitches result in hits" means 0, 1, 2, 3, 4, 5, 6, or 7 pitches result in a hit. We could add $b(9, 0; .2)$, $b(9, 1; .2)$, and so on, but it is easier to use the formula $P(E) = 1 - P(\overline{E})$. The complement of "No more than 7" is "8 or more."

$$P(\text{No more than 7}) = 1 - P \text{ (8 or more)}$$
$$= 1 - [b(9, 8; .2) + b(9, 9; .2)] = .9999$$

· ·

EXAMPLE 3

Quality Control   A machine produces light bulbs to meet certain specifications, and 80% of the bulbs produced meet these specifications. A sample of six bulbs is taken from the machine's production and placed in a box. What is the probability that three or more of them fail to meet the specifications?

SOLUTION

In this example we are looking for the probability of the event

$$E: \quad \text{At least 3 fail to meet specifications}$$

But this event is just the union of the mutually exclusive events "Exactly 3 failures," "Exactly 4 failures," "Exactly 5 failures," and "Exactly 6 failures." Hence, we use Formula (2) for $n = 6$ and $k = 3, 4, 5,$ and 6. Since the probability of failure is .20, we have

$$P(\text{Exactly 3 failures}) = b(6, 3; .20) = .0819$$
$$P(\text{Exactly 4 failures}) = b(6, 4; .20) = .0154$$
$$P(\text{Exactly 5 failures}) = b(6, 5; .20) = .0015$$
$$P(\text{Exactly 6 failures}) = b(6, 6; .20) = .0001$$

Therefore,

$$P(\text{At least 3 failures}) = P(E) = .0819 + .0154 + .0015 + .0001 = .0989$$

$\bullet$

Another way of getting this answer is to compute the probability of the complementary event

$$\overline{E}: \quad \text{Fewer than 3 failures}$$

Then

$$P(\overline{E}) = P(\text{Exactly 2 failures}) + P(\text{Exactly 1 failure}) + P(\text{Exactly 0 failures})$$
$$= b(6, 2; .20) + b(6, 1; .20) + b(6, 0; .20)$$
$$= .2458 + .3932 + .2621 = .9011$$

As a result,

Now Work Problem 35

$$P(\text{At least 3 failures}) = 1 - P(E) = 1 - .9011 = .0989$$

EXAMPLE 4

Product Testing   A man claims to be able to distinguish between two kinds of wine with 90% accuracy and presents his claim to an agency interested in promoting the consumption of one of the two kinds of wine. The following experiment is conducted to check his claim. The man is to taste the two types of wine and distinguish between them. This is to be done nine times with a 3-minute break after each taste. It is agreed that if the man is correct at least six out of the nine times, he will be hired.

The main questions to be asked are, on the one hand, whether the above procedure gives sufficient protection to the hiring agency against a person guessing and, on the other hand, whether the man is given a sufficient chance to be hired if he is really a wine connoisseur.

SOLUTION

To answer the first question, let's assume that the man is guessing. Then in each trial he has a probability of $\frac{1}{2}$ of identifying the wine correctly. Let k be the number of correct identifications. Let's compute the binomial probability for $k = 6, 7, 8, 9$, to find the likelihood of the man being hired while guessing:

$$b(9, 6; \tfrac{1}{2}) + b(9, 7; \tfrac{1}{2}) + b(9, 8; \tfrac{1}{2}) + b(9, 9; \tfrac{1}{2}) = .1641 + .0703 + .0176 + .0020$$
$$= .2540$$

Thus there is a likelihood of .254 that he will pass if he is just guessing.

To answer the second question in the case where the claim is true, we need to find the sum of the probabilities $b(9, k; .90)$ for $k = 6, 7, 8, 9$:

$$b(9, 6; .90) + b(9, 7; .90) + b(9, 8; .90) + b(9, 9; .90) = .0446 + .1722 + .3874 + .3874$$
$$= .9916$$

$\bullet$

Now Work Problem 43

Notice that the test in Example 4 is fair to the man since it practically assures him the position if his claim is true. However, the company may not like the test because 25% of the time a person who guesses will pass the test.

Application: Testing a Serum or Vaccine

Suppose that the normal rate of infection of a certain disease in cattle is 25%. To test a newly discovered serum, healthy animals are injected with it. How can we evaluate the result of the experiment?

For an absolutely worthless serum, the probability that exactly k of n test animals remain free from infection equals $b(n, k; .75)$. For $k = n = 10$, this probability is about $b(10, 10; .75) = .056$. Thus, if out of 10 test animals none is infected, this may be taken as an indication that the serum has had an effect, although it is not conclusive proof. Notice that, without serum, the probability that out of 17 animals at most 1 catches the infection is $b(17, 0; .25) + b(17, 1; .25) = .0501$. Therefore, there is *stronger evidence* in favor of the serum if out of 17 test animals at most 1 gets infected than if out of 10 all remain healthy. For $n = 23$ the probability of at most 2 animals catching the infection is about .0492 and, thus, at most 2 failures out of 23 is again better evidence for the serum than at most 1 out of 17 or 0 out of 10.

Application: Error Correction in Electronic Transmission of Data

Electronically transmitted data, be it from computer to computer or from a satellite to a ground station, is normally in the form of strings of 0s and 1s—that is, in binary form. Bursts of noise or faults in relays, for example, may at times garble the transmission and produce errors so that the message received is not the same as the one originally sent. For example,

001	⊢—⋀⋀⋀→	101
Message	Noisy	Message
sent	channel	received

is a transmission where the message received is in error since the initial 0 has been changed to a 1.

A naive way of trying to protect against such error would be to repeat the message. So instead of transmitting 001, we would send 001001. Then, were the same error to creep in as it did before, the received message would be

$$101001$$

The receiver would certainly know an error had occurred since the last half of the message is not a duplicate of the first half. But she would have no way of recovering the original message since she would not know where the error happened. For example, she would not be able to distinguish between the two messages

$$001001 \quad \text{and} \quad 101101$$

There are more sophisticated ways of coding binary data with redundancy that not only allow the detection of errors but simultaneously permit their location and correction so that the original message can be recovered. These are referred to as *error-*

correcting codes and are commonly used today in computer-implemented transmissions. One such is the (7, 4) Hamming code named after Richard Hamming, a former researcher at AT&T Laboratories. It is a code of length seven, meaning that an individual message is a string consisting of seven items, each of which is either a 0 or 1. (The four refers to the fact that the first four elements in the string can be freely chosen by the sender, while the remaining three are determined by a fixed rule and constitute the redundancy that gives the code its error correction capability.) The (7, 4) Hamming code is capable of locating and correcting a single error. That is, if during transmission a 1 has been changed to a 0 or vice versa in one of the seven locations, the Hamming code is capable of detecting and correcting this. While we will not explain how or why the Hamming code works, we will analyze the benefit obtained by its use.

By an error we mean that an individual 1 has been changed to a 0 or that a 0 has been changed to a 1. We assume that the probability of an error happening remains constant during transmission, and we designate this probability by q. Thus $p = 1 - q$ is the probability that an individual symbol remains unchanged. This is summarized in the diagram. (In practice, values of q are normally small and values of p close to 1 since we would normally be using a relatively reliable channel.)

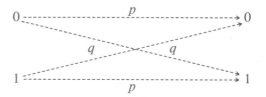

We also assume that errors occur randomly and independently. In short, we can think of the transmission of a binary string of length seven as a Bernoulli trial, with failure corresponding to a symbol being received in error.

For a message of length seven, if *no* coding were used, the probability that the receiver would get the correct message would be

$$b(7, 7; p) = p^7$$

since none of the seven symbols could have been altered. For $p = .98$, this gives $(.98)^7 = .8681$.

Using the Hamming code, the receiver will get the correct message even if one error has occurred. Hence, the corresponding probability of correct reception would be

$$\underbrace{b(7, 7; p)}_{\text{No errors}} + \underbrace{b(7, 6; p)}_{\text{1 error}} = p^7 + 7p^6q$$

For $p = .98$ this now gives $.8681 + .1240 = .9921$, which shows a considerable improvement.

There are codes in use that correct more than a single error. One such code, known by the initials of its originators as a BCH code, is a code of length 15 (messages are binary strings of length 15) that corrects up to two errors. Sending a message of length 15 with no attempt at coding would result in a probability of $b(15, 15; p) = p^{15}$ of the correct message being received. For $p = .98$ this gives $b(15, 15; .98) = .7386$. Using the BCH code that can correct two or fewer errors, the probability that a message will be correctly received becomes

$$\underbrace{b(15, 15; p)}_{\text{No errors}} + \underbrace{b(15, 14: p)}_{\text{1 error}} + \underbrace{b(15, 13; p)}_{\text{2 errors}}$$

Evaluating this for $p = .98$ we get

$$.7386 + .2261 + .0323 = .9970$$

Codes that can correct a high number of errors are clearly very desirable. Yet a basic result in the theory of codes states that as the error-correcting capability of a code increases, so, of necessity, must its length. But lengthier codes require more time for transmission and are clumsier to use. Thus, here, speed and correctness are at odds.

EXERCISE 8.2   Answers to odd-numbered problems begin on page AN-28.

In Problems 1–14 use Formula (2), page 379, and a calculator to compute each binomial probability.

1. $b(7, 5; .30)$ **2.** $b(8, 6; .40)$ **3.** $b(15, 8; .70)$

4. $b(8, 5; .60)$ **5.** $b(15, 10; \frac{1}{2})$ **6.** $b(12, 6; .90)$

7. $b(15, 3; .3) + b(15, 2; .3) + b(15, 1; .3) + b(15, 0; .3)$ **8.** $b(8, 6; .4) + b(8, 7; .4) + b(8, 8; .4)$

9. $n = 3,\ \ k = 2,\ \ p = \frac{1}{3}$ **10.** $n = 3,\ \ k = 1,\ \ p = \frac{1}{3}$ **11.** $n = 3,\ \ k = 0,\ \ p = \frac{1}{6}$

12. $n = 3,\ \ k = 3,\ \ p = \frac{1}{6}$ **13.** $n = 5,\ \ k = 3,\ \ p = \frac{2}{3}$ **14.** $n = 5,\ \ k = 0,\ \ p = \frac{2}{3}$

15. Find the probability of obtaining exactly 6 successes in 10 trials when the probability of success is .3.

16. Find the probability of obtaining exactly 5 successes in 9 trials when the probability of success is .2.

17. Find the probability of obtaining exactly 9 successes in 12 trials when the probability of success is .8.

18. Find the probability of obtaining exactly 8 successes in 15 trials when the probability of success is .75.

19. Find the probability of obtaining at least 5 successes in 8 trials when the probability of success is .25.

20. Find the probability of obtaining at most 3 successes in 7 trials when the probability of success is .15.

In Problems 21–26 a fair coin is tossed 8 times.

21. What is the probability of obtaining exactly 1 head?

22. What is the probability of obtaining exactly 2 heads?

23. What is the probability of obtaining at least 5 tails?

24. What is the probability of obtaining at most 2 tails?

25. What is the probability of obtaining exactly 2 heads if it is known that at least 1 head appeared?

26. What is the probability of obtaining exactly 3 heads if it is known that at least 1 head appeared?

27. What is the probability of obtaining 7s exactly 2 times in 5 rolls of 2 fair dice?

28. What is the probability of obtaining 11s exactly 3 times in 7 rolls of 2 fair dice?

29. Quality Control Suppose that 5% of the items produced by a factory are defective. If 8 items are chosen at random, what is the probability that

(a) Exactly 1 is defective?
(b) Exactly 2 are defective?

(c) At least 1 is defective?

(d) Fewer than 3 are defective?

30. Opinion Poll   Suppose that 60% of the voters intend to vote for a conservative candidate. What is the probability that a survey polling 8 people reveals that 3 or fewer intend to vote for a conservative candidate?

31. Family Structure   What is the probability that a family with exactly 6 children will have 3 boys and 3 girls?

32. Family Structure   What is the probability that in a family of 7 children:

(a) 4 are girls?

(b) At least 2 are girls?

(c) At least 2 and not more than 4 are girls?

33. An experiment is performed 4 times, with 2 possible outcomes, F (failure) and S (success), with probabilities $\frac{1}{4}$ and $\frac{3}{4}$, respectively.

(a) Draw a tree diagram describing the experiment.

(b) Calculate the probability of exactly 2 successes and 2 failures by using the tree diagram from Part (a).

(c) Verify your answer to Part (b) by using Formula (2).

34. Batting Averages   For a baseball player with a .250 batting average, what is the probability that the player will have at least 2 hits in 4 times at bat? What is the probability of at least 1 hit in 4 times at bat?

35. Target Shooting   If the probability of hitting a target is $\frac{1}{5}$ and 10 shots are fired independently, what is the probability of the target being hit at least twice?

36. Quality Control   A television manufacturer tests a random sample of 15 picture tubes to determine whether any are defective. The probability that a picture tube is defective has been found from past experience to be .05.

(a) What is the probability that there are no defective tubes in the sample?

(b) What is the probability that more than 2 of the tubes are defective?

37. True–False Tests   In a 15-item true–false examination, what is the probability that a student who guesses on each question will get at least 10 correct answers? If another student has .8 probability of correctly answering each question, what is the probability that this student will answer at least 12 questions correctly?

38. To screen prospective employees, a company gives a 10-question multiple-choice test. Each question has 4 possible answers, of which 1 is correct. The chance of

answering the questions correctly by just guessing is $\frac{1}{4}$ or 25%. Find the probability of answering, by chance:

(a) Exactly 3 questions correctly

(b) No questions correctly

(c) At least 8 questions correctly

(d) No more than 7 questions correctly

39. Opinion Poll   Mr. Austin and Ms. Moran are running for public office. A survey conducted just before the day of election indicates that 60% of the voters prefer Mr. Austin and 40% prefer Ms. Moran. If 8 people are chosen at random and asked their preference, find the probability that all 8 people will express a preference for Ms. Moran.

40. Working Habits   If 40% of the workers at a large factory bring their lunch each day, what is the probability that in a randomly selected sample of 8 workers

(a) Exactly 2 bring their lunch each day?

(b) At least 2 bring their lunch each day?

(c) No one brings lunch?

(d) No more than 3 bring lunch each day?

41. Heart Attack   Approximately 23% of North American deaths are due to heart attacks. What is the probability that 4 of the next 10 unrelated deaths reported in a certain community will be due to heart attacks?

42. Opinion Polls   Suppose that 60% of the voters intend to vote for a conservative candidate. What is the probability that a survey polling 8 people reveals that 3 or fewer intend to vote for a conservative candidate?

43. Product Testing   A supposed coffee connoisseur claims she can distinguish between a cup of instant coffee and a cup of drip coffee 75% of the time. You give her 6 cups of coffee and tell her that you will grant her claim if she correctly identifies at least 5 of the 6 cups.

(a) What are her chances of having her claim granted if she is in fact only guessing?

(b) What are her chances of having her claim rejected when in fact she really does have the ability she claims?

44. Opinion Poll   Opinion polls based on small samples often yield misleading results. Suppose 65% of the people in a city are opposed to a bond issue and the others favor it. If 7 people are asked for their opinion, what is the probability that a majority of them will favor the bond issue?

45. Hamming Code   There is a Hamming code of length 15 that corrects a single error. Assuming $p = .98$, find the probability that a message transmitted using this code will be correctly received.

46. Golay Code   There is a binary code of length 23 (called the Golay code) that can correct up to 3 errors. With $p = .98$, find the probability that a message transmitted using the Golay code will be correctly received.

47. What is the probability that the birthdays of 6 people fall in 2 calendar months, leaving exactly 10 months free? (Assume independent and equal probabilities for all months.)

48. A book of 500 pages contains 500 misprints. Estimate the chance that a given page contains at least 3 misprints.

Technology Exercises

Most graphing calculators have a random number function (usually RAND or RND) generating numbers between 0 and 1. Check your user's manual to see how to use this function on your graphing calculator.

Sometimes experiments are simulated using a random number function instead of actually performing the experiment. In Problems 1–4 use a random number function to simulate each experiment.

1. Tossing a Fair Coin   Consider the experiment of tossing a coin 4 times, counting the number of heads occurring in these 4 tosses. Simulate the experiment using a random number function on your calculator, considering a toss to be tails (T) if the result is less than 0.5, and considering a toss to be heads (H) if the result is more than or equal to 0.5. Record the number of heads in 4 tosses. [*Note:* Most calculators repeat the action of the last entry if you simply press the ENTER, or EXE, key again.] Repeat the experiment 10 times, thus obtaining a sequence of 10 numbers. Using these 10 numbers you can estimate the probability of k heads, $P(k)$, for each $k = 0, 1, 2, 3, 4$, by the ratio

$$\frac{\text{Number of times } k \text{ appears in your sequence}}{10}$$

Enter your estimates in the table below. Calculate the actual probabilities using the binomial probability function, and enter these numbers in the table. How close are your numbers to the actual values?

k	Your Estimate of $P(k)$	Actual Value of $P(k)$
0		
1		
2		
3		
4		

2. Tossing a Loaded Coin   Consider the experiment of tossing a loaded coin 4 times, counting the number of heads occurring in these 4 tosses. Simulate the experiment using a random number function on your calculator, con-sidering a toss to be tails (T) if the result is less than 0.80, and considering a toss to be heads (H) if the result is more than or equal to 0.80. Record the number of heads in 4 tosses. Repeat the experiment 10 times, thus obtaining a sequence of 10 numbers. Using these 10 numbers you can estimate the probability of k heads, $P(k)$, for each $k = 0, 1, 2, 3, 4$, by the ratio

$$\frac{\text{Number of times } k \text{ appears in your sequence}}{10}$$

Enter your estimates in the table below. Calculate the actual probabilities using the binomial probability function, and enter these numbers in the table. How close are your numbers to the actual values? Why is this coin loaded?

k	Your Estimate of $P(k)$	Actual Value of $P(k)$
0		
1		
2		
3		
4		

3. Tossing a Fair Coin   Consider the experiment of tossing a fair coin 8 times, counting the number of heads occur-ring in these 8 tosses. Simulate the experiment using a random number function on your calculator, considering a toss to be tails (T) if the result is less than 0.50, and consid-ering a toss to be heads (H) if the result is more than or equal to 0.50. Record the number of heads in 8 tosses. Repeat the experiment 10 times, thus obtaining a sequence of 10 numbers. Using these 10 numbers you can estimate the probability of 3 heads, $P(3)$, by the ratio

$$\frac{\text{Number of times 3 appears in your sequence}}{10}$$

Calculate the actual probability using the binomial probability function. How close is your estimate to the actual value?

4. **Tossing a Loaded Coin** Consider the experiment of tossing a loaded coin 8 times, counting the number of heads occurring in these 8 tosses. Simulate the experiment using a random number function on your calculator, consid-

ering a toss to be tails (T) if the result is less than 0.80, and considering a toss to be heads (H) if the result is more than or equal to 0.80. Record the number of heads in 8 tosses. Repeat the experiment 10 times, thus obtaining a sequence of 10 numbers. Using these 10 numbers you can estimate the probability of 3 heads, $P(3)$, by the ratio

$$\frac{\text{Number of times 3 appears in your sequence}}{10}$$

Calculate the actual probability using the binomial probability function. How close is your estimate to the actual value?

8.3 EXPECTED VALUE

An important concept, which has its origin in gambling and which uses probability, is *expected value*. For instance, gamblers are quite concerned with the *expectation,* or *expected value,* of a game. Suppose, for example, that 1000 tickets are printed and raffled off. Out of all these tickets 1 ticket has a cash value of $300, 2 are worth $100, and 100 are worth $1, while the remaining are worth $0. The *expected* (average) *value* of a ticket is then

$$E = \frac{\$300 + (\$100 + \$100) + \overbrace{(\$1 + \$1 + \cdots + \$1)}^{100 \text{ times}} + \overbrace{(\$0 + \$0 + \cdots + \$0)}^{897 \text{ times}}}{1000}$$

$$= \$300 \cdot \frac{1}{1000} + \$100 \cdot \frac{2}{1000} + \$1 \cdot \frac{100}{1000} + \$0 \cdot \frac{897}{1000}$$

$$= \frac{\$600}{\$1000} = \$0.60$$

Thus if the raffle is to be nonprofit to all, the charge for each ticket should be $0.60. The result can also be viewed as saying that if we entered such a raffle many times, $\frac{1}{1000}$ of the time we would win $300, $\frac{2}{1000}$ of the time we would win $100, and so on, with our winnings in the long run averaging $0.60 per ticket.

As another example of expectation, suppose that you are to receive $3.00 each time you obtain two heads when flipping a coin two times and $0 otherwise. Then the *expected value* is

$$E = \$3.00 \cdot \tfrac{1}{4} + \$0 \cdot \tfrac{3}{4} = \$0.75$$

This means that you should be willing to pay $0.75 each time you toss the coins if the game is to be a fair one. We arrive at the expected value E by multiplying the amount earned for a given result of the toss times the probability for that toss to occur, and adding all the products.

Another example is a game consisting of flipping a single coin. If a head shows, the player loses $1; but if a tail shows, the player wins $2. Thus half the time the player loses $1 and the other half the player wins $2. The expected value E of the game is

$$E = \$2 \cdot \tfrac{1}{2} + (-\$1) \cdot \tfrac{1}{2} = \tfrac{1}{2} = \$0.50$$

The player is expected to win an average of $0.50 on each play.

In each of the above examples we have dealt with experiments involving payoffs (numerical values) and their corresponding probabilities. In the expression for the expected value in the raffle problem, the term $\$300 \cdot \tfrac{1}{1000}$ is the pairing of the value $300 with its corresponding probability $\tfrac{1}{1000}$, namely, the probability of picking a $300 ticket. Likewise, the probability of getting a $1 ticket is $\tfrac{100}{1000}$ and this produces the term $\$1 \cdot \tfrac{100}{1000}$ in the expression for the expected value, E. So the expression for E is obtained by multiplying each ticket value by the probability of its occurrence and adding the results.

In the second example, the term $\$3.00 \cdot \tfrac{1}{4}$ is the pairing of the value $3.00 with its corresponding probability $\tfrac{1}{4}$, namely, the probability of obtaining HH. Likewise, the probability of getting HT, TH, and TT is $\tfrac{3}{4}$ with payoff 0, and this produces $\$0 \cdot \tfrac{3}{4}$ in the expression for the expected value.

Finally, in the last example the term $\$2 \cdot \tfrac{1}{2}$ is the pairing of the value $2.00 with its corresponding probability $\tfrac{1}{2}$, namely, the probability of obtaining T. Likewise, the probability of getting H is $\tfrac{1}{2}$ with payoff $-\$1$.

This leads to the following definition.

EXPECTED VALUE*

If an experiment has n partitions that are assigned the payoffs $m_1, m_2, \ldots, m_n$ occurring with probabilities $p_1, p_2, \ldots, p_n$, respectively, then the *expected value* E is given by

$$E = m_1 \cdot p_1 + m_2 \cdot p_2 + \cdots + m_n \cdot p_n$$

The term *expected value* should not be interpreted as a value that actually occurs in the experiment. For example, there was no raffle ticket costing $.60. Rather, it represents the average value per experiment were we to repeat the experiment many times.

In gambling, for instance, E is interpreted as the average winnings expected for the player in the long run. If E is positive, we say that the game is **favorable** to the player; if $E = 0$, we say the game is **fair**; and if E is negative, we say the game is **unfavorable** to the player.

When the payoff assigned to an outcome of an experiment is positive, it can be interpreted as a profit, winnings, gain, etc. When it is negative, it represents losses, penalties, deficits, etc.

The following steps outline the general procedure involved in determining the expected value.

* The assignment of numerical values to outcomes of an experiment is called a *random variable* and E is called the expected value or mean of the random variable. See Section 8.5.

> **Steps for Computing Expected Value**
>
> **Step 1** Partition the sample space S, which describes the possible outcomes when the experiment is performed.
>
> **Step 2** Determine the payoff values $m_1, m_2, \ldots, m_n$.
>
> **Step 3** Determine $p_1, p_2, \ldots, p_n$ for each payoff $m_1, m_2, \ldots, m_n$.
>
> **Step 4** Calculate $E = m_1 \cdot p_1 + m_2 \cdot p_2 + \cdots + m_n \cdot p_n$.

Now Work Problem 1

EXAMPLE 1

What is the expected number of heads in tossing a fair coin three times?

SOLUTION

Step 1 The sample space is

$$S = \{HHH, HHT, HTH, HTT, THH, THT, TTH, TTT\}$$

Each outcome is assigned the probability $\frac{1}{8}$. A coin tossed three times can have 0, 1, 2, or 3 heads, so we partition S into the four events: 0 heads, 1 head, 2 heads, 3 heads.

Step 2 A coin tossed three times can have 0, 1, 2, or 3 heads. We treat these values as payoffs.

Step 3 The probabilities corresponding to 0, 1, 2, or 3 heads in Step 2 are $\frac{1}{8}, \frac{3}{8}, \frac{3}{8}$, and $\frac{1}{8}$, respectively.

Step 4 The expected number of heads can now be found by multiplying each payoff by its corresponding probability and finding the sum of these values.

$$\text{Expected number of heads} = E = 0 \cdot \tfrac{1}{8} + 1 \cdot \tfrac{3}{8} + 2 \cdot \tfrac{3}{8} + 3 \cdot \tfrac{1}{8} = \tfrac{3}{2}$$
$$= 1.5$$

Thus, on the average, tossing a coin three times will result in 1.5 heads.

. . . .

EXAMPLE 2

Consider the experiment of rolling a fair die. The player recovers an amount of dollars equal to the number of dots on the face that turns up, except when face 5 or 6 turns up, in which case the player will lose $5 or $6, respectively. What is the expected value of the game?

SOLUTION

The sample space is $S = \{1, 2, 3, 4, 5, 6\}$. Since all six faces are equally likely to occur, we assign a probability of $\frac{1}{6}$ to each of them. The payoffs for the outcomes 1, 2, 3, 4, 5, 6 are, respectively, $1, $2, $3, $4, $- $5, $- $6. Steps 1, 2, and 3 are summarized in the table at the top of the next page.

Outcome	1	2	3	4	5	6
Probability	$\frac{1}{6}$	$\frac{1}{6}$	$\frac{1}{6}$	$\frac{1}{6}$	$\frac{1}{6}$	$\frac{1}{6}$
Payoff	\$1	\$2	\$3	\$4	− \$5	− \$6

Step 4 The expected value of the game is

$$E = \$1 \cdot \tfrac{1}{6} + \$2 \cdot \tfrac{1}{6} + \$3 \cdot \tfrac{1}{6} + \$4 \cdot \tfrac{1}{6} + (-\$5) \cdot \tfrac{1}{6} + (-\$6) \cdot \tfrac{1}{6}$$
$$= -\$\tfrac{1}{6} = -16.7¢$$

The player would expect to lose an average of 16.7¢ on each throw.

Now Work Problem 11

· ·

EXAMPLE 3

Bidding for Oil Wells An oil company may bid for only one of two contracts for oil drilling in two different locations. If oil is discovered at location I, the profit to the company will be \$3,000,000. If no oil is found, the company's loss will be \$250,000. If oil is discovered at location II, the profit will be \$4,000,000. If no oil is found, the loss will be \$500,000. The probability of discovering oil at location I is .7, and at location II it is .6. Which field should the company bid on; that is, for which location is expected profit highest?

SOLUTION

In the first field the company expects to discover oil .7 of the time at a profit of \$3,000,000. Thus it would not discover oil .3 of the time at a loss of \$250,000. The expected profit E_I is therefore

$$E_I = (\$3,000,000)(.7) + (-\$250,000)(.3) = \$2,025,000$$

Similarly, for the second field, the expected profit E_{II} is

$$E_{II} = (\$4,000,000)(.6) + (-\$500,000)(.4) = \$2,200,000$$

Since the expected profit for the second field exceeds that for the first, the oil company should bid on the second field.

Now Work Problem 19

· ·

EXAMPLE 4

Evaluating Insurance Mr. Richmond is producing an outdoor concert. He estimates that he will make \$300,000 if it does not rain and make \$60,000 if it does rain. The weather bureau predicts that the chance of rain is .34 for the day of the concert.

(a) What are Mr. Richmond's expected earnings from the concert?
(b) An insurance company is willing to insure the concert for \$150,000 against rain for a premium of \$30,000. If he buys this policy, what are his expected earnings from the concert?
(c) Based on the expected earnings, should Mr. Richmond buy an insurance policy?

SOLUTION

The sample space consists of two outcomes: R: Rain or N: No rain. The probability of rain is $P(R) = .34$; the probability of no rain is $P(N) = .66$.

(a) If it rains, the earnings are $60,000; if there is no rain, the earnings are $300,000. The expected earnings E are

$$E = 60,000 \cdot P(R) + 300,000 \cdot P(N)$$
$$= 60,000 \,(.34) + 300,000 \,(.66)$$
$$= \$218,400$$

(b) With insurance, if it rains, the earnings are

$$\underset{\substack{\text{Insurance} \\ \text{payout}}}{150,000} + \underset{\substack{\text{Concert} \\ \text{profit}}}{60,000} - \underset{\substack{\text{Cost of} \\ \text{insurance}}}{30,000} = 180,000$$

With insurance, if it does not rain, the earnings are

$$\underset{\substack{\text{No insurance} \\ \text{payout}}}{0} + \underset{\substack{\text{Concert} \\ \text{profit}}}{300,000} - \underset{\substack{\text{Cost of} \\ \text{insurance}}}{30,000} = 270,000$$

With insurance, the expected earnings E are

$$E = 180,000 \cdot P(R) + 270,000 \cdot P(N)$$
$$= 180,000 \,(.34) + 270,000 \,(.66)$$
$$= \$239,400$$

(c) Since the expected earnings are higher with insurance, it would be better to obtain the insurance.

....

EXAMPLE 5

Quality Control   A laboratory contains 10 electron microscopes, of which 2 are defective. If all microscopes are equally likely to be chosen and if 4 are chosen, what is the expected number of defective microscopes?

SOLUTION

The sample of 4 microscopes can contain 0, 1, or 2 defective microscopes. The probability p_0 that none in the sample is defective is

$$p_0 = \frac{C(2, 0) \cdot C(8, 4)}{C(10, 4)} = \frac{1}{3}$$

Similarly, the probabilities p_1 and p_2 for 1 and 2 defective microscopes are

$$p_1 = \frac{C(2, 1) \cdot C(8, 3)}{C(10, 4)} = \frac{8}{15} \qquad \text{and} \qquad p_2 = \frac{C(2, 2) \cdot C(8, 2)}{C(10, 4)} = \frac{2}{15}$$

Since we are interested in determining the expected number of defective microscopes, we assign a payoff of 0 to the outcome "0 defectives are selected," a payoff of 1 to the outcome "1 defective is chosen," and a payoff of 2 to the outcome "2 defectives are chosen." The expected value E is then

$$E = 0 \cdot p_0 + 1 \cdot p_1 + 2 \cdot p_2 = \tfrac{8}{15} + \tfrac{4}{15} = \tfrac{4}{5}$$

Of course, we cannot have $\tfrac{4}{5}$ of a defective microscope. However, we can interpret this to mean that in the long run such a sample will average just under 1 defective microscope.

....

We point out that $\frac{4}{5}$ is a reasonable answer for the expected number of defective microscopes since $\frac{1}{5}$ of the microscopes in the laboratory are defective and we are selecting a random sample consisting of 4 of these microscopes.

Expected Value for Bernoulli Trials

In 100 tosses of a coin, what is the expected number of heads? If a student guesses at random on a true–false exam with 50 questions, what is her expected grade? These are specific instances of the following more general question:

In n trials of a Bernoulli process, what is the expected number of successes?

We now compute this expected value. As before, p denotes the probability of success on any individual trial.

If $n = 1$ (there is but one trial), then the expected number of successes is

$$E = 1 \cdot p + 0 \cdot (1 - p) = p$$

If $n = 2$ (two trials), then either 0, 1, or 2 successes can occur. The expected number E of successes is

$$E = 2 \cdot p^2 + 1 \cdot 2p(1 - p) + 0 \cdot (1 - p)^2 = 2p$$

If $n = 3$ (three trials), then either 0, 1, 2, or 3 successes can occur. The expected number E of successes is

$$\begin{aligned} E &= 3 \cdot p^3 + 2 \cdot 3p^2(1 - p) + 1 \cdot 3p(1 - p)^2 + 0 \cdot (1 - p)^3 \\ &= 3p^3 + 6p^2 - 6p^3 + 3p - 6p^2 + 3p^3 \\ &= 3p \end{aligned}$$

This would seem to suggest that with n trials the expected value E would be given by $E = np$. This is indeed the case and we have the following result:

Expected Value for Bernoulli Trials

In a Bernoulli process with n trials the expected number of successes is

$$E = np$$

where p is the probability of success on any single trial.

A derivation of this result is included a little later in this section. The intuitive idea behind the result is fairly simple. Thinking of probabilities as percentages, if success results p percent of the time, then out of n attempts, p percent of them, namely, np, should be successful.

EXAMPLE 6 | In flipping a fair coin five times, there are six possibilities: 0 tails, 1 tail, 2 tails, 3 tails, 4 tails, or 5 tails, each with the respective probabilities

$$\binom{5}{0}\left(\frac{1}{2}\right)^5 \quad \binom{5}{1}\left(\frac{1}{2}\right)^5 \quad \binom{5}{2}\left(\frac{1}{2}\right)^5 \quad \binom{5}{3}\left(\frac{1}{2}\right)^5 \quad \binom{5}{4}\left(\frac{1}{2}\right)^5 \quad \binom{5}{5}\left(\frac{1}{2}\right)^5$$

The expected number of tails is

$$E = 0 \cdot \binom{5}{0}\left(\frac{1}{2}\right)^5 + 1 \cdot \binom{5}{1}\left(\frac{1}{2}\right)^5 + 2 \cdot \binom{5}{2}\left(\frac{1}{2}\right)^5$$
$$+ 3 \cdot \binom{5}{3}\left(\frac{1}{2}\right)^5 + 4 \cdot \binom{5}{4}\left(\frac{1}{2}\right)^5 + 5 \cdot \binom{5}{5}\left(\frac{1}{2}\right)^5 = \frac{5}{2}$$

Clearly, using the result $E = np$ is much easier since for $n = 5$ and $p = \frac{1}{2}$, we obtain $E = (5)(\frac{1}{2}) = \frac{5}{2}$.

. . . .

EXAMPLE 7

A multiple-choice test contains 100 questions, each with four choices. If a person guesses, what is the expected number of correct answers?

SOLUTION

This is an example of a Bernoulli trial. The probability for success (a correct answer) when guessing is $p = \frac{1}{4}$. Since there are $n = 100$ questions, the expected number of correct answers is

$$E = np = (100)(\tfrac{1}{4}) = 25$$

Now Work Problem 21

. . . .

Derivation of $E = np$

The derivation is an exercise in handling binomial coefficients and using the binomial theorem. We will make use of the following identity:

$$k\binom{n}{k} = n\binom{n-1}{k-1} \tag{1}$$

This can be established by expanding both sides using the definition of a binomial coefficient. (See Problem 21, Exercise 6.6, page 309.)

Recall that the probability of obtaining exactly k successes in n trials is given by $b(n, k; p) = \binom{n}{k}p^k q^{n-k}$. Thus the expected number of successes is

$$E = 0 \cdot \binom{n}{0}p^0 q^n + 1 \cdot \binom{n}{1}p^1 q^{n-1} + 2 \cdot \binom{n}{2}p^2 q^{n-2}$$
$$+ \cdots + k\underbrace{\binom{n}{k}p^k q^{n-k}}_{\text{}} + \cdots + n\binom{n}{n}p^n$$

No. of   Corresponding
successes   probability

Using (1) above,

$$E = n \begin{pmatrix} n-1 \\ 0 \end{pmatrix} pq^{n-1} + n \begin{pmatrix} n-1 \\ 1 \end{pmatrix} p^2 q^{n-2} + \cdots$$

$$+ n \begin{pmatrix} n-1 \\ k-1 \end{pmatrix} p^k q^{n-k} + \cdots + n \begin{pmatrix} n-1 \\ n-1 \end{pmatrix} p^n$$

Now factor out an n and a p from each of the terms on the right to get

$$E = np \left[\begin{pmatrix} n-1 \\ 0 \end{pmatrix} q^{n-1} + \begin{pmatrix} n-1 \\ 1 \end{pmatrix} pq^{n-2} + \cdots \right.$$

$$\left. + \begin{pmatrix} n-1 \\ k-1 \end{pmatrix} p^{k-1} q^{(n-1)-(k-1)} + \cdots + \begin{pmatrix} n-1 \\ n-1 \end{pmatrix} p^{n-1} \right]$$

The expression in brackets is $(q + p)^{n-1}$. To see why, use the binomial theorem. Thus

$$E = np(q + p)^{n-1}$$

Since $p + q = 1$, the result $E = np$ follows.

EXERCISE 8.3   Answers to odd-numbered problems begin on page AN-28.

1. For the data given below, compute the expected value.

Outcome	e_1	e_2	e_3	e_4
Probability	.4	.2	.1	.3
Payoff	2	3	−2	0

2. For the data below, compute the expected value.

Outcome	e_1	e_2	e_3	e_4
Probability	$\frac{1}{3}$	$\frac{1}{6}$	$\frac{1}{4}$	$\frac{1}{4}$
Payoff	1	0	4	−2

3. Attendance at a football game in a certain city results in the following pattern. If it is extremely cold, the attendance will be 35,000; if it is cold, it will be 40,000; if it is moderate, 48,000; and if it is warm, 60,000. If the probabilities for extremely cold, cold, moderate, and warm are .08, .42, .42, and .08, respectively, how many fans are expected to attend the game?

4. A player rolls a fair die and receives a number of dollars equal to the number of dots appearing on the face of the die. What is the least the player should expect to pay in order to play the game?

5. Mary will win $8 if she draws an ace from a set of 10 different cards from ace to 10. How much should she pay for one draw?

6. Thirteen playing cards, ace through king, are placed randomly with faces down on a table. The prize for guessing correctly the value of any given card is $1. What would be a fair price to pay for a guess?

7. David gets $10 if he throws a double on a single throw of a pair of dice. How much should he pay for a throw?

8. You pay $1 to toss 2 coins. If you toss 2 heads, you get $2 (including your $1); if you toss only 1 head, you get back your $1; and if you toss no heads, you lose your $1. Is this a fair game to play?

9. **Raffles** In a raffle 1000 tickets are being sold at $.60 each. The first prize is $100, and there are 3 second prizes of $50 each. By how much does the price of a ticket exceed its expected value?

10. **Raffles** In a raffle 1000 tickets are being sold at $.60 each. The first prize is $100. There are 2 second prizes of $50 each, and 5 third prizes of $10 each (there are 8 prizes in all). Jenny buys 1 ticket. How much more than the expected value of the ticket does she pay?

11. A fair coin is tossed 3 times, and a player wins $3 if 3 tails occur, wins $2 if 2 tails occur, and loses $3 if no tails occur. If 1 tail occurs, no one wins.

   (a) What is the expected value of the game?
   (b) Is the game fair?
   (c) If the answer to part (b) is "No," how much should the player win or lose for a toss of exactly 1 tail to make the game fair?

12. Colleen bets $1 on a 2-digit number. She wins $75 if she draws her number from the set of all 2-digit numbers, {00, 01, 02, . . . , 99}; otherwise, she loses her $1.

   (a) Is this game fair to the player?
   (b) How much is Colleen expected to lose in a game?

13. Two teams have played each other 14 times. Team A won 9 games, and team B won 5 games. They will play again next week. Bob offers to bet $6 on team A while you bet $4 on team B. The winner gets the $10. Is the bet fair to you in view of the past records of the two teams? Explain your answer.

14. A department store wants to sell 11 purses that cost the store $41 each and 32 purses that cost the store $9 each. If all purses are wrapped in 43 identical boxes and if each customer picks a box randomly, find

   (a) Each customer's expectation.
   (b) The department store's expected profit if it charges $13 for each box.

15. Sarah draws a card from a deck of 52 cards. She receives 40¢ for a heart, 50¢ for an ace, and 90¢ for the ace of hearts. If the cost of a draw is 15¢, should she play the game? Explain.

16. **Family Size** The following data give information about family size in the United States for a household containing a husband and wife where the husband is in the 30–34 age bracket:

Number of Children	0	1	2	3
Proportion of Families	10.2%	15.9%	31.8%	42.1%

A family is chosen at random. Find the expected number of children in the family.

17. Assume that the odds for a certain race horse to win are 7 to 5. If a bettor wins $5 when the horse wins, how much should he bet to make the game fair?

18. **Roulette** In roulette there are 38 equally likely possibilities: the numbers 1–36, 0, and 00 (double zero). See the figure. What is the expected value for a gambler who bets $1 on number 15 if she wins $35 each time the number 15 turns up and loses $1 if any other number turns up? If the gambler plays the number 15 for 200 consecutive times, what is the total expected gain?

19. **Site Selection** A company operating a chain of supermarkets plans to open a new store in 1 of 2 locations. They conducted a survey of the 2 locations and estimated that the first location will show an annual profit of $15,000 if it is successful and a $3000 loss otherwise. For the second location, the estimated annual profit is $20,000 if successful and a $6000 loss otherwise. The probability of success at each location is $\frac{1}{2}$. What location should the management decide on in order to maximize its expected profit?

20. For Problem 19 assume the probability of success at the first location is $\frac{2}{3}$ and at the second location is $\frac{1}{3}$. What location should be chosen?

21. Find the number of times the face 5 is expected to occur in a sequence of 2000 throws of a fair die.

22. What is the expected number of tails that will turn up if a fair coin is tossed 582 times?

23. A certain kind of light bulb has been found to have a .02 probability of being defective. A shop owner receives 500 light bulbs of this kind. How many of these bulbs are expected to be defective?

24. A student enrolled in a math course has a .9 probability of passing the course. In a class of 20 students, how many would you expect to fail the math course?

25. Drug Reaction A doctor has found that the probability that a patient who is given a certain drug will have unfavorable reactions to the drug is .002. If a group of 500 patients is going to be given the drug, how many of them does the doctor expect to have unfavorable reactions?

26. A true–false test consisting of 30 questions is scored by subtracting the number of wrong answers from the number of right ones. Find the expected number of correct answers of a student who just guesses on each question. What will the test score be?

27. A coin is weighted so that $P(H) = \frac{1}{4}$ and $P(T) = \frac{3}{4}$. Find the expected number of tosses of the coin required in order to obtain either a head or 4 tails.

28. A box contains 3 defective bulbs and 9 good bulbs. If 5 bulbs are drawn from the box without replacement, what is the expected number of defective bulbs?

29. Aircraft Use An airline must decide which of two aircraft it will use on a flight from New York to Los Angeles. Aircraft A has a seating capacity of 200, while aircraft B has a capacity of 300. Previous experience has

allowed the airline to estimate the number of passengers on the flight as follows:

Number of Passengers	150	180	200	250	300
Probability	.2	.3	.2	.2	.1

Regardless of aircraft used, the average cost of a ticket is $500, but there are different operating costs attached to each aircraft. There is a fixed cost (fuel, crew, etc.) of $16,000 attached to using aircraft A, while aircraft B has a fixed cost of $18,000. There is also a per passenger cost (meals, luggage, added fuel) of $200 for aircraft A and $230 for aircraft B. Which aircraft should the airline schedule so that it maximizes its profit on the flight?

30. Prove that if the numerical values assigned to the outcomes of an experiment that has expected value E are all multiplied by the constant k, then the expected value of the new experiment is $k \cdot E$. Similarly, if to all the numerical values we add the same constant k, prove that the expected value of the new experiment is $E + k$.

8.4 APPLICATIONS TO OPERATIONS RESEARCH

The field of **operations research,** the science of making optimal or best decisions, has experienced remarkable growth and development since the 1940s. Let's work some examples from operations research that utilize expected value.

EXAMPLE 1 | **Car Rentals** A national car rental agency rents cars for $16 per day (gasoline and mileage are additional expenses to the customer). The daily cost per car (for example, lease costs and overhead) is $6 per day. The daily profit to the company is $10 per car if the car is rented, and the company incurs a daily loss of $6 per car if the car is not rented. The daily profit depends on two factors: the demand for cars and the number of cars the company has available to rent. Previous rental records show that the daily demand is as given in the table:

Number of Customers	8	9	10	11	12
Probability	.10	.10	.30	.30	.20

Find the expected number of customers and determine the optimal number of cars the company should have available for rental. (This is the number that yields the largest expected profit.)

SOLUTION

The expected number of customers is

$$8(.1) + 9(.1) + 10(.3) + 11(.3) + 12(.2) = 10.4$$

If 10.4 customers are expected, how many cars should be on hand? Surely, the number should probably not exceed 11 since fewer than 11 customers are expected. However, the number may not be the integer closest to 10.4 since costs play a major role in the determination of profit. We need to compute the expected profit for each possible number of cars. The largest expected profit will tell us how many cars to have on hand.

For example, if there are 10 cars available, the expected profit for 8, 9, or 10 customers is

$$68(.1) + 84(.1) + 100(.8) = \$95.20$$

We obtain the entry 68(.1) by noting that the 10 cars cost the company $60, and 8 cars rented with probability .10 bring in $128, for a profit of $68. Similarly, we obtain the entry 84(.1) by noting that the 10 cars cost the company $60, and 9 cars rented with probability .10 bring in $144, for a profit of $84. The entry 100(.8) is obtained since for 10 or more customers (probability $.3 + .3 + .2 = .8$) the profit is $10 \times \$16 - \$60 = \$100$.

The table lists the expected profit for 8 to 12 available cars. Clearly, the optimal stock size is 11 cars, since this number of cars maximizes expected profit.

Number of Available Cars	8	9	10	11	12
Expected Profit	$80.00	$88.40	$95.20	$97.20	$94.40

. . . .

EXAMPLE 2

Figure 12

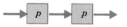

Figure 13

Quality Control   A factory produces electronic components, and each component must be tested. If the component is good, it will allow the passage of current; if the component is defective, it will block the passage of current. Let p denote the probability that a component is good. See Figure 12. With this system of testing, a large number of components requires an equal number of tests. This increases the production cost of the electronic components since it requires one test per component. To reduce the number of tests, a quality control engineer proposes, instead, a new testing procedure: Connect the components pairwise in series, as shown in Figure 13.

If the current passes two components in series, then both components are good and only one test is required. The probability that two components are good is p^2. If the current does not pass, the components must be sent individually to the quality control department, where each component is tested separately. In this case three tests are required. The probability that three tests are needed is $1 - p^2$ (1 minus probability of success p^2). The expected number of tests for a pair of components is

$$E = 1 \cdot p^2 + 3 \cdot (1 - p^2) = p^2 + 3 - 3p^2 = 3 - 2p^2$$

The number of tests saved for a pair is

$$2 - (3 - 2p^2) = 2p^2 - 1$$

The number of tests saved per component is

$$\frac{2p^2 - 1}{2} = p^2 - \frac{1}{2} \text{ tests}$$

The greater the probability p that the component is good, the greater the saving. For example, if p is almost 1, we have a saving of almost $1 - \frac{1}{2}$ or $\frac{1}{2}$, which is 50% of the original number of tests needed. Of course, if p is small, say, less than .7, we do not save anything since $(.7)^2 - \frac{1}{2}$ is less than 0, and we are wasting tests.

· ·

If the reliability of the components manufactured in Example 2 is very high, it might even be advisable to make larger groups. Suppose three components are connected in series. See Figure 14.

Figure 14

For individual testing we need three tests. For group testing we have

1   test needed with probability p^3
4   tests needed with probability $1 - p^3$

The expected number of tests is

$$E = 1 \cdot p^3 + 4 \cdot (1 - p^3) = 4 - 3p^3 \text{ tests}$$

The number of tests saved per component is

$$\frac{3p^3 - 1}{3} = p^3 - \frac{1}{3} \text{ tests}$$

In a similar way, we can show that if the components are arranged in groups of four connected in series, then the number of tests saved per component is

$$p^4 - \frac{1}{4} \text{ tests}$$

In general, for groups of n, the number of tests saved per component is

$$p^n - \frac{1}{n} \text{ tests}$$

Notice from the above formula that as n, the group size, gets very large, the number of tests saved per component gets very, very small.

To determine the optimal group size for $p = .9$, we refer to Table 1. From the table we can see that the optimal group size is four, resulting in a substantial saving of approximately 41%.

Table 1

Group Size	Expected Tests Saved per Component $p = .9$	Percent Saving
2	$p^2 - \frac{1}{2} = .81 - .50 = .31$	31
3	$p^3 - \frac{1}{3} = .729 - .333 = .396$	39.6
4	$p^4 - \frac{1}{4} = .6561 - .25 = .4061$	40.61
5	$p^5 - \frac{1}{5} = .59049 - .2 = .39049$	39.05
6	$p^6 - \frac{1}{6} = .531 - .167 = .364$	36.4
7	$p^7 - \frac{1}{7} = .478 - .143 = .335$	33.5
8	$p^8 - \frac{1}{8} = .430 - .125 = .305$	30.5

Now Work Problem 3

We also note that larger group sizes do not increase savings.

EXAMPLE 3

A \$75,000 oil detector is lowered under the sea to detect oil fields, and it becomes detached from the ship. If the instrument is not found within 24 hours, it will crack under the pressure of the sea. It is assumed that a scuba diver will find it with probability .85, but it costs \$500 to hire each diver. How many scuba divers should be hired?

SOLUTION

Let's assume that x scuba divers are hired. The probability that they will fail to discover the instrument is $.15^x$. Thus the instrument will be found with probability $1 - .15^x$.

The expected gain from hiring the scuba divers is

$$\$75,000(1 - .15^x) = \$75,000 - \$75,000(.15^x)$$

while the cost for hiring them is

$$\$500 \cdot x$$

Thus the expected net gain, denoted by $E(x)$, is

$$E(x) = \$75,000 - \$75,000(.15^x) - \$500x$$

The problem is then to choose x so that $E(x)$ is maximum.

We begin by evaluating $E(x)$ for various values of x:

$$E(1) = \$75,000 - \$75,000(.15^1) - \$500(1) = \$63,250.00$$
$$E(2) = \$75,000 - \$75,000(.15^2) - \$500(2) = \$72,312.50$$
$$E(3) = \$75,000 - \$75,000(.15^3) - \$500(3) = \$73,246.88$$
$$E(4) = \$75,000 - \$75,000(.15^4) - \$500(4) = \$72,962.03$$
$$E(5) = \$75,000 - \$75,000(.15^5) - \$500(5) = \$72,494.30$$
$$E(6) = \$75,000 - \$75,000(.15^6) - \$500(6) = \$71,999.15$$

Thus the expected net gain is optimal when three divers are hired. Note that hiring additional divers does not necessarily increase expected net gain. In fact, the expected net gain declines if more than three divers are hired.

EXERCISE 8.4 Answers to odd-numbered problems begin on page AN-29.

---

1. **Market Assessment** A car agency has fixed costs of $8 per car per day and the revenue for each car rented is $14 per day. The daily demand is given in the table:

Number of Customers	7	8	9	10	11
Probability		.10	.20	.40	.20	.10

Find the expected number of customers. Determine the optimal number of cars the company should have on hand each day. What is the expected profit in this case?

2. In Example 2 suppose $p = .8$. Show that the optimal group size is 3.

3. In Example 2 suppose $p = .95$. Show that the optimal group size is 5.

4. In Example 2 suppose $p = .99$. Compute savings for group sizes 10, 11, and 12, and thus show that 11 is the optimal group size. Determine the percent saving.

5. In Example 3 suppose the probability of the scuba divers discovering the instrument is .95. Find

   (a) An equation expressing the net expected gain.
   (b) The number x of scuba divers that maximizes the net gain.

6. In Example 2 compute the expected number of tests saved per component if on the first test the current does not pass through 2 components in series, but on the second test the current does pass through 1 of them. A third test is not made (since the other component is obviously defective).

8.5 RANDOM VARIABLES

When we perform an experiment, we are often interested not in a particular outcome, but rather in some number associated with that outcome. For example, in tossing a coin three times we may be interested in the number of heads obtained regardless of the particular sequence in which the heads appear. Similarly, the gambler throwing a pair of dice in a crap game is interested in the sum of the faces rather than the particular number on each face. When we sample mass-produced microchips, we may be interested in the number of defective microchips.

In each of these examples we are interested in numbers that are associated with experimental outcomes. This process of assigning a number to each outcome is called *random variable* assignment.

RANDOM VARIABLE

A *random variable* is a rule that assigns a number to each outcome of an experiment.

It is customary to denote the random variable by a capital letter, such as X or Y.

Table 2 summarizes the results of the experiment of flipping a fair coin three times. The first column in the table gives the sample space for this experiment. The second column shows the number of heads for each simple event.

Suppose that in this experiment we are interested only in the total number of heads. This information is given in Table 3.

Table 2

Sample Space	Number of Heads $X(e)$
e_1: *HHH*	3
e_2: *HHT*	2
e_3: *HTH*	2
e_4: *THH*	2
e_5: *HTT*	1
e_6: *THT*	1
e_7: *TTH*	1
e_8: *TTT*	0

Table 3

Number of Heads Obtained in Three Flips of a Coin	Probability
0	$\frac{1}{8}$
1	$\frac{3}{8}$
2	$\frac{3}{8}$
3	$\frac{1}{8}$

The role of the random variable is to transform the original sample space {*HHH, HHT, HTH, HTT, THH, THT, TTH, TTT*} into a new sample space that consists of the number of heads that occur: {0, 1, 2, 3}. If X denotes the random variable, then

$$X(e_1) = X(HHH) = 3 \qquad X(e_2) = X(HHT) = 2 \qquad X(e_3) = X(HTH) = 2$$
$$X(e_4) = X(THH) = 2 \qquad X(e_5) = X(HTT) = 1 \qquad X(e_6) = X(THT) = 1$$
$$X(e_7) = X(TTH) = 1 \qquad X(e_8) = X(TTT) = 0$$

We are interested in the probabilities that the random variable X assumes the values 0, 1, 2, and 3. Table 3 gives us the probabilities, so we may write

$$\text{Probability}(X = 0) = \tfrac{1}{8} \qquad \text{Probability}(X = 1) = \tfrac{3}{8}$$
$$\text{Probability}(X = 2) = \tfrac{3}{8} \qquad \text{Probability}(X = 3) = \tfrac{1}{8}$$

Probability Distribution

The discussion thus far illustrates two of the most important properties of a random variable:

1. The values it can assume.
2. The probabilities that are associated with each of these possible values.

Because values assumed by a random variable can be used to symbolize all outcomes associated with a given experiment, the probability assigned to a simple event can now be assigned as the likelihood that the random variable takes on the corresponding value.

If a random variable X has the values

$$x_1, x_2, x_3, \ldots, x_n \tag{1}$$

then the rule given by

$$p(x) = P(X = x)$$

where x assumes the values in (1) is called the *probability distribution* of X, or distribution of X.

It follows that $p(x)$ is a probability distribution of X if it satisfies the following two conditions:

1. $0 \le p(x) \le 1$ for $x = x_1, x = x_2, \ldots, x = x_n$

2. $p(x_1) + p(x_2) + \cdots + p(x_n) = 1$

where $x_1, x_2, \ldots, x_n$ are the values of X.

For example, suppose the probability distribution for the random variable X is given as

x_i	x_1	x_2	$\cdots$	x_n
p_i	p_1	p_2	$\cdots$	p_n

where $p_i = p(x_i)$. Then the **expected value of X,** denoted by $E(X)$, is

$$E(X) = x_1 p_1 + x_2 p_2 + \cdots + x_n p_n$$

If we consider again the experiment of tossing a fair coin three times, we denote the probability distribution by

$$p(x) \qquad \text{where } x = 0, 1, 2, \text{ or } 3$$

For instance, $p(3)$ is the probability of getting exactly three heads, that is,

$$p(3) = P(X = 3) = \tfrac{1}{8}$$

In a similar way, we define $p(0)$, $p(1)$, and $p(2)$. Probability distributions are also represented graphically, as shown in Figure 15. The graph of a probability distribution is often called a *histogram*.

Figure 15

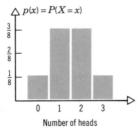

Number of heads

Binomial Probability Using Probability Distribution

We can now state the formula developed for the binomial probability model in terms of its probability distribution.

The binomial probability that assigns probabilities to the number of successes in n trials is an example of a probability distribution. For this distribution X is the random variable whose value for any outcome of the experiment is the number of successes obtained. For this distribution we may write

$$P(X = k) = b(n, k; p) = \binom{n}{k} p^k q^{n-k}$$

where $P(X = k)$ denotes the probability that the random variable equals k, that is, that exactly k successes are obtained.

EXERCISE 8.5   Answers to odd-numbered problems begin on page AN-29.

In Problems 1–6 list the values of the given random variable X together with the probability distributions.

1. A fair coin is tossed two times and X is the random variable whose value for an element in the sample space is the number of heads obtained.

2. A fair die is tossed once. The random variable X is the number showing on the top face.

3. The random variable X is the number of female children in a family with 3 children. (Assume the probability of a female birth is $\frac{1}{2}$).

4. A job applicant takes a 3-question true–false examination and guesses on each question. Let X be the number of right answers minus the number of wrong answers.

5. An urn contains 4 red balls and 6 white balls. Three balls are drawn with replacement. The random variable X is the number of red balls.

6. A couple getting married will have 3 children, and the random variable X denotes the number of boys they will have.

7. For the data given below, compute the expected value.

Outcome	e_1	e_2	e_3	e_4
Probability	.4	.2	.1	.3
x_i	2	3	-2	0

8. For the data below, compute the expected value.

Outcome	e_1	e_2	e_3	e_4
Probability	$\frac{1}{3}$	$\frac{1}{6}$	$\frac{1}{4}$	$\frac{1}{4}$
x_i	1	0	4	-2

Technology Exercises

Most graphing calculators have a random number function (usually RAND or RND) generating numbers between 0 and 1. Check your user's manual to see how to use this function on your graphing calculator.

Sometimes experiments are simulated using a random number function instead of actually performing the experiment. In Problems 1–6 use a random number function to simulate each experiment.

1. **Rolling a Fair Die**   Consider the experiment of rolling a die, and let the random variable X denote the number showing on the top face. Simulate the experiment using a random number function on your calculator, considering a roll to have the outcome k if the value of the random number function is between $(k - 1) \cdot 0.167$ and $k \cdot 0.167$. Record the outcome. Repeat the experiment 50 times, thus obtaining a sequence of 50 numbers. [*Note:* Most calculators repeat the action of the last entry if you simply press the ENTER, or EXE, key again.] Using these 50 numbers you can estimate the probability $P(X = k)$, for $k = 1, 2, 3, 4, 5, 6$, by the ratio

$$\frac{\text{Number of times } k \text{ appears in your sequence}}{50}$$

Enter your estimates in the table in the next column. Calculate the actual probabilities, and enter these numbers in the table. How close are your numbers to the actual values?

k	Your Estimate of $P(X = k)$	Actual Value of $P(X = k)$
1		
2		
3		
4		
5		
6		

2. Use a random number function to select a value for the random variable X. Repeat this experiment 50 times. Count the number of times the random variable X is between 0.6 and 0.9. Calculate the ratio

$$R = \frac{\text{Number of times the random variable } X \text{ is between 0.6 and 0.9}}{50}$$

What value of R did you obtain? Calculate the actual probability $P(0.6 \leq X < 0.9)$.

3. Use a random number function to select a value for the random variable X. Repeat this experiment 50 times. Count the number of times the random variable X is between 0.1 and 0.3. Calculate the ratio

$$R = \frac{\text{Number of times the random variable } X \text{ is between 0.1 and 0.3}}{50}$$

What value of R did you obtain? Calculate the actual probability $P(0.1 \leq X < 0.3)$.

4. **Rolling an Octahedron**   Consider the experiment of rolling an octahedron (a regular polygon with all 8 faces equal), and let the random variable X denote the number showing on the top face. Simulate the experiment using a random number function on your calculator, considering a roll to have the outcome k if the value of the random number function is between $(k - 1)/8$ and $k/8$, for $k = 1, 2, 3, 4, 5, \ldots, 8$. Record the outcome. Repeat the experiment 50 times, thus obtaining a sequence of 50 numbers. Using

k	Your Estimate of $P(X = k)$	Actual Value of $P(X = k)$
1		
2		
3		
4		
5		
6		
7		
8		

these 50 numbers you can estimate the probability $P(X = k)$, for $k = 1, 2, 3, 4, 5, 6, 7, 8$, by the ratio

$$\frac{\text{Number of times } k \text{ appears in your sequence}}{50}$$

Enter your estimates in the table shown. Calculate the actual probabilities, and enter these numbers in the table. How close are your numbers to the actual values?

5. **Rolling a Dodecahedron**   Consider the experiment of rolling a dodecahedron (a regular polygon with all 12 faces equal), and let the random variable X denote the number showing on the top face. Simulate the experiment using a random number function on your calculator, considering a roll to have the outcome k if the value of the random number function is between $(k - 1)/12$ and $k/12$, for $k = 1, 2, 3, 4, 5, \ldots, 12$. Record the outcome. Repeat the experiment 50 times, thus obtaining a sequence of 50 numbers. Using these 50 numbers you can estimate the probability $P(X = 2)$ by the ratio

$$\frac{\text{Number of times 2 appears in your sequence}}{50}$$

Calculate the actual probability, $P(X = 2)$, and compare these values. How close is your estimate to the actual value?

6. **Rolling an Icosahedron**   Consider the experiment of rolling an icosahedron (a regular polygon with all 20 faces equal), and let the random variable X denote the number showing on the top face. Simulate the experiment using a random number function on your calculator, considering a roll to have the outcome k if the value of the random number function is between $(k - 1)/20$ and $k/20$, for $k = 1, 2, 3, 4, 5, \ldots, 20$. Record the outcome. Repeat the experiment 50 times, thus obtaining a sequence of 50 numbers. Using these 50 numbers you can estimate the probability $P(X = 5)$ by the ratio

$$\frac{\text{Number of times 5 appears in your sequence}}{50}$$

Calculate the actual probability, $P(X = 5)$, and compare these values. How close is your estimate to the actual value?

CHAPTER REVIEW

IMPORTANT TERMS AND CONCEPTS

partition of a sample space 369
Bayes' formula 370
a priori/a posteriori probabilities 372

Bernoulli trials 377
formula for $b(n, k; p)$ 379
expected value 388

steps for computing expected value 389
random variable 400
probability distribution 401

IMPORTANT FORMULAS

Partitioned Sample Space

$$P(E) = P(A_1) \cdot P(E|A_1) + P(A_2) \cdot P(E|A_2)$$
$$+ P(A_3) \cdot P(E|A_3) + \cdots + P(A_n) \cdot P(E|A_n)$$

Binomial Probability Formula

$$b(n, k; p) = \binom{n}{k} p^k q^{n-k}, \quad q = 1 - p$$

Bayes' Formula

$$P(A_j|E) = \frac{P(A_j) \cdot P(E|A_j)}{P(E)}$$

Expected Value for Bernoulli Trials

$$E = np$$

TRUE–FALSE ITEMS   Answers are on page AN-29.

T_____ F_____ **1.** $b(3, 2; \frac{1}{2}) = 3 \cdot (\frac{1}{2})^3$

T_____ F_____ **2.** $P(A_1|E) = P(A_1)P(E)$

T_____ F_____ **3.** The expected value of an experiment is never negative.

T_____ F_____ **4.** Bayes' formula is useful for computing a posteriori probability.

T_____ F_____ **5.** $b(n, k; p)$ gives the probability of exactly n successes in k trials.

T_____ F_____ **6.** In flipping a fair coin 10 times, the expected number of heads is 5.

FILL IN THE BLANKS   Answers are on page AN-29.

1. The formula

$$P(A_1|E) = \frac{P(A_1) \cdot P(E|A_1)}{P(E)}$$

is called _____ _____.

2. Random experiments are called Bernoulli trials if

(a) The same experiment is repeated several times.

(b) There are only two possible outcomes, success and failure.

(c) The repeated trials are _____.

(d) The probability of each outcome remains _____ for each trial.

3. If an experiment has n outcomes that are assigned the payoffs $m_1, m_2, \ldots, m_n$, occurring with probabilities p_1, $p_2, \ldots, p_n$, then $E = m_1 p_1 + m_2 p_2 + \cdots + m_n p_n$ is the _____ _____.

4. A random variable on a sample space S is a rule that assigns _____ _____ to each element in S.

5. If X is a random variable assuming the values x_1, $x_2, \ldots x_n$, then $E(X) = x_1 p(x_1) + x_2 p(x_2) + \cdots + x_n p(x_n)$ is the _____ _____ of X.

REVIEW EXERCISES   Answers to odd-numbered problems begin on page AN-29.

In Problems 1–14 use the tree diagram below to find the indicated probability:

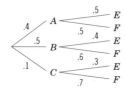

1. $P(E|A)$ **2.** $P(F|A)$ **3.** $P(E|B)$ **4.** $P(F|B)$

5. $P(E|C)$ **6.** $P(F|C)$ **7.** $P(E)$ **8.** $P(F)$

9. $P(A|E)$ **10.** $P(A|F)$ **11.** $P(B|E)$ **12.** $P(B|F)$

13. $P(C|E)$ **14.** $P(C|F)$

15. The table below indicates a survey conducted by a deodorant producer:

	Like the Deodorant	Did Not Like the Deodorant	No Opinion
Group I	180	60	20
Group II	110	85	12
Group III	55	65	7

Let the events E, F, G, H, and K be defined as follows:

E: Customer likes the deodorant

F: Customer does not like the deodorant

G: Customer is from group I

H: Customer is from group II

K: Customer is from group III

Find

(a) $P(E|G)$ (b) $P(G|E)$ (c) $P(H|E)$
(d) $P(K|E)$ (e) $P(F|G)$ (f) $P(G|F)$
(g) $P(H|F)$ (h) $P(K|F)$

16. Three machines in a factory, A_1, A_2, A_3, produce 55%, 30%, and 15% of total production, respectively. The percentage of defective output of these machines is 1%, 2%, and 3%, respectively. An item is chosen at random and it is defective. What is the probability that it came from machine A_1? From A_2? From A_3?

17. A lung cancer test has been found to have the following reliability. The test detects 85% of the people who have cancer and does not detect 15% of these people. Among the noncancerous group it detects 92% of the people not having cancer, whereas 8% of this group are detected erroneously as having lung cancer. Statistics show that about 1.8% of the population has cancer. Suppose an individual is given the test for lung cancer and it detects the disease. What is the probability that the person actually has cancer?

18. What is the expected number of girls in families having exactly 3 children?

19. Management believes that 1 out of 5 people watching a television advertisement about their new product will purchase the product. Five people who watched the advertisement are picked at random. What is the probability that 0, 1, 2, 3, 4, or 5 of these people will purchase the product?

20. Suppose that the probability of a player hitting a home run is $\frac{1}{20}$. In 5 tries what is the probability that the player hits at least 1 home run?

21. In a 12-item true–false examination, a student guesses on each question.

(a) What is the probability that the student will obtain all correct answers?
(b) If 7 correct answers constitute a passing grade, what is the probability that the student will pass?
(c) What are the odds in favor of passing?

22. In a 20-item true–false examination, a student guesses on each question.

(a) What is the probability that the student will obtain all correct answers?
(b) If 12 correct answers constitute a passing grade, what is the probability that the student will pass?
(c) What are the odds in favor of passing?

23. Find the probability of throwing an 11 at least 3 times in 5 throws of a pair of fair dice.

24. In a certain game a player has the probability $\frac{1}{4}$ of winning a prize worth $89.99 and the probability $\frac{1}{3}$ of winning another prize worth $49.99. What is the expected cost of the game for the player?

25. Frank pays $0.70 to play a certain game. He draws 2 balls (together) from a bag containing 2 red balls and 4 green balls. He receives $1 for each red ball that he draws. Has he paid too much? By how much?

26. In a lottery 1000 tickets are sold at $0.25 each. There are 3 cash prizes: $100, $50, and $30. Alice buys 5 tickets.

(a) What would have been a fair price for a ticket?
(b) How much extra did Alice pay?

27. The figure at the top of the next page shows a spinning game for which a person pays $0.30 to purchase an opportunity to spin the dial. The numbers in the figure indicate the amount of payoff and its corresponding probability. Find the expected value of this game. Is the game fair?

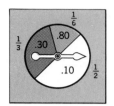

28. Consider the 3 boxes in the figure. The game is played in 2 stages. The first stage is to choose a ball from box A. If the result is a ball marked I, then we go to box I, and select a ball from there. If the ball is marked II, then we select a ball from box II. The number drawn on the second stage is the gain. Find the expected value of this game.

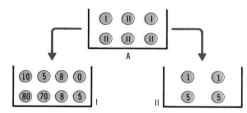

29. What is the expected number of heads that will turn up if a biased coin, $P(H) = \frac{1}{4}$, is tossed 200 times?

Mathematical Questions from Professional Exams

1. Actuary Exam — Part II   What is the probability that 10 independent tosses of an unbiased coin result in no fewer than 1 head and no more than 9 heads?

(a) $(\frac{1}{2})^9$ (b) $1 - 11(\frac{1}{2})^9$ (c) $1 - 11(\frac{1}{2})^{10}$
(d) $1 - (\frac{1}{2})^9$ (e) $1 - (\frac{1}{2})^{10}$

2. CPA Exam   The Stat Company wants more information on the demand for its products. The following data are relevant:

Units Demanded	Probability of Unit Demand	Total Cost of Units Demanded
0	.10	$0
1	.15	1.00
2	.20	2.00
3	.40	3.00
4	.10	4.00
5	.05	5.00

30. European Roulette   A European roulette wheel has only 37 compartments, 18 red, 18 black, and 1 green. A player will be paid $2 (including his $1 bet) if he picks correctly the color of the compartment in which the ball finally rests. Otherwise, he loses $1. Is the game fair to the player? Compare this answer to the answer obtained in Problem 18, Exercise 8.3.

31. The Blood Testing Problem*   A group of 1000 people is subjected to a blood test that can be administered in 2 ways: (1) each person can be tested separately (in this case 1000 tests are required) or (2) the blood samples of 30 people can be pooled and analyzed together. If we use the second way and the test is negative, then 1 test suffices for 30 people. If the test is positive, each of the 30 people can then be tested separately, and, in all, $30 + 1$ tests are required for the 30 people. Assume the probability p that the test is positive is the same for all people and that the people to be tested are independent.

(a) What is the probability that the test for a pooled sample of 30 people will be positive?
(b) What is the expected number of tests necessary under plan 2?

What is the total expected value or payoff with perfect information?

(a) $2.40   (b)$7.40   (c) $9.00   (d) $9.15

3. CPA Exam   Your client wants your advice on which of 2 alternatives he should choose. One alternative is to sell an investment now for $10,000. Another alternative is to hold the investment 3 days, after which he can sell it for a certain selling price based on the following probabilities:

Selling Price	Probability
$5,000	.4
$8,000	.2
$12,000	.3
$30,000	.1

Using probability theory, which of the following is the most reasonable statement?

* William Feller, *An Introduction to Probability Theory and Its Applications,* 3rd ed. New York: Wiley, 1968, pp. 239–240.

(a) Hold the investment 3 days because the expected value of holding exceeds the current selling price.
(b) Hold the investment 3 days because of the chance of getting $30,000 for it.
(c) Sell the investment now because the current selling price exceeds the expected value of holding.
(d) Sell the investment now because there is a 60% chance that the selling price will fall in 3 days.

4. **CPA Exam** The Polly Company wishes to determine the amount of safety stock that it should maintain for product D that will result in the lowest cost.
   The following information is available:

Stockout cost	$80 per occurrence
Carrying cost of safety stock	$2 per unit
Number of purchase orders	5 per year

The available options open to Polly are as follows:

Units of Safety Stock	10	20	30	40	50	55
Probability	50%	40%	30%	20%	10%	5%

The number of units of safety stock that will result in the lowest cost is

(a) 20   (b) 40   (c) 50   (d) 55

5. **CPA Exam** The ARC Radio Company is trying to decide whether to introduce as a new product a wrist "radiowatch" designed for shortwave reception of exact time as broadcast by the National Bureau of Standards. The "radiowatch" would be priced at $60, which is exactly twice the variable cost per unit to manufacture and sell it. The incremental fixed costs necessitated by introducing this new product would amount to $240,000 per year. Subjective estimates of the probable demand for the product are shown in the following probability distribution:

Annual Demand	6,000	8,000	10,000	12,000	14,000	16,000
Probability	.2	.2	.2	.2	.1	.1

The expected value of demand for the new product is

(a) 11,000 units   (b) 10,200 units   (c) 9000 units
(d) 10,600 units   (e) 9800 units

6. **CPA Exam** In planning its budget for the coming year, King Company prepared the following payoff probability distribution describing the relative likelihood of monthly sales volume levels and related contribution margins for product A:

Monthly Sales Volume	Contribution Margin	Probability
4,000	$ 80,000	.20
6,000	120,000	.25
8,000	160,000	.30
10,000	200,000	.15
12,000	240,000	.10

What is the expected value of the monthly contribution margin for product A?

(a) $140,000   (b) $148,000
(c) $160,000   (d) $180,000

7. **CPA Exam** A decision tree has been formulated for the possible outcomes of introducing a new product line. Branches related to alternative 1 reflect the possible payoffs from introducing the product without an advertising campaign. The branches for alternative 2 reflect the possible payoffs with an advertising campaign costing $40,000.

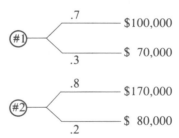

The expected values of alternatives 1 and 2, respectively, are

(a) #1: (.7 × $100,000) + (.3 × $70,000)
     #2: (.8 × $170,000) + (.2 × $80,000)
(b) #1: (.7 × $100,000) + (.3 × $70,000)
     #2: (.8 × $130,000) + (.2 × $40,000)
(c) #1: (.7 × $100,000) + (.3 × $70,000)
     #2: (.8 × $170,000) + (.2 × $80,000) − $40,000
(d) #1: (.7 × $100,000) + (.3 × $70,000) − $40,000
     #2: (.8 × $170,000) + (.2 × $80,000) − $40,000

8. **CPA Exam** A battery manufacturer warrants its automobile batteries to perform satisfactorily for as long as the owner keeps the car. Auto industry data show that only 20% of car buyers retain their cars for 3 years or more. Historical data suggest the following:

Number of Years Owned	Probability of Battery Failure	Battery Exchange Costs	Percentage of Failed Batteries Returned
Less than 3 years	.4	$50	75%
3 years or more	.6	$20	50%

If 50,000 batteries were sold this year, what is the estimated warranty cost?

(a) $375,000 (b) $435,000
(c) $500,000 (d) $660,000

Chapter 9

Statistics

S tatistics is the science of collecting, organizing, analyzing, and interpreting numerical facts. By making observations, statisticians collect **data** in the form of measurements or counts. A measurable characteristic is called a **variable.** If a variable can assume any real value between certain limits, it is called a **continuous variable.** It is called a **discrete variable** if it can assume only a finite set of values or as many values as there are whole numbers. Examples of continuous variables are weight, height, length, time, etc. Examples of discrete variables are the number of votes a candidate gets, the number of cars sold, etc. We shall discuss only discrete data in this chapter.

The **organization of data** involves the presentation of the collected measurements or counts in a form suitable for determining logical conclusions. Usually, tables or graphs are used to represent the collected data. The **analysis of data** is the process of extracting, from given measurements or counts, related and relevant information from which a brief numerical description can be formulated. In this process we use concepts known as the *mean, median, range, variance,* and *standard deviation.* By **interpretation of data** we mean the art of drawing conclusions from the analysis of the data. This involves the formulation of predictions concerning a large collection of objects based on the information available from a small collection of similar objects.

411

In collecting data concerning varied characteristics, it is often impossible or impractical to observe an entire group. Instead of examining an entire group, called the **population,** a small segment, called the **sample,** is chosen. It would be difficult, for example, to question all cigarette smokers in order to study the effects of smoking. Therefore, appropriate samples of smokers are usually selected for questioning.

The method of selecting the sample is extremely important if we want the results to be reliable. All members of the population under investigation should have an equal probability of being selected; otherwise, a **biased sample** could result. For example, if we want to study the relationship between smoking cigarettes and lung cancer, we cannot choose a sample of smokers who all live in the same location. The individuals chosen might have dozens of characteristics peculiar to their area, which would give a false impression with regard to all smokers.

Samples collected in such a way that each population item is equally likely to be chosen are called **random samples.** Of course, there are many random samples that can be chosen from a population. By combining the results of more than one random sample of a population, it is possible to obtain a more accurate representation of the population.

If a sample is representative of a population, important conclusions about the population can often be inferred from analysis of the sample. The phase of statistics dealing with conditions under which such inference is valid is called **inductive statistics** or **statistical inference.** Since such inference cannot be absolutely certain, the language of probability is often used in stating conclusions. Thus when a meteorologist makes a forecast, weather data collected over a large region are studied and, based on this study, the weather forecast is given in terms of chances. A typical forecast might be ''There is a 20% chance of rain tomorrow.''

To summarize, in statistics we are interested in four principles: collecting data or information, organizing it, analyzing it, and interpreting it.

9.1 ORGANIZATION OF DATA

Quite often a study results in a collection of large masses of data. If the data are to be understood and, at the same time, effective, they must be summarized in some manner. Two methods of presenting data are in common use. One method involves a summarized presentation of the numbers themselves according to order in a tabular form; the other involves presenting the quantitative data in pictorial form, such as graphs or diagrams.

Frequency Tables; Line Charts

EXAMPLE 1

Table 1 lists the weights of a random sample of 71 children selected from a group of 10,000.

Table 1   Weights of 71 Students, in Pounds

69	71	71	55	52	55	58	58	58	62	67	94
82	94	95	89	89	104	93	93	58	62	67	62
94	85	92	75	75	79	75	82	94	105	115	104
105	109	94	92	89	85	85	89	95	92	105	71
72	72	79	79	85	72	79	119	89	72	72	69
79	79	69	93	85	93	79	85	85	69	79	

Certain information available from the sample becomes more evident once the data are ordered according to some scheme. If the 71 measurements are written in order of magnitude, we obtain Table 2.

Table 2

52	55	55	58	58	58	58	62	62	62	67	67
69	69	69	69	71	71	71	72	72	72	72	72
75	75	75	79	79	79	79	79	79	79	79	82
82	85	85	85	85	85	85	85	89	89	89	89
89	92	92	92	93	93	93	93	94	94	94	94
94	95	95	104	104	105	105	105	109	115	119	

· ·

The data in Table 2 can be presented in a so-called **frequency table.** This is done as follows: Tally marks are used to record the occurrence of the respective weights. Then the **frequency** f with which each weight occurs can be determined. In doing this, further information may become evident. See Table 3.

Table 3

Score	Tally	Frequency, f	Score	Tally	Frequency, f
52	/	1	85	⊞⊞ //	7
55	//	2	89	⊞⊞	5
58	////	4	92	///	3
62	///	3	93	////	4
67	//	2	94	⊞⊞	5
69	////	4	95	//	2
71	///	3	104	//	2
72	⊞⊞	5	105	///	3
75	///	3	109	/	1
79	⊞⊞ ///	8	115	/	1
82	//	2	119	/	1

A graphical representation of the same data may be presented in a **line chart,** which is obtained in the following way: If we let the vertical axis (*y*-axis) denote the frequency *f* and the horizontal axis (*x*-axis) denote the weight data, we obtain the graph shown in Figure 1.

Figure 1

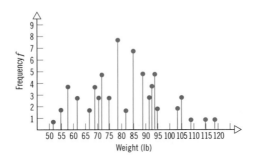

Grouping Data; Histograms

Now we show how data can be grouped.

EXAMPLE 2

Table 4 lists the monthly electric bills of 71 residential customers.

Table 4   Monthly Electric Bills

52.30	55.61	55.71	58.01	58.41	58.51	58.91	62.33	62.50	62.71
67.13	67.23	69.51	69.67	69.80	69.82	71.34	71.65	71.83	72.15
72.22	72.41	72.59	72.67	75.11	75.71	75.82	79.03	79.06	79.09
79.15	79.28	79.32	79.51	79.62	82.32	82.61	85.09	85.13	85.25
85.31	85.41	85.51	85.58	89.21	89.32	89.49	89.61	89.78	92.41
92.63	92.89	93.05	93.19	93.28	93.91	94.17	94.28	94.31	94.52
94.71	95.32	95.51	104.31	104.71	105.21	105.37	105.71	109.34	115.71
119.38									

. . . .

Notice that the data has been ordered from smallest to largest. The **range** of a set of numbers is the difference between the largest and the smallest value of the data under consideration. Thus,

$$\textbf{Range} = \textbf{(Largest value)} - \textbf{(Smallest value)}$$

For the data in Table 4 the range is

$$\text{Range} = 119.38 - 52.30 = 67.08$$

To group this data, we divide the range into intervals of equal size, called **class intervals.** Table 5 shows the data using 14 class intervals, each of size 5; Table 6 shows the same data using 7 class intervals, each of size 10.

Table 5

	Class Interval	Tally	Frequency
1	50– 54.99	/	1
2	55– 59.99	卌 /	6
3	60– 64.99	///	3
4	65– 69.99	卌 /	6
5	70– 74.99	卌 ///	8
6	75– 79.99	卌 卌 /	11
7	80– 84.99	//	2
8	85– 89.99	卌 卌 //	12
9	90– 94.99	卌 卌 //	12
10	95– 99.99	//	2
11	100–104.99	//	2
12	105–109.99	////	4
13	110–114.99		0
14	115–119.99	//	2

Table 6

	Class Interval	Tally	Frequency
1	50– 59.99	卌 //	7
2	60– 69.99	卌 ////	9
3	70– 79.99	卌 卌 卌 ////	19
4	80– 89.99	卌 卌 ////	14
5	90– 99.99	卌 卌 ////	14
6	100–109.99	卌 /	6
7	110–119.99	//	2

The class intervals shown in Tables 5 and 6 each begin at 50 and end at 119.99, so as to include all the data from Table 4. The first number in a class interval is called the **lower class limit;** the second number is called the **upper class limit.** We choose these limits so that each item in Table 4 can be assigned to one and only one class interval. The **midpoint** of a class interval is defined as

$$\text{Midpoint} = \frac{\text{Upper class limit} + \text{Lower class limit}}{2}$$

When the data are represented in the form of Table 5 (or Table 6), they are said to be **grouped data.** Notice that once raw data are converted to grouped data, it is impossible to retrieve or recover the original data from the frequency table. The best we can do is to choose the midpoint of each class interval as a representative for each class. In Table 5, for example, the actual scores of 105.21, 105.37, 105.71, and 109.34 are viewed as being represented by the midpoint of the class interval 105–109.99, namely, (105 + 109.99)/2 = 107.50.

Next we present the grouped data of Table 6 in a graph, called a **histogram.**

EXAMPLE 3

Build a histogram for the grouped data of Table 6.

SOLUTION

To build a histogram for the data in Table 6, we construct a set of rectangles having as a base the size of the class interval and as height the frequency of occurrence of data in that particular interval. The center of the base is the midpoint of each class interval. Figure 2 shows the histogram.

Figure 2

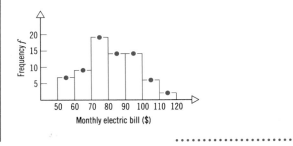

Monthly electric bill ($)

If we connect all the midpoints of the tops of the rectangles in Figure 2, we obtain a line graph called a **frequency polygon.** (In order not to leave the graph hanging, we always connect it to the horizontal axis on both sides.) See Figure 3.

Figure 3

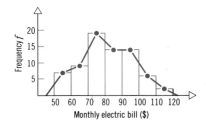

Monthly electric bill ($)

Now Work Problems
11(a)–(h)

Sometimes it is useful to learn how many cases fall below (or above) a certain value. For the data of Table 6, we convert the data as follows: Start at the lowest class interval (50–59.99) and note how many scores are in this interval. The number is 7. So we write 7 in the column labeled *cf* (cumulative frequency) of Table 7 in the row for 50–59.99. Next, we list how many scores fall in the next class interval (60–69.99), 9, and place the cumulative total $7 + 9 = 16$ in the *cf* column. The process is continued. The bottom entry of the last column should agree with the total number of scores in the sample. The numbers in this column are called the **cumulative (less than) frequencies.**

Table 7

Class Interval	Tally	*f*	*cf*
50– 59.99	ℍℍ //	7	7
60– 69.99	ℍℍ ////	9	16
70– 79.99	ℍℍ ℍℍ ℍℍ ////	19	35
80– 89.99	ℍℍ ℍℍ ////	14	49
90– 99.99	ℍℍ ℍℍ ////	14	63
100–109.99	ℍℍ /	6	69
110–119.99	//	2	71

The graph in which the horizontal axis represents class intervals and the vertical axis represents cumulative frequencies is called the **cumulative (less than) frequency distribution.** See Figure 4 for the cumulative frequency distribution for the data from Table 7. Notice that the points obtained are connected by lines to aid in visualizing the graph.

Figure 4

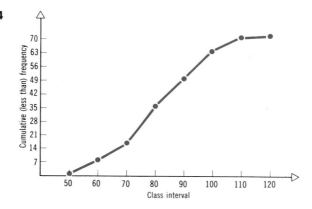

EXERCISE 9.1   Answers to odd-numbered problems begin on page AN-29.

In Problems 1–6 list some possible ways to choose random samples for each study.

1. A study to determine opinion about a certain television program.

2. A study to detect defective radio resistors.

3. A study of the opinions of people toward Medicare.

4. A study to determine opinions about an election of a U.S. president.

5. A study of the number of savings accounts per family in the United States.

6. A national study of the monthly budget for a family of four.

7. The following is an example of a biased sample: In a study of political party preferences, poorly dressed interviewers obtained a significantly greater proportion of answers favoring Democratic party candidates in their samples than did their well-dressed and wealthier-looking counterparts. Give two more examples of biased samples.

8. In a study of the number of savings accounts per family, a sample of accounts totaling less than $10,000 was taken and, from the owners of these accounts, information about the total number of accounts owned by all family members was obtained. Criticize this sample.

9. It is customary for newsreporters to sample the opinions of a few people to find out how the population at large feels about the events of the day. A reporter questions people on a downtown street corner. Is there anything wrong with such an approach?

10. In 1936 the *Literary Digest* conducted a poll to predict the presidential election. Based on its poll it predicted the election of Landon over Roosevelt. In the actual election, Roosevelt won. The sample was taken by drawing the mailing list from telephone directories and lists of car owners. What was wrong with the sample?

11. Consider the data given in the table at the top of the next page. With reference to this table, there are 12 classes, the first being the class interval 50–99. Find:

(a) The lower limit of the fifth class.
(b) The upper limit of the fourth class.
(c) The midpoint of the fifth class.
(d) The size of the fifth interval.
(e) The frequency of the third class.
(f) The class interval having the largest frequency.
(g) The number of precincts with less than 600 votes.
(h) Construct the histogram.
(i) Construct the frequency polygon.

Distribution of Cleveland Voting Precincts According to Total Vote Cast for Governor

Votes Cast	Number of Precincts
50– 99	1
100–149	4
150–199	33
200–249	120
250–299	190
300–349	150
350–399	104
400–449	67
450–499	48
500–549	26
550–599	9
600–649	1
	Total: 753

Source: Ohio Election Statistics, 1932, pp. 218–242.

12. The following scores were made on a 60-item test:

25	30	34	37	41	42	46	49	53
26	31	34	37	41	42	46	50	53
28	31	35	37	41	43	47	51	54
29	32	36	38	41	44	48	52	54
30	33	36	39	41	44	48	52	55
30	33	37	40	42	45	48	52	

(a) Set up a frequency table for the above data. What is the range?
(b) Draw a line chart for the data.
(c) Draw a histogram for the data using a class interval of size 2.
(d) Draw the frequency polygon for this histogram.
(e) Find the cumulative (less than) frequencies.
(f) Draw the cumulative (less than) frequency distribution.

13. For Table 5 in the text:

(a) Draw a histogram.
(b) Draw the frequency polygon.
(c) Find the cumulative (less than) frequencies.
(d) Draw the cumulative (less than) frequency distribution.

14. **Commercial Bank Earnings**   According to the *Fortune Directory* of June 15, 1967, the following were the earnings of the 50 largest commercial banks in the United States (as a percentage of capital funds for the year 1966):

12.2	9.9	11.2	12.5	9.8
11.5	11.8	11.1	12.3	10.1
11.4	9.2	12.8	9.8	12.6
9.9	10.2	12.6	14.4	10.9
10.2	10.3	11.6	10.2	13.1
10.4	10.9	8.4	14.6	13.4
12.3	11.4	9.2	12.8	11.0
11.2	10.9	10.1	10.9	12.9
11.2	13.2	10.2	16.0	13.6
10.9	11.4	11.6	11.7	13.0

Answer the same questions as in Problem 12, using a class interval of size 0.5 beginning with 8.0.

15. **Number of Physicians**   The following were the numbers of physicians per 100,000 population in 110 selected large U.S. cities in 1962 (*Statistical Abstract of the United States*, 1967):

131	245	145	129	155	232	256	204
296	222	185	166	198	127	153	230
175	161	240	169	169	158	116	171
111	152	126	140	218	142	141	116
176	127	156	185	207	218	153	128
176	162	100	138	129	211	178	198
132	289	165	129	137	78	146	148
145	146	161	119	119	116	245	137
95	169	131	156	136	122	194	113
184	132	172	91	110	188	185	144
105	166	154	108	144	202	212	190
165	128	131	157	115	153	127	224
171	154	149	112	134	190	130	192
123	224	131	190	136	123		

Answer the same questions as in Problem 12, using a class interval of size 10 beginning with 70.5.

16. **High School Dropouts**   The high school dropout rates in 1986 for the 50 states and the District of Columbia are given at the top of the next page. (Data are from the *1989 World Almanac*, Pharos Books, 1989.) The figures are the percentages of high school students who dropped out during the year. Make a frequency distribution for the data, using eight classes of equal width, and draw the frequency histogram.

32.7	31.7	37.0	22.0	33.3	26.9	10.2
29.3	43.2	38.0	37.3	29.2	21.0	24.2
28.3	12.5	18.5	31.4	37.3	23.5	23.4
23.3	32.2	8.6	36.7	24.4	12.8	11.9
34.8	26.7	22.4	27.7	35.8	30.0	10.3
19.6	28.4	25.9	21.5	32.7	35.5	18.5
32.6	35.7	19.7	22.4	26.1	24.8	24.8
13.7	18.8					

17. The fiscal year 1986 per capita tax burdens of the 50 states, in dollars, are given in the following list. (Data are from the *1989 World Almanac,* Pharos Books, 1989.) Make a frequency distribution for the data.

740	3490	975	770	1144	718	1202
1343	780	806	1400	743	848	810
863	777	863	807	940	1047	1314
1019	1163	731	712	755	700	1084
472	1096	989	1278	881	907	843
895	715	898	908	863	570	682
667	820	923	836	1169	964	1148
1569						

Technology Exercises

Most graphing calculators are able to draw histograms. Check your user's manual to determine the appropriate command. Use a graphing calculator to solve Problems 1–6.

1. The table at the right gives the birth rate per 1000 people in 20 states. (*Source:* U.S. Census Bureau, Statistical Abstract of the United States, 1994.) Make a frequency distribution for the data using 8 classes of equal width, and draw the frequency histogram.

State	AK	AL	AR	CA	CT	DC
Birth Rate	20.5	15.4	15.0	20.1	14.8	19.7

State	HI	IL	IN	KY	MA	ME
Birth Rate	17.6	16.8	15.3	14.6	14.7	13.6

State	MI	NH	OH	PA	RI	TN
Birth Rate	16.0	14.8	15.2	14.1	14.7	15.0

State	VT	WI
Birth Rate	14.0	14.5

2. The table at the right gives the heart disease death rate per 100,000 people in 20 states in 1991. (*Source:* U.S. Census Bureau, Statistical Abstract of the United States, 1994.) Make a frequency distribution for the data using 4 classes of equal width, and draw the frequency histogram.

State	AK	AL	AR	CA	CT
Heart Disease Rate	83	322	346	222	291

State	DC	HI	IL	IN	KY
Heart Disease Rate	312	180	309	299	322

State	MA	ME	MI	NH	OH
Heart Disease Rate	285	300	295	246	320

State	PA	RI	TN	VT	WI
Heart Disease Rate	362	323	313	259	290

3. The table below gives the cancer death rate per 100,000 people in 20 states in 1991. (*Source:* U.S. Census Bureau, Statistical Abstract of the United States, 1994.) Make a frequency distribution for the data using 5 classes of equal width, and draw the frequency histogram.

State	AK	AL	AR	CA	CT
Cancer Rate	88	216	236	165	214

State	DC	HI	IL	IN	KY
Cancer Rate	259	146	212	214	230

State	MA	ME	MI	NH	OH
Cancer Rate	230	238	206	203	222

State	PA	RI	TN	VT	WI
Cancer Rate	251	236	213	197	209

4. The table below gives the energy expenditures per person during 1991 in 20 states. (*Source:* U.S. Census Bureau, Statistical Abstract of the United States, 1994.) Make a frequency distribution for the data using 5 classes of equal width, and draw the frequency histogram.

State	AK	AL	AR	CA
Energy Expenditures	3249	2029	1975	1562

State	CT	DC	HI	IL
Energy Expenditures	1885	1899	1793	1863

State	IN	KY	MA	ME
Energy Expenditures	2125	1936	1767	2057

State	MI	NH.	OH	PA
Energy Expenditures	1786	1727	1928	1863

State	RI	TN	VT	WI
Energy Expenditures	1747	1872	1930	1645

5. The following table gives the average hourly earnings of production workers in manufacturing during 1993 in 20 states. (*Source:* U.S. Census Bureau, Statistical Abstract of the United States, 1994.) Make a frequency distribution for the data using 6 classes of equal width, and draw the frequency histogram.

State	AK	AL	AR	CA
Hourly Earnings	11.14	10.36	9.36	12.37

State	CT	DC	HI	IL
Hourly Earnings	13.01	13.18	11.98	12.04

State	IN	KY	MA	ME
Hourly Earnings	13.17	11.48	12.36	11.40

State	MI	NH	OH	PA
Hourly Earnings	15.35	11.61	14.05	12.09

State	RI	TN	VT	WI
Hourly Earings	10.22	10.33	11.81	12.17

6. The table below gives the number of hazardous waste sites in 20 states. (*Source:* U.S. Census Bureau, Statistical Abstract of the United States, 1994.) Make a frequency distribution for the data using 5 classes of equal width, and draw the frequency histogram.

State	AK	AL	AR	CA
Hazardous Waste Sites	8	14	12	95

State	CT	DC	HI	IL
Hazardous Waste Sites	15	0	3	37

State	IN	KY	MA	ME
Hazardous Waste Sites	33	20	31	10

State	MI	NH	OH	PA
Hazardous Waste Sites	76	17	36	99

State	RI	TN	VT	WI
Hazardous Waste Sites	12	15	8	40

9.2 PIE CHARTS; BAR GRAPHS

In the preceding section we discussed how line charts and histograms are often used to communicate an overall impression of data. In this section we discuss some other forms of graphs that are of great help in communicating information contained in data.

Pie Charts

The **pie chart** is another form of graphical representation of data. The total data are represented by a circle with radii drawn to divide the pie in the same proportions as the categories divide the total data. Each section is labeled with the category name, accompanied by the corresponding percentage.

Since there are 360° in a complete circle, each category is allocated a sector bounded by radii whose angle at the center of the circle is the appropriate percentage of 360°. Each percentage point is equal to an angle of 3.6° (that is, 1% of 360°).

Figure 5 illustrates, through two pie charts, the changes in the U.S. auto market from 1979 to 1989.

Figure 5

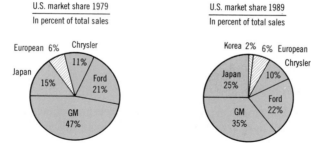

The two pie charts communicate very effectively the growth of the Japanese share of the total auto market (from 15% to 25%) and the decline of GM's share of the market (from 47% to 35%).

To illustrate the procedure of drawing pie charts, consider the data given in Table 8, which indicate the prerecorded music sales in the United States during 1988.

Table 8

Type	Amount (in billions of dollars)
Cassette tapes	3.8125
Cassette tape singles	0.0625
Compact discs	1.1875
12-inch singles	0.125
Long playing	0.875
Singles	0.1875
Total	6.25

It is more convenient to work with percentages than with amounts of money. Thus we may want to convert the amount of money spent in each category into percentages. We do this as follows:

Relative Frequency	*Angle for Sector*
$\dfrac{3.8125}{6.25} = 0.61$ or 61% of cassette tapes	$61 \times 3.6° \approx 220°$
$\dfrac{0.0625}{6.25} = 0.01$ or 1% of tape singles	$1 \times 3.6° \approx 4°$
$\dfrac{1.1875}{6.25} = 0.19$ or 19% of compact discs	$19 \times 3.6° \approx 68°$
$\dfrac{0.125}{6.25} = 0.02$ or 2% of 12-inch singles	$2 \times 3.6° \approx 7°$
$\dfrac{0.875}{6.25} = 0.14$ or 14% of long-playing	$14 \times 3.6° \approx 50°$
$\dfrac{0.1875}{6.25} = 0.03$ or 3% of singles	$3 \times 3.6° \approx 11°$

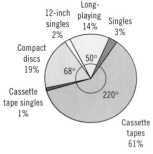

(Source Chilton Research Services.)

Figure 6

Figure 6 illustrates the above data.

Now Work Problem 5

Bar Graphs

The **bar graph** is also commonly used to graphically describe data. In such graphs **vertical bars** are usually used. The height of each bar represents the frequency of that category. Look at Figure 7(a) illustrating the gross national product (GNP) in percent change from the previous period first on an annual basis and then Figure 7(b) on a quarterly basis.

Figure 7

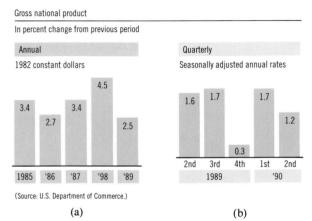

(a)   (b)

Sometimes the bars are also drawn horizontally. (See Figure 8.) Note that in Figure 8 all the bar graphs are drawn using the same scale.

Figure 8

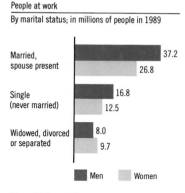

People at work

By marital status; in millions of people in 1989

Married, spouse present: 37.2 / 26.8

Single (never married): 16.8 / 12.5

Widowed, divorced or separated: 8.0 / 9.7

■ Men   ■ Women

(Source: U.S. Bureau of Labor Statistics.)

EXERCISE 9.2   Answers to odd-numbered problems begin on page AN-35.

In Problems 1 and 2 graph the data in each table using a bar graph.

1. Average home mortgage rates in percentage (*Source:* Office of Thrift Supervision, Resolution Trust Corporation).

1983	1984	1985	1986	1987	1988	1989
12.25	12.1	11.8	10.1	9	9.2	10

2. The table below gives the number of days per year the average American works to pay all federal, state, and local taxes (*Source:* Tax Foundation).

1984	1985	1986	1987	1988	1989	1990
117	121	122	124	123	123	125

3. The tables below give the average income of young families for the years 1973 and 1987 headed by a person younger than 30 years old, adjusted for inflation in thousands of 1986 dollars (*Source:* Children's Defense Fund, U.S. Census Bureau). Use a double bar graph to graph the data, that is, a bar graph for each category on the same coordinate system.

1973

White	Black	Hispanic
21	12	14

1987

White	Black	Hispanic
16	6	9

4. Graph the data given below by using a double bar graph, that is, a bar graph for each category on the same coordinate system (*Source:* U.S. Department of Commerce).

Top Bank Card Issuers
Outstanding Receivables by Institution
(in billions of dollars)

Name of Bank	1988	1989
Citibank	$21.8	$28.5
Chase Manhattan	7.2	9.3
Sears Discover Card	5.9	8.5
Bank of America	6.0	7.5
First Chicago Corp.	5.5	7.2
American Express Optima	3.8	5.2

5. Graph the data in the table below by using a pie graph: sources of U.S. personal income in percent of $4.06 trillion for 1988 (*Source:* U.S. Department of Commerce).

Wages and Salaries	Interest and Dividends	Social Security, Pensions	Self-Employed	Employee Contributions	Small Farm	Rental Income
60	17	9.5	7	5	1	0.5

6. The following data show how office workers in Chicago get to work:

Means of Transportation	Percentage
Ride alone	64
Car pool	5
Ride bus	30
Other	1

Construct a pie chart and a bar chart, and compare them to see which one seems more informative to you.

7. To study their attitudes toward a new product, 1000 people were interviewed. Their response is given in the following table:

Attitude	No. of Responses
Do not like	420
Like	360
Like very much	220

Construct a pie chart and a bar chart, and compare them to see which one seems more informative to you.

9.3 MEASURES OF CENTRAL TENDENCY

The idea of taking an *average* is familiar to practically everyone. Quite often we hear people talk about average salary, average height, average grade, and so on. The idea of averages is so commonly used it should not surprise you to learn that several kinds of averages have been introduced in statistics.

Averages are called *measures of central tendency* because they estimate the "center" of the data collected. The three most common measures of central tendency are the *arithmetic mean, median,* and *mode*.

MEAN

The *arithmetic mean,* or *mean,* of a set of real numbers $x_1, x_2, \ldots, x_n$ is denoted by $\overline{X}$ and is defined as

$$\overline{X} = \frac{x_1 + x_2 + \cdots + x_n}{n} \tag{1}$$

where n is the number of items being averaged.

EXAMPLE 1

The grades of a student on eight 100-point examinations were 70, 65, 69, 85, 94, 62, 79, and 100. Find the mean.

SOLUTION

In this example $n = 8$. The mean of this set of grades is

$$\overline{X} = \frac{70 + 65 + 69 + 85 + 94 + 62 + 79 + 100}{8} = 78$$

....

An interesting fact about the mean is that the sum of deviations of each item from the mean is zero. In Example 1 the deviation of each score from the mean $\overline{X} = 78$ is

$(100 - 78)$, $(94 - 78)$, $(85 - 78)$, $(79 - 78)$, $(70 - 78)$, $(69 - 78)$, $(65 - 78)$, and $(62 - 78)$. Table 9 lists each score, the mean, and the deviation from the mean. If we add the deviations from the mean, we obtain a sum of zero.

Table 9

Score	Mean	Deviation from Mean
62	78	-16
65	78	-13
69	78	-9
70	78	-8
79	78	1
85	78	7
94	78	16
100	78	22
		Sum of Deviations: 0

For any group of data the following result is true:

> The sum of the deviations from the mean is zero.

As a matter of fact, we could have defined the mean as that real number for which the sum of the deviations is zero.

Another interesting fact about the mean is given below.

> If Y is any guessed or assumed mean (which may be any real number) and if d_j denotes the deviation of each item of the data from the assumed mean ($d_j = x_j - Y$), then the actual mean is
>
> $$\overline{X} = Y + \frac{d_1 + d_2 + \cdots + d_n}{n} \tag{2}$$

Look again at Example 1. We know that the actual mean is 78. Suppose we had guessed the mean to be 52. Then, using Formula (2), we obtain

$$\overline{X} = 52 + \frac{(100 - 52) + (94 - 52) + (85 - 52) + (79 - 52) + (70 - 52) + (69 - 52) + (65 - 52) + (62 - 52)}{8}$$

$$= 52 + 26 = 78$$

which agrees with the mean computed in Example 1.

A method for computing the mean for grouped data given in a frequency table is illustrated by the following example.

EXAMPLE 2

Find the mean for the grouped data given in Table 6 (repeated in the first two columns of Table 10 below).

SOLUTION

1. Take the midpoint (m_i) of each of the class intervals as a reference point and enter the result in column 3 of Table 10. For example, the midpoint of the class interval 80–89.99 is 85.

Table 10

Class Interval	f_i	m_i	$f_i m_i$
50– 59.99	7	55	385
60– 69.99	9	65	585
70– 79.99	19	75	1425
80– 89.99	14	85	1190
90– 99.99	14	95	1330
100–109.99	6	105	630
110–119.99	2	115	230
$n = 71$			$5775 = $ Sum of $f_i m_i$

2. Next, multiply the entry in column 3 by the frequency f_i for that class interval and enter the product in column 4, which is labeled $f_i m_i$.
3. Add the entries in column 4.

The mean is then computed by dividing the sum by the number of entries. That is,

$$\overline{X} = \frac{\Sigma f_i m_i}{n} \tag{3}$$

where

Σ Means add the entries

$f_i = $ Number of entries in the ith class interval

$m_i = $ Midpoint of ith class interval

$n = $ Number of items

Table 10 displays the information needed to complete the example. Now we can use the data in Table 10 and Equation (3) to find the mean:

$$\overline{X} = \frac{5775}{71} = 81.34$$

· ·

When data are grouped, the original data are lost due to grouping. As a result, the number obtained by using (3) is only an approximation to the actual mean. The reason for this is that using (3) amounts to computing the weighted average midpoint of a class interval (weighted by the frequency of scores in that interval) and therefore cannot be a computation for $\overline{X}$ exactly.

MEDIAN

The *median* of a set of real numbers arranged in order of magnitude is the middle value if the number of items is odd, and it is the mean of the two middle values if the number of items is even.

EXAMPLE 3

(a) The group of data 2, 2, 3, 4, 5, 7, 7, 7, 11 has median 5.

(b) The group of data 2, 2, 3, 3, 4, 5, 7, 7, 7, 11 has median 4.5 since

$$\frac{4 + 5}{2} = 4.5$$

. . . .

Finding the median for grouped data requires more work. As with the mean, the median for grouped data only approximates the actual median that would have been obtained prior to grouping the data.

To find the median for the grouped data in Table 6, page 415, we proceed as follows: The median is that point in the data that will have 50% of the entries above it and 50% below it. Now 50% of 71 is 35.5, so we are interested in finding the point in the distribution with 35 entries above it and 35 below it.

We start by counting up from the bottom until we come as close to 35.5 as possible, but not exceeding it. This brings us through the interval 70–79.99. Thus the median must lie in the interval 80–89.99. Now the median will equal the lower limit of the interval 80–89.99, namely, 80, plus an **interpolation factor**. The interpolation factor is determined as follows:

Interpolation Factor

1. In the interval containing the median, count the number p of entries or fractional entries needed to reach the median (for the grouped data in Table 6, the number is 0.5).

2. If the frequency for the interval is q, divide the interval into q parts.

3. The interpolation factor I is

$$I = \frac{p}{q} \cdot i$$

where

p = **Number of entries or fractional entries needed to reach the median**

q = **Number of entries in the interval**

i = **Size of the interval**

The median M is then given by

$$M = \textbf{(Lower limit of interval)} + \textbf{(Interpolation factor)}$$

For the data of Table 6 the median M is

$$M = 80 + \frac{0.5}{14} \cdot 10 = 80.36$$

Again, keep in mind that this median is an approximation to the actual median since it is obtained from grouped data. If we go back to the original data listed in Table 4, we obtain $M = 82.32$.

The median is sometimes called the **fiftieth percentile** and is denoted by C_{50} to indicate that 50% of the data are below it and 50% are above it. Similarly, we can find C_{25}, or the first quartile, and C_{75}, or the third quartile.

For the data in Table 10, C_{25} is formed by first finding the class interval containing the tally equal to 25% of all the tallies. Thus the tally corresponding to C_{25} is found in the class interval 70–79.99 since

$$25\% \text{ of } 71 = 17.75$$

and 16 tallies lie in the first two class intervals. Using the interpolation factor, we find that

$$C_{25} = 70 + \tfrac{1.75}{19}(10) = 70 + 0.92 = 70.92$$

MODE

The *mode* of a set of real numbers is the value that occurs with the greatest frequency exceeding a frequency of 1.

The mode does not necessarily exist, and if it does, it is not always unique.

EXAMPLE 4   For the data listed in Table 3, page 413, the mode is 79 (8 is the highest frequency).

. . . .

EXAMPLE 5   The group of data 2, 3, 4, 5, 7, 15 has no mode.

Now Work Problem 1
. . . .

EXAMPLE 6   The group of data 2, 2, 2, 3, 3, 7, 7, 7, 11, 15 has two modes, 2 and 7, and is called **bimodal.**

. . . .

When data have been listed in a frequency table, the mode is defined as the midpoint of the interval consisting of the largest number of cases. For example, the mode for the data in Table 6, page 415, is 75 (the midpoint of the interval 70–79.99).

Of the three measures of central tendency considered so far, the mean is the most important, the most reliable, and the one most frequently used. The reason for this is that it is easy to understand, easy to compute, and uses all the data in the collection. If two samples are chosen from the same population, the two *means* corresponding to the two samples will not generally differ by as much as the two *medians* of these samples.

EXERCISE 9.3   Answers to odd-numbered problems begin on page AN-35.

In Problems 1–6 compute the mean, median, and mode of the given raw data.

1. 21, 25, 43, 36

2. 16, 18, 24, 30

3. 55, 55, 80, 92, 70

4. 65, 82, 82, 95, 70

5. 62, 71, 83, 90, 75

6. 48, 65, 80, 92, 80

In Problems 7–10 use the given graphs to determine the mean and median.

7.

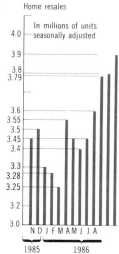

Home resales

Chicago Tribune Chart: Source:
National Association of Realtors

8.

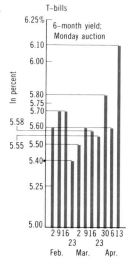

T-bills

9.

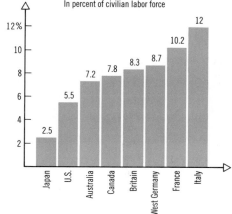

1988 Unemployment rates
In percent of civilian labor force

(Source: The Boston Company.)

10.

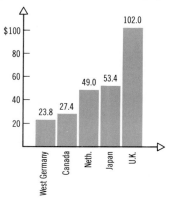

Direct investment in the United States.

(Source: U.S. Department of Commerce.)

11. If an investor purchased 50 shares of IBM stock at $85 per share, 90 shares at $105 per share, 120 shares at $110 per share, and another 75 shares at $130 per share, what is the average cost per share?

12. If a farmer sells 120 bushels of corn at $4 per bushel, 80 bushels at $4.10 per bushel, 150 bushels at $3.90 per bushel, and 120 bushels at $4.20 per bushel, what is the average income per bushel?

13. The annual salaries of five faculty members in the mathematics department at a large university are $34,000, $35,000, $36,000, $36,500, and $65,000. Compute the mean and median. Which measure describes the situation more realistically? If you were among the four lower-paid members, which measure would you use to describe the situation? What if you were the one making $65,000?

14. For the grouped data in Table 5, page 415, compute the mean, median, and mode.

15. The distribution of the monthly earnings of 1155 secretaries in May 1994 in the Chicago metropolitan area is summarized in the table. Find the mean salary and the median salary.

Monthly Earnings, $	Number of Secretaries
950–1199.99	25
1200–1449.99	55
1450–1699.99	325
1700–1949.99	410
1950–2199.99	215
2200–2449.99	75
2450–2699.99	50

16. According to an article in the October 4, 1988, edition of *The Wall Street Journal,* for companies having fewer than 5000 employees the average sales per employee were as follows:

Size of Company (Number of Employees)	Sales per Employee (Thousands of Dollars)
1–4	112
5–19	128
20–99	127
100–499	118
500–4999	120

Estimate the mean sales per employee for all firms having fewer than 5000 employees.

17. An article by Lester Thurow ("A Surge in Inequality," *Scientific American,* May 1987) presented the following data about the net worth of U.S. families:

Net Worth ($)	Percentage of Families		
	1970	1977	1983
0–4,999	38	35	33
5,000–9,999	6	5	5
10,000–24,999	14	13	12
25,000–49,999	17	17	16
50,000–99,999	13	15	17
100,000–249,999	9	12	12
250,000–499,999	2	2	3
500,000 or more	1	1	2

Estimate the mean net worth of U.S. families in each year, using a class mark of $1 million for the last class.

18. For the data given in Problems 12, 14, and 15 in Exercise 9.1 (page 418), find the mean, median, and mode.

19. Find C_{75} and C_{40} for the grouped data in Tables 5 and 6 (page 415).

20. In a labor–management wage negotiation in which the laborers are the lowest paid of the workers in the company, which measure of central tendency would labor tend to use as an argument for more pay? Which would management use? Why?

21. For the data given in Problem 11 in Exercise 9.1 (page 418), find the mean, using an assumed mean.

22. In a frequency table the score x_1 appears f_1 times, the score x_2 appears f_2 times, . . . , and the score x_n appears f_n times. Show that the mean $\overline{X}$ is given by the formula

$$\overline{X} = \frac{x_1 \cdot f_1 + x_2 \cdot f_2 + \cdots + x_n \cdot f_n}{f_1 + f_2 + \cdots + f_n}$$

9.4 MEASURES OF DISPERSION

EXAMPLE 1   Find the mean and median for each of the following sets of scores:

$$S_1: \quad 4, 6, 8, 10, 12, 14, 16$$
$$S_2: \quad 4, 7, 9, 10, 11, 13, 16$$

SOLUTION

For S_1 the mean $\overline{X}_1$ and median M_1 are

$$\overline{X}_1 = \frac{4 + 6 + 8 + 10 + 12 + 14 + 16}{7} = \frac{70}{7} = 10 \qquad M_1 = 10$$

For S_2 the mean $\overline{X}_2$ and median M_2 are

$$\overline{X}_2 = \frac{4 + 7 + 9 + 10 + 11 + 13 + 16}{7} = \frac{70}{7} = 10 \qquad M_2 = 10$$

• •

Notice that each set of scores has the same mean and the same median. As Figure 9 illustrates, the scores in S_2 seem to be more closely clustered around 10 than those in S_1.

Figure 9

We need a statistical measure to indicate the extent to which the scores in Example 1 are spread out. Such measures are called *measures of dispersion.*

Range

The simplest measure of dispersion is the **range,** which we have already defined as the difference between the largest value and the smallest value. For S_1 and S_2 the range is $16 - 4 = 12$. We can see that the range is a poor measure of dispersion since it depends on only two scores and tells us nothing about the rest of the scores.

Another measure of dispersion is the **deviation from the mean.** Recall that this measure is characterized by the fact that if the deviations from the mean of each score are all added, the result is zero. Because of this, it is not widely used as a measure of dispersion.

We need a measure that will give us an idea of how much deviation is involved without having these deviations sum to zero. By squaring each deviation from the mean, adding them, and dividing by the number of scores, we obtain an average squared deviation, which is called the **variance** of the set of scores. The formula for the variance, which is denoted by σ^2, is*

$$\sigma^2 = \frac{(x_1 - \overline{X})^2 + (x_2 - \overline{X})^2 + \cdots + (x_n - \overline{X})^2}{n}$$

where $\overline{X}$ is the mean of the scores $x_1, x_2, \ldots, x_n$ and n is the number of scores.

---

* The lowercase Greek letter sigma.

EXAMPLE 2

Calculate the variance for sets S_1 and S_2 of Example 1.

SOLUTION

For S_1, $\overline{X} = 10$ so that

$$\sigma^2 = \frac{\left[\begin{array}{c}(4 - 10)^2 + (6 - 10)^2 + (8 - 10)^2 + (10 - 10)^2 \\ + (12 - 10)^2 + (14 - 10)^2 + (16 - 10)^2\end{array}\right]}{7}$$

$$= 16$$

For S_2, $\overline{X} = 10$ so that

$$\sigma^2 = \frac{\left[\begin{array}{c}(4 - 10)^2 + (7 - 10)^2 + (9 - 10)^2 + (10 - 10)^2 \\ + (11 - 10)^2 + (13 - 10)^2 + (16 - 10)^2\end{array}\right]}{7}$$

$$= 13.14$$

· ·

Standard Deviation

In order to use the variance in practical situations (for instance, if our data represent dollars, we cannot talk about "squared dollars"), we use the square root of the variance. This is called the **standard deviation** of a set of scores. The standard deviation is denoted by σ and is given by the formula

$$\sigma = \sqrt{\frac{(x_1 - \overline{X})^2 + (x_2 - \overline{X})^2 + \cdots + (x_n - \overline{X})^2}{n}}$$

where $\overline{X}$ and the x_is are defined the same way as for the variance.

For the data in Example 1 the standard deviation for S_1 is

$$\sigma = \sqrt{\frac{36 + 16 + 4 + 0 + 4 + 16 + 36}{7}} = \sqrt{\frac{112}{7}} = \sqrt{16} = 4$$

and the standard deviation for S_2 is

$$\sigma = \sqrt{\frac{36 + 9 + 1 + 0 + 1 + 9 + 36}{7}} = \sqrt{\frac{92}{7}} = \sqrt{13.14} = 3.625$$

The fact that the standard deviation of the set S_2 is less than that for the set S_1 is an indication that the scores of S_2 are more clustered around the mean than those of S_1.

EXAMPLE 3

Find the standard deviation for the data

$$100, 90, 90, 85, 80, 75, 75, 75, 70, 70, 65, 65, 60, 40, 40, 40$$

SOLUTION

The mean is

$$\overline{X} = \frac{100 + 2 \cdot 90 + 85 + 80 + 3 \cdot 75 + 2 \cdot 70 + 2 \cdot 65 + 60 + 3 \cdot 40}{16} = 70$$

The deviations from the mean and their squares are given in Table 11. The standard deviation is

$$\sigma = \sqrt{\frac{4950}{16}} = \frac{70.4}{4} = 17.6$$

Now Work Problem 3

....

EXAMPLE 4

Find the standard deviation for the data

$$80, 80, 80, 80, 75, 75, 75, 75, 70, 70, 65, 65, 60, 60, 55, 55$$

SOLUTION

Here the mean is $\overline{X} = 70$ for the 16 scores. Table 12 gives the deviations from the mean and their squares. The standard deviation is

$$\sigma = \sqrt{\frac{1200}{16}} = \sqrt{75} = 8.7$$

....

Table 11

Scores, x	Deviation from the Mean, $x - \overline{X}$	Deviation Squared, $(x - \overline{X})^2$
40	− 30	900
40	− 30	900
40	− 30	900
60	− 10	100
65	− 5	25
65	− 5	25
70	0	0
70	0	0
75	5	25
75	5	25
75	5	25
80	10	100
85	15	225
90	20	400
90	20	400
100	30	900
Mean = 70 n = 16	Sum = 0	Sum = 4950

Table 12

Scores, x	Deviation from the Mean, $x - \overline{X}$	Deviation Squared, $(x - \overline{X})^2$
55	− 15	225
55	− 15	225
60	− 10	100
60	− 10	100
65	− 5	25
65	− 5	25
70	0	0
70	0	0
75	5	25
75	5	25
75	5	25
75	5	25
80	10	100
80	10	100
80	10	100
80	10	100
Mean = 70 n = 16	Sum = 0	Sum = 1200

These two examples show that although the samples have the same mean, 70, and the same sample size, 16, the scores in the first set deviate further from the mean than do the scores in the second set.

> In general, a relatively small standard deviation indicates that the measures tend to cluster close to the mean, and a relatively high standard deviation shows that the measures are widely scattered from the mean.

Chebychev's Theorem

Suppose we are observing an experiment with numerical outcomes and that the experiment has mean $\overline{X}$ and standard deviation σ. We wish to estimate the probability that a randomly chosen outcome lies within k units of the mean.

> **Chebychev's Theorem***
>
> For any distribution of numbers with mean $\overline{X}$ and standard deviation σ, the probability that a randomly chosen outcome lies between $\overline{X} - k$ and $\overline{X} + k$ is at least $1 - \dfrac{\sigma^2}{k^2}$.

EXAMPLE 5

Suppose that an experiment with numerical outcomes has mean 4 and standard deviation 1. Use Chebychev's theorem to estimate the probability that an outcome lies between 2 and 6.

SOLUTION

Here, $\overline{X} = 4$, $\sigma = 1$. Since we wish to estimate the probability that an outcome lies between 2 and 6, the value of k is $k = 6 - \overline{X} = 6 - 4 = 2$ (or $k = \overline{X} - 2 = 4 - 2 = 2$). Then by Chebychev's theorem, the desired probability is at least

$$1 - \frac{\sigma^2}{k^2} = 1 - \frac{1}{2^2} = 1 - \frac{1}{4} = .75$$

That is, we expect at least 75% of the outcomes of this experiment to lie between 2 and 6.

Now Work Problem 17

EXAMPLE 6

An office supply company sells boxes containing 100 paper clips. Because of the packaging procedure, not every box contains exactly 100 clips. From previous data it

* Named after the nineteenth-century Russian mathematician P. L. Chebychev.

is known that the average number of clips in a box is indeed 100 and the standard deviation is 2.8. If the company ships 10,000 boxes, estimate the number of boxes having between 94 and 106 clips, inclusive.

SOLUTION

Our experiment involves counting the number of clips in the box. For this experiment we have $\overline{X} = 100$ and $\sigma = 2.8$. Therefore, by Chebychev's theorem the fraction of boxes having between $100 - 6$ and $100 + 6$ clips ($k = 6$) should be at least

$$1 - \frac{(2.8)^2}{6^2} = 1 - .22 = .78$$

That is, we expect at least 78% of 10,000 boxes, or about 7800 boxes to have between 94 and 106 clips.

....

The importance of Chebychev's theorem stems from the fact that it applies to *any* data—only the mean and standard deviation must be known. However, the estimate is a crude one. Other results (such as the *normal distribution* given later) produce more accurate estimates about the probability of falling within k units of the mean.

Standard Deviation for Grouped Data

To find the standard deviation for grouped data, we use the formula

$$\sigma = \sqrt{\frac{(m_1 - \overline{X})^2 \cdot f_1 + (m_2 - \overline{X})^2 \cdot f_2 + \cdots + (m_k - \overline{X})^2 \cdot f_k}{n}}$$

where $m_1, m_2, \ldots, m_k$ are the class midpoints; $f_1, f_2, \ldots, f_k$ are the respective frequencies; n is the sum of the frequencies, that is, $n = f_1 + f_2 + \cdots + f_k$; and $\overline{X}$ is the mean.

EXAMPLE 7

Find the standard deviation for the grouped data given in Table 10, page 426.

SOLUTION

We have already found that the mean for the grouped data is

$$\overline{X} = 81.3$$

The class midpoints are 55, 65, 75, 85, 95, 105, and 115. The deviations of the mean from the class midpoints, their squares, and the products of the squares by the respective frequencies are listed in Table 13. The standard deviation is

$$\sigma = \sqrt{\frac{16{,}447.99}{71}} = \sqrt{231.66} = 15.22$$

Table 13

Class Midpoint	f_i	$m_i - \overline{X}$	$(m_i - \overline{X})^2$	$(m_i - \overline{X})^2 \cdot f_i$
115	2	33.7	1,135.69	2271.38
105	6	23.7	561.69	3370.14
95	14	13.7	187.69	2627.66
85	14	3.7	13.69	191.66
75	19	− 6.3	39.69	754.11
65	9	− 16.3	265.69	2391.21
55	7	− 26.3	691.69	4841.83
Sum	71			16,447.99

. . . .

A little computation shows that the sum of the deviations of the approximate mean from the class midpoints is not exactly zero. This is because we are using only an approximation to the mean. Remember, we cannot compute the exact mean for grouped data.

EXERCISE 9.4   Answers to odd-numbered problems begin on page AN-35.

---

1. Use histograms (a) and (b) to determine by inspection which distribution has the larger variance.

2. Use histograms (b) and (c) to determine by inspection which distribution has the larger variance.

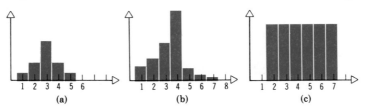

(a)   (b)   (c)

In Problems 3–8 compute the standard deviation for the given raw data.

3. 4, 5, 9, 9, 10, 14, 25

4. 6, 8, 10, 10, 11, 12, 18

5. 62, 58, 70, 70

6. 55, 65, 80, 80, 90

7. 85, 75, 62, 78, 100

8. 92, 82, 75, 75, 82

9. Find the standard deviation for the data given in Problem 11, Exercise 9.1 (page 418).

10. Find the standard deviation for the grouped data given in Table 5 (page 415).

In Problems 11 and 12 calculate the mean and the standard deviation.

11.

Class	Frequency
10–16	1
17–23	3
24–30	10
31–37	12
38–44	5
45–51	2

12.

Class	Frequency
0–3	2
4–7	5
8–11	8
12–15	6
16–19	3

13. The lifetimes of six light bulbs are 968, 893, 769, 845, 922, and 815 hours. Calculate the mean lifetime and the standard deviation.

14. A group of 25 applicants for admission to Midwestern University made the following scores on the quantitative part of an aptitude test:

591	570	425	472	555
490	415	479	517	570
606	614	542	607	441
502	506	603	488	460
550	551	420	590	482

Find the mean and standard deviation of these scores.

15. Fishing   The number of salmon caught in each of two rivers over the past 15 years is as follows:

River I		River II	
Number Caught	Years	Number Caught	Years
500–1499	4	750–1249	2
1500–2499	8	1350–1799	3
2500–3499	2	1800–2249	4
3500–4499	1	2250–2699	4
		2700–3149	2

Which river should be preferred for fishing?

16. Charge Accounts   A department store takes a sample of its customer charge accounts and finds the following:

Outstanding Balance	Number of Accounts
0–49	15
50–99	41
100–149	80
150–199	60
200–249	8

Find the mean and the standard deviation of the outstanding balances.

17. Suppose that an experiment with numerical outcomes has mean 25 and standard deviation 3. Use Chebychev's theorem to tell what percent of outcomes lie

(a) Between 19 and 31.
(b) Between 20 and 30.
(c) Between 16 and 34.
(d) Less than 19 or more than 31.
(e) Less than 16 or more than 34.

18. A watch company determines that the number of defective watches in each box averages 6 with standard deviation 2. Suppose that 1000 boxes are produced. Estimate the number of boxes having between 0 and 12 defective watches.

19. Sales   The average sale at a department store is $51.25, with a standard deviation of $8.50. Find the smallest interval such that by Chebychev's theorem at least 90% of the store's sales fall within it.

9.5 THE NORMAL DISTRIBUTION

Frequency polygons or frequency distributions can assume almost any shape or form, depending on the data. However, the data obtained from many experiments often follow a common pattern. For example, heights of adults, weights of adults, test scores, and

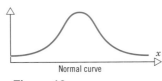

Figure 10

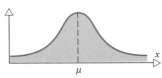

Figure 11

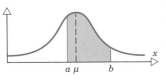

Probability between a and b
= area of the shaded region.

Figure 12

coin tossing all lead to data that have the same kind of frequency distribution. This distribution is referred to as the **normal distribution** or the **Gaussian distribution.** Because it occurs so often in practical situations, it is generally regarded as the most important distribution, and much statistical theory is based on it. The graph of the normal distribution, called the **normal curve,** is the bell-shaped curve shown in Figure 10.

The Normal Curve

Some properties of the normal curve are listed below.

1. Normal curves are bell-shaped and are symmetrical with respect to a vertical line. See Figure 11.
2. The mean μ is at the center. See Figure 11.
3. Irrespective of the shape, the area enclosed by the curve and the x-axis is always equal to 1. See the shaded region in Figure 11.
4. The probability that an outcome of a normally distributed experiment is between a and b equals the area under the associated normal curve from $x = a$ to $x = b$. See the shaded region in Figure 12.
5. The standard deviation of a normal distribution plays a major role in describing the area under the normal curve. As shown in Figure 13, the standard deviation is related to the area under the normal curve as follows:

(a) About 68.27% of the total area under the curve is within 1 standard deviation of the mean (from $\mu - \sigma$ to $\mu + \sigma$).
(b) About 95.45% of the total area under the curve is within 2 standard deviations of the mean (from $\mu - 2\sigma$ to $\mu + 2\sigma$).
(c) About 99.73% of the total area under the curve is within 3 standard deviations of the mean (from $\mu - 3\sigma$ to $\mu + 3\sigma$).

Figure 13

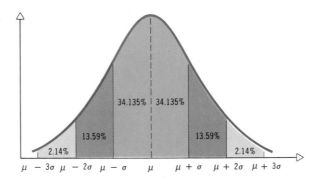

It is also important to recognize that, in theory, the normal curve will never touch the x-axis but will extend to infinity in either direction. In addition, every normal distribution has its mean, median, and mode at the same point.

Now Work Problem 1

EXAMPLE 1 | At Jefferson High School the average IQ score of the 1200 students is 100, with a standard deviation of 15. The IQ scores have a normal distribution.

(a) How many students have an IQ between 85 and 115?
(b) How many students have an IQ between 70 and 130?
(c) How many students have an IQ between 55 and 145?
(d) How many students have an IQ under 55 or over 145?
(e) How many students have an IQ over 145?

SOLUTION

See Figure 14.

Figure 14

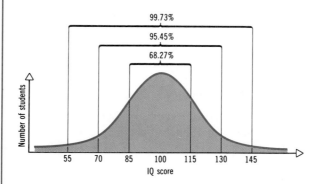

(a) The IQ scores have a normal distribution and the mean is 100. Since the standard deviation σ is 15, then 1σ either side of the mean is from 85 to 115. By Property 5(a) we know that 68.27% of 1200, or

$$(0.6827)(1200) = 819 \text{ students}$$

have IQs between 85 and 115.
(b) The scores from 70 to 130 extend $2\sigma (= 30)$ either side of the mean. By Property 5(b) we know that 95.45% of 1200, or

$$(0.9545)(1200) = 1145 \text{ students}$$

have IQs between 70 and 130.
(c) The scores from 55 to 145 extend $3\sigma (= 45)$ either side of the mean. By Property 5(c) we know that 99.73% of 1200, or

$$(0.9973)(1200) = 1197 \text{ students}$$

have IQs between 55 and 145.
(d) There are three students $(1200 - 1197)$ who have scores that are not between 55 and 145.
(e) One or two students have IQs above 145.

⋯⋯⋯⋯⋯⋯⋯⋯⋯⋯⋯

A normal distribution curve is completely determined by μ and σ. Hence, different normal distributions of data with different means and standard deviations give rise to different shapes of the normal curve. Figure 15 indicates how the normal curve changes when the standard deviation changes. For the sake of clarity, we assume all data have the same mean.

Figure 15

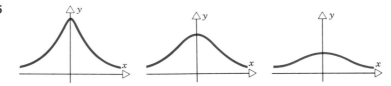

As the standard deviation increases, the normal curve flattens out. A flatter curve indicates a greater likelihood for the outcomes to be spread out. A steeper curve indicates that the outcomes are more likely to be close to the mean.

Standard Normal Curve

It would be a hopeless task to attempt to set up separate tables of normal curve areas for every conceivable value of μ and σ. Fortunately, we are able to transform all the observations to one table—the table corresponding to the so-called **standard normal curve,** which is the normal curve for which $\mu = 0$ and $\sigma = 1$. This can be achieved by introducing new score data, called **Z-scores,** defined as

$$Z = \frac{\textbf{Difference between } x \textbf{ and } \mu}{\textbf{Standard deviation}} = \frac{x - \mu}{\sigma} \qquad (1)$$

where

$$x = \text{Old score data}$$
$$\mu = \text{Mean of the old data}$$
$$\sigma = \text{Standard deviation of the old data}$$

The new score data defined by (1) will always have a *zero mean* and a *unit standard deviation.* Such data are said to be expressed in **standard units** or **standard scores.** By expressing data in terms of standard units, it becomes possible to make a comparison of distributions. Furthermore, as for all normal curves, the total area under a standard normal curve is equal to 1.

EXAMPLE 2   On a test 80 is the mean and 7 is the standard deviation. What is the Z-score of a score of

(a) 88?   (b) 62?

Interpret your results.

SOLUTION

(a) Here, 88 is the regular score. Using (1) with $x = 88$, $\mu = 80$, $\sigma = 7$, we get

$$Z = \frac{x - \mu}{\sigma} = \frac{88 - 80}{7} = \frac{8}{7} = 1.1429$$

(b) Here, 62 is the regular score. Using (1) with $x = 62$, $\mu = 80$, and $\sigma = 7$, we get

$$Z = \frac{62 - 80}{7} = \frac{-18}{7} = -2.5714$$

See Figure 16.

Figure 16

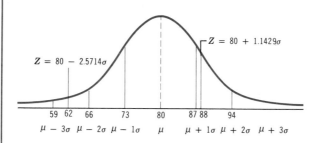

The Z-score of 1.1429 tells us that the original score of 88 is 1.1429 standard deviations *above* the mean. The Z-score of -2.5714 tells us that the original score of 62 is 2.5714 standard deviations *below* the mean. A negative Z-score always means that the score is below the mean.

Now Work Problem 5

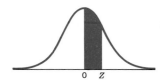

Figure 17

The curve in Figure 17 with mean $\mu = 0$ and standard deviation $\sigma = 1$ is the standard normal curve. For this curve the areas between $Z = -1$ and 1, $Z = -2$ and 2, $Z = -3$ and 3 are equal, respectively, to 68.27%, 95.45%, and 99.73% of the total area under the curve, which is 1. To find the areas cut off between other points, we proceed as in the following example.

EXAMPLE 3

(a) Find the area, that is, find the proportion of cases, included between 0 and 0.6 on a standard normal curve. Refer to the shaded area in Figure 18(a).
(b) Find the area, that is, find the proportion of cases, included between 0.6 and 1.86 on a standard normal curve. Refer to the shaded area in Figure 18(b).

Figure 18

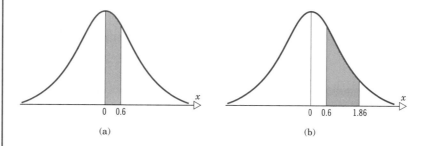

(a)   (b)

SOLUTION

We use the **standard normal curve table,** Table III in the back of the book.

(a) To find the area between 0 and 0.6, we find $Z = 0.6$ in the table. Corresponding to $Z = 0.6$ is the value 0.2257, which is the area. In other words, 22.57% of the cases will lie between 0 and 0.6.

(b) We begin by checking the table to find the area of the curve cut off between the mean and a point equivalent to a standard score of 0.6 from the mean. This value is 0.2257, as we found in Part (a). Next, we continue down the table in the left-hand column until we come to a standard score of 1.8. By looking across the row to the column below 0.06, we find that 0.4686 of the area is included between the mean and 1.86. Then the area of the curve between these two points is the difference between the two areas, $0.4686 - 0.2257$, which is 0.2429. We can then state that approximately 24.29% of the cases fall between 0.6 and 1.86, or that *the probability of a score falling between these two points is about .2429.*

Now Work Problem 11

....

In the next example we take two points that are on different sides of the mean.

EXAMPLE 4

We want to determine the area of the standard normal curve that falls between a standard score of -0.39 and one of 1.86.

SOLUTION

See Figure 19.

Figure 19

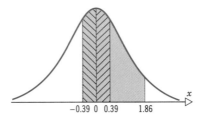

There are no values for negative standard scores in Table III. Because of the symmetry of normal curves, equal standard scores, whether positive *or* negative, give equal areas when taken from the mean. From Table III we find that a standard score of 0.39 cuts off an area of 0.1517 between it and the mean. A standard score of 1.86 includes 0.4686 of the area of the curve between it and the mean. The area included between both points is then equal to the sum of these two areas, $0.1517 + 0.4686$, which is 0.6203. Thus approximately 62.03% of the area is between -0.39 and 1.86. In other words, the probability of a score falling between these two points is about .6203.

....

EXAMPLE 5

A student receives a grade of 82 on a final examination in biology for which the mean is 73 and the standard deviation is 9. In his final examination in sociology, for which the mean grade is 81 and the standard deviation is 15, he receives an 89. In which examination is his relative standing higher?

SOLUTION

In their present form these distributions are not comparable since they have different means and, more important, different standard deviations. In order to compare the data, we transform the data to standard scores. For the biology test data the new score data for the student's examination score are

$$ Z = \frac{82 - 73}{9} = \frac{9}{9} = 1 $$

For the sociology test data, the new score data for the student's examination score are

$$ Z = \frac{89 - 81}{15} = \frac{8}{15} = 0.533 $$

This means the student's score in the biology exam is 1 standard unit above the mean, while his score in the sociology exam is 0.533 standard unit above the mean. Hence, his *relative standing* is higher in biology.

Now Work Problem 19

• •

The Normal Curve as an Approximation to the Binomial Distribution

We start with an example.

EXAMPLE 6

Consider an experiment in which a fair coin is tossed 10 times. Find the frequency distribution for the probability of tossing a head.

SOLUTION

The probability for obtaining exactly k heads is given by a binomial distribution $b(10, k; \frac{1}{2})$. Thus we obtain the distribution given in Table 14. If we graph this frequency distribution, we obtain the line chart shown in Figure 20. When we connect the tops of the lines of the line chart, we obtain a *normal curve*, as shown.

Table 14

No. of Heads	Probability $b(10, k; \frac{1}{2})$
0	.0010
1	.0098
2	.0439
3	.1172
4	.2051
5	.2461
6	.2051
7	.1172
8	.0439
9	.0098
10	.0010

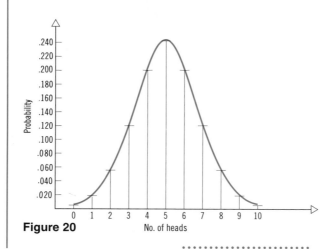

Figure 20

• •

This particular distribution for $n = 10$ and $p = \frac{1}{2}$ is not a result of the choice of n or p. As a matter of fact, the line chart for any binomial probability $b(n, k; p)$ will give an approximation to a normal curve. You should verify this for the cases in which $n = 15$, $p = .3$, and $n = 8$, $p = \frac{3}{4}$.

Probabilities associated with binomial experiments are readily obtainable from the formula $b(n, k; p)$ when n is small. If n is large, we can compute the binomial probabilities by an approximating procedure using a normal curve. It turns out that the normal distribution provides a very good approximation to the binomial distribution when n is large or p is close to $\frac{1}{2}$.

The mean μ for the binomial distribution is given by $\mu = np$ (see Expected Value for Bernoulli Trials, page 392). Moreover, it can be shown that the standard deviation is $\sigma = \sqrt{npq}$.

EXAMPLE 7

Quality Control   A company manufactures 60,000 pencils each day. Quality control studies have shown that, on the average, 4% of the pencils are defective. A random sample of 500 pencils is selected from each day's production and tested. What is the probability that in the sample there are

(a) At least 12 and no more than 24 defective pencils?
(b) 32 or more defective pencils?

SOLUTION

(a) Since $n = 500$ is very large, it is appropriate to use a normal curve approximation for the binomial distribution. Thus with $n = 500$ and $p = .04$,

$$\mu = np = 500(.04) = 20 \qquad \sigma = \sqrt{npq} = \sqrt{500(.04)(.96)} = 4.38$$

To find the approximate probability of the number of defective pencils in a sample being at least 12 and no more than 24, we find the area under a normal curve from $x = 12$ to $x = 24$. See Figure 21.

Figure 21

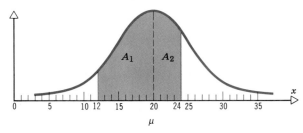

Areas A_1 and A_2 are found by converting to Z-scores and using the standard normal curve table, Table III.

$$x = 12: \quad Z_1 = \frac{x - \mu}{\sigma} = \frac{12 - 20}{4.38} = -1.83 \qquad A_1 = .4664$$

$$x = 24: \quad Z_2 = \frac{x - \mu}{\sigma} = \frac{24 - 20}{4.38} = .91 \qquad A_2 = .3186$$

Total area $= A_1 + A_2 = .4664 + .3186 = .785$

Thus the approximate probability of the number of defective pencils in the sample being at least 12 and not more than 24 is .785.

(b) We want to find the area A_2 indicated in Figure 22. We know that the area to the right of the mean is .5, and if we subtract the area A_1 from .5, we will obtain A_2. Therefore, we find the area A_1:

$$Z = \frac{x - \mu}{\sigma} = \frac{32 - 20}{4.38} = 2.74 \qquad A_1 = .4969$$

Then

$$A_2 = .5 - A_1 = .5 - .4969 = .0031$$

Thus the approximate probability of finding 32 or more defective pencils in the sample is .0031.

Figure 22

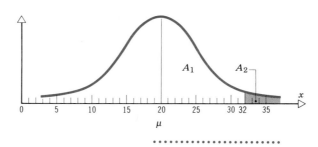

EXERCISE 9.5   Answers to odd-numbered problems begin on page AN-35.

In Problems 1–4 determine μ and σ by inspection.

1.

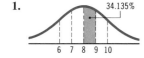

2.

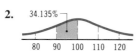

3.

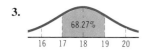

4.

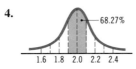

5. Given a normal distribution with a mean of 13.1 and a standard deviation of 9.3, find the Z-score equivalent of the following scores in this distribution:

7, 9, 13, 15, 29, 37, 41

6. Given a normal distribution with a mean of 15.2 and a standard deviation of 5.1, find the Z-score equivalent of the following scores in this distribution:

8, 9, 15, 16, 22, 23, 25

7. Given the following Z-scores on a standard normal distribution, find the area from the mean to each score.

(a) 0.89   (b) 1.10   (c) 2.50
(d) 3.00   (e) − 0.75   (f) − 2.31
(g) 0.80   (h) 3.03

8. An instructor assigns grades in an examination according to the following procedure:

A   if score exceeds $\mu + 1.6\sigma$
B   if score is between $\mu + 0.6\sigma$ and $\mu + 1.6\sigma$
C   if score is between $\mu - 0.3\sigma$ and $\mu + 0.6\sigma$
D   if score is between $\mu - 1.4\sigma$ and $\mu - 0.3\sigma$
F   if score is below $\mu - 1.4\sigma$

What percent of the class receives each grade, assuming that the scores are normally distributed?

In Problems 9–12 use Table III to find the area of each shaded region under the standard normal curve.

9.
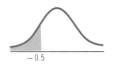
−0.5

10.

1   2

11.
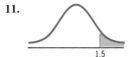
1.5

12.

−0.5 0 0.5

13. The average height of 2000 women in a random sample is 64 inches. The standard deviation is 2 inches. The heights have a normal distribution.

   (a) How many women in the sample are between 62 and 66 inches tall?
   (b) How many women in the sample are between 60 and 68 inches tall?
   (c) How many women in the sample are between 58 and 70 inches tall?

14. Corn flakes come in a box that says it holds a mean weight of 16 ounces of cereal. The standard deviation is 0.1 ounce. Suppose that the manufacturer packages 600,000 boxes with weights that have a normal distribution.

   (a) How many boxes weigh between 15.9 and 16.1 ounces?
   (b) How many boxes weigh between 15.8 and 16.2 ounces?
   (c) How many boxes weigh between 15.7 and 16.3 ounces?
   (d) How many boxes weigh under 15.7 or over 16.3 ounces?
   (e) How many boxes weigh under 15.7 ounces?

15. The weight of 100 college students closely follows a normal distribution with a mean of 130 pounds and a standard deviation of 5.2 pounds.

   (a) How many of these students would you expect to weigh at least 142 pounds?
   (b) What range of weights would you expect to include the middle 70% of the students in this group?

16. **Life Expectancy of Clothing**   If the average life of a certain make of clothing is 40 months with a standard deviation of 7 months, what percentage of these clothes can be expected to last from 28 months to 42 months? Assume that clothing lifetime follows a normal distribution.

17. **Life Expectancy of Shoes**   Records show that the average life expectancy of a pair of shoes is 2.2 years with a standard deviation of 1.7 years. A manufacturer guarantees that shoes lasting less than a year are replaced free. For every 1000 pairs sold, how many pairs should the manufacturer expect to replace free? Assume a normal distribution.

18. The attendance over a weekly period of time at a movie theater is normally distributed with a mean of 10,000 and a standard deviation of 1000 persons. Find

   (a) The number in the lowest 70% of the attendance figures.
   (b) The percent of attendance figures that falls between 8500 and 11,000 persons.
   (c) The percent of attendance figures that differs from the mean by 1500 persons or more.

19. Colleen, Mary, and Kathleen are vying for a position as editor. Colleen, who is tested with group I, gets a score of 76 on her test; Mary, who is tested with group II, gets a score of 89; and Kathleen, who is tested with group III, gets a score of 21. If the average score for group I is 82, for group II is 93, and for group III is 24, and if the standard deviation for each group is 7, 2, and 9, respectively, which person has the highest relative standing?

20. In Mathematics 135 the average final grade is 75.0 and the standard deviation is 10.0. The professor's grade distribution shows that 15 students with grades from 68.0 to 82.0 received Cs. Assuming the grades follow a normal distribution, how many students are in Mathematics 135?

21. Draw the line chart and frequency curve for the probability of a head in an experiment in which a biased coin is tossed 15 times and the probability that a head occurs is .3. [*Hint:*   Find $b(15, k; .30)$ for $k = 0, 1, \ldots, 15.$]

22. Follow the same directions as in Problem 21 for an experiment in which a biased coin is tossed 8 times and the probability that heads appears is $\frac{3}{4}$.

In Problems 23–28 suppose a binomial experiment consists of 750 trials and the probability of success for each trial is .4. Then

$$\mu = np = 300 \quad \text{and} \quad \sigma = \sqrt{npq} = \sqrt{(750)(.4)(.6)} = 13$$

Approximate the probability of obtaining the number of successes indicated by using a normal curve.

23. 285–315   **24.** 280–320   **25.** 300 or more

26. 300 or less   **27.** 325 or more   **28.** 275 or less

Technology Exercises

1. Graph the standard normal curve on a graphing calculator or a computer. For what value of x does the function assume its maximum? The equation is given by

$$y = \frac{1}{\sqrt{2\pi}} e^{-(1/2)x^2}$$

2. Graph the normal curve with $\mu = 10$ and $\sigma = 2$ on a graphing calculator or a computer. For what value of x does the function assume its maximum? The equation is given by

$$y = \frac{1}{2\sqrt{2\pi}} e^{-(1/8)(x-10)^2}$$

CHAPTER REVIEW

IMPORTANT TERMS AND CONCEPTS

frequency table 413
range 414
class interval 415
histogram 415
frequency polygon 416
pie chart 421

bar graph 422
mean 424
median 427
interpolation factor 427
mode 428
standard deviation 432

Chebychev's theorem 434
standard deviation for grouped data 435
normal distribution 438
standard normal curve 440
Z-score 440

IMPORTANT FORMULAS

Mean for Ungrouped Scores

$$\overline{X} = \frac{x_1 + x_2 + \cdots + x_n}{n}$$

Mean for Grouped Scores

$$\overline{X} = \frac{\Sigma f_i m_i}{n}$$

Standard Deviation for Ungrouped Scores

$$\sigma = \sqrt{\frac{(x_1 - \overline{X})^2 + (x_2 - \overline{X})^2 + \cdots + (x_n - \overline{X})^2}{n}}$$

Standard Deviation for Grouped Scores

$$\sigma = \sqrt{\frac{(m_1 - \overline{X})^2 \cdot f_1 + (m_2 - \overline{X})^2 \cdot f_2 + \cdots + (m_k - \overline{X})^2 \cdot f_k}{n}}$$

Z-Score

$$Z = \frac{x - \mu}{\sigma}$$

TRUE–FALSE ITEMS   Answers are on page AN-36.

T_____ F_____ 1. The range of a set of numbers is the difference between the standard deviation and the mean.

T_____ F_____ 2. Two sets of scores can have the same mean and median, yet be different.

T_____ F_____ 3. A relatively small standard deviation indicates that measures are widely scattered from the mean.

T_____ F_____ 4. The sum of the deviations from the mean is zero.

T_____ F_____ 5. For the normal distribution, 68.27% of the total area under the curve is within 2 standard deviations of the mean.

FILL IN THE BLANKS   Answers are on page AN-36.

1. The three most common measures of central tendency are

   (a) _____ (b) _____

   (c) _____ .

2. The square root of the variance is called _____ .

3. The graph of the normal distribution has a _____ shape.

4. The formula $\dfrac{x - \mu}{\sigma}$ is called the _____ .

5. The formula $1 - \dfrac{\sigma^2}{k^2}$ measures the probability that a randomly chosen variable lies between _____ and

   _____ .

REVIEW EXERCISES   Answers to odd-numbered problems begin on page AN-37.

1. The following scores were made on a math exam:

80	99	82	21	100	55	80	26	78	52
12	73	20	44	72	63	19	85	33	66
78	42	87	90	30	10	48	75	83	77
63	85	69	80	14	87	66	52	17	60
74	70	73	95	89	14	92	8	100	72

   (a) Set up a frequency table for the above data. What is the range?
   (b) Draw a line chart for the data.
   (c) Draw a histogram for the data using a class interval of size 5 beginning with 4.5.
   (d) Draw the frequency polygon for the histogram.
   (e) Find the cumulative (less than) frequencies and draw the cumulative (less than) frequency distribution.

2. The following table gives the percentage of marginal tax rates for married couples filing jointly; rates shown apply to taxable income for 1990; each portion of income is taxed at its own rate; the maximum income subject to the 33% current rate varies by exemptions claimed (*Source:* U.S. Internal Revenue Service).

Income	Current Rates in Percent
0–32,450	15
32,450–78,400	28
78,400–162,700	33
162,770+	28

   (a) Graph the data using a bar graph.
   (b) Graph the data using a pie graph.

3. Find the mean, median, and mode for each of the following sets of measurements.

   (a) 12, 10, 8, 2, 0, 4, 10, 5, 4, 4, 8, 0
   (b) 195, 5, 2, 2, 2, 2, 1, 0
   (c) 2, 5, 5, 7, 7, 7, 9, 9, 11

4. In which of the sets of data in Problem 3 is the mean a poor measure of central tendency? Why?

5. Give an example of two sets of scores for which the means are the same and the standard deviations are different.

6. Give one advantage of the standard deviation over the variance. Give an example.

7. In seven different rounds of golf, Joe scores 74, 72, 76, 81, 77, 76, and 73. What is the standard deviation of his scores?

8. A normal distribution has a mean of 25 and a standard deviation of 5.

   (a) What proportion of the scores fall between 20 and 30?
   (b) What proportion of the scores will lie above 35?

9. A set of 600 scores is normally distributed. How many scores would you expect to find:

   (a) Between $\pm 1\sigma$ of the mean?
   (b) Between 1σ and 3σ above the mean?
   (c) Between $\pm \frac{2}{3}\sigma$ of the mean?

10. **Average Life of a Dog**   The average life expectancy of a dog is 14 years, with a standard deviation of about 1.25 years. Assuming that the life spans of dogs are normally distributed, approximately how many dogs will die before reaching the age of 10 years, 4 months?

11. Use Table III to calculate the area under the normal curve between

    (a) $Z = -1.35$ and $Z = -2.75$
    (b) $Z = 1.2$ and $Z = 1.75$

12. Bob got an 89 on the final exam in mathematics and a 79 on the sociology exam. In the mathematics class the average grade was 79 with a standard deviation of 5, and in the sociology class the average grade was 72 with a standard deviation of 3.5. Assuming that the grades in both subjects were normally distributed, in which class did Bob rank higher?

13. Suppose it is known that the number of items produced in a factory has a mean of 40. If the variance of a week's production is known to equal 25, then what can be said about the probability that this week's production will be between 30 and 50?

14. From past experience a teacher knows that the test scores of students taking an examination have a mean of 75 and a variance of 25.

    (a) What can be said about the probability that a student will score between 65 and 85?
    (b) How many students would have to take the examination so as to ensure, with probability of at least .9, that the class average would be within 5 points of 75?

Mathematical Questions from Professional Exams

1. **Actuary Exam—Part II**   Under the hypothesis that a pair of dice are fair, the probability is approximately .95 that the number of 7s appearing in 180 throws of the dice will lie within $30 \pm K$. What is the value of K?

   (a) 2   (b) 4   (c) 6   (d) 8   (e) 10

2. **Actuary Exam—Part II**   If X is normally distributed with mean μ and variance μ^2 and if $P(-4 < X < 8) = .9974$, then $\mu =$

   (a) 1   (b) 2   (c) 4   (d) 6   (e) 8

3. **Actuary Exam—Part II**   A manufacturer makes golf balls whose weights average 1.62 ounces, with a standard deviation of 0.05 ounce. What is the probability that the weight of a group of 100 balls will lie in the interval 162 ± 0.5 ounces?

   (a) .18   (b) .34   (c) .68   (d) .84   (e) .96

Chapter 10

Markov Chains; Games

This chapter is divided into two parts: Markov chains (10.1 – 10.3) and games (10.4 – 10.6). These parts are completely independent of one another. Thus one may be covered and the other skipped without any difficulty. Both topics rely on probability and matrices to solve problems of interest in various fields.

A Markov chain is a probability model that consists of repeated trials of an experiment in which some change or transition is occurring, such as consumers changing their preferred choice of a soft drink. Information about how this change takes place is then used to predict the future market share.

Games provide a mathematical way to express the strategies available to people or corporations. Ways to arrive at a ''best'' strategy are discussed.

10.1 MARKOV CHAINS AND TRANSITION MATRICES

In Chapter 8 we introduced Bernoulli trials. In discussing Bernoulli trials, we made the assumption that the outcome of each trial is *independent* of the outcome of any previous trial.

Here we discuss another type of probability model, called a *Markov chain,* where there is some connection between one trial and the next. Markov chains have been shown to have applications in many areas, among them business, psychology, sociology, and biology.

Loosely speaking, a Markov chain or process is one in which what happens next is governed by what happened immediately before. At any stage a Markov experiment is in one of a finite number of states, with the next stage of the experiment consisting of movement to a possibly different state. The probability of moving to a certain state depends only on the state previously occupied and does not vary with time. Here are some situations that may be thought of as Markov chains:

1. There are yearly population shifts between a city and its surrounding suburbs. At any time, a person is either in the city or in the suburbs. So there are two states here with movement between them. As we track the population over time, we can ask: What percentage is where?
2. Several detergents compete in the market. Each year there is some shift of customer loyalty from one brand to another. At any point in time the state of a customer would correspond to the brand of detergent he or she uses.
3. A psychology experiment consists of placing a mouse in a maze composed of rooms. The mouse moves from room to room, and we think of this movement as moving from state to state.

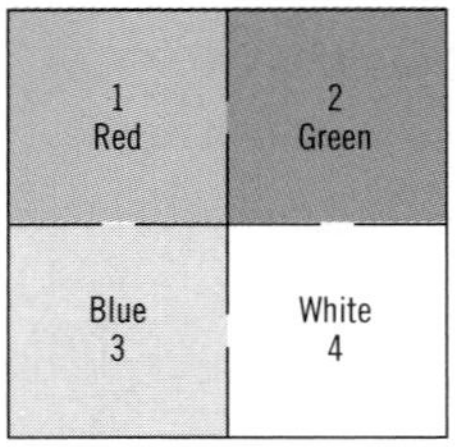

Figure 1

Consider the maze consisting of four connecting rooms shown in Figure 1. The rooms are numbered 1, 2, 3, 4 for convenience, and each room contains pulsating lights of a different color. The experiment consists of releasing a mouse in a particular room and observing its behavior.

We assume an observation is made whenever a movement occurs or after a fixed time interval, whichever comes first. Since the movement of the mouse is random in nature, we will use probabilistic terms to describe it.

We will refer to the rooms as states. For example, the probability p_{12} that the mouse moves from state 1 to state 2 might be $p_{12} = \frac{1}{2}$, while the probability of moving from state 1 to state 3 might be $p_{13} = \frac{1}{4}$. We are using p_{ij} to represent the probability of moving from state i to state j in one observation interval. So p_{11} would represent the probability that the mouse remains in room 1 for one observation interval; p_{12} represents the probability the mouse moves from room 1 to room 2 in one observation interval; and so on. Note that $p_{14} = 0$ since there is no direct passage between rooms 1 and 4.

A convenient way of writing the probabilities p_{ij} is to display them in a matrix. The resulting matrix

$$P = \begin{bmatrix} p_{11} & p_{12} & p_{13} & p_{14} \\ p_{21} & p_{22} & p_{23} & p_{24} \\ p_{31} & p_{32} & p_{33} & p_{34} \\ p_{41} & p_{42} & p_{43} & p_{44} \end{bmatrix}$$

is called the **transition matrix** of the experiment.

In our example there are four states and P is a 4×4 matrix. Were we to assign values to the remaining probabilities, a possible choice for the transition matrix P might be

$$P = \begin{array}{c} \\ 1 \\ 2 \\ 3 \\ 4 \end{array} \begin{array}{cccc} 1 & 2 & 3 & 4 \\ \begin{bmatrix} \frac{1}{4} & \frac{1}{2} & \frac{1}{4} & 0 \\ \frac{1}{6} & \frac{2}{3} & 0 & \frac{1}{6} \\ \frac{1}{3} & 0 & \frac{1}{3} & \frac{1}{3} \\ 0 & \frac{1}{4} & \frac{1}{2} & \frac{1}{4} \end{bmatrix} \end{array}$$

where the rows and columns are indexed by the four states (rooms).

We can display the entries in P using a tree diagram. See Figure 2. Thus the entries in P are conditional probabilities representing the probabilities of where the mouse will go next given that we know where it is now. This essential idea of movement from state to state with attached probabilities forms the basis of a *Markov chain*.

Figure 2

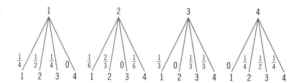

MARKOV CHAIN

A *Markov chain* is a sequence of experiments, each of which results in one of a finite number of states that we label 1, 2, . . . , *m*.

If p_{ij} is the probability of moving from state i to state j, then the **transition matrix** $P = [p_{ij}]$ is the $m \times m$ matrix

$$P = \begin{bmatrix} p_{11} & p_{12} & \cdots & p_{1m} \\ \vdots & \vdots & & \vdots \\ p_{m1} & p_{m2} & \cdots & p_{mm} \end{bmatrix}$$

Notice that the transition matrix P is a square matrix with entries that are always nonnegative since they represent probabilities. Also, the sum of the entries in every row is 1 since, as in the mouse and maze example, once the mouse is in a given room, it *Now Work Problem 1* either stays there or moves to one of the other rooms.

Computing State Distributions

The transition matrix contains the information necessary to predict what happens next, given that we know what happened before. It remains to specify what the state of affairs was at the start of the experiment. For example, what room was the mouse placed in at the start? We use a row vector for this purpose.

INITIAL PROBABILITY DISTRIBUTION

In a Markov chain with m states, the *initial probability distribution* is a $1 \times m$ row vector $v^{(0)}$ whose ith entry is the probability the experiment was in state i at the start.

For example, if the mouse was equally likely to be placed in any one of the rooms at the start, then we would have $v^{(0)} = [\frac{1}{4} \quad \frac{1}{4} \quad \frac{1}{4} \quad \frac{1}{4}]$. Whereas, if it was decided to always place the mouse initially in room 1, then we would have $v^{(0)} = [1 \quad 0 \quad 0 \quad 0]$.

PROBABILITY VECTOR

A *probability vector* is a vector whose entries are nonnegative and sum to 1.

We conclude that the initial probability distribution $v^{(0)}$ of a Markov chain is a probability row vector.

EXAMPLE 1

Look again at the maze in Figure 1. Suppose we assign the following transition probabilities:

$$\text{From room 1 to} \quad \begin{Bmatrix} 1 & 2 & 3 & 4 \\ \frac{1}{3} & \frac{1}{3} & \frac{1}{3} & 0 \end{Bmatrix}$$

Here, the mouse starts in room 1, and $\frac{1}{3}$ of the time it remains there during the time interval of observation, $\frac{1}{3}$ of the time it enters room 2, and $\frac{1}{3}$ of the time it enters room 3. Since it cannot go to room 4 directly from room 1, the probability assignment is 0. Similarly, the transition probabilities in moving from room 2, room 3, and room 4 may be given as follows:

$$\text{From room 2 to} \quad \begin{Bmatrix} 1 & 2 & 3 & 4 \\ \frac{1}{3} & \frac{1}{3} & 0 & \frac{1}{3} \end{Bmatrix}$$

$$\text{From room 3 to} \quad \begin{Bmatrix} 1 & 2 & 3 & 4 \\ \frac{1}{3} & 0 & \frac{1}{3} & \frac{1}{3} \end{Bmatrix}$$

$$\text{From room 4 to} \quad \begin{Bmatrix} 1 & 2 & 3 & 4 \\ 0 & \frac{1}{3} & \frac{1}{3} & \frac{1}{3} \end{Bmatrix}$$

(a) Find the transition matrix P.

(b) If the initial placement of the mouse is in room 4, find the initial probability distribution.

(c) What are the probabilities of being in each room after two observations?

SOLUTION

(a) The transition matrix P is

$$P = [p_{ij}] = \begin{array}{c} \\ 1 \\ 2 \\ 3 \\ 4 \end{array} \begin{array}{cccc} 1 & 2 & 3 & 4 \end{array} \\ \begin{bmatrix} \frac{1}{3} & \frac{1}{3} & \frac{1}{3} & 0 \\ \frac{1}{3} & \frac{1}{3} & 0 & \frac{1}{3} \\ \frac{1}{3} & 0 & \frac{1}{3} & \frac{1}{3} \\ 0 & \frac{1}{3} & \frac{1}{3} & \frac{1}{3} \end{bmatrix}$$

(b) Next, since the initial placement of the mouse is in room 4, the initial probability distribution, denoted by $v^{(0)}$, is

$$v^{(0)} = [p_1^{(0)} \quad p_2^{(0)} \quad p_3^{(0)} \quad p_4^{(0)}] = [0 \quad 0 \quad 0 \quad 1]$$

(c) To answer (c), we use a tree diagram. See Figure 3.

Figure 3

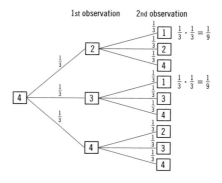

The numbers in each square refer to the room occupied. From this tree diagram we deduce, for example, that the mouse will be in state 1 after two observations with probability $\frac{1}{9} + \frac{1}{9} = \frac{2}{9}$. That is,

$$\text{Probability of moving from 4 to 1 in two stages} = \frac{2}{9}$$

Similarly,

Probability of moving from 4 to 2 in two stages $= \frac{1}{3} \cdot \frac{1}{3} + \frac{1}{3} \cdot \frac{1}{3} = \frac{2}{9}$
Probability of moving from 4 to 3 in two stages $= \frac{1}{3} \cdot \frac{1}{3} + \frac{1}{3} \cdot \frac{1}{3} = \frac{2}{9}$
Probability of moving from 4 to 4 in two stages $= \frac{1}{3} \cdot \frac{1}{3} + \frac{1}{3} \cdot \frac{1}{3} + \frac{1}{3} \cdot \frac{1}{3} = \frac{1}{3}$ (1)

Now Work Problem 3

· ·

We can record the results of Example 1(c) by writing $v^{(2)} = [\frac{2}{9} \quad \frac{2}{9} \quad \frac{2}{9} \quad \frac{1}{3}]$. Using this notation, the row vector $v^{(k)}$ would be used to record the probabilities of being in the various states after k trials. The ith entry of $v^{(k)}$ is the probability of being in state i at stage k. For example, $v^{(4)} = [\frac{1}{8} \quad \frac{1}{8} \quad 0 \quad \frac{3}{4}]$ indicates that the probability of being in state 3 at stage 4 of the experiment is 0.

Instead of following the procedure of Example 1 to find $v^{(2)}$, it is possible to compute $v^{(k)}$ directly from the transition matrix P and the initial probability distribution $v^{(0)}$.

> **Probability Distribution after k Stages**
>
> In a Markov chain the probability distribution $v^{(k)}$ after k stages is
>
> $$v^{(k)} = v^{(k-1)}P \qquad (2)$$
>
> where P is the transition matrix.

Formula (2) shows that

$$v^{(1)} = v^{(0)} P$$
$$v^{(2)} = v^{(1)} P$$
$$v^{(3)} = v^{(2)} P$$

and so on. Thus the succeeding distribution can always be derived from the previous one by multiplying by the transition matrix.

EXAMPLE 2

Using the information given in Example 1, use (2) to find the probability distribution after two observations.

SOLUTION

The initial probability distribution is $v^{(0)} = [0 \; 0 \; 0 \; 1]$, so by (2),

$$v^{(1)} = v^{(0)}P = [0 \quad 0 \quad 0 \quad 1] \begin{bmatrix} \frac{1}{3} & \frac{1}{3} & \frac{1}{3} & 0 \\ \frac{1}{3} & \frac{1}{3} & 0 & \frac{1}{3} \\ \frac{1}{3} & 0 & \frac{1}{3} & \frac{1}{3} \\ 0 & \frac{1}{3} & \frac{1}{3} & \frac{1}{3} \end{bmatrix} = [0 \quad \frac{1}{3} \quad \frac{1}{3} \quad \frac{1}{3}]$$

Using (2) again gives

$$v^{(2)} = v^{(1)}P = [0 \quad \frac{1}{3} \quad \frac{1}{3} \quad \frac{1}{3}] \begin{bmatrix} \frac{1}{3} & \frac{1}{3} & \frac{1}{3} & 0 \\ \frac{1}{3} & \frac{1}{3} & 0 & \frac{1}{3} \\ \frac{1}{3} & 0 & \frac{1}{3} & \frac{1}{3} \\ 0 & \frac{1}{3} & \frac{1}{3} & \frac{1}{3} \end{bmatrix} = [\frac{2}{9} \quad \frac{2}{9} \quad \frac{2}{9} \quad \frac{1}{3}]$$

This way of obtaining $v^{(2)}$ agrees with the results found in Example 1(c).

· ·

A few computations using (2) lead to another result.

$$v^{(1)} = v^{(0)}P$$
$$v^{(2)} = v^{(1)}P = [v^{(0)}P]P = v^{(0)}P^2$$
$$v^{(3)} = v^{(2)}P = [v^{(0)}P^2]P = v^{(0)}P^3$$
$$v^{(4)} = v^{(3)}P = [v^{(0)}P^3]P = v^{(0)}P^4$$

In each line above we have substituted the result of the preceding line. These calculations can be continued to produce the following result.

> **Probability Distribution after *k* Stages**
>
> In a Markov chain the probability distribution $v^{(k)}$ after k stages is
>
> $$v^{(k)} = v^{(0)}P^k \qquad (3)$$
>
> where P^k is the kth power of the transition matrix.

Thus the conclusion of Example 2 could also have been arrived at by squaring the transition matrix and computing

$$v^{(2)} = v^{(0)}P^2 = [0 \quad 0 \quad 0 \quad 1]\begin{bmatrix} \frac{1}{3} & \frac{2}{9} & \frac{2}{9} & \frac{2}{9} \\ \frac{2}{9} & \frac{1}{3} & \frac{2}{9} & \frac{2}{9} \\ \frac{2}{9} & \frac{2}{9} & \frac{1}{3} & \frac{2}{9} \\ \frac{2}{9} & \frac{2}{9} & \frac{2}{9} & \frac{1}{3} \end{bmatrix} = [\frac{2}{9} \quad \frac{2}{9} \quad \frac{2}{9} \quad \frac{1}{3}]$$

EXAMPLE 3

Population Movement   Suppose that the city of Oaklawn is experiencing a movement of its population to the suburbs. At present, 85% of the total population lives in the city and 15% lives in the suburbs. But each year 7% of the city people move to the suburbs, while only 1% of the suburb people move back to the city. Assuming that the total population (city and suburbs together) remains constant, what percent of the total will remain in the city after 5 years?

SOLUTION

This problem can be expressed as a sequence of experiments in which each experiment measures the proportion of people in the city and the proportion of people in the suburbs.

In the $(n + 1)$st year these proportions will depend for their value only on the proportions in the nth year and not on the proportions found in earlier years. Thus we have an experiment that can be represented as a Markov chain.

The initial probability distribution for this Markov chain is

$$\begin{array}{cc} \text{City} & \text{Suburbs} \end{array}$$
$$v^{(0)} = [.85 \qquad .15]$$

That is, initially, 85% of the people reside in the city and 15% in the suburbs.

The transition matrix P is

$$\begin{array}{cc} & \begin{array}{cc} \text{City} & \text{Suburbs} \end{array} \end{array}$$
$$P = \begin{array}{c} \text{City} \\ \text{Suburbs} \end{array}\begin{bmatrix} .93 & .07 \\ .01 & .99 \end{bmatrix}$$

That is, each year 7% of the city people move to the suburbs (so that 93% remain in the city) and 1% of the suburb people move to the city (so that 99% remain in the suburbs).

To find the probability distribution after 5 years, we need to compute $v^{(5)}$. We'll use Formula (2) five times.

$$v^{(1)} = v^{(0)}P = [.85 \quad .15] \begin{bmatrix} .93 & .07 \\ .01 & .99 \end{bmatrix} = [.792 \quad .208]$$

$$v^{(2)} = v^{(1)}P = [.792 \quad .208] \begin{bmatrix} .93 & .07 \\ .01 & .99 \end{bmatrix} = [.73864 \quad .26136]$$

$$v^{(3)} = v^{(2)}P = [.73864 \quad .26136] \begin{bmatrix} .93 & .07 \\ .01 & .99 \end{bmatrix} = [.68955 \quad .31045]$$

$$v^{(4)} = v^{(3)}P = [.68955 \quad .31045] \begin{bmatrix} .93 & .07 \\ .01 & .99 \end{bmatrix} = [.64439 \quad .35561]$$

$$v^{(5)} = v^{(4)}P = [.64439 \quad .35561] \begin{bmatrix} .93 & .07 \\ .01 & .99 \end{bmatrix} = [.60284 \quad .39716]$$

Thus, after 5 years, 60.28% of the residents live in the city and 39.72% live in the suburbs.

Now Work Problem 9

· ·

Example 3 leads us to inquire whether the situation in Oaklawn ever stabilizes. That is, after a certain number of years is an equilibrium reached? Also, does the equilibrium, if attained, depend on what the initial distribution of the population was, or is it independent of the initial state? We deal with these questions in the next section.

EXERCISE 10.1   Answers to odd-numbered problems begin on page AN-38.

1. Explain why the matrix below cannot be the transition matrix for a Markov chain.

$$\begin{bmatrix} 0 & 1 & 0 \\ \frac{1}{4} & \frac{1}{4} & \frac{1}{3} \\ \frac{1}{2} & -\frac{1}{2} & \frac{1}{2} \end{bmatrix}$$

2. Explain why the matrix below cannot be the transition matrix for a Markov chain.

$$\begin{bmatrix} 1 & \frac{1}{2} & \frac{1}{3} & \frac{1}{4} \\ 0 & 1 & 0 & 0 \\ 0 & \frac{1}{2} & \frac{1}{2} & 0 \\ 1 & 0 & 0 & 0 \end{bmatrix}$$

3. Consider a Markov chain with transition matrix

$$\begin{array}{cc} & \text{State 1} \quad \text{State 2} \\ \begin{array}{c} \text{State 1} \\ \text{State 2} \end{array} & \begin{bmatrix} \frac{1}{3} & \frac{2}{3} \\ \frac{1}{4} & \frac{3}{4} \end{bmatrix} \end{array}$$

(a) What does the entry $\frac{2}{3}$ in this matrix represent?
(b) Assuming that the system is initially in state 1, find the probability distribution one observation later. What is it two observations later?

(c) Assuming that the system is initially in state 2, find the probability distribution one observation later. What is it two observations later?

4. Consider a Markov chain with transition matrix

$$\begin{array}{cc} & \text{State 1} \quad \text{State 2} \\ \begin{array}{c} \text{State 1} \\ \text{State 2} \end{array} & \begin{bmatrix} .3 & .7 \\ .4 & .6 \end{bmatrix} \end{array}$$

(a) What does the entry .4 in the matrix represent?
(b) Assuming that the system is initially in state 1, find the probability distribution two observations later.
(c) Assuming that the system is initially in state 2, find the probability distribution two observations later.

5. Consider the transition matrix of Problem 4. If the initial probability distribution is [.25 .75], what is the probability distribution after two observations?

6. Consider a Markov chain with transition matrix

$$P = \begin{bmatrix} .7 & .2 & .1 \\ .6 & .2 & .2 \\ .4 & .1 & .5 \end{bmatrix}$$

If the initial distribution is [.25 .25 .5], what is the probability distribution in the next observation?

7. Find the values of a, b, and c that will make the following matrix a transition matrix for a Markov chain:

$$\begin{bmatrix} .2 & a & .4 \\ b & .6 & .2 \\ 0 & c & 0 \end{bmatrix}$$

8. In the maze of Figure 1 (page 452) if the initial probability distribution is $v^{(0)} = [\frac{1}{2}\ \ 0\ \ \frac{1}{2}\ \ 0]$, find the probability distribution after two observations.

9. In Example 3 if the initial probability distribution for Oaklawn is $v^{(0)} = [.7\ \ .3]$, what is the population distribution after 5 years?

10. A new rapid transit system has just been installed. It is anticipated that each week 90% of the commuters who used the rapid transit will continue to do so. Of those who traveled by car, 20% will begin to use the rapid transit instead.

    (a) Explain why the above is a Markov chain.
    (b) Set up the 2×2 matrix P with columns and rows labeled R (rapid transit) and C (car) to display these transitions.
    (c) Compute P^2 and P^3.

11. The voting pattern for a certain group of cities is such that 60% of the Democratic (D) mayors were succeeded by Democrats and 40% by Republicans (R). Also, 30% of the Republican mayors were succeeded by Democrats and 70% by Republicans.

    (a) Explain why the above is a Markov chain.
    (b) Set up the 2×2 matrix P with columns and rows labeled D and R to display these transitions.
    (c) Compute P^2 and P^3.

12. Consider the maze with nine rooms shown in the figure. The system consists of the maze and a mouse. We assume that the following learning pattern exists: If the mouse is in room 1, 2, 3, 4, or 5, it moves with equal probability to any room that the maze permits; if it is in room 8, it moves directly to room 9; if it is in room 6, 7, or 9, it remains in that room.

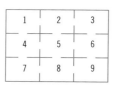

(a) Explain why the above experiment is a Markov chain.
(b) Construct the transition matrix P.

13. **Market Penetration**   A company is promoting a certain product, say brand X wine. The result of this is that 75% of the people drinking brand X wine over any given period of 1 month continue to drink it the next month; of those people drinking other brands of wine in the period of 1 month, 35% change over to the promoted wine the next month. We would like to know what fraction of wine drinkers will drink brand X after 2 months if 50% drink brand X wine now.

14. A professor either walks or drives to a university. He never drives 2 days in a row, but if he walks 1 day, he is just as likely to walk the next day as to drive his car. Show that this forms a Markov chain and give the transition matrix.

15. **Insurance**   Suppose that, during the year 1992, 45% of the drivers in a certain metropolitan area had Travelers automobile insurance, 30% had General American insurance, and 25% were insured by some other companies. Suppose also that a year later: (*i*) of those who had been insured by Travelers in 1992, 92% continued to be insured by Travelers, but 8% had switched their insurance to General American; (*ii*) of those who had been insured by General American in 1992, 90% continued to be insured by General American, but 4% had switched to Travelers and 6% had switched to some other companies; (*iii*) of those who had been insured by some other companies in 1992, 82% continued but 10% had switched to Travelers, and 8% had switched to General American. Using these data, answer the following questions:

    (a) What percentage of drivers in the metropolitan area were insured by Travelers and General American in 1993?
    (b) If these trends continued for one more year, what percentage of the drivers were insured by Travelers and General American in 1994?

16. If A is a transition matrix, what about A^2? A^3? What do you conjecture about A^n?

17. Let

$$A = \begin{bmatrix} a_{11} & a_{12} \\ a_{21} & a_{22} \end{bmatrix}$$

be a transition matrix and

$$u = [u_1\ \ u_2]$$

be a probability row vector. Prove that uA is a probability vector.

18. Let

$$P = \begin{bmatrix} p_{11} & p_{12} \\ p_{21} & p_{22} \end{bmatrix}$$

be a transition matrix. Prove that $v^{(k)} = v^{(k-1)}P$.

Technology Exercises

Most graphing calculators are able to store and multiply matrices. For example, the TI-81 can store up to three 6 × 6 (or smaller) matrices. Check your user's manual to see how to enter a matrix into the memory of your calculator.

Use a graphing calculator to solve Problems 1–4. To carry out the computations, display the vector v and enter the product ANS × P; then press the ENTER (or EXE) key repeatedly to obtain probability distributions for consecutive stages.

1. Consider the transition matrix

$$P = \begin{bmatrix} .5 & .2 & .1 & .2 \\ .3 & .3 & .2 & .2 \\ .1 & .5 & .1 & .3 \\ .25 & .25 & .25 & .25 \end{bmatrix}$$

If the initial probability distribution is $v =$ [.4 .3 .1 .2], what is the probability distribution after ten observations?

2. Consider the transition matrix

$$P = \begin{bmatrix} .7 & .15 & .05 & .1 \\ .15 & .7 & .05 & .1 \\ .1 & .05 & .7 & .15 \\ .15 & .05 & .1 & .7 \end{bmatrix}$$

If the initial probability distribution is $v =$ [.15 .15 .55 .15], what is the probability distribution after eight observations?

3. Consider the maze with six rooms shown in the figure. The system consists of the maze and a mouse. We assume the following learning pattern exists: If the mouse is in room 1, 2, 3, or 4, it moves with equal probability to any room the maze permits; if it is in room 5, it moves directly to room 6; if it is in room 6, it remains in that room.

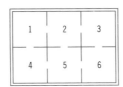

(a) Construct the transition matrix *P*.
(b) If the mouse is initially in room 2, what is the initial probability distribution?
(c) What is the probability distribution for the mouse after 10 steps?
(d) What is the most likely room for the mouse to be in after 10 steps?

4. Consider the maze with five rooms shown in the figure. The system consists of the maze and a mouse. We assume the following learning pattern exists: If the mouse is in room 1, 2, or 4, it moves with equal probability to any room the maze permits; if it is in room 3, it moves directly to room 5; if it is in room 5, it remains in that room.

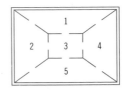

(a) Construct the transition matrix *P*.
(b) If the mouse is initially in room 2, what is the initial probability distribution?
(c) What is the probability distribution for the mouse after 10 steps?
(d) What is the most likely room for the mouse to be in after 10 steps?

10.2 REGULAR MARKOV CHAINS

One important question about Markov chains is: What happens in the long run? Is it the case that the distribution into states tends to stabilize over time? In this section we investigate the conditions under which a Markov chain produces an *equilibrium,* or *steady-state,* situation. We also give a procedure for finding this equilibrium distribution, when it exists.

Powers of the Transition Matrix

We begin by examining *powers* of the transition matrix P. Recall that the (i, j)th entry p_{ij} of P is the probability of moving from state i to state j in any one step. Are there corresponding interpretations for the entries in any *power* P^2, P^3, P^4, . . . of the transition matrix?

To motivate the answer, we use the following example.

EXAMPLE 1   **Consumer Loyalty**   A company is promoting a certain product, say, brand X wine. The result of this is that 75% of the people drinking brand X wine over any given period of 1 month continue to drink it the next month; of those people drinking other brands of wine in the period of 1 month, 35% change over to the promoted wine the next month.

The transition matrix P of this experiment is

$$P = \begin{array}{c} \\ \text{Brand X} \quad E_1 \\ \text{Other brands } E_2 \end{array} \begin{array}{c} \overset{\displaystyle \text{Brand X}}{\overset{\displaystyle E_1}{}} \quad \overset{\displaystyle \text{Other brands}}{\overset{\displaystyle E_2}{}} \\ \begin{bmatrix} .75 & .25 \\ .35 & .65 \end{bmatrix} \end{array}$$

(a) Find the probability of passing from E_1 to either E_1 or E_2 in two stages.
(b) Find the probability of passing from E_2 to either E_1 or E_2 in two stages.

SOLUTION

A tree diagram depicting two stages of this experiment is given in Figure 4.

Figure 4

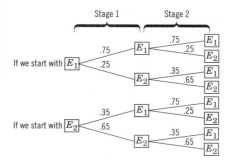

(a) The probability $p_{11}^{(2)}$ of proceeding from E_1 to E_1 in two stages is

$$p_{11}^{(2)} = (.75)(.75) + (.25)(.35) = .65$$

The probability $p_{12}^{(2)}$ from E_1 to E_2 in two stages is

$$p_{12}^{(2)} = (.75)(.25) + (.25)(.65) = .35$$

(b) The probability $p_{21}^{(2)}$ from E_2 to E_1 in two stages is

$$p_{21}^{(2)} = (.35)(.75) + (.65)(.35) = .49$$

The probability $p_{22}^{(2)}$ from E_2 to E_2 in two stages is

$$p_{22}^{(2)} = (.35)(.25) + (.65)(.65) = .51$$

....

If we square the transition matrix P, we obtain

$$P^2 = \begin{bmatrix} .75 & .25 \\ .35 & .65 \end{bmatrix} \begin{bmatrix} .75 & .25 \\ .35 & .65 \end{bmatrix}$$

$$= \begin{bmatrix} (.75)(.75) + (.25)(.35) & (.75)(.25) + (.25)(.65) \\ (.35)(.75) + (.65)(.35) & (.35)(.25) + (.65)(.65) \end{bmatrix}$$

$$= \begin{bmatrix} .65 & .35 \\ .49 & .51 \end{bmatrix}$$

and we notice that

$$P^2 = \begin{bmatrix} p_{11}^{(2)} & p_{12}^{(2)} \\ p_{21}^{(2)} & p_{22}^{(2)} \end{bmatrix}$$

Thus for Example 1 the *square* of the transition matrix gives the probabilities for moving from one state to another state in *two* stages.

This is true in general, and higher powers of the transition matrix carry similar interpretations.

Probability of Passing from State i to State j in n Stages

If P is the transition matrix of a Markov chain, then the (i, j)th entry of P^n (nth power of P) gives the probability of passing from state i to state j in n stages.

Regular Chains

We now turn to the question involving the long-term behavior of a Markov chain.

EXAMPLE 2 | Use the transition matrix P given in Example 1 and compute some of its powers.*

* A graphing calculator such as the Casio 7700fx or the TI-82 makes this easy.

SOLUTION

Some of the powers of the transition matrix P are given below.

$$P = \begin{bmatrix} .7500 & .2500 \\ .3500 & .6500 \end{bmatrix} \qquad P^7 = \begin{bmatrix} .5840 & .4160 \\ .5824 & .4176 \end{bmatrix}$$

$$P^2 = \begin{bmatrix} .6500 & .3500 \\ .4900 & .5100 \end{bmatrix} \qquad P^8 = \begin{bmatrix} .5836 & .4164 \\ .5830 & .4170 \end{bmatrix}$$

$$P^3 = \begin{bmatrix} .6100 & .3900 \\ .5460 & .4540 \end{bmatrix} \qquad P^9 = \begin{bmatrix} .5834 & .4166 \\ .5832 & .4168 \end{bmatrix}$$

$$P^4 = \begin{bmatrix} .5940 & .4060 \\ .5684 & .4316 \end{bmatrix} \qquad P^{10} = \begin{bmatrix} .5834 & .4166 \\ .5833 & .4167 \end{bmatrix}$$

$$P^5 = \begin{bmatrix} .5876 & .4124 \\ .5774 & .4226 \end{bmatrix} \qquad P^{11} = \begin{bmatrix} .5833 & .4167 \\ .5833 & .4167 \end{bmatrix} \quad \text{No change, so we stop here.}$$

$$P^6 = \begin{bmatrix} .5850 & .4150 \\ .5809 & .4191 \end{bmatrix} \qquad P^{12} = \begin{bmatrix} .5833 & .4167 \\ .5833 & .4167 \end{bmatrix}$$

We notice the interesting fact that the powers P^n seem to be converging and stabilizing around a fixed matrix. Furthermore, the rows of that matrix are equal. Also, if we let

$$\mathbf{t} = [.5833 \quad .4167]$$

and compute $\mathbf{t}P$, we find that

$$\mathbf{t}P = [.5833 \quad .4167] \begin{bmatrix} .75 & .25 \\ .35 & .65 \end{bmatrix} = [.5833 \quad .4167]$$

That is, $\mathbf{t}P = \mathbf{t}$. So, the row vector $\mathbf{t}$ is fixed by P.

Will this always happen? And what interpretation can be placed on the vector $\mathbf{t}$? Before giving the answers, we need a definition.

REGULAR MARKOV CHAIN

A Markov chain is said to be *regular* if, for some power of its transition matrix P, all of the entries are positive.

EXAMPLE 3

The transition matrix

$$P = \begin{bmatrix} \frac{1}{2} & \frac{1}{2} \\ 1 & 0 \end{bmatrix}$$

is regular since the square of P, namely,

$$P^2 = \begin{bmatrix} \frac{1}{2} & \frac{1}{2} \\ 1 & 0 \end{bmatrix}^2 = \begin{bmatrix} \frac{1}{2} & \frac{1}{2} \\ 1 & 0 \end{bmatrix} \begin{bmatrix} \frac{1}{2} & \frac{1}{2} \\ 1 & 0 \end{bmatrix} = \begin{bmatrix} \frac{3}{4} & \frac{1}{4} \\ \frac{1}{2} & \frac{1}{2} \end{bmatrix}$$

has only positive entries.

EXAMPLE 4

The matrix

$$P = \begin{bmatrix} 1 & 0 \\ \frac{3}{4} & \frac{1}{4} \end{bmatrix}$$

is not regular since every power of P will always have $p_{12} = 0$.

The matrix

$$\begin{bmatrix} 0 & 1 \\ 1 & 0 \end{bmatrix}$$

is another example of a transition matrix that is not regular.

. . . .

Fixed Probability Vector and Equilibrium

We now turn to the question involving the long-term behavior of a Markov chain. Notice that the results stated below hold only if the Markov chain is regular.

Long-Term Behavior of a Regular Markov Chain

Let P be the transition matrix of a regular Markov chain.

1. The matrices P^n approach a fixed matrix T as n gets large. We write $P^n \rightarrow T$ (as n gets large).
2. The rows of T are all identical and equal to a probability row vector **t**.
3. **t** is the unique probability vector that satisfies $\mathbf{t}P = \mathbf{t}$.
4. If $v^{(0)}$ is any initial distribution, then $v^{(n)} \rightarrow \mathbf{t}$ as n gets large.

Because of (3), **t** is called the **fixed probability vector** of the transition matrix P. Because of (4), **t** represents an equilibrium state since the state distributions $v^{(n)}$ approach **t** as time goes on. For example, the computations in Example 2 show that in the long run 58.33% of wine drinkers will be drinking brand X wine.

An important fact to note is that it is not necessary to compute higher and higher powers P^n of the transition matrix to find T and, in the process, **t**. Result (3) allows us to find **t** by solving a system of equations.

EXAMPLE 5

Find the probability vector **t** of the Markov chain whose transition matrix P is

$$P = \begin{bmatrix} .93 & .07 \\ .01 & .99 \end{bmatrix}$$

(This is the transition matrix of Example 3, page 457, which represents population movement.)

SOLUTION

First, we observe that P is regular since $P = P^1$ has all positive entries.

Because $\mathbf{t} = [t_1 \quad t_2]$ is a probability vector, we must have $t_1 + t_2 = 1$. Because of (3), we must have $\mathbf{t}P = \mathbf{t}$. Thus

$$[t_1 \quad t_2] \begin{bmatrix} .93 & .07 \\ .01 & .99 \end{bmatrix} = [t_1 \quad t_2] \qquad \text{and} \qquad t_1 + t_2 = 1$$

$$[.93t_1 + .01t_2 \quad .07t_1 + .99t_2] = [t_1 \quad t_2] \qquad \text{and} \qquad t_1 + t_2 = 1$$

$$.93t_1 + .01t_2 = t_1 \qquad .07t_1 + .99t_2 = t_2 \qquad \qquad t_1 + t_2 = 1$$

$$-.07t_1 + .01t_2 = 0 \qquad .07t_1 - .01t_2 = 0 \qquad \qquad t_1 + t_2 = 1$$

The two equations on the left are equivalent so, in fact, we have a system of two equations containing two variables, namely,

$$.07t_1 - .01t_2 = 0 \qquad \text{and} \qquad t_1 + t_2 = 1$$

or, equivalently,

$$7t_1 - t_2 = 0 \qquad \text{and} \qquad t_1 + t_2 = 1$$

Solving, we find that

$$t_1 = \tfrac{1}{8} \qquad \text{and} \qquad t_2 = \tfrac{7}{8}$$

Thus the fixed probability vector of P is $\mathbf{t} = [\tfrac{1}{8} \quad \tfrac{7}{8}]$.

In terms of Example 3, page 457, we interpret $\mathbf{t} = [\tfrac{1}{8} \quad \tfrac{7}{8}]$ as follows: in the long run, $\tfrac{1}{8}$ or 12.5% of the population will live in the city, while $\tfrac{7}{8}$ or 87.5% will live in the suburbs.

Now Work Problem 1

· ·

EXAMPLE 6

Find the fixed probability vector $\mathbf{t}$ of the transition matrix

$$P = \begin{bmatrix} \tfrac{1}{2} & 0 & \tfrac{1}{2} \\ \tfrac{1}{2} & \tfrac{1}{2} & 0 \\ \tfrac{1}{3} & \tfrac{1}{3} & \tfrac{1}{3} \end{bmatrix}$$

SOLUTION

First, P is regular, since P^2 has all positive entries (you should verify this). Let $\mathbf{t} = [t_1 \quad t_2 \quad t_3]$ be the fixed vector. Then

$$\mathbf{t}P = \mathbf{t} \qquad \text{and} \qquad t_1 + t_2 + t_3 = 1$$

$$[t_1 \quad t_2 \quad t_3] \begin{bmatrix} \tfrac{1}{2} & 0 & \tfrac{1}{2} \\ \tfrac{1}{2} & \tfrac{1}{2} & 0 \\ \tfrac{1}{3} & \tfrac{1}{3} & \tfrac{1}{3} \end{bmatrix} = [t_1 \quad t_2 \quad t_3] \qquad \text{and} \qquad t_1 + t_2 + t_3 = 1$$

$$\tfrac{1}{2}t_1 + \tfrac{1}{2}t_2 + \tfrac{1}{3}t_3 = t_1 \qquad \tfrac{1}{2}t_2 + \tfrac{1}{3}t_3 = t_2 \qquad \tfrac{1}{2}t_1 + \tfrac{1}{3}t_3 = t_3 \qquad t_1 + t_2 + t_3 = 1$$

Upon simplifying,

$$-\tfrac{1}{2}t_1 + \tfrac{1}{2}t_2 + \tfrac{1}{3}t_3 = 0 \qquad -\tfrac{1}{2}t_2 + \tfrac{1}{3}t_3 = 0 \qquad \tfrac{1}{2}t_1 - \tfrac{2}{3}t_3 = 0 \qquad t_1 + t_2 + t_3 = 1$$

Removing fractions, we obtain the following system of equations:

$$\begin{cases} t_1 + t_2 + t_3 = 1 & (1) \\ 3t_1 - 3t_2 - 2t_3 = 0 & (2) \\ \quad\;\; -3t_2 + 2t_3 = 0 & (3) \\ 3t_1 \qquad\quad - 4t_3 = 0 & (4) \end{cases}$$

Note that the sum of Equations (4) and (3) yields Equation (2), so that we really have a system of three equations containing three variables. We'll solve by substitution. From (3), $t_2 = 2t_3/3$ and from (4), $t_1 = 4t_3/3$. Substituting into (1), we obtain

$$\frac{4t_3}{3} + \frac{2t_3}{3} + t_3 = 1$$

$$3t_3 = 1$$

$$t_3 = \frac{1}{3}$$

Thus $t_1 = \frac{4}{9}$, $t_2 = \frac{2}{9}$, and $t_3 = \frac{1}{3}$. The fixed probability vector $\mathbf{t}$ of the transition matrix P is $\mathbf{t} = [\frac{4}{9} \quad \frac{2}{9} \quad \frac{1}{3}]$.

....

Now Work Problem 5

A nice feature of a regular Markov chain is that the long-run distribution $\mathbf{t}$ can be found simply by solving a system of equations. A more remarkable feature is that the same equilibrium $\mathbf{t}$ is reached no matter what the initial distribution was. We sketch the reason for this in the 2×2 case.

Suppose $v^{(0)} = [a \quad b]$ is *any* initial distribution of a regular Markov chain with transition matrix P. Let $\mathbf{t} = [t_1 \quad t_2]$ be the fixed vector. Then, by (1), we have

$$P^n \longrightarrow T = \begin{bmatrix} t_1 & t_2 \\ t_1 & t_2 \end{bmatrix}$$

so that

$$v^{(0)}P^n \longrightarrow v^{(0)}T$$

But $v^{(n)} = v^{(0)}P^n$. Thus

$$v^{(n)} \longrightarrow v^{(0)}T$$

Now

$$v^{(0)}T = [a \quad b] \begin{bmatrix} t_1 & t_2 \\ t_1 & t_2 \end{bmatrix} = [(a + b)t_1 \quad (a + b)t_2]$$

Since $v^{(0)}$ is a probability vector, it follows that $a + b = 1$.
Thus, $v^{(0)}T = [t_1 \quad t_2]$ and

$$v^{(n)} \longrightarrow v^{(0)}T = [t_1 \quad t_2] = \mathbf{t}$$

That is, the same $\mathbf{t}$ is reached regardless of the entries used in $v^{(0)}$.

EXAMPLE 7

Consumer Loyalty   Consider a certain community with three grocery stores. Within this community (we assume that the population is fixed) there always exists a shift of customers from one grocery store to another. A study was made on January 1, and it was found that $\frac{1}{4}$ of the population shopped at store I, $\frac{1}{3}$ at store II, and $\frac{5}{12}$ at store III. Each month store I retains 90% of its customers and loses 10% of them to store II. Store II retains 5% of its customers and loses 85% of them to store I and 10% of them to store III. Store III retains 40% of its customers and loses 50% of them to store I and 10% to store II. The transition matrix P is

$$
\begin{array}{cccc}
 & \text{I} & \text{II} & \text{III} \\
P = \begin{array}{c} \text{I} \\ \text{II} \\ \text{III} \end{array} & \left[\begin{array}{ccc} .90 & .10 & 0 \\ .85 & .05 & .10 \\ .50 & .10 & .40 \end{array}\right]
\end{array}
$$

We would like to answer the following questions:

(a) What proportion of customers will each store retain by February 1?
(b) By March 1?
(c) Assuming the same pattern continues, what will be the long-run distribution of customers among the three stores?

SOLUTION

(a) To answer the first question, we note that the initial probability distribution is $v^{(0)} = [\frac{1}{4} \quad \frac{1}{3} \quad \frac{5}{12}]$. By February 1, the probability distribution is

$$
v^{(1)} = v^{(0)}P = [\tfrac{1}{4} \quad \tfrac{1}{3} \quad \tfrac{5}{12}] \left[\begin{array}{ccc} .90 & .10 & .00 \\ .85 & .05 & .10 \\ .50 & .10 & .40 \end{array}\right] = [.7167 \quad .0833 \quad .2000]
$$

(b) To find the probability distribution after 2 months (March 1), we compute $v^{(2)}$:

$$
v^{(2)} = v^{(0)}P^2 = [\tfrac{1}{4} \quad \tfrac{1}{3} \quad \tfrac{5}{12}] \left[\begin{array}{ccc} .895 & .095 & .010 \\ .8575 & .0975 & .045 \\ .735 & .095 & .170 \end{array}\right] = [.8158 \quad .0958 \quad .0883]
$$

Since P^2 has all positive entries, we conclude P is regular.

(c) To find the long-run distribution, we determine the fixed probability vector $\mathbf{t}$ of the regular transition matrix P. Let $\mathbf{t} = [t_1 \quad t_2 \quad t_3]$. Then $\mathbf{t}P = \mathbf{t}$ and $t_1 + t_2 + t_3 = 1$, so that

$$
[t_1 \quad t_2 \quad t_3] \left[\begin{array}{ccc} .90 & .10 & .00 \\ .85 & .05 & .10 \\ .50 & .10 & .40 \end{array}\right] = [t_1 \quad t_2 \quad t_3] \qquad \text{and} \qquad t_1 + t_2 + t_3 = 1
$$

The solution is found to be

$$
[t_1 \quad t_2 \quad t_3] = [.8889 \quad .0952 \quad .0159]
$$

Thus in the long run store I will have about 88.9% of all customers, store II will have 9.5%, and store III will have 1.6%.

Now Work Problem 9

. . . .

EXAMPLE 8

Spread of Rumor   A U.S. senator has decided whether to vote yes or no on an important bill pending in Congress and conveys this decision to an aide. The aide then passes this news on to another individual, who passes it on to a friend, and so on, each time to a new individual. Assume p is the probability that any one person passes on the information opposite to the way he or she heard it. Then $1 - p$ is the probability a person passes on the information the same way he or she heard it. With what probability will the nth person receive the information as a yes vote?

SOLUTION

This can be viewed as a Markov chain model. Although it is not intuitively obvious, we shall find that the answer is essentially independent of p, provided $p \neq 0$.

To obtain the transition matrix, we observe that two states are possible: a yes is heard or a no is heard. The transition from a yes to a no or from a no to a yes occurs with probability p. The transition from yes to yes or from no to no occurs with probability $1 - p$. Thus the transition matrix P is

$$
\begin{array}{cc}
 & \begin{array}{cc} \text{Yes} & \text{No} \end{array} \\
\begin{array}{c} \text{Yes} \\ \text{No} \end{array} & \begin{bmatrix} 1 - p & p \\ p & 1 - p \end{bmatrix}
\end{array}
$$

The probability that the nth person will receive the information in one state or the other is given by successive powers of the matrix P, that is, by the matrices P^n. In fact, the answer in this case rapidly approaches $t_1 = \frac{1}{2}, t_2 = \frac{1}{2}$, after any considerable number of people are involved. This can easily be shown by verifying that the fixed probability vector is $\begin{bmatrix} \frac{1}{2} & \frac{1}{2} \end{bmatrix}$. Hence, no matter what the senator's initial decision is, eventually (after enough information exchange), half the people hear that the senator is going to vote yes and half hear that the senator is going to vote no.

. . . .

An interesting interpretation of this result applies to successive roll calls in Congress on the same issue. We let p represent the probability that a member changes his or her mind on an issue from one roll call to the next. Then the above example shows that if enough roll calls are forced, the Congress will eventually be near an even split on the issue. This could perhaps serve as a model for explaining a minority party's use of the parliamentary device of delaying actions.

EXERCISE 10.2   Answers to odd-numbered problems begin on page AN-39.

In Problems 1–6 determine which of the given matrices are regular. For those that are, find the fixed probability vector.

1. $\begin{bmatrix} \frac{1}{2} & \frac{1}{2} \\ 1 & 0 \end{bmatrix}$

2. $\begin{bmatrix} \frac{1}{2} & \frac{1}{2} \\ 0 & 1 \end{bmatrix}$

3. $\begin{bmatrix} 0 & 1 \\ \frac{1}{4} & \frac{3}{4} \end{bmatrix}$

4. $\begin{bmatrix} \frac{1}{3} & \frac{2}{3} \\ 1 & 0 \end{bmatrix}$

5. $\begin{bmatrix} 1 & 0 & 0 \\ \frac{1}{4} & \frac{1}{2} & \frac{1}{4} \\ 0 & 1 & 0 \end{bmatrix}$

6. $\begin{bmatrix} \frac{1}{4} & \frac{3}{4} & 0 \\ \frac{1}{2} & 0 & \frac{1}{2} \\ 0 & 1 & 0 \end{bmatrix}$

7. Show that the transition matrix P of Example 8 has the fixed probability vector $\begin{bmatrix} \frac{1}{2} & \frac{1}{2} \end{bmatrix}$.

8. Verify the result we obtained in Example 7, part (c).

9. Consumer Loyalty   A grocer stocks her store with three types of detergents, A, B, C. When brand A is sold out, the probability is .7 that she stocks up with brand A and

.15 each with brands B and C. When she sells out brand B, the probability is .8 that she will stock up again with brand B and .1 each with brands A and C. Finally, when she sells out brand C, the probability is .6 that she will stock up with brand C and .2 each with brands A and B. Find the transition matrix. In the long run what is the stock of detergents?

10. Consumer Loyalty A housekeeper buys three kinds of cereal: A, B, C. She never buys the same cereal in successive weeks. If she buys cereal A, then the next week she buys cereal B. However, if she buys either B or C, then the next week she is three times as likely to buy A as the other brand. Find the transition matrix. In the long run how often does she buy each of the three brands?

11. Voting Loyalty In England, of the sons of members of the Conservative party, 70% vote Conservative and the rest vote Labour. Of the sons of Labourites, 50% vote Labour, 40% vote Conservative, and 10% vote Socialist. Of the sons of Socialists, 40% vote Socialist, 40% vote Labour, and 20% vote Conservative. What is the probability that the grandson of a Labourite will vote Socialist? What is the membership distribution in the long run?

12. If $[\frac{1}{3} \quad 0 \quad \frac{1}{3} \quad \frac{1}{3}]$ is the fixed probability vector of a matrix P, can P be regular?

13. Family Traits The probabilities that a blond mother will have a blond, brunette, or redheaded daughter are .6, .2, and .2, respectively. The probabilities that a brunette mother will have a blond, brunette, or redheaded daughter are .1, .7, and .2, respectively. And the probabilities that a redheaded mother will have a blond, brunette, or redheaded daughter are .4, .2, and .4, respectively. What is the probability that a blond woman is the grandmother of a brunette? If the population of women is now 50% bru-

nettes, 30% blonds, and the rest redheads, what will the distribution be:

(a) After two generations?   (b) In the long run?

14. Educational Trends Use the data given in the table and assume that the indicated trends continue in order to answer the questions below.

		Maximum Education Children Achieve		
		College	H.S.	E.S.
Highest Educational Level of Parents	College	80%	18%	2%
	H.S.	40%	50%	10%
	E.S.	20%	60%	20%

(a) What is the transition matrix?
(b) What is the probability that a grandchild of a college graduate is a college graduate?
(c) What is the probability that the grandchild of a high school graduate finishes only elementary school?
(d) If at present 30%, 40%, and 30% of the population are college, high school, and elementary school graduates, respectively, what will be the distribution of the grandchildren of the present population?
(e) What will the long-run distribution be?

Technology Exercises
Use a graphing calculator to compute powers of the following transition matrices. Stop when there is no change. Use this procedure to find the fixed probability vector **t** *of each transition matrix.*

1. $\begin{bmatrix} .93 & .07 \\ .01 & .99 \end{bmatrix}$

2. $\begin{bmatrix} .90 & .10 \\ .10 & .90 \end{bmatrix}$

3. $\begin{bmatrix} \frac{1}{2} & 0 & \frac{1}{2} \\ \frac{1}{2} & \frac{1}{2} & 0 \\ \frac{1}{3} & \frac{1}{3} & \frac{1}{3} \end{bmatrix}$

4. $\begin{bmatrix} \frac{1}{4} & \frac{1}{4} & \frac{1}{2} \\ \frac{1}{2} & \frac{1}{2} & 0 \\ \frac{1}{4} & \frac{1}{2} & \frac{1}{4} \end{bmatrix}$

5. $\begin{bmatrix} .1 & 0 & .3 & .6 \\ .5 & .4 & .1 & 0 \\ .2 & .3 & .2 & .3 \\ .3 & .2 & .2 & .3 \end{bmatrix}$

6. $\begin{bmatrix} .5 & .4 & .1 & 0 \\ .3 & .3 & .3 & .1 \\ .2 & .2 & .2 & .4 \\ .1 & .1 & .4 & .4 \end{bmatrix}$

10.3 ABSORBING MARKOV CHAINS

We have already seen examples of transition matrices of Markov chains in which there are states that are impossible to leave. Such states are called *absorbing*. For example, in the transition matrix below, state 2 is absorbing since, once it is reached, it is impossible to leave.

$$
\begin{array}{cc}
 & \text{State 1} \quad \text{State 2} \\
\begin{array}{c} \text{State 1} \\ \text{State 2} \end{array} &
\begin{bmatrix} \frac{1}{4} & \frac{3}{4} \\ 0 & 1 \end{bmatrix}
\end{array}
$$

ABSORBING STATE; ABSORBING CHAIN

In a Markov chain if p_{ij} denotes the probability of going from state E_i to state E_j, then E_i is called an *absorbing state* if $p_{ii} = 1$. A Markov chain is said to be an *absorbing chain* if and only if it contains at least one absorbing state and it is possible to go from *any* nonabsorbing state to an absorbing state in one or more stages.

Now Work Problem 1

Thus an absorbing state will capture the process and will not allow any state to pass from it.

EXAMPLE 1

Consider two Markov chains with P_1 and P_2 as their transition matrices:

$$
P_1 = \begin{array}{c} E_1 \\ E_2 \\ E_3 \\ E_4 \end{array}
\begin{array}{c}
\begin{array}{cccc} E_1 & E_2 & E_3 & E_4 \end{array} \\
\begin{bmatrix}
.4 & .2 & .4 & 0 \\
0 & 1 & 0 & 0 \\
.1 & 0 & .5 & .4 \\
.1 & 0 & .3 & .6
\end{bmatrix}
\end{array}
\qquad
P_2 = \begin{array}{c} E_1 \\ E_2 \\ E_3 \end{array}
\begin{array}{c}
\begin{array}{ccc} E_1 & E_2 & E_3 \end{array} \\
\begin{bmatrix}
1 & 0 & 0 \\
0 & \frac{1}{4} & \frac{3}{4} \\
0 & \frac{1}{3} & \frac{2}{3}
\end{bmatrix}
\end{array}
$$

Determine whether either or both are absorbing chains.

SOLUTION

The chain having P_1 as its transition matrix is absorbing since the second state is an absorbing state and it is possible to pass from each of the other states to the second. Specifically, it is possible to pass from the first state directly to the second state and from either the third or the fourth state to the first state and then to the second. On the other hand, the matrix P_2 is an example of a nonabsorbing matrix since it is impossible to go from the nonabsorbing state E_2 to the absorbing state E_1.

....

When working with an absorbing Markov chain, it is convenient to rearrange the states so that the absorbing states come first and the nonabsorbing states follow.

Once this rearrangement is made, the transition matrix can be subdivided into four submatrices:

$$\left[\begin{array}{c|c} I & \mathbf{0} \\ \hline S & Q \end{array}\right]$$

Here, I is an identity matrix, $\mathbf{0}$ denotes a matrix having all 0 entries, and the matrices S and Q are the two submatrices corresponding to the absorbing and nonabsorbing states.

EXAMPLE 2

Rearrange the transition matrix given below and find the submatrices S and Q:

$$\begin{array}{c} \\ E_1 \\ E_2 \\ E_3 \\ E_4 \\ E_5 \end{array} \begin{array}{ccccc} E_1 & E_2 & E_3 & E_4 & E_5 \\ \begin{bmatrix} 0 & .5 & 0 & 0 & .5 \\ 0 & 0 & .9 & 0 & .1 \\ 0 & 0 & 0 & .7 & .3 \\ 0 & 0 & 0 & 1 & 0 \\ 0 & 0 & 0 & 0 & 1 \end{bmatrix} \end{array}$$

SOLUTION

First, we move the absorbing states E_4 and E_5 so they appear first.

$$\begin{array}{c} \\ E_4 \\ E_5 \\ E_1 \\ E_2 \\ E_3 \end{array} \begin{array}{ccccc} E_4 & E_5 & E_1 & E_2 & E_3 \\ \begin{bmatrix} 1 & 0 & 0 & 0 & 0 \\ 0 & 1 & 0 & 0 & 0 \\ 0 & .5 & 0 & .5 & 0 \\ 0 & .1 & 0 & 0 & .9 \\ .7 & .3 & 0 & 0 & 0 \end{bmatrix} \end{array}$$

Then the subdivided matrix is

$$\left[\begin{array}{cc|ccc} 1 & 0 & 0 & 0 & 0 \\ 0 & 1 & 0 & 0 & 0 \\ \hline 0 & .5 & 0 & .5 & 0 \\ 0 & .1 & 0 & 0 & .9 \\ .7 & .3 & 0 & 0 & 0 \end{array}\right] \qquad \left[\begin{array}{c|c} I & \mathbf{0} \\ \hline S & Q \end{array}\right]$$

so that

$$S = \begin{bmatrix} 0 & .5 \\ 0 & .1 \\ .7 & .3 \end{bmatrix} \qquad Q = \begin{bmatrix} 0 & .5 & 0 \\ 0 & 0 & .9 \\ 0 & 0 & 0 \end{bmatrix}$$

....

Gambler's Ruin Problem

Consider the following game involving two players, sometimes called the *gambler's ruin problem.* Player I has $3 and player II has $2. They flip a fair coin; if it is a head, player I pays player II $1, and if it is a tail, player II pays player I $1. The total amount of money in the game is, of course, $5. We would like to know how long the game will last; that is, how long it will take for one of the players to go broke or win all the money. (This game can easily be generalized by assuming that player I has M dollars and player II has N dollars.)

In this experiment how much money a player has after any given flip of the coin depends only on how much he had after the previous flip and will not depend (directly) on how much he had in the preceding stages of the game. This experiment can thus be represented by a Markov chain.

For the *gambler's ruin problem* the game does not have to involve flipping a coin. That is, the probability that player I wins does not have to equal the probability that player II wins. Also, questions can be raised as to what happens if the stakes are doubled, how long the game can be expected to last, and so on.

The coin being flipped is fair so that a probability of $\frac{1}{2}$ is assigned to each outcome. The states are the amounts of money each player has at each stage of the game. Note that each player can increase or decrease the amount of money he has by only $1 at a time.

The transition matrix P is then

$$
P = \begin{array}{c@{}c}
 & \begin{array}{cccccc} 0 & 1 & 2 & 3 & 4 & 5 \end{array} \\
\begin{array}{c} 0 \\ 1 \\ 2 \\ 3 \\ 4 \\ 5 \end{array} &
\left[\begin{array}{cccccc}
1 & 0 & 0 & 0 & 0 & 0 \\
\frac{1}{2} & 0 & \frac{1}{2} & 0 & 0 & 0 \\
0 & \frac{1}{2} & 0 & \frac{1}{2} & 0 & 0 \\
0 & 0 & \frac{1}{2} & 0 & \frac{1}{2} & 0 \\
0 & 0 & 0 & \frac{1}{2} & 0 & \frac{1}{2} \\
0 & 0 & 0 & 0 & 0 & 1
\end{array} \right]
\end{array}
$$

Here, $p_{10} = \frac{1}{2}$ is the probability of starting with $1 and losing it. Also, p_{00} is the probability of having $0 given that a player has started with $0. This is a sure event since, once a player is in state 0, he stays there forever (he is broke). Similarly, p_{55} represents the probability of having $5, given that a player started with $5, which is again a sure event (the player has won all the money).

With regard to this problem, the following questions are of interest:

(a) Given that one gambler is in a nonabsorbing state, what is the expected number of times he will hold between $1 and $4 inclusive before the termination of the game? That is, on the average, how many times will the process be in nonabsorbing states?
(b) What is the expected length of the process (game)?
(c) What is the probability that an absorbing state is reached (that is, that one gambler will eventually be wiped out)?

To answer these questions, let's look at the transition matrix P. We rearrange this matrix so that the two absorbing states will appear in the first two rows:

$$
P = \begin{array}{c@{}c}
 & \begin{array}{cccccc} 0 & 5 & 1 & 2 & 3 & 4 \end{array} \\
\begin{array}{c} 0 \\ 5 \\ 1 \\ 2 \\ 3 \\ 4 \end{array} &
\left[\begin{array}{cc:cccc}
1 & 0 & 0 & 0 & 0 & 0 \\
0 & 1 & 0 & 0 & 0 & 0 \\ \hdashline
\frac{1}{2} & 0 & 0 & \frac{1}{2} & 0 & 0 \\
0 & 0 & \frac{1}{2} & 0 & \frac{1}{2} & 0 \\
0 & 0 & 0 & \frac{1}{2} & 0 & \frac{1}{2} \\
0 & \frac{1}{2} & 0 & 0 & \frac{1}{2} & 0
\end{array} \right]
\end{array}
$$

We write P in the form

$$P = \left[\begin{array}{c|c} I_2 & \mathbf{0} \\ \hline S & Q \end{array}\right]$$

where

$$I_2 = \begin{bmatrix} 1 & 0 \\ 0 & 1 \end{bmatrix} \qquad \mathbf{0} = \begin{bmatrix} 0 & 0 & 0 & 0 \\ 0 & 0 & 0 & 0 \end{bmatrix}$$

$$S = \begin{bmatrix} \frac{1}{2} & 0 \\ 0 & 0 \\ 0 & 0 \\ 0 & \frac{1}{2} \end{bmatrix} \qquad Q = \begin{bmatrix} 0 & \frac{1}{2} & 0 & 0 \\ \frac{1}{2} & 0 & \frac{1}{2} & 0 \\ 0 & \frac{1}{2} & 0 & \frac{1}{2} \\ 0 & 0 & \frac{1}{2} & 0 \end{bmatrix}$$

As we indicated earlier, we can do this to any transition matrix of an absorbing Markov chain. If r of the states are absorbing, the transition matrix P can be written as

$$P = \left[\begin{array}{c|c} I_r & \mathbf{0} \\ \hline S & Q \end{array}\right]$$

where I_r is the $r \times r$ identity matrix, $\mathbf{0}$ is the zero matrix of dimension $r \times s$, S is of dimension $s \times r$, and Q is of dimension $s \times s$.

In order to answer the questions raised about the data of the gambler's ruin problem, we need the following result:

Fundamental Matrix of an Absorbing Markov Chain

For an absorbing Markov chain that has a transition matrix P of the form

$$P = \left[\begin{array}{c|c} I_r & \mathbf{0} \\ \hline S & Q \end{array}\right]$$

where S is of dimension $s \times r$ and Q is of dimension $s \times s$, define the matrix T, called the **fundamental matrix** of the Markov chain, to be

$$T = [I_s - Q]^{-1} \tag{1}$$

The entries of T give the expected number of times the process is in each nonabsorbing state, provided the process began in a nonabsorbing state.

In the gambler's ruin problem the fundamental matrix T is

$$T = \left[\begin{bmatrix} 1 & 0 & 0 & 0 \\ 0 & 1 & 0 & 0 \\ 0 & 0 & 1 & 0 \\ 0 & 0 & 0 & 1 \end{bmatrix} - \begin{bmatrix} 0 & \frac{1}{2} & 0 & 0 \\ \frac{1}{2} & 0 & \frac{1}{2} & 0 \\ 0 & \frac{1}{2} & 0 & \frac{1}{2} \\ 0 & 0 & \frac{1}{2} & 0 \end{bmatrix} \right]^{-1} = \begin{array}{c} 1 \\ 2 \\ 3 \\ 4 \end{array} \begin{array}{cccc} 1 & 2 & 3 & 4 \\ \begin{bmatrix} 1.6 & 1.2 & .8 & .4 \\ 1.2 & 2.4 & 1.6 & .8 \\ .8 & 1.6 & 2.4 & 1.2 \\ .4 & .8 & 1.2 & 1.6 \end{bmatrix} \end{array} \tag{2}$$

This provides the answers to question (a). The entry .8 in row 3, column 1, indicates that .8 is the expected number of times the player will have \$1 if he started with \$3. In the fundamental matrix T the column headings indicate present money, while the row headings indicate money started with.

Expected Number of Steps before Absorption

The expected number of steps before absorption for each nonabsorbing state is found by adding the entries in the corresponding row of the fundamental matrix T.

So to answer question (b), we again look at the matrix T in (2). The expected number of games before absorption (when one of the players wins or loses all the money) can be found by adding the entries in each row of T. Thus if a player starts with \$3, the expected number of games before absorption is

$$.8 + 1.6 + 2.4 + 1.2 = 6.0$$

That is, on the average, we expect this game to end on the seventh play.

If a player starts with \$1, the expected number of games before absorption is

$$1.6 + 1.2 + .8 + .4 = 4.0$$

Our last result deals with the probability of being absorbed.

Probability of Being Absorbed

The (i, j)th entry in the matrix product $T \cdot S$ gives the probability that, starting in nonabsorbing state i, we reach the absorbing state j.

Thus to answer question (c), we find the product of the matrices T and S:

$$T \cdot S = \begin{bmatrix} 1.6 & 1.2 & .8 & .4 \\ 1.2 & 2.4 & 1.6 & .8 \\ .8 & 1.6 & 2.4 & 1.2 \\ .4 & .8 & 1.2 & 1.6 \end{bmatrix} \begin{bmatrix} \frac{1}{2} & 0 \\ 0 & 0 \\ 0 & 0 \\ 0 & \frac{1}{2} \end{bmatrix} = \begin{array}{c} \\ 1 \\ 2 \\ 3 \\ 4 \end{array} \begin{array}{cc} 0 & 5 \end{array} \begin{bmatrix} .8 & .2 \\ .6 & .4 \\ .4 & .6 \\ .2 & .8 \end{bmatrix}$$

The entry in row 3, column 2, indicates the probability is .6 that the player starting with \$3 will win all the money. The entry in row 2, column 1, indicates the probability is .6 that a player starting with \$2 will lose all his money.

The above techniques are applicable in general to any transition matrix P of an absorbing Markov chain.

In the gambler's ruin problem we assumed player I started with \$3 and player II with \$2. Furthermore, we assumed the probability of player I winning \$1 was .5. Table 1 gives probabilities for ruin and expected length for other kinds of betting situations for which the bet is 1 unit.

Table 1

Probability that Player I Wins	Amount of Units Player I Starts with	Amount of Units Player II Starts with	Probability that Player I Goes Broke	Expected Length of Game	Expected Gain of Player I
.50	9	1	.1	9	0
.50	90	10	.1	900	0
.50	900	100	.1	90,000	0
.50	8000	2000	.2	16,000,000	0
.45	9	1	.210	11	−1.1
.45	90	10	.866	765.6	−76.6
.45	99	1	.182	171.8	−17.2
.40	90	10	.983	441.3	−88.3
.40	99	1	.333	161.7	−32.3

Suppose, for example, that player I starts with $90 and player II with $10, with player I having a probability of .45 of winning (the game being unfavorable to player I). If at each trial, the stake is $1, Table 1 shows that the probability is .866 that player I is ruined. If the same game is played with player I having $9 to start and player II having $1 to start, the probability that player I is ruined drops to .210.

Now Work Problem 11

Model: The Rise and Fall of Stock Prices*

This model involves using Markov chains to analyze the movement of stock prices. The stocks (shares) were those whose daily progress was recorded by the London *Financial Times*. At any point in time a given stock was classified as being in one of three states: increase (I), decrease (D), or no change (N), depending on whether its price had risen, fallen, or remained the same compared to the previous trading day. The intent of the model was to study the long-range movement of stocks through these states. Data reporting the past history of stock prices over a period of 1097 trading days were used to statistically fit a transition matrix to the model. The transition matrix used was

$$P = \begin{array}{c} \\ I \\ D \\ N \end{array} \begin{array}{c} \begin{array}{ccc} I & D & N \end{array} \\ \begin{bmatrix} .586 & .073 & .340 \\ .070 & .639 & .292 \\ .079 & .064 & .857 \end{bmatrix} \end{array}$$

Thus a share that increased one day would have probability .586 of also increasing the next day and a probability of .340 of registering no change the next day. Note that the tendency to remain in the same state is strongest for the no-change state since p_{33} is the largest of the diagonal elements.

The fixed probability vector $\mathbf{t}$ for the transition matrix can be computed to be

$$\mathbf{t} = [.156 \quad .154 \quad .687]$$

---

* Adapted from Myles M. Dryden, "Share Price Movements: A Markovian Approach," *Journal of Finance, 24* (March 1969), pp. 49–60.

Thus in the long run a stock will have increased its price 15.6% of the time, decreased its price 15.4% of the time, and remained unchanged 68.7% of the time.

How long will a stock increase before its price begins to fall? Or, in general, how long will a stock be in a given state before it moves to another one? These questions are highly reminiscent of those encountered while studying absorbing chains, yet here the transition matrix P is not absorbing. Thus the techniques used earlier do not apply.

Suppose we wanted to know on the average how long a stock would spend in states I or N before arriving in state D. We could then assume state D is an absorbing one and replace the current probability p_{22} with 1 and make all other elements in the second row of P zero. The new matrix P' we have created is

$$P' = \begin{array}{c} \\ I \\ D \\ N \end{array} \begin{array}{c} \begin{array}{ccc} I & D & N \end{array} \\ \begin{bmatrix} .586 & .073 & .340 \\ 0 & 1 & 0 \\ .079 & .064 & .857 \end{bmatrix} \end{array}$$

which is the matrix of an absorbing chain.

Subdividing this matrix, we obtain

$$\begin{array}{c} \\ D \\ I \\ N \end{array} \begin{array}{c} \begin{array}{ccc} D & I & N \end{array} \\ \begin{bmatrix} 1 & \vdots & 0 & 0 \\ \hdashline .073 & \vdots & .586 & .340 \\ .064 & \vdots & .079 & .857 \end{bmatrix} \end{array}$$

with

$$Q = \begin{array}{c} \\ I \\ N \end{array} \begin{array}{c} \begin{array}{cc} I & N \end{array} \\ \begin{bmatrix} .586 & .340 \\ .079 & .857 \end{bmatrix} \end{array}$$

The fundamental matrix of this absorbing Markov chain is

$$(I - Q)^{-1} = \begin{bmatrix} .414 & -.340 \\ -.079 & .143 \end{bmatrix}^{-1} = \begin{array}{c} \\ I \\ N \end{array} \begin{array}{c} \begin{array}{cc} I & N \end{array} \\ \begin{bmatrix} 4.42 & 10.51 \\ 2.44 & 12.80 \end{bmatrix} \end{array}$$

The entries of $(I - Q)^{-1}$ are interpreted as follows. The (i, j)th entry gives the average time spent in state j having started in state i before reaching the absorbing state D for the first time.

So, a stock currently increasing would spend an average of 4.4 days increasing and 10.5 days not changing before declining. Hence, a total of 14.9 days on the average would elapse before an increasing stock first began to decrease. Likewise, if a stock is currently exhibiting no change, the second row of the above matrix shows that an average of $2.44 + 12.80 = 15.24$ days would go by before it began to decrease.

Suppose a stock is increasing and then either levels off or declines. How long will it take before it starts rising again? In general, given we are in state i and leave, what is the average time that elapses before we return again to state i? The answer, which we state without proof, is given in the box at the top of the next page.

> **Average Time between Visits**
>
> The average amount of time elapsed between visits to state i (called **mean recurrence time**) is given by the reciprocal of the ith component of the fixed probability vector **t**.

Here

$$\mathbf{t} = [.156 \quad .154 \quad .687]$$

So we can then compute

State	Mean Recurrence Time (Days)
I	6.41
D	6.49
N	1.46

In this model, then, days on which a share's price fails to change follow each other fairly closely.

EXERCISE 10.3   Answers to odd-numbered problems begin on page AN-39.

In Problems 1–6 state which of the given matrices represent absorbing Markov chains.

1. $\begin{bmatrix} 0 & 1 \\ \frac{1}{4} & \frac{3}{4} \end{bmatrix}$

2. $\begin{bmatrix} 1 & 0 \\ \frac{1}{3} & \frac{2}{3} \end{bmatrix}$

3. $\begin{bmatrix} 1 & 0 & 0 \\ \frac{1}{8} & \frac{5}{8} & \frac{2}{8} \\ 0 & 0 & 1 \end{bmatrix}$

4. $\begin{bmatrix} 0 & 0 & 1 \\ 1 & 0 & 0 \\ 0 & 1 & 0 \end{bmatrix}$

5. $\begin{bmatrix} 0 & 1 & 0 & 0 \\ 1 & 0 & 0 & 0 \\ 0 & 0 & 1 & 0 \\ \frac{1}{4} & 0 & \frac{3}{4} & 0 \end{bmatrix}$

6. $\begin{bmatrix} \frac{1}{3} & \frac{1}{3} & 0 & \frac{1}{3} \\ 0 & \frac{1}{4} & \frac{1}{4} & \frac{1}{2} \\ 0 & 0 & 1 & 0 \\ 0 & \frac{1}{2} & 0 & \frac{1}{2} \end{bmatrix}$

7. Find the fundamental matrix T of the absorbing Markov chain in Problem 3. Also, find S and $T \cdot S$.

8. Follow the directions of Problem 7 for the matrix in Problem 6.

9. Suppose that for a certain absorbing Markov chain the fundamental matrix T is found to be

$$
\begin{array}{c c}
 & \begin{array}{ccc} \$1 & \$2 & \$3 \end{array} \\
\begin{array}{c} \$1 \\ \$2 \\ \$3 \end{array} &
\begin{bmatrix} 1.5 & .5 & .8 \\ 1.2 & 2.3 & .6 \\ .3 & 1.8 & 2.1 \end{bmatrix}
\end{array}
$$

(a) What is the expected number of times a person will have $3, given that she started with $1? With $2?

(b) If a player starts with $3, how many games can she expect to play before absorption?

10. For the data in Problem 9 suppose that we are given the following matrix S:

$$S = \begin{bmatrix} \frac{1}{2} & 0 \\ 0 & 0 \\ 0 & \frac{1}{2} \end{bmatrix}$$

What is the probability that the player will be absorbed if she started with $3?

11. **Gambler's Ruin Problem**   A person repeatedly bets $1 each day. If he wins, he wins $1 (he receives his bet of $1 plus winnings of $1). He stops playing when he goes broke, or when he accumulates $3. His probability of winning is .4 and of losing is .6. What is the probability of eventually accumulating $3 if he starts with $1? With $2?

12. The following data were obtained from the admissions office of a 2-year junior college. Of the first-year class

(F), 75% became sophomores (S) the next year and 25% dropped out (D). Of those who were sophomores during a particular year, 90% graduated (G) by the following year and 10% dropped out.

(a) Set up a Markov chain with states D, F, G, and S that describes the situation.
(b) How many states are absorbing?
(c) Determine the matrix T.
(d) Determine the probability that an entering first-year student will eventually graduate.

13. **Gambler's Ruin Problem**   Colleen wants to purchase a $4000 used car, but has only $1000 available. Not wishing to finance the purchase, she makes a series of wagers in which the winnings equal whatever is bet. The probability of winning is .4 and the probability of losing is .6. In a daring strategy Colleen decides to bet all her money or at least enough to obtain $4000 until she loses everything or has $4000. That is, if she has $1000, she bets $1000; if she has $2000, she bets $2000; if she has $3000, she bets $1000.

(a) What is the expected number of wagers placed before the game ends?
(b) What is the probability that Colleen is wiped out?
(c) What is the probability that Colleen wins the amount needed to purchase the car?

14. Answer the questions in Problem 13 if the probability of winning is .5.

15. Answer the questions in Problem 13 if the probability of winning is .6.

16. Three armored cars, A, B, and C, are engaged in a three-way battle. Armored car A has probability $\frac{1}{3}$ of destroying its target, B has probability $\frac{1}{2}$ of destroying its target, and C has probability $\frac{1}{6}$ of destroying its target. The armored cars fire at the same time and each fires at the strongest opponent not yet destroyed. Using as states the surviving cars at any round, set up a Markov chain and answer the following questions:

(a) How many states are in this chain?
(b) How many are absorbing?
(c) Find the expected number of rounds fired.
(d) Find the probability that A survives.

17. **Stock Price Model**   In the model discussed on pages 475–477 a stock currently showing no change would, on the average, stay how long in that state?

18. **Stock Price Model**   Refer to the model discussed on pages 475–477. By making state I an absorbing state, compute the average number of days elapsed before a stock currently decreasing would start to increase again.

Technology Exercises
Most graphing calculators are able to multiply and invert matrices. For example, the TI-81 can store up to three 6 × 6 (or smaller) matrices.

Use a graphing calculator to solve Problems 1 and 2.

1. Suppose that for a certain absorbing Markov chain the transition matrix P is given by

$$P = \begin{bmatrix} 1 & 0 & 0 & 0 & 0 \\ 0 & 1 & 0 & 0 & 0 \\ 0 & 0 & .05 & .95 & 0 \\ 0 & .2 & 0 & .5 & .3 \\ .75 & .25 & 0 & 0 & 0 \end{bmatrix}$$

Determine the matrices I_r, S, and Q. Use the *inverse matrix* command on a graphing calculator to find the fundamental matrix T. Determine the product $T \cdot S$.

2. Consider the maze with five rooms shown in the figure. The system consists of the maze and a mouse. We assume the following learning pattern exists: If the mouse is in room 2, 3, or 4, it moves with equal probability to any room

the maze permits; if it is in room 1, it gets caught in the trap and remains there; if it is in room 5, it remains in that room with the cheese forever.

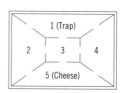

(a) Construct the transition matrix P.
(b) How many rooms represent an absorbing state for this Markov chain?
(c) If a mouse begins in room 3, what is the probability that it will find the cheese in 5 steps? Use a graphing calculator to determine the powers of matrix P.

(d) Determine the matrices I_r, S, and Q. Use the *inverse matrix* command on a graphing calculator to find the fundamental matrix T. Determine the product $T \cdot S$.

(e) How many steps, on the average, will the mouse make in the maze before it ends up in the trap?

10.4 TWO-PERSON GAMES

Game theory, as a branch of mathematics, is concerned with the analysis of human behavior in conflicts of interest. In other words, game theory gives mathematical expression to the strategies of opposing players and offers techniques for choosing the best possible strategy. In most parlor games it is relatively easy to define winning and losing and, on this basis, to quantify the best strategy for each player. However, game theory is not merely a tool for the gambler so that he or she can take advantage of the odds; nor is it merely a method for winning games like tic-tac-toe or matching pennies.

For example, when union and management sit down at the bargaining table to discuss contracts, each has definite strategies. Each will bluff, persuade, and try to discover the other's strategy, while at the same time trying to prevent the discovery of his or her own. If enough information is known, results from the theory of games can help determine what is the best possible strategy for each player. Another application of game theory can be made to politics. If two people are vying for the same political office, each has various campaign strategies. If it is possible to determine the impact of alternate strategies on the voters, the theory of games can help to find the best strategy (usually the one that gains the most votes, while losing the least votes). Thus game theory can be used in certain situations of conflict to indicate how people should behave to achieve certain goals. Of course, game theory does not tell us how people actually behave. Game theory is the study, then, of strategy in conflict situations.

TWO-PERSON GAME

Any conflict or competition between two people is called a *two-person game*.

Let's consider some examples of two-person games.

EXAMPLE 1

In a game similar to matching pennies, player I picks heads or tails and player II attempts to guess the choice. Player I will pay player II $3 if both choose heads; player I will pay player II $2 if both choose tails. If player II guesses incorrectly, he will pay player I $5.

. . . .

We use Example 1 to illustrate some terminology. First, since two players are involved, this is a two-person game. Next, notice that no matter what outcome occurs (*HH, HT, TH, TT*), whatever is lost (or gained) by player I is gained (or lost) by player II. Such games are called **two-person zero-sum games.**

If we denote the gains of player I by positive entries and his losses by negative entries, we can display this game in a 2×2 matrix as

$$\begin{array}{c} \\ H \\ T \end{array} \begin{array}{cc} H & T \\ \left[\begin{array}{cc} -3 & 5 \\ 5 & -2 \end{array} \right] \end{array}$$

This matrix is called the **game matrix** or **payoff matrix,** and each entry a_{ij} of the matrix is termed a **payoff.**

Conversely, any $m \times n$ matrix $A = [a_{ij}]$ can be regarded as the game matrix for a two-person zero-sum game in which player I chooses any one of the m rows of A and simultaneously player II chooses any one of the n columns of A. The entry in the row and column chosen is the payoff.

We will assume that the game is played repeatedly, and that the problem facing each player is what choice he should make so that he gains the most benefit. Thus, player I wishes to maximize his winnings and player II wishes to minimize his losses. By a strategy of player I for a given matrix game A, we mean the decision by player I to select rows of A in some manner.

Now Work Problem 1

EXAMPLE 2

Consider a two-person zero-sum game given by the matrix

$$\left[\begin{array}{cc} 3 & 6 \\ -2 & -3 \end{array} \right]$$

in which the entries denote the winnings of player I. The game consists of player I choosing a row and player II simultaneously choosing a column, with the intersection of row and column giving the payoff for this play in the game. For example, if player I chooses row 1 and player II chooses column 2, then player I wins $6; if player I chooses row 2 and player II chooses column 1, then player I loses $2.

It is immediately evident from the matrix that this particular game is biased in favor of player I, who will always choose row 1 since she cannot lose by doing so. Similarly, player II, recognizing that player I will choose row 1, will always choose column 1, since her losses are then minimized.

Thus the *best strategy* for player I is row 1 and the *best strategy* for player II is column 1. When both players employ their best strategy, the result is that player I wins $3. This amount is called the *value* of the game. Notice that the payoff $3 is the minimum of the entries in its row and is the maximum of the entries in its column.

· ·

STRICTLY DETERMINED GAME

A game defined by a matrix is said to be *strictly determined* if and only if there is an entry of the matrix that is the smallest element in its row and is also the largest element in its column. This entry is then called a *saddle point* and is the *value* of the game.

If a game has a positive value, the game favors player I. If a game has a negative value, the game favors player II. Any game with a value of 0 is termed a **fair game.**

If a game matrix has a saddle point, it can be shown that the row containing the saddle point is the best strategy for player I and the column containing the saddle point is the best strategy for player II. This is why such games are called *strictly determined games*. Such games are also called **games of pure strategy.**

Of course, a matrix may have more than one saddle point, in which case each player has more than one best strategy available. However, the value of the game is always the same no matter how many saddle points the matrix may have.

EXAMPLE 3

Determine whether the game defined by the matrix below is strictly determined.

$$\begin{bmatrix} 3 & 0 & -2 & -1 \\ 2 & -3 & 0 & -1 \\ 4 & 2 & 1 & 0 \end{bmatrix}$$

SOLUTION

First, we look at each row and find the smallest entry in each row:

$$\text{Row 1: } -2 \qquad \text{Row 2: } -3 \qquad \text{Row 3: } 0$$

Next, we check to see if any of the above elements are also the largest in their column. The element -2 in row 1 is not the largest entry in column 3; the element -3 in row 2 is not the largest entry in column 2; however, the element 0 in row 3 is the largest entry in column 4. Thus, this game is strictly determined. Its value is 0, so the game is fair.

....

The game of Example 3 is represented by a 3×4 matrix. This means that player I has three strategies open to her, while player II can choose from four strategies.

Now Work Problem 5

EXAMPLE 4

Two competitive franchising firms, Alpha Products and Omega Industries, are each planning to add an outlet in a certain city. It is possible for the site to be located either in the center of the city or in a large suburb of the city. If both firms decide to build in the center of the city, Alpha Products will show an annual profit of $1000 more than the profit of Omega Industries. If both firms decide to locate their outlets in the suburb, then it is determined that Alpha Products' profit will be $2000 less than the profit of Omega Industries. If Alpha locates in the suburb and Omega in the city, then Alpha will show a profit of $4000 more than Omega. Finally, if Alpha locates in the city and Omega in the suburb, then Alpha will have a profit of $3000 less than Omega's. Is there a best site for each firm to locate its outlet? In this case, by *best site* we mean the one that produces the most competition against the other firm—not the site that produces the highest gross sales.

SOLUTION

If we assign rows as Alpha strategies and columns as Omega strategies and if we use positive entries to denote the gain of Alpha over Omega and negative entries for the gain of Omega over Alpha, then the game matrix for this situation is

$$
\begin{array}{cc}
 & \text{Omega} \\
 & \begin{array}{cc} \text{City} & \text{Suburb} \end{array} \\
\text{Alpha} \begin{array}{c} \text{City} \\ \text{Suburb} \end{array} & \begin{bmatrix} 1 & -3 \\ 4 & -2 \end{bmatrix}
\end{array}
$$

where the entries are in thousands of dollars.

This game is strictly determined and the saddle point is -2, which is the value of the game. Thus, if both firms locate in the suburb, this results in the best competition. This is so since Omega will always choose to locate in the suburb, guaranteeing a larger profit than Alpha's. This being the case, Alpha, in order to minimize this larger profit of Omega, must always choose the suburb. Of course, the game is not fair since it is favorable to Omega.

....

EXERCISE 10.4   Answers to odd-numbered problems begin on page AN-39.

In Problems 1–4 write the game matrix that corresponds to each two-person conflict situation.

1. Katy and Colleen simultaneously each show one or two fingers. If they show the same number of fingers, Katy pays Colleen one dime. If they show a different number of fingers, Colleen pays Katy one dime.

2. Katy and Colleen simultaneously each show one or two fingers. If the total number of fingers shown is even, Katy pays Colleen that number of dimes. If the total number of fingers shown is odd, Colleen pays Katy that number of dimes.

3. Katy and Colleen, simultaneously and independently, each write down one of the numbers 1, 4, or 7. If the sum of the numbers is even, Katy pays Colleen that number of dimes. If the sum of the numbers is odd, Colleen pays Katy that number of dimes.

4. Katy and Colleen, simultaneously and independently, each write down one of the numbers 3, 6, or 8. If the sum of the numbers is even, Katy pays Colleen that number of dimes. If the sum of the numbers is odd, Colleen pays Katy that number of dimes.

In Problems 5–14 determine which of the two-person zero-sum games are strictly determined. For those that are, find the value of the game. All entries are the winnings of player I, who plays rows.

5. $\begin{bmatrix} -1 & 2 \\ -3 & 6 \end{bmatrix}$

6. $\begin{bmatrix} 4 & 0 \\ 0 & -1 \end{bmatrix}$

7. $\begin{bmatrix} 4 & 2 \\ 3 & 1 \end{bmatrix}$

8. $\begin{bmatrix} -6 & -1 \\ 0 & 0 \end{bmatrix}$

9. $\begin{bmatrix} 2 & 0 & -1 \\ 3 & 6 & 0 \\ 1 & 3 & 7 \end{bmatrix}$

10. $\begin{bmatrix} 2 & 3 & -2 \\ -2 & 0 & 4 \\ 0 & -3 & -2 \end{bmatrix}$

11. $\begin{bmatrix} 1 & 0 & 3 \\ -1 & 2 & 1 \\ 2 & 2 & 3 \end{bmatrix}$

12. $\begin{bmatrix} 1 & -3 & -2 \\ 2 & 5 & 4 \\ 2 & 3 & 2 \end{bmatrix}$

13. $\begin{bmatrix} 6 & 4 & -2 & 0 \\ -1 & 7 & 5 & 2 \\ 1 & 0 & 4 & 4 \end{bmatrix}$

14. $\begin{bmatrix} 8 & 6 & 4 & 0 \\ -1 & 6 & 5 & -2 \\ 0 & 1 & 3 & 3 \end{bmatrix}$

15. For what values of a is the matrix below strictly determined?

$$\begin{bmatrix} a & 8 & 3 \\ 0 & a & -9 \\ -5 & 5 & a \end{bmatrix}$$

16. Show that the matrix below is strictly determined for any choice of a, b, or c.

$$\begin{bmatrix} a & a \\ b & c \end{bmatrix}$$

17. Find necessary and sufficient conditions for the matrix below to be strictly determined.

$$\begin{bmatrix} a & 0 \\ 0 & b \end{bmatrix}$$

10.5 MIXED STRATEGIES

EXAMPLE 1

Consider a two-person zero-sum game given by the matrix

$$\begin{bmatrix} 6 & 0 \\ -2 & 3 \end{bmatrix}$$

in which the entries denote the winnings of player I. Is this game strictly determined? If so, find its value.

SOLUTION

We find that the smallest entry in each row is

$$\text{Row 1: } 0 \qquad \text{Row 2: } -2$$

The entry 0 in row 1 is not the largest element in its column; similarly, the entry -2 in row 2 is not the largest element in its column. Thus, this game is not strictly determined.

· ·

Remember that a two-person game is not usually played just once. With this in mind, player I in Example 1 might decide always to play row 1 since she may win $6 at best and win $0 at worst. Does this mean she should always employ this strategy? If she does, player II would catch on and begin to choose column 2 since this strategy limits his losses to $0. However, after a while, player I would probably start choosing row 2 to obtain a payoff of $3. Thus in a nonstrictly determined game it would be advisable for the players to *mix* their strategies rather than to use the same one all the time. That is, a random selection is desirable. Indeed, to make certain that the other player does not discover the pattern of moves, it may be best not to have any pattern at all. For instance, player I may elect to play row 1 in 40% of the plays (that is, with probability .4), while player II elects to play column 2 in 80% of the plays (that is, with probability .8). This idea of mixing strategies is important and useful in game theory. Games in which each player's strategies are **mixed** are termed **mixed-strategy games.**

Suppose we know the probability for each player to choose a certain strategy. What meaning can be given to the term *payoff of a game* if mixed strategies are used? Since the payoff has been defined for a pair of pure strategies and in a mixed-strategy situation we do not know what strategy is being used, it is not possible to define a payoff for a single game. However, in the long run we do know how often each strategy is being used, and we can use this information to compute the *expected payoff* of the game.

In Example 1 if player I chooses row 1 in 50% of the plays and row 2 in 50% of the plays, and if player II chooses column 1 in 30% of the plays and column 2 in 70% of the plays, the expected payoff of the game can be computed. For example, the strategy of row 1, column 1, is chosen $(.5)(.3) = .15$ of the time. This strategy has a payoff of $6, so that the expected payoff will be $($6)(.15) = 0.90. Table 2 summarizes the entire process. Thus, the expected payoff E of this game, when the given strategies are employed, is $1.65, which makes the game favorable to player I.

Table 2

Strategy	Payoff	Probability	Expected Payoff
Row 1—Column 1	6	.15	$0.90
Row 2—Column 1	− 2	.15	− 0.30
Row 1—Column 2	0	.35	0.00
Row 2—Column 2	3	.35	1.05
	Totals	1.00	$1.65

Now Work Problem 1

If we look very carefully at the above derivation, we get a clue as to how the expected payoff of a game that is not strictly determined should be defined.

Let's consider a game defined by the 2×2 matrix

$$A = \begin{bmatrix} a_{11} & a_{12} \\ a_{21} & a_{22} \end{bmatrix}$$

Let the strategy for player I, who plays rows, be denoted by the row vector

$$P = [p_1 \quad p_2]$$

and the strategy for player II, who plays columns, be denoted by the column vector

$$Q = \begin{bmatrix} q_1 \\ q_2 \end{bmatrix}$$

The probability that player I wins the amount a_{11} is $p_1 q_1$. Similarly, the probabilities that she wins the amounts a_{12}, a_{21}, and a_{22} are $p_1 q_2$, $p_2 q_1$, and $p_2 q_2$, respectively. If we denote by $E(P, Q)$ the expectation of player I, that is, the expected value of the amount she wins when she uses strategy P and player II uses strategy Q, then

$$E(P, Q) = p_1 a_{11} q_1 + p_1 a_{12} q_2 + p_2 a_{21} q_1 + p_2 a_{22} q_2$$

Using matrix notation, the above can be expressed as

$$E(P, Q) = PAQ$$

In general, if A is an $m \times n$ game matrix, we are led to the following definition:

EXPECTED PAYOFF

The *expected payoff* E of a two-person zero-sum game, defined by the matrix A, in which the row vector P and column vector Q define the respective strategy probabilities of player I and player II is

$$E = PAQ$$

EXAMPLE 2

Find the expected payoff of the game matrix

$$A = \begin{bmatrix} 3 & -1 \\ -2 & 1 \\ 1 & 0 \end{bmatrix}$$

if player I and player II decide on the strategies

$$P = \begin{bmatrix} \frac{1}{3} & \frac{1}{3} & \frac{1}{3} \end{bmatrix} \qquad Q = \begin{bmatrix} \frac{1}{3} \\ \frac{2}{3} \end{bmatrix}$$

SOLUTION

The expected payoff E of this game is

$$E = PAQ = \begin{bmatrix} \frac{1}{3} & \frac{1}{3} & \frac{1}{3} \end{bmatrix} \begin{bmatrix} 3 & -1 \\ -2 & 1 \\ 1 & 0 \end{bmatrix} \begin{bmatrix} \frac{1}{3} \\ \frac{2}{3} \end{bmatrix}$$

$$= \begin{bmatrix} \frac{2}{3} & 0 \end{bmatrix} \begin{bmatrix} \frac{1}{3} \\ \frac{2}{3} \end{bmatrix} = \frac{2}{9}$$

Thus the game is biased in favor of player I and has an expected payoff of $\frac{2}{9}$.

Now Work Problem 7

· ·

Most games are not strictly determined. That is, most games do not give rise to best pure strategies for each player. Examples of games that are not strictly determined are matching pennies (see Example 1, Section 10.4), bridge, poker, and so on. In the next two sections, we discuss techniques for finding optimal strategies for games that are not strictly determined.

EXERCISE 10.5   Answers to odd-numbered problems begin on page AN-40.

1. For the game of Example 1 find the expected payoff E if player I chooses row 1 in 30% of the plays and player II chooses column 1 in 40% of the plays.

2. For the game of Example 2 find the expected payoff E if player I chooses row 1 with probability .3 and row 2 with probability .4, while player II chooses column 1 half the time.

In Problems 3–6 find the expected payoff of the game matrix $\begin{bmatrix} 4 & 0 \\ 2 & 3 \end{bmatrix}$ *for the given strategies.*

3. $P = \begin{bmatrix} \frac{1}{2} & \frac{1}{2} \end{bmatrix}$ $Q = \begin{bmatrix} \frac{1}{2} \\ \frac{1}{2} \end{bmatrix}$

4. $P = \begin{bmatrix} \frac{1}{2} & \frac{1}{2} \end{bmatrix}$ $Q = \begin{bmatrix} \frac{3}{4} \\ \frac{1}{4} \end{bmatrix}$

5. $P = \begin{bmatrix} \frac{1}{4} & \frac{3}{4} \end{bmatrix}$ $Q = \begin{bmatrix} \frac{1}{2} \\ \frac{1}{2} \end{bmatrix}$

6. $P = \begin{bmatrix} 0 & 1 \end{bmatrix}$ $Q = \begin{bmatrix} 0 \\ 1 \end{bmatrix}$

In Problems 7–10 find the expected payoff of each game matrix.

7. $\begin{bmatrix} 4 & 0 \\ -3 & 6 \end{bmatrix}$; $P = \begin{bmatrix} \frac{2}{3} & \frac{1}{3} \end{bmatrix}$ $Q = \begin{bmatrix} \frac{1}{3} \\ \frac{2}{3} \end{bmatrix}$

8. $\begin{bmatrix} 1 & -1 \\ -2 & 3 \end{bmatrix}$; $P = \begin{bmatrix} \frac{1}{4} & \frac{3}{4} \end{bmatrix}$ $Q = \begin{bmatrix} \frac{1}{3} \\ \frac{2}{3} \end{bmatrix}$

9. $\begin{bmatrix} 1 & 0 & 0 \\ 0 & 1 & 0 \\ 0 & 0 & 1 \end{bmatrix}$; $P = \begin{bmatrix} \frac{1}{3} & \frac{1}{3} & \frac{1}{3} \end{bmatrix}$ $Q = \begin{bmatrix} \frac{1}{3} \\ \frac{1}{3} \\ \frac{1}{3} \end{bmatrix}$

10. $\begin{bmatrix} 4 & -1 & 0 \\ 2 & 3 & 1 \end{bmatrix}$; $P = \begin{bmatrix} \frac{1}{3} & \frac{2}{3} \end{bmatrix}$ $Q = \begin{bmatrix} \frac{2}{3} \\ \frac{1}{6} \\ \frac{1}{6} \end{bmatrix}$

11. Show that in a 2×2 game matrix $\begin{bmatrix} a_{11} & a_{12} \\ a_{21} & a_{22} \end{bmatrix}$ the only games that are not strictly determined are those for which either

(a) $a_{11} > a_{12}$ $a_{11} > a_{21}$ $a_{21} < a_{22}$ $a_{12} < a_{22}$

or

(b) $a_{11} < a_{12}$ $a_{11} < a_{21}$ $a_{21} > a_{22}$ $a_{12} > a_{22}$

10.6 OPTIMAL STRATEGY IN TWO-PERSON ZERO-SUM GAMES WITH 2 × 2 MATRICES

We have already seen that the best strategy for two-person zero-sum games that are strictly determined is found in the row and column containing the saddle point. Suppose the game is not strictly determined.

In 1927 John von Neumann, along with E. Borel, initiated research in the theory of games and proved that, even in nonstrictly determined games, there is a single course of action that represents the best strategy. In practice this means that in order to avoid always using a single strategy, a player in a game may instead choose plays randomly according to a fixed probability. This has the effect of making it impossible for the opponent to know what the player will do, since even the player will not know until the final moment. That is, by selecting a strategy randomly according to the laws of probability, the actual strategy chosen at any one time cannot be known even to the one choosing it.

For example, in the Italian game of *Morra* each player shows one, two, or three fingers and simultaneously calls out his guess as to what the sum of his and his opponent's fingers is. It can be shown that if he guesses four fingers each time and varies his own moves so that every 12 times he shows one finger 5 times, two fingers 4 times, and three fingers 3 times, he will, at worst, break even (in the long run).

EXAMPLE 1

Consider the nonstrictly determined game

$$A = \begin{bmatrix} 1 & -1 \\ -2 & 3 \end{bmatrix}$$

in which player I plays rows and player II plays columns. Determine the optimal strategy for each player.

SOLUTION

If player I chooses row 1 with probability p, then she must choose row 2 with probability $1 - p$. If player II chooses column 1, player I then expects to earn

$$E_I = p + (-2)(1 - p) = 3p - 2 \tag{1}$$

Similarly, if player II chooses column 2, player I expects to earn

$$E_I = (-1)p + 3(1 - p) = -4p + 3 \tag{2}$$

We graph these two equations using E_I as the vertical axis and p as the horizontal axis. See Figure 5.

Player I wants to maximize her expected earnings so she should maximize the minimum expected gain. This occurs when the two lines intersect since for any other choice of p, one or the other of the two expected earnings is less. Thus solving Equations (1) and (2) simultaneously, we obtain

$$3p - 2 = -4p + 3$$
$$7p = 5$$
$$p = \frac{5}{7}$$

The optimal strategy for player I is therefore

$$P = \begin{bmatrix} \frac{5}{7} & \frac{2}{7} \end{bmatrix}$$

Similarly, suppose player II chooses column 1 with probability q (and therefore column 2 with probability $1 - q$). If player I chooses row 1, player II's expected earnings are

$$E_{II} = (-1)q + (1)(1 - q) = 1 - 2q$$

If player I chooses row 2, player II's expected earnings are

$$E_{II} = (2)q + (-3)(1 - q) = 5q - 3$$

The optimal strategy for player II occurs when

$$1 - 2q = 5q - 3$$
$$-7q = -4$$
$$q = \frac{4}{7}$$

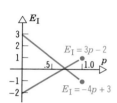

Figure 5

(figure labels: E_I; $E_I = 3p - 2$; $.5$; 1.0; p; $E_I = -4p + 3$)

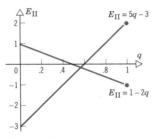

Figure 6

See Figure 6.

The optimal strategy for player II is thus

$$Q = \begin{bmatrix} \frac{4}{7} \\ \frac{3}{7} \end{bmatrix}$$

The expected payoff E corresponding to these optimal strategies is

$$E = PAQ = \begin{bmatrix} \frac{5}{7} & \frac{2}{7} \end{bmatrix} \begin{bmatrix} 1 & -1 \\ -2 & 3 \end{bmatrix} \begin{bmatrix} \frac{4}{7} \\ \frac{3}{7} \end{bmatrix} = \frac{1}{7}$$

. . . .

Value of a Game

Now consider a two-person zero-sum game given by the 2×2 matrix

$$A = \begin{bmatrix} a_{11} & a_{12} \\ a_{21} & a_{22} \end{bmatrix}$$

in which player I chooses row strategies and player II chooses column strategies.

Using the method illustrated above, it can be shown that the optimal strategy for player I is given by

$$P = \begin{bmatrix} p_1 & p_2 \end{bmatrix}$$

where

> **Optimal Strategy for Player I**
>
> $$p_1 = \frac{a_{22} - a_{21}}{a_{11} + a_{22} - a_{12} - a_{21}} \qquad p_2 = \frac{a_{11} - a_{12}}{a_{11} + a_{22} - a_{12} - a_{21}} \qquad (3)$$
>
> with
>
> $$a_{11} + a_{22} - a_{12} - a_{21} \neq 0$$

Notice that $p_1 + p_2 = 1$, as expected.

Similarly, the optimal strategy for player II is given by

$$Q = \begin{bmatrix} q_1 \\ q_2 \end{bmatrix}$$

where

> **Optimal Strategy for Player II**
>
> $$q_1 = \frac{a_{22} - a_{12}}{a_{11} + a_{22} - a_{12} - a_{21}} \qquad q_2 = \frac{a_{11} - a_{21}}{a_{11} + a_{22} - a_{12} - a_{21}} \qquad (4)$$
>
> with
>
> $$a_{11} + a_{22} - a_{12} - a_{21} \neq 0$$

Notice that $q_1 + q_2 = 1$, as expected.

The expected payoff E of the game corresponding to these optimal strategies is

---

Expected Payoff or Value of the Game

$$E = PAQ = \frac{a_{11} \cdot a_{22} - a_{12} \cdot a_{21}}{a_{11} + a_{22} - a_{12} - a_{21}}$$

When optimal strategies are used, the expected payoff E of the game is called the **value V of the game.**

EXAMPLE 2

For the game matrix

$$\begin{bmatrix} 1 & -1 \\ -2 & 3 \end{bmatrix}$$

determine the optimal strategies for player I and player II, and find the value of the game.

SOLUTION

Using Formula (3), we have

$$p_1 = \frac{3 - (-2)}{1 + 3 - (-1) - (-2)} = \frac{5}{7}$$

$$p_2 = \frac{1 - (-1)}{1 + 3 - (-1) - (-2)} = \frac{2}{7}$$

Thus player I's optimal strategy is to select row 1 with probability $\frac{5}{7}$ and row 2 with probability $\frac{2}{7}$. Also, by (4), player II's optimal strategy is

$$q_1 = \frac{3 - (-1)}{1 + 3 - (-1) - (-2)} = \frac{4}{7}$$

$$q_2 = \frac{1 - (-2)}{1 + 3 - (-1) - (-2)} = \frac{3}{7}$$

Player II's optimal strategy is to select column 1 with probability $\frac{4}{7}$ and column 2 with probability $\frac{3}{7}$. The value V of the game is

$$V = \frac{1 \cdot 3 - (-1)(-2)}{1 + 3 - (-1) - (-2)} = \frac{1}{7}$$

Thus in the long run the game is favorable to player I.

· ·

Now Work Problem 1   The results obtained in Example 2 are in agreement with those obtained earlier using the graphical technique.

EXAMPLE 3 Find the optimal strategy for each player, and determine the value of the game given by the matrix

$$\begin{bmatrix} 6 & 0 \\ -2 & 3 \end{bmatrix}$$

SOLUTION

Using the graphical technique or Formulas (3) and (4), we find player I's optimal strategy to be

$$p_1 = \tfrac{5}{11} \qquad p_2 = \tfrac{6}{11}$$

Player II's optimal strategy is

$$q_1 = \tfrac{3}{11} \qquad q_2 = \tfrac{8}{11}$$

The value V of the game is

$$V = \tfrac{18}{11} = 1.64$$

Thus the game favors player I, whose optimal strategy is $[\tfrac{5}{11} \quad \tfrac{6}{11}]$.

....

Applications

EXAMPLE 4 **Election Strategy** In a presidential campaign, there are two candidates, a Democrat (D) and a Republican (R), and two types of issues, domestic issues and foreign issues. The units assigned to each candidate's strategy are given in the table. We assume that positive entries indicate a strength for the Democrat, while negative entries indicate a weakness. We also assume that a strength of one candidate equals a weakness of the other so that the game is zero-sum. The question is, What is the best strategy for each candidate? That is, what is the value of the game?

		Republican	
		Domestic	Foreign
Democrat	Domestic	4	− 2
	Foreign	− 1	3

SOLUTION

Notice first that this game is not strictly determined. If D chooses to always play foreign issues, then R will counter with domestic issues, in which case D would also talk about domestic issues, in which case, etc., etc. There is no *single* strategy either can use. We compute that the optimal strategy for the Democrat is

$$p_1 = \frac{3 - (-1)}{4 + 3 - (-2) - (-1)} = \frac{4}{10} = .4 \qquad p_2 = \frac{4 - (-2)}{10} = .6$$

The optimal strategy for the Republican is

$$q_1 = \frac{3 - (-2)}{10} = .5 \qquad q_2 = \frac{4 - (-1)}{10} = .5$$

Thus the best strategy for the Democrat is to spend 40% of her time on domestic issues and 60% on foreign issues, while the Republican should divide his time evenly between the two issues.

The value of the game is

$$V = \frac{3 \cdot 4 - (-1)(-2)}{10} = \frac{10}{10} = 1.0$$

Thus no matter what the Republican does, the Democrat gains at least 1.0 unit by employing her best strategy.

Now Work Problem 7

....

EXAMPLE 5

War Game In a naval battle, attacking bomber planes are trying to sink ships in a fleet protected by an aircraft carrier with fighter planes. The bombers can attack either high or low, with a low attack giving more accurate results. Similarly, the aircraft carrier can send its fighters at high altitudes or low altitudes to search for the bombers. If the bombers avoid the fighters, credit the bombers with 8 points; if the bombers and fighters meet, credit the bombers with -2 points. Also, credit the bombers with 3 additional points for flying low (since this results in more accurate bombing). Find optimal strategies for the bombers and the fighters. What is the value of the game?

SOLUTION

First, we set up the game matrix. Designate the bombers as playing rows and the fighters as playing columns. Also, each entry of the matrix will denote winnings of the bombers. Then the game matrix is

$$\begin{array}{cc} & \text{Fighters} \\ & \begin{array}{cc} \text{Low} & \text{High} \end{array} \\ \text{Bombers} \begin{array}{c} \text{Low} \\ \text{High} \end{array} & \begin{bmatrix} 1 & 11 \\ 8 & -2 \end{bmatrix} \end{array}$$

The reason for a 1 in row 1, column 1, is that -2 points are credited for the planes meeting, but 3 additional points are credited to the bombers for a low flight.

Next, using Formula (3) and (4), the optimal strategies for the bombers $[p_1 \quad p_2]$ and for the fighters $[\begin{smallmatrix} q_1 \\ q_2 \end{smallmatrix}]$ are

$$p_1 = \frac{-10}{-20} = \frac{1}{2} \qquad p_2 = \frac{-10}{-20} = \frac{1}{2}$$

$$q_1 = \frac{-13}{-20} = \frac{13}{20} \qquad q_2 = \frac{-7}{-20} = \frac{7}{20}$$

The value V of the game is

$$V = \frac{-2 - 88}{-20} = \frac{-90}{-20} = 4.5$$

Thus the game is favorable to the bombers if both players employ their optimal strategies.

The bombers can decide whether to fly high or low by flipping a fair coin and flying high whenever heads appear. The fighters can decide whether to fly high or low by using a bowl with 13 black balls and 7 white balls. Each day, a ball should be selected at random and then replaced. If the ball is black, they will go low; if it is white, they will go high.

....

EXERCISE 10.6   Answers to odd-numbered problems begin on page AN-40.

In Problems 1–6 find the optimal strategy for each player and determine the value of each 2 × 2 game by using graphical techniques. Check your answers by using Formulas (3) and (4).

1. $\begin{bmatrix} 1 & 2 \\ 4 & 1 \end{bmatrix}$

2. $\begin{bmatrix} 2 & 4 \\ 3 & -2 \end{bmatrix}$

3. $\begin{bmatrix} -3 & 2 \\ 1 & 0 \end{bmatrix}$

4. $\begin{bmatrix} 3 & -2 \\ -1 & 2 \end{bmatrix}$

5. $\begin{bmatrix} 2 & -1 \\ -1 & 4 \end{bmatrix}$

6. $\begin{bmatrix} 5 & 4 \\ -3 & 7 \end{bmatrix}$

7. In Example 4 suppose the candidates are assigned the following weights for each issue:

		Republican	
		Domestic	Foreign
Democrat	Domestic	4	−1
	Foreign	0	3

What is each candidate's best strategy? What is the value of the game and whom does it favor?

8. **War Game**   For the situation described in Example 5, credit the bomber with 4 points for avoiding the fighters and with −6 points for meeting the fighters. Also, grant the bombers 2 additional points for flying low. What are the optimal strategies and the value of the game? Give instructions to the fighters and bombers as to how they should decide whether to fly high or low.

9. **Strategies for Spies**   A spy can leave an airport through two exits, one a relatively deserted exit and the other an exit heavily used by the public. His opponent, having been notified of the spy's presence in the airport, must guess which exit he will use. If the spy and opponent meet at the deserted exit, the spy will be killed; if the two meet at the heavily used exit, the spy will be arrested. Assign a payoff of 30 points to the spy if he avoids his opponent by using the deserted exit and of 10 points to the spy if he avoids his opponent by using the busy exit. Assign a payoff of −100 points to the spy if he is killed and −2 points if he is arrested. What are the optimal strategies and the value of the game?

10. Prove Formulas (3) and (4).

11. In the game matrix

$$\begin{bmatrix} a_{11} & a_{12} \\ a_{21} & a_{22} \end{bmatrix}$$

what can be said if $a_{11} + a_{22} - a_{12} - a_{21} = 0$?

CHAPTER REVIEW

IMPORTANT TERMS AND CONCEPTS

Markov chain 453
transition matrix 453
initial probability distribution 454
probability vector 454
probability distribution after
   k stages 456, 457
probability of passing from
   state *i* to state *j* in
   n stages 462

regular Markov chain 463
fixed probability vector 464
absorbing state 470
absorbing Markov chain 470
gambler's ruin problem 471
fundamental matrix of an
   absorbing Markov chain 473
expected number of steps
   before absorption 474

probability of being
   absorbed 474
two-person game 479
game matrix 480
strictly determined game 480
saddle point 480
mixed-strategy games 483
expected payoff 485, 489
optimal strategies 488

TRUE–FALSE ITEMS   Answers are on page AN-40.

T_____ F_____ **1.** The matrix $\begin{bmatrix} 1 & 0 \\ -1 & 1 \end{bmatrix}$ is a transition matrix for a Markov chain.

T_____ F_____ **2.** A Markov chain with the transition matrix $\begin{bmatrix} \frac{1}{2} & \frac{1}{2} \\ 0 & 1 \end{bmatrix}$ is regular.

T_____ F_____ **3.** A Markov chain with the transition matrix $\begin{bmatrix} 0 & 1 \\ \frac{1}{3} & \frac{2}{3} \end{bmatrix}$ is absorbing.

T_____ F_____ **4.** If P is the transition matrix of a Markov chain, then P^2 gives the probability of moving from one state to another state in two stages.

T_____ F_____ **5.** The value of a strictly determined game is unique.

T_____ F_____ **6.** In a two-person zero-sum game whatever is gained (lost) by player I is lost (gained) by player II.

T_____ F_____ **7.** In mixed-strategy games the value of the game depends on the strategy each player uses.

FILL IN THE BLANKS   Answers are on page AN-40.

1. In a Markov chain with m states the initial probability distribution is a row vector of dimension _____.

2. A probability vector is a vector whose entries are _____ and sum up to _____.

3. In a Markov chain with transition matrix P the probability distribution $v^{(k)}$ after k observations satisfies _____.

4. A Markov chain is said to be regular if, for some power of its transition matrix P, all entries are _____.

5. Each entry of a game matrix is called a _____.

6. In a nonstrictly determined game, when optimal strategies are used, the expected payoff is called the _____ of the game.

REVIEW EXERCISES   Answers to odd-numbered problems begin on page AN-40.

1. Find the fixed probability vector of

(a) $\begin{bmatrix} \frac{1}{4} & \frac{3}{4} \\ \frac{1}{2} & \frac{1}{2} \end{bmatrix}$ (b) $\begin{bmatrix} \frac{1}{3} & \frac{2}{3} \\ \frac{2}{3} & \frac{1}{3} \end{bmatrix}$ (c) $\begin{bmatrix} .7 & .1 & .2 \\ .6 & .1 & .3 \\ .4 & .2 & .4 \end{bmatrix}$

2. Define and explain in words the meaning of a *regular* transition matrix. Give an example of such a matrix and of a matrix that is not regular.

3. Customer Loyalty   Three beer distributors, A, B, and C, each presently holds $\frac{1}{3}$ of the beer market. Each wants to increase its share of the market, and to do so, each introduces a new brand. During the next year, it is learned that

(a) A keeps 50% of its business and loses 20% to B and 30% to C.

(b) B keeps 40% of its business and loses 40% to A and 20% to C.

(c) C keeps 25% of its business and loses 50% to A and 25% to B.

Assuming this trend continues, after 2 years what share of the market does each distributor have? In the long run what is each distributor's share?

4. If the current share of the market for each beer distributor A, B, and C in Problem 3 is A: 25%, B: 25%, C: 50%, and the market trend is the same, answer the same questions.

5. A representative of a book publishing company has to cover three universities, U_1, U_2, and U_3. She never sells at the same university in successive months. If she sells at university U_1, then the next month she sells at U_2. However, if she sells at either U_2 or U_3, then the next month she is three times as likely to sell at U_1 as at the other university. In the long run how often does she sell at each of the universities?

6. Family Traits   Assume that the probability of a fat father having a fat son is .7 and that of a skinny father having a skinny son is .4. What is the probability of a fat father being the great-grandfather of a fat great-grandson?

In the long run what will be the distribution? Does it depend on the initial physical state of the fathers?

7. Gambler's Ruin Problem Suppose a man has $2, which he is going to bet $1 at a time until he either loses all his money or he wins $5. Assume he wins with a probability of .45 and he loses with a probability of .55. Construct the transition matrix for this game.

(a) On the average, how many times will the process be in nonabsorbing states?

(b) What is the expected length of the game?

(c) What is the probability the man loses all his money or wins $5?

Hint: The fundamental matrix T of the transition matrix P is

$$T = \begin{bmatrix} 1.584282 & 1.062331 & .635281 & .285876 \\ 1.298405 & 2.360736 & 1.411736 & .635281 \\ .94899 & 1.725454 & 2.360736 & 1.062331 \\ .52194 & .949000 & 1.298405 & 1.584281 \end{bmatrix}$$

8. Determine which of the following two-person zero-sum games are strictly determined. For those that are, find the value of the game.

(a) $\begin{bmatrix} 5 & 3 \\ 2 & 4 \end{bmatrix}$
(b) $\begin{bmatrix} 29 & 15 \\ 79 & 3 \end{bmatrix}$
(c) $\begin{bmatrix} 50 & 75 \\ 30 & 15 \end{bmatrix}$

(d) $\begin{bmatrix} 7 & 14 \\ 9 & 13 \end{bmatrix}$
(e) $\begin{bmatrix} 0 & 2 & 4 \\ 4 & 6 & 10 \\ 16 & 14 & 12 \end{bmatrix}$

9. Find the expected payoff of the game below for the given strategies:

$$\begin{bmatrix} -1 & 1 \\ 1 & -1 \end{bmatrix}$$

(a) $P = [\frac{1}{3} \quad \frac{2}{3}] \quad Q = \begin{bmatrix} 1 \\ 0 \end{bmatrix}$

(b) $P = [0 \quad 1] \quad Q = \begin{bmatrix} \frac{1}{2} \\ \frac{1}{2} \end{bmatrix}$

(c) $P = [\frac{1}{2} \quad \frac{1}{2}] \quad Q = \begin{bmatrix} \frac{1}{2} \\ \frac{1}{2} \end{bmatrix}$

10. Investment Strategy An investor has a choice of two investments, A and B. The percentage gain of each investment over the next year depends on whether the economy is "up" or "down." This investment information is displayed in the following payoff matrix.

$$\begin{matrix} & \begin{matrix} \text{Economy} \\ \text{up} \end{matrix} & \begin{matrix} \text{Economy} \\ \text{down} \end{matrix} \end{matrix}$$
$$\begin{matrix} \text{Invest in A} \\ \text{Invest in B} \end{matrix} \begin{bmatrix} -5 & 20 \\ 18 & 0 \end{bmatrix}$$

(a) Find the best investment allocation. That is, find the row player's optimal strategy in the corresponding matrix game.

(b) Find the percentage gain the investor is assured of when using this optimal strategy. That is, find the value of the corresponding matrix game.

11. Investment Strategy There are two possible investments, A and B, and two possible states of the economy, "inflation" and "recession." The estimated percentage increases in the value of the investments over the coming year for each possible state of the economy are shown in the following payoff matrix.

$$\begin{matrix} & \text{Inflation} & \text{Recession} \end{matrix}$$
$$\begin{matrix} \text{Invest in A} \\ \text{Invest in B} \end{matrix} \begin{bmatrix} 10 & 5 \\ -5 & 20 \end{bmatrix}$$

(a) Find the investor's optimal strategy $P = [p_1 \quad p_2]$.

(b) Find the value of this game.

(c) What return can the investor expect if the optimal strategy is followed?

12. Real Estate Development A real estate developer has bought a large tract of land in Cook County. He is considering using some of the land for apartments, some for a shopping center, and some for houses. It is not certain whether the Cook County government will build a highway near his property or not. His financial advisor provides an estimate for the percentage profit to be made in each case and these percentages are given in the following table:

	Government	
Builder	Highway	No highway
Apartments	25%	5%
Shopping center	20%	15%
Houses	10%	20%

What percentage of the land should he use for each of the three categories, assuming that the government of Cook County is an active opponent?

13. **Betting Strategy**   The Pistons are going to play the Bulls in a basketball game. If you place your bet in Detroit, you can get 3 to 2 odds for a bet on the Bulls; and if you place your bet in Chicago, you can get 1 to 1 odds for a bet on the Pistons. This information is displayed in the following payoff matrix in which the entries represent the payoffs in dollars for each $1 bet.

$$
\begin{array}{cc}
 & \begin{array}{cc} \text{Bulls} & \text{Pistons} \\ \text{win} & \text{win} \end{array} \\
\begin{array}{c} \text{Bet on Bulls} \\ \text{Bet on Pistons} \end{array} &
\begin{bmatrix} 1.5 & -1 \\ -1 & 1 \end{bmatrix}
\end{array}
$$

Think of this situation as a game in which you are the row player and find your optimal strategy $P = [p_1 \quad p_2]$ and the value of the game.

Part Three

Calculus

Chapter 11

Precalculus—A Review

This chapter contains background information; much of it will be familiar to you. We begin with the role of intercepts and symmetry in graphing equations. Two sections are devoted to the important idea of a function, with special emphasis on the graph of a function. The chapter closes with a review of exponential and logarithm functions.

11.1* GRAPHS

Graphs of Equations

The **graph of an equation** in two variables x and y consists of the set of points in the xy-plane whose coordinates (x, y) satisfy the equation.

Let's look at some examples.

EXAMPLE 1

Graph the equation: $y = 2x + 5$

SOLUTION

We want to find all points (x, y) that satisfy the equation. To locate some of these points (and thus get an idea of the pattern of the graph), we assign some numbers to x and find corresponding values for y:

If	Then	Point on Graph
$x = 0$	$y = 2(0) + 5 = 5$	$(0, 5)$
$x = 5$	$y = 2(5) + 5 = 15$	$(5, 15)$
$x = -5$	$y = 2(-5) + 5 = -5$	$(-5, -5)$
$x = 10$	$y = 2(10) + 5 = 25$	$(10, 25)$

By plotting these points and then connecting them, we obtain the graph of the equation (a straight line), as shown in Figure 1.

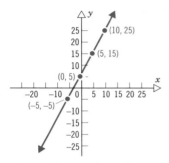

Figure 1 $y = 2x + 5$

· ·

EXAMPLE 2

Graph the equation: $y = x^2$

SOLUTION

Table 1 provides several points on the graph. In Figure 2 we plot these points and connect them with a smooth curve to obtain the graph (a *parabola*).

Table 1

x	$y = x^2$	(x, y)
-4	16	$(-4, 16)$
-3	9	$(-3, 9)$
-2	4	$(-2, 4)$
-1	1	$(-1, 1)$
0	0	$(0, 0)$
1	1	$(1, 1)$
2	4	$(2, 4)$
3	9	$(3, 9)$
4	16	$(4, 16)$

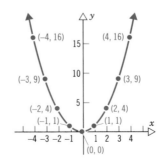

Figure 2 $y = x^2$

· ·

* Sections 11.1–11.3 are based on material from *Precalculus,* 4th ed., by Michael Sullivan. Used here with the permission of the author and Prentice-Hall, Inc.

The graphs of the equations shown in Figures 1 and 2 are necessarily incomplete. For example, in Figure 1, the point (20, 45) is a part of the graph of $y = 2x + 5$, but it is not shown. Since the graph of $y = 2x + 5$ can be extended out as far as we please, we use arrows to indicate that the pattern shown continues. Thus, it is important when illustrating a graph to present enough of the graph so that any viewer of the illustration will "see" the rest of it as an obvious continuation of what is actually there. For the most part we graph equations by plotting a sufficient number of points on the graph until a pattern becomes evident; then we connect these points with a smooth curve following the suggested pattern.

Two techniques that reduce the number of points used to graph an equation involve finding *intercepts* and checking for *symmetry*.

The points, if any, at which a graph intersects the coordinate axes are called the **intercepts.** A point at which the graph crosses or touches the x-axis is an **x-intercept,** and any point at which the graph crosses or touches the y-axis is a **y-intercept.** For example, the graph in Figure 3 has three x-intercepts, $(-3, 0)$, $(\frac{3}{2}, 0)$, $(4.5, 0)$, and three y-intercepts, $(0, -3.5)$, $(0, -\frac{4}{3})$, and $(0, 3)$.

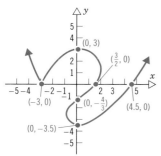

Figure 3

Procedure for Finding Intercepts

1. To find the x-intercept(s), if any, of the graph of an equation, let $y = 0$ in the equation and solve for x.

2. To find the y-intercept(s), if any, of the graph of an equation, let $x = 0$ in the equation and solve for y.

EXAMPLE 3

Find the intercepts of the graph of the equation $y = x^2 - x - 2$.

SOLUTION

To find the x-intercepts, we let $y = 0$ and solve the equation

$$x^2 - x - 2 = 0$$
$$(x - 2)(x + 1) = 0$$
$$x - 2 = 0 \quad \text{or} \quad x + 1 = 0$$
$$x = 2 \qquad\qquad x = -1$$

The x-intercepts are $(2, 0)$ and $(-1, 0)$.

To find the y-intercepts, we let $x = 0$; then $y = -2$. The intercepts are the points $(2, 0)$, $(-1, 0)$, and $(0, -2)$.

....

Another useful tool for graphing equations involves *symmetry,* particularly symmetry with respect to the x-axis, the y-axis, and the origin.

SYMMETRY WITH RESPECT TO THE *x*-AXIS

A graph is said to be *symmetric with respect to the x-axis* if, for every point (x, y) on the graph, the point $(x, -y)$ is also on the graph.

SYMMETRY WITH RESPECT TO THE *y*-AXIS

A graph is said to be *symmetric with respect to the y-axis* if, for every point (x, y) on the graph, the point $(-x, y)$ is also on the graph.

SYMMETRY WITH RESPECT TO THE ORIGIN

A graph is said to be *symmetric with respect to the origin* if, for every point (x, y) on the graph, the point $(-x, -y)$ is also on the graph.

Figure 4 illustrates these definitions. Notice that when a graph is symmetric with respect to the *x*-axis, the part of the graph above the *x*-axis is a reflection of the part below it, and vice versa. When a graph is symmetric with respect to the *y*-axis, the part of the graph to the right of the *y*-axis is a reflection of the part to the left of it, and vice versa. Notice that symmetry with respect to the origin may be viewed in two ways:

1. As a reflection about the *y*-axis, followed by a reflection about the *x*-axis.
2. As a projection along a line through the origin so that the distances from the origin are equal.

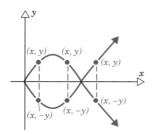

(a) Symmetric with respect to the *x*-axis

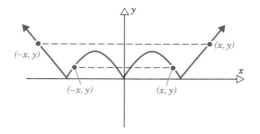

(b) Symmetric with respect to the *y*-axis

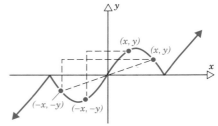

(c) Symmetric with respect to the origin

Figure 4

When the graph of an equation is symmetric, the number of points we need to plot in order to see the pattern is reduced. For example, if the graph of an equation is symmetric with respect to the *y*-axis, then, once points to the right of the *y*-axis are plotted, an equal number of points on the graph can be obtained by reflecting them about the *y*-axis. Thus, before we graph an equation, we first want to determine whether it has any symmetry. The following tests are used for that purpose.

Tests for Symmetry

To test the graph of an equation for symmetry with respect to the

***x*-axis** Replace y by $-y$ in the equation. If an equivalent equation results, the graph of the equation is symmetric with respect to the x-axis.

***y*-axis** Replace x by $-x$ in the equation. If an equivalent equation results, the graph of the equation is symmetric with respect to the y-axis.

Origin Replace x by $-x$ and y by $-y$ in the equation. If an equivalent equation results, the graph of the equation is symmetric with respect to the origin.

EXAMPLE 4

Graph the equation

$$y = x^3$$

Find any intercepts and check for symmetry before graphing.

SOLUTION

First, we seek the intercepts. When $x = 0$, then $y = 0$; and when $y = 0$, then $x = 0$. Thus, the origin $(0, 0)$ is the only intercept. Now we test for symmetry:

***x*-axis:** Replace y by $-y$. Since the result, $-y = x^3$, is not equivalent to $y = x^3$, the graph is not symmetric with respect to the x-axis.

***y*-axis:** Replace x by $-x$. Since the result, $y = -x^3$, is not equivalent to $y = x^3$, the graph is not symmetric with respect to the y-axis.

Origin: Replace x by $-x$ and y by $-y$. Since the result, $-y = -x^3$, is equivalent to $y = x^3$, the graph is symmetric with respect to the origin.

Because of the symmetry we need to locate points on the graph only for $x \geq 0$, such as $(0, 0)$, $(1, 1)$, and $(2, 8)$. Figure 5 shows the graph.

Figure 5 Symmetry with respect to the origin

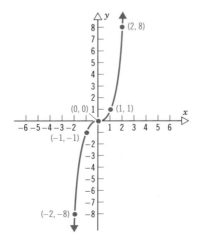

EXAMPLE 5

Graph the equation:

$$x = y^2$$

Find any intercepts and check for symmetry before graphing.

SOLUTION

The lone intercept is (0, 0). The graph is symmetric with respect to the x-axis. (Do you see why? Replace y by $-y$.) Figure 6 shows the graph.

Figure 6 Symmetry with respect to the x-axis

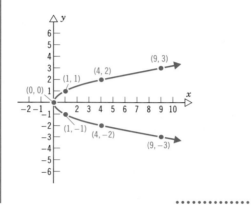

Now Work Problems 39 and 45

If we restrict y so that $y \geq 0$, the equation $x = y^2$, $y \geq 0$, may be written equivalently as $y = \sqrt{x}$. Thus, the portion of the graph of $x = y^2$ in the first quadrant is the graph of $y = \sqrt{x}$.

EXAMPLE 6

Graph the equation

$$y = \frac{1}{x}$$

Find any intercepts and check for symmetry before graphing.

SOLUTION

We check for intercepts first. If we let $x = 0$, we obtain a 0 denominator, which is not allowed. Hence, there is no y-intercept. If we let $y = 0$, we get the equation $1/x = 0$, which has no solution. Hence, there is no x-intercept. Thus, the graph of $y = 1/x$ does not cross the coordinate axes.

Replacing x by $-x$ and y by $-y$ yields $-y = -1/x$, which is equivalent to $y = 1/x$. Thus, the graph is symmetric with respect to the origin.

Finally, we set up Table 2, listing several points on the graph. Because of the symmetry with respect to the origin, we use only positive values of x. From Table 2, we infer that if x is a large and positive number, then $y = 1/x$ is a positive number close to 0. We also infer that if x is a positive number close to 0, then $y = 1/x$ is a large and positive number. Armed with this information, we can graph the equation. Figure 7 shows the graph of $y = 1/x$ and some of the points from Table 2. Notice that we made use of the fact that the graph has no intercepts and that it is symmetric with respect to the origin.

Figure 7 $y = \dfrac{1}{x}$

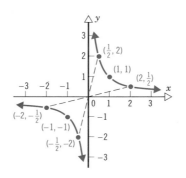

Table 2

x	$y = 1/x$	(x, y)
$\frac{1}{10}$	10	$(\frac{1}{10}, 10)$
$\frac{1}{3}$	3	$(\frac{1}{3}, 3)$
$\frac{1}{2}$	2	$(\frac{1}{2}, 2)$
1	1	$(1, 1)$
2	$\frac{1}{2}$	$(2, \frac{1}{2})$
3	$\frac{1}{3}$	$(3, \frac{1}{3})$
10	$\frac{1}{10}$	$(10, \frac{1}{10})$

EXERCISE 11.1   Answers to odd-numbered problems begin on page AN-41.

In Problems 1–8 plot each point. Then plot the point that is symmetric to it with respect to (a) the x-axis, (b) the y-axis, (c) the origin.

1. $(3, -4)$ **2.** $(5, 3)$ **3.** $(-2, 1)$ **4.** $(4, -2)$

5. $(1, 1)$ **6.** $(-1, -1)$ **7.** $(-3, -4)$ **8.** $(4, 0)$

In Problems 9–12 copy the table and fill in the missing values of the given equations. Use these points to graph each equation.

9. $y = x - 3$

x	0		2	-2	4	-4
y		0				

10. $y = -3x + 3$

x	0		2	-2	4	-4
y		0				

11. $2x - y = 6$

x	0		2	-2	4	-4
y		0				

12. $x + 3y = 9$

x	0		2	-2	4	-4
y		0				

In Problems 13–24 the graph of an equation is given.
(a) List the intercepts of the graph.
(b) Based on the graph, tell whether the graph is symmetric with respect to the
   x-axis, y-axis, origin, or none of these.

13.

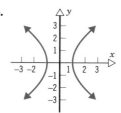

14.

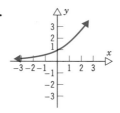

15.

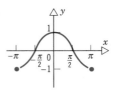

16.

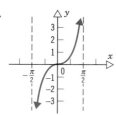

17.

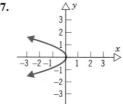

18.

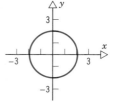

19.

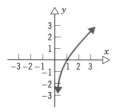

20.

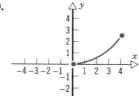

21.

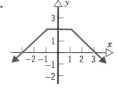

22.

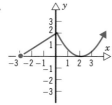

23.

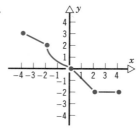

24.

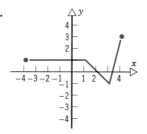

In Problems 25–38 (a) list the intercepts and (b) test for symmetry.

25. $x^2 = y$

26. $y^2 = x$

27. $y = 3x$

28. $y = -5x$

29. $x^2 + y - 9 = 0$

30. $y^2 - x - 4 = 0$

31. $4x^2 + 9y^2 = 36$

32. $x^2 + 4y^2 = 4$

33. $y = x^3 - 27$

34. $y = x^4 - 1$

35. $y = x^2 - 3x - 4$

36. $y = x^2 + 4$

37. $y = \dfrac{x}{x^2 + 9}$

38. $y = \dfrac{x^2 - 4}{x}$

In Problems 39–56 graph each equation. Find any intercepts and check for symmetry.

39. $y = 3x + 2$

40. $y = 2x - 3$

41. $3x - 2y + 6 = 0$

42. $2x - 3y + 6 = 0$

43. $y = -x^2$

44. $y = -x^2 + 3$

45. $y = x^2 + 3$

46. $y = x^2 - 3$

47. $y = x^3 - 1$

48. $y = x^3 - 8$

49. $x^2 = y + 1$

50. $x^2 = -2y$

51. $y = \sqrt{x}$

52. $x = \sqrt{y}$

53. $y = \sqrt{x - 1}$

54. $x = \sqrt{y + 2}$

55. $y = \dfrac{1}{x - 2}$

56. $y = \dfrac{1}{x - 3}$

In Problems 57–60 use the graph below.

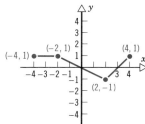

57. Extend the graph to make it symmetric with respect to the x-axis.

58. Extend the graph to make it symmetric with respect to the y-axis.

59. Extend the graph to make it symmetric with respect to the origin.

60. Extend the graph to make it symmetric with respect to the x-axis, y-axis, and origin.

61. Show that the distance between two points $P_1 = (x_1, y_1)$ and $P_2 = (x_2, y_2)$, denoted by $d(P_1, P_2)$, is given by the following **distance formula:**

$$d(P_1, P_2) = \sqrt{(x_2 - x_1)^2 + (y_2 - y_1)^2}$$

[*Hint:* Apply the Pythagorean theorem to the right triangle, whose hypotenuse is the distance d between the two points, as illustrated in the figure.]

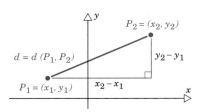

In Problems 62–66 use the formula in Problem 61 to find the distance $d(P_1, P_2)$ between the points P_1 and P_2.

62. $P_1 = (3, -4);\quad P_2 = (3, 1)$

63. $P_1 = (-3, 2);\quad P_2 = (6, 0)$

64. $P_1 = (4, -3);\quad P_2 = (6, 1)$

65. $P_1 = (2.3, 0.3);\quad P_2 = (2.3, 1.1)$

66. $P_1 = (a, b);\quad P_2 = (0, 0)$

67. Find all points having a y-coordinate of -3 whose distance from the point $(1, 2)$ is 13.

68. Find all points on the x-axis that are 5 units from the point $(2, -3)$.

Technology Exercises

In Problems 1–6 which of the suggested viewing rectangles gives the best complete graph of the indicated function.

1. $y = 2x^4$

   (a) $[-5, 5] \times [-5, 5]$
   (b) $[-10, 10] \times [-10, 10]$
   (c) $[-5, 5] \times [0, 100]$
   (d) $[-100, 100] \times [-10, 10]$
   (e) $[-10, 30] \times [-100, 0]$

2. $y = x^2$

   (a) $[0, 500] \times [0, 500]$
   (b) $[0, 20] \times [0, 400]$
   (c) $[-10, 10] \times [0, 100]$
   (d) $[-100, 100] \times [-100, 0]$
   (e) $[-20, 0] \times [0, 400]$

3. $y = \dfrac{x^3 + 1}{x}$

   (a) $[-5, 5] \times [-5, 5]$
   (b) $[-100, 100] \times [-10, 10]$
   (c) $[0, 10] \times [-10, 10]$
   (d) $[-5, 0] \times [-5, 5]$
   (e) $[-100, 100] \times [-100, 100]$

4. $y = \sqrt{x}$

   (a) $[-10, 10] \times [-10, 10]$
   (b) $[-10, 10] \times [-100, 10]$
   (c) $[0, 10] \times [0, 4]$
   (d) $[-100, 100] \times [-3000, 3000]$
   (e) $[-300, 300] \times [-100, 100]$

5. $y = -x^2$

   (a) $[0, 500] \times [0, 500]$
   (b) $[0, 10] \times [-10, 10]$
   (c) $[-10, 10] \times [-100, -50]$
   (d) $[-10, 10] \times [-10, 10]$
   (e) $[-10, 10] \times [-100, 0]$

6. $y = \dfrac{1}{x^2 + 2}$

   (a) $[0, 500] \times [0, 500]$
   (b) $[0, 100] \times [-10, 10]$
   (c) $[-100, 100] \times [-1000, 500]$
   (d) $[-10, 10] \times [-10, 10]$
   (e) $[-10, 10] \times [-1, 1]$

In Problems 7–12 use a graphing calculator to graph each function. Tell whether the graph is symmetric with respect to the x-axis, y-axis, origin, or none of these.

7. $y = 2x^4$ **8.** $y = x^2$ **9.** $y = \dfrac{x^3 + 1}{x}$

10. $y = \sqrt{x}$ **11.** $y = -x^2$ **12.** $y = \dfrac{1}{x^2 + 2}$

11.2 FUNCTIONS

In many applications a correspondence exists between two sets of numbers. For example, the revenue R resulting from the sale of x items selling for $10 each is $R = 10x$ dollars. If we know how many items have been sold, then we can calculate the revenue by using the rule $R = 10x$. This rule is an example of a *function*.

As another example, if an object is dropped from a height of 64 feet above the ground, the distance s (in feet) of the object from the ground after t seconds is given (approximately) by the formula $s = 64 - 16t^2$. When $t = 0$ seconds, the object is $s = 64$ feet above the ground. After 1 second, the object is $s = 64 - 16(1)^2 = 48$ feet above the ground. After 2 seconds, the object strikes the ground. The formula

$s = 64 - 16t^2$ provides a way of finding the distance s when the time t ($0 \le t \le 2$) is prescribed. There is a correspondence between each time t in the interval $0 \le t \le 2$ and the distance s. We say that the distance s is a *function* of the time t because

1. There is a correspondence between the set of times and the set of distances.
2. There is exactly one distance s obtained for a prescribed time t in the interval $0 \le t \le 2$.

Let's now look at the definition of a function.

Definition of a Function

FUNCTION

Let X and Y be two nonempty sets of real numbers. A *function* from X into Y is a rule or a correspondence that associates with each element of X a *unique* element of Y. The set X is called the *domain* of the function. For each element x in X, the corresponding element y in Y is called the *value* of the function at x, or the *image* of x. The set of all images of the elements of the domain is called the *range* of the function.

Refer to Figure 8. Since there may be some elements in Y that are not the image of some x in X, it follows that the range of a function may be a subset of Y.

Figure 8

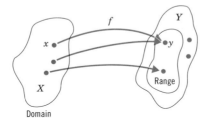

The rule (or correspondence) referred to in the definition of a function is most often given as an equation in two variables, usually denoted x and y.

EXAMPLE 1

Consider the function defined by the equation

$$y = 2x - 5 \qquad 1 \le x \le 6$$

The domain $1 \le x \le 6$ specifies that the number x is restricted to the real numbers from 1 to 6, inclusive. The rule $y = 2x - 5$ specifies that the number x is to be multiplied by 2 and then 5 is to be subtracted from the result to get y. For example, if $x = \frac{3}{2}$, then $y = 2 \cdot \frac{3}{2} - 5 = -2$. That is, the value of the function at $x = \frac{3}{2}$ is $y = -2$.

....

Functions are often denoted by letters such as f, F, g, G, and so on. If f is a function, then for each number x in its domain, the corresponding image in the range is designated by the symbol $f(x)$, read as "f of x." We refer to $f(x)$ as the **value of f at the number x.** Thus, $f(x)$ is the number that results when x is given and the rule for f is applied; $f(x)$ does *not* mean "f times x." For example, the function given in Example 1 may be written as $f(x) = 2x - 5$, $1 \leq x \leq 6$.

Figure 9 illustrates some other functions.

Figure 9

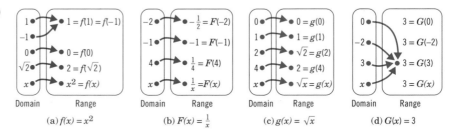

Domain   Range   Domain   Range   Domain   Range   Domain   Range

(a) $f(x) = x^2$ (b) $F(x) = \frac{1}{x}$ (c) $g(x) = \sqrt{x}$ (d) $G(x) = 3$

EXAMPLE 2

For the function

$$f(x) = x^2 + 3x - 4 \qquad -5 \leq x \leq 5$$

find the value of f at

(a) $x = 0$ (b) $x = 1$ (c) $x = -4$ (d) $x = 5$

SOLUTION

(a) The value of f at $x = 0$ is found by replacing x by 0 in the rule. Thus,

$$f(0) = 0^2 + 3(0) - 4 = -4$$

(b) $f(1) = 1^2 + 3(1) - 4 = 1 + 3 - 4 = 0$
(c) $f(-4) = (-4)^2 + 3(-4) - 4 = 16 - 12 - 4 = 0$
(d) $f(5) = 5^2 + 3(5) - 4 = 25 + 15 - 4 = 36$

Now Work Problem 1

．．．．．．．．．．．．．．．．．．．．．

A summary of some important facts to remember about a function f follows.

Summary of Important Facts about Functions

1. $f(x)$ is the image of x, or the value of f at x, when the rule f is applied to an x in the domain.

2. To each x in the domain of f, there is one and only one image $f(x)$ in the range.

3. f is the symbol we use to denote the function. It is symbolic of a domain and a rule we use to get from an x in the domain to $f(x)$ in the range.

Calculators

Most calculators have special keys that enable you to find the value of many functions. On your calculator, you should be able to find the square function, $f(x) = x^2$; the square root function, $f(x) = \sqrt{x}$; the reciprocal function, $f(x) = 1/x$; and many others that will be discussed later in this book (such as $\ln x$, $\log x$, and so on). When you enter x and then press one of these function keys, you get the value of that function at x. Try it with the functions listed in Example 3.

EXAMPLE 3

(a) $f(x) = x^2$; $f(1.234) = 1.522756$
(b) $F(x) = 1/x$; $F(1.234) = 0.8103728$
(c) $g(x) = \sqrt{x}$; $g(1.234) = 1.1108555$

....

Domain of a Function

Often the domain of a function f is not specified. Instead, only an equation defining the function is given. In such cases we agree that the domain of f is the largest set of real numbers for which the rule makes sense or, more precisely, for which the value $f(x)$ is a real number. We shall express the domain of a function using interval notation,* set notation,* or words, whichever is most convenient.

EXAMPLE 4

Find the domain of each of the following functions:

(a) $f(x) = \dfrac{3x}{x^2 - 4}$ (b) $g(x) = \sqrt{4 - 3x}$

SOLUTION

(a) The rule f tells us to divide $3x$ by $x^2 - 4$. Division by 0 is not allowed; therefore, the denominator $x^2 - 4$ can never be 0. Thus, x can never equal 2 or -2, and the domain of the function f is the set of all real numbers with -2 and 2 deleted.
(b) The rule g tells us to take the square root of $4 - 3x$. But only nonnegative numbers have real square roots. Hence, we require that

$$4 - 3x \geq 0$$
$$-3x \geq -4$$
$$x \leq \tfrac{4}{3}$$

The domain of g is $\{x | x \leq \tfrac{4}{3}\}$ or $x \leq \tfrac{4}{3}$.

Now Work Problem 43

....

* Refer to Appendix A.

If x is in the domain of a function f, we shall say that f **is defined at x,** or $f(x)$ **exists.** If x is not in the domain of f, we say that f **is not defined at x,** or $f(x)$ **does not exist.** For example, if $f(x) = x/(x^2 - 1)$, then $f(0)$ exists, but $f(1)$ and $f(-1)$ do not exist. (Do you see why?)

We have not said much about finding the range of a function. The reason is that when a function is defined by an equation, it is often difficult to find the range of a function. Therefore, we shall usually be content to find just the domain of a function when only the rule for the function is given.

When we use functions in applications, the domain may be restricted by physical or geometric considerations. For example, the domain of the function f defined by $f(x) = x^2$ is the set of all real numbers. However, if f is used as the rule for obtaining the area of a square when the length x of a side is known, then we must restrict the domain of f to the positive real numbers, since the length of a side never can be 0 or negative.

Independent Variable; Dependent Variable

Consider a function $y = f(x)$. The variable x that appears here is called the **independent variable,** because it can be assigned any of the permissible numbers from the domain. The variable y is called the **dependent variable,** because its value depends on x.

Any symbol can be used to represent the independent and dependent variables. For example, if f is the *cube function,* then f can be defined by $f(x) = x^3$ or $f(t) = t^3$ or $f(z) = z^3$. All three rules are identical—each tells us to cube the independent variable. In practice the symbols used for the independent and dependent variables are based on common usage.

EXAMPLE 5

The cost per square foot to build a house is $110.

(a) Express the cost C as a function of x, the number of square feet.
(b) What is the cost to build a 2000-square-foot house?

SOLUTION

(a) The cost C of building a house containing x square feet is $110x$ dollars. A function expressing this relationship is

$$C(x) = 110x$$

where x is the independent variable and C is the dependent variable. In this setting the domain is $\{x | x > 0\}$ since a house cannot have 0 or negative square feet.

(b) The cost to build a 2000-square-foot house is

$$C(2000) = 110(2000) = \$220{,}000$$

Now Work Problem 65

It is worth observing that, in the solution to Example 5(a), we used the symbol C in two ways: to name the function and to symbolize the dependent variable. This "double use" is common in applications.

EXAMPLE 6

The cost of eliminating a large part of the pollutants from the atmosphere (or from water) is relatively cheap. However, removing the last traces of pollutants results in a significant increase in cost. A typical relationship between the cost C, in thousands of dollars, for removal and the percent x of pollutant removed is given by the function

$$C(x) = \frac{3x}{105 - x}$$

Since x is a percentage, the domain of C consists of all real numbers x for which $0 \le x \le 100$. The cost of removing 0% of the pollutant is

$$C(0) = 0$$

The cost of removing 50% of the pollutant is

$$C(50) = \frac{150}{55} = 2.727 \text{ thousand dollars}$$

The costs of removing 60% and 70% are

$$C(60) = \frac{180}{45} = 4 \text{ thousand dollars}$$

and

$$C(70) = \frac{210}{35} = 6 \text{ thousand dollars}$$

Observe that the cost of removing an additional 10% of the pollutant after 50% had been removed is $1273, while the cost of removing an additional 10% after 60% is removed is $2000.

· ·

Construction of a Demand Function

Revenue is the amount of money derived from the sale of a product and equals the price of the product times the quantity of the product that is actually sold. But the price and the quantity sold are not independent. As the price falls, the demand for the product increases; and when the price rises, the demand decreases.

The equation that relates the price p of a quantity bought and the amount x of a quantity demanded is called the **demand equation.** If in this equation we solve for p, we have

$$p = d(x)$$

The function d is called the **price function** and $d(x)$ is the price per unit when x units are demanded. If x is the number of units sold and $d(x)$ is the price for each unit, the **revenue function** $R(x)$ is defined as

$$R(x) = x\, d(x) = xp$$

If we denote the **cost function** by $C(x)$, then the **profit function** $P(x)$ is defined as

$$P(x) = R(x) - C(x)$$

In practice a price function is found through surveys, analysis of data, history, and other sources available to the economist. The next example illustrates how a linear price function can be constructed. Observe that the fundamental assumption made here is the linear nature of the price function.

EXAMPLE 7

The manager of a toy store has observed that each week 1000 toy trucks are sold at a price of $5 per truck. When there is a special sale, the trucks sell for $4 each and 1200 per week are sold. Assuming a linear price function, construct the price function. What is the revenue function?

SOLUTION

Let p be the price of each truck and let x be the number sold. If the price function $p = d(x)$ is linear, then we know that $(1000, 5)$ and $(1200, 4)$ are two points on the line $p = d(x)$. Using the point–slope form of the equation of a line, the price function is

$$p - 4 = \frac{-1}{200}(x - 1200)$$

$$p = \frac{-1}{200}x + 10$$

The revenue function is

$$R(x) = xp = x\left(-\frac{1}{200}x + 10\right) = -\frac{1}{200}x^2 + 10x$$

· ·

The price function obtained in Example 7 is not meant to reflect extreme situations. For example, we do not expect to sell $x = 0$ trucks nor do we expect to sell too many trucks in excess of 1500, since even during a special sale only 1200 are sold. The price function does represent the relationship between price and quantity in a certain range —in this case, perhaps $500 < x < 1500$.

Our next example illustrates a constant price function.

EXAMPLE 8

No matter how much wheat a farmer can grow, it can be sold at $4 per bushel. Find the price function. What is the revenue function?

SOLUTION

Since the price per bushel is fixed at $4 per bushel, the price function is

$$p = \$4$$

The revenue function is

$$R(x) = xp = 4x$$

Now Work Problem 61

....

Elimination of Variables

In the next example the quantity you are seeking is expressed most naturally in terms of two variables. You will have to eliminate one of these variables before you can write the quantity as a function of a single variable.

EXAMPLE 9

A rectangular playpen is to be made with 180 ft of fencing material.

(a) Express the area A of the rectangle as a function of the length.
(b) Determine the domain of A (the domain will be determined by the physical constraints).
(c) Find the area for the following lengths: 20, 30, 40, 50, and 60 ft.

SOLUTION

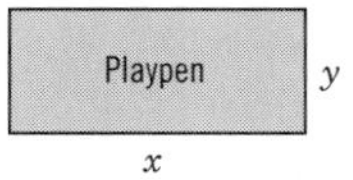

(a) Start by introducing two variables, say x and y, to denote the length and the width of the playpen. See Figure 10. The area A is equal to length times width. That is,

$$A = xy$$

The perimeter is

$$2x + 2y = 180$$

Since the goal is to express the area as a function of x alone, we must find a way to express y in terms of x. To do this, we use the fact that the perimeter is 180 ft.

$$2x + 2y = 180 \qquad \text{or} \qquad x + y = 90$$

Solve this equation for y:

$$y = 90 - x$$

and substitute the resulting expression for y into the formula for A to get

$$A(x) = x(90 - x)$$

(b) The area cannot be negative. Thus, x cannot be negative nor greater than 90. Hence, the domain is $0 \leq x \leq 90$ or [0, 90].
(c) $A(20) = 20(90 - 20) = 20 \cdot 70 = 1400 \text{ ft}^2$
   $A(30) = 30(90 - 30) = 30 \cdot 60 = 1800 \text{ ft}^2$
   $A(40) = 40(90 - 40) = 40 \cdot 50 = 2000 \text{ ft}^2$

Playpen y

x

Figure 10

$$A(50) = 50(90 - 50) = 50 \cdot 40 = 2000 \text{ ft}^2$$
$$A(60) = 60(90 - 60) = 60 \cdot 30 = 1800 \text{ ft}^2$$

In Figure 11 a graph is given showing how the area A varies with x. From the graph we see that, apparently, the maximum possible area is slightly larger than 2000 ft^2; it occurs when $x \approx 45$ ft—that is, for a playpen with dimensions 45 $\times$ 45, or an area of about 2025 ft^2. By "zooming in" on the graph near $x = 45$ (see Figure 12), we can see that a closer view suggests a maximum area of about 2025 ft^2.

For practical purposes, the results we found are probably sufficient. However, with calculus methods, we could verify that the result is indeed 2025 ft^2.

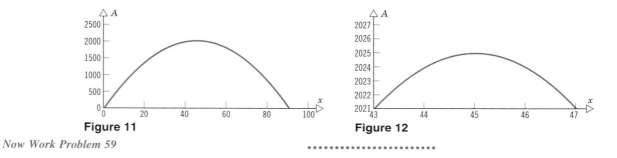

Figure 11

Figure 12

Now Work Problem 59

The Graph of a Function

In applications a graph often demonstrates more clearly the relationship between two variables than, say, an equation or table would. For example, Figure 13 shows the value of the Dow Jones average of 30 industrial stocks at the end of each day (vertical axis) from Monday, August 2, 1991, through Friday, August 23, 1991 (horizontal axis).* It is easy to see from the graph that the Dow average was falling over the few days preceding August 19 and was rising over the days from August 19 through August 23 (the days it

Figure 13

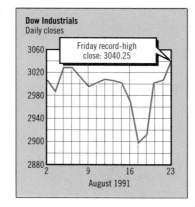

* Note that weekend dates are not included; the days marked are Fridays.

became clear that Mikhail Gorbachev would return to power in the USSR following the failed coup of August 19). The graph also shows that the lowest Dow average during this period occurred on August 19, while the highest occurred on August 23. Equations and tables, on the other hand, usually require some calculations and interpretation before this kind of information can be ''seen.''

Look again at Figure 13. The graph shows that, for each time on the horizontal axis, there is only one Dow average on the vertical axis. This is characteristic of the graph of a function.

When the rule that defines a function f is given by an equation in x and y, the **graph of f** is the graph of the equation, namely, the set of points (x, y) in the xy-plane that satisfies the equation.

Not every collection of points in the xy-plane represents the graph of a function. Remember, for a function f, each number x in the domain of f has one and only one image $f(x)$. Thus, the graph of a function f cannot contain two points with the same x-coordinate and different y-coordinates. Therefore, the graph of a function must satisfy the following **vertical line test:**

Vertical Line Test

A set of points in the xy-plane is the graph of a function if any vertical line intersects the graph in at most one point.

In other words, if any vertical line intersects a graph at more than one point, the graph is not the graph of a function.

EXAMPLE 10 | Which of the graphs in Figure 14 are graphs of functions?

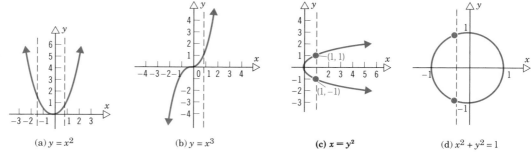

(a) $y = x^2$ (b) $y = x^3$ (c) $x = y^2$ (d) $x^2 + y^2 = 1$

Figure 14

SOLUTION

The graphs in Figures 14(a) and (b) are graphs of functions because every vertical line intersects each graph in at most one point. The graphs in Figures 14(c) and (d) are not graphs of functions because some vertical line intersects each graph in more than one point.

Ordered Pairs

The preceding discussion provides an alternative way to think of a function. We may consider a function f as a set of **ordered pairs** (x, y) or $(x, f(x))$, in which no two distinct pairs have the same first element. The set of all first elements is the domain, and the set of all second elements is the range of the function. Thus, there is associated with each element x in the domain a unique element y in the range. An example is the set of all ordered pairs (x, y) such that $y = x^2$. Some of the pairs in this set are

$$(2, 2^2) = (2, 4) \qquad\qquad (0, 0^2) = (0, 0)$$
$$(-2, (-2)^2) = (-2, 4) \qquad (\tfrac{1}{2}, (\tfrac{1}{2})^2) = (\tfrac{1}{2}, \tfrac{1}{4})$$

In this set no two pairs have the same *first* element (although there are pairs that have the same *second* element). This set is the *square function,* which associates with each real number x the number x^2. Look again at Figure 14(a).

On the other hand, the ordered pairs (x, y) for which $y^2 = x$ do not represent a function because there are ordered pairs with the same first element but different second elements. For example, $(1, 1)$ and $(1, -1)$ are ordered pairs obeying the relationship $y^2 = x$ with the same first element but different second elements. Look again at Figure 14(c).

The next example illustrates how to determine the domain and range of a function if its graph is given.

EXAMPLE 11   Let f be a function whose graph is given in Figure 15. Some points on the graph are labeled.

(a) What is the value of the function when $x = -6$, $x = -4$, $x = 0$, and $x = 6$?
(b) What is the domain of f?
(c) What is the range of f?
(d) List the intercepts. (Recall that these are the points, if any, where the graph crosses the coordinate axes.)

Figure 15

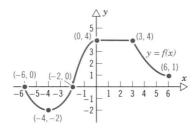

SOLUTION

(a) Since $(-6, 0)$ is on the graph of f, the y-coordinate 0 must be the value of f at the x-coordinate -6; that is, $f(-6) = 0$. In a similar way, we find that when $x = -4$, then $y = -2$, or $f(-4) = -2$; when $x = 0$, then $y = 4$, or $f(0) = 4$; and when $x = 6$, then $y = 1$, or $f(6) = 1$.

(b) To determine the domain of f, we notice that the points on the graph of f all have x-coordinates between -6 and 6, inclusive; and, for each number x between -6 and 6, there is a point $(x, f(x))$ on the graph. Thus, the domain of f is $\{x \mid -6 \leq x \leq 6\}$.

(c) The points on the graph all have y-coordinates between -2 and 4, inclusive; and for each such number y, there is at least one number x in the domain. Hence, the range of f is $\{y \mid -2 \leq y \leq 4\}$.

(d) The intercepts are $(-6, 0)$, $(-2, 0)$, and $(0, 4)$.

Now Work Problem 25

....

Summary

We list here some of the important vocabulary introduced in this section, with a brief description of each term.

Function   A rule or correspondence between two sets of real numbers so that each number x in the first set, the domain, has corresponding to it exactly one number y in the second set.

The range is the set of y values of the function for the x values in the domain.

Function Notation $y = f(x)$

f is a symbol for the rule that defines the function.

x is the independent variable.

y is the dependent variable.

$f(x)$ is the value of the function at x, or the image of x.

Graph of a Function   The collection of points (x, y) that satisfies the equation $y = f(x)$.

A collection of points is the graph of a function, provided any vertical line intersects the graph in at most one point.

EXERCISE 11.2   Answers to odd-numbered problems begin on page AN-43.

In Problems 1–8 find the following values for each function:
(a) $f(0)$ (b) $f(1)$ (c) $f(-1)$ (d) $f(2)$

1. $f(x) = -3x^2 + 2x - 4$

2. $f(x) = 2x^2 + x - 1$

3. $f(x) = \dfrac{x}{x^2 + 1}$

4. $f(x) = \dfrac{x^2 - 1}{x + 4}$

5. $f(x) = |x| + 4$

6. $f(x) = \sqrt{x^2 + x}$

7. $f(x) = \dfrac{2x + 1}{3x - 5}$

8. $f(x) = 1 - \dfrac{1}{(x + 2)^2}$

In Problems 9–20 use the given graph of the function f.

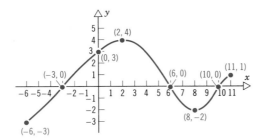

9. Find $f(0)$ and $f(2)$.

10. Find $f(8)$ and $f(-3)$.

11. Is $f(2)$ positive or negative?

12. Is $f(8)$ positive or negative?

13. For what numbers x is $f(x) = 0$?

14. For what numbers x is $f(x) > 0$?

15. What is the domain of f?

16. What is the range of f?

17. What are the x-intercepts?

18. What are the y-intercepts?

19. How often does the line $y = \frac{1}{2}$ intersect the graph?

20. How often does the line $y = 3$ intersect the graph?

In Problems 21–24 answer the questions about the given function.

21. $f(x) = \dfrac{x + 2}{x - 6}$

(a) Is the point $(3, 14)$ on the graph of f?

(b) If $x = 4$, what is $f(x)$?

(c) If $f(x) = 2$, what is x?

(d) What is the domain of f?

22. $f(x) = \dfrac{x^2 + 2}{x + 4}$

(a) Is the point $(1, \frac{3}{5})$ on the graph of f?

(b) If $x = 0$, what is $f(x)$?

(c) If $f(x) = \frac{1}{2}$, what is x?

(d) What is the domain of f?

23. $f(x) = \dfrac{2x^2}{x^4 + 1}$

(a) Is the point $(-1, 1)$ on the graph of f?

(b) If $x = 2$, what is $f(x)$?

(c) If $f(x) = 1$, what is x?

(d) What is the domain of f?

24. $f(x) = \dfrac{2x}{x - 2}$

(a) Is the point $(\frac{1}{2}, -\frac{2}{3})$ on the graph of f?

(b) If $x = 4$, what is $f(x)$?

(c) If $f(x) = 1$, what is x?

(d) What is the domain of f?

In Problems 25–38 determine whether the graph is that of a function by using the vertical line test. If it is, use the graph to find: (a) its domain and range; (b) the intercepts, if any; (c) any symmetry.

25.

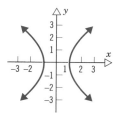

26.

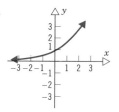

27.

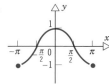

28.

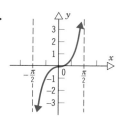

29.

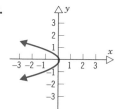

30.

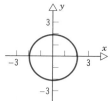

31.

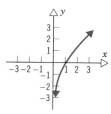

32.

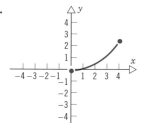

33.

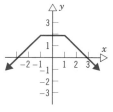

34.

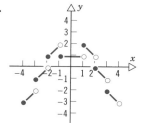

35.

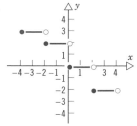

36.

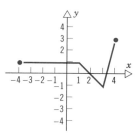

37.

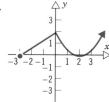

38.

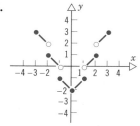

In Problems 39–52 find the domain of each function.

39. $f(x) = 2x + 1$

40. $f(x) = 3x^2 - 2$

41. $f(x) = \dfrac{x}{x^2 + 1}$

42. $f(x) = \dfrac{x^2}{x^2 + 1}$

43. $g(x) = \dfrac{x}{x^2 - 1}$

44. $h(x) = \dfrac{x}{x - 1}$

45. $F(x) = \dfrac{x - 2}{x^3 + x}$

46. $G(x) = \dfrac{x + 4}{x^3 - 4x}$

47. $h(x) = \sqrt{3x - 12}$

48. $G(x) = \sqrt{1 - x}$

49. $f(x) = \sqrt{x^2 - 9}$

50. $f(x) = \dfrac{1}{\sqrt{x^2 - 4}}$

51. $p(x) = \sqrt{\dfrac{x - 2}{x - 1}}$

52. $q(x) = \sqrt{x^2 - x - 2}$

53. If $f(x) = 2x^3 + Ax^2 + 4x - 5$ and $f(2) = 3$, what is the value of A?

54. If $f(x) = 3x^2 - Bx + 4$ and $f(-1) = 10$, what is the value of B?

55. If $f(x) = (3x + 8)/(2x - A)$ and $f(0) = 2$, what is the value of A?

56. If $f(x) = (2x - B)/(3x + 4)$ and $f(2) = \frac{1}{2}$, what is the value of B?

57. If $f(x) = (2x - A)/(x - 3)$ and $f(4) = 0$, what is the value of A? Where is f not defined?

58. If $f(x) = (x - B)/(x - A)$, $f(2) = 0$, and $f(1)$ is undefined, what are the values of A and B?

59. Area The perimeter of a rectangle is 120 ft. Express the area as a function of the width alone, and state the domain of this function.

60. Area The area of a rectangle is 30 ft². Express the perimeter as a function of the width alone, and state the domain of this function.

61. Demand Equation The price p and the quantity x sold of a certain product obey the demand equation

$$p = -\tfrac{1}{5}x + 100 \qquad 0 \le x \le 500$$

Express the revenue R as a function of x.

62. Demand Equation The price p and the quantity x sold of a certain product obey the demand equation

$$p = -\tfrac{1}{4}x + 100 \qquad 0 \le x \le 400$$

Express the revenue R as a function of x.

63. Demand Equation The price p and the quantity x sold of a certain product obey the demand equation

$$x = -20p + 100 \qquad 0 \le p \le 5$$

Express the revenue R as a function of x.

64. Demand Equation The price p and the quantity x sold of a certain product obey the demand equation

$$x = -5p + 500 \qquad 0 \le p \le 100$$

Express the revenue R as a function of x.

65. Express the gross salary G of a person who earns \$6 per hour as a function of the number x of hours worked.

66. A commissioned salesperson earns \$100 base pay plus \$10 per item sold. Express the gross salary G as a function of the number x of items sold.

67. Page Design A page with dimensions of 11 inches by 7 inches has a border of uniform width x surrounding the printed matter of the page, as shown in the figure. Write a formula for the area A of the printed part of the page as a function of the width x of the border. Give the domain and range of A.

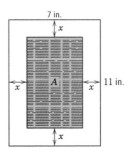

68. Cost of Flying An airplane crosses the Atlantic Ocean (3000 miles) with an airspeed of 500 miles per hour. The cost C (in dollars) per passenger is

$$C(x) = 100 + \frac{x}{10} + \frac{36{,}000}{x}$$

where x is the ground speed (airspeed $\pm$ wind).

(a) What is the cost per passenger for quiescent (no-wind) conditions?

(b) What is the cost per passenger with a head wind of 50 miles per hour?

(c) What is the cost per passenger with a tail wind of 100 miles per hour?

(d) What is the cost per passenger with a head wind of 100 miles per hour?

69. Making Boxes An open box with a square base is to be made from a square piece of cardboard 24 inches on a side by cutting out a square from each corner and turning up the sides (see the figure). Express the volume V of the box as a function of the length x of the side of the square cut from each corner.

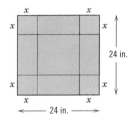

70. Making Boxes An open box with a square base is required to have a volume of 10 cubic feet. Express the amount A of material used to make such a box as a function of the length x of a side of the square base.

71. Cable Installation A cable TV company is asked to provide service to a customer whose house is located 2 miles from the road along which the cable is buried. The nearest connection box for the cable is located 5 miles down the road (see the figure).

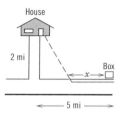

(a) If the installation cost is $100 per mile along the road and $140 per mile off the road, express the total cost C of installation as a function of the distance x (in miles) from the connection box to the point where the cable installation turns off the road.

(b) What is the domain of C?

(c) Compute the cost for $x = 1$, $x = 2$, $x = 3$, and $x = 4$.

Technology Exercises

A number x for which $f(x) = 0$ is called a zero of the function. Sketch a graph of each of the following functions and estimate the zeros. Most computer software packages contain the capability of estimating zeros. On a graphing calculator the "zoom" feature will be very helpful in Problems 1–4.

1. $f(x) = x^3 - 6x$

2. $f(x) = 6x^4 - 5x^3 - 39x^2 - 4x + 12$

3. $f(x) = 4x^5 - 10x^4 + 6x^3 - 4x^2 + 10x - 6$

4. $f(x) = x^4 + 3x^3$

5. Rework Example 9 when 187 ft of fencing is available. Using the zoom feature of your calculator, find the dimensions of the playpen with a maximum area (to three decimal places).

6. An open box is made by cutting squares of side w inches from the four corners of a sheet of cardboard that is 8.5 inches by 11 inches, and then folding up the sides.

(a) Express the volume of the box as a function of w. [*Hint*: Draw a picture.]

(b) Graph the function, and from the graph estimate the value of w that maximizes the volume of the box.

7. Use a calculator or computer graph of $f(x) = 1020 - x^3$ to determine

(a) The range of this function

(b) The number of zeros of this function

8. Use a calculator or computer graph of $f(x) = 13 - 20x - x^2 - 3x^4$ to determine

(a) The range of this function

(b) The number of zeros of this function

9. Use a graphing calculator (or a computer) to plot the graphs of $y = x^3$, $y = x^4$, and $y = x^5$ on the interval $-1 \leq x \leq 1$. Determine an appropriate range for y so that all powers will be distinguishable in the viewing rectangle. What is the y range if you plot the same graphs on the interval $-100 \leq x \leq 100$?

10. Sketch a graph of the function

$$f(x) = x^3 - 9x^2 - 48x + 52$$

(a) Start with the standard range setting $[-10, 10] \times [-10, 10]$. What do you see?

(b) Try to increase the range setting to see more of the function.

(c) Graph the function with the setting $[-10, 20] \times [-400, 200]$. What do you see now?

11.3 MORE ABOUT FUNCTIONS

Function Notation

The independent variable of a function is sometimes called the **argument** of the function. Thinking of the independent variable as an argument can sometimes make it easier to apply the rule of the function. For example, if f is the function defined by $f(x) = x^3$, then f tells us to cube the argument. Thus, $f(2)$ means to cube 2; $f(a)$ means to cube the number a; $f(x + h)$ means to cube the quantity $(x + h)$.

EXAMPLE 1   For the function G defined by $G(x) = 2x^2 - 3x$, evaluate

(a) $G(3)$ (b) $G(-x)$ (c) $-G(x)$
(d) $G(x + h)$ (e) $G(x) + G(h)$

SOLUTION

(a) We replace x by 3 in $G(x)$ to get

$$G(3) = 2(3)^2 - 3(3) = 18 - 9 = 9$$

(b) $G(-x) = 2(-x)^2 - 3(-x) = 2x^2 + 3x$

(c) $-G(x) = -(2x^2 - 3x) = -2x^2 + 3x$

(d) $G(x + h) = 2(x + h)^2 - 3(x + h)$ Notice the use of
$\qquad\qquad = 2(x^2 + 2xh + h^2) - 3x - 3h$ parentheses here.
$\qquad\qquad = 2x^2 + 4xh + 2h^2 - 3x - 3h$

(e) $G(x) + G(h) = 2x^2 - 3x + 2h^2 - 3h$

Now Work Problem 25

Example 1 illustrates certain uses of **function notation.** Another important use of function notation is to find the **difference quotient** of f:

DIFFERENCE QUOTIENT

The *difference quotient* of a function $y = f(x)$ is:

$$\frac{f(x + h) - f(x)}{h} \qquad h \neq 0 \qquad (1)$$

This expression is used often in calculus.

EXAMPLE 2

Find the difference quotient (1) of the function f defined by

$$f(x) = 2x^2 - x + 1$$

SOLUTION

It is helpful to proceed in steps.

Step 1 First, we calculate $f(x + h)$:

$$\begin{aligned} f(x + h) &= 2(x + h)^2 - (x + h) + 1 \\ &= 2(x^2 + 2xh + h^2) - x - h + 1 \\ &= 2x^2 + 4xh + 2h^2 - x - h + 1 \end{aligned}$$

Step 2 Now subtract $f(x)$ from this result:

$$f(x + h) - f(x) = 2x^2 + 4xh + 2h^2 - x - h + 1 - \underbrace{(2x^2 - x + 1)}$$

Be careful to subtract the quantity $f(x)$.

$$\begin{aligned} &= 2x^2 + 4xh + 2h^2 - x - h + 1 - 2x^2 + x - 1 \\ &= 4xh + 2h^2 - h \end{aligned}$$

Step 3 Now divide by h:

$$\begin{aligned} \frac{f(x + h) - f(x)}{h} &= \frac{4xh + 2h^2 - h}{h} \\ &= \frac{h(4x + 2h - 1)}{h} \end{aligned}$$

Factor the numerator.

$$= 4x + 2h - 1$$

Cancellation property

Now Work Problem 37

. . . .

Figure 16

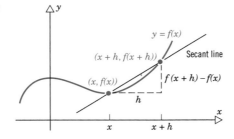

The difference quotient of a function has an important geometric interpretation. Look at the graph of $y = f(x)$ in Figure 16. Two points are labeled on the graph:

$$(x, \ f(x)) \qquad \text{and} \qquad (x + h, \ f(x + h))$$

The slope of the line (called a **secant line**) joining these two points is

$$\frac{f(x + h) - f(x)}{(x + h) - x} = \frac{f(x + h) - f(x)}{h}$$

Thus, the difference quotient of a function equals the slope of a secant line joining two points on its graph.

Increasing and Decreasing Functions

Consider the graph given in Figure 17. If you look from left to right along the graph of this function, you will notice that parts of the graph are rising, parts are falling, and parts are horizontal. In such cases the function is described as *increasing, decreasing,* and *constant,* respectively. For example, the graph is rising (increasing) on the closed interval $-4 \leq x \leq 0$, and falling (decreasing) on the closed interval $3 \leq x \leq 6$. More precise definitions follow.

Figure 17

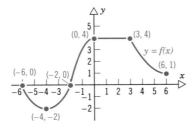

INCREASING FUNCTION

A function f is (strictly) *increasing* on an interval I if, for any choice of x_1 and x_2 in I, with $x_1 < x_2$, we have $f(x_1) < f(x_2)$.

DECREASING FUNCTION

A function f is (strictly) *decreasing* on an interval I if, for any choice of x_1 and x_2 in I, with $x_1 < x_2$, we have $f(x_1) > f(x_2)$.

CONSTANT FUNCTION

A function f is _constant_ on an interval I if, for all choices of x in I, the values $f(x)$ are equal.

Thus, the graph of an increasing function goes up from left to right; the graph of a decreasing function goes down from left to right; and the graph of a constant function remains at a fixed height. Figure 18 illustrates the definitions.

Figure 18

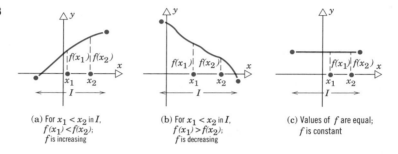

(a) For $x_1 < x_2$ in I,
$f(x_1) < f(x_2)$;
f is increasing

(b) For $x_1 < x_2$ in I,
$f(x_1) > f(x_2)$;
f is decreasing

(c) Values of f are equal;
f is constant

To answer the question of where a function is increasing, where it is decreasing, and where it is constant, we use inequalities involving the independent variable x or intervals of x-coordinates.

EXAMPLE 3

Where is the function in Figure 17 increasing? Where is it decreasing? Where is it constant?

SOLUTION

The graph is rising for $-4 \leq x \leq 0$; that is, the function is increasing on this interval. It is decreasing for $-6 \leq x \leq -4$ and for $3 \leq x \leq 6$; it is constant for $0 \leq x \leq 3$.

• •

Common Functions

We now give names to some of the functions we have encountered. In going through this list, pay special attention to the characteristics of each function, particularly to the shape of each graph.

LINEAR FUNCTION

$$f(x) = mx + b \qquad m \text{ and } b \text{ real numbers}$$

The domain of the **linear function** f consists of all real numbers. The graph of this function is a nonvertical straight line with slope m and y-intercept $(0, b)$. A linear function is increasing if $m > 0$, decreasing if $m < 0$, and constant if $m = 0$.

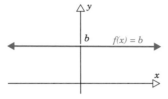

Figure 19

CONSTANT FUNCTION

$$f(x) = b \qquad b \text{ a real number}$$

See Figure 19. A **constant function** is a special linear function ($m = 0$). Its domain is the set of all real numbers; its range is the set consisting of a single number b. Its graph is a horizontal line whose y-intercept is $(0, b)$.

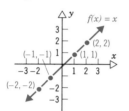

Figure 20

IDENTITY FUNCTION

$$f(x) = x$$

See Figure 20. The **identity function** is also a special linear function. Its domain and its range are the set of all real numbers. Its graph is a line whose slope is $m = 1$ and whose y-intercept is $(0, 0)$. The line consists of all points for which the x-coordinate equals the y-coordinate. The identity function is increasing over its domain. Note that the graph bisects quadrants I and III.

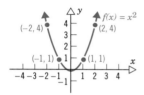

Figure 21

SQUARE FUNCTION

$$f(x) = x^2$$

See Figure 21. (Refer also to Example 2, page 500.) The domain of the **square function** f is the set of all real numbers; its range is the set of nonnegative real numbers. The graph of this function is a parabola, whose intercept is at $(0, 0)$. The square function is decreasing on the interval $(-\infty, 0]$ and increasing on the interval $[0, \infty)$.

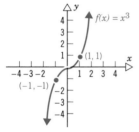

Figure 22

CUBE FUNCTION

$$f(x) = x^3$$

See Figure 22. (Refer also to Example 4, page 503.) The domain and range of the **cube function** are the set of all real numbers. The intercept of the graph is at $(0, 0)$. The cube function is increasing on the interval $(-\infty, \infty)$.

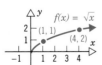

Figure 23

SQUARE ROOT FUNCTION

$$f(x) = \sqrt{x}$$

See Figure 23. (Refer also to the comments following Example 5, page 504.) The domain and range of the **square root function** are the set of nonnegative real numbers. The intercept of the graph is at $(0, 0)$. The square root function is increasing on the interval $[0, \infty)$.

RECIPROCAL FUNCTION

$$f(x) = \frac{1}{x}$$

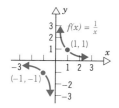

Figure 24

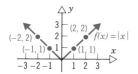

Figure 25

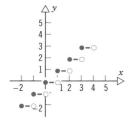

Figure 26

See Figure 24. (Refer also to Example 6, page 504.) The domain and range of the **reciprocal function** are the set of all nonzero real numbers. The graph has no intercepts. The reciprocal function is decreasing on the intervals $(-\infty, 0)$ and $(0, \infty)$.

ABSOLUTE VALUE FUNCTION

$$f(x) = |x|$$

See Figure 25. The domain of the **absolute value function** is the set of all real numbers; its range is the set of nonnegative real numbers. The intercept of the graph is at $(0, 0)$. If $x \geq 0$, then $f(x) = x$ and the graph of f is part of the line $y = x$; if $x < 0$, then $f(x) = -x$ and the graph of f is part of the line $y = -x$. The absolute value function is decreasing on the interval $(-\infty, 0]$ and increasing on the interval $[0, \infty)$.

The symbol $[\![x]\!]$, read as **"bracket x,"** stands for the largest integer less than or equal to x. For example,

$$[\![1]\!] = 1 \qquad [\![2.5]\!] = 2 \qquad [\![\tfrac{1}{2}]\!] = 0 \qquad [\![-\tfrac{3}{4}]\!] = -1 \qquad [\![\pi]\!] = 3$$

This type of correspondence occurs frequently enough in mathematics that we give it a name.

GREATEST INTEGER FUNCTION

$$f(x) = [\![x]\!] = \text{Greatest integer less than or equal to } x$$

We obtain the graph of $f(x) = [\![x]\!]$ by plotting several points. See Table 3. Figure 26 shows the graph.

Table 3

x	-1	$-\tfrac{1}{2}$	$-\tfrac{1}{4}$	0	$\tfrac{1}{4}$	$\tfrac{1}{2}$	$\tfrac{3}{4}$
$y = f(x) = [\![x]\!]$	-1	-1	-1	0	0	0	0
(x, y)	$(-1, -1)$	$(-\tfrac{1}{2}, -1)$	$(-\tfrac{1}{4}, -1)$	$(0, 0)$	$(\tfrac{1}{4}, 0)$	$(\tfrac{1}{2}, 0)$	$(\tfrac{3}{4}, 0)$

The domain of the **greatest integer function** is the set of all real numbers; its range is the set of integers. The y-intercept of the graph is at $(0, 0)$. The x-intercepts lie in the interval $[0, 1)$. The greatest integer function is constant on every interval of the form $[k, k + 1)$, for k an integer. In Figure 26, we use a solid dot to indicate, for example, that at $x = 1$, the value of f is $f(1) = 1$; we use an open circle to illustrate that the function does not assume the value 0 at $x = 1$.

From the graph of the greatest integer function, we can see why it is also called a **step function**. At $x = 0$, $x = \pm 1$, $x = \pm 2$, and so on, this function exhibits what is called a *discontinuity*. That is, at integer values, the graph suddenly "steps" from one value to another without taking on any of the intermediate values. For example, to the immediate left of $x = 3$, the y-coordinates are 2, and to the immediate right of $x = 3$, the y-coordinates are 3.

The functions we have discussed so far are basic. Whenever you encounter one of them, you should see a mental picture of its graph. For example, if you encounter the function $f(x) = x^2$, you should see in your mind's eye a picture like Figure 21.

Now Work Problem 1

Piecewise-Defined Functions

Sometimes a function is defined by a rule consisting of two or more equations. The choice of which equation to use depends on the value of the independent variable x. For example, the absolute value function $f(x) = |x|$ is actually defined by two equations: $f(x) = x$ if $x \geq 0$ and $f(x) = -x$ if $x < 0$. For convenience, we generally combine these equations into one expression as

$$f(x) = |x| = \begin{cases} x & \text{if } x \geq 0 \\ -x & \text{if } x < 0 \end{cases}$$

When functions are defined by more than one equation, they are called **piecewise-defined** functions.

Let's look at another example of a piecewise-defined function.

EXAMPLE 4

For the following function f:

$$f(x) = \begin{cases} -x + 1 & \text{if } -1 \leq x < 1 \\ 2 & \text{if } x = 1 \\ x^2 & \text{if } x > 1 \end{cases}$$

(a) Find $f(0)$, $f(1)$, and $f(2)$.   (b) Determine the domain of f.
(c) Graph f.   (d) Use the graph to find the range of f.

SOLUTION

(a) To find $f(0)$, we observe that when $x = 0$, the equation for f is given by $f(x) = -x + 1$. So we have

$$f(0) = -0 + 1 = 1$$

When $x = 1$, the equation for f is $f(x) = 2$. Thus,

$$f(1) = 2$$

When $x = 2$, the equation for f is $f(x) = x^2$. So

$$f(2) = 2^2 = 4$$

(b) To find the domain of f, we look at its definition. We conclude that the domain of f is $\{x \mid x \geq -1\}$.
(c) To graph f, we graph "each piece." Thus, we first graph the line $y = -x + 1$, and keep only the part for which $-1 \leq x < 1$. Then we plot the point $(1, 2)$, because when $x = 1$, $f(x) = 2$. Finally, we graph the parabola $y = x^2$, and keep only the part for which $x > 1$. See Figure 27.
(d) From the graph we conclude that the range of f is $\{y \mid y > 0\}$.

Figure 27 $y = f(x)$

Now Work Problem 63

EXAMPLE 5

Commonwealth Edison Company supplies electricity in the winter to residences for a monthly customer charge of $9.06 plus 10.819 cents per kilowatt-hour for the first 400 kilowatt-hours supplied in the month, and 7.093 cents per kilowatt-hour for all kilowatt-hours over 400 in the month.* If C is the monthly charge for x kilowatt-hours, express C as a function of x. What is the charge for 300 kilowatt-hours? For 700 kilowatt-hours?

SOLUTION

If $0 \le x \le 400$, the monthly charge C (in dollars) can be found by multiplying x by 0.10819 and adding the monthly customer charge of $9.06. Thus, if $0 \le x \le 400$, then

$$C(x) = 0.10819x + 9.06$$

For $x > 400$ the charge is $0.10819(400) + 9.06 + 0.07093(x - 400)$ since $x - 400$ equals usage in excess of 400 kilowatt-hours, which costs $0.07093 per kilowatt-hour. Thus, if $x > 400$, then

$$C(x) = 0.10819(400) + 9.06 + 0.07093(x - 400)$$
$$= 52.336 + 0.07093(x - 400)$$
$$= 0.07093x + 23.964$$

We follow two rules to compute C as follows:

$$C(x) = \begin{cases} 0.10819x + 9.06 & \text{if } 0 \le x \le 400 \\ 0.07093x + 23.964 & \text{if } x > 400 \end{cases}$$

The charge for $x = 300$ kilowatt-hours is

$$C(300) = 0.10819(300) + 9.06 = \$41.52$$

The charge for $x = 700$ kilowatt-hours is

$$C(700) = 0.07093(700) + 23.964 = \$73.62$$

See Figure 28 for the graph.

Figure 28

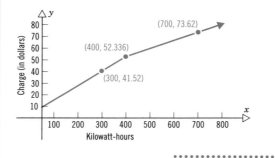

Now Work Problem 83

* *Source:* Commonwealth Edison rates for residential service, 1991.

QUADRATIC FUNCTIONS*

A *quadratic function* is a function of the form

$$f(x) = ax^2 + bx + c$$

where $a \neq 0, b$, and c are real numbers. The domain of f is the set of real numbers.

The graph of a quadratic function is called a *parabola*. To analyze the graph of a quadratic function, we complete the square on the terms involving x^2 and x.

$$y = f(x) = ax^2 + bx + c$$

$$y = a\left(x^2 + \frac{b}{a}x + \frac{c}{a}\right)$$

$$y = a\left[\left(x^2 + \frac{b}{a}x\right) + \frac{c}{a}\right]$$

Add and subtract $b^2/4a^2$ to complete the square.

$$y = a\left[\left(x^2 + \frac{b}{a}x + \frac{b^2}{4a^2}\right) + \frac{c}{a} - \frac{b^2}{4a^2}\right]$$

$$y = a\left[\left(x + \frac{b}{2a}\right)^2 + \frac{4ac - b^2}{4a^2}\right]$$

Now the value of y depends on the number x in the term $(x + b/2a)^2$. Since this term is either positive or zero, the smallest value of this term occurs when $x = -b/2a$. Thus,

The point on the parabola for which $x = -b/2a$ is called the *vertex* of the parabola. When $a > 0$, the vertex is a *minimum point* and the *graph opens upward*. See Figure 29(a). When $a < 0$, the vertex is a *maximum point* and the *graph opens downward*. See Figure 29(b).

Figure 29

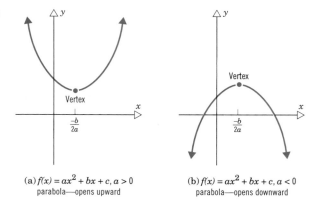

(a) $f(x) = ax^2 + bx + c, a > 0$
parabola—opens upward

(b) $f(x) = ax^2 + bx + c, a < 0$
parabola—opens downward

* Refer to Appendix A for a review of quadratic equations.

In graphing a quadratic function we proceed as follows:

Step 1 Express the given function in the form $f(x) = ax^2 + bx + c$, and identify a, b, and c. If $a > 0$, the graph opens upward. If $a < 0$, the graph opens downward.

Step 2 Locate the vertex. Its x-coordinate is $-b/2a$. The y-coordinate may be found by evaluating $f(-b/2a)$.

Step 3 Find the intercepts. The y-intercept is $(0, f(0))$. The x-intercepts, if any, obey the quadratic equation $ax^2 + bx + c = 0$. Thus,

If $b^2 - 4ac > 0$, the graph has two x-intercepts.

If $b^2 - 4ac = 0$, the graph has one x-intercept.

If $b^2 - 4ac < 0$, the graph has no x-intercepts.

Step 4 Plot the vertex, the y-intercept, and the x-intercepts, if any. If necessary, plot one or two additional points. Connect the points with a smooth curve.

Let's look at some examples.

EXAMPLE 6

Graph each quadratic function.

(a) $f(x) = x^2 + 2x - 8$ (b) $g(x) = -2x^2 + 8x$
(c) $F(x) = 2x^2 - 4x + 3$

SOLUTION

(a) 1. Since $f(x) = x^2 + 2x - 8$, we find $a = 1$, $b = 2$, $c = -8$. Since $a > 0$, the graph opens upward.
  2. $-b/2a = -2/2 = -1$; $f(-b/2a) = f(-1) = 1 - 2 - 8 = -9$. The vertex is at $(-1, -9)$.
  3. Since $f(0) = -8$, the y-intercept is $(0, -8)$. Since $b^2 - 4ac = 36 > 0$, there are two x-intercepts. They obey the equation

$$x^2 + 2x - 8 = 0$$
$$(x + 4)(x - 2) = 0$$
$$x = -4 \quad \text{or} \quad x = 2$$

  The x-intercepts are $(-4, 0)$ and $(2, 0)$.
  4. Figure 30 shows the graph.

(b) 1. Since $g(x) = -2x^2 + 8x$, we find $a = -2$, $b = 8$, $c = 0$. Since $a < 0$, the graph opens downward.
  2. $-b/2a = -8/-4 = 2$; $f(-b/2a) = f(2) = -2(4) + 8(2) = 8$. The vertex is at $(2, 8)$.
  3. Since $f(0) = 0$, the y-intercept is $(0, 0)$. Since $b^2 - 4ac = 64 > 0$, there are two x-intercepts, which obey

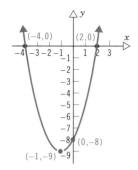

$f(x) = x^2 + 2x - 8$

Figure 30

$$-2x^2 + 8x = 0$$
$$x(-2x + 8) = 0$$
$$x = 0 \quad \text{or} \quad x = 4$$

The x-intercepts are $(0, 0)$ and $(4, 0)$.
4. Figure 31 shows the graph.
(c) 1. Since $F(x) = 2x^2 - 4x + 3$, we find $a = 2$, $b = -4$, $c = 3$. Since $a > 0$, the graph opens upward.
2. $-b/2a = -(-4)/4 = 1$; $F(-b/2a) = F(1) = 1$. The vertex is at $(1, 1)$.
3. Since $F(0) = 3$, the y-intercept is $(0, 3)$. Since $b^2 - 4ac = -8 < 0$, there are no x-intercepts.
4. Figure 32 shows the graph. Notice that one additional point $(2, 3)$ is used.

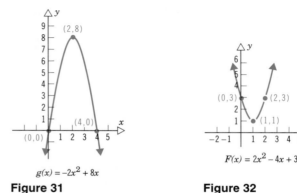

$$F(x) = 2x^2 - 4x + 3$$

Figure 31   Figure 32

Now Work Problem 69

Polynomial and Rational Functions

The linear function, the quadratic function, and the cube function are members of a larger class of functions called **polynomial functions.**

POLYNOMIAL FUNCTION

A *polynomial function* is a function of the form

$$f(x) = a_n x^n + a_{n-1} x^{n-1} + \cdots + a_2 x^2 + a_1 x + a_0$$

where $a_n \neq 0$, a_{n-1}, . . . , a_2, a_1, a_0 are real numbers and $n \geq 0$ is a nonnegative integer.

For a polynomial function the real numbers a_n, a_{n-1}, . . . , a_1, a_0 are called **coefficients** and $a_n \neq 0$ is called the **leading coefficient.** The exponent n is called the **degree of the polynomial.**

The domain of a polynomial function consists of all real numbers. The y-intercept is a_0. The x-intercepts, if any, satisfy the equation

$$a_n x^n + a_{n-1} x^{n-1} + \cdots + a_2 x^2 + a_1 x + a_0 = 0 \tag{2}$$

If $n = 1$ or $n = 2$, the solutions, if any, of equation (2) are easy to find. Although methods exist for solving this equation for $n = 3$ and $n = 4$, such methods are tedious. No general methods exist for solving this equation for $n \geq 5$. Therefore, we shall be content to identify x-intercepts only when they are easily found, or with the aid of a graphing calculator or computer.

Obtaining the graphs of most polynomials of degree 3 or higher is generally not easy with the tools we now have available. We would have to locate several well-chosen points on the graph (this is easier said than done) and then hope that by connecting them with a smooth curve, we would obtain an accurate picture. This tedious method is imprecise, and we will soon show how the power of calculus can be used to get accurate graphs without requiring a random selection of many points.

The reciprocal function $f(x) = 1/x$ is a member of the class of functions called **rational functions.**

RATIONAL FUNCTION

A *rational function* is a function of the form

$$R(x) = \frac{P(x)}{Q(x)} = \frac{a_n x^n + \cdots + a_1 x + a_0}{b_m x^m + \cdots + b_1 x + b_0}$$

where P is a polynomial function of degree n, Q is a nonzero polynomial function of degree m, and P and Q contain no common factors.

The domain of a rational function consists of all real numbers x except those for which $Q(x) = 0$.

The graphs of most rational functions, like those of most polynomial functions, require the use of calculus. We postpone a discussion of their graphs to Chapter 14. As we continue in this book, other types of functions will be encountered, classified, and discussed. For most of them, calculus will not be merely useful, but necessary, to obtain a complete description.

EXERCISE 11.3   Answers to odd-numbered problems begin on page AN-45.

In Problems 1–8 match each graph to the function whose graph most resembles the one given.

(a) *Constant function*   (b) *Linear function*   (c) *Square function*
(d) *Cube function*   (e) *Square root function*   (f) *Reciprocal function*
(g) *Absolute value function*   (h) *Greatest integer function*

1.   2.    3.    4.

5.   **6.**   **7.** 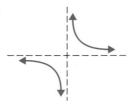   **8.**

In Problems 9–20 the graph of a function is given. Use the graph to find
(a) its domain and range; (b) the intervals on which it is increasing, decreas-
ing, or constant; (c) the intercepts, if any.

9. 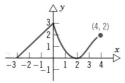   **10.** 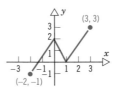   **11.**

12.   **13.** 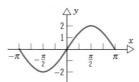   **14.**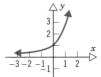

15. 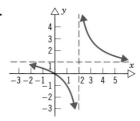   **16.** 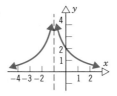   **17.**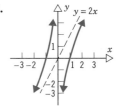

18. 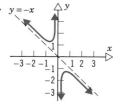   **19.** 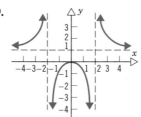   **20.**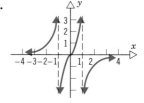

For Problems 21 and 22 determine the following for the indicated graph of a
function showing the impact of imports in the tire industry.
(a) The interval(s) on which it is increasing   *(b) The interval(s) on which it is decreasing*
(c) The highest and lowest points on the graph   *(d) The interval(s) where growth is flat*
(e) The interval(s) where growth or decline is the steepest

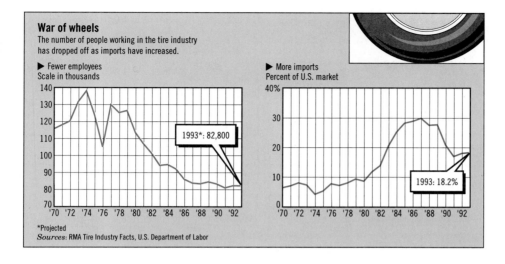

War of wheels
The number of people working in the tire industry
has dropped off as imports have increased.

▶ Fewer employees
Scale in thousands

▶ More imports
Percent of U.S. market

1993*: 82,800

1993: 18.2%

*Projected
Sources: RMA Tire Industry Facts, U.S. Department of Labor

21. The graph for fewer employees   **22.** The graph for more imports

In Problems 23–34 find the following for each function:
(a) $f(-x)$ **(b)** $-f(x)$ **(c)** $f(2x)$ **(d)** $f(x - 3)$ **(e)** $f(1/x)$ **(f)** $1/f(x)$

23. $f(x) = 2x + 3$ **24.** $f(x) = 4 - x$ **25.** $f(x) = 2x^2 - 4$

26. $f(x) = x^3 + 1$ **27.** $f(x) = x^3 - 3x$ **28.** $f(x) = x^2 + x$

29. $f(x) = \dfrac{x}{x^2 + 1}$ **30.** $f(x) = \dfrac{x^2}{x^2 + 1}$ **31.** $f(x) = |x|$

32. $f(x) = \dfrac{1}{x}$ **33.** $f(x) = 1 + \dfrac{1}{x}$ **34.** $f(x) = 4 + \dfrac{2}{x}$

In Problems 35–44 find the difference quotient,

$$\frac{f(x + h) - f(x)}{h} \qquad h \neq 0$$

for each function. Be sure to simplify.

35. $f(x) = 3$ **36.** $f(x) = 2x$ **37.** $f(x) = 1 - 3x$

38. $f(x) = x^2 + 1$ **39.** $f(x) = 3x^2 - 2x$ **40.** $f(x) = 4x - 2x^2$

41. $f(x) = x^3 - x$ **42.** $f(x) = x^3 + x$ **43.** $f(x) = \dfrac{1}{x}$

44. $f(x) = \dfrac{1}{x^2}$

*In Problems 45–68 (a) find the domain of each function; (b) locate any intercepts;
(c) graph each function; (d) based on the graph, find the range.*

45. $f(x) = 3x - 3$ **46.** $f(x) = 4 - 2x$ **47.** $g(x) = x^2 - 4$ **48.** $g(x) = x^2 + 4$

49. $h(x) = -x^2$ **50.** $F(x) = 2x^2$ **51.** $f(x) = \sqrt{x-2}$ **52.** $g(x) = \sqrt{x} + 2$

53. $h(x) = \sqrt{2-x}$ **54.** $F(x) = -\sqrt{x}$ **55.** $f(x) = |x| + 3$ **56.** $g(x) = |x+3|$

57. $h(x) = -|x|$ **58.** $F(x) = |3-x|$

59. $f(x) = \begin{cases} 2x & \text{if } x \neq 0 \\ 0 & \text{if } x = 0 \end{cases}$

60. $f(x) = \begin{cases} 3x & \text{if } x \neq 0 \\ 4 & \text{if } x = 0 \end{cases}$

61. $f(x) = \begin{cases} 1+x & \text{if } x < 0 \\ x^2 & \text{if } x \geq 0 \end{cases}$

62. $f(x) = \begin{cases} 1/x & \text{if } x < 0 \\ \sqrt{x} & \text{if } x \geq 0 \end{cases}$

63. $f(x) = \begin{cases} |x| & \text{if } -2 \leq x < 0 \\ 1 & \text{if } x = 0 \\ x^3 & \text{if } x > 0 \end{cases}$

64. $f(x) = \begin{cases} 3+x & \text{if } -3 \leq x < 0 \\ 3 & \text{if } x = 0 \\ \sqrt{x} & \text{if } x > 0 \end{cases}$

65. $g(x) = \begin{cases} 1 & \text{if } x \text{ is an integer} \\ -1 & \text{if } x \text{ is not an integer} \end{cases}$

66. $g(x) = \begin{cases} x & \text{if } x \geq 1 \\ 1 & \text{if } x < 1 \end{cases}$

67. $h(x) = 2[\![x]\!]$

68. $f(x) = [\![2x]\!]$

In Problems 69–80 determine whether the given quadratic function opens upward or downward. Find the vertex, the y-intercept, and the x-intercepts, if any. Graph each function.

69. $y = f(x) = 2x^2 + x - 3$ **70.** $y = f(x) = -2x^2 - x + 3$

71. $y = f(x) = x^2 - 4$ **72.** $y = f(x) = x^2 + 4x + 4$

73. $y = f(x) = x^2 + 1$ **74.** $y = f(x) = -3x^2 + 5x + 2$

75. $y = f(x) = -x^2 + 1$ **76.** $y = f(x) = x^2 + 2x + 1$

77. $y = f(x) = x^2 - 7x + 12$ **78.** $y = f(x) = x^2 - 10x + 25$

79. $y = f(x) = 4 - x^2$ **80.** $y = f(x) = x^2 + 4$

In Problems 81 and 82 find a piecewise-defined function whose graph is shown in the given figure. Note that each graph is made up of line segments.

81.

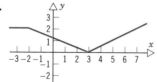

82.

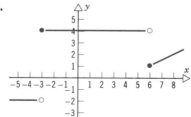

83. Discounts   A book club offers the following deal to its members: If 6 books are bought at the full price of $20.00 apiece, additional books can then be bought at $12.00 apiece. There is a limit of 10 books per customer. Express the cost of the books as a function of the number bought and draw the graph.

84. Worker's Wages   Workers at a mail-order company earn $6.00 per hour for the first 40 hours in a week, and then $8.50 per hour for additional hours. Let x be the number of hours worked in a week. Write a function W that describes a worker's pay.

85. Cost of a Taxicab Ride   The cost of a taxicab ride in a certain city is $1.90 for any ride up to, and including, 1 mile. After 1 mile, the rider pays an additional amount at the rate of 60 cents per mile (or fraction thereof). Let $C(x)$ be the total cost in dollars for a ride of x miles.

(a) Write a function C that describes the cost of a taxi ride.

(b) Compute $C(0.9)$, $C(1.1)$, and $C(1.9)$.

(c)
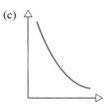

Technology Exercises

1. Each of the functions in the table decreases, but each decreases in a different way. Which of the graphs below best fits each function?

t	$f(t)$	$g(t)$	$h(t)$
1	25	35	21
2	16	31	17
3	9	25	13
4	4	20	9
5	1	9	5
6	0	0	1

(a)   (b)
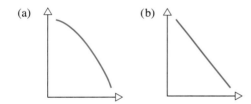

2. Graph the functions $f(x) = x^3$, $f_1(x) = x^3 + 3$, $f_2(x) = x^3 + 5$. Make a conjecture as to how the graph of $f(x)$ is moved when $f(x)$ is replaced by $f(x) + c$ where c is a constant.

3. Graph the functions $f(x) = x^3$, $f_1(x) = (x + 1)^3$, $f_2(x) = (x + 2)^3$, $f_3(x) = (x - 2)^3$. Make a conjecture as to how the graph of $f(x)$ is moved when $f(x)$ is replaced by $f(x + c)$ for c positive, c negative.

4. Compare the graphs of $y = \sqrt{x^2}$ and $y = |x|$. Explain.

5. Compare the graphs of $y = (\sqrt{x})^2$ and $y = |x|$. Explain.

11.4 THE EXPONENTIAL FUNCTION

In this section we introduce a function that plays a central role, not only in applied mathematics, but also in economics, business, and other areas of the social sciences. It is used in finance to calculate the value of investments, in demography to calculate population size, in the health sciences to study the spread of epidemics, and in archeology to date ancient artifacts. The function involves a constant raised to a variable power, such as $f(x) = 3^x$. We call such a function an *exponential function*.

EXPONENTIAL FUNCTION

An *exponential function* with a base a is a function of the form

$$f(x) = a^x$$

where a is a positive real number and $a \neq 1$. The domain of f is the set of all real numbers.

We exclude the base $a = 1$ because this function is simply the constant function $f(x) = 1^x = 1$. We also need to exclude bases that are negative because, otherwise, we would have to exclude many values of x from the domain, such as $x = \frac{1}{2}, x = \frac{3}{4}$, and so on. [Remember $(-2)^{1/2}$, $(-3)^{3/4}$, and so on, are not defined in the system of real numbers.]

Do not confuse the exponential function $f(x) = 3^x$ with the cube function $f(x) = x^3$, in which the base is the variable and the exponent is the constant.

Most scientific calculators have a $\boxed{y^x}$ key for working with exponents. To use this key, first enter the base y, then press the $\boxed{y^x}$ key, enter x, and press the $\boxed{=}$ key.

EXAMPLE 1

Using a calculator with a $\boxed{y^x}$ key, evaluate

(a) $2^{1.4}$ (b) $2^{1.41}$ (c) $2^{1.414}$ (d) $2^{1.4142}$ (e) $2^{\sqrt{2}}$

SOLUTION

(a) $2^{1.4} \approx 2.6390158$ (b) $2^{1.41} \approx 2.6573716$
(c) $2^{1.414} \approx 2.6647497$ (d) $2^{1.4142} \approx 2.6651191$
(e) $2^{\sqrt{2}} \approx 2.6651441$

Now Work Problem 17

When working with exponential functions, it may be necessary to use exponential notation and algebraic laws of exponents. These laws are as follows, where m and n are real numbers and a and b are positive.*

1. $a^n = \underbrace{a \cdot \cdots \cdot a}_{n \text{ factors}}$

2. $a^0 = 1$

3. $a^{-n} = \dfrac{1}{a^n}$

4. $a^m a^n = a^{m+n}$

5. $(a^m)^n = a^{m \cdot n}$

6. $\left(\dfrac{a}{b}\right)^m = \dfrac{a^m}{b^m}$

7. $\dfrac{a^m}{a^n} = a^{m-n}$

8. $(ab)^m = a^m b^m$

9. $a^{m/n} = \sqrt[n]{a^m} = \left(\sqrt[n]{a}\right)^m, \quad n \neq 0$

* Refer to Appendix A for a review of exponents.

Graphs of Exponential Functions

EXAMPLE 2 Graph the exponential function: $f(x) = 2^x$

SOLUTION

The domain of $f(x) = 2^x$ consists of all real numbers. We begin by locating some points on the graph of $f(x) = 2^x$, as listed in Table 4.

The data in Table 4 suggests that 2^x increases without bound as x increases without bound, and 2^x approaches zero as x decreases without bound. Since $2^x > 0$ for all x, the range of f is all positive real numbers. From this we conclude that the graph has no x-intercepts. The y-intercept is $(0, 1)$.

Using all this information, we plot some of the points from Table 4 and connect them with a smooth curve, as shown in Figure 33.

Table 4

x	$f(x) = 2^x$
-10	$2^{-10} \approx 0.00098$
-3	$2^{-3} = \frac{1}{8}$
-2	$2^{-2} = \frac{1}{4}$
-1	$2^{-1} = \frac{1}{2}$
0	$2^0 = 1$
1	$2^1 = 2$
2	$2^2 = 4$
3	$2^3 = 8$
10	$2^{10} = 1024$

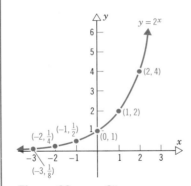

Figure 33 $y = 2^x$

Observe from Figure 33 that the graph approaches the x-axis, but never touches it. The x-axis is called a *horizontal asymptote*.

As we shall see, graphs that look like the one in Figure 33 occur very frequently in a variety of situations. For example, look at the graph in Figure 34(a), on the next page, which illustrates the amount of money paid to doctors in the Medicare program. Researchers might conclude from this graph that Medicare payments "behave exponentially"; that is, the graph exhibits "rapid, or exponential, growth." Also look at Figure 34(b), which illustrates the "rapid growth" or "exponential growth" of population in Ethiopia.

The graph of $f(x) = 2^x$ in Figure 33 is typical of all exponential functions that have a base larger than 1. Such functions are increasing functions. Their graphs lie above the x-axis, pass through the point $(0, 1)$, and thereafter rise rapidly as x increases without bound. When x decreases without bound, the x-axis is a horizontal asymptote. Finally, the graphs are smooth, with no corners or holes.

Figure 34

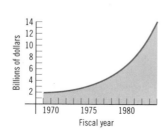

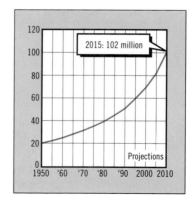

(a) Medicare payments to doctors for care of elderly
Source: Office of Management and Budget

(b) Population soars: Ethiopia, in millions of people,
with projection through 2015
Sources: News reports, World Bank, U.S. Agency
for International Development, UN Food and Agriculture Organization, Population Reference Bureau,
UNICEF, U.S. Department of Agriculture, Human
Nutrition Information Service

Figure 35 illustrates the graphs of two more exponential functions whose bases are larger than 1. Notice that for the larger base, the graph is steeper when $x > 0$ and is closer to the x-axis when $x < 0$.

Figure 35

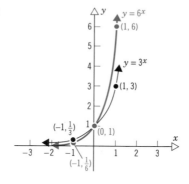

The box summarizes the information we have about $f(x) = a^x$, $a > 1$:

$$f(x) = a^x \quad a > 1$$

Domain: $(-\infty, \infty)$ **Range:** $(0, \infty)$

x-intercepts: **None**   **y-intercept:** $(0, 1)$

Horizontal asymptote: **x-axis, as x decreases without bound**

f is an increasing function passing through $(0, 1)$ and $(1, a)$

Now Work Problem 33

What happens when $0 < a < 1$? Example 3 will help us find out.

EXAMPLE 3   Graph the exponential function: $f(x) = (\frac{1}{2})^x$

SOLUTION

The domain of $f(x) = (\frac{1}{2})^x$ consists of all real numbers. As before, we locate some points on the graph, as listed in Table 5.

The data in Table 5 suggests that $(\frac{1}{2})^x$ increases without bound as x decreases without bound, and $(\frac{1}{2})^x$ approaches zero as x increases without bound. Since $(\frac{1}{2})^x > 0$ for all x, the range of f is all positive real numbers. From this we conclude that the graph has no x-intercepts. The y-intercept is $(0, 1)$.

The corresponding graph of $f(x) = (\frac{1}{2})^x$ is given in Figure 36. Observe from the figure that the graph approaches the x-axis, but never touches it. Thus, the x-axis is a horizontal asymptote.

Figure 36 $y = (\frac{1}{2})^x$

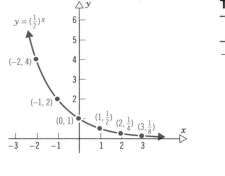

Table 5

x	$f(x) = (\frac{1}{2})^x$
-10	$(\frac{1}{2})^{-10} = 1024$
-3	$(\frac{1}{2})^{-3} = 8$
-2	$(\frac{1}{2})^{-2} = 4$
-1	$(\frac{1}{2})^{-1} = 2$
0	$(\frac{1}{2})^0 = 1$
1	$(\frac{1}{2})^1 = \frac{1}{2}$
2	$(\frac{1}{2})^2 = \frac{1}{4}$
3	$(\frac{1}{2})^3 = \frac{1}{8}$
10	$(\frac{1}{2})^{10} \approx 0.00098$

· ·

The graph of $f(x) = (\frac{1}{2})^x$ in Figure 36 is typical of all exponential functions that have a base between 0 and 1. Such functions are decreasing. Their graphs lie above the x-axis and pass through the point $(0, 1)$. The graphs rise rapidly as x decreases without bound. When x increases without bound, the x-axis is a horizontal asymptote. Figure 37 (on the next page) illustrates the graphs of two more exponential functions whose bases are between 0 and 1. Notice that the choice of a base closer to 0 results in a graph that is steeper when $x < 0$ and closer to the x-axis when $x > 0$.

The box on the following page summarizes the information we have about $f(x) = a^x$, $0 < a < 1$:

Figure 37

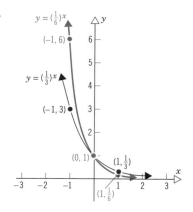

$$f(x) = a^x \qquad 0 < a < 1$$

Domain: $(-\infty, \infty)$ **Range:** $(0, \infty)$

x-intercepts: None y-intercept: (0, 1)

Horizontal asymptote: x-axis, as x increases without bound

f is a decreasing function passing through (0, 1) and (1, a)

Now Work Problem 35

The Base e

As we shall see shortly, many applied problems require the use of an exponential function whose base is a certain irrational number, symbolized by the letter e.

Let's look now at one way of arriving at this important number e.

NUMBER e

The number e is defined as the number that the expression

$$\left(1 + \frac{1}{n}\right)^n \tag{1}$$

approaches as n increases without bound. That is,

$$e \approx \left(1 + \frac{1}{n}\right)^n \approx 2.718$$

for very large n.

Table 6 illustrates what happens to the defining expression (1) as n takes on increasingly large values. The last number in the last column in the table is correct to nine decimal places and is the same as the entry given for e on your calculator (if expressed correct to nine decimal places).

The exponential function $f(x) = e^x$, whose base is the number e, occurs with such frequency in applications that it is usually referred to as *the* exponential function.

Table 6

n	$\dfrac{1}{n}$	$1 + \dfrac{1}{n}$	$\left(1 + \dfrac{1}{n}\right)^n$
1	1	2	2
2	0.5	1.5	2.25
5	0.2	1.2	2.48832
10	0.1	1.1	2.59374246
100	0.01	1.01	2.704813829
1,000	0.001	1.001	2.716923932
10,000	0.0001	1.0001	2.718145927
100,000	0.00001	1.00001	2.718268237
1,000,000	0.000001	1.000001	2.718280469
1,000,000,000	10^{-9}	$1 + 10^{-9}$	2.718281827

Indeed, many calculators have the key $\boxed{e^x}$ or $\boxed{\exp(x)}$, which may be used to evaluate the exponential function for a given value of x.* Now use your calculator to find e^x for $x = -2$, $x = -1$, $x = 0$, $x = 1$, and $x = 2$, as we have done to create Table 7 (after rounding). The graph of the exponential function $f(x) = e^x$ is given in Figure 38. Since $2 < e < 3$, the graph of $y = e^x$ lies between the graphs of $y = 2^x$ and $y = 3^x$. (Refer to Figures 33 and 35.)

Figure 38 $y = e^x$

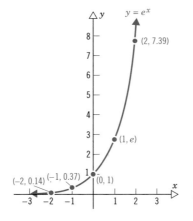

Table 7

x	e^x
-2	0.14
-1	0.37
0	1
1	e
2	7.39

* If your calculator does not have this key but does have an $\boxed{\text{Inv}}$ key and an $\boxed{\ln x}$ key, you can display the number e as follows:

Keystrokes: $\boxed{1}$ $\boxed{\text{Inv}}$

Display: $\boxed{1}$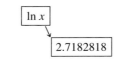

Exponential Equations

Equations that involve terms of the form a^x, $a > 0$, $a \neq 1$, are often referred to as **exponential equations.** Such equations sometimes can be solved by appropriately applying the laws of exponents and the following fact:

$$\text{If} \quad a^u = a^v \quad \text{then} \quad u = v \tag{2}$$

In other words, if two equal exponential expressions have the same base, then the exponents are equal.

EXAMPLE 4

Solve the equation: $3^{x+1} = 81$

SOLUTION

In order to apply (2), each side of the equality must be written with the same base. Thus, we need to express 81 as a power of 3:

$$3^{x+1} = 81$$
$$3^{x+1} = 3^4$$

Now we apply (2) to get

$$x + 1 = 4$$
$$x = 3$$

· ·

EXAMPLE 5

Solve the equation: $e^{-x^2} = (e^x)^2 \cdot \dfrac{1}{e^3}$

SOLUTION

We first use the laws of exponents to get the same base e on each side:

$$e^{-x^2} = (e^x)^2 \cdot \frac{1}{e^3}$$
$$e^{-x^2} = e^{2x} \cdot e^{-3}$$
$$e^{-x^2} = e^{2x-3}$$

Now we apply (2) to get

$$-x^2 = 2x - 3$$
$$x^2 + 2x - 3 = 0$$
$$(x + 3)(x - 1) = 0$$
$$x = -3 \quad \text{or} \quad x = 1$$

Now Work Problem 45

· ·

We will see how to solve exponential equations with unequal bases in Section 11.5.

EXAMPLE 6

Solve: $(\frac{1}{27})^{3x+2} = 9^{(6x-3)/2}$

SOLUTION

Since $\frac{1}{27}$ and 9 are both powers of 3, we use 3 as the base:

$$(\tfrac{1}{27}) = 3^{-3}, \qquad 9 = 3^2$$

Substituting into the equation, we get

$$(3^{-3})^{3x+2} = (3^2)^{(6x-3)/2}$$

Now we apply (2) to get

$$-3(3x + 2) = 2\left(\frac{6x-3}{2}\right)$$

$$-9x - 6 = 6x - 3$$

$$x = -\frac{1}{5} \quad \text{Solving for } x$$

· ·

Applications

Many scientific formulas dealing with exponential growth and decay are stated in terms of powers of e. In a system where the rate of growth or decay of a quantity is directly proportional to its size, the amount of the quantity present at any time can be expressed exponentially with e as the base:

$$y = Ae^{kt}$$

where the independent variable t represents time and the constant A is the amount of substance or population present at time $t = 0$.

EXAMPLE 7

The population of planet Earth is given by

$$y = Ae^{0.041t}$$

where A is the population in any given year and y is the population t years later. If the 1975 population was 4.0 billion, estimate the population in the year 2000.

SOLUTION

We let 1975 be the year $t = 0$ and measure the population in billions (that is, $A = 4$). Then

$$y = 4e^{0.041t}$$

In 2000 A.D., $t = 25$ and

$$y = 4e^{0.041(25)}$$
$$= 4e^{1.025}$$

Then

$$y = 11.148$$

Thus, in the year 2000 the population of planet Earth will be 11.148 billion.

....

EXAMPLE 8

The proportion of people responding to the advertisement of a new product after it has been on the market t days is

$$p(t) = 1 - e^{-0.2t}$$

The marketing area contains 10,000,000 potential customers, and each response to the advertisement results in profit to the company of $0.70 (on the average). This profit is exclusive of advertising cost. The fixed cost of producing the advertising is $30,000 and the variable cost is $5000 for each day the advertisement runs.

(a) Find the value of $p(t) = 1 - e^{-0.2t}$ when t is very large and interpret the answer.
(b) What percentage of customers respond after 10 days of advertising?
(c) What is the cost function $C(t)$?
(d) After 28 days of advertising, what is the net profit?
(e) Graph the function $p(t)$.

SOLUTION

(a) For large t, $e^{-0.2t} = 1/e^{0.2t}$ tends to 0. Thus, for large t, $p(t) = 1 - e^{-0.2t}$ tends to 1, so $y = 1$ is a horizontal asymptote. The interpretation is that eventually everyone will respond to the advertisement.

(b) The proportion of customers that respond after 10 days is found by substituting $t = 10$ into the response function $p(t) = 1 - e^{-0.2t}$. Thus, for $t = 10$,

$$p(10) = 1 - e^{-2} = 1 - 0.135 = 0.865$$

Thus, 86.5% of the potential customers have responded after 10 days of advertising.

(c) The cost function is

$$C(t) = \$30,000 + \$5000t$$

(d) The net profit function is the profit from sales less advertising cost. The profit from sales is the number of respondents times 0.70. Since $p(t) = 1 - e^{-0.2t}$ is the proportion of those responding, we have

$$R(t) = 10,000,000(1 - e^{-0.2t})(\$0.70) = \$7,000,000(1 - e^{-0.2t})$$

Thus, the net profit is

$$P(t) = R(t) - C(t) = \$7,000,000(1 - e^{-0.2t}) - \$30,000 - \$5000t$$

For $t = 28$,

$$P(28) = \$7,000,000(1 - e^{-5.6}) - \$30,000 - \$140,000$$
$$= \$7,000,000(0.9963) - \$170,000$$
$$= \$6,974,100 - \$170,000$$
$$= \$6,804,100$$

(e) The graph of $p(t) = 1 - e^{-0.2t}$ is given in Figure 39.

Figure 39

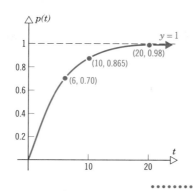

Now Work Problem 55

Continuously Compounded Interest

One use of the fact that $e \approx \left(1 + \dfrac{1}{n}\right)^n$ for large n is found in finance. Suppose a principal P is to be invested at an annual rate of interest r, which is compounded n times per year. The interest earned on a principal P at each compounding period is then $P \cdot (r/n)$. The amount A after 1 year with

1 compounding
(annually)
$$A = P + P \cdot \left(\frac{r}{1}\right) = P \cdot \left(1 + \frac{r}{1}\right)$$

2 compoundings
(semiannually)
$$A = P\left(1 + \frac{r}{2}\right) + P\left(1 + \frac{r}{2}\right)\left(\frac{r}{2}\right)$$
$$= P\left(1 + \frac{r}{2}\right)\left(1 + \frac{r}{2}\right) = P\left(1 + \frac{r}{2}\right)^2$$

4 compoundings
(quarterly)
$$A = P\left(1 + \frac{r}{4}\right)^3 + P\left(1 + \frac{r}{4}\right)^3\left(\frac{r}{4}\right)$$
$$= P\left(1 + \frac{r}{4}\right)^3\left(1 + \frac{r}{4}\right) = P\left(1 + \frac{r}{4}\right)^4$$

$\vdots$

n compoundings
per year
$$A = P\left(1 + \frac{r}{n}\right)^n$$

Compound Interest Formula

The amount A after 1 year accrued on a principal P when it is invested at an annual rate of interest r and is compounded n times per year is

$$A = P\left(1 + \frac{r}{n}\right)^n \tag{3}$$

For example, the result of investing $1000 for 1 year at an annual rate of 10% yields the amounts listed in Table 8 for various compounding periods.

Table 8

P = Principal = $1000 r = Annual rate of interest = 10% = 0.10	
n = Number of Times Compounded per Year	**A = Amount After 1 Year**
1 Annual compounding	$A = P(1 + r) = 1000(1 + 0.1) = \$1100.00$
2 Semiannual compounding	$A = P\left(1 + \dfrac{r}{2}\right)^2 = 1000(1 + 0.05)^2 = \$1102.50$
4 Quarterly compounding	$A = P\left(1 + \dfrac{r}{4}\right)^4 = 1000(1 + 0.025)^4 = \$1103.81$
12 Monthly compounding	$A = P\left(1 + \dfrac{r}{12}\right)^{12} = 1000(1 + 0.00833)^{12} = \$1104.71$
365 Daily compounding	$A = P\left(1 + \dfrac{r}{365}\right)^{365} = 1000(1 + 0.000274)^{365} = \$1105.16$

Now we ask what happens to the amount after 1 year as the number of times, n, that the interest is compounded per year gets larger and larger. The answer turns out to involve the number e.

We let

$$P\left(1 + \frac{r}{n}\right)^n = P\left[\left(1 + \frac{r}{n}\right)^{n/r}\right]^r$$

To simplify the calculation, let

$$k = \frac{n}{r} \qquad \text{so} \qquad \frac{1}{k} = \frac{r}{n}$$

We substitute to get

$$P\left[\left(1 + \frac{r}{n}\right)^{n/r}\right]^r = P\left[\left(1 + \frac{1}{k}\right)^k\right]^r$$

Since k increases without bound as n does, and since $\left(1 + \dfrac{1}{k}\right)^k$ approaches e as k increases without bound, it follows that

$$P\left[\left(1 + \frac{1}{k}\right)^k\right]^r = P[(e)]^r = Pe^r$$

Thus, no matter how often the interest is compounded during the year, the amount after 1 year has the definite ceiling Pe^r. When interest is compounded so that the amount after 1 year is Pe^r, we say that the interest is **compounded continuously.**

For example, the amount A due to investing $1000 for 1 year at an annual rate of 10% compounded continuously is

$$A = 1000e^{0.1} = \$1105.17$$

The formula $A = Pe^r$ gives the amount A after 1 year resulting from investing a principal P at the annual rate of interest r compounded continuously.

Compound Interest Formula

The amount A due to investing a principal P for a period of t years at the annual rate of interest r compounded continuously is

$$A = Pe^{rt} \qquad\qquad (4)$$

EXAMPLE 9

If $1000 is invested at 10% compounded continuously, the amount A after 3 years is

$$A = Pe^{rt} = 1000e^{(0.1)(3)} = 1000e^{0.3} = \$1349.86$$

After 5 years the amount A is

$$A = 1000e^{(0.1)(5)} = \$1648.72$$

Now Work Problem 57

・・・・・・・・・・・・・・・・・・・・・・・

The term **effective rate of interest** is often used. This is the equivalent annual rate of interest due to compounding. When interest is compounded annually, there is no difference between the annual rate and the effective rate; however, when interest is compounded more than once a year, the effective rate always exceeds the annual rate. For example, if interest of 10% per annum is compounded monthly, then after 1 year a principal of $1 is worth

$$A = \$1 \left(1 + \frac{0.10}{12} \right)^{12} = \$1.10471$$

The effective rate of interest is therefore 10.471%.

If interest of 10% per annum is compounded continuously, then $1 after 1 year is worth

$$A = \$1 \ e^{0.10} = \$1.10517$$

The effective rate of interest is therefore 10.517%.

Table 9 lists some effective rates of interest. It is worth noting that although the difference between a bank's paying interest yearly (almost none do now) versus compounding quarterly or monthly is fairly substantial, the difference between daily and continuous compounding is practically negligible.

Table 9

	Annual Rate (%)	Effective Rate (%)
Annual compounding	10	10
Semiannual compounding	10	10.25
Quarterly compounding	10	10.381
Monthly compounding	10	10.471
Daily compounding	10	10.516
Continuous compounding	10	10.517

EXERCISE 11.4 Answers to odd-numbered problems begin on page AN-48.

In Problems 1–16 evaluate each expression.

1. $27^{2/3}$

2. $8^{2/3}$

3. $9^{3/2}$

4. $16^{-3/4}$

5. $16^{-1/2}$

6. $8^{4/3}$

7. $27^{-2/3}$

8. $16^{1/4}$

9. $(\frac{1}{9})^{1/2}$

10. $(\frac{1}{8})^{-2}$

11. $(\frac{1}{8})^{-1/3}$

12. $(\frac{9}{16})^{-1/2}$

13. $(9^{1/3})(3^{1/3})$

14. $(27^{3/2})(\frac{1}{3})^{3/2}$

15. $[(8^{-1})(8^{1/3})]^3$

16. $27^{-1}(27^{1/3})^3$

In Problems 17–24 use a calculator to compute each expression, rounding off your answer to three decimal places.

17. 2^x, for $x = -\frac{1}{4}$

18. 2^x, for $x = -\frac{1}{3}$

19. 2^x, for $x = \frac{1}{3}$

20. 2^x, for $x = \frac{2}{3}$

21. 2^x, for $x = 0.1$

22. 2^x, for $x = 0.9$

23. e^3, e^{-3}, $e^{0.01}$, $e^{2/5}$, $\sqrt[3]{e}$, $e^{-0.04}$, $1/\sqrt{e}$

24. e^2, e^{-2}, $e^{0.002}$, $e^{3/5}$, $\sqrt{e}$, $e^{-0.05}$, $1/\sqrt[3]{e}$

In Problems 25–32 the graph of an exponential function is given. Match each graph to the correct function. State the reason(s) for your choice.

A. $y = 3^x$ **B.** $y = 3^{-x}$ **C.** $y = -3^x$ **D.** $y = -3^{-x}$

E. $y = 3^x - 1$ **F.** $y = 3^{x-1}$ **G.** $y = 3^{1-x}$ **H.** $y = 1 - 3^x$

25.

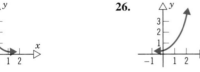

26.

27.

28.

29.

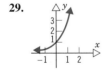

30.

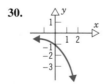

31.

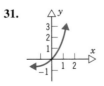

32.

In Problems 33–42 graph each function.

33. $f(x) = 3^x$

34. $f(x) = 5^x$

35. $f(x) = (\frac{1}{3})^x$

36. $f(x) = (\frac{1}{5})^x$

37. $f(x) = 4^{0.5x}$

38. $f(x) = (\frac{1}{4})^{0.5x}$

39. $f(x) = 10e^{0.3x}$, $-4 \leq x \leq 4$

40. $f(x) = 10e^{0.4x}$, $-4 \leq x \leq 4$

41. $f(x) = 100e^{-0.3x}$, $-4 \leq x \leq 4$

42. $f(x) = 100e^{-0.4x}$, $-4 \leq x \leq 4$

43. Graph $y = 3^x$ and $x = 3^y$ using the same coordinate system.

44. Graph $y = 5^x$ and $x = 5^y$ using the same coordinate system.

In Problems 45–52 use Property (2) to solve each equation.

45. $8^x = 16$ **46.** $25^x = 125$ **47.** $(\frac{1}{8})^x = 16$ **48.** $25^{-3x} = 125$

49. $5^{-3x} = \frac{1}{25}$ **50.** $9^{-x+1} = 27$ **51.** $e^{2x+1} = e^3$ **52.** $e^{-x} = (e^3)^{x+1}$

53. If $f(x) = a^x$, show that

(a) $f(x + 1) = af(x)$
(b) $f(x + 1) - f(x) = (a - 1)f(x)$
(c) $f(x + h) = a^h f(x)$

54. Graph $y = 0.4e^{-x^2/2}$.

55. Product Reaction If, in Example 8, page 548, the response function is

$$1 - e^{-0.1t}$$

and the other information remains the same, answer questions (b), (d), and (e) of the example.

56. Product Reaction If, in Example 8, the response function is

$$1 - e^{-0.05t}$$

and the profit per response is $0.10, while all other data are the same, answer questions (b), (d), and (e) of the example.

57. Compound Interest Find the amount after 1 year if $500 is invested at 6% compounded continuously for 1 year. What is the amount if the rate is $6\frac{1}{4}$% compounded quarterly? Which is better?

58. Compound Interest Find the amount after 1 year if $1000 is invested at 8% compounded continuously for 2 years. What is the amount if the rate is $8\frac{1}{2}$% compounded quarterly? Which is better?

59. Compound Interest What principal P should be invested at 6% compounded continuously in order to have $1000 after 1 year? What principal is required if the interest is compounded quarterly?

60. Compound Interest What principal P should be invested at 10% compounded continuously in order to have $2000 after 3 years? What principal is required if the interest is compounded quarterly?

61. Reliability of a Product The proportion of batteries that still maintain a charge after x years of use is

$$f(x) = (\tfrac{3}{4})^x$$

(a) What proportion of the batteries still maintain a charge after 2 years?
(b) What proportion of the batteries will fail to hold a charge between the second and third year of use?

(c) What proportion of the batteries will fail to hold a charge within the first year of use?

62. Advertising In a mass market covered by radio and television, the number of people y out of a population P who have heard a constantly repeated news headline t hours after its first announcement is

$$y = P(1 - e^{-kt})$$

If $k = 0.23$ for a given market, after how many hours will these percentages of people have heard the news?

(a) 25%   (b) 50%
(c) 75%   (d) 95%

63. Efficiency An 800 telephone number clerk at a computer terminal learning a new process can complete y transactions per hour, where

$$y = 15(1 - e^{-0.6t})$$

and t is the number of hours since beginning the job. To the nearest 0.1 transaction per hour, at what rate is the clerk working after

(a) 1 hour   (b) 2 hours
(c) 3 hours   (d) 6 hours

64. Population Growth The population of the United States is approximately

$$P = 150e^{0.0135(y - 1950)}$$

where y is the year and P is expressed in millions.

(a) What was the population in 1950?
(b) According to this formula, what will the population be in 2000?

65. Water Depletion Annual water consumption per person is estimated to be

$$W(y) = 1400(1 + e^{0.50(y - 1950)})$$

where y is the year and W is measured in gallons.

(a) What was the total water consumption in 1950, to the nearest gallon?
(b) What will the total water consumption be in 2000?

66. National Debt The national debt of the Bananian Republic, since its founding in 1964, is

$$y = Ae^{0.05t}$$

where t is measured in years and A was an initial loan of $2000. Give the values of the national debt in each year:

(a) 1974   (b) 1984   (c) 1994   (d) 2004
(e) What is the doubling time?

67. **Study Habits** In a certain mathematics class, the number of applied problems a student is expected to have solved by the end of week t of the semester is approximately
$$y = 12e^{0.5t}$$
Give the expected numbers after each week:

(a) Third   (b) Sixth
(c) Ninth   (d) Twelfth

68. **Profit Function** The annual profit of a company due to a particular item is found to be
$$P(x) = \$10,000 + \$25,000(\tfrac{1}{4})^{0.5x}$$
where x is the number of years the item has been on the market. Graph this function and estimate the profit for $x = 1, 3,$ and 5 years.

69. **Demand Function** The demand for a new product increases rapidly at first and then levels off. Suppose the percentage of actual buyers obeys
$$f(x) = 100 - 90(\tfrac{1}{3})^x$$
where x is the number of months the product is on the market. Graph the function and determine to what proportion of the market the new product sells for $x = 2, 3,$ and 5 months.

70. **Medicine** For a patient undergoing glucose infusion, the amount of glucose in his bloodstream at time t (measured in hours) is $A(t) = 10 - 8e^{-t}$ where $A(t)$ is mea-

sured in appropriate units. Plot $A(t)$ as a function of time for $t \geq 0$. Determine the equilibrium amount of glucose in the bloodstream (that is, as t gets very large).

71. **Inflation** If money loses value (because of inflation) at the rate of 6% per year, the value of $1 in t years is given by
$$y = (1 - 0.06)^t = (0.94)^t$$
(a) Use a calculator to complete the following table:

t	0	1	2	3	4	5	6	7	8	9	10
y											

(b) Graph $y = (0.94)^t$.

72. **Inflation** Use the results of Problem 71 to answer the following questions (assume that inflation is 6% per year).

(a) If a car costs $12,000 today, estimate the cost of a similar car in 10 years.
(b) Find the cost of a $4 per pound steak in 10 years.

73. **Memory Recall** The number of details that an accident witness can accurately recall is
$$y = Ae^{-kt}$$
where A is a count of the observable facts concerning the accident, t is time measured in weeks, and k depends on the person. For a certain witness whose value of k is 0.6, to an accident in which $A = 120$, determine the number of details recalled after

(a) 1 week   (b) 2 weeks
(c) 3 weeks   (d) 4 weeks

Technology Exercises

1. By trial and error, use a calculator to find (to two decimal places) the point near $x = 1$ at which $y = 2^x$ and $y = x^3$ intersect.

2. Graph the functions $y = 2^x$, $y = 3^x$, and $y = 4^x$ in the same viewing rectangle.

(a) For what values of x is it true that $2^x > 3^x > 4^x$?
(b) For what values of x is it true that $2^x < 3^x < 4^x$?
(c) For what values of x is it true that $2^x = 3^x = 4^x$?

3. For what values of x is $4^x > x^4$?

4. For what values of x is $3^x > x^3$?

5. The values of three functions are given in the table below (the numbers have been rounded to two decimal places):

t	$f(t)$	$g(t)$	$h(t)$
2.0	4.00	4.00	4.00
2.2	4.84	5.32	4.59
2.4	5.76	6.91	5.28
2.6	6.76	8.79	6.06
2.8	7.84	10.98	6.96
3.0	9.00	13.50	8.00

One function is exponential, one is of the form $y = at^2$, and one is of the form $y = bt^3$. Which is which?

6. The values of three functions are given in the tables below (the numbers have been rounded to two decimal places):

t	$f(t)$	t	$g(t)$	t	$h(t)$
3.0	27.00	3.0	27.00	4.75	26.80
3.5	46.77	3.5	36.75	5.25	36.18
4.0	81.00	4.0	48.00	5.75	47.53
4.5	130.30	4.5	60.75	6.25	61.04
5.0	243	5.0	75.00	6.75	76.89
5.5	420.89	5.5	90.75	7.25	95.27

One function is exponential, one is of the form $y = at^2$, and one is of the form $y = bt^3$. Which is which?

7. Use a graphing calculator (or a computer) to graph $y = x^4$ and $y = 3^x$. Determine the appropriate domains and ranges to give the pictures below:

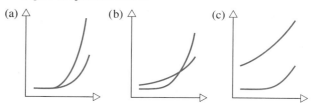

(a)   (b)   (c)

8. Using a calculator, plot $y = x^2$ and $y = 0.01e^{0.01x}$ for

(a) $0 \leq x \leq 1500, 0 \leq y \leq 225$
What do you notice? Why?
(b) $0 \leq x \leq 1500, 0 \leq y \leq 2{,}250{,}000$
What do you notice? Why?

11.5 THE LOGARITHM FUNCTION

In this section we introduce logarithm functions, which are closely related to exponential functions.

Definition of a Logarithm Function

We start by considering the exponential function defined by

$$y = 3^x \tag{1}$$

If we interchange the variables x and y in (1) we obtain an equation defined by

$$x = 3^y \tag{2}$$

Notice that the point $(1, 3)$ is part of the graph of $y = 3^x$, while $(3, 1)$ is part of the graph of $x = 3^y$. Similarly, $(0, 1)$ is part of the graph of $y = 3^x$, while $(1, 0)$ is part of the graph of $x = 3^y$. The table on the next page illustrates this relation for other selected values. If we graph each of these equations on the same coordinate system (see Figure 40), we observe that the graph of the equation $x = 3^y$ is the graph of a function. We call this function the *logarithm function to the base 3,* and write

$$y = \log_3 x \qquad \text{if and only if} \qquad x = 3^y$$

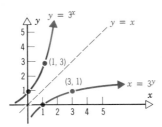

Figure 40 Note that the logarithm graph is the mirror image about $y = x$ of the exponential graph; that is, if you fold the paper along the line $y = x$, the graphs will match

Exponential Function				Logarithm Function		
x	$y = 3^x$			$x = 3^y$	y	
-3	$\frac{1}{27}$	$(-3, \frac{1}{27})$		$\frac{1}{27}$	-3	$(\frac{1}{27}, -3)$
-2	$\frac{1}{9}$	$(-2, \frac{1}{9})$		$\frac{1}{9}$	-2	$(\frac{1}{9}, -2)$
-1	$\frac{1}{3}$	$(-1, \frac{1}{3})$		$\frac{1}{3}$	-1	$(\frac{1}{3}, -1)$
0	1	$(0, 1)$		1	0	$(1, 0)$
1	3	$(1, 3)$		3	1	$(3, 1)$
2	9	$(2, 9)$		9	2	$(9, 2)$
3	27	$(3, 27)$		27	3	$(27, 3)$

In general, we define the logarithm function as follows:

LOGARITHM FUNCTION

If $x = a^y$, $a > 0$, $a \neq 1$ and, if we solve for y in terms of x, we obtain the *logarithm function*, denoted by $y = \log_a x$ and read as "y is the log to the base a of x." Thus,

$$y = \log_a x \qquad \text{means} \qquad a^y = x$$

In this sense *a logarithm of a number is an exponent.* For example,

$$\log_3 9 = 2 \qquad \text{means} \qquad 3^2 = 9$$

We say that $\log_3 9 = 2$ is the *logarithmic form* of the *exponential form* $3^2 = 9$.

EXAMPLE 1

	Exponential Form		*Logarithmic Form*
Since	$4^2 = 16$	then	$\log_4 16 = 2$
Since	$5^3 = 125$	then	$\log_5 125 = 3$
Since	$(\frac{1}{2})^3 = \frac{1}{8}$	then	$\log_{1/2} (\frac{1}{8}) = 3$
Since	$2^0 = 1$	then	$\log_2 1 = 0$

· ·

EXAMPLE 2

Logarithmic Form		*Exponential Form*		
$y = \log_2 1$	means	$2^y = 1$,	that is,	$y = 0$
$y = \log_2 4$	means	$2^y = 4$,	that is,	$y = 2$
$y = \log_3 \frac{1}{3}$	means	$3^y = \frac{1}{3}$,	that is,	$y = -1$

· ·

With the aid of Example 2, and the interpretation of $\log_a x$ as an exponent, we can now state the following general properties:

Property of $\log_a x$	Reason	Example
i. $\log_a 1 = 0$	$a^0 = 1$	$\log_2 1 = 0$
ii. $\log_a a = 1$	$a^1 = a$	$\log_5 5 = 1$
iii. $\log_a a^x = x$	$a^x = a^x$	$\log_3 9 = \log_3 3^2 = 2$
iv. $a^{\log_a x} = x$	See below	$6^{\log_6 9} = 9$

(3)

The statement $6^{\log_6 9} = 9$ follows directly from the definition of $\log_a$ since

$$\text{if} \quad y = \log_a x, \quad \text{then} \quad x = a^y; \quad \text{hence} \quad x = a^{\log_a x}$$

Now Work Problems 1 and 7

The logarithm function $y = f(x) = \log_a x$ is another form for $a^y = x$. Since $x = a^y$ is defined for $x > 0$, $-\infty < y < \infty$, it follows that the domain of a logarithm function $f(x) = \log_a x$ is the set of all positive real numbers and its range is the set of all real numbers. For $a > 1$, the graph rises from left to right; as x gets closer and closer to 0, the function values decrease without bound and the graph gets closer and closer to the y-axis. That is, the y-axis is a *vertical asymptote* of the graph. For $0 < a < 1$, the graph falls from left to right; as x gets closer and closer to 0, the function values increase without bound and the graph gets closer and closer to the y-axis. The graphs of typical logarithm functions are given in Figure 41.

Figure 41

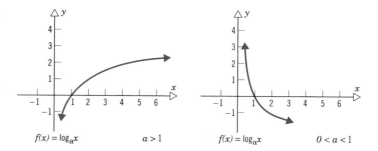

$f(x) = \log_a x$ $a > 1$ $f(x) = \log_a x$ $0 < a < 1$

We now summarize some facts about a logarithm function $f(x) = \log_a x$:

1. The x-intercept of the graph is $(1, 0)$. There is no y-intercept.
2. The y-axis is a vertical asymptote of the graph.
3. A logarithmic function is increasing if $a > 1$ and decreasing if $0 < a < 1$.
4. The graph is smooth, with no corners or gaps.

Properties of Logarithms

Logarithms have certain properties that can be derived from their definition.

> **Properties of Logarithms**
>
> If M and N represent positive real numbers and r is any real number, then
>
> (a) $\log_a(MN) = \log_a M + \log_a N$
>
> (b) $\log_a \dfrac{M}{N} = \log_a M - \log_a N$ $\qquad\qquad$ **(4)**
>
> (c) $\log_a M^r = r \log_a M$

EXAMPLE 3

(a) $\log_a[x(x + 1)] = \log_a x + \log_a(x + 1)$

(b) $\log_a \dfrac{x^2 + 4}{x + 1} = \log_a(x^2 + 4) - \log_a(x + 1)$

(c) $\log_a x^{3/2} = \frac{3}{2} \log_a x$

....

EXAMPLE 4

Using the properties listed in (4), write each of the following expressions as a single logarithm.

(a) $\log_a 7 + 3 \log_a 2$
(b) $\frac{1}{3} \log_a 75 - \log_a(5^3 - 1)$
(c) $\log_a x + \log_a 4 + \log_a(x^2 - 1) - \log_a 7$

SOLUTION

(a) $\log_a 7 + 3 \log_a 2 = \log_a 7 + \log_a 2^3 = \log_a(7)(8) = \log_a 56$
$\qquad\qquad\qquad\quad \uparrow \qquad\qquad\qquad \uparrow$
$\qquad\qquad\quad$ By (4c) $\qquad\quad$ By (4a)

(b) $\frac{1}{3} \log_a 75 - \log_a(5^3 - 1) = \log_a 75^{1/3} - \log_a 124$
$\qquad\qquad\qquad\qquad\qquad\qquad \uparrow$
$\qquad\qquad\qquad\qquad\quad$ By (4c)

$\qquad\qquad\qquad\qquad = \log_a \sqrt[3]{75} - \log_a 124$

$\qquad\qquad\qquad\qquad = \log_a \left(\dfrac{\sqrt[3]{75}}{124} \right)$
$\qquad\qquad\qquad\quad \uparrow$
$\qquad\qquad\qquad$ By (4b)

(c) $\log_a x + \log_a 4 + \log_a(x^2 - 1) - \log_a 7 = \log_a(4x) + \log_a(x^2 - 1) - \log_a 7$
$\qquad\qquad\qquad\qquad\qquad\qquad\qquad\qquad\qquad\quad \uparrow$
$\qquad\qquad\qquad\qquad\qquad\qquad\qquad\quad$ By (4a)

$\qquad\qquad\qquad\qquad\qquad\quad = \log_a[4x(x^2 - 1)] - \log_a 7$
$\qquad\qquad\qquad\qquad\qquad\qquad \uparrow$
$\qquad\qquad\qquad\qquad\qquad$ By (4a)

$\qquad\qquad\qquad\qquad\qquad\quad = \log_a \dfrac{4x(x^2 - 1)}{7}$
$\qquad\qquad\qquad\qquad\qquad\qquad \uparrow$
$\qquad\qquad\qquad\qquad\qquad$ By (4b)

Now Work Problem 25

....

Natural Logarithm

If the base of a logarithmic function is the number e, then we have the **natural logarithm function.** This function occurs so frequently in applications that it is given a special symbol, **ln** (from the Latin, *logarithmus naturalis*). Thus,

$$y = \ln x \qquad \text{if and only if} \qquad x = e^y$$

That is,

$$\ln x \qquad \text{means} \qquad \log_e x$$

For the natural logarithm function $y = \ln x$, Formulas (3i) and (3ii) are written as

$$\ln 1 = 0 \qquad \ln e = 1$$

By using a calculator with a $\boxed{\ln}$ key, we can obtain some points on the graph of $y = \ln x$. See Table 10 and Figure 42.

Figure 42 $y = \ln x$

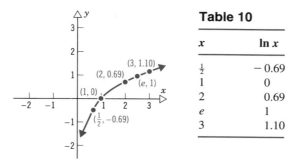

Table 10

x	$\ln x$
$\frac{1}{2}$	-0.69
1	0
2	0.69
e	1
3	1.10

Graphical View of e^x and $\ln x$

The exponential function e^x and the logarithm function $\ln x$ are *inverses*. The graphs in Figure 43 (on the next page) show what this means geometrically. The graphs of inverse functions are symmetric with respect to the line $y = x$.

Figure 43 suggests, too, why the base e is special. If you look carefully, you will see that:

At the point $(0, 1)$ on the graph of $y = e^x$ and at the point $(1, 0)$ on the graph of $y = \ln x$, we can draw two lines that are parallel to the line $y = x$.

Figure 43 Logarithm and
exponential functions
are inverses

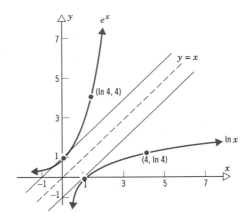

The natural logarithm can often be used to solve certain exponential equations involving e^x.

EXAMPLE 5   Solve: $2e^{0.1x} = 5$

SOLUTION

$$e^{0.1x} = \tfrac{5}{2}$$

Now we take the natural logarithm of both sides:

$$\ln e^{0.1x} = \ln \tfrac{5}{2}$$
$$0.1x \ln e = \ln \tfrac{5}{2}$$
$$0.1x = \ln \tfrac{5}{2}$$
$$x = \frac{\ln \tfrac{5}{2}}{0.1} = 9.163$$

· ·

EXAMPLE 6   Solve: $7 + (2)3^{x+1} = 10$

SOLUTION

We first isolate the terms involving the variable x on one side

$$(2)3^{x+1} = 3$$
$$3^{x+1} = \tfrac{3}{2}$$

Now take the natural logarithm of both sides

$$\ln 3^{x+1} = \ln \tfrac{3}{2}$$

Simplifying by using properties of logarithms, we get

$$(x + 1)\ln 3 = \ln \tfrac{3}{2}$$

$$x + 1 = \frac{\ln \tfrac{3}{2}}{\ln 3}$$

$$x = \frac{\ln \tfrac{3}{2}}{\ln 3} - 1$$

$$= \frac{\ln 3 - \ln 2}{\ln 3} - 1$$

$$= \frac{1.099 - 0.693}{1.099} - 1$$

$$= -0.631$$

Now Work Problem 45

·······················

EXAMPLE 7

How long will it take for a principal P to double if it is invested at 8% interest compounded continuously?

SOLUTION

We use the formula for continuously compounded interest, namely,

$$A = Pe^{rt}$$

We want to find t if $A = 2P$ and $r = 0.08$. Thus, we need to solve

$$2P = Pe^{0.08t}$$

$$2 = e^{0.08t}$$

$$0.08t = \ln 2$$

$$t = \frac{\ln 2}{0.08} = 8.66$$

It will take about 8.66 years or 8 years 8 months for a principal to double at 8% interest compounded continuously.

Now Work Problem 55

·······················

Logarithmic Equations

Equations that involve terms of the form $\ln x$ are often referred to as *logarithmic equations*. Such equations, sometimes, can be solved by applying the following fact:

If $\ln u = \ln v$ then $u = v$	(5)

EXAMPLE 8 | Solve: $\ln(2x + 3) = 2 \ln x$

SOLUTION

Here, we must assume that both $2x + 3$ and x are positive so that their logarithms are defined. We notice that for $x > 0$ both logarithms are defined. To solve the equation, we first rewrite it:

$$\ln(2x + 3) = \ln x^2$$

Using Property (5) we have

$$2x + 3 = x^2$$

or

$$x^2 - 2x - 3 = 0$$
$$(x + 1)(x - 3) = 0 \quad \text{Factoring}$$
$$x = -1, \quad x = 3$$

Because we must have $x > 0$, the only solution is 3. We verify that $x = 3$ is the solution by substituting into the original equation.

$$\ln(2 \cdot 3 + 3) = 2 \ln 3$$
$$\ln 9 = \ln 3^2$$
$$\ln 9 = \ln 9$$

Now Work Problem 43

⋯⋯⋯⋯⋯⋯⋯⋯⋯⋯⋯

EXAMPLE 9 | Plutonium-239 is a radioactive substance that decays according to the formula

$$y = Ae^{-0.000028t}$$

where y measures the amount present t years after beginning with amount A.

(a) Given 100 kilograms of plutonium-239, after how many years will the amount of substance have decayed to 95 kilograms?
(b) In how many years will the amount remaining be 50 kilograms?

SOLUTION

(a) We have $A = 100$, $y = 95$, and we solve for t.

$$95 = 100e^{-0.000028t}$$
$$0.95 = e^{-0.000028t}$$
$$\ln 0.95 = \ln e^{-0.000028t}$$
$$-0.0513 = -0.000028t \quad \text{Recall:} \quad \ln e = 1$$
$$t = \frac{-0.0513}{-0.000028} \approx 1{,}832$$

It will take about 1832 years for the substance to decay to 95 kilograms.

(b) We want to find the value of t for which $y = A/2$.

$$\frac{A}{2} = Ae^{-0.000028t}$$

$$\frac{1}{2} = e^{-0.000028t}$$

$$\ln \frac{1}{2} = -0.000028t$$

$$-\ln 2 = -0.000028t$$

Then

$$t = \frac{-\ln 2}{-0.000028} \approx \frac{-0.6931}{-0.000028} \approx 24{,}754$$

It will take almost 25,000 years for 50 kilograms of the original 100 kilograms to decay.

· ·

In Example 9(b) t was determined by the fact that the amount of substance at time t is half of the initial amount.

HALF-LIFE

In exponential decay where

$$y = Ae^{-kt}$$

the time required for half the amount present to decay is given by

$$t = \frac{\ln 2}{k}$$

This value of t is called the *half-life* of the substance.

In Example 9 the half-life of plutonium-239 was computed to be approximately 25,000 years.

EXERCISE 11.5   Answers to odd-numbered problems begin on page AN-50.

In Problems 1–6 write each logarithm expression using exponential notation.

1. $\log_3 9 = 2$

2. $\log_2 16 = 4$

3. $\log_3 \left(\frac{1}{81}\right) = -4$

4. $\log_{1/2} \left(\frac{1}{16}\right) = 4$

5. $\log_a P = Q$

6. $\log_b Y = C$

In Problems 7–10 write each exponential expression using logarithmic notation.

7. $10^3 = 1000$

8. $10^{-3} = 0.001$

9. $a^{1/2} = 3$

10. $x^3 = 6$

In Problems 11–16 evaluate each expression.

11. $\log_2 32$

12. $\log_{1/2} 32$

13. $\log_{10} 10^{-3}$

14. $\log_{10} 10^4$

15. $\log_2 24 - \log_2 12$

16. $\log_3 15 - \log_3 5$

In Problems 17–24 use $\log_{10} 2 = 0.3010$, $\log_{10} 3 = 0.4771$, and $\log_{10} 5 = 0.6990$ to compute each quantity.

17. $\log_{10} 12$

18. $\log_{10} 250$

19. $\log_{10} 7.5$

20. $\log_{10} 12.5$

21. $\log_{10} 36$

22. $\log_{10} 30$

23. $\log_{10} \left(\frac{1}{8}\right)$

24. $\log_{10} \left(\frac{1}{2}\right)$

In Problems 25–30 use the properties of logarithms to write each expression as a single logarithm.

25. $\ln 3 + \ln x$

26. $\ln x^6 - \ln x^3$

27. $\frac{1}{2} \ln 16$

28. $\ln 2 - \ln x + \ln 4$

29. $\ln(x + 1) + \ln(x + 2) + \ln(x + 3)$

30. $\ln(x + 1) - 3 \ln(x + 3)$

In Problems 31–38 find x, y, or c.

31. $\log_2 x = 5$

32. $\log_5 x = 2$

33. $\log_6 36 = y$

34. $\log_2 64 = y$

35. $\log_9 x = \frac{1}{2}$

36. $\log_{16} x = \frac{1}{2}$

37. $\log_{10} \left(\frac{1}{1000}\right) = c$

38. $\log_c 4 = \frac{1}{2}$

In Problems 39–48 use the properties of logarithms to solve for x.

39. $\log_{10}(3x + 1) = \log_{10}(x + 7)$

40. $\log_{10} x + \log_{10} 5 = \log_{10} 25$

41. $\ln x + \ln 6 = \ln 7$

42. $\ln x - \ln(x - 1) = \ln 2$

43. $\ln(-x) = \ln(x^2 - 2)$

44. $\ln(8 - 6x) = 2 \ln x$

45. $4 = e^{0.08x}$

46. $4 = 3 + 2e^{-6x}$

47. $6 + (3)2^{x+1} = 8$

48. $5 = 2 + 3^{x+1}$

In Problems 49–52 graph each function.

49. $f(x) = \log_5 x$

50. $f(x) = \log_3 x$

51. $f(x) = \log_{1/3} x$

52. $f(x) = \log_{1/5} x$

53. If $3^x = e^{cx}$, find c.

54. Doubling Time   How long will it take an amount P to double if it is invested at 4% interest compounded continuously?

55. Doubling Time   How long will it take an amount P to double if it is invested at 6% interest compounded continuously?

56. Tripling Time   How long will it take an amount P to triple if it is invested at 8% interest compounded continuously?

57. Tripling Time   How long will it take an amount P to triple if it is invested at 10% interest compounded continuously?

The commonly employed Richter scale of measuring the magnitude of an earthquake requires logarithms. On a Richter scale the magnitude R of an earthquake having intensity I is defined as

$$R = \log_{10} \frac{I}{I_0}$$

where I_0 is a fixed intensity. Apply this in Problems 58 and 59.

58. Intensity of Earthquakes   Use the formula above to find the intensity of an earthquake whose magnitude on the Richter scale is 6.

59. Intensity of Earthquakes   What was the magnitude on the Richter scale of the Los Angeles earthquake of 1971 if it had intensity of $10^{6.7}I_0$?

60. Population Growth   The formula $A = Pe^{rt}$ can also be used as a mathematical model for population growth over short periods of time. Thus, if

$$P_0 = \text{Population at time } t = 0$$
$$r = \text{Rate compounded continuously}$$
$$t = \text{Time in years}$$
$$P = \text{Population at time } t$$

then

$$P = P_0 e^{rt}$$

How long will it take the earth's population to double if it continues to grow at the rate of 2% per year (compounded continuously)?

61. Population Growth   Repeat Problem 60 assuming that the growth rate is 1% per year (compounded continuously).

62. Population Growth   If the population of a certain coun-try is doubling every 30 years, at what rate compounded continuously does the population grow?

63. Population Growth   It took the population of a certain country 100 years to double. At what rate compounded continuously did the population grow?

64. Stock Appreciation   The value of a stable stock is growing according to

$$y = 100e^{0.08t}$$

where t is measured in years. To the nearest day, what is the doubling time?

65. Radioactive Decay   A radioactive substance is decay-ing according to

$$y = 200e^{-0.15t}$$

where t is measured in weeks. What is the half-life, to the nearest hundredth of a week, if we start with 200 grams?

66. Inflation   The purchasing power of a fixed pension is decreasing as a result of inflation. The monthly purchas-ing power is given by

$$y = 1000e^{-0.05t}$$

where t is measured in years. After how many years, to the nearest tenth of a year, will the purchasing power be cut in half?

Technology Exercises

1. Graph $f(x) = \ln x$ and $g(x) = e^x$ on $[-5, 5]$ by $[-5, 5]$.

  (i)   What is the domain of $f(x)$? What is the domain of $g(x)$?
  (ii) What is the range of $f(x)$? What is the range of $g(x)$?
  (iii) From the graphs of $f(x)$ and $g(x)$ find the values for (a)–(e); if possible, check them with the built-in functions on the calculator. Find all x-values that sat-isfy $f(x)$ for (f)–(h) and all x values that satisfy $g(x)$ for (i)–(k).

(a) $\ln 1$	(b) e^0	(c) $\ln e$
(d) $\ln e^2$	(e) $e^{\ln 2}$	
(f) $f(x) > 0$	(g) $f(x) = 0$	(h) $f(x) < 0$
(i) $g(x) > 0$	(j) $g(x) = 0$	(k) $g(x) < 0$

2. Graph the functions $y = \ln(x^2)$ and $y = (\ln x)^2$. Are they the same function?

3. Graph $y = 3^x$ and $y = e^{x \ln 3}$ on the same coordinate sys-tem. You should get the same graph. Why?

4. Graph the functions $y = \ln \sqrt[3]{x^2 + 1}$ and $y = \frac{1}{3} \ln(x^2 + 1)$. Both functions have the same graph. Why?

5. Graph the function $f(t) = ke^{3t}$ for $k = 1$, 2, and 4. Graph the function $f(t) = 3e^{kt}$ for $k = 1$, 2, 4. How does the value of k affect each function?

6. Graph the function $f(x) = \ln kx$ for $k = 1$, 2, 5, and 7. How are the graphs related? Why?

7. If money loses value (because of inflation) at the rate of 6% per year, the value of $1 in t years is given by

$$y = (1 - 0.06)^t = (0.94)^t$$

(a) Use a calculator to complete the following table:

t	0	1	2	3	4	5	6	7	8	9	10
y											

(b) Graph $y = (0.94)^t$.

8. Using the definitions of $\log_b x$ and b^x approximate the following using your calculator.

(a) 2^π (b) $7 - \sqrt{2}$
(c) $(\sqrt{5})^{\sqrt{3}}$ (d) $\log_2 \pi$
(e) $\log_\pi 5$

In Problems 9–12 solve for t using logarithms. Check your answer using a graphing calculator (or computer) by plotting both sides of the equation on the same pair of axes.

9. $5^t = 7$ **10.** $2 = (1.02)^t$ **11.** $7 \cdot 3^t = 5 \cdot 2^t$ **12.** $5.02 \cdot (1.04)^t = 12.01 \cdot (1.03)^t$

CHAPTER REVIEW

IMPORTANT TERMS AND CONCEPTS

graph of an equation 500
x-intercept, y-intercept 501
finding intercepts 501
tests for symmetry 503
distance formula 507
function 509
domain of a function 509
range of a function 509
independent variable 512
dependent variable 512
demand function 513
price function 514
revenue function 514
profit function 514
vertical line test 517

ordered pairs 518
difference quotient 525
increasing function 526
decreasing function 526
constant function 526
linear function 527
identity function 528
square function 528
cube function 528
square root function 528
reciprocal function 528
absolute value function 529
greatest integer function 529
piecewise-defined function 530
quadratic function 532

parabola 532
vertex 532
polynomial function 534
rational function 535
exponential function 539
horizontal asymptote 541
the number e 544
compound interest 549
continuous compound interest 549
effective rate of interest 551
logarithm function 556
vertical asymptote 557
natural logarithm function 559
half-life 563

IMPORTANT FORMULAS

Distance Formula

$$d(P_1, P_2) = \sqrt{(x_2 - x_1)^2 + (y_2 - y_1)^2}$$

Number e

$$e \approx \left(1 + \frac{1}{n}\right)^n \approx 2.718 \quad \text{for } n \text{ large}$$

Compound Interest

$$A = P\left(1 + \frac{r}{n}\right)^n$$

Continuously Compounded Interest

$$A = Pe^{rt}$$

Properties of Logarithms

$$\log_a(MN) = \log_a M + \log_a N \qquad \log_a 1 = 0$$

$$\log_a \frac{M}{N} = \log_a M - \log_a N \qquad \log_a a = 1$$

$$\log_a M^r = r \log_a M \qquad \log_a a^x = x$$

$$a^{\log_a x} = x$$

TRUE–FALSE ITEMS   Answers are on page AN-51.

T_____ F_____ **1.** The domain of the function $f(x) = \sqrt{x}$ is the set of all real numbers.

T_____ F_____ **2.** For any function f, it follows that $f(x + h) = f(x) + f(h)$.

T_____ F_____ **3.** The graph of a function has at most one y-intercept.

T_____ F_____ **4.** If $P^R = Q$, then $\log_P R = Q$.

T_____ F_____ **5.** The solution x of $4^{x-1} = \frac{1}{8}$ is $\frac{1}{2}$.

T_____ F_____ **6.** The solution x of $\log_2 \sqrt{x} = 3$ is 64.

T_____ F_____ **7.** One of the properties of a logarithm is that $\log_a (M + N) = \log_a M + \log_a N$.

T_____ F_____ **8.** If $x = e^y$, then $y = \ln x$.

FILL IN THE BLANKS   Answers are on page AN-51.

1. If, for every point (x, y) on a graph, the point $(-x, y)$ is also on the graph, then the graph is symmetric with respect to the _____.

2. If f is a function defined by the equation $y = f(x)$, then x is called the _____ variable and y is the _____ variable.

3. A set of points in the xy-plane is the graph of a function if and only if no _____ line contains more than one point of the set.

4. A function of the form $f(x) = a^x$, $a > 0$, $a \neq 1$, is called an _____ function.

5. $\log_2 4 + \log_2 8 = $ _____.

6. The formula for continuously compounded interest involves the number _____.

7. The function $f(x) = \ln x$ is called the _____ _____ function.

REVIEW EXERCISES   Answers to odd-numbered problems begin on page AN-51.

In Problems 1–8 list the x- and y-intercepts and test for symmetry.

1. $2x = 3y^2$

2. $y = 5x$

3. $4x^2 + y^2 = 1$

4. $x^2 - 9y^2 = 9$

5. $y = x^4 + 2x^2 + 1$

6. $y = x^3 - x$

7. $x^2 + x + y^2 + 2y = 0$

8. $x^2 + 4x + y^2 - 2y = 0$

9. Given that f is a linear function, $f(4) = -2$, and $f(1) = 4$, write the equation that defines f.

10. Given that g is a linear function with slope $= -2$ and $g(-2) = 2$, write the equation that defines g.

11. A function f is defined by the equation below. If $f(1) = 4$, find A.

$$f(x) = \frac{Ax + 5}{6x - 2}$$

12. A function g is defined by the equation below. If $g(-1) = 0$, find A.

$$g(x) = \frac{A}{x} + \frac{8}{x^2}$$

13. Tell which of the graphs below are graphs of functions.

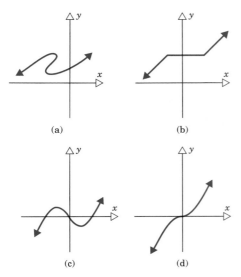

(a)   (b)

(c)   (d)

14. Use the graph of the function f shown below to find:

    (a) The domain and range of f.
    (b) The intervals on which f is increasing.
    (c) The intervals on which f is constant.
    (d) The intercepts of f.

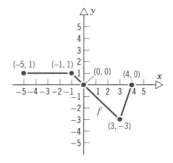

In Problems 15–20 find the following for each function:
(a) $f(-x)$ (b) $-f(x)$ (c) $f(x + 2)$ (d) $f(x - 2)$

15. $f(x) = \dfrac{x}{x^2 - 4}$

16. $f(x) = \dfrac{x^2}{x - 2}$

17. $f(x) = \sqrt{x^2 - 4}$

18. $f(x) = |x^2 - 4|$

19. $f(x) = \dfrac{x^2 - 4}{x^2}$

20. $f(x) = \dfrac{x^3}{x^2 - 4}$

In Problems 21–32 find the domain of each function.

21. $f(x) = \dfrac{x}{x^2 - 4}$

22. $f(x) = \dfrac{x^2}{x - 2}$

23. $f(x) = \sqrt{2 - x}$

24. $f(x) = \sqrt{x + 2}$

25. $h(x) = \dfrac{\sqrt{x}}{|x|}$

26. $g(x) = \dfrac{|x|}{x}$

27. $f(x) = \dfrac{x}{x^2 + 2x - 3}$

28. $F(x) = \dfrac{1}{x^2 - 3x - 4}$

29. $G(x) = \begin{cases} |x| & \text{if } -1 \le x \le 1 \\ 1/x & \text{if } x > 1 \end{cases}$

30. $H(x) = \begin{cases} 1/x & \text{if } 0 < x < 4 \\ x - 4 & \text{if } 4 \le x \le 8 \end{cases}$

31. $f(x) = \begin{cases} 1/(x - 2) & \text{if } x > 2 \\ 0 & \text{if } x = 2 \\ 3 & \text{if } 0 \le x < 2 \end{cases}$

32. $g(x) = \begin{cases} |1 - x| & \text{if } x < 1 \\ 3 & \text{if } x = 1 \\ x & \text{if } 1 < x \le 3 \end{cases}$

In Problems 33–42
(a) Find the domain of each function.   *(b) Locate any intercepts.*
(c) Graph each function.   *(d) Based on the graph, find the range.*

33. $F(x) = |x| - 4$

34. $f(x) = |x| + 4$

35. $g(x) = -|x|$

36. $g(x) = \frac{1}{2}|x|$

37. $h(x) = \sqrt{x - 1}$

38. $h(x) = \sqrt{x} - 1$

39. $f(x) = \sqrt{1 - x}$

40. $f(x) = -\sqrt{x}$

41. $F(x) = \begin{cases} x^2 + 4 & \text{if } x < 0 \\ 4 - x^2 & \text{if } x \geq 0 \end{cases}$

42. $H(x) = \begin{cases} |1 - x| & \text{if } 0 \leq x \leq 2 \\ |x - 1| & \text{if } x > 2 \end{cases}$

In Problems 43–48 graph each quadratic function. Label the vertex, the
y-intercepts, and the x-intercepts, if any.

43. $f(x) = x^2 + 2x$

44. $f(x) = x^2 - 2x$

45. $f(x) = x^2 + 2x - 8$

46. $f(x) = x^2 - 10x + 24$

47. $f(x) = x^2 + 2x + 4$

48. $f(x) = x^2 + 2x + 3$

In Problems 49–54 use a calculator to approximate each natural logarithm to five
decimal places.

49. $\ln 6.95$

50. $\ln 9.65$

51. $\ln 814$

52. $\ln 0.0814$

53. $\ln 12.5$

54. $\ln 125$

In Problems 55–58 write each expression as a single logarithm.

55. $3 \log 4 - 2 \log 2$

56. $6 \ln x + \ln z$

57. $3 \ln x - 2 \ln y + 6 \ln z$

58. $\ln x + \ln(x^2 - 1) - 4 \ln(x + 4)$

In Problems 59–68 solve for x. Express each answer to three decimal places.

59. $3^x = 4$

60. $2^{-x} = 7$

61. $4^{x-1} = 7$

62. $3^{2x+1} = 4$

63. $4^{2-x} = 10$

64. $3^{-2x} = 1.5$

65. $(1 + 0.04)^x = 100$

66. $(1 + 0.09)^x = 200$

67. $e^{2x} = 3$

68. $6e^{2x-1} - 3 = 2$

In Problems 69–78 solve each logarithmic equation.

69. $\log_{10}(3x + 1) = \log_{10}(2x + 3)$

70. $\log_{10} x = \log_{10}(1 - x)$

71. $\log_2(x + 1) = 3$

72. $\log_5(x - 1) = 1$

73. $\log_4(3x + 2) - \log_4 x = \log_4 5$

74. $\log_3 x + \log_3(x - 1) = \log_3 6$

75. $\ln 3x = \ln 6 + \ln(x - 2)$

76. $\ln x + \ln(2x - 7) = \ln 4$

77. $\ln(5x - 4) + \ln x = \ln 6$

78. $\ln 6x = \ln 2 + \ln(x - 4)$

In Problems 79–82 write each logarithmic equation in exponential form.

79. $\log_m 5 = \frac{2}{3}$

80. $\log_m 5 = -\frac{2}{3}$

81. $\log_p 13 = 3$

82. $\log_q 3 = 13$

83. Sketch the graph of $y = 4^x$ and $y = \log_4 x$.

84. Light Intensity   Suppose a light source has intensity of 100 lumens. As a beam of light passes through a window, its intensity is

$$I = 100e^{-kw}$$

where w is the thickness of the medium in centimeters and k is a constant that depends on the composition of the window. For a window 3.5 cm thick with $k = 4$, determine the intensity of the beam of light.

85. Advertising   When a news item is broadcast at frequent intervals, the number of people in a certain area who have heard the item after t hours is given by

$$f(t) = P(1 - e^{-kt})$$

where P is the total population of the area and k is a constant that depends on the "demographic nature" of the area. Suppose that, in a community of 100,000 people, half of them have heard the news flash 3 hours after it was first broadcast.

(a) Use the given information to find the value of k for this population.
(b) At what time will 90% of the population have received the news?

86. Atmospheric Pressure   The pressure of the atmosphere in pounds per square inch is

$$P(h) = 14.7e^{-0.2h}$$

where h is the altitude above sea level, measured in miles.

(a) What is the atmospheric pressure at sea level, to three significant digits?
(b) Denver is 1 mile above sea level. What is the pressure of the atmosphere there?
(c) Find the atmospheric pressure at the top of Mount McKinley, 2.2 miles above sea level.
(d) On the floor of Death Valley, 0.1 mile below sea level, find the pressure of the atmosphere.

87. Traffic   Since the completion of the Mackinac Straits Bridge in 1957, joining the upper and lower peninsulas of

Michigan, suppose that the yearly number of automobiles crossing the bridge is

$$y = 100,000 \ln(t + 1)$$

where t measures the number of years since 1957. Find the number of automobiles crossing the bridge in

(a) 1957   (b) 1967   (c) 1987   (d) 1997

88. Traffic   Given the equation of Problem 87, in what years would the number of automobiles reach

(a) 200,000?   (b) 300,000?
(c) 400,000?   (d) 1,000,000?

89. Bacteria Growth   If it takes a bacteria culture one hour to double its population, how long does it take to triple its population?

90. Bacteria Growth   If the growth of bacteria in a colony is given by

$$y = (4 \times 10^8)e^{0.27t}$$

where t is measured in days, what is the doubling time? Give the answer to the nearest hour.

91. Compound Interest   How long will it take $10,000 to double if it can be invested at 12% interest compounded continuously?

92. Demand for Oil   The demand for oil in the United States increases at a rate of 4% per year. Assuming this rate remains constant after 1993, when will the demand double? Assume compounding takes place continuously.

93. Compound Interest   At a grand opening one lucky bank customer is allowed to deposit $1,000 for one year at 100% interest. The customer may determine the length of the interest period. How much will be in the account at the end of one year if the interest periods have the following lengths?

(a) 1 year   (b) 6 months
(c) 3 months   (d) 1 month
(e) 1 week   (f) 1 day
(g) 1 hour

Mathematical Questions from Professional Exams

1. Actuary Exam—Part I   If $\log_6 2 = b$, which of the following is equal to $\log_6(4 \cdot 27)$?

(a) $3 - b$ (b) $2 + b$ (c) $5b$ (d) $3b$
(e) $2b + \log_6 27$

2. Actuary Exam—Part I   If b and c are real numbers

and there are two different real numbers x such that $(e^x)^2 + be^x + c = 0$, which of the following must be true?

I. $b^2 - 4c > 0$ II. $b < 0$ III. $c > 0$

(a) None   (b) I only   (c) I and II only
(d) I and III only   (e) I, II, and III

Chapter 12

The Limit and the Derivative of a Function

With this chapter we begin the study of calculus, which can be divided into two parts: differential calculus and integral calculus. Our main objective in studying calculus is to gain an understanding of its uses and applications. You will see that calculus is an enormously powerful branch of mathematics, with a wide range of applications, including graphing of functions, optimization of functions, and economics. The calculus allows us to analyze things that are changing, such as marginal cost or revenue, flow of a drug through the bloodstream, etc.

The centerpiece of calculus is the concept of a *limit of a function*. The concept of the limit of a function is what bridges the gap between the mathematics of algebra and geometry and the mathematics of calculus.

12.1 THE IDEA OF A LIMIT

We begin by asking a question: "What does it mean for a function f to have a limit L as x approaches some fixed number c?" To find an answer, we need to be more precise about f, L, and c. The function f must be defined in an open interval near the number c, but it does not have to be defined at c itself. The limit L is some real number. With these restrictions in mind we introduce the symbolism

$$\lim_{x \to c} f(x) = L \qquad (1)$$

which is read as "the limit of $f(x)$ as x approaches c equals the number L." This indicates that f has a limit L as x approaches c. We may describe Statement (1) in two ways:

For all x approximately equal to c, but $x \neq c$, the value $f(x)$ is approximately equal to L.

For all x sufficiently close to c, but unequal to c, the value $f(x)$ can be made as close as we please to L.

To illustrate the concept of limit, suppose you want to know what happens to the function

$$f(x) = \frac{x^2 - 1}{x - 1}$$

as the variable x approaches 1. Although f is not defined at $x = 1$, you can get a feel for the situation by evaluating f using values of x that are close to 1; some are chosen less than 1 (from the left), and some are greater than 1 (from the right). The following table summarizes the behavior of $f(x)$ for x close to 1:

	x approaches 1 from the left $\longrightarrow$						$\longleftarrow$ x approaches 1 from the right					
x	0	0.5	0.75	0.9	0.99	**0.999**	**1.001**	1.01	1.1	1.25	1.5	2
$f(x)$	1	1.5	1.75	1.9	1.99	**1.999**	**2.001**	2.01	2.1	2.25	2.5	3

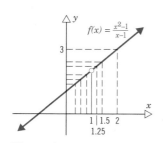

Figure 1

We see from the table that the function values of f approach the number 2 as x gets closer and closer to 1 from either side. This is also apparent from the graph of f in Figure 1. Notice there that even though the function is not defined at $x = 1$ (as indicated by the hollow dot), the function values get closer and closer to 2 as x gets closer and closer to 1. We therefore infer that the limit of f as x approaches 1 is 2, and we express this by writing

$$\lim_{x \to 1} \frac{x^2 - 1}{x - 1} = 2$$

Next, we examine the limit of a function as x approaches a number in the domain of the function. We consider the limit of $f(x) = x^3$ as x approaches 2:

$$\lim_{x \to 2} x^3$$

It is clear that as x gets closer and closer to 2, x^3 gets closer and closer to 8. This is also apparent from the table and the graph in Figure 2.

Figure 2

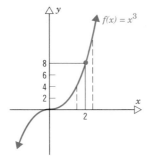

The entries in the table are obtained by using a calculator:

x	1	1.5	1.6	1.75	1.8	1.9	1.99	1.999	**1.9999**
$f(x) = x^3$	1	3.375	4.096	5.359	5.832	6.859	7.8806	7.988	**7.9988**

x approaches 2 from the left

x	3	2.5	2.4	2.25	2.2	2.1	2.01	2.001	**2.0001**
$f(x) = x^3$	27	15.625	13.824	11.3906	10.648	9.261	8.1206	8.012	**8.0012**

x approaches 2 from the right

We infer that for x ''approximately equal'' to 2, the value of $f(x) = x^3$ is approximately equal to 8. That is,

$$\lim_{x \to 2} x^3 = 8$$

Notice that in this case, the limit 8 is the value of $f(x) = x^3$ at $x = 2$, namely,

$$f(2) = 2^3 = 8$$

Now Work Problem 1   We now state the following important principle about limits:

> The limit L of a function $y = f(x)$ as x approaches the number c *does not depend on the value of f at c.*

The principle just stated is best illustrated by the graphs of the three functions in Figure 3 (at the top of the next page). For all three functions graphed, the limit of $f(x)$ as x approaches c is equal to L. Yet the functions behave quite differently at $x = c$.

Figure 3

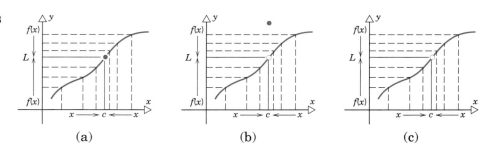

(a)   (b)   (c)

In Figure 3(a) we observe that as x gets closer to c, the value of f, as measured by its height, gets closer to the number L and in fact, $f(c)$ is equal to the limit L. In Figure 3(b) $f(c)$ is different from L (as illustrated by the "hole" in the graph of f). Figure 3(c) illustrates the case when $f(c)$ is not defined at all.

Let's look at some examples.

EXAMPLE 1

Use a graph to find: $\lim\limits_{x \to 1} (x^2 + 1)$

SOLUTION

We begin with the graph of $f(x) = x^2 + 1$. See Figure 4. Based on the illustration, for numbers x close to 1, the values of f are close to 2. In other words, as x gets closer to 1, the values of f get closer to 2. Using limit notation, these ideas are expressed by writing

$$\lim_{x \to 1} (x^2 + 1) = 2$$

Figure 4

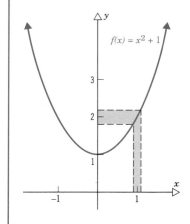

Now Work Problem 19

....

Notice in this case that the limit 2 is the value of $f(x) = x^2 + 1$ at 1, namely $f(1) = 2$.

EXAMPLE 2

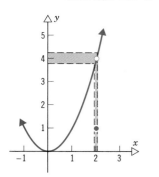

Figure 5

Use a graph to find $\lim\limits_{x\to 2} f(x)$ if

$$f(x) = \begin{cases} x^2 & \text{if} \quad x \neq 2 \\ 1 & \text{if} \quad x = 2 \end{cases}$$

SOLUTION

First we graph f. See Figure 5. Based on the illustration, we see that as x gets closer to 2, but remains unequal to 2, the values of f get closer to 4. We conclude that

$$\lim\limits_{x\to 2} f(x) = 4$$

....

Notice that $f(2) = 1$, so in this case the limit 4 does not equal the value of f at 2.

EXAMPLE 3

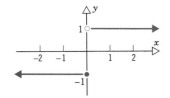

Figure 6

Now Work Problem 29

Use a graph to determine whether the function given below has a limit as x approaches 0.

$$f(x) = \begin{cases} -1 & \text{if} \quad x \leq 0 \\ 1 & \text{if} \quad x > 0 \end{cases}$$

SOLUTION

The graph of f is given in Figure 6. If x is close to 0, but remains less than 0, the values of f are -1. If x is close to 0, but remains greater than 0, the values of f are 1. Since there is no *single* number that the values of f get close to, we conclude that $\lim\limits_{x\to 0} f(x)$ does not exist.

....

If there is *no single* number that the value of f approaches, we say that f has *no limit as x approaches c* or, more simply, that the *limit does not exist at c.*

In Example 3 we showed a situation when the limit does not exist. The next example illustrates another situation when the limit of a function does not exist.

EXAMPLE 4

Use a table of values and a graph to show that

$$\lim\limits_{x\to 0} \frac{1}{x^2} \quad \text{does not exist}$$

Let $f(x) = 1/x^2$. The table at the top of the next page gives values of $f(x)$ for some values of x close to 0.

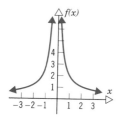

Figure 7

	x approaches 0 from the left ⟶					⟵ x approaches 0 from the right				
x	-1	$-1/2$	$-1/10$	$-1/100$	$-1/1000$	$1/1000$	$1/100$	$1/10$	$1/2$	1
$f(x)$	1	4	100	10,000	1,000,000	1,000,000	10,000	100	4	1

As x gets closer and closer to 0, the values of $f(x)$ get larger and larger without bound. This is also clear from the graph. See Figure 7. Since the values of f do not approach a number as x approaches 0, we say the limit does not exist and we write

$$\lim_{x \to 0} \frac{1}{x^2} \quad \text{does not exist}$$

· ·

Using tables of values or graphs to determine limits described in this section should be viewed as aids to understanding the idea of a limit. Each approach has its draw-backs: tables of values may require extensive calculations, and the graphical approach requires a knowledge of the graph of the function. In Section 12.2 we discuss a method for finding limits using algebra. This technique is the method we use in the study of calculus.

Left and Right Limits

We have described $\lim_{x \to c} f(x) = N$ by saying that as x gets close to c, the values of $f(x)$ get closer to the number N. The variable x can get closer to c in two ways: by approaching c from the left, through numbers less than c, or by approaching c from the right, through numbers greater than c. See Figure 8.

Figure 8

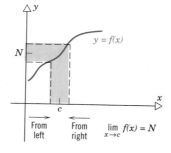

The statement

$$\lim_{x \to c^-} f(x) = L$$

read "the limit of $f(x)$ as x approaches c from the left equals L" is described as follows: As x gets closer to c, but remains less than c, the values of $f(x)$ get closer to L.

The notation $x \to c^-$ is used to remind us that x is less than c; that is, that something must be subtracted from c to yield x.

The statement

$$\lim_{x \to c^+} f(x) = R$$

read "the limit of $f(x)$ as x approaches c from the right equals R" is described as follows: As x gets closer to c, but remains greater than c, the values of $f(x)$ get closer to R.

The notation $x \to c^+$ is used to remind us that x is always larger than c.

Figure 9 illustrates these ideas by using the graph of a function.

Figure 9

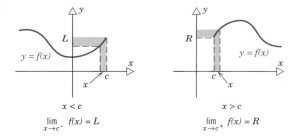

Now look at Figure 10. In Figure 10(a) we have

$$\lim_{x \to c^-} f(x) = L \qquad \lim_{x \to c^+} f(x) = R \qquad \text{and} \qquad L = R$$

In this case $\lim_{x \to c} f(x)$ exists and has the same value whether c is approached from the left or from the right. In Figure 10(b) we have

$$\lim_{x \to c^-} f(x) = L \qquad \lim_{x \to c^+} f(x) = R \qquad \text{and} \qquad L \neq R$$

Here it is apparent that $\lim_{x \to c} f(x)$ does not exist. This leads us to formulate the criterion given in the box on the next page.

Figure 10

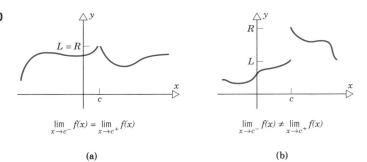

> **Criterion for $\lim\limits_{x \to c} f(x)$ to Exist**
>
> For $\lim\limits_{x \to c} f(x)$ to exist, we must have
>
> $$\lim_{x \to c^-} f(x) = \lim_{x \to c^+} f(x)$$

EXAMPLE 5

Find $\lim\limits_{x \to 1^-} f(x)$ and $\lim\limits_{x \to 1^+} f(x)$ for the function below.

$$f(x) = \begin{cases} x^2 & \text{if } x \le 1 \\ x + 3 & \text{if } x > 1 \end{cases}$$

Does $\lim\limits_{x \to 1} f(x)$ exist?

SOLUTION

To find $\lim\limits_{x \to 1^-} f(x)$, we seek the values of f for the numbers x close to 1, but less than 1. For numbers $x < 1$, $f(x) = x^2$, so that

$$\lim_{x \to 1^-} f(x) = \lim_{x \to 1^-} x^2 = 1$$

Similarly, to find $\lim\limits_{x \to 1^+} f(x)$, we want numbers x close to 1, but larger than 1. For numbers $x > 1$, $f(x) = x + 3$, so that

$$\lim_{x \to 1^+} f(x) = \lim_{x \to 1^+} (x + 3) = 4$$

Since the left limit and right limit are not equal, $\lim\limits_{x \to 1} f(x)$ does not exist. See Figure 11 for the graph.

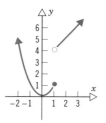

Figure 11

Now Work Problem 47

· ·

Next we give an example for which only the right limit exists.

EXAMPLE 6

For $f(x) = \sqrt{x - 2}$ find $\lim\limits_{x \to 2} f(x)$.

SOLUTION

Since f is defined only when $x \ge 2$, we may consider the limit as x approaches 2 from the right. If x is slightly greater than 2, then $x - 2$ is a positive number that is close to 0 and thus $\sqrt{x - 2}$ is close to 0. We therefore conclude that

$$\lim_{x \to 2^+} \sqrt{x - 2} = 0$$

This limit is also evident from Figure 12.

Figure 12

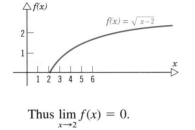

$$f(x) = \sqrt{x-2}$$

Thus $\lim\limits_{x \to 2} f(x) = 0$.

· ·

EXERCISE 12.1   Answers to odd-numbered problems begin on page AN-53.

In Problems 1–8 use a calculator to complete each table and evaluate the indicated limits.

1.

x	0.9	0.99	0.999
$f(x) = 2x$			

x	1.1	1.01	1.001
$f(x) = 2x$			

$\lim\limits_{x \to 1} f(x) = $ _____

2.

x	1.9	1.99	1.999
$f(x) = x + 3$			

x	2.1	2.01	2.001
$f(x) = x + 3$			

$\lim\limits_{x \to 2} f(x) = $ _____

3.

x	0.1	0.01	0.001
$f(x) = x^2 + 2$			

x	-0.1	-0.01	-0.001
$f(x) = x^2 + 2$			

$\lim\limits_{x \to 0} f(x) = $ _____

4.

x	-1.1	-1.01	-1.001
$f(x) = x^2 - 2$			

x	-0.9	-0.99	-0.999
$f(x) = x^2 - 2$			

$\lim\limits_{x \to -1} f(x) = $ _____

5.

x	1.9	1.99	1.999
$f(x) = \dfrac{x^2 - 4}{x - 2}$			

x	2.1	2.01	2.001
$f(x) = \dfrac{x^2 - 4}{x - 2}$			

$\lim\limits_{x \to 2} f(x) = $ _____

6.

x	-1.1	-1.01	-1.001
$f(x) = \dfrac{x^2 - 1}{x + 1}$			

x	-0.9	-0.99	-0.999
$f(x) = \dfrac{x^2 - 1}{x + 1}$			

$\lim\limits_{x \to -1} f(x) = $ _____

7.

x	-1.1	-1.01	-1.001
$f(x) = \dfrac{x^3 + 1}{x + 1}$			

x	-0.9	-0.99	-0.999
$f(x) = \dfrac{x^3 + 1}{x + 1}$			

$$\lim_{x \to -1} f(x) = \underline{\hspace{2cm}}$$

8.

x	0.9	0.99	0.999
$f(x) = \dfrac{x^3 - 1}{x - 1}$			

x	1.1	1.01	1.001
$f(x) = \dfrac{x^3 - 1}{x - 1}$			

$$\lim_{x \to 1} f(x) = \underline{\hspace{2cm}}$$

In Problems 9–18 use the graph to determine whether $\lim_{x \to c} f(x)$ exists.

9.

 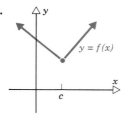

10.

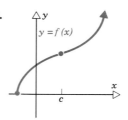

11.

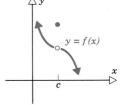

12.

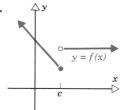

13.

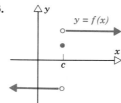

14.

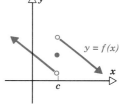

15.

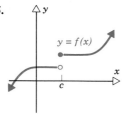

16.

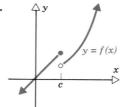

17.

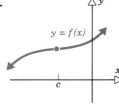

18.

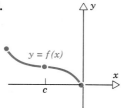

In Problems 19–36 determine whether $\lim\limits_{x \to c} f(x)$ exists by graphing the function. If it exists, find $\lim\limits_{x \to c} f(x)$.

19. $f(x) = 3x + 6, \quad c = 1$

20. $f(x) = 2x + 4, \quad c = -1$

21. $f(x) = 3x^2, \quad c = 1$

22. $f(x) = 2x^2, \quad c = 0$

23. $f(x) = x^2 + 2, \quad c = 0$

24. $f(x) = x^2 - 1, \quad c = 0$

25. $f(x) = \sqrt{x}, \quad c = 4$

26. $f(x) = \dfrac{1}{x}, \quad c = 2$

27. $f(x) = x, \quad c = 2$

28. $f(x) = 5, \quad c = 1$

29. $f(x) = \begin{cases} 2x + 5 & \text{if } x \neq 2 \\ 9 & \text{if } x = 2 \end{cases} \quad c = 2$

30. $f(x) = \begin{cases} 2x + 1 & \text{if } x \neq 0 \\ 1 & \text{if } x = 0 \end{cases} \quad c = 0$

31. $f(x) = \begin{cases} 3x - 1 & \text{if } x \neq 1 \\ 4 & \text{if } x = 1 \end{cases} \quad c = 1$

32. $f(x) = \begin{cases} 2x - 1 & \text{if } x \neq 1 \\ 2 & \text{if } x = 1 \end{cases} \quad c = 1$

33. $f(x) = \begin{cases} 3x - 1 & \text{if } x < 1 \\ \text{Not defined} & \text{if } x = 1 \\ 2x & \text{if } x > 1 \end{cases} \quad c = 1$

34. $f(x) = \begin{cases} 3x - 1 & \text{if } x < 1 \\ 2 & \text{if } x = 1 \\ 3x & \text{if } x > 1 \end{cases} \quad c = 1$

35. $f(x) = \begin{cases} x^2 & \text{if } x \leq 0 \\ 2x + 1 & \text{if } x > 0 \end{cases} \quad c = 0$

36. $f(x) = \begin{cases} x^2 & \text{if } x \leq -1 \\ -3x + 2 & \text{if } x > -1 \end{cases} \quad c = -1$

37. The graph of the function f is given below. Use the graph to find the following limits.

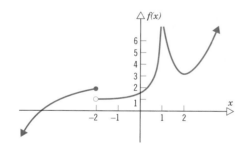

(a) $\lim\limits_{x \to -2^+} f(x)$ (b) $\lim\limits_{x \to -2^-} f(x)$ (c) $\lim\limits_{x \to -2} f(x)$

(d) $\lim\limits_{x \to 1^+} f(x)$ (e) $\lim\limits_{x \to 1^-} f(x)$ (f) $\lim\limits_{x \to 1} f(x)$

(g) $\lim\limits_{x \to 2^+} f(x)$ (h) $\lim\limits_{x \to 2^-} f(x)$ (i) $\lim\limits_{x \to 2} f(x)$

38. The graph of the function f is given below. Use the graph to find the following limits.

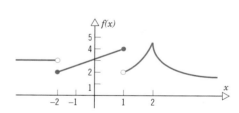

(a) $\lim\limits_{x \to -2^+} f(x)$ (b) $\lim\limits_{x \to -2^-} f(x)$ (c) $\lim\limits_{x \to -2} f(x)$

(d) $\lim\limits_{x \to 1^+} f(x)$ (e) $\lim\limits_{x \to 1^-} f(x)$ (f) $\lim\limits_{x \to 1} f(x)$

(g) $\lim\limits_{x \to 2^+} f(x)$ (h) $\lim\limits_{x \to 2^-} f(x)$ (i) $\lim\limits_{x \to 2} f(x)$

In Problems 39–46 find the limit.

39. $\lim\limits_{x \to 3^+} (x - 5)$

40. $\lim\limits_{x \to 2^+} (3 - x)$

41. $\lim\limits_{x \to 2^+} (4 - x^2)$

42. $\lim\limits_{x \to 0^+} \dfrac{3x}{x^3}$

43. $\lim\limits_{x \to 0^+} \dfrac{7}{x + 2}$

44. $\lim\limits_{x \to 3^+} \dfrac{6}{x - 3}$

45. $\lim\limits_{x \to 3^+} \sqrt{x - 3}$

46. $\lim\limits_{x \to 3^+} x\sqrt{x^2 - 9}$

47. Find

$$\lim_{x\to 2^-} f(x) \quad \text{and} \quad \lim_{x\to 2^+} f(x)$$

for the function

$$f(x) = \begin{cases} 2x + 5 & \text{if} \quad x \le 2 \\ 4x + 1 & \text{if} \quad x > 2 \end{cases}$$

Does $\lim_{x\to 2} f(x)$ exist?

48. Find

$$\lim_{x\to 1^-} f(x) \quad \text{and} \quad \lim_{x\to 1^+} f(x)$$

for the function

$$f(x) = \begin{cases} 3x - 1 & \text{if} \quad x < 1 \\ 4 & \text{if} \quad x = 1 \\ 2x & \text{if} \quad x > 1 \end{cases}$$

Does $\lim_{x\to 1} f(x)$ exist?

49. Find

$$\lim_{x\to 1^-} f(x) \quad \text{and} \quad \lim_{x\to 1^+} f(x)$$

for the function

$$f(x) = \begin{cases} 4x - 1 & \text{if} \quad x < 1 \\ \text{Not defined} & \text{if} \quad x = 1 \\ 3x & \text{if} \quad x > 1 \end{cases}$$

Does $\lim_{x\to 1} f(x)$ exist?

50. Find

$$\lim_{x\to 1^-} f(x) \quad \text{and} \quad \lim_{x\to 1^+} f(x)$$

for the function

$$f(x) = \begin{cases} 3x - 1 & \text{if} \quad x < 1 \\ 2 & \text{if} \quad x = 1 \\ 3x & \text{if} \quad x > 1 \end{cases}$$

Does $\lim_{x\to 1} f(x)$ exist?

Technology Exercises

In recent years the study of calculus was made easier by the use of computer-graphing calculators. From time to time, exercises will be given that are designed to be done with the use of a graphing calculator or a computer-graphing package such as Mathematica, Derive, Maple, Visual Calculus, or X(Plore). Using this technology has two goals. First, to direct your attention to the calculus concepts without the need for algebraic manipulation; and second, to include some exercises that would be very difficult to work without the assistance of a graphing calculator or a computer.

For example, to compute the limit

$$\lim_{x\to 1} \frac{x^3 + x^2 - x + 1}{x - 1}$$

you can graph the function

$$f(x) = \frac{x^3 + x^2 - x + 1}{x - 1}$$

and zoom in near $x = 1$.

In Problems 1–4 study the graph of the function in order to determine the given limit.

1. $\lim_{x\to 2} \dfrac{x^2 - 4}{x - 2}$ **2.** $\lim_{x\to 1} \dfrac{x^3 + 4x - 5}{x - 1}$ **3.** $\lim_{x\to .2} \dfrac{3x^3 - .12x}{x - .2}$ **4.** $\lim_{x\to 0} \dfrac{\sqrt{x + 2} - 1}{x}$

In Problems 5–9 graph the function to determine if the given limit exists.

5. $f(x) = 2 + \sqrt{x^2 - 5x + 6}$

(a) $\lim\limits_{x \to 1} f(x)$ (b) $\lim\limits_{x \to 2^-} f(x)$

(c) $\lim\limits_{x \to 2^+} f(x)$ (d) $\lim\limits_{x \to 3} f(x)$

6. $f(x) = 4 - \sqrt{27 - x^{3/2}}$

(a) $\lim\limits_{x \to 9^-} f(x)$ (b) $\lim\limits_{x \to 9^+} f(x)$

(c) $\lim\limits_{x \to 9} f(x)$

7. $f(x) = \dfrac{x^2}{x^2 - 4}$

(a) $\lim\limits_{x \to 3} f(x)$ (b) $\lim\limits_{x \to 2^-} f(x)$

(c) $\lim\limits_{x \to 2^+} f(x)$

8. $f(x) = \dfrac{x^2 - 2x}{x^2 + 3x - 10}$

(a) $\lim\limits_{x \to 2^+} f(x)$ (b) $\lim\limits_{x \to 2^-} f(x)$

(c) $\lim\limits_{x \to 4} f(x)$

9. Show that $\lim\limits_{x \to 2}(x^3 - x) = \lim\limits_{x \to 2} x^3 - \lim\limits_{x \to 2} x$ by proceeding as follows:

(a) Graph $y = x^3$ and use the graph to find $\lim\limits_{x \to 2} x^3$.

(b) Graph $y = x$ and use the graph to find $\lim\limits_{x \to 2} x$.

(c) Graph $y = x^3 - x$ and use the graph to find $\lim\limits_{x \to 2}(x^3 - x)$.

10. Complete the following table to approximate the value of
$$e = \lim\limits_{x \to \infty}\left(1 + \frac{1}{x}\right)^x:$$

x	$\left(1 + \dfrac{1}{x}\right)^x$
1	2.000000
10	
100	
1,000	
10,000	
100,000	
1,000,000	

(a) Use the graph of $y = e^x$ to approximate e.

(b) Use the graph of $y = \left(1 + \dfrac{1}{x}\right)^x$ to approximate e.

[*Note:* The notation $\lim\limits_{x \to \infty}$ is another way of saying that x gets as large as possible.]

11. Explain why, if you take n large enough (for example, $n = 10^{18}$ or more), your calculator thinks that
$$\left(1 + \frac{1}{n}\right)^n = 1$$

12.2 ALGEBRAIC TECHNIQUES FOR FINDING LIMITS

In this section we establish some useful algebraic properties of limits. These properties allow us to find limits without using numerical or graphical approaches.

Two Important Formulas

> **Limit of a Constant**
>
> For the constant function $f(x) = b$, we have
> $$\lim\limits_{x \to c} f(x) = \lim\limits_{x \to c} b = b \tag{1}$$
> for any number c.

In other words, *the limit of a constant is the constant.*

The graph of the constant function $f(x) = b$ is a horizontal straight line. Clearly, no matter what the value of x is, the height of $f(x)$ is always b. Thus, as x gets closer to c, the values of f remain fixed at b. That is, $\lim\limits_{x \to c} f(x) = \lim\limits_{x \to c} b = b$. See Figure 13.

Figure 13

EXAMPLE 1

(a) $\lim\limits_{x \to 3} 5 = 5$ (b) $\lim\limits_{x \to 0} (-2) = -2$

Now Work Problem 1

....

> **Limit of $f(x) = x$**
>
> For the identity function $f(x) = x$, we have
>
> $$\lim_{x \to c} f(x) = \lim_{x \to c} x = c \qquad (2)$$
>
> for any number c.

The graph of the identity function $f(x) = x$ is a straight line with slope 1, passing through the origin. See Figure 14 and Table 1. No matter what number c is chosen, as x gets closer to c, the values of f also get closer to c. That is,

$$\lim_{x \to c} f(x) = \lim_{x \to c} x = c$$

Figure 14

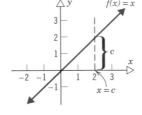

Table 1

x	$f(x)$
1	1
2	2
5	5
-2	-2
c	c

EXAMPLE 2

(a) $\lim\limits_{x \to 2} x = 2$ (b) $\lim\limits_{x \to 5} x = 5$

Now Work Problem 3

....

Properties of Limits

Various algebraic properties are needed for evaluating limits of functions that are more complicated than those just described. We state these properties and give some examples to provide a working knowledge of their many uses.

Limit of a Sum (Difference)

Let f and g be two functions whose limits as x approaches c exist; that is, suppose $\lim_{x \to c} f(x)$ and $\lim_{x \to c} g(x)$ are both known real numbers. Then

$$\lim_{x \to c} [f(x) + g(x)] = \lim_{x \to c} f(x) + \lim_{x \to c} g(x) \tag{3}$$

and

$$\lim_{x \to c} [f(x) - g(x)] = \lim_{x \to c} f(x) - \lim_{x \to c} g(x) \tag{4}$$

That is, the limit of the sum (difference) of two functions equals the sum (difference) of their limits.

EXAMPLE 3

Find: $\lim_{x \to 3} (2 + x)$

SOLUTION

We want to compute the limit of a function $h(x) = 2 + x$, which is really the sum of two other functions, $f(x) = 2$ and $g(x) = x$. From (1) and (2) we know that

$$\lim_{x \to 3} 2 = 2 \qquad \text{and} \qquad \lim_{x \to 3} x = 3$$

Then, using (3), we have

$$\lim_{x \to 3} (2 + x) = \lim_{x \to 3} 2 + \lim_{x \to 3} x \qquad \text{By (3)}$$

$$= 2 + 3 \qquad \text{By (1) and (2)}$$

Thus,

$$\lim_{x \to 3} (2 + x) = 5$$

· ·

EXAMPLE 4

Find: $\lim_{x \to -2} (x - 3)$

SOLUTION

$$\lim_{x \to -2} (x - 3) = \lim_{x \to -2} x - \lim_{x \to -2} 3 \qquad \text{By (4)}$$

$$= -2 - 3 \qquad \text{By (2) and (1)}$$

$$= -5$$

· ·

> **Limit of a Product**
>
> Let f and g be two functions whose limits as x approaches c exist; that is, suppose $\lim_{x \to c} f(x)$ and $\lim_{x \to c} g(x)$ are both known real numbers. Then
>
> $$\lim_{x \to c} [f(x) \cdot g(x)] = \left[\lim_{x \to c} f(x) \right] \cdot \left[\lim_{x \to c} g(x) \right] \qquad (5)$$
>
> Thus, the limit of the product of two functions equals the product of their limits.

EXAMPLE 5

Find: $\lim_{x \to -2} (3x)$

SOLUTION

We want to find the limit of a function $h(x) = 3x$, which is the product of two other functions, $f(x) = 3$ and $g(x) = x$. Therefore,

$$\lim_{x \to -2} (3x) = \left(\lim_{x \to -2} 3 \right) \cdot \left(\lim_{x \to -2} x \right) \qquad \text{By (5)}$$
$$= (3) \cdot (-2) \qquad \text{By (1) and (2)}$$
$$= -6$$

· ·

EXAMPLE 6

Find: $\lim_{x \to 3} (4x + 5)$

SOLUTION

$$\lim_{x \to 3} (4x + 5) = \lim_{x \to 3} (4x) + \lim_{x \to 3} 5 \qquad \text{By (3)}$$
$$= \left(\lim_{x \to 3} 4 \right) \cdot \left(\lim_{x \to 3} x \right) + \lim_{x \to 3} 5 \qquad \text{By (5)}$$
$$= (4) \cdot (3) + 5 \qquad \text{By (1) and (2)}$$
$$= 12 + 5 = 17$$

· ·

EXAMPLE 7

Find: $\lim_{x \to 4} x^2$

SOLUTION

$$\lim_{x \to 4} x^2 = \left(\lim_{x \to 4} x \right) \cdot \left(\lim_{x \to 4} x \right) \qquad \text{By (5)}$$
$$= (4) \cdot (4) \qquad \text{By (2)}$$
$$= 16$$

Now Work Problem 9

· ·

EXAMPLE 8

Suppose $n \geq 1$ is a positive integer and a is a constant. Show that

$$\lim_{x \to c} ax^n = ac^n \tag{6}$$

for any number c.

SOLUTION

$$\lim_{x \to c} ax^n = \lim_{x \to c} (\underbrace{a \cdot x \cdot x \cdot \ \cdots \ \cdot x}_{n \text{ factors}})$$

$$= \underbrace{\lim_{x \to c} a \cdot \lim_{x \to c} x \cdot \lim_{x \to c} x \cdot \ \cdots \ \cdot \lim_{x \to c} x}_{n \text{ factors}} \quad \text{By repeated use of (5)}$$

$$= \underbrace{a \cdot c \cdot c \cdot \ \cdots \ \cdot c}_{n \text{ factors}} \quad \text{By (1) and (2)}$$

$$= ac^n$$

· ·

EXAMPLE 9

Find: $\lim_{x \to 2} 3x^4$

SOLUTION

$$\lim_{x \to 2} 3x^4 = 3 \cdot 2^4 = 48$$
$$\uparrow$$
$$\text{By (6)}$$

· ·

Because of (6) and repeated applications of (3) and (4), the following result can be established.

Limit of a Polynomial

If P is a polynomial function, then

$$\lim_{x \to c} P(x) = P(c) \tag{7}$$

for any number c.

Thus, to find the limit of a polynomial function as x approaches c, merely evaluate the polynomial at c.

EXAMPLE 10

Find: $\lim\limits_{x \to 2} (3x^4 - 2x^2 + x - 5)$

SOLUTION

$$\lim\limits_{x \to 2} (3x^4 - 2x^2 + x - 5) = 3 \cdot 2^4 - 2 \cdot 2^2 + 2 - 5 \quad \text{By (7)}$$
$$= 3 \cdot 16 - 2 \cdot 4 + 2 - 5$$
$$= 48 - 8 + 2 - 5 = 37$$

· ·

The next property of limits we take up involves powers and roots of functions.

Limit of a Power or Root

Let f be a function whose limit as x approaches c exists; that is, suppose $\lim\limits_{x \to c} f(x)$ is a known real number. Then for $n \geq 2$ a positive integer, we have

(a) $\lim\limits_{x \to c}[f(x)]^n = \left[\lim\limits_{x \to c} f(x)\right]^n$ (b) $\lim\limits_{x \to c} \sqrt[n]{f(x)} = \sqrt[n]{\lim\limits_{x \to c} f(x)}$ (8)

For (8b) we also require that $\sqrt[n]{f(x)}$ and $\sqrt[n]{\lim\limits_{x \to c} f(x)}$ are defined.

EXAMPLE 11

Find:

(a) $\lim\limits_{x \to 0} (x^3 + 2)^5$ (b) $\lim\limits_{x \to 2} \sqrt{x^2 + 12}$

SOLUTION

(a) $\lim\limits_{x \to 0} (x^3 + 2)^5 = \left[\lim\limits_{x \to 0} (x^3 + 2)\right]^5 = 2^5 = 32$

$\qquad\qquad\qquad\qquad\quad \uparrow \qquad\qquad\qquad \uparrow$
$\qquad\qquad\qquad\quad \text{By (8a)} \qquad\quad \text{By (7)}$

(b) $\lim\limits_{x \to 2} \sqrt{x^2 + 12} = \sqrt{\lim\limits_{x \to 2} (x^2 + 12)} = \sqrt{16} = 4$

$\qquad\qquad\qquad\quad \uparrow \qquad\qquad\qquad \uparrow$
$\qquad\qquad\qquad \text{By (8b)} \qquad\quad \text{By (7)}$

Now Work Problem 19

· ·

The last property of limits we take up involves division of functions.

Limit of a Quotient

Let f and g be two functions whose limits as x approaches c exist; that is, suppose $\lim_{x \to c} f(x)$ and $\lim_{x \to c} g(x)$ are both known real numbers. If $\lim_{x \to c} g(x) \neq 0$, then the limit of the function f/g as x approaches c also exists and

$$\lim_{x \to c} \frac{f(x)}{g(x)} = \frac{\lim\limits_{x \to c} f(x)}{\lim\limits_{x \to c} g(x)} \qquad (9)$$

Thus, if the limit of the denominator function is not zero, the limit of the quotient of two functions equals the quotient of their limits.

EXAMPLE 12

Find: $\lim\limits_{x \to 3} \dfrac{2x^2 + 1}{x^3 - 2}$

SOLUTION

First, we look at the limit of the denominator function $g(x) = x^3 - 2$:

$$\lim_{x \to 3} (x^3 - 2) = 3^3 - 2 = 25$$

Since the limit of the denominator function is not zero, we proceed to compute the limit of the numerator function:

$$\lim_{x \to 3} (2x^2 + 1) = (2)(9) + 1 = 19$$

Thus, by (9) we have

$$\lim_{x \to 3} \frac{2x^2 + 1}{x^3 - 2} = \frac{\lim\limits_{x \to 3} (2x^2 + 1)}{\lim\limits_{x \to 3} (x^3 - 2)} = \frac{19}{25}$$

Now Work Problem 21

⋯⋯⋯⋯⋯⋯⋯⋯⋯

When the limit of the denominator function equals zero, we cannot use (9). In such cases factoring sometimes works, as the next example shows.

EXAMPLE 13

Find: $\lim\limits_{x \to -2} \dfrac{x^2 + 5x + 6}{x^2 - 4}$

SOLUTION

First, we check the limit of the denominator:

$$\lim_{x \to -2} (x^2 - 4) = 4 - 4 = 0$$

Since the limit of the denominator function is zero, we cannot use (9). However, this does not mean that the limit does not exist! Instead, we use algebraic techniques and factor:

$$\frac{x^2 + 5x + 6}{x^2 - 4} = \frac{(x + 3)(x + 2)}{(x - 2)(x + 2)}$$

Since we are interested only in the limit as x *approaches* -2, and *not* in the value when x *equals* -2, the quantity $(x + 2)$ is not zero. Hence, we can cancel the $(x + 2)$s and then apply (9):

$$\lim_{x \to -2} \frac{x^2 + 5x + 6}{x^2 - 4} = \lim_{x \to -2} \frac{(x + 3)(x + 2)}{(x - 2)(x + 2)} = \lim_{x \to -2} \frac{x + 3}{x - 2}$$

$$= \frac{\lim_{x \to -2} (x + 3)}{\lim_{x \to -2} (x - 2)} = \frac{-2 + 3}{-2 - 2} = -\frac{1}{4}$$

. . . .

Notice that, although the function $\dfrac{x^2 + 5x + 6}{x^2 - 4}$ is not defined at -2, it does have a limit as $x \to -2$.

EXAMPLE 14

Find: $\displaystyle\lim_{x \to 4} \frac{x^2 - x - 12}{x^2 - 4x}$

SOLUTION

$$\lim_{x \to 4} \frac{x^2 - x - 12}{x^2 - 4x} = \lim_{x \to 4} \frac{(x - 4)(x + 3)}{x(x - 4)} = \lim_{x \to 4} \frac{x + 3}{x}$$

$$= \frac{\lim_{x \to 4} (x + 3)}{\lim_{x \to 4} x} = \frac{7}{4}$$

Now Work Problem 23

. . . .

Let's look at some other ways to handle situations in which the limit of the denominator equals 0.

EXAMPLE 15

Find: $\displaystyle\lim_{x \to 2} \left(\frac{3x}{x - 2} - \frac{6}{x - 2} \right)$

SOLUTION

Here, we perform the indicated operation and then take the limit.

$$\lim_{x \to 2} \left(\frac{3x}{x - 2} - \frac{6}{x - 2} \right) = \lim_{x \to 2} \frac{3x - 6}{x - 2} = \lim_{x \to 2} \frac{3(x - 2)}{x - 2} = 3$$

. . . .

EXAMPLE 16

Find: $\displaystyle\lim_{x\to 9}\frac{\sqrt{x}-3}{x-9}$

SOLUTION

Here, we rationalize the numerator by multiplying the numerator and denominator by $\sqrt{x}+3$. (Because multiplying a fraction by one does not change the fraction.)

$$\lim_{x\to 9}\frac{\sqrt{x}-3}{x-9}=\lim_{x\to 9}\frac{(\sqrt{x}-3)\,(\sqrt{x}+3)}{(x-9)\,(\sqrt{x}+3)}=\lim_{x\to 9}\frac{x-9}{(x-9)\,(\sqrt{x}+3)}$$

$$=\lim_{x\to 9}\frac{1}{\sqrt{x}+3}=\frac{1}{\sqrt{9}+3}=\frac{1}{6}$$

....

Alternatively, we could have solved Example 16 by using the fact that $x-9=(\sqrt{x}-3)(\sqrt{x}+3)$. Then

$$\lim_{x\to 9}\frac{\sqrt{x}-3}{x-9}=\lim_{x\to 9}\frac{\sqrt{x}-3}{(\sqrt{x}-3)\,(\sqrt{x}+3)}=\lim_{x\to 9}\frac{1}{\sqrt{x}+3}=\frac{1}{\sqrt{9}+3}=\frac{1}{6}$$

EXAMPLE 17

For the function $f(x)=x^2-2x$, find: $\displaystyle\lim_{x\to 3}\frac{f(x)-f(3)}{x-3}$

SOLUTION

Since $f(3)=9-6=3$, we find

$$\lim_{x\to 3}\frac{f(x)-f(3)}{x-3}=\lim_{x\to 3}\frac{(x^2-2x)-3}{x-3}=\lim_{x\to 3}\frac{(x-3)\,(x+1)}{x-3}=4$$

Now Work Problem 39

....

Summary

To find $\displaystyle\lim_{x\to c}f(x)$, try the following:

1. If $f(x)$ is a polynomial, then $\displaystyle\lim_{x\to c}f(x)=f(c)$.

2. If $f(x)$ is a quotient and the limit of the denominator is not zero, then $\displaystyle\lim_{x\to c}f(x)=f(c)$.

3. If $f(x)$ is a quotient and the limit of the denominator is zero, then
  (a) try factoring (as in Examples 13 and 14); or
  (b) try performing the indicated operation (as in Example 15); or
  (c) try rationalizing (as in Example 16).

EXERCISE 12.2   Answers to odd-numbered problems begin on page AN-55.

In Problems 1–38 find the indicated limit.

1. $\lim_{x \to 0} 3$

2. $\lim_{x \to 1} 4$

3. $\lim_{x \to 0} x$

4. $\lim_{x \to -1} x$

5. $\lim_{x \to 2} (2x - 1)$

6. $\lim_{x \to -2} (3x + 2)$

7. $\lim_{x \to 4} (x^2 + x)$

8. $\lim_{x \to 2} (x^2 - x)$

9. $\lim_{x \to -3} (2x^2 - 1)$

10. $\lim_{x \to -4} (3x^2 + 4)$

11. $\lim_{x \to 1} (5x^4 - 3x^3 + 2x^2 - x + 2)$

12. $\lim_{x \to -1} (6x^5 + 5x^4 - 2x^3 + x - 4)$

13. $\lim_{x \to 2} \sqrt{x^3 + 8}$

14. $\lim_{x \to 1} \sqrt{3x^2 + 1}$

15. $\lim_{x \to 3} \dfrac{2}{x}$

16. $\lim_{x \to 4} \dfrac{3}{x^2}$

17. $\lim_{x \to 1} \dfrac{2x + 5}{x + 6}$

18. $\lim_{x \to -2} \dfrac{x + 2}{3x - 5}$

19. $\lim_{x \to 2} \sqrt{x^2 + 3x + 2}$

20. $\lim_{x \to 2} \sqrt{2x^2 - x - 3}$

21. $\lim_{x \to 2} \dfrac{x^2 + 4}{x + 3}$

22. $\lim_{x \to 3} \dfrac{x}{x^2 + 1}$

23. $\lim_{x \to 2} \dfrac{x^2 - 4}{x - 2}$

24. $\lim_{x \to -1} \dfrac{x^2 - 1}{x + 1}$

25. $\lim_{x \to -4} \dfrac{x^3 + 64}{x + 4}$

26. $\lim_{x \to 3} \dfrac{x^3 - 27}{x - 3}$

27. $\lim_{x \to 1} \dfrac{(1/x) - 1}{x - 1}$

28. $\lim_{x \to 2} \dfrac{(1/x^2) - \frac{1}{4}}{x - 2}$

29. $\lim_{x \to 1} \dfrac{\sqrt{x} - 1}{x - 1}$

30. $\lim_{x \to 4} \dfrac{\sqrt{x} - 2}{x - 4}$

31. $\lim_{x \to 2} \dfrac{x^3 - 8}{x^2 + x - 6}$

32. $\lim_{x \to 1} \dfrac{x^3 - 3x^2 + 3x - 1}{x^3 - x}$

33. $\lim_{x \to 0} \dfrac{4x^3 - 3x}{x^2 - x}$

34. $\lim_{x \to 4} \dfrac{2x^2 - 32}{x^3 - 4x^2}$

35. $\lim_{x \to 1} \dfrac{x^4 - 1}{x - 1}$

36. $\lim_{x \to -2} \dfrac{x + 2}{x^2 - 4}$

37. $\lim_{x \to -4} \left[\dfrac{x}{x + 4} + \dfrac{4}{x + 4} \right]$

38. $\lim_{x \to 3} \left[\dfrac{3}{x - 3} - \dfrac{x}{x - 3} \right]$

39. If $f(x) = 2x - 3$, find $\lim_{x \to 1} \dfrac{f(x) - f(1)}{x - 1}$.

40. If $f(x) = 3 - 4x$, find $\lim_{x \to 2} \dfrac{f(x) - f(2)}{x - 2}$.

41. If $f(x) = x^2$, find $\lim_{x \to 4} \dfrac{f(x) - f(4)}{x - 4}$.

42. If $f(x) = x^2$, find $\lim_{x \to 1} \dfrac{f(x) - f(1)}{x - 1}$.

43. If $f(x) = 3x^2 + x$, find $\lim_{x \to 2} \dfrac{f(x) - f(2)}{x - 2}$.

44. If $f(x) = 2x^2 + x$, find $\lim_{x \to 4} \dfrac{f(x) - f(1)}{x - 4}$.

45. Find $\lim_{x \to 1} f(x)$ and $f(1)$ when $f(x) = \begin{cases} 2x^2 - 3x & \text{if } x \neq 1 \\ 5 & \text{if } x = 1 \end{cases}$

46. Find $\lim_{x \to 2} f(x)$ and $f(2)$ when $f(x) = \begin{cases} 4x^3 + x & \text{if } x \neq 2 \\ 8 & \text{if } x = 2 \end{cases}$

47. Find $\lim_{x \to 4} f(x)$ and $f(4)$ when $\quad f(x) = \begin{cases} \dfrac{2x^2 - 8x}{x - 4} & \text{if } x \neq 4 \\ 0 & \text{if } x = 4 \end{cases}$

48. Find $\lim_{x \to -1} f(x)$ and $f(-1)$ when $\quad f(x) = \begin{cases} \dfrac{x^3 - x}{x + 1} & \text{if } x \neq -1 \\ 1 & \text{if } x = -1 \end{cases}$

49. Find $\lim_{x \to 1} f(x)$ and $f(1)$ when $\quad f(x) = \begin{cases} \dfrac{2x^3 - 3x + 1}{x - 1} & \text{if } x = 1 \\ 5 & \text{if } x \neq 1 \end{cases}$

50. Find $\lim_{x \to 1} f(x)$ and $f(1)$ when $\quad f(x) = \begin{cases} \dfrac{4x^3 + x - 5}{x - 1} & \text{if } x \neq 1 \\ 8 & \text{if } x = 1 \end{cases}$

In Problems 51–54 assume that $\lim\limits_{x \to c} f(x) = 5$ and $\lim\limits_{x \to c} g(x) = 2$ to find each limit.

51. $\lim\limits_{x \to c} [2f(x)]$ **52.** $\lim\limits_{x \to c} [f(x) - g(x)]$ **53.** $\lim\limits_{x \to c} g(x)^3$ **54.** $\lim\limits_{x \to c} \dfrac{f(x)}{g(x) - f(x)}$

55. If $n \geq 1$ is a positive integer, find

$$\lim_{x \to c} \frac{x^n - c^n}{x - c}$$

[*Hint:* $x^n - c^n = (x - c)(x^{n-1} + cx^{n-2} + \cdots + c^{n-2}x + c^{n-1})$.]

12.3 CONTINUOUS FUNCTIONS

Let's summarize what we have discovered so far about the limit of a function. We have seen that sometimes $\lim\limits_{x \to c} f(x)$ equals $f(c)$ and sometimes it does not. In fact, sometimes $f(c)$ is not even defined and $\lim\limits_{x \to c} f(x)$ exists. Let's look at the possibilities:

(a) $\lim\limits_{x \to c} f(x)$ exists and equals $f(c)$.

(b) $\lim\limits_{x \to c} f(x)$ exists and does not equal $f(c)$.

(c) $\lim\limits_{x \to c} f(x)$ exists and $f(c)$ is not defined.

(d) $\lim\limits_{x \to c} f(x)$ does not exist and $f(c)$ is defined.

(e) $\lim\limits_{x \to c} f(x)$ does not exist and $f(c)$ is not defined.

These situations are illustrated in Figure 15 (on the next page).

Of the five situations, the "nicest" one is that given in Figure 15(a). There, not only does $\lim\limits_{x \to c} f(x)$ exist, but it is equal to $f(c)$. Functions that have this particular quality are

Figure 15

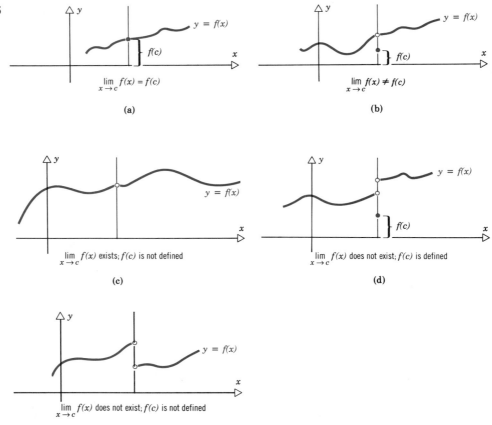

$$\lim_{x \to c} f(x) = f(c)$$

(a)

$$\lim_{x \to c} f(x) \neq f(c)$$

(b)

$\lim_{x \to c} f(x)$ exists; $f(c)$ is not defined

(c)

$\lim_{x \to c} f(x)$ does not exist; $f(c)$ is defined

(d)

$\lim_{x \to c} f(x)$ does not exist; $f(c)$ is not defined

(e)

said to be *continuous at c.* This is in agreement with the intuitive notion, usually given in elementary courses, which states ''a function is continuous if its graph can be drawn without lifting the pencil.'' The functions in Figure 15(b), (c), (d), and (e) are not continuous at c since each has a ''break'' in the graph at c. This leads us to state the following definition.

CONTINUOUS FUNCTION

Let $y = f(x)$ be a function. If

$$\lim_{x \to c} f(x) = f(c) \tag{1}$$

then the function is said to be *continuous at c.*

Thus, for continuity, the values of the function for x near c should be very close to the value of the function at c.

Conditions for a Function to Be Continuous at c

To summarize, a function f is continuous at c provided that three conditions are met:

Condition 1 $f(c)$ is defined; that is, c is in the domain of the function

Condition 2 $\lim\limits_{x \to c} f(x)$ exists

Condition 3 $\lim\limits_{x \to c} f(x) = f(c)$

If any one of these conditions is not met, then the function is said to be **discontinuous at c**. For example, in Figure 15 the graphs given in (b) through (e) are discontinuous at c: (b) does not satisfy Condition 3; (c) and (e) do not satisfy Condition 1; and (d) does not satisfy Condition 2.

Look back at equation (7) in the previous section (page 587). Based on (7) we conclude that

A polynomial function f is continuous at every real number.

For example, $f(x) = 3 - x$ and $g(x) = 3x^2 - 2x + 1$ are continuous functions since they are polynomials. Thus, the graph of a polynomial function will not contain any ''gaps'' or ''breaks.'' See Figure 16.

Figure 16

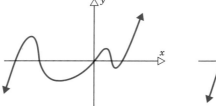

Graph of a polynomial

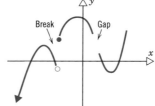

Cannot be the graph of a polynomial

EXAMPLE 1

Determine whether the function below is continuous at 3.

$$f(x) = \begin{cases} \dfrac{x^2 - 9}{x - 3} & \text{if } x \neq 3 \\ 6 & \text{if } x = 3 \end{cases}$$

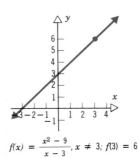

$f(x) = \dfrac{x^2 - 9}{x - 3}, x \neq 3; f(3) = 6$

Figure 17

SOLUTION

1. Since $f(3) = 6$, $f(3)$ is defined and Condition 1 is met.

2. $\lim\limits_{x \to 3} f(x) = \lim\limits_{x \to 3} \dfrac{x^2 - 9}{x - 3} = \lim\limits_{x \to 3} \dfrac{(x - 3)(x + 3)}{x - 3} = 6$

   Thus $\lim\limits_{x \to 3} f(x) = 6$ exists at 3 and Condition 2 is met.

3. Since $\lim\limits_{x \to 3} f(x) = f(3) = 6$, Condition 3 is met. Thus, f is continuous at 3.

   See Figure 17.

....

EXAMPLE 2   Determine whether the function below is continuous at 0.

$$f(x) = \begin{cases} x^2 + 1 & \text{if } x \neq 0 \\ 2 & \text{if } x = 0 \end{cases}$$

SOLUTION

1. $f(0) = 2$, so Condition 1 is met.
2. $\lim\limits_{x \to 0} f(x) = \lim\limits_{x \to 0} (x^2 + 1) = 1$, so Condition 2 is met.
3. Since $\lim\limits_{x \to 0} f(x) = 1$ and $f(0) = 2$, Condition 3 is *not* met. Thus, f is not continuous at 0.

   See Figure 18.

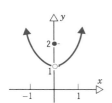

Figure 18

Now Work Problem 3

....

In Example 2 the function is discontinuous at 0 because its value at 0 does not equal 1. If we were to change the definition of f so that $f(0) = 1$, then f would be continuous at 0.

It can be shown that for rational functions (quotients of polynomial functions),

A rational function $R(x) = P(x)/Q(x)$ is continuous at every number except the numbers that make Q zero.

For example, $R(x) = \dfrac{x - 3}{x^2 - 4}$ is continuous everywhere except at $x = -2$ and $x = 2$.

EXAMPLE 3   Determine whether the function below is continuous at 2.

$$f(x) = \dfrac{x^3 - 2x^2}{x - 2}$$

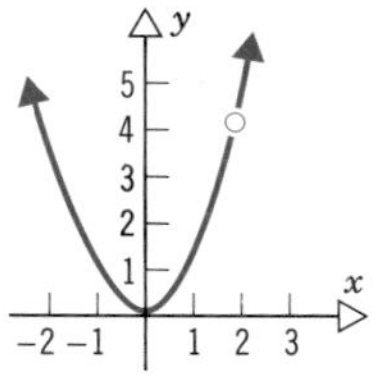

SOLUTION

Since $f(2)$ is not defined (2 is not in the domain of f), Condition 1 is not met. Thus, f is not continuous at 2. See Figure 19.

Figure 19

....

Now Work Problem 17   You should verify that if f were to be defined so that $f(2) = 4$, then f would be continuous at 2.

EXAMPLE 4

Determine the value of the constant k that will make the following function continuous for all values of x.

$$f(x) = \begin{cases} kx + 7 & \text{if } x < 3 \\ x^2 - 3x - 8 & \text{if } x \geq 3 \end{cases}$$

SOLUTION

Note that both $kx + 7$ and $x^2 - 3x - 8$ are polynomials, and therefore, $f(x)$ is continuous everywhere except possibly at $x = 3$. Also, note that

$$\lim_{x \to 3^-} f(x) = \lim_{x \to 3^-} (kx + 7) = 3k + 7$$

and

$$\lim_{x \to 3^+} f(x) = \lim_{x \to 3^+} (x^2 - 3x - 8) = -8$$

Thus, for $\lim_{x \to 3} f(x)$ to exist, we must have $3k + 7 = -8$ or $k = -5$, which means that $f(x)$ is continuous at $x = 3$ only when $k = -5$.

· ·

EXAMPLE 5

Determine whether the function below is continuous at 1.

$$f(x) = \begin{cases} 3x + 1 & \text{if } x < 1 \\ 2 & \text{if } x = 1 \\ -x + 2 & \text{if } x > 1 \end{cases}$$

SOLUTION

1. $f(1) = 2$, so Condition 1 is met.
2. To determine if $\lim_{x \to 1} f(x)$ exists, we need to look at the left limit and the right limit since the rule for f changes depending on whether $x < 1$ or $x > 1$. If $x < 1$, then $f(x) = 3x + 1$, so that

$$\lim_{x \to 1^-} f(x) = \lim_{x \to 1^-} (3x + 1) = 4$$

If $x > 1$, then $f(x) = -x + 2$, so that

$$\lim_{x \to 1^+} f(x) = \lim_{x \to 1^+} (-x + 2) = 1$$

Since the left and right limits are unequal, it follows that $\lim_{x \to 1} f(x)$ does not exist. Thus, Condition 2 is not met. As a result, f is discontinuous at 1.

See Figure 20 for the graph.

· ·

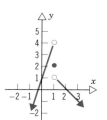

Figure 20

In Example 5 note that there is no way to define the function f at 1 to make it continuous at 1.

EXAMPLE 6 | The 1995 U.S. postage rates for first class mail weighing 8 ounces or less* are listed in the table. Graph the function describing this rate structure and determine where it is discontinuous.

Weight (ounces)	$0 < x \leq 1$	$1 < x \leq 2$	$2 < x \leq 3$	$3 < x \leq 4$	$4 < x \leq 5$	$5 < x \leq 6$	$6 < x \leq 7$	$7 < x \leq 8$
Cost (dollars)	0.32	0.55	0.78	1.01	1.24	1.47	1.70	1.93

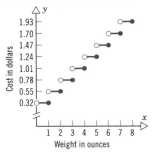

Figure 21

SOLUTION

For mail weighing up to and including 1 ounce, the charge is 32 cents. For mail weighing over 1 ounce up to and including 2 ounces, the charge is 55 cents. This pattern continues, as shown in the table. See Figure 21 for the graph.

Notice that the domain of this function is $0 < x \leq 8$, and the range is the set $\{0.32, 0.55, 0.78, \ldots, 1.93\}$. The function shown in Figure 21 is discontinuous at 1, 2, 3, 4, 5, 6, 7 and is continuous at every other number in its domain.

· ·

EXERCISE 12.3   Answers to odd-numbered problems begin on page AN-56.

In Problems 1–14 determine whether the function f is continuous at c.

1. $f(x) = \begin{cases} 3x^2 + x & \text{if } x \neq 1 \\ 4 & \text{if } x = 1 \end{cases}$ $c = 1$

2. $f(x) = \begin{cases} 1 - 3x^2 & \text{if } x \neq 0 \\ 1 & \text{if } x = 0 \end{cases}$ $c = 0$

3. $f(x) = \begin{cases} \dfrac{x^2 - 4}{x - 2} & \text{if } x \neq 2 \\ 0 & \text{if } x = 2 \end{cases}$ $c = 2$

4. $f(x) = \begin{cases} \dfrac{x^2 - 4x}{x - 4} & \text{if } x \neq 4 \\ 0 & \text{if } x = 4 \end{cases}$ $c = 4$

5. $f(x) = \begin{cases} 2x + 5 & \text{if } x \leq 2 \\ 4x + 1 & \text{if } x > 2 \end{cases}$ $c = 2$

6. $f(x) = \begin{cases} 2x + 1 & \text{if } x \leq 0 \\ 2x & \text{if } x > 0 \end{cases}$ $c = 0$

7. $f(x) = \begin{cases} 3x - 1 & \text{if } x < 1 \\ 4 & \text{if } x = 1 \\ 2x & \text{if } x > 1 \end{cases}$ $c = 1$

8. $f(x) = \begin{cases} 3x - 1 & \text{if } x < 1 \\ 2 & \text{if } x = 1 \\ 2x & \text{if } x > 1 \end{cases}$ $c = 1$

9. $f(x) = \begin{cases} 3x - 1 & \text{if } x < 1 \\ \text{Not defined} & \text{if } x = 1 \\ 2x & \text{if } x > 1 \end{cases}$ $c = 1$

10. $f(x) = \begin{cases} 3x - 1 & \text{if } x < 1 \\ 2 & \text{if } x = 1 \\ 3x & \text{if } x > 1 \end{cases}$ $c = 1$

11. $f(x) = \begin{cases} x^2 & \text{if } x \leq 0 \\ 2x & \text{if } x > 0 \end{cases}$ $c = 0$

12. $f(x) = \begin{cases} x^2 & \text{if } x < -1 \\ 2 & \text{if } x = -1 \\ -3x + 2 & \text{if } x > -1 \end{cases}$ $c = -1$

* *Source: Postal Bulletin, 1995.*

13. $f(x) = \begin{cases} 4 - 3x^2 & \text{if } x < 0 \\ 4 & \text{if } x = 0 \\ \sqrt{16 - x^2} & \text{if } 0 < x < 4 \end{cases}$ $c = 0$

14. $f(x) = \begin{cases} \sqrt{4 + x} & \text{if } x \le 4 \\ \sqrt{\dfrac{x^2 - 16}{x - 4}} & \text{if } x > 4 \end{cases}$ $c = 4$

In Problems 15 and 16 determine the value of the constant k that will make the function f continuous for all x.

15. $f(x) = \begin{cases} kx - 2 & \text{if } x \le 1 \\ -3x^2 + 2x + 7 & \text{if } 1 < x \end{cases}$

16. $f(x) = \begin{cases} 1 - 4x & \text{if } x < 2 \\ kx^2 - 3x + 2 & \text{if } 2 \le x \end{cases}$

17. Is the function f defined by $f(x) = \dfrac{x^2 - 4}{x - 2}$ continuous at 2? If not, can f be redefined at 2 to make it continuous at 2?

18. Is the function f defined by $f(x) = \dfrac{x^2 + x - 12}{x - 3}$ continuous at 3? If not, can f be redefined at 3 to make it continuous at 3?

19. Suppose that f is continuous at $x = 2$ and that $f(2) = 3$. Evaluate $\lim_{x \to 2} f(x)$.

20. Suppose that f is continuous at $x = 2$ and that $\lim_{x \to 2} f(x) = 13$. Which of the following statements is true?

(a) 2 is in the domain of f.
(b) $f(2) = 13$
(c) $\lim_{x \to 2^+} f(x) = 13$

21. Processing Oil An oil refinery has four distillation towers and operates them as they are needed to process available raw materials. Each tower has fixed costs of $500 per week whether operating or not. In addition, each tower, if in operation, will incur fixed costs of $3000 per week. The raw material cost is fixed at $1.00 per gallon of refined oil, and each tower can process at most 10,000 gallons of refined oil each week.

(a) Find the cost function $C(x)$, where x is the number of gallons of refined oil for the refinery.
(b) Find the domain of this function and graph it.
(c) Determine where the function is discontinuous.

22. Quantity Discounts The owner of a grocery store can buy bulk cashews from a particular distributor according to the following price schedule:

$3.00 per pound for 5 pounds or less

$2.50 per pound for more than 5 pounds but less than 10 pounds

$2.25 per pound for 10 or more pounds but less than 20 pounds

$2.00 per pound for 20 or more pounds

(a) Find a cost function $C(x)$, where $C(x)$ represents the cost of buying x pounds.
(b) Graph the cost function.
(c) Find those values of x where the function is discontinuous.

23. Shipping Rate The 1995 Federal Express rates for standard overnight service are given in the table at the bottom of the page. Graph the function describing the cost and determine where it is discontinuous.

24. Estimating Zeros A number r is said to be a **zero** of the function $y = f(x)$ if $f(r) = 0$. Geometrically, this is a point where the function crosses or touches the x-axis. If a continuous function changes sign on an interval, then it must *cross* the x-axis at least once on that interval. This result, which is intuitively obvious, can be used to estimate the zeros of functions.

If at $x = a$, $f(a) < 0$, and if at $x = b$, $f(b) > 0$, then there is a zero r of the function in the interval (a, b). To estimate r, we construct the line joining the two points

Table for Problem 23

Weight (pounds)	$0 < x \le 1$	$1 < x \le 2$	$2 < x \le 3$	$3 < x \le 4$	$4 < x \le 5$	$5 < x \le 6$
Cost (dollars)	15.50	16.50	17.50	18.50	19.50	21.25

$(a, f(a))$ and $(b, \ f(b))$. The place where this line crosses the x-axis will be our estimate of the zero r of the function. See the figure.

Using this result, estimate the zeros of the following functions between the values given:

(a) $f(x) = x^2 - 5$; $\quad a = 2$, $\quad b = 3$

(b) $f(x) = x^2 + \frac{3}{2}x - 1$; $\quad a = 0$, $\quad b = 1$

(c) $f(x) = x^3 + x^2 - 2x - 2$; $\quad a = 1, b = 2$

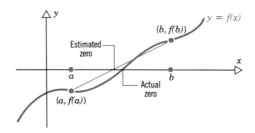

Technology Exercises

In the next set of problems we try to determine the continuity of a function using the graph of a function. Most graphing calculators and computer packages can draw graphs in two different modes: connected mode and dot mode. The connected mode works well when the function is continuous. The dot mode should be used when we suspect that the function is discontinuous. In Problems 1–6 use the graph of the given function to determine the continuity of the function.

1. $f(x) = \dfrac{x + 5}{x - 3}$

2. $f(x) = \dfrac{x^3 - 2x^2 + x - 2}{3x - 6}$

3. $f(x) = \ln x^2$

4. $f(x) = \ln(x^2 + 1)$

5. $f(x) = \dfrac{x + 3}{x^2 - 9}$

6. $f(x) = \dfrac{x - 3}{x^2 - 9}$

12.4 THE TANGENT PROBLEM; THE DERIVATIVE

Figure 22 gives the graph of a function f. Notice that we have drawn a line L_T that just *touches* the graph of f at the point P on f. The line L_T is the *tangent line* to the graph of f at P. But how should we define it?

Let's start with the notion of the tangent line to a circle. In plane geometry the line that intersects a circle in exactly one point P is defined as the **tangent line to the circle** at P. See Figure 23.

This definition, though, will not do for graphs in general. See, for example, Figure 24. Note that there are many lines, such as L_1, L_2, and L_3, which intersect the graph of f

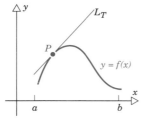

Figure 22

Figure 23

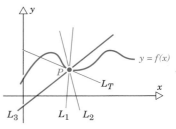

Figure 24

in one point P, but none of these lines is the tangent line. Note also that the tangent line L_T at P, in fact, intersects the graph of f elsewhere. We conclude that we require a more general definition of tangent line from the one used for circles in plane geometry.

We begin with the graph of a function f. See Figure 25. Suppose the coordinates of the point P on the graph of f are $(c, f(c))$. The tangent line L_T to the graph of f at P necessarily passes through P. To distinguish L_T from all other lines that pass through P, we need to know its slope, which we will call m_{tan}. To find m_{tan}, we locate another point, $Q = (x, f(x))$, different from P, on the graph of f. The line containing P and Q is called a **secant line**; its slope, denoted by m_{sec}, is

$$m_{sec} = \frac{f(x) - f(c)}{x - c}$$

Figure 25

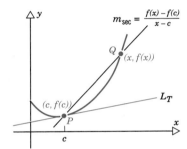

Now look at Figure 26. There we have drawn three different secant lines, L_1, L_2, and L_3. As we move along the graph of f from Q_1 to Q_2 to Q_3, the x-coordinate of these points approaches c. Furthermore, the eventual (limiting) position of these secant lines is the *tangent line to the graph of f at P*. Therefore, it follows that the limiting value of the slope of these secant lines is the slope of the tangent line. That is,

$$m_{tan} = \lim_{x \to c} m_{sec}$$

$$= \lim_{x \to c} \frac{f(x) - f(c)}{x - c}$$

Figure 26

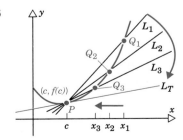

TANGENT LINE

We define the *tangent line* to the graph of f at c to be the line passing through the point $(c, f(c))$ and having the slope

$$m_{\text{tan}} = \lim_{x \to c} \frac{f(x) - f(c)}{x - c} \tag{1}$$

provided this limit exists.

To find the equation of the tangent line to the graph of $y = f(x)$ at $(c, f(c))$, we use the point–slope form for the equation of a line, namely,

$$y - y_1 = m(x - x_1)$$

Equation of the Tangent Line

An equation of the tangent line to the graph of $y = f(x)$ at c is

$$y - f(c) = m_{\text{tan}}(x - c) \tag{2}$$

provided

$$m_{\text{tan}} = \lim_{x \to c} \frac{f(x) - f(c)}{x - c}$$

exists.

EXAMPLE 1

Find the slope of the tangent line to the graph of $f(x) = 2x^2$ at $(1, 2)$. What is the equation of the tangent line? Graph the function and show its tangent line.

SOLUTION

$$m_{\text{tan}} = \lim_{x \to 1} \frac{f(x) - f(1)}{x - 1} = \lim_{x \to 1} \frac{2x^2 - 2}{x - 1}$$

$$= \lim_{x \to 1} \frac{2(x - 1)(x + 1)}{x - 1} = \lim_{x \to 1} [2(x + 1)] = 4$$

Using (2), we find an equation of this tangent line to be

$$y - 2 = 4(x - 1)$$
$$y = 4x - 2$$

The graphs of $y = 2x^2$ and its tangent line at $(1, 2)$ are shown in Figure 27.

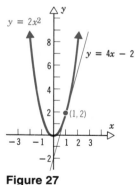

Figure 27

Now Work Problem 1

· ·

Why Do We Study Tangent Lines?

The need to study tangent lines will become more evident as we proceed with the study of calculus. For now we can illustrate this need by looking at Figure 28. The graph illustrates a total profit function. The largest (maximum) profit occurs at a point when the graph has a horizontal tangent line; that is, where the tangent line has slope $m_{\text{tan}} = 0$.

Figure 28

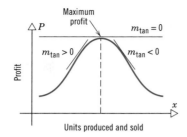

Note that profit is increasing up to the point where the tangent line has slope 0; on this interval, the tangent lines all have positive slope. From then on, the profit is decreasing and the tangent lines have negative slope.

Derivative of a Function

The limit in (1) has an important generalization:

DERIVATIVE OF A FUNCTION

Let $y = f(x)$ be a function and let c be in the domain of f. The *derivative of f at c*, denoted by $f'(c)$, and read "f prime of c," is the number

$$f'(c) = \lim_{x \to c} \frac{f(x) - f(c)}{x - c} \tag{3}$$

provided this limit exists.

The calculation of the derivative of a function f is done by following the steps outlined below:

Steps to Compute Derivatives

Step 1 Find $f(c)$.

Step 2 Subtract $f(c)$ from $f(x)$ to get $f(x) - f(c)$ and form the quotient

$$\frac{f(x) - f(c)}{x - c}$$

Step 3 Find the limit (if it exists) of the quotient found in Step 2 as $x \to c$:

$$f'(c) = \lim_{x \to c} \frac{f(x) - f(c)}{x - c}$$

EXAMPLE 2

Find $f'(2)$ if $f(x) = x^2 + 2x$.

SOLUTION

Step 1 $f(2) = 2^2 + 2(2) = 8$

Step 2 $\dfrac{f(x) - f(2)}{x - 2} = \dfrac{x^2 + 2x - 8}{x - 2}$

Step 3 $f'(2) = \lim\limits_{x \to 2} \dfrac{f(x) - f(2)}{x - 2} = \lim\limits_{x \to 2} \dfrac{x^2 + 2x - 8}{x - 2} = \lim\limits_{x \to 2} \dfrac{(x - 2)(x + 4)}{x - 2} = 6$

Hence, $f'(2) = 6$.

· ·

EXAMPLE 3

Find the derivative of $f(x) = x^3$ at 2.

SOLUTION

We go directly to Step 3: Since $f(2) = 8$, we have

$$f'(2) = \lim_{x \to 2} \frac{f(x) - f(2)}{x - 2} = \lim_{x \to 2} \frac{x^3 - 8}{x - 2} = \lim_{x \to 2} \frac{(x - 2)(x^2 + 2x + 4)}{x - 2} = 12$$

Now Work Problem 11

· ·

In Example 3 we calculated the derivative of $f(x) = x^3$ at 2. It is often just as easy to find the derivative at an arbitrary number c, as the next example illustrates.

EXAMPLE 4

Find the derivative of $f(x) = x^3$ at c.

SOLUTION

Since $f(c) = c^3$, we have

$$f'(c) = \lim_{x \to c} \frac{f(x) - f(c)}{x - c} = \lim_{x \to c} \frac{x^3 - c^3}{x - c} = \lim_{x \to c} \frac{(x - c)(x^2 + cx + c^2)}{x - c}$$

$$= \lim_{x \to c} (x^2 + cx + c^2) = 3c^2$$

· ·

Notice that for $c = 2$, $f'(c) = 3c^2 = (3)(2^2) = 12$, which agrees with the answer found in Example 3.

Since the limit

$$f'(c) = \lim_{x \to c} \frac{f(x) - f(c)}{x - c} = 3c^2$$

exists for any choice of c, it is convenient to replace c by x and write $f'(x) = 3x^2$. Thus, the derivative $f'(x)$ of a function f at x is itself a function. This function f' is called the **derivative of f**. We also say that f is "differentiable." Thus, "differentiate f" means the same as "find the derivative of f."

Formula (3) for the derivative of a function f at c, namely,

$$f'(c) = \lim_{x \to c} \frac{f(x) - f(c)}{x - c}$$

may be written in another form. See Figure 29.

Figure 29

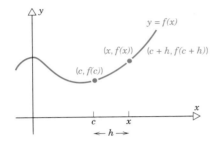

If we let h be the distance from c to x, then $h = x - c$, from which we find $x = c + h$. The point $(x, f(x))$ on the graph of f takes the form $(c + h, f(c + h))$. The slope of the secant line may be written as

$$m_{\text{sec}} = \frac{f(x) - f(c)}{x - c} \qquad \text{or as} \qquad m_{\text{sec}} = \frac{f(c + h) - f(c)}{h}$$

Since $x \to c$ is equivalent to $h \to 0$, we have

$$m_{\text{tan}} = \lim_{x \to c} \frac{f(x) - f(c)}{x - c} \qquad \text{or} \qquad m_{\text{tan}} = \lim_{h \to 0} \frac{f(c + h) - f(c)}{h}$$

Since $m_{\text{tan}} = f'(c)$, it follows that

$$f'(c) = \lim_{h \to 0} \frac{f(c + h) - f(c)}{h} \tag{4}$$

Now replace c by x in (4). This gives us the following formula for finding the derivative of f at any number x.

Formula for the Derivative of f at x

$$f'(x) = \lim_{h \to 0} \frac{f(x + h) - f(x)}{h} \tag{5}$$

Based on (5), the derivative of a function is the limit of a difference quotient.

EXAMPLE 5

(a) Use Formula (5) to find the derivative of $f(x) = x^2 + 2x$.

(b) Find $f'(0), f'(-1), f'(3)$.

SOLUTION

(a) First, we compute the difference quotient.

$$\frac{f(x+h)-f(x)}{h} = \frac{(x+h)^2 + 2(x+h) - (x^2 + 2x)}{h}$$

$$= \frac{[x^2 + 2xh + h^2 + 2x + 2h] - x^2 - 2x}{h}$$

$$= \frac{2xh + h^2 + 2h}{h}$$

$$= \frac{h(2x + h + 2)}{h}$$

$$= 2x + h + 2$$

The derivative of f is the limit of the difference quotient as $h \to 0$. That is,

$$f'(x) = \lim_{h \to 0} \frac{f(x+h)-f(x)}{h} = \lim_{h \to 0} (2x + h + 2) = 2x + 2$$

Thus, $f'(x) = 2x + 2$

(b) The result of (a) gives the derivative of f at any x, namely,

$$f'(x) = 2x + 2$$

Thus,

$$f'(0) = 2 \cdot 0 + 2 = 2$$
$$f'(-1) = 2(-1) + 2 = 0$$
$$f'(3) = 2(3) + 2 = 8$$

Now Work Problem 21

Example 5 illustrates again that the derivative at a point is the slope of the tangent line to the function at that point. Hence, when the point changes, so does the derivative.

Cases for Which $f'(x)$ Does Not Exist

In the preceding discussion we defined differentiability of $f(x)$ at $x = c$ in terms of a limit. If this limit does not exist, then we say that the function fails to have a derivative at $x = c$, or the *derivative does not exist at $x = c$*.

Let's look at two cases for which the derivative does not exist at a number. Suppose we try to find the derivative of $f(x) = |x|$ at $x = 0$. Note that $f(0) = |0| = 0$, so the function is defined at $x = 0$.

Since

$$f'(x) = \lim_{h \to 0} \frac{f(x+h)-f(x)}{h}$$

$$= \lim_{h \to 0} \frac{|x+h| - |x|}{h}$$

at $x = 0$ we have

$$f'(0) = \lim_{h \to 0} \frac{|0 + h| - |0|}{h}$$

$$= \lim_{h \to 0} \frac{|h|}{h}$$

Table 2 summarizes the behavior of $|h|/h$ close to 0:

Table 2

				From the left						From the right	
h	-2	-1	-0.1	-0.01	-0.001	h	2	1	0.1	0.01	0.001
$\dfrac{|h|}{h}$	$\dfrac{2}{-2} = -1$	$\dfrac{1}{-1} = -1$	-1	-1	-1	$\dfrac{|h|}{h}$	$\dfrac{2}{2} = 1$	$\dfrac{1}{1} = 1$	1	1	1

Note that as h approaches 0 from the right, $|h|/h$ approaches 1. As h approaches 0 from the left, $|h|/h$ approaches -1. Thus,

$$\lim_{h \to 0} \frac{|h|}{h} \quad \text{does not exist}$$

so

$$f'(0) \quad \text{does not exist}$$

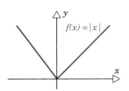

Figure 30

See Figure 30 for the graph of $f(x) = |x|$ at $x = 0$. Note that the graph has a **corner** there.

As a second example, let us find the derivative of $f(x) = x^{2/3}$ at $x = 0$. Note that $f(0) = 0^{2/3} = 0$, so the function is defined at $x = 0$. Since

$$f'(x) = \lim_{h \to 0} \frac{(x + h)^{2/3} - x^{2/3}}{h}$$

$$f'(0) = \lim_{h \to 0} \frac{(0 + h)^{2/3} - 0^{2/3}}{h} = \lim_{h \to 0} \frac{h^{2/3}}{h} = \lim_{h \to 0} \frac{1}{h^{1/3}}$$

But as $h \to 0$, $h^{1/3} \to 0$ and $\dfrac{1}{h^{1/3}}$ gets very large. Thus,

$$\lim_{h \to 0} \frac{1}{h^{1/3}} \quad \text{does not exist}$$

so

$$f'(0) \quad \text{does not exist}$$

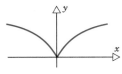

Figure 31 $f(x) = x^{2/3}$

Look at the graph of $f(x) = x^{2/3}$. Notice how the graph of $f(x)$ comes to a **sharp point** at $x = 0$. See Figure 31.

Figure 32 (on the next page) illustrates the graphs of some functions that do not have tangent lines at $x = c$, and, hence, the derivative does not exist at $x = c$.

(i)   The function is not defined, as in Figure 32(a).
(ii) The graph of the function has a "sharp point" or "corner," as in Figure 32(b), (c), and (d).

(iii) The graph has a vertical tangent line (recall that slope is not defined for vertical lines), as in Figure 32(c) and (e).

Figure 32

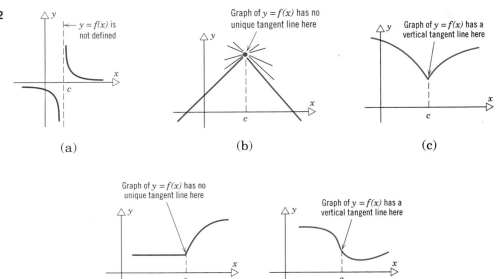

(a)   (b)   (c)

(d)   (e)

Differentiability and Continuity

Note that the functions whose graphs are given in Figure 32(b), (c), (d), and (e) are all continuous at $x = c$, yet the derivative does not exist at $x = c$. That is:

Continuity does not imply differentiability.

However, it can be shown that:

If the derivative of a function exists at $x = c$, then $f(x)$ is continuous at $x = c$.

Intuitively, this means that a graph can have a derivative at a point only if the graph is "smooth" in the vicinity of the point.

EXERCISE 12.4   Answers to odd-numbered problems begin on page AN-56.

In Problems 1–10 find the slope of the tangent line to the graph of f at the given point. Then find an equation of this tangent line. For Problems 1–4 also graph f and the tangent line.

1. $f(x) = 2x^2$ at $(-1, 2)$

2. $f(x) = x^2 + 4$ at $(1, 5)$

3. $f(x) = x^2 - 3$ at $(-2, 1)$

4. $f(x) = 3x^2$ at $(-1, 3)$

5. $f(x) = x^2 + x$ at $(2, 6)$

6. $f(x) = x^2 + 3x$ at $(2, 10)$

7. $f(x) = -x^2 + 4x$ at $(1, 3)$

8. $f(x) = -x^2 - x$ at $(-2, -2)$

9. $f(x) = 2x^2 + x - 1$ at $(2, 9)$

10. $f(x) = 3x^2 - 2x + 1$ at $(0, 1)$

In Problems 11–20 find the derivative of f at the given number. Use Formula (3).

11. $f(x) = 3x^2$ at 2

12. $f(x) = 2x^2$ at -1

13. $f(x) = 2x^2 + 1$ at 1

14. $f(x) = 3x^2 + 4$ at 2

15. $f(x) = -x^2 + 2x - 1$ at -1

16. $f(x) = -2x^2 - x + 1$ at 1

17. $f(x) = x^3 + 1$ at 2

18. $f(x) = 2x^3 + 1$ at -1

19. $f(x) = x^4 + 2x + 1$ at 0

20. $f(x) = x^5 - 3x + 1$ at 0

In Problems 21–30 find the derivative of f. Use formula (5).

21. $f(x) = 2x - 4$

22. $f(x) = 3x + 5$

23. $f(x) = x^2 + 2$

24. $f(x) = 2x^2 - 3$

25. $f(x) = 3x^2 - 2x + 1$

26. $f(x) = 2x^2 + x + 1$

27. $f(x) = x^3 - 1$

28. $f(x) = x^3 + 1$

29. $f(x) = mx + b$

30. $f(x) = ax^2 + bx + c$

31. Does the tangent line to the graph of $y = x^2$ at $(1, 1)$ pass through the point $(2, 5)$?

32. Does the tangent line to the graph of $y = x^3$ at $(1, 1)$ pass through the point $(2, 5)$?

33. A dive bomber is flying from right to left along the graph of $y = x^2$. When a rocket bomb is released, it follows a path that approximately follows the tangent line. Where should the pilot release the bomb if the target is at $(1, 0)$?

34. Answer the question in Problem 33 if the plane is flying from right to left along the graph of $y = x^3$.

35. Use the graph of the function f below to determine the points at which the derivative does not exist.

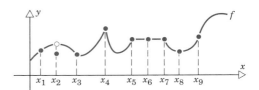

Technology Exercises

1. Graph the function $f(x) = \sqrt{x^2 + 0.01}$ within the x range $[-20, 20]$. Does $f(x)$ have a derivative at $x = 0$? What if you zoom in within $[-0.01, 0.01]$?

2. Graph the functions $f(x) = x^3$, $g(x) = f(x) - 3$, and $h(x) = f(x) + 3$ on the same set of axes. What can you say about the slopes of the tangent lines to the three graphs at the point $x = 0$? $x = -1$? Any point?

3. Use a graphing calculator to estimate the derivative of $f(x) = x^x$ at $x = 2$.

4. For $f(x) = 3x^{3/2} - x$, use a graphing calculator to construct a graph of $f(x)$ for x between 0 and 2. From your graph, estimate $f'(0)$ and $f'(1)$.

5. (a) Using a tangent line to e^x at $x = 0$, show that

$$e^x \geq 1 + x$$

for all values of x.

(b) Let $x = 1/n$, and show that for every positive integer n,

$$e > \left(1 + \frac{1}{n}\right)^n$$

(c) Consider $x = -\dfrac{1}{n+1}$, and show that

$$e < \left(1 + \frac{1}{n}\right)^{n+1}$$

for all positive integers n.

(d) Use a calculator with some specific choices of n to prove that $2.7 < e < 2.8$.

12.5 THE DERIVATIVE AS AN INSTANTANEOUS VELOCITY

We have given a geometrical interpretation of the derivative as being the slope of the tangent line to the graph at a point. The purpose of this section is to introduce an important physical interpretation of the derivative. The discussion begins with a familiar practical situation, which will serve as a model for the general case.

Average Velocity

EXAMPLE 1

Mr. Jones and his family leave on a car trip Saturday at 5 A.M. and arrive at their destination at 11 A.M. When they began the trip, the car's odometer read 26,700 kilometers, and when they arrived, it read 27,000 kilometers. What was the average velocity for the trip?

SOLUTION

Recall the formula $s = vt$, which can be written in the form

$$v = \frac{s}{t} = \frac{\text{Total distance}}{\text{Elapsed time}} = \text{Average rate}$$

The distance in this case is $27,000 - 26,700 = 300$ kilometers, and the elapsed time is $11 - 5 = 6$ hours. Thus, the average velocity is

$$\text{Average velocity} = \frac{300}{6} = 50 \text{ kilometers per hour}$$

Sometimes Mr. Jones will be traveling faster and sometimes slower, but the *average* rate is 50 kilometers per hour.

· ·

AVERAGE VELOCITY

The *average velocity* is the ratio of the change in distance to the change in time. If s denotes distance and t denotes time, we have

$$\text{Average velocity} = \frac{\text{Total distance}}{\text{Elapsed time}} = \frac{\Delta s}{\Delta t}$$

Instantaneous Velocity

More important than the average velocity is the instantaneous velocity. In order to see how this definition arises naturally, we use the following example involving the velocity of a falling body.

The function $s = f(t) = 16t^2$ can be used to measure the distance s (in feet) an object dropped from a tall building has traveled after a time t (in seconds). When $0 \leq t \leq 6$, we calculate the average velocity of the object over the entire time interval of 6 seconds to be

$$\frac{\Delta s}{\Delta t} = \frac{f(6) - f(0)}{6 - 0} = \frac{16(6)^2 - 16(0)^2}{6} = 96 \text{ feet per second}$$

This average accurately describes the velocity of the object over the 6-second interval, but it gives no information at all about the actual velocity at any particular instant of time. We now want to find the velocity of the object at a particular instant.

To see how this might be done, we will seek the exact velocity of the object at the instant when $t = 3$. This velocity is called the *instantaneous velocity at 3*.

So far, we have no *mathematical* method for finding instantaneous velocities. However, we can *estimate* the instantaneous velocity at $t = 3$ seconds by computing the average velocity for some intervals of time beginning at $t = 3$. For example, let's compute the average velocity for the 1-second interval beginning at $t = 3$ and ending at $t = 4$. The distance of the object from the starting position is

$$s = f(3) = 16(9) = 144 \text{ feet}$$

At $t = 4$,

$$s = f(4) = 16(16) = 256 \text{ feet}$$

Thus, over the 1-second interval from $t = 3$ to $t = 4$,

$$\text{Average velocity} = \frac{\Delta s}{\Delta t} = \frac{f(4) - f(3)}{4 - 3} = \frac{256 - 144}{1} = 112 \text{ feet per second}$$

The average velocity for the smaller intervals of time $\Delta t = 0.5, 0.1, 0.01, 0.0001$ may be found similarly. Table 3 gives the five estimates obtained for the instantaneous velocity.

Table 3

Start	End	Δt	Average Velocity (feet per second) $\frac{\Delta s}{\Delta t} = \frac{f(t) - f(3)}{t - 3}$
3	4	1	$\frac{\Delta s}{\Delta t} = \frac{f(4) - f(3)}{1} = \frac{16(4)^2 - 16(9)}{1} = 112$
3	3.5	0.5	$\frac{\Delta s}{\Delta t} = \frac{f(3.5) - f(3)}{0.5} = \frac{16(3.5)^2 - 16(9)}{0.5} = 104$
3	3.1	0.1	$\frac{\Delta s}{\Delta t} = \frac{f(3.1) - f(3)}{0.1} = \frac{16(3.1)^2 - 16(9)}{0.1} = 97.6$
3	3.01	0.01	$\frac{\Delta s}{\Delta t} = \frac{f(3.01) - f(3)}{0.01} = \frac{16(3.01)^2 - 16(9)}{0.01} = 96.16$
3	3.0001	0.0001	$\frac{\Delta s}{\Delta t} = \frac{f(3.0001) - f(3)}{0.0001} = \frac{16(3.0001)^2 - 16(9)}{0.0001} = 96.0016$

We can see from the table that for this example, the larger the time interval Δt, the larger the average velocity $\Delta s/\Delta t$. The most accurate estimates for instantaneous velocity will correspond to very small time intervals Δt. For example, over the interval $\Delta t = 0.0001$ second, we would not expect the velocity of the object to change very much. Thus, the average velocity of 96.0016 feet per second during the very short time interval $\Delta t = 0.0001$ should be very close to the instantaneous velocity at $t = 3$.

But what is the exact or instantaneous velocity at $t = 3$? It must be close to 96.0016, but is it 96.0016? Or is it 96.0001? Or what?

To obtain the precise answer, we first use some algebra. Specifically, we find the average velocity for the object in the time interval that begins at 3 and ends at t, where $t \neq 3$ is close to 3:

$$f(3) = 16(9) = 144 \qquad f(t) = 16t^2$$

Thus,

$$\text{Average velocity} = \frac{\Delta s}{\Delta t} = \frac{f(t) - f(3)}{t - 3} = \frac{16t^2 - 144}{t - 3} = \frac{16(t^2 - 9)}{t - 3}$$

$$= \frac{16(t - 3)(t + 3)}{t - 3}$$

Since $t \neq 3$, we may cancel $(t - 3)$ in the numerator and denominator to get

$$\frac{\Delta s}{\Delta t} = 16(t + 3)$$

We are now at the important step in this procedure. As t gets closer and closer to 3, but not equal to 3, the values of $\Delta s/\Delta t = 16(t + 3)$ get closer to 96. This is apparent since we can make $\Delta s/\Delta t = 16(t + 3)$ as close as we please to 96 by taking t sufficiently close to 3.

Using the terminology we introduced earlier in this chapter, 96 is called the *limit* of the average velocity $\Delta s/\Delta t$ as Δt approaches 0, or equivalently, as $t \to 3$. In symbols we write

$$\lim_{\Delta t \to 0} \frac{\Delta s}{\Delta t} = \lim_{t \to 3} \frac{f(t) - f(3)}{t - 3} = \lim_{t \to 3} [16(t + 3)] = 96$$

Intuition tells us that the limit 96 feet per second is what we mean by the (instantaneous) velocity of the object at time $t = 3$ seconds. The limit on the right is simply the derivative of f with respect to t at $t_0 = 3$. Thus, the instantaneous velocity of the object at $t_0 = 3$ is just $f'(t_0)$ and we may write

$$v = f'(t_0) = \lim_{t \to t_0} \frac{f(t) - f(t_0)}{t - t_0}$$

INSTANTANEOUS VELOCITY

In general if $s = f(t)$ is a function that describes the distance s a particle travels in time t, the *(instantaneous) velocity* v of the particle at time t_0 is defined as the limit

of the average rate of change $\Delta s/\Delta t$ as Δt approaches 0. Specifically, the velocity v at time t_0 is

$$v = f'(t_0) = \lim_{\Delta t \to 0} \frac{\Delta s}{\Delta t} = \lim_{t \to t_0} \frac{f(t) - f(t_0)}{t - t_0}$$

provided this limit exists.

EXERCISE 12.5   Answers to odd-numbered problems begin on page AN-57.

---

1. The function $s = f(t) = 6t(t + 1)$ relates the distance s in kilometers a car travels in time t (in hours). Compute the car's average velocity from t_0 to t_1 for

   (a) $t_0 = 2, \quad t_1 = 3$ (b) $t_0 = 2, \quad t_1 = 2.5$

2. Follow the directions in Problem 1 for

   (a) $t_0 = 2, \quad t_1 = 2.1$ (b) $t_0 = 2, \quad t_1 = 2.01$

3. **Velocity**   Suppose the function $s = f(t) = 16t^2$ relates the distance s (in feet) an object travels in time t (in seconds). Compute the average velocity, $\Delta s/\Delta t$, from $t = 3$ to

   (a) $t = 3.5$ (b) $t = 3.1$

4. **Velocity**   Follow the directions in Problem 3 if the change in time is from $t = 3$ to

   (a) $t = 3.01$ (b) $t = 3.0001$

5. **Velocity**   The distance s (in meters) that a particle moves in time t (in seconds) is given by $s = f(t) = 3t^2 + 4t$. Find the velocity at $t = 0$. At $t = 2$. At any time t.

6. **Velocity**   The distance s (in meters) that a particle moves in time t (in seconds) is $s = f(t) = t^2 - 4t$. Find the velocity at $t = 0$. At $t = 3$. At any time t.

7. **Velocity**   The distance a man can walk in time t obeys the formula

$$s = \sqrt{t}$$

where s is measured in kilometers and t is measured in hours. What is his average velocity from $t = 0$ to $t = 16$ hours? What is his average velocity from $t = 1$ to $t = 4$ hours? From $t = 1$ to $t = 2$ hours?

8. **Velocity**   The figure below shows the position versus time path for a particle moving along a straight line. Estimate each of the following from the graph.

   (a) The average velocity over the interval $0 \le t \le 4$
   (b) The values of t at which the instantaneous velocity is either maximum or minimum
   (c) The values of t at which the instantaneous velocity is zero
   (d) The instantaneous velocity when $t = 4$ seconds

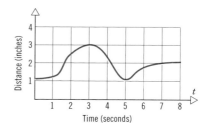

CHAPTER REVIEW

IMPORTANT TERMS AND CONCEPTS

IMPORTANT FORMULAS

$\lim_{x \to c} b = b$

$\lim_{x \to c} x = c$

$\lim_{x \to c} [f(x) \pm g(x)] = \lim_{x \to c} f(x) \pm \lim_{x \to c} g(x)$

$\lim_{x \to c} f(x) \cdot g(x) = \lim_{x \to c} f(x) \cdot \lim_{x \to c} g(x)$

$\lim_{x \to c} ax^n = ac^n$

$\lim_{x \to c} P(x) = P(c)$ P a polynomial function

$\lim_{x \to c} [f(x)]^n = \left[\lim_{x \to c} f(x) \right]^n$

$\lim_{x \to c} \sqrt[n]{f(x)} = \sqrt[n]{\lim_{x \to c} f(x)}$

$\lim_{x \to c} \dfrac{f(x)}{g(x)} = \dfrac{\lim_{x \to c} f(x)}{\lim_{x \to c} g(x)}, \quad \lim_{x \to c} g(x) \neq 0$

A Function f Is Continuous at c if

1. $f(c)$ is defined   **2.** $\lim_{x \to c} f(x)$ exists

3. $\lim_{x \to c} f(x) = f(c)$

Equation of Tangent Line

$y - f(c) = m_{\tan}(x - c)$

$f'(c) = \lim_{x \to c} \dfrac{f(x) - f(c)}{x - c}$

Slope of Tangent Line

$m_{\tan} = \lim_{x \to c} \dfrac{f(x) - f(c)}{x - c}$

$f'(x) = \lim_{h \to 0} \dfrac{f(x + h) - f(x)}{h}$

TRUE–FALSE ITEMS   Answers are on page AN-57.

T_____ F_____ **1.** The limit of the sum of two functions equals the sum of their limits, provided each limit exists.

T_____ F_____ **2.** The limit of a function f as x approaches c always equals $f(c)$.

T_____ F_____ **3.** $\lim_{x \to 4} \dfrac{x^2 - 16}{x - 4} = 8$

T_____ F_____ **4.** The function $f(x) = \dfrac{x}{x^2 + 4}$ is continuous at $x = -2$.

T_____ F_____ **5.** The limit of a quotient of two functions equals the quotient of their limits, provided each limit exists and the limit of the denominator is not zero.

T_____ F_____ **6.** The derivative of a function is the limit of a certain difference quotient.

FILL IN THE BLANKS   Answers are on page AN-57.

1. The notation _____ may be described by saying "For x approximately equal to c, but $x \neq c$, the value $f(x)$ is approximately equal to L."

2. If $\lim_{x \to c} f(x) = L$ and f is continuous at c, then $f(c)$ _____ L.

3. If there is no single number that the value of f approaches when x is close to c, then $\lim_{x \to c} f(x)$ does _____ .

4. When $\lim_{x \to c} f(x) = f(c)$, we say f is _____ at c.

5. $\lim_{x \to c} \dfrac{f(x)}{g(x)} = \dfrac{\lim_{x \to c} f(x)}{\lim_{x \to c} g(x)}$ provided $\lim_{x \to c} f(x)$ and $\lim_{x \to c} g(x)$ each exist and $\lim_{x \to c} g(x)$ _____ 0.

6. If $\lim_{x \to c^-} f(x) = L$ and $\lim_{x \to c^+} f(x) = R$, then $\lim_{x \to c} f(x)$ exists provided L _____ R.

7. The derivative of f at c equals the slope of the _____ line to the graph of f at $(c, f(c))$.

REVIEW EXERCISES   Answers to odd-numbered problems begin on page AN-57.

In Problems 1–22 find each limit.

1. $\lim\limits_{x \to 3} (3x - 4)$

2. $\lim\limits_{x \to 2} (2x + 5)$

3. $\lim\limits_{x \to 2} (x^2 + 2)$

4. $\lim\limits_{x \to 1} (x^2 - 2x)$

5. $\lim\limits_{x \to -2} (3 - 2x)$

6. $\lim\limits_{x \to 2}(8 - x)$

7. $\lim\limits_{x \to 1} \dfrac{x^2 - 1}{x - 1}$

8. $\lim\limits_{x \to -3} \dfrac{x^2 - 9}{x + 3}$

9. $\lim\limits_{x \to 1} \dfrac{x^3 - 1}{x - 1}$

10. $\lim\limits_{x \to -1} \dfrac{x^3 - 1}{x + 1}$

11. $\lim\limits_{x \to 4} \dfrac{(1/x) - \frac{1}{4}}{x - 4}$

12. $\lim\limits_{x \to 2} \dfrac{(1/x - \frac{1}{2})}{x - 2}$

13. $\lim\limits_{x \to 1} \sqrt{3x^3 - 4x + 5}$

14. $\lim\limits_{x \to 2} \sqrt{x^3 - 4}$

15. $\lim\limits_{x \to -2} \dfrac{2x^2 + 3x - 2}{x^2 + 5x + 6}$

16. $\lim\limits_{x \to -3} \dfrac{3x^2 + 10x + 3}{x^2 + 2x - 3}$

17. $\lim\limits_{x \to 4} \dfrac{\sqrt{x} - 2}{x^2 - 16}$

18. $\lim\limits_{x \to 9} \dfrac{x^2 - 9x}{\sqrt{x} - 3}$

19. $\lim\limits_{x \to 1} \left(\dfrac{3x^2}{x - 1} - \dfrac{3x}{x - 1} \right)$

20. $\lim\limits_{x \to -2} \left(\dfrac{3x}{x + 2} + \dfrac{6}{x + 2} \right)$

21. $\lim\limits_{x \to 2} f(x)$; $f(x) = \begin{cases} \dfrac{x^2 - 4x + 4}{x^2 - 4} & \text{if } x \neq 2 \\ 5 & \text{if } x = 2 \end{cases}$

22. $\lim\limits_{x \to -1} f(x)$; $f(x) = \begin{cases} \dfrac{x^4 - 1}{x^2 + x} & \text{if } x \neq -1 \\ 0 & \text{if } x = -1 \end{cases}$

In Problems 23–26 find the limit given below for each function.

$$\lim_{x \to 4} \frac{f(x) - f(4)}{x - 4}$$

23. $f(x) = 4x$

24. $f(x) = 2x + 1$

25. $f(x) = x^2 + x$

26. $f(x) = x^2 - x$

In Problems 27–32 determine whether the function f is continuous at c.

27. $f(x) = \begin{cases} \dfrac{x^2 - 16}{x - 4} & \text{if } x \neq 4 \\ 8 & \text{if } x = 4 \end{cases}$ $c = 4$

28. $f(x) = \begin{cases} \dfrac{x^2 + x}{x} & \text{if } x \neq 0 \\ 1 & \text{if } x = 0 \end{cases}$ $c = 0$

29. $f(x) = \begin{cases} \dfrac{x^3 - 8}{x - 2} & \text{if } x \neq 2 \\ 4 & \text{if } x = 2 \end{cases}$ $c = 2$

30. $f(x) = \begin{cases} \dfrac{x^3 + 1}{x + 1} & \text{if } x \neq -1 \\ 0 & \text{if } x = -1 \end{cases}$ $c = -1$

31. $f(x) = \begin{cases} 2x^2 + 1 & \text{if } x < 0 \\ 1 & \text{if } x = 0 \\ x^2 + 1 & \text{if } x > 0 \end{cases}$ $c = 0$

32. $f(x) = \begin{cases} 3x + 4 & \text{if } x < 1 \\ 2 & \text{if } x = 1 \\ x^2 - x + 7 & \text{if } x > 1 \end{cases}$ $c = 1$

33. Is the function f defined by

$$f(x) = \frac{x^2 - 3x}{x^2 - 9}$$

continuous at 3? If not, can f be redefined at 3 to make it continuous at 3?

34. Is the function f defined by

$$f(x) = \frac{2x + 4}{x^2 - 4}$$

continuous at -2? If not, can f be redefined at -2 to make it continuous at -2?

In Problems 35–38 find the slope of the tangent line to the graph of f at the given point. Also find an equation of this tangent line.

35. $f(x) = 2x^2 + 1$ at $(1, 3)$

36. $f(x) = -x^2 + 2x$ at $(-1, -3)$

37. $f(x) = x^2 - 2x + 4$ at $(-2, 12)$

38. $f(x) = -2x^2 + x + 1$ at $(2, -5)$

In Problems 39–42 find the derivative of f at the given number.

39. $f(x) = 2x^2 + 1$ at 2

40. $f(x) = -x^2 + 2x$ at 1

41. $f(x) = x^2 - 2x + 4$ at -1

42. $f(x) = -2x^2 + x + 1$ at 3

In Problems 43–46 find the derivative of f.

43. $f(x) = 2x^2 + 1$

44. $f(x) = -x^2 + 2x$

45. $f(x) = x^2 - 2x + 4$

46. $f(x) = -2x^2 + x + 1$

47. Use the graph of the function f below to determine the points at which the derivative does not exist.

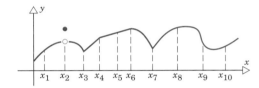

Mathematical Question from Professional Exam

1. Actuary Exam—Part I $\lim\limits_{x \to 3} \dfrac{x^2 - x - 6}{x^2 - 9}$

(a) 0   (b) $\frac{5}{6}$ (c) 1   (d) $\frac{5}{3}$ (e) Undefined

Chapter 13

Derivative Formulas

$\mathbf{I}$n Chapter 12 we solved the tangent problem: the problem of defining the slope of the tangent line to the graph of a function f. This led to the definition of the derivative of a function $y = f(x)$ as the limit of its difference quotient:

$$f'(x) = \lim_{h \to 0} \frac{f(x + h) - f(x)}{h} \tag{1}$$

In this chapter we will use this definition to derive formulas for finding derivatives.

13.1 THE SIMPLE POWER RULE; SUM AND DIFFERENCE FORMULAS

We begin by considering the constant function $f(x) = b$, where b is a real number. Since the graph of the constant function f is a horizontal line (see Figure 1), the tangent line to f at any point is also a horizontal line (the slope equals 0). Since the derivative equals the slope of the tangent line to the graph of a function f at a point, then the derivative of f should be 0. Algebraically, the derivative is obtained by using Formula (1).

The difference quotient of $f(x) = b$ is

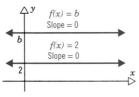

Figure 1

$$\frac{f(x + h) - f(x)}{h} = \frac{b - b}{h} = \frac{0}{h} = 0$$

Thus,

$$f'(x) = \lim_{h \to 0} \frac{f(x + h) - f(x)}{h} = \lim_{h \to 0} 0 = 0$$

Derivative of the Constant Function

For the constant function $f(x) = b$, the derivative is $f'(x) = 0$. In other words, the derivative of a constant is 0.

Besides the **prime notation** f', there are several other ways to denote the derivative of a function $y = f(x)$. The most common ones are

$$y' \quad \text{and} \quad \frac{dy}{dx}$$

The notation dy/dx, often referred to as the **Leibniz notation,** may also be written as

$$\frac{dy}{dx} = \frac{d}{dx}(y) = \frac{d}{dx}f(x)$$

where $\dfrac{d}{dx}f(x)$ is an instruction to compute the derivative of the function f with respect to its independent variable x. A change in the symbol used for the independent variable does not affect the meaning. Thus, if $s = f(t)$ is a function of t, then ds/dt is an instruction to differentiate f with respect to t.

In terms of the Leibniz notation, if b is a constant, then

$$\frac{d}{dx}b = 0 \qquad\qquad (2)$$

EXAMPLE 1

(a) If $f(x) = 5$, then $f'(x) = 0$.
(b) If $y = -1.7$, then $y' = 0$.
(c) If $y = \frac{2}{3}$, then $dy/dx = 0$.
(d) If $s = f(t) = \sqrt{5}$, then $ds/dt = f'(t) = 0$.

Now Work Problem 1

$\cdots\cdots\cdots\cdots\cdots\cdots\cdots\cdots$

In subsequent work with derivatives we shall use the prime notation or the Leibniz notation, or sometimes a mixture of the two, depending on which is more convenient.

Simple Power Rule

We now investigate the derivative of the power function $f(x) = x^n$ for some positive integers n to see if a pattern appears.

For $f(x) = x$, $n = 1$, we have

$$f'(x) = \lim_{h \to 0} \frac{f(x + h) - f(x)}{h} = \lim_{h \to 0} \frac{(x + h) - x}{h} = \lim_{h \to 0} \frac{h}{h} = \lim_{h \to 0} 1 = 1$$

For $f(x) = x^2$, $n = 2$, we have

$$f'(x) = \lim_{h \to 0} \frac{f(x + h) - f(x)}{h} = \lim_{h \to 0} \frac{(x + h)^2 - x^2}{h} = \lim_{h \to 0} \frac{x^2 + 2xh + h^2 - x^2}{h}$$

$$= \lim_{h \to 0} \frac{2xh + h^2}{h} = \lim_{h \to 0} \frac{h(2x + h)}{h}$$

$$= \lim_{h \to 0} (2x + h) = 2x$$

For $f(x) = x^3$, $n = 3$, we have

$$f'(x) = \lim_{h \to 0} \frac{f(x + h) - f(x)}{h} = \lim_{h \to 0} \frac{(x + h)^3 - x^3}{h}$$

$$= \lim_{h \to 0} \frac{x^3 + 3x^2h + 3xh^2 + h^3 - x^3}{h}$$

$$= \lim_{h \to 0} \frac{3x^2h + 3xh^2 + h^3}{h}$$

$$= \lim_{h \to 0} \frac{h(3x^2 + 3xh + h^2)}{h}$$

$$= \lim_{h \to 0} (3x^2 + 3xh + h^2) = 3x^2$$

In the Leibniz notation these results take the form

$$\frac{d}{dx} x = 1 \qquad \frac{d}{dx} x^2 = 2x \qquad \frac{d}{dx} x^3 = 3x^2$$

This pattern suggests the following formula:

Simple Power Rule: Derivative of x^n

For the function $f(x) = x^n$, n a positive integer, the derivative is $f'(x) = nx^{n-1}$. That is,

$$\frac{d}{dx} x^n = nx^{n-1} \qquad (3)$$

The Simple Power Rule may be restated as

The derivative with respect to x of x raised to the power n, where n is a positive integer, is n times x raised to the power $n - 1$.

Problems 110 and 111 outline proofs of the Simple Power Rule.

EXAMPLE 2

(a) If $f(x) = x^6$, then $f'(x) = 6x^{6-1} = 6x^5$

(b) $\dfrac{d}{dt} t^5 = 5t^4$

(c) $\dfrac{d}{dx} x = 1$

Now Work Problem 3

....

So far, Formula (3) has only been shown to be valid when n is a positive integer. However, it turns out that this formula is true if the exponent is any real number. That is,

---

Simple Power Rule

$$\frac{d}{dx} x^n = nx^{n-1} \qquad \text{for any real number } n \qquad (4)$$

The general proof of this rule will be given in Section 13.5.

EXAMPLE 3

(a) If $f(x) = x^{-9}$, then $f'(x) = -9x^{-9-1} = -\dfrac{9}{x^{10}}$.

(b) If $f(x) = x^{4/3}$, then $f'(x) = \dfrac{4}{3} x^{(4/3)-1} = \dfrac{4}{3} x^{1/3}$.

(c) If $f(x) = x^{1/3}$, then $f'(x) = \dfrac{1}{3} x^{(1/3)-1} = \dfrac{1}{3} x^{-2/3} = \dfrac{1}{3x^{2/3}}$

....

When we apply Rule (4) to find the derivative of a function, sometimes the function must be rewritten so that it is in the form of the rule. Some examples of the Simple Power Rule follow.

$f(x)$	$f'(x)$
$\sqrt{x} = x^{1/2}$	$\dfrac{1}{2}x^{-1/2} = \dfrac{1}{2x^{1/2}} = \dfrac{1}{2\sqrt{x}}$
$\sqrt[4]{x^3} = x^{3/4}$	$\dfrac{3}{4}x^{-1/4} = \dfrac{3}{4x^{1/4}} = \dfrac{3}{4\sqrt[4]{x}}$
$\sqrt[3]{x^4} = x^{4/3}$	$\dfrac{4}{3}x^{1/3} = \dfrac{4}{3}\sqrt[3]{x}$
$\dfrac{1}{x^{1/5}} = x^{-1/5}$	$-\dfrac{1}{5}x^{-6/5} = -\dfrac{1}{5x^{6/5}}$
$\dfrac{1}{x^{4/3}} = x^{-4/3}$	$-\dfrac{4}{3}x^{-7/3} = -\dfrac{4}{3x^{7/3}}$

EXAMPLE 4 Find $f'(4)$ if $f(x) = x^3$.

SOLUTION
$$f'(x) = 3x^2$$
$$f'(4) = 3(4)^2 = 48$$

· ·

Formula (4) allows us to compute some derivatives with ease. However, do not forget that a derivative is, in actuality, the limit of a difference quotient. Next, we establish another differentiation formula.

Derivative of a Constant Times a Function

The derivative of a constant times a function equals the constant times the derivative of the function. That is, if C is a constant and f is a differentiable function, then

$$\frac{d}{dx}[Cf(x)] = C\frac{d}{dx}f(x) \qquad (5)$$

Proof We verify Formula (5) as follows.

$$\frac{d}{dx}Cf(x) = \lim_{h \to 0} \frac{Cf(x+h) - Cf(x)}{h}$$

$$= \lim_{h \to 0} C\frac{f(x+h) - f(x)}{h}$$

$$\frac{d}{dx} Cf(x) = C \lim_{h \to 0} \frac{f(x + h) - f(x)}{h}$$

$$= C \frac{d}{dx} f(x)$$

∎

The usefulness and versatility of this formula are often overlooked, especially when the constant appears in the denominator. Note that

$$\frac{d}{dx}\left[\frac{f(x)}{C}\right] = \frac{d}{dx}\left[\frac{1}{C}f(x)\right] = \frac{1}{C}\frac{d}{dx}[f(x)]$$

Always be on the lookout for constant factors *before* differentiating.

EXAMPLE 5

Differentiate the following functions:

(a) $f(x) = 10x^3$ (b) $g(x) = \frac{7x}{3}$ (c) $f(x) = \frac{2\sqrt{3}}{3}x^3$

SOLUTION

(a) Here, f is a constant (10) times a function (x^3).

$$\frac{d}{dx}(10x^3) = 10 \frac{d}{dx}(x^3) \qquad \text{By (5)}$$

$$= 10(3x^{3-1}) = 30x^2 \quad \text{By (3)}$$

(b) Here, we first rewrite g as a constant times a function and then differentiate, using Rule 3. The constant is $\frac{7}{3}$ and the function is x.

$$g(x) = \frac{7x}{3} = \frac{7}{3}x$$

$$g'(x) = \frac{d}{dx}\left(\frac{7}{3}x\right) = \frac{7}{3}\frac{d}{dx}(x) \quad \text{By (5)}$$

$$= \frac{7}{3}(1) = \frac{7}{3} \qquad \text{By (3)}$$

(c) Here, f is a constant $\left(\frac{2\sqrt{3}}{3}\right)$ times a function (x^3).

$$\frac{d}{dx}\frac{2\sqrt{3}}{3}x^3 = \frac{2\sqrt{3}}{3}\frac{d}{dx}(x^3) \qquad \text{By (5)}$$

$$= \frac{2\sqrt{3}}{3}(3x^2) = 2\sqrt{3}x^2 \quad \text{By (3)}$$

· ·

> **Sum and Difference Formulas**
>
> If f and g are two differentiable functions, then
>
> $$\frac{d}{dx}[f(x) + g(x)] = \frac{d}{dx}f(x) + \frac{d}{dx}g(x) \qquad (6)$$
>
> and
>
> $$\frac{d}{dx}[f(x) - g(x)] = \frac{d}{dx}f(x) - \frac{d}{dx}g(x) \qquad (7)$$

That is, the derivative of the sum (or difference) of two differentiable functions equals the sum (or difference) of their derivatives.

Proof We verify Formula (6) as follows. To compute

$$\frac{d}{dx}[f(x) + g(x)]$$

we need to find the limit of the difference quotient of $f(x) + g(x)$.

$$\frac{d}{dx}[f(x) + g(x)] = \lim_{h \to 0} \frac{[f(x + h) + g(x + h)] - [f(x) + g(x)]}{h}$$

$$= \lim_{h \to 0} \frac{[f(x + h) - f(x)] + [g(x + h) - g(x)]}{h}$$

$$= \lim_{h \to 0} \left[\frac{f(x + h) - f(x)}{h} + \frac{g(x + h) - g(x)}{h} \right]$$

$$= \lim_{h \to 0} \left[\frac{f(x + h) - f(x)}{h} \right] + \lim_{h \to 0} \left[\frac{g(x + h) - g(x)}{h} \right]$$

$$= \frac{d}{dx}f(x) + \frac{d}{dx}g(x) \qquad \blacksquare$$

This formula for differentiating states that functions that are sums can be differentiated "term by term." The proof of (7) is similar to that of (6).

EXAMPLE 6   Find the derivative of $f(x) = x^2 + 4x$.

SOLUTION

The function f is the sum of the two functions x^2 and $4x$. Thus, we can differentiate term by term:

$$\frac{d}{dx}f(x) = \frac{d}{dx}(x^2 + 4x) = \frac{d}{dx}x^2 + \frac{d}{dx}4x = 2x + 4\frac{d}{dx}x = 2x + 4$$

······················

EXAMPLE 7   Find the derivative of $f(x) = x^2 + \sqrt{x}$.

SOLUTION

The function f is the sum of the two functions x^2 and $\sqrt{x}$. Thus, we can differentiate term by term:

$$\frac{d}{dx}f(x) = \frac{d}{dx}(x^2 + \sqrt{x}) = \frac{d}{dx}x^2 + \frac{d}{dx}\sqrt{x} = 2x + \frac{1}{2\sqrt{x}}$$

······················

Formulas (6) and (7) extend to sums and differences of more than two functions. Thus, with the differentiation rules stated so far, we can compute the derivative of any polynomial function as well as a variety of other functions.

EXAMPLE 8   Find the derivative of $f(x) = 6x^4 - 3x^2 + 10x - 8$.

SOLUTION

$$f'(x) = \frac{d}{dx}(6x^4 - 3x^2 + 10x - 8)$$

$$= \frac{d}{dx}(6x^4) - \frac{d}{dx}(3x^2) + \frac{d}{dx}(10x) - \frac{d}{dx}8$$

$$= 24x^3 - 6x + 10$$

Now Work Problem 15

······················

EXAMPLE 9   If $f(x) = -\dfrac{x^4}{2} - 2x + 3$, find: (a) $f'(x)$ (b) $f'(-1)$

SOLUTION

(a) $f'(x) = -\dfrac{4x^3}{2} - 2 + 0 = -2x^3 - 2$

(b) $f'(-1) = -2(-1)^3 - 2 = 0$

······················

EXAMPLE 10   If $h(t) = 3 - \dfrac{1}{2t^4}$, find: (a) $h'(t)$ (b) $h'(2)$

SOLUTION

(a) We first write $h(t)$ as

$$h(t) = 3 - \frac{1}{2} t^{-4}$$

Then

$$h'(t) = 0 - \left(-\frac{4}{2} t^{-5} \right) = 2t^{-5} = \frac{2}{t^5}$$

(b) $h'(2) = \dfrac{2}{2^5} = \dfrac{1}{16}$

· ·

EXAMPLE 11

Find the value of dy/dx at the point $(2, \frac{1}{2})$ on the graph of $y = 4x^{-3}$.

SOLUTION

$$\frac{dy}{dx} = \frac{d}{dx} 4x^{-3} = -12x^{-4} = \frac{-12}{x^4}$$

At the point $(2, \frac{1}{2})$ we have $x = 2$, so

$$\frac{dy}{dx} = \frac{-12}{2^4} = -\frac{12}{16} = -\frac{3}{4}$$

Now Work Problem 77

· ·

We have already seen that the slope of the tangent line to the graph of a function f equals the derivative. In particular, if the tangent line is horizontal, then its slope is zero, so the derivative of f will also be zero. See Figure 2.

Figure 2

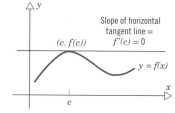

The following result is a consequence of this fact.

Criterion for a Horizontal Tangent Line

The tangent line to the graph of f at a point $(c, f(c))$ on the graph of f is horizontal if and only if $f'(c) = 0$.

Thus, to determine where the tangent line to the graph of $y = f(x)$ is horizontal, we need to solve the equation $f'(x) = 0$.

EXAMPLE 12

At what point(s) is the tangent line to the graph of

$$f(x) = x^3 + 3x^2 - 24x$$

horizontal?

SOLUTION

We first find $f'(x)$:

$$f'(x) = 3x^2 + 6x - 24$$

Horizontal tangent lines occur where $f'(x) = 0$. Thus, we solve the equation

$$3x^2 + 6x - 24 = 0$$
$$3(x^2 + 2x - 8) = 0$$
$$3(x + 4)(x - 2) = 0$$
$$x = -4 \quad \text{and} \quad x = 2$$

The tangent line is horizontal at the points $(-4, f(-4)) = (-4, 80)$ and $(2, f(2)) = (2, -28)$ on the graph of f.

• •

Figure 3 shows the graph of the function discussed in Example 12. Notice that we substitute $x = -4$ and $x = 2$ into the function $y = f(x)$ to find the y-coordinate of the point on the graph of f at which the tangent line is horizontal. Be careful not to substitute these values of x into the derivative $f'(x)$ because this gives the value of the derivative *Now Work Problem 95* rather than the value of the function.

Figure 3

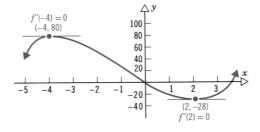

EXERCISE 13.1   Answers to odd-numbered problems begin on page AN-57.

In Problems 1–66 find the derivative of each function.

1. $f(x) = 4$

2. $f(x) = -2$

3. $f(x) = x^3$

4. $f(x) = x^4$

5. $f(x) = 6x^2$

6. $f(x) = -8x^3$

7. $f(t) = \dfrac{t^4}{4}$

8. $f(t) = \dfrac{t^3}{6}$

9. $f(x) = x^2 + x$

10. $f(x) = x^2 - x$

11. $f(x) = x^3 - x^2 + 1$

12. $f(x) = x^4 - x^3 + x$

13. $f(t) = 2t^2 - t + 4$

14. $f(t) = 3t^3 - t^2 + t$

15. $f(x) = \frac{1}{2}x^8 + 3x + \frac{2}{3}$

16. $f(x) = \frac{2}{3}x^6 - \frac{1}{2}x^4 + 2$

17. $f(x) = \frac{1}{3}(x^5 - 8)$

18. $f(x) = \dfrac{x^3 + 2}{5}$

19. $f(x) = \frac{2}{3}x^6 + \frac{2}{5}x^5$

20. $q(x) = \dfrac{x^8}{8} + \dfrac{3}{5}x$

21. $f(x) = \dfrac{3 - x^7}{4}$

22. $f(x) = \dfrac{6(x^3 - 2)}{5}$

23. $h(t) = \sqrt{3}\,t^3 + \dfrac{t^2}{\sqrt{2}}$

24. $r(t) = \dfrac{3 - t}{5}$

25. $f(x) = x^{2/3}$

26. $f(x) = x^{1/3}$

27. $y = \frac{2}{3}x^{1/2}$

28. $y = 3u^{-7/17}$

29. $y = x^{1.2}$

30. $y = x^{-3.5}$

31. $y = x^{\sqrt{5}}$

32. $y = x^{\sqrt{3}}$

33. $y = x^{2/3} + x^{3/2}$

34. $y = -2x^{1/2} + 3x^{-3/2}$

35. $f(x) = \sqrt{x}$

36. $f(x) = \sqrt[4]{x^3}$

37. $f(x) = 4\sqrt{x^3}$

38. $f(x) = 8\sqrt[6]{x^3}$

39. $f(x) = x^{-3}$

40. $v(t) = 5t^{-4}$

41. $y = x^{-3} + x^{-2} + 2x^2$

42. $y = 15x^{-2} - 75x^{-1/2}$

43. $y = \dfrac{1}{x^2}$

44. $f(x) = \dfrac{1}{x^3}$

45. $f(x) = \dfrac{3}{x^3}$

46. $y = \dfrac{1}{3x^3}$

47. $f(x) = ax^2 + bx + c$

48. $f(x) = ax^3 + bx^2 + cx + d$

49. $f(x) = \dfrac{10}{x^4} + \dfrac{3}{x^2}$

50. $f(x) = \dfrac{2}{x^5} - \dfrac{3}{x^3}$

51. $f(x) = \dfrac{x - 1}{2x}$

52. $f(z) = \dfrac{z + 1}{2z}$

53. $f(t) = 3t + \dfrac{1}{3t}$

54. $f(u) = 4u - \dfrac{1}{4u}$

55. $f(x) = 3x^3 - \dfrac{1}{3x^2}$

56. $f(x) = x^5 - \dfrac{5}{x^5}$

57. $f(t) = \dfrac{1}{t} - \dfrac{1}{t^2} + \dfrac{1}{t^3}$

58. $f(x) = \dfrac{3}{x^{1/2}} + \dfrac{1}{x^{1/3}} + \dfrac{1}{x^{1/4}}$

59. $f(x) = \dfrac{3}{2x^2}$

60. $f(x) = \dfrac{3}{(2x)^2}$

61. $f(x) = \dfrac{3}{2x^{-2}}$

62. $f(x) = \dfrac{3}{(2x)^{-2}}$

63. $f(x) = \dfrac{1}{\sqrt[4]{x^3}}$

64. $f(x) = \dfrac{5}{\sqrt[4]{x}}$

65. $f(x) = \dfrac{2}{\sqrt{x}} - 3x^{-3} + x$

66. $f(x) = \dfrac{3}{\sqrt[3]{x^2}} - 3x^{-2} + x^2 + 1$

In Problems 67–72 find the indicated derivative.

67. $\dfrac{d}{dx}\left(\sqrt{3}x + \frac{1}{2}\right)$

68. $\dfrac{d}{dx}\left(\dfrac{2x^4 - 5}{8}\right)$

69. $\dfrac{dA}{dR}$ if $A = \pi R^2$

70. $\dfrac{dC}{dR}$ if $C = 2\pi R$

71. $\dfrac{dV}{dR}$ if $V = \frac{4}{3}\pi R^3$

72. $\dfrac{dP}{dT}$ if $P = 0.2T$

In Problems 73–82 find the value of the derivative at the indicated point.

73. $y = x^4$, at $(1, 1)$

74. $y = x^4$, at $(2, 16)$

75. $y = \sqrt{x}$, at $(4, 2)$

76. $y = 1/x^2$, at $(3, \frac{1}{9})$

77. $y = 1/\sqrt[3]{x}$, at $(-8, -\frac{1}{2})$

78. $y = \sqrt[3]{x}$, at $(-8, -2)$

79. $y = \frac{1}{2}x^2$, at $(1, \frac{1}{2})$

80. $y = 1/(2x)^2$, at $(1, \frac{1}{4})$

81. $y = 2 - 2/x^3$, at $(2, \frac{7}{4})$

82. $y = 2x^2 - \frac{1}{2}x + 3$, at $(0, 3)$

In Problems 83 and 84 find the slope of the tangent line to the graph of the function f at the indicated point. What is an equation of the tangent line?

83. $f(x) = x^3 + 3x - 1$, at $(0, -1)$

84. $f(x) = x^4 + 2x - 1$, at $(1, 2)$

In Problems 85–90 find those x, if any, at which f'(x) = 0.

85. $f(x) = 3x^2 - 12x + 4$

86. $f(x) = x^2 + 4x - 3$

87. $f(x) = x^3 - 3x + 2$

88. $f(x) = x^4 - 4x^3$

89. $f(x) = x^3 + x$

90. $f(x) = x^5 - 5x^4 + 1$

In Problems 91–102 find any points at which the graph of f has a horizontal tangent line.

91. $f(x) = x^2 - 4x$

92. $f(x) = x^2 + 2x$

93. $f(x) = -x^2 + 8x$

94. $f(x) = -x^2 - 12x + 1$

95. $f(x) = -2x^2 + 8x + 1$

96. $f(x) = -3x^2 + 12x$

97. $f(x) = -x^3 + 3x + 1$

98. $f(x) = -2x^3 + 6x^2 + 1$

99. $f(x) = x^5 - 10x^4$

100. $f(x) = x^5 + 5x$

101. $f(x) = 3x^5 + 20x^3 - 1$

102. $f(x) = 3x^5 - 5x^3 - 1$

103. Find all points on the graph of $f(x) = x^2 - 6x + 8$ where the slope of the tangent line is 3.

104. Find all points on the graph of $f(x) = \dfrac{x^3}{3} - 8x + 4$ where the slope of the tangent line is 1.

105. Find the point(s), if any, on the graph of the function $y = 9x^3$ at which the tangent line is parallel to the line $3x - y + 2 = 0$.

106. Find the point(s), if any, on the graph of the function $y = 4x^2$ at which the tangent line is parallel to the line $2x - y - 6 = 0$.

107. Two lines through the point $(1, -3)$ are tangent to the graph of the function $y = 2x^2 - 4x + 1$. Find the equations of these two lines.

108. Two lines through the point $(0, 2)$ are tangent to the graph of the function $y = 1 - x^2$. Find the equations of these two lines.

109. The graph of the derivative of a function f appears at the top of the next column. (Note that the graph of f is not given.)

(a) Suppose that $f(2) = 3$. Find an equation of the line tangent to the graph of f at $(2, 3)$.

(b) Suppose that $f(-2) = -4$. Find an equation of the line tangent to the graph of f at $(-2, -4)$.

110. Use the following factoring rule to prove the Simple Power Rule:

$$f(x) = x^n - c^n$$
$$= (x - c)(x^{n-1} + x^{n-2}c + x^{n-3}c^2 + \cdots + c^{n-1})$$

Now apply Formula (3), page 620, to find $f'(c)$.

111. Use the binomial theorem to prove the Simple Power Rule.

[*Hint:* $(x + h)^n - x^n = x^n + nx^{n-1}h + \dfrac{n(n-1)}{2}x^{n-2}h^2 + \cdots + h^n - x^n = nx^{n-1}h + h^2 \cdot$ (Terms involving x and h)

Now apply Formula (1), page 617.]

Technology Exercises

1. Let $f(x) = x^2 - 1$.

(a) Graph $f(x) = x^2 - 1$ using a calculator. Describe the graph of $f(x)$.
(b) Find $f'(x)$. This is a new function with the same domain as $f(x)$.
(c) Complete the following table [use your calculator to evaluate $f(x)$ and $f'(x)$] and plot the points $(x, f'(x))$ on a suitable range.

x	$f(x)$	$f'(x)$
-4	15	-8
-2		
-1		
0		
1		
2		
4		

(d) Describe the graph of $f'(x)$.
(e) Graph $f(x)$ and $f'(x)$ on your calculator on the same screen. When $f'(x) = 0$, what happens to the graph of $f(x)$?

2. Graph the function $y_1 = x^3 + 1$ and the lines $y_2 = 3x - 1$, $y_3 = 5x + 2$, and $y_4 = 4x + 3$. Which line appears to be the tangent line to the graph at $(1, 2)$?

3. Confirm by graphing that $y = -2x + 3$ is the tangent line to the graph of $y = 1/x^2$ at $(1, 1)$.

4. From the graph of the function

$$f(x) = \frac{9}{(x^2 + 2x - 7)}$$

determine whether the slope of the tangent line would be positive, negative, or zero at

(a) $x = 1$ (b) $x = 0$
(c) $x = 2$ (d) $x = 3$

5. Graph the function $y = x^2$ and the line $y = 4.01x - 4$ in the same viewing rectangle $[0, 5] \times [0, 10]$.

(a) Does the line $y = 4.01x - 4$ appear to be tangent to the graph of $y = x^2$ at the point $(2, 4)$?
(b) Zoom in on the point $(2, 4)$ several times. What happens? Why?
(c) What is the tangent line to the function $y = x^2$ at $(2, 4)$?

6. Endangered Species Prior to the arrival of Europeans, more than 60 million bison wandered over the plains of North America. By 1899, there were just 550 plains bison alive. The bison population in North America during 1700–1899 can be approximated by the function

$$P(t) = \frac{60,000,000}{1 + e^{-438 + 0.23673t}}$$

with t denoting the year. Use a graphing calculator (or a computer) to:

(a) Graph P within a $[1700, 1900] \times [0, 60000000]$ viewing rectangle.
(b) Estimate how many bison there would have been in 1925 if nothing was changed in our policy toward bison.
(c) Using the same model, estimate how many bison there were in 1850.
(d) Estimate when the bison population was declining most rapidly. How large was the bison population at that time?

13.2 RATES OF CHANGE; APPLICATION TO ECONOMICS: MARGINAL ANALYSIS

Earlier we learned that the derivative of a function f equals the slope of the tangent line to the graph of f. In this section we look at other uses for the derivative.

If the independent variable x of a function $y = f(x)$ changes from c to d, there will be a corresponding change in the dependent variable, from $f(c)$ to $f(d)$. If we denote the change in x by Δx (read "delta x") and the change in y by Δy, then **the average rate of change of y with respect to x is**

$$\begin{pmatrix} \text{Average rate of} \\ \text{change of } y \text{ with} \\ \text{respect to } x \end{pmatrix} = \frac{\Delta y}{\Delta x} = \frac{f(d) - f(c)}{d - c} \tag{1}$$

EXAMPLE 1

At Dan's Toy Store the revenue R, in dollars, derived from selling x electric trucks is

$$R(x) = -\frac{1}{200} x^2 + 20x$$

(a) After 1000 trucks are sold, what is the average rate of change in revenue derived from selling an additional 100 trucks?

(b) After 1000 trucks are sold, what is the average rate of change in revenue derived from selling an additional 50 trucks?

SOLUTION

(a) The additional revenue derived from selling an additional 100 trucks after 1000 have been sold requires a change in x from 1000 to 1100. The corresponding change in revenue is

$$\Delta R = R(1100) - R(1000) = 15{,}950 - 15{,}000 = 950$$

Thus, the average rate of change in revenue due to the additional sales is

$$\frac{\Delta R}{\Delta x} = \frac{R(1100) - R(1000)}{1100 - 1000} = \frac{950}{100} = \$9.50 \text{ per truck}$$

(b) Here, the change in x is from 1000 to 1050, so the corresponding change in revenue is

$$\Delta R = R(1050) - R(1000) = 15{,}487.5 - 15{,}000 = 487.5$$

The average rate of change in revenue is

$$\frac{\Delta R}{\Delta x} = \frac{R(1050) - R(1000)}{1050 - 1000} = \frac{487.5}{50} = \$9.75 \text{ per truck}$$

Now Work Problem 11

Look at Figure 4.

Figure 4

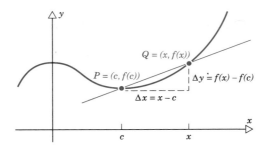

The average rate of change of f, $\Delta y/\Delta x$, equals the slope of the secant line joining P and Q. That is,

$$\frac{\Delta y}{\Delta x} = \frac{f(x) - f(c)}{x - c}$$

Now let $x \to c$. Then $\Delta x = x - c \to 0$. The limit of the average rate of change as $\Delta x \to 0$ equals the limit of the slope of the secant line as $x \to c$. But this is the derivative of f at c. Since $\Delta x = x - c$, it follows that $x = c + \Delta x$. Thus,

$$f'(c) = \lim_{\Delta x \to 0} \frac{\Delta y}{\Delta x} = \lim_{\Delta x \to 0} \frac{f(c + \Delta x) - f(c)}{\Delta x} \qquad (2)$$

In this setting, we call $f'(c)$ the **(instantaneous) rate of change of y with respect to x at c.**

EXAMPLE 2

During a month-long advertising campaign, the total sales of a magazine are given by

$$S(x) = 5x^2 + 100x + 10{,}000$$

where x represents the number of days of the campaign, $0 \le x \le 30$.

(a) What is the average rate of change of sales from $x = 10$ to $x = 20$ days?
(b) What is the instantaneous rate of change of sales when $x = 10$ days?

SOLUTION

(a) From $x = 10$ to $x = 20$ the average rate of change of sales is

$$\frac{\Delta S}{\Delta x} = \frac{S(20) - S(10)}{20 - 10} = \frac{14{,}000 - 11{,}500}{10} = 250 \text{ magazines per day}$$

(b) The instantaneous rate of change of sales is

$$S'(x) = \frac{d}{dx}(5x^2 + 100x + 10{,}000) = 10x + 100$$

When $x = 10$,

$$S'(10) = 10(10) + 100 = 200 \text{ magazines per day}$$

· ·

We interpret the results of Example 2 as follows: The fact that the average rate of sales from $x = 10$ to $x = 20$ is $\Delta S/\Delta x = 250$ magazines per day indicates that on the 10th day of the campaign, we can expect to average 250 magazines per day of additional sales if we continue the campaign for 10 more days. The fact that $S'(10) = 200$ magazines per day indicates that on the 10th day of the campaign, one more day of advertising will result in additional sales of approximately 200 magazines per day.

Now Work Problem 15

Application to Economics: Marginal Analysis

Economics is one of the many fields in which calculus has been used to great advantage. Economists have a special name for the application of derivatives to problems in economics—it is **marginal analysis.** Thus, whenever the term *marginal* appears in a discussion, it signals the presence of derivatives in the background.

MARGINAL COST

Suppose $C = C(x)$ is the cost of producing x units. Then the derivative $C'(x)$ is called the *marginal cost.*

Let's look at a way to interpret the marginal cost. Since

$$C'(x) = \lim_{\Delta x \to 0} \frac{C(x + \Delta x) - C(x)}{\Delta x}$$

it follows for small values of Δx that

$$C'(x) \approx \frac{C(x + \Delta x) - C(x)}{\Delta x}$$

That is to say,

$$C'(x) \approx \frac{\text{Cost of increasing production from } x \text{ to } x + \Delta x}{\text{Increment } \Delta x}$$

In many practical situations we deal with large x. Because of this, many economists let $\Delta x = 1$, which is small compared to large x. Then marginal cost may be interpreted as

$$C'(x) \approx C(x + 1) - C(x) = \textbf{Cost of increasing production by one unit}$$

EXAMPLE 3   Suppose that the cost in dollars for a weekly production of x tons of steel is given by the formula:

$$C = \tfrac{1}{10} x^2 + 5x + 1000$$

(a) Find the marginal cost.
(b) Find the cost and marginal cost when $x = 1000$ tons.
(c) Interpret $C'(1000)$.

SOLUTION

(a) $C'(x) = \dfrac{d}{dx} (\tfrac{1}{10} x^2 + 5x + 1000) = \tfrac{1}{5}x + 5$

(b) $C(1000) = \tfrac{1}{10}(1000)^2 + 5 \cdot 1000 + 1000 = \$106{,}000$
   $C'(1000) = \tfrac{1}{5} \cdot 1000 + 5 = \$205/\text{ton}$

(c) $C'(1000) = \$205$ per ton means that the cost of producing one additional ton of steel after 1000 tons have been produced is approximately \$205.

. . . .

Note that the average cost of producing one more ton of steel after the 1000th ton is

$$\frac{\Delta C}{\Delta x} = \frac{C(1001) - C(1000)}{1001 - 1000}$$

$$= (\tfrac{1}{10} \cdot 1001^2 + 5 \cdot 1001 + 1000) - (\tfrac{1}{10} \cdot 1000^2 + 5 \cdot 1000 + 1000)$$

$$= \$205.10/\text{ton}$$

We observe that the average cost differs from the marginal cost by only 0.1 dollar/ton, which is less than $\tfrac{1}{20}$th of 1% of the marginal cost. Note, too, that the marginal cost is easier to compute than the actual average cost.

The money received by the steel producer for selling steel is called revenue. Specifically, let $R = R(x)$ be the total revenue received from selling x tons of steel. Then the derivative $R'(x)$ is called the **marginal revenue.** For this example, marginal revenue, like marginal cost, is measured in dollars per ton. An approximate value for $R'(x)$ is obtained by noting again that

$$R'(x) \approx \frac{R(x + \Delta x) - R(x)}{\Delta x}$$

When x is large, then $\Delta x = 1$ is small by comparison, so that

$R'(x) \approx R(x + 1) - R(x) =$ Revenue resulting from the sale of one additional unit

This is the interpretation many economists give to marginal revenue.

EXAMPLE 4   Suppose that the revenue R for a weekly sale of x tons of steel is given by the formula

$$R(x) = x^2 + 5x$$

(a) Find the marginal revenue.

(b) Find the revenue and marginal revenue when $x = 1000$ tons.

(c) Interpret $R'(1000)$.

SOLUTION

(a) $R'(x) = \dfrac{d}{dx}(x^2 + 5x) = 2x + 5$

(b) The revenue when $x = 1000$ tons is

$$R(1000) = (1000)^2 + 5(1000) = \$1{,}005{,}000$$

The marginal revenue when $x = 1000$ tons is

$$R'(1000) = 2(1000) + 5 = \$2005/\text{ton}$$

(c) $R'(1000) = \$2005/\text{ton}$ means that the revenue due to selling one additional ton of steel after 1000 tons have been sold is approximately \$2005.

· ·

Note that the average revenue derived from selling one additional ton after 1000 tons have been sold is

$$\frac{\Delta R}{\Delta x} = \frac{R(1001) - R(1000)}{1001 - 1000} = 1{,}007{,}006 - 1{,}005{,}000 = \$2006/\text{ton}$$

Now Work Problem 13   Observe that the actual average revenue differs from the marginal revenue by only \$1/ton, or 0.05% of the average revenue.

Relative and Percentage Rates of Change

So far, we have used $f'(x)$ as a rate of change of a quantity $f(x)$. The rate of change $f'(x)$ by itself may not give a clear picture of how much $f(x)$ is changing. Thus, we use the *relative* rate of change and the *percentage* rate of change to compare $f'(x)$ to $f(x)$. For example, a monthly rate of change of 1500 unit sales in a company that sells 6 million units a month would be insignificant. However, the same monthly rate of change would have a much larger impact for a company with only 50,000 unit sales per month. By considering the ratio

$$\frac{\text{Rate of change of quantity}}{\text{Size of quantity}}$$

we have a way of comparing the rate of change of a quantity with the quantity itself. This ratio is called the **relative rate of change.** Therefore, a rate of change of 1500 units of sales per month in a company that sells 6,000,000 is

$$\frac{1500}{6{,}000{,}000} = 0.00025$$

and the same rate of change in a company that sells 50,000 is

$$\frac{1500}{50,000} = 0.03$$

The second relative rate of change is much larger than the first. By multiplying relative rates by 100, we obtain **percentage rates of change**. The percentage rate for 6,000,000 is $0.00025 \times 100 = 0.025\%$, and for 50,000 is $0.03 \times 100 = 3\%$. Here are formulas for relative rate of change and percentage rate of change written in terms of the derivative.

RELATIVE RATE OF CHANGE; PERCENTAGE RATE OF CHANGE

The *relative rate of change* of $f(x)$ is

$$\frac{f'(x)}{f(x)}$$

The *percentage rate of change* of $f(x)$ is

$$\frac{f'(x)}{f(x)} \cdot 100$$

EXAMPLE 5

Determine the relative and percentage rate of change of

$$y = f(x) = 2x^2 - 3x + 10$$

when $x = 2$.

SOLUTION

Here,

$$f'(x) = 4x - 3$$

Since $f'(2) = 4 \cdot 2 - 3 = 5$ and $f(2) = 2 \cdot 2^2 - 3 \cdot 2 + 10 = 12$, the relative rate of change of y when $x = 2$ is

$$\frac{f'(2)}{f(2)} = \frac{5}{12} \approx 0.417$$

Multiplying 0.417 by 100 gives a percentage rate of change of 41.7%.

Now Work Problem 33

··················

EXERCISE 13.2   Answers to odd-numbered problems begin on page AN-58.

In Problems 1–10 find (a) the average rate of change as x changes from 1 to 3;
(b) the (instantaneous) rate of change at 1.

1. $f(x) = 3x + 4$ **2.** $f(x) = 2x - 6$ **3.** $f(x) = 3x^2 + 1$

4. $f(x) = 2x^2 + 1$

5. $f(x) = x^2 + 2x$

6. $f(x) = x^2 - 4x$

7. $f(x) = 2x^2 - x + 1$

8. $f(x) = 2x^2 + 3x - 2$

9. $f(x) = x^3 - 1$

10. $f(x) = x^3 + 4$

11. Ticket Sales   The cumulative ticket sales for the 10 days preceding a popular concert is given by

$$S = 4x^2 + 50x + 5000$$

where x represents the 10 days leading up to the concert, $1 \le x \le 10$.

(a) What is the average rate of change in sales from day 1 to day 5?

(b) What is the average rate of change in sales from day 1 to day 10?

(c) What is the average rate of change in sales from day 5 to day 10?

12. Marginal Cost   The cost per day, $C(x)$, in dollars, of producing x pairs of eyeglasses is

$$C(x) = 0.2x^2 + 3x + 1000$$

(a) Find the average cost due to producing 10 additional pairs of eyeglasses after 100 have been produced.

(b) Find the marginal cost.

(c) Find the marginal cost at $x = 100$.

(d) Interpret $C'(100)$.

13. Toy Truck Sales   At Dan's Toy Store the revenue R, in dollars, derived from selling x electric trucks is

$$R(x) = -0.005x^2 + 20x$$

(a) What is the average rate of change in revenue due to selling 10 additional trucks after 1000 have been sold?

(b) What is the marginal revenue?

(c) What is the marginal revenue at $x = 1000$?

(d) Interpret $R'(1000)$.

(e) For what value of x is $R'(x) = 0$?

14. Typewriter Sales   The weekly revenue R, in dollars, due to selling x typewriters is

$$R(x) = -20x^2 + 1000x$$

(a) Find the average rate of change in revenue due to selling 5 additional typewriters after the 20th has been sold.

(b) Find the marginal revenue.

(c) Find the marginal revenue at $x = 20$.

(d) Interpret the answers found in (a) and (c).

(e) For what value of x is $R'(x) = 0$?

15. Supply and Demand   Suppose $S(x) = 50x^2 - 50x$ is the supply function describing the number of crates of grapefruit a farmer is willing to supply to the market for x dollars per crate.

(a) How many crates is the farmer willing to supply for $10 per crate?

(b) How many crates is the farmer willing to supply for $13 per crate?

(c) Find the average rate of change in supply from $10 per crate to $13 per crate.

(d) Find the instantaneous rate of change in supply at $x = 10$.

(e) Interpret the answers found in (c) and (d).

16. Medicine   In a metabolic experiment the mass M of glucose decreases over time t according to the formula

$$M = 4.5 - 0.03t^2$$

(a) Find the average rate of change of the mass from $t = 0$ to $t = 2$.

(b) Find the instantaneous rate of change of mass at $t = 0$.

(c) Interpret the answers found in (a) and (b).

17. Cost and Revenue Functions   For a certain production facility the cost function is

$$C(x) = 2x + 5$$

and the revenue function is

$$R(x) = 8x - x^2$$

where x is the number of units produced (in thousands) and R and C are measured in millions of dollars. Find

(a) The marginal revenue

(b) The marginal cost

(c) The break-even point(s) [the number(s) x for which $R(x) = C(x)$]

(d) The number x for which marginal revenue equals marginal cost

(e) Graph $C(x)$ and $R(x)$ on the same set of axes.

18. Cost and Revenue Functions   For a certain production facility the cost function is

$$C(x) = x + 5$$

and the revenue function is

$$R(x) = 12x - 2x^2$$

where x is the number of units produced (in thousands) and R and C are measured in millions of dollars. Find

(a) The marginal revenue
(b) The marginal cost
(c) The break-even point(s) [the number(s) x for which $R(x) = C(x)$]
(d) The number x for which marginal revenue equals marginal cost
(e) Graph $C(x)$ and $R(x)$ on the same set of axes.

19. **Demand Equation**   The price p per ton of cement when x tons of cement are demanded is given by the equation

$$p = -10x + 2000$$

dollars. Find:

(a) The revenue function [*Hint*: $R = xp$, where p is the unit price.]
(b) The marginal revenue
(c) The marginal revenue at $x = 100$ tons
(d) The average rate of change in revenue from $x = 100$ to $x = 101$ tons

20. **Demand Equation**   The cost function and demand equation for a certain product are

$$C(x) = 50x + 40{,}000 \qquad \text{and} \qquad p = 100 - 0.01x$$

Find:

(a) The revenue function
(b) The marginal revenue
(c) The marginal cost
(d) The break-even point(s)
(e) The number x for which marginal revenue equals marginal cost

21. **Demand Equation**   A certain item can be produced at a cost of \$10 per unit. The demand equation for this item is

$$p = 90 - 0.02x$$

where p is the price in dollars and x is the number of units. Find

(a) The revenue function
(b) The marginal revenue
(c) The marginal cost
(d) The break-even point(s)
(e) The number x for which marginal revenue equals marginal cost

22. **Geometry**   A circle of radius r has area $A = \pi r^2$ and circumference $C = 2\pi r$. If the radius changes from r to $(r + \Delta r)$, find the

(a) Change in area
(b) Change in circumference
(c) Average rate of change of area with respect to the radius
(d) Average rate of change of the circumference with respect to the radius
(e) Rate of change of area with respect to the radius
(f) Rate of change of the circumference with respect to the radius

23. **Respiration Rate**   A human being's respiration rate R (in breaths per minute) is given by

$$R = -10.35p + 0.59p^2$$

where p is the partial pressure of carbon dioxide in the lungs. Find the rate of change in respiration rate when $p = 50$.

24. **Work Output**   The relationship* between the amount $A(t)$ of work output and the elapsed time t, $t \geq 0$, was found through empirical means to be

$$A(t) = a_3 t^3 + a_2 t^2 + a_1 t + a_0$$

where a_0, a_1, a_2, a_3 are constants. Find the instantaneous rate of change of work output at time t.

25. **Consumer Price Index**   The consumer price index (CPI) of an economy is described by the function

$$I(t) = -0.2t^2 + 3t + 200 \qquad (0 \leq t \leq 10)$$

where $t = 0$ corresponds to the year 1990.

(a) What was the average rate of increase in the CPI over the period from 1990 to 1993?
(b) At what rate was the CPI of the economy changing in 1990? in 1993? in 1995?

26. **Height of a Person**   The figure on the next page shows the graph of the height H (in inches) versus the age t (in years) of an individual from a little after birth to age 25.

(a) When is the growth rate greatest?
(b) Estimate the growth rate at age 10.
(c) At what approximate age between 10 and 20 is the growth rate greatest?

---

* M. R. Neifeld and A. T. Poffenberger, ''A Mathematical Analysis of Curves,'' *Journal of General Psychology,* **1** (1928), pp. 448–456.

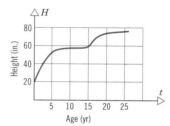

Figure for Problem 26

27. **Economics**   The following are statements about a function and its derivative. Interpret each case, indicate clearly what the function is, what each variable means, the appropriate units, and the derivative.

(a) The cost of a product decreases as more of it is produced.

(b) The price of a product decreases as more of it is produced.

(c) The increase in demand for a new product decreases over time.

(d) The U.S. population is growing more slowly now (in 1995) than it was 10 years ago.

(e) Health care costs continue to rise, but at a slower rate than 5 years ago.

(f) The world consumption of oil continued to decline in the past 5 years.

28. **Cost Function**   An airplane crosses the Atlantic Ocean (3000 miles) with an air speed of 500 miles per hour. The cost C (in dollars) per person is

$$C(x) = 100 + \frac{x}{10} + \frac{36000}{x}$$

where x is the ground speed (air speed $\pm$ wind). Find

(a) The rate of change of cost $C'(x)$

(b) $C'(50)$, $C'(100)$

In Problems 29–32, find (a) the rate of change of y with respect to x, (b) the relative rate of change of y at the given value, (c) the percentage rate of change of y.

29. $y = f(x) = 3x + 5$; $x = 4$

30. $y = f(x) = 6 - 3x$; $x = 2$

31. $y = f(x) = 3x^2 - 5x + 2$; $x = 3$

32. $y = f(x) = 4 - x^2$; $x = 1$

33. **Gross National Product**   The gross national product (GNP) of a certain country is given by $G(t) = t^2 + 3t + 102$ billion dollars t years after 1994.

(a) At what rate was the GNP changing with respect to time in 1994?

(b) At what relative and percentage rates was the GNP changing with respect to time in 1994?

34. **Revenue Function**   The revenue R obtained from the sale of q units of a product is given by

$$R(q) = 20q - 0.2q^2$$

(a) Find the rate of change of R with respect to q, when $q = 4$.

(b) Find the relative rate of change of R.

(c) Find the percentage rate of change of R.

35. **Cost Fluctuation**   Due to seasonal fluctuation, the price of oranges changes. If $P(t) = -\frac{1}{3}t^2 + 16t + 120$ represents the price in dollars per ton as a function of time, where t is measured in weeks,

(a) Find the rate of change of P with respect to t, when $t = 5$.

(b) Find the percentage rate of change.

36. **Cheese Production**   A dairy produces Q pounds of cottage cheese in t days, where $Q = 1000t + 0.2t^2$. Find the relative rate of change of Q with respect to t.

37. Consider the function $f(x) = 2x - 1$ and its derivative, $f'(x) = 2$.

(a) Graph $f(x)$ and $f'(x)$.

(b) Use the graphs to explain—in terms of rates of change—why the derivative is the constant function.

Technology Exercises

1. Graph the following cost and revenue functions for an airline company (in millions of dollars):

$$C(x) = 8 - 5x + x^2 \qquad R(x) = 5x - x^2$$

where $0 \le x \le 6$. From the graphs, estimate the following:

(a) For what values of x does the airline make a profit?

(b) When is the value of marginal cost zero?

2. The cost (in dollars) for a weekly production of x tons of steel is $C(x) = 0.1x^2 + 5x + 1000$.

(a) Find the marginal cost $C'(x)$.
(b) Find the cost when $x = 1000$ tons, that is, $C(1000)$.
(c) Find the marginal cost when $x = 1000$ tons, that is, $C'(1000)$.
(d) Find the equation of the tangent line to $C(x)$ at $x = 1000$:

$$y_{tan} = C'(1000)(x - 1000) + C(1000)$$

(e) Find the average cost of producing one more ton of steel after 1000 have been produced, $\Delta c / \Delta x$.

(f) Find the equation of the secant line through the points $(1000, C(1000))$ and $(1001, C(1001))$,

$$y_{sec} = \frac{\Delta c}{\Delta x} (x - 1000) + C(1000)$$

(g) Graph $C(x)$, y_{tan}, and y_{sec} on the range $[0, 2000]$ Scl 500 by $[0, 150000]$ Scl 50000.

(h) Trace to $x = 1000$ (or as close as your calculator allows). What is the difference between y_{tan} and y_{sec} at this point? Why? Is the marginal cost or the average cost easier to compute? What restriction must be made on Δx for marginal cost to be a good approximation to average cost?

13.3 PRODUCT AND QUOTIENT FORMULAS

In this section we develop formulas for the derivative of products and quotients of functions.

The Derivative of a Product

In Section 13.1 we noted that the derivative of the sum or the difference of two functions is simply the sum or the difference of their derivatives. The natural inclination at this point may be to assume that differentiating a product or quotient of two functions is as simple. But this is not the case, as illustrated for the case of a product of two functions. Consider

$$F(x) = f(x) \cdot g(x) = (3x^2 - 3)(2x^3 - x) \tag{a}$$

where $f(x) = 3x^2 - 3$ and $g(x) = 2x^3 - x$. The derivative of $f(x)$ is $f'(x) = 6x$ and the derivative of $g(x)$ is $g'(x) = 6x^2 - 1$. The product of these derivatives is

$$f'(x) \cdot g'(x) = 6x(6x^2 - 1) = 36x^3 - 6x \tag{b}$$

To see if this is equal to the derivative of the product, we first multiply the right side of (a) and then differentiate, using the rules of differentiation of the preceding section:

$$F(x) = (3x^2 - 3)(2x^3 - x) = 6x^5 - 9x^3 + 3x$$

so that

$$F'(x) = 30x^4 - 27x^2 + 3 \tag{c}$$

Since (b) and (c) are not equal, we conclude that the derivative of a product *is not* equal to the product of the derivatives. The formula for finding the derivative of the product of two functions is given in the next box.

> **Derivative of a Product**
>
> The derivative of the product of two differentiable functions equals the first function times the derivative of the second plus the second function times the derivative of the first. That is,
>
> $$\frac{d}{dx}[f(x)g(x)] = f(x)\frac{d}{dx}g(x) + g(x)\frac{d}{dx}f(x) \tag{1}$$

A proof is given at the end of the section.

The following version of Formula (1) may help you remember it.

> $$\frac{d}{dx}(\text{First} \cdot \text{Second}) = \text{First} \cdot \frac{d}{dx}\text{Second} + \text{Second} \cdot \frac{d}{dx}\text{First}$$

EXAMPLE 1

Find the derivative of $F(x) = (x^2 + 2x - 5)(x^3 - 1)$.

SOLUTION

The function F is the product of the two functions $f(x) = x^2 + 2x - 5$ and $g(x) = x^3 - 1$. We compute

$$\frac{d}{dx}f(x) = 2x + 2 \qquad \text{and} \qquad \frac{d}{dx}g(x) = 3x^2$$

By (1), we have

$$F'(x) = (x^2 + 2x - 5)\left[\frac{d}{dx}(x^3 - 1)\right] + (x^3 - 1)\left[\frac{d}{dx}(x^2 + 2x - 5)\right]$$

$$= (x^2 + 2x - 5)(3x^2) + (x^3 - 1)(2x + 2)$$

$$= 5x^4 + 8x^3 - 15x^2 - 2x - 2$$

Now Work Problem 1

· ·

Now that you know the formula for the derivative of a product, be careful not to use it unnecessarily. When one of the factors is a constant, you should use the formula for the derivative of a constant times a function. For example, it is easier to work

$$\frac{d}{dx}[5(x^2 + 1)] = 5\frac{d}{dx}(x^2 + 1) = (5)(2x) = 10x$$

than it is to work

$$\frac{d}{dx}[5(x^2 + 1)] = 5\left[\frac{d}{dx}(x^2 + 1)\right] + (x^2 + 1)\left(\frac{d}{dx}5\right)$$

$$= (5)(2x) + (x^2 + 1)(0) = 10x$$

EXAMPLE 2

Find the derivative of $g(x) = \left(1 - \dfrac{1}{x^2}\right)(x + 1)$.

SOLUTION

We apply the formula for the derivative of a product:

$$g'(x) = \left(1 - \frac{1}{x^2}\right)\frac{d}{dx}(x + 1) + (x + 1)\frac{d}{dx}\left(1 - \frac{1}{x^2}\right)$$

$$= \left(1 - \frac{1}{x^2}\right)(1) + (x + 1)\frac{d}{dx}(1 - x^{-2})$$

$$= 1 - \frac{1}{x^2} + (x + 1)(2x^{-3}) = 1 - \frac{1}{x^2} + \frac{2(x + 1)}{x^3}$$

$$= 1 - \frac{1}{x^2} + \frac{2x}{x^3} + \frac{2}{x^3} = 1 + \frac{1}{x^2} + \frac{2}{x^3}$$

· ·

Alternatively, we could have solved Example 2 by multiplying the factors first. Then

$$g(x) = \left(1 - \frac{1}{x^2}\right)(x + 1) = x + 1 - \frac{1}{x} - \frac{1}{x^2}$$

so

$$g'(x) = 1 + \frac{1}{x^2} + \frac{2}{x^3}$$

EXAMPLE 3

Find the derivative of $f(x) = (2\sqrt{x} + 1)(\sqrt[3]{x} - 2)$.

SOLUTION

First, we change each radical to its fractional exponent equivalent:

$$f(x) = (2x^{1/2} + 1)(x^{1/3} - 2)$$

By the product rule,

$$f'(x) = (2x^{1/2} + 1)(\tfrac{1}{3}x^{-2/3}) + (x^{1/3} - 2)(x^{-1/2})$$

$$= \frac{2x^{1/2} + 1}{3x^{2/3}} + \frac{x^{1/3} - 2}{x^{1/2}} = \frac{5x + x^{1/2} - 6x^{2/3}}{3x^{7/6}}$$

· ·

The Derivative of a Quotient

As in the case with a product, the derivative of a quotient *is not* the quotient of the derivatives.

Derivative of a Quotient

The derivative of the quotient of two differentiable functions is equal to the denominator times the derivative of the numerator minus the numerator times the derivative of the denominator, all divided by the square of the denominator.

$$\frac{d}{dx}\left[\frac{f(x)}{g(x)}\right] = \frac{g(x)\dfrac{d}{dx}f(x) - f(x)\dfrac{d}{dx}g(x)}{[g(x)]^2} \qquad \text{where } g(x) \neq 0 \qquad (2)$$

You may want to memorize the following version of Formula (2):

$$\frac{d}{dx}\frac{\textbf{Numerator}}{\textbf{Denominator}} = \frac{(\textbf{Denominator})\dfrac{d}{dx}(\textbf{Numerator}) - (\textbf{Numerator})\dfrac{d}{dx}(\textbf{Denominator})}{(\textbf{Denominator})^2}$$

EXAMPLE 4

Find the derivative of $F(x) = \dfrac{x^2 + 1}{x - 3}$.

SOLUTION

Here, the function F is the quotient of $f(x) = x^2 + 1$ and $g(x) = x - 3$. Thus, we use (2) to get

$$\begin{aligned}
\frac{d}{dx}\left(\frac{x^2 + 1}{x - 3}\right) &= \frac{(x - 3)\dfrac{d}{dx}(x^2 + 1) - (x^2 + 1)\dfrac{d}{dx}(x - 3)}{(x - 3)^2} \\
&= \frac{(x - 3)(2x) - (x^2 + 1)(1)}{(x - 3)^2} \\
&= \frac{2x^2 - 6x - x^2 - 1}{(x - 3)^2} = \frac{x^2 - 6x - 1}{(x - 3)^2}
\end{aligned}$$

Now Work Problem 15

EXAMPLE 5

Find the derivative of $y = \dfrac{(1 - 3x)(2x + 1)}{3x - 2}$.

SOLUTION

We shall solve the problem in two ways.

Method 1   Use the formula for the derivative of a quotient right away.

$$y' = \frac{d}{dx} \frac{(1 - 3x)(2x + 1)}{3x - 2}$$

$$= \frac{(3x - 2)\dfrac{d}{dx}[(1 - 3x)(2x + 1)] - (1 - 3x)(2x + 1)\dfrac{d}{dx}(3x - 2)}{(3x - 2)^2}$$

$$= \frac{(3x - 2)[(1 - 3x)(2) + (2x + 1)(-3)] - (1 - 3x)(2x + 1)(3)}{(3x - 2)^2}$$

$$= \frac{(3x - 2)[2 - 6x - 6x - 3] - (-18x^2 - 3x + 3)}{(3x - 2)^2}$$

$$= \frac{-36x^2 + 21x + 2 + 18x^2 + 3x - 3}{(3x - 2)^2}$$

$$= \frac{-18x^2 + 24x - 1}{(3x - 2)^2}$$

Method 2 First, multiply the factors in the numerator and then apply the formula for the derivative of a quotient.

$$y = \frac{(1 - 3x)(2x + 1)}{3x - 2} = \frac{-6x^2 - x + 1}{3x - 2}$$

Thus,

$$y' = \frac{d}{dx} \frac{-6x^2 - x + 1}{3x - 2}$$

$$= \frac{(3x - 2)\dfrac{d}{dx}(-6x^2 - x + 1) - (-6x^2 - x + 1)\dfrac{d}{dx}(3x - 2)}{(3x - 2)^2}$$

$$= \frac{(3x - 2)(-12x - 1) - (-6x^2 - x + 1)(3)}{(3x - 2)^2}$$

$$= \frac{-36x^2 + 21x + 2 + 18x^2 + 3x - 3}{(3x - 2)^2} = \frac{-18x^2 + 24x - 1}{(3x - 2)^2}$$

················

As you can see from this example, looking at alternative methods may make the differentiation easier.

EXAMPLE 6

Find the derivative of $f(x) = \dfrac{x^{1/2} - 2}{x^{1/2}}$.

(a) By using the quotient rule first and then simplifying.
(b) By first simplifying $f(x)$ and then differentiating.

SOLUTION

(a) If we apply the quotient rule before doing any simplification, we get

$$f'(x) = \frac{x^{1/2}(\frac{1}{2}x^{-1/2}) - (x^{1/2} - 2)\frac{1}{2}x^{-1/2}}{(x^{1/2})^2} = \frac{\frac{1}{2} - \frac{1}{2} + x^{-1/2}}{x} = \frac{x^{-1/2}}{x} = \frac{1}{x^{3/2}}$$

(b) If we simplify first, the problem becomes to find the derivative of

$$f(x) = \frac{x^{1/2} - 2}{x^{1/2}} = \frac{x^{1/2}}{x^{1/2}} - \frac{2}{x^{1/2}} = 1 - \frac{2}{x^{1/2}} = 1 - 2x^{-1/2}$$

Thus,

$$f'(x) = \frac{d}{dx}(1 - 2x^{-1/2}) = 0 - 2\left(-\frac{1}{2}\right)x^{-3/2} = x^{-3/2} = \frac{1}{x^{3/2}}$$

· ·

Application

EXAMPLE 7

The value $v(t)$, in dollars, of a car t years after its purchase is given by the equation

$$v(t) = \frac{8000}{t} + 5000 \qquad 1 \le t \le 5$$

Graph the function $v = v(t)$. Then find

(a) The average rate of change in value from $t = 1$ to $t = 4$
(b) The instantaneous rate of change in value
(c) The instantaneous rate of change in value after 1 year
(d) The instantaneous rate of change in value after 3 years
(e) Interpret the answers to (c) and (d).

SOLUTION

The graph of $v = v(t)$ is given in Figure 5.

Figure 5

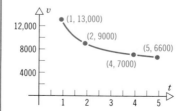

(a) The average rate of change in value from $t = 1$ to $t = 4$ is given by

$$\frac{v(4) - v(1)}{4 - 1} = \frac{7000 - 13{,}000}{3} = -2000$$

So the average rate of change in value from $t = 1$ to $t = 4$ is $-\$2000$ per year.

(b) The derivative $v'(t)$ of $v(t)$ equals the instantaneous rate of change in the value of the car. Thus,

$$v'(t) = \frac{-8000}{t^2}$$

Notice that $v'(t) < 0$; we interpret this to mean that the value of the car is decreasing over time.

(c) After 1 year, $v'(1) = -\frac{8000}{1} = -\$8000$.

(d) After 3 years, $v'(3) = -\frac{8000}{9} = -\$888.89$.

(e) $v'(1) = -\$8000$ means that the value of the car after 1 year will decline by approximately $\$8000$ over the next year; $v'(3) = -\$888.89$ means that the value of the car after 3 years will decline by approximately $\$888.89$ over the next year.

Now Work Problem 51

· ·

Summary

Each of the derivative formulas given thus far can be written without reference to the independent variable of the function. If f and g are differentiable functions, we have the following formulas:

Derivative of a constant times a function	$(cf)' = cf'$
Derivative of a sum	$(f + g)' = f' + g'$
Derivative of a difference	$(f - g)' = f' - g'$
Derivative of a product	$(f \cdot g)' = f \cdot g' + g \cdot f'$
Derivative of a quotient	$\left(\dfrac{f}{g}\right)' = \dfrac{g \cdot f' - f \cdot g'}{g^2}$

Proofs

Derivative of a Product   Let $F(x) = f(x)g(x)$; then

$$F'(x) = \lim_{h \to 0} \frac{F(x + h) - F(x)}{h} = \lim_{h \to 0} \frac{f(x + h)g(x + h) - f(x)g(x)}{h}$$

Now subtract and add $f(x + h)\, g(x)$ in the numerator:

$$F'(x) = \lim_{h \to 0} \frac{f(x + h)g(x + h) - f(x + h)g(x) + f(x + h)g(x) - f(x)g(x)}{h}$$

$$= \lim_{h \to 0} \left[f(x + h) \frac{g(x + h) - g(x)}{h} + g(x) \frac{f(x + h) - f(x)}{h} \right]$$

$$= \lim_{h \to 0} f(x + h) \frac{g(x + h) - g(x)}{h} + \lim_{h \to 0} g(x) \frac{f(x + h) - f(x)}{h}$$

$$= \left[\lim_{h \to 0} f(x + h) \right] \left[\lim_{h \to 0} \frac{g(x + h) - g(x)}{h} \right] + \left[\lim_{h \to 0} g(x) \right] \left[\lim_{h \to 0} \frac{f(x + h) - f(x)}{h} \right]$$

From the differentiability of f and g, we obtain the following limits:

$$\lim_{h \to 0} f(x + h) = f(x) \qquad \lim_{h \to 0} \frac{g(x + h) - g(x)}{h} = g'(x)$$

$$\lim_{h \to 0} g(x) = g(x) \qquad \lim_{h \to 0} \frac{f(x + h) - f(x)}{h} = f'(x)$$

Therefore, we have $F'(x) = f(x)g'(x) + g(x)f'(x)$ and

$$\frac{d}{dx} [f(x)g(x)] = f(x) \frac{d}{dx} g(x) + g(x) \frac{d}{dx} f(x)$$

∎

Derivative of a Quotient   Let

$$F(x) = \frac{f(x)}{g(x)}$$

Then

$$F'(x) = \lim_{h \to 0} \frac{F(x + h) - F(x)}{h} = \lim_{h \to 0} \frac{\dfrac{f(x + h)}{g(x + h)} - \dfrac{f(x)}{g(x)}}{h}$$

$$= \lim_{h \to 0} \frac{g(x)f(x + h) - f(x)g(x + h)}{hg(x)g(x + h)}$$

Now by adding and subtracting $f(x)g(x)$ in the numerator, we have

$$F'(x) = \lim_{h \to 0} \frac{g(x)f(x + h) - f(x)g(x) + f(x)g(x) - f(x)g(x + h)}{hg(x)g(x + h)}$$

$$= \frac{\displaystyle\lim_{h \to 0} \frac{g(x)[f(x + h) - f(x)]}{h} - \lim_{h \to 0} \frac{f(x)[g(x + h) - g(x)]}{h}}{\displaystyle\lim_{h \to 0} [g(x)g(x + h)]}$$

$$= \frac{g(x) \left[\displaystyle\lim_{h \to 0} \frac{f(x + h) - f(x)}{h} \right] - f(x) \left[\displaystyle\lim_{h \to 0} \frac{g(x + h) - g(x)}{h} \right]}{\displaystyle\lim_{h \to 0} [g(x)g(x + h)]}$$

$$= \frac{g(x)f'(x) - f(x)g'(x)}{[g(x)]^2}$$

Thus,

$$\frac{d}{dx}[f(x)g(x)] = \frac{g(x)\dfrac{d}{dx}f(x) - f(x)\dfrac{d}{dx}g(x)}{[g(x)]^2}$$

∎

EXERCISE 13.3   Answers to odd-numbered problems begin on page AN-59.

In Problems 1–14 find the derivative of each function by using the formula for the derivative of a product.

1. $f(x) = (2x + 1)(4x - 3)$

2. $f(x) = (3x - 4)(2x + 5)$

3. $f(t) = (t^2 + 1)(t^2 - 4)$

4. $f(t) = (t^2 - 3)(t^2 + 4)$

5. $f(x) = (3x - 5)(2x^2 + 1)$

6. $f(x) = (3x^2 - 1)(4x + 1)$

7. $f(x) = (x^5 + 1)(3x^3 + 8)$

8. $f(x) = (x^6 - 2)(4x^2 + 1)$

9. $f(x) = x^3(x^2 - 8)$

10. $f(x) = 5x^3(x^2 - 9)$

11. $f(x) = (x^3 - 1)(3x^2 - 2x + 1) + 3(x^2 - 2)$

12. $f(x) = 3(x^2 - 1) - 4(x^3 - 2)(x^2 - 5)$

13. $y = (5 - t^{1/2})(t^2 + 1)$

14. $y = \frac{5}{3}(\sqrt{u} - 2)(3u + 2)$

In Problems 15–30 find the derivative of each function.

15. $f(x) = \dfrac{x}{x + 1}$

16. $f(x) = \dfrac{x + 4}{x^2}$

17. $f(x) = \dfrac{3x + 4}{2x - 1}$

18. $f(x) = \dfrac{3x - 5}{4x + 1}$

19. $f(x) = \dfrac{x^2}{x - 4}$

20. $f(x) = \dfrac{x}{x^2 - 4}$

21. $f(x) = \dfrac{2x + 1}{3x^2 + 4}$

22. $f(x) = \dfrac{2x^2 - 1}{5x + 2}$

23. $f(t) = \dfrac{-2}{t^2}$

24. $f(t) = \dfrac{4}{t^3}$

25. $f(x) = 1 + \dfrac{1}{x} + \dfrac{1}{x^2}$

26. $f(x) = 1 - \dfrac{1}{x} + \dfrac{1}{x^2}$

27. $f(v) = \dfrac{v^3 - 7}{v}$

28. $y = \dfrac{x - 3}{\sqrt{x}}$

29. $y = \dfrac{3x^2 - 2x + 1}{\sqrt{x}}$

30. $y = \left(\dfrac{1 - v}{v}\right)(1 - v^2)$

In Problems 31–34 find the slope of the tangent line to the graph of the function f at the indicated point. What is an equation of the tangent line?

31. $f(x) = (x^3 - 2x + 2)(x + 1)$ at $(1, 2)$

32. $f(x) = (2x^2 - 5x + 1)(x - 3)$ at $(1, 4)$

33. $f(x) = \dfrac{x^3}{x + 1}$ at $(1, \frac{1}{2})$

34. $f(x) = \dfrac{x^2}{x - 1}$ at $(-1, -\frac{1}{2})$

In Problems 35–38 find those x, if any, at which f'(x) = 0.

35. $f(x) = (x^2 - 2)(2x - 1)$

36. $f(x) = (3x^2 - 3)(2x^3 - x)$

37. $f(x) = \dfrac{x^2}{x + 1}$

38. $f(x) = \dfrac{x^2 + 1}{x}$

In Problems 39–50 find y′.

39. $y = x^2(3x - 2)$

40. $y = (x^2 + 2)(x - 1)$

41. $y = (x^{-2} + 4)(4x^2 + 3)$

42. $y = (2x^{-1} + 3)(x^{-3} + x^{-2})$

43. $y = \dfrac{(2x + 3)(x - 4)}{3x + 5}$

44. $y = \dfrac{(3x - 2)(x^2 + 1)}{4x - 3}$

45. $y = \dfrac{3x + 1}{(x - 2)(x + 2)}$

46. $y = \dfrac{2x - 5}{(1 - x)(1 + x)}$

47. $y = \dfrac{(3x + 4)(2x - 3)}{(2x + 1)(3x - 2)}$

48. $y = \dfrac{(2 - 3x)(1 - x)}{(x + 2)(3x + 1)}$

49. $y = \dfrac{x^{-2} - x^{-1}}{x^{-2} + x^{-1}}$

50. $y = \dfrac{3x^{-4} - x^{-2}}{x^{-3} + x^{-1}}$

51. Value of a Car The value v of a luxury car after t years is

$$v(t) = \frac{10{,}000}{t} + 6000 \qquad 1 \le t \le 6$$

Graph $v = v(t)$.

(a) What is the average rate of change in value from $t = 2$ to $t = 5$?

(b) What is the instantaneous rate of change in value?

(c) What is the instantaneous rate of change after 2 years?

(d) What is the instantaneous rate of change after 5 years?

(e) Interpret the answers found in (c) and (d).

52. Value of a Painting The value v of a famous painting t years after it is purchased is

$$v(t) = \frac{100t^2 + 50}{t} + 400 \qquad 1 \le t \le 5$$

Graph $v = v(t)$.

(a) What is the average rate of change in value from $t = 1$ to $t = 3$?

(b) What is the instantaneous rate of change in value?

(c) What is the instantaneous rate of change after 1 year?

(d) What is the instantaneous rate of change after 3 years?

(e) Interpret the answers found in (c) and (d).

53. Demand Equation The demand equation for a certain commodity is

$$p = 10 + \frac{40}{x} \qquad 1 \le x \le 10$$

where p is the price in dollars when x units are demanded. Find

(a) The revenue function

(b) The marginal revenue

(c) The marginal revenue for $x = 4$

(d) The marginal revenue for $x = 6$

54. Satisfaction and Reward The relationship* between satisfaction S and total reward r has been found to be

$$S(r) = \frac{ar}{g - r}$$

where $g \ge 0$ is the predetermined goal level and $a > 0$ is the perceived justice per unit of reward. Show that the instantaneous rate of change of satisfaction with respect to reward is inversely proportional to the square of the difference between the personal goal of the individual and the amount of reward received.

55. Cost Function The cost of fuel in operating a luxury yacht is given by the equation

$$C(s) = \frac{-3s^2 + 1200}{s}$$

where s is the speed of the yacht. Find the rate at which the cost is changing when $s = 10$.

56. Price–Demand Function The price–demand function for calculators is given by

$$D(p) = \frac{100{,}000}{p^2 + 10p + 50} \qquad 5 \le p \le 20$$

where D is the quantity demanded per week and p is the unit price in dollars.

(a) Find $D'(p)$, the rate of change of demand with respect to price.

(b) Find $D'(5)$, $D'(10)$, and $D'(15)$ and interpret your results.

---

* R. Carzo and J. N. Yanouzas, *Formal Organization: A Systems Approach.* Homewood, Ill.: Richard D. Irwin Press, 1967.

57. Rising Object   The height, in kilometers, that a balloon will rise in t hours is given by the formula

$$s = \frac{t^2}{2 + t}$$

Find the rate at which the balloon is rising after (a) 10 minutes, (b) 20 minutes.

58. Population Growth   A population of 1000 bacteria is introduced into a culture and grows in number according to the formula

$$P(t) = 1000 \left(1 + \frac{4t}{100 + t^2} \right)$$

where t is measured in hours. Find the rate at which the population is growing when

(a) $t = 1$ (b) $t = 2$ (c) $t = 3$ (d) $t = 4$

59. Drug Concentration   The concentration of a certain drug in a patient's bloodstream t hours after injection is given by

$$C(t) = \frac{0.4t}{2t^2 + 1}$$

Find the rate at which the concentration of the drug is changing with respect to time. At what rate is the concentration changing

(a) 10 minutes after the injection?
(b) 30 minutes after the injection?
(c) 1 hour after the injection?
(d) 3 hours after the injection?

60. Intensity of Illumination   The intensity of illumination I on a surface is inversely proportional to the square of the distance r from the surface to the source of light. If the intensity is 1000 units when the distance is 1 meter, find the rate of change of the intensity with respect to the distance when the distance is 10 meters.

61. Cost Function   The cost C, in thousands of dollars, for removal of pollution from a certain lake is

$$C(x) = \frac{5x}{110 - x}$$

where x is the percent of pollutant removed.

(a) Find $C'(x)$, the rate of change of cost with respect to the amount of pollutant removed.
(b) Compute $C'(10)$, $C'(20)$, $C'(70)$, $C'(90)$.

62. Cost Function   An airplane crosses the Atlantic Ocean (3000 miles) with an air speed of 500 miles per hour. The cost C (in dollars) per person is

$$C(x) = 100 + \frac{x}{10} + \frac{36,000}{x}$$

where x is the ground speed (air speed $\pm$ wind). Find:

(a) The marginal cost
(b) The marginal cost at a ground speed of 500 mph
(c) The marginal cost at a ground speed of 550 mph
(d) The marginal cost at a ground speed of 450 mph

63. Average Cost Function   If C is the total cost function, then $\overline{C} = C/x$ is defined as the **average cost function,** that is, the cost per unit produced. Typically, the graph of the average cost function has a U shape. This is so since we expect higher average costs because of plant inefficiency at low output levels and also at high output levels near plant capacity. Suppose a company estimates that the total cost of producing x units of a certain product is given by

$$C(x) = 400 + 0.02x + 0.0001x^2$$

Then the average cost is given by

$$\overline{C}(x) = \frac{C(x)}{x} = \frac{400}{x} + 0.02 + 0.0001x$$

(a) Find the marginal average cost $\overline{C}'(x)$.
(b) Find the marginal average cost at $x = 200$, 300, and 400, and interpret your results.

64. Prove that if f, g, and h are differentiable functions, then

$$\frac{d}{dx} [f(x)g(x)h(x)]$$
$$= f(x)g(x)h'(x) + f(x)h(x)g'(x) + h(x)g(x)f'(x)$$

From this, deduce that

$$\frac{d}{dx} [f(x)]^3 = 3[f(x)]^2 f'(x)$$

In Problems 65–67 use the formula for finding the derivative of the product of three functions, as given in Problem 64, to find dy/dx.

65. $y = (x^2 + 1)(x - 1)(x + 5)$

66. $y = (x - 1)(x^2 + 5)(x^3 - 1)$

67. $y = (x^4 + 1)^3$

68. Prove that if $g \neq 0$ is a differentiable function, then

$$\frac{d}{dx}\left[\frac{1}{g(x)}\right] = \frac{-g'(x)}{[g(x)]^2}$$

13.4 THE CHAIN RULE; THE EXTENDED POWER RULE

The Chain Rule allows us to take the derivative of a function within a function. To be able to use the Chain Rule formula, we first have to understand what we mean by a *composite function.*

Consider the function $y = (2x + 3)^2$. If we write $y = f(u) = u^2$ and $u = g(x) = 2x + 3$, then, by a substitution process, we can obtain the original function, namely, $y = f(u) = f(g(x)) = (2x + 3)^2$. This process is called **composition**, and the function $y = (2x + 3)^2$ is called the **composite function** of $y = f(u) = u^2$ and $u = g(x) = 2x + 3$.

EXAMPLE 1

Find the composite function of

$$y = f(u) = \sqrt{u} \qquad \text{and} \qquad u = g(x) = x^2 + 4$$

SOLUTION

The composite function is

$$y = f(u) = \sqrt{u} = \sqrt{g(x)} = \sqrt{x^2 + 4}$$

. . . .

Additional examples of composite functions are shown in Example 2.

EXAMPLE 2

(a) If $y = (5x + 1)^3$, then $y = u^3$ and $u = 5x + 1$.

(b) If $y = (x^2 + 1)^{-2}$, then $y = u^{-2}$ and $u = x^2 + 1$.

(c) If $y = \dfrac{5}{(2x + 3)^3}$, then $y = \dfrac{5}{u^3}$ and $u = 2x + 3$.

(d) If $y = \sqrt{x^2 + 1}$, then $y = \sqrt{u}$ and $u = x^2 + 1$.

(e) If $y = \sqrt[3]{2x + 5}$, then $y = \sqrt[3]{u}$ and $u = 2x + 5$.

. . . .

In the above examples the composite function was "broken up" into simpler functions. Now we turn our attention to the Chain Rule.

We would have a difficult time using the differentiation formulas derived so far to compute the derivative of the function

$$f(x) = (x^3 - 2x - 1)^{100}$$

(Our only option would be to expand the polynomial.) In this section we derive a result that will enable us to compute the derivative of this function as well as a large number of other functions. The idea behind the result is illustrated below.

Suppose that $y = f(u)$ is a function of u, and in turn $u = g(x)$ is a function of x. That is, y is a composite function $y = f(g(x))$. What then is the derivative of $f(g(x))$? It turns out that the derivative of the composite function $f(g(x))$, where $u = g(x)$ can be written as the product of

$$\frac{dy}{du} \cdot \frac{du}{dx}$$

The formula for this important derivative is called the **Chain Rule**.

Another way to motivate the Chain Rule is to interpret derivatives as rates of change. Suppose the total manufacturing cost of a certain product is a function of the number of units produced, which in turn is a function of the number of hours it took to produce the units. If we let c denote the cost (in dollars), q the number of units, and t the time (in hours), then

$$\frac{dc}{dq} = \text{Rate of change of cost with respect to output} \quad \text{Dollars per unit}$$

and

$$\frac{dq}{dt} = \text{Rate of change of output with respect to time} \quad \text{Units per hour}$$

The product of these two rates is the rate of change of cost with respect to time, dc/dt. That is,

$$\frac{dc}{dt} = \frac{dc}{dq} \cdot \frac{dq}{dt} = \text{Rate of change of cost with respect to time} \quad \text{Dollars per hour}$$

This product,

$$\frac{dc}{dt} = \frac{dc}{dq} \cdot \frac{dq}{dt}$$

is an example of the Chain Rule, which is used to find the derivative of a composite function.

Chain Rule

Suppose y is a differentiable function of u and u is a differentiable function of x. Then y is regarded as a composite function of x and

$$\frac{dy}{dx} = \frac{dy}{du} \cdot \frac{du}{dx} \qquad (1)$$

The **Chain Rule** states that the derivative of y with respect to x is the derivative of y with respect to u times the derivative of u with respect to x.

One way to remember the chain rule is to treat dy/du and du/dx as fractions; when we multiply, the denominator in dy/du cancels the numerator in du/dx. The complete proof of this rule, which requires advanced topics, will not be given in this book. A partial proof is given at the end of this section.

EXAMPLE 3 | Use the Chain Rule to find the derivative of $y = (5x + 1)^3$.

SOLUTION

We break up y into simpler functions: If $y = (5x + 1)^3$, then $y = u^3$ and $u = 5x + 1$. To find dy/dx, we first find dy/du and du/dx:

$$\frac{dy}{du} = \frac{d}{du}(u^3) = 3u^2 \qquad \text{and} \qquad \frac{du}{dx} = \frac{d}{dx}(5x + 1) = 5$$

By the Chain Rule,

$$\frac{dy}{dx} = \frac{dy}{du}\frac{du}{dx} = (3u^2)(5) = 15u^2 = 15(5x + 1)^2$$
$$\underset{u\,=\,5x\,+\,1}{\uparrow}$$

Now Work Problem 1 | $\cdots\cdots\cdots\cdots\cdots\cdots\cdots\cdots$

Notice that when using the Chain Rule, we must substitute for u in the expression for dy/du so that we obtain a function of x.

EXAMPLE 4 | Find the derivative of $y = (4x^3 - 6x^2 + 5x - 2)^4$.

SOLUTION

Here, $y = u^4$ and $u = 4x^3 - 6x^2 + 5x - 2$. Thus,

$$\frac{dy}{dx} = \frac{dy}{du}\frac{du}{dx}$$

$$= (4u^3)(12x^2 - 12x + 5) = 4(4x^3 - 6x^2 + 5x - 2)^3(12x^2 - 12x + 5)$$

$$\uparrow$$
$$u = 4x^3 - 6x^2 + 5x - 2$$

. . . .

The Extended Power Rule

In Example 4 we used the Chain Rule to find the derivative of

$$y = (4x^3 - 6x^2 + 5x - 2)^4$$

which is a function raised to a power 4. In cases like this, where the function we deal with takes the form $(g(x))^r$ and r is a real number, we can use the **Extended Power Rule.** It is a special case of the Chain Rule and can be used to find derivatives of functions that take the form $(g(x))^r$.

The Extended Power Rule

If g is a differentiable function and r is any real number, then

$$\frac{d}{dx}[g(x)]^r = r[g(x)]^{r-1}g'(x) \qquad (2)$$

Notice the similarity between the Extended Power Rule and the formula

$$\frac{d}{dx}x^r = rx^{r-1}$$

The main difference between these formulas is the factor $g'(x)$.

EXAMPLE 5   Find the derivative of the function $f(x) = (x^2 + 1)^3$.

SOLUTION

We could, of course, expand the right-hand side and proceed according to techniques discussed earlier. However, the significance of the Extended Power Rule is that it enables us to find derivatives of functions like this without resorting to tedious (and sometimes impossible) computation.

The function $f(x) = (x^2 + 1)^3$ is the function $g(x) = x^2 + 1$ raised to the power 3. By the Extended Power Rule,

$$\frac{d}{dx} f(x) = \frac{d}{dx}(x^2 + 1)^3 = 3(x^2 + 1)^2 \frac{d}{dx}(x^2 + 1)$$

$$= 3(x^2 + 1)^2(2x) = 6x(x^2 + 1)^2$$

Now Work Problem 11

∙∙∙∙∙∙∙∙∙∙∙∙∙∙∙∙∙∙∙∙∙∙∙

EXAMPLE 6

Find $f'(x)$ if $f(x) = \dfrac{1}{(x^3 + 4)^5}$.

SOLUTION

We write $f(x)$ as $f(x) = (x^3 + 4)^{-5}$. Then we use the Extended Power Rule:

$$f'(x) = \frac{d}{dx}(x^3 + 4)^{-5} = -5(x^3 + 4)^{-6}\frac{d}{dx}(x^3 + 4)$$

$$= -5(x^3 + 4)^{-6}(3x^2) = \frac{-15x^2}{(x^3 + 4)^6}$$

∙∙∙∙∙∙∙∙∙∙∙∙∙∙∙∙∙∙∙∙∙∙∙

EXAMPLE 7

Additional examples of the Extended Power Rule are

(a) $f(x) = (3 - x^3)^{-5}$
   $f'(x) = (-5)(3 - x^3)^{-6}(-3x^2) = 15x^2(3 - x^3)^{-6}$
(b) $f(x) = (2x + 3)^{3/2}$
   $f'(x) = (\frac{3}{2})(2x + 3)^{1/2}(2) = 3(2x + 3)^{1/2}$
(c) $f(x) = \sqrt[3]{(x^3 - 3x^2 + 1)} = (x^3 - 3x^2 + 1)^{1/3}$
   $f'(x) = (\frac{1}{3})(x^3 - 3x^2 + 1)^{-2/3}(3x^2 - 6x) = (x^2 - 2x)(x^3 - 3x^2 + 1)^{-2/3}$

∙∙∙∙∙∙∙∙∙∙∙∙∙∙∙∙∙∙∙∙∙∙∙

Often, we must use at least one other formula along with the Extended Power Rule to differentiate a function. Here are two examples.

EXAMPLE 8

Find the derivative of the function $f(x) = x(x^2 + 1)^3$.

SOLUTION

The function f is the product of x and $(x^2 + 1)^3$. We begin by using the formula for the derivative of a product. That is,

$$f'(x) = x\frac{d}{dx}(x^2 + 1)^3 + (x^2 + 1)^3 \frac{d}{dx}x$$

We continue by using the Extended Power Rule:

$$f'(x) = x\left[3(x^2 + 1)^2 \frac{d}{dx}(x^2 + 1)\right] + (x^2 + 1)^3 \cdot 1$$

$$f'(x) = (x)(3)(x^2 + 1)^2(2x) + (x^2 + 1)^3$$
$$= (x^2 + 1)^2(6x^2) + (x^2 + 1)^2(x^2 + 1)$$
$$= (x^2 + 1)^2[6x^2 + (x^2 + 1)] = (x^2 + 1)^2(7x^2 + 1)$$

Now Work Problem 17

....

EXAMPLE 9

Find the derivative of the function $f(x) = \left(\dfrac{3x + 2}{4x^2 - 5}\right)^5$.

SOLUTION

Here, f is the quotient $(3x + 2)/(4x^2 - 5)$ raised to the power 5. Thus, we begin by using the Extended Power Rule and then use the formula for the derivative of a quotient:

$$f'(x) = (5)\left(\frac{3x + 2}{4x^2 - 5}\right)^4\left[\frac{d}{dx}\left(\frac{3x + 2}{4x^2 - 5}\right)\right]$$
↑
Apply Extended Power Rule

$$= (5)\left(\frac{3x + 2}{4x^2 - 5}\right)^4\left[\frac{(4x^2 - 5)\dfrac{d}{dx}(3x + 2) - (3x + 2)\dfrac{d}{dx}(4x^2 - 5)}{(4x^2 - 5)^2}\right]$$
↑
Apply quotient formula

$$= (5)\left(\frac{3x + 2}{4x^2 - 5}\right)^4\left[\frac{(4x^2 - 5)(3) - (3x + 2)(8x)}{(4x^2 - 5)^2}\right]$$

$$= \frac{5(3x + 2)^4(-12x^2 - 16x - 15)}{(4x^2 - 5)^6}$$

or

$$= \frac{-5(3x + 2)^4(12x^2 + 16x + 15)}{(4x^2 - 5)^6}$$

Now Work Problem 37

....

As this example illustrates, whenever a problem involves finding the derivative of a function raised to a power, either the Extended Power Rule or the Chain Rule may be used. However, the Extended Power Rule, as a special case of the Chain Rule, will not always work—especially when exponential and logarithmic functions are involved (see Chapter 11).

Application

The revenue R derived from selling x units of a product at a price p per unit is

$$R = xp$$

where $p = d(x)$ is the demand equation, namely, the equation that gives the price p when the number x of units demanded is known. The marginal revenue is then the derivative of R with respect to x:

$$R'(x) = \frac{d}{dx} xp = p + x \frac{dp}{dx} \qquad (3)$$

It is sometimes easier to find the marginal revenue by using (3) instead of differentiating the revenue function directly.

EXAMPLE 10

Suppose the price p per ton when x tons of polished aluminum are demanded is given by the equation

$$p = \frac{2000}{x + 20} - 10 \qquad 0 < x < 90$$

Find

(a) The rate of change of price with respect to x
(b) The revenue function
(c) The marginal revenue
(d) The marginal revenue at $x = 20$ and $x = 80$

SOLUTION

(a) The rate of change of price with respect to x is the derivative dp/dx.

$$\frac{dp}{dx} = \frac{d}{dx}\left(\frac{2000}{x + 20} - 10\right) = \frac{d}{dx} 2000(x + 20)^{-1} - \frac{d}{dx} 10 = \frac{-2000}{(x + 20)^2}$$

(b) The revenue function is

$$R(x) = xp = \frac{2000x}{x + 20} - 10x$$

(c) The marginal revenue is

$$R'(x) = p + x \frac{dp}{dx} = \frac{2000}{x + 20} - 10 + x \left(\frac{-2000}{(x + 20)^2}\right)$$

$$= \frac{2000}{x + 20} - 10 - \frac{2000x}{(x + 20)^2}$$

(d) $R'(20) = \dfrac{2000}{40} - 10 - \dfrac{2000(20)}{(40)^2} = \$15/\text{ton}$

$ R'(80) = \dfrac{2000}{100} - 10 - \dfrac{2000(80)}{(100)^2} = -\$6/\text{ton}$

Now Work Problem 60

....

EXAMPLE 11

A destroyer is traveling due north on a straight course at a constant velocity of 15 kilometers per hour. At 2 P.M. the destroyer is 50 kilometers due south of a tanker that is moving west at 5 kilometers per hour. At what velocity are the ships approaching each other 2 hours later? At 6 P.M., are the ships approaching each other or receding from each other? When are the ships closest?

SOLUTION

First, we construct Figure 6, which shows the relative position of the destroyer and the tanker at an arbitrary time t after 2 P.M.

Figure 6

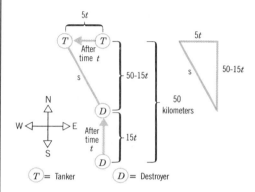

T = Tanker D = Destroyer

From the Pythagorean theorem, the distance s separating the two ships at time t is

$$s(t) = \sqrt{(5t)^2 + (50 - 15t)^2} = \sqrt{250t^2 - 1500t + 2500}$$

Note that when $t = 0$, the ships are 50 kilometers apart as stipulated. The velocity at which the ships are approaching 2 hours later is given by $s'(2)$. Using the Extended Power Rule, we have

$$s'(t) = \frac{1}{2}(250t^2 - 1500t + 2500)^{-1/2}(500t - 1500)$$

$$= \frac{250(t - 3)}{\sqrt{250t^2 - 1500t + 2500}} = \frac{50(t - 3)}{\sqrt{10t^2 - 60t + 100}}$$

When $t = 2$,

$$s'(2) = \frac{-50}{\sqrt{40 - 120 + 100}} = \frac{-50}{2\sqrt{5}} = -5\sqrt{5}$$

The fact that $s'(2) < 0$ means the ships are approaching at the rate of $5\sqrt{5}$ kilometers per hour.

At 6 P.M. (when $t = 4$), we have

$$s'(4) = \frac{50}{\sqrt{160 - 240 + 100}} = \frac{50}{2\sqrt{5}} = 5\sqrt{5}$$

Thus, the ships are moving apart at the rate of $5\sqrt{5}$ kilometers per hour at 6 P.M.
Since for $t < 3$, $s'(t) < 0$ (the ships are approaching) and for $t > 3$, $s'(t) > 0$ (the ships are receding), it follows that at $t = 3$ the ships are closest.

Now Work Problem 69

....

Proofs

Partial Proof of the Chain Rule To find the derivative dy/dx when $y = f(u)$ and $u = g(x)$, we look at the difference quotient $\Delta y/\Delta x$. If $\Delta u \neq 0$,* we can write

$$\frac{\Delta y}{\Delta x} = \frac{\Delta y}{\Delta u}\frac{\Delta u}{\Delta x}$$

Thus,†

$$\frac{dy}{dx} = \lim_{\Delta x \to 0}\frac{\Delta y}{\Delta x} = \lim_{\Delta x \to 0}\left(\frac{\Delta y}{\Delta u}\cdot\frac{\Delta u}{\Delta x}\right) = \left(\lim_{\Delta x \to 0}\frac{\Delta y}{\Delta u}\right)\cdot\left(\lim_{\Delta x \to 0}\frac{\Delta u}{\Delta x}\right)$$

$$= \left(\lim_{\Delta u \to 0}\frac{\Delta y}{\Delta u}\right)\left(\frac{du}{dx}\right) = \frac{dy}{du}\frac{du}{dx} \qquad\blacksquare$$

Proof of the Extended Power Rule Using the Chain Rule Let $y = u^n$ and $u = g(x)$.
Then $y = [g(x)]^n$ and

$$\frac{dy}{dx} = \frac{dy}{du}\frac{du}{dx} = (nu^{n-1})(g'(x)) = n[g(x)]^{n-1}g'(x)$$
$$\underset{u = g(x)}{\uparrow} \qquad\qquad\qquad\qquad\blacksquare$$

EXERCISE 13.4   Answers to odd-numbered problems begin on page AN-60.

In Problems 1–10 find dy/dx using the Chain Rule.

1. $y = u^5$, $u = x^3 + 1$

2. $y = u^3$, $u = 2x + 5$

3. $y = \dfrac{u}{u+1}$, $u = x^2 + 1$

4. $y = \dfrac{u-1}{u}$, $u = x^2 - 1$

5. $y = (u+1)^2$, $u = \dfrac{1}{x}$

6. $y = (u^2 - 1)^3$, $u = \dfrac{1}{x+2}$

7. $y = (u^3 - 1)^5$, $u = x^{-2}$

8. $y = (u^2 + 4)^4$, $u = x^{-2}$

9. $y = u^3$, $u = \dfrac{1}{x^3 + 1}$

10. $y = 4u^2$, $u = \dfrac{3}{x^2 + 1}$

---

* The case for $\Delta u = 0$ is more complicated. Interested readers should consult books on calculus.

† Since $u = g(x)$ is continuous, when $\Delta x \to 0$, then $\Delta u \to 0$. For, as $\Delta x \to 0$, we see that

$$\lim_{\Delta x \to 0}\Delta u = \lim_{\Delta x \to 0}\left(\frac{\Delta u}{\Delta x}\cdot\Delta x\right) = g'(x)\cdot 0 = 0.$$

In Problems 11–52 find f'(x) using the Extended Power Rule.

11. $f(x) = (2x - 3)^4$

12. $f(x) = (5x + 4)^3$

13. $f(x) = (x^2 + 4)^3$

14. $f(x) = (x^2 - 1)^4$

15. $f(x) = (3x^2 + 4)^2$

16. $f(x) = (9x^2 + 1)^2$

17. $f(x) = x(x + 1)^3$

18. $f(x) = x(x - 4)^2$

19. $f(x) = 4x^2(2x + 1)^4$

20. $f(x) = 3x^2(x^2 + 1)^3$

21. $f(x) = [x(x - 1)]^3$

22. $f(x) = [x(x + 4)]^4$

23. $f(x) = (3x - 1)^{-2}$

24. $f(x) = (2x + 3)^{-3}$

25. $f(x) = \sqrt{x^2 + 1}$

26. $f(x) = \sqrt{x^2 - 4}$

27. $f(x) = \sqrt[3]{3x - 1}$

28. $f(x) = \sqrt[3]{5x + 2}$

29. $f(x) = x\sqrt{x^2 + 1}$

30. $f(x) = x\sqrt{x^2 - 1}$

31. $f(x) = x^2\sqrt{3x + 1}$

32. $f(x) = x^2\sqrt{4x - 1}$

33. $f(x) = \dfrac{4}{x^2 + 4}$

34. $f(x) = \dfrac{3}{x^2 - 9}$

35. $f(x) = \dfrac{-4}{(x^2 - 9)^3}$

36. $f(x) = \dfrac{-2}{(x^2 + 2)^4}$

37. $f(x) = \left(\dfrac{x}{x + 1}\right)^3$

38. $f(x) = \left(\dfrac{x^2}{x + 5}\right)^4$

39. $f(x) = \dfrac{(2x + 1)^4}{3x^2}$

40. $f(x) = \dfrac{(3x + 4)^3}{9x}$

41. $f(x) = \dfrac{(x^2 + 1)^3}{x}$

42. $f(x) = \dfrac{(3x^2 + 4)^2}{2x}$

43. $f(x) = \left(x + \dfrac{1}{x}\right)^3$

44. $f(x) = \left(x - \dfrac{1}{x}\right)^4$

45. $f(x) = \dfrac{3x^2}{(x^2 + 1)^2}$

46. $g(x) = \dfrac{2x^3}{(x^2 - 4)^2}$

47. $f(x) = \sqrt{2x}\,(x + 1)^3$

48. $f(x) = \sqrt{3x}\,(x - 3)^3$

49. $f(x) = \dfrac{1}{\sqrt{x + 3}}$

50. $f(x) = \dfrac{2}{\sqrt[3]{x^2 - 1}}$

51. $f(x) = \dfrac{2x}{\sqrt{x^2 - 2x + 1}}$

52. $f(x) = \dfrac{2x^2}{\sqrt{x^2 + 2x - 3}}$

In Problems 53 and 54 find the slope of the tangent line to the graph of the function f at the indicated point. What is an equation of the tangent line?

53. $f(x) = \sqrt{2x - 5}$ at $(3, 1)$

54. $f(x) = x\sqrt{x^3 + 1}$ at $(2, 6)$

55. Find the derivative y' of $y = (x^3 + 1)^2$ by

  (a) Using the Chain Rule

  (b) Using the Extended Power Rule

  (c) Expanding and then differentiating

56. Follow the directions in Problem 55 for the function $y = (x^2 - 2)^3$.

57. If $f(x) = x\sqrt{1 - x^2}$, find the numbers x at which $f'(x) = 0$. Are there any numbers x for which $f'(x)$ does not exist?

58. Follow the directions of Problem 57 for $f(x) = x^2\sqrt{4 - x}$.

59. Car Depreciation   A certain car depreciates according to the formula

$$V = \dfrac{9000}{1 + 0.4t + 0.1t^2}$$

where t represents the time of purchase (in years). The derivative $V'(t)$ gives the rate at which the car depreciates. Find the rate at which the car is depreciating

  (a) 1 year after purchase

  (b) 2 years after purchase

  (c) 3 years after purchase

  (d) 4 years after purchase

60. Demand Function   The demand function for a certain calculator is given by

$$d(x) = \dfrac{100}{0.02x^2 + 1} \qquad 0 \le x \le 20$$

where x (measured in units of a thousand) is the quantity demanded per week and $d(x)$ is the unit price in dollars.

(a) Find $d'(x)$.

(b) Find $d'(10)$, $d'(15)$, and $d'(20)$ and interpret your results.

(c) Find the revenue function.

(d) Find the marginal revenue.

61. Demand Equation   The price p per pound when x pounds of a certain commodity are demanded is

$$p = \frac{10,000}{5x + 100} - 5 \qquad 0 < x < 90$$

Find

(a) The rate of change of price with respect to x

(b) The revenue function

(c) The marginal revenue

(d) The marginal revenue at $x = 10$ and at $x = 40$

(e) Interpret the answer to (d).

62. Revenue Function   The weekly revenue R in dollars resulting from the sale of x typewriters is

$$R(x) = \frac{100x^5}{(x^2 + 1)^2} \qquad 0 \le x \le 100$$

Find

(a) The marginal revenue

(b) The marginal revenue at $x = 40$

(c) The marginal revenue at $x = 60$

(d) Interpret the answers to (b) and (c).

63. Amino Acids   A protein disintegrates into amino acids according to the formula

$$M = \frac{28}{t + 2}$$

where M, the mass of the protein, is measured in grams and t is time measured in hours.

(a) Find the average rate of change in mass from $t = 0$ to $t = 2$ hours.

(b) Find $M'(0)$.

(c) Interpret the answers to (a) and (b).

64. Pollution   The amount of pollution in a certain lake is found to be

$$A(t) = (t^{1/4} + 3)^3$$

where t is measured in years and $A(t)$ is measured in appropriate units. What is the instantaneous rate of change of the amount of pollution? At what rate is the amount of pollution changing after 16 years?

65. Enrollment Projection   The Office of Admissions estimates that the total student enrollment in the University Division will be given by

$$N(t) = -\frac{1000}{\sqrt{1 + 0.1t}} + 11,000$$

where $N(t)$ denotes the number of students enrolled in the division t years from now. Find an expression for $N'(t)$. How fast is the student enrollment increasing currently? 10 years from now?

66. Learning Curve   The psychologist L. L. Thurstone suggested the following relationship between the learning time T it takes to memorize a list of n words:

$$T = f(n) = Cn \sqrt{n - b}$$

where C and b are constants depending upon the person and the task.

(a) Compute dT/dn and interpret your results.

(b) Suppose that for a certain person and a certain task, $C = 2$ and $b = 2$. Compute $f'(10)$ and $f'(30)$ and interpret your results.

67. Production Function   The production of commodities sometimes requires several resources, such as land, labor, machinery, and the like. If there are two inputs that require the amounts x and y, then the output z is given by a function of two variables: $z = f(x, y)$. Here z is called a **production function.** For example, if we use x to represent land and y to represent capital, and z to be the amount of a particular commodity produced, a possible production function is

$$z = x^{0.5}y^{0.4}$$

Set z equal to a fixed amount produced and show that $dy/dx = -5y/4x$. Thus, show that the rate of change of capital with respect to land is always negative when the amount produced is fixed.

68. Price Function   It is estimated that t months from now, the average price (in dollars) of a personal computer will be given by

$$P(t) = \frac{300}{1 + \frac{1}{6}\sqrt{t}} + 100 \qquad 0 \le t \le 60$$

(a) Find an expression for $P'(t)$.

(b) Compute $P'(0)$, $P'(10)$, and $P'(49)$ and interpret your results.

69. Distance between Moving Objects   At 6 P.M. one ship, traveling 8 kilometers per hour in an easterly direction, is 52 kilometers due west of a second ship that is moving at a velocity of 12 kilometers per hour due north. At what rate are the ships approaching each other at 7 P.M.? Are they approaching or receding from each other at 10 P.M.? At what time are they closest?

70. Distance between Moving Objects   A submarine, submerged at 200 feet, is moving along at the rate of 15 feet per second and passes under a stationary destroyer. How fast is the distance from the submarine to the destroyer changing after 1 minute?

71. Distance between Moving Objects   At noon a carrier is 10 kilometers due west of a destroyer. The carrier is traveling east at 2 kilometers per hour and the destroyer is going south at 4 kilometers per hour. When are the ships closest?

13.5 DERIVATIVE OF THE EXPONENTIAL AND LOGARITHM FUNCTIONS

Before getting started on this section, you may find it helpful to review Sections 11.4 and 11.5. In these two sections we review the exponential and the logarithm functions. You are also encouraged to review Section 12.4, where the derivative is first introduced. Such a review will serve to reinforce the important ideas of calculus, and it will enable you to follow the discussion in this section more easily.

The Derivative of e^x

We begin the discussion of the derivative of e^x by considering the function

$$f(x) = a^x \qquad a > 0, a \neq 1$$

To find the derivative of $f(x) = a^x$, we use the formula for finding the derivative of f at x, namely,

$$f'(x) = \lim_{h \to 0} \frac{f(x + h) - f(x)}{h}$$

Thus, for $f(x) = a^x$, we have

$$f'(x) = \frac{d}{dx} a^x = \lim_{h \to 0} \frac{a^{x+h} - a^x}{h} = \lim_{h \to 0} a^x \left(\frac{a^h - 1}{h} \right) = a^x \lim_{h \to 0} \frac{a^h - 1}{h}$$

Suppose we seek $f'(0)$. Assuming the limit on the right exists and equals some number, it follows (since $a^0 = 1$) that the derivative of $f(x) = a^x$ at 0 is

$$f'(0) = \lim_{h \to 0} \frac{a^h - 1}{h}$$

This limit equals the slope of the tangent line to the graph of $f(x) = a^x$ at the point $(0, 1)$. The value of this limit depends upon the choice of a. Observe in Figure 7 (on the next page) that the slope of the tangent line to the graph of $f(x) = 2^x$ at $(0, 1)$ is less than 1, and that the slope of the tangent line to the graph of $f(x) = 3^x$ at $(0, 1)$ is greater than 1.

Figure 7

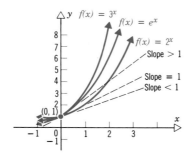

From this we conclude there is a number a, $2 < a < 3$, for which the slope of the tangent line to the graph of $f(x) = a^x$ at $(0, 1)$ is exactly 1. The function $f(x) = a^x$ for which $f'(0) = 1$ is $f(x) = e^x$, where $e = 2.71828$. . . is the number we discussed in Section 11.4. Thus, the number e has the property that

$$\lim_{h \to 0} \frac{e^h - 1}{h} = 1$$

Using this result, we find that

$$\frac{d}{dx} e^x = \lim_{h \to 0} \frac{e^{x+h} - e^x}{h} = \lim_{h \to 0} \frac{e^x(e^h - 1)}{h} = e^x \lim_{h \to 0} \frac{e^h - 1}{h} = e^x(1) = e^x$$

> **Derivative of $f(x) = e^x$**
>
> The derivative of the exponential function $f(x) = e^x$ is e^x. That is,
>
> $$\frac{d}{dx} e^x = e^x$$

The function $f(x) = e^x$ behaves in the same way as the functions $y = 2^x$ and $y = 3^x$ except that computing its derivative is much easier. This is why the function $f(x) = e^x$ appears in many applications.

EXAMPLE 1   Find the derivative of each function:

(a) $f(x) = x^2 + e^x$ (b) $f(x) = xe^x$ (c) $f(x) = \dfrac{e^x}{x}$

SOLUTION

(a) $f'(x) = \dfrac{d}{dx} (x^2 + e^x) = 2x + e^x$

(b) We use the formula for the derivative of a product:

$$f'(x) = \frac{d}{dx} x e^x = x \frac{d}{dx} e^x + e^x \frac{d}{dx} x = x e^x + e^x(1) = e^x(x + 1)$$

(c) We write $f(x) = e^x/x = e^x x^{-1}$ in order to use the formula for the derivative of a product:

$$f'(x) = \frac{d}{dx} (e^x x^{-1}) = e^x \frac{d}{dx} x^{-1} + x^{-1} \frac{d}{dx} e^x = e^x(-x^{-2}) + x^{-1} e^x$$

$$= \frac{e^x}{x} - \frac{e^x}{x^2} = e^x \left(\frac{1}{x} - \frac{1}{x^2} \right)$$

Now Work Problem 1

· ·

The Derivative of $e^{g(x)}$

In Section 13.4 the Chain Rule was used to find the derivative of a composite function. We now form composite functions by using the exponential function e^x.

For example, the composite function $y = e^{x^2+1}$ is composed of the two functions $y = e^{g(x)}$ and $g(x) = x^2 + 1$. As another example, $y = e^{x^3}$ is composed of $y = e^{g(x)}$ and $g(x) = x^3$.

Derivative of $e^{g(x)}$

The formula for finding the derivative of a composite function $y = e^{g(x)}$, where g is a differentiable function, is

$$\frac{d}{dx} e^{g(x)} = e^{g(x)} \frac{d}{dx} g(x) \tag{1}$$

Proof Recall that if

$$y = f(u) \qquad \text{and} \qquad u = g(x)$$

then by the Chain Rule,

$$\frac{dy}{dx} = \frac{dy}{du} \cdot \frac{du}{dx}$$

In particular, if

$$y = e^u \qquad u = g(x)$$

then by the Chain Rule,

$$\frac{d}{dx} e^{g(x)} = \frac{d}{du} e^u \cdot \frac{du}{dx} = e^u \frac{du}{dx} = e^{g(x)} \cdot \frac{d}{dx} g(x)$$

$$\uparrow$$
$$u = g(x)$$

∎

EXAMPLE 2

Find the derivative of each function:

(a) $f(x) = 4e^{5x}$　　(b) $f(x) = e^{x^2+1}$　　(c) $f(x) = e^{2x^2-(1/x)}$

SOLUTION

(a) Using (1), where $g(x) = 5x$, we find

$$f'(x) = \frac{d}{dx}(4e^{5x}) = 4\frac{d}{dx}e^{5x} = 4 \cdot e^{5x}\frac{d}{dx}5x = 4e^{5x}(5) = 20e^{5x}$$

(b) Using (1), where $g(x) = x^2 + 1$, we find

$$f'(x) = \frac{d}{dx}e^{x^2+1} = e^{x^2+1}\frac{d}{dx}(x^2+1) = e^{x^2+1}(2x)$$

(c) Here, $g(x) = 2x^2 - (1/x)$, so that $g'(x) = 4x + (1/x^2)$. By (1),

$$f'(x) = \frac{d}{dx}e^{2x^2-(1/x)} = e^{2x^2-(1/x)}\frac{d}{dx}\left(2x^2 - \frac{1}{x}\right) = e^{2x^2-(1/x)}\left(4x + \frac{1}{x^2}\right)$$

Now Work Problem 23

· ·

The Derivative of ln x

To find the derivative of $y = \ln x$, we observe that if $y = \ln x$, then $e^y = x$. That is,

$$e^{\ln x} = x$$

If we differentiate both sides with respect to x, we obtain

$$\frac{d}{dx}e^{\ln x} = \frac{d}{dx}x$$

$$e^{\ln x}\frac{d}{dx}\ln x = 1$$

$$\frac{d}{dx}\ln x = \frac{1}{e^{\ln x}}$$

But $e^{\ln x} = x$. Thus, we have the following formula:

Derivative of ln x

$$\frac{d}{dx}\ln x = \frac{1}{x} \qquad\qquad (2)$$

EXAMPLE 3

Find the derivative of each function.

(a) $f(x) = x^2 + \ln x$　　(b) $f(x) = x \ln x$

SOLUTION

(a) $f'(x) = \dfrac{d}{dx}(x^2 + \ln x) = \dfrac{d}{dx}x^2 + \dfrac{d}{dx}\ln x = 2x + \dfrac{1}{x}$

(b) $f'(x) = \dfrac{d}{dx}x \ln x = x\dfrac{d}{dx}\ln x + \ln x \dfrac{d}{dx}x$

$$= (x)\left(\dfrac{1}{x}\right) + (\ln x)(1) = 1 + \ln x$$

Now Work Problem 31

· ·

Sometimes it is necessary to differentiate the natural logarithm of a composite function, such as $\ln g(x)$.

The Derivative of ln $g(x)$

The formula for finding the derivative of the composite function $f(x) = \ln g(x)$, where g is a differentiable function, is

$$\dfrac{d}{dx}\ln g(x) = \dfrac{\dfrac{d}{dx}g(x)}{g(x)} \qquad (3)$$

Proof   If

$$y = f(u) \qquad \text{and} \qquad u = g(x)$$

then by the Chain Rule,

$$\dfrac{dy}{dx} = \dfrac{dy}{du} \cdot \dfrac{du}{dx}$$

In particular, if

$$y = \ln u \qquad \text{and} \qquad u = g(x)$$

then

$$\dfrac{d}{dx}\ln g(x) = \dfrac{d}{du}\ln u \cdot \dfrac{du}{dx} = \dfrac{1}{u}\cdot\dfrac{du}{dx} = \dfrac{1}{g(x)}\cdot\dfrac{d}{dx}g(x) = \dfrac{\dfrac{d}{dx}g(x)}{g(x)}$$

$$\underset{u = g(x)}{\uparrow}$$

■

EXAMPLE 4   Find the derivative of each function:

(a) $f(x) = \ln(x^2 + 1)$ (b) $f(x) = \ln\sqrt{x^2 + 1}$ (c) $f(x) = (\ln x)^2$

SOLUTION

(a) Here, $g(x) = x^2 + 1$ and $g'(x) = 2x$, so that

$$f'(x) = \frac{d}{dx} \ln(x^2 + 1) = \frac{2x}{x^2 + 1}$$

(b) Always simplify first if possible! Here, we may use (4c), properties of logarithms (page 557), to write $\ln\sqrt{x^2 + 1}$ as $\frac{1}{2} \ln(x^2 + 1)$. Then we may use the solution to Part (a) above to get

$$f'(x) = \frac{1}{2} \frac{2x}{x^2 + 1} = \frac{x}{x^2 + 1}$$

(c) We use the power rule with

$$f(x) = [g(x)]^2 \qquad \text{and} \qquad g(x) = \ln x$$

Then

$$f'(x) = 2 \ln x \left(\frac{d}{dx} \ln x \right) = \frac{2 \ln x}{x}$$

Now Work Problem 39

· ·

Logarithmic Differentiation

The properties of logarithms (page 557) can sometimes be used to simplify the work needed to find the derivative of certain algebraic functions.

EXAMPLE 5

Find the derivative of $f(x) = \ln[(2x - 1)^3(2x + 1)^5]$.

SOLUTION

Rather than attempt to use (3), we first use the fact that a logarithm transforms products into sums. That is, we may write

$$f(x) = \ln(2x - 1)^3 + \ln(2x + 1)^5 = 3 \ln(2x - 1) + 5 \ln(2x + 1)$$

$\uparrow$ $\uparrow$

By (4a), page 557   By (4c), page 557

Now we differentiate using (3):

$$f'(x) = (3) \left(\frac{2}{2x - 1} \right) + (5) \left(\frac{2}{2x + 1} \right) = \frac{6}{2x - 1} + \frac{10}{2x + 1} = \frac{4(8x - 1)}{4x^2 - 1}$$

· ·

As Example 5 illustrates, some thought should be given to the possibility of simplification before differentiating. The next example illustrates a somewhat more subtle procedure.

EXAMPLE 6

Find the derivative of $f(x) = \dfrac{x^2\sqrt{5x + 1}}{(3x - 2)^3}$.

SOLUTION

As you will see, it is easier to take the natural logarithm of both sides before differentiating. That is, look instead at

$$\ln f(x) = \ln \frac{x^2 \sqrt{5x + 1}}{(3x - 2)^3}$$

Using the properties of logarithms, we may write the above expression as

$$\ln f(x) = \ln x^2 + \ln \sqrt{5x + 1} - \ln(3x - 2)^3$$

$$= 2 \ln x + \frac{1}{2} \ln(5x + 1) - 3 \ln(3x - 2)$$

We now use (3) to find $f'(x)$:

$$\frac{d}{dx} \ln f(x) = \frac{d}{dx} \left[2 \ln x + \frac{1}{2} \ln(5x + 1) - 3 \ln(3x - 2) \right]$$

$$\frac{f'(x)}{f(x)} = 2 \cdot \frac{1}{x} + \frac{1}{2} \cdot \frac{5}{5x + 1} - 3 \cdot \frac{3}{3x - 2}$$

$$f'(x) = f(x) \left[\frac{2}{x} + \frac{5}{2(5x + 1)} - \frac{9}{3x - 2} \right]$$

$$f'(x) = \frac{x^2 \sqrt{5x + 1}}{(3x - 2)^3} \left[\frac{2}{x} + \frac{5}{2(5x + 1)} - \frac{9}{3x - 2} \right]$$

· ·

We refer to the procedure used in Example 6 as **logarithmic differentiation.** Let's do another example to illustrate the procedure.

EXAMPLE 7

Find the derivative of $f(x) = \dfrac{\sqrt{4x + 3}}{(2x - 5)^3}$.

SOLUTION

We take the natural logarithm of both sides:

$$\ln f(x) = \ln \frac{(4x + 3)^{1/2}}{(2x - 5)^3} = \ln(4x + 3)^{1/2} - \ln(2x - 5)^3$$

$$= \frac{1}{2} \ln(4x + 3) - 3 \ln(2x - 5)$$

$$\frac{f'(x)}{f(x)} = \frac{1}{2} \left(\frac{4}{4x + 3} \right) - 3 \left(\frac{2}{2x - 5} \right) = \frac{2}{4x + 3} - \frac{6}{2x - 5}$$

$$f'(x) = \frac{\sqrt{4x + 3}}{(2x - 5)^3} \left(\frac{2}{4x + 3} - \frac{6}{2x - 5} \right)$$

Now Work Problem 53

· ·

We are now in a position to use logarithmic differentiation to finally establish the power rule:

$$\frac{d}{dx}(x^r) = rx^{r-1} \qquad r \text{ a real number}$$

Let $f(x) = x^r$. Then

$$\ln f(x) = \ln x^r = r \ln x$$

Differentiation of this equation yields

$$\frac{f'(x)}{f(x)} = r \cdot \frac{1}{x}$$

$$f'(x) = r \cdot \frac{1}{x} \cdot f(x) = r \cdot \frac{1}{x} \cdot x^r = rx^{r-1}$$

and the power rule is proved.

Derivative of $\log_a x$ and a^x

The problem of determining the derivative of the logarithm function $f(x) = \log_a x$ for any base a is solved by using the change-of-base formula:

Change-of-Base Formula*

$$\log_a M = \frac{\log_b M}{\log_b a}$$

To develop the formula for the derivative of $f(x) = \log_a x$, we change over to the natural base e. Then

$$f(x) = \log_a x = \frac{\log_e x}{\log_e a} = \frac{\ln x}{\ln a}$$

Since $\ln a$ is a constant, we have

$$f'(x) = \frac{d}{dx} \log_a x = \frac{d}{dx} \frac{\ln x}{\ln a} = \frac{1}{\ln a} \frac{d}{dx} \ln x = \frac{1}{\ln a} \frac{1}{x} = \frac{1}{x \ln a}$$

Thus, we have the formula

Derivative of $\log_a x$

$$\frac{d}{dx} \log_a x = \frac{1}{x \ln a} \tag{4}$$

---

* Refer to Appendix A.

EXAMPLE 8

Find the derivative of $f(x) = \log_2 x$.

SOLUTION

Using (4), we have

$$f(x) = \frac{d}{dx} \log_2 x = \frac{1}{x \ln 2}$$

Now Work Problem 61

$\cdots\cdots\cdots\cdots\cdots\cdots\cdots\cdots$

Finally, we want to find the derivative of $f(x) = a^x$, where $a > 0$, $a \neq 1$, is any real constant.

To solve this problem, we use the definition of a logarithm and the change-of-base formula. Then, if $y = a^x$, we have

$$x = \log_a y$$

$$x = \frac{\ln y}{\ln a}$$

$$x = \frac{\ln a^x}{\ln a}$$

Now we differentiate both sides with respect to x and use (3), where $g(x) = a^x$:

$$\frac{d}{dx} x = \frac{d}{dx} \frac{\ln a^x}{\ln a}$$

$$1 = \frac{1}{\ln a} \frac{d}{dx} \ln a^x$$

$$1 = \frac{1}{\ln a} \frac{\frac{d}{dx} a^x}{a^x}$$

$$1 = \frac{\frac{d}{dx} a^x}{a^x \ln a}$$

$$\frac{d}{dx} a^x = a^x \ln a$$

Thus, we have the following formula:

Derivative of a^x

$$\frac{d}{dx} a^x = a^x \ln a \qquad (5)$$

EXAMPLE 9   Find the derivative of $f(x) = 2^x$.

SOLUTION

Using (5), we have

$$f'(x) = \frac{d}{dx} 2^x = 2^x \ln 2$$

· ·

EXERCISE 13.5   Answers to odd-numbered problems begin on page AN-61.

In Problems 1–52 find the derivative of each function.

1. $f(x) = 5e^x$

2. $f(x) = 2e^x$

3. $f(x) = e^{5x}$

4. $f(x) = e^{-2x}$

5. $f(x) = 8e^{-x/2}$

6. $f(x) = \dfrac{1}{e^{-4x}}$

7. $f(x) = xe^x$

8. $f(x) = x^2 e^x$

9. $f(x) = e^{x^2}$

10. $f(x) = e^{x^3}$

11. $f(x) = e^{\sqrt{x}}$

12. $f(x) = e^{-\sqrt{x}}$

13. $f(x) = \sqrt{e^x}$

14. $f(x) = (e^x)^{1/2}$

15. $f(x) = 1 - e^x$

16. $f(x) = 1 - e^{-2x}$

17. $f(x) = \dfrac{e^x + e^{-x}}{2}$

18. $f(x) = \dfrac{e^x - e^{-x}}{2}$

19. $f(x) = e^{-3x} - 3x$

20. $f(x) = e^{-2x} - 2x^2 + x$

21. $f(x) = \dfrac{e^{3x} - e^{-x}}{e^x}$

22. $f(x) = \dfrac{e^x + e^{-x}}{e^x}$

23. $f(x) = e^{2x^2 + x + 1}$

24. $f(x) = e^{x^3 + x - (1/x)}$

25. $f(x) = \dfrac{e^x}{x}$

26. $f(x) = \dfrac{e^{-x}}{x}$

27. $f(x) = e^{x - (1/x)}$

28. $f(x) = e^{x + (1/x)}$

29. $f(x) = \sqrt{1 + e^x}$

30. $f(x) = \sqrt{1 - e^{-x}}$

31. $f(x) = 6 \ln x$

32. $f(x) = -2 \ln x$

33. $f(x) = \ln 3x$

34. $f(x) = \ln 5x$

35. $f(x) = 8 \ln \dfrac{x}{2}$

36. $f(x) = \frac{1}{2} \ln 4x$

37. $f(x) = x \ln x$

38. $f(x) = x^2 \ln x$

39. $f(x) = \ln x^2$

40. $f(x) = \ln x^3$

41. $f(x) = \ln \sqrt{x}$

42. $f(x) = \ln \sqrt[3]{x}$

43. $f(x) = \sqrt{\ln x}$

44. $f(x) = \sqrt[3]{\ln x}$

45. $f(x) = \dfrac{1}{x} \ln x$

46. $f(x) = e^x + x \ln x$

47. $f(x) = e^{\ln x}$

48. $f(x) = \ln(\ln x)$

49. $f(x) = x \ln(x^2 + 4)$

50. $f(x) = x \ln(x^2 + 5x + 1)$

51. $f(x) = x \ln \sqrt{x^2 + 1}$

52. $f(x) = x \ln \sqrt[3]{3x + 1}$

In Problems 53–58 use logarithmic differentiation to find the derivative of each function.

53. $f(x) = (x^2 + 1)^2(2x^3 - 1)^4$

54. $f(x) = (3x^2 + 4)^3(x^2 + 1)^4$

55. $f(x) = (x^3 + 1)(x - 1)(x^4 + 5)$

56. $f(x) = \sqrt{x^2 + 1}(x^3 - 5)(3x + 4)$

57. $f(x) = \dfrac{x^2(x^3 + 1)}{\sqrt{x^2 + 1}}$

58. $f(x) = \dfrac{\sqrt{x}(x^3 + 2)^2}{\sqrt[3]{3x + 4}}$

In Problems 59–62 find the derivative of each function.

59. $f(x) = 3^x$

60. $f(x) = 5^x$

61. $f(x) = \log_3 x$

62. $f(x) = \log_5 x$

63. Find an equation of the tangent line to the graph of $y = e^{3x-2}$ at the point $(\frac{2}{3}, 1)$.

64. Find an equation of the tangent line to the graph of $y = e^{-x^2}$ at the point $(1, 1/e)$.

65. Find the equation of the tangent line to $y = e^x$ that is parallel to the line $y = x$.

66. Find the equation of the tangent line to $y = e^{3x}$ that is perpendicular to the line $y = -\frac{1}{2}x$.

67. Find the equation of the tangent line to the graph of $y = \ln x$ at the point $(2, \ln 2)$.

68. If $f(x) = \log_a x$, show that

$$f(x + h) - f(x) = \log_a\left(1 + \frac{h}{x}\right)$$

69. Marginal Cost   The cost (in dollars) of producing x units (measured in thousands) of a certain product is found to be

$$C(x) = 20 + \ln(x + 1)$$

Find the marginal cost.

70. Weber–Fechner Law   When a certain drug is administered, the reaction R to the dose x is given by the **Weber–Fechner law:**

$$R = 5.5 \ln x + 10$$

Find the reaction rate for a dose of 5 units.

71. Atmospheric Pressure   The atmospheric pressure at a height of x meters above sea level is $P(x) = 10^4 e^{-0.00012x}$ kilograms per square meter. What is the rate of change of the pressure with respect to the height at $x = 500$ meters? At $x = 700$ meters?

72. Revenue   Revenue sales analysis of a new toy by Toys Inc. indicates that the relationship between the unit price p and the monthly sales q for its new toy is given by the equation

$$p = 10e^{-0.04q}$$

Find

(a) The revenue function $R(q)$

(b) The rate of change of R when $q = 200$

73. Market Penetration   The function

$$A(t) = 102 - 90e^{-0.21t}$$

expresses the relationship between A, the percentage of the market penetrated by video sets, and t, the time in years, where $t = 0$ corresponds to the year 1960.

(a) Find the rate of change A' with respect to time.

(b) Evaluate A' at $t = 5$ and interpret your result.

(c) Evaluate A' at $t = 10$ and interpret your result.

(d) Evaluate A' at $t = 30$ and interpret your result.

74. Sales   Because of lack of promotion, the yearly sales of a product decline according to the equation

$$f(t) = 3000e^{-0.80t}$$

where $f(t)$ represents yearly sales at time t. Find

(a) The rate of change of sales

(b) The rate of change of sales at $t = 0.5$

(c) The rate of change of sales at $t = 2$

75. Advertising   The equation

$$S(x) = 100,000 + 400,000 \ln x$$

expresses the relation between sales (in dollars) of a product and the advertising for the product, where x is in thousands of dollars. Find

(a) The rate of change of S with respect to x

(b) $S'(10)$ and interpret

(c) $S'(20)$ and interpret

76. Water Consumption Annual water consumption per person is estimated to be $W(t) = 1400(1 + e^{0.05(t-1950)})$, where t is the year and W is measured in gallons. Find

(a) The rate of change of W with respect to t
(b) The rate of change at $t = 1950$
(c) The rate of change at $t = 2000$

Technology Exercises
In Problems 1–4, graph the given function and its derivative on the same coordinate system.

1. $f(x) = 3^x$ **2.** $f(x) = (\frac{1}{3})^x$

3. $f(x) = 10^x$ **4.** $f(x) = e^x$

5. The temperature T of a pizza at room temperature (75°F) that is put into a hot oven (maintained at 400°F) is given as a function of time by

$$T = 330(1 - e^{-kx}) + 70$$

where T is in degrees Fahrenheit and x is time in minutes.

(a) Graph the function T with the x range $[0, 60]$ for $k = 1$, $k = 0.1$, and $k = 0.01$.
(b) Which of the three functions looks the most realistic? Explain.
(c) How fast is the temperature of the pizza rising initially for your choice in (b)? After 10 minutes? After 50 minutes?

6. Graph the function $f(x) = \ln x$ between 0 and 4. Estimate the derivative $f'(1)$.

7. Advertising The equation

$$S(x) = 100,000 + 400,000 \ln x$$

expresses the relation between sales (in dollars) of a product and the advertising for the product, where x is in thousands of dollars (as in Problem 75).

(a) Graph on a graphing calculator (or a computer) the rate of change of S with respect to x.
(b) Interpret the graph in terms of advertising.

8. Sales Because of lack of promotion, the yearly sales of a product decline according to the equation

$$f(t) = 3000e^{-0.80t}$$

where $f(t)$ represents yearly sales at time t (as in Problem 74).

(a) Graph on a graphing calculator (or a computer) the rate of change of f with respect to t.
(b) Interpret the graph in terms of sales.

13.6 IMPLICIT DIFFERENTIATION

So far, we have considered only functions whose law of correspondence is expressed in the form $y = f(x)$. This expression of the relationship between x and y is said to be in *explicit form* because we have solved for the dependent variable y. For example, the equations

$$y = 7x - 2, \qquad s = -16t^2 + 10t + 100, \qquad v = 4h^2 - h$$

are all written in explicit form, and we say that y, s, and v are functions of x, t, and h, respectively.

If the functional relationship between the independent variable x and the dependent variable y is not of this form, we say that x and y are related *implicitly*. For example, x and y are related implicitly in the expression

$$x^4 + 2x^2y^2 + y^4 = 9x^2 - 9y^2$$

In this equation we would be hard pressed to find y as a function of x. How, then, do we go about computing dy/dx in such a case? Figure 8 illustrates the graph of this equation and the tangent line at $(\sqrt{5}, -1)$. (The graph is called a lemniscate.)

Figure 8

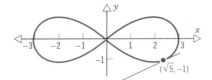

The procedure for finding the derivative of y with respect to x when the functional relationship between x and y is given implicitly is called **implicit differentiation;** however, keep in mind that we still want to find dy/dx, which means that we are still differentiating y with respect to x. The procedure is to think of y as a function f of x, without actually expressing y in terms of x. If this requires differentiating terms like y^4, which we think of as $(f(x))^4$, then we use the Extended Power Rule. Thus, the derivative of y^4 or $(f(x))^4$ is

$$4(f(x))^3 f'(x) \qquad \text{or} \qquad 4y^3 \frac{dy}{dx}$$

In the next example we differentiate a function implicitly.

EXAMPLE 1

Find dy/dx if

$$3x + 4y - 5 = 0$$

SOLUTION

We begin by assuming that there is a differentiable function $y = f(x)$ implied by the above relationship. That is, that

$$3x + 4f(x) - 5 = 0$$

We differentiate both sides of the equality (an identity) with respect to x. Then

$$\frac{d}{dx}[3x + 4f(x) - 5] = \frac{d}{dx}(0)$$

$$\frac{d}{dx}(3x) + \frac{d}{dx}[4f(x)] - \frac{d}{dx}(5) = \frac{d}{dx}(0)$$

$$3 + 4 \cdot \frac{d}{dx}f(x) = 0$$

Solving for $\frac{d}{dx}f(x)$, we find

$$\frac{d}{dx}f(x) = -\frac{3}{4}$$

Replacing $\dfrac{d}{dx} f(x)$ by $\dfrac{dy}{dx}$, we have

$$\frac{dy}{dx} = -\frac{3}{4}$$

....

For the function in Example 1 it is possible to solve for y as a function of x by algebraically solving for y. In this case

$$y = \frac{1}{4}(5 - 3x)$$

Then

$$\frac{dy}{dx} = -\frac{3}{4}$$

which agrees with the result obtained using implicit differentiation. Often, though, it is very difficult, or even impossible, to actually solve for y in terms of x.

EXAMPLE 2

Find dy/dx if $3x^2 + 4y^2 = 2x$.

SOLUTION

We again assume that y is a differentiable function $y = f(x)$ and that when y is replaced by $f(x)$ in the expression $3x^2 + 4y^2 = 2x$, we obtain an identity in x. We proceed to differentiate both sides of this identity with respect to x:

$$\frac{d}{dx}(3x^2 + 4y^2) = \frac{d}{dx}(2x)$$

$$\frac{d}{dx}(3x^2) + \frac{d}{dx}(4y^2) = 2$$

$$6x + 4\left[\frac{d}{dx}(y^2)\right] = 2$$

Since $\dfrac{d}{dx} y^2 = 2y \dfrac{dy}{dx}$, we obtain

$$6x + 4\left(2y\frac{dy}{dx}\right) = 2 \qquad \text{or} \qquad 6x + 8y\frac{dy}{dx} = 2$$

Collect terms involving dy/dx on the left. Solving for dy/dx, we have

$$8y\frac{dy}{dx} = 2 - 6x$$

$$\frac{dy}{dx} = \frac{1 - 3x}{4y} \qquad \text{provided} \quad y \neq 0$$

....

We have learned two ways of finding dy/dx of an equation: explicitly and implicitly. To find dy/dx explicitly, we rewrite the equation and solve for y and then find dy/dx. When an equation defines y implicitly, as a function of x, we find dy/dx as follows:

Steps to Differentiate Implicitly

Step 1 Differentiate both sides of the equation with respect to x.

Step 2 Collect all terms involving dy/dx on one side of the equation, and collect all other terms on the other side.

Step 3 Factor out dy/dx from the side involving the dy/dx terms.

Step 4 Solve for dy/dx by dividing both sides of the equation by the expression "beside" dy/dx.

EXAMPLE 3

Find dy/dx if $x^4 + 3xy^3 - 16 = y^3$.

SOLUTION

Since y is not given explicitly in terms of x, we proceed to differentiate implicitly.

Step 1
$$\frac{d}{dx}(x^4 + 3xy^3 - 16) = \frac{d}{dx}(y^3)$$

$$\frac{d}{dx}(x^4) + \frac{d}{dx}(3xy^3) - \frac{d}{dx}(16) = \frac{d}{dx}(y^3)$$

To find $\dfrac{d}{dx}(3xy^3)$, we use the product rule:

$$4x^3 + 3\left[x\frac{d}{dx}(y^3) + y^3\frac{d}{dx}(x)\right] - 0 = 3y^2\frac{dy}{dx}$$

$$4x^3 + 3\left[x\left(3y^2\frac{dy}{dx}\right) + y^3(1)\right] = 3y^2\frac{dy}{dx}$$

$$4x^3 + 9xy^2\frac{dy}{dx} + 3y^3 = 3y^2\frac{dy}{dx}$$

Step 2 Collect terms involving dy/dx on one side of the equation, and collect all other terms on the other side:

$$9xy^2\frac{dy}{dx} - 3y^2\frac{dy}{dx} = -4x^3 - 3y^3$$

Step 3 Factor out dy/dx:

$$\frac{dy}{dx}(9xy^2 - 3y^2) = -4x^3 - 3y^3$$

Step 4 Solve for dy/dx:

$$\frac{dy}{dx} = \frac{-4x^3 - 3y^3}{9xy^2 - 3y^2} \qquad \text{provided} \quad 9xy^2 - 3y^2 \neq 0$$

Now Work Problem 1

· ·

EXAMPLE 4

Find the equation of the tangent line to the graph of $x^3 + xy + y^3 = 5$ at the point $(-1, 2)$.

SOLUTION

To find the equation of the tangent line, we need (a) the slope of the tangent line (that is, the derivative), and (b) a point on that line.

(a) To find the slope of the tangent line, we differentiate the equation implicitly:

$$\frac{d}{dx}(x^3 + xy + y^3) = \frac{d}{dx}5$$

$$\frac{d}{dx}x^3 + \frac{d}{dx}(xy) + \frac{d}{dx}y^3 = 0$$

$$3x^2 + \left(x\frac{dy}{dx} + y\right) + 3y^2\frac{dy}{dx} = 0$$

$$(3x^2 + y) + (x + 3y^2)\frac{dy}{dx} = 0$$

Solving for dy/dx, we find

$$\frac{dy}{dx} = \frac{-(3x^2 + y)}{x + 3y^2} \qquad \text{provided} \quad x + 3y^2 \neq 0$$

(b) The derivative dy/dx equals the slope of the tangent line to the graph at any point (x, y) for which $x + 3y^2 \neq 0$. In particular, for $x = -1$ and $y = 2$, we find the slope of the tangent line to the graph at $(-1, 2)$ to be

$$\frac{dy}{dx} = \frac{-(3 + 2)}{-1 + 12} = -\frac{5}{11}$$

The equation of the tangent line at the point $(-1, 2)$ is

$$y - y_1 = m(x - x_1)$$

$$y - 2 = \frac{-5}{11}(x + 1)$$

$$y - 2 = \frac{-5}{11}x + \frac{-5}{11}$$

$$y = \frac{-5}{11}x + \frac{17}{11}$$

Now Work Problem 33

· ·

EXAMPLE 5

Find dy/dx if $e^{xy} = x^2 + y$.

SOLUTION

Since y is not given explicitly in terms of x, we proceed to differentiate implicitly.

$$\frac{d}{dx}(e^{xy}) = \frac{d}{dx}(x^2 + y)$$

$$e^{xy}\frac{d}{dx}(xy) = 2x + \frac{dy}{dx}$$

$$e^{xy}\left[x\frac{dy}{dx} + y\frac{d}{dx}(x)\right] = 2x + \frac{dy}{dx} \qquad \text{Product rule}$$

$$e^{xy}\left(x\frac{dy}{dx} + y\right) = 2x + \frac{dy}{dx}$$

$$xe^{xy}\frac{dy}{dx} + ye^{xy} = 2x + \frac{dy}{dx}$$

$$xe^{xy}\frac{dy}{dx} - \frac{dy}{dx} = 2x - ye^{xy}$$

$$\frac{dy}{dx}(xe^{xy} - 1) = 2x - ye^{xy}$$

$$\frac{dy}{dx} = \frac{2x - ye^{xy}}{xe^{xy} - 1} \qquad \text{provided} \quad xe^{xy} - 1 \neq 0$$

· ·

EXAMPLE 6

Find dy/dx if $y + x = \ln y + \ln x$.

SOLUTION

Since y is not given explicitly in terms of x, we proceed to differentiate implicitly.

$$\frac{d}{dx}(y) + \frac{d}{dx}(x) = \frac{d}{dx}(\ln y) + \frac{d}{dx}(\ln x)$$

$$\frac{dy}{dx} + 1 = \frac{1}{y} \cdot \frac{dy}{dx} + \frac{1}{x}$$

$$\frac{dy}{dx} - \frac{1}{y}\frac{dy}{dx} = \frac{1}{x} - 1$$

$$\frac{dy}{dx}\left(1 - \frac{1}{y}\right) = \frac{1}{x} - 1$$

Simplifying,

$$\frac{dy}{dx}\left(\frac{y-1}{y}\right) = \frac{1-x}{x}$$

$$\frac{dy}{dx} = \frac{1-x}{x} \cdot \frac{y}{y-1} = \frac{(1-x)y}{x(y-1)} \qquad \text{provided} \quad x(y-1) \neq 0$$

....

EXAMPLE 7

For a particular commodity, the demand equation is

$$3x^2 + 4p^2 = 1200 \qquad 0 < x < 20, \quad 0 < p < 10\sqrt{3}$$

where x is the amount demanded and p is the price (in dollars). Find the marginal revenue when $x = 8$.

SOLUTION

The revenue function is

$$R = xp$$

We could solve for p in the price demand function and then compute dR/dx. However, the technique we introduced earlier is easier.

We differentiate $R = xp$ with respect to x, remembering that p is a function of x. Then, by the rule for differentiating a product, we obtain the marginal revenue.

$$R'(x) = p + x\frac{dp}{dx}$$

To find dp/dx, we differentiate the demand function implicitly:

$$3x^2 + 4p^2 = 1200$$

$$6x + 8p\frac{dp}{dx} = 0$$

Solving for dp/dx, we have

$$\frac{dp}{dx} = \frac{-6x}{8p} = \frac{-3x}{4p}$$

Then $R'(x)$ becomes

$$R'(x) = p + x\left(\frac{-3x}{4p}\right) = \frac{4p^2 - 3x^2}{4p}$$

When $x = 8$, $4p^2 = 1200 - 192 = 1008$; $p = \sqrt{252} = 15.87$, so that the marginal revenue at $x = 8$ is

$$R'(8) = \frac{1008 - 192}{4(15.87)} = \$12.85 \text{ per unit}$$

....

EXERCISE 13.6   Answers to odd-numbered problems begin on page AN-62.

In Problems 1–32 find dy/dx by using implicit differentiation.

1. $x^2 + y^2 = 4$ **2.** $3x^2 - 2y^2 = 6$ **3.** $x^2y = 8$

4. $x^3 y = 5$

5. $x^2 + y^2 - xy = 2$

6. $x^2 y + xy^2 = x + 1$

7. $x^2 + 4xy + y^2 = y$

8. $x^2 + 2xy + y^2 = x$

9. $3x^2 + y^3 = 1$

10. $y^4 - 4x^2 = 5$

11. $4x^3 + 2y^3 = x^2$

12. $5x^2 + xy - y^2 = 0$

13. $\dfrac{1}{x^2} - \dfrac{1}{y^2} = 4$

14. $\dfrac{1}{x^2} + \dfrac{1}{y^2} = 6$

15. $\dfrac{1}{x} + \dfrac{1}{y} = 2$

16. $\dfrac{1}{x} - \dfrac{1}{y} = 4$

17. $\dfrac{x}{y} + \dfrac{y}{x} = 6$

18. $x^2 + y^2 = \dfrac{2y}{x}$

19. $x^2 = \dfrac{y^2}{y^2 - 1}$

20. $x^2 + y^2 = \dfrac{2y^2}{x^2}$

21. $(2x + 3y)^2 = x^2 + y^2$

22. $x^2 + y^2 = (3x - 4y)^2$

23. $(x^2 + y^2)^2 = (x - y)^3$

24. $(x^2 - y^2)^2 = (x + y)^3$

25. $(x^3 + y^3)^2 = x^2 y^2$

26. $(x^3 - y^3)^2 = xy^2$

27. $e^y + e^x = x$

28. $e^y - e^x = x$

29. $e^{xy} = x + 4$

30. $e^{xy} = x - y$

31. $y - x = \ln y - \ln x$

32. $x^2 = \ln(x + y)$

In Problems 33 and 34 find the slope of the tangent line at the indicated point. Write an equation for this tangent line.

33. $x^2 + y^2 = 5$,   at $(1, 2)$

34. $x^2 - y^2 = 8$,   at $(3, 1)$

35. Use implicit differentiation to show that the tangent line to a circle $x^2 + y^2 = r^2$ at any point P on the circle is perpendicular to OP, where O is the center of the circle.

In Problems 36–38 find those points (x, y) (if there are any) where the tangent line is horizontal $(dy/dx = 0)$.

36. $x^2 + y^2 = 4$

37. $xy + y^2 - x^2 = 4x$

38. $y^2 + 4x + 6 = 0$

39. Given the equation $x + xy + 2y^2 = 6$:

(a) Find an expression for the slope of the tangent line at any point (x, y) on the graph.

(b) Write an equation for the line tangent to the graph at the point $(2, 1)$.

(c) Find the coordinates of all points (x, y) on the graph at which the slope of the tangent line equals the slope of the tangent line at $(2, 1)$.

40. The graph of the function $(x^2 + y^2)^2 = x^2 - y^2$ contains exactly four points at which the tangent line is horizontal. Find them.

41. Gas Pressure   For ideal gases **Boyle's law** states that pressure is inversely proportional to volume. A more realistic relationship between pressure P and volume V is given by **van der Waals' equation**

$$P + \frac{a}{V^2} = \frac{C}{V - b}$$

where C is the constant of proportionality, a is a constant that depends on molecular attraction, and b is a constant that depends on the size of the molecules. Find the compressibility of the gas, which is measured by dV/dP.

42. Cost Function   If the relationship between cost C per unit (in dollars) and the number x (in thousands) of units produced is

$$\frac{x^2}{9} - C^2 - 1 = 0 \qquad x > 0, \quad C > 0$$

find the marginal cost by using implicit differentiation. For what number of units produced (approximately) does $C'(x) = 1$?

43. Pollution   The amount of pollution in a certain lake is found to be

$$A(t) = (t^{1/4} + 3)^3$$

where t is measured in years and $A(t)$ is measured in appropriate units. What is the instantaneous rate of change of the amount of pollution? At what rate is the amount of pollution changing after 16 years?

44. A large container is being filled with water. After t hours there are $8t - 4t^{1/2}$ liters of water in the container. At what rate is the water filling the container (in liters per hour) when $t = 4$?

45. A young child travels s feet down a slide in t seconds, where $s = t^{3/2}$. What is the child's velocity after 1 second? If the slide is 8 feet long, with what velocity does the child strike the ground?

13.7 HIGHER-ORDER DERIVATIVES; VELOCITY AND ACCELERATION

Earlier we concluded that the derivative of a function $y = f(x)$ is also a function called the derivative function $f'(x)$.

For example, if

$$f(x) = 6x^3 - 3x^2 + 2x - 5$$

then

$$f'(x) = 18x^2 - 6x + 2$$

The derivative of the function $f'(x)$ is called the **second derivative of f** and is denoted by $f''(x)$. For the function f above,

$$f''(x) = \frac{d}{dx} f'(x) = \frac{d}{dx} (18x^2 - 6x + 2) = 36x - 6$$

By continuing in this fashion, we can find the third derivative $f'''(x)$, the fourth derivative $f^{(4)}(x)$, and so on, provided that these derivatives exist.*

For example, the first, second, and third derivatives of the function

$$f(x) = x^4 + 3x^3 - 2x^2 + 5x - 6$$

are

$$f'(x) = 4x^3 + 9x^2 - 4x + 5$$

---

* The symbols $f'(x)$, $f''(x)$, and so on for **higher-order derivatives** have several parallel notations. If $y = f(x)$, we may write

$$y' = f'(x) = \frac{dy}{dx} = \frac{d}{dx} f(x)$$

$$y'' = f''(x) = \frac{d^2y}{dx^2} = \frac{d^2}{dx^2} f(x)$$

$$y''' = f'''(x) = \frac{d^3y}{dx^3} = \frac{d^3}{dx^3} f(x)$$

$$\vdots$$

$$y^{(n)} = f^{(n)} = \frac{d^n y}{dx^n} = \frac{d^n}{dx^n} f(x)$$

$$f''(x) = \frac{d}{dx}f'(x) = 12x^2 + 18x - 4$$

$$f'''(x) = \frac{d}{dx}f''(x) = 24x + 18$$

For this function f, observe that $f^{(4)}(x) = 24$ and that all derivatives of order 5 or more equal 0. The result obtained in this example can be generalized:

For a polynomial function f of degree n, we have

$$f(x) = a_n x^n + a_{n-1} x^{n-1} + \cdots + a_1 x + a_0$$
$$f'(x) = na_n x^{n-1} + (n-1)a_{n-1} x^{n-2} + \cdots + a_1$$

Thus, the first derivative of a polynomial function of degree n is a polynomial function of degree $n - 1$. By continuing the differentiation process, it follows that the nth-order derivative of f is

$$f^{(n)}(x) = n(n-1)(n-2) \cdot \cdots \cdot (3)(2)(1)a_n = n!a_n$$

a polynomial of degree 0—a constant. Therefore, all derivatives of order greater than n will equal 0.

Now Work Problems 3 and 25

In some applications it is important to find both the first and second derivatives of a function and to solve for those numbers x that make these derivatives equal 0.

EXAMPLE 1

For $f(x) = 4x^3 - 12x^2 + 2$, find those numbers x, if any, at which the derivative $f'(x) = 0$. For what numbers x will $f''(x) = 0$?

SOLUTION

$$f'(x) = 12x^2 - 24x = 12x(x - 2) = 0 \qquad \text{when} \qquad x = 0 \qquad \text{or} \qquad x = 2$$
$$f''(x) = 24x - 24 = 24(x - 1) = 0 \qquad \text{when} \qquad x = 1$$

·······················

The prime notation y', y'', and so on, is usually used in finding higher-order derivatives for implicitly defined functions.

EXAMPLE 2

Using implicit differentiation, find y' and y'' in terms of x and y if

$$xy + y^2 - x^2 = 5$$

SOLUTION

$$\frac{d}{dx}(xy + y^2 - x^2) = \frac{d}{dx}5$$

$$\frac{d}{dx}(xy) + \frac{d}{dx}y^2 - \frac{d}{dx}x^2 = 0$$

$$(y + xy') + 2yy' - 2x = 0 \tag{1}$$

$$y'(x + 2y) = 2x - y$$

$$y' = \frac{2x - y}{x + 2y} \qquad \text{provided} \quad x + 2y \neq 0 \tag{2}$$

It is easier to find y'' by differentiating (1) than by using (2):

$$y' + y' + xy'' + 2y'(y') + 2yy'' - 2 = 0$$

$$y''(x + 2y) = 2 - 2y' - 2(y')^2$$

$$y'' = \frac{2 - 2y' - 2(y')^2}{x + 2y} \qquad \begin{array}{l} \text{provided} \\ x + 2y \neq 0 \end{array}$$

To express y'' in terms of x and y, use (2). Then

$$y'' = \frac{2 - 2\left(\dfrac{2x - y}{x + 2y}\right) - 2\left(\dfrac{2x - y}{x + 2y}\right)^2}{x + 2y}$$

$$= \frac{-10(x^2 - xy - y^2)}{(x + 2y)^3} = \frac{50}{(x + 2y)^3}$$

$$\uparrow$$
$$x^2 - xy - y^2 = -5$$

Now Work Problem 27

⋯⋯⋯⋯⋯⋯⋯⋯⋯

Velocity and Acceleration

We first review the definitions of *average velocity* and *instantaneous velocity*.

AVERAGE VELOCITY

The *average velocity* is the ratio of the change in distance to the change in time. If s denotes distance and t denotes time, we have

$$\text{Average velocity} = \frac{\text{Total distance}}{\text{Elapsed time}} = \frac{\Delta s}{\Delta t}$$

INSTANTANEOUS VELOCITY

The rate of change of distance with respect to time is called *(instantaneous) velocity*. Thus, if $s = f(t)$ is a function that describes the position s of a particle at time t, the velocity of the particle at time t is

$$v = \frac{ds}{dt} = f'(t)$$

EXAMPLE 3

An object is propelled vertically upward from ground level with an initial velocity of 80 feet per second. The distance s in feet of the object from the ground after t seconds is given by the formula

$$s = f(t) = -16t^2 + 80t$$

What is the velocity of the object after 2 seconds?

SOLUTION

The velocity v at any time t is

$$v = f'(t) = -32t + 80$$

After 2 seconds the velocity is

$$v = f'(2) = -64 + 80 = 16 \text{ feet per second}$$

....

ACCELERATION

The *acceleration* a of a particle is defined as the rate of change of velocity with respect to time. That is,

$$a = \frac{dv}{dt} = \frac{d}{dt} v = \frac{d}{dt}\left(\frac{ds}{dt}\right) = \frac{d^2s}{dt^2}$$

In other words, acceleration is the second derivative of the function $s = f(t)$ with respect to time.

EXAMPLE 4

A ball is thrown vertically upward from ground level with an initial velocity of 19.6 meters per second. The distance s (in meters) of the ball above the ground is $s = -4.9t^2 + 19.6t$, where t is the number of seconds elapsed from the moment the ball is thrown.

(a) What is the velocity of the ball at the end of 1 second?
(b) When will the ball reach its highest point?
(c) What is the maximum height the ball reaches?
(d) What is the acceleration of the ball at any time t?
(e) How long is the ball in the air?
(f) What is the velocity of the ball upon impact?
(g) What is the total distance traveled by the ball?

SOLUTION

(a) The velocity is

$$v = \frac{ds}{dt} = \frac{d}{dt}(-4.9t^2 + 19.6t) = -9.8t + 19.6$$

At $t = 1$, $v = 9.8$ meters per second.

(b) The ball will reach its highest point when it is stationary; that is, when $v = 0$.

$$v = -9.8t + 19.6 = 0 \qquad \text{when } t = 2 \text{ seconds}$$

(c) At $t = 2$, $s = -4.9(4) + 19.6(2) = 19.6$ meters.

(d) $a = \dfrac{d^2s}{dt^2} = \dfrac{dv}{dt} = -9.8$ meters per second per second

(e) We can answer this question in two ways. First, since the ball starts at ground level and it takes 2 seconds for the ball to reach its maximum height, it follows that it will take another 2 seconds to reach the ground, for a total time of 4 seconds in the air. The second way is to set $s = 0$ and solve for t:

$$-4.9t^2 + 19.6t = 0$$

$$t = 0 \qquad \text{or} \qquad t = \frac{19.6}{4.9} = 4$$

The ball is at ground level when $t = 0$ and when $t = 4$ seconds.

(f) Upon impact, $t = 4$. Hence, when $t = 4$,

$$v = (-9.8)(4) + 19.6 = -19.6 \text{ meters per second}$$

The minus sign here indicates that the direction of velocity is downward.

(g) The total distance traveled is

$$\text{Distance up} + \text{Distance down} = 19.6 + 19.6 = 39.2 \text{ meters}$$

See Figure 9 for an illustration.

Figure 9

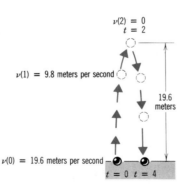

Now Work Problem 41

• •

In Example 4 the acceleration of the ball is constant. This is approximately true for all falling bodies provided air resistance is ignored. In fact, the constant is the same for all falling bodies, as Galileo (1564–1642) discovered in the sixteenth century. We can use calculus to see this. Galileo found by experimentation that all falling bodies obey the law that the distance they fall when they are dropped is proportional to the square of the time t it takes to fall that distance. Of great importance is the fact that the constant of proportionality c is the same for all bodies. Thus, Galileo's law states that the distance s a body falls in time t is given by

$$s = -ct^2$$

The reason for the minus sign is that the body is falling and we have chosen our coordinate system so that the positive direction is up, along the vertical axis.

The velocity v of this freely falling body is

$$v = \frac{ds}{dt} = -2ct$$

and its acceleration a is

$$a = \frac{dv}{dt} = \frac{d^2s}{dt^2} = -2c$$

Thus, the acceleration of a freely falling body is a constant. Usually, we denote this constant by $-g$ so that

$$a = -g$$

The number g is called the **acceleration of gravity.** For our planet g may be approximated by 32 feet per second per second or 980 centimeters per second per second.* On the planet Jupiter, $g \approx 2600$ centimeters per second per second, and on our moon, $g \approx 160$ centimeters per second per second.

EXAMPLE 5   Could a major league pitcher throw a ball to the 250-foot ceiling of the King Dome (home of the Seattle Mariners)?

SOLUTION

To answer this question, we have to make the following assumptions:

(a) The equation

$$h(t) = h_0 + v_0 t - 16t^2 \tag{3}$$

is a formula for a free fall describing the path of the ball, where $h(t)$ is the altitude (in feet), and h_0 is the height of the object at time $t = 0$ ("ground level").

(b) The derivative of $h(t)$,

$$h'(t) = v(t) = v_0 - 32t \tag{4}$$

is the velocity (in feet per second), where v_0 is the initial velocity of the ball.

(c) On the average, fastball pitchers throw the ball with an initial velocity v_0 of around 95 miles per hour ≈ 140 feet per second.

* The Earth, as you know, is not perfectly round; it bulges slightly at the equator. But neither is it perfectly oval, and its mass is not distributed uniformly. As a result, the acceleration of any freely falling body varies slightly from these constants.

(d) The ball is thrown upward with an initial height of 6 feet (h_0) (average height of a pitcher).

Making these assumptions, we will show that the answer to this question is that a pitcher can throw the ball to the 250-foot ceiling. We have

$$v_0 = 140 \text{ feet per second} \qquad h_0 = 6 \text{ feet}$$

Replacing these figures in the formulas above, we get

$$h(t) = 6 + 140t - 16t^2 \qquad v(t) = 140 - 32t$$

Height is greatest at the instant when upward velocity is zero, that is, when

$$v(t) = 140 - 32t = 0$$

or

$$t = \tfrac{140}{32} \text{ seconds}$$

At that time,

$$h(\tfrac{140}{32}) = 6 + 140(\tfrac{140}{32}) - 16(\tfrac{140}{32})^2$$
$$= 312.25 \text{ feet}$$

which is higher than the King Dome ceiling. The graphs of $h(t)$ and $v(t)$ are given in Figure 10.

Figure 10

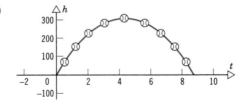
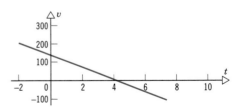

· ·

EXERCISE 13.7 Answers to odd-numbered problems begin on page AN-62.

In Problems 1–16 find f' and f".

1. $f(x) = 2x + 5$

2. $f(x) = 3x + 2$

3. $f(x) = 3x^2 + x - 2$

4. $f(x) = 5x^2 + 1$

5. $f(x) = -3x^4 + 2x^2$

6. $f(x) = -4x^3 + x^2 - 1$

7. $f(x) = \dfrac{1}{x}$

8. $f(x) = \dfrac{1}{x^2}$

9. $f(x) = x + \dfrac{1}{x}$

10. $f(x) = x - \dfrac{1}{x}$

11. $f(x) = \dfrac{x}{x + 1}$

12. $f(x) = \dfrac{x + 1}{x}$

13. $f(x) = \dfrac{x^2}{x + 1}$

14. $f(x) = \dfrac{x + 1}{x^2}$

15. $f(x) = (x^2 + 4)^3$

16. $f(x) = (x^2 - 1)^4$

In Problems 17 and 18 find y', y'', *and* y''', *where a is a constant.*

17. $y = e^{ax}$

18. $y = \ln(ax)$

19. If $y = e^{2x}$, show that $y'' - 4y = 0$.

20. If $y = e^{-2x}$, show that $y'' - 4y = 0$.

In Problems 21–26 find the indicated derivative.

21. $f^{(4)}(x)$ if $f(x) = x^3 - 3x^2 + 2x - 5$

22. $f^{(5)}(x)$ if $f(x) = 4x^3 + x^2 - 1$

23. $\dfrac{d^{20}}{dx^{20}}(8x^{19} - 2x^{14} + 2x^5)$

24. $\dfrac{d^{14}}{dx^{14}}(x^{13} - 2x^{10} + 5x^3 - 1)$

25. $\dfrac{d^8}{dx^8}(\frac{1}{8}x^8 - \frac{1}{7}x^7 + x^5 - x^3)$

26. $\dfrac{d^6}{dx^6}(x^6 + 5x^5 - 2x + 4)$

In Problems 27–30 find y' *and* y'' *in terms of x and y.*

27. $x^2 + y^2 = 4$

28. $x^2 - y^2 = 1$

29. $xy + yx^2 = 2$

30. $4xy = x^2 + y^2$

In Problems 31–34 find the velocity v and acceleration a of an object whose position s at time t is given.

31. $s = 16t^2 + 20t$

32. $s = 16t^2 + 10t + 1$

33. $s = 4.9t^2 + 4t + 4$

34. $s = 4.9t^2 + 5t$

35. Find the nth-order derivative of $f(x) = (2x + 3)^n$.

36. Find the nth-order derivative of $f(x) = \dfrac{1}{3x - 4}$.

In Problems 37 and 38 find the derivative of order n of f.

37. $f(x) = e^{ax}$, where a is a constant

38. $f(x) = \ln x$

39. Find the second derivative of $f(x) = x^2 g(x)$,   where g' and g'' exist.

40. Find the second derivative of $f(x) = g(x)/x$,   where g' and g'' exist.

41. Falling Body   A rock is dropped from a height of 88.2 meters and falls toward the ground in a straight line. In t seconds the rock falls $9.8t^2$ meters.

   (a) How long does it take for the rock to hit the ground?

   (b) What is the average velocity of the rock during the time it is falling?

   (c) What is the average velocity of the rock for the first 3 seconds?

   (d) What is the velocity of the rock when it hits the ground?

42. Falling Body   A ball is thrown upward. Let the height in feet of the ball be given by $s(t) = 100t - 16t^2$, where t is the time elapsed in seconds. What is the velocity

when $t = 0$, $t = 1$, and $t = 4$? At what time does the ball strike the ground? At what time does the ball reach its highest point?

43. Falling Body   A ball is thrown vertically upward with an initial velocity of 80 feet per second. The distance s (in feet) of the ball from the ground after t seconds is given by $s = 6 + 80t - 16t^2$.

   (a) What is the velocity of the ball after 2 seconds?

   (b) When will the ball reach its highest point?

   (c) What is the maximum height the ball reaches?

   (d) What is the acceleration of the ball at any time t?

   (e) How long is the ball in the air?

   (f) What is the velocity of the ball upon impact?

   (g) What is the total distance traveled by the ball?

44. Falling Body   An object is propelled vertically upward with an initial velocity of 39.2 meters per second. The distance s (in meters) of the object from the ground after t seconds is $s = -4.9t^2 + 39.2t$.

(a) What is the velocity of the object at any time t?
(b) When will the object reach its highest point?
(c) What is the maximum height?
(d) What is the acceleration of the object at any time t?
(e) How long is the object in the air?
(f) What is the velocity of the object upon impact?
(g) What is the total distance traveled by the object?

45. Ballistics   A bullet is fired horizontally into a bale of paper. The distance s (in meters) the bullet travels in the bale of paper in t seconds is given by $s = 8 - (2 - t)^3$ for $0 \le t \le 2$. Find the velocity of the bullet after 1 second. Find the acceleration of the bullet at any time t.

46. Falling Rocks on Jupiter   If a rock falls from a height of 20 meters on the planet Jupiter, then its height H after t seconds is approximately

$$H(t) = 20 - 13t^2$$

(a) What is the average velocity of the rock from $t = 2$ to $t = 3$?
(b) What is the instantaneous velocity at time $t = 3$?
(c) What is the acceleration of the rock?

47. Population Growth   A population grows from an initial size of 50,000 to an amount $A(t)$, given by

$$A(t) = 50{,}000(1 + 0.2t + t^2)$$

What is the acceleration in the size of the population? Interpret your result.

48. Population Growth   A population grows from an initial size of 50,000 to an amount $A(t)$, given by

$$A(t) = 50{,}000(1 + 0.4t + t^2)$$

What is the acceleration in the size of the population? Interpret your result.

49. Marginal Cost   If $C(q) = 0.2q^2 + 3q + 600$ is a cost function, how fast is the marginal cost changing when $q = 50$?

50. Marginal Revenue   If $P(q) = 500 - 30q - q^2$ is a demand equation, how fast is marginal revenue changing when $q = 20$?

Technology Exercises

1. The height s (in feet) of a ball thrown upward from ground level is given by

$$s = -16t^2 + 16t$$

where t is time (in seconds).

(a) Graph the distance function.

(b) From the graph, estimate (to the nearest tenth of a second) when the velocity will be zero.
(c) Estimate (to the nearest foot) the maximum height the ball will reach.
(d) Estimate (to the nearest second) when the ball will hit the ground.

The equations in Problems 2–6 give the position $s = f(t)$ of a particle moving along the x-axis, where s is measured in feet and t in seconds. Graph simultaneously both the position of the particle and its velocity function $v(t) = s'(t)$.
(a) When is the particle at rest?
(c) Where is the particle after 4 seconds?
(b) When is the particle moving the fastest?
(d) Where is the particle initially?

2. $s = f(t) = 3t, \quad 0 \le t \le 5$

3. $s = f(t) = t^2 - 4, \quad 0 \le t \le 5$

4. $s = f(t) = t^3 - 12t, \quad 0 \le t \le 3$

5. $s = f(t) = \sqrt{t}, \quad 1 \le t \le 9$

6. $s = f(t) = \dfrac{t}{t + 1}, \quad 0 \le t \le 16$

7. Find the quadratic function $g(x) = ax^2 + bx + c$ that best fits the function $f(x) = e^x$ at $x = 0$, in the sense that $f(0) = g(0)$, and $f'(0) = g'(0)$, and $f''(0) = g''(0)$. Using a graphing calculator (or a computer), sketch graphs of f and g in the same viewing rectangle. What do you notice?

CHAPTER REVIEW

IMPORTANT TERMS AND CONCEPTS

prime notation 618
Leibniz notation 618
Simple Power Rule 619
derivative of x^n 620
derivative of a sum (difference) 623
criterion for a horizontal
   tangent line 625
average rate of change 630
instantaneous rate of change 631
marginal analysis 632
marginal cost 632
marginal revenue 633

relative rate of change 635
percentage rate of
   change 635
derivative of a product 640
derivative of a quotient 642
composition 650
composite function 650
Chain Rule 652
Extended Power Rule 653
derivative of e^x 662
derivative of $e^{g(x)}$ 663
derivative of $\ln x$ 664

derivative of $\ln g(x)$ 665
logarithmic
   differentiation 667
derivative of $\log_a x$ 668
derivative of a^x 669
implicit differentiation 675
second derivative 680
higher-order derivatives 680
average velocity 682
instantaneous velocity 682
acceleration 682

IMPORTANT FORMULAS

Derivative of a Constant

$$\frac{d}{dx} b = 0$$

Simple Power Rule

$$\frac{d}{dx} x^n = nx^{n-1}$$

Constant Times a Function

$$\frac{d}{dx} [Cf(x)] = C \frac{d}{dx} f(x)$$

Sum (Difference) of Two Functions

$$\frac{d}{dx} [f(x) \pm g(x)] = \frac{d}{dx} f(x) \pm \frac{d}{dx} g(x)$$

Average Rate of Change

$$\frac{\Delta y}{\Delta x} = \frac{f(d) - f(c)}{d - c}$$

Instantaneous Rate of Change

$$f'(c) = \lim_{\Delta x \to 0} \frac{\Delta y}{\Delta x} = \lim_{\Delta x \to 0} \frac{f(c + \Delta x) - f(c)}{\Delta x}$$

Marginal Cost

$$C'(x) \approx C(x + 1) - C(x)$$

Marginal Revenue

$$R'(x) \approx R(x + 1) - R(x)$$

Relative Rate of Change of y with Respect to x

$$\frac{f'(x)}{f(x)}$$

Percentage Rate of Change of y with Respect to x

$$100 \frac{f'(x)}{f(x)}$$

Product Rule

$$\frac{d}{dx} [f(x) \cdot g(x)] = f(x) \frac{d}{dx} g(x) + g(x) \frac{d}{dx} f(x)$$

Quotient Rule

$$\frac{d}{dx} \left[\frac{f(x)}{g(x)} \right] = \frac{g(x) \dfrac{d}{dx} f(x) - f(x) \dfrac{d}{dx} g(x)}{[g(x)]^2}, \quad g(x) \neq 0$$

Chain Rule

$$\frac{dy}{dx} = \frac{dy}{du} \cdot \frac{du}{dx}$$

Extended Power Rule

$$\frac{d}{dx}[g(x)]^r = r[g(x)]^{r-1} \cdot g'(x)$$

Second Derivative

$$\frac{d^2y}{dx^2} = f''(x) = y''$$

Velocity

$$v = \frac{ds}{dt} = s'(t)$$

Acceleration

$$a = \frac{dv}{dt} = \frac{d^2s}{dt^2} = s''(t)$$

$$\frac{d}{dx} e^x = e^x$$

$$\frac{d}{dx} e^{g(x)} = e^{g(x)} \frac{d}{dx} g(x)$$

$$\frac{d}{dx} \ln x = \frac{1}{x}$$

$$\frac{d}{dx} \ln g(x) = \frac{\frac{d}{dx} g(x)}{g(x)}$$

$$\frac{d}{dx} \log_a x = \frac{1}{x \ln a}$$

$$\frac{d}{dx} a^x = a^x \ln a$$

TRUE–FALSE ITEMS   Answers are on page AN-63.

T_____ F_____ **1.** The derivative of a function is the limit of a difference quotient.

T_____ F_____ **2.** The derivative of a product equals the product of the derivatives.

T_____ F_____ **3.** If $f(x) = 1/x$, then $f'(x) = -1/x^2$.

T_____ F_____ **4.** The expression "rate of change of a function" means the derivative of the function.

T_____ F_____ **5.** Every function has a derivative at each number in its domain.

T_____ F_____ **6.** If $x^3 - y^3 = 1$, then $dy/dx = 3x^2 - 3y^2$.

T_____ F_____ **7.** The derivative of e^x is e^x.

FILL IN THE BLANKS   Answers are on page AN-63.

1. The derivative of f at c equals the slope of the _____ line to f at c.

2. If $C = C(x)$ denotes the cost C of producing x items, then $C'(x)$ is called the _____ _____.

3. The derivative of $f(x) = (x^2 + 1)^{3/2}$ may be obtained by using either the _____ _____ or the _____ _____.

4. The acceleration of an object equals the rate of change of _____ with respect to time.

5. The fifth-order derivative of a polynomial of degree 4 equals _____.

6. The derivative of $x^3 - y^4x + 3y = 5$ is obtained using _____ differentiation.

7. If $y = e^{x^2}$, then $y' = $ _____.

REVIEW EXERCISES   Answers to odd-numbered problems begin on page AN-63.

In Problems 1–4 find the average rate of change from x = 0 to x = 1 for each function.

1. $f(x) = 2x^2 + 1$

2. $f(x) = 2x^3 - 3$

3. $f(x) = x^2 + 2x - 3$

4. $f(x) = 3x^2 - x + 4$

In Problems 5–40 find the derivative of each function.

5. $f(x) = 8x^3$

6. $f(x) = -2x^4$

7. $f(x) = -3x^2 + 5$

8. $f(x) = 4 - 2x^3$

9. $f(x) = 3x^4 - 2x^2 + 5x - 2$

10. $f(x) = 5x^3 + 2x^2 - 6x - 9$

11. $f(x) = 8\sqrt{x}$

12. $f(x) = 6\sqrt[3]{x}$

13. $g(x) = \dfrac{4}{x}$

14. $h(x) = -\dfrac{2}{x^2}$

15. $f(t) = (t^2 + 1)^3$

16. $f(w) = (w^3 - 3)^2$

17. $f(x) = \dfrac{3}{x} - \dfrac{x}{3}$

18. $f(u) = \dfrac{4}{u} + \dfrac{u}{4}$

19. $f(x) = (3x^2 + x + 1)(4x^3 - x + 2)$

20. $f(x) = (5x^3 - x + 1)(2x^2 + x + 3)$

21. $g(t) = \dfrac{t}{t^2 + 1}$

22. $f(u) = \dfrac{u}{u^2 - 1}$

23. $f(w) = \tfrac{2}{5}w^{5/2} - 2w^{3/2}$

24. $g(u) = \tfrac{3}{5}u^{5/3} + 3u^{4/3}$

25. $f(x) = (3x - 2)^3 + 5(3x - 2)^2$

26. $g(x) = (2x - 3)^4 - 6(2x - 3)^2$

27. $f(x) = \sqrt{x + \sqrt{x}}$

28. $f(x) = \sqrt{x - \sqrt{x}}$

29. $f(t) = t^2\sqrt{t - 1}$

30. $f(u) = u\sqrt{u^3 + 1}$

31. $f(x) = 4e^{5x}$

32. $f(x) = 2e^{-3x}$

33. $f(x) = 15 \ln \dfrac{x}{3}$

34. $f(x) = \tfrac{1}{2} \ln(6x)$

35. $f(x) = e^{2x^2 + 5}$

36. $f(x) = e^{x^2 + 1}$

37. $f(x) = \ln(2x^2 + 5)$

38. $f(x) = \ln(x^3 + 1)$

39. $f(x) = (x^2 + 1)^2(x^2 - 1)^3$

40. $f(x) = (2x + 1)^2(x^2 + 4)^3$

In Problems 41–46 find f′(1) and f″(1) for each function.

41. $f(x) = 3x^4 - 2x^2 + 4$

42. $f(x) = 6x^3 - 3x^2 + 2$

43. $f(x) = x^{2/3}$

44. $f(x) = x^{1/3}$

45. $f(x) = 3/x$

46. $f(x) = 4/x$

In Problems 47–50 find f′(x) and f″(x). Find all numbers x for which f′(x) = 0, and calculate f″(x) at these numbers.

47. $f(x) = 2x^3 + 3x^2 - 12x + 6$

48. $f(x) = 2x^3 - 15x^2 + 36x - 2$

49. $f(x) = (x^2 - 1)^{3/2}$ **50.** $f(x) = (x^2 + 1)^{3/2}$

In Problems 51–56 find dy/dx by using implicit differentiation.

51. $3xy^2 - 2x^2y = x$ **52.** $4xy^3 - 8xy = 5$ **53.** $4x^2 - y^2 = x + y$

54. $3y^2 - 5x^2 = 2x - y$ **55.** $\sqrt{x} + \sqrt{y} = xy$ **56.** $\sqrt{x^2 + y^2} = 2x^2$

57. The graph of a function f is shown below. Rank the values of $f'(-4), f'(-2), f'(0), f'(1), f'(2)$, and $f'(3)$ in increasing order.

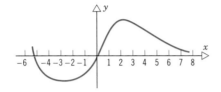

Mathematical Question from Professional Exam

1. Actuary Exam—Part I $\dfrac{d}{dx}(x^2 e^{x^2}) = ?$

(a) $2xe^{x^2}$

(b) $\dfrac{x^3}{3} e^{x^2}$

(c) $4x^2 e^{x^2}$

(d) $2xe^{x^2} + x^4 e^{x^2 - 1}$

(e) $2x^3 e^{x^2} + 2xe^{x^2}$

Chapter 14

Applications: Graphing Functions; Optimization

In this chapter we discuss applications of the derivative. We shall use the derivative as a tool for obtaining the graph of a function as well as for finding the maximum and minimum values of a function. Then we shall see how the derivative can be used to solve certain problems in which related variables vary over time. Finally, we shall use the derivative as an approximation tool.

14.1 INCREASING AND DECREASING FUNCTIONS; THE FIRST DERIVATIVE TEST

Consider the graph of the function $y = f(x)$ given in Figure 1. Notice that as we proceed from left to right along this graph, the function is increasing on the intervals (a, b) and (c, d) while the function is decreasing on the intervals (b, c) and (d, e). (Refer to Chapter 11, Section 11.5, to review the definition of increasing and decreasing functions.) Notice also that the points $(b, f(b))$ and $(d, f(d))$ (where the graph changes direction from increasing to decreasing) are higher than nearby points. The point $(c, f(c))$ is lower than nearby points. We describe this characteristic by saying the function f has a **local maximum at b and at d** and a **local minimum at c.**

Figure 1

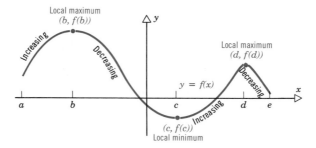

Now Work Problems 3 and 7

In this section we will give a test for differentiable functions that provides a straightforward way to determine the intervals on which the function is increasing or decreasing.

Look at the graph of the function f given in Figure 2. On the interval (a, b), where f is increasing, we have drawn several tangent lines. Notice that each tangent line has a positive slope. This is characteristic of tangent lines of increasing functions. Since the derivative of f equals the slope of the tangent line, it follows that wherever the derivative is positive, the function f will be increasing.

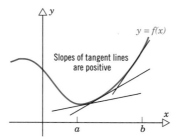

Figure 2   *Note:* Slope lines rise to the right, denoting positive slopes (positive derivatives).

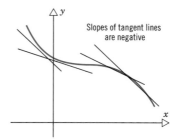

Figure 3   *Note:* Slope lines drop to the right, denoting negative slopes (negative derivatives).

Similarly, as Figure 3 illustrates, the tangent lines of a decreasing function have negative slope. Whenever the derivative of f is negative, the function f is decreasing. This leads us to the following test for an increasing or decreasing function.

Test for Increasing and Decreasing Function

A differentiable function is

(a) Increasing on the interval (a, b) if $f'(x) > 0$ for all x in (a, b).

(b) Decreasing on the interval (c, d) if $f'(x) < 0$ for all x in (c, d).

Thus, to determine where the graph of a function is increasing or decreasing, follow these steps:

Steps for Finding Where a Function Is Increasing and Decreasing

Step 1 Find the derivative $f'(x)$ of the function $y = f(x)$.

Step 2 Set up a table that solves the two inequalities:

$$f'(x) > 0 \qquad \text{and} \qquad f'(x) < 0$$

If you wish to review how to solve inequalities, see Appendix A.

EXAMPLE 1

Determine where the function

$$f(x) = x^3 - 6x^2 + 9x - 2$$

is increasing and where it is decreasing.

SOLUTION

We follow the steps given above.

Step 1 $f'(x) = 3x^2 - 12x + 9$

Step 2 To set up the table, we need to factor $f'(x)$ and solve the equation $f'(x) = 0$.

$$f'(x) = 3x^2 - 12x + 9 = 3(x^2 - 4x + 3) = 3(x - 1)(x - 3)$$

The solutions of the equation $f'(x) = 3(x - 1)(x - 3) = 0$ are 1 and 3. These numbers separate the real number line into three intervals,

$$x < 1 \qquad 1 < x < 3 \qquad 3 < x$$

as shown below:

We determine the sign of the derivative on each interval as follows. If $f'(x)$ is positive for one value of x in the interval, then it will be positive for all x in the interval. Similarly, if it is negative for one value of x in the interval, it will be negative for all x in the interval. Therefore, to check the sign in each interval, we merely choose any test value of x in each interval; we call it a test value and make a substitution. We usually choose as test values numbers for which it is easy to compute the derivative. For this example we choose as test value for the interval $x < 1$, the number $x = 0$. For the interval $1 < x < 3$ we choose $x = 2$. For the interval $3 < x$ we choose $x = 4$.

Test 0: $f'(0) = 3 \cdot 0^2 - 12 \cdot 0 + 9 = 9 > 0$

Test 2: $f'(2) = 3 \cdot 2^2 - 12 \cdot 2 + 9 = -3 < 0$

Test 4: $f'(4) = 3 \cdot 4^2 - 12 \cdot 4 + 9 = 9 > 0$

We construct a table showing the sign of $f'(x) = 3(x - 1)(x - 3)$ on each of the intervals.

Interval	$x < 1$	$1 < x < 3$	$3 < x$
Test Value	0	2	4
f' **at Test Value**	$f'(0) = 9 > 0$	$f'(2) = -3 < 0$	$f'(4) = 9 > 0$
Sign of f'	+	−	+
Graph of f	Increasing	Decreasing	Increasing

We conclude that the function f is increasing on the intervals $x < 1$ and $3 < x$, and that it is decreasing on the interval $1 < x < 3$.

. . . .

Local Maximum and Local Minimum

We have already observed that if a function f is increasing to the left of a point A on the graph of f and is decreasing to the right of A, then A is a local maximum since the graph of f is higher at A than at nearby points. Similarly, if f is decreasing to the left of a point B on the graph of f and is increasing to the right of B, then B is a local minimum. Since the first derivative of a function supplies information about where the function is increasing or decreasing, we call the test for locating local maxima* and local minima the **First Derivative Test.**

First Derivative Test

Let f denote a differentiable function. Find the derivative of f and set up a table to determine where f is increasing and decreasing.

1. If f is increasing to the left of a point A on the graph of f and is decreasing to the right of A, then the point A is a local maximum.
2. If f is decreasing to the left of a point B on the graph of f and is increasing to the right of B, then the point B is a local minimum.

EXAMPLE 2

Use the First Derivative Test to locate the local maxima and local minima, if any, of

$$f(x) = x^3 - 6x^2 + 9x - 2$$

SOLUTION

This is the same function discussed in Example 1, where f is increasing for $x < 1$ and is decreasing for $1 < x < 3$. When $x = 1$, we have $y = f(1) = 2$. Thus, by the First Derivative Test, f has a local maximum at the point $(1, 2)$. Similarly, f is decreasing for $1 < x < 3$ and is increasing for $3 < x$. When $x = 3$, we have $y = f(3) = -2$. Thus, by the First Derivative Test, f has a local minimum at the point $(3, -2)$. See Figure 4 for the graph of f.

Figure 4

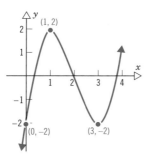

· ·

Graphing Functions

We shall use the First Derivative Test to graph functions. In graphing a function $y = f(x)$, we follow the steps outlined in the box on the next page.

* *Maxima is the plural of maximum.*

Steps for Graphing Functions

Step 1 Find the domain of f.

Step 2 Locate the intercepts of f (skip the x-intercepts if they are too hard to find).

Step 3 Determine where the graph of f is increasing or decreasing.

Step 4 Find any local maxima or local minima of f by using the First Derivative Test.

Step 5 Locate all points on the graph of f at which the tangent line is horizontal or for which the derivative does not exist.

EXAMPLE 3

Graph the function $f(x) = x^3 - 12x$.

SOLUTION

Step 1 The domain of f is all real numbers.

Step 2 Let $x = 0$. Then $y = f(0) = 0$. Thus, the y-intercept is $(0, 0)$. To find the x-intercepts, if any, let $y = 0$. Then

$$x^3 - 12x = 0$$
$$x(x^2 - 12) = 0$$
$$x(x + 2\sqrt{3})(x - 2\sqrt{3}) = 0$$

The x-intercepts are thus $(0, 0)$, $(-2\sqrt{3}, 0)$, and $(2\sqrt{3}, 0)$.

Step 3 To determine where the graph of f is increasing or decreasing, we first need to find $f'(x)$.

$$f(x) = x^3 - 12x$$
$$f'(x) = 3x^2 - 12 = 3(x^2 - 4) = 3(x + 2)(x - 2)$$

The solutions of the equation $f'(x) = 3(x + 2)(x - 2) = 0$ are -2 and 2. These numbers separate the number line into three parts:

$$x < -2 \qquad -2 < x < 2 \qquad 2 < x$$

For this case we choose the test values to be -3, 0, and 3.

Test -3: $f'(-3) = 3 \cdot (-3)^2 - 12 = 15 > 0$

Test 0: $f'(0) = 3(0)^2 - 12 = -12 < 0$

Test 3: $f'(3) = 3(3)^2 - 12 = 15 > 0$

We construct a table showing the sign of $f'(x) = 3(x + 2)(x - 2)$ on each of the intervals.

Interval	$x < -2$	$-2 < x < 2$	$2 < x$
Test Value	-3	0	3
f' at Test Value	$f'(-3) = 15 > 0$	$f'(0) = -12 < 0$	$f'(3) = 15 > 0$
Sign of f'	$+$	$-$	$+$
Graph of f	Increasing	Decreasing	Increasing

We conclude that the graph of f is increasing on the intervals $x < -2$ and $2 < x$. It is decreasing on the interval $-2 < x < 2$.

Step 4 Because the graph of f is increasing for $x < -2$ and is decreasing for $-2 < x < 2$, therefore, at $x = -2$, the graph changes from increasing to decreasing. Consequently, the point $(-2, f(-2)) = (-2, 16)$ is a local maximum. Similarly, at $x = 2$, the graph changes from decreasing to increasing. Thus, the point $(2, f(2)) = (2, -16)$ is a local minimum.

Step 5 The derivative of f is $f'(x) = 3x^2 - 12 = 3(x + 2)(x - 2)$. We see that $f'(x) = 0$ if $x = -2$ or if $x = 2$. Thus, the graph of f has a horizontal tangent line at the points $(-2, f(-2)) = (-2, 16)$ and $(2, f(2)) = (2, -16)$. Since $f(x)$ is a polynomial, the derivative exists everywhere.

To graph f, we plot the intercepts, the local maximum, and the local minimum (the points at which the tangent line is horizontal), and connect these points with a smooth curve. See Figure 5.

Figure 5

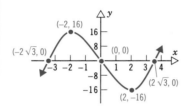

Now Work Problem 11

EXAMPLE 4

Graph the function $f(x) = x^4 - 4x^3$.

SOLUTION

Step 1 The domain of f is all real numbers.

Step 2 Let $x = 0$. Then $y = f(0) = 0$. Thus, the y-intercept is $(0, 0)$. To find the x-intercepts, if any, let $y = 0$. Then

$$x^4 - 4x^3 = 0$$
$$x^3(x - 4) = 0$$

The x-intercepts are $(0, 0)$ and $(4, 0)$.

Step 3 To determine where the graph of f is increasing or decreasing, we first need to find $f'(x)$:

$$f(x) = x^4 - 4x^3$$
$$f'(x) = 4x^3 - 12x^2 = 4x^2(x - 3)$$

The solutions of the equation $f'(x) = 4x^2(x - 3) = 0$ are 0 and 3. These numbers separate the number line into three parts:

$$x < 0 \qquad 0 < x < 3 \qquad 3 < x$$

The test values we select for these intervals are -1, 1, and 4, respectively. We summarize the results in the following table.

Interval	$x < 0$	$0 < x < 3$	$3 < x$
Test Value	-1	1	4
f' **at Test Value**	$f'(-1) = -16 < 0$	$f'(1) = -8 < 0$	$f'(4) = 64 > 0$
Sign of f'	$-$	$-$	$+$
Graph of f	Decreasing	Decreasing	Increasing

We conclude that the graph of f is decreasing on the intervals $x < 0$ and $0 < x < 3$. It is increasing on the interval $3 < x$.

Step 4 Since the graph of f is decreasing for $x < 3$ and is increasing for $3 < x$, therefore, at the point $(3, f(3)) = (3, -27)$, the graph changes from decreasing to increasing. Consequently, the point $(3, -27)$ is a local minimum.

Step 5 The derivative of f is $f'(x) = 4x^3 - 12x^2 = 4x^2(x - 3)$. We see that $f'(x) = 0$ if $x = 0$ or if $x = 3$. Thus, the graph of f has a horizontal tangent line at the points $(0, f(0)) = (0, 0)$ and $(3, f(3)) = (3, -27)$. Since $f(x)$ is a polynomial, the derivative exists everywhere.

To graph f, we plot the intercepts, the local minimum, and the point $(0, 0)$ (the points at which the tangent line is horizontal), and connect these points with a smooth curve. See Figure 6.

Figure 6

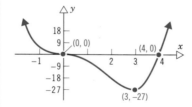

Now Work Problem 23

EXAMPLE 5 Graph the function $f(x) = 2x^{5/3} - 5x^{2/3}$.

SOLUTION

Step 1 The domain of f is all real numbers.

Step 2 Let $x = 0$. Then $y = f(0) = 0$. Thus, the y-intercept is $(0, 0)$. To find the x-intercepts, if any, let $y = 0$. Then

$$2x^{5/3} - 5x^{2/3} = 0$$
$$x^{2/3}(2x - 5) = 0$$

The x-intercepts are $(0, 0)$ and $(\frac{5}{2}, 0)$.

Step 3 To determine where the graph of f is increasing or decreasing, we first need to find $f'(x)$:

$$f(x) = 2x^{5/3} - 5x^{2/3}$$

$$f'(x) = \frac{10x^{2/3}}{3} - \frac{10x^{-1/3}}{3} = \frac{10x^{2/3}}{3} - \frac{10}{3x^{1/3}}$$

$$= \frac{10x - 10}{3x^{1/3}} = \frac{10(x - 1)}{3x^{1/3}}$$

Now, $f'(x) = 0$ when $x = 1$, and $f'(x)$ is undefined when $x = 0$. Thus, we use the numbers 0 and 1 to separate the number line into three parts:

$$x < 0 \qquad 0 < x < 1 \qquad 1 < x$$

The test values we select for these intervals are -1, $\frac{1}{2}$, and 2, respectively. We summarize the results in the following table.

Interval	$x < 0$	$0 < x < 1$	$1 < x$
Test Value	-1	$\frac{1}{2}$	2
f' at Test Value	$f'(-1) = \frac{20}{3} > 0$	$f'(\frac{1}{2}) \approx -2.1$	$f'(2) \approx 2.6$
Sign of f'	$+$	$-$	$+$
Graph of f	Increasing	Decreasing	Increasing

We conclude that the graph of f is increasing on the intervals $x < 0$ and $1 < x$; it is decreasing on the interval $0 < x < 1$.

Step 4 From the table, we see that the graph of f is increasing for $x < 0$ and is decreasing for $0 < x < 1$. Therefore, at the point $(0, f(0)) = (0, 0)$, the graph changes from increasing to decreasing. Consequently, the point $(0, 0)$ is a local maximum. Similarly, for $0 < x < 1$, the graph of f is decreasing, and for $1 < x$, the graph of f is increasing. Thus, the point $(1, f(1)) = (1, -3)$ is a local minimum.

Step 5 The derivative of f is $f'(x) = 10(x - 1)/3x^{1/3}$. We see that $f'(x) = 0$ if $x = 1$. Thus, the graph of f has a horizontal tangent line at the point $(1, -3)$. Also, $f'(x)$ does not exist at $x = 0$. There is a vertical tangent line at the point $(0, 0)$.

To graph f, we plot the intercepts, the local maximum, and the local minimum (the points at which the tangent line is horizontal or vertical), and connect these points. See Figure 7. Notice how the vertical tangent line and local maximum at $(0, 0)$ are shown in the graph.

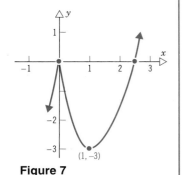

Figure 7

Now Work Problem 29

In looking back over Examples 2–5, you may have noticed that the local maximum and local minimum, when they existed, always occurred at a number c for which either $f'(c) = 0$ or $f'(c)$ does not exist.

Necessary Condition for a Local Maximum or Local Minimum

If a function f has a local maximum or a local minimum at c, then either $f'(c) = 0$ or $f'(c)$ does not exist.

Because of this result, we call the numbers c for which $f'(c) = 0$ or $f'(c)$ does not exist **critical numbers of f.** The corresponding points $(c, f(c))$ on the graph of f are called **critical points.**

Be careful! The critical points only provide a list of potential local maxima and local minima for a graph. It is possible for a critical point to be neither a local maximum nor a local minimum. See Figure 8 for some illustrations.

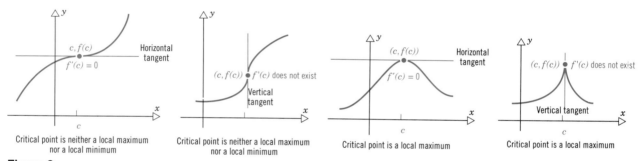

Figure 8

EXAMPLE 6

It is estimated that the profit realized by a company for the manufacture and sale of x units of digital watches per week is

$$P(x) = -0.001x^2 + 8x - 10{,}000$$

dollars. Determine the level of production for which the profit is maximum. What is the maximum profit?

SOLUTION

We find the derivative $P'(x)$:

$$P'(x) = -0.002x + 8 = -0.002(x - 4000)$$

If $x < 4000$, then $P'(x) > 0$, and the graph of P is increasing. If $x > 4000$, then $P'(x) < 0$, and the graph of P is decreasing.

Thus, at 4000 there is a local maximum. Consequently, a production level of 4000 units will yield the maximum profit. The maximum profit is $P(4000) = \$6000$. See Figure 9 for the graph.

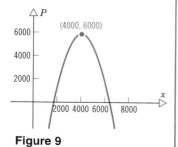

Figure 9

Now Work Problem 43

Summary

The box below summarizes what the first derivative tells us about the **graph of a** function.

Derivative of f	Graph of f
$= 0$ at c	Horizontal tangent line at the point $(c, f(c))$
does not exist at c	Possible vertical tangent line at the point $(c, f(c))$
> 0 for $a < x < b$	Increasing on the interval (a, b)
< 0 for $a < x < b$	Decreasing on the interval (a, b)

EXERCISE 14.1   Answers to odd-numbered problems begin on page AN-64.

In Problems 1–10 use the given graph of $y = f(x)$.

1. What is the domain of f?

2. List the intercepts of f.

3. On what intervals, if any, is the graph of f increasing?

4. On what intervals, if any, is the graph of f decreasing?

5. For what numbers does $f'(x) = 0$?

6. For what numbers does $f'(x)$ not exist?

7. List the point(s) at which f has a local maximum.

8. List the point(s) at which f has a local minimum.

9. List the critical numbers of f.

10. List the critical points of f.

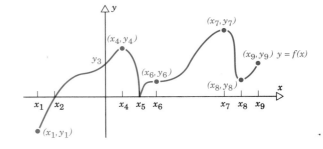

In Problems 11–34 follow the five steps on page 698 to graph f.

11. $f(x) = x^2 - 4x$

12. $f(x) = x^2 + 2x$

13. $f(x) = -x^2 + 6x + 7$

14. $f(x) = -x^2 - 4x + 5$

15. $f(x) = -2x^2 + 4x - 2$

16. $f(x) = -3x^2 + 12x$

17. $f(x) = x^3 - 9x^2 + 27x - 27$

18. $f(x) = x^3 - 6x^2 + 12x - 8$

19. $f(x) = 2x^3 - 15x^2 + 36x$

20. $f(x) = 2x^3 + 6x^2 + 6x$

21. $f(x) = -x^3 + 3x - 1$

22. $f(x) = -2x^3 + 6x^2 + 1$

23. $f(x) = 3x^4 - 12x^3 + 2$

24. $f(x) = x^4 - 4x + 2$

25. $f(x) = x^5 - 5x + 1$

26. $f(x) = x^5 + 5x^4 + 1$

27. $f(x) = 3x^5 - 20x^3 + 1$

28. $f(x) = 3x^5 - 5x^3 - 1$

29. $f(x) = x^{2/3} + 2x^{1/3}$

30. $f(x) = x^{2/3} - 2x^{1/3}$

31. $f(x) = (x^2 - 1)^{2/3}$

32. $f(x) = (x^2 - 1)^{4/3}$

33. $f(x) = \dfrac{x}{x^2 - 4}$

34. $f(x) = \dfrac{x^2}{x^2 - 1}$

Determine the intervals where the function $f(x)$ in Problems 35 and 36 is increasing and where it is decreasing.

35. $f(x) = \ln x^2$

36. $f(x) = \ln 2x$

37. Graphs (a)–(e), on the next page, are the graphs of functions. Graphs (i)–(v) are the graphs of the associated derivative functions. Match each function with its derivative.

(a)
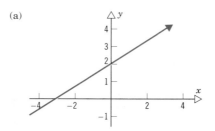

(b)
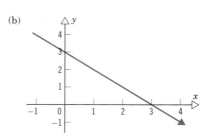

(c)
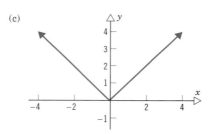

(d)
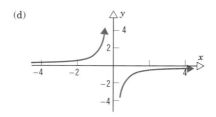

(e)
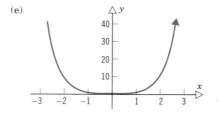

(i)
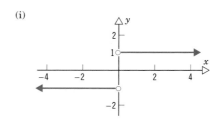

(ii)
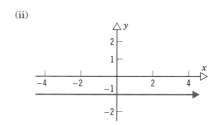

(iii)
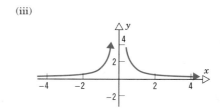

(iv)
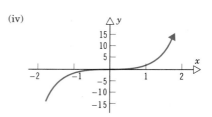

(v)
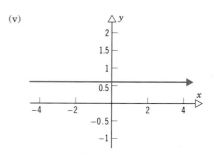

38. Cost Function   The cost C per day, in dollars, of producing x pairs of eyeglasses is

$$C(x) = 0.2x^2 + 3x + 1000$$

Show that C is an increasing function.

39. Sales of Toy Trucks   At a toy store the revenue R, in dollars, derived from selling x electric trucks is

$$R(x) = -0.005x^2 + 20x$$

(a) Determine where the graph of R is increasing and where it is decreasing.
(b) How many trucks need to be sold to maximize revenue?
(c) What is the maximum revenue?
(d) Graph the function R.

40. Sales of Typewriters   The weekly revenue R, in dollars, from selling x typewriters is

$$R(x) = -20x^2 + 1000x$$

(a) Determine where the graph of R is increasing and where it is decreasing.
(b) How many typewriters need to be sold to maximize revenue?
(c) What is the maximum revenue?
(d) Graph the function R.

41. Supply and Demand   Suppose

$$S(x) = -50x^2 + 700x$$

is the supply function describing the number of crates of grapefruit a farmer is willing to supply to the market for x dollars per crate.

(a) Determine where the graph of S is increasing and where it is decreasing.
(b) For what price will the number of crates supplied be a maximum?
(c) What is the maximum number of crates supplied?
(d) Graph the function S.

42. Minimum Cost   Suppose the cost C for producing x units of electronic pocket calculators is

$$C(x) = 0.2x^2 - 1.6x + 1200$$

where C is measured in hundreds of dollars. Determine

(a) When cost is increasing
(b) When cost is decreasing
(c) When cost is a minimum

43. Maximum Profit   A firm can sell as many units as it can produce at $30 per unit. The total cost to the firm of producing x units is

$$C(x) = 25 + 2x + 0.01x^2$$

The profit function P is given by $P(x) = 30x - C(x)$. Find the number of units the firm should produce to obtain a maximum profit. What is the maximum profit?

44. Transatlantic Crossing   An airplane crosses the Atlantic Ocean (3000 miles) with an air speed of 500 mph.

(a) Find the time saved with a 25 mile per hour tail wind.
(b) Find the time lost with a 50 mile per hour head wind.
(c) If the cost per passenger is

$$C(x) = 100 + \frac{x}{10} + \frac{36,000}{x}$$

where x is the ground speed and $C(x)$ is the cost in dollars, what is the cost per passenger when there is no wind?
(d) What is the cost with a tail wind of 25 miles per hour?
(e) What is the cost with a head wind of 50 miles per hour?
(f) What ground speed minimizes the cost?
(g) What is the minimum cost per passenger?

45. Drug Concentration   The concentration C of a certain drug in the bloodstream t hours after injection into muscle tissue is given by

$$C(t) = \frac{2t}{16 + t^3}$$

When is the concentration greatest?

46. Profit Function   The cost C in dollars of producing n machines is $C(n) = 10n + 0.05n^2 + 150,000$. If a machine is priced at $\$m$, the estimated number that would be sold is $n = 2000 - m/5$. At what price would the profit be maximum?

47. Marginal Cost   If $C(q) = 4q^3 + 2q^2 - 8q + 12$ is a cost function, when is marginal cost increasing?

48. Revenue   The monthly revenue R, in dollars, from selling q toys is

$$R(q) = 80q + 19q^2 - \tfrac{1}{3}q^3$$

Determine the output for maximum revenue.

49. Cost Function   Show that for the cost function $C(q) = \sqrt{q}$, the marginal cost and the average cost are always decreasing for $q > 0$. See Problem 63, Section 13.3, for a definition of average cost.

50. Profit   Given the revenue function $R(x) = 4x^3 - x^4 - 500x^2 + 4000x$ and the cost function $C(x) = 1000x + 1000$, find the number x that should be produced in order to maximize the profit.

In Problems 51–54 use the discussion that follows.

Rolle's Theorem   *If a function $y = f(x)$ has the following three properties:*

1. *It is continuous on the closed interval [a, b]*
2. *It is differentiable on the open interval (a, b)*
3. $f(a) = f(b)$

then there is at least one number c, $a < c < b$, at which $f'(c) = 0$.
This result is called **Rolle's** *theorem. See the figure.*

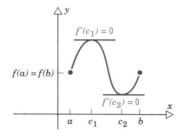

Verify Rolle's theorem by finding the number(s) c for each function on the interval indicated.

51. $f(x) = 2x^2 - 2x$ on [0, 1]   **52.** $f(x) = x^2 + 2x$ on [−2, 0]

53. $f(x) = x^4 - 1$ on [−1, 1]   **54.** $f(x) = x^4 - 2x^2 - 8$ on [−2, 2]

In Problems 55–58 use the discussion that follows.

Mean Value Theorem   *If $y = f(x)$ is a continuous function on an interval [a, b] and is differentiable on (a, b), there is at least one number in (a, b) at which the slope of the tangent line equals the slope of the line joining $(a, f(a))$ and $(b, f(b))$. That is, there is a number c, $a < c < b$, at which*

$$f'(c) = \frac{f(b) - f(a)}{b - a}$$

This result is called the **mean value theorem.** *See the figure.*

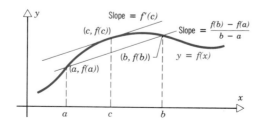

Verify the mean value theorem by finding the number(s) c for each function on the interval indicated.

55. $f(x) = x^2$ on [0, 3]   **56.** $f(x) = x^3$ on [0, 2]   **57.** $f(x) = \dfrac{1}{x^2}$ on [1, 2]

58. $f(x) = x^{3/2}$ on [0, 1]

Technology Exercises
In Problems 1 and 2 graph the given function to determine where it is increasing and where it is decreasing.

1. $f(x) = \dfrac{10}{x^2 + 9}$

2. $f(x) = \dfrac{x^2}{4 - x^2}$

3. Graph $f(x) = x^4 - 8x^3 + 24x^2 - 32x + 17$ and estimate (to the nearest integer) the coordinates of all relative maximum and relative minimum points.

4. Graph $f'(x) = 3x^2 - 6x + 3$ and use the graph to conclude where $f(x)$ has a relative maximum and/or a relative minimum.

5. Let $f(x) = x^4 + 8x^3 + 18x^2 - 8$.

(a) Find $f'(x)$.
(b) Graph $f'(x)$.

(c) From the graph of $f'(x)$, find the interval of x values that make $f'(x) = 0$; $f'(x) > 0$; $f'(x) < 0$.
(d) Find the intervals where $f(x)$ is increasing.
(e) Find the intervals where $f(x)$ is decreasing.
(f) Graph $f(x)$ and $f'(x)$ on the same screen and check your results.

6. Graph the function

$$f(x) = \dfrac{x^2 - 5 \ln x}{4x}$$

Use the graph to determine if $f'(3)$ is positive, negative, zero, or undefined.

14.2 CONCAVITY; THE SECOND DERIVATIVE TEST

Consider the graphs of $y = x^2$ ($x \geq 0$) and $y = \sqrt{x}$ as shown in Figure 10. Each graph starts at (0, 0), is increasing, and passes through the point (1, 1). However, the graph of $y = x^2$ increases rapidly while the graph of $y = \sqrt{x}$ increases slowly. We use the terms *concave up* or *concave down* to describe these characteristics of graphs.

Figure 10

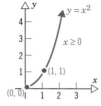
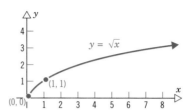

Now look at Figure 11 (at the top of the next page). The graph of the function in (a) is concave up on the interval (a, b), while the graph in (b) is concave down on the interval (a, b). Notice that the tangent lines to the graph in (a) lie below the graph, while the tangent lines to the graph in (b) lie above the graph. This observation is the basis of the following definition.

Figure 11

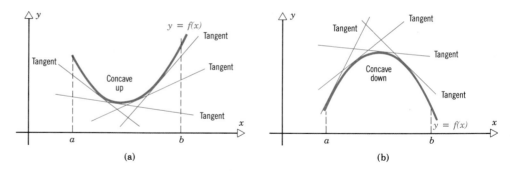

(a)   (b)

CONCAVE UP; CONCAVE DOWN

Let f denote a function that is differentiable on the interval (a, b).

1. **The graph of f is *concave up* on (a, b) if throughout (a, b) the tangent lines to the graph of f lie below the graph.**
2. **The graph of f is *concave down* on (a, b) if throughout (a, b) the tangent lines to the graph of f lie above the graph.**

Now Work Problem 9

Figure 11 also provides a test for determining whether a graph is concave up or concave down. Notice that in Figure 11(a), as one proceeds from left to right along the graph of f, the slopes of the tangent lines are increasing, starting off very negative and ending up very positive. Since the derivative $f'(x)$ equals the slope of a tangent line and these slopes are increasing, it follows that $f'(x)$ is an increasing function. Thus, its derivative $f''(x)$ must be positive on (a, b). Similarly, in Figure 11(b), the slopes of the tangent lines to the graph of f are decreasing. Now it follows that $f'(x)$ is a decreasing function and so $f''(x)$ must be negative on (a, b). Thus, the second derivative provides information about the concavity of a function.

Test for Concavity

Let $y = f(x)$ be a function and let $f''(x)$ be its second derivative.

1. If $f''(x) > 0$ for all x in the interval (a, b), then the graph of f is concave up on (a, b).
2. If $f''(x) < 0$ for all x in the interval (a, b), then the graph of f is concave down on (a, b).

EXAMPLE 1

Determine where the graph of $f(x) = x^3 - 12x^2$ is concave up or concave down.

SOLUTION

We proceed to find $f''(x)$:

$$f(x) = x^3 - 12x^2$$
$$f'(x) = 3x^2 - 24x$$
$$f''(x) = 6x - 24 = 6(x - 4)$$

Now we need to solve the inequalities $f''(x) > 0$ and $f''(x) < 0$. We see that if $x < 4$, then $f''(x) < 0$. Thus, the graph of f is concave down on the interval $x < 4$. If $x > 4$, then $f''(x) > 0$. Thus, the graph of f is concave up on the interval $x > 4$.

. . . .

EXAMPLE 2   Graph the function $f(x) = x^3 - 12x^2$.

SOLUTION

Step 1 The domain of f is all real numbers.

Step 2 Let $x = 0$. Then $y = f(0) = 0$. Thus, the y-intercept is $(0, 0)$. To find the x-intercepts, if any, let $y = 0$. Then

$$x^3 - 12x^2 = 0$$
$$x^2(x - 12) = 0$$

The x-intercepts are $(0, 0)$ and $(12, 0)$.

Step 3 To determine where the graph of f is increasing or decreasing, we first need to find $f'(x)$:

$$f(x) = x^3 - 12x^2$$
$$f'(x) = 3x^2 - 24x = 3x(x - 8)$$

The solutions of the equation $f'(x) = 3x(x - 8) = 0$ are 0 and 8. These numbers separate the number line into three parts:

$$x < 0 \qquad 0 < x < 8 \qquad 8 < x$$

The test values we select for these intervals are -1, 1, and 9, respectively. We summarize the results in the following table:

Interval	$x < 0$	$0 < x < 8$	$8 < x$
Test Values	-1	1	9
f' at Test Values	$f'(-1) = 27 > 0$	$f'(1) = -21 < 0$	$f'(9) = 27 > 0$
Sign of f'	$+$	$-$	$+$
Graph of f	Increasing	Decreasing	Increasing

Step 4 From the table, the graph of f is increasing for $x < 0$ and is decreasing for $0 < x < 8$. Thus, at $x = 0$, the graph changes from increasing to decreasing. Consequently, the point $(0, f(0)) = (0, 0)$ is a local maximum. Similarly, at $x = 8$, the graph changes from decreasing to increasing. Thus, the point $(8, f(8)) = (8, -256)$ is a local minimum.

Step 5 The derivative of f is $f'(x) = 3x(x - 8)$. We see that $f'(x) = 0$ if $x = 0$ or $x = 8$. Thus, the graph of f has a horizontal tangent line at the points $(0, 0)$ and $(8, -256)$. There are no vertical tangent lines.

Finally, we use the result of Example 1, namely, that the graph of f is concave down on $x < 4$ and is concave up on $4 < x$.

To graph f, we plot the intercepts, the local maximum, and the local minimum (the points at which the graph has a horizontal tangent line), and connect these points with a smooth curve, keeping the concavity of the graph in mind. See Figure 12.

Figure 12

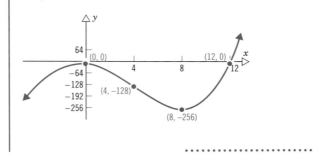

The point $(4, -128)$ on the graph of $f(x) = x^3 - 12x^2$ is the point at which the concavity of f changed from down to up. This point is called an *inflection point*.

INFLECTION POINT

An *inflection point* of a continuous function f is a point on the graph of f at which the concavity of f changes.

Figure 13 illustrates two graphs with points of inflections. In Figure 13(a) $f''(x) < 0$ for $x < b$, so the graph of f is concave down to the left of b; and $f''(x) > 0$ for $x > b$, so the graph of f is concave up to the right of b. Therefore, the point $(b, f(b))$ is an inflection point. For this case $f''(b) = 0$. In Figure 13(b) the graph of f is also concave down to the left of b and concave up to the right of b, so $(b, f(b))$ is also an inflection point. In this case $f''(b)$ is not defined.

To locate inflection points, find the intervals where $f''(x) > 0$, and the intervals where $f''(x) < 0$. Points where $f''(x)$ changes sign are points of inflection. At a point of inflection $(b, f(b))$

$$f''(b) = 0 \qquad \text{or} \qquad f''(b) \text{ is not defined}$$

Figure 13

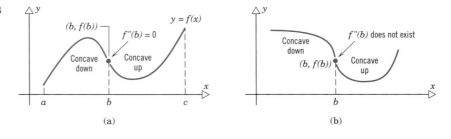

EXAMPLE 3

Let

$$f(x) = x^3 - 6x^2 + 9x + 3$$

Find all inflection points.

SOLUTION

We compute $f'(x)$ and $f''(x)$:

$$f'(x) = 3x^2 - 12x + 9$$
$$f''(x) = 6x - 12 = 6(x - 2)$$

When $x > 2$, we see that $f''(x) > 0$. Thus, the function is concave up for $x > 2$. Similarly, it is concave down for $x < 2$.

To find the inflection points, set

$$f''(x) = 6(x - 2) = 0$$

The only candidate for an inflection point is $(2, 5)$. Since for $x < 2$, the function is concave down and for $x > 2$, it is concave up, the point $(2, 5)$ is an inflection point.

Now Work Problem 11

....

The next example illustrates the case where $f''(x)$ does not exist at an inflection point.

EXAMPLE 4

Find all inflection points of the function $f(x) = x^{5/3}$.

$$f'(x) = \frac{5}{3} x^{2/3} \qquad f''(x) = \frac{10}{9} x^{-1/3} = \frac{10}{9x^{1/3}}$$

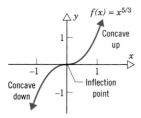

Figure 14

SOLUTION

Here, $f''(x)$ is never zero. But we observe that $f''(x)$ does not exist at $x = 0$, which is in the domain. Since the function is concave down for $x < 0$ and concave up for $x > 0$, the point $(0, 0)$ is an inflection point. See Figure 14.

....

Figure 15

The concavity of a function can change at a point, yet the point is not necessarily an inflection point. For example, consider the function $f(x) = 1/x$ and its graph as given in Figure 15. As we see from the graph, the function is concave down for x negative and is concave up for x positive, yet $x = 0$ is not an inflection point because the function is not continuous at $x = 0$.

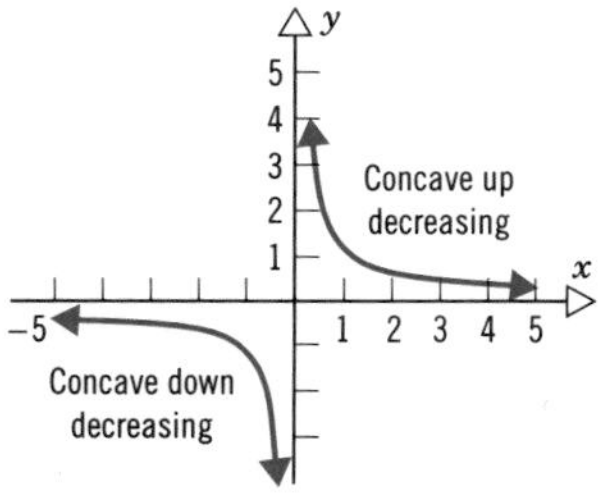

To the five steps we used earlier to graph a function, we now add Step 6:

> **Steps for Graphing Functions**
>
> **Step 1** Find the domain of f.
>
> **Step 2** Locate the intercepts of f. (Skip the x-intercepts if they are too hard to find.)
>
> **Step 3** Determine where the graph of f is increasing or decreasing.
>
> **Step 4** Find any local maxima or local minima of f by using the First Derivative Test.
>
> **Step 5** Locate all points on the graph of f at which the tangent line is either horizontal or where $f'(x)$ does not exist.
>
> **Step 6** Use $f''(x)$ to determine where the concavity of the graph changes. Locate the inflection points, if any.

EXAMPLE 5   Graph the function $f(x) = 3x^5 - 5x^4$.

SOLUTION

Step 1 The domain of f is all real numbers.

Step 2 Let $x = 0$. Then $y = f(0) = 0$. Thus, the y-intercept is $(0, 0)$. To find the x-intercepts, if any, let $y = 0$. Then

$$3x^5 - 5x^4 = 0$$
$$x^4(3x - 5) = 0$$

The x-intercepts are $(0, 0)$ and $(\frac{5}{3}, 0)$.

Step 3 To determine where the graph of f is increasing or decreasing, we first need to find $f'(x)$:

$$f(x) = 3x^5 - 5x^4$$
$$f'(x) = 15x^4 - 20x^3 = 5x^3(3x - 4)$$

The solutions of the equation $f'(x) = 5x^3(3x - 4) = 0$ are 0 and $\frac{4}{3}$. These numbers separate the number line into three parts:

$$x < 0 \qquad 0 < x < \tfrac{4}{3} \qquad \tfrac{4}{3} < x$$

The test values we select for these intervals are -1, 1, and 2, respectively. We summarize the results in the following table:

Interval	$x < 0$	$0 < x < \frac{4}{3}$	$\frac{4}{3} < x$
Test Values	-1	1	2
f' at Test Value	$f'(-1) = 35 > 0$	$f'(1) = -5 < 0$	$f'(2) = 80 > 0$
Sign of f'	$+$	$-$	$+$
Graph of f	Increasing	Decreasing	Increasing

We conclude that the graph of f is increasing on the intervals $x < 0$ and $\frac{4}{3} < x$. It is decreasing on the interval $0 < x < \frac{4}{3}$.

Step 4 From the table, the graph of f is increasing for $x < 0$ and is decreasing for $0 < x < \frac{4}{3}$. Therefore, at $x = 0$, the graph changes from increasing to decreasing. Consequently, the point $(0, f(0)) = (0, 0)$ is a local maximum. Similarly, at $x = \frac{4}{3}$, the graph changes from decreasing to increasing. Thus, the point $(\frac{4}{3}, f(\frac{4}{3})) = (\frac{4}{3}, -\frac{256}{81})$ is a local minimum.

Step 5 The derivative of f is $f'(x) = 5x^3(3x - 4)$. We see that $f'(x) = 0$ if $x = 0$ or $x = \frac{4}{3}$. Thus, the graph of f has a horizontal tangent line at the points $(0, 0)$ and $(\frac{4}{3}, -\frac{256}{81})$. Since f is a polynomial, $f'(x)$ exists everywhere.

Step 6 To locate the inflection points, if any, of f, we need to find $f''(x)$:

$$f(x) = 3x^5 - 5x^4$$
$$f'(x) = 15x^4 - 20x^3$$
$$f''(x) = 60x^3 - 60x^2 = 60x^2(x - 1)$$

We use the numbers 0 and 1 to separate the number line into three parts:

$$x < 0 \qquad 0 < x < 1 \qquad 1 < x$$

To determine the sign of $f''(x)$, we select the following test values for these intervals: -1, $\frac{1}{2}$, and 2, respectively. We summarize the results in the following table:

Interval	$x < 0$	$0 < x < 1$	$1 < x$
Test Values	-1	$\frac{1}{2}$	2
f'' at Test Values	$f''(-1) = -120 < 0$	$f''(\frac{1}{2}) = -\frac{15}{2} < 0$	$f''(2) = 240 > 0$
Sign of f''	$-$	$-$	$+$
Concavity of f	Down	Down	Up

We conclude that the graph of f is concave down on $x < 1$ and is concave up on $1 < x$. Further, since the concavity changes at the point $(1, f(1)) = (1, -2)$, we conclude that $(1, -2)$ is an inflection point.

To graph f, we plot the intercepts, the local maximum, the local minimum, and the inflection point, and connect these points with a smooth curve. See Figure 16.

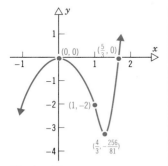

Figure 16

Now Work Problem 27

EXAMPLE 6

Graph the function $f(x) = xe^{-x}$.

SOLUTION

Step 1 The domain of f is all real numbers and f is continuous on its domain.

Step 2 Let $x = 0$. Then $y = f(0) = 0$, so $(0, 0)$ is the y-intercept. Let $y = 0$. Then $xe^{-x} = 0$. Since $e^{-x} > 0$ for all x, the only solution is $x = 0$. Thus, the only x-intercept is $(0, 0)$.

Step 3 To find where the graph of f is increasing or decreasing, we need to find $f'(x)$:

$$f'(x) = \frac{d}{dx} xe^{-x} = x\frac{d}{dx} e^{-x} + e^{-x}\frac{d}{dx} x = xe^{-x}(-1) + e^{-x}(1)$$

$$= e^{-x}(-x + 1) = (1 - x)e^{-x}$$

Since $e^{-x} > 0$ for all x, it follows that $f'(x) > 0$ if $x < 1$, and $f'(x) < 0$ if $1 < x$. Thus, the graph of f is increasing if $x < 1$ and is decreasing if $1 < x$.

Step 4 Using the First Derivative Test, we conclude that the point $(1, e^{-1}) = (1, 0.368)$ is a local maximum.

Step 5 Since $f'(1) = 0$, the tangent line is horizontal at $(1, 0.368)$.

Step 6 To locate any inflection points, we need to find $f''(x)$.

$$f''(x) = \frac{d}{dx} f'(x) = \frac{d}{dx} (1 - x)e^{-x} = (1 - x)\frac{d}{dx} e^{-x} + e^{-x}\frac{d}{dx} (1 - x)$$

$$= (1 - x) \cdot e^{-x}(-1) + e^{-x}(-1) = e^{-x}[(1 - x)(-1) + (-1)]$$
$$= (x - 2)e^{-x}$$

It follows that $f''(x) < 0$ if $x < 2$, and $f''(x) > 0$ if $2 < x$. The graph of f is concave down if $x < 2$ and is concave up if $2 < x$. The point $(2, 2e^{-2}) = (2, 0.27)$ is an inflection point. Put all this information together and draw the graph. See Figure 17.

Figure 17

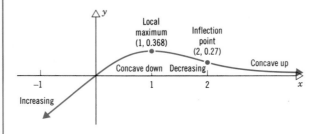

Now Work Problem 45

· ·

EXAMPLE 7

The total sales $S(x)$, in hundreds of thousands of dollars, of the Big Apple, a manufacturer of microcomputers, is related to the amount of money that the company spends on advertising by the function

$$S(x) = -0.02x^3 + 1.2x^2 + 1000 \qquad 0 \le x \le 60$$

Find the inflection point of the function S and discuss its significance.

SOLUTION

The first and second derivatives of S are

$$S'(x) = -0.06x^2 + 2.4x = -0.06x(x - 40)$$
$$S''(x) = -0.12x + 2.4 = -0.12(x - 20)$$

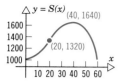

Figure 18

Now

$$S''(x) > 0 \quad \text{for} \quad x < 20 \quad \text{and} \quad S''(x) < 0 \quad \text{for} \quad x > 20$$

Thus, the point $(20, 1320)$ is an inflection point of the function S. Figure 18 illustrates the graph of S.

Notice that the total sales of the company increase very slowly at the beginning, but as advertising money increases, total sales increase rapidly. This rapid sales growth indicates that consumers are responding to the advertisement. However, there is a point on the graph at which the *rate of growth* in sales changes from positive to negative—where the rate of growth changes from increasing to decreasing—resulting in a slower rate of increased sales. This point, commonly known as the *point of diminishing returns,* is the point of inflection of S.

Now Work Problem 69

· ·

The Second Derivative Test

There is another test for finding the local maximum and local minimum that is sometimes easier to use than the First Derivative Test. This test is based on the geometric observation that the graph of a function has a local maximum at a point where the tangent line is horizontal and the graph is concave down. See Figure 19(a).

Figure 19

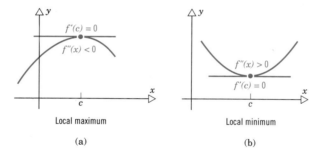

(a)   (b)

Similarly, the graph of a function has a local minimum at a point where the tangent line is horizontal and the graph is concave up. See Figure 19(b). This leads us to formulate the Second Derivative Test.

Second Derivative Test

Let $y = f(x)$ be a function that is differentiable on an open interval I and suppose that the second derivative $f''(x)$ exists on I. Also suppose c is a number for which $f'(c) = 0$.

1. If $f''(c) < 0$, then the point $(c, f(c))$ is a local maximum.
2. If $f''(c) > 0$, then the point $(c, f(c))$ is a local minimum.
3. If $f''(c) = 0$, the test is inconclusive and the First Derivative Test must be used.

Again the reader is reminded that maxima and minima occur at $(c, f(c))$ and not at $(c, f'(c))$.

Notice that the Second Derivative Test is used to locate those local maxima and local minima that occur where the tangent line is horizontal. Since local maxima and local minima can also occur at points on the graph at which the derivative does not exist, this test is usually used only for functions that are differentiable. Also note that the test provides no information if the second derivative is zero at c. Whenever either of these situations arises, you must use the First Derivative Test.

EXAMPLE 8

Use the Second Derivative Test to find the local maxima and local minima of

$$f(x) = x^3 - 12x^2$$

SOLUTION

First, we find those numbers at which $f'(x) = 0$:

$$f(x) = x^3 - 12x^2$$
$$f'(x) = 3x^2 - 24x = 3x(x - 8)$$

Thus, $f'(x) = 0$ at $x = 0$ and $x = 8$.

Next, we evaluate $f''(x)$ at these numbers:

$$f''(x) = 6x - 24$$

At $x = 0$, $f''(0) = -24 < 0$, so the graph of f is concave down. Thus, f has a local maximum at $(0, 0)$. At $x = 8$, $f''(8) = 24 > 0$, so the graph of f is concave up. Thus, f has a local minimum at $(8, -256)$. [The graph of $f(x) = x^3 - 12x^2$ was given earlier in Figure 12.]

Now Work Problem 47

· ·

EXAMPLE 9

Sketch the graph of a function

$$y = f(x) \qquad 0 \le x \le 6$$

which is continuous and has the following properties:

1. The points $(0, 4)$, $(1, 3)$, $(3, 5)$, $(5, 7)$, and $(6, 6)$ are on the graph.
2. $f'(1) = 0$ and $f'(5) = 0$; $f'(x)$ is not 0 anywhere else.
3. $f''(x) > 0$ for $x < 3$ and $f''(x) < 0$ for $x > 3$.

SOLUTION

First, we plot the points $(0, 4)$, $(1, 3)$, $(3, 5)$, $(5, 7)$, and $(6, 6)$. Since we are told that $f'(1) = 0$ and $f'(5) = 0$, we know the tangent lines to the graph at $(1, 3)$ and at $(5, 7)$ are horizontal. See Figure 20(a).

Since $f''(x) > 0$ if $x < 3$, it follows that $f''(1) > 0$. By the Second Derivative Test, the point $(1, 3)$ is a local minimum. Similarly, since $f''(x) < 0$ if $x > 3$, it follows that $f''(5) < 0$, so the point $(5, 7)$ is a local maximum. See Figure 20(b).

Finally, since the graph is concave up for $x < 3$ and concave down for $x > 3$, it follows that the point $(3, 5)$ is an inflection point. See Figure 20(c).

Figure 20

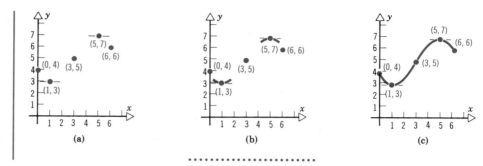

(a)   (b)   (c)

Now Work Problem 55

Application: Average Cost

If the cost C of producing x units is given by the cost function $C = C(x)$, the **average cost $\overline{C}(x)$** of producing x units is defined as

$$\overline{C}(x) = \frac{C(x)}{x}$$

EXAMPLE 10

Consider the cost function

$$C(x) = 1000 + \tfrac{1}{10}x^2$$

where x is the number of units manufactured.

(a) What is the average cost?
(b) What is the minimum average cost?
(c) Find the marginal cost.
(d) Graph the average cost function and the marginal cost function, using the same coordinate system.

SOLUTION

(a) The average cost $\overline{C}(x)$ is

$$\overline{C}(x) = \frac{C(x)}{x} = \frac{1000 + \tfrac{1}{10}x^2}{x} = \frac{1000}{x} + \frac{1}{10}x$$

(b) To find the minimum average cost, we use the Second Derivative Test. Thus, we begin by finding $\overline{C}'(x)$:

$$\overline{C}'(x) = \frac{-1000}{x^2} + \frac{1}{10} = \frac{-10,000 + x^2}{10x^2}$$

$$\overline{C}'(x) = 0 \quad \text{when } x^2 - 10,000 = 0 \quad \text{or} \quad x = \pm\,100$$

We disregard $x = -100$, since the number of units must be positive. Next, we find the second derivative of $\overline{C}(x)$. Since

$$\overline{C}'(x) = \frac{-1000}{x^2} + \frac{1}{10}$$

we have

$$\overline{C}''(x) = \frac{2000}{x^3}$$

Since $\overline{C}''(100) = 2000/1{,}000{,}000 > 0$, it follows by the Second Derivative Test that $\overline{C}(x)$ has a local minimum at $x = 100$. The minimum average cost is therefore

$$\overline{C}(100) = \frac{C(100)}{100} = \frac{2000}{100} = 20$$

(c) The cost function is $C(x) = 1000 + \frac{1}{10}x^2$. Thus, the marginal cost is

$$C'(x) = \tfrac{1}{5}x$$

(d) See Figure 21 for the graph.

Figure 21

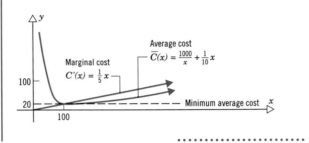

Notice that the minimum average cost occurs at the intersection of the average cost and the marginal cost. In fact, this is always the case.

> The minimum average cost occurs when the average cost and marginal cost are equal.

EXERCISE 14.2 Answers to odd-numbered problems begin on page AN-67.

In Problems 1–10 use the given graph of y = f(x).

1. What is the domain of f?

2. List the intercepts of f.

3. On what intervals, if any, is the graph of f increasing?

4. On what intervals, if any, is the graph of f decreasing?

5. For what numbers does $f'(x) = 0$?

6. For what numbers does $f'(x)$ not exist?

7. List the point(s) at which f has a local maximum.

8. List the point(s) at which f has a local minimum.

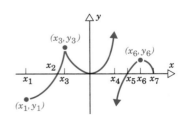

9. On what intervals, if any, is the graph of f concave up?

10. On what intervals, if any, is the graph of f concave down?

In Problems 11–26 determine the intervals on which the graph of f is concave up and concave down. List any inflection points.

11. $f(x) = x^3 - 6x^2 + 1$

12. $f(x) = x^3 + 3x^2 + 2$

13. $f(x) = x^4 - 2x^3 + 6x - 1$

14. $f(x) = x^4 + 2x^3 - 8x + 8$

15. $f(x) = 3x^5 - 5x^4 + 60x + 10$

16. $f(x) = 3x^5 + 5x^4 + 20x - 4$

17. $f(x) = 3x^5 - 10x^3 + 10x + 10$

18. $f(x) = 3x^5 - 10x^3 - 8x + 8$

19. $f(x) = x^5 - 10x^2 + 4$

20. $f(x) = x^5 + 10x^2 - 8x + 4$

21. $f(x) = 3x^{1/3} + 9x + 2$

22. $f(x) = 3x^{2/3} - 8x + 4$

23. $f(x) = x^{2/3}(x - 10)$

24. $f(x) = x^{2/3}(x - 15)$

25. $f(x) = x^{2/3}(x^2 - 16)$

26. $f(x) = x^{2/3}(x^2 - 4)$

In Problems 27–46 follow the six steps on page 712 to graph f.

27. $f(x) = x^3 - 6x^2 + 1$

28. $f(x) = x^3 + 6x^2 + 2$

29. $f(x) = x^4 - 2x^2 + 1$

30. $f(x) = 2x^4 - 4x^2 + 2$

31. $f(x) = x^5 - 10x^4$

32. $f(x) = x^5 + 5x$

33. $f(x) = x^6 - 3x^5$

34. $f(x) = x^6 + 3x^5$

35. $f(x) = 3x^4 - 12x^3$

36. $f(x) = 3x^4 + 12x^3$

37. $f(x) = x^5 - 10x^2 + 4$

38. $f(x) = x^5 + 10x^2 + 2$

39. $f(x) = x^{2/3}(x - 10)$

40. $f(x) = x^{2/3}(x - 15)$

41. $f(x) = x^{2/3}(x^2 - 16)$

42. $f(x) = x^{2/3}(x - 4)$

43. $f(x) = x \ln x$

44. $f(x) = \dfrac{1}{x} \ln x$

45. $f(x) = xe^x$

46. $f(x) = x^2 e^x$

In Problems 47–54 determine where $f'(x) = 0$. Use the Second Derivative Test to determine the local maxima and local minima of each function.

47. $f(x) = x^3 - 3x + 2$

48. $f(x) = x^3 - 12x - 4$

49. $f(x) = 3x^4 + 4x^3 - 3$

50. $f(x) = 3x^4 - 6x^2 + 4$

51. $f(x) = x^5 - 5x^4 + 2$

52. $f(x) = 3x^5 - 20x^3$

53. $f(x) = x + \dfrac{1}{x}$

54. $f(x) = 2x + \dfrac{1}{x^2}$

55. Sketch the graph of a function $y = f(x)$ that is continuous for all x and has the following properties:

1. $(0, 10)$, $(6, 15)$, and $(10, 0)$ are on the graph.
2. $f'(6) = 0$ and $f'(10) = 0$; $f'(x)$ is not 0 anywhere else.
3. $f''(x) < 0$ for $x < 9$, $f''(9) = 0$, and $f''(x) > 0$ for $x > 9$.

56. Sketch the graph of a function $y = f(x)$ that is continuous for all x and has the following properties:

1. $(-1, 3)$, $(1, 5)$, and $(3, 7)$ are on the graph.
2. $f'(3) = 0$ and $f'(-1) = 0$; $f'(x)$ is not 0 anywhere else.
3. $f''(x) > 0$ for $x < 1$, $f''(1) = 0$, and $f''(x) > 0$ for $x > 1$.

57. Sketch the graph of a function $y = f(x)$ that is continuous for all x and has the following properties:

1. $(1, 5)$, $(2, 3)$, and $(3, 1)$ are on the graph.
2. $f'(1) = 0$ and $f'(3) = 0$; $f'(x)$ is not 0 anywhere else.
3. $f''(x) < 0$ for $x < 2$, $f''(2) = 0$, and $f''(x) > 0$ for $x > 2$.

58. Sketch the graph of a function $y = f(x)$ that is continuous for all x and has the following properties:

1. Domain of $f(x)$ is $x \geq 0$.
2. $(0, 0)$ and $(6, 7)$ are on the graph.
3. $f'(x) > 0$ for $x > 0$.
4. $f''(x) < 0$ for $x < 6$, $f''(6) = 0$, and $f''(x) > 0$ for $x > 6$.

59. For the function $f(x) = ax^3 + bx^2$, determine a and b so that the point $(1, 6)$ is a point of inflection of $f(x)$.

60. Let $f(x) = ax^2 + bx + c$, where $a \neq 0$, b and c are real numbers. Is it possible for $f(x)$ to have an inflection point? Explain your answer.

61. Cost and Revenue Functions   For a certain production facility the cost function is

$$C(x) = 2x + 5$$

and the revenue function is

$$R(x) = 8x - x^2$$

where x is the number of units produced (in thousands) and R and C are measured in millions of dollars.

(a) Find the profit function $P(x) = R(x) - C(x)$.
(b) Where is the profit a maximum?
(c) What is the maximum profit?
(d) Where is the revenue a maximum?
(e) What is the maximum revenue?

62. Cost and Revenue Functions   For a certain production facility, the cost function is

$$C(x) = x^2 + 5$$

and the revenue function is

$$R(x) = 12x - 2x^2$$

where x is the number of units produced (in thousands) and R and C are measured in millions of dollars.

(a) Find the profit function $P(x) = R(x) - C(x)$.
(b) Where is the profit a maximum?
(c) What is the maximum profit?
(d) Where is the revenue a maximum?
(e) What is the maximum revenue?

63. Demand Equation   The cost function and demand equation for a certain product are

$$C(x) = 50x + 40,000$$
$$p = d(x) = 100 - 0.01x$$

Find

(a) The revenue function
(b) The maximum revenue
(c) The profit function
(d) The maximum profit

64. Demand Equation   A certain item can be produced at a cost of $10 per unit. The demand equation for this item is

$$p = d(x) = 90 - 0.02x$$

where p is the price in dollars and x is the number of units. Find

(a) The revenue function
(b) The maximum revenue
(c) The profit function
(d) The maximum profit

65. Demand Equation   The demand equation for a certain commodity is

$$p = d(x) = 10 + \frac{40}{x}   \quad 1 \le x \le 10$$

where p is the price in dollars when x units are demanded. Find

(a) The revenue function
(b) The number x of units demanded that maximizes revenue
(c) The maximum revenue

66. Demand Equation   The price p per pound when x pounds of a certain commodity are demanded is

$$p = d(x) = \frac{10,000}{5x + 100} - 5   \quad 0 \le x \le 90$$

Find

(a) The revenue function
(b) The number x of units demanded that maximizes revenue
(c) The maximum revenue

67. Let $f(x) = ax^3 + bx^2 + cx + d$, where $a \ne 0, b, c$, and d are real numbers. Is it possible for $f(x)$ to have an inflection point? Explain your answer.

68. Advertisement Cost   The total sales $S(x)$, in thousands of dollars, of AM Company is related to the amount of money x that the company spends on advertising its products by the function

$$S(x) = -0.003x^3 + 0.04x^2 + 500   \quad 0 \le x \le 200$$

where x is measured in thousands of dollars. Find the inflection point and interpret its significance.

69. The cost (in dollars) of producing x units is given by the function

$$C(x) = 0.001x^3 - 0.3x^2 + 30x + 42$$

Find where this cost function is concave up and where it is concave down, and find the inflection point. Interpret the information.

70. Average Cost   The cost function for producing x items is

$$C(x) = 500 + 10x + \frac{x^2}{500}$$

(a) What is the average cost $\overline{C}(x)$?
(b) What is the minimum average cost?
(c) What is the marginal cost?
(d) Graph the average cost and the marginal cost function using the same coordinate system.

71. Repeat Problem 70 for the cost function

$$C(x) = 800 + 0.04x + 0.0002x^2$$

72. Profit Function   A company estimates that the profit $P(x)$ is related to the selling price x of an item by

$$P(x) = 15x^2 - 100 - \tfrac{1}{3}x^3$$

(a) Determine where profit is increasing.
(b) What selling price results in maximum profit?

73. Blood Velocity   The nineteenth century physician Poise-ville discovered that the velocity (in centimeters per second) of blood flowing r centimeters from the central axis of an artery is given by

$$v(r) = k(R^2 - r^2)$$

where k is a constant and R is the radius of the artery. Show that velocity of blood is greatest along the central axis.

74. Profit Function   A farmer has a crop of 3000 pounds of oranges that can be sold for 20 cents per pound. If he waits to pick the crop, it will increase in weight at the rate of 200 pounds per week but the price will fall 1 cent per pound per week. When should he pick the fruit so as to maximize his profit?

75. Minimum Profit   The cost (in dollars) of producing x units (measured in thousands) of a certain product is found to be

$$C(x) = 20 + \ln(x + 1)$$

The revenue (in dollars) derived from the sale of x units (measured in thousands) of the product is

$$R(x) = 0.1x$$

Find the profit function $P(x) = R(x) - C(x)$, and the number of units that will minimize profit.

Technology Exercises

1. Graph the function

$$f(x) = \frac{0.9x^2}{x^2 - 4}$$

and then determine where it is concave up, concave down, and the coordinates of the point(s) of inflection.

2. Let $f(x) = x^3 - 2x^2 - 3x + 2$.

(a) Find $f''(x)$.   (b) Graph $f''(x)$.
(c) From the graph of $f''(x)$, find the values of x that make $f''(x) = 0$; $f''(x) > 0$; $f''(x) < 0$.
(d) Find the intervals where the graph of $f(x)$ is concave up.
(e) Find the intervals where the graph of $f(x)$ is concave down.

(f) Find the points of inflection.
(g) Graph $f(x)$ and check your results.

3. Graph the function $f(x) = x - e^{x-2}$.

(a) Use the graph to determine the local maxima or minima.
(b) Use the graph to determine the interval when it is concave up or down.

4. Graph the function $f(x) = 6.4 + 1.2x - 4 \ln x$.

(a) Use the graph to find the local maxima or minima.
(b) Use the graph to determine the interval when it is concave up or down.

14.3 ASYMPTOTES

In Chapter 12 we described $\lim_{x \to c} f(x) = L$ by saying that the value of f can be made as close as we please to L by choosing numbers x sufficiently close to c. It was understood that L and c were real numbers. In this section we extend the language of limits to allow c to be ∞ or $-\infty$ *(limits at infinity)* and to allow L to be ∞ or $-\infty$ *(infinite limits).** These

* Remember that the symbols ∞ (infinity) and $-\infty$ (minus infinity) are *not numbers*. Infinity expresses the idea of unboundedness in the positive direction; minus infinity expresses the idea of unboundedness in the negative direction.

limits, it turns out, are useful for locating *asymptotes* and hence aid in obtaining the graph of certain functions.

We begin with limits at infinity.

Limits at Infinity

Let's look at the function $f(x) = 1/x$, whose domain is $x \neq 0$. See Figure 22. This function has the property that the values of f can be made as close as we please to 0 as x becomes unbounded in the positive direction. The table illustrates this fact for selected numbers x:

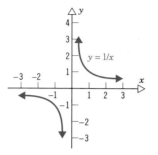

Figure 22

x	1	10	100	1000	10,000	100,000
$f(x) = 1/x$	1	0.1	0.01	0.001	0.0001	0.00001

This characteristic is expressed by saying that $f(x) = 1/x$ has the limit 0 as x approaches ∞ and is symbolized by writing

$$\lim_{x \to \infty} \frac{1}{x} = 0 \tag{1}$$

In the same way, we can write

$$\lim_{x \to -\infty} \frac{1}{x} = 0 \tag{2}$$

to indicate that $1/x$ can be made as close as we please to 0 as x becomes unbounded in the negative direction. We summarize statements (1) and (2) by saying that $f(x) = 1/x$ has **limits at infinity.**

EXAMPLE 1

Find: (a) $\displaystyle\lim_{x \to \infty} \frac{3x - 2}{4x - 1}$ (b) $\displaystyle\lim_{x \to \infty} \frac{5x^3 - 3x^2 + 2}{x^3 + 5}$

SOLUTION

(a) We evaluate this limit by first dividing each term of both the numerator and the denominator by the highest power of x that appears in the denominator (in this case, x). Then

$$\lim_{x \to \infty} \frac{3x - 2}{4x - 1} = \lim_{x \to \infty} \frac{3 - (2/x)}{4 - (1/x)} = \frac{\displaystyle\lim_{x \to \infty} [3 - (2/x)]}{\displaystyle\lim_{x \to \infty} [4 - (1/x)]}$$

$$= \frac{\displaystyle\lim_{x \to \infty} 3 - \lim_{x \to \infty} (2/x)}{\displaystyle\lim_{x \to \infty} 4 - \lim_{x \to \infty} (1/x)} = \frac{3 - 0}{4 - 0} = \frac{3}{4}$$

(b) We follow the same procedure as in Part (a):

$$\lim_{x \to \infty} \frac{5x^3 - 3x^2 + 2}{x^3 + 5} = \lim_{\substack{\uparrow \\ x \to \infty}} \frac{5 - (3/x) + (2/x^3)}{1 + (5/x^3)} = 5$$

Divide by x^3

Now Work Problem 1

....

Infinite Limits

We again use the function $f(x) = 1/x$, whose graph was given in Figure 22, to introduce the idea of **infinite limits.** We construct a table that gives values of f for selected numbers x that are close to 0:

x	1	0.1	0.01	0.001	0.0001	0.00001
$f(x) = 1/x$	1	10	100	1000	10,000	100,000

Here, we see that as x gets closer to 0 from the right, the value of $f(x) = 1/x$ can be made as positive as we please—that is, $1/x$ becomes unbounded in the positive direction. We express this fact by writing

$$\lim_{x \to 0^+} \frac{1}{x} = \infty \qquad (3)$$

Similarly, we use the notation

$$\lim_{x \to 0^-} \frac{1}{x} = -\infty \qquad (4)$$

to indicate that $1/x$ can be made as negative as we please by selecting numbers x sufficiently close to 0, but less than 0. We summarize (3) and (4) by saying that $f(x) = 1/x$ has **one-sided infinite limits** at 0.

We now apply the ideas of limits at infinity and infinite limits to the problem of finding *horizontal* and *vertical asymptotes.*

Horizontal Asymptotes

Limits at infinity have the following important geometric interpretation. When $\lim_{x \to \infty} f(x) = N$, it means that as x becomes unbounded in the positive direction, the value of f can be made as close as we please to N. That is, the graph of $y = f(x)$ for x sufficiently positive is as close as we please to the horizontal line $y = N$. Similarly, $\lim_{x \to -\infty} f(x) = M$ means that the graph of $y = f(x)$ for x sufficiently negative is as close as we please to the horizontal line $y = M$. These lines are called **horizontal asymptotes** of the graph of f. See Figure 23.

Figure 23

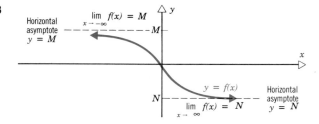

EXAMPLE 2

Find the horizontal asymptotes, if any, of the graph of

$$f(x) = \frac{4x^2}{x^2 + 2}$$

SOLUTION

To find any horizontal asymptotes, we need to find two limits:
$\lim\limits_{x \to \infty} f(x)$ and $\lim\limits_{x \to -\infty} f(x)$.

$$\lim_{x \to \infty} f(x) = \lim_{x \to \infty} \frac{4x^2}{x^2 + 2} = \lim_{x \to \infty} \frac{4}{1 + \dfrac{2}{x^2}} = 4$$

We conclude that the line $y = 4$ is a horizontal asymptote of the graph when x is sufficiently positive.

$$\lim_{x \to -\infty} f(x) = \lim_{x \to -\infty} \frac{4x^2}{x^2 + 2} = \lim_{x \to -\infty} \frac{4}{1 + \dfrac{2}{x^2}} = 4$$

We conclude that the line $y = 4$ is a horizontal asymptote of the graph when x is sufficiently negative.

See Figure 24 for the graph.

Figure 24

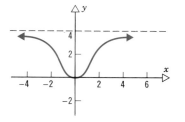

Vertical Asymptotes

Infinite limits are used to find vertical asymptotes. Figure 25 illustrates some of the possibilities that can occur when a function has an infinite limit.

Whenever

$$\lim_{x \to c^-} f(x) = \infty \text{ or } -\infty \qquad \text{or} \qquad \lim_{x \to c^+} f(x) = \infty \text{ or } -\infty$$

we call the line $x = c$ a **vertical asymptote** of the graph of f.

Figure 25

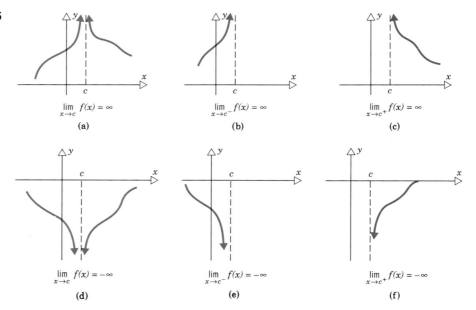

$$\lim_{x \to c} f(x) = \infty$$
(a)

$$\lim_{x \to c^-} f(x) = \infty$$
(b)

$$\lim_{x \to c^+} f(x) = \infty$$
(c)

$$\lim_{x \to c} f(x) = -\infty$$
(d)

$$\lim_{x \to c^-} f(x) = -\infty$$
(e)

$$\lim_{x \to c^+} f(x) = -\infty$$
(f)

EXAMPLE 3

Find the vertical asymptotes, if any, of

$$f(x) = \frac{x^2}{x - 4}$$

SOLUTION

Look at the denominator. The function f has an infinite limit when $x = 4$. Thus, the line $x = 4$ is a vertical asymptote. To verify this and to see how the graph behaves near the vertical asymptote, we find the two one-sided limits at 4:

$\lim\limits_{x \to 4^-} f(x)$: Since $x \to 4^-$, we know $x < 4$, so $x - 4 < 0$. Since $x^2 \geq 0$, it follows that the expression $x^2/(x - 4)$ is negative and becomes unbounded as $x \to 4^-$. Thus,

$$\lim_{x \to 4^-} f(x) = \lim_{x \to 4^-} \frac{x^2}{x - 4} = -\infty$$

$\lim\limits_{x \to 4^+} f(x)$:   Since $x \to 4^+$, we know $x > 4$, so $x - 4 > 0$. Since $x^2 \geq 0$, it follows that the expression $x^2/(x - 4)$ is positive and becomes unbounded as $x \to 4^+$. Thus,

$$\lim_{x \to 4^+} f(x) = \lim_{x \to 4^+} \frac{x^2}{x - 4} = \infty$$

· ·

The graph of $f(x) = x^2/(x - 4)$, based on the information found in Example 3, will exhibit the behavior shown in Figure 26 near $x = 4$.

Figure 26

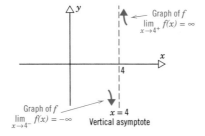

To the steps for graphing a function, page 712, we add another step:

Step 7 Find any horizontal or vertical asymptotes of the graph.

Now Work Problem 9

EXAMPLE 4

Graph: $f(x) = \dfrac{x^2}{x^2 - 1}$

SOLUTION

Step 1 The domain of f is $\{x \mid x \neq -1, x \neq 1\}$.

Step 2 Let $x = 0$. Then $y = f(0) = 0$. The y-intercept is $(0, 0)$. Now let $y = 0$. Then $x^2/(x^2 - 1) = 0$, so $x = 0$. The x-intercept is also $(0, 0)$.

Step 3 To find where the graph is increasing or decreasing, we need to find $f'(x)$:

$$f'(x) = \frac{(x^2 - 1)\dfrac{d}{dx}x^2 - (x^2)\dfrac{d}{dx}(x^2 - 1)}{(x^2 - 1)^2}$$

$$= \frac{(x^2 - 1)(2x) - x^2(2x)}{(x^2 - 1)^2} = \frac{-2x}{(x^2 - 1)^2}$$

We use the numbers $-1, 0$, and 1 to separate the number line into four parts: the intervals $x < -1$, $-1 < x < 0$, $0 < x < 1$, and $1 < x$. The test values we select for these intervals are -2, $-\frac{1}{2}$, $\frac{1}{2}$, and 2, respectively. We summarize the results in the following table:

Intervals	$x < -1$	$-1 < x < 0$	$0 < x < 1$	$1 < x$
Test Value	-2	$-1/2$	$1/2$	2
f' at Test Value	$f'(-2) = \frac{4}{9} > 0$	$f'(-\frac{1}{2}) = \frac{16}{9} > 0$	$f'(\frac{1}{2}) = -\frac{16}{9} < 0$	$f'(2) = -\frac{4}{9} < 0$
Sign of f'	$+$	$+$	$-$	$-$
Graph of f	Increasing	Increasing	Decreasing	Decreasing

Step 4 Using the First Derivative Test and the table, we conclude that the point $(0, 0)$ is a local maximum.

Step 5 At $x = 0$, we have $f'(0) = 0$, so at $(0, 0)$, the tangent line is horizontal. There are no vertical tangent lines ($x = -1$ and $x = 1$ are not in the domain of f).

Step 6 To locate any inflection points, we need to find $f''(x)$:

$$f'(x) = \frac{-2x}{(x^2 - 1)^2}$$

$$f''(x) = -2\left[\frac{(x^2 - 1)^2 \frac{d}{dx} x - x \frac{d}{dx}(x^2 - 1)^2}{(x^2 - 1)^4}\right]$$

$$= -2\left[\frac{(x^2 - 1)^2 \cdot 1 - x \cdot 2(x^2 - 1) \cdot 2x}{(x^2 - 1)^4}\right]$$

$$= \frac{-2(x^2 - 1)[(x^2 - 1) - 4x^2]}{(x^2 - 1)^4} = \frac{-2[-3x^2 - 1]}{(x^2 - 1)^3}$$

$$= \frac{2(3x^2 + 1)}{(x^2 - 1)^3} = \frac{2(3x^2 + 1)}{(x + 1)^3(x - 1)^3}$$

Note that $3x^2 + 1$ is the sum of squares; so $f''(x) \neq 0$.

We use the numbers -1 and 1 to form the intervals $x < -1$, $-1 < x < 1$, and $1 < x$. The test values we select for these intervals are -2, 0, and 2, respectively. We summarize the results in the following table:

Interval	$x < -1$	$-1 < x < 1$	$1 < x$
Test Value	-2	0	2
f'' at Test Value	$f''(-2) = \frac{26}{27} > 0$	$f''(0) = -2 < 0$	$f''(2) = \frac{26}{27} > 0$
Sign of f''	$+$	$-$	$+$
Concavity of f	Up	Down	Up

The concavity changes at $x = -1$ and at $x = 1$, but these are not in the domain of f. Thus, the graph has no inflection points.

Step 7 To find any horizontal asymptotes, we find $\lim\limits_{x \to \infty} f(x)$ and $\lim\limits_{x \to -\infty} f(x)$:

$$\lim_{x \to \infty} f(x) = \lim_{x \to \infty} \frac{x^2}{x^2 - 1} = \lim_{x \to \infty} \frac{1}{1 - \dfrac{1}{x^2}} = 1$$

and

$$\lim_{x \to -\infty} f(x) = \lim_{x \to -\infty} \frac{x^2}{x^2 - 1} = \lim_{x \to -\infty} \frac{1}{1 - \dfrac{1}{x^2}} = 1$$

Thus, the line $y = 1$ is a horizontal asymptote for x sufficiently positive and for x sufficiently negative.

To find any vertical asymptotes, we first determine where f is unbounded. Since this happens when $x = -1$ and when $x = 1$, it follows that the lines $x = -1$ and $x = 1$ are vertical asymptotes. To find the behavior of the graph near the vertical asymptotes, we find the one-sided limits:

$$\lim_{x \to -1^-} \frac{x^2}{x^2 - 1} = \infty \qquad \text{and} \qquad \lim_{x \to -1^+} \frac{x^2}{x^2 - 1} = -\infty$$

$$\lim_{x \to 1^-} \frac{x^2}{x^2 - 1} = -\infty \qquad \text{and} \qquad \lim_{x \to 1^+} \frac{x^2}{x^2 - 1} = \infty$$

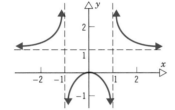

Figure 27

Now Work Problem 15

Putting all these facts together, we obtain the graph. See Figure 27.

· ·

Application

EXAMPLE 5

A company estimates that the fixed costs for producing a new toy are $50,000 and the variable costs per unit are $3. The cost function of producing x toys is therefore

$$C(x) = 50{,}000 + 3x$$

The average cost function $\overline{C}(x)$ is

$$\overline{C}(x) = \frac{C(x)}{x} = \frac{50{,}000 + 3x}{x} = \frac{50{,}000}{x} + 3$$

Since

$$\lim_{x \to \infty} \overline{C}(x) = \lim_{x \to \infty} \left(\frac{50{,}000}{x} + 3 \right) = 3$$

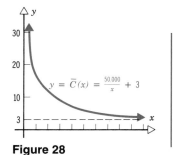

Figure 28

the line $y = 3$ is a horizontal asymptote for the average cost function. Also, notice that $3 is the variable unit cost. Next,

$$\lim_{x \to 0^+} \overline{C}(x) = \lim_{x \to 0^+} \left(\frac{50{,}000}{x} + 3 \right) = \infty$$

Thus, $\overline{C}(x)$ has a vertical asymptote at $x = 0$. (Since $C(x)$ is defined only for $x \geq 0$, we calculate only the right-hand limit.) See Figure 28.

· ·

As the graph illustrates, the more units that are produced, the closer the average cost will get to the unit cost, and the less important the fixed cost becomes.

EXERCISE 14.3   Answers to odd-numbered problems begin on page AN-70.

In Problems 1–8 find the indicated limit.

1. $\lim\limits_{x \to \infty} \dfrac{x^3 + x^2 + 2x - 1}{x^3 + x + 1}$

2. $\lim\limits_{x \to \infty} \dfrac{2x^2 - 5x + 2}{5x^2 + 7x - 1}$

3. $\lim\limits_{x \to \infty} \dfrac{2x + 4}{x - 1}$

4. $\lim\limits_{x \to \infty} \dfrac{x + 1}{x}$

5. $\lim\limits_{x \to \infty} \dfrac{3x^2 - 1}{x^2 + 4}$

6. $\lim\limits_{x \to -\infty} \dfrac{x^2 - 2x + 1}{x^3 + 5x + 4}$

7. $\lim\limits_{x \to -\infty} \dfrac{5x^3 - 1}{x^2 + 1}$

8. $\lim\limits_{x \to -\infty} \dfrac{x^2 + 1}{x^3 - 1}$

In Problems 9–14 locate all horizontal and vertical asymptotes, if any, of the function f.

9. $f(x) = 3 + \dfrac{1}{x}$

10. $f(x) = 2 - \dfrac{1}{x^2}$

11. $f(x) = \dfrac{2}{(x - 1)^2}$

12. $f(x) = \dfrac{3x - 1}{x + 1}$

13. $f(x) = \dfrac{x^2}{x^2 - 4}$

14. $f(x) = \dfrac{x}{x^2 - 1}$

In Problems 15–24 graph each function.

15. $f(x) = \dfrac{2}{x^2 - 4}$

16. $f(x) = \dfrac{1}{x^2 - 1}$

17. $f(x) = \dfrac{2x - 1}{x + 1}$

18. $f(x) = \dfrac{x - 2}{x}$

19. $f(x) = \dfrac{x}{x^2 + 1}$

20. $f(x) = \dfrac{2x}{x^2 - 4}$

21. $f(x) = \dfrac{8}{x^2 - 16}$

22. $f(x) = \dfrac{x^2}{4 - x^2}$

23. $f(x) = x + \dfrac{1}{x}$

24. $f(x) = 2x + \dfrac{1}{x^2}$

25. Sketch the graph of a function f that is defined and continuous for $-1 \leq x \leq 2$ and that satisfies the following conditions:

$$f(-1) = 1 \quad f(1) = 2 \quad f(2) = 3 \quad f(0) = 0 \quad f(\tfrac{1}{2}) = 3$$

$$\lim_{x \to -1^+} f'(x) = -\infty \quad \lim_{x \to 1^-} f'(x) = -1 \quad \lim_{x \to 1^+} f'(x) = \infty$$

f has a local minimum at 0 and f has a local maximum at $\tfrac{1}{2}$.

26. Average Cost   A business has a cost function of $C = 5000 + 1.5x$, where C is measured in dollars and x is the number of units produced.

(a) Find the average cost $\overline{C}(x)$.

(b) Compute $\overline{C}(1000)$, $\overline{C}(10{,}000)$, $\overline{C}(100{,}000)$.

(c) Find $\lim\limits_{x \to \infty} \overline{C}(x)$, $\lim\limits_{x \to 0^+} \overline{C}(x)$.

(d) Graph $\overline{C}(x)$.

27. Average Profit   The cost and revenue functions for a particular item are given by

$$C(x) = 10{,}000 + 18.5x \quad \text{and} \quad R(x) = 76.5x$$

(a) Find the average profit function

$$\overline{P}(x) = \frac{R(x) - C(x)}{x}$$

(b) Find $\overline{P}(1000)$, $\overline{P}(10{,}000)$, $\overline{P}(100{,}000)$.
(c) Evaluate $\lim\limits_{x \to \infty} \overline{P}(x)$.

28. Marginal Cost   The total cost function for printing x greeting cards is

$$C(x) = 1000 + 2x + \frac{200}{x}$$

(a) Find the marginal cost function $C'(x)$.
(b) Evaluate $\lim\limits_{x \to \infty} C'(x)$.

29. Pollution Control   The cost C, in thousands of dollars, for removal of a pollutant from a certain lake is

$$C(x) = \frac{5x}{100 - x} \qquad 0 \le x < 100$$

where x is the percent of pollutant removed. Find $\lim\limits_{x \to 100^-} C(x)$. Is it possible to remove 100% of the pollutant?

30. Drug Concentration   The concentration C of a certain drug in a patient's bloodstream t hours after injection is given by

$$C(t) = \frac{0.3t}{t^2 + 2}$$

milligrams per cubic centimeter.

(a) Find the horizontal asymptote of $C(t)$.
(b) Interpret your answer.

31. Advertising   The sales in units of a new product over a period of time are expected to follow the relationship

$$S(x) = \frac{2000x^2}{3.5x^2 + 1000}$$

where x is the amount of money spent on advertising. Evaluate $\lim\limits_{x \to \infty} S(x)$ and interpret your answer.

32. Learning   The function

$$T(x) = 5\left(1 + \frac{1}{\sqrt{x}}\right)$$

describes the time T in minutes it takes a worker on an assembly line to perform an operation after performing it x number of times. Evaluate $\lim\limits_{x \to \infty} T(x)$ and interpret your answer.

Technology Exercises

1. Investment Revenue   The weekly revenue, R (in hundred thousand dollars), from investing x dollars in a project is

$$R(x) = xe^{-0.1x}$$

where x is in thousands of dollars.

(a) Graph this function on your graphing calculator (or a computer). Estimate the limit $\lim\limits_{x \to \infty} f(x)$.

(b) Interpret your answer.

In Problems 2–7 locate all horizontal and vertical asymptotes, if any, of the function f. Check your work by looking at the actual graph of each on a graphing calculator or a computer.

2. $f(x) = 3 + \dfrac{1}{x}$

3. $f(x) = 2 - \dfrac{1}{x^2}$

4. $f(x) = \dfrac{2}{(x - 1)^2}$

5. $f(x) = \dfrac{3x - 1}{x + 1}$

6. $f(x) = \dfrac{x^2}{x^2 - 4}$

7. $f(x) = \dfrac{x}{x^2 - 1}$

14.4 OPTIMIZATION

Absolute Maximum and Absolute Minimum

Earlier we described the local maximum (and local minimum) of a function as a point on the graph that is higher (or lower) than nearby points on the graph. However, the value of the function at such points may not be the largest (or smallest) value of the function on its domain.

The largest value, if one exists, of a function on its domain is called the **absolute maximum** of the function. The smallest value, if it exists, of a function on its domain is called the **absolute minimum** of the function.

Let's look at some of the possibilities. See Figure 29.

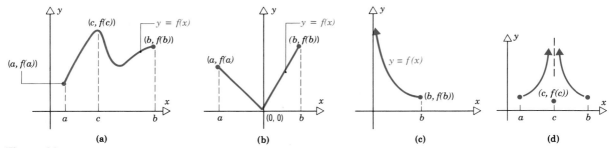

Figure 29

In Figure 29(a) the function f is continuous on the closed interval $[a, b]$. The absolute maximum of f is $f(c)$. Note that c is a critical number since $f'(c) = 0$. The absolute minimum of f is $f(a)$.

In Figure 29(b) the function f is continuous on the closed interval $[a, b]$. The absolute maximum of f is $f(b)$. The absolute minimum of f is $f(0) = 0$. Note that 0 is a critical number since $f'(0)$ does not exist.

In Figure 29(c) the function f is continuous on the interval $(0, b]$. The function f has no absolute maximum since there is no largest value of f on $(0, b]$. The absolute minimum is $f(b)$.

In Figure 29(d) the function f, whose domain is the closed interval $[a, b]$, is discontinuous at c. The function f has no absolute maximum since there is no largest value of f on $[a, b]$. The absolute minimum is $f(c)$.

The following result, which we state without proof, gives a condition under which a function will have an absolute maximum and an absolute minimum.

> **Condition for a Function to Have an Absolute Maximum and an Absolute Minimum**
>
> If a continuous function has as its domain a closed interval, the absolute maximum and absolute minimum exist.

Thus, continuous functions defined on a closed interval will have an absolute maximum and an absolute minimum. Look again at Figures 29(a) and (b). We see that each function f is continuous on a closed interval $[a, b]$. Note that the absolute maximum and the absolute minimum occur either at a critical number or at an endpoint. This leads us to formulate the following test for finding the absolute maximum and the absolute minimum.

Test for Absolute Maximum and Absolute Minimum

If a continuous function $y = f(x)$ has a closed interval $[a, b]$ as its domain, we can find the absolute maximum (minimum) by choosing the largest (smallest) value from among the following:

1. Values of f at the critical numbers in (a, b)
2. $f(a)$
3. $f(b)$

If critical numbers of $y = f(x)$ are found that are not in the interval $[a, b]$, these critical numbers should be ignored since we are concerned only with the function on the interval $[a, b]$.

EXAMPLE 1

Consider the function $f(x) = x^3 - 3x$. If the domain of f is $[0, 2]$, find the absolute maximum and absolute minimum.

SOLUTION

$$f'(x) = 3x^2 - 3 = 3(x^2 - 1) = 3(x + 1)(x - 1)$$

There are no numbers at which $f'(x)$ does not exist. So the critical numbers obey

$$f'(x) = 3(x + 1)(x - 1) = 0$$
$$x = -1 \quad \text{and} \quad x = 1$$

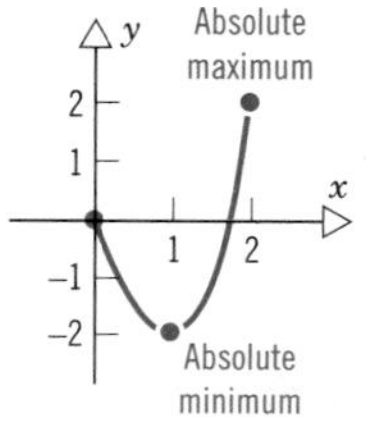

We ignore the critical number $x = -1$ since it is not in the domain, $0 \le x \le 2$. For the critical number $x = 1$, we have

$$f(1) = -2$$

Figure 30

See Figure 30.

The values of f at the endpoints 0 and 2 of the interval $[0, 2]$ are

$$f(0) = 0 \quad \text{and} \quad f(2) = 2$$

Thus, the absolute maximum of f on $[0, 2]$ is 2 and the absolute minimum is -2.

Now Work Problem 1

· ·

EXAMPLE 2

Find the absolute maximum and absolute minimum of the function $f(x) = x^{2/3}$ on the interval $[-1, 8]$.

SOLUTION

First, we locate the critical numbers, if any:

$$f'(x) = \frac{2}{3} x^{-1/3} = \frac{2}{3x^{1/3}}$$

The only critical numbers are those for which $f'(x)$ does not exist. Thus, the only critical number is $x = 0$. If $x = 0$, we have $f(0) = 0$.

Next, we find the values of $f(x)$ at the endpoints:

$$f(-1) = 1 \qquad \text{and} \qquad f(8) = 4$$

The absolute maximum of $f(x) = x^{2/3}$ on $[-1, 8]$ is 4 and the absolute minimum is 0. Figure 31 illustrates this situation.

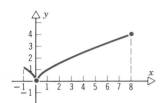

Figure 31

······················

EXAMPLE 3

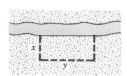

Figure 32

A farmer with 4000 meters of fencing wants to enclose a rectangular plot that borders a straight river. If the farmer does not fence the side along the river, what is the largest area that can be enclosed? See Figure 32.

SOLUTION

The quantity to be maximized is the area. We denote it by A and denote the dimensions of the rectangle by x and y. The area A is then

$$A = xy$$

But x and y are related since the length of fence to be used is 4000 meters. That is,

$$x + y + x = 4000$$
$$y = 4000 - 2x$$

Thus, the area A is

$$A = A(x) = x(4000 - 2x) = 4000x - 2x^2$$

The restrictions on x are $x > 0$ and $x < 2000$ (since if $x > 2000$, the length of fence used, $x + y + x$, would be greater than 4000 meters).

The problem is to maximize $A(x) = 4000x - 2x^2$ on $[0, 2000]$. The critical numbers obey

$$A'(x) = 4000 - 4x = 0$$
$$x = 1000$$

Since this is the only critical number, we use the test for absolute maximum and calculate the values of A at this critical number and at the endpoints of the interval:

$$A(1000) = 2,000,000 \qquad A(0) = 0 \qquad A(2000) = 0$$

The maximum area that can be enclosed is 2,000,000 square meters.

Now Work Problem 25

······················

Optimization Problems

In general each type of problem we will discuss requires some quantity to be minimized or maximized. We assume that the quantity we want to optimize can be represented by a function. Once this function is determined, the problem can be reduced to the question of determining at what number the function assumes its absolute maximum or absolute minimum.

Even though each applied problem has its unique features, it is possible to outline in a rough way a procedure for obtaining a solution. This five-step procedure is shown in the following box:

Steps for Solving Applied Problems

Step 1 Identify the quantity for which a maximum or a minimum value is to be found.

Step 2 Assign symbols to represent other variables in the problem. If possible, use an illustration to assist you.

Step 3 Determine the relationships among these variables.

Step 4 Express the quantity to be optimized as a function of one of these variables. Be sure to state the domain.

Step 5 Apply the test for absolute maximum and absolute minimum to this function.

The following examples illustrate this procedure.

EXAMPLE 4

From each corner of a square piece of sheet metal 18 centimeters on a side, remove a small square of side x centimeters and turn up the edges to form an open box. What should be the dimensions of the box so as to maximize the volume?

SOLUTION

The quantity to be maximized is the volume. Therefore, let's denote it by V and denote the dimension of the sides of the small square by x, as shown in Figure 33. Although the area of the sheet metal is fixed, the sides of the square can be changed and thus are treated as variables. Let y denote the portion left to make the square after cutting out the x^2 parts. We have

$$y = 18 - 2x$$

The height of the box is x, while the area of the base of the box is y^2. The volume V is therefore

$$V = xy^2$$

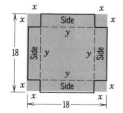

Figure 33

Since we have a function of two variables, we need to reduce it to a function in one variable. We can do this by substituting for y in the formula for volume. This gives

$$V = V(x) = x(18 - 2x)^2$$

This is the function to be maximized. Its domain is the set of real numbers. However, physically, the only numbers x that make sense are those between 0 and 9. Thus, we want to find the absolute maximum of

$$V(x) = x(18 - 2x)^2 \qquad 0 \leq x \leq 9$$

To find the number x that maximizes V, we differentiate and find the critical numbers, if any:

$$V'(x) = (18 - 2x)^2 + 2x(18 - 2x)(-2)$$
$$= (18 - 2x)(18 - 6x)$$

Now we set $V'(x) = 0$ and solve for x:

$$(18 - 2x)(18 - 6x) = 0$$
$$18 - 2x = 0 \quad \text{or} \quad 18 - 6x = 0$$
$$x = 9 \quad \text{or} \quad x = 3$$

The only critical number in $(0, 9)$ is $x = 3$. Thus, we calculate the values of $V(x)$ at this critical number and at the endpoints of the interval:

$$V(0) = 0 \qquad V(3) = 3(18 - 6)^2 = 432 \qquad V(9) = 0$$

The maximum volume is 432 cubic centimeters, and the dimensions of the box that yield the maximum volume are $x = 3$ centimeters deep by $y = 18 - 2(3) = 12$ centimeters on each side.

Now Work Problem 33

· ·

EXAMPLE 5

Figure 34

Playpen Problem*   A certain manufacturer makes a playpen with flexible construction that permits the linkage of its sides (each side is of unit length, and the playpen is normally square) to be attached at right angles to a wall (the side of a house, for example). See Figure 34. When the playpen is placed as in Figure 34, the area enclosed is 2 square units, which doubles the child's play area. Is there a configuration that will do better than double the child's play area?

SOLUTION

Since the playpen must be attached at right angles to the wall, the possible configurations depend on the amount of wall that is used as a fifth side for the playpen. See

---

* Adapted from *Proceedings, Summer Conference for College Teachers on Applied Mathematics,* University of Missouri–Rolla, 1971.

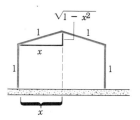

Figure 35

Figure 35. Let x represent half the length of the wall used as a fifth side. The area A is a function of x and is the sum of two rectangles (with sides 1 and x) and two right triangles (hypotenuse 1 and base x). Thus, the quantity to be maximized is

$$A(x) = 2x + x\sqrt{1 - x^2} \qquad 0 \le x \le 1$$

Now

$$A'(x) = 2 + \sqrt{1 - x^2} + x(\tfrac{1}{2})(1 - x^2)^{-1/2}(-2x)$$

$$= 2 + \sqrt{1 - x^2} - \frac{x^2}{\sqrt{1 - x^2}}$$

$$= \frac{2\sqrt{1 - x^2} + 1 - 2x^2}{\sqrt{1 - x^2}}$$

Set $A'(x) = 0$. Then

$$\frac{2\sqrt{1 - x^2} + 1 - 2x^2}{\sqrt{1 - x^2}} = 0$$

Ratios equal 0 when the numerator equals 0; thus, $A'(x) = 0$ if

$$2\sqrt{1 - x^2} + 1 - 2x^2 = 0$$
$$2\sqrt{1 - x^2} = 2x^2 - 1$$
$$4(1 - x^2) = 4x^4 - 4x^2 + 1$$
$$4x^4 = 3$$
$$x = \sqrt[4]{\tfrac{3}{4}} \approx 0.931$$

$A'(x)$ does not exist at $x = -1$ and $x = 1$, but these values are not in the interval $(0, 1)$. Thus, the only critical number in $(0, 1)$ is 0.931.

Compute $A(x)$ at the endpoints $x = 0$ and $x = 1$, and at the critical number $x = 0.931$. The results are

$$A(0) = 0 \qquad A(1) = 2 \qquad A(0.931) = 2.20$$

Thus, a wall of length $2x = 2(0.931) = 1.862$ will maximize the area, and a configuration like the one in Figure 35 increases the play area by about 10% (from 2 to 2.20).

....

The next two examples illustrate how to solve optimization problems in which the function to be optimized has a domain that is not a closed interval.

EXAMPLE 6   A can company wants to produce a cylindrical container with a capacity of 1000 cubic centimeters. The top and bottom of the container must be made of material that costs $0.05 per square centimeter, while the sides of the container can be made of material costing $0.03 per square centimeter. Find the dimensions that will minimize the total cost of the container.

SOLUTION

Figure 36 shows a cylindrical container and the area of its top, bottom, and lateral surfaces.

Figure 36

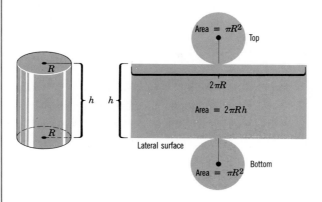

As indicated in the figure, if we let h stand for the height of the can and let R stand for the radius, then the total area of the bottom and top is $2\pi R^2$ and the area of the lateral surface of the can is $2\pi Rh$. The total cost C of manufacturing the can is

$$C = (\$0.05)(2\pi R^2) + (\$0.03)(2\pi Rh) = 0.1\pi R^2 + 0.06\pi Rh$$

This is the function we want to minimize.

The cost function is a function of two variables, h and R. But there is a relationship between h and R since the volume of the cylinder is fixed at 1000 cubic centimeters. That is,

$$V = 1000 = \pi R^2 h$$

$$h = \frac{1000}{\pi R^2}$$

Substituting this expression for h into the cost function C, we obtain

$$C = C(R) = 0.1\pi R^2 + 0.06\pi R\left(\frac{1000}{\pi R^2}\right) = 0.1\pi R^2 + \frac{60}{R}$$

The domain of C is $\{R\,|\,R > 0\}$. The derivative of C with respect to R is

$$C'(R) = 0.2\pi R - \frac{60}{R^2} = \frac{0.2\pi R^3 - 60}{R^2}$$

Set $C'(R) = 0$. Then

$$0.2\pi R^3 - 60 = 0$$

$$R^3 = \frac{300}{\pi}$$

$$R = \sqrt[3]{300/\pi} \approx 4.57$$

Since $R > 0$, the only critical number is $R \approx 4.57$. Using the Second Derivative Test, we obtain

$$C''(R) = 0.2\pi + \frac{120}{R^3}$$

and

$$C''(\sqrt[3]{300/\pi}) = 0.2\pi + \frac{120\pi}{300} > 0$$

Thus, for $R = \sqrt[3]{300/\pi} = 4.57$ centimeters, the cost is a local minimum.

Since the only physical constraint is that R be positive, this local minimum value is the absolute minimum. The corresponding height of this can is

$$h = \frac{1000}{\pi R^2} = \frac{1000}{20.89\pi} = 15.24 \text{ centimeters}$$

These are the dimensions that will minimize the cost of the material.

Now Work Problem 37

· ·

If the cost of the material is the same for the top, bottom, and lateral surfaces of a cylindrical container, then the minimum cost occurs when the surface area is minimum. It can be shown (see Problem 49) that for any fixed volume, the minimum surface area is obtained when the height equals twice the radius.

EXAMPLE 7

A company charges $200 for each box of tools on orders of 150 or fewer boxes. The cost to the buyer on every box is reduced by $1 for each ordered in excess of 150. For what size order is revenue maximum?

SOLUTION

For an order of exactly 150 boxes, the company's revenue is

$$\$200(150) = \$30,000$$

For an order of 160 boxes (which is 10 in excess of 150), the per box charge is $200 - 10(1) = 190$ and the revenue is

$$\$190(160) = \$30,400$$

To solve the problem, let x denote the number of boxes sold. The revenue R is

$$R = (\text{Number of boxes})(\text{Cost per box}) = x(\text{Cost per box})$$

If $x \geq 150$, the charge per box is

$$200 - 1 \left(\begin{array}{c} \text{Number of boxes} \\ \text{in excess of 150} \end{array} \right) = 200 - 1(x - 150) = 350 - x$$

Hence, the revenue R is

$$R = x(350 - x) = 350x - x^2$$

To find the number of boxes leading to maximum revenue, we find the critical numbers of R:

$$R'(x) = 350 - 2x$$
$$R'(x) = 0 \quad \text{when} \quad x = 175$$

Since $R''(x) = -2 < 0$ for all x, there is an absolute maximum at $x = 175$. Thus, a purchase of 175 boxes maximizes the company's revenue. The maximum revenue is

$$R = (350)(175) - (175)^2 = \$30,625$$

Of course, the company would set this figure as the most it would allow anyone to purchase on this plan, since revenue to the company starts to decrease for orders in excess of 175.

....

Maximum Profit

The revenue derived from selling x units is $R(x) = x\, d(x)$, where $p = d(x)$ is the price function. If $C(x)$ is the cost of producing x units, the **profit function** P, assuming whatever is produced can be sold, is

$$P(x) = R(x) - C(x)$$

What quantity x will maximize profit?

To maximize $P(x)$, we find the critical numbers that obey

$$\frac{d}{dx} P(x) = \frac{d}{dx} [R(x) - C(x)] = \frac{d}{dx} R(x) - \frac{d}{dx} C(x) = 0$$

or

$$R'(x) - C'(x) = 0$$

We apply the Second Derivative Test to the function P:

$$P''(x) = R''(x) - C''(x)$$

The profit P has a local maximum at a number x if $P''(x) < 0$. This will occur at a number x for which the marginal revenue function equals the marginal cost and $R''(x) < C''(x)$.

Maximizing Profit

The equality

$$R'(x) = C'(x)$$

is the basis for the classical economic criterion for maximum profit—that marginal revenue and marginal cost be equal.

EXAMPLE 8

Suppose that a wheat farmer can sell wheat at a fixed price of $4 per bushel and can produce anywhere from 0 to 25,000 bushels. Suppose the cost function (in dollars) is

$$C = \frac{x^2}{10,000} + 1500$$

where x represents the number of bushels produced. We interpret this cost function as consisting of total fixed costs of $1500 and total variable costs of $x^2/10,000$ dollars. The total fixed cost of $1500 is due to costs of land, equipment, and other costs that will not vary with the number of bushels produced. The total variable cost represents the cost of planting, fertilizing, and harvesting the crop. For example, the cost of producing 25,000 bushels is

$$C = \frac{(25,000)^2}{10,000} + 1500 = 62,500 + 1500 = \$64,000$$

Of particular interest is the marginal cost, which is

$$C'(x) = \frac{x}{5000}$$

For example, for $x = 5000$ bushels, the marginal cost is

$$C'(5000) = 1$$

and for $x = 6000$ bushels, the marginal cost is

$$C'(6000) = 1.2$$

The difference in marginal cost is an indication that the cost is increasing from $1 per bushel to $1.20 per bushel. This increase is acceptable provided it is not detrimental to the total profit picture. That is, the combination of cost and revenue is what is critical — not cost alone. Thus, we need to ask how the revenue function changes relative to the quantity produced. Comparing this to the marginal cost will provide valuable information to the farmer.

Assuming the farmer can sell all the wheat that is grown, how much wheat should be produced to maximize profit?

The maximum profit occurs when marginal cost equals marginal revenue. Since $R(x) = 4x$, we have

$$C'(x) = \frac{x}{5000} \qquad R'(x) = 4$$

These are equal when

$$\frac{x}{5000} = 4$$

$$x = 20,000$$

Since

$$P(x) = R(x) - C(x) = 4x - \left(\frac{x^2}{10,000} + 1500 \right) \qquad 0 \le x \le 25,000$$

we have

$$P(0) = -1500$$

$$P(20,000) = 4(20,000) - \left[\frac{(20,000)^2}{10,000} + 1500 \right]$$

$$= 80,000 - 41,500 = \$38,500$$

$$P(25,000) = 4(25,000) - 64,000 = \$36,000$$

Thus, the maximum profit occurs for production that is 5000 bushels under the maximum output of 25,000 bushels—that is, at 20,000 bushels.

· ·

EXAMPLE 9

Let us return to a problem discussed earlier (Section 11.4, Example 8). The proportion of people responding to the advertisement of a new product after it has been on the market t days is

$$p(t) = 1 - e^{-0.2t}$$

The marketing area contains 10,000,000 potential customers, and each response to the advertisement results in profit to the company of \$0.70 (on the average). The profit is exclusive of advertising cost. The fixed cost of producing the advertising is \$30,000, and the variable cost is \$5000 for each day the advertisement runs.

(a) What is the cost function $C(t)$?
(b) After 28 days of advertising, what is the net profit?
(c) How many days should the advertisement be run to maximize profits?
(d) Graph the function $p(t)$.

SOLUTION

(a) The cost function is

$$C(t) = \$30,000 + \$5000t$$

(b) The net profit function is the profit from sales less advertising cost. The profit from sales is the number of respondents times 0.70. Since $p(t) = 1 - e^{-0.2t}$ is the proportion of those responding, we have

$$R(t) = 10,000,000(1 - e^{-0.2t})(\$0.70) = \$7,000,000(1 - e^{-0.2t})$$

Thus, the net profit is

$$P(t) = R(t) - C(t) = \$7,000,000(1 - e^{-0.2t}) - \$30,000 - \$5000t$$

For $t = 28$,

$$P(28) = \$7,000,000(1 - e^{-5.6}) - \$30,000 - \$140,000$$
$$= \$7,000,000(0.9963) - \$170,000$$
$$= \$6,974,100 - \$170,000$$
$$= \$6,804,100$$

(c) Profit is maximum when marginal revenue is equal to marginal cost. From Part (a), we know that $C(t) = 30,000 + 5000t$; hence, by taking the derivative, we find

$$C'(t) = 5000$$

From Part (b), we know that $R(t) = 7,000,000(1 - e^{-0.2t})$; then, by taking the derivative, we find

$$R'(t) = 1,400,000e^{-0.2t}$$

Setting $C'(t) = R'(t)$, we have

$$5000 = 1,400,000e^{-0.2t}$$
$$e^{0.2t} = 280$$
$$0.2t = \ln 280$$
$$t = \frac{\ln 280}{0.2} = 28.17$$

Also note that

$$R''(t) = -280,000e^{-0.2t} \quad \text{and} \quad C''(t) = 0$$

Thus, $R''(t) < 0$ for every t, in particular at $t = 28.17$, and therefore the profit is maximum at $t = 28.17$. The company should advertise the product for approximately 28 days to maximize profit.

(d) The graph of $p(t) = 1 - e^{-0.2t}$ is given in Figure 37.

Figure 37

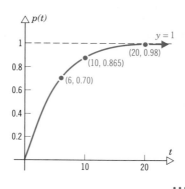

Now Work Problem 53

Application: Maximizing Tax Revenue

In determining the tax rate on cars, telephones, and other consumer goods, the government is always faced with the following problem: How large should the tax be so that the tax revenue will be as large as possible? Let's examine this situation. When the government places a tax on a product, the price of this product for the consumer may increase and the quantity demanded may decrease accordingly. A very large tax may cause the quantity demanded to diminish to zero with the result that no tax revenue is collected. On the other hand, if no tax is levied, there will be no tax revenue at all. Thus, the problem is to find the tax rate that optimizes tax revenue. (Tax revenue is the product of the tax per unit times the actual market quantity consumed.)

Let's assume that because of long-time experience in levying taxes, the government is able to determine that the relationship between the market quantity consumed on a certain product and the related tax is

$$t = \sqrt{27 - 3q^2}$$

where t denotes the amount of tax per unit of a product and q is the market quantity consumed (measured in appropriate units). (It must be pointed out that the relationship between tax rate and consumption is derived by government economists and is subject to both change and criticism.)

Notice that the relationship between tax rate and quantity consumed conforms to the restrictions discussed earlier. For example, when the tax rate $t = 0$, the quantity consumed is $q = 3$; when the tax is at a maximum ($t = 5.2\%$), the quantity consumed is zero. Figure 38 illustrates the graph of this relationship.

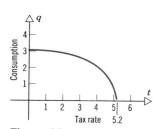

Figure 38

The revenue R due to the tax rate t is the product of tax rate per unit and the market quantity consumed:

$$R(q) = qt = q(27 - 3q^2)^{1/2}$$

where R is measured in dollars. Since both q and t are assumed to be nonnegative, the domain of R is $0 \le q \le 3$. Also, $R = 0$ at both $q = 0$ and $q = 3$, so that for some positive number q between 0 and 3, R attains its absolute maximum.

To find the absolute maximum, we take the derivative of R with respect to q, using the formula for the derivative of a product and the Extended Power Rule:

$$R'(q) = (27 - 3q^2)^{1/2} + \frac{1}{2}q(27 - 3q^2)^{-1/2}(-6q)$$

$$= \frac{27 - 3q^2 - 3q^2}{(27 - 3q^2)^{1/2}} = \frac{27 - 6q^2}{(27 - 3q^2)^{1/2}}$$

The critical numbers obey

$$27 - 6q^2 = 0 \qquad \text{and} \qquad 27 - 3q^2 = 0$$
$$q^2 = 4.5 \qquad\qquad\qquad q^2 = 9$$
$$q = \pm\sqrt{4.5} \qquad\qquad\quad q = \pm 3$$

The only critical number in the interval $(0, 3)$ is $q = \sqrt{4.5} \approx 2.12$.

To find the absolute maximum, we compare the value of R at the endpoints $q = 0$ and $q = 3$ with its value at $q = \sqrt{4.5} \approx 2.12$:

$$R(0) = 0 \qquad R(\sqrt{4.5}) = \sqrt{4.5}\sqrt{13.5} = (2.12)(3.67) = 7.79 \qquad R(3) = 0$$

Thus, the revenue is maximized at $q = 2.12$. The tax rate corresponding to maximum revenue is

$$t = \sqrt{27 - 3q^2} = \sqrt{13.5} = 3.67$$

This means that, for a tax rate of 3.67%, a maximum revenue $R = 7.79$ is generated.

Now Work Problem 47

Application: Contraction of Windpipe While Coughing

Coughing is caused by an increase in pressure in the lungs and is accompanied by a decrease in the diameter of the windpipe. From physics, one can show that the amount V

of air flowing through a cross section of the windpipe per second is related to the radius r of the windpipe and the pressure difference p at each end by the equation $V = kpr^4$, where k is a constant. The radius r will decrease with increased pressure p according to the formula $r_0 - r = cp$, where r_0 is the radius of the windpipe when there is no difference in pressure and c is a positive constant. We wish to find the radius r that allows the most air to flow through the windpipe.

We shall restrict r so that

$$0 < \frac{r_0}{2} \le r \le r_0$$

The amount V of air flowing through the windpipe is given by

$$V = kpr^4$$

where $cp = r_0 - r$. Hence, we wish to maximize

$$V(r) = k\left(\frac{r_0 - r}{c}\right)r^4 = \frac{k}{c}r_0 r^4 - \frac{k}{c}r^5 \qquad \frac{r_0}{2} \le r \le r_0$$

The derivative is

$$V'(r) = 4\frac{k}{c}r_0 r^3 - 5\frac{k}{c}r^4 = \frac{k}{c}r^3(4r_0 - 5r)$$

The only critical number is $r = 4r_0/5$. (We exclude $r = 0$ because $0 < r_0/2 \le r \le r_0$.) Using the test for an absolute maximum, we find

$$V\left(\frac{r_0}{2}\right) = \frac{kr_0^5}{32c} \qquad V(r_0) = 0 \qquad V\left(\frac{4r_0}{5}\right) = \frac{256\, kr_0^5}{3125c}$$

Since $(256kr_0^5)/(3125c) > (kr_0^5)/(32c)$, the maximum air flow is obtained when the radius of the windpipe is $\frac{4}{5}r_0$; that is, the windpipe contracts by a factor of $\frac{1}{5}$.

Application: Growth Curves; Decay Curves

Many models in business and economics require the use of exponential functions. The graphs of these functions are called **growth curves** or **decay curves,** depending on whether they are increasing or decreasing functions.

For example, we could use a growth curve to measure the increase in gross national product per unit time; sometimes, the relationship between price and demand is a decay curve.

Such curves can be used to reflect accurately situations in which the values of the function increase without bound (or decrease to zero). However, curves of this type cannot be used if the growth (or decay), after a fast start, begins to level off, approaching a maximum (or minimum) value. For example, the value of a car depreciates very fast initially, but then levels off and, of course, for tax purposes, can never depreciate more than its original cost. The demand for a new product, such as video games, hula-hoops, big wheels, etc., starts high, increases rapidly, and then levels off. Curves of this kind are called **modified growth curves,** and are described by functions of the form

$$f(x) = C(1 - b^{-ax})$$

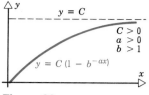

Figure 39

where a and C are positive real numbers and $b > 1$ is a real number. Figure 39 illustrates a modified growth curve. Note that the line $y = C$ is a horizontal asymptote.

If the modified growth curve is used as a model for the growth in sales of a new product, it may not accurately reflect the real situation. Sometimes, the initial rate of growth is slow, and, as time progresses, the rate increases to a maximum value and then begins to decline. In other words, the modified growth curve may be a good model only for what happens at the end of the sales cycle. Curves that describe a situation in which the rate of growth is slow at first, increases to a maximum rate, and then decreases are called **logistic curves,** or **saturation curves.** These curves are best characterized by their "S" shape. Figure 40 illustrates a typical general logistic curve and a graph showing the acceptance years for TVs and VCRs.

Figure 40

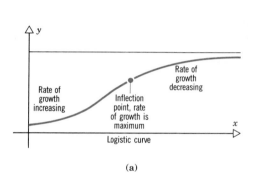

(a)

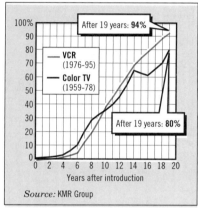

(b) It took years for color TVs and VCRs to become fixtures in many U.S. households

EXAMPLE 10

The sales of a new line of color televisions over a period of time is expected to follow the logistic curve

$$f(x) = \frac{10,000}{1 + 100e^{-x}}$$

where x is measured in years. Analyze the graph of this function, determine the year in which a maximum sales rate is achieved, and find the upper limit to sales in any year.

SOLUTION

Step 1 The domain of this function is $x \geq 0$ and the function is continuous on its domain.

Step 2 If $x = 0$,

$$y = f(0) = \frac{10,000}{1 + 100} = 99.01$$

The y-intercept is $(0, 99.01)$. There is no x-intercept. The y-intercept represents the predicted number of television sets sold when production begins.

Step 3 The derivative of the function is

$$f'(x) = \frac{d}{dx} \frac{10,000}{1 + 100e^{-x}} = \frac{d}{dx} 10,000(1 + 100e^{-x})^{-1}$$

$$= 10,000(-1)(1 + 100e^{-x})^{-2} \frac{d}{dx}(1 + 100e^{-x})$$

$$= \frac{-10,000}{(1 + 100e^{-x})^2}[100e^{-x}(-1)]$$

$$= \frac{1,000,000\ e^{-x}}{(1 + 100e^{-x})^2}$$

Since $e^{-x} > 0$ for all x, it follows that $f'(x) > 0$ for $x \geq 0$. Thus, the function is increasing, which means that sales are increasing each year.

Step 4 There are no local maxima since the graph is increasing.

Step 5 There are no vertical or horizontal tangent lines.

Step 6 To find any inflection points, we compute $f''(x)$.

$$f''(x) = \frac{d}{dx} f'(x) = \frac{d}{dx} \frac{1,000,000e^{-x}}{(1 + 100e^{-x})^2}$$

$$= 1,000,000 \left[\frac{(1 + 100e^{-x})^2 e^{-x}(-1) - e^{-x} \cdot 2(1 + 100e^{-x}) \frac{d}{dx}(1 + 100e^{-x})}{(1 + 100e^{-x})^4} \right]$$

$$= 1,000,000 \left\{ \frac{-e^{-x}(1 + 100e^{-x})^2 - 2e^{-x}(1 + 100e^{-x})[100e^{-x}(-1)]}{(1 + 100e^{-x})^4} \right\}$$

$$= 1,000,000 \left\{ \frac{-e^{-x}(1 + 100e^{-x})[(1 + 100e^{-x}) + 2(-100e^{-x})]}{(1 + 100e^{-x})^4} \right\}$$

$$= 1,000,000 \left[\frac{(-e^{-x})(1 - 100e^{-x})}{(1 + 100e^{-x})^3} \right]$$

$$= 1,000,000e^{-x} \left[\frac{100e^{-x} - 1}{(1 + 100e^{-x})^3} \right]$$

The sign of $f''(x)$ is controlled by the numerator since $1 + 100e^{-x} > 0$ for all x. Now $100e^{-x} - 1 = 0$ if $e^x = 100$, which happens if $x = \ln 100 = 4.6$. Thus, if $x < 4.6$, we have $100e^{-x} - 1 > 0$, so $f''(x) > 0$, and the graph of f is concave up. If $x > 4.6$, we have $100e^{-x} - 1 < 0$, so $f''(x) < 0$, and the graph of f is concave down. Further, at $(4.6, 5000)$ there is an inflection point. At this point, the first derivative, f', achieves its maximum value. Thus, at 4.6 years, the rate of growth in sales is a maximum.

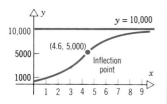

Figure 41

Step 7 As $x \to \infty$, we have $e^{-x} \to 0$. Thus,

$$\lim_{x \to \infty} f(x) = \lim_{x \to \infty} \frac{10,000}{1 + 100e^{-x}} = 10,000$$

Thus, $y = 10,000$ is a horizontal asymptote as x becomes unbounded in the positive direction. This number represents the upper estimate for sales. The graph is given in Figure 41.

· ·

EXERCISE 14.4   Answers to odd-numbered problems begin on page AN-71.

In Problems 1–24 find the absolute maximum and absolute minimum of f on the indicated interval.

1. $f(x) = x^2 + 2x$;   on $[-3, 3]$

2. $f(x) = x^2 - 8x$;   on $[-1, 10]$

3. $f(x) = 1 - 6x - x^2$;   on $[0, 4]$

4. $f(x) = 4 - 2x - x^2$;   on $[-2, 2]$

5. $f(x) = x^3 - 3x^2$;   on $[1, 4]$

6. $f(x) = x^3 - 6x$;   on $[-2, 2]$

7. $f(x) = x^4 - 2x^2 + 1$;   on $[0, 1]$

8. $f(x) = 3x^4 - 4x^3$;   on $[-2, 0]$

9. $f(x) = x^{2/3}$;   on $[-1, 1]$

10. $f(x) = x^{1/3}$;   on $[-1, 1]$

11. $f(x) = 2\sqrt{x}$;   on $[1, 4]$

12. $f(x) = 4 - \sqrt{x}$;   on $[0, 4]$

13. $f(x) = x\sqrt{1 - x^2}$;   on $[-1, 1]$

14. $f(x) = x^2\sqrt{2 - x}$;   on $[0, 2]$

15. $f(x) = \dfrac{x^2}{x - 1}$;   on $[-1, \frac{1}{2}]$

16. $f(x) = \dfrac{x}{x^2 - 1}$;   on $[-\frac{1}{2}, \frac{1}{2}]$

17. $f(x) = (x + 2)^2(x - 1)^{2/3}$;   on $[-4, 5]$

18. $f(x) = (x - 1)^2(x + 1)^3$;   on $[-2, 7]$

19. $f(x) = \dfrac{(x - 4)^{1/3}}{x - 1}$;   on $[2, 12]$

20. $f(x) = \dfrac{(x + 3)^{2/3}}{x + 1}$;   on $[-4, -2]$

21. $f(x) = \dfrac{\ln x}{x}$;   on $[1, 3]$

22. $f(x) = xe^x$;   on $[0, 4]$

23. $f(x) = xe^x$;   on $[-10, 10]$

24. $f(x) = xe^{-x}$;   on $[-10, 10]$

25. Cost of Fencing   A farmer with 3000 meters of fencing wants to enclose a rectangular plot that borders on a straight highway. If the farmer does not fence the side along the highway, what is the largest area that can be enclosed?

26. Cost of Fencing   If the farmer in Problem 25 also decides to fence the side along the highway, what is the largest area that can be enclosed?

27. Cost of Fencing   Find the dimensions of the rectangle of the largest area that can be enclosed by 200 meters of fencing.

28. Cost of Fencing   A builder wants to fence in 135,000 square meters of land in a rectangular shape. Because of security reasons, the fence in the front will cost $2 per meter, while the fence for the other three sides will cost $1 per meter. How much of each type of fence will be needed in order to minimize the cost? What is the minimum cost?

29. Best Dimensions for a Field   A farmer with 30,000 meters of fencing wants to enclose a rectangular field and then divide it into two plots with a fence parallel to one of the sides. See the figure. What is the largest area that can be enclosed?

30. Best Dimensions for a Field   A farmer wants to enclose 6000 square meters of land in a rectangular plot and then divide it into two plots with a fence parallel to one of the sides. What are the dimensions of the rectangular plot that require the least amount of fence?

31. A rectangle has a perimeter of length L. What should the dimensions of the rectangle be if its area is to be a maximum?

32. Best Dimensions for a Box   An open box with a square base is to be made from a square piece of cardboard 12 centimeters on a side by cutting out a square from each corner and turning up the sides. Find the dimensions of the box that yield the maximum volume.

33. Best Dimensions for a Box   An open box with a square base is to be made from a square piece of cardboard 24 centimeters on a side by cutting out a square from each corner and turning up the sides. Find the dimensions of the box that yield the maximum volume.

34. Best Dimensions for a Box   A box, open at the top with a square base, is to have a volume of 8000 cubic centimeters. What should the dimensions of the box be if the amount of material used is to be a minimum?

35. If the box in Problem 34 is to be closed on top, what should the dimensions of the box be if the amount of material used is to be a minimum?

36. Best Dimensions for a Can   A cylindrical container is to be produced that will have a capacity of 4000 cubic centimeters. The top and bottom of the container are to be made of material that costs $0.50 per square centimeter, while the side of the container is to be made of material costing $0.40 per square centimeter. Find the dimensions that will minimize the total cost of the container.

37. Best Dimensions for a Can   A cylindrical container is to be produced that will have a capacity of 10 cubic meters. The top and bottom of the container are to be made of a material that costs $2 per square meter, while the side of the container is to be made of material costing $1.50 per square meter. Find the dimensions that will minimize the total cost of the container.

38. Setting Rental Prices   A car rental agency has 24 identical cars. The owner of the agency finds that at a price of $10 per day, all the cars can be rented. However, for each $1 increase in rental, one of the cars is not rented. What should be charged to maximize income?

39. Charter-Flight Charges   A charter-flight club charges its members $200 per year. But for each new member in excess of 60, the charge for every member is re-

duced by $2. What number of members leads to a maximum revenue?

40. Placing Telephone Boxes   A telephone company is asked to provide telephone service to a customer whose house is located 2 kilometers away from the road along which the telephone lines run. The nearest telephone box is located 5 kilometers down this road. If the cost to connect the telephone line is $50 per kilometer along the road and $60 per kilometer away from the road, where along the road from the box should the company connect the telephone line so as to minimize construction cost? [*Hint:* Let x denote the distance from the box to the connection so that $5 - x$ is the distance from this point to the point on the road closest to the house.]

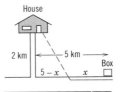

41. Minimizing Travel Time   A small island is 3 kilometers from the nearest point P on the straight shoreline of a large lake. If a woman on the island can row her boat 2.5 kilometers per hour and can walk 4 kilometers per hour, where should she land her boat in order to arrive in the shortest time at a town 12 kilometers down the shore from P?

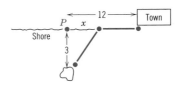

42. Maximum Profit   The relationship between profit P of a firm and the selling price x of its goods is

$$P = 1000x - 25x^2 \qquad 0 \le x \le 40$$

Over what interval of selling prices is profit increasing? For what selling price is profit maximized?

43. Most Economical Speed   A truck has a top speed of 75 miles per hour and, when traveling at the rate of x miles per hour, consumes gasoline at the rate of $\frac{1}{200}[(1600/x) + x]$ gallon per mile. If the length of the trip is 200 miles and the price of gasoline is $1.60 per gallon, the cost is

$$C(x) = 1.60 \left(\frac{1600}{x} + x \right)$$

where $C(x)$ is measured in dollars. What is the most economical speed for the truck? Use the interval $[10, 75]$.

44. If the driver of the truck in Problem 43 is paid \$8 per hour, what is the most economical speed for the truck?

45. Page Layout   A printer plans on having 50 square inches of printed matter per page and is required to allow for margins of 1 inch on each side and 2 inches on the top and bottom. What are the most economical dimensions for each page if the cost per page depends on the area of the page?

46. Dimensions for a Window   A window is to be made in the shape of a rectangle surmounted by a semicircle with diameter equal to the width of the rectangle. See the figure. If the perimeter of the window is 22 feet, what dimensions will let in the most light?

47. Tax Revenue   On a particular product, government economists determine that the relationship between tax rate t and the quantity q consumed is

$$t + 3q^2 = 18$$

Graph this relationship and explain how it could be justified. Find the optimal tax rate and the revenue generated by this tax rate.

48. Most Economical Speed   A truck has a top speed of 75 miles per hour and, when traveling at the rate of x miles per hour, consumes gasoline at the rate of $\frac{1}{200}[(1600/x) + x]$ gallon per mile. This truck is to be taken on a 200-mile trip by a driver who is to be paid at the rate of b dollars per hour plus a commission of c dollars. Since the time required for this trip at x miles per hour is $200/x$, the total cost, if gasoline costs a dollars per gallon, is

$$C(x) = \left(\frac{1600}{x} + x\right)a + \frac{200}{x}b + c$$

Find the most economical possible speed under each of the following sets of conditions:

(a) $b = 0$, $c = 0$
(b) $a = 1.50$, $b = 8.00$, $c = 500$
(c) $a = 1.60$, $b = 10.00$, $c = 0$

49. Prove that a cylindrical container of fixed volume V requires the least material (minimum surface area) when its height is twice its radius.

50. Spread of Rumor   In a city of 50,000 people, the number of people at time t who have heard a certain rumor obeys

$$N(t) = \frac{50,000}{1 + 49,999e^{-t}}$$

At what time t is the rate of spreading of the rumor greatest? Graph the function $N(t)$.

51. Maximum Sales Rate   An advertising company conducts a special campaign to promote sales of a certain product. They estimate that the benefits of the campaign will result in extra sales and, when the campaign is over, the extra sales will obey a curve of the form

$$S = 4000e^{-0.3t}$$

where S is the amount of extra sales and t is the time in days after the advertising campaign is over. How many extra sales are obtained 10 days after the close of the advertising campaign? What is the rate of extra sales at $t = 10$? What is the maximum rate of extra sales in the 10 days?

52. Normal Density   The function

$$f(x) = \frac{1}{\sqrt{2\pi}}e^{-x^2/2}$$

is often encountered in probability theory and is called the **normal density function.** Determine where this function is increasing and decreasing, find all local maxima and local minima, find all inflection points, and determine intervals of concavity. Graph the function.

53. Find the number x that maximizes the function

$$E(x) = 75,000(1 - e^{-0.405x}) - 500x$$

54. Maximum Sales Rate   The sales of a new stereo system over a period of time are expected to follow the relationship

$$f(x) = \frac{5000}{1 + 5e^{-x}}$$

where x is measured in years. Determine the year in which the sales rate is a maximum. Graph the function.

55. Maximum Sales Rate   The sales of a new car model over a period of time are expected to follow the relationship

$$f(x) = \frac{20,000}{1 + 50e^{-x}}$$

where x is measured in months. Determine the month in which the sales rate is a maximum. Graph the function.

56. In general a logistic curve is of the form

$$f(x) = \frac{M}{1 + ae^{-bx}}$$

where $a > 0$, $M > 0$, $b > 0$ are real numbers. At what number x is the derivative $f'(x)$ a maximum?

57. Spread of Disease   In a town of 50,000 people the number of people at time t who have influenza is

$$N(t) = \frac{10,000}{1 + 9999e^{-t}}$$

where t is measured in days. Note that the flu is spread by the one person who has it at $t = 0$. At what time t is the rate of spreading the greatest? Graph the function.

58. Drug Reaction   The reactions, as a function of time (measured in hours), to two drugs are $r_1(t) = te^{-t}$ and $r_2(t) = t^2e^{-t}$. Which drug has the larger maximum reaction? Which drug is slower in its action?

59. Bacteria Growth Rate   A bacteria population grows from an initial population of 800 to a population $p(t)$ at time t (measured in days) according to the equation

$$p(t) = \frac{800e^t}{1 + \frac{1}{10}(e^t - 1)}$$

Determine the growth rate $p'(t)$. When is the growth rate a maximum? Determine $\lim_{t \to \infty} p(t)$, the equilibrium population.

60. Drug Concentration   A drug is injected into the bloodstream at time $t = 0$. The concentration of this drug in the bloodstream at time t is described by $c(t) = d(e^{-at} - e^{-bt})$, where a, b, and d are positive constants with $a < b$.

(a) Show that $c(0) = 0$ and that $c(t) > 0$ for $t > 0$.
(b) What is the maximum value of $c(t)$ and when does it occur?
(c) Draw a graph of the concentration $c(t)$ as a function of time when $a = 1$, $b = 2$, and $d = 1$.

Technology Exercises

1. Let $f(x) = 4x^{1/3}$, domain $[-1, 1]$.

(a) Find the critical values of $f(x)$.
(b) Evaluate $f(x)$ at $x = -1$.
(c) Evaluate $f(x)$ at $x = 1$.
(d) Evaluate $f(x)$ at its critical points.
(e) Find the absolute maximum of $f(x)$.
(f) Find the absolute minimum of $f(x)$.
(g) Graph $f(x)$ on $-1 \le x \le 1$ and a suitable y range.
(h) Verify your answers to Parts (e) and (f).

2. Let $C(x) = (3 + x)^2$ be the cost function, $R(x) = 60x$ be the revenue function, and $P(x) = R(x) - C(x)$ be the profit function.

(a) Graph $C(x)$, $R(x)$, and $P(x)$.
(b) Graph $C'(x)$, $R'(x)$, and $P'(x)$.
(c) Show from the graph that profit is maximum when $R'(x) = C'(x)$, that is, when $P'(x) = 0$.

3. The sales of a new car model over a period of time are expected to follow the relationship (also called a logistic curve)

$$f(x) = \frac{20,000}{1 + 50e^{-x}}$$

when x is measured in months. Graph the function. Use the graph to determine the month in which the sales rate is

maximum. Determine where $f(x)$ is concave up or down, and give it a business interpretation.

4. The function

$$f(x) = \frac{1}{\sqrt{2\pi}} e^{-x^2/2}$$

is often encountered in probability theory and is called the **normal density function.** Graph the function. Determine where this function is increasing and decreasing, find all local maxima and local minima, find all inflection points, and determine where $f(x)$ is concave up or down.

5. Graph the function $f(x) = 4x - \sqrt{x^3 + 1}$. Use the graph to estimate the absolute maximum and the absolute minimum on the interval $[-1, 5]$.

6. The cost of producing x units of a product is

$$C(x) = x^3 - 17x^2 + 85x - 65 \qquad 2 \le x \le 10$$

(a) Graph the marginal cost function and the average cost function $C(x)/x$ on the same coordinate system.
(b) Find the point of intersection of $C'(x)$ and $C(x)/x$. Does the point of intersection have any economic significance?

7. Graph the function $f(x) = x^3 - 0.3x^2 + 0.3x + 0.1$. Use the graph to estimate (to the nearest tenth) the absolute maximum and the absolute minimum of the function on the interval $[1, 4]$.

14.5 ELASTICITY OF DEMAND

In this section we will study how economists describe the effect that changes in price have on demand and revenue. Recall that a demand equation expresses the market price p that will generate a demand of exactly x. Suppose the price p and the quantity x demanded for a certain product are related by the following demand equation:

$$p = 200 - 0.02x \qquad (1)$$

The equation states that in order to sell x units, the price must be set at $200 - 0.02x$ dollars. For example, to sell 5000 units, the price must be set at

$$200 - 0.02(5000) = \$100 \text{ per unit}$$

In problems involving revenue, sales, and profit, it is customary to use the demand equation to express price as a function of demand. Since we are now interested in the effects that changes in price have on demand, it is more practical to express demand as a function of price. Equation (1) may be solved for x in terms of p to yield

$$x = \frac{1}{0.02}(200 - p) \qquad \text{or} \qquad x = f(p) = 50(200 - p)$$

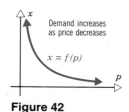

Demand increases as price decreases

$x = f(p)$

Figure 42

This equation expresses quantity x as a function of the price. Since x and p must be nonnegative, we must restrict p so that $0 \le p \le 200$.

Usually, increasing the price of a commodity lowers the quantity demanded, while decreasing the price results in higher demand. Therefore, the typical demand function $x = f(p)$ is decreasing and has a negative slope everywhere. See Figure 42.

A differentiable demand function $x = f(p)$ relates the quantity demanded to price. Therefore, the derivative $f'(p)$ equals the rate of change in quantity demanded with respect to price. **Elasticity,** on the other hand, measures the relative rate of change of the quantity demanded with the relative rate of change of price. Specifically, consider a particular demand function $x = f(p)$ and a particular price p. Then,

$$\textbf{Relative change in quantity demanded} = \frac{\Delta x}{x} = \frac{f(p + \Delta p) - f(p)}{f(p)}$$
$$\textbf{Relative change in price} = \frac{\Delta p}{p} \qquad (2)$$

$$\textbf{Elasticity} = \frac{\textbf{Relative change in quantity demanded}}{\textbf{Relative change in price}} = \frac{\dfrac{\Delta x}{x}}{\dfrac{\Delta p}{p}} \qquad (3)$$

Economists use elasticity to study the effect of price change on quantity demanded. Since elasticity depends on p and Δp, if we let $\Delta p \to 0$, we obtain an expression for the elasticity of demand at price p, denoted by $E(p)$:

$$E(p) = \lim_{\Delta p \to 0} \dfrac{\dfrac{\Delta x}{x}}{\dfrac{\Delta p}{p}}$$

$$= \lim_{\Delta p \to 0} \dfrac{\dfrac{f(p + \Delta p) - f(p)}{f(p)}}{\dfrac{\Delta p}{p}}$$

$$= \lim_{\Delta p \to 0} \dfrac{f(p + \Delta p) - f(p)}{f(p)} \cdot \dfrac{p}{\Delta p}$$

$$= \dfrac{p}{f(p)} \lim_{\Delta p \to 0} \dfrac{f(p + \Delta p) - f(p)}{\Delta p}$$

$$= \dfrac{p}{f(p)} \cdot f'(p)$$

ELASTICITY OF DEMAND

The *elasticity of demand $E(p)$ at price p* for the demand function $x = f(p)$ is defined as

$$E(p) = \dfrac{p f'(p)}{f(p)}$$

Since $f'(p)$ is always negative for a typical demand function, the quantity $pf'(p)/f(p)$ will be negative for all values of p. For convenience, economists prefer to work with positive numbers. Therefore, the *price elasticity of demand* is taken to be $|E(p)|$. For a given price p, if $|E(p)| > 1$, the demand is said to be *elastic*. If $|E(p)| < 1$, the demand is said to be *inelastic*.

EXAMPLE 1   Suppose $x = f(p) = 5000 - 30p^2$ is the demand function for a certain commodity, where p is the price per pound and x is the quantity demanded in pounds.

(a) What quantity can be sold at $10 per pound?
(b) Determine the function $E(p)$.
(c) Determine and interpret the elasticity of demand at $p = 5$.
(d) Determine and interpret the elasticity of demand at $p = 10$.

SOLUTION

(a) At $p = 10$, $f(10) = 5000 - 30(10)^2 = 2000$. Therefore, 2000 pounds of the commodity can be sold at a price of $10.

(b) $E(p) = \dfrac{p f'(p)}{f(p)} = \dfrac{p(-60p)}{5000 - 30p^2} = \dfrac{-60p^2}{5000 - 30p^2}$

(c) The elasticity of demand at price $p = 5$ is $E(5)$.

$$E(5) = \frac{-60(5)^2}{5000 - 30(5)^2} = -\frac{1500}{4250} = -0.353$$

or $|E(5)| = 0.353$

When the price is set at $5 per pound, a small increase in price will result in a relative rate of decrease in quantity demanded of about 0.353 times the relative rate of increase in price. For instance, if the price is increased from $5 by 10%, then the quantity demanded will decrease by $(0.353)(10\%) = 0.0353 = 3.53\%$.

(d) When $p = 10$, we have

$$E(10) = \frac{-60(10)^2}{5000 - 30(10)^2} = -3$$

or $|E(10)| = 3$

When the price is set at 10, a small increase in price will result in a relative rate of decrease in quantity demanded of 3 times the relative rate of increase of price. For instance, a 10% price increase will result in a decrease in quantity demanded of approximately $3(10\%) = 30\%$.

Now Work Problem 1

Revenue and Elasticity of Demand

The concept of elasticity has an interesting relationship to the total revenue $R(p)$:

(a) If the demand is elastic, then an increase in the price per unit will result in a decrease in total revenue.

(b) If the demand is inelastic, then an increase in the price per unit will result in an increase in total revenue.

Now we establish these relationships between elasticity of demand and total revenue.

Since total revenue is given by $R(p) = xp = f(p)p$, we calculate the marginal revenue to be

$$R'(p) = f(p) \cdot 1 + f'(p) \cdot p = f(p)\left[1 + \frac{pf'(p)}{f(p)}\right] = f(p)\,[1 + E(p)]$$

We know that $f(p) > 0$. If the demand is elastic, then $|E(p)| > 1$. Since $E(p) < 0$, this means that $E(p) < -1$, so that $1 + E(p) < 0$. As a result, $R'(p) < 0$. In other words, by the First Derivative Test, $R(p)$ is decreasing at p. Thus, an increase in price will result in a decrease in total revenue when the demand is elastic.

Similarly, if the demand is inelastic, then $|E(p)| < 1$, so that $E(p) > -1$ or $1 + E(p) > 0$. Hence, $R'(p) > 0$. This implies that an increase in price will result in an increase in total revenue when the demand is inelastic.

EXERCISE 14.5 Answers to odd-numbered problems begin on page AN-73.

1. Given the demand equation $p + \frac{1}{100}x = 40$:

(a) Express the demand x as a function of p.

(b) Find the elasticity of demand $E(p)$.

(c) What is the elasticity of demand when $p = \$5$? If the price is increased by 10%, what is the approximate change in demand?

(d) What is the elasticity of demand when $p = \$15$? If the price is increased by 10%, what is the approximate change in demand?

(e) What is the elasticity of demand when $p = \$20$? If the price is increased by 10%, what is the approximate change in demand?

2. Repeat Problem 1 for the demand equation

$$p + \frac{1}{200}x = 80$$

3. Given the demand equation $p + \frac{1}{200}x = 50$:

(a) Express the demand x as a function of p.

(b) Find the elasticity of demand $E(p)$.

(c) What is the elasticity of demand when $p = \$10$? If the price is increased by 5%, what is the approximate change in demand?

(d) What is the elasticity of demand when $p = \$25$? If the price is increased by 5%, what is the approximate change in demand?

(e) What is the elasticity of demand when $p = \$35$? If the price is increased by 5%, what is the approximate change in demand?

4. Repeat Problem 3 for the demand function

$$p + \frac{1}{200}x = 100$$

In Problems 5–16 a demand function is given. Find E(p) and determine if demand is elastic or inelastic (or neither) at the indicated price.

5. $x = f(p) = 600 - 3p; \quad p = 50$

6. $x = f(p) = 700 - 4p; \quad p = 40$

7. $x = f(p) = \dfrac{600}{p + 4}; \quad p = 10$

8. $x = f(p) = \dfrac{500}{p + 6}; \quad p = 10$

9. $x = f(p) = 10{,}000 - 10p^2; \quad p = 10$

10. $x = f(p) = 2250 - p^2; \quad p = 15$

11. $x = f(p) = \sqrt{100 - p}; \quad p = 10$

12. $x = f(p) = \sqrt{2500 - 2p^2}; \quad p = 25$

13. $x = f(p) = 3(p - 4)^2; \quad p = 2$

14. $x = f(p) = 20(p - 2)^2; \quad p = 4$

15. $x = f(p) = 20 - 3\sqrt{p}; \quad p = 4$

16. $x = f(p) = 30 - 4\sqrt{p}; \quad p = 20$

In Problems 17–20 use implicit differentiation to find the elasticity of demand at the indicated values of x and p.

17. $x^{1/2} + 2px + p^2 = 84; \quad x = 16, \quad p = 4$

18. $x^{3/2} + 2px + p^3 = 1088; \quad x = 4, \quad p = 10$

19. $2x^2 + 3px + 10p^2 = 600; \quad x = 10, \quad p = 5$

20. $3x^3 + x^2p^2 + 10p^3 = 3480; \quad x = 10, \quad p = 2$

The discussion about elasticity assumed that the demand x is a function of price p. However, in economics it is common to use x as the independent variable. Thus, if $p = F(x)$ is the demand equation, then it can be shown the elasticity of demand is given by

$$E(x) = \frac{F(x)}{xF'(x)}$$

In Problems 21–26 use this formula to find the elasticity of demand at the given value of x.

21. $p = F(x) = 10 - \frac{1}{20}x; \quad x = 5$

22. $p = F(x) = 40 - \frac{1}{10}x; \quad x = 4$

23. $p = F(x) = 10 - 2x^2; \quad x = 2$

25. $p = F(x) = 50 - 2\sqrt{x}; \quad x = 100$

24. $p = F(x) = 20 - 4x^2; \quad x = 4$

26. $p = F(x) = 20 - 4\sqrt{x}; \quad x = 400$

27. Revenue and Elasticity A movie theater has a capacity of 1000 people. The number of people attending the show at a price of $\$p$ per ticket is $x = f(p) = 6000/p - 500$. Currently the price is $4 per ticket.

(a) Determine whether the demand is elastic or inelastic at $4.

(b) If the price is increased, will revenue increase or decrease?

28. Revenue and Elasticity The demand function for a rechargeable hand-held vacuum cleaner is given by

$$x = \tfrac{1}{3}(300 - p^2)$$

where x (measured in units of 100) is the quantity demanded per week and p is the unit price in dollars. The manufacturer would like to increase revenue.

(a) Is demand elastic or inelastic at $p = \$15$?

(b) Should the price of the unit be raised or lowered?

29. Revenue and Elasticity The demand function for digital watches is

$$x = \sqrt{300 - 6p}$$

where x is measured in units of hundreds.

(a) Is demand elastic or inelastic at $p = \$10$?

(b) If the price is lowered slightly, will revenue increase or decrease?

30. Revenue and Elasticity A company wishes to increase its revenue by lowering the price of its product. The demand function for this product is

$$x = \frac{10{,}000}{p^2}$$

(a) Compute $E(p)$.

(b) Will the company succeed in raising its revenue?

31. Revenue and Elasticity When a wholesaler sold a certain product at $15 per unit, sales were 2000 units each week. However, after a price rise of $3, the average number of units sold decreased to 1800 per week. Assume that the demand function is linear.

(a) Determine the demand function.

(b) Find the elasticity of demand at the new price.

(c) Approximate the change in demand if the price is increased by 5%.

(d) Will the price increase cause the revenue to increase or decrease?

14.6 RELATED RATES

In all of the natural sciences and many of the social and behavioral sciences, quantities that are related, but vary with time, are encountered. For example, the pressure of an ideal gas of fixed volume is proportional to temperature, yet each of these quantities may change over a period of time. Problems involving rates of related variables are referred to as **related rate problems.** In such problems we normally want to find the rate at which one of the variables is changing at a certain time, while the rates at which the other variables are changing are known. Let's look at an example.

EXAMPLE 1

A child throws a stone into a still millpond, causing a circular ripple to spread. If the radius of the circle increases at the constant rate of 0.5 feet per second, how fast is the area of the ripple increasing when the radius of the ripple is 30 feet? See Figure 43.

SOLUTION

The variables involved are

$t = $ Time (in seconds) elapsed from the time of the throw

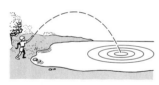

Figure 43

$$r = r(t) = \text{Radius of the ripple (in feet) after } t \text{ seconds}$$
$$A = A(t) = \text{Area of the ripple (in square feet) after } t \text{ seconds}$$

The rates involved are

$$\frac{dr}{dt} = \text{Rate at which the radius is increasing at each instant}$$

$$\frac{dA}{dt} = \text{Rate at which the area is increasing with time}$$

We wish to find dA/dt when $r = 30$; that is, the rate at which the area of the ripple is increasing at the instant when $r = 30$. The relationship between A and r is given by the formula for the area of a circle:

$$A = \pi r^2 \tag{1}$$

Since A and r are functions of t, we use implicit differentiation to differentiate both sides of (1) with respect to t to obtain

$$\frac{dA}{dt} = 2\pi r \frac{dr}{dt} \tag{2}$$

Since the radius increases at the rate of 0.5 feet per second, we know that

$$\frac{dr}{dt} = 0.5 \tag{3}$$

By substituting (3) into (2), we get

$$\frac{dA}{dt} = 2\pi r(0.5) = \pi r$$

Thus, when $r = 30$, the area of the ripple is increasing at the rate

$$\frac{dA}{dt} = \pi(30) = 30\pi = 94.25 \text{ square feet per second}$$

$$\cdots\cdots\cdots\cdots\cdots\cdots\cdots\cdots\cdots\cdots$$

Example 1 illustrates some general guidelines that will prove helpful for solving related rate problems:

Steps for Solving Related Rate Problems

Step 1 If possible, draw a picture illustrating the problem.

Step 2 Identify the variables and assign symbols to them.

Step 3 Identify and interpret rates of change as derivatives.

Step 4 Express all relationships among the variables by equations.

Step 5 Obtain additional relationships among the variables and their derivatives by differentiating.

Step 6 Substitute numerical values for the variables and the derivatives. Solve for the unknown rate.

[*Note:*   It is important to remember that the substitution of numerical values must occur after the differentiation process (Step 5).]

EXAMPLE 2

A racing car at the Indianapolis 500 travels along the ellipse $x^2 + 2y^2 = 9$, where x and y are measured in miles. See Figure 44. The position of the car at any time t is described by its coordinates x and y, which change as the car moves. When the car is at the point $(1, 2)$, the x-coordinate of the point is changing at the rate (speed) of 3 miles per minute. How fast is the y-coordinate changing at that moment?

SOLUTION

Draw a figure.
    The variables of the problem are

t = Time (in minutes) measured from the moment the car started traveling

$x = x(t)$ = The x-coordinate of the car at time t

$y = y(t)$ = The y-coordinate of the car at time t

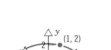

Figure 44

The rates of change are

$$\frac{dx}{dt} = \text{Rate of change of the } x\text{-coordinate with respect to time } t$$

$$\frac{dy}{dt} = \text{Rate of change of the } y\text{-coordinate with respect to time } t$$

The variables x and y are related by $x^2 + 2y^2 = 9$. Our problem is to find dy/dt, given $x = 1$, $y = 2$, and $dx/dt = 3$. Differentiating each term in the equation with respect to t, we obtain

$$\frac{d}{dt}(x^2) + \frac{d}{dt}(2y^2) = \frac{d}{dt}(9)$$

$$2x\frac{dx}{dt} + 4y\frac{dy}{dt} = 0$$

using implicit differentiation.
    Substituting $x = 1$, $y = 2$, and $dx/dt = 3$, and solving for dy/dt, we obtain

$$2 \cdot 1 \cdot 3 + 4 \cdot 2\frac{dy}{dt} = 0$$

$$6 + 8\frac{dy}{dt} = 0$$

$$\frac{dy}{dt} = -\frac{6}{8} = \frac{-3}{4} \text{ mile per minute}$$

This means that the y-coordinate of the car is decreasing at the rate of $\frac{3}{4}$ mile per minute when the car passes through the point $(1, 2)$.

Now Work Problem 9

· ·

EXAMPLE 3

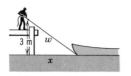

Figure 45

A person is standing on a pier and pulling a boat inward by pulling a rope at the rate of 2 meters per second. The end of the rope is 3 meters above water level. (See Figure 45.) How fast is the boat approaching the base of the pier when 5 meters of rope are left to pull in? Disregard sagging of the rope and assume the rope is attached to the boat at water level.

SOLUTION

The variables of the problem are

$$t = \text{Time (in seconds)}$$
$$x = x(t) = \text{Distance (in meters) from the boat to the base of the pier}$$
$$w = w(t) = \text{Distance (in meters) from the boat to the person (that is,}$$
$$\text{the length of rope)}$$

The rates of change are

$$\frac{dx}{dt} = \text{Rate at which the boat is approaching the pier}$$

$$\frac{dw}{dt} = \text{Rate at which the rope is being pulled}$$

From Figure 45 we see that 3, x, and w form the sides of a right triangle. By the Pythagorean theorem,

$$w^2 = 9 + x^2 \tag{4}$$

The variables w and x are functions of time t. By differentiating (4) with respect to time t, we obtain

$$2w\frac{dw}{dt} = 2x\frac{dx}{dt}$$

$$w\frac{dw}{dt} = x\frac{dx}{dt} \tag{5}$$

We seek to find dx/dt, the rate at which the boat is approaching the pier, when $w = 5$ meters and $dw/dt = -2$ meters per second. (The negative sign is used to indicate that the length of the rope is *decreasing* at the rate of 2 meters per second.)

The value of x when $w = 5$ is found from Equation (4) to be $x = \sqrt{25 - 9} = 4$ meters. By substituting these values into (5), we find

$$5(-2) = 4\frac{dx}{dt}$$

$$\frac{dx}{dt} = -2.5 \text{ meters per second}$$

Thus, the boat is approaching the pier at the rate of 2.5 meters per second.

. . . .

EXAMPLE 4

Suppose that for a company manufacturing digital watches, the cost, revenue, and profit functions are given by

Cost function: $C(x) = 10,000 + 3x$

Revenue function: $R(x) = 5x - \dfrac{x^2}{2000}$

Profit function: $P(x) = R(x) - C(x)$

where x is the daily production of digital watches. If production is increasing at the rate of 50 watches per day when production is 1000 watches, find the rate of increase in (a) cost, (b) revenue, and (c) profit.

SOLUTION

The variables involved are

$$t = \text{Time in days}$$
$$x = x(t) = \text{Production } x \text{ as a function of time}$$
$$C = C(t) = \text{Cost as a function of time}$$
$$R = R(t) = \text{Revenue as a function of time}$$
$$P = P(t) = \text{Profit as a function of time}$$

The rates involved are

$$\frac{dx}{dt} = 50 = \text{Rate at which production is increasing when } x = 1000$$

$$\frac{dC}{dt} = \text{Rate at which cost is increasing}$$

$$\frac{dR}{dt} = \text{Rate at which revenue is increasing}$$

$$\frac{dP}{dt} = \text{Rate at which profit is increasing}$$

(a)
$$C(x) = 10,000 + 3x$$
$$\frac{dC}{dt} = \frac{d}{dt}(10,000) + \frac{d}{dt}(3x) = 3\frac{dx}{dt}$$

Since

$$\frac{dx}{dt} = 50 \qquad \text{when} \qquad x = 1000$$

then

$$\frac{dC}{dt} = 3(50) = \$150 \text{ per day}$$

Thus, the cost is increasing at the rate of $150 per day.

(b)
$$R(x) = 5x - \frac{x^2}{2000}$$

$$\frac{dR}{dt} = \frac{d}{dt}(5x) - \frac{d}{dt}\left(\frac{x^2}{2000}\right) = 5\frac{dx}{dt} - \frac{x}{1000}\frac{dx}{dt}$$

Since

$$\frac{dx}{dt} = 50 \qquad \text{when} \qquad x = 1000$$

then

$$\frac{dR}{dt} = 5(50) - \frac{1000}{1000}(50) = \$200 \text{ per day}$$

Revenue is increasing at the rate of $200 per day.

(c)
$$P = R - C$$
$$\frac{dP}{dt} = \frac{dR}{dt} - \frac{dC}{dt} = \$200 - \$150 = \$50 \text{ per day}$$

Profit is increasing at the rate of $50 per day.

· ·

EXERCISE 14.6 Answers to odd-numbered problems begin on page AN-74.

In Problems 1–4 assume x and y are differentiable functions of t. Find dx/dt when
x = 2, y = 3, and dy/dt = 2.

1. $x^2 + y^2 = 13$ **2.** $x^2 - y^2 = -5$ **3.** $x^3 y^2 = 72$ **4.** $x^2 y^3 = 108$

5. Suppose h is a differentiable function of t and suppose that when $h = 3$, $dh/dt = \frac{1}{12}$. Find dV/dt if $V = 80h^2$.

6. Suppose x is a differentiable function of t and suppose that when $x = 15$, $dx/dt = 3$. Find dy/dt if $y^2 = 625 - x^2$.

7. Suppose h is a differentiable function of t and suppose that $dh/dt = \frac{5}{16}\pi$ when $h = 8$. Find dV/dt if $V = \frac{1}{12}\pi h^3$.

8. Suppose x and y are differentiable functions of t and suppose that when $t = 20$, $dx/dt = 5$, $dy/dt = 4$, $x = 150$, and $y = 80$. Find ds/dt if $s^2 = x^2 + y^2$.

9. A particle is traveling along the graph of $x^2 + y^2 = 5$. When the particle is at $(1, 2)$, its y-coordinate is increasing at the rate of 3 units per minute. How fast is the x-coordinate changing at that point?

10. A point is moving along the graph $y = 3x^2$. When the point is at $(2, 12)$, its x-coordinate is decreasing at the rate

of 2 units per second. How fast is the y-coordinate changing at that point?

11. You are standing on top of a 10-foot ladder that is leaning against a vertical wall. See the figure. The foot of the ladder is slipping away from the wall at the rate of 0.5 foot per second.

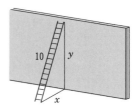

(a) Use the Pythagorean theorem to find an equation relating x and y.
(b) At what rate is the top of the ladder coming down when it is 5 feet off the ground?

12. When a metal plate is heated, it expands. If the shape of the metal is circular and if its radius, as a result of expansion, increases at the rate of 0.02 centimeter per second, at what rate is the area of the top surface increasing when the radius is 3 centimeters?

13. If each edge of a cube is increasing at the constant rate of 3 centimeters per second, how fast is the volume increasing when x, the length of an edge, is 10 centimeters long?

14. If the radius of a sphere is increasing at 1 centimeter per second, find the rate of change of its volume when the radius is 6 centimeters. $(V = \frac{4}{3}\pi r^3)$

15. Consider a right triangle with hypotenuse of (fixed) length 45 centimeters and variable legs of lengths x and y, respectively. If the leg of length x increases at the rate of 2 centimeters per minute, how fast is y changing when x is 4 centimeters long?

16. Air is pumped into a balloon with a spherical shape at the rate of 80 cubic centimeters per second. How fast is the surface area of the balloon increasing when the radius is 10 centimeters? $(S = 4\pi r^2)$

17. A spherical balloon filled with gas has a leak that permits the gas to escape at a rate of 1.5 cubic meters per minute. How fast is the surface area of the balloon shrinking when the radius is 4 meters?

18. **Cost, Revenue, Profit Functions** Suppose that for a company manufacturing microcomputers, the cost, revenue, and profit functions are given by

$$C(x) = 85{,}000 + 300x \qquad R(x) = 400x - \frac{x^2}{20}$$

$$P(x) = R(x) - C(x)$$

where x is the weekly production of microcomputers. If production is increasing at the rate of 400 microcom-

puters per week when production output is 5000 microcomputers, find the rate of increase in

(a) Cost   (b) Revenue   (c) Profit

19. **Pollution** Assume that oil spilled from a ruptured tanker forms a circular oil slick whose radius increases at a constant rate of 0.42 feet per minute $(dr/dt = 0.42)$. Estimate the rate dA/dt (in square feet per minute) at which the area of the spill is increasing when the radius of the spill is 120 feet $(r = 120)$. Use $\pi = 3.14$.

20. **Demand, Revenue Functions** The marketing department of a manufacturing company estimates that the demand q (in thousands of units per year) for an electric typewriter is related to price by the demand equation $q = 200 - 0.9p$. Because of efficiency and technological advances, the prices are falling at a rate of $30 per year $(dp/dt = -30)$. The current price of a typewriter is $650. At what rate (dR/dt) are revenues changing?

21. **Cost, Revenue, Profit** The cost C and revenue R of a company are given by

$$C(x) = 5x + 5000 \qquad R(x) = 15x - \frac{x^2}{10{,}000}$$

where x is the daily production. If production is increasing at the rate of 100 units per day at a production level of 1000 units, find

(a) The rate of change in daily cost
(b) The rate of change in daily revenue
(c) Whether revenue is increasing or decreasing
(d) The profit function
(e) The rate of change in profit

22. Rework Problem 21 if production is decreasing at the rate of 40 units per day at a production level of 2000 units.

14.7 THE DIFFERENTIAL AND LINEAR APPROXIMATIONS

The Differential

In studying the derivative of a function $y = f(x)$, we use the notation dy/dx to represent the derivative. The symbols dy and dx, called *differentials,* which appear in this notation may also be given their own meanings. To pursue this, recall that for a differentiable function f, the derivative is defined as

$$\frac{dy}{dx} = f'(x) = \lim_{\Delta x \to 0} \frac{\Delta y}{\Delta x} = \lim_{\Delta x \to 0} \frac{f(x + \Delta x) - f(x)}{\Delta x}$$

That is, the derivative f' is the limit of the ratio of the change in y to the change in x as Δx tends to 0, but $\Delta x \neq 0$. In other words, for Δx sufficiently close to 0, we can make $\Delta y/\Delta x$ as close as we please to $f'(x)$. We express this fact by writing

$$\frac{\Delta y}{\Delta x} \approx f'(x) \qquad \text{when} \qquad \Delta x \approx 0 \, (\Delta x \neq 0) \tag{1}$$

Another way of writing (1) is

$$\Delta y \approx f'(x) \, \Delta x \qquad \text{when} \qquad \Delta x \approx 0 \, (\Delta x \neq 0)$$

The quantity $f'(x) \, \Delta x$ is given a special name, the *differential of y*.

DIFFERENTIAL

Let f denote a differentiable function and let Δx denote a change in x.

(a) The *differential of y*, denoted by dy, is defined as $dy = f'(x) \, \Delta x$.
(b) The *differential of x*, denoted by dx, is defined as $dx = \Delta x \neq 0$.

Thus, using the notation of differentials, we have

$$dy = f'(x) \, dx \tag{2}$$

Since $dx \neq 0$, (2) can be written as

$$\frac{dy}{dx} = f'(x) \tag{3}$$

The expression in (3) should look very familiar. Interestingly enough, we have given an independent meaning to the symbols dy and dx in such a way that, when dy is divided by dx, their quotient will be equal to the derivative. That is, the differential of y divided by the differential of x is equal to the derivative $f'(x)$. For this reason, *we may formally regard the derivative as a quotient of differentials.*

Note that the differential dy is a function of both x and dx. For example, the differential dy of the function $y = x^3$ is

$$dy = 3x^2 \, dx$$

so that

$$\begin{array}{llll}
\text{if } x = 1 & \text{and} & dx = 2, & \text{then} & dy = 3(1)^2(2) = 6 \\
\text{if } x = 0.5 & \text{and} & dx = 0.1, & \text{then} & dy = 3(0.5)^2(0.1) = 0.075 \\
\text{if } x = 2 & \text{and} & dx = -5, & \text{then} & dy = 3(2)^2(-5) = -60
\end{array}$$

EXAMPLE 1

(a) If $y = x^2 + 3x - 5$, then $dy = (2x + 3) \, dx$.

(b) If $y = \sqrt{x^2 + 4}$, then $dy = \dfrac{x}{\sqrt{x^2 + 4}} \, dx$.

Now Work Problem 1

....

Geometric Interpretation

We use Figure 46 to arrive at a geometric interpretation of the differentials dx and dy and their relationship to Δx and Δy. From the definition, the differential dx and the change Δx are equal. Therefore, we concentrate on the relationship between dy and Δy.

In Figure 46(a), $P = (x, y)$ is a point on the graph of $y = f(x)$, and the nearby point $Q = (x + \Delta x, y + \Delta y)$ is also on the graph of f. The slope of the tangent line to the graph of f at P is $f'(x)$. From the figure, it follows that

$$f'(x) = \frac{dy}{\Delta x} = \frac{dy}{dx} \quad \text{or} \quad dy = f'(x)\, dx$$

Figure 46(a) illustrates the case for which $dy < \Delta y$ and $\Delta x > 0$. The case for which $dy > \Delta y$ and $\Delta x > 0$ is illustrated in Figure 46(b). The remaining cases, in which $\Delta x = dx < 0$, have similar graphical representations.

Figure 46

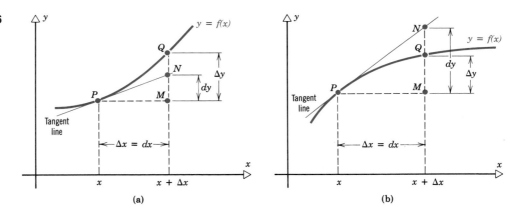

(a)   (b)

Linear Approximations

Let us now establish a relationship between Δy and dy. In Figure 46(a), $\Delta x = dx$ and the increment Δy is represented by the length of the line segment $|MQ|$. Thus, $\Delta y - dy$ is just the length of the line segment $|NQ|$. The size of $|NQ|$ equals the amount by which the graph departs from its tangent line. In fact, for $dx = \Delta x$ sufficiently small, the graph does not depart very much from its tangent line. As a result, the function whose graph is this tangent line is referred to as the *linear approximation to f near P*.

Approximating Δy

Thus, for $dx = \Delta x$ sufficiently small, the differential dy is a good approximation to Δy, in the sense that dy differs from Δy by a small percentage of dx. That is,

$$\Delta y \approx dy \quad \text{if} \quad \Delta x \approx 0 \tag{4}$$

We can use (4) to obtain the linear approximation to a function f near a point $P = (x_0, y_0)$ on f. Since

$$dy = f'(x_0)\,dx = f'(x_0)\,\Delta x = f'(x_0)(x - x_0)$$

we find from (4) that

$$\Delta y \approx dy$$
$$f(x) - f(x_0) \approx f'(x_0)(x - x_0)$$
$$f(x) \approx f(x_0) + f'(x_0)(x - x_0)$$

Linear Approximation to f Near x_0

For $dx = \Delta x$ sufficiently small, that is, for x close to x_0,

$$f(x) \approx f(x_0) + f'(x_0)(x - x_0) \qquad (5)$$

The function $y = f(x_0) + f'(x_0)(x - x_0)$ is called the **linear approximation to f near x_0.**

EXAMPLE 2

Find the linear approximation to $f(x) = x^2 + 2x$ near $x = 1$. Graph f and the linear approximation.

SOLUTION

First, $f(1) = 3$. Next, $f'(x) = 2x + 2$ so that $f'(1) = 2(1) + 2 = 4$. By (5), the linear approximation to f near $x = 1$ is

$$f(x) \approx f(1) + f'(1)(x - 1) = 3 + 4(x - 1) = 4x - 1$$

Figure 47 illustrates the graph of f and the linear approximation $y = 4x - 1$ to f near 1.

Figure 47

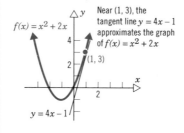

Now Work Problem 15

$\cdots\cdots\cdots\cdots\cdots\cdots\cdots\cdots\cdots$

Although the practicality of the next example is lessened due to the availability of hand calculators, whenever $f(x)$ is difficult to compute and $f(x_0)$ and $f'(x_0)$ are easy to compute, (5) may be used to obtain a numerical approximation.

EXAMPLE 3

Use differentials to approximate $\sqrt{123}$.

SOLUTION

The closest perfect square to 123 is 121, and $\sqrt{121} = 11$. What we wish to know is the value of $y = f(x) = \sqrt{x}$ when $x = 123$. The change in x from a square root we know ($x_0 = 121$) to the square root we seek ($x = 123$) is $x - x_0 = 2$. At $x_0 = 121$ [we know $f'(x) = 1/(2x^{1/2})$],

$$f(x_0) = \sqrt{121} = 11 \qquad \text{and} \qquad f'(x_0) = \frac{1}{2\sqrt{121}} = \frac{1}{22}$$

From (5),

$$f(x) = \sqrt{123} \approx f(x_0) + f'(x_0)(x - x_0) = 11 + \tfrac{1}{22}(2) = 11 + 0.0909 = 11.0909$$

(On a calculator, $\sqrt{123} = 11.090537$.)

$\cdots\cdots\cdots\cdots\cdots\cdots\cdots$

The next two examples use (4).

EXAMPLE 4

A bearing with a spherical shape has a radius of 3 centimeters when it is new. Find the approximate volume of the metal lost after it wears down to a radius of 2.971 centimeters.

SOLUTION

The exact volume of metal lost equals the change, ΔV, in volume V of the sphere, where $V = \tfrac{4}{3}\pi r^3$. The change in radius r is $\Delta r = -0.029$ centimeter. Since the change Δr is small, we can use the differential dV of volume to approximate the change ΔV in volume. Therefore,

$$\Delta V \approx dV = 4\pi r^2\, dr = (4\pi)(9)(-0.029) \approx -3.28$$

$$\uparrow$$
$$r = 3$$
$$dr = \Delta r = -0.029$$

The approximate loss in volume is 3.28 cubic centimeters.

Now Work Problem 23

$\cdots\cdots\cdots\cdots\cdots\cdots\cdots$

The use of dy to approximate Δy when dx is small may also be helpful in approximating errors.

If Q is the quantity to be measured and if ΔQ is the change in Q, we define

$$\textbf{Relative error in } Q = \frac{|\Delta Q|}{Q} \qquad \textbf{Percentage error in } Q = \frac{|\Delta Q|}{Q}(100\%)$$

For example, if $Q = 50$ units and the change ΔQ in Q is measured to be 5 units, then

$$\text{Relative error in } Q = \frac{5}{50} = 0.10$$

$$\text{Percentage error in } Q = 10\%$$

EXAMPLE 5

Suppose a company manufactures spherical ball bearings with radius 3 centimeters, and the percentage error in the radius must be no more than 1%. What is the approximate percentage error for the surface area of the ball bearing?

SOLUTION

If S is the surface area of a sphere of radius r, then $S = 4\pi r^2$. The actual error $\Delta S/S$ we seek may be approximated by the use of differentials. That is,

$$\frac{\Delta S}{S} \approx \frac{dS}{S} = \frac{8\pi r \, dr}{4\pi r^2} = \frac{2dr}{r} = \frac{2\Delta r}{r} = 2(0.01) = 0.02$$

The percentage error in the surface area is 2%.

· ·

In Example 5 the percentage error of 1% in the radius of the sphere means the radius will lie somewhere between 2.97 and 3.03 centimeters. But the percentage error of 2% in the surface area means the surface area lies within a factor of $\pm(0.02)$ of $S = 4\pi r^2 = 36\pi$; that is, it lies between $35.28\pi = 110.84$ and $36.72\pi = 115.36$ square centimeters. A rather small error in the radius results in a more significant range of possibilities for the surface area!

Now Work Problem 29

Differential Formulas

All the formulas derived earlier for finding derivatives carry over to differentials. We use the symbol df to indicate the differential of the function f. Thus, if $f(x) = x^2 + 2$, then $d(x^2 + 2) = 2x \, dx$. The list below gives formulas for differentials next to the corresponding derivative formulas.

Derivative	*Differential*
1. $\dfrac{d}{dx} c = 0$	1'. $dc = 0$ if c is constant
2. $\dfrac{d}{dx} (kx) = k$	2'. $d(kx) = k \, dx$ if k is constant
3. $\dfrac{d}{dx} (u + v) = \dfrac{du}{dx} + \dfrac{dv}{dx}$	3'. $d(u + v) = du + dv$
4. $\dfrac{d}{dx} (uv) = u\dfrac{dv}{dx} + v\dfrac{du}{dx}$	4'. $d(uv) = u\,dv + v\,du$
5. $\dfrac{d}{dx}\left(\dfrac{u}{v}\right) = \dfrac{v\dfrac{du}{dx} - u\dfrac{dv}{dx}}{v^2}$	5'. $d\left(\dfrac{u}{v}\right) = \dfrac{v\,du - u\,dv}{v^2}$
6. $\dfrac{d}{dx} x^r = rx^{r-1}$	6'. $d(x^r) = rx^{r-1} \, dx$ r a real number

From now on, to find the differential of a function $y = f(x)$, either find the derivative dy/dx and then multiply by dx or employ Formulas (1') through (6'). For example, if $y = x^3 + 2x + 1$,   then

$$\frac{dy}{dx} = 3x^2 + 2 \qquad \text{so that} \qquad dy = (3x^2 + 2)\, dx$$

By using the differential formulas,

$$dy = d(x^3 + 2x + 1) = d(x^3) + d(2x) + d(1)$$
$$= 3x^2\, dx + 2dx + 0 = (3x^2 + 2)\, dx$$

But be careful! The use of dy on the left side of an equation requires dx on the right side. Thus, $dy = 3x^2 + 2$ is incorrect.

The symbol d is an instruction to take the differential!

EXAMPLE 6

(a) $d(x^2 - 3x) = (2x - 3)\, dx$ (b) $d(3y^4 - 2y + 4) = (12y^3 - 2)\, dy$

(c) $d(\sqrt{z^2 + 1}) = \dfrac{z}{\sqrt{z^2 + 1}}\, dz$

....

The differential can be used to find the derivative of a function that is defined implicitly.

EXAMPLE 7

Find dy/dx and dx/dy if $x^2 + y^2 = 2xy^2$.

SOLUTION

We take the differential of each side:

$$d(x^2 + y^2) = d(2xy^2)$$
$$2x\, dx + 2y\, dy = 2(y^2\, dx + 2xy\, dy)$$
$$(y - 2xy)\, dy = (y^2 - x)\, dx$$
$$\frac{dy}{dx} = \frac{y^2 - x}{y - 2xy} \qquad \text{provided} \qquad y - 2xy \neq 0$$
$$\frac{dx}{dy} = \frac{y - 2xy}{y^2 - x} \qquad \text{provided} \qquad y^2 - x \neq 0$$

....

EXERCISE 14.7   Answers to odd-numbered problems begin on page AN-74.

In Problems 1–4 find the differential dy.

1. $y = x^3 - 2x + 1$ **2.** $y = 4(x^2 + 1)^{3/2}$ **3.** $y = \dfrac{x - 1}{x^2 + 2x - 8}$ **4.** $y = \sqrt{x^2 - 1}$

In Problems 5–10 find dy/dx and dx/dy by means of differentials.

5. $xy = 6$

6. $3x^2y + 2x - 10 = 0$

7. $x^2 + y^2 = 16$

8. $4xy^2 + yx^2 + 6 = 0$

9. $x^3 + y^3 = 3x^2y$

10. $2x^2 + y^3 = xy^2$

In Problems 11–14 find the indicated differential.

11. $d(\sqrt{x} - 2)$

12. $d\left(\dfrac{1 - x}{1 + x}\right)$

13. $d(x^3 - x - 4)$

14. $d(x^2 + 5)^{2/3}$

In Problems 15–18 find the linear approximation to f near x_0. Graph f and the linear approximation.

15. $f(x) = x^2 - 2x + 1; \quad x_0 = 2$

16. $f(x) = x^3 - 1; \quad x_0 = 0$

17. $f(x) = \sqrt{x}; \quad x_0 = 4$

18. $f(x) = x^{2/3}; \quad x_0 = 1$

19. Use (5) to approximate

(a) $\sqrt{35}$ (b) $\sqrt{26.2}$ (c) $\dfrac{1}{\sqrt{1.2}}$

20. Use (5) to approximate

(a) $\sqrt[3]{126}$ (b) $\sqrt[3]{123}$ (c) $\sqrt[4]{15}$

21. Use (5) to find the approximate change in

(a) $y = f(x) = x^2$ as x changes from 3 to 3.001
(b) $y = f(x) = 1/(x + 2)$ as x changes from 2 to 1.98

22. Use (5) to find the approximate change in

(a) $y = x^3$ as x changes from 3 to 3.01
(b) $y = 1/(x - 1)$ as x changes from 2 to 1.98

23. A circular plate is heated and expands. If the radius of the plate increases from $r = 10$ centimeters to $r = 10.1$ centimeters, find the approximate increase in area of the top surface.

24. In a wooden block 3 centimeters thick, an existing circular hole with a radius of 2 centimeters is enlarged to a hole with a radius of 2.2 centimeters. Approximately what volume of wood is removed?

25. Find the approximate change in volume of a spherical balloon of radius 3 meters as the balloon swells to a radius of 3.1 meters.

26. A bee flies around the circumference of a circle traced on a ball with a radius of 7 centimeters at a constant distance of 2 centimeters from the ball. An ant travels along the circumference of the same circle on the ball. Approximately how many more centimeters does the bee travel in one trip around than does the ant?

27. If the percentage error in measuring the edge of a cube is 2%, what is the percentage error in computing its volume?

28. The radius of a spherical ball is computed by measuring the volume of the sphere (by finding how much water it displaced). The volume is found to be 40 cubic centimeters, with a percentage error of 1%. Compute the corresponding percentage error in the radius (due to the error in measuring the volume).

29. A manufacturer produces paper cups in the shape of a right circular cone with radius equal to one-fourth its height. Specifications call for the cups to have a diameter of 4 centimeters. After production, it is discovered that the diameters measure only 3 centimeters. Assuming that the radius is still one-fourth of the height, what is the approximate loss in the capacity of the cup?

30. The oil pan of a car is shaped in the form of a hemisphere with a radius of 8 centimeters. The depth h of the oil is found to be 3 centimeters, with a percentage error of 10%. Approximate the percentage error in the volume. [*Hint:* The volume V for a spherical segment is $V = \frac{1}{3}\pi h^2(3r - h)$, where r is the radius.]

31. To find the height of a building, the length of the shadow of a 3-meter pole placed 9 meters from the building is measured. This measurement is found to be 1 meter, with a percentage error of 1%. What is the estimated height of the building? What is the percentage error in the estimate? See the figure below.

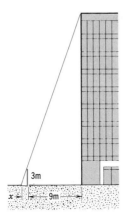

3m

x   9m

Technology Exercises

1. Find the linear approximation to $f(x) = 6x^2$ at $x_0 = \frac{1}{2}$.

   (a) Find $f'(x)$.
   (b) Write the equation for the tangent line, $g(x)$, to $f(x)$ at $x = x_0 = \frac{1}{2}$.
   (c) Graph $f(x)$ and $g(x)$ on $[-2, 2]$ by $[-5, 15]$.
   (d) From the graph, estimate an interval on the x-axis where $f(x)$ can be approximated by $g(x)$.
   (e) Trace to the point of intersection of the curve and its tangent. Zoom in on this point several times. What happens to the curve as you zoom in?
   (f) Use your calculator to evaluate $f(x)$ and $g(x)$ and complete the table.

x	$f(x)$	$g(x)$	$|f(x) - g(x)|$
0.2			
0.3			
0.4			
0.5			
0.6			
0.7			
0.8			

   (g) For what range of values for x is $|f(x) - g(x)| < 0.2$?
   (h) For what range of values for x is $|f(x) - g(x)| < 0.1$?

32. The period of the pendulum of a grandfather clock is $T = 2\pi\sqrt{l/g}$, where l is the length (in meters) of the pendulum, T is the period (in seconds), and g is the acceleration due to gravity (9.8 meters per second per second). Suppose the length of the pendulum, a thin wire, increases by 1% due to an increase in temperature. What is the corresponding percentage error in the period? How much time will the clock lose each day?

33. Refer to Problem 32. If the pendulum of a grandfather clock is normally 1 meter long and the length is increased by 10 centimeters, how many minutes will the clock lose each day?

34. What is the approximate volume enclosed by a hollow sphere if its inner radius is 2 meters and its outer radius is 2.1 meters?

2. Use a calculator to see how good the estimate $\sqrt{4.006} \approx 2.0015$ is. How large is the error?

3. Given the following data about a function, f:

x	4.5	5	5.5	6
$f(x)$	22.3	16.9	9.5	4.2

   (a) Estimate an equation of the tangent line to $y = f(x)$ at $x = 5$.
   (b) Using this equation, estimate $f(5.25)$, $f(6.5)$, and $f(7)$. Which of these estimates do you feel most confident about? Why?

4. Given the following data about a function, f:

x	1	1.5	2	2.5
$f(x)$	2.33	6.99	14.5	24

   (a) Estimate an equation of the tangent line to $y = f(x)$ at $x = 1.5$.
   (b) Using this equation, estimate $f(1.75)$, $f(2.75)$, and $f(3)$. Which of these estimates do you feel most confident about? Why?

CHAPTER REVIEW

IMPORTANT TERMS AND CONCEPTS

local maximum/minimum 694
test for increasing/decreasing
    function 695
First Derivative Test 697
critical number 702
critical point 702
concave up; concave down 708
test for concavity 708
inflection point 710
steps for graphing functions 712
Second Derivative Test 715

average cost 717
limits at infinity 722
infinite limits 723
horizontal asymptotes 723
vertical asymptotes 725
absolute maximum/
    minimum 731
test for absolute maximum/
    minimum 732
growth/decay curves 744

modified growth curves 744
logistic curves 745
elasticity of demand 752
steps for solving related
    rate problems 756
differential 762
linear approximation to f
    near x_0 764
relative error; percentage
    error 765

IMPORTANT FORMULAS

Average Cost

$$\overline{C}(x) = \frac{C(x)}{x}$$

Profit

$$P(x) = R(x) - C(x)$$

Maximizing Profit

$$R'(x) = C'(x)$$

Relative Change in Quantity Demanded

$$\frac{\Delta x}{x} = \frac{f(p + \Delta p) - f(p)}{f(p)}$$

Relative Change in Price

$$\frac{\Delta p}{p}$$

Elasticity

$$\frac{\text{Relative change in demand}}{\text{Relative change in price}} = \frac{\Delta x/x}{\Delta p/p}$$

Elasticity of Demand

$$E(p) = \frac{pf'(p)}{f(p)}$$

Differential of y

$$dy = f'(x)\,\Delta x$$

Differential of x

$$dx = \Delta x \neq 0$$

Linear Approximation of f

$$f(x) \approx f(x_0) + f'(x_0)(x - x_0)$$

Relative Error in Q

$$\frac{|\Delta Q|}{Q}$$

Percentage Error in Q

$$\frac{|\Delta Q|}{Q}(100\%)$$

TRUE–FALSE ITEMS   Answers are on page AN-75.

T_____ F_____ **1.** If the derivative of a function f does not exist at c, then f has a vertical tangent line at c.

T_____ F_____ **2.** A differentiable function f is increasing on (a, b) if $f'(x) > 0$ throughout (a, b).

T_____ F_____ **3.** The absolute maximum of a function equals the value of the function at a critical number.

T_____ F_____ **4.** If $f''(x) > 0$ for all x in (a, b), then the graph of f is concave down on (a, b).

T_____ F_____ **5.** If $\lim\limits_{x \to \infty} f(x) = N$, then the line $y = N$ is a horizontal asymptote of the graph of f.

T_____ F_____ **6.** If x and y are two differentiable functions of time t, then after differentiating $x^2 - y^3 + 4y - x = 100$,   we obtain

$$2x \frac{dx}{dt} - 3y^2 \frac{dy}{dt} + 4 \frac{dy}{dt} - \frac{dx}{dt} = 0$$

T_____ F_____ **7.** If $y = x^2 + 2x$, then $dy = 2x + 2$.

T_____ F_____ **8.** The graph of $f(x) = e^x$ is increasing.

T_____ F_____ **9.** The graph of $f(x) = \ln x$ is decreasing.

FILL IN THE BLANKS   Answers are on page AN-75.

1. The function $f(x)$ is _____ on (a, b) if $f'(x) < 0$ for all x in (a, b).

2. A point $(c, f(c))$ is a local minimum if the graph of the function is _____ to the left of c and _____ to the right of c.

3. A differentiable function is _____ _____ on (a, b) if the tangent lines to the graph at every point lie below its graph.

4. At a critical point the graph of a differentiable function f has a _____ tangent line.

5. At an inflection point the graph exhibits a change in _____.

6. If $y = f(x)$, f a differential function, then the differential of y is _____.

7. The function $y = f(x_0) + f'(x_0)(x - x_0)$ is called the _____ _____ to f at x_0.

REVIEW EXERCISES   Answers to odd-numbered problems begin on page AN-75.

1. Sketch the graph of a continuous function having all the given properties:

1. $(1, 34)$, $(2, 32)$, and $(3, 30)$ are on the graph
2. $f'(1) = 0$ and $f'(3) = 0$
3. $f''(x) < 0$ if $x < 2$ and $f''(x) > 0$ if $x > 2$

2. Refer to the given graph of $y = f(x)$ to identify the points or intervals on the x-axis that exhibit the indicated behavior.

(a) f is increasing
(b) $f'(x) < 0$
(c) Graph of f is concave down
(d) Local minima
(e) Absolute maximum
(f) $f'(x) = 0$
(g) Inflection points
(h) $f'(x)$ does not exist

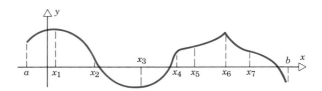

In Problems 3–14
(a) Find the domain of f.
(b) Locate the intercepts of f (skip the x-intercepts if they are too hard to find).
(c) Determine where the graph of f is increasing or decreasing.
(d) Find any local maxima or local minima.
(e) Find the slope of the tangent line at each local maximum and local minimum.
(f) Find the inflection points, if any.
(g) Find the horizontal and vertical asymptotes, if any.
(h) Graph f.

3. $f(x) = x^3 - 3x^2 + 3x - 1$

4. $f(x) = 2x^3 - x^2 + 2$

5. $f(x) = x^4 - 2x^2$

6. $f(x) = x^4 + 2x^2$

7. $f(x) = x^5 - 5x$

8. $f(x) = x^5 + 5x^4$

9. $f(x) = x^4 - 4x^3 + 4x^2$

10. $f(x) = x^4 + \frac{4}{3}x^3 - 4x^2$

11. $f(x) = x^{4/3} + 4x^{1/3}$

12. $f(x) = x^{4/3} - 4x^{1/3}$

13. $f(x) = \dfrac{2x}{x^2 + 1}$

14. $f(x) = \dfrac{4x}{x^2 + 4}$

In Problems 15–20 find the absolute maximum and absolute minimum for each function on the domain indicated.

15. $f(x) = x^3 - 3x^2 + 3x - 1$;   on $[0, 1]$

16. $f(x) = 2x^3 - x^2 + 2$;   on $[0, 1]$

17. $f(x) = x^4 - 2x^2$;   on $[-1, 1]$

18. $f(x) = x^4 + 2x^4$;   on $[-1, 1]$

19. $f(x) = x^{4/3} + 4x^{1/3}$;   on $[-1, 1]$

20. $f(x) = x^{4/3} - 4x^{1/3}$;   on $[-1, 1]$

In Problems 21–24 suppose that x and y are both functions of t and are related by the given equation. Calculate dx/dt for the given values of x and y.

21. $x^2 + y^2 = 8$; $x = 2, y = 2, dy/dt = 3$

22. $x^2 - y^2 = 5$; $x = 4, y = 3, dy/dt = 2$

23. $y^2 - 6x^2 = 3$; $x = 1, y = 3, dy/dt = -2$

24. $xy + 6x + y^3 = -2$; $x = 2, y = -3, dy/dt = 3$

In Problems 25–28 find dy.

25. $y = 3x^4 - 2x^3 + x$

26. $y = 3(x^2 - 1)^5$

27. $y = \dfrac{3 - 2x}{1 + x}$

28. $y = \sqrt[3]{2 + x^4}$

29. A child throws a stone into a still millpond, causing a circular ripple to spread. If the radius of the circle increases at the constant rate of 0.5 meter per second, how fast is the area of the ripple increasing when the radius of the ripple is 20 meters? [*Hint:* The area of a circle is $A = \pi r^2$.]

30. A balloon in the form of a sphere is being inflated at the rate of 10 cubic meters per minute. Find the rate at which the surface area of the sphere is increasing at the instant when the radius of the sphere is 3 meters. [*Hint:* The volume of a sphere is $V = \frac{4}{3}\pi r^3$; the surface area is $S = 4\pi r^2$.]

31. Maximizing Profit   A company's history shows that profits increase, as a result of advertising, according to

$$P(x) = 150 + 120x - 3x^2$$

where x is the number of dollars spent on advertising (measured in thousands). How much should be spent on advertising to maximize profits?

32. Best Dimensions of a Can   A beer can is cylindrical and holds 500 cubic centimeters of beer. If the cost of the material used to make the sides, top, and bottom is the same, what dimensions should the can have to minimize cost?

33. If a function f is continuous for all x and if f has a local maximum at $(-1, 4)$ and a local minimum at $(3, -2)$, which of the following statements must be true?

(a) The graph of f has a point of inflection somewhere between $x = -1$ and $x = 3$.
(b) $f'(-1) = 0$
(c) The graph of f has a horizontal asymptote.
(d) The graph of f has a horizontal tangent line at $(3, -2)$.
(e) The graph of f intersects both axes.

34. Let $f(x)$ be a function with derivative

$$f'(x) = \frac{1}{1 + x^2}$$

Show that the graph of $f(x)$ has an inflection point at $x = 0$. Note that $f'(x) > 0$ for all x.

35. Maximizing Profit   The price function of a certain mobile home producer is

$$p(x) = 2402.50 - 0.5x^2$$

where p is the price (in dollars) and x is the number of units sold. The cost of production for x units is

$$C(x) = 1802.5x + 1500$$

How many units need to be sold to maximize profit?

36. Setting Refrigerator Prices   A distributor of refrigerators has average monthly sales of 1500 refrigerators, each selling for $300. From past experience, the distributor knows that a special month-long promotion will enable them to sell 200 additional refrigerators for each $15 decrease in price. What should be charged for each refrigerator during the month of promotion in order to maximize revenue?

37. Size of a Burn   A burn on a person's skin is in the shape of a circle, so that if r is the radius of the burn and A is the area of the burn, then $A = \pi r^2$. Use the differential to approximate the decrease in the area of the burn when the radius decreases from 10 to 8 millimeters.

38. A spherical ball is being inflated. Find the approximate change in volume if the radius increases from 3 to 3.1 cm.

39. Size of a Tumor   A tumor is approximately spherical in shape. If the radius of the tumor changes from 11 to 13 millimeters, find the approximate change in volume.

40. Wound Healing   A wound on a person's skin is in the shape of a circle and is healing at the rate of 30 square millimeters per day. How fast is the radius r of the wound decreasing when $r = 10$ millimeters? [*Hint:* $A = \pi r^2$.]

41. Demand Function   The demand for peanuts (in hundreds of pounds) at a price of x dollars is

$$D(x) = -4x^3 - 3x^2 + 2000$$

Approximate the change in demand as the price changes from

(a) $1.50 to $2.00   (b) $2.50 to $3.50

42. Drug Concentration   The concentration of a certain drug in the bloodstream x hours after being administered is

$$c(x) = \frac{3x}{4 + 2x^2}$$

Approximate the change in concentration as x changes from

(a) 1.2 to 1.3   (b) 2 to 2.25

Mathematical Questions from Professional Exams

1. CPA Exam   The mathematical notation for the total cost for a business is $2X^3 + 4X^2 + 3X + 5$ where X equals production volume. Which of the following is the mathematical notation for the marginal cost function for this business?

(a) $2(X^3 + 2X^2 + 1.5X + 2.5)$
(b) $6X^2 + 8X + 3$
(c) $2X^3 + 4X^2 + 3X$
(d) $3X + 5$

2. CPA Exam   The mathematical notation for the total cost function for a business is $4X^3 + 6X^2 + 2X + 10$, where X equals production volume. Which of the following is the mathematical notation for the average cost function for that business?

(a) $2(2X^2 + 3X + 2)$
(b) $2X^3 + 3X^2 + X + 5$
(c) $0.4X^3 + 0.6X^2 + 0.2X + 1$
(d) $4X^2 + 6X + 2 + \dfrac{10}{X}$

3. **CPA Exam** The mathematical notation for the average cost function for a business is $6X^3 + 4X^2 + 2X + 8 + 2/X$, where X equals production volume. What would be the mathematical notation for the total cost function for the business?

(a) The average cost function multiplied by X.
(b) The average cost function divided by X.
(c) The average cost function divided by $X/2$.
(d) The first derivative of the average cost function.

4. **CPA—Review** To find a minimum cost point, given a total cost equation, the initial steps are to find the first derivative, set this derivative equal to zero, and solve the equation. Using the solution(s) so derived, what additional steps must be taken, and what result indicates a minimum?

(a) Substitute the solution(s) into the first derivative equation, and a positive solution indicates a minimum.
(b) Substitute the solution(s) into the first derivative equation, and a negative solution indicates a minimum.

(c) Substitute the solution(s) into the second derivative equation, and a positive solution indicates a minimum.
(d) Substitute the solution(s) into the second derivative equation, and a negative solution indicates a minimum.

5. **Actuary Exam—Part I** Figure A could represent the graph of which of the following?

(a) $y = x^3 e^{x^2}$
(b) $y = xe^x$
(c) $y = xe^{-x}$
(d) $y = x^2 e^x$
(e) $y = x^2 e^{x^2}$

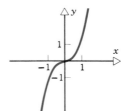

Figure A

Chapter 15

The Integral Calculus

With this chapter we begin the study of *integral calculus,* which may be subdivided into two parts: the indefinite integral and the definite integral. The *indefinite integral* concerns the inverse process of differentiation, usually referred to as finding the *antiderivative of a function.* The *definite integral* plays a major role in applications to geometry (finding the area under a graph), to business and economics (marginal analysis, consumer's surplus, maximizing profit over time), and to probability.

15.1 ANTIDERIVATIVES; THE INDEFINITE INTEGRAL

Antiderivatives

We start this section by asking the following question: "Is there a mathematical process that reverses differentiation, and if so, what is it?" We have already learned that to each differentiable function f there corresponds a derivative function f'. It is also possible to ask the following question: "If a function f is given, can we find a function F whose derivative is f? That is, is it possible to find a function F so that $F'(x) = f(x)$?" If such a function F can be found, it is called *an antiderivative of f.*

ANTIDERIVATIVE

A function F is called an *antiderivative* of the function f if

$$F'(x) = f(x)$$

EXAMPLE 1

An antiderivative of $f(x) = 2x$ is x^2 since

$$\frac{d}{dx} x^2 = 2x$$

Another antiderivative of $2x$ is $x^2 + 3$ since

$$\frac{d}{dx} (x^2 + 3) = 2x$$

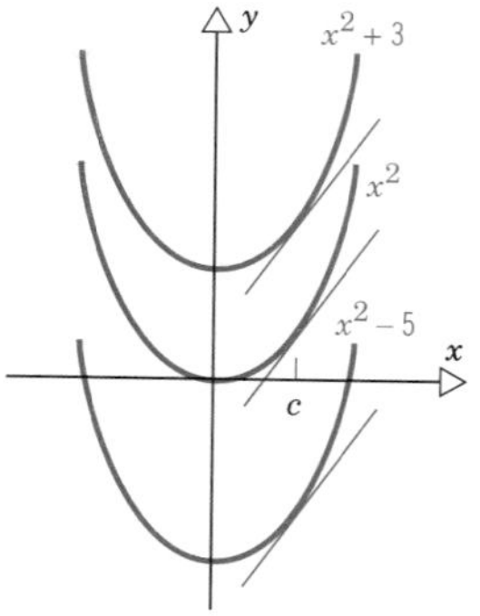

. . . .

Figure 1   Slopes of the tangent lines at $x = c$ are the same.

This example leads us to suspect that the function $f(x) = 2x$ has an unlimited number of antiderivatives. Indeed, any of the functions x^2, $x^2 + \frac{1}{2}$, $x^2 + 2$, $x^2 + \sqrt{5}$, $x^2 - \pi$, $x^2 - 1$, $x^2 + K$, where K is any constant, has the property that its derivative is $2x$. The fact that all these functions have the same derivative, $f(x) = 2x$, means that their slopes at any particular value of x are the same, as shown in Figure 1 where $x = c$. So all functions of the form $F(x) = x^2 + K$, where K is any constant, are antiderivatives of $f(x) = 2x$. Are there others? The answer is no! As the following result tells us, *all* the antiderivatives of $f(x) = 2x$ are of the form $F(x) = x^2 + K$, where K is any constant.

Antiderivatives of f

If F is any antiderivative of f, then all the antiderivatives of f are of the form

$$F(x) + K$$

where K is any constant.

EXAMPLE 2

All the antiderivatives of $f(x) = x^5$ are of the form

$$\frac{x^6}{6} + K$$

where K is any constant.

We check the answer by finding its derivative:

$$\frac{d}{dx}\left(\frac{x^6}{6} + K\right) = x^5$$

· ·

EXAMPLE 3

Find all the antiderivatives of $f(x) = x^{1/2}$.

SOLUTION

The derivative of the function $\frac{2}{3}x^{3/2}$ is

$$\frac{d}{dx}\frac{2}{3}x^{3/2} = \frac{2}{3}\left(\frac{3}{2}x^{1/2}\right) = x^{1/2}$$

So $\frac{2}{3}x^{3/2}$ is an antiderivative of $f(x) = x^{1/2}$. Thus, all the antiderivatives of $f(x) = x^{1/2}$ are of the form

$$\tfrac{2}{3}x^{3/2} + K$$

where K is any constant.

· ·

In Example 3 you may ask how we knew that we should choose the function $\frac{2}{3}x^{3/2}$. First, we know that

$$\frac{d}{dx}x^r = rx^{r-1}$$

That is, differentiation of a power of x reduces the exponent by 1. Antidifferentiation is the inverse process, so it should increase the exponent by 1. This is how we obtained the $x^{3/2}$ part of $\frac{2}{3}x^{3/2}$. Second, the $\frac{2}{3}$ factor is needed so that, when we differentiate, we get $x^{1/2}$ and not $\frac{3}{2}x^{1/2}$.

Thus, because

$$\frac{d}{dx}x^{r+1} = (r+1)x^r$$

for any real number r, it follows that

Antiderivatives of x^r

All the antiderivatives of x^r are of the form

$$\frac{x^{r+1}}{r+1} + K$$

where r is any real number, except -1, and K is any constant.

Notice that $r = -1$ is excluded from the formula. This case requires special attention and is considered a little later.

Indefinite Integrals

We use a special symbol to represent all the antiderivatives of a function—the **integral sign, $\int$.**

INDEFINITE INTEGRAL

Let F be an antiderivative of the function f. The *indefinite integral of f*, denoted by $\int f(x)\, dx$, is defined as

$$\int f(x)\, dx = F(x) + K$$

where K is any constant.

Thus, the indefinite integral of f is a symbol for all the antiderivatives of f.

In the expression $\int f(x)\, dx$, the integral sign $\int$ indicates that the operation of antidifferentiation is to be performed on the function f, and the dx reinforces the fact that the operation is to be performed with respect to the variable x. The function f is called the **integrand,** and the process of antidifferentiation is called **integration.**

Basic Integration Formulas

Based on the relationship between the process of differentiation and that of indefinite integration, or antidifferentiation, we can construct a list of formulas. These formulas may be verified by differentiating the right-hand side.

Basic Integration Formulas

If c is a real number and K is any constant,

$$\int c\, dx = cx + K \tag{1}$$

$$\int x^r\, dx = \frac{x^{r+1}}{r+1} + K \qquad r \neq -1 \tag{2}$$

$$\int [f(x) + g(x)]\, dx = \int f(x)\, dx + \int g(x)\, dx \tag{3}$$

$$\int [f(x) - g(x)]\, dx = \int f(x)\, dx - \int g(x)\, dx \tag{4}$$

$$\int cf(x)\, dx = c \int f(x)\, dx \tag{5}$$

As a special case of Formula (1), let $c = 1$. Then we find that

$$\int dx = \int 1 \cdot dx = 1 \cdot x + K = x + K$$

where K is any constant.

Formulas (3) and (4) state that the integral of a sum or a difference equals the sum or difference of the integrals.

Formula (5) states that a constant factor can be moved across an integral sign. Be careful! A variable factor cannot be moved across an integral sign.

EXAMPLE 4

(a) $\displaystyle\int 5 \, dx = 5x + K$
   ↑
   Formula (1)

(b) $\displaystyle\int x^5 \, dx = \frac{x^{5+1}}{5+1} + K = \frac{x^6}{6} + K$
   ↑ Formula (2)

(c) $\displaystyle\int 3x^4 \, dx = 3 \int x^4 \, dx = 3\,\frac{x^5}{5} + K = \frac{3}{5} x^5 + K$
   ↑ Formula (5)   ↑ Formula (2)

(d) $\displaystyle\int (x^2 + x^3) \, dx = \int x^2 \, dx + \int x^3 \, dx = \frac{x^3}{3} + K_1 + \frac{x^4}{4} + K_2$
   ↑ Formula (3)   ↑ Formula (2)

$$= \frac{x^3}{3} + \frac{x^4}{4} + K$$

where $K = K_1 + K_2$.

Now Work Problems 1 and 5

· ·

Formulas (3), (4), and (5) can be combined and used for sums and differences of three or more functions.

EXAMPLE 5

$$\int \left(7x^5 + \frac{1}{2} x^2 - x\right) dx = \int 7x^5 \, dx + \int \frac{1}{2} x^2 \, dx - \int x \, dx$$

$$= 7 \int x^5 \, dx + \frac{1}{2} \int x^2 \, dx - \int x \, dx$$

$$= \left(\frac{7x^6}{6} + K_1\right) + \left(\frac{1}{2}\frac{x^3}{3} + K_2\right) - \left(\frac{x^2}{2} + K_3\right)$$

$$= \frac{7}{6} x^6 + \frac{1}{6} x^3 - \frac{1}{2} x^2 + K$$

where $K = K_1 + K_2 - K_3$.

Now Work Problem 15

· ·

As Example 5 illustrates, we can now integrate any polynomial function.

Sometimes it is necessary to use algebra to put the integrand in a proper form before applying integration formulas.

EXAMPLE 6

(a) $\displaystyle\int \frac{1}{\sqrt{x}}\,dx = \int x^{-1/2}\,dx = \frac{x^{(-1/2)+1}}{-\frac{1}{2}+1} + K = \frac{x^{1/2}}{\frac{1}{2}} + K = 2x^{1/2} + K$

$\uparrow\uparrow$
Algebra$$(2)

(b) $\displaystyle\int 4\sqrt[3]{x^5}\,dx = 4\int \sqrt[3]{x^5}\,dx = 4\int x^{5/3}\,dx = 4\,\frac{x^{5/3+1}}{\frac{5}{3}+1} + K = \frac{3}{2}x^{8/3} + K$

$\uparrow\uparrow\uparrow$
(5)Algebra$$(2)

(c) $\displaystyle\int \frac{15\,dx}{x^5} = 15\int \frac{1}{x^5}\,dx = 15\int x^{-5}\,dx = \frac{15x^{-4}}{-4} + K = \frac{-15}{4x^4} + K$

$\uparrow\uparrow\uparrow$
(5)Algebra$$(2)

(d) $\displaystyle\int \frac{x^{3/2} - 2x}{\sqrt{x}}\,dx = \int \frac{x^{3/2} - 2x}{x^{1/2}}\,dx = \int \left(\frac{x^{3/2}}{x^{1/2}} - \frac{2x}{x^{1/2}} \right)\,dx$

$\uparrow$
Algebra

$= \displaystyle\int (x - 2x^{1/2})\,dx = \int x\,dx - 2\int x^{1/2}\,dx$

$\uparrow$
(4)

$= \dfrac{x^2}{2} - \dfrac{2x^{3/2}}{\frac{3}{2}} + K = \dfrac{x^2}{2} - \dfrac{4}{3}x^{3/2} + K$

Now Work Problem 21

· ·

Indefinite Integrals Involving Exponential and Logarithmic Functions

The next three integration formulas involve the exponential and logarithmic functions. Each one is a direct result of formulas developed in Chapter 13.

$$\int e^x\,dx = e^x + K \tag{6}$$

$$\int e^{ax}\,dx = \frac{1}{a}e^{ax} + K \qquad a \neq 0 \tag{7}$$

$$\int \frac{1}{x}\,dx = \ln |x| + K \qquad x \neq 0 \tag{8}$$

where K is any constant.

Proof Formulas (6) and (7) follow from the facts that

$$\frac{d}{dx}e^x = e^x \qquad \text{and} \qquad \frac{d}{dx}\frac{1}{a}e^{ax} = e^{ax}$$

Formula (8) requires more attention. To prove it, we need to show that

$$\frac{d}{dx} \ln |x| = \frac{1}{x}$$

Noting that $x \neq 0$ (do you see why?), we consider two cases: $x > 0$ and $x < 0$.

Case 1 $x > 0$:

$$\frac{d}{dx} \ln |x| = \frac{d}{dx} \ln x = \frac{1}{x}$$

$$\uparrow$$
$$|x| = x$$
$$\text{since } x > 0$$

Case 2 $x < 0$:

$$\frac{d}{dx} \ln |x| = \frac{d}{dx} \ln(-x) = \frac{1}{-x}\frac{d}{dx}(-x) = \frac{1}{-x}(-1) = \frac{1}{x}$$

$$\uparrow \qquad\qquad \uparrow$$
$$|x| = -x \qquad \text{Chain Rule}$$
$$\text{since } x < 0$$

Thus, in each case,

$$\frac{d}{dx} \ln |x| = \frac{1}{x} \qquad x \neq 0$$

Hence,

$$\int \frac{1}{x}\, dx = \ln |x| + K \qquad x \neq 0$$

∎

Formula (8) takes care of finding $\int \frac{1}{x}\, dx = \int x^{-1}\, dx$. Now we can find $\int x^r\, dx$ for any real number r.

We have still not discussed the indefinite integral of $\ln x$. We postpone a discussion of $\int \ln x\, dx$ until Section 15.2, where we introduce *integration by parts*.

EXAMPLE 7

Evaluate each of the following indefinite integrals:

(a) $\displaystyle\int \frac{e^x + e^{-x}}{2}\, dx$ (b) $\displaystyle\int \frac{3x^7 - 4x}{x^2}\, dx$

SOLUTION

(a) $\displaystyle\int \frac{e^x + e^{-x}}{2}\, dx = \frac{1}{2}\int (e^x + e^{-x})\, dx$

$$= \frac{1}{2}\left[\int e^x\, dx + \int e^{-x}\, dx\right]$$

$$= \frac{1}{2}[e^x + (-e^{-x})] + K$$

$$= \frac{e^x - e^{-x}}{2} + K$$

(b) $\displaystyle\int \frac{3x^7 - 4x}{x^2}\, dx = \int \left(\frac{3x^7}{x^2} - \frac{4x}{x^2}\right) dx$

$\displaystyle = \int 3x^5\, dx - \int \frac{4}{x}\, dx$

$\displaystyle = 3\int x^5\, dx - 4\int \frac{1}{x}\, dx$

$\displaystyle = \frac{3x^6}{6} - 4\ln|x| + K$

$\displaystyle = \frac{x^6}{2} - 4\ln|x| + K$

Now Work Problems 27 and 29

One of the most important steps to follow in finding antiderivatives is to *rewrite the integrand* in a form that fits one of the basic integration Formulas (1) through (8). To further illustrate, we give some additional examples.

EXAMPLE 8

Given	Rewrite	Integrate	Simplify
(a) $\displaystyle\int \sqrt[3]{x}\, dx$	$\displaystyle\int x^{1/3}\, dx$	$\dfrac{x^{4/3}}{\frac{4}{3}} + K$	$\frac{3}{4}x^{4/3} + K$
(b) $\displaystyle\int \frac{3}{\sqrt{x}}\, dx$	$\displaystyle 3\int x^{-1/2}\, dx$	$\dfrac{3x^{1/2}}{\frac{1}{2}} + K$	$6x^{1/2} + K$
(c) $\displaystyle\int \frac{x^2 + 4}{x}\, dx$	$\displaystyle\int \left(x + \frac{4}{x}\right) dx$	$\frac{1}{2}x^2 + 4\ln|x| + K$	$\frac{1}{2}x^2 + 4\ln|x| + K$
(d) $\displaystyle\int \sqrt{x}(x^3 - 1)\, dx$	$\displaystyle\int (x^{7/2} - x^{1/2})\, dx$	$\dfrac{x^{9/2}}{\frac{9}{2}} - \dfrac{x^{3/2}}{\frac{3}{2}} + K$	$\frac{2}{9}x^{9/2} - \frac{2}{3}x^{3/2} + K$
(e) $\displaystyle\int \frac{e^x - e^{2x}}{e^x}\, dx$	$\displaystyle\int (1 - e^x)\, dx$	$x - e^x + K$	$x - e^x + K$

Applications

Sometimes in applications the rate of change of a quantity is known and the goal is to find an expression for the quantity itself. For example, marginal revenue and marginal cost are defined as the derivative of the revenue function and cost function, respectively. As a result, if the marginal revenue (or marginal cost) is a known function, the revenue function R (or cost function C) may be found by using the process of antidifferentiation. These and other examples are shown in the following table.

Derivative		Integration (Antiderivative)	
Name	**Symbol**	**Name**	**Symbol**
Marginal cost	$C'(x)$	Cost	$\int C'(x)\, dx = C(x) + K$
Marginal profit	$P'(x)$	Profit	$\int P'(x)\, dx = P(x) + K$
Marginal revenue	$R'(x)$	Revenue	$\int R'(x)\, dx = R(x) + K$
Velocity	$v(t) = s'(t)$	Distance	$\int s'(t)\, dt = s(t) + K$
Acceleration	$a(t) = v'(t)$	Velocity	$\int v'(t)\, dt = v(t) + K$

The presence of a constant K in the antiderivative of a function plays an important role in applications since the undetermined constant K can assume any value that a particular practical situation demands.

EXAMPLE 9

By experimenting with various production techniques, a manufacturer finds that the marginal cost of production is given by the function

$$C'(x) = 2x + 6$$

where x is the number of units produced and C' is the marginal cost in dollars. The fixed cost of production is known to be \$9. Find the cost of production.

SOLUTION

The antiderivative of the marginal cost of production C' is the cost C of production. That is,

$$C(x) = \int C'(x)\, dx = \int (2x + 6)\, dx = x^2 + 6x + K$$

where K is a constant. We can find the value of the constant K by observing that of all the cost functions with derivative $2x + 6$, only one has a fixed cost of production of \$9, namely, the one whose cost is 9 when $x = 0$. We use this requirement to find the constant K.

$$C(x) = x^2 + 6x + K$$
$$C(0) = 0^2 + 6\cdot 0 + K = 9$$
$$K = 9$$

Thus,

$$C(x) = x^2 + 6x + 9$$

· ·

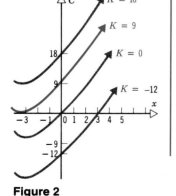

$K = 18$
$K = 9$
$K = 0$
$K = -12$

Figure 2

Now Work Problem 39

Figure 2 illustrates various cost functions whose marginal cost is $2x + 6$. The one having a fixed cost of \$9 is shown in color.

EXAMPLE 10

Suppose the manufacturer in Example 9 receives a price of $60 per unit. This means the marginal revenue is $60. That is,

$$R'(x) = 60$$

(a) Find the revenue function R.
(b) Find the profit function P.
(c) Find the sales volume that yields maximum profit.
(d) What is the profit at this sales volume?

SOLUTION

(a) The revenue function R is the antiderivative of the marginal revenue function R'. That is,

$$R(x) = \int R'(x)\,dx = \int 60\,dx = 60x + K$$

Now, of all these revenue functions, there is only one for which revenue equals zero for $x = 0$ units sold. To find it, we need to find K. We find K as follows:

$$R(0) = 60(0) + K = 0$$
$$K = 0$$

This means the revenue function is

$$R(x) = 60x$$

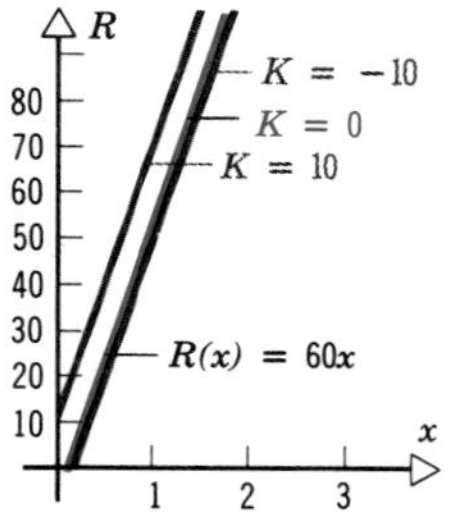

See Figure 3.

Figure 3

(b) The profit function P is the difference between revenue and cost. Since $R(x) = 60x$ and $C(x) = x^2 + 6x + 9$ (from Example 9), we have

$$P(x) = R(x) - C(x) = 60x - (x^2 + 6x + 9) = -x^2 + 54x - 9$$

(c) The maximum profit is obtained when marginal revenue equals marginal cost:

$$R'(x) = C'(x)$$
$$60 = 2x + 6$$
$$2x = 54$$
$$x = 27$$

Thus, when sales total 27 units, a maximum profit is obtained.

(d) The profit for sales of 27 units is

$$P(27) = -(27)^2 + 54(27) - 9 = \$720$$

....

EXAMPLE 11

The size of the mosquito population is changing t months from now, at the rate of $432t^2 - 5t^4$ per month. If the current population is 40, what will the population size be 5 months from now?

SOLUTION

If $P(t)$ is the population of mosquitoes now, then $P'(t)$ represents the rate of change of population with respect to time, that is,

$$P'(t) = 432t^2 - 5t^4$$

It follows that $P(t)$ is the antiderivative of $432t^2 - 5t^4$, or

$$P(t) = \int (432t^2 - 5t^4)\, dt = 144t^3 - t^5 + K$$

The constant K can be determined from the condition that at $t = 0$ the population is 40. That is,

$$40 = 144(0^3) - 0^5 + K \qquad \text{or} \qquad K = 40$$

Hence,

$$P(t) = 144t^3 - t^5 + 40$$

The population 5 months from now will be

$$P(5) = 144(5^3) - 5^5 + 40 = 14{,}915$$

See Figure 4.

Figure 4
The population function
$P(t) = 144t^3 - t^5 + 40$

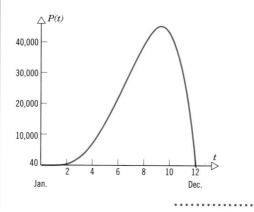

· ·

EXAMPLE 12

Determine the function $f(x)$ whose graph has a tangent with slope $3x^2 + 1$ at each value of x, and whose graph passes through the point $(1, 4)$.

SOLUTION

The slope of the tangent is the derivative of f. Thus,

$$f'(x) = 3x^2 + 1$$

The integral

$$\int (3x^2 + 1)\, dx = x^3 + x + K$$

gives a whole family of functions $f(x) = x^3 + x + K$, each of which has a slope given by $3x^2 + 1$. To find the one whose graph passes through $(1, 4)$—that is, to determine K—we substitute $x = 1$ and $f(1) = 4$ into the equation.

$$4 = 1^3 + 1 + K \qquad \text{or} \qquad K = 2$$

The desired function is $f(x) = x^3 + x + 2$.

Now Work Problem 47

....

EXERCISE 15.1   Answers to odd-numbered problems begin on page AN-77.

In Problems 1–30 evaluate each indefinite integral.

1. $\displaystyle\int 3\, dx$

2. $\displaystyle\int -4\, dx$

3. $\displaystyle\int x\, dx$

4. $\displaystyle\int x^2\, dx$

5. $\displaystyle\int x^{1/3}\, dx$

6. $\displaystyle\int x^{4/3}\, dx$

7. $\displaystyle\int x^{-2}\, dx$

8. $\displaystyle\int x^{-3}\, dx$

9. $\displaystyle\int x^{-1/2}\, dx$

10. $\displaystyle\int x^{-2/3}\, dx$

11. $\displaystyle\int (2x^3 + 5x)\, dx$

12. $\displaystyle\int (3x^2 - 4x)\, dx$

13. $\displaystyle\int (x^2 + 2e^x)\, dx$

14. $\displaystyle\int (3x + 5e^x)\, dx$

15. $\displaystyle\int (x^3 - 2x^2 + x - 1)\, dx$

16. $\displaystyle\int (2x^4 + x^2 - 5)\, dx$

17. $\displaystyle\int \left(\frac{x-1}{x}\right) dx$

18. $\displaystyle\int \left(\frac{x+1}{x}\right) dx$

19. $\displaystyle\int \left(2e^x - \frac{3}{x}\right) dx$

20. $\displaystyle\int \left(\frac{8}{x} - e^{-x}\right) dx$

21. $\displaystyle\int \left(\frac{3\sqrt{x}+1}{\sqrt{x}}\right) dx$

22. $\displaystyle\int \left(\frac{2\sqrt{x}-4}{\sqrt{x}}\right) dx$

23. $\displaystyle\int \frac{x^2-4}{x+2}\, dx$

24. $\displaystyle\int \frac{x^2-1}{x-1}\, dx$

25. $\displaystyle\int x(x-1)\, dx$

26. $\displaystyle\int x(x+2)\, dx$

27. $\displaystyle\int \frac{3x^5+2}{x}\, dx$

28. $\displaystyle\int \frac{x^6+x^2+1}{x^3}\, dx$

29. $\displaystyle\int \frac{4e^x+e^{2x}}{e^x}\, dx$

30. $\displaystyle\int \frac{3e^x+xe^{2x}}{xe^x}\, dx$

In Problems 31–34 find the revenue function. Assume that revenue is zero when zero units are sold.

31. $R'(x) = 600$

32. $R'(x) = 350$

33. $R'(x) = 20x + 5$

34. $R'(x) = 50x - x^2$

In Problems 35–38 find the cost function and determine where the cost is a minimum.

35. $C'(x) = 14x - 2800$
    Fixed cost = $4300

36. $C'(x) = 6x - 2400$
    Fixed cost = $800

37. $C'(x) = 20x - 8000$
    Fixed cost = $500

38. $C'(x) = 15x - 3000$
    Fixed cost = $1000

39. Marginal Cost   The marginal cost of production is found to be

$$C'(x) = 1000 - 20x + x^2$$

where x is the number of units produced. The fixed cost of production is $9000. Find the cost function.

40. Profit Function   In Problem 39 the manufacturer fixes the price per unit at $3400.

(a) Find the revenue function.
(b) Find the profit function.
(c) Find the sales volume that yields maximum profit.
(d) What is the profit at this sales volume?
(e) Graph the revenue, cost, and profit functions.

41. Cost Function   A company determines that the marginal cost of producing x units of a particular commodity during 1 day of operation is $C'(x) = 6x - 141$, where the production cost is in dollars. The selling price of the commodity is fixed at $9 per unit, and the fixed cost is $1800 per day.

(a) Find the cost function.
(b) Find the revenue function.
(c) Find the profit function.
(d) What is the maximum profit that can be obtained in 1 day of operation?
(e) Graph the revenue, cost, and profit functions.

42. Population Growth   It is estimated that the size of a population of a certain town changes at the rate of $2 + t^{4/5}$ people per month. If the current population is 20,000, what will the population be in 10 months?

43. Resource Depletion   The water currently used from a lake is estimated to amount to 150 million gallons a month. Water usage (in millions of gallons) is expected to increase at the rate of $3 + 0.01x$ after x months. What will the water usage be a year from now?

44. Population Growth   There are currently 20,000 citizens of voting age in a small town. Demographics indicate that the voting population will change at the rate of $2.2t - 0.8t^2$ (in thousands of voting citizens), and t denotes time in years. How many citizens of voting age will there be 3 years from now?

45. Air Pollution   An environmental study of a certain town suggests that t years from now the level of carbon monoxide in the air will be changing at the rate of $0.2t + 0.2$ parts per million per year. If the current level of carbon monoxide in the air is 3.8 parts per million, what will the level be 5 years from now?

46. Chemical Reaction   The end product of a chemical reaction is produced at the rate of $(\sqrt{t} - 2)/t$ milligrams per minute. If the reaction started at time $t = 1$, determine the amount produced during the first 4 minutes.

47. Geometry   Determine the function whose graph has a tangent with a slope $2x + 1$ for each value of x, and whose graph passes through the point $(1, 3)$.

48. Geometry   Determine the function whose graph has a tangent with a slope $3x^2 - 2x + 1$ for each value of x, and whose graph passes through the point $(1, 6)$.

49. Free Fall   An object dropped from an airplane 3200 feet above the ground falls at the rate of

$$v(t) = 32t \text{ (feet per second)}$$

where t is given in seconds. If the object has fallen 576 feet after 6 seconds, how long after it is dropped will the object hit the ground?

50. Water Depletion   A water reservoir is being filled at the rate of 15,000 gallons per hour. Due to increased consumption, the water in the reservoir is decreasing at the rate of $\frac{5}{2}t$ gallons per hour at time t. When will the reservoir be empty if the initial water volume was 100,000 gallons?

51. Verify the following statements:

(a) $\displaystyle\int (x \cdot \sqrt{x})\, dx \neq \int x\, dx \cdot \int \sqrt{x}\, dx$

(b) $\displaystyle\int x(x^2 + 1)\, dx \neq x \int (x^2 + 1)\, dx$

(c) $\displaystyle\int \frac{x^2 - 1}{x - 1}\, dx \neq \frac{\displaystyle\int (x^2 - 1)\, dx}{\displaystyle\int (x - 1)\, dx}$

15.2 INTEGRATION BY SUBSTITUTION; INTEGRATION BY PARTS

Integration By Substitution

Indefinite integrals that cannot be evaluated by using Formulas (1)–(8) of Section 15.1 may sometimes be evaluated by the **substitution method.** This method involves the introduction of a function that changes the integrand into a form to which the formulas of Section 15.1 apply.

The basic idea behind integration by substitution is the Chain Rule. To see how integration by substitution works, consider the following example: Recall the Chain Rule for the derivatives as used in

$$\frac{d}{dx}(x^2 + 5)^4 = 4(x^2 + 5)^3(2x)$$

Note that the result in this example is a product of two functions, $4(x^2 + 5)^3$ and $2x$. This is often the result when we use the Chain Rule, where we get a product of two functions. Because of this, whenever the integral is formed by a product of two functions, we can sometimes integrate by using the Chain Rule in reverse. In the example above, working backwards from the derivatives gives

$$\int 4(x^2 + 5)^3\, 2x\, dx = (x^2 + 5)^4 + K$$

We may simplify the procedure used above by *changing the variables.* Introduce the variable u, defined as

$$u = x^2 + 5$$

The derivative $du/dx = 2x$ can be written in differential notation as

$$du = 2x\, dx$$

Now we express the integrand as a function of u and the differential dx as a function of u and du.

$$\int 4(x^2 + 5)^3 2x\, dx = \int 4u^3\, du = \frac{4}{4}u^4 + K = (x^2 + 5)^4 + K$$

$$\uparrow$$
$$u = x^2 + 5 \qquad\qquad u = x^2 + 5$$
$$du = 2x\, dx$$

Although we could have done the above integration by multiplying out the $(x^2 + 5)^3$, the substitution technique is easier. Sometimes substitution is the only method available.

EXAMPLE 1

Evaluate: $\int \dfrac{dx}{2x + 1}$

SOLUTION

We try the substitution $u = 2x + 1$ to see if it simplifies the integral. Then

$$\frac{du}{dx} = 2 \qquad \text{so} \qquad du = 2\, dx \qquad \text{and} \qquad \frac{du}{2} = dx$$

$$\int \frac{dx}{2x+1} = \int \frac{du/2}{u} = \frac{1}{2}\int \frac{du}{u} = \frac{1}{2}\ln|u| + K = \frac{1}{2}\ln|2x+1| + K$$

$$u = 2x + 1$$
$$\frac{du}{2} = dx$$
$$u = 2x + 1$$

Now Work Problem 1

· ·

EXAMPLE 2

Evaluate: $\int x\sqrt{x^2 + 1}\, dx$

SOLUTION

We try the substitution $u = x^2 + 1$ to see if it simplifies the integral. Then

$$\frac{du}{dx} = 2x \qquad \text{so} \qquad du = 2x\, dx \qquad \text{and} \qquad \frac{du}{2} = x\, dx$$

Now

$$\int x\sqrt{x^2 + 1}\, dx = \int \sqrt{x^2 + 1}\, x\, dx$$

$$= \int \sqrt{u}\,\frac{du}{2}$$

$$= \frac{1}{2}\int u^{1/2}\, du$$

$$= \frac{1}{2}\frac{u^{3/2}}{\frac{3}{2}} + K$$

$$= \frac{(x^2 + 1)^{3/2}}{3} + K$$

· ·

Note that in using the substitution $u = x^2 + 1$ in Example 2 we must substitute not only for the integrand $x\sqrt{x^2 + 1}$ but also for dx. In fact, it is the existence of x as part of the integrand that makes the substitution $u = x^2 + 1$ work. For example, if we try this same substitution to evaluate $\int \sqrt{x^2 + 1}\, dx$, we obtain

$$\int \sqrt{x^2 + 1}\, dx = \int \sqrt{u}\,\frac{du}{2\sqrt{u-1}} = \int \frac{\sqrt{u}}{2\sqrt{u-1}}\, du$$

In this case the substitution results in an integrand that is *more complicated* than the original one.

Thus, the idea behind the substitution method is to obtain an integral $\int h(u)\, du$ that is simpler than the original integral $\int f(x)\, dx$. When a substitution does not simplify the integral, other substitutions should be tried. If these do not work, other integration methods must be applied. Since integration, unlike differentiation, has no prescribed method, a lot of practice in integration is required.

Sometimes more than one substitution will work. In the next example we use two different substitutions to evaluate an integral.

EXAMPLE 3

Evaluate: $\displaystyle\int x\sqrt{4+x}\,dx$

SOLUTION

Substitution I   Let $u = 4 + x$. Then $du = dx$ and $x = u - 4$, so that

$$\int x\sqrt{4+x}\,dx = \int (u-4)\sqrt{u}\,du$$

$$= \int (u^{3/2} - 4u^{1/2})\,du$$

$$= \frac{u^{5/2}}{\frac{5}{2}} - \frac{4u^{3/2}}{\frac{3}{2}} + K$$

$$= \frac{2(4+x)^{5/2}}{5} - \frac{8(4+x)^{3/2}}{3} + K$$

Substitution II   Let $u^2 = 4 + x$. Then $2u\dfrac{du}{dx} = 1$,   so $2u\,du = dx$.   Since $x = u^2 - 4$, we have

$$\int x\sqrt{4+x}\,dx = \int (u^2-4)u\,2u\,du$$

$$= 2\int (u^4 - 4u^2)\,du$$

$$= \frac{2u^5}{5} - \frac{8u^3}{3} + K$$

$$= \frac{2(4+x)^{5/2}}{5} - \frac{8(4+x)^{3/2}}{3} + K$$

Now Work Problem 17

· ·

EXAMPLE 4

Evaluate: $\displaystyle\int \frac{dx}{2\sqrt{x}(1+\sqrt{x})^3}$

SOLUTION

We try the substitution

$$u = 1 + \sqrt{x}$$

Then

$$\frac{du}{dx} = \frac{d}{dx}(1+\sqrt{x}) = \frac{1}{2}x^{-1/2} = \frac{1}{2\sqrt{x}}$$

so that

$$du = \frac{dx}{2\sqrt{x}}$$

Now

$$\int \frac{dx}{2\sqrt{x}(1 + \sqrt{x})^3} = \int \frac{du}{u^3} = \int u^{-3}\, du = \frac{u^{-2}}{-2} + K$$

$$= \frac{-1}{2u^2} + K = \frac{-1}{2(1 + \sqrt{x})^2} + K$$

Now Work Problem 27

....

EXAMPLE 5

Evaluate: $\displaystyle\int \frac{dx}{x \ln x}$

SOLUTION

Let $u = \ln x$. Then $du = (1/x)\, dx$ and

$$\int \frac{dx}{x \ln x} = \int \frac{du}{u} = \ln |u| + K = \ln |\ln x| + K$$

....

A summary of the method of integration by substitution is given in the flowchart in Figure 5.

Figure 5

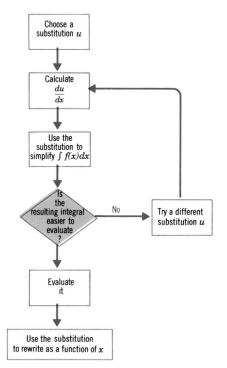

Integration By Parts

Next, we discuss a method for evaluating indefinite integrals such as

$$\int xe^x \, dx \qquad \text{and} \qquad \int \ln x \, dx$$

where the substitution technique does not work.

This method, called **integration by parts,** is based on the product rule for differentiation. Recall that if u and v are differentiable functions of x, then

$$\frac{d}{dx}(uv) = u\frac{dv}{dx} + v\frac{du}{dx}$$

By integrating both sides, we obtain

$$uv = \int u\frac{dv}{dx}\,dx + \int v\frac{du}{dx}\,dx$$

so that

$$\int u\frac{dv}{dx}\,dx = uv - \int v\frac{du}{dx}\,dx$$

In abbreviated form this formula may be written in the following way.

Integration by Parts Formula

$$\int u \, dv = uv - \int v \, du \qquad\qquad (1)$$

To use the integration by parts formula, we separate the integrand into two parts. We call one u and the other dv. We differentiate u to obtain du and integrate dv to obtain v. If we can then integrate $\int v \, du$, the problem is solved. The goal of this procedure, then, is to choose u and dv so that the term $\int v \, du$ is easier to solve than the original problem, $\int u \, dv$. As the examples will illustrate, this usually happens when u is simplified by differentiation.

EXAMPLE 6

Evaluate: $\displaystyle\int xe^x \, dx$

SOLUTION

To use the integration by parts formula, we choose u and dv so that

$$\int u \, dv = \int xe^x \, dx$$

and $\int v \, du$ is easier to evaluate than $\int u \, dv$. In this example we decide to choose

$$u = x \qquad \text{and} \qquad dv = e^x \, dx$$

As a result of this choice,

$$du = dx \quad \text{and} \quad v = \int dv = \int e^x \, dx = e^x$$

Note that we require only a particular antiderivative of dv at this state; we will add the constant of integration later. Substitution into (1) results in

$$\int \overbrace{x}^{u} \overbrace{e^x \, dx}^{dv} = \overbrace{x}^{u} \overbrace{e^x}^{v} - \int \overbrace{e^x}^{v} \overbrace{dx}^{du}$$

$$= xe^x - e^x + K$$
$$= e^x(x - 1) + K$$

Now Work Problem 35

• •

Let's look once more at Example 6. Suppose we had chosen u and dv differently:

$$u = e^x \quad \text{and} \quad dv = x \, dx$$

This choice would have resulted in

$$du = e^x \, dx \quad \text{and} \quad v = \frac{x^2}{2}$$

and Equation (1) would have yielded

$$\int xe^x \, dx = \frac{x^2}{2} e^x - \int \frac{x^2 e^x}{2} \, dx$$

As you can see, instead of obtaining an integral that is easier to evaluate, we obtain one that is more complicated than the original. This means that an unwise choice of u and dv has been made.

Unfortunately, there are no general directions for choosing u and dv except these:

Steps to Integrate by Parts

Step 1 dx is always a part of dv.

Step 2 It must be possible to integrate dv.

Step 3 u and dv are chosen so that $\int v \, du$ is easier to evaluate than the original integral $\int u \, dv$; this often happens when u is simplified by differentiation.

In making an initial choice for u and dv, a certain amount of trial and error is used. If a selection appears to hold little promise, abandon it and try some other choice. If no choices work, it may be that some other technique of integration should be tried.

Let's look at some more examples.

EXAMPLE 7

Evaluate: $\displaystyle\int x \ln x \, dx$

SOLUTION

We choose

$$u = \ln x \qquad \text{and} \qquad dv = x \, dx$$

Then

$$du = \frac{1}{x} \, dx \qquad \text{and} \qquad v = \frac{x^2}{2}$$

Now

$$\int x \ln x \, dx = \frac{x^2}{2} \ln x - \int \frac{x^2}{2} \frac{1}{x} \, dx$$

$$= \frac{1}{2} x^2 \ln x - \frac{1}{2} \int x \, dx$$

$$= \frac{1}{2} x^2 \ln x - \frac{1}{4} x^2 + K$$

$$= \frac{x^2}{2} \left(\ln x - \frac{1}{2} \right) + K$$

Now Work Problem 39

· ·

Sometimes it is necessary to integrate by parts more than once to solve a particular problem, as illustrated by the next example.

EXAMPLE 8

Evaluate: $\displaystyle\int x^2 e^x \, dx$

SOLUTION

Let

$$u = x^2 \qquad \text{and} \qquad dv = e^x \, dx$$

so that

$$du = 2x \, dx \qquad \text{and} \qquad v = \int e^x \, dx = e^x$$

Then

$$\int x^2 e^x \, dx = x^2 e^x - \int 2x e^x \, dx$$

Although we must still evaluate $\int xe^x \, dx$, we can see that the original integral has been replaced by a simpler one. In fact, in Example 6 we found (by using integration by parts) that

$$\int xe^x \, dx = e^x(x - 1) + K$$

Thus,

$$\int x^2 e^x \, dx = x^2 e^x - 2[e^x(x - 1)] + K$$
$$= x^2 e^x - 2xe^x + 2e^x + K$$
$$= e^x(x^2 - 2x + 2) + K$$

. . . .

EXAMPLE 9   Find all the antiderivatives of ln x.

SOLUTION

To find all the antiderivatives of ln x, namely ∫ ln x dx, we use the integration by parts formula. Taking $u = \ln x$ and $dv = dx$, we find $du = (1/x) \, dx$ and $v = x$. Then

$$\int \ln x \, dx = x \ln x - \int x \frac{1}{x} \, dx$$
$$= x \ln x - x + K$$
$$= x(\ln x - 1) + K$$

. . . .

The integration by parts formula is useful for the evaluation of indefinite integrals that have integrands composed of e^x times a polynomial function of x or ln x times a polynomial function of x. It can also be used for other types of indefinite integrals that will not be discussed in this book.

EXERCISE 15.2   Answers to odd-numbered problems begin on page AN-77.

In Problems 1–34 evaluate each indefinite integral. Use the substitution method.

1. $\int (2x + 1)^5 \, dx$

2. $\int (3x - 5)^4 \, dx$

3. $\int e^{2x-3} \, dx$

4. $\int e^{3x+4} \, dx$

5. $\int (-2x + 3)^{-2} \, dx$

6. $\int (5 - 2x)^{-3} \, dx$

7. $\int (x^2 + 4)^2 \, x \, dx$

8. $\int (x^2 - 2)^3 \, x \, dx$

9. $\int e^{x^3+1} x^2 \, dx$

10. $\int e^{2x^2+1} x \, dx$

11. $\int (e^x + e^{-x}) \, dx$

12. $\int (e^x - e^{-x}) \, dx$

13. $\int (x^3 + 2)^6 x^2 \, dx$

14. $\int (x^3 - 1)^4 x^2 \, dx$

15. $\int \frac{x}{\sqrt[3]{1 + x^2}} \, dx$

16. $\int \frac{x}{\sqrt[5]{1 - x^2}} \, dx$

17. $\int x\sqrt{x + 3} \, dx$

18. $\int x\sqrt{x - 3} \, dx$

19. $\displaystyle\int \frac{e^x}{e^x + 1}\,dx$

20. $\displaystyle\int \frac{e^{-x}}{e^{-x} + 4}\,dx$

21. $\displaystyle\int \frac{e^{\sqrt{x}}\,dx}{\sqrt{x}}$

22. $\displaystyle\int \frac{e^{\sqrt[3]{x}}\,dx}{x^{2/3}}$

23. $\displaystyle\int \frac{(x^{1/3} - 1)^6\,dx}{x^{2/3}}$

24. $\displaystyle\int \frac{(x^{1/3} + 2)^3}{x^{2/3}}\,dx$

25. $\displaystyle\int \frac{(x + 1)\,dx}{(x^2 + 2x + 3)^2}$

26. $\displaystyle\int \frac{(x + 4)\,dx}{(x^2 + 8x + 2)^3}$

27. $\displaystyle\int \frac{dx}{\sqrt{x}(1 + \sqrt{x})^4}$

28. $\displaystyle\int \frac{(3 - 2\sqrt{x})^2}{\sqrt{x}}\,dx$

29. $\displaystyle\int \frac{dx}{2x + 3}$

30. $\displaystyle\int \frac{dx}{3x - 5}$

31. $\displaystyle\int \frac{x\,dx}{4x^2 + 1}$

32. $\displaystyle\int \frac{x\,dx}{5x^2 - 2}$

33. $\displaystyle\int \frac{x + 1}{x^2 + 2x + 2}\,dx$

34. $\displaystyle\int \frac{2x - 1}{x^2 - x + 4}\,dx$

In Problems 35–48 evaluate each indefinite integral. Use integration by parts.

35. $\displaystyle\int xe^{2x}\,dx$

36. $\displaystyle\int xe^{-3x}\,dx$

37. $\displaystyle\int x^2 e^{-x}\,dx$

38. $\displaystyle\int x^2 e^{2x}\,dx$

39. $\displaystyle\int \sqrt{x}\ln x\,dx$

40. $\displaystyle\int x(\ln x)^2\,dx$

41. $\displaystyle\int (\ln x)^2\,dx$

42. $\displaystyle\int \frac{\ln x}{x^2}\,dx$

43. $\displaystyle\int x^2 \ln 3x\,dx$

44. $\displaystyle\int x^2 \ln 5x\,dx$

45. $\displaystyle\int x^2(\ln x)^2\,dx$

46. $\displaystyle\int x^3(\ln x)^2\,dx$

47. $\displaystyle\int \frac{\ln x}{x}\,dx$

48. $\displaystyle\int \sqrt{x}\,(\ln \sqrt{x})^2\,dx$

49. Use a substitution to verify the formula
$$\int (ax + b)^n\,dx = \frac{(ax + b)^{n+1}}{a(n + 1)} + K \qquad a \neq 0, \quad n \neq -1$$

50. Cost The marginal cost of production is found to be
$$C'(x) = 2x(x^2 + 20)^2$$
where x is the number of units produced. The fixed cost is $15,000. Find the cost function.

51. Revenue The marginal revenue (in thousands of dollars) from the sale of x tractors is
$$R'(x) = \frac{2x(x^2 + 10)^2}{1000}$$
Find the total revenue if the revenue from the sale of 4 tractors is $198,000.

52. Growth Work Force The number of employees $N(t)$ at the Ajex Steel Company is growing at a rate given by the equation
$$N'(t) = 20e^{0.01t} \quad \text{(people per year)}$$
The number of employees currently is 400.

(a) Find an equation for $N(t)$.
(b) How long will it take the work force to reach 800 employees?

53. Pollution An oil tanker is leaking oil and producing a circular oil slick that is growing at the rate of
$$R'(t) = \frac{30}{\sqrt{t + 4}} \qquad t \geq 0$$
where R is the radius in meters of the slick after t minutes. Find the radius of the slick after 21 minutes if the radius is 0 when $t = 0$.

15.3 THE DEFINITE INTEGRAL

We begin with an example illustrating the general idea of a *definite integral*.

EXAMPLE 1

The marginal cost of a certain firm is given by the equation

$$C'(x) = 4 - 0.2x \qquad 0 \le x \le 10$$

where C' is in units of thousands of dollars and the quantity x produced is in hundreds of units per day. If the number of units produced in a given day changes from 200 to 500 units, what is the change in cost?

SOLUTION

If C is the cost function, the change in cost from $x = 2$ to $x = 5$ is

$$C(5) - C(2)$$

This is the number we seek. The cost C is an antiderivative of $C'(x) = 4 - 0.2x$. Thus,

$$C(x) = \int C'(x)\, dx = \int (4 - 0.2x)\, dx = 4x - 0.1x^2 + K$$

We use this to compute $C(5) - C(2)$:

$$C(5) - C(2) = [4(5) - (0.1)(25) + K] - [4(2) - (0.1)(4) + K] = 9.9 \quad (1)$$

Thus, the change in cost is 9.9 thousand dollars.

· ·

In this example the change in C was computed by using an antiderivative of C', which is symbolized by $\int C'(x)\, dx$. To indicate that the change is from $x = 2$ to $x = 5$, we add to this notation as follows:

$$\text{Change in } C \text{ from 2 to 5} = \int_2^5 C'(x)\, dx$$

This form is called a **definite integral**.

DEFINITE INTEGRAL

The *definite integral* of a continuous function f from a to b is the difference

$$\int_a^b f(x)\, dx = F(b) - F(a) \tag{2}$$

where F is an antiderivative of f. That is, the definite integral is the net change in the antiderivative between $x = a$ and $x = b$.

Equation (2) shows the connection between antiderivatives and definite integrals. This definition provides us with a method for evaluating definite integrals.* In $\int_a^b f(x)\,dx$, the numbers a and b are called the **lower** and **upper limits of integration,** respectively.

EXAMPLE 2

The definite integral from 2 to 3 of $f(x) = x^2$ is computed by first finding an antiderivative of f. One such antiderivative is $F(x) = x^3/3$. Thus, the definite integral from 2 to 3 of x^2 is

$$F(3) - F(2) = \frac{27}{3} - \frac{8}{3} = \frac{19}{3}$$

That is,

$$\int_2^3 x^2\,dx = \frac{19}{3}$$

Now Work Problem 1

In computing $\int_a^b f(x)\,dx$, we find that the choice of an antiderivative of f does not matter. Look back at Equation (1) in the solution to Example 1. The constant K drops out. Now look at Example 2. If we had used $F(x) = x^3/3 + K$ as the antiderivative of x^2, we would have found that

$$\int_2^3 x^2\,dx = F(3) - F(2) = \left(\frac{27}{3} + K\right) - \left(\frac{8}{3} + K\right) = \frac{19}{3}$$

Again the constant K drops out. This will always be the case.

Any antiderivative of f can be used to evaluate $\int_a^b f(x)\,dx$.

For convenience we introduce the notation

$$F(x)\Big|_a^b = F(b) - F(a)$$

In terms of this new notation, to calculate $F(x)|_a^b$, first replace x by the upper limit b to obtain $F(b)$, and from this subtract $F(a)$, obtained by letting $x = a$.

EXAMPLE 3

(a) $\displaystyle\int_{-1}^{5} 6x\,dx = 3x^2\Big|_{-1}^{5} = 3(5)^2 - 3(-1)^2 = 75 - 3 = 72$

(b) $\displaystyle\int_{1}^{2} x^3\,dx = \frac{x^4}{4}\Big|_{1}^{2} = \frac{(2)^4}{4} - \frac{(1)^4}{4} = \frac{16}{4} - \frac{1}{4} = \frac{15}{4}$

*The justification for this statement is provided in Section 15.5.

It is important to distinguish between the indefinite integral and the definite integral. The indefinite integral, a symbol for all the antiderivatives of a function, is a function. On the other hand, the definite integral is a number.

EXAMPLE 4

Evaluate: $\displaystyle\int_{1}^{4} \sqrt{x}\, dx$

SOLUTION

An antiderivative of $\sqrt{x} = x^{1/2}$ is

$$\frac{x^{3/2}}{\frac{3}{2}} = \frac{2}{3}\, x^{3/2}$$

Thus,

$$\int_{1}^{4} \sqrt{x}\, dx = \frac{2}{3}\, x^{3/2}\,\Big|_{1}^{4} = \frac{2}{3}\,(4)^{3/2} - \frac{2}{3}\,(1)^{3/2} = \frac{16}{3} - \frac{2}{3} = \frac{14}{3}$$

Now Work Problem 5

· ·

Properties of the Definite Integral

We list some properties of definite integrals below.

Properties of Definite Integrals

If f is a continuous function that has an antiderivative on the interval $[a, b]$, then

$$\int_{a}^{b} f(x)\, dx = -\int_{b}^{a} f(x)\, dx \tag{3}$$

$$\int_{a}^{a} f(x)\, dx = 0 \tag{4}$$

EXAMPLE 5

(a) $\displaystyle\int_{4}^{1} \sqrt{x}\, dx = -\int_{1}^{4} \sqrt{x}\, dx = -\frac{14}{3}$ (b) $\displaystyle\int_{1}^{1} x\, dx = 0$

· ·

Formulas (3) and (4) are an immediate consequence of the definition of a definite integral. Specifically, if F is an antiderivative of f, then

$$\int_{a}^{b} f(x)\, dx = F(b) - F(a) = -[F(a) - F(b)] = -\int_{b}^{a} f(x)\, dx$$

and

$$\int_{a}^{a} f(x)\, dx = F(a) - F(a) = 0$$

Properties of Definite Integrals

If f is a continuous function that has an antiderivative on the interval $[a, b]$, and if c is between a and b, then

$$\int_a^b f(x)\, dx = \int_a^c f(x)\, dx + \int_c^b f(x)\, dx \tag{5}$$

If f is a continuous function that has an antiderivative on the interval $[a, b]$ and if c is a constant, then

$$\int_a^b cf(x)\, dx = c \int_a^b f(x)\, dx \tag{6}$$

EXAMPLE 6

(a) If $\int_1^3 f(x)\, dx = 5$ and $\int_3^6 f(x)\, dx = 7$, then

$$\int_1^6 f(x)\, dx = \int_1^3 f(x)\, dx + \int_3^6 f(x)\, dx = 5 + 7 = 12$$

(b) $\displaystyle\int_1^2 16x^2\, dx = 16 \int_1^2 x^2\, dx = 16 \left(\frac{x^3}{3} \Big|_1^2 \right)$

$$= 16 \left(\frac{8}{3} - \frac{1}{3} \right) = (16)\left(\frac{7}{3} \right) = \frac{112}{3}$$

· ·

Properties of Definite Integrals

If f and g are continuous functions that have antiderivatives on the interval $[a, b]$, then

$$\int_a^b [f(x) \pm g(x)]\, dx = \int_a^b f(x)\, dx \pm \int_a^b g(x)\, dx \tag{7}$$

EXAMPLE 7

$$\int_1^2 (x^2 + \sqrt{x})\, dx = \int_1^2 x^2\, dx + \int_1^2 \sqrt{x}\, dx$$

$$= \frac{x^3}{3} \Big|_1^2 + \frac{2}{3} x^{3/2} \Big|_1^2$$

$$= \frac{7}{3} + \frac{2}{3} (2\sqrt{2} - 1)$$

$$= \frac{4\sqrt{2}}{3} + \frac{5}{3}$$

· ·

EXAMPLE 8

Evaluate: $\displaystyle\int_1^2 3x(x^2 - 1)\,dx$

SOLUTION

$$\int_1^2 3x(x^2 - 1)\,dx = \int_1^2 (3x^3 - 3x)\,dx$$

$$= \int_1^2 3x^3\,dx - \int_1^2 3x\,dx$$

$$= 3\int_1^2 x^3\,dx - 3\int_1^2 x\,dx$$

$$= 3\left(\frac{x^4}{4}\Big|_1^2\right) - 3\left(\frac{x^2}{2}\Big|_1^2\right)$$

$$= 3\left(4 - \frac{1}{4}\right) - 3\left(2 - \frac{1}{2}\right)$$

$$= 3\left(\frac{15}{4}\right) - 3\left(\frac{3}{2}\right) = \frac{27}{4}$$

Now Work Problem 11

....

EXAMPLE 9

The marginal cost function for producing x units is $3x^2 - 200x + 1500$ dollars. Find the increase in cost if production is increased from 90 to 100 units.

SOLUTION

If C equals the cost of producing x units, then

$$C'(x) = 3x^2 - 200x + 1500$$

The increase in cost due to a production increase from 90 to 100 units is

$$C(100) - C(90) = \int_{90}^{100} C'(x)\,dx$$

$$= \int_{90}^{100} (3x^2 - 200x + 1500)\,dx$$

$$= \left(x^3 - 200\frac{x^2}{2} + 1500x\right)\Big|_{90}^{100}$$

$$= [1{,}000{,}000 - 100(10{,}000) + 1500(100)]$$
$$\quad - [(90)^3 - 100(8100) + 1500(90)]$$

$$= 96{,}000 \text{ dollars}$$

....

Application: The Learning Curve

Quite often, the managerial planning and control component of a production industry is faced with the problem of predicting labor time requirements and cost per unit of product. The tool used to achieve such predictions is the **learning curve.** The basic assumption made here is that, in certain production industries such as the assembling of televisions and cars, the worker learns from experience. As a result, the more often a worker repeats an operation, the more efficiently the job is performed. Hence, direct labor input per unit of product declines. If the *rate* of improvement is regular enough, the learning curve can be used to predict future reductions in labor requirements.

One function that might be used to model such a situation is

$$f(x) = cx^k$$

where $f(x)$ is the number of hours of direct labor required to produce the xth unit, $-1 \le k < 0$, and $c > 0$. The choice of x^k, with $-1 \le k < 0$, guarantees that, as the number x of units produced increases, the direct labor input decreases. See Figure 6.

The function $f(x) = cx^k$ describes a rate of learning per unit produced. This rate is measured in terms of labor-hours per unit. As Figure 6 illustrates, the number of direct labor-hours declines as more items are produced.

Once a learning curve has been determined for a gross production process, it can be used as a predictor to determine the number of production hours for future work.

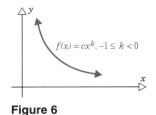

Figure 6

Learning Curves

For a learning curve $f(x) = cx^k$, the total number of labor-hours required to produce units numbered a through b is

$$N = \int_a^b f(x)\, dx = \int_a^b cx^k\, dx$$

EXAMPLE 10

The Ace Air Conditioning Company manufactures air conditioners on an assembly line. From experience, it was determined that the first 100 air conditioners required 1272 labor-hours. For each subsequent 100 air conditioners (1 unit), fewer labor-hours were required according to the learning curve

$$f(x) = 1272x^{-0.25}$$

where $f(x)$ is the rate of labor-hours required to assemble the xth unit (each unit being 100 air conditioners). This curve was determined after 30 units had been manufactured.

The company is in the process of bidding for a large contract involving 5000 additional air conditioners, or 50 additional units. The company can estimate the labor-hours required to assemble these units by evaluating

$$N = \int_{30}^{80} 1272x^{-0.25}\, dx$$
$$= \frac{1272x^{0.75}}{0.75}\bigg|_{30}^{80}$$
$$= 1696(80^{0.75} - 30^{0.75})$$
$$= 1696(26.75 - 12.82)$$
$$= 23{,}625$$

Now Work Problem 39

Thus, the company can bid, estimating the total labor-hours needed as 23,625.

••••••••••••••••••••••

Application: Annuities

An **annuity** is a sequence of equal periodic payments. For example, a 20-payment life insurance policy is an annuity in which 20 equal premiums, or payments, are paid on an annual basis, earning a fixed interest rate. In this section we discuss the situation in which the interest earned is compounded continuously. The **amount of an annuity** is the sum of all payments made, plus all interest accumulated.

> **Amount of Annuity with Continuous Compounding of Equal Annual Payments**
>
> If an annuity consists of equal annual payments P in which an interest rate of $r\%$ per annum is compounded continuously, the amount A of the annuity after N payments is
>
> $$A = \int_0^N Pe^{rt}\, dt$$

EXAMPLE 11

A savings and loan association pays 6% per annum compounded continuously. If a person places $1000 in a savings account at the beginning of each year, how much will be in the account after 3 years?

SOLUTION

Here, $P = 1000$, $N = 3$, and $r = 0.06$. The amount A after 3 years is

$$A = \int_0^3 1000e^{0.06t}\, dt$$

$$= \frac{1000}{0.06} e^{0.06t}\Big|_0^3$$

$$= \frac{1000}{0.06}(e^{0.18} - 1)$$

$$= \$3286.96$$

Now Work Problem 41

•••••••••••••••••••••••

Application: Rate of Sales

> **Total Sales over Time**
>
> When the rate of sales of a product is a known function, say $f(t)$, where t is the time, the total sales of this product over a time period T are
>
> $$\text{Total sales over time } T = \int_0^T f(t)\, dt$$

For example, suppose the rate of sales of a new product is given by

$$f(t) = 100 - 90e^{-t}$$

where t is the number of days the product is on the market. The total sales during the first 4 days are

$$
\begin{aligned}
\int_0^4 f(t)\, dt &= \int_0^4 (100 - 90e^{-t})\, dt \\
&= (100t + 90e^{-t})\Big|_0^4 \\
&= 400 + 90e^{-4} - 90 \\
&= 311.6 \text{ units}
\end{aligned}
$$

EXAMPLE 12

A company has current sales of $1,000,000 per month, and profit to the company averages 10% of sales. The company's past experience with a certain advertising strategy is that sales will increase by 2% per month over the length of the advertising campaign (12 months). The company now needs to decide whether to embark on a similar campaign that will have a total cost of $130,000. The decision will be yes, provided the increase in sales due to the campaign results in profits that exceed $13,000. (This is a 10% return on the advertising investment of $130,000.)

SOLUTION

The monthly rate of sales during the advertising campaign obeys a growth curve of the form

$$\$1,000,000 e^{0.02t}$$

where t is measured in months. The total sales after 12 months (the length of the campaign) are

$$
\text{Total sales} = \int_0^{12} 1,000,000 e^{0.02t}\, dt = \frac{1,000,000 e^{0.02t}}{0.02}\bigg|_0^{12}
$$

$$= 50,000,000(e^{0.24} - 1) = \$13,562,458$$

The profit to the company is 10% of sales, so that the profit due to the increase in sales is

$$0.10(13,562,458 - 12,000,000) = \$156,246$$

This $156,246 profit was achieved through the expenditure of $130,000 in advertising. Thus, the advertising yielded a true profit of

$$\$156,246 - \$130,000 = \$26,246$$

Since this represents more than a 10% return on the cost of the advertising, the company should proceed with the advertising campaign.

....

EXERCISE 15.3   Answers to odd-numbered problems begin on page AN-78.

In Problems 1–34 evaluate each definite integral.

1. $\displaystyle\int_{1}^{2} (3x - 1)\, dx$

2. $\displaystyle\int_{1}^{2} (2x + 1)\, dx$

3. $\displaystyle\int_{0}^{1} (3x^2 + e^x)\, dx$

4. $\displaystyle\int_{-2}^{0} (e^x + x^2)\, dx$

5. $\displaystyle\int_{0}^{1} \sqrt{u}\, du$

6. $\displaystyle\int_{1}^{4} \sqrt{u}\, du$

7. $\displaystyle\int_{0}^{1} (t^2 - t^{3/2})\, dt$

8. $\displaystyle\int_{1}^{4} (\sqrt{x} - 4x)\, dx$

9. $\displaystyle\int_{-2}^{3} (x - 1)(x + 3)\, dx$

10. $\displaystyle\int_{0}^{1} (z^2 + 1)^2\, dz$

11. $\displaystyle\int_{1}^{2} \frac{x^2 - 1}{x^4}\, dx$

12. $\displaystyle\int_{1}^{3} \frac{2 - x^2}{x^4}\, dx$

13. $\displaystyle\int_{1}^{4} \left(\sqrt[5]{t^2} + \frac{1}{t}\right) dt$

14. $\displaystyle\int_{1}^{4} \left(\sqrt{u} + \frac{1}{u}\right) du$

15. $\displaystyle\int_{1}^{4} \frac{x + 1}{\sqrt{x}}\, dx$

16. $\displaystyle\int_{1}^{9} \frac{\sqrt{x} + 1}{x^2}\, dx$

17. $\displaystyle\int_{3}^{3} (5x^4 + 1)^{3/2}\, dx$

18. $\displaystyle\int_{-1}^{1} (x + 1)^3\, dx$

19. $\displaystyle\int_{-1}^{1} (x + 1)^2\, dx$

20. $\displaystyle\int_{-1}^{-1} \sqrt[3]{x^2 + 4}\, dx$

21. $\displaystyle\int_{1}^{e} \left(x - \frac{1}{x}\right) dx$

22. $\displaystyle\int_{1}^{e} \left(x + \frac{1}{x}\right) dx$

23. $\displaystyle\int_{0}^{1} e^{-x}\, dx$

24. $\displaystyle\int_{0}^{1} x^2 e^{x^3}\, dx$

25. $\displaystyle\int_{1}^{3} \frac{dx}{x + 1}$

26. $\displaystyle\int_{-2}^{2} e^{-7x/2}\, dx$

27. $\displaystyle\int_{0}^{1} \frac{\sqrt{x}}{x^{3/2} + 1}\, dx$

28. $\displaystyle\int_{2}^{3} \frac{dx}{x \ln x}$

29. $\displaystyle\int_{1}^{3} x e^{2x}\, dx$

30. $\displaystyle\int_{0}^{4} (1 + x e^{-x})\, dx$

31. $\displaystyle\int_{1}^{2} x e^{-3x}\, dx$

32. $\displaystyle\int_{1}^{3} x^2 \ln x\, dx$

33. $\displaystyle\int_{1}^{5} \ln x\, dx$

34. $\displaystyle\int_{1}^{2} x \ln x\, dx$

35. A continuous function f is an **even function** if $f(-x) = f(x)$. It can be shown that if f is an even function, then

$$\int_{-a}^{a} f(x)\, dx = 2 \int_{0}^{a} f(x)\, dx \qquad a > 0$$

Verify the above formula by evaluating the following definite integrals:

(a) $\displaystyle\int_{-1}^{1} x^2\, dx$

(b) $\displaystyle\int_{-1}^{1} (x^4 + x^2)\, dx$

36. A continuous function f is an **odd function** if $f(-x) = -f(x)$. It can be shown that if f is an odd function, then

$$\int_{-a}^{a} f(x)\, dx = 0 \qquad a > 0$$

Verify the above formula by evaluating the following definite integrals:

(a) $\displaystyle\int_{-1}^{1} x\, dx$

(b) $\displaystyle\int_{-1}^{1} x^3\, dx$

37. The marginal cost function for producing x units is $6x^2 - 100x + 1000$ dollars. Find the increase in cost if production is increased from 100 to 110 units.

38. The marginal revenue function for selling x units is $10 - 4x$. Find the increase in revenue if selling is increased from 10 to 12 units.

39. Learning Curve

(a) Rework Example 10 for the learning curve $f(x) = 1272x^{-0.35}$.

(b) Rework Example 10 for the learning curve $f(x) = 1272x^{-0.15}$.

(c) Based on the answers to (a) and (b), explain the role of k in the learning curve $f(x) = cx^k$.

40. Learning Curve

(a) Rework Example 10 for the learning curve $f(x) = 1500x^{-0.25}$. [This means it was determined that the first 100 air conditioners (1 unit) required 1500 labor-hours.]

(b) Rework Example 10 for the learning curve $f(x) = 1000x^{-0.25}$.

(c) Based on the answers to (a) and (b), explain the role of c in the learning curve $f(x) = cx^k$.

41. Annuity If \$500 is deposited each year in a savings account paying 5.5% per annum compounded continuously, how much is in the account after 4 years?

42. Annuity If \$1200 is deposited each year in a savings account paying 5% per annum compounded continuously, how much is in the account after 3 years?

43. Annuity How much needs to be saved each year in a savings account paying 6% per annum compounded continuously in order to accumulate \$6000 in 3 years?

44. Annuity Answer Problem 43 if the rate of interest is 8%.

45. Rate of Sales The rate of sales of a new product is given by

$$f(x) = 1200 - 950e^{-x}$$

where x is the number of months the product is on the market. Find the total sales during the first year.

46. Rate of Sales A company whose annual sales are currently \$300,000 has been experiencing sales increases of 10% per year. Assuming this rate of growth continues, what will annual sales be in 4 years?

47. Rate of Sales In Example 12 what decision should the company make if sales due to advertising increase by only 1.5% per month?

48. Rate of Sales In Example 12 what decision should the company make if sales due to advertising increase by only 1% per month?

49. Learning Curve After producing 35 units, a company determines that its production facility is following a learning curve of the form $f(x) = 1000x^{-0.5}$, where $f(x)$ is the rate of labor-hours required to assemble the xth unit. How many total labor-hours should the company estimate are required to produce an additional 25 units?

50. Learning Curve Tina's Auto Shop has found that, after tuning up 50 cars, a learning curve of the form $f(x) = 1000x^{-1}$ is being followed. How many total labor-hours should the shop estimate are required to tune up an additional 50 cars?

15.4 AREA UNDER A GRAPH

The development of the integral, like that of the derivative, was originally motivated to a large extent by attempts to solve a basic problem in geometry—namely, the **area problem.** The question is "Given a nonnegative function f, whose domain is the closed interval $[a, b]$, what is the area enclosed by the graph of f, the x-axis, and the vertical lines $x = a$ and $x = b$?" Figure 7 illustrates the area to be found.

Figure 7

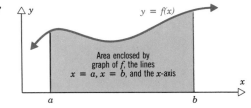

In plane geometry we learn how to find the area of certain geometric figures, such as squares, rectangles, and circles. For example, the area of a square with a side of length

3 feet is 9 square feet. The reason is that the square can be subdivided into nine smaller squares, each having sides of length 1 foot.

We also know that the area of a rectangle with length a units and width b units is ab square units.

Properties of Area

All area problems have certain features in common. For example, whenever the area of an object is computed, it is expressed as a number of square units; this number is never negative. *Thus, one property of area is that it is nonnegative.*

Consider the trapezoid shown in Figure 8. This trapezoid has been decomposed into two nonoverlapping geometric figures, a triangle (with area A_1) and a rectangle (with area A_2). Clearly, the area of the trapezoid is the sum $A_1 + A_2$ of the two component areas. Thus, as long as two regions do not overlap (except perhaps for a common boundary), the total area can be found by adding the component areas. We sometimes call this the **additive property of area.**

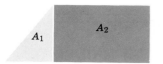

Figure 8

Properties of Area

Two properties of area are

I. Area ≥ 0

II. If A and B are two nonoverlapping regions with areas that are known, then

Total area of A and B = Area of A + Area of B

The above two properties enable us to compute the areas of a wide variety of regions. However, we still are not able to calculate the area of a region enclosed by an arbitrary graph. For example, the problem of determining the area "under the graph of $f(x) = x^2$ from $x = 0$ to $x = 1$," that is, the area of the region enclosed by $f(x) = x^2$, the x-axis, and the vertical lines $x = 0$ and $x = 1$, cannot be solved by using the methods of plane geometry. See Figure 9.

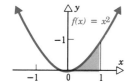

Figure 9

Evaluating Area

The next result gives a technique for evaluating areas such as the shaded region shown in Figure 9.

Area under a Graph

Suppose $y = f(x)$ is a continuous function defined on a closed interval I and $f(x) \geq 0$ for all points x in I. Then, for $a < b$ in I, the definite integral

$$\int_a^b f(x)\, dx \qquad (1)$$

is the area under the graph of $y = f(x)$ and above the x-axis between the lines $x = a$ and $x = b$.

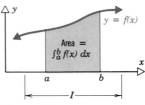

Figure 10

Figure 10 illustrates the above statement. A proof is given at the end of this section.

We are now able to find the area under the graph of $y = f(x)$, provided three conditions are met:

1. f is continuous on $[a, b]$.
2. f is nonnegative on $[a, b]$, that is, $f(x) \geq 0$ for $a \leq x \leq b$.
3. An antiderivative for f can be found.

We can now solve the area problem illustrated in Figure 9.

EXAMPLE 1

Find the area enclosed by $f(x) = x^2$, the x-axis, $x = 0$, and $x = 1$.

SOLUTION

The area we seek is given by the definite integral

$$\int_0^1 x^2 \, dx = \left.\frac{x^3}{3}\right|_0^1 = \frac{1}{3}$$

Thus, the area illustrated in Figure 9 is $\frac{1}{3}$ square unit.

Now Work Problem 1

· ·

Suppose a function f is continuous on the interval I, $a \leq x \leq b$, and has an antiderivative on I, but is negative for $a \leq x \leq c$ and is positive for $c \leq x \leq b$. How, in this situation, do we compute the area enclosed by $y = f(x)$, the x-axis, $x = a$, and $x = b$? See Figure 11.

Figure 11

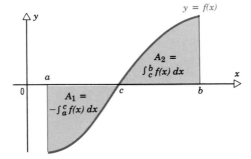

Notice in Figure 11 that the area A in question is composed of two nonoverlapping areas, A_1 and A_2, so that, by the additive property of area,

$$A = A_1 + A_2$$

Also, we know that on the interval $[c, b]$, the function is nonnegative, so that

$$A_2 = \int_c^b f(x) \, dx$$

To find the area A_1, we note that, since $f(x) \leq 0$ on $a \leq x \leq c$, then $-f(x) \geq 0$, and, by symmetry, the area A_1 equals

$$A_1 = \int_a^c [-f(x)] \, dx = - \int_a^c f(x) \, dx$$

The total area A we seek is, therefore,

$$A = A_1 + A_2 = - \int_a^c f(x) \, dx + \int_c^b f(x) \, dx$$

The next example illustrates this procedure for calculating area.

EXAMPLE 2

Find the area enclosed by $f(x) = x^3$, the x-axis, $x = -1$, and $x = \frac{1}{2}$.

SOLUTION

The desired area is indicated by the shaded region in Figure 12. Notice that it is composed of two regions: A_1, in which $f(x) < 0$ over the interval $-1 \leq x < 0$; and A_2, in which $f(x) \geq 0$ over the interval $0 \leq x \leq \frac{1}{2}$. To solve the problem, we use the additive property of area. Since $f(x) < 0$ for $-1 \leq x < 0$,

$$A_1 = - \int_{-1}^0 x^3 \, dx = - \frac{x^4}{4} \Big|_{-1}^0 = \frac{1}{4}$$

For the area A_2 we have

$$A_2 = \int_0^{1/2} x^3 \, dx = \frac{x^4}{4} \Big|_0^{1/2} = \frac{1}{64}$$

Figure 12

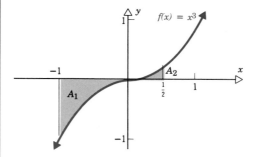

The total area A (since the regions do not overlap) is

$$A = A_1 + A_2 = \frac{1}{4} + \frac{1}{64} = \frac{17}{64}$$

Example 2 illustrates the necessity of graphing the function before any attempt is made to compute the area. In subsequent examples we shall always graph the function before doing anything else.

EXAMPLE 3

Find the area enclosed by $f(x) = x^2 - 4$ and the x-axis from $x = 0$ to $x = 4$.

SOLUTION

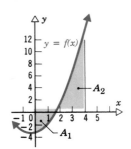

Figure 13

On the interval $0 \leq x \leq 4$, the graph crosses the x-axis at $x = 2$ since $f(2) = 0$. Also, $f(x) < 0$ from $x = 0$ to $x = 2$, and $f(x) > 0$ from $x = 2$ to $x = 4$. The areas A_1 and A_2 as depicted in Figure 13 are

$$A_1 = -\int_0^2 (x^2 - 4)\, dx = -\left(\frac{x^3}{3} - 4x\right)\Big|_0^2 = -\left(\frac{8}{3} - 8\right) = \frac{16}{3}$$

$$A_2 = \int_2^4 (x^2 - 4)\, dx = \left(\frac{x^3}{3} - 4x\right)\Big|_2^4 = \left(\frac{64}{3} - 16\right) - \left(\frac{8}{3} - 8\right)$$

$$= \frac{56}{3} - 8 = \frac{32}{3}$$

The total area A is therefore

$$A = A_1 + A_2 = \frac{16}{3} + \frac{32}{3} = \frac{48}{3} = 16$$

....

Area Enclosed by Two Graphs

The next example illustrates how to find the area enclosed by the graphs of two functions.

EXAMPLE 4

Find the area enclosed by the graphs of the functions

$$f(x) = 2x^2 \quad \text{and} \quad g(x) = 2x + 4$$

SOLUTION

First, we graph each of the functions, as shown in Figure 14.

The area to be calculated (the shaded portion of Figure 14) lies under the graph of the line $g(x) = 2x + 4$ and above the graph of $f(x) = 2x^2$. To find this area, we first need to find the numbers x at which the graphs intersect, that is, all numbers x for which $2x^2 = 2x + 4$. The solutions of this equation are obtained as follows:

$$2x^2 - 2x - 4 = 0$$
$$x^2 - x - 2 = 0$$
$$(x + 1)(x - 2) = 0$$
$$x + 1 = 0 \qquad x - 2 = 0$$
$$x = -1 \qquad x = 2$$

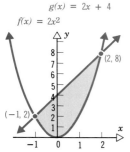

$g(x) = 2x + 4$

$f(x) = 2x^2$

$(2, 8)$

$(-1, 2)$

Figure 14

Now Work Problem 11

Thus, the points of intersection of the two graphs are $(-1, 2)$ and $(2, 8)$, as shown in Figure 14.

From Figure 14 we can see that if we subtract the area under $f(x) = 2x^2$, between $x = -1$ and $x = 2$, from the area under $g(x) = 2x + 4$, between $x = -1$ and $x = 2$, we will have the area A we seek. Thus,

$$A = \int_{-1}^{2} g(x)\, dx - \int_{-1}^{2} f(x)\, dx$$

$$= \int_{-1}^{2} [g(x) - f(x)]\, dx$$

$$= \int_{-1}^{2} [(2x + 4) - 2x^2]\, dx$$

$$= \left(x^2 + 4x - \frac{2x^3}{3} \right)\Big|_{-1}^{2}$$

$$= (4 + 8 - \tfrac{16}{3}) - (1 - 4 + \tfrac{2}{3}) = 9$$

The technique used in Example 4 can be used whenever we are asked to determine the area enclosed by the graphs of two continuous nonnegative functions f and g from $x = a$ to $x = b$.

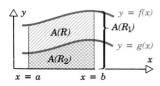

$y = f(x)$

$A(R_1)$

$A(R)$

$y = g(x)$

$A(R_2)$

$x = a$ $x = b$

Figure 15

Suppose, as depicted in Figure 15, $f(x) \geq g(x) \geq 0$ for x in $[a, b]$, and we wish to determine the area enclosed by f, g, $x = a$, and $x = b$. If we denote this area by $A(R)$, the area under f by $A(R_1)$, and the area under g by $A(R_2)$, then

$$A(R) = A(R_1) - A(R_2)$$

$$= \int_{a}^{b} f(x)\, dx - \int_{a}^{b} g(x)\, dx$$

$$= \int_{a}^{b} [f(x) - g(x)]\, dx$$

The next example illustrates this formula.

EXAMPLE 5

Find the area enclosed by the graphs of the functions

$$f(x) = 10x - x^2 \qquad \text{and} \qquad g(x) = 30 - 3x$$

SOLUTION

First, we graph the two functions. See Figure 16 (on the next page). The points of intersection of the two graphs were obtained by finding all numbers x for which

$$g(x) = f(x)$$

$$30 - 3x = 10x - x^2$$

$$x^2 - 13x + 30 = 0$$

$$(x - 3)(x - 10) = 0$$

$$x - 3 = 0 \qquad\qquad x - 10 = 0$$

$$x = 3 \qquad\qquad x = 10$$

Figure 16

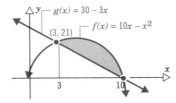

Thus, the points where the two curves meet are $(3, 21)$ and $(10, 0)$. We also see that for $3 \le x \le 10$,

$$f(x) \ge g(x) \ge 0$$

Thus, the required area, indicated by the shaded portion in Figure 16, is

$$\int_3^{10} [(10x - x^2) - (30 - 3x)] \, dx = \int_3^{10} [-x^2 + 13x - 30] \, dx$$

$$= \left(\frac{-x^3}{3} + \frac{13x^2}{2} - 30x \right) \Bigg|_3^{10} = \frac{343}{6}$$

· ·

Remember, when computing area by using the formula $\int_a^b [f(x) - g(x)] \, dx$, it must be true that $f(x) \ge g(x)$ on $[a, b]$. If this condition is not met, break the area up into pieces on which the inequality does hold and compute each one separately.

EXAMPLE 6

Find the area enclosed by the graphs of the functions

$$f(x) = x^3 \qquad \text{and} \qquad g(x) = -x^2 + 2x$$

SOLUTION

First, we graph the two functions. See Figure 17.
  The points of intersection of the two graphs are found by solving the equation

$$f(x) = g(x)$$
$$x^3 = -x^2 + 2x$$
$$x^3 + x^2 - 2x = 0$$
$$x(x^2 + x - 2) = 0$$
$$x(x + 2)(x - 1) = 0$$
$$x = 0 \qquad x = -2 \qquad x = 1$$

The points of intersection are $(0, 0)$, $(-2, -8)$, and $(1, 1)$. Notice that $f(x) \ge g(x)$ on the interval $-2 \le x \le 0$, while $g(x) \ge f(x)$ on the interval $0 \le x \le 1$. To find the

Figure 17

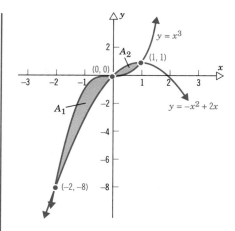

area A enclosed by the graphs of f and g, we compute the areas A_1 and A_2. Thus,

$$A = A_1 + A_2 = \int_{-2}^{0} [x^3 - (-x^2 + 2x)] \, dx + \int_{0}^{1} [(-x^2 + 2x) - x^3] \, dx$$

$$= \int_{-2}^{0} (x^3 + x^2 - 2x) \, dx + \int_{0}^{1} (-x^3 - x^2 + 2x) \, dx$$

$$= \left(\frac{x^4}{4} + \frac{x^3}{3} - x^2 \right) \Big|_{-2}^{0} + \left(\frac{-x^4}{4} - \frac{x^3}{3} + x^2 \right) \Big|_{0}^{1}$$

$$= 0 - \left(4 - \frac{8}{3} - 4 \right) + \left(-\frac{1}{4} - \frac{1}{3} + 1 \right) - 0$$

$$= \frac{8}{3} + \frac{5}{12} = \frac{37}{12}$$

Now Work Problem 25

· ·

Justification of Area as a Definite Integral

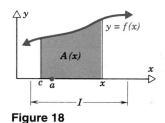

Figure 18

Look at Figure 18. Choose a number c in I so that $c < a$. Suppose x in I is an arbitrary number for which $x > c$. Let $A(x)$ denote the area enclosed by $y = f(x)$ and the x-axis from c to x. We want to show that $A'(x) = f(x)$ for all x in I, $x > c$.

Now choose $h > 0$ so that $x + h$ is in I. Then $A(x + h)$ is the area enclosed by $y = f(x)$ and the x-axis from c to $x + h$. See Figure 19 (on the next page). The difference $A(x + h) - A(x)$ is just the area enclosed by $y = f(x)$ and the x-axis from x to $x + h$. See Figure 20.

Next, we construct a rectangle with base h and area $A(x + h) - A(x)$. The height of the rectangle is then

$$\frac{A(x + h) - A(x)}{h}$$

Now we superimpose this rectangle on Figure 20 to obtain Figure 21.

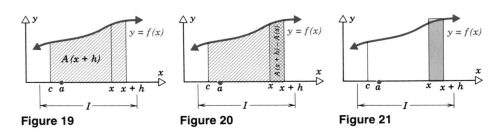

Figure 19   **Figure 20**   **Figure 21**

Since $y = f(x)$ is assumed to be a continuous function, and since both the rectangle and the shaded area have the same base and the same area, the upper edge of the rectangle must cross the graph of $y = f(x)$.

As we let $h \to 0^+$, the height of the rectangle tends to $f(x)$; that is,

$$\frac{A(x + h) - A(x)}{h} \to f(x) \qquad \text{as} \qquad h \to 0^+$$

A similar argument applies if we choose $h < 0$ and let $h \to 0^-$. Thus,

$$\lim_{h \to 0} \frac{A(x + h) - A(x)}{h} = f(x)$$

The limit above is the derivative of A. Thus,

$$A'(x) = f(x)$$

Since the choice of x is arbitrary (except for the condition that $x > c$), it follows that

$$A'(x) = f(x) \qquad \text{for all } x \text{ in } I, x > c$$

In other words, we have shown that the area A is an antiderivative of f on I. Hence,

$$\int_a^b f(x)\, dx = A(x) \Big|_a^b = A(b) - A(a)$$

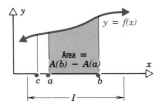

Figure 22

But the area we want to find is the area enclosed by $y = f(x)$ and the x-axis from a to b. Since a and b are in I, this is the quantity $A(b) - A(a)$. See Figure 22.

Hence,

$$\text{Area enclosed by } y = f(x) \text{ and the } x\text{-axis from } a \text{ to } b \text{ is} \quad \int_a^b f(x)\, dx$$

Application: Consumer's Surplus; Producer's Surplus

Suppose the price p a consumer is willing to pay for a quantity x of a particular commodity is governed by the demand curve

$$p = D(x)$$

In general the demand function D is a decreasing function, indicating that, as the price of the commodity increases, the quantity the consumer is willing to buy declines.

Suppose the price p that a producer is willing to charge for a quantity x of a particular commodity is governed by the supply curve

$$p = S(x)$$

In general the supply function S is an increasing function since, as the price p of a commodity increases, the more the producer is willing to supply the commodity.

The point of intersection of the demand curve and the supply curve is called the **equilibrium point** E. If the coordinates of E are (q, p), then p, the **market price,** is the price a consumer is willing to pay for—and a producer is willing to sell at—a quantity q, the **demand level,** of the commodity. See Figure 23.

Figure 23

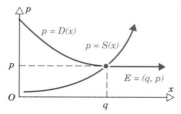

The total revenue of the producer at a market price p and a demand level q is pq (the price per unit times the number of units). This revenue can be interpreted geometrically as the area of the rectangle $OpEq$ in Figure 23.

In a free market economy there are times when some consumers would be willing to pay more for a commodity than the market price p that they actually do pay. The benefit of this to consumers—that is, the difference between what consumers *actually* paid and what they were *willing* to pay—is called the **consumer's surplus CS.** To obtain a formula for consumer's surplus CS, we use Figure 24 as a guide.

The quantity $\int_0^q D(x)\, dx$ is the area under the demand curve $D(x)$ from $x = 0$ to $x = q$ and represents the total revenue that would have been generated by the willingness of some consumers to pay more. By subtracting pq (the revenue actually achieved), the result is a surplus CS to the consumer. Thus, we have the formula

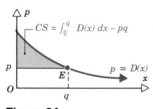

Figure 24

$$CS = \int_0^q D(x)\, dx - pq \tag{2}$$

In a free market economy there are also times when some producers would be willing to sell at a price below the market price p that the consumer actually pays. The benefit of this to the producer—that is, the difference between the revenue producers *actually* receive and what they would have been willing to receive—is called the **producer's surplus PS.** To obtain a formula for PS, we use Figure 25 as a guide.

The quantity $\int_0^q S(x)\, dx$ is the area under the supply curve $S(x)$ from $x = 0$ to $x = q$ and represents the total revenue that would have been generated by some producer's willingness to sell at a lower price. If we subtract this amount from pq (the revenue actually achieved), the result is a surplus to the producer, PS. Thus, the formula for PS is

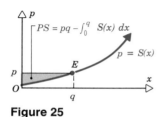

Figure 25

$$PS = pq - \int_0^q S(x)\, dx \tag{3}$$

Example 7 illustrates a situation in which both the supply and demand curves are linear.

EXAMPLE 7 Find CS and PS for the demand curve

$$D(q) = 18 - 3q$$

and the supply curve

$$S(q) = 3q + 6$$

where

$$p = D(q) = S(q)$$

SOLUTION

We first determine the equilibrium point E by solving the equation

$$D(q) = S(q)$$
$$18 - 3q = 3q + 6$$
$$6q = 12$$
$$q = 2$$

To find p, we compute $D(q)$ or $S(q)$:

$$p = D(q) = D(2) = 18 - 6 = 12$$

To find CS and PS, we use formulas (2) and (3):

$$CS = \int_0^2 (18 - 3q)\, dq - (2)(12) = \left(18q - \frac{3q^2}{2}\right)\bigg|_0^2 - 24$$
$$= 36 - 6 - 24 = 6$$

$$PS = (2)(12) - \int_0^2 (3q + 6)\, dq = 24 - \left(\frac{3q^2}{2} + 6q\right)\bigg|_0^2$$
$$= 24 - (6 + 12) = 6$$

Thus, in this example the consumer's surplus and producer's surplus each equal \$6.

Now Work Problem 29

· ·

Application: Maximizing Profit Over Time

The model introduced here concerns business operations of a special character. In oil drilling, mining, and other depletion operations, the initial revenue rate is generally higher than the revenue rate after a period of time has passed. That is, revenue rate, as a function of time, is a decreasing function (this is because depletion is occurring).

The cost rate of such operations generally increases with time because of inflation and other reasons. That is, cost rate, as a function of time, is an increasing function. The problem that management faces is to determine the time t_{max} that maximizes the profit function $P = P(t)$.

To construct a model, we denote the cost function by $C = C(t)$ and the revenue function by $R = R(t)$, where t denotes time. This representation of cost and revenue deviates from the usual economic definitions of cost per unit times number of units, and price per unit times number of units. The derivatives C' and R', taken with respect to time, represent cost and revenue as time rates. Furthermore, we make the natural assumption that the revenue rate, say, dollars per week, is greater than the cost rate at the beginning of the business operation under consideration. Also, as time goes on, we assume that the cost rate increases to the revenue rate and thereafter exceeds it. The optimum time at which the business operation should terminate is that point in time where the rates are equal. That is, the optimum time t_{max} obeys

$$C'(t_{max}) = R'(t_{max})$$

The profit rate P' is the difference between the revenue rate and the cost rate. That is,

$$P'(t) = R'(t) - C'(t)$$

Hence,

$$P(t) - P(0) = \int_0^t [R'(t) - C'(t)]\, dt$$

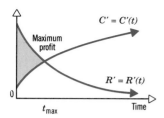

Figure 26

Since $P'(t_{max}) = R'(t_{max}) - C'(t_{max}) = 0$, the maximum profit is obtained when $t = t_{max}$. Thus, the maximum profit is $P(t_{max})$. Geometrically, the maximum profit $P(t_{max})$ is the area enclosed by the graphs of C' and R' from $t = 0$ to $t = t_{max}$. See Figure 26.

Notice that in Figure 26 the revenue rate function obeys the assumptions made in constructing the model: it is decreasing and it is very high initially. Also, the cost rate function is increasing and is concave down, indicating that the cost rate eventually levels off.

EXAMPLE 8

The G-B Oil Company's revenue rate (in millions of dollars per year) at time t years is

$$R'(t) = 9 - t^{1/3}$$

and the corresponding cost rate function (also in millions of dollars) is

$$C'(t) = 1 + 3t^{1/3}$$

Determine how long the oil company should continue to operate and what the total profit will be at the end of the operation.

SOLUTION

Recall that the time t_{max} of optimal termination is found when

$$R'(t) = C'(t)$$
$$9 - t^{1/3} = 1 + 3t^{1/3}$$
$$8 = 4t^{1/3}$$
$$2 = t^{1/3}$$
$$t_{max} = 8 \text{ years}$$

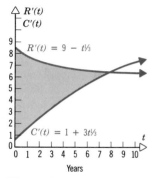

Figure 27

The revenue and cost rate functions are given in Figure 27.

At $t_{max} = 8$, both revenue and cost rates are 7 million dollars per year. The profit $P(t_{max})$ is

$$P(t_{max}) = \int_0^8 [R'(t) - C'(t)] \, dt$$

$$= \int_0^8 [(9 - t^{1/3}) - (1 + 3t^{1/3})] \, dt$$

$$= (8t - 3t^{4/3}) \Big|_0^8 = 16 \text{ million dollars}$$

· ·

In Example 8 we were forced to overlook the *fixed* cost for the cost function at time $t = 0$. This is because if C contains a constant ʼ(the fixed cost), then it becomes zero when we take the derivative C'. Thus, in the final analysis of the problem, total profit should be reduced by the amount corresponding to the fixed cost.

EXERCISE 15.4   Answers to odd-numbered problems begin on page AN-78.

In Problems 1–10 find the area described. Be sure to sketch the graph first.

1. Enclosed by $f(x) = 3x + 2$, the x-axis, and the lines $x = 2$ and $x = 6$

2. Enclosed by $f(x) = 3 - x$, the x-axis, and the lines $x = 0$ and $x = 3$

3. Enclosed by $f(x) = x^2$, the x-axis, and the lines $x = 0$ and $x = 2$

4. Enclosed by $f(x) = x^2$, the x-axis, and the lines $x = -2$ and $x = 1$

5. Enclosed by $f(x) = x^2 + 2$, the x-axis, and the lines $x = -2$ and $x = 1$

6. Enclosed by $f(x) = x^2 - 4$, the x-axis, and the lines $x = 2$ and $x = 4$

7. Enclosed by $f(x) = x$, the x-axis, and the lines $x = 1$ and $x = 2$

8. Enclosed by $f(x) = 1/x$, the x-axis, and the lines $x = 1$ and $x = 2$

9. Enclosed by $f(x) = e^x$, the x-axis, and the lines $x = 0$ and $x = 1$

10. Enclosed by $f(x) = x^3$, the x-axis, and the lines $x = 0$ and $x = 1$

In Problems 11–28 find the area enclosed by the graphs of the given functions and lines. Draw a sketch first.

11. $f(x) = x$, $g(x) = 2x$, $x = 0$, $x = 1$

12. $f(x) = x$, $g(x) = 3x$, $x = 0$, $x = 3$

13. $f(x) = x^2$, $g(x) = x$

14. $f(x) = x^2$, $g(x) = 4x$

15. $f(x) = x^2 + 1$, $g(x) = x + 1$

16. $f(x) = x^2 + 1$, $g(x) = 4x + 1$

17. $f(x) = \sqrt{x}$, $g(x) = x^3$

18. $f(x) = x^2$, $g(x) = x^3$

19. $f(x) = x^2$, $g(x) = x^4$

20. $f(x) = \sqrt{x}$, $g(x) = x^2$

21. $f(x) = x^2 - 4x$, $g(x) = -x^2$

22. $f(x) = x^2 - 8x$, $g(x) = -x^2$

23. $f(x) = 4 - x^2$, $g(x) = x + 2$

24. $f(x) = 2 + x - x^2$, $g(x) = -x - 1$

25. $f(x) = x^3$, $g(x) = 4x$

26. $f(x) = x^3$, $g(x) = 16x$

27. $y = x^2$, $y = x$, $y = -x$

28. $y = x^2 - 1$, $y = x - 1$, $y = -x - 1$

29. Consumer's and Producer's Surplus Find the consumer's surplus and the producer's surplus for the demand curve

$$D(x) = -5x + 20$$

and the supply curve

$$S(x) = 4x + 8$$

Sketch the graphs.

30. Consumer's and Producer's Surplus Follow the same directions as in Problem 29 if

$$D(x) = -0.4x + 15 \quad \text{and} \quad S(x) = 0.8x + 0.5$$

31. Profit Function The revenue and the cost rate of Gold Star mining operation are, respectively,

$$R'(t) = 19 - t^{1/2} \quad \text{and} \quad C'(t) = 3 + 3t^{1/2}$$

where t is measured in years and R and C are measured in millions of dollars. Determine how long the operation should continue and the profit that can be generated during this period. Ignore any fixed costs.

32. Consumer's Surplus Find the consumer's surplus for the demand curve

$$D(q) = 50 - 0.025q^2$$

if it is known that the demand level q is 20 units.

33. Mean Value Theorem for Integrals If $y = f(x)$ is continuous on an interval I, $a \leq x \leq b$, then there is a number c, $a < c < b$, so that

$$\int_a^b f(x)\, dx = f(c)(b - a)$$

The interpretation of this result is that there is a rectangle with base $b - a$ and height $f(c)$, whose area is numerically equal to the area $\int_a^b f(x)\, dx$. See the illustration at the top of the next column.

Technology Exercises

1. Find the area enclosed by the graphs of $f(x) = x^2 + 1$ and $g(x) = x + 1$.

(a) Graph $f(x)$ and $g(x)$.
(b) Use your calculator to find the intersections of $f(x)$ and $g(x)$.
(c) Change the range so that the intersections of $f(x)$ and $g(x)$ are in the lower left and upper right corners of the screen, and regraph $f(x)$ and $g(x)$.

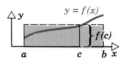

Verify this result by finding c for the functions below. Graph each function.

(a) $f(x) = x^2$, $a = 0$, $b = 1$
(b) $f(x) = 1/x^2$, $a = 1$, $b = 4$

34. Show that the shaded area in the figure is $\frac{2}{3}$ of the area of the parallelogram $ABCD$. (This illustrates a result due to Archimedes concerning sectors of parabolas.)

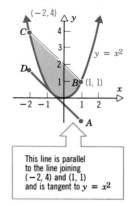

This line is parallel to the line joining $(-2, 4)$ and $(1, 1)$ and is tangent to $y = x^2$

35. If $y = f(x)$ is continuous on the interval I and if it has an antiderivative on I, then for some a in I,

$$\frac{d}{dx} \int_a^x f(t)\, dt = f(x) \qquad \text{for } x > a \text{ in } I$$

This result gives us a technique for finding the derivative of a definite integral in which the lower limit is fixed and the upper limit is variable. Use this result to find

(a) $\dfrac{d}{dx} \displaystyle\int_1^x t^2\, dt$ (b) $\dfrac{d}{dx} \displaystyle\int_2^x \sqrt{t^2 - 2}\, dt$

(c) $\dfrac{d}{dx} \displaystyle\int_5^x \sqrt{t^2 + 2t}\, dt$

(d) Evaluate $A = \displaystyle\int_a^b [f(x) - g(x)]\, dx$.

2. Find the area enclosed by $f(x) = x^2$, the x-axis, $x = 0$, and $x = 1$.

(a) Graph $f(x)$ on $[0, 1]$ by $[f(0), f(1)]$.
(b) Evaluate $\displaystyle\int_0^1 x^2\, dx$ by using the definition of the definite integral or your calculator.

3. **Oil Well Profits**   The figure below is a graph of the monthly yield, $y(t)$ (in gallons per month), from an oil well after t months of pumping.

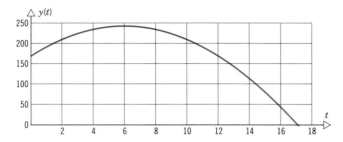

(a) During which month is the well most profitable?
(b) Estimate (from the graph) the total yield in the first year of pumping. [*Hint:* Total yield is the area under the graph of $y(t)$—count the squares.]
(c) What is the total yield from this well?

4. **Investment Profit**   The figure below is a graph of the daily fluctuation of value of an investment in a particular stock t days after investing $200,000.

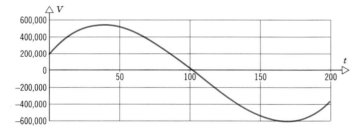

(a) When is the stock most profitable?
(b) Estimate (from the graph) the total profit during the first 100 days.
(c) Estimate (from the graph) the total profit through 200 days.
(d) What would be the best strategy for the investor? When should the investor sell the stock to maximize profit?

15.5 APPROXIMATING DEFINITE INTEGRALS

Thus far, the evaluation of a definite integral

$$\int_a^b f(x)\, dx$$

has required that we find an antiderivative F of f so that

$$\int_a^b f(x)\, dx = F(x)\Big|_a^b = F(b) - F(a) \qquad \text{where} \qquad F'(x) = f(x)$$

But what if we can't find an antiderivative? In fact, sometimes it is impossible to find an antiderivative. In such situations it is necessary to *approximate* the definite integral. One way is to use rectangles.

Approximating the Definite Integral Using Rectangles

We have already discussed the fact that when f is a continuous nonnegative function defined on the closed interval $[a, b]$, then the definite integral $\int_a^b f(x) \, dx$ equals the area under the graph of f from a to b. We will use this idea to obtain an approximation to $\int_a^b f(x) \, dx$.

Consider the graph of the function f in Figure 28(a) The area under the graph of f from a to b is $\int_a^b f(x) \, dx$. Pick a number u in the interval $[a, b]$ and form the rectangle whose height is $f(u)$ and whose base is $b - a$. See Figure 28(b). The area of this rectangle provides a rough approximation to

$$\int_a^b f(x) \, dx$$

That is,

$$\int_a^b f(x) \, dx \approx f(u)(b - a)$$

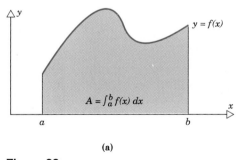

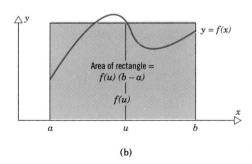

(a)   (b)

Figure 28

A better approximation to $\int_a^b f(x) \, dx$ can be obtained by dividing the interval $[a, b]$ into two subintervals of the same length. See Figure 29(a), where x_1 is the midpoint of $[a, b]$. Now pick a number u_1 between a and x_1, and a number u_2 between x_1 and b, and form two rectangles: One whose height is $f(u_1)$ and whose base is $x_1 - a$ and the other whose height is $f(u_2)$ and whose base is $b - x_1$. See Figure 29(b). The sum of the areas of these two rectangles provides an approximation to $\int_a^b f(x) \, dx$. That is,

$$\int_a^b f(x) \, dx \approx f(u_1)(x_1 - a) + f(u_2)(b - x_1)$$

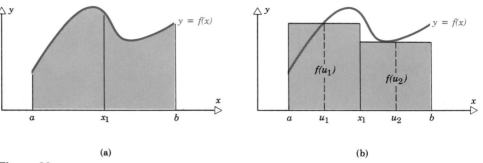

Figure 29

To get an even better approximation, divide the interval $[a, b]$ into three subintervals, each of the same length Δx. See Figure 30(a), where x_1 and x_2 are chosen so that

$$\Delta x = x_1 - a = x_2 - x_1 = b - x_2$$

Now pick numbers u_1, u_2, u_3 in each subinterval and form three rectangles of heights $f(u_1)$, $f(u_2)$, $f(u_3)$, each with the same base Δx. See Figure 30(b). The sum of the areas of these rectangles approximates

$$\int_a^b f(x)\, dx$$

That is,

$$\int_a^b f(x)\, dx \approx f(u_1)\, \Delta x + f(u_2)\, \Delta x + f(x_3)\, \Delta x$$

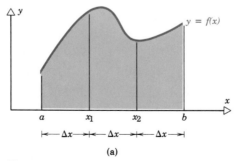

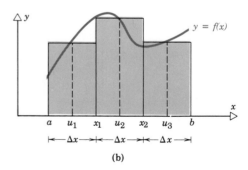

Figure 30

It should be apparent that the more subintervals we divide $[a, b]$ into, the better the approximation. Suppose we divide $[a, b]$ into n subintervals

$$[a, x_1], [x_1, x_2], \ldots, [x_{k-1}, x_k], \ldots, [x_{n-1}, b]$$

each of length $\Delta x = (b - a)/n$. See Figure 31(a). Now pick numbers $u_1, u_2, \ldots, u_n$ in each subinterval and form n rectangles having base Δx and with heights

$f(u_1), f(u_2), \ldots, f(u_n)$. See Figure 31(b). The sum of the areas of these rectangles approximates $\int_a^b f(x)\, dx$. That is,

$$\int_a^b f(x)\, dx \approx f(u_1)\, \Delta x + f(u_2)\, \Delta x + \cdots + f(u_n)\, \Delta x$$

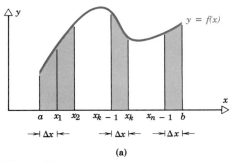
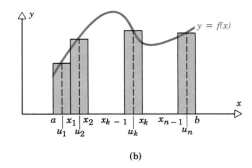

(a)   (b)

Figure 31

Let's summarize this process.

Steps for Approximating an Integral

To approximate $\int_a^b f(x)\, dx$, follow these steps:

Step 1 Divide the interval $[a, b]$ into n subintervals of equal length $\Delta x = (b - a)/n$. The larger n is, the better your approximation will usually be.

Step 2 Pick a number u in each subinterval and evaluate $f(u)$.

Step 3 The sum

$$f(u_1)\, \Delta x + f(u_2)\, \Delta x + \cdots + f(u_n)\, \Delta x$$

approximates

$$\int_a^b f(x)\, dx$$

EXAMPLE 1

Approximate $\int_0^4 (3 + 2x)\, dx$ by dividing the interval $[0, 4]$ into four subintervals of equal length. Pick u_i as the left endpoint of each subinterval.

SOLUTION

Figure 32 illustrates the graph of $f(x) = 3 + 2x$ on $[0, 4]$.

Step 1 Divide $[0, 4]$ into four subintervals of equal length

$$[0, 1], [1, 2], [2, 3], [3, 4]$$

Figure 32

$f(x) = 3 + 2x$

Step 2 Pick the left endpoint of each of these subintervals and evaluate f there: $f(0) = 3, f(1) = 5, f(2) = 7, f(3) = 9$.

Step 3 $\displaystyle\int_0^4 (3 + 2x)\,dx \approx f(0) \cdot 1 + f(1) \cdot 1 + f(2) \cdot 1 + f(3) \cdot 1$

$$= 3 + 5 + 7 + 9 = 24$$

The exact answer is

$$\int_0^4 (3 + 2x)\,dx = (3x + x^2)\Big|_0^4 = 12 + 16 = 28$$

Now Work Problem 1

$\cdots\cdots\cdots\cdots\cdots\cdots\cdots\cdots$

EXAMPLE 2

Approximate: $\displaystyle\int_0^8 x^2\,dx$

(a) By dividing the interval $[0, 8]$ into eight subintervals of equal length and picking u_i as the right endpoint of each subinterval.

(b) By dividing the interval $[0, 8]$ into four subintervals of equal length and picking u_i as the midpoint of each subinterval.

SOLUTION

Figure 33 illustrates the graph of $f(x) = x^2$ on the interval $[0, 8]$.

(a) We divide $[0, 8]$ into eight subintervals of equal length:

$$[0, 1], [1, 2], [2, 3], [3, 4], [4, 5], [5, 6], [6, 7], [7, 8]$$

Now we evaluate f at the right endpoint of each of these subintervals.

$$f(1) = 1, \quad f(2) = 4, \quad f(3) = 9, \quad f(4) = 16, \quad f(5) = 25,$$
$$f(6) = 36, \quad f(7) = 49, \quad f(8) = 64$$

Since each subinterval is of length $\Delta x = 1$, we find

$$\int_0^8 x^2\,dx \approx f(1) + f(2) + f(3) + f(4) + f(5) + f(6) + f(7) + f(8) = 204$$

(b) We divide $[0, 8]$ into four subintervals of equal length:

$$[0, 2], [2, 4], [4, 6], [6, 8]$$

Now we evaluate f at the midpoint of each subinterval:

$$f(1) = 1, \quad f(3) = 9, \quad f(5) = 25, \quad f(7) = 49$$

Since each subinterval is of length $\Delta x = 2$, we find

$$\int_0^8 x^2\,dx \approx f(1) \cdot 2 + f(3) \cdot 2 + f(5) \cdot 2 + f(7) \cdot 2$$

$$= 2 + 18 + 50 + 98 = 168$$

$\cdots\cdots\cdots\cdots\cdots\cdots\cdots\cdots$

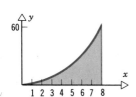

Figure 33

In this case we know that $\frac{1}{3}x^3$ is an antiderivative of x^2, so we can find the *exact* value of $\int_0^8 x^2\,dx$, namely,

$$\int_0^8 x^2\,dx = \frac{1}{3}x^3 \bigg|_0^8 = \frac{512}{3} = 170.7$$

A natural question is to ask whether we could somehow determine in advance that the second approximation would end up better than the first. This question is not easy to answer—the area of mathematics called *numerical analysis* makes attempts at an-

Now Work Problem 7   swering it.

Riemann Sums

Let's review the procedure for approximating a definite integral except now when we subdivide each interval, we won't require that each length be the same. Suppose f is a continuous function defined on the interval $[a, b]$. Suppose we partition, or divide, this interval into n subintervals. The point $a = x_0$ is the initial point, the first point of the subdivision is x_1, the second is x_2, . . . , and the nth point is $b = x_n$. See Figure 34.

Figure 34

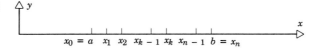

The original interval $[a, b]$ now consists of n subintervals, and the length of each one is

First, Second, Third, . . . ,

$$\Delta x_1 = x_1 - x_0, \qquad \Delta x_2 = x_2 - x_1, \qquad \Delta x_3 = x_3 - x_2, \ldots,$$

kth, . . . , nth

$$\Delta x_k = x_k - x_{k-1}, \ldots, \qquad \Delta x_n = x_n - x_{n-1}$$

We use the symbol Δ to denote the largest such length. (The value of Δ, of course, depends on how the partition itself has been chosen.) We call Δ the **norm of the partition.**

Next, we concentrate on the function. Pick a number in each subinterval (you may select a number in the interval or either endpoint, if you wish) and evaluate the function at this number. To fix our ideas, let u_k denote the chosen number. The corresponding value of the function is $f(u_k)$. This represents the height of the graph of the function at u_k.

Multiply $f(u_1)$ times $\Delta x_1 = x_1 - x_0, f(u_2)$ times $\Delta x_2 = x_2 - x_1, \ldots, f(u_n)$ times $\Delta x_n = x_n - x_{n-1}$, and add these products. The result is the sum

$$f(u_1)\,\Delta x_1 + f(u_2)\,\Delta x_2 + \cdots + f(u_n)\,\Delta x_n$$

This is called a **Riemann sum** for the function f on $[a, b]$.

Finally, we take the limit of this sum as the norm $\Delta \rightarrow 0$, that is, as all the intervals get smaller. If this limit exists, it is the definite integral of $f(x)$ from a to b. Thus,

$$\int_a^b f(x)\,dx = \lim_{\Delta \rightarrow 0} [f(u_1)\,\Delta x_1 + f(u_2)\,\Delta x_2 + \cdots + f(u_n)\,\Delta x_n] \tag{1}$$

The above formula, in a more formal course in calculus, is taken as the definition of a definite integral. Then it can be proven as a theorem—the **Fundamental Theorem of Calculus**—that

$$\int_a^b f(x)\, dx = F(b) - F(a)$$

where F is an antiderivative of f.

EXERCISE 15.5   Answers to odd-numbered problems begin on page AN-80.

In Problems 1–6 approximate

$$\int_0^{10} (x^2 - 5x)\, dx$$

1. By dividing [0, 10] into two subintervals of equal length; always pick u_i as the left endpoint of each subinterval.

2. By dividing [0, 10] into two subintervals of equal length; always pick u_i as the right endpoint of each subinterval.

3. By dividing [0, 10] into two subintervals of equal length; always pick u_i as the midpoint of each subinterval.

4. By dividing [0, 10] into five subintervals of equal length; always pick u_i as the left endpoint of each subinterval.

5. By dividing [0, 10] into five subintervals of equal length; always pick u_i as the right endpoint of each subinterval.

6. By dividing [0, 10] into five subintervals of equal length; always pick u_i as the midpoint of each subinterval.

In Problems 7–12 approximate

$$\int_0^{10} (x^3 - x^2)\, dx$$

7. By dividing [0, 10] into two subintervals of equal length; always pick u_i as the left endpoint of each subinterval.

8. By dividing [0, 10] into two subintervals of equal length; always pick u_i as the right endpoint of each subinterval.

9. By dividing [0, 10] into two subintervals of equal length; always pick u_i as the midpoint of each subinterval.

10. By dividing [0, 10] into five subintervals of equal length; always pick u_i as the left endpoint of each subinterval.

11. By dividing [0, 10] into five subintervals of equal length; always pick u_i as the right endpoint of each subinterval.

12. By dividing [0, 10] into five subintervals of equal length; always pick u_i as the midpoint of each subinterval.

In Problems 13–18 approximate each definite integral by dividing the interval [a, b] into subintervals of equal length 1. Always pick u_i as the midpoint of each subinterval.

13. $\int_0^3 (1 - 2x)\, dx$

14. $\int_0^5 (3 + 4x)\, dx$

15. $\int_1^4 4x^2\, dx$

16. $\int_1^5 x^3\, dx$

17. $\int_{-2}^1 e^x\, dx$

18. $\int_1^5 \ln x\, dx$

Technology Exercises

1. Evaluate $\int_2^4 (x - 1)\, dx$.

   (a) Graph $f(x) = x - 1$ on [2, 4] by [−2, 4].
   (b) This figure is a trapezoid and the area can be found from geometry, using the formula given at the right.

$A = \frac{1}{2}(h_1 + h_2)w$ where $h_1 = f(2)$, $h_2 = f(4)$, and $w = b - a$

Find the area A from this formula.

(c) Use a calculator to find A.
(d) Find the antiderivative $F(x)$ of $f(x)$.
(e) Find $A = F(b) - F(a)$.

In Problems 2–7 evaluate the definite integrals in the following two ways:
(a) *Find the antiderivative $F(x)$ and (you may use your calculator) evaluate*
     $F(b) - F(a)$.
(b) *Use your calculator to approximate the integral directly.*

2. $\displaystyle\int_0^3 (x^3 - 4x + 1)\, dx$ 　　　**3.** $\displaystyle\int_{-1}^2 |x|\, dx$ 　　　**4.** $\displaystyle\int_0^2 2x(x^2 + 1)^3\, dx$

5. $\displaystyle\int_0^1 \frac{\sqrt{x}}{x^{3/2} + 1}\, dx$ 　　　**6.** $\displaystyle\int_1^2 xe^{-3x}\, dx$ 　　　**7.** $\displaystyle\int_1^2 x^2 \ln x\, dx$

15.6 IMPROPER INTEGRALS

Recall that in defining $\int_a^b f(x)\, dx$ two basic assumptions are made:

1. The limits of integration a and b are both finite.
2. The function is continuous on $[a, b]$.

In many situations one or more of these assumptions are not met. For example, one of the limits of integration might be infinity; or the function $y = f(x)$ might be discontinuous on $[a, b]$. If either of the Conditions (1) and (2) are not satisfied, then $\int_a^b f(x)\, dx$ is called an **improper integral.**

One Limit of Integration is Infinite

We begin with an example.

EXAMPLE 1   Find the area between the graph of $f(x) = 1/x^2$, the x-axis, and to the right of $x = 1$.

SOLUTION

First, we graph $f(x) = 1/x^2$. See Figure 35. The area to the right of $x = 1$ is shaded. To find this area, we pick a number b to the right of $x = 1$. The area between the graph of $f(x) = 1/x^2$, the x-axis, $x = 1$, and $x = b$ is

$$\int_1^b \frac{1}{x^2}\, dx = \left(-\frac{1}{x}\right)\bigg|_1^b = -\frac{1}{b} + 1$$

Figure 35

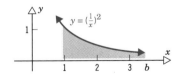

As we should expect, this area depends on the choice of b. Now the area we seek is the one obtained by letting $b \to \infty$. Since

$$\lim_{b \to \infty} \left(-\frac{1}{b} + 1 \right) = 1$$

we conclude that the area between the graph of $f(x) = x^2$, the x-axis, and to the right of $x = 1$ is 1.

....

The area we found in Example 1 can be represented symbolically by the *improper integral*

$$\int_1^\infty \frac{1}{x^2}\, dx$$

and it can be evaluated by finding

$$\lim_{b \to \infty} \int_1^b \frac{1}{x^2}\, dx$$

This leads us to formulate the following definition.

IMPROPER INTEGRAL

The *improper integral $\int_a^\infty f(x)\, dx$* is defined as

$$\int_a^\infty f(x)\, dx = \lim_{b \to \infty} \int_a^b f(x)\, dx$$

provided this limit exists and is a real number. If this limit does not exist or if it is infinite, the improper integral has no value.

The *improper integral $\int_{-\infty}^b f(x)\, dx$* is defined as

$$\int_{-\infty}^b f(x)\, dx = \lim_{a \to -\infty} \int_a^b f(x)\, dx$$

provided that this limit exists and is a real number. If this limit does not exist or if it is infinite, the improper integral has no value.

EXAMPLE 2

Find the value of: (a) $\displaystyle\int_{-\infty}^0 e^x\, dx$ (b) $\displaystyle\int_1^\infty \frac{1}{x}\, dx$

SOLUTION

(a) $\displaystyle\int_{-\infty}^0 e^x\, dx = \lim_{a \to -\infty} \int_a^0 e^x\, dx = \lim_{a \to -\infty} [e^x] \Big|_a^0$

$= \lim_{a \to -\infty} [1 - e^a] = 1 - \lim_{a \to -\infty} e^a = 1$

(b) $\displaystyle\int_1^\infty \frac{1}{x}\,dx = \lim_{b\to\infty}\int_1^b \frac{1}{x}\,dx = \lim_{b\to\infty}\ln x \Big|_1^b = \lim_{b\to\infty}\ln b = \infty$

Thus, since the limit is infinite, $\displaystyle\int_1^\infty \frac{1}{x}\,dx$ has no value.

Now Work Problem 7

The Integrand is Discontinuous

The improper integrals studied in the above examples had infinity either as a lower limit of integration or as an upper limit of integration. A second type of improper integral occurs when the integrand f in $\int_a^b f(x)\,dx$ is discontinuous at either a or b (here a and b are both real numbers). Again, we illustrate the technique for evaluating this type of improper integral by an example.

EXAMPLE 3

The integrand of $\int_0^4 1/\sqrt{x}\,dx$ is discontinuous at $x = 0$. To evaluate this improper integral, we proceed as follows:

$$\int_0^4 \frac{1}{\sqrt{x}}\,dx = \lim_{t\to 0^+}\int_t^4 \frac{1}{\sqrt{x}}\,dx = \lim_{t\to 0^+}\frac{x^{1/2}}{\frac{1}{2}}\Big|_t^4$$

$$= \lim_{t\to 0^+}(2\sqrt{4} - 2\sqrt{t}) = 4 - \lim_{t\to 0^+}2\sqrt{t} = 4$$

IMPROPER INTEGRAL

If f is continuous on $[a, b)$, but is discontinuous at $x = b$, we define the *improper integral* $\int_a^b f(x)\,dx$ to be

$$\int_a^b f(x)\,dx = \lim_{t\to b^-}\int_a^t f(x)\,dx$$

provided that this limit exists and is a real number. If this limit does not exist or if it is infinite, $\int_a^b f(x)\,dx$ has no value.

If f is continuous on $(a, b]$, but is discontinuous at $x = a$, we define the *improper integral* $\int_a^b f(x)\,dx$ to be

$$\int_a^b f(x)\,dx = \lim_{t\to a^+}\int_t^b f(x)\,dx$$

provided that this limit exists and is a real number. If this limit does not exist or if it is infinite, $\int_a^b f(x)\,dx$ has no value.

[*Note:* When the upper limit is the point of discontinuity, the limit is left-handed; when the lower limit is the point of discontinuity, the limit is right-handed.]

EXAMPLE 4

Find the area between the graph of $f(x) = 1/\sqrt[3]{x}$ and the x-axis from $x = -1$ to $x = 0$.

SOLUTION

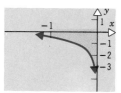

Figure 36

We begin by graphing the function. See Figure 36. Since the area lies below the x-axis, the area is $-\int_{-1}^{0} 1/\sqrt[3]{x}\, dx$.

We observe that the integrand $1/\sqrt[3]{x}$ is discontinuous at $x = 0$ (the upper limit of integration). Hence,

$$\int_{-1}^{0} \frac{1}{\sqrt[3]{x}}\, dx = \lim_{t \to 0^-} \int_{-1}^{t} \frac{1}{\sqrt[3]{x}}\, dx = \lim_{t \to 0^-} \int_{-1}^{t} x^{-1/3}\, dx$$

$$= \lim_{t \to 0^-} \frac{x^{2/3}}{\frac{2}{3}} \Big|_{-1}^{t} = \lim_{t \to 0^-} \frac{3}{2}(t^{2/3} - 1) = -\frac{3}{2}$$

The area is thus $\frac{3}{2}$ square units.

Now Work Problem 13

• •

There are other types of improper integrals besides those discussed here. Such types are discussed in Problems 18 and 19 below.

EXERCISE 15.6   Answers to odd-numbered problems begin on page AN-80.

In Problems 1–6 determine which of the integrals are improper and give a reason why they are, or are not, improper.

1. $\int_{0}^{\infty} x^2\, dx$

2. $\int_{2}^{3} \frac{dx}{x-1}$

3. $\int_{0}^{1} \frac{1}{x}\, dx$

4. $\int_{-1}^{1} \frac{x}{x^2+1}\, dx$

5. $\int_{1}^{2} \frac{dx}{x-1}$

6. $\int_{0}^{1} \frac{x}{x^2-1}\, dx$

In Problems 7–14 determine the value, if any, of each improper integral.

7. $\int_{1}^{\infty} e^{-4x}\, dx$

8. $\int_{-\infty}^{-1} \frac{1}{x^3}\, dx$

9. $\int_{0}^{\infty} \sqrt{x}\, dx$

10. $\int_{0}^{\infty} x\, e^{-x}\, dx$

11. $\int_{-1}^{0} \frac{1}{\sqrt[5]{x}}\, dx$

12. $\int_{2}^{4} \frac{x\, dx}{\sqrt{x^2-4}}$

13. $\int_{0}^{1} \frac{1}{x}\, dx$

14. $\int_{0}^{1} \frac{\ln x}{x}\, dx$

15. Find the area, if it exists, between the graph of

$$f(x) = \frac{1}{\sqrt{x}}$$

and the x-axis from $x = 0$ to $x = 1$.

16. Find the area, if it exists, between the graph of $f(x) = \sqrt{x}$ and the x-axis to the right of $x = 0$.

17. Drug Reaction   The rate of reaction to a given dose of a drug at time t hours after administration is given by $r(t) = te^{-t^2}$ (measured in appropriate units). Why is it reasonable to define the **total reaction** as the area under the curve $y = r(t)$ from $t = 0$ to $t = \infty$? Evaluate the total reaction to the given dose of the drug.

18. If $y = f(x)$ is continuous, the improper integral $\int_{-\infty}^{\infty} f(x)\, dx$ is defined as

$$\int_{-\infty}^{\infty} f(x)\, dx = \int_{-\infty}^{0} f(x)\, dx + \int_{0}^{\infty} f(x)\, dx$$

provided that each of the improper integrals on the right has a value. Use this definition to find the value, if it exists, of

(a) $\displaystyle\int_{-\infty}^{\infty} e^x\, dx$

(b) $\displaystyle\int_{-\infty}^{\infty} \frac{x\, dx}{(x^2 + 1)^2}$

19. If $y = f(x)$ is continuous on $[a, b]$, except at a point c, $a < c < b$, the integral $\int_{a}^{b} f(x)\, dx$ is improper and is defined by

$$\int_{a}^{b} f(x)\, dx = \int_{a}^{c} f(x)\, dx + \int_{c}^{b} f(x)\, dx$$

provided that each of the improper integrals on the right has a value. Use this definition to evaluate

(a) $\displaystyle\int_{-1}^{1} \frac{1}{x^2}\, dx$

(b) $\displaystyle\int_{0}^{4} \frac{x\, dx}{\sqrt[3]{x^2 - 4}}$

Technology Exercises

In Problems 1–6 determine the value, if any, of the improper integral $\int_a^\infty f(x)\, dx$ by using a graphing calculator (or a computer).

(a) *First determine the antiderivative, $F(x)$, of the function $f(x)$.*

(b) *Then graph the function $F(x)$ in the x ranges $[a, 1]$, $[a, 10]$, and $[a, 100]$ to estimate the limit $\lim_{x \to \infty} F(x)$ and the value of the improper integral $\int_a^\infty f(x)\, dx$.*

(c) *Determine analytically the exact value, if any, of the improper integral, and compare with your estimate from Part (b).*

1. $\displaystyle\int_{0}^{\infty} x^2\, dx$

2. $\displaystyle\int_{0}^{\infty} e^{-4x}\, dx$

3. $\displaystyle\int_{0}^{\infty} xe^{-x}\, dx$

4. $\displaystyle\int_{1}^{\infty} \frac{1}{x^2}\, dx$

5. $\displaystyle\int_{0}^{\infty} xe^{-x^2}\, dx$

6. $\displaystyle\int_{1}^{\infty} \frac{1}{\sqrt{x}}\, dx$

CHAPTER REVIEW

IMPORTANT TERMS AND CONCEPTS

antiderivative 776
antiderivative of x^r 777
indefinite integral 778
basic integration formulas 778, 780
integration by substitution 787
integration by parts 792
steps to integrate by parts 793
definite integral 797
lower and upper limits of
  integration 798

properties of definite
  integrals 799, 800
learning curve 801, 802
annuity 803
rate of sales 803
properties of area 807
area under a graph 807
consumer's surplus;
  producer's surplus 815

steps for approximating an
  integral 823
Riemann sum 825
Fundamental Theorem
  of Calculus 826
improper integral: infinite limits
  of integration 828
improper integral: discontinuous
  integrand 829

IMPORTANT FORMULAS

Indefinite Integrals

$$\int f(x)\, dx = F(x) + K$$

Properties of Indefinite Integrals

$$\int c\, dx = cx + K$$

$$\int x^r\, dx = \frac{x^{r+1}}{r + 1} + K \qquad r \neq -1$$

$$\int [f(x) \pm g(x)]\, dx = \int f(x)\, dx \pm \int g(x)\, dx$$

$$\int cf(x)\, dx = c \int f(x)\, dx$$

$$\int e^x\, dx = e^x + K$$

$$\int e^{ax}\, dx = \frac{1}{a} e^{ax} + K$$

$$\int \frac{1}{x}\, dx = \ln |x| + K$$

Integration by Parts

$$\int u\, dv = uv - \int v\, du$$

Fundamental Theorem of Calculus

$$\int_a^b f(x)\, dx = F(b) - F(a)$$

Properties of Definite Integrals

$$\int_a^b f(x)\, dx = - \int_b^a f(x)\, dx$$

$$\int_a^a f(x)\, dx = 0$$

$$\int_a^b f(x)\, dx = \int_a^c f(x)\, dx + \int_c^b f(x)\, dx$$

$$\int_a^b cf(x)\, dx = c \int_a^b f(x)\, dx$$

$$\int_a^b [f(x) \pm g(x)]\, dx = \int_a^b f(x)\, dx \pm \int_a^b g(x)\, dx$$

Definite Integral as a Sum

$$\int_a^b f(x)\, dx = \lim_{\Delta \to 0} [\, f(u_1)\, \Delta x_1 + f(u_2)\, \Delta x_2 + \cdots + f(u_n)\, \Delta x_n]$$

Improper Integrals

$$\int_a^b f(x)\, dx = \lim_{t \to b^-} \int_a^t f(x)\, dx \qquad \int_a^b f(x)\, dx = \lim_{t \to a^+} \int_t^b f(x)\, dx$$

TRUE–FALSE ITEMS   Answers are on page AN-82.

T_____ F_____   **1.** The integral of the sum of two functions equals the sum of their integrals.

T_____ F_____   **2.** $\int \left(\frac{1}{2} x^3 + x^{3/2} - 1 \right) dx$

$$= \frac{x^4}{6} + \frac{2x^{7/2}}{7} - x + K$$

T_____ F_____   **3.** $\int \frac{x^2 + 4}{x}\, dx = \dfrac{\frac{x^3}{3} + 4x}{\frac{x^2}{2}} + K$

T_____ F_____   **4.** $\int \frac{\sqrt{x^2 + 1}}{x}\, dx = \frac{1}{x} \int \sqrt{x^2 + 1}\, dx$

T_____ F_____   **5.** $\int \ln x\, dx = \frac{1}{x} + K$

T_____ F_____   **6.** Any antiderivative of $f(x)$ can be used to evaluate $\int_a^b f(x)\, dx$.

T_____ F_____   **7.** The definite integral $\int_a^b f(x)\, dx$, if it exists, is a unique number.

T_____ F_____   **8.** $\int_a^b f(x)\, dx + \int_b^a f(x)\, dx = 0$

T_____ F_____   **9.** $\int_0^1 x^2\, dx = 3$

T_____ F_____   **10.** $\int_0^1 \frac{1}{x - 1}\, dx$ is improper.

FILL IN THE BLANKS   Answers are on page AN-82.

1. A function F is called an antiderivative of the function f if

   _____.

2. The symbol _____ represents all the antiderivatives of a function f.

3. The formula $\int u\, dv = uv - \int v\, du$ is referred to as the _____ _____ _____ formula.

4. In $\int_a^b f(x)\, dx$ the numbers a and b are called the

   _____ and _____ _____ of

   _____, respectively.

5. $\int_a^a f(x)\, dx =$ _____.

6. If f is continuous on $[a, b]$ and $F' = f$, then $\int_a^b f(x)\, dx =$

   _____.

7. If f is discontinuous at 2, but continuous elsewhere, then

   $\int_0^2 f(x)\, dx =$ _____, provided this limit exists.

REVIEW EXERCISES   Answers to odd-numbered problems begin on page AN-82.

In Problems 1–18 evaluate each indefinite integral.

1. $\displaystyle\int (x^3 - 3x + 1)\, dx$

2. $\displaystyle\int (x^3 + 4x - 2)\, dx$

3. $\displaystyle\int (x^{1/3} - 4x^{1/2})\, dx$

4. $\displaystyle\int (x^{3/2} + 5x^{1/2})\, dx$

5. $\displaystyle\int (1 + e^{-x})\, dx$

6. $\displaystyle\int (1 - e^x)\, dx$

7. $\displaystyle\int x\sqrt{x^2 - 1}\, dx$

8. $\displaystyle\int x\sqrt{3x^2 - 1}\, dx$

9. $\displaystyle\int \sqrt{3x - 2}\, dx$

10. $\displaystyle\int \sqrt{2 - 3x}\, dx$

11. $\displaystyle\int \frac{x^3\, dx}{(x^4 + 1)^{3/2}}$

12. $\displaystyle\int \frac{x^2\, dx}{(x^3 - 1)^{1/2}}$

13. $\displaystyle\int \frac{5x\, dx}{x^2 + 1}$

14. $\displaystyle\int \frac{5x^2}{x^3 - 1}\, dx$

15. $\displaystyle\int xe^{x/2}\, dx$

16. $\displaystyle\int x \ln 3x\, dx$

17. $\displaystyle\int x^2 e^{x^3}\, dx$

18. $\displaystyle\int \frac{x^{3/2}\, dx}{x^{5/2} + 2}$

In Problems 19–28 evaluate each definite integral.

19. $\displaystyle\int_0^1 (3x^4 - 8x + 2)\, dx$

20. $\displaystyle\int_{-2}^2 (4x^3 - 8x)\, dx$

21. $\displaystyle\int_{-1}^2 e^{2x}\, dx$

22. $\displaystyle\int_{-1}^0 e^{-x}\, dx$

23. $\displaystyle\int_1^2 \frac{1}{x}\, dx$

24. $\displaystyle\int_1^2 \frac{1}{x^2}\, dx$

25. $\displaystyle\int_0^1 3x(x^2 + 4)^2\, dx$

26. $\displaystyle\int_1^4 x \ln x\, dx$

27. $\displaystyle\int_0^1 xe^{-x}\, dx$

28. $\displaystyle\int_1^2 x^2 \ln x\, dx$

29. Find the area enclosed by $f(x) = x^2$ and $g(x) = x^3$.

30. Find the area enclosed by $f(x) = -x^2 + 2x + 2$ and $g(x) = x^2 - 4x + 2$.

31. Find the area enclosed by $f(x) = x^2$ and $g(x) = \sqrt{x}$.

32. Find the area enclosed by $f(x) = x^3 - x$ and the x-axis from $x = -1$ to $x = 2$.

33. Find the area enclosed by $f(x) = 2x/(x^2 + 1)$ and the x-axis from $x = 0$ to $x = 2$.

34. Find the area enclosed by $f(x) = xe^{3x^2}$ and the x-axis from $x = 0$ to $x = 1$.

35. Find the value, if it exists, of

$$\int_0^2 \frac{1}{(x-2)^2}\, dx$$

36. Find the value, if it exists, of

$$\int_0^{-1} \frac{1}{(x+1)^2}\, dx$$

37. Find the value, if it exists, of

$$\int_0^\infty 2e^{-3x}\, dx$$

38. Find the value, if it exists, of

$$\int_1^\infty \frac{\ln x}{x}\, dx$$

In Problems 39–46 approximate each definite integral by using a partition in which each subinterval is of length 1. Choose u_i as the left endpoint of each subinterval.

39. $\displaystyle\int_2^4 x^2\, dx$

40. $\displaystyle\int_0^3 x^3\, dx$

41. $\displaystyle\int_{-1}^1 (x^2 + x - 1)\, dx$

42. $\displaystyle\int_0^2 (2x^2 - x + 4)\, dx$

43. $\displaystyle\int_0^4 e^x\, dx$

44. $\displaystyle\int_1^2 \frac{1}{x}\, dx$

45. $\displaystyle\int_0^2 x^2\sqrt{x^3 + 1}\, dx$

46. $\displaystyle\int_0^1 x\sqrt{x^2 + 1}\, dx$

47. Savings   If a person saves $2000 per year for 5 years in a savings account that pays 6% per annum compounded continuously, will there be enough in the account to make a 20% down payment on a $100,000 house?

48. Savings   How much should the person in Problem 47 save each year in order to have the required down payment?

49. Total Sales   The rate of sales of a certain product obeys

$$f(x) = 1340 - 850e^{-x}$$

where x is the number of years the product is on the market. Find the total sales during the first 5 years.

50. Consumer's Surplus   Find the consumer's surplus and the producer's surplus for the demand curve $D(x) = 12 - (x/50)$ and the supply curve $S(x) = (x/20) + 5$.

Mathematical Questions from Professional Exams

1. Actuary Exam—Part I $\displaystyle\int_1^e \frac{1}{x} \ln x\, dx = ?$

(a) $1/e$ (b) $\frac{1}{2}$ (c) 1
(d) e (e) e^2

2. Actuary Exam—Part I $\displaystyle\int_0^1 x \ln x\, dx = ?$

(a) $-\infty$ (b) -2 (c) -1
(d) $-\frac{1}{4}$ (e) $-\frac{2}{9}$

3. Actuary Exam—Part I   If $\int_1^b f(x)\, dx = b^2 e^b - e$ for all $b > 0$, then for all $x > 0$, $f(x) = ?$

(a) $x^2 e^x$ (b) $\dfrac{x^3}{3} e^x$ (c) $x^2 e^x + 2xe^x$
(d) $2xe^x$ (e) $x^2 e^x - e^{x-1}$

4. Actuary Exam—Part I   If the area of the region bounded by $y = f(x)$, the x-axis, and the lines $x = a$ and $x = b$ is given by $\int_a^b f(x)\, dx$, which of the following must be true?

(a) $a < b$ and $f(x) > 0$ (b) $a < b$ and $f(x) < 0$
(c) $a > b$ and $f(x) > 0$ (d) $a > b$ and $f(x) < 0$
(e) None of the above

Chapter 16

More Applications of the Integral

In this chapter we provide three additional applications of the integral: finding the average value of a function (Section 16.1); investigating probability distributions (Sections 16.2 and 16.3); and solving differential equations (Section 16.4). These applications are independent of one another and may be covered in any order, or omitted altogether, without any loss of continuity.

Section 16.1, ''Average Value of a Function,'' examines how the definite integral can be used to calculate averages.

Sections 16.2 and 16.3 deal with probability. We shall extend the concept of sample space to include *infinite* sample spaces as well as investigate the use of certain *probability distributions* (the Poisson, uniform, and exponential distributions).

Section 16.4, ''Differential Equations,'' discusses applications of the integral in biology and the natural sciences.

16.1 AVERAGE VALUE OF A FUNCTION

At the U.S. Weather Bureau, a continuous reading of the temperature over a 24-hour period is taken daily. To obtain the average daily temperature, 12 readings may be taken at 2-hour intervals beginning at midnight: $f(0)$, $f(2)$, $f(4)$, . . . , $f(20)$, $f(22)$. The average temperature is then calculated as

$$\frac{f(0) + f(2) + f(4) + \cdots + f(20) + f(22)}{12}$$

This number represents a good approximation to the true average as long as there were no drastic temperature changes over the short periods of time. To improve the approximation, readings may be taken every hour. The average in this case would be

$$\frac{f(0) + f(1) + \cdots + f(22) + f(23)}{24}$$

An even better approximation would be obtained if readings were recorded every half hour.

In general, if $y = f(x)$ is a continuous function defined on the closed interval $[a, b]$, we can obtain the *average of f on* $[a, b]$ as follows: Partition the interval $[a, b]$ into n subintervals

$$[a, x_1], \quad [x_1, x_2], \quad . . . , \quad [x_{k-1}, x_k], \quad . . . , \quad [x_{n-1}, b]$$

each of equal length $\Delta x = (b - a)/n$. This is the norm Δ of the partition. Pick a number in each subinterval and let these numbers be $u_1, u_2, . . . , u_n$. An approximation of the average value of f over the interval $[a, b]$ is then the sum

$$\frac{f(u_1) + f(u_2) + \cdots + f(u_n)}{n} \tag{1}$$

If we multiply and divide the expression in (1) by $b - a$, we get

$$\frac{f(u_1) + f(u_2) + \cdots + f(u_n)}{n} = \frac{1}{b - a}\left[f(u_1)\frac{b - a}{n} + f(u_2)\frac{b - a}{n} + \cdots + f(u_n)\frac{b - a}{n} \right]$$

$$= \frac{1}{b - a}[f(u_1)\,\Delta x + f(u_2)\,\Delta x + \cdots + f(u_n)\,\Delta x]$$

The sum obtained gives an approximation to the average value. As the norm $\Delta \to 0$, this sum is a better and better approximation to the average value of f on $[a, b]$. However, this sum is a Riemann sum,* so its limit is a definite integral. This suggests the following definition:

AVERAGE VALUE OF A FUNCTION OVER AN INTERVAL

The *average value, AV,* of a continuous function f over the interval $[a, b]$ is

$$AV = \frac{1}{b - a}\int_a^b f(x)\,dx \tag{2}$$

---

* See Section 15.5 for a discussion of Riemann sums.

EXAMPLE 1

The average value of $f(x) = x^3$ over the interval $[0, 2]$ is

$$AV = \frac{1}{2 - 0} \int_0^2 x^3 \, dx = \frac{1}{2}(4) = 2$$

Now Work Problem 1

Geometric Interpretation

The average value AV of a function f, as defined in (2), has an interesting geometric interpretation. If we rearrange the formula for AV, we obtain

$$AV(b - a) = \int_a^b f(x) \, dx \tag{3}$$

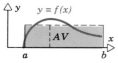

If $f(x) \geq 0$ on $[a, b]$, then the right side of (3) represents the area enclosed by the graph of $y = f(x)$, the x-axis, the line $x = a$, and the line $x = b$. The left side of the equation can be interpreted as the area of a rectangle of height AV and base $b - a$. Hence, (3) asserts that the average value of the function is the height of a rectangle with base $b - a$ and area equal to the area under the graph of f. See Figure 1.

Figure 1

Application

EXAMPLE 2

Suppose the current world population is 5×10^9 and the population in t years is assumed to grow exponentially according to the law

$$P(t) = (5 \times 10^9)e^{0.02t}$$

What will be the average world population during the next 20 years? (This number is helpful for long-range planning of agricultural and industrial output.)

SOLUTION

The average value of the population function $P(t)$ from $t = 0$ to $t = 20$ is

$$AV = \frac{1}{20 - 0} \int_0^{20} P(t) \, dt$$

$$= \frac{1}{20} \int_0^{20} (5 \times 10^9)e^{0.02t} \, dt$$

$$= \frac{5 \times 10^9}{20} \int_0^{20} e^{0.02t} \, dt$$

$$= \left(\frac{10^9}{4}\right) \frac{e^{0.02t}}{0.02} \Big|_0^{20}$$

$$= \left(\frac{10^9}{0.08}\right)(e^{0.4} - 1)$$

$$\approx (6.15 \times 10^9)$$

Now Work Problem 11

EXERCISE 16.1   Answers to odd-numbered problems begin on page AN-82.

In Problems 1–10 find the average value of each function f over the given interval.

1. $f(x) = x^2$,   over $[0, 1]$

2. $f(x) = 2x^2$,   over $[4, 2]$

3. $f(x) = 1 - x^2$,   over $[-1, 1]$

4. $f(x) = 16 - x^2$,   over $[-4, 4]$

5. $f(x) = 3x$,   over $[1, 5]$

6. $f(x) = 4x$,   over $[-5, 5]$

7. $f(x) = -5x^4 + 4x - 10$,   over $[-2, 2]$

8. $f(x) = 10x^4 - 2x + 7$,   over $[-1, 2]$

9. $f(x) = e^x$,   over $[0, 1]$

10. $f(x) = e^{-x}$,   over $[0, 1]$

11. Population Growth   Rework Example 2 if the population function is given by $P(t) = (5 \times 10^9)e^{0.03t}$. (This is a 3% growth rate.)

12. Population Growth   Rework Example 2 if the growth rate is 1%.

13. Average Temperature   A rod 3 meters long is heated to $25x$ degrees Celsius, where x is the distance (in meters) from one end of the rod. Calculate the average temperature of the rod.

14. Average Daily Rainfall   The rainfall per day, measured in centimeters, x days after the beginning of the year, is $0.00002(6511 + 366x - x^2)$. By integration, estimate the average daily rainfall for the first 180 days of the year.

15. Average Speed   A car starting from rest accelerates at the rate of 3 meters per second per second. Find its average speed over the first 8 seconds.

16. Average Area of a Circle   What is the average area of all circles with radii between 1 and 3 meters?

Technology Exercises
In Problems 1–4 find the average value,

$$AV = \frac{1}{b - a} \int_a^b f(x) \, dx$$

of each function over the given interval. Use a graphing calculator to graph the functions $y = f(x)$ and $y = AV$ together on the same pair of axes within the x range $[a, b]$.

1. $f(x) = e^x$,   over $[0, 1]$

2. $f(x) = -10 + 4x - 5x^4$,   over $[-2, 2]$

3. $f(x) = 7 - 2x + 10x^4$,   over $[-1, 2]$

4. $f(x) = x^4 - 8x^3 + 19x^2 - 12x$,   over $[-1, 5]$

16.2 DISCRETE PROBABILITY FUNCTIONS

We begin with the concept of a random variable. Intuitively, a *random variable* is a quantity that is measured in connection with a random experiment. For example, if the random experiment involves weighing individuals, then the weights of the individuals would be random variables. As another example, if the random experiment is to determine the time between arrivals of customers at a gas station, then the time between arrivals of customers at the gas station would be a random variable.

Let's consider some examples that demonstrate how to obtain random variables from random experiments.

When we perform a simple experiment, we are often interested not in a particular outcome, but rather in some number associated with that outcome. For example, in tossing a coin three times, we may be interested in the number of heads obtained, regardless of the particular sequence in which the heads appear. Similarly, the gambler throwing a pair of dice is generally more interested in the sum of the faces than in the particular number on each face.

Table 1 summarizes the results of the simple experiment of flipping a fair coin three times. The first column in the table gives a sample space of this experiment. The second column shows the number of heads for each outcome, and the third column shows the probability associated with each outcome.

Table 1

	Number of Heads	Probability
HHH	3	$\frac{1}{8}$
HHT	2	$\frac{1}{8}$
HTH	2	$\frac{1}{8}$
THH	2	$\frac{1}{8}$
HTT	1	$\frac{1}{8}$
THT	1	$\frac{1}{8}$
TTH	1	$\frac{1}{8}$
TTT	0	$\frac{1}{8}$

Suppose that in this experiment we are interested only in the total number of heads. This information is given in Table 2.

Table 2

Number of Heads Obtained in Three Flips of a Coin	Probability
0	$\frac{1}{8}$
1	$\frac{3}{8}$
2	$\frac{3}{8}$
3	$\frac{1}{8}$

The role of the random variable is to transform the original sample space {HHH, HHT, HTH, HTT, THH, THT, TTH, TTT} into a new sample space that consists of the number of heads that occur: {0, 1, 2, 3}. If X denotes the random variable, then X may take on any of the values 0, 1, 2, 3. From Table 2, the probability that the random variable X assumes the value 2 is

$$\text{Probability}(X = 2) = \tfrac{3}{8}$$

Also,

$$\text{Probability}(X = 5) = 0$$

We see that a random variable indicates the rule of correspondence between any member of a sample space and a number assigned to it. Thus, a random variable is a function.

RANDOM VARIABLE*

A *random variable* is a function that assigns a numerical value to each outcome of a sample space S.

We shall use the capital letter X to represent a random variable. In the coin-flipping example, the random variable X is

$$X(HHH) = 3 \quad X(HHT) = 2 \quad X(HTH) = 2 \quad X(THH) = 2$$
$$X(HTT) = 1 \quad X(THT) = 1 \quad X(TTH) = 1 \quad X(TTT) = 0$$

The random variable X indicates a relationship between the first two columns of Table 1; it pairs each outcome of the experiment with one of the real numbers 0, 1, 2, or 3.

EXAMPLE 1

Consider the experiment of a player rolling a fair die. He wins \$2 if the outcome is 1, 2, or 3, and he loses \$2 if the outcome is 4, 5, or 6. The random variable X is

$$X(1) = X(2) = X(3) = +2$$
$$X(4) = X(5) = X(6) = -2$$

What is the probability for the random variable to assume the value 2? In other words, what is $P(X = 2)$?

SOLUTION

Since the random variable takes on the value 2 at the outcomes 1, 2, and 3, we have

$$P(X = 2) = \tfrac{1}{2}$$

....

Random variables fall into two classes: those related to *discrete sample spaces* and those associated with *continuous sample spaces*.

DISCRETE SAMPLE SPACE; DISCRETE RANDOM VARIABLE

A sample space is *discrete* if it contains a finite number of outcomes or as many outcomes as there are whole numbers. A random variable is said to be *discrete* if it is defined over a discrete sample space.

* See Section 8.5 for a review of this topic.

EXAMPLE 2

The random experiment of flipping a fair coin until a head appears is represented by an infinite sample space that is discrete:

$$\{H, \quad TH, \quad TTH, TTTH, TTTTH, \ldots\}$$

1 flip 2 flips 3 flips 4 flips   5 flips

· ·

The sample space associated with the random experiment of measuring the height of each citizen of the United States is, of course, finite. Because of the proximity of all these heights, we usually allow the random variable to assume any real number so that the number of possible heights is infinite and the random variable is *continuous*. Thus, whenever a random variable has values that consist of an entire interval of real numbers, it is called a **continuous random variable.** In such cases we also say the sample space

Now Work Problem 5   is **continuous.**

Any practical problem that *measures* such dimensions as height, weight, time, and age will utilize a continuous random variable. As a result, the sample space associated with such experiments is also taken as continuous.

Look again at Table 2. This table contains two columns: the first lists the values the random variable X can take on and the second lists the corresponding probability. The relationship between the probability and the random variable is called a **probability function.** Such functions tell us how the total probability is distributed among the various values of the random variable.

EXAMPLE 3

In the experiment of one toss of two fair dice, compute the value of the probability function at $x = 7$ (that is, the two dice add up to 7).

SOLUTION

$$f(7) = P(X = 7) = P\{(1, 6), (2, 5), (3, 4), (4, 3), (5, 2), (6, 1)\} = \tfrac{6}{36} = \tfrac{1}{6}$$

· ·

We can compute the probabilities $f(2), f(3), \ldots, f(12)$ in a similar way. These values are summarized in the following table:

Values of x	2	3	4	5	6	7	8	9	10	11	12
Probability Function of x, $f(x)$	$\frac{1}{36}$	$\frac{2}{36}$	$\frac{3}{36}$	$\frac{4}{36}$	$\frac{5}{36}$	$\frac{6}{36}$	$\frac{5}{36}$	$\frac{4}{36}$	$\frac{3}{36}$	$\frac{2}{36}$	$\frac{1}{36}$

A **discrete probability function** is one for which the random variable is discrete. A **continuous probability function** is one for which the random variable is continuous.

In this section we limit our discussion to discrete probability functions. In Section 16.3 we discuss continuous probability functions.

The Binomial Probability Function

The binomial probability model was discussed earlier in Chapter 8. You may wish to review this material now.

The binomial probability function that assigns probabilities to the number of successes in n independent trials, where the probability of success p is the same from trial to trial, is an example of a discrete probability function. For this function, X is the random variable whose value for any outcome of the experiment is the number of successes obtained. For the binomial probability function, we may write

$$P(X = k) = b(n, k; p) = \binom{n}{k} p^k q^{n-k}$$

where $P(X = k)$ denotes the probability that the random variable equals k, that is, that exactly k successes are obtained, p is the probability of success, and $q = 1 - p$.

For example, consider the experiment of flipping a fair coin three times. If we let $f(x)$ denote the probability that the random variable X assumes the value x, then, for this experiment, we may write

$$f(0) = P(X = 0) = \tfrac{1}{8} \qquad f(1) = P(X = 1) = \tfrac{3}{8}$$
$$f(2) = P(X = 2) = \tfrac{3}{8} \qquad f(3) = P(X = 3) = \tfrac{1}{8}$$

Whenever possible, we try to express probability functions by means of formulas that enable us to calculate probabilities associated with the various values that the random variable assumes.

For example, in the experiment of tossing a fair coin three times, the probability function is

$$f(x) = \binom{3}{x} \left(\frac{1}{2}\right)^3 = \frac{3!}{(3-x)!x!} \frac{1}{8} = \frac{3}{4x!(3-x)!} \qquad x = 0, 1, 2, 3$$

where x denotes the number of heads obtained. Then,

$$f(0) = \frac{3}{(4)(0!)(3!)} = \frac{3}{(4)(1)(6)} = \frac{1}{8} \qquad f(1) = \frac{3}{(4)(1!)(2!)} = \frac{3}{(4)(1)(2)} = \frac{3}{8}$$

$$f(2) = \frac{3}{(4)(2!)(1!)} = \frac{3}{(4)(2)(1)} = \frac{3}{8} \qquad f(3) = \frac{3}{(4)(3!)(0!)} = \frac{3}{(4)(6)(1)} = \frac{1}{8}$$

as before.

Poisson Probability Function

The binomial probability function involves experiments consisting of repeated trials that are independent. Although this function has many applications, some natural phenomena and everyday applications do not conform to it. For instance, when n (the number of trials) is very large or p (the probability of success) is very small, say .05, another probability function, the **Poisson probability function,** is easier to use. We conclude this section with a discussion of this well-known and widely applied probability function.

Experience has shown that the Poisson probability function is an excellent model to use for computing probabilities associated with the following random experiments:

The number of cars arriving at a toll gate during a fixed period of time

The number of phone calls occurring within a certain time interval

The study of bacteria distribution in a culture

The arrival of customers at the checkout counter of a supermarket in a given time interval

The number of defects in a manufactured product

The number of earthquakes in a period of time

The number of typing errors on a given printed page

For everyday random phenomena, such as those mentioned above, we use the following result, which we state without proof.

THE POISSON PROBABILITY FUNCTION

Let X be a random variable and let the probability that exactly x successes occur in a given interval be given by

$$P(X = x) = f(x) = \frac{(np)^x e^{-np}}{x!} \qquad x = 0, 1, 2, \ldots$$

where n is the number of trials and p is the probability of success. Then the random variable X is known as a *Poisson variable* and f is the *Poisson probability function*. (Here, np is the average number of successes occurring in the given time interval.)

EXAMPLE 4

A department store has found that the daily demand for color television sets averages 3 in 100 customers. On a given day, 50 appliances are sold. What is the probability that more than 3 of the 50 sales are requests for television sets?

SOLUTION

We assume the problem to be a Poisson model since the probability that a television set is requested is small (.03). The random variable here is the occurrence of customers wanting to buy televisions. Also, n, the number of appliances sold, is 50, so that $np = 50(.03) = 1.5$. If X is a Poisson random variable, then the probability that more than 3 of the sales are for televisions is

$$P(X > 3) = 1 - P(X \leq 3) = 1 - [f(0) + f(1) + f(2) + f(3)]$$

$$= 1 - \left[\frac{(1.5)^0}{0!} e^{-1.5} + \frac{(1.5)^1}{1!} e^{-1.5} + \frac{(1.5)^2}{2!} e^{-1.5} + \frac{(1.5)^3}{3!} e^{-1.5} \right]$$

$$= 1 - [.223 + .335 + .251 + .126] = .065$$

Now Work Problem 13

EXAMPLE 5

Weather records show that, of the 30 days in November, on the average 3 days are snowy. What is the probability that November of next year will have at most 4 snowy days? Use a Poisson model.

SOLUTION

Let X be a Poisson random variable. From the information given,

$$n = 30 \qquad p = \frac{3}{30} \qquad np = 30\left(\frac{3}{30}\right) = 3$$

The probability that at most 4 days are snowy is

$$P(X \le 4) = f(0) + f(1) + f(2) + f(3) + f(4)$$

$$= e^{-3} + 3e^{-3} + \frac{9e^{-3}}{2} + \frac{27e^{-3}}{6} + \frac{81e^{-3}}{24}$$

$$= .050 + .149 + .224 + .224 + .168 = .815$$

....

Although the Poisson probability function is a limiting case of the binomial probability function, it has many applications to problems that are not directly related to the binomial probability function. The next example illustrates this.

EXAMPLE 6

During a certain period of time, people arrive at a ticket counter at an average rate of once every 2 minutes. What is the probability that, during a given 1-minute period, no people arrive at the ticket counter? What is the probability that at least two people arrive in this 1-minute period?

SOLUTION

Before solving the problem, we observe that no information is given about the probability that exactly one person will arrive in a 1-minute interval. However, we do know that this probability is small, and we know that the average rate of arrival in a 1-minute period is $\frac{1}{2}$, so that $np = \frac{1}{2}$.

If we use a Poisson probability function to solve the problem, the probability that no people arrive in the 1-minute interval is

$$f(0) = \frac{(\frac{1}{2})^0 e^{-1/2}}{0!} = e^{-1/2} = .607$$

Thus, there is a 61% probability that no people will arrive in a 1-minute interval.

To answer the second question, we need to compute $f(2) + f(3) + \cdots$. It is much easier to consider the complementary event: "At most one person arrives." To do this, we need to compute $f(0)$ and $f(1)$. We already have $f(0)$ from above, so now we compute $f(1)$:

$$f(1) = \frac{(np)e^{-np}}{1!} = \left(\frac{1}{2}\right)e^{-1/2} = .303$$

The probability that at most one person arrives in the 1-minute interval is

$$P(X \leq 1) = f(0) + f(1) = .607 + .303 = .910$$

Thus, the probability that two or more people arrive during a 1-minute period is

$$1 - .910 = .090$$

....

EXERCISE 16.2   Answers to odd-numbered problems begin on page AN-82.

In Problems 1–4 list the values of the given random variable X together with the probabilities of occurrence.

1. A fair coin is tossed two times and X is the random variable whose value for an outcome in the sample space is the number of heads obtained.

2. A fair die is tossed once. The random variable X is the number showing on the top face.

3. The random variable X is the number of female children in a family with three children. (Assume the probability of a female birth is $\frac{1}{2}$.)

4. A job applicant takes a three question true–false examination and guesses on each question. Let X be the number of right answers minus the number of wrong answers.

In Problems 5–8 classify the given random variable as discrete or continuous. If it is discrete, state whether the sample space is infinite or finite.

5. X is the number of tosses of a coin in the experiment of tossing a fair coin repeatedly until a head occurs.

6. X is the number of defective items in a lot of 10,000 items.

7. X is the length of time a person must wait in line at a checkout counter.

8. X denotes the time elapsed in minutes between the arrival of airplanes at an airport.

In Problems 9–12 find each probability. Assume X is a Poisson random variable with np = 6.

9. $P(X \leq 5)$

10. $P(X > 5)$

11. $P(X = 5)$

12. $P(1 < X < 4)$

13. Consumer Arrival   At a supermarket, customers arrive at a checkout counter at the rate of 60 per hour. What is the probability that 8 or fewer will arrive in a period of 10 minutes?

14. Insurance Policy   An insurance company insures 5000 people against the loss of both eyes in a car accident. Based on previous data, the rates were computed on the assumption that on the average 8 people in 100,000 will have car accidents each year that result in this type of injury. What is the probability that more than 3 of the insured will collect on their policy in a given year?

15. Defective Parts   A machine produces parts to meet certain specifications, and the probability that a part is defec-

tive is .05. A sample of 50 parts is taken. What is the probability that it will have 2 or more defective parts? Compute this probability, using both the Poisson and the binomial probability functions.

16. Frequency of Tornadoes   From past data it has been shown that the number of tornadoes hitting the Midwest each year is a random variable whose probability function can be approximated by a Poisson probability function with $np = 7$. Find the following:

(a) The probability that, in a given year, fewer than 5 tornadoes will hit the Midwest.

(b) The probability that, in a given year, no more than 7 tornadoes will hit the area.

Technology Exercises
Most graphing calculators have a **random number** *function (usually RAND or RND) generating numbers between 0 and 1. Every time you use a random number function, a different number is selected. Check your user's manual to see how to use this function on your graphing calculator.*

Sometimes probabilities are found by experimentation, that is, by performing an experiment. In Problems 1–6 use a random number function to perform an experiment.

1. **Tossing a Fair Coin**   Consider an experiment of tossing a coin four times and a random variable X counting the number of heads occurring in these four tosses. Simulate the experiment using the random number function on your calculator, considering a toss to be tails (T) if the result is less than 0.5 and considering a toss to be heads (H) if the result is more than or equal to 0.5. Record the number of heads in four tosses. [*Note:* Most calculators repeat the action of the last entry if you simply press the ENTER, or EXE, key again.] Repeat the experiment ten times, thus obtaining a sequence of ten numbers. Using these ten numbers you can estimate the probability $P(X = k)$ for each $k = 0, 1, 2, 3, 4$ by the ratio

$$\frac{\text{Number of times } k \text{ appears in your sequence}}{10}$$

Enter your estimates in the first column of the table below. Calculate the actual probabilities using the binomial probability function, and enter these numbers in the second column of the table. How close are your numbers to the actual values?

k	Your estimate of $P(X = k)$	Actual value of $P(X = k)$
0		
1		
2		
3		
4		

2. **Rolling a Fair Die**   Consider an experiment of rolling a die and a random variable X denoting the number showing on the top face. Simulate the experiment using the random number function on your calculator, considering a roll to have the outcome k if the value of the random number function is between $(k - 1) \cdot 0.167$ and $k \cdot 0.167$. Record the outcome. Repeat the experiment 50 times, thus obtaining a sequence of 50 numbers. [*Note:* Most calculators repeat the action of the last entry if you simply press the ENTER, or EXE, key again.] Using these 50 numbers you can estimate the probability $P(X = k)$ for each $k = 1, 2, 3, 4, 5, 6$ by the ratio

$$\frac{\text{Number of times } k \text{ appears in your sequence}}{50}$$

Enter your estimates in the first column of the table below. Calculate the actual probabilities, and enter these numbers in the second column of the table. How close are your numbers to the actual values?

k	Your estimate of $P(X = k)$	Actual value of $P(X = k)$
1		
2		
3		
4		
5		
6		

3. **Tossing a Loaded Coin**   Consider an experiment of tossing a loaded coin four times and a random variable X counting the number of heads occurring in these four tosses. Simulate the experiment using the random number function on your calculator, considering a toss to be tails (T) if the result is less than 0.80 and considering a toss to

be heads (H) if the result is more than or equal to 0.80. Record the number of heads in four tosses. [*Note:* Most calculators repeat the action of the last entry if you simply press the ENTER, or EXE, key again.] Repeat the experiment ten times, thus obtaining a sequence of ten numbers. Using these ten numbers you can estimate the probability $P(X = k)$ for each $k = 0, 1, 2, 3, 4$ by the ratio

$$\frac{\text{Number of times } k \text{ appears in your sequence}}{10}$$

Enter your estimates in the first column of the table below. Calculate the actual probabilities using the binomial probability function, and enter these numbers in the second column of the table. How close are your numbers to the actual values? Why is this coin considered to be loaded?

k	Your estimate of $P(X = k)$	Actual value of $P(X = k)$
0		
1		
2		
3		
4		

4. **Rolling an Octahedron**   Consider an experiment of rolling an octahedron (a regular polygon with all eight faces equal) and a random variable X denoting the number showing on the top face. Simulate the experiment using the random number function of your calculator, considering a roll to have the outcome k if the value of the random number function is between $(k - 1)/8$ and $k/8$ for $k = 1, 2, 3, \ldots, 8$. Record the outcome. Repeat the experiment 50 times, thus obtaining a sequence of 50 numbers. [*Note:* Most calculators repeat the action of the last entry if you simply press the ENTER, or EXE, key again.] Using these 50 numbers you can estimate the probability $P(X = k)$ for each $k = 1, 2, 3, 4, 5, 6, 7, 8$ by the ratio

$$\frac{\text{Number of times } k \text{ appears in your sequence}}{50}$$

Enter your estimates in the first column of the table at the top of the next column. Calculate the actual probabilities, and enter these numbers in the second column of the table. How close are your numbers to the actual values?

k	Your estimate of $P(X = k)$	Actual value of $P(X = k)$
1		
2		
3		
4		
5		
6		
7		
8		

5. **Rolling a Dodecahedron**   Consider an experiment of rolling a dodecahedron (a regular polygon with all twelve faces equal) and a random variable X denoting the number showing on the top face. Simulate the experiment using the random number function on your calculator, considering a roll to have the outcome k if the value of the random number function is between $(k - 1)/12$ and $k/12$ for $k = 1, 2, 3, \ldots, 12$. Record the outcome. Repeat the experiment 50 times, thus obtaining a sequence of 50 numbers. [*Note:* Most calculators repeat the action of the last entry if you simply press the ENTER, or EXE, key again.] Using these 50 numbers you can estimate the probability $P(X = 2)$ by the ratio

$$\frac{\text{Number of times } 2 \text{ appears in your sequence}}{50}$$

Calculate the actual probability $P(X = 2)$, and compare these values. How close is your estimate to the actual value?

6. **Rolling an Icosahedron**   Consider an experiment of rolling an icosahedron (a regular polygon with all twenty faces equal) and a random variable X denoting the number showing on the top face. Simulate the experiment using the random number function on your calculator, considering a roll to have the outcome k if the value of the random number function is between $(k - 1)/20$ and $k/20$ for $k = 1, 2, 3, \ldots, 20$. Record the outcome. Repeat the experiment 50 times, thus obtaining a sequence of 50 numbers. [*Note:* Most calculators repeat the action of the last entry if you simply press the ENTER, or EXE, key again.]

Using these 50 numbers you can estimate the probability $P(X = 5)$ by the ratio

$$\frac{\text{Number of times 5 appears in your sequence}}{50}$$

Calculate the actual probability $P(X = 5)$, and compare these values. How close is your estimate to the actual value?

16.3 PROBABILITY DENSITY FUNCTIONS

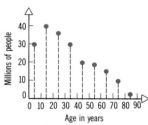

Figure 2

The random experiments we have encountered so far have resulted in probability models with sample spaces that are either finite or discrete. However, we often have to deal with random experiments that are neither finite nor discrete. Such random experiments may occur physically as approximations to discrete spaces with a very large number of outcomes.

For example, if the random experiment involves weighing individuals, the weight of each individual is treated as a continuous random variable. Likewise, in the random experiment of determining the time between arrival of customers at a gas station, the time between arrivals is treated as a continuous random variable.

Suppose the distribution of the population of the United States by ages is given by data grouped in 10-year intervals. See Figure 2.

In this illustration there are 30 million people in the age group between 0 and 10, 40 million between 10 and 20, and so on. We also know that the total population is 200 million.

Thus, since there are 40 million people in the age group $10-20$, the probability that a person is in this age group is $\frac{40}{200} = .20$. Figure 3 illustrates the distribution of probabilities for each age group. The function constructed by connecting the probability values by a smooth curve is an example of a **probability density function.**

When probabilities are associated with intervals, it is reasonable to assume that their values depend not only on the lengths of the intervals but also on their locations. For instance, there is no reason why the probability that a person is in the age group $10-20$ should equal the probability that a person is in the age group $60-70$, even though the two intervals have the same length. If we assume that there exists a function f with the values $f(x)$, then the probability that a person will be in a certain age group on a small interval from x to $x + \Delta x$ is approximately $f(x)\Delta x$, and the size of Δx determines how good the approximation is. This approximation is given by the area of the shaded rectangle shown in Figure 4, where the height of the rectangle is $f(x)$ and the base is Δx.

In a similar manner, we obtain the probabilities of other age groups by computing the areas corresponding to different subintervals. The desired probability for the whole

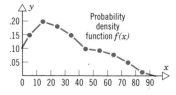

Figure 3

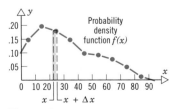

Figure 4

interval is approximately the sum of the areas

$$f(x_1)\,\Delta x + f(x_2)\,\Delta x + \cdots + f(x_n)\,\Delta x$$

for $x_1, x_2, \ldots, x_n$ in the interval.

Using the fact that the exact area under the graph of f is given by a definite integral, the exact probability that a person is between the ages c and d, denoted by $P(c \le X \le d)$, is given by

$$P(c \le X \le d) = \int_c^d f(x)\,dx$$

Thus, for the density function illustrated in Figure 3, the probability that a person is between 22 and 24 years of age is

$$\int_{22}^{24} f(x)\,dx$$

The argument we have presented here leads to the following definition of probability:

PROBABILITY DENSITY FUNCTION

The probability that the outcome of an experiment results in a value of a random variable X between c and d is given by

$$P(c \le X \le d) = \int_c^d f(x)\,dx \tag{1}$$

where $f(x)$ denotes the values of an appropriate function f called a *probability density function*.

Thus, we have reduced the problem of finding probabilities to that of finding the area enclosed by the graph of a probability density function and the x-axis on the interval from a to b. When probability density functions are integrated between any two values, they yield probabilities, so they possess the properties given in the following definition:

Properties of a Probability Density Function

A function f is a **probability density function** on the interval $[a, b]$ if it has two properties:

I $\displaystyle\int_a^b f(x)\,dx = 1$

where the interval $[a, b]$, possibly the interval $(-\infty, \infty)$, contains all values that the random variable X can assume.

II $f(x) \ge 0$

With (1) in mind, the rationale behind Property I becomes apparent. Since the interval $[a, b]$ contains all the values the random variable can assume, the probability that a random variable lies between a and b must equal 1.

EXAMPLE 1

Show that the function $f(x) = \frac{3}{56}(5x - x^2)$ is a probability density function over the interval $[0, 4]$.

SOLUTION

If f is indeed a probability density function, it has to satisfy Properties I and II. To verify Property I, we evaluate

$$\int_0^4 \frac{3}{56}(5x - x^2)\, dx = \frac{3}{56}\left(\frac{5x^2}{2} - \frac{x^3}{3}\right)\Bigg|_0^4$$

$$= \frac{3}{56}\left[\frac{(5)16}{2} - \frac{64}{3}\right] = \frac{3}{56}\left(\frac{56}{3}\right) = 1$$

Hence, Property I is satisfied.
Property II is also satisfied since

$$f(x) = \frac{3}{56}(5x - x^2) = \frac{3}{56}x\,(5 - x) \geq 0$$

for all x in the interval $[0, 4]$.
Thus, $f(x) = \frac{3}{56}(5x - x^2)$, for x in $0 \leq x \leq 4$, is a probability density function.

Now Work Problem 1

....

EXAMPLE 2

Compute the probability that the random variable X with probability density function $\frac{3}{56}(5x - x^2)$ assumes values between 1 and 2.

SOLUTION

To compute $P(1 \leq X \leq 2)$, we use (1):

$$P(1 \leq X \leq 2) = \int_1^2 \frac{3}{56}(5x - x^2)\, dx = \frac{3}{56}\left(\frac{5x^2}{2} - \frac{x^3}{3}\right)\Bigg|_1^2$$

$$= \frac{3}{56}\left[\frac{(5)(4)}{2} - \frac{8}{3}\right] - \frac{3}{56}\left(\frac{5}{2} - \frac{1}{3}\right) = \frac{31}{112}$$

Now Work Problem 25

....

How do we obtain a probability density function f? For individual random experiments, it is possible to construct them as we indicated in the example on age probabilities. However, the construction of a probability density function is usually a tedious and difficult task and often depends on the nature of the problem. Fortunately, several relatively simple probability density functions are available that can be used to fit most random experiments. In every example we discuss, the probability density function is given.

Uniform Density Function

The **uniform density function,** or the **uniform distribution,** the simplest of probability density functions, is one in which the random variable assumes all its values with equal probability. The probability density function for this random variable is

Figure 5

$$f(x) = \begin{cases} \dfrac{1}{b-a} & \text{if} \quad a \le x \le b \\ 0 & \text{if} \quad x < a \text{ or } x > b \end{cases}$$

The graph of this function is given in Figure 5.

Notice that the function has the value 0 outside the interval $a \le x \le b$.

To verify that the uniform density function is a probability density function, we need to show that Properties I and II are satisfied.

I $$\int_a^b f(x)\, dx = \int_a^b \frac{1}{b-a}\, dx = \frac{x}{b-a}\Big|_a^b = \frac{b-a}{b-a} = 1$$

II Since $b > a$, $f(x) = 1/(b-a) > 0$, for $a \le x \le b$.

EXAMPLE 3

Trains leave a terminal every 40 minutes. What is the probability that a passenger arriving at a random time to catch a train will have to wait at least 10 minutes?

SOLUTION

Let T (time) be a random variable and assume it is uniformly distributed for time $0 \le T \le 40$. The probability that the passenger must wait at least 10 minutes is

$$P(T \ge 10) = \int_{10}^{40} \frac{1}{40}\, dt = \frac{t}{40}\Big|_{10}^{40} = \frac{40}{40} - \frac{10}{40} = \frac{3}{4}$$

· ·

Exponential Density Function

EXPONENTIAL DENSITY FUNCTION

Let X be a continuous random variable. Then X is said to be *exponentially distributed* if X has the probability density function

$$f(x) = \begin{cases} \lambda e^{-\lambda x} & \text{if} \quad x \ge 0 \\ 0 & \text{if} \quad x < 0 \end{cases}$$

where λ* is a positive constant. The function f is called the *exponential density function*.

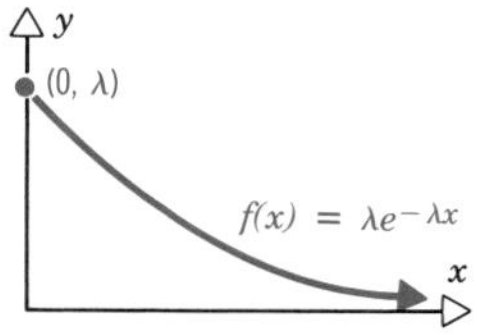

The graph of the exponential density function is given in Figure 6.

Figure 6

* The Greek letter *lambda*.

The exponential density function is nonnegative for all x. Also,

$$\int_0^\infty f(x)\, dx = \int_0^\infty \lambda e^{-\lambda x}\, dx = \lim_{b\to\infty} \int_0^b \lambda e^{-\lambda x}\, dx$$

$$= \lim_{b\to\infty} \lambda \left(-\frac{1}{\lambda}\right) e^{-\lambda x}\Big|_0^b = \lim_{b\to\infty} -e^{-\lambda b} + 1 = 1$$

Thus, the exponential density function satisfies Properties I and II and is a probability density function.

In general any situation that deals with *waiting time* between successive events will lead to an exponential density function. In the exponential density function the constant λ plays the role of the average number of arrivals per unit time.

EXAMPLE 4

Airplanes arriving at an airport follow a pattern similar to the exponential density function with an average of $\lambda = 15$ arrivals per hour. Determine the probability of an arrival within 0.1 hour (6 minutes). Use an exponential density function.

SOLUTION

The probability is

$$P(t \le 0.1) = \int_0^{0.1} 15e^{-15t}\, dt = -e^{-15t}\Big|_0^{0.1} = -e^{-1.5} + 1$$

$$= 1 - .223 = .777$$

Thus, the probability of an arrival within 0.1 hour is .777.

Now Work Problem 27

. . . .

The next example illustrates how the average λ of an exponential density function can be calculated.

EXAMPLE 5

From past data it is known that a certain machine normally produces 1 defective product in every 200. To detect defective products, an inspector tests a continuous stream of products and records the interval in which a defective product appears. Use an exponential density function to find the probability that, after a defective product is found, the next 200 are nondefective.

SOLUTION

If we start from a defective item, the probability that the $(x + 1)$st product will be the next defective is

$$f(x) = \lambda e^{-\lambda x}$$

The average defective rate, $\frac{1}{200}$, equals λ, so the exponential density function is

$$f(x) = \frac{1}{200} e^{-x/200}$$

The probability that a defective product will be found within the first 200 items following a defective one is

$$\int_0^{200} \frac{1}{200} e^{-x/200} \, dx = -e^{-x/200} \Big|_0^{200} = -e^{-1} + 1 = .632$$

Thus, the probability that the next defective product will not be within the next 200 is $1 - .632 = .368$.

························

Expectation

We begin by referring to the definition of expectation for finite sample spaces as discussed in Chapter 8. There we defined expectation by associating both a probability and a payoff to each outcome of the sample space. These payoffs are, in fact, the values of a discrete random variable, and the corresponding probability is the value of a probability function.

EXPECTED VALUE

If X is a discrete (finite) random variable with the probability function listed below,

x	x_1	x_2	$\cdots$	x_n
$P(X = x)$	$f(x_1)$	$f(x_2)$	$\cdots$	$f(x_n)$

the *expected value of X* is

$$E(X) = x_1 f(x_1) + x_2 f(x_2) + \cdots + x_n f(x_n)$$

We now state the following definition for a continuous random variable.

EXPECTED VALUE FOR A CONTINUOUS RANDOM VARIABLE

If X is a continuous random variable with the probability density function $f(x)$, $a \le x \le b$, the *expected value of X* is

$$E(X) = \int_a^b x f(x) \, dx$$

Once again the expected value of the random variable X is interpreted as the average (mean) value of the random variable. For example, if X is a random variable measuring heights, then $E(X)$ is the average (mean) height of the population.

EXAMPLE 6   A passenger arrives at a train terminal where trains arrive every 40 minutes. Determine the expected waiting time by using a uniform density function.

SOLUTION

Let the random variable T measure waiting time with uniform density function $f(t) = \frac{1}{40}$, where $0 \le T \le 40$. The expected value $E(T)$ is then

$$E(T) = \int_0^{40} t \frac{1}{40}\, dt = \frac{t^2}{80}\bigg|_0^{40} = \frac{1}{80}(40)^2 = \frac{1600}{80} = 20 \text{ minutes}$$

Now Work Problem 17

EXAMPLE 7   Show that the expected value of the random variable with a uniform density function $f(x) = 1/(b - a)$, $a \le x \le b$, is the midpoint of the interval $[a, b]$.

SOLUTION

Let X be a random variable with density function $f(x) = 1/(b - a)$, $a \le x \le b$. Then

$$E(X) = \int_a^b x \frac{1}{b - a}\, dx = \frac{1}{2}\frac{x^2}{b - a}\bigg|_a^b = \frac{1}{2}\frac{b^2 - a^2}{b - a}$$

$$= \frac{1}{2}\frac{(b - a)(b + a)}{b - a} = \frac{a + b}{2}$$

EXERCISE 16.3   Answers to odd-numbered problems begin on page AN-83.

In Problems 1–8 verify that each function is a probability density function over the indicated interval.

1. $f(x) = \frac{1}{2}$,   over $[0, 2]$

2. $f(x) = \frac{1}{5}$,   over $[0, 5]$

3. $f(x) = 2x$,   over $[0, 1]$

4. $f(x) = \frac{1}{8}x$,   over $[0, 4]$

5. $f(x) = \frac{3}{250}(10x - x^2)$,   over $[0, 5]$

6. $f(x) = \frac{6}{27}(3x - x^2)$,   over $[0, 3]$

7. $f(x) = \frac{1}{x}$,   over $[1, e]$

8. $f(x) = \frac{4}{3(x + 1)^2}$,   over $[0, 3]$

If $f(x) \ge 0$ is not a probability density function, we can find a constant k such that $kf(x)$ satisfies the condition $\int_a^b kf(x)\, dx = 1$. For the functions in Problems 9–16 determine the constant k that will make each one a probability density function over the interval indicated.

9. $f(x) = 1$,   over $[0, 3]$

10. $f(x) = 1$,   over $[0, 4]$

11. $f(x) = x$,   over $[0, 2]$

12. $f(x) = x$,   over $[0, \frac{1}{2}]$

13. $f(x) = 10x - x^2$,   over $[0, 5]$

14. $f(x) = 10x - x^2$,   over $[0, 8]$

15. $f(x) = \frac{1}{x}$,   over $[1, 2]$

16. $f(x) = \frac{1}{(x + 1)^3}$,   over $[3, 7]$

In Problems 17–24 compute the expected value for each probability density function.

17. $f(x) = \frac{1}{2}$,   over $[0, 2]$

18. $f(x) = \frac{1}{5}$,   over $[0, 5]$

19. $f(x) = 2x$,   over $[0, 1]$

20. $f(x) = \frac{1}{8}x$,   over $[0, 4]$

21. $f(x) = \frac{3}{250}(10x - x^2)$,   over $[0, 5]$

22. $f(x) = \frac{6}{27}(3x - x^2)$,   over $[0, 3]$

23. $f(x) = \dfrac{1}{x}$,   over $[1, e]$

24. $f(x) = \dfrac{4}{3(x + 1)^2}$,   over $[0, 3]$

25. A number x is selected at random from the interval $[0, 5]$. The probability density function for x is

$$f(x) = \tfrac{1}{5} \quad \text{for } 0 \le x \le 5$$

Find the probability that a number is selected in the subinterval $[1, 3]$.

26. A number x is selected at random from the interval $[0, 10]$. The probability density function for x is

$$f(x) = \tfrac{1}{10} \quad \text{for } 0 \le x \le 10$$

Find the probability that a number is selected in the subinterval $[6, 9]$.

27. Frequency of Telephone Calls The time between incoming telephone calls at a hotel switchboard has an exponential density function with $\lambda = 0.5$ minute. What is the probability that there is an interval of at least 6 minutes between incoming calls?

28. Inventory Control The demand for an inventory item has a probability density function given by

$$f(x) = 0.2e^{-0.2x}$$

where $f(x)$ is the probability that x items will be in demand over a 1-week period. What is the probability that fewer than 5 items will be in demand? Fewer than 100? More than 10?

29. Psychological Testing Let T be the random variable that a subject in a psychological testing program will make a certain choice after t seconds. If the probability density function is

$$f(t) = 0.4e^{-0.4t}$$

what is the probability that the subject will make the choice in less than 5 seconds?

30. Waiting Time Buses on a certain route run every 50 minutes. What is the probability that a person arriving at a random stop along the route will have to wait at least 30

minutes? Assume that the random variable T is the time the person will have to wait and assume that T is uniformly distributed.

31. Learning Time A manufacturer of educational games for children finds through extensive psychological research that the average time it takes for a child in a certain age group to learn the rules of the game is predicted by a **beta probability density function,**

$$f(x) = \begin{cases} \dfrac{1}{4500}(30x - x^2) & \text{if } 0 \le x \le 30 \\ 0 & \text{if } x < 0 \text{ or } x > 30 \end{cases}$$

where x is the time in minutes. What is the probability a child will learn how to play the game within 10 minutes? What is the probability a child will learn the game after 20 minutes? What is the probability the game is learned in at least 10 minutes, but no more than 20 minutes?

32. Waiting Time A passenger arrives at a train station where trains arrive every 20 minutes. Determine the expected waiting time by using a uniform density function.

33. Cost Estimate The probability density function that gives the probability that an electrical contractor's cost estimate is off by x percent is

$$f(x) = \tfrac{3}{56}(5x - x^2)$$

for x in the interval $[0, 4]$. On average, by what percent can the contractor be expected to be off?

34. The **variance** σ^2 associated with the probability density function f on $[a, b]$ is defined as

$$\sigma^2 = \int_a^b x^2 f(x)\, dx - [E(x)]^2$$

Verify that for the uniform density function,

$$\sigma^2 = \frac{(b - a)^2}{12}$$

Technology Exercises
Triangular probability distributions are often used to model business situations.
The graph of a possible triangular probability function, p(x), is shown below. Such
a distribution is used when we know only two pieces of information: the interval of
possible values of the random variable X, that is, a ≤ X ≤ b, and the most likely
value for the random variable X, that is, the value X = c with the largest
probability of occurring.

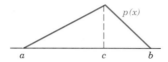

The probability of X = c can be calculated geometrically using Property I:

$$\int_a^b p(x)\, dx = 1$$

That is, the area of the triangle above must be 1, and since the probability p(X = c)
is the height of the triangle, we have

$$\frac{(b - a) \cdot p(X = c)}{2} = 1$$

Using this information, we can write the probability density function as a
piecewise-defined function consisting of two line segments:

$$p(x) = \begin{cases} m_1 x + b_1 & \text{if } a \le x \le c \\ m_2 x + b_2 & \text{if } c \le x \le b \end{cases}$$

Use this model in Problems 1 and 2.

1. **Price of a New Car**   A business analyst predicts that a new car will cost between $10,000 and $20,000, with the most likely price being $17,000. Using the triangular model above, find the equation of the probability density function $p(x)$.

   (a) What is the probability that the car will cost less than $15,000?

   (b) Graph the function $p(x)$ using a graphing calculator (or a computer).

   (c) Estimate the expected price of the car from the graph.

   (d) Evaluate the expected price of the car. Why is the expected price different from $17,000?

2. **Stock Market Analysis**   A stock market analyst predicts that a stock will be worth between $15 and $19, with the most likely value being $16. Using the triangular model above, find the equation of the probability density function $p(x)$.

   (a) What is the probability that the stock will be worth more than $18?

   (b) Graph the function $p(x)$ using a graphing calculator (or a computer).

   (c) Estimate the expected value of the stock from the graph.

   (d) Evaluate the expected value of the stock. Why is the expected value of the stock different from $16?

*Most graphing calculators have a **random number** function (usually RAND or RND) generating numbers between 0 and 1. Every time you use a random number function, a different number is selected. Check your user's manual to see how to use this function on your graphing calculator.*

Sometimes probabilities are found by experimentation, that is, by performing an experiment. In Problems 3 and 4 use a random number function to perform an experiment.

3. Use a random number function to select a value for the random variable X. Repeat this experiment 50 times. [*Note:* Most calculators repeat the action of the last entry if you simply press the ENTER, or EXE, key again.] Count the number of times the random variable X is between 0.6 and 0.9. Calculate the ratio

$$R = \frac{\text{the random variable } X \text{ is between 0.6 and 0.9}}{50}$$

What is your result? Calculate the actual probability $P(0.6 \le X < 0.9)$.

4. Use a random number function to select a value for the random variable X. Repeat this experiment 50 times. [*Note:* Most calculators repeat the action of the last entry if you simply press the ENTER, or EXE, key again.] Count the number of times the random variable X is between 0.1 and 0.3. Calculate the ratio

$$R = \frac{\text{the random variable } X \text{ is between 0.1 and 0.3}}{50}$$

What is your result? Calculate the actual probability $P(0.1 \le X < 0.3)$.

16.4 DIFFERENTIAL EQUATIONS

In studies of physical, chemical, biological, and other natural phenomena, scientists attempt, on the basis of long observation, to deduce mathematical laws that will describe and predict nature's behavior. Such laws often involve the derivatives of some unknown function F, and it is required to find this unknown function F. For example, it may be required to find all functions $y = F(x)$ so that

$$\frac{dy}{dx} = f(x) \tag{1}$$

This equation is an example of what is called a **differential equation.** A function $y = F(x)$ for which $dy/dx = f(x)$ is a **solution** of the differential equation. The **general solution** of $dy/dx = f(x)$ consists of all the antiderivatives of f.

EXAMPLE 1 | The general solution of the differential equation

$$\frac{dy}{dx} = 5x^2 + 2 \tag{2}$$

is

$$y = \frac{5x^3}{3} + 2x + K$$

. . . .

A **particular solution** of the differential equation $dy/dx = f(x)$ occurs when K is assigned a particular value. When a particular solution is required, we use a **boundary condition.**

EXAMPLE 2

In the differential Equation (2) we require the general solution to obey the boundary condition that $y = 5$ when $x = 3$. Find the particular solution.

SOLUTION

In the general solution

$$y = \frac{5x^3}{3} + 2x + K$$

let $x = 3$ and $y = 5$. Then

$$5 = \frac{5(27)}{3} + (2)(3) + K$$

$$K = -46$$

The particular solution of (2) with the boundary condition that $y = 5$ when $x = 3$ is, therefore,

$$y = \frac{5x^3}{3} + 2x - 46$$

· ·

EXAMPLE 3

Solve the differential equation below with the boundary condition that $y = -1$ when $x = 3$.

$$\frac{dy}{dx} = x^2 + 2x + 1$$

SOLUTION

The general solution of the differential equation is

$$y = \frac{x^3}{3} + x^2 + x + K$$

To determine the number K, we use the boundary condition. Then

$$-1 = \frac{3^3}{3} + 3^2 + 3 + K$$

$$K = -22$$

The particular solution of the differential equation with the boundary condition that $y = -1$ when $x = 3$ is

$$y = \frac{x^3}{3} + x^2 + x - 22$$

Now Work Problem 1

· ·

Applications

The statement below describes many situations in physics, chemistry, and biology.

> The amount A of a substance varies with time t in such a way that the time rate of change of A is proportional to A itself.

We may state this in the form of the differential equation

$$\frac{dA}{dt} = kA \tag{3}$$

where $k \neq 0$ is a real number. If $k > 0$, then (3) asserts that the time rate of change of A is positive, so that the amount A of the substance is increasing; if $k < 0$, then (3) asserts that the time rate of change of A is negative, so that the amount A of the substance is decreasing.

We seek a solution to (3). We begin by rewriting the equation in terms of differentials as

$$\frac{dA}{A} = k \, dt$$

Then we integrate each side to get

$$\int \frac{dA}{A} = \int k \, dt$$

from which we obtain

$$\ln A = kt + K$$

To determine the constant K, we use the boundary condition that when $t = 0$, the initial amount present is A_0. As a result, $\ln A_0 = K$, so that

$$\ln A = kt + \ln A_0$$
$$\ln A - \ln A_0 = kt$$
$$\ln \frac{A}{A_0} = kt \qquad \text{Properties of logarithms, page 558}$$
$$\frac{A}{A_0} = e^{kt} \qquad \text{Definition of natural logarithm, page 559}$$

The solution of the differential Equation (3) is therefore

$$A = A_0 e^{kt} \tag{4}$$

When a function $A = A(t)$ varies according to the law given by (3), or its equivalent (4), it is said to follow the **exponential law,** or the **law of uninhibited growth or decay,** or the **law of continuously compounded interest.**

As previously noted, the sign of k determines whether A is increasing (if $k > 0$) or decreasing (if $k < 0$). Figure 7 illustrates the graphs of (4) for $k > 0$ and for $k < 0$.

Figure 7

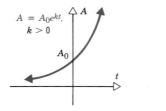
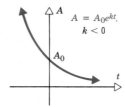

Bacterial Growth

Our first application is to bacterial growth.

EXAMPLE 4

Assume that the population of a colony of bacteria *increases at a rate proportional to the number present.** If the number of bacteria doubles in 5 hours, how long will it take for the bacteria to triple?

SOLUTION

Let $N(t)$ be the number of bacteria present at time t. Then the assumption that this colony of bacteria increases at a rate proportional to the number present can be mathematically written as

$$\frac{dN}{dt} = kN \tag{5}$$

where k is a positive constant of proportionality. By (4), the solution of (5) is

$$N(t) = N_0 e^{kt}$$

where N_0 is the initial number of bacteria in this colony. Since the number of bacteria doubles in 5 hours, we have

$$N(5) = 2N_0$$

But $N(5) = N_0 e^{k(5)}$, so that

$$N_0 e^{5k} = 2N_0$$
$$e^{5k} = 2$$
$$5k = \ln 2$$
$$k = (\tfrac{1}{5}) \ln 2 = 0.1386$$

---

* This is a model of uninhibited growth. However, after enough time has passed, growth will not continue at a rate proportional to the number present. Other factors, such as lack of living space and dwindling food supply, will start to affect the rate of growth. The model presented accurately reflects the way growth occurs in the early stages.

The function that gives the number N of bacteria present at time t is

$$N(t) = N_0 e^{0.1386t}$$

The time t that is required for this colony to triple must satisfy the equation

$$N(t) = 3N_0$$
$$N_0 e^{0.1386t} = 3N_0$$
$$e^{0.1386t} = 3$$
$$0.1386t = \ln 3$$
$$t = \frac{\ln 3}{0.1386} = 7.926 \text{ hours}$$

Now Work Problem 11

$\cdots\cdots\cdots\cdots\cdots\cdots\cdots\cdots$

Radioactive Decay

Our second application is to radioactive decay, and in particular its use in carbon dating. For a radioactive substance, *the rate of decay is proportional to the amount present at a given time t.* That is, if $A(t)$ represents the amount of a radioactive substance at time t, we have

$$\frac{dA}{dt} = kA$$

where the constant k is negative and depends on the radioactive substance. The **half-life** of a radioactive substance is the time required for half of the substance to decay.

In carbon dating we use the fact that all living organisms contain two kinds of carbon, carbon-12 (a stable carbon) and carbon-14 (a radioactive carbon). As a result, when an organism dies, the amount of carbon-12 present remains unchanged, while the amount of carbon-14 begins to decrease. This change in the amount of carbon-14 present relative to the amount of carbon-12 present makes it possible to calculate the time at which the organism lived.

EXAMPLE 5

In the skull of an animal found in an archaeological dig, it was determined that about 20% of the original amount of carbon-14 was still present. If the half-life of carbon-14 is 5600 years, find the approximate age of the animal.

SOLUTION

Let $A(t)$ be the amount of carbon-14 present in the skull at time t. Then $A(t)$ satisfies the differential equation $dA/dt = kA$, whose solution is

$$A(t) = A_0 e^{kt}$$

where A_0 is the amount of carbon-14 present at time $t = 0$. To determine the constant k, we use the fact that when $t = 5600$, half of the original amount A_0 will remain. Thus,

$$\tfrac{1}{2}A_0 = A_0e^{5600k}$$
$$\tfrac{1}{2} = e^{5600k}$$
$$5600k = \ln \tfrac{1}{2}$$
$$k = -0.000124$$

The relationship between the amount $A(t)$ of carbon-14 and time t is therefore

$$A(t) = A_0e^{-(0.000124)t}$$

If the amount $A(t)$ of carbon-14 is 20% of the original amount A_0, we have

$$0.2A_0 = A_0e^{-(0.000124)t}$$
$$0.2 = e^{-(0.000124)t}$$
$$-(0.000124)t = \ln 0.2$$
$$t = \frac{-1.6094}{-0.000124} = 12{,}979 \text{ years}$$

Thus, the animal lived approximately 13,000 years ago.

Now Work Problem 15   . . . .

EXERCISE 16.4   Answers to odd-numbered problems begin on page AN-83.

In Problems 1–10 solve each differential equation, using the indicated boundary condition.

1. $\dfrac{dy}{dx} = x^2 - 1$

   $y = 0$ when $x = 0$

2. $\dfrac{dy}{dx} = x^2 + 4$

   $y = 1$ when $x = 0$

3. $\dfrac{dy}{dx} = x^2 - x$

   $y = 3$ when $x = 3$

4. $\dfrac{dy}{dx} = x^2 + x$

   $y = 5$ when $x = 3$

5. $\dfrac{dy}{dx} = x^3 - x + 2$

   $y = 1$ when $x = -2$

6. $\dfrac{dy}{dx} = x^3 + x - 5$

   $y = 1$ when $x = -2$

7. $\dfrac{dy}{dx} = e^x$

   $y = 4$ when $x = 0$

8. $\dfrac{dy}{dx} = \dfrac{1}{x}$

   $y = 0$ when $x = 1$

9. $\dfrac{dy}{dx} = \dfrac{x^2 + x + 1}{x}$

   $y = 0$ when $x = 1$

10. $\dfrac{dy}{dx} = x + e^x$

    $y = 4$ when $x = 0$

11. **Bacterial Growth** The rate of growth of bacteria is proportional to the amount present. If initially there are 100 bacteria and 5 minutes later there are 150 bacteria, how many bacteria will be present after 1 hour? How many are present after 90 minutes? How long will it take for the number of bacteria to reach 1,000,000?

12. Answer the questions posed in Problem 11 if after 8 minutes the number of bacteria present grows from 100 to 150.

13. **Radioactive Decay** The half-life of radium is 1690 years. If 8 grams of radium are present now, how many grams will be present in 100 years?

14. **Radioactive Decay** If 25% of a radioactive substance disappears in 10 years, what is the half-life of the substance?

15. **Age of a Tree** A piece of charcoal is found to contain 30% of the carbon-14 it originally had. When did the tree from which the charcoal came die? Use 5600 years as the half-life of carbon-14.

16. **Age of a Fossil** A fossilized leaf contains 70% of a normal amount of carbon-14. How old is the fossil?

17. **Population Growth** The population growth of a colony of mosquitoes obeys the uninhibited growth equation. If there are 1500 mosquitoes initially, and there are 2500 mosquitoes after 24 hours, what is the size of the mosquito population after 3 days?

18. **Population Growth** The population of a suburb doubled in size in an 18-month period. If this growth continues and the current population is 8000, what will the population be in 4 years?

19. **Bacterial Growth** The number of bacteria in a culture is growing at a rate of $3000e^{2t/5}$ per unit of time t. At $t = 0$, the number of bacteria present was 7500. Find the number present at $t = 5$.

20. **Bacterial Growth** At any time t, the rate of increase in the area of a culture of bacteria is twice the area of the culture. If the initial area of the culture is 10, then what is the area at time t?

21. **Bacterial Growth** The rate of change in the number of bacteria in a culture is proportional to the number present. In a certain laboratory experiment, a culture had 10,000 bacteria initially, 20,000 bacteria at time t_1 minutes, and 100,000 bacteria at $(t_1 + 10)$ minutes.
    (a) In terms of t only, find the number of bacteria in the culture at any time t minutes ($t \geq 0$).
    (b) How many bacteria were there after 20 minutes?
    (c) At what time were 20,000 bacteria observed? That is, find the value of t_1.

22. **Chemistry** Salt (NaCl) decomposes in water into sodium (Na^+) and chloride (Cl^-) ions at a rate proportional to its mass. If the initial amount of salt is 25 kilograms, and after 10 hours, 15 kilograms are left:
    (a) How much salt would be left after 1 day?
    (b) After how many hours would there be less than $\frac{1}{2}$ kilogram of salt left?

23. **Age of a Fossil** Radioactive beryllium is sometimes used to date fossils found in deep-sea sediment. The decay of radioactive beryllium satisfies the equation $dA/dt = -\alpha A$, where $\alpha = 1.5 \times 10^{-7}$, and t is measured in years. What is the half-life of radioactive beryllium?

24. **Pressure** Atmospheric pressure is a function of altitude above sea level and is given by the equation $dP/da = \beta P$, where β is a constant. The pressure is measured in millibars (mb). At sea level ($a = 0$), $P(0)$ is 1013.25 mb, which means that the atmosphere at sea level will support a column of mercury 1013.25 millimeters high at a standard temperature of 15°C. At an altitude of $a = 1500$ meters, the pressure is 845.6 mb.
    (a) What is the pressure at $a = 4000$ meters?
    (b) What is the pressure at 10 kilometers?
    (c) In California the highest and lowest points are Mount Whitney (4418 meters) and Death Valley (86 meters below sea level). What is the difference in their atmospheric pressures?
    (d) What is the atmospheric pressure at Mount Everest (elevation 8848 meters)?
    (e) At what elevation is the atmospheric pressure equal to 1 mb?

CHAPTER REVIEW

IMPORTANT TERMS AND CONCEPTS

IMPORTANT FORMULAS

Average Value

$$AV = \frac{1}{b-a} \int_a^b f(x)\, dx$$

Binomial Probability Function

$$\binom{n}{k} p^k q^{n-k}$$

Poisson Probability Function

$$f(x) = \frac{(np)^x e^{-np}}{x!} \quad x = 0, 1, 2, \ldots$$

Probability Density Function

$$P(c \le X \le d) = \int_c^d f(x)\, dx$$

Uniform Density Function

$$f(x) = \begin{cases} \dfrac{1}{b-a} & \text{if} \quad a \le x \le b \\ 0 & \text{if} \quad x < a \text{ or } x > b \end{cases}$$

Exponential Density Function

$$f(x) = \begin{cases} \lambda e^{-\lambda x} & \text{if} \quad x \ge 0 \\ 0 & \text{if} \quad x < 0 \end{cases}$$

Expectation

$$E(X) = x_1 f(x_1) + x_2 f(x_2) + \cdots + x_n f(x_n)$$

$$E(X) = \int_a^b x f(x)\, dx$$

TRUE–FALSE ITEMS   Answers are on page AN-83.

T_____ F_____ **1.** The average value of a function f over the interval $[a, b]$ equals $\dfrac{f(b) - f(a)}{b - a}$.

T_____ F_____ **2.** The Poisson probability function is a good model for computing probabilities associated with experiments involving the number of phone calls occurring in a time interval.

T_____ F_____ **3.** A function f is a probability density function if $\int_a^b f(x)\, dx = 1$.

T_____ F_____ **4.** The general solution of the differential equation $dy/dx = f(x)$ is $\int f(x)\, dx$.

FILL IN THE BLANKS   Answers are on page AN-83.

1. For a continuous function f defined on $[a, b]$, the number $\dfrac{1}{b-a} \int_a^b f(x)\, dx$ is called the _____ _____ of f.

2. A function that assigns a numerical value to each outcome of a sample space is called a _____ _____.

3. The function constructed by connecting probability values by a smooth curve is called a _____ _____ function.

4. A continuous random variable X is _____ _____ if X has the density function

$$f(x) = \begin{cases} \lambda e^{-\lambda x} & \text{if} \quad x \ge 0 \\ 0 & \text{if} \quad x < 0 \end{cases} \quad \text{where } \lambda > 0 \text{ is a constant}$$

5. The equation $dy/dx = f(x)$ is an example of what is called a _____ _____.

6. A particular solution of $dy/dx = f(x)$ occurs when K is assigned a _____ _____.

REVIEW EXERCISES   Answers to odd-numbered problems begin on page AN-83.

1. Find the average value of $f(x) = x^3$ over the interval $[0, 1]$.

2. Find the average value of $f(x) = x^n$, $n \geq 1$ an integer, over the interval $[-1, 1]$.

3. Classify the following random variables as discrete or continuous. If the random variable is discrete, state whether the sample space is infinite or finite.

   (a) X is the number of defective automobiles that come off an assembly line on a given day.
   (b) X is the age in years of voters in a local election.

4. **Defective Items**   A shipment of 1000 light bulbs from a certain manufacturer contains 5% that are defective or broken. Use a Poisson distribution to find the probability that 3 or more are defective or broken in a shipment of 360.

5. **Frequency of Calls**   A fire department in a medium-sized city receives an average of 2.5 calls each minute. Use a Poisson distribution to find the probability that in a given minute no calls occur. What is the probability that exactly 4 calls will occur in a given minute?

6. **Probability Density Function**   Show that the function

$$f(x) = \frac{3}{688,000} (-x^2 + 200x - 5000)$$

is a probability density function over the interval $[20, 100]$.

7. **Life Insurance**   A man who is currently 20 years old wants to purchase life insurance. The insurance company is interested in determining at what age X (in years) he is likely to die. If the probability density function given in

Problem 6 measures this likelihood, find the probability that the man is likely to die at or before age 40. What is the probability he will die at or before age 60?

8. An experiment has the probability density function $f(x) = 6(x - x^2)$ and outcomes lying between 0 and 1. Determine the probability that an outcome

   (a) Lies between $\frac{1}{3}$ and $\frac{1}{2}$.
   (b) Lies between 0 and $\frac{3}{4}$.

9. Suppose the outcome X of an experiment lies between 0 and 2, and the probability density function for X is $f(x) = \frac{1}{2}x$. Find

   (a) $P(X \leq 1)$ (b) $P(1 \leq X \leq 1.5)$
   (c) $P(1.5 \leq X)$

10. **Customer Arrival**   At a fast-food counter it takes an average of 3 minutes to get served. Suppose that the service time X for a customer has an exponential probability density function.

   (a) What fraction of the customers are served within 2 minutes?
   (b) What is the probability that a customer will have to wait at least 3 minutes?

11. **Waiting Time**   A toy machine produces a toy every 2 minutes. An inspector arrives at a random time and must wait X minutes for a toy.

   (a) Find the probability density function for X.
   (b) Find the probability that the inspector has to wait at least 1 minute.
   (c) Find the probability that the inspector has to wait no more than 1 minute.

In Problems 12–15 solve each differential equation, using the indicated boundary condition.

12. $\dfrac{dy}{dx} = \dfrac{x + 1}{x}$, $y = 1$ when $x = 1$

13. $\dfrac{dy}{dx} = e^x + x$, $y = 5$ when $x = 0$

14. $\dfrac{dy}{dx} = x\sqrt{x^2 + 1}$, $y = 6$ when $x = 0$

15. $\dfrac{dy}{dx} = x\sqrt{x^2 - 16}$, $y = 0$ when $x = 5$

16. Find the general solution of the differential equation

$$\frac{dy}{dx} = 4x^3 - 6x^2 + x - 4$$

Find a particular solution if $y = 3$ when $x = 1$.

17. Find the general solution of the differential equation

$$\frac{dy}{dx} = 3e^{5x} - 2x + 1$$

Find a particular solution if $y = -2$ when $x = 0$.

18. **Bacterial Growth**   Bacteria grown in a certain culture increase at a rate proportional to the amount present. If there are 2000 bacteria present initially and the amount triples in 2 hours, how many bacteria will there be in $4\frac{1}{2}$ hours?

19. **Population Growth**   In 1990 the population of Glenwood was 3000. By 1995 the population doubled. Assuming that this growth continues and assuming that the rate of increase is proportional to the population, what will the population be in 2000?

20. **Radioactive Decay**   The half-life of a certain radioactive material is 1000 years. How long will it take for the amount of radioactive material present to decay to 20% of its original amount?

21. **Age of an Animal**   The skeleton of an animal is found to contain 35% of the original amount of carbon-14. What is the approximate age of this animal? (The half-life of carbon-14 is 5600 years.)

Mathematical Questions from Professional Exams

1. **Actuary Exam—Part I**   For a Poisson distribution $f(x)$ in which $f(0) = 2f(1)$, what is $Pr(X \geq 2)$?

(a) $\frac{2}{3}$ (b) $\frac{2}{3}e^{-1/3}$ (c) $1 - \frac{2}{3}e^{-1/2}$
(d) $1 - \frac{3}{2}e^{-1/2}$ (e) $1 - 3e^{-2}$

2. **Actuary Exam—Part II**   Two men patronize the same barber shop. If they both arrive independently between 3 P.M. and 4 P.M. on the same day and both stay for 15 minutes, what is the probability that they are in the barber shop for part or all of the same time?

(a) $\frac{3}{8}$ (b) $\frac{7}{16}$ (c) $\frac{1}{2}$
(d) $\frac{9}{16}$ (e) $\frac{5}{8}$

3. **Actuary Exam—Part II**   In a certain process for enameling copper wire, small bare spots occur at random with an average frequency of two such spots per 1000 feet of wire. What is the probability that a 5000-foot roll of this copper wire will contain no more than one such bare spot?

(a) $10e^{-10}$ (b) $11e^{-10}$ (c) $61e^{-10}$
(d) $1 - 50e^{-10}$ (e) $1 - 10e^{-10}$

4. **Actuary Exam—Part II**   Each day X arrives at a point A between 8:00 and 9:00 A.M., his times of arrival being uniformly distributed. Y arrives independently at A between 8:30 and 9:00 A.M., his times of arrival also being uniformly distributed. What is the probability that Y arrives before X?

(a) $\frac{1}{8}$ (b) $\frac{1}{6}$ (c) $\frac{2}{9}$
(d) $\frac{1}{4}$ (e) $\frac{1}{2}$

Chapter 17

Functions of Two or More Variables

In this chapter we discuss functions that have more than one independent variable. Many of the ideas introduced in Chapters 1, 11, and 12 for functions of one variable have a parallel representation for functions of two or more variables.

17.1 FUNCTIONS AND THEIR GRAPHS

So far, we have considered only functions of one independent variable, usually expressed explicitly by an equation $y = f(x)$. Quite often, models require functions of more than one variable. Consider the following example:

Suppose a firm specializes in producing only one item at a profit of $45 per item. If x is the number of items produced, then the total profit $P(x)$ is given by

$$P(x) = 45x$$

$P(x)$ is an example of a function of *one variable*. Suppose the firm specializes in producing two items x and y at a profit of $45 and $55 per item, respectively. Then its total profit is now a function of two variables x and y, and the profit is given by

$$P(x, y) = 45x + 55y$$

$P(x, y)$ is an example of a function of *two variables x and y*.

Another example that gives rise to functions of two variables is the cost of producing a certain item, which may depend on variables such as labor and material. In economic theory, supply and demand of a commodity often depend not only on the commodity's own price, but also on the prices of related commodities and some other factors (such as income level, time of year, etc.).

Note that for the profit function $P(x, y) = 45x + 55y$, the function assigns for each pair (x, y) a unique number $45x + 55y$. For example, if $x = 3$ and $y = 4$, we get

$$P(3, 4) = 45(3) + 55(4) = 355$$

Definition of a Function of Two Variables

FUNCTION OF TWO VARIABLES

A *function f of two variables x and y* is a rule $z = f(x, y)$ that assigns a unique real number z to each ordered pair (x, y) of real numbers in a set D of the xy-plane. The set D is called the *domain* of the function f.

In the equation $z = f(x, y)$ we refer to z as the **dependent variable** and to x and y as the **independent variables**. The **range** of f consists of all real numbers $f(x, y)$, where (x, y) is in D. Figure 1 illustrates one way of depicting $z = f(x, y)$.

Figure 1

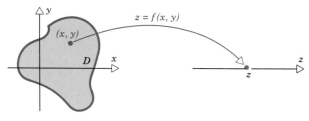

D is the domain of f

EXAMPLE 1

If $z = f(x, y) = x\sqrt{y} + xy^2$, find

(a) $f(0, 0)$ (b) $f(2, 4)$ (c) $f(1, 1)$ (d) $f(x + \Delta x, y)$
(e) $f(x, y + \Delta y)$

SOLUTION

(a) $f(0, 0) = 0\sqrt{0} + 0 \cdot 0^2 = 0$ (b) $f(2, 4) = 2\sqrt{4} + 2 \cdot 4^2 = 36$
(c) $f(1, 1) = 1\sqrt{1} + 1 \cdot 1^2 = 2$
(d) $f(x + \Delta x, y) = (x + \Delta x)\sqrt{y} + (x + \Delta x) \cdot y^2$
(e) $f(x, y + \Delta y) = x\sqrt{y + \Delta y} + x(y + \Delta y)^2$

Now Work Problem 19

....

We can similarly define functions of three independent variables, $w = f(x, y, z)$, or a function of four independent variables. In this chapter we will primarily concern ourselves with functions of two or three variables.

As with functions of one variable, a function of two variables is usually given by a rule or an equation and, unless otherwise stated, *the domain is taken to be the largest set of points in the plane for which this rule or equation makes sense in the real number system.*

EXAMPLE 2

Find the domain of each function:

(a) $z = f(x, y) = \dfrac{2x^2 + 5y}{x - 2y}$ (b) $z = g(x, y) = \ln(y - x^2)$

SOLUTION

(a) Since division by zero is not allowed, the only ordered pairs (x, y) for which f cannot be evaluated are those for which $x = 2y$. Hence, the domain of f consists of all ordered pairs (x, y) of real numbers for which $x \neq 2y$.

(b) The rule for g is the natural logarithm of the expression $y - x^2$. Since only logarithms of positive numbers are allowed, we must have

$$y - x^2 > 0$$

from which

$$y > x^2$$

This inequality describes the domain of g, namely, the set of points (x, y) "inside" the parabola $y = x^2$. See Figure 2. Note that we use a dashed rule to show that the points on the parabola $y = x^2$ are not part of the domain.

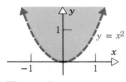

Figure 2

Now Work Problem 35

....

Three-Dimensional Coordinate Systems

To graph functions of two variables, we need a **three-dimensional rectangular coordinate system.** In Chapter 1 we established a correspondence between points on a line and real numbers. Then we showed that each point in a plane can be associated with an

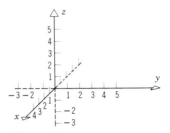

Figure 3

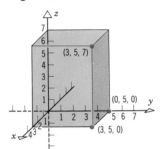

Figure 4

Now Work Problems 1
and 13

ordered pair of real numbers. Here, we show that each point in (three-dimensional) space can be associated with an **ordered triple** of real numbers.

First, select a fixed point called the **origin.** Through the origin, we draw three mutually perpendicular lines. These are called the **coordinate axes,** and are labeled the **x-axis, y-axis,** and **z-axis.** On each of the three lines we choose one direction as positive and select an appropriate scale on each axis. Figure 3 shows one possible way to draw the three axes.

Just as we did in one and two dimensions, we assign coordinates to each point P in space. Specifically, we identify each point P with an ordered triple of real numbers (x, y, z), and we refer to it as "the point (x, y, z)." Thus, "the point $(3, 5, 7)$" is the point for which $x = 3$, $y = 5$, $z = 7$. So, starting at the origin, we reach P by moving 3 units along the positive x-axis, then 5 units in the direction of the positive y-axis, and, finally, 7 units in the direction of the positive z-axis. Figure 4 illustrates the location of the point $(3, 5, 7)$, as well as the points $(3, 5, 0)$ and $(0, 5, 0)$. Observe that any point on the x-axis will have the form $(x, 0, 0)$. Similarly, $(0, y, 0)$ and $(0, 0, z)$ represent points on the y-axis and the z-axis, respectively.

In addition, all points of the form $(x, y, 0)$ constitute a plane called the **xy-plane.** This plane is perpendicular to the z-axis. Similarly, the points $(0, y, z)$ form the **yz-plane,** which is perpendicular to the x-axis; and the points $(x, 0, z)$ form the **xz-plane,** which is perpendicular to the y-axis (see Figure 5). Figure 5 also illustrates that points of the form (x, y, z), where $z = 5$, lie in a plane parallel to the xy-plane. Similarly, points (x, y, z), where $y = 7$, lie in a plane parallel to the xz-plane.

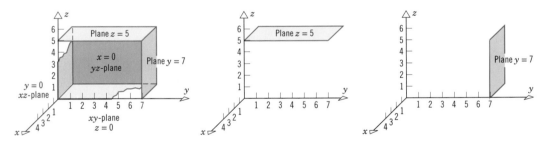

Figure 5

Graphing Functions of Two Variables

The graph of a function $z = f(x, y)$ of two variables, called a **surface,** consists of all points (x, y, z) for which $z = f(x, y)$ and (x, y) is in the domain of f. See Figure 6.

Figure 6

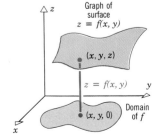

EXAMPLE 3

Describe the graph of the function: $z = f(x, y) = \sqrt{4 - x^2 - y^2}$

SOLUTION

First, we observe that the domain of f consists of all points (x, y) for which $x^2 + y^2 \leq 4$. Since $z = f(x, y)$ is defined as a square root, it follows that $z \geq 0$. Thus, the graph will lie above the xy-plane. By squaring both sides of $z = \sqrt{4 - x^2 - y^2}$, we find that

$$z^2 = 4 - x^2 - y^2 \qquad z \geq 0$$
$$x^2 + y^2 + z^2 = 4 \qquad z \geq 0$$

The reader might recognize this as the equation of a hemisphere, center at $(0, 0, 0)$ and radius 2. Figure 7 illustrates the graph.

· ·

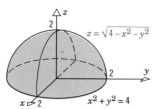

Figure 7

Figure 8 depicts some three-dimensional surfaces in space that were generated using a computer. Obtaining the shape, or even a rough sketch, of the graph of most functions

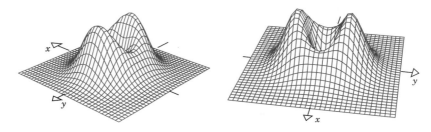

(a) Two views of the surface
$z = f(x, y) = (x^2 + 2y^2)\,e^{1 - x^2 - y^2}$

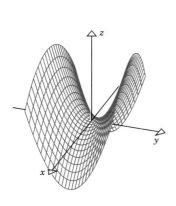

(b) The surface of
$z = f(x, y) = 4y^2 - x^2$

Figure 8

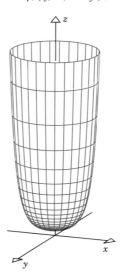

(c) The surface of
$z = f(x, y) = e^{x^2 + y^2}$

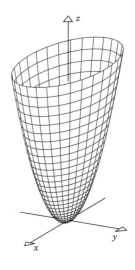

(d) The surface of
$z = f(x, y) = x^2 + 4y^2$

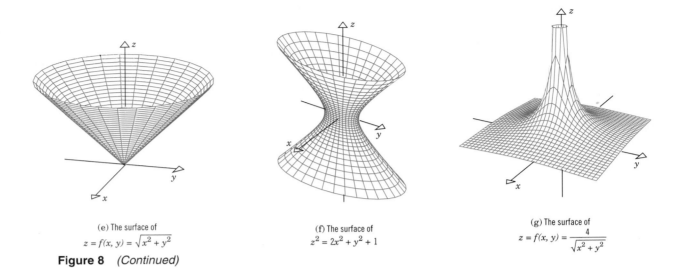

(e) The surface of
$z = f(x, y) = \sqrt{x^2 + y^2}$

(f) The surface of
$z^2 = 2x^2 + y^2 + 1$

(g) The surface of
$z = f(x, y) = \dfrac{4}{\sqrt{x^2 + y^2}}$

Figure 8   *(Continued)*

of two variables is a difficult task without the use of computer graphics, and is not taken up in this book.

Functions of Three Variables

The function f defined by $w = f(x, y, z)$ is a function of the three independent variables x, y, and z. For each ordered triple (x, y, z) in the domain, the rule f assigns a value to w, the dependent variable. In this case the domain is a collection of points in space. Figure 9 illustrates a way of depicting the function $w = f(x, y, z)$.

Figure 9

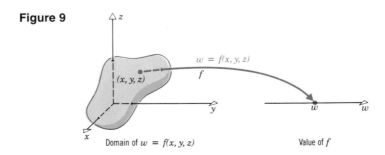

Domain of $w = f(x, y, z)$

Value of f

As with functions of two variables, a function of three variables is usually given by an equation, and, unless otherwise stated, the domain is taken to be the largest set of points in space for which this rule or formula makes sense in the real number system.

EXAMPLE 4

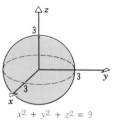

$x^2 + y^2 + z^2 = 9$

Figure 10

Find the domain of the function $w = f(x, y, z) = \sqrt{9 - x^2 - y^2 - z^2}$.

SOLUTION

Since square roots of negative numbers are not permitted, the domain of this function consists of all points for which $x^2 + y^2 + z^2 - 9 \leq 0$. Therefore, the domain consists of all points inside and on the sphere with center at $(0, 0, 0)$ and radius 3. See Figure 10 for the graph of the domain.

· ·

The graph of a function $w = f(x, y, z)$ of three variables consists of all points (x, y, z, w) for which (x, y, z) is in the domain of f and $w = f(x, y, z)$. Because this requires locating points in a four-dimensional space, it is impossible for us to draw such a graph.

EXERCISE 17.1   Answers to odd-numbered problems begin on page AN-84.

In Problems 1–6 plot each point.

1. $(1, 1, 1)$ **2.** $(0, 0, 1)$ **3.** $(0, 2, 5)$ **4.** $(-1, 5, 0)$

5. $(-3, 1, 0)$ **6.** $(4, -1, -3)$

In Problems 7–12 opposite vertices of a rectangular box whose edges are parallel to the coordinate axes are given. List the coordinates of the other six vertices of the box.

7. $(0, 0, 0); (2, 1, 3)$ **8.** $(0, 0, 0); (4, 2, 2)$ **9.** $(1, 2, 3); (3, 4, 5)$

10. $(5, 6, 1); (3, 8, 2)$ **11.** $(-1, 0, 2); (4, 2, 5)$ **12.** $(-2, -3, 0); (-6, 7, 1)$

In Problems 13–18 describe in words the set of all points (x, y, z) that satisfy the given conditions.

13. $y = 3$ **14.** $z = -3$ **15.** $x = 0$

16. $x = 1$ and $y = 0$ **17.** $z = 5$ **18.** $x = y$ and $z = 0$

In Problems 19–28 evaluate $f(2, 1)$.

19. $f(x, y) = x^2 + y$ **20.** $f(x, y) = x - y^2$ **21.** $f(x, y) = \sqrt{xy}$

22. $f(x, y) = x\sqrt{y}$ **23.** $f(x, y) = \dfrac{1}{2x + y}$ **24.** $f(x, y) = \dfrac{x}{x - 3y}$

25. $f(x, y) = \dfrac{x^2 - y}{x - y}$ **26.** $f(x, y) = \dfrac{x + y^2}{x^2 - y^2}$ **27.** $f(x, y) = \sqrt{4 - x^2 y^2}$

28. $f(x, y) = \sqrt{9 - x^2 y^2}$

29. Let $f(x, y) = 3x + 2y + xy$. Find

   (a) $f(1, 0)$ (b) $f(0, 1)$
   (c) $f(2, 1)$ (d) $f(x + \Delta x, y)$
   (e) $f(x, y + \Delta y)$

30. Let $f(x, y) = x^2y + x + 1$. Find

   (a) $f(0, 0)$ (b) $f(0, 1)$
   (c) $f(2, 1)$ (d) $f(x + \Delta x, y)$
   (e) $f(x, y + \Delta y)$

31. Let $f(x, y) = \sqrt{xy} + x$. Find

   (a) $f(0, 0)$ (b) $f(0, 1)$
   (c) $f(a^2, t^2)$; $a > 0, t > 0$
   (d) $f(x + \Delta x, y)$ (e) $f(x, y + \Delta y)$

32. Let $f(x, y) = e^{x+y}$. Find

   (a) $f(0, 0)$ (b) $f(1, -1)$
   (c) $f(x + \Delta x, y)$ (d) $f(x, y + \Delta y)$

33. Let $f(x, y, z) = x^2y + y^2z$. Find

   (a) $f(1, 2, 3)$ (b) $f(0, 1, 2)$
   (c) $f(-1, -2, -3)$

34. Let $f(x, y, z) = 3x^2 + y^2 - 2z^2$. Find

   (a) $f(1, 2, 3)$ (b) $f(0, 1, 2)$
   (c) $f(-1, -2, -3)$

In Problems 35–48 find the domain of each function.

35. $z = f(x, y) = \sqrt{x}\,\sqrt{y}$

36. $z = f(x, y) = \sqrt{xy}$

37. $z = f(x, y) = \sqrt{9 - x^2 - y^2}$

38. $z = f(x, y) = \sqrt{x^2 + y^2 - 16}$

39. $z = f(x, y) = \dfrac{\ln x}{\ln y}$

40. $z = f(x, y) = \ln \dfrac{x}{y}$

41. $z = f(x, y) = \dfrac{3}{x^2 + y^2 - 4}$

42. $z = f(x, y) = \dfrac{4}{9 - x^2 - y^2}$

43. $z = f(x, y) = \ln (x^2 + y^2)$

44. $z = f(x, y) = \ln(4x - y^2)$

45. $w = f(x, y, z) = \sqrt{x^2 + y^2 + z^2 - 16}$

46. $w = f(x, y, z) = \sqrt{9 - (x^2 + y^2 + z^2)}$

47. $w = f(x, y, z) = \dfrac{4}{x^2 + y^2 + z^2}$

48. $w = f(x, y, z) = \ln(x^2 + y^2 + z^2)$

49. For the function $z = f(x, y) = 3x + 4y$, find

   (a) $f(x + \Delta x, y)$
   (b) $f(x + \Delta x, y) - f(x, y)$
   (c) $\dfrac{f(x + \Delta x, y) - f(x, y)}{\Delta x}$, $\Delta x \neq 0$
   (d) $\displaystyle\lim_{\Delta x \to 0} \dfrac{f(x + \Delta x, y) - f(x, y)}{\Delta x}$, $\Delta x \neq 0$

50. For the function $z = f(x, y) = 4x + 5y$, find

   (a) $f(x, y + \Delta y)$
   (b) $f(x, y + \Delta y) - f(x, y)$
   (c) $\dfrac{f(x, y + \Delta y) - f(x, y)}{\Delta y}$, $\Delta y \neq 0$
   (d) $\displaystyle\lim_{\Delta y \to 0} \dfrac{f(x, y + \Delta y) - f(x, y)}{\Delta y}$, $\Delta y \neq 0$

51. Cost of Construction The cost of the bottom and top of a cylindrical tank is \$300 per square meter, and the cost of the sides is \$500 per square meter. Write the total cost of constructing such a tank as a function of the radius r and height h (both in meters).

52. Cost of Construction The cost per square centimeter of the material to be used for an open rectangular box is \$4 for the bottom and \$2 for the other sides. Write the total cost of constructing such a box as a function of its bottom and side dimensions.

53. Baseball A pitcher's earned run average is given by

$$A(N, I) = 9 \left(\frac{N}{I}\right)$$

where N is the total number of earned runs given up in I innings of pitching. Find

   (a) $A(3, 4)$ (b) $A(6, 3)$ (c) $A(2, 9)$

54. Intelligence Quotient In psychology, intelligence quotient (IQ) is measured by

$$IQ = f(M, C) = 100 \frac{M}{C}$$

where M is a person's mental age and C is the person's chronological or actual age.

(a) Find the IQ of a 12-year-old child whose mental age is 10.
(b) Find the IQ of a 10-year-old child whose mental age is 12.
(c) If a 10-year-old girl has an IQ of 120, what is her mental age?

55. Production Function The production function for a toy manufacturer is given by the equation

$$Q(L, M) = 400L^{0.3}M^{0.7}$$

where Q is the output in units, L is the labor in hours, and M is the number of machine hours. Find

(a) $Q(19, 21)$ (b) $Q(21, 20)$

17.2 PARTIAL DERIVATIVES

For a function $y = f(x)$ of one independent variable, we introduced the idea of a derivative f'. For a function $z = f(x, y)$ of two independent variables, we introduce the idea of a *partial derivative*. A function $z = f(x, y)$ of two variables x and y will have two partial derivatives: f_x, the partial derivative of f with respect to x; and f_y, the partial derivative of f with respect to y. The partial derivative of f with respect to x is found by differentiating f with respect to x while treating y as if it were a constant. The partial derivative of f with respect to y is found by differentiating f with respect to y while treating x as if it were a constant.

For example, if $z = f(x, y) = 2xy + 3xy^2$, then we can find f_x by differentiating $z = 2xy + 3xy^2$ with respect to x, while treating y as if it were a constant. The result is

$$f_x = 2y + 3y^2$$

Similarly, by treating x as if it were constant and differentiating with respect to y, we obtain

$$f_y = 2x + 6xy$$

We define these partial derivatives as limits of certain difference quotients.

PARTIAL DERIVATIVES OF $z = f(x, y)$

Let $z = f(x, y)$ be a function of two variables. Then the partial derivatives of f with respect to x and with respect to y are functions f_x and f_y defined as follows:

$$f_x(x, y) = \lim_{\Delta x \to 0} \frac{f(x + \Delta x, y) - f(x, y)}{\Delta x}$$

$$f_y(x, y) = \lim_{\Delta y \to 0} \frac{f(x, y + \Delta y) - f(x, y)}{\Delta y} \tag{1}$$

provided these limits exist.

Observe the similarity between the above definitions and the definition of a derivative given in Chapter 12. Observe also that in $f_x(x, y)$, an increment Δx is given to x, while y is fixed; in $f_y(x, y)$, an increment Δy is given to y, while x is fixed.

> **Finding Partial Derivatives of $z = f(x, y)$**
>
> To find $f_x(x, y)$:   Differentiate f with respect to x while treating y as a constant.
> To find $f_y(x, y)$:   Differentiate f with respect to y while treating x as a constant.

EXAMPLE 1

Find f_x and f_y for

$$z = f(x, y) = 4x^3 + 2x^2y + y^2$$

SOLUTION

To find f_x, we treat y as a constant and differentiate $z = 4x^3 + 2x^2y + y^2$ with respect to x. The result is

$$f_x(x, y) = 12x^2 + 4xy$$

To find f_y, we treat x as a constant and differentiate $z = 4x^3 + 2x^2y + y^2$ with respect to y. The result is

$$f_y(x, y) = 2x^2 + 2y$$

· ·

For the function in the preceding example, let us evaluate f_x and f_y at $(3, -1)$:

$$f_x(3, -1) = 12(3^2) + 4(3)(-1) = 96 \qquad f_y(3, -1) = 2(3^2) + 2(-1) = 16$$

EXAMPLE 2

(a) Find f_x and f_y for $z = f(x, y) = x^2 \ln y + ye^x$.
(b) Evaluate $f_x(2, 1)$ and $f_y(2, 1)$.

SOLUTION

(a) $f_x(x, y) = 2x \ln y + ye^x \qquad\qquad f_y(x, y) = \dfrac{x^2}{y} + e^x$

(b) $f_x(2, 1) = 2 \cdot 2 \cdot \ln 1 + 1 \cdot e^2 = e^2 \qquad f_y(2, 1) = \dfrac{2^2}{1} + e^2 = 4 + e^2$

· ·

Another Notation

There is another notation used for the partial derivatives f_x and f_y of a function $z = f(x, y)$, which we introduce here:

$$f_x(x, y) = \frac{\partial f}{\partial x} = \frac{\partial z}{\partial x} \qquad f_y(x, y) = \frac{\partial f}{\partial y} = \frac{\partial z}{\partial y}$$

The symbols $\partial/\partial x$ and $\partial/\partial y$ denote operations performed on a function to obtain the partial derivatives with respect to x in the case of $\partial/\partial x$ and with respect to y in the case of $\partial/\partial y$.

EXAMPLE 3

(a) $\dfrac{\partial}{\partial x}(3e^x y^2 - xy) = \dfrac{\partial}{\partial x}3e^x y^2 - \dfrac{\partial}{\partial x}xy = 3y^2\dfrac{\partial}{\partial x}e^x - y\dfrac{\partial}{\partial x}x = 3e^x y^2 - y$

(b) $\dfrac{\partial}{\partial y}(3e^x y^2 - xy) = \dfrac{\partial}{\partial y}3e^x y^2 - \dfrac{\partial}{\partial y}xy = 3e^x\dfrac{\partial}{\partial y}y^2 - x\dfrac{\partial}{\partial y}y = 6e^x y - x$

· ·

Geometric Interpretation of Partial Derivatives

For a geometric interpretation of the partial derivatives of $z = f(x, y)$, we look at the graph of the surface $z = f(x, y)$. See Figure 11. In computing f_x we hold y fixed, say, at $y = y_0$, and then differentiate with respect to x. But holding y fixed at y_0 is equivalent to intersecting the surface $z = f(x, y)$ with the plane $y = y_0$, the result being the curve $z = f(x, y_0)$. Thus, the partial derivative f_x is the slope of the tangent line to this curve. In particular:

Geometric Interpretation of $f_x(x_0, y_0)$

The slope of the tangent line to the curve of intersection of the surface $z = f(x, y)$ and the plane $y = y_0$ at the point (x_0, y_0, z_0) on the surface equals $f_x(x_0, y_0)$.

Geometric Interpretation of $f_y(x_0, y_0)$

The slope of the tangent line to the curve of intersection of the surface $z = f(x, y)$ and the plane $x = x_0$ at the point (x_0, y_0, z_0) on the surface equals $f_y(x_0, y_0)$.

See Figure 12 (p. 878).

Figure 11

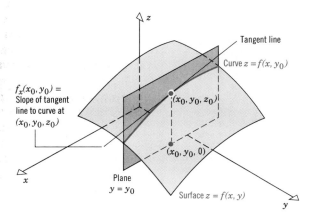

Figure 12

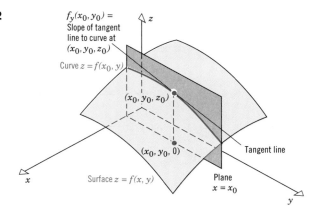

EXAMPLE 4

Find the slope of the tangent line to the curve of intersection of the surface $z = f(x, y) = 16 - x^2 - y^2$:

(a) With the plane $y = 2$ at the point $(1, 2, 11)$.
(b) With the plane $x = 1$ at the point $(1, 2, 11)$.

SOLUTION

(a) The slope of the tangent line to the curve of intersection of the surface $z = 16 - x^2 - y^2$ and the plane $y = 2$ at any point is $f_x(x, y) = -2x$. At the point $(1, 2, 11)$, the slope is $f_x(1, 2) = -2(1) = -2$. See Figure 13(a).

Figure 13

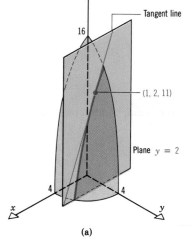

(a)

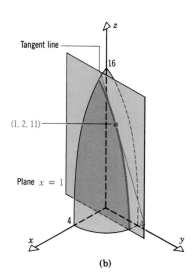

(b)

(b) The slope of the tangent line to the curve of intersection of the surface $z = 16 - x^2 - y^2$ and the plane $x = 1$ at any point is $f_y(x, y) = -2y$. At the point $(1, 2, 11)$, the slope is $f_y(1, 2) = -2(2) = -4$. See Figure 13(b).

Now Work Problem 31

· ·

Application to Marginal Analysis

We have already noted the similarity of the definition of the partial derivatives of $z = f(x, y)$ and the definition of the derivatives of a function of one variable, and we have given a geometric interpretation. Let's look at how partial derivatives are interpreted in a business environment.

EXAMPLE 5

In many production processes the total cost of manufacturing consists of a fixed cost and two variable costs: the cost of raw materials and the cost of labor. If the total cost is given by

$$C(x, y) = 180 + 18x + 40y$$

where x is the cost (in dollars) of raw materials and y is the cost (in dollars) of labor, find C_x and C_y and give an interpretation to your answer.

SOLUTION

We interpret $C_x(x, y) = 18$ to mean that when the cost of labor y is held fixed, an increase of $1 in the cost of raw materials causes an increase of $18 in the total cost of the product. The partial derivative C_x thus measures the incremental cost due to an increase of $1 in the cost of raw material, while labor costs are held fixed. The partial derivative $C_y(x, y) = 40$ means that when the cost of raw materials x is held fixed, an increase of $1 in the cost of labor y causes a $40 increase in the total cost of the product. The partial derivative C_x is called the **marginal cost of raw material** and C_y is the **marginal cost of labor.**

Now Work Problem 45

· ·

Production Functions

The production of most commodities requires the use of at least two factors of production—for example, labor, land, capital, materials, or machines. If the quantity z of a commodity is produced by using the amounts x and y, respectively, of two factors of production, then the **production function**

$$z = f(x, y)$$

gives the amount of **output** z when the amounts x and y, respectively, of the **inputs** are used simultaneously. For such a representation to be economically meaningful, it is assumed that the amounts of the inputs can be varied without restriction, at least in the range of interest, and that the production function is continuous.

MARGINAL PRODUCTIVITY

If the production function is given by $z = f(x, y)$, then the partial derivative $\partial z/\partial x$ of z with respect to x (with y held constant) is the *marginal productivity of x* or the *marginal product of x*. The partial derivative $\partial z/\partial y$ of z with respect to y (with x held constant) is the *marginal productivity of y* or the *marginal product of y*.

Note that the marginal productivity of either input is the rate of increase of the total product as that input is increased, assuming that the amount of the other input remains constant.

Usually, for a considerable range marginal productivity is positive—that is, as the amount of one input increases (with the amount of the other input held constant), the output also increases. However, as the input of one factor increases, the output usually increases at a decreasing rate until the point is reached at which there is no further increase in output but, in fact, a decrease in total output occurs with additional inputs of the particular factor under analysis. This characteristic behavior of production functions is known as the *law of eventually diminishing marginal productivity.*

EXAMPLE 6

Let $z = 2x^{1/2}y^{1/2}$ be a production function. Find the marginal productivity with respect to x and the marginal productivity with respect to y.

SOLUTION

The marginal productivity with respect to x is

$$\frac{\partial z}{\partial x} = x^{-1/2}y^{1/2} = \frac{y^{1/2}}{x^{1/2}}$$

and the marginal productivity with respect to y is

$$\frac{\partial z}{\partial y} = x^{1/2}y^{-1/2} = \frac{x^{1/2}}{y^{1/2}}$$

Note that $\partial z/\partial x$ is always positive, but decreases as x increases; similarly, $\partial z/\partial y$ is always positive, but decreases as y increases.

Now Work Problem 47

Higher-Order Partial Derivatives

For a function $z = f(x, y)$ of two variables for which the Limits (1) exist, there are two **first-order partial derivatives:** f_x and f_y. If it is possible to differentiate each of these partially with respect to x or y, there will result four **second-order partial derivatives,** namely,

$$f_{xx}(x, y) = \frac{\partial}{\partial x} f_x(x, y) = \frac{\partial}{\partial x}\frac{\partial z}{\partial x} = \frac{\partial^2 z}{\partial x^2} \qquad f_{xy}(x, y) = \frac{\partial}{\partial y} f_x(x, y) = \frac{\partial}{\partial y}\frac{\partial z}{\partial x} = \frac{\partial^2 z}{\partial y \partial x}$$

$$f_{yx}(x, y) = \frac{\partial}{\partial x} f_y(x, y) = \frac{\partial}{\partial x}\frac{\partial z}{\partial y} = \frac{\partial^2 z}{\partial x \partial y} \qquad f_{yy}(x, y) = \frac{\partial}{\partial y} f_y(x, y) = \frac{\partial}{\partial y}\frac{\partial z}{\partial y} = \frac{\partial^2 z}{\partial y^2}$$

The two second-order partial derivatives

$$\frac{\partial^2 z}{\partial x \, \partial y} = f_{yx}(x, y) \qquad \text{and} \qquad \frac{\partial^2 z}{\partial y \, \partial x} = f_{xy}(x, y)$$

are called **mixed partials.** Observe the differences in these two equations. The notation f_{yx} means that first we should differentiate f partially with respect to y and then differentiate the result partially with respect to x—in that order! On the other hand, f_{xy} means we should differentiate with respect to x and then with respect to y. As it turns out, for most functions the two mixed partials are equal. Although there are functions for which the mixed partials are unequal, they are rare and will not be encountered in this book.

EXAMPLE 7

Find all second-order partial derivatives of: $f(x, y) = x^2 y + x^3 y^2$

SOLUTION

Since

$$f_x = 2xy + 3x^2 y^2$$

we have

$$f_{xx} = \frac{\partial}{\partial x}(f_x) = \frac{\partial}{\partial x}(2xy + 3x^2 y^2) = 2y + 6xy^2$$

and

$$f_{xy} = \frac{\partial}{\partial y}(f_x) = \frac{\partial}{\partial y}(2xy + 3x^2 y^2) = 2x + 6x^2 y$$

Also, since

$$f_y = x^2 + 2x^3 y$$

we have

$$f_{yy} = \frac{\partial}{\partial y}(f_y) = \frac{\partial}{\partial y}(x^2 + 2x^3 y) = 2x^3$$

and

$$f_{yx} = \frac{\partial}{\partial x}(f_y) = \frac{\partial}{\partial x}(x^2 + 2x^3 y) = 2x + 6x^2 y$$

·······················

EXAMPLE 8

Find all second-order partial derivatives of: $z = f(x, y) = x \ln y + ye^x$

SOLUTION

$$f_x = \ln y + ye^x \qquad f_y = \frac{x}{y} + e^x$$

Therefore,

$$f_{xx} = \frac{\partial}{\partial x}(f_x) = \frac{\partial}{\partial x}(\ln y + ye^x) = ye^x$$

$$f_{xy} = \frac{\partial}{\partial y}(f_x) = \frac{\partial}{\partial y}(\ln y + ye^x) = \frac{1}{y} + e^x$$

$$f_{yx} = \frac{\partial}{\partial x}(f_y) = \frac{\partial}{\partial x}\left(\frac{x}{y} + e^x\right) = \frac{1}{y} + e^x$$

$$f_{yy} = \frac{\partial}{\partial y}(f_y) = \frac{\partial}{\partial y}\left(\frac{x}{y} + e^x\right) = \frac{-x}{y^2}$$

Now Work Problem 7

....

Functions of Three Variables

The idea of partial differentiation may be extended to a function of three variables. Thus, if $w = f(x, y, z)$ is a function of three variables, there will be three partial derivatives: the partial derivative with respect to x is f_x; the partial derivative with respect to y is f_y; and the partial derivative with respect to z is f_z. Each of these is calculated by differentiating with respect to the indicated variable, while treating the other two as constants.

EXAMPLE 9   Find f_x, f_y, f_z, if $f(x, y, z) = 10x^2y^3z^4$.

SOLUTION

$$f_x(x, y, z) = 10(2x)y^3z^4 = 20xy^3z^4$$
$$f_y(x, y, z) = 10x^2(3y^2)z^4 = 30x^2y^2z^4$$
$$f_z(x, y, z) = 10x^2y^3(4z^3) = 40x^2y^3z^3$$

....

EXERCISE 17.2   Answers to odd-numbered problems begin on page AN-85.

In Problems 1–6 find f_x, f_y, $f_x(2, -1)$, and $f_y(-2, 3)$.

1. $f(x, y) = 3x - 2y + 3y^3$ **2.** $f(x, y) = 2x^3 - 3y + x^2$ **3.** $f(x, y) = (x - y)^2$

4. $f(x, y) = (x - y)^3$ **5.** $f(x, y) = \sqrt{x^2 + y^2}$ **6.** $f(x, y) = \sqrt{x^2 - y^2}$

In Problems 7–16 find f_x, f_y, f_{xx}, f_{yy}, f_{yx}, and f_{xy}.

7. $f(x, y) = y^3 - 2xy + y^2 - 12x^2$ **8.** $f(x, y) = x^3 - xy + 10y^2x$

9. $f(x, y) = xe^y + ye^x + x$ **10.** $f(x, y) = xe^x + xe^y + y$

11. $f(x, y) = \dfrac{x}{y}$ **12.** $f(x, y) = \dfrac{y}{x}$

13. $f(x, y) = \ln(x^2 + y^2)$ **14.** $f(x, y) = \ln(x^2 - y^2)$

15. $f(x, y) = \dfrac{10 - x + 2y}{xy}$ **16.** $f(x, y) = \dfrac{5 + 3x - 2y}{xy}$

In Problems 17–22 verify that $f_{xy} = f_{yx}$.

17. $f(x, y) = x^3 + y^2$

18. $f(x, y) = x^2 - y^3$

19. $f(x, y) = 3x^4y^2 + 7x^2y$

20. $f(x, y) = 5x^3y - 8xy^2$

21. $f(x, y) = \dfrac{y}{x^2}$

22. $f(x, y) = \dfrac{x}{y^2}$

In Problems 23–30 find f_x, f_y, f_z.

23. $f(x, y, z) = x^2y - 3xyz + z^3$

24. $f(x, y, z) = 3xy + 4yz + 8z^2$

25. $f(x, y, z) = xe^y + ye^z$

26. $f(x, y, z) = x \ln y + y \ln z$

27. $f(x, y, z) = x \ln(yz) + y \ln(xz)$

28. $f(x, y, z) = e^{(3x + 4y + 5z)}$

29. $f(x, y, z) = \ln(x^2 + y^2 + z^2)$

30. $f(x, y, z) = e^{(x^2 + y^2 + z^2)}$

In Problems 31–38 find the slope of the tangent line to the curve of intersection of the surface $z = f(x, y)$ with the given plane at the indicated point.

31. $z = f(x, y) = 5x^2 + 3y^2$;   plane: $y = 3$;
point: $(2, 3, 47)$

32. $z = f(x, y) = 2x^2 - 4y^2$;   plane: $x = 2$;
point: $(2, 3, -28)$

33. $z = f(x, y) = \sqrt{16 - x^2 - y^2}$;   plane: $x = 1$;
point: $(1, 2, \sqrt{11})$

34. $z = f(x, y) = \sqrt{x^2 - y^2}$;   plane: $y = 0$;   point: $(4, 0, 4)$

35. $z = f(x, y) = e^x \ln y$;   plane: $x = 0$;   point: $(0, 1, 0)$

36. $z = f(x, y) = e^{2x + 3y}$;   plane: $y = 0$;   point: $(0, 0, 1)$

37. $z = f(x, y) = 2 \ln \sqrt{x^2 + y^2}$;   plane: $x = 1$;
point: $(1, 1, 2 \ln 2)$

38. $z = f(x, y) = e^{x^2 + y^2}$;   plane: $y = 0$;   point: $(1, 0, e)$

39. If $z = x^2 + 4y^2$, show that $x \dfrac{\partial z}{\partial x} + y \dfrac{\partial z}{\partial y} = 2z$.

40. If $z = x^2y$, show that $x \dfrac{\partial z}{\partial x} + y \dfrac{\partial z}{\partial y} = 3z$.

41. If $z = xy^2$, show that $x \dfrac{\partial z}{\partial x} + y \dfrac{\partial z}{\partial y} = 3z$.

42. If $z = \ln \sqrt{x^2 + y^2}$, show that $\dfrac{\partial^2 z}{\partial x^2} + \dfrac{\partial^2 z}{\partial y^2} = 0$.

43. If $z = e^{x \ln y}$, find $\dfrac{\partial^2 z}{\partial x^2}, \dfrac{\partial^2 z}{\partial y^2},$ and $\dfrac{\partial^2 z}{\partial x\, \partial y}$.

44. If you are told that a function $z = f(x, y)$ has the two partial derivatives $f_x(x, y) = 3x - y$ and $f_y(x, y) = x - 3y$, should you believe it? Explain.

45. Demand for Butter   In a large town the demand for butter (measured in pounds) is given by the formula

$$z = 2400 - 50x + 90y$$

where x is the average price per pound (in cents) of butter and y is the average price per pound (in cents) of margarine. Find the two first-order partial derivatives of z and interpret your answer.

46. Marginal Productivity   The production function of a certain commodity is given by

$$P = 8I - I^2 + 3Ik + 50k - k^2$$

where I and k are the labor and capital inputs, respectively. Find the marginal productivities of I and k at $I = 2$ and $k = 5$.

47. Marginal Productivity   A production function is given by

$$z = 2xy - x^2 - 3y^2$$

Find the marginal productivity of x and the marginal productivity of y. Interpret your answer.

48. Marginal Productivity   The production function for the commodities x and y is given by

$$z = 4xy + 2x^2 + 2y^2$$

Find the marginal productivity when $x = 1, y = 1$. Interpret your answer.

49. Marginal Productivity   The production function for the commodities x and y is given by

$$z = 40 - \frac{1}{x} - \frac{1}{y}$$

Find the marginal productivity when $x = 1$ and $y = 1$. Interpret your answer.

50. The **Cobb-Douglas production** for the economy as a whole is given by

$$z = ax^b y^{1-b}$$

where z is total product, x is quantity of labor, y is quantity of capital, and a and b are constants. Find the marginal productivity.

17.3 LOCAL MAXIMA AND LOCAL MINIMA

We saw in Chapter 14 that an important application of the derivative is finding the local maxima and local minima of a function of one variable. In this section we learn that the partial derivatives of a function of two variables are used in a similar way to find the local maxima and minima of a function of two variables.

LOCAL MAXIMUM; LOCAL MINIMUM

Let $z = f(x, y)$ denote a function of two variables:

1. f has a *local maximum* at (x_0, y_0) if

$$f(x, y) \leq f(x_0, y_0)$$

for all points (x, y) close to the point (x_0, y_0).
2. f has a *local minimum* at (x_0, y_0) if

$$f(x_0, y_0) \leq f(x, y)$$

for all points (x, y) close to the point (x_0, y_0).

Figure 14 illustrates the graph of a function $z = f(x, y)$ that has several local maxima and minima.

Figure 14

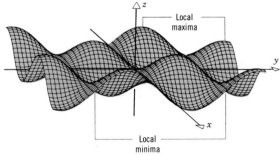

Figure 15

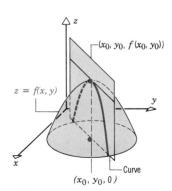

Suppose f has a local maximum at the point (x_0, y_0). Then the curve that results from the intersection of the surface $z = f(x, y)$ and any plane through the point $(x_0, y_0, 0)$ that is perpendicular to the xy-plane will have a local maximum at (x_0, y_0). See Figure 15.

In particular, the curve that results from intersecting $z = f(x, y)$ with the plane $x = x_0$ has this property. This means that if f_y exists at (x_0, y_0), then $f_y(x_0, y_0) = 0$. By a similar argument, we must also have $f_x(x_0, y_0) = 0$. This leads us to formulate the following necessary condition for local extrema:

Condition for Local Maxima and Local Minima

Let (x_0, y_0) be a point in the domain of $z = f(x, y)$. If f has a local maximum or a local minimum at (x_0, y_0), then

$$f_x(x_0, y_0) = 0 \quad \text{and} \quad f_y(x_0, y_0) = 0$$

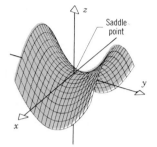

Figure 16

From this theorem we see the importance of those points at which the partial derivatives exist and are zero simultaneously. Such points are called **critical points.** Thus, we say that f has a **critical point** at (x_0, y_0) if

$$f_x(x_0, y_0) = 0 \quad \text{and} \quad f_y(x_0, y_0) = 0$$

It can happen that f has a critical point at (x_0, y_0) but that the point is neither a local maximum nor a local minimum. Such points are called **saddle points.** Figure 16 illustrates a saddle point—and suggests justification for the name. Note that this saddle point is a maximum when viewed relative to x and a minimum when viewed relative to y.

EXAMPLE 1

Find all the critical points of

$$z = f(x, y) = x^2 + y^4 - 2y^2 + 6$$

SOLUTION

We compute the partial derivatives f_x and f_y, set each one equal to zero, and solve the resulting system of equations.

$$f_x = 2x = 0$$
$$f_y = 4y^3 - 4y = 0$$

From the first equation we get $x = 0$; from the second we get

$$4y^3 - 4y = 0$$
$$4y(y^2 - 1) = 0$$
$$y = 0 \qquad y = -1 \qquad y = 1$$

Thus, f has three critical points: $(0, 0)$, $(0, -1)$, and $(0, 1)$.

Now Work Problem 1

....

Tests for Local Maxima, Local Minima, Saddle Points

We still don't know, though, the character of these critical points. What we need is a test, or tests, to tell us whether these critical points are in fact local maxima, local minima, or saddle points. For functions that possess both first- and second-order partial derivatives (this is true of all the ones we will encounter in this book), the following tests may be used.

Test for a Local Maximum

If (x_0, y_0) is a critical point of $z = f(x, y)$ and if

1. $f_{xx}(x_0, y_0) < 0$
2. $D = f_{xx}(x_0, y_0) \cdot f_{yy}(x_0, y_0) - [f_{xy}(x_0, y_0)]^2 > 0$

then the function has a local maximum at (x_0, y_0).

Test for a Local Minimum

If (x_0, y_0) is a critical point of $z = f(x, y)$ and if

1. $f_{xx}(x_0, y_0) > 0$
2. $D = f_{xx}(x_0, y_0) \cdot f_{yy}(x_0, y_0) - [f_{xy}(x_0, y_0)]^2 > 0$

then the function f has a local minimum at (x_0, y_0).

Test for a Saddle Point

If (x_0, y_0) is a critical point of $z = f(x, y)$ and if

$$D = f_{xx}(x_0, y_0) \cdot f_{yy}(x_0, y_0) - [f_{xy}(x_0, y_0)]^2 < 0$$

then the point (x_0, y_0) results in a saddle point of f.

Two comments about these tests:

1. If $D > 0$, then $f_{xx}(x_0, y_0)$ and $f_{yy}(x_0, y_0)$ are each of the same sign.
2. If $D = 0$, no information results.

EXAMPLE 2 Return to the function discussed in Example 1, $f(x, y) = x^2 + y^4 - 2y^2 + 6$, and determine the character of each critical point.

SOLUTION

We need to compute the second-order partial derivatives. Since

$$f_x(x, y) = 2x \qquad \text{and} \qquad f_y(x, y) = 4y^3 - 4y$$

we find

$$f_{xx}(x, y) = 2 \qquad f_{xy}(x, y) = f_{yx}(x, y) = 0 \qquad f_{yy}(x, y) = 12y^2 - 4$$

Next, we evaluate these partial derivatives at each critical point:

$(0, 0)$: $\qquad f_{xx}(0, 0) = 2 \qquad f_{xy}(0, 0) = 0 \qquad f_{yy}(0, 0) = -4$

$\qquad\qquad D = f_{xx}(0, 0) \cdot f_{yy}(0, 0) - [f_{xy}(0, 0)]^2 = -8 < 0$

Since $D < 0$, the point $(0, 0)$ results in a saddle point of f.

$(0, -1)$: $f_{xx}(0, -1) = 2$ $f_{xy}(0, -1) = 0$ $f_{yy}(0, -1) = 8$

$$D = 2 \cdot 8 - 0^2 = 16 > 0$$

Since $f_{xx}(0, -1) = 2 > 0$, and $D > 0$, f has a local minimum at the point $(0, -1)$.

$(0, 1)$: $f_{xx}(0, 1) = 2$ $f_{xy}(0, 1) = 0$ $f_{yy}(0, 1) = 8$

$$D = 2 \cdot 8 - 0^2 = 16 > 0$$

Since $f_{xx}(0, 1) = 2 > 0$ and $D > 0$, f has a local minimum at the point $(0, 1)$.

Now Work Problem 7

· ·

EXAMPLE 3

For the function

$$z = f(x, y) = x^2 + xy + y^2 - 6x + 6$$

find all critical points. Determine the character of each one.

SOLUTION

First, we compute the first-order partial derivatives of $z = f(x, y)$.

$$f_x = 2x + y - 6 \qquad f_y = x + 2y$$

The critical points, if there are any, obey the system of equations

$$f_x = 0 \qquad \text{and} \qquad f_y = 0$$
$$2x + y - 6 = 0 \qquad \text{and} \qquad x + 2y = 0$$

Solving these equations simultaneously, we find that $x = 4$, $y = -2$ is the only critical point. Now

$$f_{xx}(x, y) = 2 \qquad f_{xy}(x, y) = f_{yx}(x, y) = 1 \qquad f_{yy}(x, y) = 2$$

so that

$$f_{xx}(4, -2) = 2 \qquad f_{xy}(4, -2) = f_{yx}(4, -2) = 1 \qquad f_{yy}(4, -2) = 2$$

Hence,

$$D = 2 \cdot 2 - 1^2 = 3$$

Since $f_{xx}(4, -2) = 2 > 0$ and $D > 0$, f has a local minimum at the point $(4, -2)$.

· ·

Applications

EXAMPLE 4

The demand functions for two products are

$$p = 12 - 2x \qquad \text{and} \qquad q = 20 - y$$

where p and q are the respective prices (in thousands of dollars) for each product, and x and y are the respective amounts (in thousands of units) of each sold. Suppose the joint cost function is

$$C(x, y) = x^2 + 2xy + 2y^2$$

Find the revenue function and the profit function. Determine the prices and amounts that will maximize profit. What is the maximum profit?

SOLUTION

The revenue function R is the sum of the revenues due to each product. Thus,

$$R = xp + yq = x(12 - 2x) + y(20 - y)$$

The profit function P is

$$
\begin{aligned}
P &= P(x, y) \\
&= R - C \\
&= x(12 - 2x) + y(20 - y) - (x^2 + 2xy + 2y^2) \\
&= -3x^2 - 3y^2 - 2xy + 12x + 20y
\end{aligned}
$$

The first-order partial derivatives of P are

$$P_x = -6x - 2y + 12 \qquad P_y = -6y - 2x + 20$$

The critical points obey

$$-6x - 2y + 12 = 0 \quad \text{and} \quad -6y - 2x + 20 = 0$$

Solving these equations, we find that the only critical point is $(1, 3)$.
   The second-order partial derivatives of P are

$$P_{xx}(x, y) = -6 \qquad P_{xy}(x, y) = P_{yx}(x, y) = -2 \qquad P_{yy}(x, y) = -6$$

for any value (x, y). At the critical point $(1, 3)$, we see that

$$P_{xx}(1, 3) = -6 < 0$$

$$D = P_{xx}(1, 3) \cdot P_{yy}(1, 3) - [P_{xy}(1, 3)]^2 = (-6)(-6) - (-2)^2 = 32 > 0$$

Thus, P has a local maximum at $(1, 3)$. For these quantities sold, namely, $x = 1000$ units and $y = 3000$ units, the corresponding prices p and q are $p = \$10,000$ and $q = \$17,000$. The maximum profit is $P(1, 3) = \$36,000$.

· ·

EXERCISE 17.3   Answers to odd-numbered problems begin on page AN-86.

In Problems 1–6 find all the critical points of each function.

1. $f(x, y) = x^4 - 2x^2 + y^2 + 15$

2. $f(x, y) = x^2 - y^2 + 6x - 2y + 14$

3. $f(x, y) = 4xy - x^4 - y^4 + 12$

4. $f(x, y) = x^3 + 6xy + 3y^2 + 8$

5. $f(x, y) = x^4 + y^4$

6. $f(x, y) = xy + \dfrac{2}{x} + \dfrac{4}{y}$

In Problems 7–24 find all critical points and determine whether they are a local maximum, a local minimum, or a saddle point.

7. $f(x, y) = 3x^2 - 2xy + y^2$

8. $f(x, y) = x^2 - 2xy + 3y^2$

9. $f(x, y) = x^2 + y^2 - 3x + 12$

10. $f(x, y) = x^2 + y^2 - 6y + 10$

11. $f(x, y) = x^2 - y^2 + 4x + 8y$

12. $f(x, y) = x^2 - y^2 - 2x + 4y$

13. $f(x, y) = x^2 + 4y^2 - 4x + 8y - 1$

14. $f(x, y) = x^2 + y^2 - 4x + 2y - 4$

15. $f(x, y) = x^2 + y^2 + xy - 6x + 6$

16. $f(x, y) = x^2 + y^2 + xy - 8y$

17. $f(x, y) = 2 + x^2 - y^2 + xy$

18. $f(x, y) = x^2 - y^2 + 2xy$

19. $f(x, y) = x^3 - 6xy + y^3$

20. $f(x, y) = x^3 - 3xy - y^3$

21. $f(x, y) = x^3 + x^2y + y^2$

22. $f(x, y) = 3y^3 - x^2y + x$

23. $f(x, y) = \dfrac{y}{x + y}$

24. $f(x, y) = \dfrac{x}{x + y}$

25. Maximizing Profits The demand functions for two products are $p = 12 - x$ and $q = 8 - y$, where p and q are the respective prices (in thousands of dollars), and x and y are the respective amounts (in thousands of units) of each product sold. If the joint cost function is $C(x, y) = x^2 + 2xy + 3y^2$, determine the quantities x, y and prices p, q that maximize profit. What is the maximum profit?

26. Minimizing Cost The labor cost of a firm is given by the function

$$Q(x, y) = x^2 + y^3 - 6xy + 3x + 6y - 5$$

where x is the number of days required by a skilled worker and y is the number of days required by a semiskilled worker. Find the values of x and y for which the labor cost is a minimum.

27. A certain mountain is in the shape of the surface

$$z = 2xy - 2x^2 - y^2 - 8x + 6y + 4$$

(The unit of distance is 1000 feet.) If sea level is the xy-plane, how high is the mountain?

28. Reaction to Drugs Two drugs are used simultaneously as a treatment for a certain disease. The reaction R (measured in appropriate units) to x units of the first drug and y units of the second drug is

$$R(x, y) = x^2y^2(a - x)(b - y)$$

$$0 \le x \le a, \quad 0 \le y \le b$$

For a fixed amount x of the first drug, what amount y of the second drug produces the maximum reaction? For a fixed amount y of the second drug, what amount x of the first drug produces the maximum reaction? If x and y are both variable, what amount of each maximizes the reaction?

29. Reaction to Drugs The reaction R to x units of a drug t hours after the drug has been administered is given by

$$R(x, t) = x^2(a - x)t^2e^{-t} \qquad 0 \le x \le a$$

For what amount x is the reaction as large as possible? When does the maximum reaction occur?

30. Reaction to Drugs The reaction y to an injection of x units of a certain drug, t hours after the injection, is given by

$$y = x^2(a - x)t \qquad a \text{ constant}$$

Find the values of x and t, if any, that will maximize y.

31. Production Function A steel manufacturer produces two grades of steel, x tons of grade A and y tons of grade B. His cost C and revenue R are given in dollars by the formulas

$$C = \frac{1}{20}x^2 + 700x + y^2 - 150y - \frac{1}{2}xy$$

$$R = 2700x - \frac{3}{20}x^2 + 1000y - y^2 + \frac{1}{2}xy + 10,000$$

If $P = \text{Profit} = R - C$, find the production (in tons) of grades A and B that maximizes the manufacturer's profit.

32. Expansion of Gas The volume of a fixed amount of gas varies proportionally with the temperature and inversely with the pressure. Therefore, $V = k \cdot (T/P)$, where $k > 0$ is a constant and V, T, and P are the volume, temperature, and pressure, respectively. Calculate $\partial V/\partial T$ and $\partial V/\partial P$. Show that

$$P \cdot \frac{\partial V}{\partial P} + T \cdot \frac{\partial V}{\partial T} = 0$$

33. Postage Pricing The U.S. Post Office regulations state that the combined (sum) length and girth of a parcel post package being sent to a first-class post office in the United States may not exceed 84 inches. If this combined length and girth exceeds 84 inches, extra postage will be charged according to weight. Find

(a) The length, width, and height of the rectangular box of maximum volume that can be mailed, subject to the 84-inch restriction.

(b) The dimensions of a circular tube of maximum volume that can be mailed, subject to the 84-inch restriction.

34. Metal Detector A metal detector is used to locate an underground pipe. After several readings of the detector are taken, it is determined that the reading at an arbitrary point (x, y) is given by

$$D = y(x - x^2) - x^2 \quad \text{volts} \qquad x \geq 0, y \geq 0$$

Determine the point (x, y) where the reading is largest.

35. Waste Management A car manufacturer used in its production x tons of steel at the rate of y tons per week. It is found that waste due to storage and interplant distribution amounts to

$$W = \frac{1}{100} \left[\frac{1}{20} x^2 + 25y^2 - x(y + 4) \right] \text{ tons}$$

Determine the value of x and y for which waste is minimum.

17.4 LAGRANGE MULTIPLIERS

In the previous section we introduced a method to find the local maximum and local minimum of a function of two variables without any constraints or conditions on the function or the variables. However, in many practical problems we are faced with maximizing or minimizing a function subject to conditions or constraints on the variables involved.

For example, a consumer may want to maximize utility derived from the consumption of commodities, subject to budget constraints, or a manufacturer may want to produce a box with a fixed volume so that the least amount of material is used.

Let's look again at an example we solved earlier.

EXAMPLE 1

A farmer wants to enclose a rectangular plot that borders on a straight river with a fence. He will not fence in the side along the river. If the farmer has 4000 meters of fencing, what is the largest area that can be enclosed?

SOLUTION

Refer to Figure 17. If A is the area to be enclosed, then the problem we face is to find the maximum value of $A = xy$, subject to the condition that $2x + y = 4000$—that is, subject to the 4000 meters of fence that are available. To express the problem in terms of a single variable, we solve for y in the equation $2x + y = 4000$. Then the area A can be expressed in terms of x alone as

$$A = xy = x(4000 - 2x) = 4000x - 2x^2$$

This equation for A is easy to differentiate. Thus, we can find the critical numbers of A:

$$A'(x) = 4000 - 4x = 0$$
$$x = 1000$$

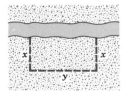

Figure 17

Since $A''(x) = -4 < 0$, this critical number yields a maximum value for the area A, namely,

$$A = 4000(1000) - 2(1000)^2 = 2{,}000{,}000 \text{ square meters}$$

· ·

In the preceding example the problem required that we maximize $A = A(x, y) = xy$, subject to a side condition or *constraint* involving x and y, namely, $2x + y - 4000 = 0$. We were able to solve this problem by using earlier techniques for two reasons:

1. In the equation $2x + y - 4000 = 0$ it was easy to solve for y in terms of x.
2. After substituting into $A = xy$, the area A became a function of the single variable x, which was easy to differentiate.

Suppose we want to maximize or minimize a function $z = f(x, y)$, subject to a constraint $g(x, y) = 0$ in which

1. It is *not* easy to solve the equation $g(x, y) = 0$ for x or for y, or
2. After substitution, the resulting function z of a single variable is not easy to differentiate.

In such cases we can instead use the *method of Lagrange multipliers.* We describe the method below:

Consider a function $z = f(x, y)$ of two variables x and y, subject to a single constraint $g(x, y) = 0$. We introduce a new variable λ, called the **Lagrange multiplier,** and construct the function

$$F(x, y, \lambda) = f(x, y) + \lambda g(x, y)$$

This new function F is a function of three variables x, y, and λ. The following result establishes the connection between the function F and the local maxima and the local minima of $z = f(x, y)$.

Method of Lagrange Multipliers

Suppose that, subject to the constraint $g(x, y) = 0$, the function $z = f(x, y)$ has a local maximum or a local minimum at the point (x_0, y_0). Form the function

$$F(x, y, \lambda) = f(x, y) + \lambda g(x, y)$$

Then there is a value of λ such that (x_0, y_0, λ) is a solution of the system of equations

$$\frac{\partial F}{\partial x} = \frac{\partial f}{\partial x} + \lambda \frac{\partial g}{\partial x} = 0$$

$$\frac{\partial F}{\partial y} = \frac{\partial f}{\partial y} + \lambda \frac{\partial g}{\partial y} = 0 \tag{1}$$

$$\frac{\partial F}{\partial \lambda} = g(x, y) = 0$$

provided all the partial derivatives exist.

In other words, the above result tells us that if we find all the solutions of the system of Equations (1), then among the solutions we find the points at which $z = f(x, y)$ may have a local maximum or a local minimum subject to the condition $g(x, y) = 0$.

EXAMPLE 2

Find the maximum value of

$$z = f(x, y) = xy$$

subject to the constraint

$$g(x, y) = x + y - 16 = 0$$

SOLUTION

First, we construct the function F:

$$F(x, y, \lambda) = f(x, y) + \lambda g(x, y) = xy + \lambda(x + y - 16)$$

The system of Equations (1) is

$$\frac{\partial F}{\partial x} = \frac{\partial f}{\partial x} + \lambda \frac{\partial g}{\partial x} = 0 \qquad \frac{\partial F}{\partial y} = \frac{\partial f}{\partial y} + \lambda \frac{\partial g}{\partial y} = 0 \qquad g(x, y) = x + y - 16 = 0$$

$$y + \lambda = 0 \qquad\qquad\qquad x + \lambda = 0 \qquad\qquad\qquad x + y - 16 = 0$$

Using the solutions of the first two equations, namely, $y = -\lambda, x = -\lambda$, in the third equation, we get

$$-\lambda - \lambda - 16 = 0 \qquad \text{or} \qquad \lambda = -8$$

Since $x = -\lambda$ and $y = -\lambda$, the only solution of the system is

$$x = 8 \qquad y = 8 \qquad \lambda = -8$$

Thus, $z = f(x, y) = xy$ has a local maximum at $(8, 8)$; the maximum value is $z = f(8, 8) = 64$.

Now Work Problem 1

$\cdots\cdots\cdots\cdots\cdots\cdots\cdots\cdots$

EXAMPLE 3

Find the minimum value of

$$z = f(x, y) = xy$$

subject to the constraint

$$g(x, y) = x^2 + y^2 - 4 = 0$$

SOLUTION

First, we construct the function F:

$$F(x, y, \lambda) = f(x, y) + \lambda g(x, y) = xy + \lambda(x^2 + y^2 - 4)$$

The system of Equations (1) is

$$\frac{\partial F}{\partial x} = y + \lambda \cdot 2x = 0$$

$$\frac{\partial F}{\partial y} = x + \lambda \cdot 2y = 0 \qquad (2)$$

$$\frac{\partial F}{\partial \lambda} = x^2 + y^2 - 4 = 0$$

From the first of these, we find that

$$y = -2x\lambda \qquad (3)$$

Substituting into the second equation of (2), we obtain

$$x - 4x\lambda^2 = 0$$
$$x(1 - 4\lambda^2) = 0$$

Thus, either

$$x = 0 \qquad \text{or} \qquad 1 - 4\lambda^2 = 0$$

From (3), $x = 0$ implies that $y = 0$, and these two numbers do not satisfy the constraint $g(x, y) = 0$. Hence, we must have

$$1 - 4\lambda^2 = 0$$
$$\lambda = \pm\tfrac{1}{2}$$

Substituting these values for λ into Equation (3), we find

$$y = x \qquad \text{or} \qquad y = -x$$

Since x and y are subject to the constraint $g(x, y) = 0$, we must have

$$x^2 + y^2 - 4 = x^2 + x^2 - 4 = 0$$
$$2x^2 = 4$$
$$x = \pm\sqrt{2}$$

Similarly,

$$y = \pm\sqrt{2}$$

Thus, the solutions of the system are

$$(\sqrt{2}, \sqrt{2}), \quad (\sqrt{2}, -\sqrt{2}), \quad (-\sqrt{2}, \sqrt{2}), \quad (-\sqrt{2}, -\sqrt{2})$$

Checking the value of $z = f(x, y)$ at each of these points, we get

$$f(\sqrt{2}, \sqrt{2}) = \sqrt{2}\,\sqrt{2} = 2$$
$$f(\sqrt{2}, -\sqrt{2}) = \sqrt{2}(-\sqrt{2}) = -2$$
$$f(-\sqrt{2}, \sqrt{2}) = -\sqrt{2}\,\sqrt{2} = -2$$
$$f(-\sqrt{2}, -\sqrt{2}) = (-\sqrt{2})(-\sqrt{2}) = 2$$

Hence, $z = f(x, y)$ attains its minimum value at the two points $(-\sqrt{2}, \sqrt{2})$ and $(\sqrt{2}, -\sqrt{2})$. The minimum value is -2.

Now Work Problem 3

· ·

Applications

EXAMPLE 4

A manufacturer produces two types of engines, x and y, and the joint profit function is given by

$$P(x, y) = x^2 + 3xy - 6y$$

To maximize profit, how many engines of each type should be produced if there must be a total of 42 engines?

SOLUTION

The condition of a total of 42 engines constitutes the constraint of the problem. Thus, the constraint is

$$g(x, y) = x + y - 42 = 0$$

The function F is

$$F(x, y, \lambda) = P(x, y) + \lambda g(x, y) = x^2 + 3xy - 6y + \lambda(x + y - 42)$$

The system of Equations (1) is

$$\frac{\partial F}{\partial x} = 2x + 3y + \lambda = 0$$

$$\frac{\partial F}{\partial y} = 3x - 6 + \lambda = 0$$

$$\frac{\partial F}{\partial \lambda} = x + y - 42 = 0$$

The only solution of this system is

$$x = 33 \qquad y = 9 \qquad \lambda = -93$$

Thus, maximum profit is achieved for $x = 33$ and $y = 9$.

Now Work Problem 17

· ·

Functions of Three Variables

One of the advantages of the method of Lagrange multipliers is that it extends easily to functions of three variables.

Method of Lagrange Multipliers for Three Variables

Suppose that, subject to the constraint $g(x, y, z) = 0$, the function $w = f(x, y, z)$ has a local maximum or a local minimum at the point (x_0, y_0, z_0). Form the function

$$F(x, y, z) = f(x, y, z) + \lambda g(x, y, z)$$

Then there is a value of λ so that (x_0, y_0, z_0, λ) is a solution of the following system of equations:

$$F_x(x, y, z, \lambda) = f_x(x, y, z) + \lambda g_x(x, y, z) = 0$$
$$F_y(x, y, z, \lambda) = f_y(x, y, z) + \lambda g_y(x, y, z) = 0$$
$$F_z(x, y, z, \lambda) = f_z(x, y, z) + \lambda g_z(x, y, z) = 0$$
$$F_\lambda(x, y, z, \lambda) = g(x, y, z) = 0$$

provided each of the partial derivatives exists.

EXAMPLE 5

The material for a rectangular container costs $3 per square foot for the bottom and $2 per square foot for the sides and top. Find the dimensions of the container so that its volume is 12 cubic feet and the cost is minimum.

SOLUTION

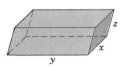

Figure 18

Refer to Figure 18. If x and y (in feet) equal the length and width of the container and z (in feet) equals its height, then our problem is to minimize

$$C(x, y, z) = \underbrace{3xy}_{\text{Bottom}} + \underbrace{2xy}_{\text{Top}} + \underbrace{4yz + 4xz}_{\text{Sides}} = 5xy + 4yz + 4xz$$

subject to

$$g(x, y, z) = xyz - 12 = 0$$

Form the function

$$F(x, y, z, \lambda) = C(x, y, z) + \lambda g(x, y, z)$$
$$= 5xy + 4yz + 4xz + \lambda(xyz - 12)$$

The system of equations to be solved is

$$F_x(x, y, z, \lambda) = 5y + 4z + \lambda yz = 0$$
$$F_y(x, y, z, \lambda) = 5x + 4z + \lambda xz = 0$$
$$F_z(x, y, z, \lambda) = 4y + 4x + \lambda xy = 0$$
$$F_\lambda(x, y, z, \lambda) = xyz - 12 = 0$$

Since $x > 0$, $y > 0$, $z > 0$, we can solve for λ in the first three equations to get

$$\lambda = \frac{-(5y + 4z)}{yz} \qquad \lambda = \frac{-(5x + 4z)}{xz} \qquad \lambda = \frac{-4(y + x)}{xy}$$

From the first two of these we find that

$$\frac{5y + 4z}{yz} = \frac{5x + 4z}{xz}$$

$$xz(5y + 4z) = yz(5x + 4z)$$

$$5xyz + 4xz^2 = 5xyz + 4yz^2$$

$$4xz^2 = 4yz^2$$

$$x = y$$

From the second two of these we find that

$$\frac{5x + 4z}{xz} = \frac{4y + 4x}{xy}$$

$$5x^2y + 4xyz = 4xyz + 4x^2z$$

$$5y = 4z$$

$$y = \frac{4}{5}z$$

Using these results in the fourth equation, we get

$$xyz - 12 = 0$$

$$\tfrac{4}{5}z \cdot \tfrac{4}{5}z \cdot z = 12$$

$$z^3 = \frac{75}{4}$$

$$z = 2.657$$

The only solution is $x = y = 2.126$, $z = 2.657$ feet. These are the dimensions that minimize the cost of the container.

· ·

EXERCISE 17.4   Answers to odd-numbered problems begin on page AN-86.

In Problems 1–12 use the method of Lagrange multipliers.

1. Find the maximum value of $z = f(x, y) = 3x + 4y$ subject to the constraint $g(x, y) = x^2 + y^2 - 9 = 0$.

2. Find the maximum value of $z = f(x, y) = 3xy$ subject to the constraint $g(x, y) = x^2 + y^2 - 4 = 0$.

3. Find the minimum value of $z = f(x, y) = x^2 + y^2$ subject to the constraint $g(x, y) = x + y - 1 = 0$.

4. Find the minimum value of $z = f(x, y) = 3x + 4y$ subject to the constraint $g(x, y) = x^2 + y^2 - 9 = 0$.

5. Find the maximum value of $z = f(x, y) = 12xy - 3y^2 - x^2$ subject to the constraint $g(x, y) = x + y - 16 = 0$.

6. Find the maximum value of $z = f(x, y) = xy$ subject to the constraint $g(x, y) = x + y - 8 = 0$.

7. Find the minimum value of $z = f(x, y) = 5x^2 + 6y^2 - xy$ subject to the constraint $g(x, y) = x + 2y - 24 = 0$.

8. Find the minimum value of $z = f(x, y) = x^2 + y^2$ subject to the constraint $g(x, y) = 2x + 3y - 4 = 0$.

9. Find the maximum value of $w = f(x, y, z) = xyz$ subject to the constraint $g(x, y, z) = x + 2y + 2z - 120 = 0$.

10. Find the maximum value of $w = f(x, y, z) = x + y + z$ subject to the constraint $g(x, y, z) = x^2 + y^2 + z^2 - 12 = 0$.

11. Find the minimum value of $w = f(x, y, z) = x^2 + y^2 + z^2 - x - 3y - 5z$ subject to the constraint $g(x, y, z) = x + y + 2z - 20 = 0$.

12. Find the minimum value of $w = f(x, y, z) = 4x + 4y + 2z$ subject to the constraint $g(x, y, z) = x^2 + y^2 + z^2 - 9 = 0$.

13. Find two numbers x and y so that their product is a maximum while their sum is 100.

14. Find two numbers x and y so that the sum of their squares is a minimum while their sum is 100.

15. Find three numbers x, y, and z so that their sum is a maximum while the sum of their squares is 25.

16. Find three numbers x, y, and z so that their sum is a minimum while the sum of their squares is 25.

17. **Joint Cost Function**   Let x and y be two types of items produced by a factory, and let

$$C = 18x^2 + 9y^2$$

be the joint cost of production of x and y. If $x + y = 54$, find x and y that minimize cost.

18. **Production Function**   The production function of ABC Manufacturing is

$$P(x, y) = x^2 + 3xy - 6x$$

where x and y represent two different types of input. Find the amounts of x and y that maximize production if $x + y = 40$.

19. **Minimizing Materials**   A container producer wants to build a closed rectangular box with a volume of 175 cubic feet. Determine what the dimensions of the container should be so as to use the least amount of material in construction.

20. **Best Dimensions of a Box**   A rectangular box, open at the top, is to be made from material costing $2 per square foot. If the volume is to be 12 cubic feet, what dimensions will minimize the cost?

21. **Best Dimensions of a Box**   A rectangular box is to have a bottom made from material costing $2 per square foot while the top and sides are made from material costing $1 per square foot. If the volume of the box is to be 18 cubic feet, what dimensions will minimize the cost?

17.5 THE DOUBLE INTEGRAL

The definite integral of a function of a single variable can be extended to functions of two variables. Integrals of a function of two variables are called **double integrals.** Recall that the definite integral of a function of one variable is defined over an interval. Double integrals, on the other hand, involve integration over a region in the plane.

An example of a double integral is

$$\int_0^3 \int_1^4 x^2 y \, dx \, dy$$

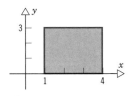

Figure 19

where the integrand is the function $f(x, y) = x^2 y$. We will integrate this function over the region determined by the limits of integration: $1 \le x \le 4$ and $0 \le y \le 3$. Figure 19 illustrates the region in the plane, which is referred to as a rectangular region. We first show how to evaluate double integrals over rectangular regions.

The evaluation of a double integral of a function f of two variables whose domain is a rectangle is equivalent to the evaluation of a pair of definite integrals in which one of the integrations is performed *partially*. Partial integration is merely the reverse of partial differentiation. The symbol $\int_a^b f(x, y) \, dx$ is an instruction to hold y fixed and integrate with respect to x. The result will be a function of y alone. Similarly, $\int_c^d f(x, y) \, dy$ is an instruction to hold x fixed and integrate with respect to y. The result here is a function of x alone.

EXAMPLE 1

Evaluate: (a) $\displaystyle\int_1^2 2x^2y\,dx$ (b) $\displaystyle\int_0^4 2x^2y\,dy$

SOLUTION

(a) The dx tells us to integrate with respect to x, holding y as a constant. Then by the Fundamental Theorem of Calculus, we substitute 2 for x, and then 1 for x.

$$\int_1^2 2x^2y\,dx = 2y\int_1^2 x^2\,dx = 2y\left.\frac{x^3}{3}\right|_1^2 = 2y\left(\frac{8}{3}-\frac{1}{3}\right) = 2y\cdot\frac{7}{3} = \frac{14y}{3}$$

(b) The dy tells us to integrate with respect to y, holding x as a constant. Then by the Fundamental Theorem of Calculus, we substitute 4 for y, and then 0 for y.

$$\int_0^4 2x^2y\,dy = 2x^2\int_0^4 y\,dy = 2x^2\left.\frac{y^2}{2}\right|_0^4 = x^2(16-0) = 16x^2$$

························

EXAMPLE 2

Evaluate: (a) $\displaystyle\int_1^2 (6x^2y + 3y^2)\,dy$ (b) $\displaystyle\int_2^3 (6x^2y + 3y^2)\,dx$

SOLUTION

(a) The dy tells us to integrate with respect to y, holding x as a constant. Then by the Fundamental Theorem of Calculus, we substitute 2 for y, and then 1 for y.

$$\int_1^2 (6x^2y + 3y^2)\,dy = (3x^2y^2 + y^3)\Big|_1^2 = 3x^2(2)^2 + 2^3 - (3x^2\cdot 1 + 1^3)$$

$$= 9x^2 + 7$$

(b) The dx tells us to integrate with respect to x, holding y as a constant. Then by the Fundamental Theorem of Calculus, we substitute 3 for x, and then 2 for x.

$$\int_2^3 (6x^2y + 3y^2)\,dx = (2x^3y + 3y^2x)\Big|_2^3$$

$$= 2(3)^3y + 3y^2(3) - [2\cdot 2^3y + 3y^2(2)]$$
$$= 54y + 9y^2 - 16y - 6y^2$$
$$= 38y + 3y^2$$

Now Work Problem 3

························

In the case of the double integral we perform partial integrals one at a time.

EXAMPLE 3

Evaluate: (a) $\displaystyle\int_0^4\left[\int_1^2 2x^2y\,dx\right]dy$ (b) $\displaystyle\int_1^2\left[\int_0^4 2x^2y\,dy\right]dx$

SOLUTION

(a) $\displaystyle\int_0^4 \left[\int_1^2 2x^2y\ dx\right] dy = \int_0^4 \dfrac{14y}{3}\ dy = \dfrac{7y^2}{3}\bigg|_0^4 = \dfrac{112}{3}$

↑

From Example 1, Part (a)

(b) $\displaystyle\int_1^2 \left[\int_0^4 2x^2y\ dy\right] dx = \int_1^2 16x^2\ dx = \dfrac{16x^3}{3}\bigg|_1^2 = \dfrac{112}{3}$

↑

From Example 1, Part (b)

· ·

Double integrals of the form

$$\int_a^b \left[\int_c^d f(x,\ y)\ dy\right] dx \qquad \text{and} \qquad \int_c^d \left[\int_a^b f(x,\ y)\ dx\right] dy$$

are called **iterated integrals.** In the iterated integral on the left, the function f is integrated partially with respect to y from c to d, resulting in a function of x that is integrated from a to b. In the iterated integral on the right, the function f is integrated partially with respect to x from a to b, resulting in a function of y that is integrated from c to d.

EXAMPLE 4

Evaluate:

(a) $\displaystyle\int_1^2 \left[\int_0^1 (6x^2y + 8y^3)\ dy\right] dx$ (b) $\displaystyle\int_0^1 \left[\int_1^2 (6x^2y + 8y^3)\ dx\right] dy$

SOLUTION

(a) $\displaystyle\int_1^2 \left[\int_0^1 (6x^2y + 8y^3)\ dy\right] dx = \int_1^2 (3x^2y^2 + 2y^4)\bigg|_0^1\ dx$

$\displaystyle = \int_1^2 (3x^2 + 2)\ dx = (x^3 + 2x)\bigg|_1^2$

$= (8 + 4) - (1 + 2) = 9$

(b) Note that this is the same integrand with the same limits of integration as in Part (a), but the order of integration is reversed.

$\displaystyle\int_0^1 \left[\int_1^2 (6x^2y + 8y^3)\ dx\right] dy = \int_0^1 (2x^3y + 8y^3x)\bigg|_1^2\ dy$

$\displaystyle = \int_0^1 [16y + 16y^3 - (2y + 8y^3)]\ dy$

$\displaystyle = \int_0^1 (14y + 8y^3)\ dy$

$= (7y^2 + 2y^4)\big|_0^1 = 7 + 2 = 9$

Now Work Problem 21

· ·

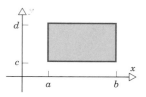

Figure 20

The answer in Part (a) is equal to the answer in Part (b). It can be shown that for a large class of functions, the following equation holds true. If R is the rectangular region given by $a \le x \le b$, $c \le y \le d$ as shown in Figure 20, and if $f(x, y)$ is continuous on R, then

$$\iint_R f(x, y)\, dx\, dy = \int_a^b \left[\int_c^d f(x, y)\, dy \right] dx = \int_c^d \left[\int_a^b f(x, y)\, dx \right] dy \qquad (1)$$

Note that the equation states that either form of the iterated integral in (1) may be used to evaluate a double integral over a rectangular region.

Since the two double integrals are equal, the brackets are not needed, and either integral is given by

$$\int_a^b \int_c^d f(x, y)\, dx\, dy$$

EXAMPLE 5

Evaluate $\displaystyle\iint_R 2xy\, dx\, dy$ if R is the rectangular region $1 \le x \le 2$, $0 \le y \le 1$.

SOLUTION

We choose to evaluate the following double integral:

$$\int_0^1 \int_1^2 2xy\, dx\, dy = \int_0^1 x^2 y \Big|_1^2 dy = \int_0^1 (4y - y)\, dy = \left(2y^2 - \frac{y^2}{2} \right)\Big|_0^1$$

$$= 2 - \frac{1}{2} = \frac{3}{2}$$

Now Work Problem 27

....

Until now, we have been evaluating double integrals of a function defined on a rectangular region. In the next example we show how to evaluate double integrals with variable limits of integration.

EXAMPLE 6

Evaluate: $\displaystyle\int_1^2 \int_y^{y^2} 6x^2 y\, dx\, dy$

SOLUTION

Integrate first with respect to x, then with respect to y.

$$\int_1^2 \int_y^{y^2} 6x^2 y\, dx\, dy = \int_1^2 \left[\int_y^{y^2} 6x^2 y\, dx \right] dy = \int_1^2 (2x^3 y) \Big|_y^{y^2} dy$$

Replace x first with y^2 and then with y and subtract.

$$= \int_{1}^{2} [2(y^2)^3\, y - 2y^3 y]\, dy$$

$$= \int_{1}^{2} (2y^7 - 2y^4)\, dy$$

$$= \left(\frac{y^8}{4} - \frac{2y^5}{5} \right) \Big|_{1}^{2}$$

$$= \left(64 - \frac{64}{5} \right) - \left(\frac{1}{4} - \frac{2}{5} \right) = \frac{1027}{20}$$

Notice in the above example that variable limits are in the inner integral. This is always true when we deal with variable limits.

EXAMPLE 7

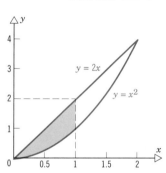

Figure 21

Now Work Problem 31

Evaluate $\displaystyle\iint_{R} xy\, dy\, dx$, where R is the region shown in Figure 21.

SOLUTION

The region over which integration takes place is all points (x, y) such that $x^2 \le y \le 2x$ and $0 \le x \le 1$. The double integral is evaluated by first treating x as a constant and integrating xy with respect to y between x^2 and $2x$. Then we integrate with respect to x between 0 and 1.

$$\int_{0}^{1} \int_{x^2}^{2x} xy\, dy\, dx = \int_{0}^{1} x \frac{y^2}{2} \Big|_{x^2}^{2x} dx = \int_{0}^{1} \left[x \frac{(2x)^2}{2} - x \frac{(x^2)^2}{2} \right] dx$$

$$= \int_{0}^{1} \left(2x^3 - \frac{x^5}{2} \right) dx = \frac{x^4}{2} - \frac{x^6}{12} \Big|_{0}^{1}$$

$$= \frac{1}{2} - \frac{1}{12} = \frac{5}{12}$$

Finding Volume by Using Double Integrals

One application of the definite integral $\displaystyle\int_{a}^{b} f(x)\, dx$ is to find the area under a curve. In a similar manner, the double integral is used to find the volume of a solid bounded above by the surface $z = f(x, y)$, below by the xy-plane, and on the sides by the vertical walls defined by the region R, as illustrated in Figure 22 (on the next page).

Figure 22 Volume of a
solid under a surface

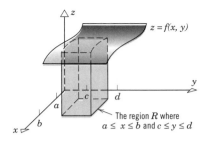

The region R where
$a \leq x \leq b$ and $c \leq y \leq d$

VOLUME

If $f(x, y) \geq 0$ over a rectangular region R: $a \leq x \leq b$, $c \leq y \leq d$, then the *volume*
of the solid under the graph of f and over the region R is

$$V = \iint_R f(x, y) \, dx \, dy$$

EXAMPLE 8

Find the volume V under $f(x, y) = x^2 + y^2$ and over the rectangular region
$0 \leq x \leq 2$, $0 \leq y \leq 1$.

SOLUTION

Figure 23 illustrates the volume we seek. The volume V is

$$\iint_R (x^2 + y^2) \, dy \, dx = \int_0^2 \left[\int_0^1 (x^2 + y^2) \, dy \right] dx$$

$$= \int_0^2 \left(x^2 y + \frac{y^3}{3} \right) \Big|_0^1 dx$$

$$= \int_0^2 \left(x^2 + \frac{1}{3} \right) dx$$

$$= \left(\frac{x^3}{3} + \frac{x}{3} \right) \Big|_0^2$$

$$= \frac{8}{3} + \frac{2}{3} = \frac{10}{3} \text{ cubic units}$$

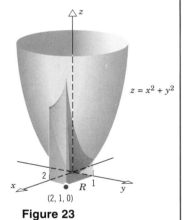

$z = x^2 + y^2$

$(2, 1, 0)$

Figure 23

Now Work Problem 39

EXERCISE 17.5   Answers to odd-numbered problems begin on page AN-86.

In Problems 1–16 evaluate each partial integral.

1. $\displaystyle\int_0^2 (xy^3 + x^2) \, dx$

2. $\displaystyle\int_1^3 (xy^3 - x^2) \, dx$

3. $\displaystyle\int_2^4 (3x^2 y + 2x) \, dy$

4. $\displaystyle\int_0^1 (6xy^2 - 2xy + 3) \, dy$

5. $\displaystyle\int_2^3 (x + 3y) \, dx$

6. $\displaystyle\int_1^3 (6xy + 12x^2 y^3) \, dy$

7. $\displaystyle\int_2^4 (4x - 6y + 7)\, dy$

8. $\displaystyle\int_1^3 (4x - 6y + 7)\, dx$

9. $\displaystyle\int_0^1 \frac{x^2}{\sqrt{1 + y^2}}\, dx$

10. $\displaystyle\int_0^2 \frac{y^3}{\sqrt{1 + x^2}}\, dy$

11. $\displaystyle\int_0^2 e^{x+y}\, dx$

12. $\displaystyle\int_0^2 e^{x+6y}\, dx$

13. $\displaystyle\int_0^4 xe^{x^2 - 4y}\, dx$

14. $\displaystyle\int_2^4 ye^{x+y^2}\, dy$

15. $\displaystyle\int_0^2 \frac{x}{\sqrt{y + x^2}}\, dx$

16. $\displaystyle\int_1^4 \frac{y}{\sqrt{x^2 + y^2}}\, dy$

In Problems 17–26 evaluate each iterated integral.

17. $\displaystyle\int_0^2 \left[\int_0^4 y\, dx \right] dy$

18. $\displaystyle\int_1^2 \left[\int_3^4 x\, dy \right] dx$

19. $\displaystyle\int_1^2 \left[\int_1^3 (x^2 + y)\, dx \right] dy$

20. $\displaystyle\int_0^1 \left[\int_2^3 (x + y)\, dy \right] dx$

21. $\displaystyle\int_0^1 \left[\int_1^2 (x^2 + y)\, dx \right] dy$

22. $\displaystyle\int_0^3 \left[\int_1^2 (x - y^2)\, dy \right] dx$

23. $\displaystyle\int_1^2 \left[\int_3^4 (4x + 2y + 5)\, dx \right] dy$

24. $\displaystyle\int_1^2 \left[\int_3^4 (6x + 4y + 7)\, dy \right] dx$

25. $\displaystyle\int_2^4 \left[\int_0^1 (6xy^2 - 2xy + 3)\, dy \right] dx$

26. $\displaystyle\int_0^2 \left[\int_1^3 (6xy + 12x^2y^3)\, dy \right] dx$

In Problems 27–30 evaluate each double integral over the indicated rectangular region R.

27. $\displaystyle\iint_R (y + 3x^2)\, dx\, dy \quad 0 \le x \le 2, \quad 1 \le y \le 3$

28. $\displaystyle\iint_R (x + 3y^2)\, dx\, dy \quad 0 \le x \le 3, \quad 0 \le y \le 4$

29. $\displaystyle\iint_R (x + y)\, dy\, dx \quad 0 \le x \le 2, \quad 1 \le y \le 4$

30. $\displaystyle\iint_R (x^2 - 2xy)\, dy\, dx \quad 0 \le x \le 2, \quad 1 \le y \le 4$

In Problems 31–38 evaluate each double integral.

31. $\displaystyle\int_0^2 \int_0^x (x^2 + y^2)\, dy\, dx$

32. $\displaystyle\int_0^4 \int_0^{2x} y\, dy\, dx$

33. $\displaystyle\int_2^4 \int_2^{3x} (3x + 4y)\, dy\, dx$

34. $\displaystyle\int_2^4 \int_1^{1+x} (2x + 4y)\, dy\, dx$

35. $\displaystyle\int_0^2 \int_1^{2y+1} (x + y)\, dx\, dy$

36. $\displaystyle\int_0^2 \int_{y^2}^{2y} x\, dx\, dy$

37. $\displaystyle\int_0^1 \int_0^y e^{x+y}\, dx\, dy$

38. $\displaystyle\int_1^2 \int_2^3 e^{x-y}\, dx\, dy$

In Problems 39 and 40 find the volume under the surface of f(x, y) and above the indicated rectangle.

39. $f(x, y) = 2x + 3y + 4 \quad 1 \le x \le 2, \quad 3 \le y \le 4$

40. $f(x, y) = x + y - 1 \quad 0 \le x \le 1, \quad 0 \le y \le 1$

CHAPTER REVIEW

IMPORTANT TERMS AND CONCEPTS

function of two variables 868
three-dimensional coordinate
  system 869
surface 870
partial derivative 875
geometric interpretation of
  $f_x(x_0, y_0)$ and $f_y(x_0, y_0)$ 877
production function 879
marginal productivity 880

second-order partial derivatives 880
function of three variables 872,
  882
local maximum; local minimum
  884
necessary condition for local
  maximum and local minimum
  885
critical point 885

tests for local maximum,
  local minimum, and
  saddle points 886
method of Lagrange
  multipliers 891
double integral 897
iterated integral 899
volume of a solid under
  a surface 902

IMPORTANT FORMULAS

$$\frac{\partial f}{\partial x} = f_x(x, y) = \lim_{\Delta x \to 0} \frac{f(x + \Delta x, y) - f(x, y)}{\Delta x}$$

$$\frac{\partial f}{\partial y} = f_y(x, y) = \lim_{\Delta y \to 0} \frac{f(x, y + \Delta y) - f(x, y)}{\Delta y}$$

Tests for Local Maxima, Local Minima, and Saddle Points:

(x_0, y_0) is a critical point if $f_x = f_y = 0$. Let

$$D = f_{xx} \cdot f_{yy} - (f_{xy})^2$$

If $D > 0$ and $f_{xx} > 0$, f has local minima.
If $D > 0$ and $f_{xx} < 0$, f has local maxima.
If $D < 0$, f has a saddle point.

Method of Lagrange Multipliers
To find a local maximum or a local minimum for $z = f(x, y)$, subject to $g(x, y) = 0$, solve the equations

$$\frac{\partial f}{\partial x} + \lambda \frac{\partial g}{\partial x} = 0 \qquad \frac{\partial f}{\partial y} + \lambda \frac{\partial g}{\partial y} = 0$$

$$g(x, y) = 0$$

Double Integrals

$$\iint_R f(x, y) \, dx \, dy = \int_a^b \left[\int_c^d f(x, y) \, dy \right] dx$$

$$= \int_c^d \left[\int_a^b f(x, y) \, dx \right] dy$$

$$\text{Volume} = V = \iint_R f(x, y) \, dx \, dy$$

$$R: a \le x \le b, \, c \le y \le d$$

TRUE–FALSE ITEMS   Answers are on page AN-86.

T_____ F_____ **1.** The domain of a function of two variables is a set of points in the xy-plane.

T_____ F_____ **2.** The partial derivative $f_x(x, y)$ of $z = f(x, y)$ is

$$f_x(x, y) = \lim_{\Delta x \to 0} \frac{f(x + \Delta x, y + \Delta y) - f(x, y)}{\Delta x}$$

provided the limit exists.

T_____ F_____ **3.** For most functions in this book, $f_{xy} \ne f_{yx}$.

T_____ F_____ **4.** If (x_0, y_0) is a critical point of $z = f(x, y)$ and if $f_{xx}(x_0, y_0) > 0$ and

$$D = f_{xx}(x_0, y_0) \cdot f_{yy}(x_0, y_0) - [f_{xy}(x_0, y_0)]^2 < 0$$

then f has a local minimum at (x_0, y_0).

FILL IN THE BLANKS   Answers are on page AN-86.

1. The graph of a function of two variables is called a _____.

2. If $f(x, y) = x^2 y - \sqrt{xy}$, then $f_x(1, 2) =$ _____.

3. The partial derivative $f_y(x_0, y_0)$ equals the slope of the tan-

gent line to the curve of intersection of the surface $z = f(x, y)$ and the plane _____ at the point (x_0, y_0, z_0) on the surface.

4. A critical point that is neither a local maximum nor a local minimum is a _____ _____.

REVIEW EXERCISES   Answers to odd-numbered problems begin on page AN-86.

In Problems 1–4 find $f(1, -3)$ and $f(4, -2)$ for each function.

1. $f(x, y) = 2x^2 + 6xy - y^3$

2. $f(x, y) = 3x^2 y - x^2 + y^2$

3. $f(x, y) = \dfrac{x + 2y}{x - 3y}$

4. $f(x, y) = \dfrac{2x + y}{x^2 - y}$

5. For the function $z = f(x, y) = x^2 + xy$, find

(a) $f(x + \Delta x, y)$

(b) $f(x + \Delta x, y) - f(x, y)$

(c) $\dfrac{f(x + \Delta x, y) - f(x, y)}{\Delta x}$, $\Delta x \neq 0$

(d) $\lim\limits_{\Delta x \to 0} \dfrac{f(x + \Delta x, y) - f(x, y)}{\Delta x}$

6. Find f_x, f_y, f_{xx}, f_{xy}, and f_{yy} for each function:

(a) $z = f(x, y) = x^2 y + 4x$

(b) $z = f(x, y) = x^2 + y^2 + 2xy$

(c) $z = f(x, y) = y^2 e^x + x \ln y$

7. For each of the following surfaces, find all local maxima, local minima, and saddle points:

(a) $z = f(x, y) = xy - 6x - x^2 - y^2$

(b) $z = f(x, y) = x^2 + 2x + y^2 + 4y + 10$

(c) $z = f(x, y) = 2x - x^2 + 4y - y^2 + 10$

(d) $z = f(x, y) = xy$

(e) $z = f(x, y) = x^2 - 9y + y^3$

8. Show that the plane $z = 3x + 4y - 2$ has no critical points.

9. Use the method of Lagrange multipliers to find the maximum value of each of the following functions $z = f(x, y)$, subject to the constraint $g(x, y) = 0$:

(a) $f(x, y) = 5x^2 - 3y^2 + xy$

$g(x, y) = 2x - y - 20 = 0$

(b) $f(x, y) = x\sqrt{y}$

$g(x, y) = 2x + y - 3000 = 0$

10. Use the method of Lagrange multipliers to find the minimum value of each of the following functions $z = f(x, y)$, subject to the constraint $g(x, y) = 0$:

(a) $f(x, y) = x^2 + y^2$ $g(x, y) = 2x + y - 4 = 0$

(b) $f(x, y) = xy^2$ $g(x, y) = x^2 + y^2 - 1 = 0$

11. Evaluate: $\displaystyle\int (4x^2 y - 12y)\, dx$

12. Evaluate: $\displaystyle\int (4x^2 y + 2y)\, dx$

13. Evaluate: $\displaystyle\int_1^2 \left[\int_0^3 (6x^2 y + 2x)\, dy \right] dx$

14. Evaluate: $\displaystyle\iint_R (2x + 4y)\, dy\, dx$;

$R: -1 \leq x \leq 1,\quad 1 \leq y \leq 3$

15. Evaluate: $\displaystyle\iint_R (3x + 2)\, dx\, dy$;

$R: 0 \leq y \leq 1,\quad y \leq x \leq 3 + y$

16. Find the volume under the given surface:

$z = f(x, y) = 3x + 2y + 1;\quad 1 \leq x \leq 8,\quad 0 \leq y \leq 6$

17. Evaluate: $\displaystyle\int_0^2 \int_0^{3x} 2xy\, dy\, dx$

18. Production Function   The Cobb-Douglas production function for a certain factory is

$$z = f(K, L) = 80K^{1/4}L^{3/4}$$

units, where K is the capital investment measured in units of $1,000,000, and L is the size of the labor force measured in worker-hours.

(a) Find: $\dfrac{\partial z}{\partial K}$ and $\dfrac{\partial z}{\partial L}$

(b) Evaluate $\dfrac{\partial z}{\partial K}$ and $\dfrac{\partial z}{\partial L}$ when $K = \$800,000$ and $L = 20,000$ worker-hours.

19. Joint Profit Function A company produces two products at a total cost

$$C(x, y) = x^2 + 200x + y^2 + 100y - xy$$

where x and y represent the units produced of each product. The revenue function is

$$R(x, y) = 2000x - 2x^2 + 100y - y^2 + xy$$

Find the number of units of each product that will maximize profit.

20. Minimizing Material A rectangular cardboard box with an open top is to have a volume of 96 cubic feet. Find the dimensions of the box so that the amount of cardboard used is minimized.

Appendix A

Review Topics from Algebra and Geometry*

A.1 BASIC ALGEBRA

Classification of Numbers

It is helpful to classify the various kinds of numbers. The **counting numbers,** or **natural numbers,** are the numbers 1, 2, 3, 4, ... (The three dots, called an **ellipsis,** indicate that the pattern continues indefinitely.) As their name implies, these numbers are often used to count things. For example, there are 26 letters in our alphabet; there are 100 cents in a dollar. The **whole numbers** are the numbers 0, 1, 2, 3, . . . , that is, the counting numbers together with 0.

INTEGERS

The *integers* are the numbers . . . , − 3, − 2, − 1, 0, 1, 2, 3, ...

These numbers prove useful in many situations. For example, if your checking account has $10 in it and you write a check for $15, you can represent the current balance as − $5.

---

* Based on material from *College Algebra,* 4th ed., by Michael Sullivan. Used here with the permission of the author and Prentice-Hall, Inc.

Notice that the counting numbers are included among the whole numbers. Each time we expand a number system, such as from the whole numbers to the integers, we do so in order to be able to handle new, and usually more complicated, problems. Thus the integers allow us to solve problems requiring both positive and negative counting numbers, such as profit/loss, height above/below sea level, temperature above/below 0°F, and so on.

But integers alone are not sufficient for *all* problems. For example, they do not answer the questions, "What part of a dollar is 38 cents?" or "What part of a pound is 5 ounces?" To answer such questions, we enlarge our number system to include *rational numbers*. For example, $\frac{38}{100}$ answers the question, "What part of a dollar is 38 cents?" and $\frac{5}{16}$ answers the question, "What part of a pound is 5 ounces?"

RATIONAL NUMBER

A *rational number* is a number that can be expressed as a quotient a/b of two integers. The integer a is called the *numerator,* and the integer b, which cannot be 0, is called the *denominator.*

Examples of rational numbers are $\frac{3}{4}, \frac{5}{2}, \frac{0}{4}, -\frac{2}{3}$, and $\frac{100}{3}$. Since $a/1 = a$ for any integer a, it follows that the rational numbers contain the integers as a special case.

Real Numbers

Rational numbers may be represented as **decimals.** For example, the rational numbers $\frac{3}{4}$, $\frac{5}{2}, -\frac{2}{3}$, and $\frac{7}{66}$ may be represented as decimals by merely carrying out the indicated division:

$$\frac{3}{4} = 0.75 \qquad \frac{5}{2} = 2.5 \qquad -\frac{2}{3} = -0.666\ldots \qquad \frac{7}{66} = 0.1060606\ldots$$

Notice that the decimal representations of $\frac{3}{4}$ and $\frac{5}{2}$ terminate, or end. The decimal representations of $-\frac{2}{3}$ and $\frac{7}{66}$ do not terminate, but they do exhibit a pattern of repetition. For $-\frac{2}{3}$ the 6 repeats indefinitely; for $\frac{7}{66}$ the block 06 repeats indefinitely. It can be shown that every rational number may be represented by a decimal that either terminates or is nonterminating with a repeating block of digits, and vice versa.

On the other hand, there are decimals that do not fit into either of these categories. Such decimals represent **irrational numbers.** For example, the decimal 0.1234567891011121314. . . , in which we write the positive integers successively one after the other, will neither repeat (think about it) nor terminate, and so, represents an irrational number. Thus every irrational number may be represented by a decimal that neither repeats nor terminates.

Irrational numbers occur naturally. For example, consider the isosceles right triangle whose legs are each of length 1; see Figure 1. The number that equals the length of the hypotenuse is the positive number whose square is 2. It can be shown that this number, symbolized by $\sqrt{2}$, is an irrational number.

Also, the number that equals the ratio of the circumference C to the diameter d of any circle, denoted by the symbol π (the Greek letter pi), is an irrational number. See Figure 2.

Figure 1

Figure 2 $\pi = C/d$

The irrational numbers $\sqrt{2}$ and π have decimal representations that begin as follows:

$$\sqrt{2} = 1.414213. . . \qquad \pi = 3.14159. . .$$

In practice irrational numbers are generally represented by approximations. For example, using the symbol $\approx$ (read as ''approximately equal to''), we can write

$$\sqrt{2} \approx 1.4142 \qquad \pi \approx 3.1416$$

Other examples of irrational numbers are $\sqrt{3}$, $\sqrt{5}$, $\sqrt{7}$, $\sqrt[3]{2}$, $\sqrt[3]{3}$, and so on.

REAL NUMBERS

Together, the rational numbers and irrational numbers form the *real numbers*.

Thus every decimal may be represented by a real number (either rational or irrational) and, conversely, every real number may be represented by a decimal. It is this feature of real numbers that gives them their practicality. In the physical world many changing quantities such as the length of a heated rod, the velocity of a falling object, and so on, are assumed to pass through every possible magnitude from the initial one to the final one as they change. Real numbers in the form of decimals provide a convenient way to measure such quantities as they change. Figure 3 shows the relationship of various types of numbers.

Figure 3

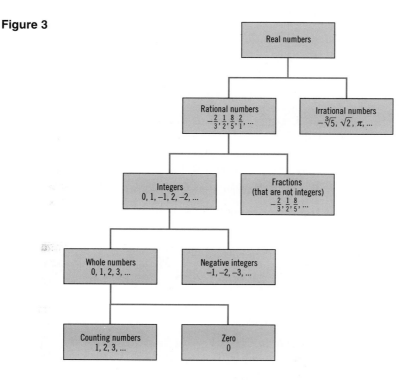

Properties of Real Numbers

As an aid to your review of real numbers, we list below several important properties and notations. The letters a, b, c, d represent real numbers; any exceptions will be noted as they occur.

1. *Commutative Properties*
   (a) $a + b = b + a$ (b) $a \cdot b = b \cdot a$

2. *Associative Properties*
   (a) $a + b + c = (a + b) + c = a + (b + c)$
   (b) $a \cdot b \cdot c = (a \cdot b) \cdot c = a \cdot (b \cdot c)$

3. *Distributive Property*
   $a \cdot (b + c) = (a \cdot b) + (a \cdot c)$

4. *Arithmetic of Ratios*

   (a) $\dfrac{a}{b} = a \cdot \dfrac{1}{b}$ $b \neq 0$

   (b) $\dfrac{a}{b} + \dfrac{c}{d} = \dfrac{(a \cdot d) + (b \cdot c)}{b \cdot d}$ $b \neq 0, \quad d \neq 0$

   (c) $\dfrac{a}{b} \cdot \dfrac{c}{d} = \dfrac{a \cdot c}{b \cdot d}$ $b \neq 0, \quad d \neq 0$

   (d) $\dfrac{a}{b} \div \dfrac{c}{d} = \dfrac{a}{b} \cdot \dfrac{d}{c} = \dfrac{a \cdot d}{b \cdot c}$ $b \neq 0, \quad c \neq 0, \quad d \neq 0$

5. *Rules for Division*

   $0 \div a = 0$ or $\dfrac{0}{a} = 0$ for any real number a different from 0

   $a \div a = 1$ or $\dfrac{a}{a} = 1$ for any real number a different from 0

   Note: Division by zero is not allowed. One reason is to avoid the following difficulty: $2/0 = x$ means to find x such that $0 \cdot x = 2$. But $0 \cdot x = 0$ for all x, so there is no number x such that $2/0 = x$.

6. *Cancellation Properties*
   (a) If $a \cdot c = b \cdot c$ and c is not 0, then $a = b$.

   (b) If b and c are not 0, then $\dfrac{a \cdot c}{b \cdot c} = \dfrac{a}{b}$.

7. *Product Law*
   If $a \cdot b = 0$, then either $a = 0$ or $b = 0$.

8. *Rules of Signs*
   (a) $a \cdot (-b) = -(a \cdot b)$ (b) $(-a) \cdot b = -(a \cdot b)$
   (c) $(-a) \cdot (-b) = a \cdot b$ (d) $-(-a) = a$

9. *Agreements: Notations*
   (a) Given $a \cdot b + c$ or $c + a \cdot b$, we agree to multiply $a \cdot b$ first, and then add c.
   (b) A *mixed number* $3\frac{5}{8}$ means $3 + \frac{5}{8} = 3.625$; 3 *times* $\frac{5}{8}$ is written as $3(\frac{5}{8})$ or $(3)(\frac{5}{8})$ or $3 \cdot \frac{5}{8} = \frac{15}{8} = 1\frac{7}{8}$.

10. *Exponents*

For any positive integer n and any real number x we define

$$x^1 = x, \quad x^2 = x \cdot x, \quad \ldots, \quad x^n = \underbrace{x \cdot x \cdot \cdots \cdot x}_{n \text{ factors}}$$

For any real number $x \neq 0$ we define

$$x^0 = 1, \quad x^{-1} = \frac{1}{x}, \quad x^{-2} = \frac{1}{x^2}, \quad \ldots, \quad x^{-n} = \frac{1}{x^n}$$

A review of exponents and logarithms may be found in Section A.2.

The Real Number Line

It can be shown that there is a one-to-one correspondence between real numbers and points on a line. That is, every real number corresponds to a point on the line and, conversely, each point on the line has a unique real number associated with it. We establish this correspondence of real numbers with points on a line in the following manner.

We start with a line that is, for convenience, drawn horizontally. We pick a point on the line and label it O, for **origin.** Then we pick another point some fixed distance to the right of O and label it U, for unit, as shown in Figure 4.

Figure 4   Horizontal line

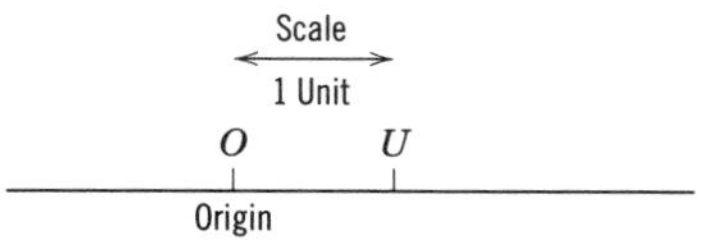

Scale
1 Unit
O U
Origin

The fixed distance, which may be 1 inch, 1 centimeter, 1 light-year, or any unit distance, determines the **scale.** We associate the real number 0 with the origin O and the number 1 with the point U. Refer now to Figure 5. The point to the right of U that is twice as far from O as U is associated with the number 2. The point to the right of U that is three times as far from O as U is associated with the number 3. The point midway between O and U is assigned the number 0.5, or $\frac{1}{2}$. Corresponding points to the left of the origin O are assigned the numbers $-\frac{1}{2}, -1, -2, -3$, and so on, depending on how far they are from O. Notice in Figure 5 that we placed an arrowhead on the right end of the line to indicate the direction in which the assigned numbers increase. Figure 5 also shows the points associated with the irrational numbers $\sqrt{2}$ and π.

Figure 5   The real number line

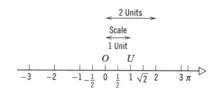

COORDINATE; REAL NUMBER LINE

The real number associated with a point P is called the *coordinate* of P, and the line whose points have been assigned coordinates is called the *real number line*.

The real number line divides the real numbers into three classes, as shown in Figure 6.

Figure 6
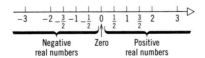

1. The **negative real numbers** are the coordinates of points to the left of the origin O.
2. The real number **zero** is the coordinate of the origin O.
3. The **positive real numbers** are the coordinates of points to the right of the origin O.

Negative and positive numbers have the following multiplication properties:

1. The product of two positive numbers is a positive number.
2. The product of two negative numbers is a positive number.
3. The product of a positive number and a negative number is a negative number.

Inequality Symbols

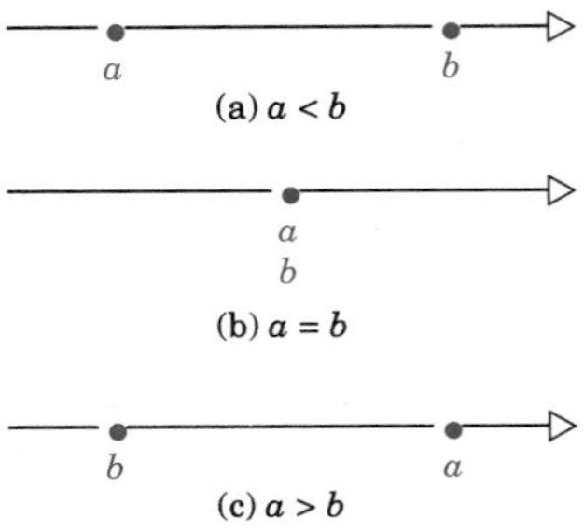

An important property of the real number line follows from the fact that, given two numbers (points) a and b, either a is to the left of b, a equals b, or a is to the right of b. See Figure 7.

If a is to the left of b, we say "a is less than b" and write $a < b$. If a is to the right of b, we say "a is greater than b" and write $a > b$. Of course, if a equals b, we write $a = b$. If a is either less than or equal to b, we write $a \leq b$. Similarly, $a \geq b$ means a is either greater than or equal to b. Collectively, the symbols $<$, $>$, $\leq$, $\geq$ are called **inequality symbols.** A formal definition of *less than* and *greater than* follows.

(a) $a < b$

(b) $a = b$

(c) $a > b$

Figure 7

Let a and b be two real numbers. We say that a is less than b, written $a < b$, or, equivalently, that b is greater than a, written $b > a$, if there is a positive number p so that $a + p = b$.

Note that $a < b$ and $b > a$ mean the same thing. Thus it does not matter whether we write $2 < 3$ or $3 > 2$.

EXAMPLE 1

(a) $3 < 7$ (b) $-5 > -16$ (c) $-6 < 0$
(d) $-8 < -4$ (e) $4 > -1$ (f) $8 > 0$

....

In Example 1(a) we conclude that $3 < 7$ either because 3 is to the left of 7 on the real number line or because $3 + 4 = 7$.

Similarly, we conclude in Example 1(b) that $-5 > -16$ either because -5 lies to the right of -16 on the real number line or because $-16 + 11 = -5$.

Now Work Problem 3

Look again at Example 1. You may find it useful to observe that the inequality symbol always points in the direction of the smaller number.

Statements of the form $a < b$ or $b > a$ are called **strict inequalities,** while statements of the form $a \le b$ or $b \ge a$ are called **nonstrict inequalities. An inequality** is a statement in which two expressions are related by an inequality symbol. The expressions are referred to as the sides of the inequality.

Based on the discussion thus far, we conclude that

$a > 0$ is equivalent to a is positive

$a < 0$ is equivalent to a is negative

Thus we sometimes read $a > 0$ by saying that ''a is positive.'' If $a \ge 0$, then either $a > 0$ or $a = 0$, and we may read this as ''a is nonnegative.''

Suppose a and b are two real numbers and $a < b$. We shall use the notation $a < x < b$ to mean that x is a number *between* a and b. Thus the expression $a < x < b$ is equivalent to the two inequalities $a < x$ and $x < b$. Similarly, the expression $a \le x \le b$ is equivalent to the two inequalities $a \le x$ and $x \le b$. The remaining two possibilities, $a \le x < b$ and $a < x \le b$, are defined analogously.

Although it is acceptable to write $3 \ge x \ge 2$, it is preferable to reverse the inequality symbols and write instead $2 \le x \le 3$, so that, as you read from left to right, the values go from smaller to larger.

A statement such as $2 \le x \le 1$ is false because there is no number x for which $2 \le x$ and $x \le 1$. Finally, we never mix inequality symbols, as in $2 \le x \ge 3$.

Inequalities are useful in representing certain subsets of real numbers. In so doing, though, other variations of the inequality notation may be used.

EXAMPLE 2

(a) In the inequality $x > 4$, x is any number greater than 4. In Figure 8 we use an open circle at 4 to indicate that the number 4 is not part of the graph.

(b) In the inequality $4 < x \le 6$, x is any number between 4 and 6, including 6 but excluding 4. In Figure 9 we use a solid dot at 6 to indicate that 6 is part of the graph.

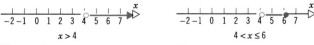

$x > 4$

Figure 8

$4 < x \le 6$

Figure 9

Now Work Problem 7

Inequalities have the following properties:

1. *Addition Property*
   If $a \leq b$, then $a + c \leq b + c$ for any choice of c. That is, the addition of a number to each side of an inequality will not affect the sense or direction of the inequality.
2. *Multiplication Properties*
   (a) If $a \leq b$ and $c > 0$, then $a \cdot c \leq b \cdot c$.
   (b) If $a \leq b$ and $c < 0$, then $a \cdot c \geq b \cdot c$.

   When multiplying each side of an inequality by a number, the sense or direction of the inequality remains the same if we multiply by a positive number; it is reversed if we multiply by a negative number.

EXAMPLE 3

(a) Since $2 < 3$, then $2 + 5 < 3 + 5$, or $7 < 8$.
(b) Since $2 < 3$ and $6 > 0$, then $2 \cdot 6 < 3 \cdot 6$, or $12 < 18$.
(c) Since $2 < 3$ and $-4 < 0$, then $2 \cdot (-4) > 3 \cdot (-4)$, or $-8 > -12$.

....

Constants and Variables

In algebra we use letters such as x, y, a, b, and c to represent numbers. If the letter used is to represent *any* number from a given set of numbers, it is called a **variable**. A **constant** is either a fixed number, such as 5, $\sqrt{3}$, etc., or a letter that represents a fixed (possibly unspecified) number.

An **equation** is a statement involving one or more variables and an "equals" sign ($=$). To **solve** an equation means to find all possible numbers that the variables can assume to make the statement true. The set of all such numbers is called the **solution.** Two equations with the same solution are called **equivalent equations.**

EXAMPLE 4

Solve the equation: $2x + 3(x + 2) = 1$

SOLUTION

$$2x + 3(x + 2) = 1$$
$$2x + (3x + 6) = 1 \qquad \text{Remove parentheses using the Distributive Property.}$$
$$(2x + 3x) + 6 = 1 \qquad \text{Use the Associative Property.}$$
$$5x + 6 = 1 \qquad \text{Combine } 2x + 3x = (2 + 3)x = 5x.$$
$$5x = -5 \qquad \text{Subtract 6 from each side.}$$
$$x = -1 \qquad \text{Divide each side by 5.}$$

Check: $2(-1) + 3(-1 + 2) = -2 + 3(1) = -2 + 3 = 1$

....

An **inequality** is a statement involving one or more variables and one of the inequality symbols ($<$, $\leq$, $>$, $\geq$). To **solve** an inequality means to find all possible numbers that the variables can assume to make the statement true. The set of all such numbers is called the **solution.** Two inequalities with the same solution are called **equivalent inequalities.**

To find the solution of an inequality, we apply the properties for inequalities.

EXAMPLE 5

Solve the inequality: $x + 2 \leq 3x - 5$. Graph the solution.

SOLUTION

$$x + 2 \leq 3x - 5$$
$$-2x \leq -7 \qquad \text{Subtract 2 and then } 3x \text{ from each side.}$$
$$x \geq \tfrac{7}{2} \qquad \text{Multiply by } -\tfrac{1}{2} \text{ and remember to reverse the inequality because we multiplied by a negative number.}$$

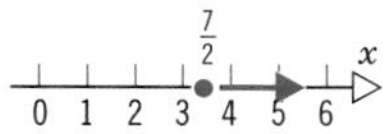

The solution is all real numbers to the right of $\tfrac{7}{2}$, including $\tfrac{7}{2}$. See Figure 10 for the graph of the solution.

Figure 10

. . . .

EXERCISE A.1   Answers to odd-numbered problems begin on page AN-87.

In Problems 1–6 write each statement as an inequality.

1. x is positive

2. z is negative

3. x is less than 2

4. y is greater than -5

5. x is less than or equal to 1

6. x is greater than or equal to 2

In Problems 7–16 graph the numbers x, if any, on the real number line.

7. $x \geq -1$

8. $x < 1$

9. $x \geq 4$ and $x < 6$

10. $x > 3$ and $x \leq 7$

11. $x \leq 0$ or $x < 6$

12. $x > 0$ or $x \geq 5$

13. $x \leq -2$ and $x > 1$

14. $x \geq 4$ and $x < -2$

15. $x \leq -2$ or $x > 1$

16. $x \geq 4$ or $x < -2$

In Problems 17–24 solve each equation.

17. $2x + 5 = 7$

18. $x + 6 = 2$

19. $6 - x = 0$

20. $6 + x = 0$

21. $3(2 - x) = 9$

22. $5(x + 1) = 10$

23. $4x + 3 = 2x - 5$

24. $5x - 8 = 2x + 1$

In Problems 25–32 solve each inequality. Graph the solutions.

25. $3x + 5 \leq 2$

26. $14x - 21x + 16 \leq 3x - 2$

27. $3x + 5 \geq 2$

28. $4 - 5x \geq 3$

29. $-3x + 5 \leq 2$

30. $8 - 2x \leq 5x - 6$

31. $6x - 3 \geq 8x + 5$

32. $-3x \leq 2x + 5$

A.2 EXPONENTS AND LOGARITHMS

Exponents

Integer exponents provide a convenient notation for repeated multiplication of a real number.

If a is a real number and n is a positive integer, then the **symbol a^n** represents the product of n factors of a. That is,

$$a^n = \underbrace{a \cdot a \cdot \cdots \cdot a}_{n \text{ factors}}$$

where it is understood that $a^1 = a$.

In the expression a^n, read "a raised to the power n," a is called the **base** and n is called the **exponent** or **power.**

If n is a positive integer and $a \neq 0$, then we define

$$a^{-n} = \frac{1}{a^n} \qquad a \neq 0$$

Also, we define

$$a^0 = 1 \qquad a \neq 0$$

EXAMPLE 1

(a) $2^3 = 2 \cdot 2 \cdot 2 = 8$ (b) $3^{-2} = \dfrac{1}{3^2} = \dfrac{1}{9}$ (c) $5^0 = 1$

Now Work Problem 1

The **principal nth root of a real number a,** written $\sqrt[n]{a}$, is defined as follows:

(a) If $a > 0$ and n is even, then $\sqrt[n]{a}$ is the positive number x for which $x^n = a$, $x > 0$.
(b) If $a < 0$ and n is even, then $\sqrt[n]{a}$ does not exist.
(c) If n is odd, then $\sqrt[n]{a}$ is the number x for which $x^n = a$.
(d) $\sqrt[n]{0} = 0$ for any n.

If $n = 2$, we write $\sqrt{a}$ in place of $\sqrt[2]{a}$.

EXAMPLE 2

(a) $\sqrt[3]{8} = 2$ because $2^3 = 8$
(b) $\sqrt{64} = 8$ because $8^2 = 64$ and $8 > 0$
(c) $\sqrt[3]{-27} = -3$ because $(-3)^3 = -27$

Now Work Problem 7

We use principal roots to define exponents that are rational. If a is a real number and $n \geq 2$ is an integer, then

$$a^{1/n} = \sqrt[n]{a}$$

provided $\sqrt[n]{a}$ exists.

If a is a real number and m, n are integers containing no common factors with $n \geq 2$, then

$$a^{m/n} = \sqrt[n]{a^m} = (\sqrt[n]{a})^m$$

provided $\sqrt[n]{a}$ exists.

EXAMPLE 3

(a) $4^{3/2} = (\sqrt{4})^3 = 2^3 = 8$ (b) $(-8)^{4/3} = (\sqrt[3]{-8})^4 = (-2)^4 = 16$
(c) $(32)^{-2/5} = (\sqrt[5]{32})^{-2} = 2^{-2} = \frac{1}{4}$

Now Work Problem 13

· ·

The Laws of Exponents establish some rules for working with exponents.

LAWS OF EXPONENTS

If a and b are real numbers and r and s are rational numbers, then

$$a^r \cdot a^s = a^{r+s} \qquad (a^r)^s = a^{rs} \qquad (ab)^r = a^r b^r \qquad a^{-r} = \frac{1}{a^r}, \quad a \neq 0$$

But what is the meaning of a^x, where the base a is a positive real number and the exponent x is an irrational number? Although a rigorous definition requires methods discussed in calculus, the basis for the definition is easy to follow: Select a rational number r that is formed by truncating (removing) all but a finite number of digits from the irrational number x. Then it is reasonable to expect that

$$a^x \approx a^r$$

For example, take the irrational number $\pi \approx 3.14159. \ldots$ An approximation to a^π is

$$a^\pi \approx a^{3.14}$$

where the digits after the hundredths position have been removed from the value for π. A better approximation would be

$$a^\pi \approx a^{3.14159}$$

where the digits after the hundred-thousandths position have been removed. Continuing in this way, we can obtain approximations to a^π to any desired degree of accuracy.

Most scientific calculators have an $\boxed{x^y}$ key (or a $\boxed{y^x}$ key) for working with exponents. To use this key, first enter the base x, then press the $\boxed{x^y}$ key, enter y, and press the $\boxed{=}$ key.

EXAMPLE 4

Using a calculator with an $\boxed{x^y}$ key, evaluate

(a) $2^{1.4}$ (b) $2^{1.41}$ (c) $2.^{1.414}$ (d) $2^{1.4142}$ (e) $2^{\sqrt{2}}$

SOLUTION

(a) $2^{1.4} \approx 2.6390158$ (b) $2^{1.41} \approx 2.6573716$
(c) $2^{1.414} \approx 2.6647497$ (d) $2.^{1.4142} \approx 2.6651191$
(e) $2^{\sqrt{2}} \approx 2.6651441$

Now Work Problem 19

· ·

It can be shown that the Laws of Exponents hold for real exponents.

Logarithms

We have given meaning to a number a raised to the power x. Suppose $N = a^x$, $a > 0$, $a \neq 1$, x real. The definition of a logarithm is based on this exponential relationship. The **logarithm to the base a of N** is symbolized by $\log_a N$ and is defined by

$$\log_a N = x \quad \text{if and only if} \quad N = a^x$$

EXAMPLE 5

(a) $\log_2 8 = x$ means $2^x = 8 = 2^3$, so $x = 3$
(b) $\log_3 9 = x$ means $3^x = 9 = 3^2$, so $x = 2$
(c) $\log_{10} N = 3$ means $N = 10^3 = 1000$
(d) $\log_4 N = 2$ means $N = 4^2 = 16$
(e) $\log_a 16 = 2$ means $16 = a^2$, so $a = 4$
(f) $\log_a 3 = -1$ means $3 = a^{-1}$, so $a = \frac{1}{3}$

· ·

Now Work Problems 25, 29, and 33

Thus logarithms are another name for exponents.

EXAMPLE 6

(a) If $100 = (1.1)^n$, then $n = \log_{1.1} 100$
(b) If $350 = (1.25)^n$, then $n = \log_{1.25} 350$
(c) If $1000 = (1.005)^n$, then $n = \log_{1.005} 1000$

· ·

To evaluate the expressions shown in Example 6, we use a scientific calculator. Such calculators have a key labeled $\boxed{\log}$ to evaluate logarithms to the base 10. To use the calculator, we need to know the following formula involving changing the base.

Change-of-Base Formula

$$\log_a M = \frac{\log_{10} M}{\log_{10} a}$$

To use your calculator to evaluate $\log_{10} M$, enter $\boxed{M}$ and press $\boxed{\log}$.

EXAMPLE 7

(a) $\log_{1.1} 100 = \dfrac{\log_{10} 100}{\log_{10} 1.1} = \dfrac{2}{0.0414} = 48.3177$

(b) $\log_{1.25} 350 = \dfrac{\log_{10} 350}{\log_{10} 1.25} = \dfrac{2.5441}{0.0969} = 26.2519$

(c) $\log_{1.005} 1000 = \dfrac{\log_{10} 1000}{\log_{10} 1.005} = \dfrac{3}{0.002166} = 1385.0021$

· ·

Comparing Examples 6 and 7, we find that:

(a) $100 = (1.1)^n$ means $n = \log_{1.1} 100 = 48.3177$

(b) $350 = (1.25)^n$ means $n = \log_{1.25} 350 = 26.2519$

Now Work Problem 37   (c) $1000 = (1.005)^n$ means $n = \log_{1.005} 1000 = 1385.0021$

EXERCISE A.2   Answers to odd-numbered problems begin on page AN-87.

In Problems 1–18 evaluate each expression.

1. 4^3 **2.** 8^2 **3.** 2^{-3} **4.** 4^{-2} **5.** 8^0

6. $(-3)^0$ **7.** $\sqrt{16}$ **8.** $\sqrt{9}$ **9.** $\sqrt[3]{27}$ **10.** $\sqrt[3]{-1}$

11. $\sqrt[4]{16}$ **12.** $\sqrt{0}$ **13.** $8^{2/3}$ **14.** $9^{3/2}$ **15.** $16^{-3/2}$

16. $8^{-2/3}$ **17.** $(-8)^{-2/3}$ **18.** $(-27)^{-4/3}$

In Problems 19–24 approximate each number using a calculator.

19. (a) $3^{2.2}$ (b) $3^{2.23}$ (c) $3^{2.236}$ (d) $3^{\sqrt{5}}$

20. (a) $5^{1.7}$ (b) $5^{1.73}$ (c) $5^{1.732}$ (d) $5^{\sqrt{3}}$

21. (a) $2^{3.14}$ (b) $2^{3.141}$ (c) $2^{3.1415}$ (d) 2^{π}

22. (a) $2^{2.7}$ (b) $2^{2.71}$ (c) $2^{2.718}$ (d) 2^{e}

23. (a) $3.1^{2.7}$ (b) $3.14^{2.71}$ (c) $3.141^{2.718}$ (d) π^{e}

24. (a) $2.7^{3.1}$ (b) $2.71^{3.14}$ (c) $2.718^{3.141}$ (d) e^{π}

In Problems 25–28 evaluate each expression.

25. $\log_3 27$ **26.** $\log_2 16$ **27.** $\log_2 \frac{1}{2}$ **28.** $\log_3 \frac{1}{9}$

In Problems 29–32 find N.

29. $\log_2 N = 3$ **30.** $\log_2 N = -2$ **31.** $\log_3 N = -1$ **32.** $\log_8 N = \frac{1}{3}$

In Problems 33–36 find a.

33. $\log_a 8 = 3$ **34.** $\log_a 4 = -1$ **35.** $\log_a 9 = 2$ **36.** $\log_a 16 = -2$

In Problems 37–46 use a calculator to evaluate each logarithm.

37. $\log_{1.1} 200$

38. $\log_{1.05} 200$

39. $\log_{1.005} 1000$

40. $\log_{1.001} 10$

41. $\log_{1.002} 20$

42. $\log_{1.005} 30$

43. $\log_{1.0005} 500$

44. $\log_{1.0001} 1000$

45. $\log_{1.003} 500$

46. $\log_{1.006} 500$

A.3 GEOMETRIC SEQUENCES

A **sequence** is a rule that assigns a real number to each positive integer.

A sequence is often represented by listing its values in order. For example, the sequence whose rule is to assign to each positive integer its reciprocal may be represented as

$$s_1 = 1, \qquad s_2 = \frac{1}{2}, \qquad s_3 = \frac{1}{3}, \qquad s_4 = \frac{1}{4}, \ldots, \qquad s_n = \frac{1}{n}, \ldots$$

or merely as the list

$$1, \quad \frac{1}{2}, \quad \frac{1}{3}, \quad \frac{1}{4}, \ldots, \quad \frac{1}{n}, \ldots$$

The list never ends, as the dots indicate. The real numbers in this ordered list are called the **terms** of the sequence. For the sequence above we have a rule for the nth term, namely, $s_n = 1/n$, so it is easy to find any term of the sequence. In this case we represent the sequence by placing braces around the formula for the nth term, writing $\{s_n\} = \{1/n\}$.

EXAMPLE 1

Write the first six terms of the sequence below.

$$\{a_n\} = \left\{\frac{n-1}{n}\right\}$$

SOLUTION

$$a_1 = 0$$
$$a_2 = \tfrac{1}{2}$$
$$a_3 = \tfrac{2}{3}$$
$$a_4 = \tfrac{3}{4}$$
$$a_5 = \tfrac{4}{5}$$
$$a_6 = \tfrac{5}{6}$$

······················

EXAMPLE 2

Write the first six terms of the sequence below.

$$\{b_n\} = \left\{(-1)^{n-1}\left(\frac{2}{n}\right)\right\}$$

SOLUTION

$$b_1 = 2$$
$$b_2 = -1$$
$$b_3 = \tfrac{2}{3}$$
$$b_4 = -\tfrac{1}{2}$$
$$b_5 = \tfrac{2}{5}$$
$$b_6 = -\tfrac{1}{3}$$

Now Work Problem 5

. . . .

A second way of defining a sequence is to assign a value to the first term (or the first few terms) and specify the nth term by a formula or equation that involves one or more of the terms preceding it. Sequences defined this way are said to be defined **recursively,** and the rule or formula is called a **recursive formula.**

EXAMPLE 3

Write the first five terms of the recursively defined sequence given below.

$$s_1 = 1, \qquad s_n = 4s_{n-1}$$

SOLUTION

The first term is given as $s_1 = 1$. To get the second term, we use $n = 2$ in the formula to get $s_2 = 4s_1 = 4 \cdot 1 = 4$. To get the third term, we use $n = 3$ in the formula to get $s_3 = 4s_2 = 4 \cdot 4 = 16$. To get a new term requires that we know the value of the preceding term. The first five terms are

$$s_1 = 1$$
$$s_2 = 4 \cdot 1 = 4$$
$$s_3 = 4 \cdot 4 = 16$$
$$s_4 = 4 \cdot 16 = 64$$
$$s_5 = 4 \cdot 64 = 256$$

. . . .

EXAMPLE 4

Write the first five terms of the recursively defined sequence given below.

$$u_1 = 1, \qquad u_2 = 1, \qquad u_{n+2} = u_n + u_{n+1}$$

SOLUTION

We are given the first two terms. To get the third term requires that we know each of the previous two terms. Thus

$$u_1 = 1$$
$$u_2 = 1$$
$$u_3 = u_1 + u_2 = 2$$
$$u_4 = u_2 + u_3 = 1 + 2 = 3$$
$$u_5 = u_3 + u_4 = 2 + 3 = 5$$

. . . .

The sequence defined in Example 4 is called a **Fibonacci sequence,** and the terms of this sequence are called **Fibonacci numbers.** These numbers appear in a wide variety of applications.

EXAMPLE 5

Write the first five terms of the recursively defined sequence given below.
$$f_1 = 1, \qquad f_{n+1} = (n + 1)f_n$$

SOLUTION

Here,
$$
\begin{aligned}
f_1 &= 1 \\
f_2 &= 2f_1 = 2 \cdot 1 = 2 \\
f_3 &= 3f_2 = 3 \cdot 2 = 6 \\
f_4 &= 4f_3 = 4 \cdot 6 = 24 \\
f_5 &= 5f_4 = 5 \cdot 24 = 120
\end{aligned}
$$

. . . .

Now Work Problems 13 and 21

The nth term of the sequence in Example 5 is n **factorial.**

Geometric Sequences

When the ratio of successive terms of a sequence is always the same nonzero number, the sequence is called **geometric.** Thus a **geometric sequence*** may be defined recursively as $a_1 = a$, $a_{n+1}/a_n = r$, or as:

> **Geometric Sequence**
>
> $$a_1 = a, \qquad a_{n+1} = ra_n$$
>
> where $a = a_1$ and $r \neq 0$ are real numbers. The number a is the **first term,** and the nonzero number r is called the **common ratio.**

Thus the terms of a geometric sequence with first term a and common ratio r follow the pattern
$$a, \quad ar, \quad ar^2, \quad ar^3, \ldots$$

EXAMPLE 6

The sequence
$$2, \quad 6, \quad 18, \quad 54, \quad 162, \ldots$$
is geometric since the ratio of successive terms is 3. The first term is 2, and the common ratio is 3.

. . . .

* Sometimes called a **geometric progression.**

EXAMPLE 7

Show that the sequence below is geometric. Find the first term and the common ratio.

$$\{s_n\} = \{2^{-n}\}$$

SOLUTION

The first term is $s_1 = 2^{-1} = \frac{1}{2}$. The $(n + 1)$st and nth terms of the sequence $\{s_n\}$ are

$$s_{n+1} = 2^{-(n+1)} \qquad \text{and} \qquad s_n = 2^{-n}$$

Their ratio is

$$\frac{s_{n+1}}{s_n} = \frac{2^{-(n+1)}}{2^{-n}} = 2^{-n-1+n} = 2^{-1} = \frac{1}{2}$$

Because the ratio of successive terms is a nonzero number independent of n, the sequence $\{s_n\}$ is geometric with common ratio $\frac{1}{2}$.

....

EXAMPLE 8

Show that the sequence below is geometric. Find the first term and the common ratio.

$$\{t_n\} = \{4^n\}$$

SOLUTION

The first term is $t_1 = 4^1 = 4$. The $(n + 1)$st and nth terms are

$$t_{n+1} = 4^{n+1} \qquad \text{and} \qquad t_n = 4^n$$

Their ratio is

$$\frac{t_{n+1}}{t_n} = \frac{4^{n+1}}{4^n} = 4$$

Thus $\{t_n\}$ is a geometric sequence with common ratio 4.

....

Adding the First n Terms of a Geometric Sequence

The next result gives us a formula for finding the sum of the first n terms of a geometric sequence.

> **Sum of the First n Terms of a Geometric Sequence**
>
> Let $\{a_n\}$ be a geometric sequence with first term a and common ratio r. The sum S_n of the first n terms of $\{a_n\}$ is
>
> $$S_n = a\left(\frac{1 - r^n}{1 - r}\right) \qquad r \neq 0, 1 \qquad (1)$$

Formula (1) may be derived as follows:

$$S_n = a + ar + \cdots + ar^{n-1} \qquad (2)$$

Multiply each side by r to obtain

$$rS_n = ar + ar^2 + \cdots + ar^n \tag{3}$$

Now subtract (3) from (2). The result is

$$S_n - rS_n = a - ar^n$$
$$(1 - r)S_n = a(1 - r^n)$$

Since $r \neq 1$, we can solve for S_n:

$$S_n = a\left(\frac{1 - r^n}{1 - r}\right)$$

EXAMPLE 9   Find the sum S_n of the first n terms of the sequence $\{(\frac{1}{2})^n\}$; that is, find

$$\frac{1}{2} + \frac{1}{4} + \frac{1}{8} + \cdots + \left(\frac{1}{2}\right)^n$$

SOLUTION

The sequence $\{(\frac{1}{2})^n\}$ is a geometric sequence with $a = \frac{1}{2}$ and $r = \frac{1}{2}$. The sum S_n we seek is the sum of the first n terms of the sequence, so we use Formula (1) to get

$$S_n = \frac{1}{2} + \frac{1}{4} + \frac{1}{8} + \cdots + \left(\frac{1}{2}\right)^n$$
$$= \frac{1}{2}\left[\frac{1 - (\frac{1}{2})^n}{1 - \frac{1}{2}}\right]$$
$$= \frac{1}{2}\left[\frac{1 - (\frac{1}{2})^n}{\frac{1}{2}}\right]$$
$$= 1 - \left(\frac{1}{2}\right)^n$$

....

EXERCISE A.3   Answers to odd-numbered problems begin on page AN-87.

In Problems 1–12 write the first five terms of each sequence.

1. $\{n\}$

2. $\{n^2 + 1\}$

3. $\left\{\dfrac{n}{n + 1}\right\}$

4. $\left\{\dfrac{2n + 3}{2n - 1}\right\}$

5. $\{(-1)^{n+1}n^2\}$

6. $\left\{(-1)^{n-1}\left(\dfrac{n}{2n - 1}\right)\right\}$

7. $\left\{\dfrac{2^n}{3^n + 1}\right\}$

8. $\left\{\left(\dfrac{4}{3}\right)^n\right\}$

9. $\left\{\dfrac{(-1)^n}{(n + 1)(n + 2)}\right\}$

10. $\left\{\dfrac{3^n}{n}\right\}$

11. $\left\{\dfrac{n}{e^n}\right\}$

12. $\left\{\dfrac{n^2}{2^n}\right\}$

In Problems 13–26 a sequence is defined recursively. Write the first five terms.

13. $a_1 = 1;\quad a_{n+1} = 2 + a_n$

14. $a_1 = 3;\quad a_{n+1} = 5 - a_n$

15. $a_1 = -2;\quad a_{n+1} = n + a_n$

16. $a_1 = 1;\quad a_{n+1} = n - a_n$

17. $a_1 = 5;\quad a_{n+1} = 2a_n$

18. $a_1 = 2;\quad a_{n+1} = -a_n$

19. $a_1 = 3;\quad a_{n+1} = \dfrac{a_n}{n}$

20. $a_1 = -2;\quad a_{n+1} = n + 3a_n$

21. $a_1 = 1;\quad a_2 = 2;\quad a_{n+2} = a_n a_{n+1}$

22. $a_1 = -1;\quad a_2 = 1;\quad a_{n+2} = a_{n+1} + na_n$

23. $a_1 = A;\quad a_{n+1} = a_n + d$

24. $a_1 = A;\quad a_{n+1} = ra_n,\quad r \neq 0$

25. $a_1 = \sqrt{2};\quad a_{n+1} = \sqrt{2 + a_n}$

26. $a_1 = \sqrt{2};\quad a_{n+1} = \sqrt{a_n/2}$

In Problems 27–36 a geometric sequence is given. Find the common ratio and write out the first four terms. Also, find the sum of the first n terms.

27. $\{2^n\}$

28. $\{(-4)^n\}$

29. $\left\{ -3\left(\dfrac{1}{2}\right)^n \right\}$

30. $\left\{ \left(\dfrac{5}{2}\right)^n \right\}$

31. $\left\{ \dfrac{2^{n-1}}{4} \right\}$

32. $\left\{ \dfrac{3^n}{9} \right\}$

33. $\{2^{n/3}\}$

34. $\{3^{2n}\}$

35. $\left\{ \dfrac{3^{n-1}}{2^n} \right\}$

36. $\left\{ \dfrac{2^n}{3^{n-1}} \right\}$

A.4 FACTORING POLYNOMIALS

Consider the following product:

$$(2x + 3)(x - 4) = 2x^2 - 5x - 12$$

The two polynomials on the left side are called **factors** of the polynomial on the right side. Expressing a given polynomial as a product of other polynomials is called **factoring.**

We shall restrict our discussion here to factoring polynomials in one variable into products of polynomials in one variable, where all coefficients are integers. We call this **factoring over the integers.** There will be times, though, when we will want to **factor over the rational numbers** and even **factor over the real numbers.** Factoring over the rational numbers means to write a given polynomial, whose coefficients are rational numbers, as a product of polynomials whose coefficients are also rational numbers. Factoring over the real numbers means to write a given polynomial, whose coefficients are real numbers, as a product of polynomials whose coefficients are also real numbers. Unless specified otherwise, we will be factoring over the integers.

Any polynomial can be written as the product of 1 times itself or as -1 times its additive inverse. If a polynomial cannot be written as the product of two other polynomials (excluding 1 and -1), then the polynomial is said to be **prime.** When a polynomial has been written as a product consisting only of prime factors, then it is said to be **factored completely.** Examples of prime polynomials are

$$2, \quad 3, \quad 5, \quad x, \quad x + 1, \quad x - 1, \quad 3x + 4$$

The first factor to look for in a factoring problem is a common monomial factor present in each term of the polynomial. If one is present, use the distributive property to factor it out. For example:

Polynomial	Common Monomial Factor	Remaining Factor	Factored Form
$2x + 4$	2	$x + 2$	$2x + 4 = 2(x + 2)$
$3x - 6$	3	$x - 2$	$3x - 6 = 3(x - 2)$
$2x^2 - 4x + 8$	2	$x^2 - 2x + 4$	$2x^2 - 4x + 8 = 2(x^2 - 2x + 4)$
$8x - 12$	4	$2x - 3$	$8x - 12 = 4(2x - 3)$
$x^2 + x$	x	$x + 1$	$x^2 + x = x(x + 1)$
$x^3 - 3x^2$	x^2	$x - 3$	$x^3 - 3x^2 = x^2(x - 3)$
$6x^2 + 9x$	$3x$	$2x + 3$	$6x^2 + 9x = 3x(2x + 3)$

Notice that, once all common monomial factors have been removed from a polynomial, the remaining factor is either a prime polynomial of degree 1 or a polynomial of degree 2 or higher. (Do you see why?) Thus we concentrate on techniques for factoring polynomials of degree 2 or higher that contain no monomial factors.

Now Work Problem 1

Special Factors

Equations (1) through (6) below provide a list of factoring formulas. For example, Equation (1) states that if the polynomial is the difference of two squares, $x^2 - a^2$, it can be factored into $(x - a)(x + a)$.

Difference of Two Squares

$$x^2 - a^2 = (x - a)(x + a) \tag{1}$$

Squares of Binomials, or Perfect Squares

$$x^2 + 2ax + a^2 = (x + a)^2 \tag{2a}$$

$$x^2 - 2ax + a^2 = (x - a)^2 \tag{2b}$$

Miscellaneous Trinomials

$$x^2 + (a + b)x + ab = (x + a)(x + b) \tag{3a}$$

$$acx^2 + (ad + bc)x + bd = (ax + b)(cx + d) \tag{3b}$$

Cubes of Binomials, or Perfect Cubes

$$x^3 + 3ax^2 + 3a^2x + a^3 = (x + a)^3 \tag{4a}$$

$$x^3 - 3ax^2 + 3a^2x - a^3 = (x - a)^3 \tag{4b}$$

Difference of Two Cubes

$$x^3 - a^3 = (x - a)(x^2 + ax + a^2) \tag{5}$$

Sum of Two Cubes

$$x^3 + a^3 = (x + a)(x^2 - ax + a^2) \tag{6}$$

EXAMPLE 1

Factor completely: $x^2 - 4$

SOLUTION

We notice that $x^2 - 4$ is the difference of two squares, x^2 and 2^2. Thus, using Equation (1), we find
$$x^2 - 4 = (x - 2)(x + 2)$$
· ·

EXAMPLE 2

Factor completely: $x^3 - 1$

SOLUTION

Equation (5) states that the difference of two cubes, $x^3 - a^3$, can be factored as $(x - a)(x^2 + ax + a^2)$. Because $x^3 - 1$ is the difference of two cubes, x^3 and 1^3, we find
$$x^3 - 1 = (x - 1)(x^2 + x + 1)$$
· ·

EXAMPLE 3

Factor completely: $x^3 + 8$

SOLUTION

Equation (6) states that the sum of two cubes, $x^3 + a^3$, can be factored as $(x + a)(x^2 - ax + a^2)$. Because $x^3 + 8$ is the sum of two cubes, x^3 and 2^3, we have
$$x^3 + 8 = (x + 2)(x^2 - 2x + 4)$$
· ·

EXAMPLE 4

Factor completely: $x^4 - 16$

SOLUTION

Using Equation (1) for the difference of two squares, $x^4 = (x^2)^2$ and $16 = 4^2$, we have
$$x^4 - 16 = (x^2 - 4)(x^2 + 4)$$
But $x^2 - 4$ is also the difference of two squares. Thus
$$x^4 - 16 = (x^2 - 4)(x^2 + 4) = (x - 2)(x + 2)(x^2 + 4)$$

Now Work Problem 11

· ·

Whenever the first term and third term of a trinomial are both positive and are perfect squares, such as x^2, $9x^2$, 1, 4, and so on, check to see whether either of the special products (2a) or (2b) applies.

EXAMPLE 5

Factor completely: $x^2 + 4x + 4$

SOLUTION

The first term, x^2, and the third term, $4 = 2^2$, are perfect squares. Because the middle term is twice the product of x and 2, we use Equation (2a) to obtain

$$x^2 + 4x + 4 = (x + 2)^2$$

· ·

EXAMPLE 6   Factor completely: $9x^2 - 6x + 1$

SOLUTION

The first term, $9x^2 = (3x)^2$, and the third term, $1 = 1^2$, are perfect squares. Because the middle term is twice the product of $3x$ and 1, we use Equation (2b) to obtain

$$9x^2 - 6x + 1 = (3x - 1)^2$$

· ·

EXAMPLE 7   Factor completely: $25x^2 + 30x + 9$

SOLUTION

The first term, $25x^2 = (5x)^2$, and the third term, $9 = 3^2$, are perfect squares. Because the middle term is twice the product of $5x$ and 3, we use Equation (2a) to obtain

$$25x^2 + 30x + 9 = (5x + 3)^2$$

Now Work Problem 21

· ·

If a trinomial is not a perfect square, it may be possible to factor it using the technique discussed next.

Factoring Second-Degree Polynomials

Factoring a second-degree polynomial, $Ax^2 + Bx + C$, where A, B, and C are integers, is a matter of skill, experience, and often some trial and error. The idea behind factoring $Ax^2 + Bx + C$ is to see whether it can be made equal to the product of two, possibly equal, first-degree polynomials. Thus we want to see whether there are integers a, b, c, and d so that

$$Ax^2 + Bx + C = (ax + b)(cx + d)$$

We start with second-degree polynomials that have a leading coefficient of 1. Such a polynomial, if it can be factored, must follow the form of the special product (3a).

$$x^2 + Bx + C = (x + a)(x + b) = x^2 + (a + b)x + ab \quad B = a + b, C = ab$$

Note the pattern in this formula: The correct choice of the factors a and b of the constant term C must sum to the coefficient of the middle term B.

EXAMPLE 8   Factor completely: $x^2 + 7x + 12$

SOLUTION

First, we determine all possible integral factors of the constant term 12 and compute their sums:

Factors of 12	1, 12	$-1, -12$	2, 6	$-2, -6$	3, 4	$-3, -4$
Sum	13	-13	8	-8	7	-7

The factors of 12 that sum to 7, the coefficient of the middle term, are 3 and 4. Thus

$$x^2 + 7x + 12 = (x + 3)(x + 4)$$

. . . .

EXAMPLE 9

Factor completely: $x^2 - 6x + 8$

SOLUTION

First, we determine all possible integral factors of the constant term 8 and compute each sum:

Factors of 8	1, 8	$-1, -8$	2, 4	$-2, -4$
Sum	9	-9	6	-6

Since -6 is the coefficient of the middle term, we have

$$x^2 - 6x + 8 = (x - 2)(x - 4)$$

. . . .

EXAMPLE 10

Factor completely: $x^2 - x - 12$

SOLUTION

First, we determine all possible integral factors of -12 and compute each sum:

Factors of -12	1, -12	$-1, 12$	2, -6	$-2, 6$	3, -4	$-3, 4$
Sum	-11	11	-4	4	-1	1

Since -1 is the coefficient of the middle term, we have

$$x^2 - x - 12 = (x + 3)(x - 4)$$

. . . .

EXAMPLE 11

Factor completely: $x^2 + 4x - 12$

SOLUTION

The factors -2 and 6 of -12 have the sum 4. Thus

$$x^2 + 4x - 12 = (x - 2)(x + 6)$$

. . . .

To avoid errors in factoring, always check your answer by multiplying it out to see if the result equals the original expression.

When none of the possibilities works, the polynomial is prime.

EXAMPLE 12

Show that $x^2 + 9$ is prime.

SOLUTION

First, we list the integral factors of 9, and compute their sums:

Factors of 9	1, 9	$-1, -9$	3, 3	$-3, -3$
Sum	10	-10	6	-6

Since the coefficient of the middle term in $x^2 + 9$ is 0 and none of the sums above equals 0, we conclude that $x^2 + 9$ is prime.

Now Work Problem 19

....

When the leading coefficient is not 1, a somewhat longer list of possibilities may be required. Observe the pattern in the following formula:

$$Ax^2 + Bx + C = (ax + b)(cx + d) = acx^2 + (ad + bc)x + bd$$
$$A = ac, \quad B = ad + bc, \quad C = bd$$

Our task is to find factors a and c of the leading coefficient A and factors b and d of the constant term C so that the expression $ad + bc$ equals the coefficient B of the middle term.

EXAMPLE 13

Factor completely: $2x^2 + 5x + 3$

SOLUTION

The positive integral factors of the leading coefficient $ac = 2$ are $a = 2, c = 1$. We begin the factorization by writing

$$2x^2 + 5x + 3 = (2x \quad)(x \quad)$$

The positive integral factors of the constant term $bd = 3$ are $b = 1, d = 3$, or $b = 3, d = 1$. This suggests the following possibilities:

$$2x^2 + 5x + 3 \begin{cases} (2x \quad 1)(x \quad 3) \\ (2x \quad 3)(x \quad 1) \end{cases}$$

Next, we select the signs to be placed inside the factors. Since the constant term is positive and the coefficient of the middle term is positive, the only signs that can possibly work are $+$ signs. (Do you see why?) Thus

$$2x^2 + 5x + 3 \begin{cases} (2x + 1)(x + 3) = 2x^2 + 7x + 3 \\ (2x + 3)(x + 1) = 2x^2 + 5x + 3 \end{cases}$$

Thus we conclude that

$$2x^2 + 5x + 3 = (2x + 3)(x + 1)$$

· ·

EXAMPLE 14

Factor completely: $2x^2 - x - 6$

SOLUTION

The positive integral factors of 6 are 1, 6, and 2, 3. This suggests the following possibilities:

$$2x^2 - x - 6 \begin{cases} (2x \quad 1)(x \quad 6) \\ (2x \quad 6)(x \quad 1) \\ (2x \quad 2)(x \quad 3) \\ (2x \quad 3)(x \quad 2) \end{cases}$$

Since the constant term is negative, the signs chosen for each of the possible factors must be opposite. This leads to the possibilities

$$(2x - 1)(x + 6) = 2x^2 + 11x - 6 \qquad (2x - 2)(x + 3) = 2x^2 + 4x - 6$$
$$(2x + 1)(x - 6) = 2x^2 - 11x - 6 \qquad (2x + 2)(x - 3) = 2x^2 - 4x - 6$$
$$(2x - 6)(x + 1) = 2x^2 - 4x - 6 \qquad (2x - 3)(x + 2) = 2x^2 + x - 6$$
$$(2x + 6)(x - 1) = 2x^2 + 4x - 6 \qquad (2x + 3)(x - 2) = 2x^2 - x - 6$$

Thus $2x^2 - x - 6 = (2x + 3)(x - 2)$.

· ·

[*Note:* $(2x \pm 3)(x \mp 2)$ is a notation that can be used to save space. It actually represents two products: $(2x + 3)(x - 2)$ and $(2x - 3)(x + 2)$.]

Study the patterns illustrated in these examples carefully. Practice will give you the experience needed to use this factoring technique skillfully and efficiently.

Now Work Problem 59

Summary

Type of Polynomial	Method	Example
Any polynomial	Look for common monomial factors. (Always do this first!)	$6x^2 + 9x = 3x(2x + 3)$
Binomials of degree 2 or higher	Check for a special product: Difference of two squares, $x^2 - a^2$	Examples 1, 4
	Difference of two cubes, $x^3 - a^3$	Example 2
	Sum of two cubes, $x^3 + a^3$	Example 3
Trinomials of degree 2	Check for a perfect square $(x \pm a)^2$.	Examples 5–7
	List possibilities.	Examples 8–14

EXERCISE A.4   Answers to odd-numbered problems begin on page AN-87.

In Problems 1–10 factor each polynomial by removing the common monomial factor.

1. $3x + 6$

2. $7x - 14$

3. $ax^2 + a$

4. $ax - a$

5. $x^3 + x^2 + x$

6. $x^3 - x^2 + x$

7. $2x^2 + 2x + 2$

8. $3x^2 - 3x + 3$

9. $3x^2y - 6xy^2 + 12xy$

10. $60x^2y - 48xy^2 + 72x^3y$

In Problems 11–68 factor completely each polynomial. If the polynomial cannot be factored, say it is prime.

11. $x^2 - 1$

12. $x^2 - 4$

13. $4x^2 - 1$

14. $9x^2 - 1$

15. $x^2 + 7x + 10$

16. $x^2 + 3x - 4$

17. $x^2 - 10x + 21$

18. $x^2 - 4x - 21$

19. $x^2 - 7x - 8$

20. $x^2 - 6x + 5$

21. $x^2 + 2x + 1$

22. $x^2 - 4x + 4$

23. $x^2 + 4x + 4$

24. $x^2 - 2x - 1$

25. $x^2 - 2x - 15$

26. $x^2 - 6x - 14$

27. $3x^2 - 12x - 36$

28. $x^3 + 8x^2 - 20x$

29. $y^4 + 11y^3 + 30y^2$

30. $3y^3 - 18y^2 - 48y$

31. $16x^2 + 8x + 1$

32. $25x^2 + 10x + 1$

33. $4x^2 + 12x + 9$

34. $9x^2 - 12x + 4$

35. $ax^2 - 4a^2x - 45a^3$

36. $bx^2 + 14b^2x + 45b^3$

37. $x^3 - 27$

38. $x^3 + 27$

39. $8x^3 + 27$

40. $27 - 8x^3$

41. $2x^4 + 16x$

42. $3x^5 - 3x^2$

43. $3x^2 + 4x + 1$

44. $4x^2 + 3x - 1$

45. $x^4 - 81$

46. $x^4 - 1$

47. $x^6 - 2x^3 + 1$

48. $x^6 + 2x^3 + 1$

49. $x^7 - x^5$

50. $x^8 - x^5$

51. $2z^2 + 5z + 3$

52. $6z^2 - z - 1$

53. $16x^2 - 24x + 9$

54. $9x^2 + 24x + 16$

55. $16x^2 - 16x - 5$

56. $16x^2 - 11x - 5$

57. $4y^2 + 16y + 15$

58. $9y^2 + 9y - 4$

59. $18x^2 - 9x - 27$

60. $8x^2 - 6x - 2$

61. $8x^2 + 2x + 6$

62. $9x^2 - 3x + 3$

63. $x^2 - x + 4$

64. $x^2 + 6x + 9$

65. $4x^3 - 10x^2 - 6x$

66. $27x^3 - 9x^2 - 6x$

67. $x^2 + 4$

68. $x^2 + x + 1$

A.5 QUADRATIC EQUATIONS

Quadratic equations are equations equivalent to one written in the **standard form**

$$ax^2 + bx + c = 0 \tag{1}$$

where a, b, and c are real numbers and $a \neq 0$.

When a quadratic equation is written in standard form, $ax^2 + bx + c = 0$, it may be possible to factor the expression on the left side as the product of two first-degree polynomials.

EXAMPLE 1 Solve the equation: $x^2 = 12 - x$

SOLUTION

We put the equation in standard form by adding $x - 12$ to each side:

$$x^2 = 12 - x$$
$$x^2 + x - 12 = 0$$

The left side may now be factored as

$$(x + 4)(x - 3) = 0 \qquad \text{The factors 4 and } -3 \text{ of } -12$$
$$\text{have the sum 1.}$$

By the product law* we can set each factor equal to 0:

$$x + 4 = 0 \qquad \text{or} \qquad x - 3 = 0$$
$$x = -4 \qquad\qquad\qquad x = 3$$

The solution set is $\{-4, 3\}$.

· ·

When the left side factors into two linear equations with the same solution, the quadratic equation is said to have a **repeated solution.** We also call this solution a **root of multiplicity 2,** or a **double root.**

EXAMPLE 2 Solve the equation: $x^2 - 6x + 9 = 0$

SOLUTION

This equation is already in standard form, and the left side can be factored:

$$x^2 - 6x + 9 = 0$$
$$(x - 3)(x - 3) = 0$$

so that

$$x = 3 \qquad \text{or} \qquad x = 3$$

The equation has only the repeated solution 3.

· ·

Quadratic equations also can be solved by using the *quadratic formula,* given at the top of the next page.

---

* The product law states that if the product of two expressions is 0, then one or both expressions must equal 0. That is, if $AB = 0$, then $A = 0$ or $B = 0$ or both equal 0.

Quadratic Formula

If $b^2 - 4ac \geq 0$, the real solution(s) of the quadratic equation

$$ax^2 + bx + c = 0 \qquad a \neq 0 \qquad \text{(2)}$$

is (are) given by the **quadratic formula:**

$$x = \frac{-b \pm \sqrt{b^2 - 4ac}}{2a} \qquad \text{(3)}$$

The quantity $b^2 - 4ac$ is called the **discriminant** of the quadratic equation. It tells us whether the equation has real solutions and how many solutions to expect.

Discriminant of a Quadratic Equation

For a quadratic equation $ax^2 + bx + c = 0$:

1. If $b^2 - 4ac > 0$, there are two unequal real solutions.
2. If $b^2 - 4ac = 0$, there is a repeated real solution—a root of multiplicity 2.
3. If $b^2 - 4ac < 0$, there is no real solution.

EXAMPLE 3

Use the quadratic formula to find the real solutions, if any, of the equation:

$$3x^2 - 5x + 1 = 0$$

SOLUTION

The equation is in standard form, so we compare it to $ax^2 + bx + c = 0$ to find a, b, and c:

$$3x^2 - 5x + 1 = 0$$
$$ax^2 + bx + c = 0$$

With $a = 3$, $b = -5$, and $c = 1$, we evaluate the discriminant $b^2 - 4ac$:

$$b^2 - 4ac = (-5)^2 - 4(3)(1) = 25 - 12 = 13$$

Since $b^2 - 4ac > 0$, there are two real solutions, which can be found using the quadratic formula:

$$x = \frac{-b \pm \sqrt{b^2 - 4ac}}{2a} = \frac{5 \pm \sqrt{13}}{6}$$

The solution set is $\{(5 - \sqrt{13})/6, (5 + \sqrt{13})/6\}$.

Now Work Problem 1

· ·

EXAMPLE 4

Use the quadratic formula to find the real solutions, if any, of the equation:

$$3x^2 + 2 = 4x$$

SOLUTION

The equation, as given, is not in standard form.

$$3x^2 + 2 = 4x$$

$$3x^2 - 4x + 2 = 0 \qquad \text{Put in standard form.}$$

$$ax^2 + bx + c = 0 \qquad \text{Compare to standard form.}$$

With $a = 3$, $b = -4$, and $c = 2$, we find

$$b^2 - 4ac = 16 - 24 = -8$$

Since $b^2 - 4ac < 0$, the equation has no real solution.

••••••••••••••••••••

EXERCISE A.5   Answers to odd-numbered problems begin on page AN-88.

In Problems 1–10 find the real solutions, if any, of each equation. Use the quadratic formula.

1. $x^2 - 4x + 2 = 0$

2. $x^2 + 4x + 2 = 0$

3. $x^2 - 5x - 1 = 0$

4. $x^2 + 5x + 3 = 0$

5. $2x^2 - 5x + 3 = 0$

6. $2x^2 + 5x + 3 = 0$

7. $4y^2 - y + 2 = 0$

8. $4t^2 + t + 1 = 0$

9. $4x^2 = 1 - 2x$

10. $2x^2 = 1 - 2x$

In Problems 11–16 use the discriminant to determine whether each quadratic equation has two unequal real solutions, a repeated real solution, or no real solution, without solving the equation.

11. $x^2 - 5x + 7 = 0$

12. $x^2 + 5x + 7 = 0$

13. $9x^2 - 30x + 25 = 0$

14. $25x^2 - 20x + 4 = 0$

15. $3x^2 + 5x - 2 = 0$

16. $2x^2 - 3x - 4 = 0$

A.6 SOLVING INEQUALITIES

Intervals

Often we write the solution of an inequality using *intervals*. Let a and b represent two real numbers with $a < b$:

A **closed interval,** denoted by **[a, b],** consists of all real numbers x for which $a \leq x \leq b$.

An **open interval,** denoted by **(a, b),** consists of all real numbers x for which $a < x < b$.

The **half-open,** or **half-closed intervals** are **(a, b],** consisting of all real numbers x for which $a < x \leq b$, and **[a, b),** consisting of all real numbers x for which $a \leq x < b$.

In each of these definitions, a is called the **left endpoint** and b the **right endpoint** of the interval. Figure 11 illustrates each type of interval.

Figure 11

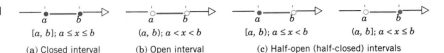

$[a, b]; a \leq x \leq b$ $(a, b); a < x < b$ $[a, b); a \leq x < b$ $(a, b]; a < x \leq b$

(a) Closed interval   (b) Open interval   (c) Half-open (half-closed) intervals

The symbol ∞ (read as "infinity") is not a real number but a notational device used to indicate unboundedness in the positive direction. The symbol $-\infty$ (read as "minus infinity") also is not a real number but a notational device used to indicate unboundedness in the negative direction. Using the symbols ∞ and $-\infty$, we can define four other kinds of intervals:

$[a, \infty)$ consists of all real numbers x for which $x \geq a$

(a, ∞) consists of all real numbers x for which $x > a$

$(-\infty, a]$ consists of all real numbers x for which $x \leq a$

$(-\infty, a)$ consists of all real numbers x for which $x < a$

Figure 12 illustrates these types of intervals.

Figure 12

$[a, \infty); x \geq a$ $(a, \infty); x > a$ $(-\infty, a]; x \leq a$ $(-\infty, a); x < a$

EXAMPLE 1

Write each inequality using interval notation.

(a) $1 \leq x \leq 3$ (b) $-4 < x < 0$ (c) $x > 5$ (d) $x \leq 1$

SOLUTION

(a) $1 \leq x \leq 3$ describes all numbers x between 1 and 3, inclusive. In interval notation we write $[1, 3]$.

(b) In interval notation $-4 < x < 0$ is written $(-4, 0)$.

(c) $x > 5$ consists of all numbers x greater than 5. In interval notation we write $(5, \infty)$.

(d) In interval notation $x \leq 1$ is written $(-\infty, 1]$.

Now Work Problem 1

· ·

EXAMPLE 2

Write each interval as an inequality involving x.

(a) $[1, 4)$ (b) $(2, \infty)$ (c) $[2, 3]$ (d) $(-\infty, -3]$

SOLUTION

(a) $[1, 4)$ consists of all numbers x for which $1 \leq x < 4$.

(b) $(2, \infty)$ consists of all numbers x for which $x > 2$.

(c) [2, 3] consists of all numbers x for which $2 \leq x \leq 3$.

(d) $(-\infty, -3]$ consists of all numbers x for which $x \leq -3$.

Now Work Problem 9

. . . .

Two inequalities having exactly the same solution set are called **equivalent in-equalities.** As with equations, one method for solving an inequality is to replace it by a series of equivalent inequalities until an inequality with an obvious solution, such as $x < 3$, is obtained. We obtain equivalent inequalities by applying some of the same operations as those used to find equivalent equations. The procedures listed below form the basis for solving inequalities.

Procedures that Leave the Inequality Symbol Unchanged

1. Add or subtract the same expression on both sides of the inequality.

2. Multiply or divide both sides of the inequality by the same *positive* expression.

Procedures that Reverse the Sense or Direction of the Inequality Symbol

1. Interchange the two sides of the inequality.

2. Multiply or divide both sides of the inequality by the same *negative* expression.

EXAMPLE 3

Solve the inequality $2x - 3 \leq 4x + 7$, and graph the solution set.

SOLUTION

$$2x - 3 \leq 4x + 7$$
$$(2x - 3) + 3 \leq 4x + 7 + 3 \qquad \text{Add 3 to both sides.}$$
$$2x \leq 4x + 10 \qquad \text{Simplify.}$$
$$2x - 4x \leq 4x + 10 - 4x \qquad \text{Subtract } 4x \text{ from both sides.}$$
$$-2x \leq 10 \qquad \text{Simplify.}$$
$$\frac{-2x}{-2} \geq \frac{10}{-2} \qquad \text{Divide both sides by } -2. \text{ (The sense of the inequality is reversed.)}$$
$$x \geq -5 \qquad \text{Simplify.}$$

The solution set is $\{x \mid x \geq -5\}$, or, using interval notation, all numbers in the interval $[-5, \infty)$.

See Figure 13 for the graph.

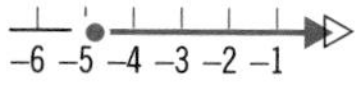

Now Work Problem 17

. . . .

Figure 13

$x \geq -5$ or $[-5, \infty)$

To solve inequalities that contain polynomials of degree 2 and higher as well as some that contain rational expressions, we rearrange them so that the polynomial or rational expression is on the left side and 0 is on the right side. An example will show you why.

EXAMPLE 4

Solve the inequality $x^2 + x - 12 > 0$, and graph the solution set.

SOLUTION

We factor the left side, obtaining

$$x^2 + x - 12 > 0$$
$$(x + 4)(x - 3) > 0$$

We then construct a graph that uses the solutions to the equation

$$x^2 + x - 12 = (x + 4)(x - 3) = 0$$

namely, $x = -4$ and $x = 3$. These numbers separate the real number line into three parts:

$$x < -4 \qquad -4 < x < 3 \qquad x > 3$$

or, in interval notation, into $(-\infty, -4)$, $(-4, 3)$, and $(3, \infty)$. See Figure 14(a). Now if $x < -4$, then $x + 4 < 0$. We indicate this fact about the expression $x + 4$ by placing minus signs $(- - -)$ to the left of -4. If $x > -4$, then $x + 4 > 0$. We indicate this fact about $x + 4$ by placing plus signs $(+ + +)$ to the right of -4.

Similarly, if $x < 3$, then $x - 3 < 0$. We indicate this fact about $x - 3$ by placing minus signs to the left of 3. If $x > 3$, then $x - 3 > 0$. We indicate this fact about $x - 3$ by placing plus signs to the right of 3. See Figure 14(b).

Figure 14

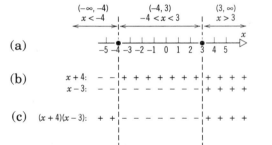

(a)

(b)

(c)

(d)

Now we prepare Figure 14(c). Since we know that the expressions $x + 4$ and $x - 3$ are both negative for $x < -4$, it follows that their product will be positive for $x < -4$. Since we know that $x + 4$ is positive and $x - 3$ is negative for $-4 < x < 3$, it follows that their product is negative for $-4 < x < 3$. Finally, since both expressions are positive for $x > 3$, their product is positive for $x > 3$. Thus the product $(x + 4)(x - 3) = x^2 + x - 12$ is positive when $x < -4$ or when $x > 3$.

The solution set is $\{x \mid x < -4 \text{ or } x > 3\}$. In interval notation we write the solution as $(-\infty, -4) \cup (3, \infty)$. See Figure 14(d).

·····················

The preceding discussion demonstrates that the sign of each factor of an expression is the same on each interval the real number line was divided into. Consequently, an alternative, and simpler, approach to obtaining Figure 14(b) would be to select a **test number** in each interval and use it to evaluate each factor to see if it is positive or negative. You may choose any number in the interval as a test number. See Figure 15.

Figure 15

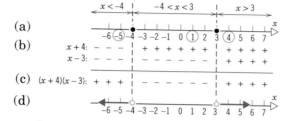

(a)
(b)
(c)
(d)

In Figure 15(a) the test numbers we selected, $-5, 1, 4$, have been circled. For $x + 4$, we insert $---$ under $x < -4$ because, for the test number -5, the value of $x + 4$ is $-5 + 4 = -1$, a negative number. Continuing across, we insert $+++$ under $-4 < x < 3$ because, for the test number 1, the value of $x + 4$ is $1 + 4 = 5$, a positive number. This process is continued for each factor and for each interval to obtain Figure 15(b). The rest of Figure 15 is obtained as before.

We shall employ the method of using a test number to solve inequalities. Here is another example showing all the details.

EXAMPLE 5

Solving Quadratic Inequalities   Solve the inequality $x^2 \le 4x + 12$, and graph the solution set.

SOLUTION

First, we rearrange the inequality so that 0 is on the right side:

$$x^2 \le 4x + 12$$
$$x^2 - 4x - 12 \le 0$$
$$(x + 2)(x - 6) \le 0 \qquad \text{Factor.}$$

Next, we set the left side equal to 0 and solve the resulting equation:

$$(x + 2)(x - 6) = 0$$

The solutions of the equation are -2 and 6, and they separate the real number line into three intervals:

$$x < -2 \qquad -2 < x < 6 \qquad x > 6$$

See Figure 16(a).

Now for the test number -3 we find that $x + 2 = -3 + 2 = -1$, a negative number, so we place minus signs under the interval $x < -2$. For the test number 1 we find $x + 2 = 1 + 2 = 3$, a positive number, so we place plus signs under the interval $-2 < x < 6$. Continuing in this fashion, we obtain Figure 16(b).

Next, we enter the signs of the product $(x + 2)(x - 6)$ in Figure 16(c). Since we want to know where the product $(x + 2)(x - 6)$ is negative, we conclude that the solutions are numbers x for which $-2 < x < 6$. However, because the original inequality is nonstrict, numbers x that satisfy the equation $x^2 = 4x + 12$ are also solutions of the inequality $x^2 \leq 4x + 12$. Thus we include -2 and 6, and the solution set of the given inequality is $\{x | -2 \leq x \leq 6\}$; that is, all x in $[-2, 6]$. See Figure 16(d).

Figure 16

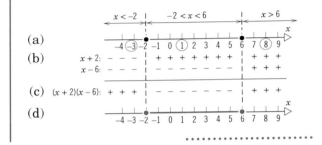

(a)
(b)
(c)
(d)

Now Work Problem 35

We have been solving inequalities by rearranging the inequality so that 0 is on the right side, setting the left side equal to 0, and solving the resulting equation. The solutions are then used to separate the real number line into intervals. But what if the resulting equation has no real solution? In this case we rely on the following result.

Theorem

If a polynomial equation has no real solutions, the polynomial is either always positive or always negative.

For example, the equation

$$x^2 + 5x + 8 = 0$$

has no real solutions. (Do you see why? Its discriminant, $b^2 - 4ac = 25 - 32 = -7$, is negative.) The value of $x^2 + 5x + 8$ is therefore always positive or always negative. To see which is true, we test its value at some number (0 is the easiest). Because $0^2 + 5(0) + 8 = 8$ is positive, we conclude that $x^2 + 5x + 8 > 0$ for all x.

EXAMPLE 6

Solving a Polynomial Inequality Solve the inequality $x^4 \leq x$, and graph the solution set.

SOLUTION

We rewrite the inequality so that 0 is on the right side:

$$x^4 \leq x$$
$$x^4 - x \leq 0$$

Then we proceed to factor the left side:

$$x^4 - x \leq 0$$
$$x(x^3 - 1) \leq 0$$
$$x(x - 1)(x^2 + x + 1) \leq 0$$

The solutions of the equation

$$x(x - 1)(x^2 + x + 1) = 0$$

are just 0 and 1, since the equation $x^2 + x + 1 = 0$ has no real solutions. We note that the expression $x^2 + x + 1$ is always positive. (Do you see why?) Next, we use 0 and 1 to separate the real number line into three intervals:

$$x < 0 \qquad 0 < x < 1 \qquad x > 1$$

Then we construct Figure 17 showing the signs of x, $x - 1$, and $x^2 + x + 1$. Note that since $x^2 + x + 1 > 0$ for all x, we enter plus signs next to it.

Figure 17

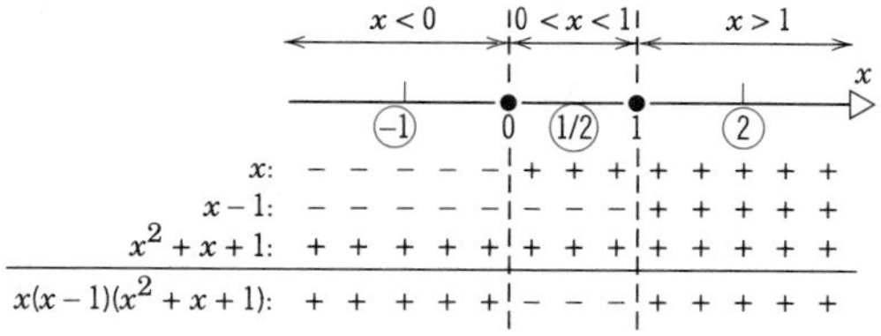

Since we want to know where $x^4 \leq x$ or where $x(x - 1)(x^2 + x + 1) \leq 0$, we conclude from Figure 17 that the solution set is $\{x | 0 \leq x \leq 1\}$, that is, all x in $[0, 1]$. See Figure 18.

Figure 18
$0 \leq x \leq 1$ or $[0, 1]$

Now Work Problem 49

········

Let's solve a rational inequality.

EXAMPLE 7

Solving a Rational Inequality   Solve the inequality $\dfrac{4x + 5}{x + 2} \geq 3$, and graph the solution set.

SOLUTION

We first note that the domain of the variable consists of all real numbers except -2. We rearrange terms so that 0 is on the right side:

$$\frac{4x + 5}{x + 2} \geq 3$$

$$\frac{4x + 5}{x + 2} - 3 \geq 0$$

$$\frac{4x + 5 - 3(x + 2)}{x + 2} \geq 0 \qquad \text{Rewrite using } x + 2 \text{ as the denominator.}$$

$$\frac{x - 1}{x + 2} \geq 0 \qquad \text{Simplify.}$$

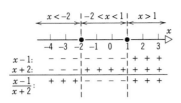

Figure 19

Figure 20
$x < -2$ or
$(-\infty, -2) \cup [1, \infty)$

The sign of a rational expression depends on the sign of its numerator and the sign of its denominator. Thus, for a rational expression, we separate the real number line into intervals using the numbers obtained by setting the numerator and the denominator equal to 0. For this example, they are -2 and 1. See Figure 19.

The bottom line in Figure 19 reveals the numbers x for which $(x - 1)/(x + 2)$ is positive. However, we want to know where the expression $(x - 1)/(x + 2)$ is positive or 0. Since $(x - 1)/(x + 2) = 0$ only if $x = 1$, we conclude that the solution set is $\{x | x < -2 \text{ or } x \geq 1\}$; that is, all x in $(-\infty, -2) \cup [1, \infty)$. See Figure 20.

In Example 7 you may wonder why we did not first multiply both sides of the inequality by $x + 2$ to clear the denominator. The reason is that we do not know whether $x + 2$ is positive or negative and, as a result, we do not know whether to reverse the sense of the inequality symbol after multiplying by $x + 2$. However, there is nothing to prevent us from multiplying both sides by $(x + 2)^2$, which is always positive, since $x \neq -2$. (Do you see why?)

$$\frac{4x + 5}{x + 2} \geq 3 \qquad\qquad x \neq -2$$

$$\frac{4x + 5}{x + 2}(x + 2)^2 \geq 3(x + 2)^2$$

$$(4x + 5)(x + 2) \geq 3(x^2 + 4x + 4)$$

$$4x^2 + 13x + 10 \geq 3x^2 + 12x + 12$$

$$x^2 + x - 2 \geq 0$$

$$(x + 2)(x - 1) \geq 0 \qquad\qquad x \neq -2$$

This last expression leads to the same solution set obtained in Example 7.

EXERCISE A.6   Answers to odd-numbered problems begin on page AN-88.

In Problems 1–8 write each inequality using interval notation, and illustrate each inequality using the real number line.

1. $0 \leq x \leq 4$ **2.** $-1 < x < 5$ **3.** $4 \leq x < 6$ **4.** $-2 < x \leq 0$

5. $x \geq 2$ **6.** $x \leq 2$ **7.** $x < -4$ **8.** $x > 1$

In Problems 9–16 write each interval as an inequality involving x, and illustrate each inequality using the real number line.

9. $[2, 5]$ **10.** $(1, 2)$ **11.** $(-3, -2)$ **12.** $[0, 1)$

13. $[4, \infty)$ **14.** $(-\infty, 2]$ **15.** $(-\infty, -1)$ **16.** $(-6, \infty)$

In Problems 17–60 solve each inequality. Graph the solution set.

17. $3x - 1 \geq 3 + x$

18. $2x - 2 \geq 3 + x$

19. $-2(x + 3) < 6$

20. $-3(1 - x) < 9$

21. $4 - 3(1 - x) \leq 3$

22. $8 - 4(2 - x) \leq -2x$

23. $\frac{1}{2}(x - 4) > x + 8$

24. $3x + 4 > \frac{1}{2}(x - 2)$

25. $\frac{x}{2} \geq 1 - \frac{x}{4}$

26. $\frac{x}{3} \geq 2 + \frac{x}{6}$

27. $0 \leq 2x - 6 \leq 4$

28. $4 \leq 2x + 2 \leq 10$

29. $-6 \leq 1 - 3x \leq 2$

30. $-3 \leq 2 - 2x \leq 4$

31. $(x - 3)(x + 1) < 0$

32. $(x - 1)(x + 2) < 0$

33. $-x^2 + 9 > 0$

34. $-x^2 + 1 > 0$

35. $x^2 + x > 12$

36. $x^2 + 7x < -12$

37. $x(x - 7) > -12$

38. $x(x + 1) > 12$

39. $4x^2 + 9 < 6x$

40. $25x^2 + 16 < 40x$

41. $(x - 1)(x^2 + x + 1) > 0$

42. $(x + 2)(x^2 - x + 1) > 0$

43. $(x - 1)(x - 2)(x - 3) < 0$

44. $(x + 1)(x + 2)(x + 3) < 0$

45. $-x^3 + 2x^2 + 8x < 0$

46. $-x^3 - 2x^2 + 8x < 0$

47. $x^3 > x$

48. $x^3 < 4x$

49. $x^3 > x^2$

50. $x^3 < 3x^2$

51. $\frac{x + 1}{1 - x} < 0$

52. $\frac{3 - x}{x + 1} < 0$

53. $\frac{(x - 1)(x + 1)}{x} < 0$

54. $\frac{(x - 3)(x + 2)}{x - 1} < 0$

55. $\frac{x - 2}{x^2 - 1} \geq 0$

56. $\frac{x + 5}{x^2 - 4} \geq 0$

57. $\frac{x + 4}{x - 2} \leq 1$

58. $\frac{x + 2}{x - 4} \geq 1$

59. $\frac{2x + 5}{x + 1} > \frac{x + 1}{x - 1}$

60. $\frac{1}{x + 2} > \frac{3}{x + 1}$

A.7 GEOMETRY TOPICS

Pythagorean Theorem

A **right angle** is an angle of 90°. A **right triangle** is one that contains a right angle. The side of the triangle opposite the 90° angle is called the **hypotenuse;** the remaining two sides are called the **legs.** The *Pythagorean theorem* is a statement about right triangles relating the length of the hypotenuse to the length of the legs.

In Figure 21, we have used c to represent the length of the hypotenuse and a and b to represent the lengths of the legs. Notice the use of the symbol $\ulcorner$ to show the 90° angle. We now state the Pythagorean theorem.

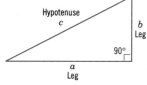

Figure 21

Pythagorean Theorem

In a right triangle, the square of the length of the hypotenuse is equal to the sum of the squares of the lengths of the legs. That is, in the right triangle shown in Figure 21,

$$c^2 = a^2 + b^2 \qquad (1)$$

EXAMPLE 1

In a right triangle one leg is of length 4 and the other is of length 3. What is the length of the hypotenuse?

SOLUTION

Since the triangle is a right triangle, we use the Pythagorean theorem with $a = 4$ and $b = 3$ to find the length c of the hypotenuse. Thus from Equation (1) we have

$$c^2 = a^2 + b^2$$
$$c^2 = 4^2 + 3^2 = 16 + 9 = 25$$
$$c = 5$$

Now Work Problem 1

....

The converse of the Pythagorean theorem is also true.

Converse of the Pythagorean Theorem

In a triangle if the square of the length of one side equals the sum of the squares of the lengths of the other two sides, then the triangle is a right triangle. The 90° angle is opposite the longest side.

EXAMPLE 2

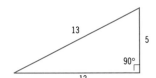

Figure 22

Now Work Problem 7

Show that a triangle whose sides are of lengths 5, 12, and 13 is a right triangle. Identify the hypotenuse.

SOLUTION

We square the lengths of the sides:

$$25, \quad 144, \quad 169$$

Notice that the sum of the first two squares (25 and 144) equals the third square (169). Hence, the triangle is a right triangle. The longest side, 13, is the hypotenuse. See Figure 22.

....

Geometry Formulas

For a **rectangle** of length l and width w:

$$\textbf{Area} = \textbf{\textit{lw}} \qquad \textbf{Perimeter} = \textbf{2\textit{l} + 2\textit{w}}$$

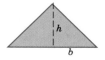

For a **triangle** with base b and altitude h:

$$\textbf{Area} = \tfrac{1}{2}\textbf{\textit{bh}}$$

For a **circle** of radius r:

$$\text{Area} = \pi r^2 \qquad \text{Circumference} = 2\pi r$$

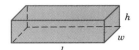

For a **rectangular box** of length l, width w, and height h:

$$\text{Volume} = lwh \qquad \text{Surface area} = 2lw + 2lh + 2wh$$

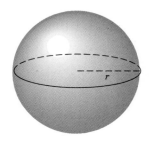

For a **sphere** with radius r:

$$\text{Volume} = \frac{4\pi r^3}{3} \qquad \text{Surface area} = 4\pi r^2$$

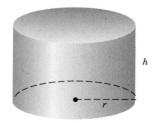

For a **right circular cylinder** with radius r and height h:

$$\text{Volume} = \pi r^2 h \qquad \text{Surface area} = 2\pi r^2 + 2\pi rh$$

For a **right circular cone** with radius r and height h,

$$\text{Volume} = \frac{\pi r^2 h}{3}$$

EXAMPLE 3

The perimeter of a rectangular hall is 34 meters. If its width is 4 meters, what is its length?

SOLUTION

If l represents the length and w the width, the perimeter is $2l + 2w$. Thus

$$2l + 2w = 34$$
$$2l + 2(4) = 34$$
$$2l = 34 - 8 = 26$$
$$l = 13 \text{ meters}$$

. . . .

EXERCISE A.7 Answers to odd-numbered problems begin on page AN-89.

In Problems 1–6 the lengths of the legs of a right triangle are given. Find the hypotenuse.

1. $a = 5, b = 12$

2. $a = 6, b = 8$

3. $a = 10, b = 24$

4. $a = 4, b = 3$

5. $a = 7, b = 24$

6. $a = 14, b = 48$

In Problems 7–14 the lengths of the sides of a triangle are given. Determine which are right triangles. For those that are, identify the hypotenuse.

7. 3, 4, 5

8. 6, 8, 10

9. 4, 5, 6

10. 2, 2, 3

11. 7, 24, 25

12. 10, 24, 26

13. 6, 4, 3

14. 5, 4, 7

In Problems 15–18 find the area and the perimeter of a rectangle with length l and width w.

15. $l = 2, w = 3$

16. $l = 5, w = 2$

17. $l = \frac{1}{2}, w = \frac{1}{3}$

18. $l = \frac{3}{4}, w = \frac{4}{3}$

In Problems 19–22 find the area of a triangle with base b and altitude h.

19. $b = 2, h = 1$

20. $b = 3, h = 4$

21. $b = \frac{1}{2}, h = \frac{3}{4}$

22. $b = \frac{3}{4}, h = \frac{4}{3}$

In Problems 23–26 find the area and circumference of a circle with radius r (use $\pi \approx 3.14$).

23. $r = 1$

24. $r = 2$

25. $r = \frac{1}{2}$

26. $r = \frac{3}{4}$

In Problems 27–30 find the volume and surface area of a rectangular box with length l, width w, and height h.

27. $l = 1, w = 1, h = 1$

28. $l = 2, w = 3, h = 4$

29. $l = \frac{1}{2}, w = \frac{3}{2}, h = \frac{4}{3}$

30. $l = \frac{3}{4}, w = \frac{1}{2}, h = \frac{5}{3}$

31. Find the volume and surface area of a sphere of radius 2 (use $\pi \approx 3.14$).

32. Find the volume of a right circular cylinder with radius 2 and height 3 (use $\pi \approx 3.14$).

33. The tallest inhabited building in North America is the Sears Tower in Chicago.* If the observation tower is 1454 feet above ground level, use the figure to determine how far a person standing in the observation tower can see (with the aid of a telescope). Use 3960 miles for the radius of the Earth. [*Note:* 1 mile = 5280 feet]

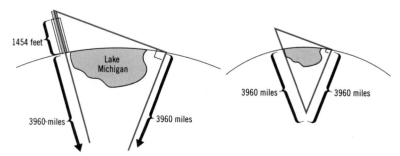

* Source: *Guinness Book of World Records.*

34. The conning tower of the USS *Silversides,* a World War II submarine now permanently stationed in Muskegon, Michigan, is approximately 20 feet above sea level. How far can one see from the conning tower?

35. A person who is 6 feet tall is standing on the beach in Fort Lauderdale, Florida, and looks out onto the Atlantic Ocean. Suddenly, a ship appears on the horizon. How far is the ship from shore?

36. The deck of a destroyer is 100 feet above sea level. How far can a person see from the deck? How far can a person see from the bridge, which is 150 feet above sea level?

Appendix B

Using LINDO to Solve Linear Programming Problems

A number of available software packages can be used to solve linear programming problems. Three of the more popular ones are these:

GAMS—General Algebraic Modeling System.   This is a general-purpose optimization system specifically designed for modeling and solving large, complex linear, nonlinear, and mixed-integer programming problems. It is available for PC, 386/486, workstation, or mainframe.

GINO—General INteractive Optimizer.   This system can solve linear and nonlinear constrained optimization problems on a PC, Mac, 386, or mainframe.

LINDO—Linear INteractive and Discrete Optimizer.   This is the most popular interactive linear, integer, and quadratic programming system on the market. It is available for PC and Mac.

Because LINDO is the most popular software package, we have chosen to include a brief section illustrating how it can be used to solve some linear programming problems. LINDO was designed to allow users to do simple problems in an easy, cost-efficient manner. At the other extreme, LINDO has been used to solve real industrial programs involving more than 10,000 rows and several thousand variables. In solving such problems, LINDO uses a *revised* simplex method—one that exploits the special character of large linear programming problems. As a result, the intermediate tableaus obtained using LINDO may differ from those obtained manually.

EXAMPLE 1   Use LINDO to maximize

$$P = 3x_1 + 4x_2$$

subject to

$$2x_1 + 4x_2 \le 120$$
$$2x_1 + 2x_2 \le 80$$
$$x_1 \ge 0 \qquad x_2 \ge 0$$

LINDO SOLUTION

The objective function

```
MAX      3 X1 + 4 X2
```

The constraints: numbered 2 and 3

```
SUBJECT TO
        2)   2 X1 + 4 X2 <=   120
        3)   2 X1 + 2 X2 <=    80
END
```

Instruction to begin

```
: tabl
```

The initial tableau
SLK 2 is the slack variable associated with constraint number 2.
ROW 1 ART is the objective row.
The column for P is not written since pivoting does not affect it.
The column on the far right is RHS.
The bottom row ART is added by LINDO to represent the phase I objective of minimizing the sum of the infeasibilities.

```
THE TABLEAU
    ROW (BASIS)        X1        X2     SLK   2    SLK   3
      1 ART          -3.000    -4.000      .000       .000       .000
      2 SLK    2      2.000     4.000     1.000       .000    120.000
      3 SLK    3      2.000     2.000      .000      1.000     80.000
ART   3 ART          -3.000    -4.000      .000       .000       .000
```

Instruction to pivot
The pivot element is in row 3, column X1.
X1 will become a basic variable.

```
: piv
    X1 ENTERS AT VALUE    40.000   IN  ROW   3  OBJ. VALUE=   120.00
```

After pivoting, SLK 2 and X1 are basic variables. $P = 120$ at this stage; SLK 2 = 40; X1 = 40; SLK 3 = 0; X2 = 0.

```
: tabl
THE TABLEAU
    ROW (BASIS)        X1        X2     SLK   2    SLK   3
      1 ART            .000    -1.000      .000      1.500    120.000
      2 SLK    2       .000     2.000     1.000     -1.000     40.000
      3      X1       1.000     1.000      .000       .500     40.000
```

The pivot element is in row 2, column X2.
After pivoting, X2 and X1 are basic variables.

```
: piv
    X2 ENTERS AT VALUE    20.000   IN  ROW   2  OBJ. VALUE=   140.00
: tabl
THE TABLEAU
    ROW (BASIS)        X1        X2     SLK   2    SLK   3
      1 ART            .000      .000      .500      1.000    140.000
      2      X2        .000     1.000      .500      -.500     20.000
      3      X1       1.000      .000     -.500      1.000     20.000
: piv
```

This is a final tableau. The solution is $P = 140$, X1 = 20, X2 = 20.

```
LP OPTIMUM FOUND AT STEP     2

        OBJECTIVE FUNCTION VALUE

    1)      140.00000

VARIABLE           VALUE          REDUCED COST
     X1          20.000000           .000000
     X2          20.000000           .000000
```

The value of the slack variables are SLK 2 = 0, SLK 3 = 0.

```
    ROW     SLACK OR SURPLUS      DUAL PRICES
     2)           .000000           .500000
     3)           .000000          1.000000
```

Two pivots were used.

```
NO. ITERATIONS=        2
```

....

Compare the steps in Example 1, using LINDO, with the steps given in Example 1, Section 4.2.

EXAMPLE 2 | Use LINDO to maximize

$$P = 20x_1 + 15x_2$$

subject to

$$x_1 + x_2 \geq 7$$
$$9x_1 + 5x_2 \leq 45$$
$$2x_1 + x_2 \geq 8$$
$$x_1 \geq 0 \qquad x_2 \geq 0$$

SOLUTION

```
MAX      20 X1 + 15 X2
SUBJECT TO
      2)    X1 + X2 >=    7
      3)    9 X1 + 5 X2 <=    45
      4)    2 X1 + X2 >=    8
END

: tabl

THE TABLEAU
      ROW  (BASIS)        X1        X2   SLK    2   SLK    3   SLK    4
        1 ART         -20.000   -15.000     .000       .000       .000       .000
        2 SLK    2     -1.000    -1.000    1.000       .000       .000     -7.000
        3 SLK    3      9.000     5.000     .000      1.000       .000     45.000
        4 SLK    4     -2.000    -1.000     .000       .000      1.000     -8.000
ART     4 ART          -3.000    -2.000    1.000       .000      1.000    -15.000

: piv
      X1 ENTERS AT VALUE    5.0000      IN ROW    3 OBJ. VALUE=    100.00

: tabl

THE TABLEAU
      ROW  (BASIS)        X1        X2   SLK    2   SLK    3   SLK    4
        1 ART            .000    -3.889     .000      2.222       .000    100.000
        2 SLK    2       .000     -.444    1.000       .111       .000     -2.000
        3         X1    1.000      .556     .000       .111       .000      5.000
        4 SLK    4       .000      .111     .000       .222      1.000      2.000
ART     4 ART            .000     -.444    1.000       .111       .000     -2.000

: piv
      X2 ENTERS AT VALUE    4.5000      IN ROW    2 OBJ. VALUE=    117.50

: tabl

THE TABLEAU
      ROW  (BASIS)        X1        X2   SLK    2   SLK    3   SLK    4
        1 ART            .000      .000   -8.750      1.250       .000    117.500
        2         X2      .000     1.000   -2.250      -.250       .000      4.500
        3         X1    1.000      .000    1.250       .250       .000      2.500
        4 SLK    4       .000      .000     .250       .250      1.000      1.500
ART     4 ART            .000      .000   -8.750      1.250

: piv
SLK    2 ENTERS AT VALUE    2.0000      IN ROW    3 OBJ. VALUE=    117.50
```

```
: tabl
THE TABLEAU
    ROW  (BASIS)          X1        X2    SLK    2  SLK    3   SLK    4
      1  ART           7.000      .000      .000   3.000     .000   135.000
      2     X2         1.800     1.000      .000    .200     .000     9.000
      3  SLK   2         .800      .000     1.000    .200     .000     2.000
      4  SLK   4        -.200      .000      .000    .200    1.000     1.000
: piv
LP OPTIMUM FOUND AT STEP       3

        OBJECTIVE FUNCTION VALUE

    1)      135.00000

VARIABLE         VALUE          REDUCED COST
    X1            .000000         7.000000
    X2           9.000000          .000000

    ROW    SLACK OR SURPLUS      DUAL PRICES
    2)         2.000000           .000000
    3)          .000000          3.000000
    4)         1.000000           .000000

NO. ITERATIONS=      3
```

· ·

Compare the final tableau found using LINDO with the final tableau of Example 1, Section 4.4, namely,

$$
\begin{array}{c|cccccc|c}
\text{BV} & P & x_1 & x_2 & s_1 & s_2 & s_3 & \text{RHS} \\
\hline
P & 1 & 7 & 0 & 0 & 3 & 0 & 135 \\
\hline
x_2 & 0 & \frac{9}{5} & 1 & 0 & \frac{1}{5} & 0 & 9 \\
s_3 & 0 & -\frac{1}{5} & 0 & 0 & \frac{1}{5} & 1 & 1 \\
s_1 & 0 & \frac{4}{5} & 0 & 1 & \frac{1}{5} & 0 & 2
\end{array}
$$

The solution, in both cases, is $P = 135$, $x_2 = 9$, $x_1 = 0$, $s_1 = $ SLK 2 $= 2$, $s_2 = $ SLK 3 $= 0$, $s_3 = $ SLK 4 $= 1$. Notice also that the intermediate tableaus are different. This is because LINDO uses a *revised* simplex method that requires additional computations not easily performed manually.

EXAMPLE 3   Use LINDO to minimize

$$C = 5x_1 + 6x_2$$

subject to

$$x_1 + x_2 \le 10$$
$$x_1 + 2x_2 \ge 12$$
$$2x_1 + x_2 \ge 12$$
$$x_1 \ge 3$$
$$x_1 \ge 0 \qquad x_2 \ge 0$$

| SOLUTION

```
MIN      5 X1 + 6 X2
SUBJECT TO
        2)    X1 + X2 <=    10
        3)    X1 + 2 X2 >=    12
        4)    2 X1 + X2 >=    12
        5)    X1 >=    3
END

: tabl

THE TABLEAU
     ROW  (BASIS)        X1      X2    SLK    2   SLK    3   SLK    4   SLK    5
       1 ART           5.000   6.000    .000       .000       .000       .000       .000
       2 SLK    2      1.000   1.000   1.000       .000       .000       .000     10.000
       3 SLK    3     -1.000  -2.000    .000      1.000       .000       .000    -12.000
       4 SLK    4     -2.000  -1.000    .000       .000      1.000       .000    -12.000
       5 SLK    5     -1.000    .000    .000       .000       .000      1.000     -3.000
ART    5 ART         -4.000  -3.000    .000      1.000      1.000      1.000    -27.000

: piv
    X1 ENTERS AT VALUE    6.0000     IN ROW     4 OBJ. VALUE= -30.000

: tabl

THE TABLEAU
     ROW  (BASIS)        X1      X2    SLK    2   SLK    3   SLK    4   SLK    5
       1 ART            .000   3.500    .000       .000      2.500       .000    -30.000
       2 SLK    2       .000    .500   1.000       .000       .500       .000      4.000
       3 SLK    3       .000  -1.500    .000      1.000      -.500       .000     -6.000
       4        X1     1.000    .500    .000       .000      -.500       .000      6.000
       5 SLK    5       .000    .500    .000       .000      -.500      1.000      3.000
ART    5 ART           .000  -1.500    .000      1.000      -.500       .000     -6.000

 : piv
    X2 ENTERS AT VALUE    4.0000     IN ROW     3 OBJ. VALUE= -44.000

: tabl

THE TABLEAU
     ROW  (BASIS)        X1      X2    SLK    2   SLK    3   SLK    4   SLK    5
       1 ART            .000    .000    .000      2.333      1.333       .000    -44.000
       2 SLK    2       .000    .000   1.000       .333       .333       .000      2.000
       3        X2      .000   1.000    .000      -.667       .333       .000      4.000
       4        X1     1.000    .000    .000       .333      -.667       .000      4.000
       5 SLK    5       .000    .000    .000       .333      -.667      1.000      1.000
ART    5 ART           .000    .000    .000      2.333      1.333

: piv
LP OPTIMUM FOUND AT STEP      2

OBJECTIVE FUNCTION VALUE

       1)     44.000000

    VARIABLE         VALUE         REDUCED COST
        X1         4.000000           .000000
        X2         4.000000           .000000

       ROW    SLACK OR SURPLUS      DUAL PRICES
        2)        2.000000           .000000
        3)         .000000         -2.333333
        4)         .000000         -1.333333
        5)        1.000000           .000000

NO. ITERATIONS=        2
```

....

Compare this result, using LINDO, to the result obtained in Example 2, Section 4.4. Note again that the same solution is found: $C = 44$, $x_1 = 4$, $x_2 = 4$, $s_1 = $ SLK 2 $= 2$, $s_2 = $ SLK 3 $= 0$, $s_3 = $ SLK 4 $= 0$, $s_4 = $ SLK 5 $= 1$. Also, note again the different intermediate tableaus.

EXERCISE B.1   Answers to odd-numbered problems begin on page AN-89.

In Problems 1–24 use LINDO (or any other software package) to solve each linear programming problem.

1. Maximize
$$P = 3x_1 + 2x_2 + x_3$$
subject to
$$3x_1 + x_2 + x_3 \le 30$$
$$5x_1 + 2x_2 + x_3 \le 24$$
$$x_1 + x_2 + 4x_3 \le 20$$
$$x_1 \ge 0 \quad x_2 \ge 0 \quad x_3 \ge 0$$

2. Maximize
$$P = x_1 + 4x_2 + 3x_3 + x_4$$
subject to
$$2x_1 + x_2 \le 10$$
$$3x_1 + x_2 + x_3 + 2x_4 \le 18$$
$$x_1 + x_2 + x_3 + x_4 \le 14$$
$$x_1 \ge 0 \quad x_2 \ge 0 \quad x_3 \ge 0 \quad x_4 \ge 0$$

3. Maximize
$$P = 3x_1 + x_2 + x_3$$
subject to
$$x_1 + x_2 + x_3 \le 6$$
$$2x_1 + 3x_2 + 4x_3 \le 10$$
$$x_1 \ge 0 \quad x_2 \ge 0 \quad x_3 \ge 0$$

4. Maximize
$$P = 3x_1 + x_2 + x_3$$
subject to
$$x_1 + x_2 + x_3 \le 8$$
$$2x_1 + x_2 + 4x_3 \ge 6$$
$$x_1 \ge 0 \quad x_2 \ge 0$$

5. Maximize
$$P = 2x_1 + x_2 + 3x_3$$
subject to
$$x_1 + x_2 - x_3 \le 10$$
$$x_2 + x_3 \le 4$$
$$x_1 \ge 0 \quad x_2 \ge 0 \quad x_3 \ge 0$$

6. Maximize
$$P = 2x_1 + 2x_2 + 3x_3$$
subject to
$$x_1 - x_2 + x_3 \le 6$$
$$x_1 \le 4$$
$$x_1 \ge 0 \quad x_2 \ge 0 \quad x_3 \ge 0$$

7. Maximize
$$P = x_1 + x_2 + x_3$$
subject to
$$x_1 + x_2 + x_3 \le 6$$
$$4x_1 + x_2 \ge 12$$
$$x_1 \ge 0 \quad x_2 \ge 0 \quad x_3 \ge 0$$

8. Maximize
$$P = 2x_1 + x_2 + 3x_3$$
subject to
$$-x_1 + x_2 + x_3 \ge -6$$
$$2x_1 - 3x_2 \ge -12$$
$$x_1 \ge 0 \quad x_2 \ge 0 \quad x_3 \ge 0$$

9. Maximize
$$P = 2x_1 + x_2 + 3x_3$$
subject to
$$5x_1 + 2x_2 + x_3 \le 20$$
$$6x_1 + x_2 + 4x_3 \le 24$$
$$x_1 + x_2 + 4x_3 \le 16$$
$$x_1 \ge 0 \quad x_2 \ge 0 \quad x_3 \ge 0$$

10. Maximize
$$P = 3x_1 + 2x_2 + x_3$$
subject to
$$3x_1 + 2x_2 - x_3 \le 10$$
$$x_1 - x_2 + 3x_3 \le 12$$
$$2x_1 + x_2 + x_3 \le 6$$
$$x_1 \ge 0 \quad x_2 \ge 0 \quad x_3 \ge 0$$

11. Maximize

$$P = 2x_1 + 3x_2 + x_3$$

subject to

$$x_1 + x_2 + x_3 \leq 50$$
$$3x_1 + 2x_2 + x_3 \leq 10$$
$$x_1 \geq 0 \qquad x_2 \geq 0 \qquad x_3 \geq 0$$

12. Maximize

$$P = 4x_1 + 4x_2 + 2x_3$$

subject to

$$3x_1 + x_2 + x_3 \leq 10$$
$$x_1 + x_2 + 3x_3 \leq 5$$
$$x_1 \geq 0 \qquad x_2 \geq 0 \qquad x_3 \geq 0$$

13. Maximize

$$P = 2x_1 + x_2 + x_3$$

subject to

$$-2x_1 + x_2 - 2x_3 \leq 4$$
$$x_1 - 2x_2 + x_3 \leq 2$$
$$x_1 \geq 0 \qquad x_2 \geq 0 \qquad x_3 \geq 0$$

14. Maximize

$$P = 4x_1 + 2x_2 + 5x_3$$

subject to

$$x_1 + 3x_2 + 2x_3 \leq 30$$
$$2x_1 + x_2 + 3x_3 \leq 12$$
$$x_1 \geq 0 \qquad x_2 \geq 0 \qquad x_3 \geq 0$$

15. Maximize

$$P = 2x_1 + x_2 + 3x_3$$

subject to

$$x_1 + 2x_2 + x_3 \leq 25$$
$$3x_1 + 2x_2 + 3x_3 \leq 30$$
$$x_1 \geq 0 \qquad x_2 \geq 0 \qquad x_3 \geq 0$$

16. Maximize

$$P = 6x_1 + 3x_2 + 2x_3$$

subject to

$$2x_1 + 2x_2 + 3x_3 \leq 30$$
$$2x_1 + 2x_2 + x_3 \leq 12$$
$$x_1 \geq 0 \qquad x_2 \geq 0 \qquad x_3 \geq 0$$

17. Maximize

$$P = 2x_1 + 4x_2 + x_3 + x_4$$

subject to

$$2x_1 + x_2 + 2x_3 + 3x_4 \leq 12$$
$$2x_2 + x_3 + 2x_4 \leq 20$$
$$2x_1 + x_2 + 4x_3 \leq 16$$
$$x_1 \geq 0 \qquad x_2 \geq 0 \qquad x_3 \geq 0 \qquad x_4 \geq 0$$

18. Maximize

$$P = 2x_1 + 4x_2 + x_3$$

subject to

$$-x_1 + 2x_2 + 3x_3 \leq 6$$
$$-x_1 + 4x_2 + 5x_3 \leq 5$$
$$-x_1 + 5x_2 + 7x_3 \leq 7$$
$$x_1 \geq 0 \qquad x_2 \geq 0 \qquad x_3 \geq 0$$

19. Maximize

$$P = 2x_1 + x_2 + x_3$$

subject to

$$x_1 + 2x_2 + 4x_3 \leq 20$$
$$2x_1 + 4x_2 + 4x_3 \leq 60$$
$$3x_1 + 4x_2 + x_3 \leq 90$$
$$x_1 \geq 0 \qquad x_2 \geq 0 \qquad x_3 \geq 0$$

20. Maximize

$$P = x_1 + 2x_2 + 4x_3$$

subject to

$$8x_1 + 5x_2 - 4x_3 \leq 30$$
$$-2x_1 + 6x_2 + x_3 \leq 5$$
$$-2x_1 + 2x_2 + x_3 \leq 15$$
$$x_1 \geq 0 \qquad x_2 \geq 0 \qquad x_3 \geq 0$$

21. Maximize

$$P = x_1 + 2x_2 + 4x_3 - x_4$$

subject to

$$5x_1 + 4x_3 + 6x_4 \leq 20$$
$$4x_1 + 2x_2 + 2x_3 + 8x_4 \leq 40$$
$$x_1 \geq 0 \qquad x_2 \geq 0 \qquad x_3 \geq 0 \qquad x_4 \geq 0$$

22. Maximize

$$P = x_1 + 2x_2 - x_3 + 3x_4$$

subject to

$$2x_1 + 4x_2 + 5x_3 + 6x_4 \leq 24$$
$$4x_1 + 4x_2 + 2x_3 + 2x_4 \leq 4$$
$$x_1 \geq 0 \qquad x_2 \geq 0 \qquad x_3 \geq 0 \qquad x_4 \geq 0$$

23. Minimize

$$C = x_1 + x_2 + x_3 + x_4 + x_5 + x_6 + x_7$$

subject to

$$4x_1 + 2x_2 + x_3 \quad\quad + 2x_5 + x_6 \quad\quad \geq 75$$
$$x_2 + 2x_3 + 3x_4 \quad\quad + x_6 \quad\quad \geq 110$$
$$x_5 + x_6 + 2x_7 \geq 50$$
$$x_1 \geq 0 \quad x_2 \geq 0 \quad x_3 \geq 0 \quad x_4 \geq 0$$
$$x_5 \geq 0 \quad x_6 \geq 0 \quad x_7 \geq 0$$

24. Minimize

$$C = x_1 + x_2 + x_3 + x_4 + x_5 + x_6 + x_7$$

subject to

$$4x_1 + x_2 + 2x_3 \quad\quad + 2x_5 \quad\quad + x_7 \geq 75$$
$$2x_2 + x_3 + 3x_4 \quad\quad + x_7 \geq 180$$
$$x_5 + 2x_6 + x_7 \geq 50$$
$$x_1 \geq 0 \quad x_2 \geq 0 \quad x_3 \geq 0 \quad x_4 \geq 0$$
$$x_5 \geq 0 \quad x_6 \geq 0 \quad x_7 \geq 0$$

Tables

Table I Amount of an Annuity

(a) Annual Compounding

No. of Periods n	8% per annum $\dfrac{(1 + 0.08)^n - 1}{0.08}$	$\left[\dfrac{(1 + 0.08)^n - 1}{0.08}\right]^{-1}$	10% per annum $\dfrac{(1 + 0.10)^n - 1}{0.10}$	$\left[\dfrac{(1 + 0.10)^n - 1}{0.10}\right]^{-1}$	12% per annum $\dfrac{(1 + 0.12)^n - 1}{0.12}$	$\left[\dfrac{(1 + 0.12)^n - 1}{0.12}\right]^{-1}$
1	1.00000000	1.00000000	1.00000000	1.00000000	1.00000000	1.00000000
2	2.08000000	0.48076923	2.10000000	0.47619048	2.12000000	0.47169811
3	3.24640000	0.30803351	3.31000000	0.30211480	3.37440000	0.29634898
4	4.50611200	0.22192080	4.64100000	0.21547080	4.77932800	0.20923444
5	5.86660096	0.17045645	6.10510000	0.16379748	6.35284736	0.15740973
6	7.33592904	0.13631539	7.71561000	0.12960738	8.11518904	0.12322572
7	8.92280336	0.11207240	9.48717100	0.10540550	10.0890117	0.09911774
8	10.6366276	0.09401476	11.4358881	0.08744402	12.2996931	0.08130284
9	12.4875578	0.08007971	13.5794769	0.07364054	14.7756563	0.06767889
10	14.4865625	0.06902949	15.9374246	0.06274539	17.5487351	0.05698416
11	16.6454875	0.06007634	18.5311671	0.05396314	20.6545833	0.04841540
12	18.9771265	0.05269502	21.3842838	0.04676332	24.1331333	0.04143681
13	21.4952966	0.04652181	24.5227121	0.04077852	28.0291093	0.03567720
14	24.2149203	0.04129685	27.9749834	0.03574622	32.3926024	0.03087125
15	27.1521139	0.03682954	31.7724817	0.03147378	37.2797147	0.02682424
16	30.3242830	0.03297687	35.9497299	0.02781662	42.7532804	0.02339002
17	33.7502257	0.02962943	40.5447029	0.02466413	48.8836741	0.02045673
18	37.4502437	0.02670210	45.5991731	0.02193022	55.7497150	0.01793731
19	41.4462632	0.02412763	51.1590904	0.01954687	63.4396808	0.01576300
20	45.7619643	0.02185221	57.2749995	0.01745962	72.0524424	0.01387878
21	50.4229214	0.01983225	64.0024994	0.01562439	81.6987355	0.01224009
22	55.4567552	0.01803207	71.4027494	0.01400506	92.5025838	0.01081051
23	60.8932956	0.01642217	79.5430243	0.01257181	104.602894	0.00955996
24	66.7647592	0.01497796	88.4973268	0.01129978	118.155241	0.00846344
25	73.1059400	0.01367878	98.3470594	0.01016807	133.333870	0.00749997
26	79.9544151	0.01250713	109.181765	0.00915904	150.333934	0.00665186

Table I *(Continued)*

(a) Annual Compounding

No. of Periods n	8% per annum $\dfrac{(1 + 0.08)^n - 1}{0.08}$	8% per annum $\left[\dfrac{(1 + 0.08)^n - 1}{0.08}\right]^{-1}$	10% per annum $\dfrac{(1 + 0.10)^n - 1}{0.10}$	10% per annum $\left[\dfrac{(1 + 0.10)^n - 1}{0.10}\right]^{-1}$	12% per annum $\dfrac{(1 + 0.12)^n - 1}{0.12}$	12% per annum $\left[\dfrac{(1 + 0.12)^n - 1}{0.12}\right]^{-1}$
27	87.3507684	0.01144810	121.099942	0.00825764	169.374007	0.00590409
28	95.3388298	0.01048891	134.209936	0.00745101	190.698887	0.00524387
29	103.965936	0.00961854	148.630930	0.00672807	214.582754	0.00466021
30	113.283211	0.00882743	164.494023	0.00607925	241.332684	0.00414366
31	123.345868	0.00810728	181.943425	0.00549621	271.292606	0.00368606
32	134.213537	0.00745081	201.137767	0.00497172	304.847719	0.00328033
33	145.950620	0.00685163	222.251442	0.00449941	342.429446	0.00292031
34	158.626670	0.00630411	245.476699	0.00407371	384.520979	0.00260064
35	172.316804	0.00580326	271.024368	0.00368971	431.663496	0.00231662
36	187.102148	0.00534467	299.126805	0.00334306	484.463116	0.00206414
37	203.070320	0.00492440	330.039486	0.00302994	543.598690	0.00183959
38	220.315945	0.00453894	364.043434	0.00274692	609.830533	0.00163980
39	238.941221	0.00418513	401.447778	0.00249098	684.010197	0.00146197
40	259.056519	0.00386016	442.592556	0.00225941	767.091420	0.00130363

(b) Monthly Compounding

No. of Periods n	8% per annum $\dfrac{(1 + \frac{0.08}{12})^n - 1}{\frac{0.08}{12}}$	8% per annum $\left[\dfrac{(1 + \frac{0.08}{12})^n - 1}{\frac{0.08}{12}}\right]^{-1}$	10% per annum $\dfrac{(1 + \frac{0.10}{12})^n - 1}{\frac{0.10}{12}}$	10% per annum $\left[\dfrac{(1 + \frac{0.10}{12})^n - 1}{\frac{0.10}{12}}\right]^{-1}$	12% per annum $\dfrac{(1 + \frac{0.12}{12})^n - 1}{\frac{0.12}{12}}$	12% per annum $\left[\dfrac{(1 + \frac{0.12}{12})^n - 1}{\frac{0.12}{12}}\right]^{-1}$
12	12.4499260	0.08032176	12.5655681	0.07958255	12.6825030	0.07884879
24	25.9331897	0.03856062	26.4469154	0.03781159	26.9734649	0.03707347
36	40.5355577	0.02466970	41.7818211	0.02393385	43.0768784	0.02321431
48	56.3499150	0.01774626	58.7224919	0.01702925	61.2226078	0.01633384
60	73.4768562	0.01360973	77.4370723	0.01291371	81.6696699	0.01224445
72	92.0253250	0.01086657	98.1113137	0.01019250	104.709931	0.00955019
84	112.113307	0.00891955	120.950418	0.00826785	130.672274	0.00765273
96	133.868583	0.00747001	146.181076	0.00684083	159.927293	0.00625284
108	157.429535	0.00635205	174.053713	0.00574535	192.892579	0.00518423
120	182.946035	0.00546609	204.844979	0.00488174	230.038689	0.00434709
132	210.580392	0.00474878	238.860493	0.00418654	271.895856	0.00367788
144	240.508386	0.00415786	276.437876	0.00361745	319.061559	0.00313419
156	272.920390	0.00366407	317.950103	0.00314515	372.209054	0.00268666
168	308.022573	0.00324652	363.809201	0.00274869	432.096982	0.00231430
180	346.038221	0.00288985	414.470347	0.00241272	499.580198	0.00200168
192	387.209149	0.00258258	470.436376	0.00212569	575.621974	0.00173725
204	431.797243	0.00231590	532.262781	0.00187877	661.307751	0.00151216
216	480.086127	0.00208296	600.563217	0.00166510	757.860630	0.00131950
228	532.382965	0.00187835	676.015602	0.00147926	866.658830	0.00115386
240	589.020414	0.00169773	759.368837	0.00131688	989.255365	0.00101086
252	650.358744	0.00153761	851.450246	0.00117447	1127.40021	0.00088700
264	716.788125	0.00139511	953.173781	0.00104913	1283.06528	0.00077938
276	788.731112	0.00126786	1065.54910	0.00093848	1458.47257	0.00068565
288	866.645331	0.00115387	1189.69158	0.00084055	1656.12591	0.00060382
300	951.026392	0.00105150	1326.83341	0.00075367	1878.84663	0.00053224

TABLE I *(Continued)*

(b) Monthly Compounding

No. of Periods n	8% per annum		10% per annum		12% per annum	
	$\dfrac{(1 + \frac{0.08}{12})^n - 1}{\frac{0.08}{12}}$	$\left[\dfrac{(1 + \frac{0.08}{12})^n - 1}{\frac{0.08}{12}}\right]^{-1}$	$\dfrac{(1 + \frac{0.10}{12})^n - 1}{\frac{0.10}{12}}$	$\left[\dfrac{(1 + \frac{0.10}{12})^n - 1}{\frac{0.10}{12}}\right]^{-1}$	$\dfrac{(1 + \frac{0.12}{12})^n - 1}{\frac{0.12}{12}}$	$\left[\dfrac{(1 + \frac{0.12}{12})^n - 1}{\frac{0.12}{12}}\right]^{-1}$
312	1042.41104	0.00095931	1478.33577	0.00067644	2129.81391	0.00046952
324	1141.38057	0.00087613	1645.70241	0.00060764	2412.61013	0.00041449
336	1248.56452	0.00080092	1830.59453	0.00054627	2731.27198	0.00036613
348	1364.64468	0.00073279	2034.84726	0.00049144	3090.34813	0.00032359
360	1490.35944	0.00067098	2260.48793	0.00044238	3494.96413	0.00028613

TABLE II Present Value of an Annuity

(a) Annual Compounding

No. of Periods n	8% per annum		10% per annum		12% per annum	
	$\dfrac{1 - (1 + 0.08)^{-n}}{0.08}$	$\left[\dfrac{1 - (1 + 0.08)^{-n}}{0.08}\right]^{-1}$	$\dfrac{1 - (1 + 0.10)^{-n}}{0.10}$	$\left[\dfrac{1 - (1 + 0.10)^{-n}}{0.10}\right]^{-1}$	$\dfrac{1 - (1 + 0.12)^{-n}}{0.12}$	$\left[\dfrac{1 - (1 + 0.12)^{-n}}{0.12}\right]^{-1}$
1	0.92592593	1.08000000	0.90909091	1.10000000	0.89285714	1.12000000
2	1.78326475	0.56076923	1.73553719	0.57619048	1.69005102	0.59169811
3	2.57709699	0.38803351	2.48685199	0.40211480	2.40183127	0.41634898
4	3.31212684	0.30192080	3.16986545	0.31547080	3.03734935	0.32923444
5	3.99271004	0.25045645	3.79078677	0.26379748	3.60477620	0.27740973
6	4.62287966	0.21631539	4.35526070	0.22960738	4.11140732	0.24322572
7	5.20637006	0.19207240	4.86841882	0.20540550	4.56375654	0.21911774
8	5.74663894	0.17401476	5.33492620	0.18744402	4.96763977	0.20130284
9	6.24688791	0.16007971	5.75902382	0.17364054	5.32824979	0.18767889
10	6.71008140	0.14902949	6.14456711	0.16274539	5.65022303	0.17698416
11	7.13896426	0.14007634	6.49506101	0.15396314	5.93769913	0.16841540
12	7.53607802	0.13269502	6.81369182	0.14676332	6.19437423	0.16143681
13	7.90377594	0.12652181	7.10335620	0.14077852	6.42354842	0.15567720
14	8.24423698	0.12129685	7.36668746	0.13574622	6.62816823	0.15087125
15	8.55947869	0.11682954	7.60607951	0.13147378	6.81086449	0.14682424
16	8.85136916	0.11297687	7.82370864	0.12781662	6.97398615	0.14339002
17	9.12163811	0.10962943	8.02155331	0.12466413	7.11963049	0.14045673
18	9.37188714	0.10670210	8.20141210	0.12193022	7.24967008	0.13793731
19	9.60359920	0.10412763	8.36492009	0.11954687	7.36577686	0.13576300
20	9.81814741	0.10185221	8.51356372	0.11745962	7.46944362	0.13387878
21	10.0168032	0.09983225	8.64869429	0.11562439	7.56200324	0.13224009
22	10.2007437	0.09803207	8.77154026	0.11400506	7.64464575	0.13081051
23	10.3710589	0.09642217	8.88321842	0.11257181	7.71843370	0.12955997
24	10.5287583	0.09497796	8.98474402	0.11129978	7.78431581	0.12846344
25	10.6747762	0.09367878	9.07704002	0.11016807	7.84313911	0.12749997
26	10.8099780	0.09250713	9.16094547	0.10915904	7.89565992	0.12665186
27	10.9351648	0.09144810	9.23722316	0.10825764	7.94255350	0.12590409
28	11.0510785	0.09048891	9.30656651	0.10745101	7.98442277	0.12524387
29	11.1584060	0.08961854	9.36960591	0.10672807	8.02180604	0.12466021
30	11.2577833	0.08882743	9.42691447	0.10607925	8.05518397	0.12414366
31	11.3497994	0.08810728	9.47901315	0.10549621	8.08498569	0.12368606
32	11.4349994	0.08745081	9.52637559	0.10497172	8.11159436	0.12328033

TABLE II *(Continued)*

(a) Annual Compounding

No. of Periods n	8% per annum $\dfrac{1-(1+0.08)^{-n}}{0.08}$	$\left[\dfrac{1-(1+0.08)^{-n}}{0.08}\right]^{-1}$	10% per annum $\dfrac{1-(1+0.10)^{-n}}{0.10}$	$\left[\dfrac{1-(1+0.10)^{-n}}{0.10}\right]^{-1}$	12% per annum $\dfrac{1-(1+0.12)^{-n}}{0.12}$	$\left[\dfrac{1-(1+0.12)^{-n}}{0.12}\right]^{-1}$
33	11.5138884	0.08685163	9.56943236	0.10449941	8.13535211	0.12292031
34	11.5869337	0.08630411	9.60857487	0.10407371	8.15656438	0.12260064
35	11.6545682	0.08580326	9.64415897	0.10368971	8.17550391	0.12231662
36	11.7171928	0.08534467	9.67650816	0.10334306	8.19241421	0.12206414
37	11.7751785	0.08492440	9.70591651	0.10302994	8.20751269	0.12183959
38	11.8288690	0.08453894	9.73265137	0.10274692	8.22099347	0.12163980
39	11.8785824	0.08418513	9.75695579	0.10249098	8.23302988	0.12146197
40	11.9246133	0.08386016	9.77905072	0.10225941	8.24377668	0.12130363

(b) Monthly Compounding

No. of Periods n	8% per annum $\dfrac{1-(1+\frac{0.08}{12})^{-n}}{\frac{0.08}{12}}$	$\left[\dfrac{1-(1+\frac{0.08}{12})^{-n}}{\frac{0.08}{12}}\right]^{-1}$	10% per annum $\dfrac{1-(1+\frac{0.10}{12})^{-n}}{\frac{0.10}{12}}$	$\left[\dfrac{1-(1+\frac{0.10}{12})^{-n}}{\frac{0.10}{12}}\right]^{-1}$	12% per annum $\dfrac{1-(1+\frac{0.12}{12})^{-n}}{\frac{0.12}{12}}$	$\left[\dfrac{1-(1+\frac{0.12}{12})^{-n}}{\frac{0.12}{12}}\right]^{-1}$
12	11.4957818	0.08698843	11.3745084	0.08791589	11.2550775	0.08884879
24	22.1105436	0.04522729	21.6708548	0.04614493	21.2433873	0.04707347
36	31.9118055	0.03133637	30.9912356	0.03226719	30.1075050	0.03321431
48	40.9619129	0.02441292	39.4281601	0.02536258	37.9739595	0.02633384
60	49.3184333	0.02027639	47.0653691	0.02124704	44.9550384	0.02224445
72	57.0345221	0.01753324	53.9786655	0.01852584	51.1503915	0.01955019
84	64.1592611	0.01558621	60.2366674	0.01660118	56.6484528	0.01765273
96	70.7379704	0.01413668	65.9014885	0.01517416	61.5277030	0.01625284
108	76.8124971	0.01301871	71.0293549	0.01407869	65.8577898	0.01518423
120	82.4214808	0.01213276	75.6711634	0.01321507	69.7005220	0.01434709
132	87.6006002	0.01141545	79.8729861	0.01251988	73.1107518	0.01367788
144	92.3827995	0.01082453	83.6765283	0.01195078	76.1371575	0.01313419
156	96.7984979	0.01033074	87.1195419	0.01147848	78.8229389	0.01268666
168	100.875784	0.00991318	90.2362006	0.01108203	81.2064335	0.01231430
180	104.640592	0.00955652	93.0574389	0.01074605	83.3216640	0.01200168
192	108.116871	0.00924925	95.6112588	0.01045902	85.1988236	0.01173725
204	111.326733	0.00898257	97.9230083	0.01021210	86.8647075	0.01151216
216	114.290596	0.00874963	100.015633	0.00999843	88.3430948	0.01131950
228	117.027313	0.00854501	101.909902	0.00981259	89.6550886	0.01115386
240	119.554292	0.00836400	103.624619	0.00965022	90.8194164	0.01101086
252	121.887607	0.00820428	105.176801	0.00950780	91.8526982	0.01088700
264	124.042099	0.00806178	106.581856	0.00938246	92.7696833	0.01077938
276	126.031475	0.00793453	107.853730	0.00927182	93.5834610	0.01068565
288	127.868388	0.00782054	109.005045	0.00917389	94.0356475	0.01060382
300	129.564523	0.00771816	110.047230	0.00908701	94.9465513	0.01053224
312	131.130668	0.00762598	110.990629	0.00900977	95.5153208	0.01046952
324	132.576786	0.00754280	111.844605	0.00894098	96.0200749	0.01041449
336	133.912076	0.00746759	112.617635	0.00887960	96.4680186	0.01036613
348	135.145031	0.00739946	113.317392	0.00882477	96.8655458	0.01032359
360	136.283494	0.00733765	113.950820	0.00877572	97.2183311	0.01028613

TABLE III   Normal Curve Table
Z = Z-Score

An entry in the table is the area under the curve between $Z = 0$ and a positive value of Z. Areas for negative values of Z are obtained by symmetry.

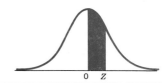

Z	0.00	0.01	0.02	0.03	0.04	0.05	0.06	0.07	0.08	0.09
0.0	0.0000	0.0040	0.0080	0.0120	0.0160	0.0199	0.0239	0.0279	0.0319	0.0359
0.1	0.0398	0.0438	0.0478	0.0517	0.0557	0.0596	0.0636	0.0675	0.0714	0.0753
0.2	0.0793	0.0832	0.0871	0.0910	0.0948	0.0987	0.1026	0.1064	0.1103	0.1141
0.3	0.1179	0.1217	0.1255	0.1293	0.1331	0.1368	0.1406	0.1433	0.1480	0.1517
0.4	0.1554	0.1591	0.1628	0.1664	0.1700	0.1736	0.1772	0.1808	0.1844	0.1879
0.5	0.1915	0.1950	0.1985	0.2019	0.2054	0.2088	0.2123	0.2157	0.2190	0.2224
0.6	0.2257	0.2291	0.2324	0.2357	0.2389	0.2422	0.2454	0.2486	0.2517	0.2549
0.7	0.2580	0.2611	0.2642	0.2673	0.2703	0.2734	0.2764	0.2794	0.2823	0.2852
0.8	0.2881	0.2910	0.2939	0.2967	0.2995	0.3023	0.3051	0.3078	0.3106	0.3133
0.9	0.3159	0.3186	0.3212	0.3238	0.3264	0.3289	0.3315	0.3340	0.3365	0.3389
1.0	0.3413	0.3438	0.3461	0.3485	0.3508	0.3531	0.3554	0.3577	0.3599	0.3621
1.1	0.3642	0.3665	0.3686	0.3708	0.3729	0.3749	0.3770	0.3790	0.3810	0.3830
1.2	0.3849	0.3869	0.3888	0.3907	0.3925	0.3944	0.3962	0.3980	0.3997	0.4015
1.3	0.4032	0.4049	0.4066	0.4082	0.4099	0.4115	0.4131	0.4147	0.4162	0.4177
1.4	0.4192	0.4207	0.4222	0.4236	0.4251	0.4265	0.4279	0.4292	0.4306	0.4319
1.5	0.4332	0.4345	0.4357	0.4370	0.4382	0.4394	0.4406	0.4418	0.4429	0.4441
1.6	0.4452	0.4463	0.4474	0.4484	0.4495	0.4505	0.4515	0.4525	0.4535	0.4545
1.7	0.4554	0.4564	0.4573	0.4582	0.4591	0.4599	0.4608	0.4616	0.4625	0.4633
1.8	0.4641	0.4649	0.4656	0.4664	0.4671	0.4678	0.4686	0.4693	0.4699	0.4706
1.9	0.4713	0.4719	0.4726	0.4732	0.4738	0.4744	0.4750	0.4756	0.4761	0.4767
2.0	0.4772	0.4778	0.4783	0.4788	0.4793	0.4798	0.4803	0.4808	0.4812	0.4817
2.1	0.4821	0.4826	0.4830	0.4834	0.4838	0.4842	0.4846	0.4850	0.4854	0.4857
2.2	0.4861	0.4864	0.4868	0.4871	0.4875	0.4878	0.4881	0.4884	0.4887	0.4890
2.3	0.4893	0.4896	0.4898	0.4901	0.4904	0.4906	0.4909	0.4911	0.4913	0.4916
2.4	0.4918	0.4920	0.4922	0.4925	0.4927	0.4929	0.4931	0.4932	0.4934	0.4936
2.5	0.4938	0.4940	0.4941	0.4943	0.4945	0.4946	0.4948	0.4949	0.4951	0.4952
2.6	0.4953	0.4955	0.4956	0.4957	0.4959	0.4960	0.4961	0.4962	0.4963	0.4964
2.7	0.4965	0.4966	0.4967	0.4968	0.4969	0.4970	0.4971	0.4972	0.4973	0.4974
2.8	0.4974	0.4975	0.4976	0.4977	0.4977	0.4978	0.4979	0.4979	0.4980	0.4981
2.9	0.4981	0.4982	0.4982	0.4983	0.4984	0.4984	0.4985	0.4985	0.4986	0.4986
3.0	0.4987	0.4987	0.4987	0.4988	0.4988	0.4989	0.4989	0.4989	0.4990	0.4990

Answers To Odd-Numbered Problems

CHAPTER 1

Exercise 1.1 (page 14)

1. $A = (4, 2)$, $B = (6, 2)$, $C = (5, 3)$, $D = (-2, 1)$, $E = (-2, -3)$, $F = (3, -2)$, $G = (6, -2)$, $H = (5, 0)$

3.

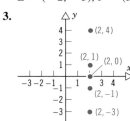

The set of points of the form $(2, y)$, where y is a real number, is a vertical line passing through 2 on the x-axis.

5.

x	0	3	2	-2	4	-4
y	-3	0	-1	-5	1	-7

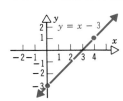

7.

x	0	3	2	-2	4	-4
y	-6	0	-2	-10	2	-14

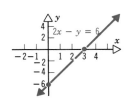

9. $\frac{1}{2}$ **11.** -1

13. Slope $= 3$

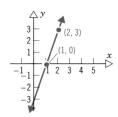

15. Slope $= \dfrac{-1}{2}$

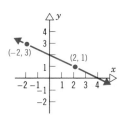

17. Slope $= 0$

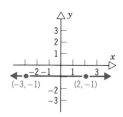

19. Undefined slope (vertical line)

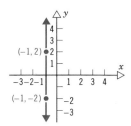

21. Slope $= \dfrac{\sqrt{3} - 3}{1 - \sqrt{2}} \approx 3.0611$

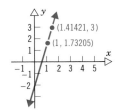

AN-1

23.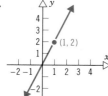

25.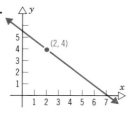

51. Slope $= \frac{2}{3}$; y-intercept $= (0, -2)$

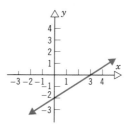

27.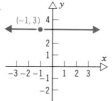

29. (y-axis)

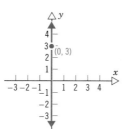

53. Slope $= -1$; y-intercept $= (0, 1)$

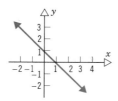

55. Slope is undefined; no y-intercept

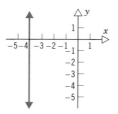

31. $x - 2y = 0$ **33.** $x + y = 2$ **35.** $2x - y = -7$

37. $2x + 3y = -1$ **39.** $x - 2y = -5$

41. $3x + y = 3$ **43.** $x - 2y = 2$ **45.** $x = 1$

47. Slope $= 2$; y-intercept $= (0, 3)$

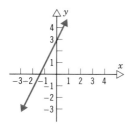

57. Slope $= 0$; y-intercept $= (0, 5)$

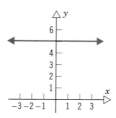

49. Slope $= 2$; y-intercept $= (0, -2)$

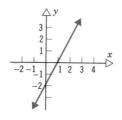

59. Slope $= 1$; y-intercept $= (0, 0)$

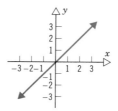

61. Slope $= \frac{3}{2}$; y-intercept $= (0, 0)$ **63.** $y = 0$

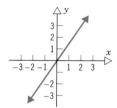

65. $9C - 5F = -160$ (where C represents degrees Celsius and F represents degrees Fahrenheit). When $F = 70$, $C = 21\frac{1}{9} \approx 21.111$.

67. $P = 0.50x - 100$ dollars, or $2P - x = 100$

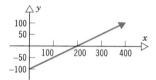

69. $C = 0.10189x + 9.06$;   when $x = 300$, $C = \$39.63$; when $x = 900$, $C = \$100.76$.

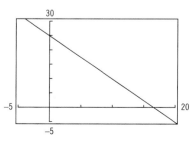

71. $C = 0.2x + 280$ dollars   **73.** $w = 5h - 190$

75. (a) $w = -4.75d + 347.5$ (where d represents the day of the month and w represents million gallons of water)
(b) 281 million gallons

Technology Exercises (page 17)

1. x-intercept $= (16.66, 0)$, y-intercept $= (0, 25.00)$

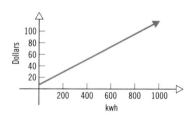

3. x-intercept $= (.24, 0)$, y-intercept $= (0, -530.00)$

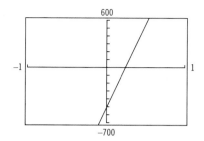

5. x-intercept $= (2.83, 0)$, y-intercept $= (0, 2.55)$

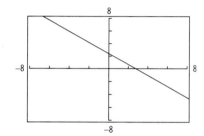

7. x-intercept $= (.77, 0)$, y-intercept $= (0, -1.41)$

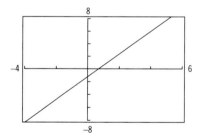

9.

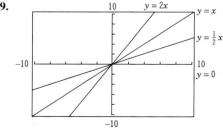

11. Lines having the same slope and different *y*-intercepts are parallel.

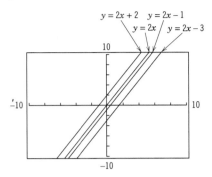

21. $m_1 = \frac{1}{3}$, $m_2 = -3$, $m_1 m_2 = -1$

23. $m_1 = -\frac{1}{2}$, $m_2 = 2$, $m_1 m_2 = -1$

25. $m_1 = -\frac{1}{4}$, $m_2 = 4$, $m_1 m_2 = -1$

27. $y = 2x - 3$ or $2x - y = 3$

29. $y = -\frac{1}{2}x + \frac{9}{2}$ or $x + 2y = 9$

31. $y = 3x + 5$ or $3x - y = -5$

33. $y = 2x$ or $2x - y = 0$ **35.** $x = 4$

37. $y = -\frac{1}{2}x - \frac{5}{2}$ or $x + 2y = -5$

39. $y = -\frac{1}{2}x + \frac{19}{30}$ or $15x + 30y = 19$ **41.** $t = 4$

43. $y = \frac{5}{19}x - \frac{85}{19}$ or $5x - 19y = 85$ **45.** $y = -3$

Exercise 1.3 (page 29)

1. (a) $80,000 (b) $95,000 (c) $105,000 (d) $120,000

3. $C = 500\,t + 7000$ (where $t = 0$ corresponds to the year 1992 and C represents the average cost)
In 1997, $t = 5$ and $C = 9500.

5. (a) SAT $= -7t + 592$ (where $t = 0$ corresponds to the year 1990)
(b) 536

7. $x = 30$ **9.** $x = 500$

Exercise 1.2 (page 22)

1. Parallel   **3.** Intersecting   **5.** Coincident

7. Parallel   **9.** Intersecting

11. (3, 2)   **13.** (3, 1)

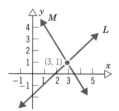

15. (1, 0)   **17.** (2, 1)

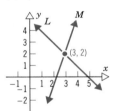

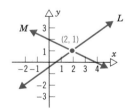

19. (−1, 1)

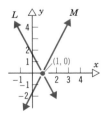

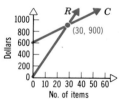

11. 1200 items   **13.** 200 newspapers

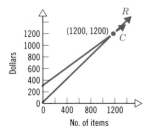

15. 20 caramels and 30 creams; increase the number of caramels to obtain a profit

17. $50,000 in AA bonds and $100,000 in S&L Certificates

19. 1358 adults (and 1242 children)

21. 30 cc of 15% solution and 70 cc of 5% solution

23. $p = \$1$ **25.** $p = \$10$

27. Market price is 1; supply at the market price is 1.1; the point of intersection occurs at the market price, i.e., the price at which the amount of sugar that sellers will supply matches the amount of sugar that consumers will purchase.

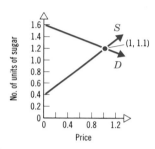

29. $D = 23 - 4p$

Chapter Review: True–False Items *(page 32)*

1. F   **2.** T   **3.** T   **4.** F   **5.** F   **6.** T

7. F   **8.** T   **9.** F   **10.** F

Fill in the Blanks *(page 32)*

1. x-coordinate, y-coordinate   **2.** undefined, zero

3. negative   **4.** parallel   **5.** coincident

6. perpendicular   **7.** intersecting

Review Exercises *(page 33)*

1.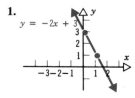

3.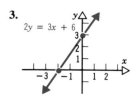

5. $x + 2y = 5$ **7.** $3x + 2y = 0$

9. $y = -2x + 3$ or $2x + y = 3$

11. $x = -3$ **13.** $y = -\frac{1}{5}x - 2$ or $x + 5y = -10$

15. $y = \frac{2}{3}x + \frac{19}{3}$ or $2x - 3y = -19$

17. $y = -\frac{3}{2}x - \frac{9}{2}$ or $3x + 2y = -9$

19. Slope $= -4\frac{1}{2}$; y-intercept: $(0, 9)$

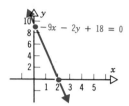

21. Slope $= -2$; y-intercept: $(0, 4\frac{1}{2})$

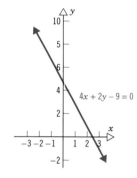

23. Parallel   **25.** Intersecting   **27.** Coincident

29. $(5, 1)$ **31.** $(1, 3)$

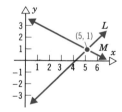   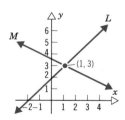

33. $(-2, 1)$

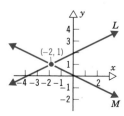

35. $78,571.43 in B-rated bonds and $11,428.57 in the bank

37. (a) & (b)

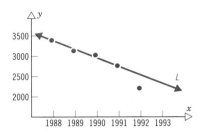

(c) $y - 3400 = -200(x - 1988)$, using the points (1988, 3400) and (1991, 2800)

(d) 2400 thousand units (Answer varies with choice of L.)

Mathematical Questions from Professional Exams (page 34)

1. b   **2.** d   **3.** d   **4.** b

5. c   **6.** c   **7.** b   **8.** b

CHAPTER 2

Exercise 2.1 (page 49)

1. $4 + 1 = 5$; $10 - 2 = 8$: solution

3. $6 - 2 = 4$; $2 - \frac{3}{2} = \frac{1}{2}$: solution

5. $3 - 3 + 4 = 4$; $1 + 1 - 2 = 0$; $-2 - 6 = -8$: solution

7. $x = 6, y = 2$ **9.** $x = 3, y = 2$

11. $x = 8, y = -4$ **13.** $x = 4, y = -2$

15. Inconsistent   **17.** $x = \frac{1}{2}, y = \frac{3}{4}$

19. x is any real number, $y = 2 - \frac{1}{2}x$ (*Alternatively:* $x = 4 - 2y$, y is any real number)

21. $x = \frac{1}{10}, y = \frac{2}{5}$ **23.** $x = \frac{3}{2}, y = 1$

25. x is any real number, $y = \frac{5}{3} - \frac{2}{3}x$ (*Alternatively:* $x = \frac{5}{2} - \frac{3}{2}y$, y is any real number)

27. $x = \frac{4}{3}, y = \frac{1}{5}$ **29.** $x = 8, y = 2, z = 0$

31. $x = 2, y = -1, z = 1$ **33.** Inconsistent

35. x is any real number, $y = \frac{4}{5}x - \frac{7}{5}, z = \frac{1}{5}x + \frac{2}{5}$ (*Alternatively:* $x = \frac{5}{4}y + \frac{7}{4}, z = \frac{1}{4}y + \frac{3}{4}$, y is any real number or $x = 5z - 2, y = 4z - 3$, z is any real number)

37. Inconsistent   **39.** $x = 1, y = 3, z = -2$

41. $x = -3, y = \frac{1}{2}, z = 1$

43. 61 and 20 (or $\frac{181}{3}$ and $\frac{224}{3}$)

45. Length $= 30$ ft, width $= 15$ ft

47. Cheeseburger: $1.55; shake: $0.85

49. $31,250 in AA bonds; $18,750 in S&L Certificates

51. 8 nickels   **53.** 60 cc of 10% acid; 40 cc of 30% acid

55. 3260 adults

57. $37,500 in the first (10%) investment; $12,500 in the second (12%) investment

59. 100 acres of corn; 900 acres of soybeans

61. 40 of the $20 sets; 160 of the $15 sets

Technology Exercises (page 51)

1. $x = -2.54, y = 28.81$

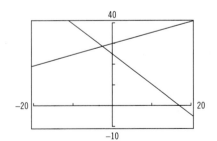

3. $x = 3.07, y = -.21$

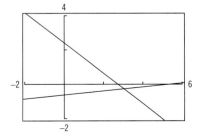

Exercise 2.2 (page 62)

1. $\begin{bmatrix} 2 & -3 & | & 5 \\ 1 & -1 & | & 3 \end{bmatrix}$

3. $\begin{bmatrix} 2 & 1 & | & -6 \\ 1 & 1 & | & -1 \end{bmatrix}$

5. $\begin{bmatrix} 2 & -1 & -1 & | & 0 \\ 1 & -1 & -1 & | & 1 \\ 3 & -1 & 0 & | & 2 \end{bmatrix}$

7. $\begin{bmatrix} 2 & -3 & 1 & | & 7 \\ 1 & 1 & -1 & | & 1 \\ 2 & 2 & -3 & | & -4 \end{bmatrix}$

9. $\begin{bmatrix} 4 & -1 & 2 & -1 & | & 4 \\ 1 & 1 & 0 & 0 & | & -6 \\ 0 & 2 & -1 & 1 & | & 5 \end{bmatrix}$

11. (a) $\begin{bmatrix} 1 & -3 & -5 & | & -2 \\ 0 & 1 & 6 & | & 9 \\ -3 & 5 & 4 & | & 6 \end{bmatrix}$

(b) $\begin{bmatrix} 1 & -3 & -5 & | & -2 \\ 0 & 1 & 6 & | & 9 \\ 0 & -4 & -11 & | & 0 \end{bmatrix}$

(c) $\begin{bmatrix} 1 & -3 & -5 & | & -2 \\ 0 & 1 & 6 & | & 9 \\ 0 & 0 & 13 & | & 36 \end{bmatrix}$

13. (a) $\begin{bmatrix} 1 & -3 & 4 & | & 3 \\ 0 & 1 & -2 & | & 0 \\ -3 & 3 & 4 & | & 6 \end{bmatrix}$

(b) $\begin{bmatrix} 1 & -3 & 4 & | & 3 \\ 0 & 1 & -2 & | & 0 \\ 0 & -6 & 16 & | & 15 \end{bmatrix}$

(c) $\begin{bmatrix} 1 & -3 & 4 & | & 3 \\ 0 & 1 & -2 & | & 0 \\ 0 & 0 & 4 & | & 15 \end{bmatrix}$

15. (a) $\begin{bmatrix} 1 & -3 & 2 & | & -6 \\ 0 & 1 & -1 & | & 8 \\ -3 & -6 & 4 & | & 6 \end{bmatrix}$

(b) $\begin{bmatrix} 1 & -3 & 2 & | & -6 \\ 0 & 1 & -1 & | & 8 \\ 0 & -15 & 10 & | & -12 \end{bmatrix}$

(c) $\begin{bmatrix} 1 & -3 & 2 & | & -6 \\ 0 & 1 & -1 & | & 8 \\ 0 & 0 & -5 & | & 108 \end{bmatrix}$

17. (a) $\begin{bmatrix} 1 & -3 & 1 & | & -2 \\ 0 & 1 & 4 & | & 2 \\ -3 & 1 & 4 & | & 6 \end{bmatrix}$

(b) $\begin{bmatrix} 1 & -3 & 1 & | & -2 \\ 0 & 1 & 4 & | & 2 \\ 0 & -8 & 7 & | & 0 \end{bmatrix}$

(c) $\begin{bmatrix} 1 & -3 & 1 & | & -2 \\ 0 & 1 & 4 & | & 2 \\ 0 & 0 & 39 & | & 16 \end{bmatrix}$

19. (a) $\begin{bmatrix} 1 & -3 & -2 & | & 3 \\ 0 & 1 & 6 & | & -7 \\ -3 & -2 & 4 & | & 6 \end{bmatrix}$

(b) $\begin{bmatrix} 1 & -3 & -2 & | & 3 \\ 0 & 1 & 6 & | & -7 \\ 0 & -11 & -2 & | & 15 \end{bmatrix}$

(c) $\begin{bmatrix} 1 & -3 & -2 & | & 3 \\ 0 & 1 & 6 & | & -7 \\ 0 & 0 & 64 & | & -62 \end{bmatrix}$

21. $x = 2, y = 4$ **23.** $x = 2, y = 1$

25. $x = 2, y = 1$ **27.** $x = 2, y = -3$

29. $x = \frac{1}{2}, y = \frac{1}{3}$ **31.** $x = \frac{3}{2}, y = \frac{2}{3}$

33. $x = 2, y = 3$ **35.** $x = \frac{2}{3}, y = \frac{1}{3}$

37. $x = 1, y = 4, z = 0$ **39.** $x = -1, y = 1, z = 2$

41. $x = 2, y = -1, z = 1$

43. $x = 0.5, y = 0.25, z = 0.75$

45. $x = \frac{1}{3}, y = \frac{2}{3}, z = 1$ **47.** No solution

49. Infinitely many solutions $(2x - 3y = 6)$

51. No solution   **53.** Unique solution $(x = 1, y = 1)$

55. 22.5 lb of cashews

57. 10 two-student work stations and 6 three-student work stations

59. $5.56

61. 40 orchestra seats, 260 main seats, and 200 balcony seats

63. $3000 invested at 6%, $1500 invested at 8%, and $2000 invested at 9%

65. 20 cases of orange juice, 12 cases of tomato juice, 6 cases of pineapple juice

67. 2 packages of first type, 10 packages of second type, 4 packages of third type

69. 5 units of food I, 5 units of food II, $\frac{5}{3}$ units of food III

71. 4 assorted cartons, 8 mixed cartons, 5 single cartons

73. 3 large cans, 2 mammoth cans, 4 giant cans

Technology Exercises (page 65)

1. $\begin{bmatrix} -2 & 0 & -8 & | & 0 \\ -2 & 1 & -4 & | & 1 \\ 3 & -1 & 0 & | & 1 \end{bmatrix}$

3. $\begin{bmatrix} 1 & 0 & 4 & | & 0 \\ 1 & 1 & 8 & | & 1 \\ 3 & -1 & 0 & | & 1 \end{bmatrix}$

5. $x = \frac{2}{9}, y = -\frac{2}{3}, z = \frac{2}{9}$ **7.** $x = 2, y = -1, z = 3$

9. $x_1 = 20, x_2 = -13, x_3 = 17, x_4 = -4$

Exercise 2.3 (page 75)

1. Not in reduced row-echelon form

3. Not in reduced row-echelon form

5. Not in reduced row-echelon form

7. In reduced row-echelon form

9. In reduced row-echelon form

11. Infinitely many solutions

13. One solution   **15.** Infinitely many solutions

17. Infinitely many solutions

19. Infinitely many solutions

21. $\begin{bmatrix} 1 & 0 & | & 2 \\ 0 & 1 & | & 1 \end{bmatrix}$; $x = 2, y = 1$

23. $\begin{bmatrix} 1 & 0 & | & 0 \\ 0 & 1 & | & 0 \\ 0 & 0 & | & 1 \end{bmatrix}$; inconsistent

25. $\begin{bmatrix} 1 & -2 & | & 4 \\ 0 & 0 & | & 0 \\ 0 & 0 & | & 0 \end{bmatrix}$; infinitely many solutions: $x = 4 + 2y, y$ is any real number

27. $\begin{bmatrix} 1 & 0 & 0 & | & -3 \\ 0 & 1 & 3 & | & 5 \\ 0 & 0 & 0 & | & 0 \end{bmatrix}$; infinitely many solutions: $x = -3, y = 5 - 3z, z$ is any real number

29. $\begin{bmatrix} 1 & 0 & 0 & | & 3 \\ 0 & 1 & 0 & | & 2 \\ 0 & 0 & 1 & | & -4 \end{bmatrix}$; $x = 3, y = 2, z = -4$

31. $\begin{bmatrix} 1 & 0 & 0 & 0 & | & -17 \\ 0 & 1 & 0 & 0 & | & 24 \\ 0 & 0 & 1 & 0 & | & 33 \\ 0 & 0 & 0 & 1 & | & 14 \end{bmatrix}$; $x_1 = -17, x_2 = 24, x_3 = 33, x_4 = 14$

33. $\begin{bmatrix} 1 & 0 & 0 & -\frac{1}{3} & | & \frac{4}{3} \\ 0 & 1 & 0 & -\frac{11}{15} & | & \frac{14}{15} \\ 0 & 0 & 1 & \frac{4}{15} & | & -\frac{16}{15} \end{bmatrix}$;

infinitely many solutions; we can solve for x_1, x_2, and x_3 in terms of x_4: $x_1 = \frac{4}{3} + \frac{1}{3} x_4$, $x_2 = \frac{14}{15} + \frac{11}{15} x_4$, $x_3 = -\frac{16}{15} - \frac{4}{15} x_4$

35. $\begin{bmatrix} 1 & -1 & 1 & | & 0 \\ 0 & 0 & 0 & | & 1 \end{bmatrix}$; inconsistent

37. $\begin{bmatrix} 1 & 0 & \frac{7}{12} & | & 0 \\ 0 & 1 & -\frac{1}{4} & | & 0 \\ 0 & 0 & 0 & | & 1 \end{bmatrix}$; inconsistent

39. $\begin{bmatrix} 1 & 0 & 0 & 0 & | & 1 \\ 0 & 1 & 0 & 0 & | & 2 \\ 0 & 0 & 1 & 0 & | & 0 \\ 0 & 0 & 0 & 1 & | & 1 \end{bmatrix}$; $x_1 = 1, x_2 = 2, x_3 = 0, x_4 = 1$

41. $\begin{bmatrix} 1 & 0 & 0 & | & 0 \\ 0 & 1 & 1 & | & 0 \\ 0 & 0 & 0 & | & 1 \end{bmatrix}$; inconsistent

43. $\begin{bmatrix} 1 & 0 & 0 & | & 0 \\ 0 & 1 & -1 & | & -6 \\ 0 & 0 & 0 & | & 0 \end{bmatrix}$; infinitely many solutions: $x = 0, y = -6 + z, z$ is any real number

45.

No. of Liters 10% Solution	No. of Liters 30% Solution	No. of Liters 50% Solution
55	15	30
50	25	25
45	35	20
40	45	15
35	55	10
30	65	5

47. The information is not sufficient to determine the prices; possibilities include:

Hamburger	Fries	Cola
$2.13	$0.89	$0.62
$2.10	$0.90	$0.65
$2.07	$0.91	$0.68
$2.04	$0.92	$0.71
$2.01	$0.93	$0.74
$1.98	$0.94	$0.77
$1.95	$0.95	$0.80
$1.92	$0.96	$0.83
$1.89	$0.97	$0.86
$1.86	$0.98	$0.89

49.

Treasury Bills	Corporate Bonds	Junk Bonds
$12,500	$12,500	$0
$13,500	$10,500	$1,000
$14,500	$8,500	$2,000
$15,500	$6,500	$3,000
$16,500	$4,500	$4,000
$17,500	$2,500	$5,000
$18,500	$500	$6,000

51. (a) If they invest all $25,000 in Treasury bills, their re-
turn will be $1750 per year, greater than the $1500
they require.

(b)

Treasury Bills	Corporate Bonds	Junk Bonds
$500	$11,500	$13,000
$1,500	$9,500	$14,000
$2,500	$7,500	$15,000
$3,500	$5,500	$16,000
$4,500	$3,500	$17,000
$5,500	$1,500	$18,000

(c) As the required return increases, the amount they can
invest in Treasury bills diminishes, and the amount
they must invest in junk bonds increases.

Technology Exercises (page 77)

1. $\begin{bmatrix} 1 & 0 & 0 & 0 & 3 \\ 0 & 1 & 0 & 0 & -5 \\ 0 & 0 & 1 & 0 & 2 \\ 0 & 0 & 0 & 1 & 3 \end{bmatrix}$; $x_1 = 3, x_2 = -5,$ $x_3 = 2, x_4 = 3$

3. $\begin{bmatrix} 1 & 0 & 0 & 0 & -24 \\ 0 & 1 & 0 & 0 & 20 \\ 0 & 0 & 1 & 0 & 9 \\ 0 & 0 & 0 & 1 & -19 \end{bmatrix}$; $x_1 = -24, x_2 = 20,$ $x_3 = 9, x_4 = -19$

5. $\begin{bmatrix} 1 & 0 & 0 & 0 & 80 \\ 0 & 1 & 0 & 0 & -10 \\ 0 & 0 & 1 & 0 & 25 \\ 0 & 0 & 0 & 1 & -10 \end{bmatrix}$; $x_1 = 80, x_2 = -10,$ $x_3 = 25, x_4 = -10$

Exercise 2.4 (page 86)

1. 2×2 **3.** 2×3 **5.** 2×1 **7.** 1×1

9. False. Two equal matrices must have the same dimensions.

11. True   **13.** True   **15.** True

17. $\begin{bmatrix} 1 & 1 \\ 5 & 5 \end{bmatrix}$ **19.** $\begin{bmatrix} 6 & 18 & 0 \\ 12 & -6 & 3 \end{bmatrix}$

21. $\begin{bmatrix} 2 & 4 & -8 \\ 1 & -4 & -1 \end{bmatrix}$ **23.** $\begin{bmatrix} 13a & 54 \\ -2b & -7 \\ -2c & -6 \end{bmatrix}$

25. $\begin{bmatrix} 3 & -5 & 4 \\ 5 & 3 & 3 \end{bmatrix}$ **27.** $\begin{bmatrix} 13 & -6 & -7 \\ -6 & 1 & -7 \end{bmatrix}$

29. $\begin{bmatrix} 9 & -5 & -6 \\ 1 & 1 & -3 \end{bmatrix}$ **31.** $\begin{bmatrix} -2 & -17 & 32 \\ 28 & 14 & 23 \end{bmatrix}$

33. $\begin{bmatrix} 5 & -2 & 3 \\ -12 & 1 & -5 \end{bmatrix}$

35. $A + B = \begin{bmatrix} 2+1 & -3+(-2) & 4+0 \\ 0+5 & 2+1 & 1+2 \end{bmatrix}$

$= \begin{bmatrix} 3 & -5 & 4 \\ 5 & 3 & 3 \end{bmatrix}$;

$B + A = \begin{bmatrix} 1+2 & -2+(-3) & 0+4 \\ 5+0 & 1+2 & 2+1 \end{bmatrix}$

$= \begin{bmatrix} 3 & -5 & 4 \\ 5 & 3 & 3 \end{bmatrix}$; so

$A + B = \begin{bmatrix} 3 & -5 & 4 \\ 5 & 3 & 3 \end{bmatrix} = B + A$

37. $A + (-A) = \begin{bmatrix} 2 & -3 & 4 \\ 0 & 2 & 1 \end{bmatrix} + \begin{bmatrix} -2 & 3 & -4 \\ 0 & -2 & -1 \end{bmatrix}$

$= \begin{bmatrix} 2+(-2) & -3+3 & 4+(-4) \\ 0+0 & 2+(-2) & 1+(-1) \end{bmatrix}$

$= \begin{bmatrix} 0 & 0 & 0 \\ 0 & 0 & 0 \end{bmatrix} = 0$

39. $2B + 3B = \begin{bmatrix} 2 & -4 & 0 \\ 10 & 2 & 4 \end{bmatrix} + \begin{bmatrix} 3 & -6 & 0 \\ 15 & 3 & 6 \end{bmatrix}$

$= \begin{bmatrix} 5 & -10 & 0 \\ 25 & 5 & 10 \end{bmatrix} = 5B$

41. $x = 4, z = 3$ **43.** $x = 5, y = 1$

45. $x = 4, y = -11, z = 6$

47.

	$\frac{1}{2}''$	$1''$	$2''$
Steel	25	45	35
Aluminum	13	20	23

or

	Steel	Aluminum
$\frac{1}{2}''$	25	13
$1''$	45	20
$2''$	35	23

49.

	< $25,000	≥ $25,000
Dem.	351	203
Rep.	271	215
Ind.	73	55

or

	Dem.	Rep.	Ind.
< $25,000	351	271	73
≥ $25,000	203	215	55

51.

	LAS	ENG	EDUC
Male	200	225	120
Female	200	75	180

Technology Exercises (page 88)

1.
$$\begin{bmatrix} -2 & 1 & 7 & 5 \\ 4 & 6 & 7 & 5 \\ -3.5 & 8 & -4 & 13 \\ 12 & -1 & 7 & 6 \end{bmatrix}$$

3.
$$\begin{bmatrix} 19 & -11 & -14 & -15 \\ -12 & -13 & -21 & -17 \\ 15.5 & -24 & 19 & -39 \\ -29 & 10 & -14 & -11 \end{bmatrix}$$

5.
$$\begin{bmatrix} 37 & -17 & 30 & 15 \\ 4 & 9 & 13 & 1 \\ 20.5 & 16 & -3 & 31 \\ 29 & 18 & 22 & 43 \end{bmatrix}$$

Exercise 2.5 (page 99)

1. [14]   **3.** [14 −6]   **5.** $\begin{bmatrix} 4 \\ 2 \end{bmatrix}$ **7.** $\begin{bmatrix} 4 & 2 \\ 2 & 8 \end{bmatrix}$

9. $\begin{bmatrix} 4 & 2 \\ 2 & 8 \\ 9 & 8 \end{bmatrix}$ **11.** BA is defined; dimension is 3×4.

13. AB is not defined.   **15.** $(BA)C$ is not defined.

17. $BA + A$ is defined; dimension is 3×4.

19. $DC + B$ is defined; dimension is 3×3.

21. $\begin{bmatrix} -1 & 10 & -1 \\ -4 & 16 & -8 \end{bmatrix}$ **23.** $\begin{bmatrix} 11 & 5 \\ 13 & -9 \end{bmatrix}$

25. $\begin{bmatrix} 6 & 10 \\ 8 & 2 \\ -4 & 5 \end{bmatrix}$ **27.** $\begin{bmatrix} 3 & -1 \\ 4 & 2 \end{bmatrix}$

29. $\begin{bmatrix} 8 & 4 & 22 \\ 4 & 32 & 16 \end{bmatrix}$ **31.** $\begin{bmatrix} -14 & 7 \\ -20 & -6 \end{bmatrix}$

33. $\begin{bmatrix} 10 & 30 & 37 \\ 15 & 16 & 50 \\ -6 & 20 & -8 \end{bmatrix}$

35. $D(CB) =$

$$\begin{bmatrix} 1 & 0 & 4 \\ 0 & 1 & 2 \\ 0 & -1 & 1 \end{bmatrix}\left(\begin{bmatrix} 3 & 1 \\ 4 & -1 \\ 0 & 2 \end{bmatrix}\begin{bmatrix} 1 & 2 & 3 \\ -1 & 4 & -2 \end{bmatrix}\right)$$

$$= \begin{bmatrix} 1 & 0 & 4 \\ 0 & 1 & 2 \\ 0 & -1 & 1 \end{bmatrix}\begin{bmatrix} 2 & 10 & 7 \\ 5 & 4 & 14 \\ -2 & 8 & -4 \end{bmatrix}$$

$$= \begin{bmatrix} -6 & 42 & -9 \\ 1 & 20 & 6 \\ -7 & 4 & -18 \end{bmatrix};$$

$(DC)B =$

$$\left(\begin{bmatrix} 1 & 0 & 4 \\ 0 & 1 & 2 \\ 0 & -1 & 1 \end{bmatrix}\begin{bmatrix} 3 & 1 \\ 4 & -1 \\ 0 & 2 \end{bmatrix}\right)\begin{bmatrix} 1 & 2 & 3 \\ -1 & 4 & -2 \end{bmatrix}$$

$$= \begin{bmatrix} 3 & 9 \\ 4 & 3 \\ -4 & 3 \end{bmatrix}\begin{bmatrix} 1 & 2 & 3 \\ -1 & 4 & -2 \end{bmatrix}$$

$$= \begin{bmatrix} -6 & 42 & -9 \\ 1 & 20 & 6 \\ -7 & 4 & -18 \end{bmatrix}; \text{ so } D(CB) = (DC)B$$

37. $AB = \begin{bmatrix} 4 & -2 \\ 6 & 4 \end{bmatrix}$; $BA = \begin{bmatrix} 7 & -3 \\ 7 & 1 \end{bmatrix}$

39. $A = \begin{bmatrix} 2 & 1 \\ -\frac{1}{2} & -\frac{1}{2} \end{bmatrix}$

41. $\begin{bmatrix} 1 & 2 & 5 \\ 2 & 4 & 10 \\ -1 & -2 & -5 \end{bmatrix}\begin{bmatrix} 1 & 2 & 5 \\ 2 & 4 & 10 \\ -1 & -2 & -5 \end{bmatrix}$

$$= \begin{bmatrix} 0 & 0 & 0 \\ 0 & 0 & 0 \\ 0 & 0 & 0 \end{bmatrix}$$

43. The possibilities are: $a = 0$, $b = 0$; $a = -1$, $b = 0$; $a = -\frac{1}{2}$, $b = \frac{1}{2}$; $a = -\frac{1}{2}$, $b = -\frac{1}{2}$.

45. $[\frac{1}{3} \quad \frac{2}{3}]$

47. (a) PQ represents the matrix of raw materials needed to fill order: $PQ = [138 \quad 189 \quad 97 \quad 399]$
  (b) QC represents the matrix of costs to produce each product:
  $$QC = \begin{bmatrix} 311 \\ 653 \\ 614 \end{bmatrix}$$
  (c) PQC represents the total cost to produce the order: $PQC = \$13,083$

49. $\begin{bmatrix} 4 & 3 \\ 1 & 1 \\ 2 & 0 \end{bmatrix}$ **51.** $\begin{bmatrix} 1 & 0 & 1 \\ 11 & 12 & 4 \end{bmatrix}$ **53.** $[8 \quad 6 \quad 3]$

55. (a) $y = \frac{54}{35}x + \frac{27}{5}$ (b) 17.743 thousand units

57. $y = \frac{334}{223}x + \frac{8058}{223}$

59. (1) *Not* symmetric (2) Symmetric (3) *Not* symmetric Yes; a symmetric matrix *must* be square.

Technology Exercises (page 102)

1. $\begin{bmatrix} .5 & 16 & -30 & 25 \\ 21 & 14 & 28 & 64 \\ 19.5 & 8 & -23 & 33 \\ -9.5 & 45 & -9 & 83 \end{bmatrix}$

3. $\begin{bmatrix} 31.5 & 251 & -31.5 & 143 \\ 861 & 350 & 791 & 420 \\ 369.5 & 115 & 206.5 & 215 \\ 412.5 & 882 & 451.5 & 491 \end{bmatrix}$

5. $\begin{bmatrix} 66 & 74 & 94 & 38 \\ 71 & 13 & 106 & 28 \\ 165 & 124.5 & 79 & 52 \\ 158 & -3 & 152 & 46 \end{bmatrix}$

7. $\begin{bmatrix} -5 & 23 & -102 & 44 \\ -108 & -56 & -70 & 122 \\ 108 & -152 & -67 & 36 \\ -346 & 279 & -249 & 187 \end{bmatrix}$

9. (a) $A^2 = \begin{bmatrix} -2.95 & -.5 & -1.21 & 1.6 & -.49 \\ -2.8 & 2.56 & 2.54 & 1.4 & 1.44 \\ 5.2 & 8.6 & 1.11 & -2.6 & -.26 \\ -1.5 & -1 & .3 & 1 & .3 \\ 3.6 & -1.8 & -.63 & -1.8 & -.18 \end{bmatrix};$

$A^{10} = \begin{bmatrix} 433.0971 & -1583.5617 & -369.8169 & -216.5481 & -141.6383 \\ 1207.6949 & 5998.4376 & 2563.2274 & -603.8474 & 1045.7047 \\ -1423.0228 & 8065.0704 & 4271.7984 & 711.5114 & 2023.3641 \\ 479.4764 & -246.5381 & -231.5431 & -239.7372 & -140.5573 \\ -899.3405 & -3064.3130 & -1627.5015 & 449.6703 & -698.6086 \end{bmatrix};$

(Rounded to four decimal places)

$A^{15} = \begin{bmatrix} -2247.8448 & -118449.3176 & -69122.3644 & 1123.9224 & -32134.3035 \\ -11094.8095 & 542433.5133 & 249928.4018 & 5547.4047 & 110820.2098 \\ 100366.7830 & 831934.5635 & 386083.7906 & -50183.3915 & 165597.1358 \\ -18782.2970 & -38432.0936 & -18654.0856 & 9391.1485 & -7395.9599 \\ -1633.6655 & -306291.0222 & -130154.5685 & 816.8327 & -56015.4536 \end{bmatrix}$

(Rounded to four decimal places)

(b) $A^2 = \begin{bmatrix} 1.16 & 1.42 & -.408 & .29 \\ 1.25 & -1.26 & .506 & .91 \\ 13.9 & -5.29 & .632 & 7.8 \\ -2.74 & -.53 & .6 & -.788 \end{bmatrix};$

$A^{10} = \begin{bmatrix} -31.1428 & 14.0580 & 1.0561 & -16.1157 \\ 28.4909 & -.3398 & -3.2689 & 14.6394 \\ -50.8954 & 149.5593 & -30.8932 & -34.3845 \\ 52.3686 & -47.5375 & 5.6205 & 29.5703 \end{bmatrix};$

(Rounded to four decimal places)

$A^{15} = \begin{bmatrix} 111.7124 & -169.5588 & 31.0626 & 68.5252 \\ -155.7116 & 145.0638 & -15.6641 & -86.5465 \\ -621.4964 & -218.2318 & 157.5564 & -283.4745 \\ -1.3722 & 252.7427 & -69.4323 & -20.2157 \end{bmatrix}$

(Rounded to four decimal places)

(c) $A^2 = \begin{bmatrix} 1 & 0 & 1 \\ 1 & 2 & 1 \\ 2 & 2 & 2 \end{bmatrix}$; $A^{10} = \begin{bmatrix} 171 & 170 & 171 \\ 341 & 342 & 341 \\ 512 & 512 & 512 \end{bmatrix}$;

$A^{15} = \begin{bmatrix} 5461 & 5462 & 5461 \\ 10923 & 10922 & 10923 \\ 16384 & 16384 & 16384 \end{bmatrix}$

Exercise 2.6 (page 110)

1. $\begin{bmatrix} 1 & 2 \\ 2 & 3 \end{bmatrix}\begin{bmatrix} -3 & 2 \\ 2 & -1 \end{bmatrix} = \begin{bmatrix} 1 & 0 \\ 0 & 1 \end{bmatrix} = I_2$

3. $\begin{bmatrix} -1 & -2 \\ 3 & 4 \end{bmatrix}\begin{bmatrix} 2 & 1 \\ -\frac{3}{2} & -\frac{1}{2} \end{bmatrix} = \begin{bmatrix} 1 & 0 \\ 0 & 1 \end{bmatrix} = I_2$

5. $\begin{bmatrix} 1 & 2 & 3 \\ 2 & 3 & 4 \\ 1 & 2 & 1 \end{bmatrix}\begin{bmatrix} -\frac{5}{2} & 2 & -\frac{1}{2} \\ 1 & -1 & 1 \\ \frac{1}{2} & 0 & -\frac{1}{2} \end{bmatrix} = \begin{bmatrix} 1 & 0 & 0 \\ 0 & 1 & 0 \\ 0 & 0 & 1 \end{bmatrix} = I_3$

7. $\begin{bmatrix} 3 & -5 \\ -1 & 2 \end{bmatrix}$ **9.** $\begin{bmatrix} 4 & -1 \\ 3 & -1 \end{bmatrix}$

11. $\begin{bmatrix} 1.5 & -0.5 \\ -2 & 1 \end{bmatrix}$ **13.** $\begin{bmatrix} 0 & 0 & 1 \\ 0 & 1 & 0 \\ 1 & 0 & 0 \end{bmatrix}$

15. $\begin{bmatrix} \frac{4}{9} & \frac{1}{9} & \frac{1}{9} \\ \frac{4}{3} & -\frac{2}{3} & \frac{1}{3} \\ \frac{7}{9} & -\frac{5}{9} & \frac{4}{9} \end{bmatrix}$ **17.** $\begin{bmatrix} 1 & -1 & 2 \\ -1 & 2 & -3 \\ -1 & 1 & -1 \end{bmatrix}$

19. $\begin{bmatrix} 2 & -1 & -1 & -2 \\ -1 & 1 & 1 & 2 \\ -2 & 1 & 2 & 3 \\ -1 & 1 & 1 & 1 \end{bmatrix}$

21. $\begin{bmatrix} 4 & 6 & | & 1 & 0 \\ 2 & 3 & | & 0 & 1 \end{bmatrix} \rightarrow \begin{bmatrix} 4 & 6 & | & 1 & 0 \\ \boxed{0 & 0} & | & -\frac{1}{2} & 0 \end{bmatrix}$

23. $\begin{bmatrix} -8 & 4 & | & 1 & 0 \\ -4 & 2 & | & 0 & 1 \end{bmatrix} \rightarrow \begin{bmatrix} -8 & 4 & | & 1 & 0 \\ \boxed{0 & 0} & | & -\frac{1}{2} & 0 \end{bmatrix}$

25. $\begin{bmatrix} 1 & 1 & 1 & | & 1 & 0 & 0 \\ 3 & -4 & 2 & | & 0 & 1 & 0 \\ \boxed{0 & 0 & 0} & | & 0 & 0 & 1 \end{bmatrix}$

27. $\begin{bmatrix} 2 & -1 \\ -1 & 1 \end{bmatrix}$ **29.** $\begin{bmatrix} \frac{1}{3} & \frac{1}{3} \\ 0 & \frac{1}{2} \end{bmatrix}$ **31.** No inverse

33. $\begin{bmatrix} \frac{7}{2} & -1 & -\frac{3}{2} \\ 1 & -1 & -1 \\ \frac{1}{2} & 1 & \frac{1}{2} \end{bmatrix}$

35. $A^{-1} = \begin{bmatrix} \frac{1}{5} & \frac{2}{5} \\ \frac{2}{5} & -\frac{1}{5} \end{bmatrix}$; $B^{-1} = \begin{bmatrix} -\frac{1}{11} & \frac{4}{11} \\ \frac{3}{11} & -\frac{1}{11} \end{bmatrix}$;

$A^{-1} - B^{-1} = \begin{bmatrix} \frac{16}{55} & \frac{2}{55} \\ \frac{7}{55} & -\frac{6}{55} \end{bmatrix}$

37. $A = \begin{bmatrix} 1 & 3 & 2 \\ 2 & 7 & 3 \\ 1 & 0 & 6 \end{bmatrix}$; $B = \begin{bmatrix} 2 \\ 1 \\ 3 \end{bmatrix}$;

$A^{-1}B = \begin{bmatrix} 42 & -18 & -5 \\ -9 & 4 & 1 \\ -7 & 3 & 1 \end{bmatrix}\begin{bmatrix} 2 \\ 1 \\ 3 \end{bmatrix}$

39. $x = 1, y = 1$ **41.** $x = 2, y = 1$

43. $x = 88, y = -36$ **45.** $x = \frac{14}{9}, y = \frac{26}{3}, z = \frac{65}{9}$

47. $x = \frac{20}{3}, y = \frac{72}{3}, z = \frac{56}{3}$ **49.** $x = \frac{22}{9}, y = \frac{46}{3}, z = \frac{79}{9}$

51. $\begin{bmatrix} a & b \\ c & d \end{bmatrix}\begin{bmatrix} d & -b \\ -c & a \end{bmatrix} = \begin{bmatrix} ad - bc & 0 \\ 0 & -bc + ad \end{bmatrix} =$

$\begin{bmatrix} \Delta & 0 \\ 0 & \Delta \end{bmatrix}$; thus, $\begin{bmatrix} a & b \\ c & d \end{bmatrix}\begin{bmatrix} \frac{d}{\Delta} & \frac{-b}{\Delta} \\ \frac{-c}{\Delta} & \frac{a}{\Delta} \end{bmatrix} = \begin{bmatrix} 1 & 0 \\ 0 & 1 \end{bmatrix}$

53. $\begin{bmatrix} 1 & 5 \\ 2 & 0 \end{bmatrix}^{-1} = -\frac{1}{10}\begin{bmatrix} 0 & -5 \\ -2 & 1 \end{bmatrix} = \begin{bmatrix} 0 & \frac{1}{2} \\ \frac{1}{5} & -\frac{1}{10} \end{bmatrix}$

Technology Exercises (page 111)

1. $\begin{bmatrix} .0054 & .0509 & -.0066 \\ .0104 & -.0186 & .0095 \\ -.0193 & .0116 & .0344 \end{bmatrix}$

3. $\begin{bmatrix} .0249 & -.0360 & -.0057 & .0059 \\ -.0171 & .0521 & .0292 & -.0305 \\ .0206 & .0081 & -.0421 & 4.8764 \\ -.0175 & .0570 & .0657 & .0619 \end{bmatrix}$

5. $x = 4.5666, y = -6.4436, z = -24.0747$

7. $x = -1.1874, y = 2.4568, z = 8.2650$

9. $\begin{bmatrix} .25 & -.0625 & -.28125 & .3125 & .09375 \\ -1.5 & .125 & 3.0625 & -2.625 & .3125 \\ 1.75 & -.1875 & -2.84375 & 2.9375 & -.71875 \\ -.5 & .375 & .6875 & -.875 & .4375 \\ -1.25 & .3125 & 2.40625 & -2.5625 & .53125 \end{bmatrix}$;

$\begin{bmatrix} 1 & 0 & 0 & 0 & 0 \\ 0 & 1 & 0 & 0 & 0 \\ 0 & 0 & 1 & 0 & 0 \\ 0 & 0 & 0 & 1 & 0 \\ 0 & 0 & 0 & 0 & 1 \end{bmatrix}$

Exercise 2.7 (*page 124*)

1. A's wages = C's wages = \$30,000; B's wages = $\frac{3}{4}$(C's wages) = \$22,500

3. A's wages = $\frac{2}{5}$(C's wages) = \$12,000; B's wages = $\frac{11}{15}$(C's wages) = \$22,000; C's wages = \$30,000

5. $\begin{bmatrix} 203.282 \\ 166.977 \\ 137.847 \end{bmatrix}$

7. Farmer's wages = $\frac{4}{5}$(Rancher's wages) = \$20,000; Builder's wages = $\frac{18}{25}$(Rancher's wages) = \$18,000; Tailor's wages = $\frac{12}{25}$(Rancher's wages) = \$12,000; Rancher's wages = \$25,000

9. $X = \begin{bmatrix} 160 \\ 75.38 \end{bmatrix}$

11. (a) LIFE IS HARD   (b) ET TU BRUTE

13. A CORRECT ANSWER

15.

Dept.	Total Costs	Direct Costs	Indirect Costs	
			S_1	S_2
S_1	\$3109.09	\$2000	\$345.45	\$763.64
S_2	\$2290.91	\$1000	\$1036.36	\$254.55
P_1	\$3354.54	\$2500	\$345.45	\$509.09
P_2	\$2790.91	\$1500	\$1036.36	\$254.55
P_3	\$3854.54	\$3000	\$345.45	\$509.09
Totals			\$3109.07	\$2290.92

Technology Exercises (*page 126*)

1. $\begin{bmatrix} 7.5806451613 \\ 9.27419354838 \\ 16.0080645161 \\ 4.51612903227 \end{bmatrix}$ 3. $\begin{bmatrix} 7.70007770007 \\ 10.1165501166 \\ 3.64413364413 \\ 1.04118104118 \end{bmatrix}$

5. $\begin{bmatrix} 495.857760806 \\ 463.86087336 \\ 679.19467876 \\ 512.929924508 \\ 468.155972489 \end{bmatrix}$

Chapter Review: True–False Items (*page 127*)

1. T   2. F   3. F   4. T   5. F   6. F   7. F

Fill in the Blanks (*page 127*)

1. 3×2 2. one; infinitely many   3. rows; columns

4. inverse   5. 3×3 6. 5×5

Review Exercises (*page 127*)

1. $\begin{bmatrix} -1 & 3 & 16 \\ 3 & 15 & 8 \\ 5 & 10 & 29 \end{bmatrix}$ 3. $\begin{bmatrix} -3 & 9 & 48 \\ 9 & 45 & 24 \\ 15 & 30 & 87 \end{bmatrix}$

5. $\begin{bmatrix} -9 & -9 & -6 \\ -3 & 3 & -6 \\ -3 & -6 & 39 \end{bmatrix}$ 7. $\begin{bmatrix} -20 & 0 & 70 \\ 10 & 80 & 30 \\ 20 & 40 & 210 \end{bmatrix}$

9. $\begin{bmatrix} -\frac{7}{2} & -\frac{3}{2} & \frac{25}{2} \\ 3 & \frac{9}{2} & \frac{11}{2} \\ -\frac{37}{2} & -10 & 19 \end{bmatrix}$ 11. $\begin{bmatrix} 19 & 36 & 38 \\ 26 & 77 & 73 \\ 73 & 160 & 206 \end{bmatrix}$

13. $\begin{bmatrix} 16 & 32 & 27 \\ 16 & 10 & 19 \\ -104 & -80 & -113 \end{bmatrix}$ 15. $\begin{bmatrix} \frac{1}{3} & 0 \\ \frac{2}{3} & 1 \end{bmatrix}$

17. $\begin{bmatrix} 1 & -3 & 2 \\ -3 & 3 & -1 \\ 2 & -1 & 0 \end{bmatrix}$ 19. $\begin{bmatrix} \frac{1}{16} & \frac{1}{32} & \frac{1}{4} \\ \frac{3}{16} & \frac{3}{32} & -\frac{1}{4} \\ -\frac{3}{16} & \frac{13}{32} & \frac{1}{4} \end{bmatrix}$

21. $\begin{bmatrix} 1 & 2 & -3 \\ 0 & -2 & 14 \\ \boxed{0 \quad 0 \quad 0} \end{bmatrix} \begin{bmatrix} 1 & 0 & 0 \\ -4 & 1 & 0 \\ 3 & 0 & 1 \end{bmatrix}$ The matrix has no inverse.

23. $x = \frac{14}{9}, y = \frac{26}{9}$ 25. $x = -103, y = 32, z = 9$

27. $x = 29, y = -10, z = -1$

29. $\begin{bmatrix} 1 & 1 & -1 & | & 2 \\ 0 & 1 & -1 & | & 1 \\ \boxed{0 \quad 0 \quad 0} & | & \frac{1}{2} \end{bmatrix}$ There is no solution.

31. $x = 9, y = -\frac{56}{3}, z = -\frac{37}{3}$

33. $x = 29, y = 8, z = -24$

35. $x = \frac{10}{7} + \frac{3}{7}z, y = -\frac{9}{7} + \frac{5}{7}z$; three sample solutions: $x = 1, y = -2, z = -1; x = 4, y = 3, z = 6; x = 7, y = 8, z = 13$

37. $x = 1 - \frac{3}{5}z, y = 2 + \frac{4}{5}z$; three sample solutions: $x = 4, y = -2, z = -5; x = 1, y = 2, z = 0; x = -2, y = 6, z = 5$

39. $\begin{bmatrix} 1 & 0 & | & 4 \\ 0 & 1 & | & 2 \\ \boxed{0 \quad 0} & | & \frac{7}{2} \end{bmatrix}$ There is no solution.

41. $y = -z, x = w$

43. 20 caramels, 30 creams; increase the number of caramels to obtain a profit.

45.

Pounds of Almonds	Pounds of Cashews	Pounds of Peanuts
5	60	35
20	40	40
35	20	45
50	0	50

47. (a)

Treasury Bills	Corporate Bonds	Junk Bonds
$36,000	$3,000	$1,000
$36,500	$2,000	$1,500
$37,000	$1,000	$2,000
$37,500	$0	$2,500

(b)

Treasury Bills	Corporate Bonds	Junk Bonds
$10,000	$30,000	$0
$11,000	$28,000	$1,000
$12,000	$26,000	$2,000
$13,000	$24,000	$3,000
$14,000	$22,000	$4,000
$25,000	$0	$15,000

(c)

Treasury Bills	Corporate Bonds	Junk Bonds
$0	$25,000	$15,000
$1,000	$23,000	$16,000
$2,000	$21,000	$17,000
$3,000	$19,000	$18,000
$10,000	$5,000	$25,000
$12,500	$0	$27,500

Mathematical Questions from Professional Exams **(page 129)**

1. b   **2.** b   **3.** d   **4.** c

CHAPTER 3

Exercise 3.1 **(page 140)**

1. $x \geq 0$

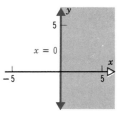

3. $x \geq 0, \ y \geq 0$

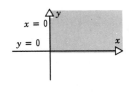

5. $2x - 3y \leq -6$

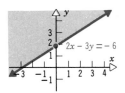

7. $5x + y \leq -10$

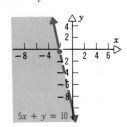

9. $x \geq 5$

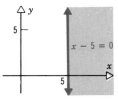

11. Only $P_1 = (3, 8)$ is part of the system's graph.
13. Only $P_3 = (5, 1)$ is part of the system's graph.
15. None of the points are part of the system's graph.
17. *b*   **19.** *b*   **21.** *b*   **23.** *c*
25. Bounded; corner points:
(0, 0), (0, 2), (2, 0)

27. Bounded; corner points:
(0, 2), (3, 0), (2, 0)

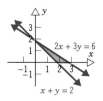

29. Bounded; corner points:
(0, 2), (0, 8), (2, 6), (5, 0), (2, 0)

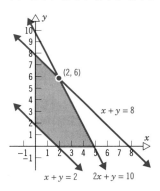

31. Bounded; corner points:
(0, 2), (0, 4), $(\frac{24}{7}, \frac{12}{7})$, (4, 0), (2, 0)

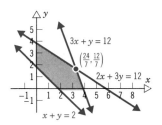

33. Bounded; corner points:
$(0, \frac{1}{2})$, (0, 5), (10, 0), (1, 0)

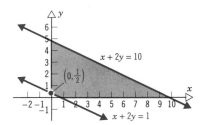

35. (a) $\begin{cases} 4x + 8y \le 960 \\ 12x + 8y \le 1440 \\ x \ge 0, y \ge 0 \end{cases}$ or $\begin{cases} x + 2y \le 240 \\ 3x + 2y \le 360 \\ x \ge 0, y \ge 0 \end{cases}$

(b) Corner points: (0, 0), (0, 120), (120, 0), (60, 90)

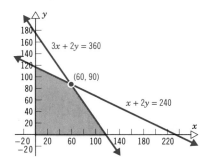

37. (a) $\begin{cases} 3x + 2y \le 80 \\ 4x + 3y \le 120 \\ x \ge 0, y \ge 0 \end{cases}$

(b) Corner points: (0, 40), (0, 0), $(\frac{80}{3}, 0)$

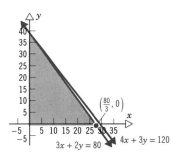

39. (a) $\begin{cases} x \ge 15 \\ y \le 10 \\ x + y \le 25 \\ y \ge 0 \end{cases}$ Where x thousand dollars are invested in Treasury bills and y thousand dollars are invested in corporate bonds; the monetary unit is $1000.

(b) Corner points: (15, 0), (15, 10), (25, 0)

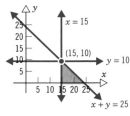

(c) (15, 0): invest $15,000 in Treasury bills and do not invest the remaining $10,000.
(15,10): invest $15,000 in Treasury bills and $10,000 in corporate bonds.
(25, 0): invest all $25,000 in Treasury bills.

41. (a) $\begin{cases} x + 2y \geq 5 \\ 5x + y \geq 16 \\ x \geq 0, y \geq 0 \end{cases}$ Where x = no. of units of first grain and y = no. of sec-ond grain.

(b) Corner points: (0, 16), (3, 1), (5, 0)

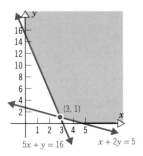

43. (a) $\begin{cases} 5x + 4y \geq 85 \\ 3x + 3y \geq 70 \\ 2x + 3y \geq 50 \\ x \geq 0, y \geq 0 \end{cases}$ Where x = no. of ounces of food A and y = no. of ounces of food B.

(b) Corner points: $(0, \frac{70}{3})$, $(20, \frac{10}{3})$, $(25, 0)$

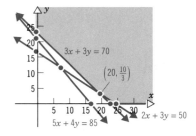

Technology Exercises **(*page 143*)**

1. Corner points: $(-1, 2)$, $(2, -1)$, $(2, 2)$

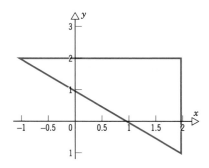

3. Corner points: (1, 0), (1, 1.33), (3, 0)

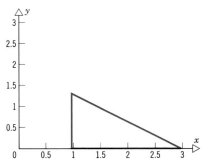

5. Corner points: $(0, -1)$, (0, 5), (3, 2)

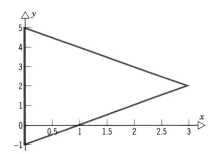

Exercise 3.2 **(*page 156*)**

1. The maximum is 38 at (7, 8); the minimum is 10 at (2, 2).

3. The maximum is 71 at (7, 8); the minimum is 16 at (8, 1).

5. The maximum is 55 at (7, 8); the minimum is 14 at any point on the line segment joining (2, 2) and (8, 1).

7. (13, 0), (3, 0), (0, 4) [*Note:* The set of feasible points is unbounded above.]

9. (0, 10), (5, 10), (15, 0), (0, 0)

11. (0, 8), (10, 8), (10, 0), (3, 0), (0, 4)

13. The maximum is 14 at (0, 2).

15. The maximum is 15 at (3, 0).

17. The maximum is 56 at (0, 8).

19. The maximum is 58 at (6, 4).

21. The minimum is 4 at (2, 0).

23. The minimum is 4 at (2, 0).

25. The minimum is 0 at (0, 0).

27. The maximum is 10 at any point on the line segment joining (0, 10) and (10, 0). The minimum is $\frac{20}{3}$ at $(\frac{10}{3}, \frac{10}{3})$.

29. The maximum is 50 at (10, 0); the minimum is 20 at (0, 10).

31. The maximum is 40 at $(0, 10)$; the minimum is $\frac{70}{3}$ at $\left(\frac{10}{3}, \frac{10}{3}\right)$.

33. The maximum is 204 at $(4, 4)$; the minimum is 54 at $(3, 0)$.

35. The maximum is 208 at $(4, 5)$; the minimum is 21 at $(3, 0)$.

37. The maximum is 240 at $(3, 10)$; the minimum is -300 at $(15, 0)$.

39. The maximum is 216 at $(2, 10)$; the minimum is -180 at $(15, 0)$.

41. Make 90 packages of the low-grade mixture and 105 packages of the high-grade mixture (for a maximum profit of $69).

43. Plant x acres of soybeans, where $24 \le x \le 30$, and $y = 60 - 2x$ acres of corn (for a maximum profit of $9000).

45. Manufacture 15 units of the first product and 25 units of the second product (for a maximum profit of $2100).

47. Invest $7500 in the first security and $22,500 in the second security (for a maximum income of $2550).

49. Employ the first repairman for 2 weeks and the second repairman for 3 weeks (for a minimum cost of $1160).

51. Produce 700 units of item A and 400 units of item B (for a maximum legal profit of $1450).

53. Add 5 oz of supplement I and 1 oz of supplement II to each 100 oz of feed (for a minimum cost of $0.19 for the supplements used per 100 oz of feed).

55. The price per roll of high-grade carpet should be between $520 and $620.

Technology Exercises (page 160)

1. Corner points: $(3, 0)$, $(1, 2)$, $(4, 5)$, $(6, 3)$ Minimum is $z = 6$ at $x = 1$, $y = 2$. Maximum is $z = 24.75$ at $x = 6$, $y = 3$.

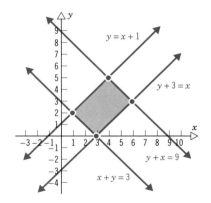

3. Corner points: $(.54, 5.27)$, $(4, 7)$, $(5.33, 4.33)$, $(1.5, .5)$ Minimum is $z = 5.875$ at $x = 1.5$, $y = .5$. Maximum is $z = 24.0675$ at $x = 5.33$, $y = 4.33$.

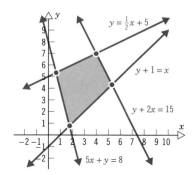

5. Corner points: $(.54, 5.27)$, $(4, 7)$, $(6.66, 1.66)$, $(5, 0)$, $(1.60, 0)$ Minimum is $z = 5.6$ at $x = 1.60$, $y = 0$. Maximum is $z = 25.385$ at $x = 6.66$, $y = 1.66$.

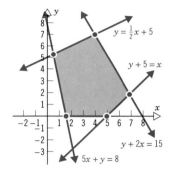

Chapter Review: True–False Items (page 160)

1. T   **2.** F   **3.** T   **4.** T   **5.** T   **6.** F

Fill in the Blanks (page 161)

1. half-plane   **2.** objective   **3.** feasible
4. bounded   **5.** corner point

Review Exercises (page 161)

1.

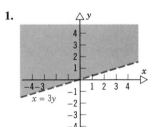

3.

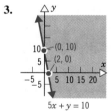

5. Only $P_2 = (2, -6)$ is part of the system's graph.

7. a

9. Corner points: $(0, 1)$, $(0, 6)$, $(4, 0)$, $(1, 0)$; bounded graph

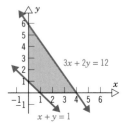

11. Corner points: $(0, 6)$, $(\frac{8}{5}, \frac{6}{5})$, $(4, 0)$; unbounded graph

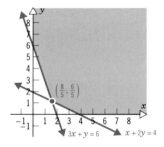

13. Corner points: $(0, 3)$, $(0, 4)$, $(2, 3)$, $(4, 0)$, $(2, 0)$; bounded graph

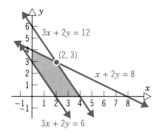

15–21. Corner points: $(0, 10)$, $(0, 20)$, $(\frac{40}{3}, \frac{40}{3})$, $(20, 0)$, $(10, 0)$

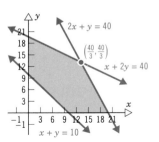

15. Maximum $z = \frac{80}{3}$ at $(x, y) = (\frac{40}{3}, \frac{40}{3})$

17. Minimum $z = 20$ at $(x, y) = (0, 10)$

19. Maximum $z = 40$ at any point (x, y) on the line segment joining $(\frac{40}{3}, \frac{40}{3})$ and $(20, 0)$

21. Minimum $z = 20$ at $(x, y) = (10, 0)$

23. Maximum $z = 235$ at $(x, y) = (5, 8)$; Minimum $z = 60$ at any point (x, y) on the line segment joining $(0, 3)$ and $(4, 0)$

25. Maximum $z = 155$ at $(x, y) = (5, 4)$; Minimum $z = 0$ at $(x, y) = (0, 0)$

27. Maximum $z = 25.2$ at $(x, y) = (9, 6)$

29. Maximum $z = 18.36$ at $(x, y) = (8, 7)$

31. Katy should buy $7\frac{1}{2}$ lb of food A and $11\frac{1}{4}$ lb of food B each month (for a minimum bill of \$18.75).

33. Make 8 downhill skis and 24 cross-county skis for a maximum profit of \$1760.

35. The maximum weekly profit is \$7200.

37. The maximum profit is \$160 (by producing 20 dozen cans of food A and 10 dozen cans of food B).

Mathematical Questions from Professional Exams (page 163)

1. b   **2.** a   **3.** c   **4.** c   **5.** d   **6.** c   **7.** c

8. b   **9.** b   **10.** a   **11.** b   **12.** e   **13.** c

14. b

CHAPTER 4

Exercise 4.1 (page 176)

1. Standard form   **3.** Not in standard form

5. Not in standard form

7. Not in standard form   **9.** Standard form

11. Cannot be modified to be in standard form

13. Cannot be modified to be in standard form

15. Maximize $P = 2x_1 + x_2 + 3x_3$ subject to the constraints $x_1 - x_2 - x_3 \le 6$, $-2x_1 + 3x_2 \le 12$, $x_3 \le 2$, $x_1 \ge 0$, $x_2 \ge 0$, and $x_3 \ge 0$.

17.

BV	P	x_1	x_2	x_3	s_1	s_2	s_3	RHS
P	1	-2	-1	-3	0	0	0	0
s_1	0	5	2	1	1	0	0	20
s_2	0	6	1	4	0	1	0	24
s_3	0	1	1	4	0	0	1	16

19.

BV	P	x_1	x_2	s_1	s_2	s_3	RHS
P	1	-3	-5	0	0	0	0
s_1	0	2.2	-1.8	1	0	0	5
s_2	0	0.8	1.2	0	1	0	2.5
s_3	0	1	1	0	0	1	0.1

21.

BV	P	x_1	x_2	x_3	s_1	s_2	RHS
P	1	-2	-3	-1	0	0	0
s_1	0	1	1	1	1	0	50
s_2	0	3	2	1	0	1	10

23.

BV	P	x_1	x_2	x_3	s_1	s_2	s_3	RHS
P	1	-3	-4	-2	0	0	0	0
s_1	0	3	1	4	1	0	0	5
s_2	0	1	1	0	0	1	0	5
s_3	0	2	-1	1	0	0	1	6

25. New tableau:

BV	P	x_1	x_2	s_1	s_2	RHS
P	1	0	0	1	0	300
x_2	0	$\frac{1}{2}$	1	$\frac{1}{2}$	0	150
s_2	0	2	0	-1	1	180

New system: $P = \qquad\qquad - s_1 + 300$
$$x_2 = -\tfrac{1}{2}x_1 - \tfrac{1}{2}s_1 + 150$$
$$s_2 = -2x_1 + \qquad s_1 + 180$$

Current values: $P = 300,\ x_2 = 150,\ s_2 = 180$

27. New tableau:

BV	P	x_1	x_2	x_3	s_1	s_2	s_3	RHS
P	1	$\frac{5}{4}$	$-\frac{1}{2}$	0	0	0	$\frac{3}{4}$	$\frac{27}{2}$
s_1	0	-2	0	0	1	0	-1	6
s_2	0	$\frac{5}{4}$	$-\frac{3}{2}$	0	0	1	$-\frac{1}{4}$	$\frac{55}{2}$
x_3	0	$\frac{3}{4}$	$\frac{1}{2}$	1	0	0	$\frac{1}{4}$	$\frac{9}{2}$

New system: $P = -\frac{5}{4}x_1 + \frac{1}{2}x_2 - \frac{3}{4}s_3 + \frac{27}{2}$
$$s_1 = 2x_1 + s_3 + 6$$
$$s_2 = -\tfrac{5}{4}x_1 + \tfrac{3}{2}x_2 + \tfrac{1}{4}s_3 + \tfrac{55}{2}$$
$$x_3 = -\tfrac{3}{4}x_1 - \tfrac{1}{2}x_2 - \tfrac{1}{4}s_3 + \tfrac{9}{2}$$

Current values: $P = \frac{27}{2},\ s_1 = 6,\ s_2 = \frac{55}{2},\ x_3 = \frac{9}{2}$

29. New tableau:

BV	P	x_1	x_2	x_3	x_4	s_1	s_2	s_3	s_4	RHS
P	1	7	-2	-3	0	0	4	0	0	96
s_1	0	-3	0	1	0	1	0	0	0	20
x_4	0	2	0	0	1	0	1	0	0	24
s_3	0	0	-3	1	0	0	0	1	0	28
s_4	0	-2	-3	0	0	0	-1	0	1	0

New system: $P = -7x_1 + 2x_2 + 3x_3 - 4s_2 + 96$
$s_1 = 3x_1 - x_3 + 20$
$x_4 = -2x_1 - s_2 + 24$
$s_3 = 3x_2 - x_3 + 28$
$s_4 = 2x_1 + 2x_3 + s_2$

Current values: $P = 96$, $s_1 = 20$, $x_4 = 24$, $s_3 = 28$,
$s_4 = 0$

Technology Exercises (page 179)

1.

BV	P	x_1	x_2	s_1	s_2	RHS
P	1	-1	0	0	1	150
s_1	0	1.67	0	1	$-.67$	100
s_2	0	.33	1	0	.33	50

3.

BV	P	x_1	x_2	s_1	s_2	RHS
P	1	-1.25	0	.75	0	75
s_1	0	.25	1	.25	0	25
s_2	0	.75	0	-1.25	1	10

Exercise 4.2 (page 194)

1. (b); the pivot element is 1 in row 2, column 2

3. (a); the solution is $P = \frac{256}{7}$, $x_1 = \frac{32}{7}$, $x_2 = 0$ **5.** (c)

7. The maximum is $P = \frac{204}{7} = 29\frac{1}{7}$ when $x_1 = \frac{24}{7}$, $x_2 = \frac{12}{7}$.

9. The maximum is $P = 8$ when $x_1 = \frac{2}{3}$, $x_2 = \frac{2}{3}$.

11. The maximum is $P = 6$ when $x_1 = 2$, $x_2 = 0$.

13. There is no maximum for P; the feasible region is unbounded.

15. The maximum is $P = 30$ when $x_1 = 0$, $x_2 = 0$, $x_3 = 10$.

17. The maximum is $P = 42$ when $x_1 = 1$, $x_2 = 10$, $x_3 = 0$, $x_4 = 0$.

19. The maximum is $P = 40$ when $x_1 = 20$, $x_2 = 0$, $x_3 = 0$.

21. The maximum is $P = 50$ when $x_1 = 0$, $x_2 = 15$, $x_3 = 5$, $x_4 = 0$.

23. The maximum profit is $1500 when the manufacturer makes 400 of Jean I, 0 of Jean II, and 50 of Jean III.

25. The maximum profit is $190 from the sale of 0 of product A, 40 of product B, and 75 of product C.

27. The maximum revenue is $275,000 when 200,000 gal of regular, 0 gal of premium, and 25,000 gal of super premium are mixed.

29. The maximum return is $7500 when she invests $45,000 in stocks, $15,000 in corporate bonds, and $30,000 in municipal bonds.

31. The maximum profit is $14,400 when 180 acres of crop A, 20 acres of crop B, and 0 acres of crop C are planted.

33. The maximum revenue is $2800 for 50 cans of can I, no cans of can II, and 70 cans of can III. (The revenue is also $2800 for no cans of can I, 100 cans of can II, and 20 cans of can III.)

35. The maximum profit is $12,000 from 1200 television cabinets and no stereo or radio cabinets.

37. The maximum profit is $30,000 when no TVs are shipped from Chicago, 375 TVs are shipped from New York, and no TVs are shipped from Denver. (The profit is also $30,000 when no TVs are shipped from Chicago, 350 TVs are shipped from New York, and 50 TVs are shipped from Denver.)

Technology Exercises (page 197)

1. Maximum is $P = 167.86$ at $x_1 = 14.29$, $x_2 = 32.14$.

3. Maximum is $P = 8$ at $x_1 = .67$, $x_2 = .67$.

Exercise 4.3 (page 205)

1. Standard form   **3.** Not in standard form

5. Not in standard form

7. Maximize $P = 2y_1 + 6y_2$ subject to $y_1 + 2y_2 \le 2$, $y_1 + 3y_2 \le 3$, $y_1 \ge 0$, $y_2 \ge 0$.

9. Maximize $P = 5y_1 + 4y_2$ subject to $y_1 + 2y_2 \le 3$, $y_1 + y_2 \le 1$, $y_1 \le 1$, $y_1 \ge 0$, $y_2 \ge 0$.

11. $C = 6$, $x_1 = 0$, $x_2 = 2$ **13.** $C = 12$, $x_1 = 0$, $x_2 = 4$

15. $C = \frac{21}{5}$, $x_1 = \frac{8}{5}$, $x_2 = 0$, $x_3 = \frac{13}{5}$

17. $C = 5$, $x_1 = 1$, $x_2 = 1$, $x_3 = 0$, $x_4 = 0$

19. The minimum cost is $0.22 when 2 P pills and 4 Q pills are taken as supplements.

21. The minimum cost is $290 when $A = 20$ units, $B = 30$ units, and $C = 150$ units.

23. The minimum cost is $65.20 when 4 lunch #1, 3 lunch #2, and 2 lunch #3 are ordered.

Technology Exercises (*page 207*)

1. Minimum is $C = 167.86$ at $x_1 = 14.29$, $x_2 = 32.14$.

3. Minimum is $C = 8$ at $x_1 = .67$, $x_2 = .67$.

Exercise 4.4 (*page 217*)

1. The maximum is $P = 44$ when $x_1 = 4$, $x_2 = 8$.

3. The maximum is $P = 27$ when $x_1 = 9$, $x_2 = x_3 = 0$.

5. The minimum is $z = \frac{20}{3}$ when $x_1 = x_2 = 0$, $x_3 = \frac{20}{3}$.

7. The maximum is $P = 7$ when $x_1 = 1$, $x_2 = 2$.

9. The minimum total shipping charge is $150,000; ship 100 engines from M1 to A1, 300 engines from M1 to A2, 400 engines from M2 to A1, and no engines from M2 to A2.

11. The minimum preparation cost is $7.50, using $\frac{5}{8}$ unit of food I, $\frac{25}{4}$ units of food II, and no units of food III. (The cost is also $7.50 using no units of food I or food III, and $\frac{15}{2}$ units of food II.)

13. The minimum cost is $965 when 55 sets are shipped from W_1 to R_1 and 75 sets are shipped from W_2 to R_2.

15. The minimum cost is $120 when 0 units of A and 15 units of B are used. (The cost is also $120 when 2 units of A and 14 units of B are used.)

Technology Exercises (*page 219*)

1. Maximum is $P = 3.255$ at $x_1 = .47$, $x_2 = .85$.

3. Minimum is $C = 13.796$ at $x_1 = 2.27$, $x_2 = .55$.

5. Maximum is $P = 44.805$ at $x_1 = 3.25$, $x_2 = 6.20$, $x_3 = 15.2$, $x_4 = 0$.

Chapter Review: True–False Items (*page 220*)

1. T   **2.** F   **3.** T   **4.** F   **5.** T   **6.** T

Fill in the Blanks (*page 221*)

1. slack variables   **2.** column   **3.** $\geq$

4. Von Neumann duality   **5.** Phase I/Phase II

Review Exercises (*page 221*)

1. The maximum is $P = 22,500$ when $x_1 = 0$, $x_2 = 100$, $x_3 = 50$.

3. The maximum is $P = 352$ when $x_1 = 0$, $x_2 = \frac{6}{5}$, $x_3 = \frac{28}{5}$.

5. The minimum is $z = 0$ when $x_1 = 0$, $x_2 = 0$.

7. The minimum is $z = 350$ when $x_1 = 0$, $x_2 = 50$, $x_3 = 50$. ($z = 350$ also when $x_1 = 25$, $x_2 = 0$, and $x_3 = 75$).

9. The maximum is $P = 12,250$ when $x_1 = 0$, $x_2 = 5$, $x_3 = 25$.

11. Make $83\frac{1}{3}$ lb of hamburger patties and 500 lb of picnic patties (for a maximum of $583\frac{1}{3}$ lb of meat used).

13. The maximum profit is approx. $5714.29 for cultivation of 0 acres of corn, 0 acres of wheat, and $\frac{1000}{7} \approx 143$ acres of soybeans.

Mathematical Questions from Professional Exams (*page 222*)

1. c   **2.** d   **3.** c   **4.** a   **5.** b   **6.** a

7. c   **8.** d   **9.** d   **10.** a   **11.** d

CHAPTER 5

Exercise 5.1 (*page 230*)

1. 45%   **3.** 112%   **5.** 6%   **7.** 0.25%

9. 0.42   **11.** 0.002   **13.** 0.00001   **15.** 0.734

17. 150   **19.** 18   **21.** 105   **23.** 5%

25. 160%   **27.** 250   **29.** $333\frac{1}{3}$ **31.** $10

33. $45   **35.** $150   **37.** 10%   **39.** $33\frac{1}{3}$%

41. $13\frac{1}{3}$%   **43.** $1140   **45.** $1680

47. $1263.16   **49.** $2380.95

51. The discounted loan at 9% has less interest for 6 months.

53. $471.70   **55.** $3\frac{1}{4}$ yr

57. She should choose the simple interest loan at 12.3%.

59. 8.58%

Exercise 5.2 (*page 238*)

1. $1348.18   **3.** $545   **5.** $854.36   **7.** $95.14

9. $456.97

11. (a) Amount = $1295.03; Interest earned = $295.03
(b) Amount = $1302.26; Interest earned = $302.26
(c) Amount = $1306.05; Interest earned = $306.05
(d) Amount = $1308.65; Interest earned = $308.65

13. (a) $1266.77 (b) $1425.76 (c) $1604.71

15. (a) $3947.05 (b) $3115.83

17. (a) 8.16% (b) 12.68%

19. 25.99%

21. Approx. $11\frac{1}{2}$ yr

23. The 10% loan compounded monthly has less interest.

25. $1759.11 **27.** 5.35% **29.** 6.82%

31. $6\frac{1}{4}$% compounded annually

33. 9% compounded monthly **35.** $109,400

37. $656.07 **39.** $29,137.83 **41.** $42,640.10

43. 19,918 **45.** Yes, the stock returns 7.26% quarterly.

47. $18,508.09 **49.** $10,810.76 **51.** 9.07%

53. Approx. $15\frac{1}{4}$ yr

Exercise 5.3 (*page 247*)

1. $1593.74 **3.** $5073 **5.** $7867.22

7. $147.05 **9.** $1868.68 **11.** $4121.33

13. $62,822.56 **15.** $9126.56 **17.** $524.04

19. $22,192.08 **21.** $205,367.98 **23.** $1655.57

25. Approx. 34 yr

Exercise 5.4 (*page 254*)

1. $15,495.62 **3.** $856.60 **5.** $85,135.64

7. $229,100; $25,239.51 **9.** $470.73 **11.** $530.76

13. $2008.18 **15.** $25,906.15

17.

Loan	Monthly Payment
8%, 20 yr	~ $794.62
9%, 25 yr	~ $797.24

The 9% loan for 25 yr has the larger monthly payment. Clearly, the interest is greater on the 9% loan for 25 yr, since the monthly payment is larger, and there are more monthly payments.

Loan	Equity after 10 Years
8%, 20 yr	~ $54,506.24
9%, 25 yr	~ $41,397.39

After 10 yr, the equity from the 8%, 20 yr, loan is greater.

19. $55.82 **21.** (a) $1207.64 (b) $36.24

23. (a) $15,200 (b) $60,800 (c) $489.21 (d) $115,315.60
(e) 199 mo = 16 yr 7 mo (f) $56,452.79

25. $332.79

27. Monthly payment: $474.01; Total interest paid: $4752.48

29. For a 30-yr mortgage: Monthly payment: $966.37; Total interest paid: $235,893.20
For a 15-yr mortgage: Monthly payment: $1189.89; Total interest paid: $102,180.20

31. Approx. 4 yr, 8 mo

Exercise 5.5 (*page 259*)

1. Leasing is preferable.

3. Machine A is preferable. (Machine A's annual cost is $125.56 *less* than its labor savings, but machine B costs $36.86 *more* each year than its labor savings.)

5. $1086.46

Chapter Review: True–False Items (*page 260*)

1. T **2.** T **3.** F **4.** F

Fill in the Blanks (*page 260*)

1. proceeds **2.** present value **3.** annuity

4. amortized

Review Exercises (*page 260*)

1. $I = \$36, A = \$436$ **3.** $125.12

5. The interest from loan (a) is $1080, while the interest from loan (b) is $1044.55. The 10% loan compounded monthly costs Mike less.

7. $71.36 **9.** $404.81

11. (a) $545.22 (b) $103,566 (c) $23,501.79

13. Monthly payment: $1049; Equity after 10 yr: $21,575.51

15. $119,431.77 **17.** $108,003.59 **19.** $10,078.44

21. $37.98 **23.** $141.22 **25.** 9.38%

27. $1156.60 **29.** $2087.09 **31.** $330.74

Mathematical Questions from Professional
Exams **(page 262)**

1. b   **2.** c   **3.** b   **4.** b   **5.** d

6. a.   **7.** c

CHAPTER 6

Exercise 6.1 **(page 278)**

1. True   **3.** False   **5.** False   **7.** True

9. True   **11.** {2, 3}   **13.** {1, 2, 3, 4, 5}

15. ∅   **17.** {*a, b, d, e, f, q*}   **19.** 4   **21.** 10

23. (a) {0, 1, 2, 3, 5, 7, 8}   (b) {5}   (c) {5}
(d) {0, 1, 2, 3, 4, 6, 7, 8, 9}   (e) {4, 6, 9}
(f) {0, 1, 5, 7}   (g) ∅   (h) {5}

25. (a) {*b, c, d, e, f, g*}   (b) {*c*}
(c) {*a, h, i, j, k, l, m, . . . , x, y, z*}
(d) {*a, b, d, e, f, g, h, i, . . . , x, y, z*}

27. (a) $\overline{A} \cap B$ (b) $(\overline{A} \cap \overline{B}) \cup C$

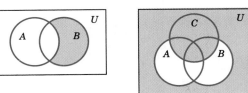

(c) and (d) $A \cap (A \cup B) = A \cup (A \cap B)$

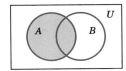

(e) and (f) $(A \cup B) \cap (A \cup C) = A \cup (B \cap C)$

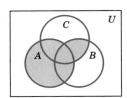

(g) $(A \cap B)$ $(A \cap \overline{B})$

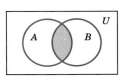

$A = (A \cap B) \cup (A \cap \overline{B})$

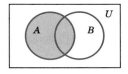

(h) $(A \cap B)$ $(\overline{A} \cap B)$

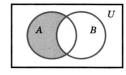

$B = (A \cap B) \cup (\overline{A} \cap B)$

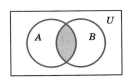

29. {*x*|*x* is both a customer of IBM and a member of the board of directors of IBM}

31. {*x*|*x* is either a customer of IBM or a stockholder of IBM}

33. $M \cap S$ = {All male students who smoke}

35. $\overline{M} \cap \overline{S}$ = {All female students who do not smoke}

37. 3   **39.** 6   **41.** 5   **43.** 2   **45.** 10

47. 452   **49.** 36   **51.** 46   **53.** 24   **55.** 63

57. 3   **59.** (a) 536 (b) 317 (c) 134

61. (a) 259 (b) 455 (c) 227 (d) 76 (e) 118
(f) 93 (g) 912

63. (a) 40 (b) 35 (c) 40 (d) 205 (e) 155

65. There are eight possible blood types:

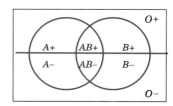

67. 46

69. $\varnothing$, {a}, {b}, {c}, {d}, {a, b}, {a, c}, {a, d}, {b, c}, {b, d},
{c, d}, {a, b, c}, {a, b, d}, {a, c, d}, {b, c, d}, {a, b, c, d}
There are 16 (= 2^4) subsets of {a, b, c, d}.

Exercise 6.2 *(page 284)*

1. 8   **3.** 24   **5.** 864   **7.** 36   **9.** 1320

11. 1200

13. 6 people: 720 ways; 8 people: 40,320 ways

15. No repeated letters: 360 code words; repeated letters:
1296 code words

17. 5040   **19.** $4^{10} \cdot 2^{15} = 2^{35} = 34,359,738,368$

21. (a) 6,760,000 (b) 3,407,040 (c) 3,276,000

23. 16   **25.** 60   **27.** $2^8 = 256$

29. $50^3 = 125,000$ **31.** 8

Exercise 6.3 *(page 290)*

1. 60   **3.** 120   **5.** 90   **7.** 9   **9.** 28

11. 42   **13.** 8   **15.** 1   **17.** 56   **19.** 1

21. (a) 6! = 720 (b) 5! = 120 (c) 4! = 24

23. P(10, 4) = 5040   **25.** P(9, 5) = 15,120

27. P(12, 8) = 19,958,400

29. P(1500, 3) = 3,368,253,000

31. P(15, 4) = 32,760   **33.** 5! = 120

Exercise 6.4 *(page 295)*

1. 15   **3.** 21   **5.** 5   **7.** 28   **9.** 56

11. 2380   **13.** 2300   **15.** 15   **17.** 60

19. 26,046,720   **21.** 10   **23.** 1,192,052,400

25. 75,287,520   **27.** 1,217,566,350   **29.** 60

31. 10,626

33. $P(50, 15) = \frac{50!}{35!} = 2,943,352,142,120,754,524,160,000$

35. 5040

Exercise 6.5 *(page 302)*

1. (a) 1024 (b) 210 (c) 56 (d) 968

3. (a) 63 (b) 35 (c) 1   **5.** 70   **7.** 1260

9. 4,989,600   **11.** 27,720

13. (a) 280 (b) 280 (c) 640

15. $\dfrac{30!}{(6!)^5} \approx 1.37 \times 10^{18}$ **17.** 93   **19.** 826

21. Even number of 1s: 128; Odd number of 1s: 128

Exercise 6.6 *(page 309)*

1. $x^5 + 5x^4y + 10x^3y^2 + 10x^2y^3 + 5xy^4 + y^5$

3. $x^3 + 9x^2y + 27xy^2 + 27y^3$

5. $16x^4 - 32x^3y + 24x^2y^2 - 8xy^3 + y^4$

7. 10   **9.** 405   **11.** 32   **13.** 1023   **15.** 512

17. $\binom{10}{7} = \binom{9}{7} + \binom{9}{6}$ and $\binom{9}{7} = \binom{8}{7} + \binom{8}{6}$,
so $\binom{10}{7} = \binom{8}{7} + \binom{8}{6} + \binom{9}{6}$
Also, $\binom{8}{7} = \binom{7}{7} + \binom{7}{6}$ and $\binom{7}{7} = 1 = \binom{6}{6}$,
so $\binom{10}{7} = \binom{6}{6} + \binom{7}{6} + \binom{8}{6} + \binom{9}{6}$

19. $\binom{12}{6}$

21. $k \cdot \binom{n}{k} = k \cdot \dfrac{n!}{k!(n-k)!} = k \cdot \dfrac{n!}{k \cdot (k-1)!(n-k)!}$

$= \dfrac{n \cdot (n-1)!}{(k-1)!(n-k)!}$

$= n \cdot \dfrac{(n-1)!}{(k-1)!((n-1)-(k-1))!}$

$= n \cdot \binom{n-1}{k-1}$

Alternatively, suppose we wish to select a team of k people from a class of n students, and designate one of the
team members as "team leader." We could do this by
selecting the team [in $\binom{n}{k}$ ways] and then designate the
leader (k ways), so there are $k \cdot \binom{n}{k}$ possible results, Or,
we could first choose the leader (n choices) and then

choose the remaining $k - 1$ team members [$\binom{n-1}{k-1}$ ways], so there are $n \cdot \binom{n-1}{k-1}$ possible results. Since these two methods count the same set of results, the two answers must be the same, i.e., $k \cdot \binom{n}{k} = n \cdot \binom{n-1}{k-1}$.

Chapter Review: True–False Items *(page 310)*

1. T   **2.** T   **3.** F   **4.** F   **5.** T   **6.** F   **7.** F

Fill in the Blanks *(page 310)*

1. disjoint   **2.** permutation   **3.** combination

4. Pascal's   **5.** binomial coefficients

6. Binomial Theorem   **7.** $\binom{5}{3}$ $2^2 = 40$

Review Exercises *(page 310)*

1. None of these.   **3.** None of these.

5. None of these.   **7.** $\subset$, $\subseteq$

9. $\subseteq$, $=$ **11.** None of these.

13. $\subset$, $\subseteq$ **15.** None of these.

17. (a) $\{3, 6, 8, 9\}$ (b) $\{6\}$ (c) $\{2, 3, 6, 7\}$

19. 3   **21.** (a) 45   (b) 33   (c) 50   **23.** 120

25. 10   **27.** 6   **29.** 72

31. The maximum number of words is 12; 6 words are possible.

33. 218,400   **35.** (a) 525   (b) 1715

37. $(4! \cdot 5! \cdot 6!) \cdot 3! = 12,441,600$ **39.** 20,790

41. 240   **43.** 30   **45.** 24

47. (a) 4845   (b) 5700   (c) 7805

49. 924   **51.** $x^4 + 8x^3 + 24x^2 + 32x + 16$

53. 560

CHAPTER 7

Exercise 7.1 *(page 323)*

1. (a) $\{H, T\}$ (b) $\{0, 1, 2\}$ (c) $\{M, D\}$

3. $S = \{HH, HT, TH, TT\}$

5. $S = \{HHH, HHT, HTH, HTT, THH, THT, TTH, TTT\}$

7. $S = \{HH1, HH2, HH3, HH4, HH5, HH6, HT1, HT2, HT3, HT4, HT5, HT6, TH1, TH2, TH3, TH4, TH5, TH6, TT1, TT2, TT3, TT4, TT5, TT6\}$

9. $S = \{RA, RB, RC, GA, GB, GC\}$

11. $S = \{RR, RG, GR, GG\}$

13. $S = \{AA1, AA2, AA3, AA4, AB1, AB2, AB3, AB4, BA1, BA2, BA3, BA4, BB1, BB2, BB3, BB4, AC1, AC2, AC3, AC4, CA1, CA2, CA3, CA4, BC1, BC2, BC3, BC4, CB1, CB2, CB3, CB4, CC1, CC2, CC3, CC4\}$

15. $S = \{RA1, RA2, RA3, RA4, RB1, RB2, RB3, RB4, RC1, RC2, RC3, RC4, GA1, GA2, GA3, GA4, GB1, GB2, GB3, GB4, GC1, GC2, GC3, GC4\}$

17. $2^4 = 16$ **19.** $6^3 = 216$ **21.** $\dfrac{52 \cdot 51}{2} = 1326$

23. 1, 2, 3, and 6   **25.** 2   **27.** $P(H) = \frac{3}{4}$, $P(T) = \frac{1}{4}$

29. $P(1) = P(3) = P(5) = \frac{2}{9}$, $P(2) = P(4) = P(6) = \frac{1}{9}$

31–35. The sample space is $S = \{(x, y) | x = 1, \ldots, 6, y = 1, \ldots, 6\}$, for which the probability of each event is $\frac{1}{36}$.

31. $\frac{2}{36} = \frac{1}{18}$ **33.** $\frac{4}{36} = \frac{1}{9}$ **35.** $\frac{6}{36} = \frac{1}{6}$

37–41. The sample space is $S = \{(x, y) | x = 1, \ldots, 6, y = H \text{ or } T\}$, for which the probability of each event is $\frac{1}{12}$.

37. $\frac{1}{2}$ **39.** $\frac{5}{6}$ **41.** $\frac{1}{2}$

43.

Outcome	HH	HT	TH	TT
Probability	$\frac{1}{4}$	$\frac{1}{4}$	$\frac{1}{4}$	$\frac{1}{4}$

45.

Outcome	$1H$	$1T$	$2H$	$2T$	$3H$	$3T$	$4H$	$4T$	$5H$	$5T$	$6H$	$6T$
Probability	$\frac{1}{12}$	$\frac{1}{12}$	$\frac{1}{12}$	$\frac{1}{12}$	$\frac{1}{12}$	$\frac{1}{12}$	$\frac{1}{12}$	$\frac{1}{12}$	$\frac{1}{12}$	$\frac{1}{12}$	$\frac{1}{12}$	$\frac{1}{12}$

47.

Outcome	*HHHH*	*HHHT*	*HHTH*	*HHTT*	*HTHH*	*HTHT*	*HTTH*	*HTTT*
Probability	$\frac{1}{16}$	$\frac{1}{16}$	$\frac{1}{16}$	$\frac{1}{16}$	$\frac{1}{16}$	$\frac{1}{16}$	$\frac{1}{16}$	$\frac{1}{16}$

Outcome	*THHH*	*THHT*	*THTH*	*THTT*	*TTHH*	*TTHT*	*TTTH*	*TTTT*
Probability	$\frac{1}{16}$	$\frac{1}{16}$	$\frac{1}{16}$	$\frac{1}{16}$	$\frac{1}{16}$	$\frac{1}{16}$	$\frac{1}{16}$	$\frac{1}{16}$

49. $\{HTTT, TTTT\}$

51. $\{HHHT, HHTH, HHTT, HTHH, HTHT, HTTH, THHH, THHT, THTH, TTHH\}$

53.

Outcome	*RRR*	*RRL*	*RLR*	*LRR*	*RLL*	*LRL*	*LLR*	*LLL*
Probability	$\frac{1}{12}$	$\frac{1}{6}$	$\frac{1}{12}$	$\frac{1}{12}$	$\frac{1}{6}$	$\frac{1}{6}$	$\frac{1}{12}$	$\frac{1}{6}$

(a) $\frac{1}{3}$ (b) $\frac{1}{6}$ (c) $\frac{1}{2}$ (d) $\frac{1}{2}$

55. $P(C_1) = \frac{1}{4}$, $P(C_2) = \frac{1}{2}$, $P(C_3) = \frac{1}{4}$

Exercise 7.2 *(page 332)*

1. .8   **3.** .5   **5.** .35   **7.** $\frac{1}{52}$ **9.** $\frac{13}{52} = \frac{1}{4}$

11. $\frac{12}{52} = \frac{3}{13}$ **13.** $\frac{20}{52} = \frac{5}{13}$ **15.** $\frac{48}{52} = \frac{12}{13}$

17. Yes, the events "Sum is 2" and "Sum is 12" are mutually exclusive. $P(\text{"Sum is 2" or "Sum is 12"}) = \frac{2}{36} = \frac{1}{18}$

19. $\frac{3}{23}$ **21.** $\frac{7}{23}$ **23.** $\frac{8}{23}$ **25.** $\frac{11}{23}$ **27.** $\frac{6}{36} = \frac{1}{6}$

29. .30   **31.** .2   **33.** (a) .7 (b) .4 (c) .2 (d) .3

35. (a) .68 (b) .58 (c) .32

37. (a) .57 (b) .95 (c) .83 (d) .38 (e) .29 (f) .05 (g) .78 (h) .71

39. $\frac{65}{150} = \frac{13}{30} \approx .433$ **41.** $\frac{3}{4}$ **43.** $\frac{5}{12}$ **45.** $\frac{1}{2}$

47. The odds are 7 to 3 for *E*; 3 to 7 against *E*.

49. The odds are 4 to 1 for *F*; 1 to 4 against *F*.

51. 1 to 5; 1 to 17; 2 to 7   **53.** 23 to 27

55. $P(A \text{ or } B \text{ wins}) = \frac{1}{3} + \frac{2}{5} = \frac{11}{15}$; the odds that either *A* or *B* wins are 11 to 4.

57. $P(E \cup F) = P(E) + P(\overline{E} \cap F) = P(E) + P(F) - P(E \cap F)$

59.
$$P(E) = \frac{a}{b}(1 - P(E)) = \frac{a}{b} - \frac{a}{b}P(E)$$
$$P(E) + \frac{a}{b}P(E) = \frac{a}{b}$$
$$\left(1 + \frac{a}{b}\right)P(E) = \frac{a}{b}$$
$$\frac{b+a}{b}P(E) = \frac{a}{b}$$
$$P(E) = \frac{a}{b} \cdot \frac{b}{a+b}$$
$$P(E) = \frac{a}{a+b}$$

Technology Exercises *(page 335)*

1. $P(H) = .50$, $P(T) = .50$

3. $P(H) = .75$, $P(T) = .25$

5. $P(R) = .33$, $P(Y) = .14$, $P(W) = .53$

Exercise 7.3 *(page 341)*

1. .940   **3.** .664   **5.** (a) .3125 (b) .03125

7. (a) .00463 (b) .126

9. $P(\text{All 5 are defective}) = \frac{6 \cdot 5 \cdot 4 \cdot 3 \cdot 2}{50 \cdot 49 \cdot 48 \cdot 47 \cdot 46} \approx .00000283$; $P(\text{At least 2 are defective}) = \frac{218,246}{2,118,760} \approx .103$

11. $\frac{4}{37} \approx .108$ **13.** $1 - \frac{1320}{1728} = \frac{17}{72} \approx .236$

15. $1 - \frac{970,200}{1,000,000} = \frac{149}{5000} = .0298$

17. $1 - \frac{365 \cdot 364 \cdot 363 \cdot \cdots \cdot 266}{365} \approx .999999692751$

19. $\frac{1}{2}$ **21.** $\frac{1}{5}$ **23.** $\frac{11}{13} \approx .846$

25. $\frac{342,132,219}{63,501,355,960} \approx .00539$ **27.** $\frac{105}{512} \approx .205$

Exercise 7.4 *(page 349)*

1. .5   **3.** .75   **5.** .3   **7.** .5

9. $P(E|F) = .5$; $P(F|E) = .25$ **11.** .5   **13.** $\frac{4}{13}$

15. (a) $\frac{1}{2}$ (b) $\frac{2}{3}$ **17.** .69   **19.** .9   **21.** .2

23. .1   **25.** .2   **27.** $\frac{2}{3} \approx .67$ **29.** $\frac{1}{2}$

31. $P(HHHH) = \frac{1}{16}$; yes: $P(?H??) = \frac{1}{8}$ **33.** $\frac{25}{204}$

35. $\frac{1}{13}$ **37.** (a) $\frac{1}{26}$ (b) $\frac{1}{2}$ (c) $\frac{1}{13}$

39. $\frac{32}{75}$, or approx. 42.67%   **41.** .40   **43.** .24

45. .10   **47.** .08   **49.** $\frac{5}{12}$ **51.** $\frac{1}{3}$

53. (a) $\frac{5}{11}$ (b) $\frac{6}{23}$ (c) $\frac{5}{11}$ (d) $\frac{5}{12}$ (e) $\frac{14}{23}$ (f) $\frac{1}{4}$

55. (a) .7175 (b) .3300 (c) .0714 (d) .1939 (e) .2362
(f) .3574 (g) .3796 (h) .3137

57. $\frac{2}{3}$, or approx. 66.7%   **59.** .5   **61.** 71.5%

63. $P(E|E) = \dfrac{P(E \cap E)}{P(E)} = \dfrac{P(E)}{P(E)} = 1$ when $P(E) \neq 0$

65. $P(E|S) = \dfrac{P(E \cap S)}{P(S)} = \dfrac{P(E)}{1} = P(E)$

Exercise 7.5 *(page 357)*

1. .15   **3.** .125   **5.** No

7. (a) .3 (b) .5 (c) .15 (d) .65   **9.** $\frac{4}{147}$

11. $P(E|F) = .5$; no   **13.** No   **15.** Yes

17. (a) $\frac{1}{4}$ (b) $\frac{1}{4}$ (c) $\frac{1}{16}$ **19.** No

21. $P(A) = \frac{1}{4} + \frac{1}{4} = \frac{1}{2}$; $P(B) = \frac{1}{4} + \frac{1}{4} = \frac{1}{2}$;
$P(C) = \frac{1}{4} + \frac{1}{4} = \frac{1}{2}$
$P(A \cap B) = \frac{1}{4} = P(A) \cdot P(B)$, so A and B are
independent.
$P(A \cap C) = \frac{1}{4} = P(A) \cdot P(C)$, so A and C are
independent.
$P(B \cap C) = \frac{1}{4} = P(B) \cdot P(C)$, so B and C are
independent.

23. (a) $\frac{9}{16}$ (b) $\frac{1}{16}$ (c) $\frac{3}{8}$

25. (a) $\frac{27}{64}$ (b) $\frac{9}{64}$ **27.** (a) .4096 (b) .1536 (c) .9728

29. (a) $\frac{25}{81}$ (b) $\frac{40}{81}$

31. (a) $\frac{8}{65} \approx .1231$ (b) $\frac{1}{103} \approx .0097$ (c) No
(d) No (e) No (f) No

33. (a) $\frac{16}{25}$ (b) $\frac{1}{25}$ (c) $\frac{8}{25}$

35. $P(\text{Event (a)}) \approx .5177$; $P(\text{Event (b)}) \approx .491$; event (a) is
more likely to occur.

37. $P(E) \cdot P(F) = P(E \cap F) = 0$, so either $P(E) = 0$,
$P(F) = 0$, or $P(E) = P(F) = 0$.

39. $P(E) \cdot P(F) = P(E \cap F)$; $P(\overline{E}) \cdot P(\overline{F}) =$
$[1 - P(E)][1 - P(F)] = 1 - P(E) - P(F) +$
$P(E) \cdot P(F) = 1 - P(E \cup F) = P(\overline{E \cup F}) =$
$P(\overline{E} \cap \overline{F})$

41. See the hint.

Chapter Review: True–False Items *(page 361)*

1. T   **2.** F   **3.** F   **4.** F   **5.** T   **6.** F
7. T   **8.** T

Fill in the Blanks *(page 361)*

1. $\frac{1}{2}$ **2.** 32   **3.** 1; 0   **4.** .8   **5.** for
6. equally likely   **7.** mutually exclusive

Review Exercises *(page 361)*

1. $S = \{MM, MF, FM, FF\}$

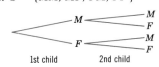

1st child   2nd child

3. (a) $\frac{10}{91}$ (b) $\frac{45}{91}$ (c) $\frac{55}{91}$

5. (a) .21 (b) .23 (c) .1825

7. (a) $\frac{1}{2}$ (b) $\frac{11}{24}$ (c) $\frac{13}{24}$ **9.** (a) No (b) 000 (c) $\frac{9}{64}$

11. $\frac{7}{13}$ **13.** (a) $\frac{1}{3} \approx .333$ (b) $\frac{9}{38} \approx .237$ (c) $\frac{14}{25} = .56$

15. (a) .025 (b) .0625 (c) .4

17. (a) $\frac{69}{200}$ (b) $\frac{21}{50}$ (c) $\frac{13}{50}$ (d) $\frac{2}{5}$ (e) $\frac{147}{400}$
(f) $\frac{138}{400} \cdot \frac{85}{400} = \frac{1173}{16,000} \neq \frac{4}{400}$

19. $\frac{1}{4}$

Mathematical Questions from Professional Exams *(page 363)*

1. b   **2.** e   **3.** b   **4.** d   **5.** c   **6.** d   **7.** b
8. a   **9.** c   **10.** b

CHAPTER 8

Exercise 8.1 *(page 374)*

1. .4   **3.** .2   **5.** .7   **7.** .31   **9.** $\frac{12}{31}$ **11.** $\frac{7}{31}$

13. $\frac{12}{31}$ **15.** .017   **17.** .018

19. $P(A_1|E) = \frac{3}{17} \approx .176$; $P(A_2|E) = \frac{14}{17} \approx .824$

21. $P(A_1|E) = \frac{5}{18} \approx .278$; $P(A_2|E) = \frac{9}{18} = .5$; $P(A_3|E) = \frac{4}{18} \approx .222$

23. $P(A_1|E) = \frac{8}{29} \approx .276$; $P(A_2|E) = \frac{20}{29} \approx .690$; $P(A_3|E) = \frac{1}{29} \approx .034$

25. $P(A_2|E) = \frac{0}{31} = 0$; $P(A_3|E) = \frac{2}{31} \approx .065$; $P(A_4|E) = \frac{0}{31} = 0$; $P(A_5|E) = \frac{2}{31} \approx .065$

27. $P(U_I|E) = \frac{5}{15} \approx .333$; $P(U_{II}|E) = \frac{3}{15} = .2$; $P(U_{III}|E) = \frac{7}{15} \approx .467$

29. A priori: $\frac{2}{3} \approx .667$; a posteriori: .8   **31.** .945

33. $P(\text{Democrat}|\text{Voted}) = .385$; $P(\text{Republican}|\text{Voted}) = .39$; $P(\text{Independent}|\text{Voted}) = .225$

35. $P(\text{Rock}|\text{Positive test}) \approx .385$; $P(\text{Clay}|\text{Positive test}) \approx .209$; $P(\text{Sand}|\text{Positive test}) \approx .405$

37. $P(\text{Republican}) = .466$; $P(\text{Northeasterner}|\text{Republican}) \approx .343$

39. .217   **41.** (a) .858   (b) .503   (c) .961

43. Since F is a subset of E, $E \cap F = F$ and $P(E \cap F) = P(F)$. So, $P(E|F) = \dfrac{P(E \cap F)}{P(F)} = \dfrac{P(F)}{P(F)} = 1$.

Exercise 8.2 *(page 384)*

1. .0250   **3.** .0811   **5.** $\frac{3003}{32,768} \approx .0916$ **7.** .2969

9. $\frac{2}{9} \approx .2222$ **11.** $\frac{125}{216} \approx .5787$ **13.** $\frac{80}{243} \approx .3292$

15. .0368   **17.** .2362   **19.** .0273   **21.** $\frac{1}{32} = .03125$

23. $\frac{93}{256} \approx .3633$ **25.** $\frac{28}{255} \approx .1098$ **27.** $\frac{625}{3888} \approx .1608$

29. (a) .2793   (b) .0515   (c) .3366   (d) .9942

31. $\frac{5}{16} = .3125$

33. (a)

(b) $(\frac{1}{4} \cdot \frac{1}{4} \cdot \frac{3}{4} \cdot \frac{3}{4}) + (\frac{1}{4} \cdot \frac{3}{4} \cdot \frac{1}{4} \cdot \frac{3}{4}) + (\frac{1}{4} \cdot \frac{3}{4} \cdot \frac{3}{4} \cdot \frac{1}{4}) +$
$(\frac{3}{4} \cdot \frac{1}{4} \cdot \frac{1}{4} \cdot \frac{3}{4}) + (\frac{3}{4} \cdot \frac{1}{4} \cdot \frac{3}{4} \cdot \frac{1}{4}) + (\frac{3}{4} \cdot \frac{3}{4} \cdot \frac{1}{4} \cdot \frac{1}{4}) =$
$\frac{54}{256} \approx .2109$

(c) $b(4, 2; \frac{1}{4}) = 6 \cdot (\frac{1}{4})^2 \cdot (\frac{3}{4})^2 = \frac{54}{256}$

35. .6242

37. $\sum\limits_{k=10}^{15} b(15, k; \frac{1}{2}) = \frac{309}{2048} \approx .1509$; $\sum\limits_{k=12}^{15} b(15, k; .8) \approx .6482$

39. .0007

41. .1225   **43.** (a) $\frac{7}{64} \approx .1094$ (b) .4661   **45.** .9647

47. $\dfrac{341}{12^5} \approx .00137$

Technology Exercises *(page 386)*

1.

k	Actual Value of $P(k)$
0	.0625
1	.25
2	.375
3	.25
4	.0625

3. .21875

Exercise 8.3 *(page 394)*

1. 1.2   **3.** 44,560

5. She should pay $0.80 for a fair game.

7. He should pay $1.67 for a fair game.   **9.** 35¢

11. (a) $0.75   (b) No   (c) Lose $2

13. It is not a fair bet; your expected loss is $\$\frac{3}{7} \approx \$0.43$.

15. No, she should not play the game; her expected loss is $\frac{15}{13}$¢ ≈ 1.2¢.

17. $7   **19.** The second site has the higher expected profit.

21. $\frac{2000}{6} = 333\frac{1}{3}$ **23.** 10   **25.** 1

27. $\frac{175}{64} \approx 2.734$ tosses   **29.** Aircraft A

Exercise 8.4 *(page 400)*

1. The expected number of customers is 9; the optimal number of cars is 9; the expected daily profit is $48.40.

3. For $p = .95$:

Group Size	2	3	4	5	6	7
Expected Tests Saved per Component	.4025	.524	.565	.574	.568	.555

The optimal group size is 5.

5. (a) Expected net gain in dollars = $75,000 - 75,000(.05^x) - 500x$, where x = Number of divers hired.

   (b) Hiring 2 divers will maximize the net gain.

Exercise 8.5 *(page 403)*

1. $P(X = 0) = \frac{1}{4}$; $P(X = 1) = \frac{1}{2}$; $P(X = 2) = \frac{1}{4}$

3. $P(X = 0) = \frac{1}{8}$; $P(X = 1) = \frac{3}{8}$;
   $P(X = 2) = \frac{3}{8}$; $P(X = 3) = \frac{1}{8}$

5. $P(X = 0) = \frac{27}{125} = .216$; $P(X = 1) = \frac{54}{125} = .432$;
   $P(X = 2) = \frac{36}{125} = .288$; $P(X = 3) = \frac{8}{125} = .064$

7. 1.2

Technology Exercises *(page 403)*

1. $P(X = k) = .167$, for $k = 1, 2, 3, 4, 5, 6$

3. .2   **5.** .083

Chapter Review: True–False Items *(page 405)*

1. T   **2.** F   **3.** F   **4.** T   **5.** F   **6.** T

Fill in the Blanks *(page 405)*

1. Bayes' formula

2. (c) independent   (d) unchanged (or the same)

3. expected value   **4.** a number   **7.** expected value

Review Exercises *(page 405)*

1. .5   **3.** .4   **5.** .3   **7.** .43   **9.** $\frac{20}{43} \approx .4651$

11. $\frac{20}{43} \approx .4651$ **13.** $\frac{3}{43} \approx .0698$

15. (a) $\frac{180}{260} \approx .6923$ (b) $\frac{180}{345} \approx .5217$ (c) $\frac{110}{345} \approx .3188$
   (d) $\frac{55}{345} \approx .1594$ (e) $\frac{60}{260} \approx .2308$ (f) $\frac{60}{210} \approx .2857$
   (g) $\frac{85}{210} \approx .4048$ (h) $\frac{65}{210} \approx .3095$

17. .163

19. 1; individual probabilities are:

k	0	1	2	3	4	5
Probability of k purchasers	.3277	.4096	.2048	.0512	.0064	.0003

21. (a) $\frac{1}{4096}$ (b) $\frac{793}{2048} \approx .3872$ (c) 793 to 1255

23. $\frac{31}{19,683} \approx .001575$ **25.** He paid $3\frac{1}{3}$¢ too much.

27. The expected value is $0.2833 - $0.30 = -$0.0167$. The game is not fair.

29. 50   **31.** (a) $1 - q^{30}$ (b) $31 - 30q^{30}$ tests

Mathematical Questions from Professional Exams *(page 407)*

1. d   **2.** a   **3.** a   **4.** d   **5.** b   **6.** b

7. b *or* c (they're equal!)   **8.** d

CHAPTER 9

Exercise 9.1 *(page 417)*

1. A poll should be taken either door-to-door or by means of the telephone.

3. A poll should be taken door-to-door in which people are asked to fill out a questionnaire.

5. The data should be gathered from all different kinds of banks.

7. (1) Asking a group of children if they like candy to determine what percentage of people like candy.
(2) Asking a group of people over 65 their opinion toward Medicare to determine the opinion of people in general about Medicare.

9. By taking a poll downtown, you would question mostly people who are either shopping or working downtown. For instance, you would question few students.

11. (a) 250   (b) 249   (c) 274.5   (d) 49   (e) 33
(f) The fifth class: 250–299   (g) 752
(h) Histogram:

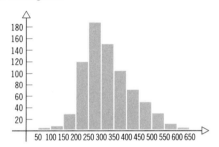

(i) Frequency polygon:

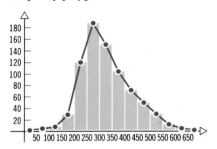

13. (a) Histogram:

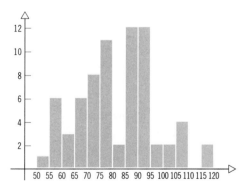

(b) Frequency polygon:

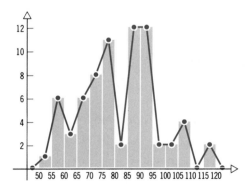

(c) Cumulative frequencies:

x	50	55	60	65	70	75	80	85	90	95	100	105	110	115	120
Number < x	0	1	7	10	16	24	35	37	49	61	63	65	69	69	71

(d) Cumulative frequency distribution:

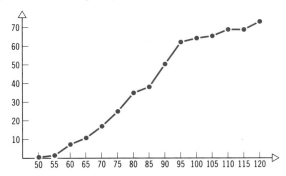

15. (a) Frequency table:

Physicians	Freq.	Physicians	Freq.
78	1	122	1
91	1	123	2
95	1	126	1
100	1	127	3
105	1	128	2
108	1	129	3
110	1	130	1
111	1	131	4
112	1	132	2
113	1	134	1
115	1	136	2
116	3	137	2
119	2	138	1

Physicians	Freq.	Physicians	Freq.
140	1	157	1
141	1	158	1
142	1	161	2
144	2	162	1
145	2	165	2
146	2	166	2
148	1	169	3
149	1	171	2
152	1	172	1
153	3	175	1
154	2	176	2
155	1	178	1
156	2	184	1

Physicians	Freq.	Physicians	Freq.
185	3	224	2
188	1	230	1
190	3	232	1
192	1	240	1
194	1	245	2
198	2	256	1
202	1	289	1
204	1	296	1
207	1		
211	1		
212	1		
218	2		
222	1		

Range: $296 - 78 = 218$

(b) Line chart:

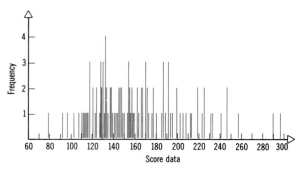

(c) Histogram:

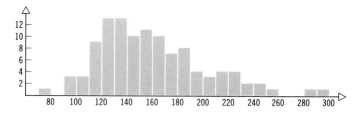

(d) Frequency polygon:

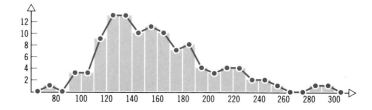

(e) Cumulative (less than) frequencies:

Class Interval	Freq.	< Cum. Freq.	Class Interval	Freq.	< Cum. Freq.
70.5–80	1	1	150.5–160	11	63
80.5–90	0	1	160.5–170	10	73
90.5–100	3	4	170.5–180	7	80
100.5–110	3	7	180.5–190	8	88
110.5–120	9	16	190.5–200	4	92
120.5–130	13	29	200.5–210	3	95
130.5–140	13	42	210.5–220	4	99
140.5–150	10	52	220.5–230	4	103

Class Interval	Freq.	< Cum. Freq.
230.5–240	2	105
240.5–250	2	107
250.5–260	1	108
260.5–270	0	108
270.5–280	0	108
280.5–290	1	109
290.5–300	1	110

(f) Cumulative (less than) frequency distribution:

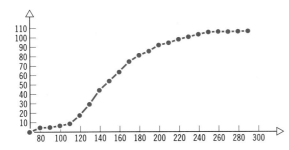

	Class Interval	Tally	Freq.	< Cum. Freq.
17.	400–499	I	1	1
	500–599	I	1	2
	600–699	II	2	4
	700–799	⊮ ⊮ I	11	15
	800–899	⊮ ⊮ III	13	28
	900–999	⊮ II	7	35
	1000–1099	IIII	4	39
	1100–1199	IIII	4	43
	1200–1299	II	2	45
	1300–1399	II	2	47
	1400–1499	I	1	48
	1500–3500	II	2	50

Technology Exercises (*page 419*)

1.

	Class Interval	Tally	Frequency, f
1	13–13.9	I	1
2	14–14.9	⫴⫴ III	8
3	15–15.9	⫴⫴	5
4	16–16.9	II	2
5	17–17.9	I	1
6	18–18.9		0
7	19–19.9	I	1
8	20–20.9	II	2

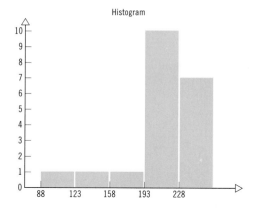

3.

	Class Interval	Tally	Frequency, f
1	88–122	I	1
2	123–157	I	1
3	158–192	I	1
4	193–227	⫴⫴ ⫴⫴	10
5	228–262	⫴⫴ II	7

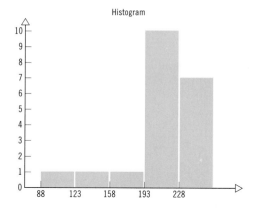

5.

	Class Interval	Tally	Frequency, f
1	9.36–10.35	III	3
2	10.36–11.35	II	2
3	11.36–12.35	⫴⫴ III	8
4	12.36–13.35	⫴⫴	5
5	13.36–14.35	I	1
6	14.36–15.35	I	1

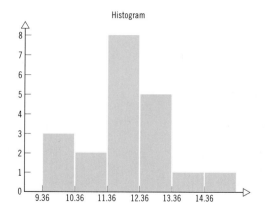

Exercise 9.2 (*page 423*)

1.

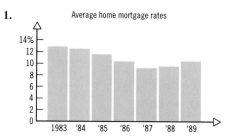

Average home mortgage rates

Sources: Office of Thrift Supervision, Resolution Trust, Corp.

3.

Median income of young families
Headed by a person younger than 30, adjusted
for inflation in thousands of 1986 dollars

Sources: Children's Defense Fund, U.S. Census Bureau

5.

Sources of U.S. personal income
in percent of $4.06 trillion for 1988

Rental income .5
Small farm 1
Employee contributions to pensions
Self-employed
Social Security, pensions
Wages and salaries 60%
5
7
9.5
17
Interest and dividends

Source: U.S. Department of Commerce

7.

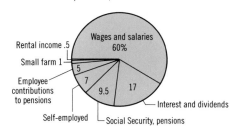

Not like 42%
Like very much 22%
Like 36%

Exercise 9.3 (*page 429*)

1. Mean = 31.25; median = 30.5; no mode

3. Mean = 70.4; median = 70; mode = 55

5. Mean = 76.2; median = 75; no mode

7. Mean = 3.52 million units; median = 3.45 million units

9. Mean = 7.8%; median = 8.05%   **11.** $109.40

13. Mean = $41,300; median = $36,000. The median describes the situation more realistically since it is closer to the salary of most of the faculty in this sample.

15. Mean = $1826.08; median = $1805.18

17. 1970 mean = $50,725; 1977 mean = $57,150; 1983 mean = $69,300

19. For Table 5, $C_{75} \approx 91.77$, $C_{40} \approx 77$. For Table 6, $C_{75} \approx 93.04$, $C_{40} \approx 76.53$.

21. With assumed mean = 325, $n = 753$,
$$\sum_{i=1}^{12} f_i(m_i - 325) = -650, \quad \bar{X} = 325 + \tfrac{(-650)}{753} \approx 324.14$$
votes.

Exercise 9.4 (*page 436*)

1. (b) has the larger variance.   **3.** $\sigma \approx 6.534$

5. $\sigma \approx 5.196$ **7.** $\sigma \approx 12.474$ **9.** $\sigma \approx 91.77$ votes

11. $\bar{X} = 31.9$; $\sigma = 7.80$

13. Mean lifetime ≈ 868.67 hr; $\sigma \approx 66.68$ hr

15. Since the mean number of salmon caught in river I is 2000 and the mean number of salmon caught in river II is 2038, river II is preferred.

17. (a) 75%   (b) 64%   (c) $88\frac{8}{9}$%   (d) 25%   (e) $11\frac{1}{9}$%

19. ($24.37, $78.13)

Exercise 9.5 (*page 445*)

1. $\mu = 8$; $\sigma = 1$ **3.** $\mu = 18$; $\sigma = 1$

5.

x	7	9	13	15	29	37	41
Z-score	−0.6559	−0.4409	−0.0108	0.2043	1.7097	2.5699	3.0

7. (a) .3133 (b) .3642 (c) .4938 (d) .4987
(e) .2734 (f) .4896 (g) .2881 (h) .4988

9. .3085   **11.** .0668

13. (a) Approx. 1365 women (b) Approx. 1909 women
(c) Approx. 1995 women

15. (a) Approx. 1 student (b) Between 124.6 and 135.4 lb

17. 239 pairs   **19.** Kathleen

21.

k	0	1	2	3	4	5	6	7
b(15, k; .30)	.0047	.0305	.0916	.1700	.2186	.2061	.1472	.0811
k	8	9	10	11	12	13	14	15
b(15, k; .30)	.0348	.0116	.0030	.0006	.0001	.0000	.0000	.0000

Line chart:

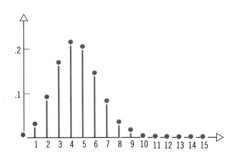

Frequency curve:

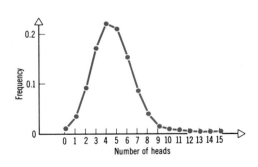

23. .7498   **25.** .5   **27.** .0274

Technology Exercises **(*page 447*)**

1. Maximum for $x = 0$.

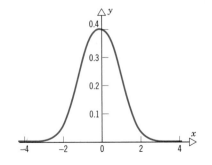

Chapter Review: True – False Items **(*page 448*)**

1. F   **2.** T   **3.** F   **4.** T   **5.** F

Fill in the Blanks **(*page 448*)**

1. mean, median, mode   **2.** standard deviation

3. bell   **4.** Z-score   **5.** $\bar{x} - k$, $\bar{x} + k$

Review Exercises (page 448)

1. (a) Frequency table:

Score	Freq.	Score	Freq.
8	1	33	1
10	1	42	1
12	1	44	1
14	2	48	1
17	1	52	2
19	1	55	1
20	1	60	1
21	1	63	2
26	1	66	2
30	1	69	1

Score	Freq.	Score	Freq.
70	1	85	2
72	2	87	2
73	2	89	1
74	1	90	1
75	1	92	1
77	1	95	1
78	2	99	1
80	3	100	2
82	1		
83	1		

Range $= 100 - 8 = 92$

(b) Line chart:

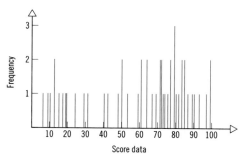

(c) Histogram:

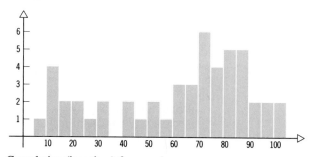

(d) Frequency polygon:

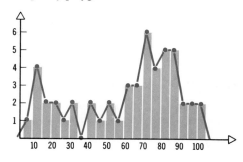

(e) Cumulative (less than) frequencies:

Class Interval	< Cum. Freq.	Class Interval	< Cum. Freq.
4.5–9	1	29.5–34	12
9.5–14	5	34.5–39	12
14.5–19	7	39.5–44	14
19.5–24	9	44.5–49	15
24.5–29	10	49.5–54	17

Class Interval	< Cum. Freq.	Class Interval	< Cum. Freq.
54.5–59	18	79.5–84	39
59.5–64	21	84.5–89	44
64.5–69	24	89.5–94	46
69.5–74	30	94.5–99	48
74.5–79	34	99.5–104	50

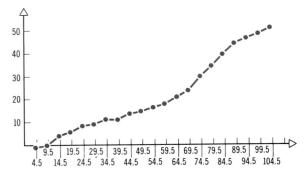

3. (a) Mean $= \frac{67}{12} \approx 5.58$; Median $= 4.5$; Mode $= 4$
(b) Mean $= \frac{209}{8} = 26.125$; Median $= 2$; Mode $= 2$
(c) Mean $= \frac{62}{9} \approx 6.89$; Median $= 7$; Mode $= 7$

5. $A = \{-1, 1\}$; $B = \{-10, 10\}$. Both sets have mean 0, but the standard deviation for A is 1, while the standard deviation for B is 10.

7. ~ 2.7701

9. (a) Approx. 410   (b) Approx. 94   (c) Approx. 298

11. (a) .0855   (b) .075

13. At least .75 (using Chebychev's theorem)

Mathematical Questions from Professional Exams (page 449)

1. e   **2.** b   **3.** c

CHAPTER 10

Exercise 10.1 (page 458)

1. The sum of the entries in row 3 is not equal to 1; there is a negative entry in row 3, column 2.

3. (a) The probability of a change from state 1 to state 2 is $\frac{2}{3}$.
(b) $[\frac{1}{3} \quad \frac{2}{3}]$; $[\frac{5}{18} \quad \frac{13}{18}]$
(c) $[\frac{1}{4} \quad \frac{3}{4}]$; $[\frac{13}{48} \quad \frac{35}{48}]$

5. $[.3625 \quad .6375]$ **7.** $a = .4, b = .2, c = 1$

9. $[.50397 \quad .49603]$

11. (a) Each experiment measures the proportion of mayors in these cities who are Democrats and the proportion who are Republicans during a given term of office.

These proportions depend only on the proportions during the preceding mayoral terms, so the sequence of experiments can be represented as a Markov chain.

(b) $P = \begin{array}{c} D \\ R \end{array} \begin{bmatrix} .6 & .4 \\ .3 & .7 \end{bmatrix}$ with column headings $D \quad R$

(c) $P^2 = \begin{bmatrix} .48 & .52 \\ .39 & .61 \end{bmatrix}$; $P^3 = \begin{bmatrix} .444 & .556 \\ .417 & .583 \end{bmatrix}$

13. 57%

15. (a) 45.1% by Travelers and 32.6% by General American
(b) 45.026% by Travelers and 34.732% by General American

17. $uA = [u_1a_{11} + u_2a_{21} \quad u_1a_{12} + u_2a_{22}]$. Note that since the entries of u and A are nonnegative, the entries of uA (being sums of products of nonnegative numbers) are nonnegative. Also, $(u_1a_{11} + u_2a_{21}) + (u_1a_{12} + u_2a_{22}) = (u_1a_{11} + u_1a_{12}) + (u_2a_{21} + u_2a_{22}) = u_1(a_{11} + a_{12}) + u_2(a_{21} + a_{22}) = u_1(1) + u_2(1) = u_1 + u_2 = 1$. Since uA is a row vector whose entries are all nonnegative and sum to 1, it is a probability vector.

*Technology Exercises (**page 460**)*

1. $v^{(10)} = [.32 \quad .29 \quad .16 \quad .23]$

3. (a) $P = \begin{bmatrix} 0 & \frac{1}{2} & 0 & \frac{1}{2} & 0 & 0 \\ \frac{1}{3} & 0 & \frac{1}{3} & 0 & \frac{1}{3} & 0 \\ 0 & \frac{1}{2} & 0 & 0 & 0 & \frac{1}{2} \\ \frac{1}{2} & 0 & 0 & 0 & \frac{1}{2} & 0 \\ 0 & 0 & 0 & 0 & 0 & 1 \\ 0 & 0 & 0 & 0 & 0 & 1 \end{bmatrix}$

(b) $v^{(0)} = [0 \quad 1 \quad 0 \quad 0 \quad 0 \quad 0]$

(c) $v^{(10)} = [0 \quad .0187 \quad 0 \quad .0125 \quad 0 \quad .9688]$

(d) Room 6, with 96.88% probability

*Exercise 10.2 (**page 468**)*

1. Regular; fixed probability vector $= [\frac{2}{3} \quad \frac{1}{3}]$

3. Regular; fixed probability vector $= [\frac{1}{5} \quad \frac{4}{5}]$

5. Not regular

7. $\begin{bmatrix} \frac{1}{2} & \frac{1}{2} \\ 2 & 2 \end{bmatrix} \begin{bmatrix} 1-p & p \\ p & 1-p \end{bmatrix}$

$= \begin{bmatrix} \dfrac{1-p}{2} + \dfrac{p}{2} & \dfrac{p}{2} + \dfrac{1-p}{2} \end{bmatrix} = \begin{bmatrix} \dfrac{1}{2} & \dfrac{1}{2} \end{bmatrix}$

9. $P = \begin{bmatrix} .7 & .15 & .15 \\ .1 & .8 & .1 \\ .2 & .2 & .6 \end{bmatrix}$. In the long run, $\frac{4}{13}$ ($\approx 30.8\%$) of the detergent stock is brand A, $\frac{6}{13}$ ($\approx 46.2\%$) of the stock is brand B, and $\frac{3}{13}$ ($\approx 23.1\%$) of the stock is brand C.

11. The probability that the grandson of a Labourite will vote Socialist is .09 (or 9%). The membership distribution in the long run is $\frac{26}{47}$ ($\approx 55.3\%$) Conservative, $\frac{18}{47}$ ($\approx 38.3\%$) Labourite, and $\frac{3}{47}$ ($\approx 6.4\%$) Socialist.

13. The probability that a blond is the grandmother of a brunette is .3 (or 30%).

(a) 32.7% blonds, 42.5% brunettes, 24.8% redheads

(b) 35% blonds, 40% brunettes, 25% redheads

*Technology Exercises (**page 469**)*

1. $t = [.125 \quad .875]$

3. $t = [.444444 \quad .222222 \quad .333333] = [\frac{4}{9} \quad \frac{2}{9} \quad \frac{1}{3}]$

5. $t = [.267647058824 \quad .208823529412 \quad .205882352941 \quad .317647058824]$

*Exercise 10.3 (**page 477**)*

1. Not absorbing   **3.** Absorbing

5. Not absorbing

7. $T = [\frac{8}{3}]$; $S = [\frac{1}{8} \quad \frac{2}{8}]$; $T \cdot S = [\frac{1}{3} \quad \frac{2}{3}]$

9. (a) .8; .6   (b) 4.2

11. Starting with \$1: $\frac{4}{19} \approx .2105$; starting with \$2: $\frac{10}{19} \approx .5263$

13. (a) 1.4 wagers   (b) .84   (c) .16

15. (a) 1.6 wagers   (b) .64   (c) .36   **17.** 6.993 days

*Technology Exercises (**page 478**)*

1. $I_r = \begin{bmatrix} 1 & 0 \\ 0 & 1 \end{bmatrix}$, $S = \begin{bmatrix} 0 & 0 \\ 0 & .2 \\ .75 & .25 \end{bmatrix}$,

$Q = \begin{bmatrix} .05 & .95 & 0 \\ 0 & .5 & .3 \\ 0 & 0 & 0 \end{bmatrix}$, $T = \begin{bmatrix} 1.05 & 2 & .6 \\ 0 & 2 & .6 \\ 0 & 0 & 1 \end{bmatrix}$,

$T \cdot S = \begin{bmatrix} .45 & .55 \\ .45 & .55 \\ .75 & .25 \end{bmatrix}$

*Exercise 10.4 (**page 482**)*

1. Katy's payoff matrix: $\begin{matrix} & 1 & 2 \\ & \begin{bmatrix} -1 & 1 \\ 1 & -1 \end{bmatrix} & \begin{matrix} 1 \text{ finger} \\ 2 \text{ fingers} \end{matrix} \end{matrix}$

3. Katy's payoff matrix: $\begin{matrix} & 1 & 4 & 7 \\ & \begin{bmatrix} -2 & 5 & -8 \\ 5 & -8 & 11 \\ -8 & 11 & -14 \end{bmatrix} & \begin{matrix} 1 \\ 4 \\ 7 \end{matrix} \end{matrix}$

5. Strictly determined; value $= -1$

7. Strictly determined; value $= 2$

9. Not strictly determined

11. Strictly determined; value $= 2$

13. Not strictly determined

15. $0 \le a \le 3$ **17.** $ab \le 0$

Exercise 10.5 (*page 485*)

1. 1.42   **3.** $\frac{9}{4} = 2.25$ **5.** $\frac{19}{8} = 2.375$

7. $\frac{17}{9} \approx 1.889$ **9.** $\frac{1}{3}$

11. If the game is not strictly determined, then none of the entries $a_{11}, a_{12}, a_{21}, a_{22}$ can be a saddle point. Comparing a_{11} and a_{12}, either $a_{11} < a_{12}$, $a_{11} = a_{12}$, or $a_{11} > a_{12}$, If $a_{11} < a_{12}$, then $a_{11} < a_{21}$, for otherwise a_{11} would be a saddle point. Then $a_{21} > a_{22}$, else a_{21} would be a saddle point. Then $a_{12} > a_{22}$, so that a_{22} won't be a saddle point. Thus, if $a_{11} < a_{12}$ and the game is not strictly determined, then the inequalities listed in (b) follow. Note that $a_{11} \ne a_{12}$, or there would be a saddle point in the matrix. Finally, if $a_{11} > a_{12}$, then $a_{12} < a_{22}$, as a_{12} isn't a saddle point. Then $a_{21} < a_{22}$ and $a_{11} > a_{21}$, to prevent a_{22} and a_{21} from being saddle points. Thus, if $a_{11} > a_{12}$ and the game is not strictly determined, then the inequalities listed in (a) follow.

Exercise 10.6 (*page 492*)

1. Player I: optimal strategy is to choose row 1 with probability $\frac{3}{4}$; $p_1 = \frac{3}{4}$, $p_2 = \frac{1}{4}$. Player II: optimal strategy is to choose column 1 with probability $\frac{1}{4}$; $q_1 = \frac{1}{4}$, $q_2 = \frac{3}{4}$. Game value $V = \frac{7}{4}$.

3. Player I: optimal strategy is to choose row 1 with probability $\frac{1}{6}$; $p_1 = \frac{1}{6}$, $p_2 = \frac{5}{6}$. Player II: optimal strategy is to choose row 1 with probability $\frac{1}{3}$; $q_1 = \frac{1}{3}$, $q_2 = \frac{2}{3}$. Game value $V = \frac{1}{3}$.

5. Player I: optimal strategy is to choose row 1 with probability $\frac{5}{8}$; $p_1 = \frac{5}{8}$, $p_2 = \frac{3}{8}$. Player II: optimal strategy is to choose row 1 with probability $\frac{5}{8}$; $q_1 = \frac{5}{8}$, $q_2 = \frac{3}{8}$. Game value $V = \frac{7}{8}$.

7. The Democrat should spend $\frac{3}{8}$ ($= 37.5\%$) of her/his time on domestic issues and $\frac{5}{8}$ ($= 62.5\%$) on foreign issues. The Republican should divide her/his time evenly between the two issues. The value of the game is $\frac{3}{2}$, favoring the Democrat.

9. The spy should choose the deserted exit $\frac{6}{71}$ ($\approx 8.5\%$) of the time and the heavily used exit $\frac{65}{71}$ ($\approx 91.5\%$) of the time. The spy's opponent should choose the deserted exit $\frac{16}{71}$

($\approx 22.5\%$) of the time and the heavily used exit $\frac{55}{71}$ ($\approx 77.5\%$) of the time. The game's value is $\frac{50}{71}$ ($\approx .7042$), favoring the spy.

11. The game must be strictly determined.

Chapter Review: True – False Items (*page 493*)

1. F   **2.** F   **3.** F   **4.** T   **5.** T   **6.** T   **7.** F

Fill in the Blanks (*page 493*)

1. m **2.** nonnegative, one   **3.** $v^{(k)} = v(0)P^k$

4. positive   **5.** payoff   **6.** value

Review Exercises (*page 493*)

1. (a) $[\frac{2}{5} \quad \frac{3}{5}]$ (b) $[\frac{1}{2} \quad \frac{1}{2}]$ (c) $[\frac{48}{79} \quad \frac{10}{79} \quad \frac{21}{79}]$

3. After 2 years: A holds $\frac{283}{600} \approx 47.17\%$, B holds $\frac{323}{1200} \approx 26.92\%$, and C holds $\frac{311}{1200} \approx 25.92\%$ of the beer market. In the long run: A will hold $\frac{80}{169} \approx 47.34\%$, B will hold $\frac{45}{169} \approx 26.63\%$, and C will hold $\frac{44}{169} \approx 26.04\%$ of the beer market.

5. She will sell at U_1 $\frac{3}{7}$ ($\approx 42.86\%$) of the time, at U_2 $\frac{16}{35}$ ($\approx 45.71\%$) of the time, and at U_3 $\frac{4}{35}$ ($\approx 11.43\%$) of the time.

7. (a)

(Nonabsorbing) State, $x	Expected No. of Times Process Is in State $x
$1	1.298405
$2	2.360736
$3	1.411736
$4	0.635281

(b) The expected number of bets before absorption is 5.706158.

(c) The probability that he loses all his money is .714123. (The probability that he wins $5 is $1 - .714123 = .285877$.)

9. (a) $\frac{1}{3}$ (b) 0   (c) 0

11. (a) $[\frac{5}{6} \quad \frac{1}{6}]$ (b) 7.5   (c) 7.5% increase

13. $[\frac{4}{9} \quad \frac{5}{9}]$; $\frac{1}{9}$

CHAPTER 11

Exercise 11.1 (page 505)

1.

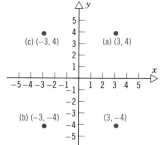

3.

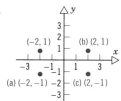

5.

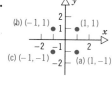

7.

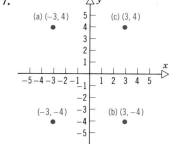

9.

x	0	3	2	-2	4	-4
y	-3	0	-1	-5	1	-7

11.

x	0	3	2	-2	4	-4
y	-6	0	-2	-10	2	-14

13. (a) Intercepts: $(-1, 0)$, $(1, 0)$
　　(b) Symmetric with respect to x-axis, y-axis, and origin

15. (a) Intercepts: $(-\pi/2, 0)$, $(\pi/2, 0)$, $(0, 1)$
　　(b) Symmetric with respect to y-axis

17. (a) Intercept: $(0, 0)$
　　(b) Symmetric with respect to x-axis

19. (a) Intercept: $(1, 0)$
　　(b) *Not* symmetric with respect to x-axis, y-axis, *nor* origin ("none of these")

21. (a) Intercepts: $(-3, 0)$, $(3, 0)$, $(0, 2)$
　　(b) Symmetric with respect to y-axis

23. (a) Intercept: $(0, 0)$
　　(b) *Not* symmetric with respect to x-axis, y-axis, *nor* origin ("none of these")

25. (a) Intercept: $(0, 0)$
　　(b) Symmetric with respect to y-axis

27. (a) Intercept: $(0, 0)$
　　(b) Symmetric with respect to origin

29. (a) Intercepts: $(-3, 0)$, $(3, 0)$, $(0, 9)$
　　(b) Symmetric with respect to y-axis

31. (a) Intercepts: $(-3, 0)$, $(3, 0)$, $(0, -2)$, $(0, 2)$
　　(b) Symmetric with respect to x-axis, y-axis, and origin

33. (a) Intercepts: $(3, 0)$, $(0, -27)$
　　(b) *Not* symmetric with respect to x-axis, y-axis, *nor* origin (but it *is* symmetric with respect to its y-intercept)

35. (a) Intercepts: $(-1, 0)$, $(4, 0)$, $(0, -4)$
　　(b) *Not* symmetric with respect to x-axis, y-axis, *nor* origin (but it *is* symmetric with respect to the line $x = 1.5$)

37. (a) Intercept: $(0, 0)$
　　(b) Symmetric with respect to origin

39. Intercepts: $(-\frac{2}{3}, 0)$ and $(0, 2)$; *not* symmetric with respect to x-axis, y-axis, *nor* origin

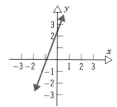

41. Intercepts: $(-2, 0)$ and $(0, 3)$; *not* symmetric with respect to x-axis, y-axis, *nor* origin

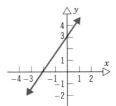

43. Intercept: $(0, 0)$; symmetric with respect to y-axis

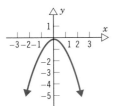

45. Intercept $(0, 3)$; symmetric with respect to y-axis

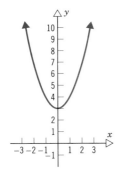

47. Intercepts: $(1, 0)$ and $(0, -1)$; *not* symmetric with respect to x-axis, y-axis, *nor* origin

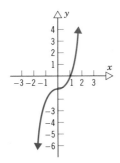

49. Intercepts: $(-1, 0)$, $(1, 0)$, and $(0, -1)$; symmetric with respect to y-axis

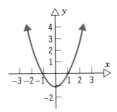

51. Intercept: $(0, 0)$; *not* symmetric with respect to x-axis, y-axis, *nor* origin

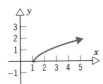

53. Intercept: $(1, 0)$; *not* symmetric with respect to x-axis, y-axis, *nor* origin

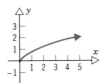

55. Intercept: $(0, -\frac{1}{2})$; *not* symmetric with respect to x-axis, y-axis, *nor* origin

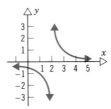

57.

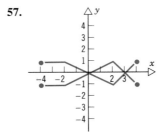

59.

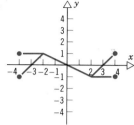

61. $[d(P_1, P_2)]^2 = (x_2 - x_1)^2 + (y_2 - y_1)^2$; thus, $d(P_1, P_2) = \sqrt{(x_2 - x_1)^2 + (y_2 - y_1)^2}$

63. $\sqrt{85} \approx 9.21954$ **65.** 0.8

67. $(-11, -3)$; $(13, -3)$

Technology Exercises *(page 508)*

1. (c)   **3.** (a)   **5.** (e)

7. Symmetric with respect to the y-axis

9. No symmetry

11. Symmetric with respect to the y-axis

Exercise 11.2 *(page 519)*

1. (a) -4 (b) -5 (c) -9 (d) -12

3. (a) 0   (b) $\frac{1}{2}$ (c) $-\frac{1}{2}$ (d) $\frac{2}{3}$

5. (a) 4   (b) 5   (c) 5   (d) 6

7. (a) $-\frac{1}{3}$ (b) $-1\frac{1}{2}$ (c) $\frac{1}{8}$ (d) 5

9. $f(0) = 3$; $f(2) = 4$ **11.** Positive

13. $x = -3, 6, 10$ **15.** $[-6, 11]$ or $\{x|-6 \leq x \leq 11\}$

17. $(-3, 0)$, $(6, 0)$, and $(10, 0)$ **19.** Three times

21. (a) No   (b) -3 (c) 14
(d) $(-\infty, 6) \cup (6, \infty)$ or $\{x|x \neq 6\}$

23. (a) Yes   (b) $\frac{8}{17} \approx 0.470588$ (c) $x = -1$ or 1
(d) All real numbers

25. (a) Domain $= (-\infty, -1] \cup [1, \infty)$ or $\{x|x \leq -1$ or $1 \leq x\}$; range $= (-\infty, \infty) =$ all real numbers
(b) x-intercepts: $(-1, 0)$ and $(1, 0)$; no y-intercept
(c) Symmetric with respect to x-axis, y-axis, and with respect to origin

27. (a) Domain $= [-\pi, \pi] = \{x|-\pi \leq x \leq \pi\}$;
range $= [-1, 1] = \{y|-1 \leq y \leq 1\}$

(b) x-intercepts: $(-\pi/2, 0)$ and $(\pi/2, 0)$; y-intercept: $(0, 1)$
(c) Symmetric with respect to y-axis

29. (a) Domain $= (-\infty, 0]$ or $\{x|x \leq 0\}$; range $= (-\infty, \infty) =$ all real numbers
(b) x-intercept and y-intercept: $(0, 0)$
(c) Symmetric with respect to x-axis

31. (a) Domain $= (0, \infty) = \{x|0 < x\}$; range $=$ all real numbers
(b) x-intercept: $(1, 0)$; no y-intercept
(c) *No* standard symmetries

33. (a) Domain $=$ all real numbers; range $= (-\infty, 2] = \{y|y \leq 2\}$
(b) x-intercepts: $(-3, 0)$ and $(3, 0)$; y-intercept: $(0, 2)$
(c) Symmetric with respect to the y-axis

35. (a) Domain $= [-4, 4) = \{x|-4 \leq x < 4\}$; range $= \{-2, 0, 2, 3\}$
(b) x-intercepts: $\{(x, 0)|0 \leq x < 2\}$; y-intercept: $(0, 0)$
(c) *No* standard symmetries

37. (a) Domain $= [-4, 4) = \{x|-4 \leq x < 4\}$; range $= [-3, -2] \cup [-1, 0] \cup [1, 2] = \{y|-3 \leq y \leq -2$ or $-1 \leq y \leq 0$ or $1 \leq y \leq 2\}$
(b) x-intercept: $(2, 0)$; y-intercept: $(0, 1)$
(c) *No* standard symmetries

39. All real numbers   **41.** All real numbers

43. $\{x|x \neq -1$ and $x \neq 1\} = \{x|(x < -1)$ or $(-1 < x < 1)$ or $(1 < x)\} = (-\infty, -1) \cup (-1, 1) \cup (1, \infty)$

45. $\{x|x \neq 0\} = \{x|x < 0$ or $0 < x\} = (-\infty, 0) \cup (0, \infty)$

47. $\{x|4 \leq x\} = [4, \infty)$

49. $\{x|x \leq -3$ or $3 \leq x\} = (-\infty, -3] \cup [3, \infty)$

51. $\{x|x < 1$ or $2 \leq x\} = (-\infty, 1) \cup [2, \infty)$

53. -4 **55.** -4 **57.** 8; $f(x)$ not defined for $x = 3$

59. $A(y) = (60 - y)y$, where $y =$ width; the domain is $\{y|0 \leq y \leq 60\} = [0, 60]$.

61. $R(x) = x(-\frac{1}{5}x + 100) = -\frac{1}{5}x^2 + 100x, 0 \leq x \leq 500$

63. $R(x) = x\left(\dfrac{x - 100}{-20}\right) = -\dfrac{1}{20}x^2 + 5x, 0 \leq x \leq 100$

65. $G(x) = 6x, 0 \leq x$

67. $A(x) = (7 - 2x)(11 - 2x) = 4x^2 - 36x + 77$;
domain $= \{x|0 \leq x \leq 3\frac{1}{2}\} = [0, 3\frac{1}{2}]$; range $= \{y|0 \leq y \leq 77\} = [0, 77]$

69. $V(x) = x(24 - 2x)(24 - 2x)$
     $= 4x^3 - 96x^2 + 576x$ in.³, $0 \le x \le 12$

71. (a) $C(x) = 100x + 140 \sqrt{(5 - x)^2 + 4}$
     $= 100x + 140 \sqrt{x^2 - 10x + 29}$ dollars

(b) All real numbers (A more practical domain for the cable TV company is $\{x | 0 \le x \le 5\} = [0, 5]$.)

(c) $C(1) = \$726.10$, $C(2) = \$704.78$, $C(3) = \$695.98$, $C(4) = \$713.05$

Technology Exercises (*page 523*)

1. $f(x) = x^3 - 6x$
   Zeros are: $x = 0$, $x = \sqrt{6}$, $x = -\sqrt{6}$

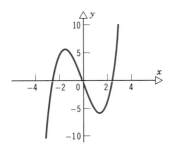

3. $f(x) = 4x^5 - 10x^4 + 6x^3 - 4x^2 + 10x - 6$
   Zeros are: $x = \frac{3}{2}$, $x = 1$

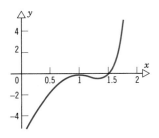

5. (a) Area $A(l) = l \cdot \dfrac{187 - 2l}{2}$

(b) Domain: $0 \le l \le 93.5$

(c) $A(20) = 1470$, $A(30) = 1905$, $A(40) = 2140$,
    $A(50) = 2175$, $A(60) = 2010$
Area in $[0, 93.5]$ by $[0, 2200]$ window:

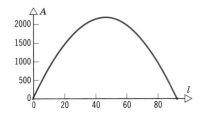

Zoomed in to $[46.5, 47]$:

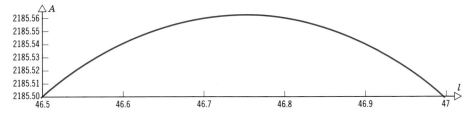

The area is maximum for length $l = 46.75$ and width $w = 46.75$.

7. (a) $-\infty < y < +\infty$ (b) 1 zero

9. Viewing rectangles $[-1, \ 1] \times [-0.5, \ 0.5]$ and $[-100, \ 100] \times [-1000000, \ 1000000]$ give distinguishable view

Exercise 11.3 (*page 535*)

1. c   **3.** e   **5.** b   **7.** f

9. (a) Domain $= \{x| -3 \le x \le 4\} = [-3, \ 4]$; range $= \{y|0 \le y \le 3\} = [0, 3]$
(b) Increasing on $[-3, \ 0]$ and on $[2, 4]$; decreasing on $[0, 2]$
(c) x-intercepts: $(-3, 0)$; $(2, 0)$; y-intercept: $(0, 3)$

11. (a) Domain $=$ all real numbers $= (-\infty, \ \infty)$; range $= \{y|0 < y\} = (0, \infty)$
(b) Increasing on $(-\infty, \infty)$
(c) No x-intercepts; y-intercept: $(0, 1)$

13. (a) Domain $= \{x| -\pi \le x \le \pi\} = [-\pi, \ \pi]$; range $= \{y|-2 \le y \le 2\} = [-2, 2]$
(b) Increasing on $[-\pi/2, \ \pi/2]$; decreasing on $[-\pi, \ -\pi/2]$ and on $[\pi/2, \ \pi]$
(c) x-intercepts: $(-\pi, 0), (0, 0), (\pi, 0)$; y-intercept: $(0, 0)$

15. (a) Domain $= \{x|x \ne 2\} = (-\infty, 2) \cup (2, \infty)$; range $= \{y|y \ne 1\} = (-\infty, 1) \cup (1, \infty)$
(b) Decreasing on $(-\infty, 2)$ and on $(2, \infty)$
(c) x-intercept: $(0, 0)$; y-intercept: $(0, 0)$

17. (a) Domain $= \{x|x \ne 0\} = (-\infty, 0) \cup (0, \infty)$; range $=$ all real numbers $= (-\infty, \infty)$
(b) Increasing on $(-\infty, 0)$ and on $(0, \infty)$
(c) x-intercepts: $(-1, 0), (1, 0)$; no y-intercepts

19. (a) Domain $= \{x|x \ne -2 \text{ and } x \ne 2\} = (-\infty, \ -2) \cup (-2, 2) \cup (2, \infty)$; range $= \{y|y \le 0 \text{ or } 1 < y\} = (-\infty, 0] \cup (1, \infty)$
(b) Increasing on $(-\infty, \ -2)$ and on $(-2, 0]$; decreasing on $[0, 2)$ and on $(2, \infty)$
(c) x-intercept: $(0, 0)$; y-intercept: $(0, 0)$

21. (a) $(70, 74), (76, 77), (78, 79), (88, 89),$ and $(91, 93)$
(b) $(74, 76), (77, 78), (79, 83), (84, 88),$ and $(89, 91)$
(c) Highest, 74; lowest, 91.
(d) $(83, 84)$
(e) $(76, 77)$

23. (a) $-2x + 3$ (b) $-2x - 3$
(c) $4x + 3$ (d) $2x - 3$ (e) $\dfrac{2}{x} + 3$ (f) $\dfrac{1}{2x + 3}$

25. (a) $2x^2 - 4$ (b) $-2x^2 + 4$ (c) $8x^2 - 4$
(d) $2x^2 - 12x + 14$ (e) $\dfrac{2}{x^2} - 4$ (f) $\dfrac{1}{2x^2 - 4}$

27. (a) $-x^3 + 3x$ (b) $-x^3 + 3x$
(c) $8x^3 - 6x$ (d) $x^3 - 9x^2 + 24x - 18$
(e) $\dfrac{1}{x^3} - \dfrac{3}{x}$ (f) $\dfrac{1}{x^3 - 3x}$

29. (a) $\dfrac{-x}{x^2 + 1}$ (b) $\dfrac{-x}{x^2 + 1}$ (c) $\dfrac{2x}{4x^2 + 1}$
(d) $\dfrac{x - 3}{x^2 - 6x + 10}$ (e) $\dfrac{x}{x^2 + 1}$ (f) $x + \dfrac{1}{x}$

31. (a) $|x|$ (b) $-|x|$ (c) $|2x|$
(d) $|x - 3|$ (e) $\dfrac{1}{|x|}$ (f) $\dfrac{1}{|x|}$

33. (a) $1 - \dfrac{1}{x}$ (b) $-1 - \dfrac{1}{x}$ (c) $1 + \dfrac{1}{2x}$
(d) $\dfrac{x - 2}{x - 3}$ (e) $1 + x$ (f) $\dfrac{x}{x + 1}$

35. 0, if $h \ne 0$ **37.** -3, if $h \ne 0$

39. $6x + 3h - 2$, if $h \ne 0$

41. $3x^2 + 3xh + h^2 - 1$, if $h \ne 0$

43. $\dfrac{-1}{x^2 + xh}$, if $h \ne 0$

45. (a) All real numbers
(b) x-intercept: $(1, 0)$; y-intercept: $(0, -3)$
(c)

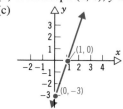

(d) All real numbers

47. (a) All real numbers
(b) x-intercepts: $(-2, 0)$ and $(2, 0)$; y-intercept: $(0, -4)$
(c)

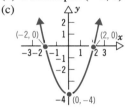

(d) $\{y|-4 \le y\} = [-4, \infty)$

49. (a) All real numbers
(b) x-intercept: $(0, 0)$; y-intercept: $(0, 0)$
(c)
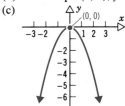
(d) $\{y | y \leq 0\} = (-\infty, 0]$

51. (a) $\{x | 2 \leq x\} = [2, \infty)$
(b) x-intercepts: $(2, 0)$; no y-intercept
(c)
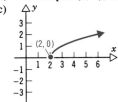
(d) $\{y | 0 \leq y\} = [0, \infty)$

53. (a) $\{x | x \leq 2\} = (-\infty, 2]$
(b) x-intercept: $(2, 0)$; y-intercept: $(0, \sqrt{2}) \approx (0, 1.41416)$
(c)
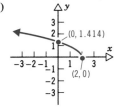
(d) $\{y | 0 \leq y\} = [0, \infty)$

55. (a) All real numbers
(b) No x-intercept; y-intercept: $(0, 3)$
(c)
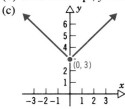
(d) $\{y | 3 \leq y\} = [3, \infty)$

57. (a) All real numbers
(b) x-intercept: $(0, 0)$; y-intercept: $(0, 0)$
(c)
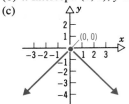
(d) $\{y | y \leq 0\} = (-\infty, 0]$

59. (a) All real numbers
(b) x-intercept: $(0, 0)$; y-intercept: $(0, 0)$
(c)
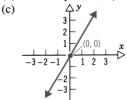
(d) All real numbers

61. (a) All real numbers
(b) x-intercepts: $(-1, 0)$ and $(0, 0)$; y-intercept: $(0, 0)$
(c)
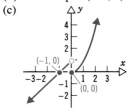
(d) All real numbers

63. (a) $[-2, \infty) = \{x | -2 \leq x\}$
(b) No x-intercept; y-intercept: $(0, 1)$
(c)
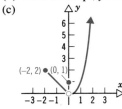
(d) $(0, \infty)$

65. (a) All real numbers

(b) No *x*-intercept; *y*-intercept: (0, 1)

(c)

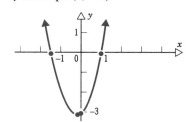

(d) $\{-1, 1\}$

67. (a) All real numbers

(b) *x*-intercepts: $\{(x, 0) | 0 \le x < 1\}$; *y*-intercept: (0, 0)

(c)

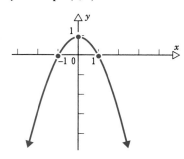

(d) All even integers

69. Upward

Vertex: $\left(-\frac{1}{4}, -\frac{25}{8}\right)$

x-intercepts: $\left(-\frac{3}{2}, 0\right)$, (1, 0)

y-intercept: (0, −3)

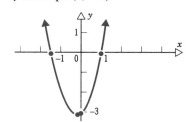

71. Upward

Vertex: (0, −4)

x-intercepts: (−2, 0), (2, 0)

y-intercept: (0, −4)

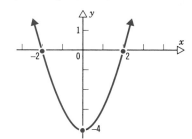

73. Upward

Vertex: (0, 1)

y-intercept: (0, 1)

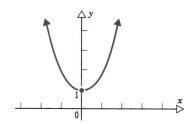

75. Downward

Vertex: (0, 1)

x-intercepts: (−1, 0), (1, 0)

y-intercept: (0, 1)

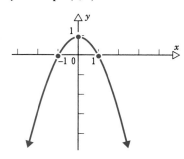

77. Upward

Vertex: $\left(\frac{7}{2}, -\frac{1}{4}\right)$

x-intercepts: (3, 0), (4, 0)

y-intercept: (0, 12)

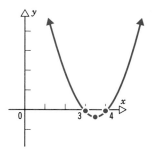

79. Downward
Vertex: $(0, 4)$
x-intercepts: $(-2, 0)$, $(2, 0)$
y-intercept: $(0, 4)$

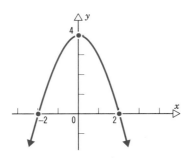

81. $f(x) = \begin{cases} 2 & \text{if} & x < -2 \\ -\frac{2}{3}(x - 3) & \text{if} & -2 \le x < 3 \\ \frac{2}{3}(x - 3) & \text{if} & 3 \le x \end{cases}$

83. $C(x) = \begin{cases} 20x & \text{if} & 0 \le x \le 6 \\ 12x + 48 & \text{if} & 6 < x \le 10 \end{cases}$

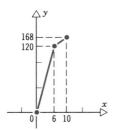

85. (a) $C(x) = \begin{cases} 1.90 & \text{if} & 0 \le x \le 1 \\ 1.90 + 0.60(x - 1) & \text{if} & 1 < x \end{cases}$

(b) $C(0.9) = \$1.9$,
$C(1.1) = \$1.96$,
$C(1.9) = \$2.44$

Technology Exercises **(*page 539*)**

1. (a) $g(t)$
   (b) $h(t)$
   (c) $f(t)$

3. $f(x) = x^3$
$f_1(x) = (x + 1)^3$
$f_2(x) = (x + 2)^3$
$f_3(x) = (x - 2)^3$
If c is positive, graph
shifts to the left. If c is
negative, graph shifts
to the right

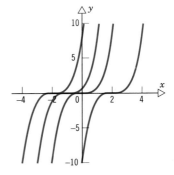

5. $(\sqrt{x})^2 = |x|$ for $x \ge 0$, so the graphs are the same on the
positive side of the x-axis.
$(\sqrt{x})^2$ is undefined for negative xs.

Exercise 11.4 **(*page 552*)**

1. 9   **3.** 27   **5.** $\frac{1}{4}$ **7.** $\frac{1}{9}$ **9.** $\frac{1}{3}$

11. 2   **13.** 3   **15.** $\frac{1}{64}$ **17.** 0.841

19. 1.260   **21.** 1.072

23. 20.086, 0.0498, 1.010, 1.492, 1.396, 0.961, 0.607

25. B; $y = 3^{-x}$, since $3^{-0} = 1$; $3^{-1} = \frac{1}{3}$; $3^{-(-1)} = 3$

27. D; $y = -3^{-x}$, since $-3^{-0} = -1$; $-3^{-1} = -\frac{1}{3}$;
$-3^{-(-1)} = -3$

29. A; $y = 3^x$, since $3^0 = 1$; $3^1 = 3$; $3^{-1} = \frac{1}{3}$

31. E; $y = 3^x - 1$, since $3^0 - 1 = 0$; $3^1 - 1 = 2$;
$3^{-1} - 1 = -\frac{2}{3}$

33. $f(x) = 3^x$

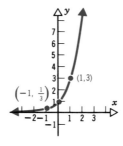

35. $f(x) = (\frac{1}{3})^x$

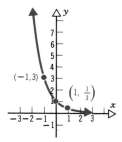

37. $f(x) = 4^{0.5x}$

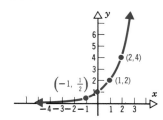

39. $f(x) = 10e^{0.3x}, -4 \le x \le 4$

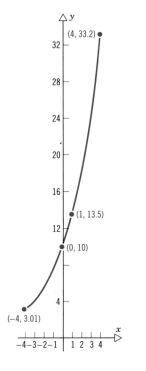

41. $f(x) = 100e^{-0.3x}, -4 \le x \le 4$

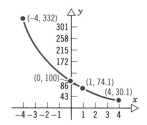

43.

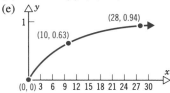

45. $x = \frac{4}{3}$ **47.** $x = -\frac{4}{3}$ **49.** $x = \frac{2}{3}$

51. $x = 1$

53. (a) $f(x + 1) = a^{(x+1)} = a^x a^1 = f(x) \cdot a = af(x)$
(b) $f(x + 1) - f(x) = af(x) - f(x) = (a - 1)f(x)$
(c) $f(x + h) = a^{(x+h)} = a^x a^h = f(x) \cdot a^h = a^h f(x)$

55. (b) 63.2% (d) \$6,404,329.56
(e)

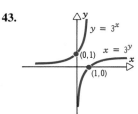

57. At 6%, compounded continuously: $\$500e^{0.06} \approx \$530.92$.
At $6\frac{1}{4}\%$, compounded quarterly:
$$\$500\left(1 + \frac{0.0625}{4}\right)^4 \approx \$531.99.$$
$6\frac{1}{4}\%$ compounded quarterly yields a greater return than 6% compounded continuously.

59. Compounded continuously, $P = \$941.77$; compounded quarterly, $P = \$942.19$.

61. (a) $\frac{9}{16}$ (b) $\frac{9}{64}$ (c) $\frac{1}{4}$

63. (a) 6.8 (b) 10.5 (c) 12.5 (d) 14.6

65. (a) 2800 (b) $1.008068591 \times 10^{14}$

67. (a) 53.78 (b) 241.03 (c) 1080.21 (d) 4841.15

69. $f(2) = 90\%, f(3) \approx 96.7\%, f(5) \approx 99.6\%$

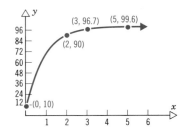

71. (a)

t	0	1	2	3	4	5
y	1	0.94	0.884	0.831	0.781	0.734

t	6	7	8	9	10
y	0.690	0.648	0.610	0.573	0.539

(b)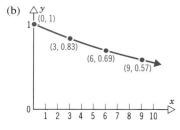

73. (a) 65.86   (b) 36.14   (c) 19.84   (d) 10.89

Technology Exercises (*page 554*)

1. $x = 1.37$

3. $x > 2$

5. $f(t) = t^2$, $g(t) = \frac{1}{2}t^3$, $h(t) = 2^t$

7. (a) $[0, 5] \times [0, 500]$ (b) $[0, 2] \times [0, 20]$
(c) $[0, 1] \times [0, 3]$

Exercise 11.5 (*page 563*)

1. $3^2 = 9$ **3.** $3^{-4} = \frac{1}{81}$ **5.** $a^Q = P$

7. $\log_{10} 1000 = 3$ **9.** $\log_a 3 = \frac{1}{2}$ **11.** 5

13. -3 **15.** 1   **17.** 1.0791   **19.** 0.8751

21. 1.5562   **23.** -0.9030 **25.** $\ln(3x)$ **27.** $\ln 4$

29. $\ln[(x + 1)(x + 2)(x + 3)]$ **31.** $x = 32$

33. $y = 2$ **35.** $x = 3$ **37.** $c = -3$

39. $x = 3$ **41.** $x = \frac{7}{6}$ **43.** $x = -2$

45. $x = 17.33$ **47.** $x = -1.58$

49.
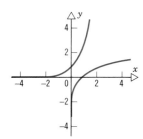

51.

53. $c = \ln 3$ **55.** $\dfrac{\ln 2}{0.06} \approx 11.55$ years

57. $10 \ln(3) \approx 10.99$ years   **59.** $R = 6.7$

61. $100 \ln(2) \approx 69.3$ years

63. $\dfrac{\ln(2)}{100} \times 100\% \approx 0.69\%$ **65.** $t = 4.62$ weeks

Technology Exercises (*page 565*)

1. $f(x) = \ln x$
$g(x) = e^x$

(i)   Domain of f: $x > 0$; domain of g: all $x \in R$
(ii) Range of f: all $y \in R$; range of g: all $y > 0$
(iii) (a) $\ln 1 = 0$ (b) $e^0 = 1$ (c) $\ln e = 1$
      (d) $\ln e^2 = 2$ (e) $e^{\ln 2} = 2$
      (f) $f(x) > 0$ for $x < 1$ (g) $f(x) = 0$ for $x = 1$
      (h) $f(x) < 0$ for $1 < x$ (i) $g(x) > 0$ for all $x \in R$
      (j) $g(x) = 0$ for no $x \in R$
      (k) $g(x) < 0$ for no $x \in R$

3. $f(x) = 3^x$
$f(x) = e^{x \ln 3}$
Because $3^x = e^{x \ln 3}$

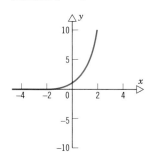

5. $f(t) = e^{3t}$
$f(t) = 2e^{3t}$
$f(t) = 4e^{3t}$

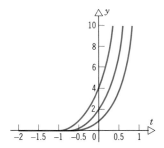

$f(t) = 3e^t$
$f(t) = 3e^{2t}$
$f(t) = 3e^{4t}$

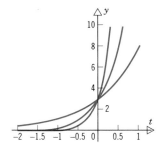

7. (a)

t	0	1	2	3	4
y	1	0.94	0.8836	0.83058	0.78075

t	5	6	7
y	0.7339	0.68987	0.64848

t	8	9	10
y	0.60957	0.57299	0.53862

(b)

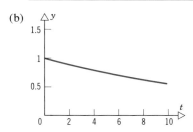

9. $t = 1.209$

11. $t = -0.8298$

Chapter Review: True–False Items (*page 567*)

1. F   **2.** F   **3.** T   **4.** F   **5.** F   **6.** T
7. F   **8.** T

Fill in the Blanks (*page 567*)

1. y-axis   **2.** independent; dependent   **3.** vertical
4. exponential   **5.** 5   **6.** e **7.** natural logarithm

Review Exercises (*page 567*)

1. x-intercept: $(0, 0)$; y-intercept: $(0, 0)$; symmetric with respect to the x-axis

3. x-intercepts: $\left(-\frac{1}{2}, 0\right)$ and $\left(\frac{1}{2}, 0\right)$; y-intercepts: $(0, -1)$ and $(0, 1)$; symmetric with respect to x-axis, to y-axis, and to origin

5. No x-intercepts; y-intercept: $(0, 1)$; symmetric with respect to y-axis

7. x-intercepts: $(-1, 0)$ and $(0, 0)$; y-intercepts: $(0, -2)$ and $(0, 0)$; no "standard" symmetries [though the graph is a circle with center $\left(-\frac{1}{2}, -1\right)$]

9. $f(x) = -2x + 6$ **11.** 11   **13.** b, c, and d

15. (a) $\dfrac{-x}{x^2 - 4}$ (b) $\dfrac{-x}{x^2 - 4}$ (c) $\dfrac{x + 2}{x^2 + 4x}$ (d) $\dfrac{x - 2}{x^2 - 4x}$

17. (a) $\sqrt{x^2 - 4}$ (b) $-\sqrt{x^2 - 4}$ (c) $\sqrt{x^2 + 4x}$
(d) $\sqrt{x^2 - 4x}$

19. (a) $\dfrac{x^2 - 4}{x^2}$ (b) $\dfrac{-x^2 + 4}{x^2}$ (c) $\dfrac{x^2 + 4x}{x^2 + 4x + 4}$
(d) $\dfrac{x^2 - 4x}{x^2 - 4x + 4}$

21. $\{x | x \neq -2 \text{ and } x \neq 2\} = (-\infty, -2) \cup (-2, 2) \cup (2, \infty)$

23. $\{x | x \leq 2\} = (-\infty, 2]$ **25.** $\{x | 0 < x\} = (0, \infty)$

27. $\{x | x \neq -3 \text{ and } x \neq 1\} = (-\infty, -3) \cup (-3, 1) \cup (1, \infty)$

29. $\{x | -1 \leq x\} = [-1, \infty)$ **31.** $\{x | 0 \leq x\} = [0, \infty)$

33. (a) All real numbers
(b) x-intercepts: $(-4, 0)$ and $(4, 0)$; y-intercept: $(0, -4)$
(c)
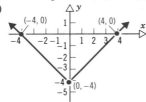

(d) $\{y | -4 \leq y\} = [-4, \infty)$

35. (a) All real numbers
(b) x-intercept: $(0, 0)$; y-intercept: $(0, 0)$
(c)
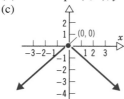

(d) $\{y | y \leq 0\} = (-\infty, 0]$

37. (a) $\{x | 1 \leq x\} = [1, \infty)$
(b) x-intercept: $(1, 0)$; no y-intercept
(c)
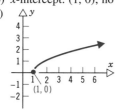

(d) $\{y | 0 \leq y\} = [0, \infty)$

39. (a) $\{x | x \leq 1\} = (-\infty, 1]$
(b) x-intercept: $(1, 0)$; y-intercept: $(0, 1)$
(c)
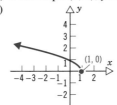

(d) $\{y | 0 \leq y\} = [0, \infty)$

41. (a) All real numbers
(b) x-intercept: $(2, 0)$; y-intercept: $(0, 4)$
(c)
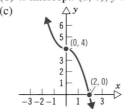

(d) All real numbers

43.
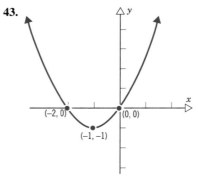

45.
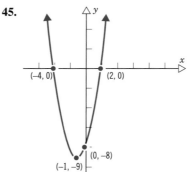

47.

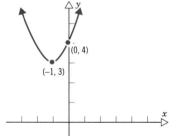

(0, 4)

(−1, 3)

49. 1.93874   **51.** 6.70196   **53.** 2.52573

55. 4 log 2   **57.** $\ln \dfrac{x^3 z^6}{y^2}$ **59.** $x = 1.262$

61. $x = 2.404$ **63.** $x = 0.339$ **65.** $x = 117.417$

67. $x = 0.549$ **69.** $x = 2$ **71.** $x = 7$ **73.** $x = 1$

75. $x = 4$ **77.** $x = \dfrac{2}{5} + \dfrac{\sqrt{34}}{5} \approx 1.56$

79. $m^{2/3} = 5$ **81.** $p^3 = 13$

83.

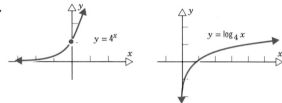

$y = 4^x$ $y = \log_4 x$

85. (a) $k = \dfrac{\ln 2}{3} \approx 0.231$ (b) $t = \dfrac{3 \ln 10}{\ln 2} \approx 9.9657$ hours

87. (a) 0   (b) 239,789.53   (c) 343,398.72
(d) 371,357.21

89. $t = \dfrac{\ln 3}{\ln 2} \approx 1.585$ hours

91. $t = \dfrac{\ln 2}{0.12} \approx 5.776$ years

93. (a) $2,000   (b) $2,250   (c) $2,441.41   (d) $2,613.04
(e) $2,692.60   (f) $2,714.57   (g) $2,718.13

Mathematical Questions from Professional Exams (*page 570*)

1. e

CHAPTER 12

Exercise 12.1 (*page 579*)

1.

x	0.9	0.99	0.999
$f(x) = 2x$	1.8	1.98	1.998

x	1.1	1.01	1.001
$f(x) = 2x$	2.2	2.02	2.002

$\lim\limits_{x \to 1} f(x) = 2$

3.

x	0.1	0.01	0.001
$f(x) = x^2 + 2$	2.01	2.0001	2.000001

x	-0.1	-0.01	-0.001
$f(x) = x^2 + 2$	2.01	2.0001	2.000001

$\lim\limits_{x \to 0} f(x) = 2$

5.

x	1.9	1.99	1.999
$f(x) = \dfrac{x^2 - 4}{x - 2}$	3.9	3.99	3.999

x	2.1	2.01	2.001
$f(x) = \dfrac{x^2 - 4}{x - 2}$	4.1	4.01	4.001

$\lim\limits_{x \to 2} f(x) = 4$

7.

x	-1.1	-1.01	-1.001
$f(x) = \dfrac{x^3 + 1}{x + 1}$	3.31	3.0301	3.003001

x	-0.9	-0.99	-0.999
$f(x) = \dfrac{x^3 + 1}{x + 1}$	2.71	2.9701	2.997001

$\lim\limits_{x \to -1} f(x) = 3$

9. Exists   **11.** Exists   **13.** Does not exist

15. Does not exist   **17.** Exists

19. $\lim\limits_{x \to 1} f(x) = 9$

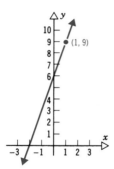

21. $\lim\limits_{x \to 1} f(x) = 3$

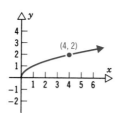

23. $\lim\limits_{x \to 0} f(x) = 2$

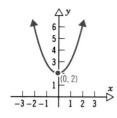

25. $\lim\limits_{x \to 4} f(x) = 2$

27. $\lim\limits_{x \to 2} f(x) = 2$

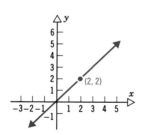

29. $\lim\limits_{x \to 2} f(x) = 9$

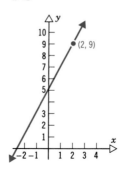

31. $\lim\limits_{x \to 1} f(x) = 2$

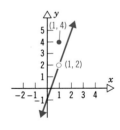

33. $\lim\limits_{x \to 1} f(x) = 2$

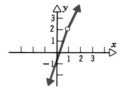

35. $\lim\limits_{x \to 0} f(x)$ does not exist

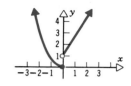

37. (a) 1   (b) 2   (c) Does not exist
(d) Does not exist   (e) Does not exist
(f) Does not exist   (g) 3   (h) 3   (i) 3

39. -2 **41.** 0   **43.** $\frac{7}{2}$ **45.** 0

47. $\lim\limits_{x \to 2^-} f(x) = 9$; $\lim\limits_{x \to 2^+} f(x) = 9$; thus, $\lim\limits_{x \to 2} f(x)$ exists.

49. $\lim\limits_{x \to 1^-} f(x) = 2$; $\lim\limits_{x \to 1^+} f(x) = 2$; thus, $\lim\limits_{x \to 1} f(x)$ exists.

Technology Exercises (page 582)

1. $\lim\limits_{x \to 2} \dfrac{x^2 - 4}{x - 2} = 4$

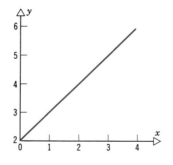

3. $\lim\limits_{x \to .2} \dfrac{3x^3 - .12x}{x - .2} = .24$

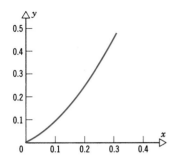

5. $f(x) = 2 + \sqrt{x^2 - 5x + 6}$

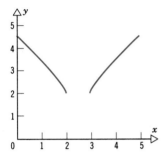

(a) $\lim\limits_{x \to 1} f(x) = 2 + \sqrt{2}$

(b) $\lim\limits_{x \to 2^-} f(x) = 2$

(c) $\lim\limits_{x \to 2^+} f(x)$ does not exist

(d) $\lim\limits_{x \to 3} f(x)$ does not exist

7. $f(x) = \dfrac{x^2}{x^2 - 4}$

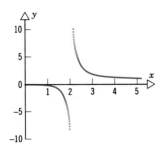

(a) $\lim\limits_{x \to 3} f(x) = \frac{9}{5}$

(b) $\lim\limits_{x \to 2^-} f(x)$ does not exist

(c) $\lim\limits_{x \to 2^+} f(x)$ does not exist

9. (a) $\lim\limits_{x \to 2^-} x^3 = 8$

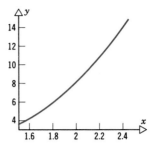

(b) $\lim\limits_{x \to 2} x = 2$

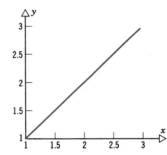

(c) $\lim\limits_{x \to 2} (x^3 - x) = 6$

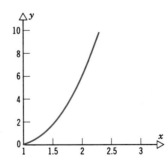

11. Because the calculator thinks $1/10^{18} = 0$; it exceeds the precision for the calculator.

Exercise 12.2 *(page 592)*

1. 3   **3.** 0   **5.** 3   **7.** 20   **9.** 17   **11.** 5

13. 4   **15.** $\frac{2}{3}$ **17.** 1   **19.** $2\sqrt{3}$ **21.** $\frac{8}{5}$

23. 4   **25.** 48   **27.** -1 **29.** $\frac{1}{2}$ **31.** $\frac{12}{5}$

33. 3   **35.** 4   **37.** 1   **39.** 2   **41.** 8   **43.** 13

45. $\lim\limits_{x \to 1} f(x) = -1; f(1) = 5$

47. $\lim\limits_{x \to 4} f(x) = 8; f(4) = 0$

49. $\lim\limits_{x\to 1} f(x) = 3; f(1) = 5$ **51.** 10   **53.** 8

55. $\underbrace{c^{n-1} + cc^{n-2} + \cdots + c^{n-1}}_{n \text{ terms}} = nc^{n-1}$

Exercise 12.3 *(page 598)*

1. Continuous   **3.** Not continuous   **5.** Continuous
7. Not continuous   **9.** Not continuous
11. Continuous   **13.** Continuous   **15.** $k = 8$
17. Not continuous;

$$f(x) = \begin{cases} \dfrac{x^2 - 4}{x - 2} & \text{if } x \neq 2 \\ 4 & \text{if } x = 2 \end{cases} \quad \text{is continuous at } x = 2$$

19. 3

21. (a) $C(x) = \begin{cases} 2000 & \text{if } x = 0 \\ 5000 + x & \text{if } 0 < x \leq 10{,}000 \\ 8000 + x & \text{if } 10{,}000 < x \leq 20{,}000 \\ 11{,}000 + x & \text{if } 20{,}000 < x \leq 30{,}000 \\ 14{,}000 + x & \text{if } 30{,}000 < x \leq 40{,}000 \end{cases}$

(b) Domain = $\{x | 0 \leq x \leq 40{,}000\}$

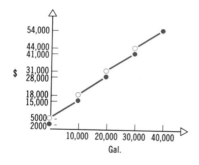

(c) $C(x)$ is discontinuous at $x = 0$, $x = 10{,}000$, $x = 20{,}000$, and $x = 30{,}000$

23. The cost function is discontinuous at $x = 1, x = 2, x = 3$, $x = 4$, and $x = 5$.

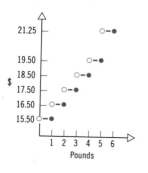

Technology Exercises *(page 600)*

1. Discontinuous at $x = 3$
3. Discontinuous at $x = 0$
5. Discontinuous at $x = -3$ and $x = 3$

Exercise 12.4 *(page 608)*

1. $y = -4x - 2$

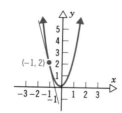

3. $y = -4x - 7$

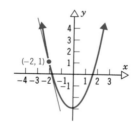

5. $y = 5x - 4$ **7.** $y = 2x + 1$ **9.** $y = 9x - 9$
11. 12   **13.** 4   **15.** 4   **17.** 12   **19.** 2
21. 2   **23.** $2x$ **25.** $6x - 2$ **27.** $3x^2$
29. m **31.** No
33. Release the bomb at the point $(2, 4)$ **35.** x_4, x_5, x_7, x_9

Technology Exercises *(page 609)*

1.
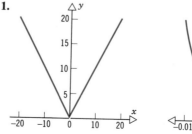 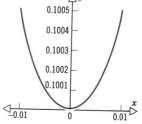

It does have a derivative at $x = 0$.

3. $f'(2) = 6.7726$

5. (a) Tangent line is below the function, so $e^x \geq 1 + x$, and $e^x = 1 + x$ only for $x = 0$.

(b) $e^{1/n} > 1 + \dfrac{1}{n}$, so $e > \left(1 + \dfrac{1}{n}\right)^n$.

(c) $e^{-1/(n+1)} > 1 - \dfrac{1}{n+1}$, or $\left(\dfrac{1}{e}\right)^{1/(n+1)} > \dfrac{n}{n+1}$,

so $\left(1 + \dfrac{1}{n}\right)^{n+1} > e$.

(d) Using Parts (b) and (c) for $n = 100$,

$\left(1 + \dfrac{1}{n}\right)^n = 2.7 < e < 2.73 = \left(1 + \dfrac{1}{n}\right)^{n+1}$

Exercise 12.5 (*page 613*)

1. (a) $\Delta s/\Delta t = 36$ km/hr (b) $\Delta s/\Delta t = 33$ km/hr

3. (a) $\Delta s/\Delta t = 104$ ft/sec (b) $\Delta s/\Delta t = 97.6$ ft/sec

5. 4 m/sec; 16 m/sec; $f'(t) = 6t + 4$

7. $\frac{1}{4}$ km/hr; $\frac{1}{3}$ km/hr; $(\sqrt{2} - 1)$ km/hr ≈ 0.41 km/hr

Chapter Review: True–False Items (*page 614*)

1. T **2.** F **3.** T **4.** T **5.** T **6.** T

Fill in the Blanks (*page 614*)

1. $\lim\limits_{x \to c} f(x) = L$ **2.** equals **3.** not exist

4. continuous **5.** $\neq$ **6.** equals **7.** tangent

Review Exercises (*page 615*)

1. 5 **3.** 6 **5.** 7 **7.** 2 **9.** 3

11. $-\frac{1}{16}$ **13.** 2 **15.** -5 **17.** $\frac{1}{32}$ **19.** 3

21. 0 **23.** 4 **25.** 9 **27.** Continuous

29. Not continuous **31.** Continuous

33. Not continuous at $x = 3$;

$$f(x) = \begin{cases} \dfrac{x^2 - 3x}{x^2 - 9} & \text{if } x \neq 3 \\ \dfrac{1}{2} & \text{if } x = 3 \end{cases} \quad \text{is continuous at } x = 3$$

35. Slope $= 4$; $y = 4x - 1$

37. Slope $= -6$; $y = -6x$ **39.** 8 **41.** -4

43. $4x$ **45.** $2x - 2$ **47.** x_3, x_4, x_7, x_9

Mathematical Question from Professional Exam (*page 616*)

1. b

CHAPTER 13

Exercise 13.1 (*page 626*)

1. 0 **3.** $3x^2$ **5.** $12x$ **7.** t^3 **9.** $2x + 1$

11. $3x^2 - 2x$ **13.** $4t - 1$ **15.** $4x^7 + 3$

17. $\frac{5}{3}x^4$ **19.** $4x^5 + 2x^4$ **21.** $-\frac{7}{4}x^6$

23. $3\sqrt{3}t^2 + \dfrac{2}{\sqrt{2}}t$ **25.** $\frac{2}{3}x^{-1/3}$ **27.** $\frac{1}{3}x^{-1/2}$

29. $1.2x^{0.2}$ **31.** $\sqrt{5}x^{\sqrt{5} - 1}$ **33.** $\frac{2}{3}x^{-1/3} + \frac{3}{2}x^{1/2}$

35. $\dfrac{1}{2\sqrt{x}}$ **37.** $6\sqrt{x}$ **39.** $-3x^{-4}$

41. $-3x^{-4} - 2x^{-3} + 4x$ **43.** $\dfrac{-2}{x^3}$ **45.** $\dfrac{-9}{x^4}$

47. $2ax + b$ **49.** $-\dfrac{40}{x^5} - \dfrac{6}{x^3}$ **51.** $\dfrac{1}{2x^2}$

53. $3 - \dfrac{1}{3t^2}$ **55.** $9x^2 + \dfrac{2}{3x^3}$

57. $-\dfrac{1}{t^2} + \dfrac{2}{t^3} - \dfrac{3}{t^4}$ **59.** $-\dfrac{3}{x^3}$ **61.** $\dfrac{3}{x^{-1}}$

63. $\dfrac{-3}{4x\sqrt[4]{x^3}}$ **65.** $-\dfrac{1}{x\sqrt{x}} + 9x^{-4} + 1$

67. $\sqrt{3}$ **69.** $2\pi R$ **71.** $4\pi R^2$ **73.** 4 **75.** $\frac{1}{4}$

77. $-\frac{1}{48}$ **79.** 1 **81.** $\frac{3}{8}$ **83.** $y = 3x - 1$ **85.** 2

87. $1, -1$ **89.** No real number solution

91. Horizontal tangent at $x = 2$

93. Horizontal tangent at $x = 4$

95. Horizontal tangent at $x = 2$

97. Horizontal tangents at $x = -1$ and at $x = 1$

99. Horizontal tangents at $x = 0$ and at $x = 8$

101. Horizontal tangents at $x = -4$ and at $x = 0$

103. $x = \frac{9}{2}$ **105.** $(\frac{1}{3}, \frac{1}{3})$ and $(-\frac{1}{3}, -\frac{1}{3})$

107. $y = -4x + 1$ and $y = 4x - 7$

109. (a) $f'(2) \approx 4$, so the tangent line is $y - 3 = 4(x - 2)$.
(b) $f'(-2) \approx 4.7$, so the tangent line is $y + 4 = 4.7(x + 2)$.

111. $\dfrac{d}{dx}(x^n) = \lim\limits_{h \to 0} \dfrac{(x + h)^n - x^n}{h}$

$$= \lim_{h \to 0} \frac{(x^n + nx^{n-1}h + n(n - 1)x^{n-2}h^2 + \cdots + h^n) - x^n}{h}$$

$$= \lim_{h \to 0} \frac{nx^{n-1}h + n(n - 1)x^{n-2}h^2 + \cdots + h^n}{h}$$

$$= \lim_{h \to 0} [nx^{n-1} + n(n - 1)x^{n-2}h + \cdots + h^{n-1}]$$

$$= nx^{n-1}$$

Technology Exercises (*page 629*)

1. (a)

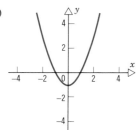

(b) $f'(x) = 2x$

(c)

x	$f(x)$	$f'(x)$
-4	15	-8
-2	3	-4
-1	0	-2
0	-1	0
1	0	2
2	3	4
4	15	8

(e)

3.

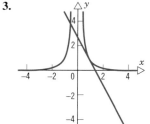

5. (a) Yes.
(b) The line is not tangent.
(c) $y = 4x - 4$

Exercise 13.2 (*page 635*)

1. (a) 3   (b) 3   **3.** (a) 12   (b) 6

5. (a) 6   (b) 4   **7.** (a) 7   (b) 3   **9.** (a) 13   (b) 3

11. (a) 74   (b) 94   (c) 110

13. (a) $9.95 per truck   (b) $-0.01x + 20$
(c) $10 per truck
(d) If 1001 trucks are sold, the store should expect a revenue increase of approximately $10 over the revenue for selling 1000 trucks.
(e) 2000 trucks

15. (a) 4500   (b) 7800   (c) 1100   (d) 950
(e) If the price rises from $10 to $13 per crate, the farmer will be willing to increase the supply of crates by an average of 1100 crates for each $1 increase in price. Assuming a price of $10 per crate, the farmer would supply approximately 950 more crates if the price were to rise to $11.

17. (a) $8 - 2x$ (b) 2   (c) $x = 1$ and $x = 5$ (d) $x = 3$
(e)

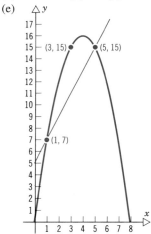

19. (a) $-10x^2 + 2000x$ (b) $-20x + 2000$ (c) 0
(d) -10

21. (a) $90x - 0.02x^2$ (b) $90 - 0.04x$ (c) 10
(d) $x = 0$ or $x = 4000$ (e) $x = 2000$

23. 48.65   **25.** (a) 2.4   (b) 3; 1.8; 1

27. (a) Function: $f(x)$ is the cost of a product when x units are produced. Variable: x is the number of products produced. Derivative: $f'(x)$ is the rate of change of the cost, and is a negative function; $f'(x) < 0$.

(b) Function: $f(x)$ is the price of a product when x units are produced. Variable: x is the number of products produced. Derivative: $f'(x)$ is the rate of change of the price, and is a negative function; $f'(x) < 0$.

(c) Function: $f(t)$ is the demand for a product. Variable: t is the time (days). Derivative: $f'(t)$ is the rate of change of the demand per day; $f'(t) < 0$.

(d) Function: $f(t)$ is the size of the U.S. population. Variable: t is the time (years). Derivative: $f'(t)$ is the rate of change of the size of the population; $f'(1985) > f'(1995) > 0$.

(e) Function: $f(t)$ are the health care costs. Variable: t is the time (years). Derivative: $f'(t)$ is the rate of change of the health care costs per year; $f'(1990) > f'(1995) > 0$.

(f) Function: $f(t)$ is the world consumption of oil (number of gallons). Variable: t is the time (years). Derivative: $f'(t)$ is the rate of change of the world consumption of oil per year (number of gallons/year); $f'(t) < 0$, for $1990 \leq t \leq 1995$.

29. (a) 3   (b) $\frac{3}{17}$ (c) 17.647%

31. (a) 13   (b) $\frac{13}{14}$ (c) 92.857%

33. (a) 3   (b) $\frac{3}{102} \approx 2.9412 \times 10^{-2}$; 2.9412%

35. (a) 14   (b) 7.1795%

37. (a) f: 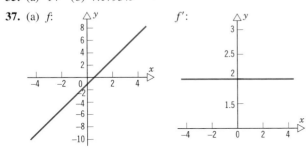 f':

(b) The rate of change of $f(x)$ is a constant positive value.

Technology Exercises (page 638)

1. (a) $1 \leq x \leq 4$ (b) $x = 2.5$

Exercise 13.3 (page 647)

1. $16x - 2$ **3.** $4t^3 - 6t$ **5.** $18x^2 - 20x + 3$

7. $24x^7 + 40x^4 + 9x^2$

9. $5x^4 - 24x^2$ **11.** $15x^4 - 8x^3 + 3x^2 + 2$

13. $-\frac{5}{2}t^{3/2} + 10t - \frac{1}{2}t^{-1/2}$ **15.** $\dfrac{1}{(1 + x)^2}$

17. $-\dfrac{11}{(2x - 1)^2}$ **19.** $\dfrac{x^2 - 8x}{(x - 4)^2}$ **21.** $\dfrac{-6x^2 - 6x + 8}{(3x^2 + 4)^2}$

23. $\dfrac{4}{t^3}$ **25.** $-\dfrac{1}{x^2} - \dfrac{2}{x^3}$ **27.** $\dfrac{2v^3 + 7}{v^2}$

29. $\dfrac{9}{2}\sqrt{x} - \dfrac{1}{\sqrt{x}} - \dfrac{1}{2x\sqrt{x}}$

31. Slope $= 3$; $y = 3x - 1$ **33.** Slope $= \frac{5}{4}$; $y = \frac{5}{4}x - \frac{3}{4}$

35. $-\frac{2}{3}$ and 1   **37.** -2 and 0   **39.** $9x^2 - 4x$

41. $32x - \dfrac{6}{x^3}$ **43.** $\dfrac{6x^2 + 20x + 11}{(3x + 5)^2}$

45. $\dfrac{3x^2 - 2x - 12}{(x - 2)^2(x + 2)^2}$ **47.** $\dfrac{120x - 10}{(2x + 1)^2(3x - 2)^2}$

49. $-\dfrac{2}{(x + 1)^2}$

51. Graph:

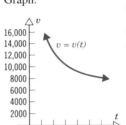

(a) -1000 (b) $-\dfrac{10{,}000}{t^2}$

(c) -2500 (d) -400

(e) The value of the car will drop approximately \$2500 during its third year. The value of the car will drop approximately \$400 during its sixth year.

53. (a) $10x + 40$, $1 \leq x \leq 10$ (b) 10, $1 \leq x \leq 10$
(c) 10   (d) 10

55. -15

57. (a) $\frac{25}{169}$ (b) $\frac{13}{49}$

59. $C'(t) = \dfrac{-8t^2 + 2}{5(1 + 2t^2)^2}$ (a) $\dfrac{612}{1805}$ (b) $\dfrac{4}{45}$ (c) $-\dfrac{2}{45}$
(d) $-\dfrac{34}{1805}$

61. (a) $C'(x) = \dfrac{550}{(x - 110)^2}$ (b) $\frac{11}{200}, \frac{11}{162}, \frac{11}{32}, \frac{11}{8}$

63. (a) $\overline{C}'(x) = -\dfrac{400}{x^2} + 0.0001$

(b) -0.0099; -0.004344; -0.0024
Since the marginal average cost is negative across this range of values, the average cost is decreasing; thus, increasing production will continue to decrease the average cost. However, since the marginal average cost is decreasing in magnitude across this range of values, increasing production will have less and less effect on the average cost.

65. $(x^2 + 1)(x - 1) + (x^2 + 1)(x + 5) +$
$(x - 1)(x + 5)(2x) = 4x^3 + 12x^2 - 8x + 4$

67. $(x^4 + 1)^2(4x^3) + (x^4 + 1)^2(4x^3) + (x^4 + 1)^2(4x^3) =$
$12x^3(x^4 + 1)^2$

Exercise 13.4 *(page 658)*

1. $5(x^3 + 1)^4(3x^2) = 15x^2(x^3 + 1)^4$ **3.** $\dfrac{2x}{(x^2 + 2)^2}$

5. $-2\left(\dfrac{1}{x} + 1\right)\left(\dfrac{1}{x^2}\right)$ **7.** $-30(x^{-6} - 1)^4 x^{-7}$

9. $-\dfrac{9x^2}{(x^3 + 1)^4}$ **11.** $8(2x - 3)^3$ **13.** $6x(x^2 + 4)^2$

15. $12x(3x^2 + 4)$

17. $(x + 1)^3 + 3x(x + 1)^2 = (x + 1)^2(4x + 1)$

19. $32x^2(2x + 1)^3 + (2x + 1)^4(8x) = 8x(2x + 1)^3(6x + 1)$

21. $3(x(x - 1))^3(2x - 1)$

23. $-6(3x - 1)^{-3} = -\dfrac{6}{(3x - 1)^3}$ **25.** $\dfrac{x}{\sqrt{x^2 + 1}}$

27. $\dfrac{1}{\sqrt[3]{(3x - 1)^2}}$ **29.** $\sqrt{x^2 + 1} + \dfrac{x^2}{\sqrt{x^2 + 1}}$

31. $2x\sqrt{3x + 1} + \dfrac{3x^2}{2\sqrt{3x + 1}}$

33. $-8x(x^2 + 4)^{-2} = -\dfrac{8x}{(x^2 + 4)^2}$

35. $24x(x^2 - 9)^{-4} = \dfrac{24x}{(x^2 - 9)^4}$

37. $3\left(\dfrac{x}{x + 1}\right)^2\left(\dfrac{1}{(x + 1)^2}\right) = \dfrac{3x^2}{(x + 1)^4}$

39. $\dfrac{24x^2(2x + 1)^3 - (2x + 1)^4(6x)}{9x^4} = \dfrac{2(2x + 1)^3(2x - 1)}{3x^3}$

41. $\dfrac{6x^2(x^2 + 1)^2 - (x^2 + 1)^3}{x^2} = \dfrac{(x^2 + 1)^2(5x^2 - 1)}{x^2}$

43. $3\left(x + \dfrac{1}{x}\right)^2\left(1 - \dfrac{1}{x^2}\right)$

45. $\dfrac{(x^2 + 1)^2(6x) - 12x^3(x^2 + 1)}{(x^2 + 1)^4} = \dfrac{6x(1 - x^2)}{(x^2 + 1)^3}$

47. $\dfrac{(x + 1)^3}{2\sqrt{2x}} + 6\sqrt{x}(x + 1)^2$ **49.** $-\dfrac{1}{2(x + 3)\sqrt{x + 3}}$

51. $\dfrac{2}{x - 1} - \dfrac{2x}{(x - 1)^2}$ **53.** $y = x - 2$

55. (a) $2(x^3 + 1)(3x^2)$ (b) $2(x^3 + 1)(3x^2)$

(c) $\dfrac{d}{dx}(x^6 + 2x^3 + 1) = 6x^5 + 6x^2$

57. $f'(x) = 0$ for $x = \pm 1/\sqrt{2}$; $f'(x)$ is undefined when $x = \pm 1$ [neither $f(x)$ nor $f'(x)$ is defined when $x < -1$ or when $x > 1$]

59. (a) $V'(1) = -2400$ (b) $V'(2) \approx -1487.60$
(c) $V'(3) \approx -936.52$ (d) $V'(4) \approx -612.24$

61. (a) $-\dfrac{50{,}000}{(5x + 100)^2}$, $0 < x < 90$

(b) $\dfrac{10{,}000x}{5x + 100} - 5x$, $0 < x < 90$

(c) $\dfrac{1{,}000{,}000}{(5x + 100)^2} - 5$, $0 < x < 90$ (d) $39\frac{4}{9}$; $6\frac{1}{9}$

(e) Both values are positive, so revenue will increase as more pounds of the commodity are demanded. However, note that an increase in demand from 10 lb to 11 lb would produce a significantly larger increase in revenue than a change in demand from 40 to 41 lb.

63. (a) $-\frac{7}{2}$ (b) -7
(c) Since $dM/dt < 0$, the mass of the protein is decreasing; that is, it is disintegrating into amino acids. The magnitude of $M'(0)$ is twice that of the average rate of change of the mass from $t = 0$ to $t = 2$ hr, indicating that the protein disintegrates more rapidly initially.

65. $N'(t) = \dfrac{500}{(1 + 0.1t)^{3/2}}$; $N'(0) = 500$ students/yr;

$N'(10) = \dfrac{500}{2^{3/2}} \approx 176.78$ students/yr

67. Since z is fixed, $\dfrac{dz}{dx} = 0$. Implicitly differentiating $z = x^{0.5}y^{0.4}$ yields $0 = 0.5x^{-0.5}y^{0.4} + 0.4x^{0.5}y^{-0.6}\dfrac{dy}{dx}$, so

$\dfrac{dy}{dx} = \dfrac{-0.5x^{-0.5}y^{0.4}}{0.4x^{0.5}y^{-0.6}} = -\dfrac{5y}{4x}$.

69. At 7 P.M. they approach each other at the rate of $-\sqrt{\frac{104}{5}} \approx 4.5607$ kph. They are receding at 10 P.M. They are closest at 8 P.M.

71. 1 hr after noon

Exercise 13.5 (page 670)

1. $5e^x$ **3.** $5e^{5x}$ **5.** $-4e^{-x/2}$ **7.** $e^x + xe^x$

9. $2xe^{x^2}$ **11.** $\dfrac{1}{2\sqrt{x}}\,e^{\sqrt{x}}$ **13.** $\dfrac{e^x}{2\sqrt{e^x}} = \dfrac{1}{2}e^{x/2}$ **15.** $-e^x$

17. $\dfrac{e^x - e^{-x}}{2}$ **19.** $-3e^{-3x} - 3$ **21.** $2e^{2x} + 2e^{-2x}$

23. $(4x + 1)e^{2x^2 + x + 1}$ **25.** $\dfrac{xe^x - e^x}{x^2} = \dfrac{(x - 1)e^x}{x^2}$

27. $\left(1 + \dfrac{1}{x^2}\right)e^{x - (1/x)}$ **29.** $\dfrac{e^x}{2\sqrt{1 + e^x}}$ **31.** $\dfrac{6}{x}$

33. $\dfrac{1}{x}$ **35.** $\dfrac{8}{x}$ **37.** $1 + \ln x$ **39.** $\dfrac{2}{x}$ **41.** $\dfrac{1}{2x}$

43. $\dfrac{1}{2x\sqrt{\ln x}}$ **45.** $\dfrac{1 - \ln x}{x^2}$ **47.** 1

49. $\dfrac{2x^2}{x^2 + 4} + \ln(x^2 + 4)$ **51.** $\dfrac{x^2}{x^2 + 1} + \ln\sqrt{x^2 + 1}$

53. $f'(x) = (x^2 + 1)^2(2x^3 - 1)^4\left(\dfrac{4x}{x^2 + 1} + \dfrac{24x^2}{2x^3 - 1}\right)$

$= 4x(x^2 + 1)(2x^3 - 1)^4 + 24x^2(x^2 + 1)^2(2x^3 - 1)^3$

$= 4x(x^2 + 1)(2x^3 - 1)^3(8x^3 + 6x - 1)$

55. $f'(x) = (x^3 + 1)(x - 1)(x^4 + 5) \times$

$\left(\dfrac{3x^2}{x^3 + 1} + \dfrac{1}{x - 1} + \dfrac{4x^3}{x^4 + 5}\right)$

57. $f'(x) = \dfrac{x^2(x^3 + 1)}{\sqrt{x^2 + 1}}\left(\dfrac{2}{x} + \dfrac{3x^2}{x^3 + 1} - \dfrac{x}{x^2 + 1}\right)$

59. $3^x \ln 3$ **61.** $\dfrac{1}{x \ln 3}$ **63.** $y = 3x - 1$

65. $y = x + 1$

67. $y = \frac{1}{2}x - 1 + \ln 2 \approx \frac{1}{2}x - 0.306853$

69. $C'(x) = \dfrac{1}{x + 1}$ **71.** $-1.13; -1.1$

73. (a) $A'(t) = 18.9e^{-0.21t}$
  (b) $A'(5) \approx 6.6138$. In 1965 an additional 6.6138% of the market was penetrated by video sets.
  (c) $A'(10) \approx 2.3144$. In 1970 an additional 2.3144% of the market was penetrated by video sets.
  (d) $A'(30) \approx 0.0347$. In 1990 an additional 0.0347% of the market was penetrated by video sets.

75. (a) $S'(x) = 400,000/x$
  (b) $S'(10) = 40,000$. If we spend $1000 over $10,000, the sales increase will be $40,000.
  (c) $S'(20) = 20,000$. If we spend $1000 over $20,000, the sales increase will be $20,000.

Technology Exercises (page 672)

1. $f(x) = 3^x$
  $f'(x) = 3^x \ln 3$

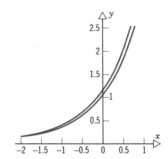

3. $f(x) = 10^x$
  $f'(x) = 10^x \ln 10$

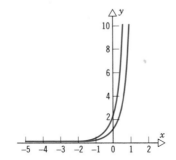

5. (a)

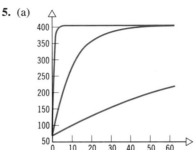

  (b) $k = 0.1$
  (c) $T'(x) = 330k(e^{-kx})$; $T'(0) = 33°F$; $T'(10) = 12.1°F$; $T'(50) = .22°F$

7. (a)

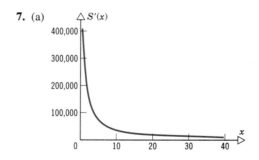

(b) The rate at which the sales increase after investing in advertising is the strongest at the beginning of the advertising campaign. It is decreasing, and is negligible, after investing more than $20,000.

27. $\dfrac{dy}{dx} = \dfrac{1 - e^x}{e^y}$ **29.** $\dfrac{dy}{dx} = \dfrac{1 - ye^{xy}}{xe^{xy}}$

31. $\dfrac{dy}{dx} = \dfrac{y(x - 1)}{x(y - 1)}$

33. Slope $= -\frac{1}{2}$; $y = -\frac{1}{2}x + \frac{5}{2}$

35. If $(x, y) = (\pm R, 0)$ or $(0, \pm R)$, the result is clear. Otherwise, the tangent line at $P(x, y)$ has slope $dy/dx = -x/y$ and OP has slope y/x. Since these two slopes have a product equal to -1, the tangent line at P is perpendicular to the radius OP.

37. The tangent line is never horizontal.

39. (a) $-\dfrac{1 + y}{x + 4y}$ (b) $y = -\dfrac{1}{3}x + \dfrac{5}{3}$ (c) $(6, -3)$

41. $\dfrac{dV}{dP} = \dfrac{V^3(V - b)^2}{2a(V - b)^2 - CV^3}$

43. $A'(t) = \frac{3}{4}(t^{1/4} + 3)^2 t^{-3/4}$; $A'(16) = \frac{75}{32} = 2.34375$

45. 1.5 ft/sec; 3 ft/sec

Exercise 13.6 (page 678)

1. $\dfrac{dy}{dx} = -\dfrac{x}{y}$ **3.** $\dfrac{dy}{dx} = -\dfrac{2y}{x}$ **5.** $\dfrac{dy}{dx} = \dfrac{2x - y}{2y - x}$

7. $\dfrac{dy}{dx} = \dfrac{2x + 4y}{1 - 4x - 2y}$ **9.** $\dfrac{dy}{dx} = -\dfrac{2x}{y^2}$

11. $\dfrac{dy}{dx} = \dfrac{x - 6x^2}{3y^2}$ **13.** $\dfrac{dy}{dx} = \dfrac{y^3}{x^3}$ or $\dfrac{dy}{dx} = \dfrac{x(1 + 4y^2)}{y(1 - 4x^2)}$

15. $\dfrac{dy}{dx} = -\dfrac{y^2}{x^2}$ or $\dfrac{dy}{dx} = \dfrac{2y - 1}{1 - 2x}$

17. $\dfrac{dy}{dx} = \dfrac{y^3 - x^2y}{xy^2 - x^3} = \dfrac{y}{x}$ or $\dfrac{dy}{dx} = \dfrac{3y - x}{y - 3x}$

19. $\dfrac{dy}{dx} = -\dfrac{x}{y}(1 - y^2)^2$ or $\dfrac{dy}{dx} = \dfrac{x(1 - y^2)}{y(x^2 - 1)}$

21. $\dfrac{dy}{dx} = -\dfrac{3(x + 2y)}{2(3x + 4y)}$

23. $\dfrac{dy}{dx} = \dfrac{3(x - y)^2 - 4x(x^2 + y^2)}{4y(x^2 + y^2) + 3(x - y)^2}$

$= \dfrac{3x^2 - 4x^3 - 6xy + 3y^2 - 4xy^2}{3x^2 - 6xy + 4x^2y + 3y^2 + 4y^3}$

25. $\dfrac{dy}{dx} = \dfrac{xy^2 - 3x^2(x^3 + y^3)}{3y^2(x^3 + y^3) - x^2y} = \dfrac{-3x^5 + xy^2 - 3x^2y^3}{-x^2y + 3x^3y^2 + 3y^5}$

Exercise 13.7 (page 686)

1. 2; 0   **3.** $6x + 1$; 6   **5.** $-12x^3 + 4x$; $-36x^2 + 4$

7. $-\dfrac{1}{x^2}$; $\dfrac{2}{x^3}$ **9.** $1 - \dfrac{1}{x^2}$; $\dfrac{2}{x^3}$

11. $\dfrac{1}{(x + 1)^2}$; $-\dfrac{2}{(x + 1)^3}$ **13.** $\dfrac{x^2 + 2x}{(x + 1)^2}$; $\dfrac{2}{(x + 1)^3}$

15. $6x(x^2 + 4)^2$;
$24x^2(x^2 + 4) + 6(x^2 + 4)^2 = 6(5x^2 + 4)(x^2 + 4)$

17. $y' = ae^{ax}$; $y'' = a^2e^{ax}$; $y''' = a^3e^{ax}$

19. $y'' = 4e^{2x}$, so $y'' - 4y = 4e^{2x} - 4e^{2x} = 0$

21. 0   **23.** 0   **25.** $7 \cdot 6 \cdot 5 \cdot 4 \cdot 3 \cdot 2 \cdot 1 = 5040$

27. $y' = -\dfrac{x}{y}$; $y'' = -\dfrac{x^2 + y^2}{y^3}$

29. $y' = -\dfrac{y(1 + 2x)}{x + x^2}$; $y'' = \dfrac{2y(1 + 3x + 3x^2)}{(x + x^2)^2}$

31. $v = 32t + 20$; $a = 32$ **33.** $v = 9.8t + 4$; $a = 9.8$

35. $n! \cdot 2^n = [n \cdot (n - 1) \cdot (n - 2) \cdot \cdots \cdot 2 \cdot 1] \cdot 2^n$

37. $a^n e^{ax}$ **39.** $2g(x) + 4xg'(x) + x^2 g''(x)$

41. (a) 3 sec   (b) -29.4 m/sec
  (c) -29.4 m/sec   (d) -58.8 m/sec

43. (a) 16 ft/sec   (b) $2\frac{1}{2}$ sec   (c) 106 ft   (d) -32 ft/sec²
  (e) $\dfrac{10 + \sqrt{106}}{4} \approx 5.07391$ sec
  (f) $-8\sqrt{106} \approx -82.365$ ft/sec   (g) 206 ft

45. 3 m/sec; $-6(2 - t) = -12 + 6t$ m/sec², $0 \le t \le 2$

47. The growth of the population is increasing by 100,000 per unit of time.

49. $C''(50) = 0.4$

Technology Exercises (*page 688*)

1. (a)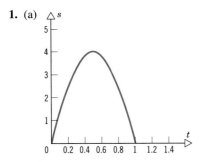

  (b) $t = .5$ sec
  (c) $s = 4$ ft
  (d) $t = 1$ sec

3.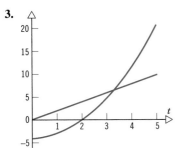

  (a) $t = 0$
  (b) $t = 5$
  (c) 12 units to the right of the origin
  (d) 4 units to the left of the origin

5.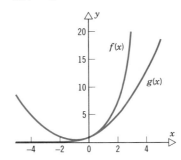

  (a) $t = 1$
  (b) $t = 9$
  (c) 2 units to the right of the origin
  (d) 1 unit to the right of the origin

7. $g(x) = \frac{1}{2}x^2 + x + 1$

Functions overlap around $x = 0$, have same shape and slope, but differ away from 0.

Chapter Review: True–False Items (*page 690*)

1. T   **2.** F   **3.** T   **4.** T   **5.** F   **6.** F   **7.** T

Fill in the Blanks (*page 690*)

1. tangent   **2.** marginal cost
3. power rule; chain rule   **4.** velocity   **5.** 0
6. implicit   **7.** $2xe^{x^2}$

Review Exercises (*page 691*)

1. 2   **3.** 3   **5.** $24x^2$ **7.** $-6x$

9. $12x^3 - 4x + 5$ **11.** $\dfrac{4}{\sqrt{x}}$ **13.** $-\dfrac{4}{x^2}$

15. $6t(t^2 + 1)^2$ **17.** $-\dfrac{3}{x^2} - \dfrac{1}{3}$

19. $(3x^2 + 10x + 1)(12x^2 - 1) + (4x^3 - x + 2)(6x + 1)$

21. $\dfrac{1 - t^2}{(t^2 + 1)^2}$ **23.** $w^{3/2} - 3w^{1/2}$

25. $9(3x - 2)^2 + 30(3x - 2)$

27. $\dfrac{1 + \dfrac{1}{2\sqrt{x}}}{2\sqrt{x} + \sqrt{x}} = \dfrac{1 + 2\sqrt{x}}{4\sqrt{x}\sqrt{x + \sqrt{x}}}$

29. $2t\sqrt{t - 1} + \dfrac{t^2}{2\sqrt{t - 1}} = \dfrac{5t^2 - 4t}{2\sqrt{t - 1}}$ **31.** $20e^{5x}$

33. $\dfrac{15}{x}$ **35.** $4xe^{2x^2 + 5}$ **37.** $\dfrac{4x}{2x^2 + 5}$

39. $(x^2 + 1)^2(x^2 - 1)^3 \left(\dfrac{4x}{x^2 + 1} + \dfrac{6x}{x^2 - 1} \right) =$

   $(x^2 + 1)(x^2 - 1)^2(2x)(5x^2 + 1)$

41. $f'(1) = 8; f''(1) = 32$ **43.** $f'(1) = \frac{2}{3}; f''(1) = -\frac{2}{9}$

45. $f'(1) = -3; f''(1) = 6$

47. $f'(x) = 6x^2 + 6x - 12; f'(x) = 0$ when $x = -2$ or $x = 1; f''(x) = 12x + 6; f''(-2) = -18; f''(1) = 18$

49. $f'(x) = 3x\sqrt{x^2 - 1}; f'(x) = 0$ when $x = \pm 1$ (0 is not in the domain of f); $f''(x) = 3\sqrt{x^2 - 1} + \dfrac{3x^2}{\sqrt{x^2 - 1}}; f''(x)$ does not exist when $x = \pm 1$

51. $\dfrac{dy}{dx} = \dfrac{1 + 4xy - 3y^2}{6xy - 2x^2}$ **53.** $\dfrac{dy}{dx} = \dfrac{8x - 1}{2y + 1}$

55. $\dfrac{dy}{dx} = \dfrac{\dfrac{1}{2\sqrt{x}} - y}{x - \dfrac{1}{2\sqrt{y}}} = \dfrac{\sqrt{y}\,(1 - 2y\sqrt{x})}{\sqrt{x}\,(2x\sqrt{y} - 1)}$

57. $f'(-4), f'(3), f'(2), f'(-2), f'(1), f'(0)$

Mathematical Question from Professional Exam **(page 692)**

1. e

CHAPTER 14

Exercise 14.1 (*page 703*)

1. $[x_1, x_9] = \{x \mid x_1 \le x \le x_9\}$

3. Increasing on (x_1, x_4), (x_5, x_7), and (x_8, x_9)

5. x_4, x_6, x_7, and x_8

7. $(x_4, y_4), (x_7, y_7)$, and (x_9, y_9)

9. x_4, x_5, x_6, x_7, and x_8

11. Domain: all real numbers
   Increasing on $(2, \infty)$
   Local minimum at $(2, -4)$
   Intercepts: $(0, 0)$ and $(4, 0)$
   Decreasing on $(-\infty, 2)$
   Horizontal tangent at $x = 2$

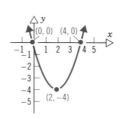

13. Domain: all real numbers
   Increasing on $(-\infty, 3)$
   Local maximum at $(3, 16)$
   Intercepts: $(0, 7), (-1, 0)$,
   and $(7, 0)$

Decreasing on $(3, \infty)$
Horizontal tangent at $x = 3$

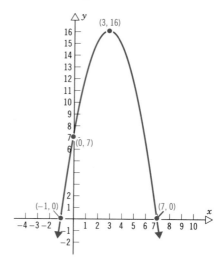

15. Domain: all real numbers
Increasing on $(-\infty, 1)$
Local maximum
at $(1, 0)$
Intercepts: $(0, -2)$
and $(1, 0)$
Decreasing on $(1, \infty)$
Horizontal tangent
at $x = 1$

17. Domain: all real numbers
Increasing on $(-\infty, \infty)$
No local maximum or
local minimun
Intercepts: $(0, -27)$
and $(3, 0)$
Not decreasing
Horizontal tangent
at $x = 3$

19. Domain: all real
numbers
Increasing on
$(-\infty, 2)$
and $(3, \infty)$
Local maximum
at $(2, 28)$
Local minimum
at $(3, 27)$
Intercept: $(0, 0)$
Decreasing
on $(2, 3)$
Horizontal tangents
at $x = 2$
and $x = 3$

21. Domain: all real numbers
Increasing on $(-1, 1)$
Local maximum at $(1, 1)$
Local minimum
at $(-1, -3)$
Intercepts: $(0, -1)$ and
three x-intercepts
as shown
Decreasing on
$(-\infty, -1)$
and $(1, \infty)$
Horizontal tangents at
$x = -1$ and $x = 1$

23. Domain: all real numbers
Not increasing
No local maximum or
local minimum
Intercepts: $(0, 2)$ and one
x-intercept as shown
Decreasing on $(-\infty, \infty)$
Horizontal tangent at $x = 0$

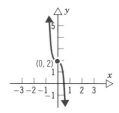

25. Domain: all real numbers
Increasing on $(-\infty, -1)$
and $(1, \infty)$
Local maximum at $(-1, 5)$
Local minimum at $(1, -3)$
Intercepts: $(0, 1)$ and three
x-intercepts as shown
Decreasing on $(-1, 1)$
Horizontal tangent at $x = -1$
and $x = 1$

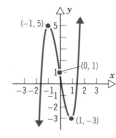

27. Domain: all real
numbers
Increasing on
$(-\infty, -2)$
and $(2, \infty)$
Local maximum
at $(-2, 65)$
Local minimum
at $(2, -63)$
(*Note:* Different scales
for x- and y-axes.)
Intercepts: $(0, 1)$ and three x-intercepts as shown
Decreasing on $(-2, 2)$
Horizontal tangent at $x = -2$, $x = 0$, and $x = 2$

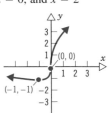

29. Domain; all real numbers
Increasing on $(-1, \infty)$
Local minimum at $(-1, -1)$
Intercept: $(0, 0)$
Decreasing on $(-\infty, -1)$
Horizontal tangent at $x = -1$
Vertical tangent at $x = 0$

31. Domain: all real numbers
Increasing on $(-1, 0)$
and $(1, \infty)$
Local maximum at $(0, 1)$
Local minimum at $(-1, 0)$
and $(1, 0)$
Intercepts: $(0, 1)$, $(-1, 0)$,
and $(1, 0)$
Decreasing on $(-\infty, -1)$ and $(0, 1)$

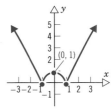

(Continued)

31. *(Continued)*

Horizontal tangent at $x = 0$

No tangent line at $x = -1$
or at $x = 1$

33. Domain: $\{x|x \neq -2 \text{ and } x \neq 2\}$

Not increasing

No local maximum
or local minimum

Intercept: $(0, 0)$

Decreasing on $(-\infty, -2)$,
$(-2, 2)$, and $(2, \infty)$

No horizontal or
vertical tangent lines

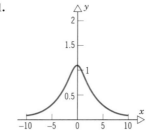

35. $f(x)$ is increasing when $x > 0$; $f(x)$ is decreasing when $x < 0$

37. (a) v   (b) ii   (c) i   (d) iii   (e) iv

39. (a) Increasing on $(0, 2000)$; decreasing on $(2000, \infty)$

   (b) 2000 trucks   (c) $20,000

   (d)

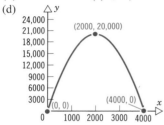

41. (a) Increasing on $(0, 7)$; decreasing on $(7, \infty)$

   (b) $7 per crate   (c) 2450 crates

   (d)

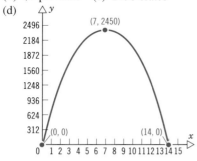

43. 1400 units; $19,575   **45.** 2 hr after injection

47. The function $C'(q)$ is increasing for $q > -\frac{1}{6}$. Since q is taking only positive values, the marginal cost is increasing for every q.

49. $C'(q) = \dfrac{1}{2\sqrt{q}}$; $C''(q) = -\dfrac{1}{4q\sqrt{q}}$; $-\dfrac{1}{4q\sqrt{q}} < 0$

(for $q > 0$); $C''(q) = (C'(q))' < 0$ (for $q > 0$). Thus, the function $C'(q)$ is always decreasing for $q > 0$.

$\overline{C}(q) = \dfrac{\sqrt{q}}{q}$; $\overline{C}'(q) = -\dfrac{1}{2q\sqrt{q}}$; $-\dfrac{1}{2q\sqrt{q}} < 0$ (for $q > 0$);

$\overline{C}'(q) < 0$ (for $q > 0$). Thus, the function $\overline{C}(q)$ is always decreasing for $q > 0$.

51. $c = \frac{1}{2}$ **53.** $c = 0$ **55.** $c = \frac{3}{2}$

57. $c = 2/\sqrt[3]{3} \approx 1.38672$

Technology Exercises (page 707)

1.

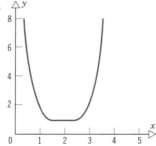

Increasing: $x < 0$; decreasing: $x > 0$

3.

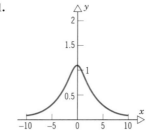

Relative minimum: $(2, 1)$

5. (a) $f'(x) = 4x^3 + 24x^2 + 36x$

   (b)

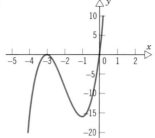

   (c) $f'(x) = 0$ for $x = -3$, $x = 0$;
   $f'(x) > 0$ for $x > 0$; $f'(x) < 0$ for $x < -3$ and
   $-3 < x < 0$

   (d) $x > 0$ (e) $x < -3$ and $-3 < x < 0$

(f)

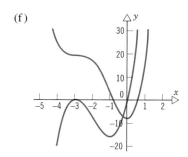

Exercise 14.2 *(page 718)*

1. $\{x \mid x_1 \le x < x_4 \text{ or } x_4 < x \le x_7\} =$
$[x_1, x_4) \cup (x_4, x_7]$

3. Increasing on (x_1, x_3), $(0, x_4)$, and (x_4, x_6)

5. $x = 0$ and $x = x_6$

7. Local maximum at (x_3, y_3) and (x_6, y_6)

9. Concave up on (x_1, x_3) and (x_3, x_4)

11. Concave down on $(-\infty, 2)$
Concave up on $(2, \infty)$
Inflection point: $(2, -15)$

13. Concave down on $(0, 1)$
Concave up on $(-\infty, 0)$ and $(1, \infty)$
Inflection points: $(0, -1)$ and $(1, 4)$

15. Concave down on $(-\infty, 1)$
Concave up on $(1, \infty)$
Inflection point: $(1, 68)$

17. Concave down on $(-\infty, -1)$ and $(0, 1)$
Concave up on $(-1, 0)$ and $(1, \infty)$
Inflection points: $(-1, 7)$, $(1, 13)$, and $(0, 10)$

19. Concave down on $(-\infty, 1)$
Concave up on $(1, \infty)$
Inflection point: $(1, -5)$

21. Concave down on $(-\infty, 0)$
Concave up on $(0, \infty)$
Inflection point: $(0, 2)$

23. Concave down on $(-\infty, -2)$
Concave up on $(-2, \infty)$
Inflection point:
$(-2, -12 \cdot 2^{2/3}) \approx (-2, -19.0488)$

25. Concave up on $(-\infty, \infty)$
No inflection points

27. Domain: all real numbers
Increasing on $(-\infty, 0)$ and $(4, \infty)$
Local maximum at $(0, 1)$
Local minimum at $(4, -31)$
Intercepts: $(0, 1)$ and three
x-intercepts as shown
Decreasing on $(0, 4)$
Horizontal tangent at $x = 0$
and $x = 4$
Inflection point at $(2, -15)$

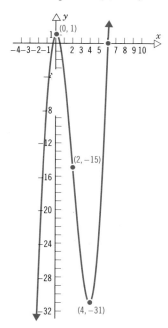

29. Domain: all real numbers
Increasing on $(-1, 0)$
and $(1, \infty)$
Local maximum at $(0, 1)$
Local minima at $(-1, 0)$ and
$(1, 0)$
Intercepts: $(0, 1)$, $(-1, 0)$, and
$(1, 0)$
Decreasing on $(-\infty, -1)$ and
$(0, 1)$
Horizontal tangent at $x = -1$,
$x = 0$, and $x = 1$
Inflection points at
$\left(\pm \dfrac{1}{\sqrt{3}}, \dfrac{4}{9}\right) \approx (\pm 0.577, 0.444)$

31. Domain: all real numbers
Increasing on $(-\infty, 0)$
and $(8, \infty)$
Local maximum at $(0, 0)$
Local minimum at
$(8, -8192)$
Intercepts: $(0, 0)$ and
$(10, 0)$
Decreasing on $(0, 8)$
Horizontal tangent
at $x = 0$ and $x = 8$
Inflection point at
$(6, -5184)$

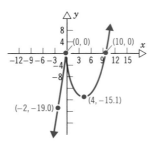

33. Domain: all real numbers
Increasing on $(\frac{5}{2}, \infty)$
Local minimum
at $(\frac{5}{2}, -\frac{3125}{64}) \approx$
$(2.5, -48.828)$
Intercepts: $(0, 0)$ and $(3, 0)$
Decreasing on $(-\infty, \frac{5}{2})$
Horizontal tangent at $x = 0$
and $x = \frac{5}{2}$
Inflection point at $(2, -32)$

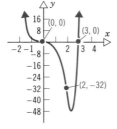

35. Domain: all real numbers
Increasing on $(3, \infty)$
Local minimum
at $(3, -81)$
Intercepts: $(0, 0)$ and $(4, 0)$
Decreasing on $(-\infty, 3)$
Horizontal tangent
at $x = 0$ and $x = 3$
Inflection point
at $(2, -48)$

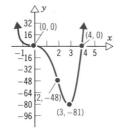

37. Domain: all real numbers
Increasing on $(-\infty, 0)$ and
$(\sqrt[3]{4}, \infty)$
Local maximum at $(0, 4)$
Local minimum at $(\sqrt[3]{4}, f(\sqrt[3]{4})) \approx$
$(1.587, -11.119)$
Intercepts: $(0, 4)$ and three
x-intercepts as shown
Decreasing on $(0, \sqrt[3]{4})$
Horizontal tangent at $x = 0$
and $x = \sqrt[3]{4}$
Inflection point at $(1, -5)$

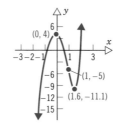

39. Domain: all real numbers
Increasing on $(-\infty, 0)$ and $(4, \infty)$
Local maximum at $(0, 0)$
Local minimum at $(4, f(4)) \approx (4, -15.119)$

Intercepts: $(0, 0)$ and $(10, 0)$
Decreasing on $(0, 4)$
Horizontal tangent at $x = 4$
Vertical tangent at $x = 0$
Inflection point at $(-2, f(-2)) \approx (-2, -19.049)$

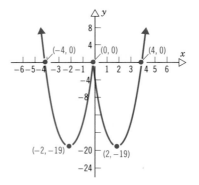

41. Domain: all real numbers
Increasing on $(-2, 0)$ and $(2, \infty)$
Local maximum at $(0, 0)$
Local minima at $(\pm 2, -12\sqrt[3]{4}) \approx (\pm 2, -19.049)$
Intercepts: $(0, 0)$, $(-4, 0)$, and $(4, 0)$
Decreasing on $(-\infty, -2)$ and $(0, 2)$
Horizontal tangent at $x = -2$ and $x = 2$
Vertical tangent at $x = 0$
No inflection points

43. Domain: $\{x | x > 0\}$
Increasing on $(1/e, \infty)$
Local minimum at $(1/e, -1/e)$
Intercept: $(1, 0)$
Decreasing on $(0, 1/e)$
Horizontal tangent at $x = 1/e$
No inflection points

45. Domain: all real numbers
Increasing on $(-1, \infty)$
Local minimum at $(-1, -1/e)$
Intercept: $(0, 0)$

Decreasing on $(-\infty, -1)$
Horizontal tangent at $x = -1$
Inflection point at $(-2, -2/e^2)$

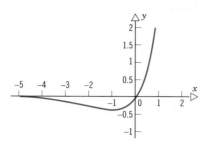

47. Local minimum at $(1, 0)$; local maximum at $(-1, 4)$

49. Local minimum at $(-1, -4)$
No conclusion from Second Derivative Test at $x = 0$;
$(0, -3)$ is neither a maximum nor a minimum.

51. Local minimum at $(4, -254)$
No conclusion from Second Derivative Test at $x = 0$;
$(0, 2)$ is a local maximum.

53. Local minimum at $(1, 2)$; local maximum at $(-1, -2)$

55. 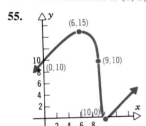   **57.**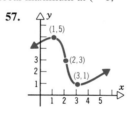

59. $a = -3, b = 9; f(x) = -3x^3 + 9x^2$

61. (a) $P(x) = -x^2 + 6x - 5$
(b) $x = 3$ units   (c) $P(3) = \$4$ million
(d) $x = 4$ (e) $R(4) = \$16$ million

63. (a) Revenue $= -0.01x^2 + 100x$
(b) $250,000$ (when $x = 5000$)
(c) Profit $= -0.01x^2 + 50x - 40,000$
(d) $22,500$ (when $x = 2500$)

65. (a) Revenue $= 10x + 40, 1 < x < 10$
(b) x near 10 (but 10 isn't in the domain!)
(c) $\lim\limits_{x \to 10^-}$ Revenue $= \$140$

67. $f(x)$ *must* have an inflection point: Since $a \neq 0, f''(x) = 6ax + 2b = 0$ when $x = -b/3a$ and $f''(x)$ changes sign across $x = -b/3a$. Thus, $(-b/3a, f(-b/3a))$ is an inflection point for $f(x)$.

69. Assuming the domain of $C(x)$ to be $[0, \infty)$, $C(x)$ is concave up on $(100, \infty)$ and concave down on $(0, 100)$. The inflection point is $(100, 1042)$. The cost increases rapidly at the beginning, but its increase is slowing down until the 100th unit is produced. After that, the increase in cost of production is faster and faster.

71. (a) $\overline{C}(x) = \dfrac{800}{x} + 0.04 + 0.0002x$

(b) 0.84 (when $x = 2000$)
(c) $C'(x) = 0.04 + 0.0004x$
(d)

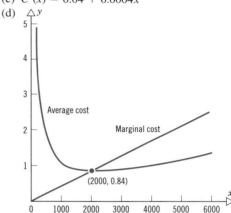

73. $v'(r) = -2kr$, assuming $k > 0, v'(r) < 0$ for $0 < r < R$. Thus, $v(r)$ is a decreasing function throughout its domain (i.e., on $[0, R]$), and the maximum value for $v(r)$ occurs at the left endpoint, $r = 0$. That is, the maximum velocity of blood occurs along the central axis.

75. $P(x) = 0.1x - 20 - \ln(x + 1)$; 9 units

Technology Exercises (*page 721*)

1.

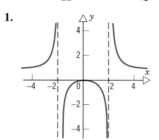

Concave up: $x < -2$ and $x > 2$
Concave down: $-2 < x < 2$
Inflection points: none

3.

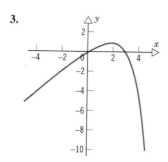

(a) Maximum at $(2, 1)$
(b) Concave down for $-\infty < x < +\infty$

Exercise 14.3 (page 729)

1. 1   **3.** 2   **5.** 3   **7.** $-\infty$

9. Horizontal asymptote: $y = 3$; vertical asymptote: $x = 0$

11. Horizontal asymptote: $y = 0$; vertical asymptote: $x = 1$

13. Horizontal asymptote: $y = 1$; vertical asymptotes: $x = -2$ and $x = 2$

15. Domain: $\{x | x \neq \pm 2\}$
Local maximum at $(0, -\frac{1}{2})$
Vertical asymptotes: $x = -2$ and $x = 2$
Intercept: $(0, -\frac{1}{2})$
No inflection points
Horizontal asymptote: $y = 0$

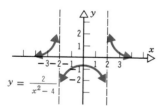

17. Domain: $\{x | x \neq -1\}$
No local extrema
Vertical asymptote: $x = -1$
Intercepts: $(0, -1)$ and $(\frac{1}{2}, 0)$
No inflection points
Horizontal asymptote: $y = 2$

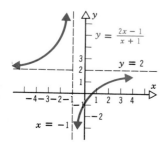

19. Domain: all real numbers
Local maximum at $(1, \frac{1}{2})$
Local minimum at $(-1, -\frac{1}{2})$
No vertical asymptote
Intercept: $(0, 0)$
Inflection points:
$(-\sqrt{3}, \sqrt{3}/4) \approx (-1.732, -0.433)$,
$(0, 0)$, and $(\sqrt{3}, \sqrt{3}/4) \approx (1.732, 0.433)$
Horizontal asymptote: $y = 0$

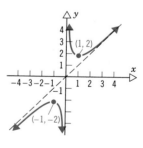

21. Domain: $\{x | x \neq \pm 4\}$
Local maximum at $(0, -\frac{1}{2})$
Vertical asymptotes: $x = -4$ and $x = 4$
Horizontal asymptote: $y = 0$
Intercept: $(0, -\frac{1}{2})$
No inflection points
Horizontal asymptote: $y = 0$

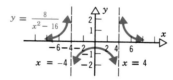

23. Domain: $\{x | x \neq 0\}$
Local maximum at $(-1, -2)$
No inflection points
Vertical asymptote: $x = 0$
No intercepts
Local minimum at $(1, 2)$
"Tilted" asymptote: $y = x$

25. One possible graph appears here. (Dashed lines indicate tangent rays.)

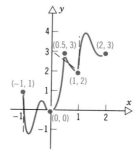

27. (a) $\bar{P}(x) = 58 + \dfrac{10{,}000}{x}$

(b) $\bar{P}(1000) = 48; \bar{P}(10{,}000) = 57; \bar{P}(100{,}000) = 57.9$

(c) $\lim\limits_{x \to \infty} \bar{P}(x) = 58$

29. $\lim\limits_{x \to 100^-} C(x) = \infty$. Thus, obtaining an entirely pollution-free lake is prohibitively expensive.

31. $\lim\limits_{x \to \infty} S(x) = \frac{2000}{3.5} = 571\frac{3}{7} \approx 571.429$. Note that $S(x)$ approaches 571.429 from below as $x \to \infty$. Spending more money on advertising will increase sales, but sales will never exceed 571.429 units.

Technology Exercises (page 730)

1. (a)

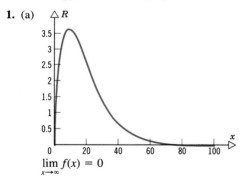

$\lim\limits_{x \to \infty} f(x) = 0$

(b) After investing more than $80,000 the revenue is closer and closer to zero.

3. Horizontal asymptote: $y = 2$

Vertical asymptote: $x = 0$

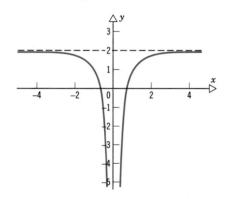

5. Horizontal asymptote: $y = 3$

Vertical asymptote: $x = -1$

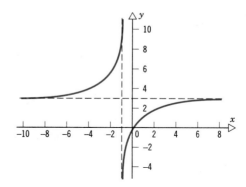

7. Horizontal asymptote: $y = 0$

Vertical asymptotes: $x = -1$ and $x = 1$

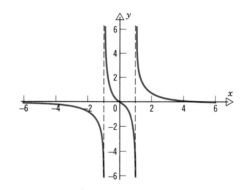

Exercise 14.4 (page 747)

1. Absolute maximum: $f(3) = 15$; absolute minimum: $f(-1) = -1$

3. Absolute maximum: $f(0) = 1$; absolute minimum: $f(4) = -39$

5. Absolute maximum: $f(4) = 16$; absolute minimum: $f(2) = -4$

7. Absolute maximum: $f(0) = 1$; absolute minimum: $f(1) = 0$

9. Absolute maximum: $f(-1) = f(1) = 1$; absolute minimum: $f(0) = 0$

11. Absolute maximum: $f(1) = 2$; absolute minimum: $f(4) = 4$

13. Absolute maximum: $f(1/\sqrt{2}) = \frac{1}{2}$; absolute minimum: $f(-1/\sqrt{2}) = -\frac{1}{2}$

15. Absolute maximum: $f(0) = 0$; absolute minimum: $f(-1) = f(\frac{1}{2}) = -\frac{1}{2}$

17. Absolute maximum: $f(5) = 98\sqrt[3]{2} \approx 123.472$; absolute minimum: $f(-2) = f(1) = 0$

19. Absolute maximum: $f(\frac{11}{2}) = \frac{2}{9}\sqrt[3]{\frac{3}{2}} \approx 0.254$; absolute minimum: $f(2) = -\sqrt[3]{2} \approx -1.260$

21. Absolute maximum: $f(e) = 1/e \approx 0.367879$; absolute minimum: $f(1) = 0$

23. Absolute maximum: $f(10) = 10e^{10} \approx 220{,}265$; absolute minimum: $f(-1) = -e^{-1} \approx -0.367879$

25. (750 m perpendicular to road) × (1500 m parallel to road) = 1,123,000 m²

27. 50 m × 50 m

29. 37,500,000 m²: three 5000 m fences and two 7500 m fences

31. $\dfrac{L}{4} \times \dfrac{L}{4}$

33. (4 cm high) × (16 cm wide) × (16 cm deep)

35. (20 cm) × (20 cm) × (20 cm)

37. Radius $= \sqrt[3]{\dfrac{15}{4\pi}} \approx 1.061$ m; height $= \sqrt[3]{\dfrac{640}{9\pi}} \approx 2.829$ m

39. 80 members

41. $5\sqrt{\frac{3}{13}} \approx 2.402$ km down shore from P $(12 - 5\sqrt{\frac{3}{13}} \approx 9.598$ km from town)

43. 40 mph   **45.** (7 in. wide) × (14 in. tall)

47. Graph of $t = 18 - 3q^2$:

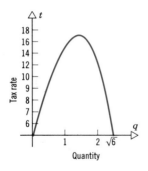

Quantity

Graph of $q = \sqrt{\dfrac{18 - t}{3}}$:

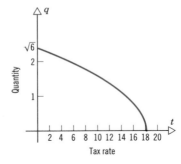

Tax rate

In the second graph, note that a higher tax rate (predictably) diminishes consumption. Revenue $= tq$; the optimal tax rate is $t = 12$, yielding revenue of $12\sqrt{2} \approx 16.971$.

49. Let h be the height of the container and r be its radius. Then $V = \pi r^2 h$, and the surface area is given by $S = 2\pi r^2 + 2\pi rh = 2\pi r^2 + 2\pi r\left(\dfrac{V}{\pi r^2}\right) = 2\pi r^2 + \dfrac{2V}{r}$

(for $r > 0$). $\dfrac{dS}{dr} = 4\pi r - \dfrac{2V}{r^2} = 0$ when $r^3 = \dfrac{V}{2\pi}$, or $r = \sqrt[3]{\dfrac{V}{2\pi}}$. Note that $S''(r) = \dfrac{8V}{r^3} > 0$ for $r > 0$, $\lim\limits_{r \to 0^+} S(r) = \infty$, and $\lim\limits_{r \to \infty} S(r) = \infty$, so the surface area is minimized when $r = \sqrt[3]{\dfrac{V}{2\pi}}$. Here the height is

$$\dfrac{V}{\pi\left(\dfrac{V}{2\pi}\right)^{2/3}} = 2\sqrt[3]{\dfrac{V}{2\pi}} = 2r.$$

51. 199; -59.7; -59.7 **53.** $x = \dfrac{\ln(60.75)}{0.405} \approx 10.1402$

55. At $x = \ln(50) \approx 3.912$ months, the sales rate is maximum.

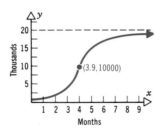

Months

57. At $t = \ln(9999) \approx 9.210$ days, the flu is spreading fastest.

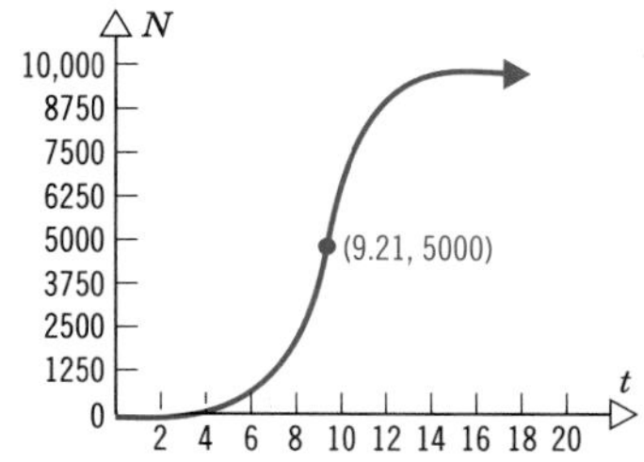

59. $p'(t) = \dfrac{72{,}000\, e^t}{(e^t + 9)^2}$. The growth rate is maximum when $t = \ln 9 \approx 2.19722$ days; $\lim_{t\to\infty} p(t) = 8000$.

Technology Exercises (page 750)

1. (a) $x = 0$
(b) $y = -4$
(c) $y = 4$
(d) $f(0) = 0$
(e) $f(1) = 4$
(f) $f(-1) = -4$
(g)

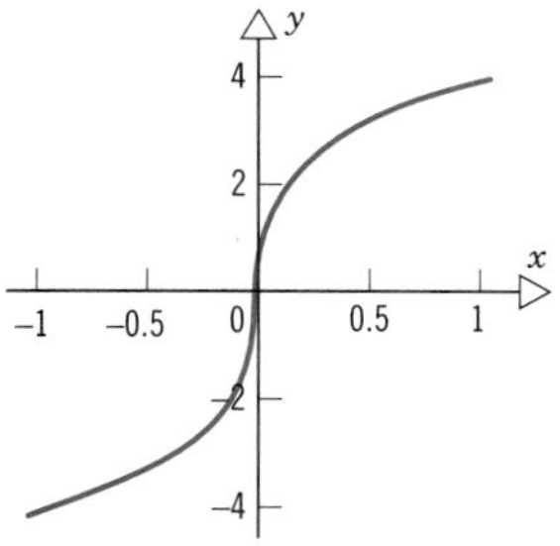

3. Sales rate is maximum at $x = \ln 50 \approx 3.912$, or at the end of the 4th month.
Concave up: $x < \ln 50$
Concave down: $\ln 50 < x$
The increase in sales is growing for the first 4 months.
The increase in sales is falling for the last 8 months.

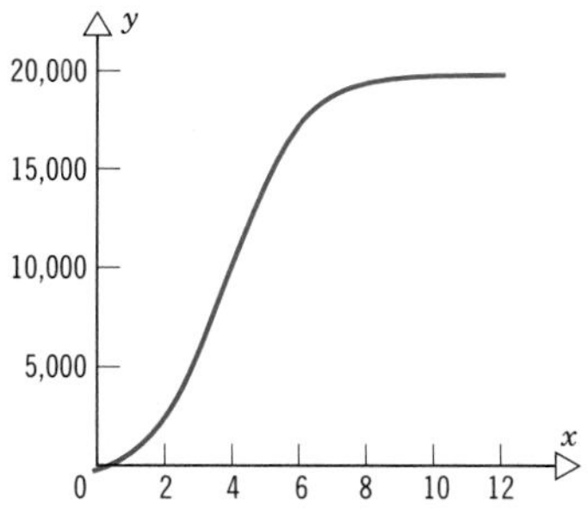

5.

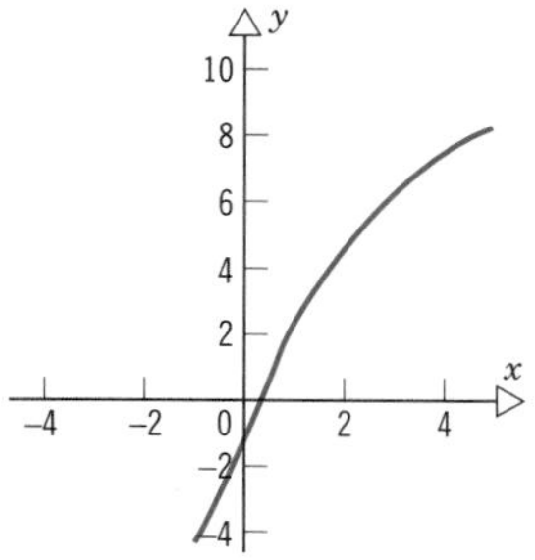

Maximum: $f(5) = 8.775$
Minimum: $f(-1) = -4$

7.

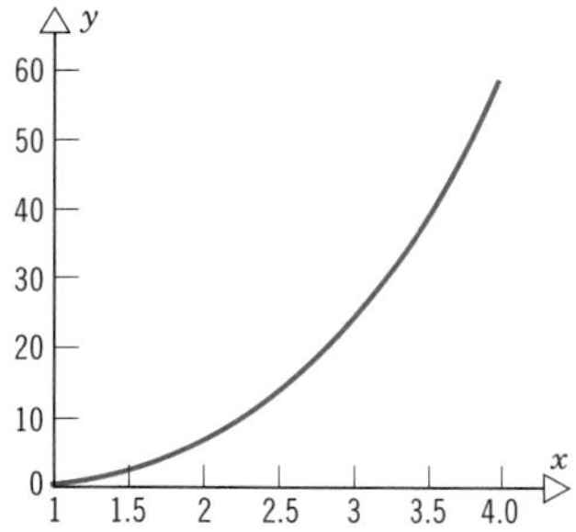

Maximum: $f(4) = 60.5$
Minimum: $f(1) = 1.1$

Exercise 14.5 (page 754)

1. (a) $x = 4000 - 100p$
(b) $E(p) = \dfrac{-100p}{4000 - 100p} = \dfrac{p}{p - 40}$
(c) $E(5) = -\frac{1}{7} \approx 0.143$; 1.43% decrease in quantity demanded
(d) $E(15) = -\frac{3}{5} = -0.6\%$; 6% decrease in quantity demanded
(e) $E(20) = -1$; 10% decrease in quantity demanded

3. (a) $x = 10{,}000 - 200p$ (b) $E(p) = \dfrac{p}{p - 50}$
(c) $E(10) = -\frac{1}{4} = -0.25$; 1.25% decrease in quantity demanded
(d) $E(25) = -1$; 5% decrease in quantity demanded
(e) $E(35) = -\frac{7}{3} \approx -2.333$; 11.67% decrease in quantity demanded

5. $E(p) = \dfrac{p}{p - 200}$; $E(50) = -\dfrac{1}{3}$; inelastic

7. $E(p) = \dfrac{-p}{p + 4}$; $E(10) = -\dfrac{5}{7}$; inelastic

9. $E(p) = \dfrac{2p^2}{p^2 - 1000}$; $E(10) = -\dfrac{2}{9}$; inelastic

11. $E(p) = \dfrac{p}{2p - 200}$; $E(10) = -\dfrac{1}{18}$; inelastic

13. $E(p) = \dfrac{2p}{p - 4}$; $E(2) = -2$; elastic

15. $E(p) = \dfrac{-3\sqrt{p}}{2(20 - 3\sqrt{p})}$; $E(4) = -\dfrac{3}{14}$; inelastic

17. $E(4) = -\frac{64}{65} \approx -0.985$

19. $E(5) = -\frac{15}{17} \approx -0.882$ **21.** $E(5) = -39$

23. $E(2) = -\frac{1}{8} = -0.125$ **25.** $E(100) = -3$

27. (a) Elastic   (b) Decrease

29. (a) Inelastic   (b) Decrease

31. (a) $3000 - \dfrac{200p}{3}$ (b) $E(18) = -\frac{2}{3} \approx -0.667$

(c) $-\frac{10}{3} \approx -3.33\%$ (d) Increase

Exercise 14.6 (*page 760*)

1. $dx/dt = -3$ **3.** $dx/dt = -\frac{8}{9}$ **5.** $dV/dt = 40$

7. $dV/dt = 5\pi^2$

9. x-coordinate is decreasing at the rate of 6 units per minute.

11. (a) $x^2 + y^2 = 100$

(b) $\sqrt{3}/2 \approx 0.866$ ft/sec

13. 900 cm³/sec   **15.** $dy/dt = -8/\sqrt{29}$ cm/min

17. $\frac{3}{4}$ m²/sec   **19.** $dA/dt \approx 317$ ft²/min

21. (a) 500 per day   (b) 1480 per day   (c) Increasing

(d) $P(x) = 10x - \dfrac{x^2}{10,000} - 5000$ (e) 980 per day

Exercise 14.7 (*page 767*)

1. $dy = (3x^2 - 2)\,dx$ **3.** $dy = \left[\dfrac{-x^2 + 2x - 6}{(x^2 + 2x - 8)^2}\right] dx$

5. $\dfrac{dy}{dx} = -\dfrac{y}{x}$; $\dfrac{dx}{dy} = -\dfrac{x}{y}$ **7.** $\dfrac{dy}{dx} = -\dfrac{x}{y}$; $\dfrac{dx}{dy} = -\dfrac{y}{x}$

9. $3x^2 + 3y^2\,dy = 6xy + 3x^2\,dy$;

$3x^2\,dx + 3y^2 = 6xy\,dx + 3x^2$; $\dfrac{dy}{dx} = \dfrac{x^2 - 2xy}{x^2 - y^2}$;

$\dfrac{dx}{dy} = \dfrac{x^2 - y^2}{x^2 - 2xy}$

11. $\dfrac{1}{2\sqrt{x - 2}}\,dx$ **13.** $(3x^2 - 1)\,dx$

15. $f(x) \approx 2x - 3$ near $x = 2$

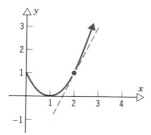

17. $f(x) \approx \dfrac{x}{4} + 1$ near $x = 4$

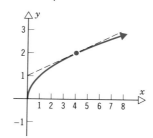

19. (a) $\dfrac{71}{12} \approx 5.9167$ (b) 5.12

(c) Using $f(x) = \sqrt{x}$ and $x_0 = 1$, $1/\sqrt{1.2} \approx 0.91667$.
Using $f(x) = \sqrt{x}$ and $x_0 = 1/1.21$, $1/\sqrt{1.2} \approx 0.91288$.
Using $f(x) = 1/\sqrt{x}$ and $x_0 = 1$, $1/\sqrt{1.2} \approx 0.9$.
Using $f(x) = 1/\sqrt{x}$ and $x_0 = 1.21$, $1/\sqrt{1.2} \approx 0.91285$.

21. (a) 0.006   (b) 0.00125   **23.** $2\pi \approx 6.28319$ cm²

25. $3.6\pi \approx 11.3097$ m³   **27.** 6%   **29.** 8π cm³

31. 30 m; 0.9% error   **33.** Approx. 72 min

Technology Exercises (*page 769*)

1. (a) $f'(x) = 12x$
(b) $g(x) = 6x - \frac{3}{2}$
(c)

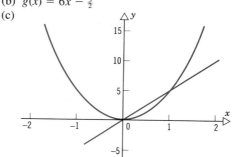

(d) $0 < x < 1$

(e)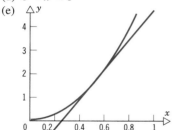

(f)

x	$f(x)$	$g(x)$	$\lvert f(x) - g(x)\rvert$
0.2	0.24	-0.3	0.54
0.3	0.54	0.3	0.24
0.4	0.96	0.9	0.06
0.5	1.5	1.5	0
0.6	2.16	2.1	0.06
0.7	2.94	2.7	0.24
0.8	3.84	3.3	0.54

(g) $0.31743 < x < 0.68257$

(h) $0.3709 < x < 0.6291$

3. (a) $y - 16.9 = \dfrac{9.5 - 22.3}{5.5 - 4.5}(x - 5)$, so $y = 80.9 - 12.8x$

(b) $f(5.25) \approx 13.7$, $f(6.5) \approx -2.3$, $f(7) \approx -8.7$; 5.25 is the closest to $x = 5$, so we are the most confident about the estimate $f(5.25) \approx 13.7$.

Chapter Review: True–False Items (*page 771*)

1. F   **2.** T   **3.** F   **4.** F   **5.** T   **6.** T

7. F $[dy = (2x + 2)\,dx]$ **8.** T   **9.** F

Fill in the Blanks (*page 771*)

1. decreasing   **2.** decreasing; increasing

3. concave up   **4.** horizontal   **5.** concavity

6. $f'(x)\,dx$ **7.** tangent line

Review Exercises (*page 771*)

1.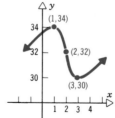

3. (a) Domain: all real numbers
   (b) Intercept: $(1, 0)$
   (c) Increasing on $(-\infty, \infty)$
   (d), (e) No local maxima or minima
   (f) Inflection point at $(1, 0)$
   (g) No horizontal or vertical asymptotes
   (h) Graph:

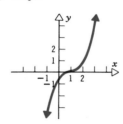

5. (a) Domain: all real numbers
   (b) Intercepts: $(-\sqrt{2}, 0)$, $(0, 0)$, and $(\sqrt{2}, 0)$
   (c) Increasing on $(-1, 0)$ and $(1, \infty)$; decreasing on $(-\infty, -1)$ and $(0, 1)$
   (d) Local maximum at $(0, 0)$; local minima at $(-1, -1)$ and $(1, -1)$
   (e) Tangent line has slope 0 at $(0, 0)$, $(-1, -1)$, and $(1, -1)$
   (f) Inflection points at $\left(-1/\sqrt{3}, -\frac{5}{9}\right)$ and $\left(1/\sqrt{3}, -\frac{5}{9}\right)$
   (g) No horizontal or vertical asymptotes
   (h) Graph:

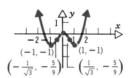

7. (a) Domain: all real numbers
   (b) Intercepts: $(\pm \sqrt[4]{5}, 0) \approx (\pm 1.495, 0)$ and $(0, 0)$
   (c) Increasing on $(-\infty, -1)$ and $(1, \infty)$; decreasing on $(-1, 1)$

(d) Local maximum at $(-1, 4)$; local minimum at $(1, -4)$

(e) Tangent line has slope 0 at $(-1, 4)$ and $(1, -4)$

(f) Inflection point at $(0, 0)$

(g) No horizontal or vertical asymptotes

(h) Graph:

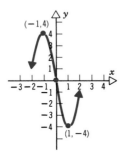

9. (a) Domain: all real numbers

(b) Intercepts: $(0, 0)$ and $(2, 0)$

(c) Increasing on $(0, 1)$ and $(2, \infty)$; decreasing on $(-\infty, 0)$ and $(1, 2)$

(d) Local maximum at $(1, 1)$; local minima at $(0, 0)$ and $(2, 0)$

(e) Tangent line has slope 0 at $(1, 1)$, $(0, 0)$, and $(2, 0)$

(f) Inflection points at $\left(\dfrac{3 - \sqrt{3}}{2}, f\left(\dfrac{3 - \sqrt{3}}{2}\right)\right) \approx$ $(0.423, 0.444)$

and at $\left(\dfrac{3 + \sqrt{3}}{2}, f\left(\dfrac{3 + \sqrt{3}}{2}\right)\right) \approx (1.577, 0.444)$

(g) No horizontal or vertical asymptotes

(h) Graph:

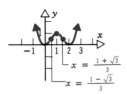

11. (a) Domain: all real numbers

(b) Intercepts: $(-4, 0)$ and $(0, 0)$

(c) Increasing on $(-1, \infty)$; decreasing on $(-\infty, -1)$

(d) Local minimum at $(-1, -3)$

(e) Tangent line has slope 0 at $(-1, -3)$

(f) Inflection points at $(0, 0)$ and $(2, f(2)) \approx (2, 7.560)$

(g) No horizontal or vertical asymptotes

(h) Graph:

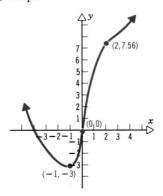

13. (a) Domain: all real numbers

(b) Intercept: $(0, 0)$

(c) Increasing on $(-1, 1)$; decreasing on $(-\infty, -1)$ and $(1, \infty)$

(d) Local maximum at $(1, 1)$; local minimum at $(-1, -1)$

(e) Tangent line has slope 0 at $(1, 1)$ and $(-1, -1)$

(f) Inflection points at $(-\sqrt{3}, -\sqrt{3}/2) \approx$ $(-1.732, -0.866)$, at $(0, 0)$, and at $(\sqrt{3}, \sqrt{3}/2) \approx$ $(1.732, 0.866)$

(g) Horizontal asymptote: $y = 0$

(h) Graph:

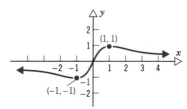

15. Absolute maximum: $f(0) = -1$; absolute minimum: $f(1) = 0$

17. Absolute maximum: $f(-1) = f(1) = -1$; absolute minimum: $f(0) = 0$

19. Absolute maximum: $f(-1) = 5$; absolute minimum: $f(1) = -3$

21. $dx/dt = -3$ 23. $dx/dt = -1$

25. $(12x^3 - 6x^2 + 1)\,dx$ 27. $\left(\dfrac{-5}{(1 + x)^2}\right)dx$

29. $20\pi \approx 62.832$ m²/sec   31. $20 thousand

33. Only statement (e) *must* be true. [If $f(x)$ is assumed to be *differentiable* for all x, then statements (a), (b), (d), and (e) must be true.]

35. 20 units   **37.** $40\pi \approx 125.7$ mm²

39. $968\pi \approx 3041$ mm³

41. (a) Demand decreases by 18 hundred pounds
(b) Demand decreases by 90 hundred pounds

CHAPTER 15

Exercise 15.1 *(page 786)*

1. $3x + K$ **3.** $\dfrac{x^2}{2} + K$ **5.** $\frac{3}{4}x^{4/3} + K$

7. $-x^{-1} + K$ **9.** $2x^{1/2} + K$ **11.** $\dfrac{x^4}{2} + \dfrac{5}{2}x^2 + K$

13. $\dfrac{x^3}{3} + 2e^x + K$ **15.** $\dfrac{x^4}{4} - \dfrac{2}{3}x^3 + \dfrac{x^2}{2} - x + K$

17. $x - \ln|x| + K$ **19.** $2e^x - 3 \ln|x| + K$

21. $3x + 2\sqrt{x} + K$ **23.** $\dfrac{x^2}{2} - 2x + K$

25. $\dfrac{x^3}{3} - \dfrac{x^2}{2} + K$ **27.** $\frac{3}{5}x^5 + 2 \ln|x| + K$

29. $4x + e^x + K$ **31.** $R(x) = 600x$

33. $R(x) = 10x^2 + 5x$

35. $C(x) = 7x^2 - 2800x + 4300$; cost is minimum when $x = 200$

37. $C(x) = 10x^2 - 8000x + 500$; cost is minimum when $x = 400$

39. $C(x) = 3400 + 1000x - 10x^2 + \frac{1}{3}x^3$

41. (a) $C(x) = 8x^2 - 1591x + 1800$ (b) $R(x) = 9x$
(c) $P(x) = R(x) - C(x) = -8x^2 + 1600x - 1800$
(d) Maximum profit is $P(100) = \$78{,}200$
(e)

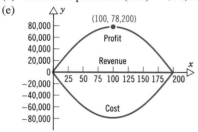

43. 187.76 million gallons a month

45. 7.3 ppm   **47.** $f(x) = x^2 + x + 1$ **49.** 14.14 sec

51. (a) If $y = \displaystyle\int x \, dx \cdot \int \sqrt{x} \, dx = \left(\dfrac{x^2}{2} + K_1 \right) \times$
$\left(\dfrac{2x^{3/2}}{3} + K_2 \right) = \dfrac{x^{7/2}}{3} + \dfrac{K_2 x^2}{2} + \dfrac{2K_1 x^{3/2}}{3} + K_1 K_2,$
then $\dfrac{dy}{dx} = \dfrac{7}{6} x^{5/2} + K_2 x + K_2 \sqrt{x} \neq x\sqrt{x}$. So
$\displaystyle\int x \, dx \cdot \int \sqrt{x} \, dx \neq \int (x \cdot \sqrt{x}) \, dx.$

(b) If $y = x \displaystyle\int (x^2 + 1) \, dx = x\left(\dfrac{x^3}{3} + x + K \right)$
$= \dfrac{x^4}{3} + x^2 + Kx$, then $\dfrac{dy}{dx} = \dfrac{4}{3} x^3 + 2x + K$
$\neq x(x^2 + 1)$. So $x\displaystyle\int (x^2 + 1) \, dx \neq \int x(x^2 + 1) \, dx.$

(c) If $y = \dfrac{\displaystyle\int (x^2 - 1) dx}{\displaystyle\int (x - 1) dx} = \dfrac{\dfrac{x^3}{3} - x + K_1}{\dfrac{x^2}{2} - x + K_2}$, then $\dfrac{dy}{dx}$

$= \dfrac{\left(\dfrac{x^2}{2} - x + K_2 \right)(x^2 - 1) - \left(\dfrac{x^3}{3} - x + K_1 \right)(x - 1)}{\left(\dfrac{x^2}{2} - x + K_2 \right)^2}$

$\neq \dfrac{x^2 - 1}{x - 1}$. So $\dfrac{\displaystyle\int (x^2 - 1) \, dx}{\displaystyle\int (x - 1) \, dx} \neq \int \dfrac{x^2 - 1}{x - 1} \, dx.$

Exercise 15.2 *(page 795)*

1. $\frac{1}{12}(2x + 1)^6 + K$ **3.** $\frac{1}{2}e^{(2x-3)} + K$

5. $\frac{1}{2}(-2x + 3)^{-1} + K$ **7.** $\frac{1}{6}(x^2 + 4)^3 + K$

1. b   **3.** a   **5.** a

9. $\frac{1}{3}e^{(x^3+1)} + K$ **11.** $e^x - e^{-x} + K$

13. $\frac{1}{21}(x^3 + 2)^7 + K$ **15.** $\frac{3}{4}(1 + x^2)^{2/3} + K$

17. $\frac{2}{5}(x + 3)^{5/2} - 2(x + 3)^{3/2} + K$ **19.** $\ln(e^x + 1) + K$

21. $2e^{\sqrt{x}} + K$ **23.** $\frac{3}{7}(x^{1/3} - 1)^7 + K$

25. $\dfrac{-1}{2(x^2 + 2x + 3)} + K$ **27.** $\dfrac{-2}{3(1 + \sqrt{x})^3} + K$

29. $\frac{1}{2}\ln|2x + 3| + K$ **31.** $\frac{1}{8}\ln(4x^2 + 1) + K$

33. $\frac{1}{2}\ln(x^2 + 2x + 2) + K$ **35.** $\frac{1}{2}xe^{2x} - \frac{1}{4}e^{2x} + K$

37. $-x^2e^{-x} + 2(-xe^{-x} - e^{-x}) + K =$
$-1(x^2 + 2x + 1)e^{-x} + K$

39. $\frac{2}{3}x^{3/2}(\ln x) - \frac{4}{9}x^{3/2} + K$

41. $x(\ln x)^2 - 2x \ln x + 2x + K$

43. $\frac{1}{3}x^3 \ln(3x) - \frac{1}{9}x^3 + K$

45. $\frac{1}{3}x^3(\ln x)^2 - \frac{2}{3}(\frac{1}{3}x^3 \ln x - \frac{1}{9}x^3) + K =$
$\frac{1}{3}x^3(\ln x)^2 - \frac{2}{9}x^3 \ln x + \frac{2}{27}x^3 + K$

47. $\frac{1}{2}(\ln x)^2 + K$

49. Set $u = ax + b$, so $du = a\,dx$, or $\dfrac{1}{a}\,du = dx$. Then

(assuming $a \neq 0$ and $n \neq -1$): $\displaystyle\int (ax + b)^n\,dx =$

$\dfrac{1}{a}\displaystyle\int u^n\,du = \dfrac{1}{a(n + 1)}u^{n+1} + K = \dfrac{(ax + b)^{n+1}}{a(n + 1)} + K$

51. $R(x) = \dfrac{(x^2 + 10)^3}{3000} + 192.14$ thousand dollars

53. 180 m

Exercise 15.3 (*page 805*)

1. $\frac{7}{2}$ **3.** e **5.** $\frac{2}{3}$ **7.** $-\frac{1}{15}$ **9.** $\frac{5}{3}$ **11.** $\frac{5}{24}$

13. $\frac{5}{7} \cdot 4^{7/5} - \frac{5}{7} + \ln 4 \approx 5.64658$ **15.** $\frac{20}{3}$ **17.** 0

19. $\frac{8}{3}$ **21.** $\dfrac{e^2}{2} - \dfrac{3}{2}$ **23.** $-\dfrac{1}{e} + 1$ **25.** $\ln 2$

27. $\frac{2}{3}(\ln 2)$ **29.** $\dfrac{5e^6}{4} - \dfrac{e^2}{4}$ **31.** $-\frac{7}{9}e^{-6} + \frac{4}{9}e^{-3}$

33. $5 \ln 5 - 4$

35. (a) $\displaystyle\int_{-1}^{1} x^2\,dx = \frac{2}{3}; \quad 2 \cdot \int_{0}^{1} x^2\,dx = 2 \cdot \frac{1}{3} = \frac{2}{3}$

(b) $\displaystyle\int_{-1}^{1} (x^4 + x^2)\,dx = \frac{16}{15}$,

$2 \cdot \displaystyle\int_{0}^{1} (x^4 + x^2)\,dx = 2 \cdot \frac{8}{15} = \frac{16}{15}$

37. $567{,}000$

39. (a) $\displaystyle\int_{30}^{80} 1272x^{-0.35}\,dx \approx 15{,}921.34$

(b) $\displaystyle\int_{30}^{80} 1272x^{-0.15}\,dx \approx 35{,}089.39$

(c) In the learning function $f(x) = cx^k$, $-1 \le k < 0$, the nearer k is to 0, the flatter the learning curve (see graph), and hence the less influence the number of units produced has on the time required to produce the next item. Parts (a), (b), and Example 10 illustrate this, since the total number of labor-hours required to produce units 30 through 80, $\displaystyle\int_{30}^{80} 1272x^k\,dx$, increases as k goes from -0.35 to -0.15.

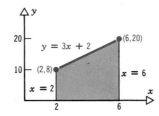

41. $2237.06   **43.** $1825.40

45. $13{,}450 + \dfrac{950}{e^{12}} \approx 13{,}450$

47. Don't advertise; the additional profit from advertising is $15{,}218 *less than* the advertising cost.

49. Approx. 3660 labor-hours

Exercise 15.4 (*page 818*)

1. Area = 56

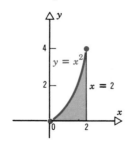

3. Area = $\frac{8}{3}$ **5.** Area = 9

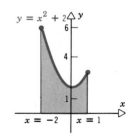

7. Area $= \frac{3}{2}$

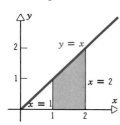

9. Area $= e - 1$

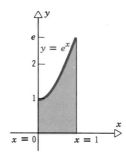

21. Area $= \frac{8}{3}$

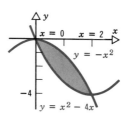

23. Area $= \frac{9}{2}$

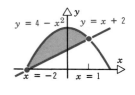

11. Area $= \frac{1}{2}$

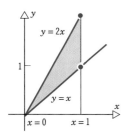

13. Area $= \frac{1}{6}$

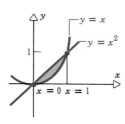

25. Area $= 8$

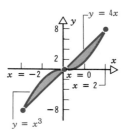

27. Area $= \frac{1}{3}$

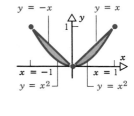

15. Area $= \frac{1}{6}$

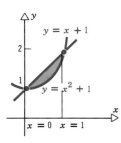

17. Area $= \frac{5}{12}$

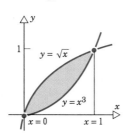

29. Consumer's surplus $= \frac{40}{9}$ Producer's surplus $= \frac{32}{9}$

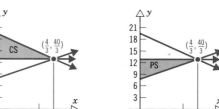

31. The operation should continue for 16 years, generating a profit of $\$85\frac{1}{3}$ million.

33. (a) $c = 1/\sqrt{3}$

(b) $c = 2$

19. Area $= \frac{4}{15}$

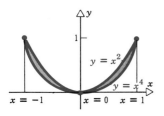

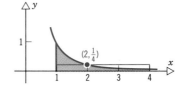

35. (a) x^2 (b) $\sqrt{x^2 - 2}$ (c) $\sqrt{x^2 + 2x}$

Technology Exercises (page 819)

1. (a)

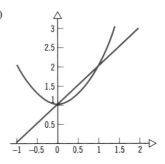

(b) $(0, 1)$ and $(1, 2)$

(c)

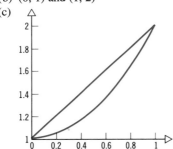

(d) $A = \int_0^1 [(x + 1) - (x^2 + 1)]\, dx = \frac{1}{6}$

3. (a) The 6th month   (b) 2640 gal   (c) 3116.67 gal

Exercise 15.5 (page 826)

1. 0   **3.** 62.5   **5.** 140   **7.** 500   **9.** 1875

11. 3160   **13.** -6 **15.** 83   **17.** 2.47838

Technology Exercises (page 826)

1. (a)

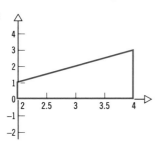

(b) $A = \frac{1}{2}(h_1 + h_2)w = \frac{1}{2}(1 + 3)2 = 4$

(c) $A = \int_2^4 (x - 1)\, dx = 4.0$

(d) $F(x) = \frac{1}{2}x^2 - x$

(e) $A = F(4) - F(2) = 4$

3. (a) $\int_{-1}^{2} |x|\, dx = \int_{-1}^{0} (-x)\, dx + \int_{0}^{2} (x)\, dx$, so $F_1(x) = -\frac{1}{2}x^2$ and $F_2(x) = \frac{1}{2}x^2$. The area is $F_1(0) - F_1(-1) + F_2(2) - F_2(0) = \frac{5}{2}$.

(b) $\int_{-1}^{0} (-x)\, dx + \int_{0}^{2} (x)\, dx = 2.5$

5. (a) $F(x) = \frac{2}{3} \ln|x^{3/2} + 1| + K$. The area is $F(1) - F(0) = \frac{2}{3} \ln 2$.

(b) $\int_0^1 \frac{\sqrt{x}}{x^{3/2} + 1}\, dx = .4621$

7. (a) $F(x) = \frac{1}{3}x^3 \ln x - \frac{1}{9}x^3$. The area is $F(2) - F(1) = \frac{8}{3} \ln 2 - \frac{7}{9}$.

(b) $\int_1^2 x^2 \ln x\, dx = 1.0706$

Exercise 15.6 (page 830)

1. Improper; upper limit is ∞.

3. Improper; integrand $1/x$ is discontinuous at lower limit 0.

5. Improper; integrand $\dfrac{1}{x - 1}$ is discontinuous at lower limit 1.

7. $\frac{1}{4}e^{-4}$ **9.** Does not exist   **11.** $-\frac{5}{4}$

13. Does not exist   **15.** 2 square units

17. The reaction for the first b hours after administration is $\int_0^b r(t)\, dt$. Thus, the *total reaction* is approximated by $\lim\limits_{b \to \infty} \int_0^b r(t)\, dt = \int_0^\infty r(t)\, dt$. Then, $\int_0^\infty te^{-t^2}\, dt = \dfrac{1}{2}$.

19. (a) Does not exist   (b) $\frac{3}{2}(-\sqrt[3]{2} + \sqrt[3]{18})$

Technology Exercises (page 831)

1. (a) $F(x) = \frac{1}{3}x^3$

(b)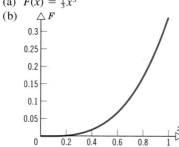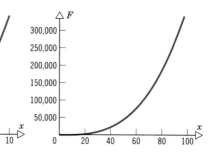

$\lim\limits_{x \to \infty} F(x)$ does not exist (infinite)

(c) Does not exist (infinite)

3. (a) $F(x) = -xe^{-x} - e^{-x}$

(b)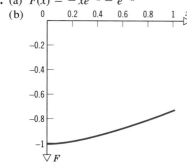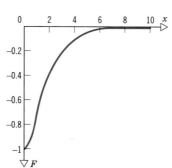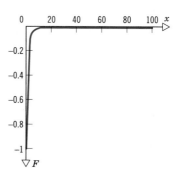

$\lim\limits_{x \to \infty} F(x) = 0$

(c) 1

5. (a) $F(x) = -\frac{1}{2}e^{-x^2}$

(b)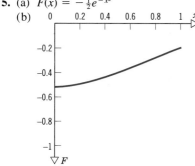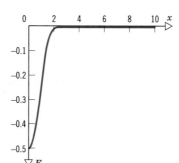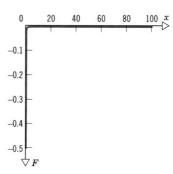

$\lim\limits_{x \to \infty} F(x) = 0$

(c) $\frac{1}{2}$

Chapter Review: True–False Items (*page 832*)

1. T   **2.** F   **3.** F   **4.** F   **5.** F   **6.** T

7. T   **8.** T   **9.** F   **10.** T

Fill in the Blanks (*page 833*)

1. $F'(x) = f(x)$ **2.** $\int$ **3.** integration by parts

4. lower; upper limits; integration   **5.** 0

6. $F(b) - F(a)$ **7.** $\lim\limits_{b \to 2^-} \int_0^b f(x)\ dx$

Review Exercises (*page 833*)

1. $\frac{1}{4}x^4 - \frac{3}{2}x^2 + x + K$ **3.** $\frac{3}{4}x^{4/3} - \frac{8}{5}x^{3/2} + K$

5. $x - e^{-x} + K$ **7.** $\frac{1}{3}(x^2 - 1)^{3/2} + K$

9. $\frac{2}{9}(3x - 2)^{3/2} + K$ **11.** $-\frac{1}{2}(x^4 + 1)^{-1/2} + K$

13. $\frac{5}{2}\ln(x^2 + 1) + K$ **15.** $2x\ e^{x/2} - 4e^{x/2} + K$

17. $\frac{1}{3}e^{x^3} + K$ **19.** $-\frac{7}{5}$ **21.** $\frac{1}{2}(e^4 - 1)$

23. $\ln 2$ **25.** $\frac{61}{2}$ **27.** $-2e^{-1} + 1$ **29.** $\frac{1}{12}$

31. $\frac{1}{3}$ **33.** $\ln 5$ **35.** Does not exist   **37.** $\frac{2}{3}$

39. 13   **41.** -2 **43.** $1 + e + e^2 + e^3 \approx 31.1929$

45. $\sqrt{2} \approx 1.41421$

47. No, the account balance will be approx. $8338 less than the down payment.

49. 5855.73

Mathematical Questions from Professional Exams (*page 834*)

1. b   **3.** c

CHAPTER 16

Exercise 16.1 (*page 838*)

1. $\frac{1}{3}$ **3.** $\frac{2}{3}$ **5.** 9   **7.** -26 **9.** $e - 1 \approx 1.71828$

11. 6.85099×10^9 **13.** $37.5°C$ **15.** 12 m/sec

3. $A = 28$

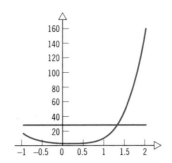

Technology Exercises (*page 838*)

1. $A = 1.7182818$

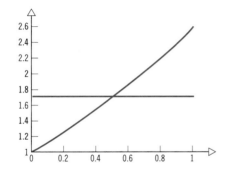

Exercise 16.2 (*page 845*)

1.

No. Heads in 2 Coin Flips	0	1	2
Probability	$\frac{1}{4}$	$\frac{1}{2}$	$\frac{1}{4}$

3.

No. Females among 3 Children	0	1	2	3
Probability	$\frac{1}{8}$	$\frac{3}{8}$	$\frac{3}{8}$	$\frac{1}{8}$

5. Discrete; sample space is infinite

7. Continuous   **9.** $179.8e^{-6} \approx .44568$

11. $64.8e^{-6} \approx .160623$ **13.** $.33282$

15. Poisson: $.712703$; binomial: $.720568$

Technology Exercises (page 846)

1. $P(X = 0) = 0.0625$, $P(X = 1) = 0.25$, $P(X = 2) = 0.375$, $P(X = 3) = 0.25$, $P(X = 4) = 0.0625$

3. $P(X = 0) = 0.4096$, $P(X = 1) = 0.4096$, $P(X = 2) = 0.1536$, $P(X = 3) = 0.0256$, $P(X = 4) = 0.0016$

5. $P(X = 2) = \frac{1}{12} = 0.0834$

Exercise 16.3 (page 854)

1. $\int_0^2 \frac{1}{2}\, dx = 1$; $f(x) = \frac{1}{2} \geq 0$ on $[0, 2]$

3. $\int_0^1 2x\, dx = x^2\big|_0^1 = 1$; $f(x) = 2x \geq 0$ on $[0, 1]$

5. $\int_0^5 \frac{3}{250}\,(10x - x^2)\, dx = \frac{3}{250}\left(125 - \frac{125}{3}\right) = 1$;

$f(x) = \frac{3}{250}\,(10 - x)x \geq 0$ on $[0, 5]$

7. $\int_1^e \frac{1}{x}\, dx = \ln e - \ln 1 = 1$; $f(x) = \frac{1}{x} \geq 0$ on $[1, e]$

9. $\frac{1}{3}$ **11.** $\frac{1}{2}$ **13.** $\frac{3}{250}$ **15.** $\frac{1}{\ln 2}$

17. 1 **19.** $\frac{2}{3}$ **21.** $\frac{25}{8} = 3.125$ **23.** $e - 1$

25. $\frac{2}{5}$ **27.** $e^{-3} \approx .0497871$ **29.** $1 - e^2 \approx .864665$

31. $\frac{7}{27} \approx .259259$; $\frac{7}{27} \approx .259259$; $\frac{13}{27} \approx .481481$

33. $\frac{16}{7} \approx 2.28571\%$

Technology Exercises (page 856)

1. $p(x) = \begin{cases} \frac{1}{35,000,000}\,x - \frac{1}{3500} & \text{if} \quad 10,000 \leq x \leq 17,000 \\ -\frac{1}{15,000,000}\,x + \frac{1}{750} & \text{if} \quad 17,000 < x \leq 20,000 \end{cases}$

(a) 0.357

(b)

(d) $E(X) = \$15,666.67$; because the two line segments are not of the same length

3. 0.3

Exercise 16.4 (page 862)

1. $y = \frac{1}{3}x^3 - x$ **3.** $y = \frac{1}{3}x^3 - \frac{1}{2}x^2 - \frac{3}{2}$

5. $y = \frac{1}{4}x^4 - \frac{1}{2}x^2 + 2x + 3$ **7.** $y = e^x + 3$

9. $y = \frac{1}{2}x^2 + x + \ln x - \frac{3}{2}$

11. $12{,}975$ bacteria; $147{,}789$ bacteria; approx. 113.58 min

13. 7.67852 g   **15.** The charcoal is about 9727 years old.

17. 6944 mosquitoes   **19.** $7500e^2 \approx 55{,}418$ bacteria

21. (a) $10{,}000 \cdot 5^{t/10}$ bacteria
(b) $250{,}000$ bacteria   (c) $t_1 \approx 4.3068$

23. $4{,}620{,}981$ yr

Chapter Review: True–False Items (page 864)

1. F   **2.** T   **3.** F   **4.** T

Fill in the Blanks (page 864)

1. average value   **2.** random value

3. probability density   **4.** exponentially distributed

5. differential equation   **6.** particular value

Review Exercises (page 865)

1. $\frac{1}{4}$ **3.** (a) Discrete and finite   (b) Discrete and finite

5. $.082085$; $.133602$ **7.** $\frac{1}{172} \approx .00581$; $\frac{19}{86} \approx .22093$

9. (a) $.25$ (b) $.3125$ (c) $.4375$

11. (a) $f(x) = \dfrac{x}{2}, 0 \le x \le 2$ (b) $\frac{1}{2}$ (c) $\frac{1}{2}$

13. $y = \frac{1}{2}x^2 + e^x + 4$ **15.** $y = \frac{1}{3}(x^2 - 16)^{3/2} - 9$

17. $y = \frac{3}{5}e^{5x} - x^2 + x + K; y = \frac{3}{5}e^{5x} - x^2 + x - \frac{13}{5}$

19. 12,000   **21.** 8482 yr

*Mathematical Questions from Professional
Exams (page 866)*

1. d   **3.** b

CHAPTER 17

Exercise 17.1 (page 873)

1.
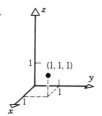

3.
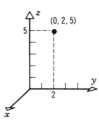

5.
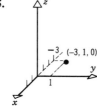

7. (2, 0, 0), (2, 1, 0), (0, 1, 0), (0, 0, 3), (2, 0, 3), (0, 1, 3)

9. (3, 2, 3), (3, 4, 3), (1, 4, 3), (1, 2, 5), (3, 2, 5), (1, 4, 5)

11. (4, 0, 2), (4, 2, 2), (4, 0, 5), (− 1, 0, 5), (− 1, 2, 2), (− 1, 2, 5)

13. "Vertical" plane parallel to, and 3 units to the right of, the xz-plane

15. The yz-plane

17. "Horizontal" plane parallel to, and 5 units above, the xy-plane

19. 5   **21.** $\sqrt{2}$ **23.** $\frac{1}{5}$ **25.** 3   **27.** 0

29. (a) 0   (b) 2   (c) 10
(d) $3x + 2y + xy + (3 + y)\,\Delta x$
(e) $3x + 2y + xy + (2 + x)\,\Delta y$

31. (a) 0   (b) 0   (c) $at + a^2$
(d) $\sqrt{xy} + y\,\Delta x + x + \Delta x$ (e) $\sqrt{xy} + x\,\Delta x + x$

33. (a) 14   (b) 2   (c) − 14

35. $\{(x, y)\,|\,x \ge 0 \text{ and } y \ge 0\}$
(or, the first quadrant of the xy-plane, including its boundary)

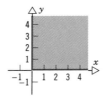

37. $\{(x, y)\,|\,x^2 + y^2 \le 9\}$
(or, the disk of radius 3 centered at the origin)

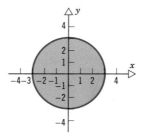

39. $\{(x, y)\,|\,x > 0 \text{ and } (0 < y < 1 \text{ or } 1 < y)\}$
(or, the first quadrant of the xy-plane, exclusive of the axes and of the horizontal line $y = 1$)

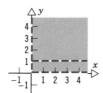

41. $\{(x, y)\mid x^2 + y^2 \neq 4\}$
(or, the xy-plane exclusive of the circle
of radius 2 centered at the origin)

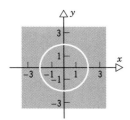

43. $\{(x, y)\mid x \neq 0 \text{ and } y \neq 0\}$
(or, the xy-plane exclusive of the origin)

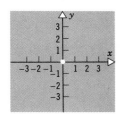

45. $\{(x, y, z)\mid x^2 + y^2 + z^2 \geq 16\}$
[or, outside or on the sphere of radius 4 centered at $(0, 0, 0)$]

47. $\{(x, y, z)\mid x \neq 0 \text{ and } y \neq 0 \text{ and } z \neq 0\}$ (or, xyz-space exclusive of the origin)

49. (a) $3x + 4y + 3\,\Delta x$ (b) $3\,\Delta x$ (c) 3   (d) 3

51. Total cost $= 600\pi r^2 + 1000\pi rh$ dollars

53. (a) 6.75   (b) 18   (c) 2

55. (a) $Q(19, 21) = 8151.54$ (b) $Q(21, 20) = 8117.96$

Exercise 17.2 (page 882)

1. $f_x = 3; f_y = -2 + 9y^2; f_x(2, -1) = 3; f_y(-2, 3) = 79$

3. $f_x = 2x - 2y; f_y = 2y - 2x; f_x(2, -1) = 6;$
$f_y(-2, 3) = 10$

5. $f_x = \dfrac{x}{\sqrt{x^2 + y^2}}; f_y = \dfrac{y}{\sqrt{x^2 + y^2}};$

$f_x(2, -1) = \dfrac{2}{\sqrt{5}}; f_y(-2, 3) = \dfrac{3}{\sqrt{13}}$

7. $f_x = -2y - 24x$ $f_y = 3y^2 - 2x + 2y$
$f_{xx} = -24$ $f_{yy} = 6y + 2$
$f_{yx} = -2$ $f_{xy} = -2$

9. $f_x = e^y + ye^x + 1$ $f_y = xe^y + e^x$
$f_{xx} = ye^x$ $f_{yy} = xe^y$
$f_{yx} = e^y + e^x$ $f_{xy} = e^y + e^x$

11. $f_x = \dfrac{1}{y}$ $f_y = -\dfrac{x}{y^2}$

$f_{xx} = 0$ $f_{yy} = \dfrac{2x}{y^3}$

$f_{yx} = -\dfrac{1}{y^2}$ $f_{xy} = -\dfrac{1}{y^2}$

13. $f_x = \dfrac{2x}{x^2 + y^2}$ $f_y = \dfrac{2y}{x^2 + y^2}$

$f_{xx} = \dfrac{2y^2 - 2x^2}{(x^2 + y^2)^2}$ $f_{yy} = \dfrac{2x^2 - 2y^2}{(x^2 + y^2)^2}$

$f_{yx} = \dfrac{-4xy}{(x^2 + y^2)^2}$ $f_{xy} = \dfrac{-4xy}{(x^2 + y^2)^2}$

15. $f_x = \dfrac{-(10 + 2y)}{x^2 y}$ $f_y = \dfrac{x - 10}{xy^2}$

$f_{xx} = \dfrac{4y + 20}{x^3 y}$ $f_{yy} = \dfrac{-2x + 20}{xy^3}$

$f_{yx} = \dfrac{10}{x^2 y^2}$ $f_{xy} = \dfrac{10}{x^2 y^2}$

17. $f_x = 3x^2$ **19.** $f_x = 12x^3y^2 + 14xy$
$f_y = 2y$ $f_y = 6x^4y + 7x^2$
$f_{xy} = 0$ $f_{xy} = 24x^3y + 14x$
$f_{yx} = 0$ $f_{yx} = 24x^3y + 14x$

21. $f_x = -\dfrac{2y}{x^3}$ $f_{xy} = -\dfrac{2}{x^3}$

$f_y = \dfrac{1}{x^2}$ $f_{yx} = -\dfrac{2}{x^3}$

23. $f_x = 2xy - 3yz; f_y = x^2 - 3xz; f_z = -3xy + 3z^2$

25. $f_x = e^y; f_y = xe^y + e^z; f_z = ye^z$

27. $f_x = \ln(yz) + \dfrac{y}{x}; f_y = \dfrac{x}{y} + \ln(xz); f_z = \dfrac{x}{z} + \dfrac{y}{z}$

29. $f_x = \dfrac{2x}{x^2 + y^2 + z^2}$; $f_y = \dfrac{2y}{x^2 + y^2 + z^2}$;

$f_z = \dfrac{2z}{x^2 + y^2 + z^2}$

31. 20   **33.** $-2/\sqrt{11}$ **35.** 1   **37.** 2

39. $x\dfrac{\partial z}{\partial x} + y\dfrac{\partial z}{\partial y} = x(2x) + y(8y) = 2(x^2 + 4y^2) = 2z$

41. $x\dfrac{\partial z}{\partial x} + y\dfrac{\partial z}{\partial y} = x(y^2) + y(2xy) = 3xy^2 = 3z$

43. $\dfrac{\partial^2 z}{\partial x^2} = (\ln y)^2 e^{x \ln y}$; $\dfrac{\partial^2 z}{\partial y^2} = \left(\dfrac{x^2 - x}{y^2}\right) e^{x \ln y}$;

$\dfrac{\partial^2 z}{\partial x \, \partial y} = \left(\dfrac{x \ln y + 1}{y}\right) e^{x \ln y}$

45. $\dfrac{\partial z}{\partial x} = -50$: If the cost of margarine is held constant, then an increase of 1¢ in the price of butter will cause the demand for butter to decrease by about 50 lb.

$\dfrac{\partial z}{\partial y} = 90$: If the cost of butter is held constant, then an increase of 1¢ in the price of margarine will cause the demand for butter to increase by about 90 lb.

47. $\dfrac{\partial z}{\partial x} = 2y - 2x$; $\dfrac{\partial z}{\partial y} = 2x - 6y$

If y is held constant, then an increase of x by 1 unit will cause the productivity to change by $2y - 2x$.
If x is held constant, then an increase of y by 1 unit will cause the productivity to change by $2x - 6y$.

49. $\dfrac{\partial z}{\partial x} = \dfrac{1}{x^2}$; $\dfrac{\partial z}{\partial y} = \dfrac{1}{y^2}$

If y is held constant, then an increase of commodity x from 1 to 2 will cause the productivity to increase by 1.
If x is held constant, then an increase of commodity y from 1 to 2 will cause the productivity to increase by 1.

Exercise 17.3 (*page 888*)

1. $(-1, 0)$, $(0, 0)$, $(1, 0)$ **3.** $(-1, -1)$, $(0, 0)$, $(1, 1)$

5. $(0, 0)$ **7.** Local minimum at $(0, 0)$

9. Local minimum at $\left(\frac{3}{2}, 0\right)$ **11.** Saddle point at $(-2, 4)$

13. Local minimum at $(2, -1)$

15. Local minimum at $(4, -2)$ **17.** Saddle point at $(0, 0)$

19. Saddle point at $(0, 0)$; local minimum at $(2, 2)$

21. Saddle point at $(0, 0)$ (with more of a "seat" shape than a "saddle" shape: note that $f_y(x, 0) = x^2 > 0$ for $x \neq 0$, so f is increasing along the x-axis); local minimum at $(2, 2)$

23. No critical points and no local extrema

25. The maximum profit is $\frac{128}{7} \approx 18.286$ thousand dollars when $x = \frac{20}{7} \approx 2.857$ thousand units, $y = \frac{2}{7} \approx 0.286$ thousand units, $p = \frac{64}{7} \approx 9.143$ thousand dollars, and $q = \frac{54}{7} \approx 7.714$ thousand dollars.

27. 14 thousand ft

29. The maximum reaction occurs to $x = \left(\frac{2}{3}a\right)$ units of the drug, $t = 2$ hours after it is administered.

31. Produce $x = 15{,}250$ tons of grade A steel and $y = 4100$ tons of grade B steel for maximum profit.

33. (a) Length = 28 in., width = 14 in., height = 14 in.
(b) Radius = $28/\pi \approx 8.913$ in., length = 28 in.

35. The waste is minimum if 50 tons of steel are used at the rate of 1 ton per week.

Exercise 17.4 (*page 896*)

1. 15   **3.** $\frac{1}{2}$ **5.** 528   **7.** 612   **9.** 16,000

11. $\frac{233}{12} \approx 19.4167$ **13.** $x = y = 50$

15. $x = y = z = \dfrac{5}{\sqrt{3}} \approx 2.88675$

17. $x = 18$, $y = 36$

19. Length = width = height = $\sqrt[3]{175} \approx 5.59344$ ft

21. Length = width = $\sqrt[3]{12} \approx 2.28943$ ft; height = $\frac{3}{2} \cdot \sqrt[3]{12} \approx 3.43414$ ft

Exercise 17.5 (*page 902*)

1. $2y^3 + \frac{8}{3}$ **3.** $56y + 12$ **5.** $\frac{5}{2} + 3y$ **7.** $8x - 22$

9. $\dfrac{1}{\sqrt[3]{1 + y^2}}$ **11.** $e^{2+y} - e^y$

13. $\frac{1}{2} e^{16-4y} - \frac{1}{2} e^{-4y}$ **15.** $\sqrt{y + 4} - \sqrt{y}$ **17.** 8

19. $\frac{35}{3}$ **21.** $\frac{17}{6}$ **23.** 22   **25.** 12   **27.** 24

29. 21   **31.** $\frac{16}{3}$ **33.** 452   **35.** $\frac{44}{3}$

37. $\frac{1}{2} e^2 - e + \frac{1}{2}$ **39.** $\frac{35}{2}$

Chapter Review: True–False Items (*page 904*)

1. T   **2.** F   **3.** F   **4.** F

Fill in the Blanks (*page 905*)

1. surface   **2.** $4 - \dfrac{1}{\sqrt{2}}$ **3.** $x = x_0$ **4.** saddle point

Review Exercises (*page 905*)

1. $f(1, -3) = 11$; $f(4, -2) = -8$

3. $f(1, -3) = -0.5$; $f(4, -2) = 0$

5. (a) $x^2 + 2x\,\Delta x + (\Delta x)^2 + xy + y\,\Delta x$
   (b) $2x\,\Delta x + (\Delta x)^2 + y\,\Delta x$
   (c) $2x + \Delta x + y$ (d) $2x + y$

7. (a) Local maximum $z = 12$ at $(x, y) = (-4, 2)$
   (b) Local minimum $z = 5$ at $(x, y) = (-1, -2)$
   (c) Local maximum $z = 15$ at $(x, y) = (1, 2)$
   (d) Saddle point $z = 0$ at $(x, y) = (0, 0)$

   (e) Saddle point $z = 6\sqrt{3}$ at $(x, y) = (0, -\sqrt{3})$; local minimum $z = -6\sqrt{3}$ at $(x, y) = (0, \sqrt{3})$

9. (a) $f(22, 24) = 1220$ (b) $f(1000, 1000) = 10,000\,\sqrt{10}$

11. $\frac{4}{3}x^3y - 12yx$ **13.** 90   **15.** 24   **17.** 36

19. Maximum profit is 324,000 for $x = 360$ units and $y = 180$ units.

APPENDIX A

Exercise A.1 (page A-9)

1. $x > 0$ **3.** $x < 2$ **5.** $x \le 1$

7.   **9.**

11.   **13.**

15.   **17.** $x = 1$ **19.** $x = 6$

21. $x = -1$ **23.** $x = -4$

25. $x \le -1$

27. $x \ge -1$

29. $x \ge 1$

31. $x \le -4$

Exercise A.2 (page A-13)

1. 64   **3.** $\frac{1}{8}$ **5.** 1   **7.** 4   **9.** 3   **11.** 2

13. 4   **15.** $\frac{1}{64}$ **17.** $\frac{1}{4}$

19. (a) 11.211578   (b) 11.587251   (c) 11.663882
   (d) 11.664753

21. (a) 8.8152409   (b) 8.8213533   (c) 8.8244111
   (d) 8.8249778

23. (a) 21.216638   (b) 22.216690   (c) 22.440403
   (d) 22.459158

25. 3   **27.** -1 **29.** 8   **31.** $\frac{1}{3}$ **33.** 2   **35.** 3

37. 55.5903   **39.** 1385.0021   **41.** 1499.3635

43. 12,432.323   **45.** 2074.6418

Exercise A.3 (page A-18)

1. 1, 2, 3, 4, 5   **3.** $\frac{1}{2}, \frac{2}{3}, \frac{3}{4}, \frac{4}{5}, \frac{5}{6}$

5. $1, -4, 9, -16, 25$ **7.** $\frac{1}{2}, \frac{2}{5}, \frac{3}{7}, \frac{4}{41}, \frac{8}{61}$

9. $-\frac{1}{6}, \frac{1}{12}, -\frac{1}{20}, \frac{1}{30}, -\frac{1}{42}$

11. $\dfrac{1}{e}, \dfrac{2}{e^2}, \dfrac{3}{e^3}, \dfrac{4}{e^4}, \dfrac{5}{e^5}$ **13.** 1, 3, 5, 7, 9

15. $-2, -1, 1, 4, 8$ **17.** 5, 10, 20, 40, 80

19. $3, 3, \frac{3}{2}, \frac{1}{2}, \frac{1}{8}$ **21.** 1, 2, 2, 4, 8

23. $A, A + d, A + 2d, A + 3d, A + 4d$

25. $\sqrt{2}, \; \sqrt{2 + \sqrt{2}}, \; \sqrt{2 + \sqrt{2 + \sqrt{2}}}, \; \sqrt{2 + \sqrt{2 + \sqrt{2 + \sqrt{2}}}},$
   $\sqrt{2 + \sqrt{2 + \sqrt{2 + \sqrt{2 + \sqrt{2}}}}}$

27. $r = 2$; 2, 4, 8, 16; $S_n = -2(1 - 2^n)$

29. $r = \frac{1}{2}$; $-\frac{3}{2}, -\frac{3}{4}, -\frac{3}{8}, -\frac{3}{16}$; $S_n = -3[1 - (\frac{1}{2})^n]$

31. $r = 2$; $\frac{1}{4}, \frac{1}{2}, 1, 2$; $S_n = -\frac{1}{4}(1 - 2^n)$

33. $r = 2^{1/3}$; $2^{1/3}, 2^{2/3}, 2, 2^{4/3}$; $S_n = 2^{1/3}\left(\dfrac{1 - 2^{n/3}}{1 - 2^{1/3}}\right)$

35. $r = \frac{3}{2}$; $\frac{1}{2}, \frac{3}{4}, \frac{9}{8}, \frac{27}{16}$; $S_n = -[1 - (\frac{3}{2})^n]$

Exercise A.4 (page A-26)

1. $3(x + 2)$ **3.** $a(x^2 + 1)$ **5.** $x(x^2 + x + 1)$

7. $2(x^2 + x + 1)$ **9.** $3xy(x - 2y + 4)$

11. $(x - 1)(x + 1)$ **13.** $(2x - 1)(2x + 1)$

15. $(x + 2)(x + 5)$ **17.** $(x - 3)(x - 7)$

19. $(x + 1)(x - 8)$ **21.** $(x + 1)^2$ **23.** $(x + 2)^2$

25. $(x + 3)(x - 5)$ **27.** $3(x + 2)(x - 6)$

29. $y^2(y + 5)(y + 6)$ **31.** $(4x + 1)^2$ **33.** $(2x + 3)^2$

35. $a(x + 5a)(x - 9a)$ **37.** $(x - 3)(x^2 + 3x + 9)$

39. $(2x + 3)(4x^2 - 6x + 9)$

41. $2x(x + 2)(x^2 - 2x + 4)$ **43.** $(x + 1)(3x + 1)$

45. $(x - 3)(x + 3)(x^2 + 9)$ **47.** $(x - 1)^2(x^2 + x + 1)^2$

49. $x^5(x - 1)(x + 1)$ **51.** $(z + 1)(2z + 3)$

53. $(4x - 3)^2$ **55.** $(4x + 1)(4x - 5)$

57. $(2y + 3)(2y + 5)$ **59.** $9(x + 1)(2x - 3)$

61. $2(4x^2 + x + 3)$ **63.** $x^2 - x + 4$ is prime

65. $2x(x - 3)(2x + 1)$

67. No pair of factors of 4 sums to 0 (the x coefficient of $x^2 + 4$).

Exercise A.5 (page A-29)

1. $2 + \sqrt{2}, 2 - \sqrt{2}$ **3.** $\dfrac{5 - \sqrt{29}}{2}, \dfrac{5 + \sqrt{29}}{2}$

5. $1, \frac{3}{2}$ **7.** $b^2 - 4ac = -31 < 0$; no real solutions

9. $\dfrac{-1 + \sqrt{5}}{4}, \dfrac{-1 - \sqrt{5}}{4}$

11. $b^2 - 4ac = -3 < 0$; no real solutions

13. $b^2 - 4ac = 0$; repeated solution

15. $b^2 - 4ac = 49 > 0$; two unequal real solutions

Exercise A.6 (page A-36)

1. $[0, 4]$

3. $[4, 6)$

5. $[2, \infty)$

7. $(-\infty, -4)$

9. $2 \le x \le 5$

11. $-3 < x < -2$

13. $4 \le x$

15. $x < -1$

17. $2 \le x$

19. $-6 < x$

21. $x \le \frac{2}{3}$

23. $x < -20$

25. $\frac{4}{3} \le x$

27. $3 \le x \le 5$

29. $-\frac{1}{3} \le x \le \frac{7}{3}$

31. $-1 < x < 3$

33. $-3 < x < 3$

35. $x < -4$ or $3 < x$

37. $x < 3$ or $4 < x$

39. No real solutions

41. $1 < x$

43. $x < 1$ or $2 < x < 3$

45. $-2 < x < 0$ or $4 < x$

47. $-1 < x < 0$ or $1 < x$

49. $1 < x$

51. $x < -1$ or $1 < x$

53. $x < -1$ or $0 < x < 1$

55. $-1 < x < 1$ or $2 \le x$

57. $x < 2$

$$-2\ -1\ \ 0\ \ 1\ \ 2\ \ 3\ \ 4\ \ 5$$

59. $x < -3$ or $-1 < x < 1$ or $2 < x$

$$-4\ -3\ -2\ -1\ \ 0\ \ 1\ \ 2\ \ 3$$

Exercise A.7 (page A-40)

1. 13   **3.** 26   **5.** 25

7. Right triangle: hypotenuse has length 5

9. Not a right triangle

11. Right triangle: hypotenuse has length 25

13. Not a right triangle

15. Area = 6; Perimeter = 10

17. Area = $\frac{1}{6} \approx 0.167$; Perimeter = $\frac{5}{3} \approx 1.667$

19. 1   **21.** $\frac{3}{16} = 0.1875$

23. Area = $\pi \approx 3.14$; Circumference = $2\pi \approx 6.18$

25. Area = $\pi/4 \approx 0.785$; Circumference = $\pi \approx 3.14$

27. Volume = 1; Surface area = 6

29. Volume = 1; Surface area = $\frac{41}{6} \approx 6.833$

31. Volume = $32\pi/3 \approx 33.49$; Surface area = $16\pi \approx 50.24$ [*Note:* These approximations use $\pi \approx 3.14$; a better (typical calculator) approximation for π gives Volume $\approx$ 33.51 and Surface area $\approx$ 50.27.]

33. Approx. 246586.5 ft, or 46.70 mi

35. Approx. 15840 ft, or 3 mi

APPENDIX B

Exercise B.1 (page A-48)

1. The maximum is $P = 24$ when $x_1 = 0$, $x_2 = \frac{76}{7}$, $x_3 = \frac{16}{7}$. ($P = 24$ also when $x_1 = 0$, $x_2 = 12$, $x_3 = 0$.)

3. The maximum is $P = 15$ when $x_1 = 5$, $x_2 = 0$, $x_3 = 0$.

5. The maximum is $P = 40$ when $x_1 = 14$, $x_2 = 0$, $x_3 = 4$.

7. The maximum is $P = 6$ when $x_1 = 2$, $x_2 = 4$, $x_3 = 0$. ($P = 6$ also when $x_1 = 3$, $x_2 = 0$, $x_3 = 3$.)

9. The maximum is $P = \frac{76}{5}$ when $x_1 = \frac{8}{5}$, $x_2 = \frac{24}{5}$, $x_3 = \frac{12}{5}$.

11. The maximum is $P = 15$ when $x_1 = 0$, $x_2 = 5$, $x_3 = 0$.

13. There is no maximum for P; the feasible region is unbounded.

15. The maximum is $P = 30$ when $x_1 = 0$, $x_2 = 0$, $x_3 = 10$.

17. The maximum is $P = 42$ when $x_1 = 1$, $x_2 = 10$, $x_3 = 0$, $x_4 = 0$.

19. The maximum is $P = 40$ when $x_1 = 20$, $x_2 = 0$, $x_3 = 0$.

21. The maximum is $P = 50$ when $x_1 = 0$, $x_2 = 15$, $x_3 = 5$, $x_4 = 0$.

23. The minimum is $C = \frac{305}{4}$ when $x_1 = \frac{25}{4}$, $x_2 = 0$, $x_3 = 0$, $x_4 = 20$, $x_5 = 0$, $x_6 = 50$, $x_7 = 0$.

Index

COUNTING/PERMUTATIONS/COMBINATIONS

Counting Formula
$$c(A \cup B) = c(A) + c(B) - c(A \cap B)$$

Permutation
$$P(n, r) = \frac{n!}{(n - r)!}$$

Combination
$$C(n, r) = \binom{n}{r} = \frac{n!}{(n - r)!r!} = \frac{P(n, r)}{r!}$$

Binomial Theorem
$$(x + y)^n = \binom{n}{0}x^n + \binom{n}{1}yx^{n-1} + \cdots + \binom{n}{k}y^k x^{n-k} + \cdots$$
$$+ \binom{n}{n}y^n$$

PROBABILITY

Equally Likely Outcomes
$$P(E) = \frac{c(E)}{c(S)}$$

Additive Rule
$$P(E \cup F) = P(E) + P(F) - P(E \cap F)$$

Complement of an Event
$$P(\overline{E}) = 1 - P(E)$$

Product Rule
$$P(E \cap F) = P(F) \cdot P(E|F)$$

Binomial Probability
$$b(n, k; p) = \binom{n}{k}p^k(1 - p)^{n-k}$$

Bayes' Theorem
$$P(A_j|E) = \frac{P(A_j)P(E|A_j)}{P(E)}$$
$$= \frac{P(A_j)P(E|A_j)}{P(A_1)P(E|A_1) + P(A_2)P(E|A_2) + \cdots + P(A_n)P(E|A_n)}$$

Expectation
$$E = m_1 p_1 + m_2 p_2 + \cdots + m_n p_n$$

STATISTICS

Mean
$$\overline{X} = \frac{x_1 + x_2 + \cdots + x_n}{n}$$

Standard Deviation
$$\sigma = \sqrt{\frac{(x_1 - \overline{X})^2 + (x_2 - \overline{X})^2 + \cdots + (x_n - \overline{X})^2}{n}}$$

FINANCE

Simple Interest
$$I = Prt$$

Compound Interest
$$A_n = P(1 + i)^n$$

Amount of an Annuity
$$A = P\frac{(1 + i)^n - 1}{i}$$

Present Value of an Annuity
$$V = P\frac{1 - (1 + i)^{-n}}{i}$$